Arithmetic Operations:

$$ab + ac = a(b + c)$$

$$\frac{a}{b} + \frac{c}{d} = \frac{ad + bc}{bd}$$

$$\frac{a + b}{c} = \frac{a}{c} + \frac{b}{c}$$

$$\frac{\left(\frac{a}{b}\right)}{\left(\frac{c}{d}\right)} = \frac{ad}{bc}$$

$$a\left(\frac{b}{c}\right) = \frac{ab}{c}$$

$$\frac{a - b}{c - d} = \frac{b - a}{d - c}$$

$$\frac{ab + ac}{a} = b + c, \; a \neq 0$$

$$\frac{\left(\frac{a}{b}\right)}{c} = \frac{a}{bc}$$

$$\frac{a}{\left(\frac{b}{c}\right)} = \frac{ac}{b}$$

Exponents and Radicals:

$$a^0 = 1, \; a \neq 0$$

$$\frac{a^x}{a^y} = a^{x-y}$$

$$\left(\frac{a}{b}\right)^x = \frac{a^x}{b^x}$$

$$\sqrt[n]{a^m} = a^{m/n} = \left(\sqrt[n]{a}\right)^m$$

$$a^{-x} = \frac{1}{a^x}$$

$$(a^x)^y = a^{xy}$$

$$\sqrt{a} = a^{1/2}$$

$$\sqrt[n]{ab} = \sqrt[n]{a}\sqrt[n]{b}$$

$$a^x a^y = a^{x+y}$$

$$(ab)^x = a^x b^x$$

$$\sqrt[n]{a} = a^{1/n}$$

$$\sqrt[n]{\left(\frac{a}{b}\right)} = \frac{\sqrt[n]{a}}{\sqrt[n]{b}}$$

Algebraic Errors to Avoid:

$$\frac{a}{x + b} \neq \frac{a}{x} + \frac{a}{b}$$

(To see this error, let $a = b = x = 1$.)

$$\sqrt{x^2 + a^2} \neq x + a$$

(To see this error, let $x = 3$ and $a = 4$.)

$$a - b(x - 1) \neq a - bx - b$$

[Remember to distribute negative signs. The equation should be $a - b(x - 1) = a - bx + b$.]

$$\frac{\left(\frac{x}{a}\right)}{b} \neq \frac{bx}{a}$$

[To divide fractions, invert and multiply. The equation should be

$$\frac{\left(\frac{x}{a}\right)}{b} = \frac{\left(\frac{x}{a}\right)}{\left(\frac{b}{1}\right)} = \left(\frac{x}{a}\right)\left(\frac{1}{b}\right) = \frac{x}{ab}.]$$

$$\sqrt{-x^2 + a^2} \neq -\sqrt{x^2 - a^2}$$

(The negative sign cannot be factored out of the square root.)

$$\frac{a + bx}{a} \neq 1 + bx$$

(This is one of many examples of incorrect cancellation. The equation should be $\frac{a + bx}{a} = \frac{a}{a} + \frac{bx}{a} = 1 + \frac{bx}{a}$.)

$$\frac{1}{x^{1/2} - x^{1/3}} \neq x^{-1/2} - x^{-1/3}$$

(This error is a more complex version of the first error.)

$$(x^2)^3 \neq x^5$$

[This equation should be $(x^2)^3 = x^2 x^2 x^2 = x^6$.]

Conversion Table:

1 centimeter ≈ 0.394 inch
1 meter ≈ 39.370 inches
≈ 3.281 feet
1 kilometer ≈ 0.621 mile
1 liter ≈ 0.264 gallon
1 newton ≈ 0.225 pound

1 joule ≈ 0.738 foot-pound
1 gram ≈ 0.035 ounce
1 kilogram ≈ 2.205 pounds
1 inch ≈ 2.540 centimeters
1 foot ≈ 30.480 centimeters
≈ 0.305 meter

1 mile ≈ 1.609 kilometers
1 gallon ≈ 3.785 liters
1 pound ≈ 4.448 newtons
1 foot-lb ≈ 1.356 joules
1 ounce ≈ 28.350 grams
1 pound ≈ 0.454 kilogram

College Algebra

Fifth Edition

College Algebra

Fifth Edition

Ron Larson
▶ **Robert P. Hostetler**

The Pennsylvania State University
The Behrend College

▶ **With the assistance of David C. Falvo**

The Pennsylvania State University
The Behrend College

Houghton Mifflin Company   Boston   New York

Sponsoring Editor: Jack Shira
Managing Editor: Cathy Cantin
Senior Associate Editor: Maureen Ross
Associate Editor: Laura Wheel
Assistant Editor: Carolyn Johnson
Supervising Editor: Karen Carter
Project Editor: Patty Bergin
Editorial Assistant: Kate Hartke
Art Supervisor: Gary Crespo
Marketing Manager: Michael Busnach
Senior Manufacturing Coordinator: Sally Culler
Composition and Art: Meridian Creative Group

Printed in the U.S.A.

Library of Congress Catalog Card Number: 00-131896

ISBN: 0-618-05284-4

56789–DOW–05 04 03 02

Contents

A Word from the Authors

Welcome to *College Algebra*, Fifth Edition. In this revision we focused on student success, accessibility, and flexibility.

Student Success: During the past 30 years of teaching and writing, we have learned many things about the teaching and learning of mathematics. We found that students are most successful when they know what they are expected to learn and why it is important to learn. With that in mind, we have restructured the Fifth Edition to include a thematic study thread in every chapter.

Each chapter begins with a study guide called *How to Study This Chapter*, which includes a comprehensive overview of the chapter concepts (*The Big Picture*), a list of *Important Vocabulary* that is integral to learning *The Big Picture* concepts, a list of study resources, and a general study tip. The study guide allows students to get organized and prepare for the chapter.

An old pedagogical recipe goes something like this, "First I'm going to tell you what I'm going to teach you, then I will teach it to you, and finally I will go over what I taught you." Following this recipe, we have also included a set of learning objectives in every section that outlines what students are expected to learn, followed by an interesting real-life application that illustrates why it is important to learn the concepts in that section. Finally, the chapter summary (*What did you learn?*), which reinforces the section objectives, and the chapter *Review Exercises*, which are correlated to the chapter summary, provide additional study support at the conclusion of each chapter.

Our new *Student Success Organizer* supplement takes this study thread one step further, providing a content-based study aid.

Accessibility: Over the years we have taken care to write a text for the student. We have paid careful attention to the presentation, using precise mathematical language and clear writing, to create an effective learning tool. We believe that every student can learn mathematics and we are committed to providing a text that makes the mathematics within it accessible to all students. In the Fifth Edition, we have revised and improved upon many text features designed for this purpose. The *Technology, Exploration,* and *Study Tip* features have been expanded. *Chapter Tests,* which give students an opportunity for self-assessment, now follow every chapter in the Fifth Edition. The exercise sets now include both *Synthesis* exercises, which check students' conceptual understanding, and *Review* exercises, which reinforce skills learned in previous sections and chapters. Also, students have access to several media resources that accompany this text— videotapes, *Interactive College Algebra* CD-ROM, and a *College Algebra* website—that provide additional text-specific support.

Flexibility: From the time we first began writing in the early 1970s, we have always viewed part of our authoring role as that of providing instructors with flexible teaching programs. The optional features within the text allow instructors with different pedagogical approaches to design their courses to meet both their instructional needs and the needs of their students. Instructors who stress applications and problem solving, or exploration and technology, or more traditional methods, will be able to use this text successfully. In addition, we provide several print and media resources to support instructors, including a new *Instructor Success Organizer*.

We hope you enjoy the Fifth Edition.

Ron Larson

Ron Larson

Robert P. Hostetler

Robert P. Hostetler

Acknowledgments

We would like to thank the many people who have helped us at various stages of this project to prepare the text and supplements package. Their encouragement, criticisms, and suggestions have been invaluable to us.

Fifth Edition Reviewers

James Alsobrook, Southern Union State Community College; Sherry Biggers, Clemson University; Charles Biles, Humboldt State University; Randall Boan, Aims Community College; Jeremy Carr, Pensacola Junior College; D. J. Clark, Portland Community College; Donald Clayton, Madisonville Community College; Linda Crabtree, Metropolitan Community College; David DeLatte, University of North Texas; Gregory Dlabach, Northeastern Oklahoma A & M College; Joseph Lloyd Harris, Gulf Coast Community College; Jeff Heiking, St. Petersburg Junior College; Celeste Hernandez, Richland College; Heidi Howard, Florida Community College at Jacksonville; Wanda Long, St. Charles County Community College; Wayne F. Mackey, University of Arkansas; Rhonda MacLeod, Florida State University; M. Maheswaran, University of Wisconsin–Marathon County; Valerie Miller, Georgia State University; Katharine Muller, Cisco Junior College; Bonnie Oppenheimer, Mississippi University for Women; James Pohl, Florida Atlantic University; Hari Pulapaka, Valdosta State University; Michael Russo, Suffolk County Community College; Cynthia Floyd Sikes, Georgia Southern University; Susan Schindler, Baruch College–CUNY; Stanley Smith, Black Hills State University. In addition, we would like to thank all the college algebra instructors who took the time to respond to our survey.

We would like to extend a special thanks to Hari Pulapaka for his contributions to this revision.

We would like to thank the staff of Larson Texts, Inc. and the staff of Meridian Creative Group, who assisted in proofreading the manuscript, preparing and proofreading the art package, and typesetting the supplements.

On a personal level, we are grateful to our wives, Deanna Gilbert Larson and Eloise Hostetler, for their love, patience, and support. Also, a special thanks goes to R. Scott O'Neil.

If you have suggestions for improving this text, please feel free to write to us. Over the past two decades we have received many useful comments from both instructors and students, and we value these comments very much.

Ron Larson
Robert P. Hostetler

Features Highlights

Student Success Tools

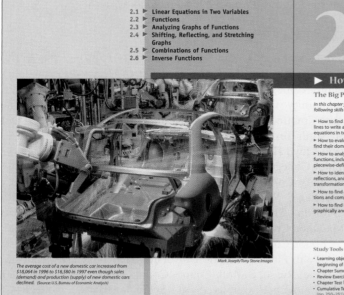

▶ **How to Study This Chapter**

The Big Picture

In this chapter you will learn the following skills and concepts.

▶ How to find and use the slopes of lines to write and graph linear equations in two variables

▶ How to evaluate functions and find their domains

▶ How to analyze graphs of functions, including linear, step, and piecewise-defined functions

▶ How to identify and graph shifts, reflections, and nonrigid transformations of functions

▶ How to find arithmetic combinations and compositions of functions

▶ How to find inverses of functions graphically and algebraically

Important Vocabulary

As you encounter each new vocabulary term in this chapter, add the term and its definition to your notebook glossary.

Linear equation in two variables (p. 172)
Slope (p. 172)
Slope-intercept form (p. 172)
Ratio (p. 174)
Rate of change (p. 174)
Point-slope form (p. 177)
Two-point form (p. 177)
General form (p. 178)
Parallel (p. 179)
Perpendicular (p. 179)
Function (p. 187)
Domain (p. 187)
Range (p. 187)
Independent variable (p. 188)
Dependent variable (p. 188)
Function notation (p. 189)
Piecewise-defined function (p. 190)
Implied domain (p. 191)
Graph of a function (p. 201)
Vertical Line Test (p. 202)

Zeros of a function (p. 203)
Increasing function (p. 204)
Decreasing function (p. 204)
Constant function (p. 204)
Relative minimum (p. 205)
Relative maximum (p. 205)
Linear function (p. 206)
Greatest integer function (p. 207)
Step function (p. 207)
Even function (p. 208)
Odd function (p. 208)
Vertical and horizontal shifts (p. 215)
Reflection (p. 217)
Vertical stretch (p. 219)
Vertical shrink (p. 219)
Arithmetic combination of functions (p. 225)
Composition of functions (p. 227)
Inverse (p. 233)
Horizontal Line Test (p. 236)

Study Tools

· Learning objectives at the beginning of each section
· Chapter Summary (p. 243)
· Review Exercises (pp. 244–247)
· Chapter Test (p. 249)
· Cumulative Test for Chapters P–2 (pp. 250–251)

Additional Resources

· Study and Solutions Guide
· Interactive College Algebra
· Videotapes for Chapter 2
· College Algebra Website
· Student Success Organizer

STUDY TIP

During class, take notes on definitions, examples, concepts, and rules—whatever it is you identify as important to the instructor. Then, as soon after class as possible, review your notes.

▶ **"How to Study This Chapter"**

The new chapter-opening study guide includes:

• *The Big Picture*—an objective-based overview of the main concepts of the chapter

• *Important Vocabulary*—mathematical terms integral to learning *The Big Picture* concepts

• *Study Tools*

• *Additional Resources*

• *Study Tip*

New Section Openers include:

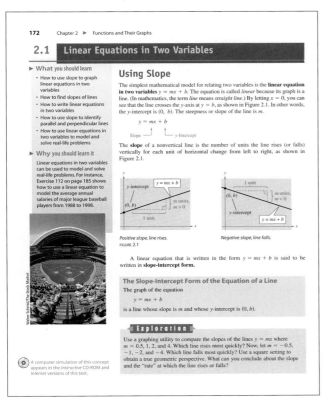

2.1   Linear Equations in Two Variables

▶ **What you should learn**
- How to use slope to graph linear equations in two variables
- How to find slopes of lines
- How to write linear equations in two variables
- How to use slope to identify parallel and perpendicular lines
- How to use linear equations in two variables to model and solve real-life problems

▶ **Why you should learn it**
Linear equations in two variables can be used to model and solve real-life problems. For instance, Exercise 112 on page 185 shows how to use a linear equation to model the average annual salaries of major league baseball players from 1988 to 1998.

A computer simulation of this concept appears in the Interactive CD-ROM and Internet versions of this text.

Using Slope

The simplest mathematical model for relating two variables is the **linear equation in two variables** $y = mx + b$. The equation is called *linear* because its graph is a line. (In mathematics, the term *line* means *straight line*.) By letting $x = 0$, you can see that the line crosses the y-axis at $y = b$, as shown in Figure 2.1. In other words, the y-intercept is $(0, b)$. The steepness or slope of the line is m.

$$y = mx + b$$

Slope ⎯⎯   ⎯⎯ y-Intercept

The **slope** of a nonvertical line is the number of units the line rises (or falls) vertically for each unit of horizontal change from left to right, as shown in Figure 2.1.

Positive slope, line rises.   *Negative slope, line falls.*
FIGURE 2.1

A linear equation that is written in the form $y = mx + b$ is said to be written in **slope-intercept form.**

The Slope-Intercept Form of the Equation of a Line

The graph of the equation

$$y = mx + b$$

is a line whose slope is m and whose y-intercept is $(0, b)$.

Exploration

Use a graphing utility to compare the slopes of the lines $y = mx$ where $m = 0.5, 1, 2,$ and 4. Which line rises most quickly? Now, let $m = -0.5, -1, -2,$ and -4. Which line falls most quickly? Use a square setting to obtain a true geometric perspective. What can you conclude about the slope and the "rate" at which the line rises or falls?

▶ **"What you should learn"**

Objectives outline the main concepts and help keep students focused on *The Big Picture*.

▶ **"Why you should learn it"**

A real-life application or a reference to other branches of mathematics illustrates the relevance of the section's content.

▶ **"What did you learn?" Summary**

This chapter summary provides a concise, section-by-section review of the section objectives. These objectives are correlated to the chapter Review Exercises.

Chapter Summary

What did you learn?

	Review Exercises
Section 2.1	
☐ How to use slope to graph linear equations in two variables	1–14
☐ How to find slopes of lines	15–18
☐ How to write linear equations in two variables	19–26
☐ How to use slope to identify parallel and perpendicular lines	27–28
☐ How to use linear equations in two variables to model and solve real-life problems	29–32
Section 2.2	
☐ How to decide whether relations between two variables are functions	33–38
☐ How to use function notation and evaluate functions	39, 40
☐ How to find the domains of functions	41–46
☐ How to use functions to model and solve real-life problems	47–50
Section 2.3	
☐ How to use the Vertical Line Test for functions	51–54
☐ How to find the zeros of functions	55–58
☐ How to determine intervals on which functions are increasing or decreasing	59–62
☐ How to identify and graph linear functions	63–66
☐ How to identify and graph step functions and other piecewise-defined functions	67, 68
☐ How to identify even and odd functions	69–72
Section 2.4	
☐ How to recognize graphs of common functions	73–76
☐ How to use vertical and horizontal shifts to sketch graphs of functions	77–80
☐ How to use reflections to sketch graphs of functions	81–84
☐ How to use nonrigid transformations to sketch graphs of functions	85–88
Section 2.5	
☐ How to add, subtract, multiply, and divide functions	89–92
☐ How to find compositions of one function with another function	93–96
☐ How to use combinations of functions to model and solve real-life problems	97, 98
Section 2.6	
☐ How to find inverse functions informally and verify that two functions are inverses of each other	99–102
☐ How to use graphs of functions to decide whether functions have inverses	103–108
☐ How to find inverse functions algebraically	109–114

FEATURES

Revised Exercises and Applications

Section 2.2   ▶   Functions   **197**

42. $h(x) = \begin{cases} 9 - x^2, & x < 3 \\ x - 3, & x \geq 3 \end{cases}$

x	1	2	3	4	5
$h(x)$					

In Exercises 43–50, find all real values of x such that $f(x) = 0$.

43. $f(x) = 15 - 3x$ 44. $f(x) = 5x + 1$

45. $f(x) = \dfrac{3x - 4}{5}$ 46. $f(x) = \dfrac{12 - x^2}{5}$

47. $f(x) = x^2 - 9$ 48. $f(x) = x^2 - 8x + 15$

49. $f(x) = x^3 - x$

50. $f(x) = x^3 - x^2 - 4x + 4$

In Exercises 51–54, find the value(s) of x for which $f(x) = g(x)$.

51. $f(x) = x^2$, $g(x) = x + 2$

52. $f(x) = x^2 + 2x + 1$, $g(x) = 3x + 3$

53. $f(x) = \sqrt{3x} + 1$, $g(x) = x + 1$

54. $f(x) = x^4 - 2x^2$, $g(x) = 2x^2$

In Exercises 55–68, find the domain of the function.

55. $f(x) = 5x^2 + 2x - 1$ 56. $g(x) = 1 - 2x^2$

57. $h(t) = \dfrac{4}{t}$ 58. $s(y) = \dfrac{3y}{y + 5}$

59. $g(y) = \sqrt{y - 10}$ 60. $f(t) = \sqrt[3]{t + 4}$

61. $f(x) = \sqrt[4]{1 - x^2}$ 62. $f(x) = \sqrt[4]{x^2 + 3x}$

63. $g(x) = \dfrac{1}{x} - \dfrac{3}{x + 2}$ 64. $h(x) = \dfrac{10}{x^2 - 2x}$

65. $f(s) = \dfrac{\sqrt{s - 1}}{s - 4}$ 66. $f(x) = \dfrac{\sqrt{x + 6}}{6 + x}$

67. $f(x) = \dfrac{\sqrt[4]{x - 4}}{x}$ 68. $f(x) = \dfrac{x - 5}{x^2 - 9}$

In Exercises 69–72, assume that the domain of f is the set $A = \{-2, -1, 0, 1, 2\}$. Determine the set of ordered pairs that represents the function f.

69. $f(x) = x^2$ 70. $f(x) = \dfrac{2x}{x^2 + 1}$

71. $f(x) = \sqrt{x + 2}$ 72. $f(x) = |x + 1|$

Exploration **In Exercises 73–76, determine the function from**

$f(x) = cx$, $g(x) = cx^2$, $h(x) = c\sqrt{|x|}$, and $r(x) = \dfrac{c}{x}$

and the value of the constant c that will make the function fit the data in the table.

73.

x	-4	-1	0	1	4
y	-32	-2	0	-2	-32

74.

x	-4	-1	0	1	4
y	-1	$-\frac{1}{4}$	0	$\frac{1}{4}$	1

75.

x	-4	-1	0	1	4
y	-8	-32	Undef.	32	8

76.

x	-4	-1	0	1	4
y	6	3	0	3	6

🔷 **In Exercises 77–84, find the difference quotient and simplify your answer.**

77. $f(x) = x^2 - x + 1$, $\dfrac{f(2 + h) - f(2)}{h}$, $h \neq 0$

78. $f(x) = 5x - x^2$, $\dfrac{f(5 + h) - f(5)}{h}$, $h \neq 0$

79. $f(x) = x^3$, $\dfrac{f(x + c) - f(x)}{c}$, $c \neq 0$

80. $f(x) = 2x$, $\dfrac{f(x + c) - f(x)}{c}$, $c \neq 0$

81. $g(x) = 3x - 1$, $\dfrac{g(x) - g(3)}{x - 3}$, $x \neq 3$

82. $f(t) = \dfrac{1}{t}$, $\dfrac{f(t) - f(1)}{t - 1}$, $t \neq 1$

83. $f(x) = \sqrt{5x}$, $\dfrac{f(x) - f(5)}{x - 5}$, $x \neq 5$

84. $f(x) = x^{2/3} + 1$, $\dfrac{f(x) - f(8)}{x - 8}$, $x \neq 8$

85. *Geometry* Express the area A of a square as a function of its perimeter P.

86. *Geometry* Express the area A of a circle as a function of its circumference C.

The symbol 🔷 indicates an example or exercise that highlights algebraic techniques specifically used in calculus.

▶ Exercises

- Each exercise set contains a variety of computational, conceptual, and applied problems.
- Each exercise set is carefully graded in difficulty to allow students to gain confidence as they progress.
- Each exercise set now concludes with two new types of exercises:
  - *Synthesis* exercises promote further exploration of mathematical concepts, critical thinking skills, and writing about mathematics. These exercises require students to synthesize the main concepts presented in the section and chapter.
  - *Review* exercises reinforce previously learned skills and concepts.

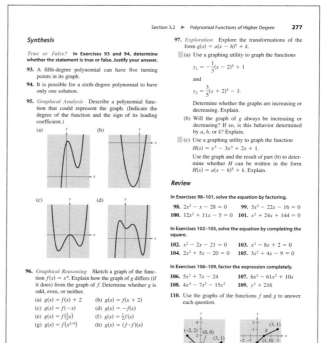

Section 3.2   ▶   Polynomial Functions of Higher Degree   **277**

Synthesis

True or False? **In Exercises 93 and 94, determine whether the statement is true or false. Justify your answer.**

93. A fifth-degree polynomial can have five turning points in its graph.

94. It is possible for a sixth-degree polynomial to have only one solution.

95. *Graphical Analysis* Describe a polynomial function that could represent the graph. (Indicate the degree of the function and the sign of its leading coefficient.)

(a)   (b)

(c)   (d)

96. *Graphical Reasoning* Sketch a graph of the function $f(x) = x^4$. Explain how the graph of g differs (if it does) from the graph of f. Determine whether g is odd, even, or neither.

(a) $g(x) = f(x) + 2$ (b) $g(x) = f(x + 2)$
(c) $g(x) = f(-x)$ (d) $g(x) = -f(x)$
(e) $g(x) = f\left(\frac{1}{2}x\right)$ (f) $g(x) = \frac{1}{2}f(x)$
(g) $g(x) = f\left(x^{3/4}\right)$ (h) $g(x) = (f \circ f)(x)$

97. *Exploration* Explore the transformations of the form $g(x) = a(x - h)^5 + k$.

(a) Use a graphing utility to graph the functions

$$y_1 = -\frac{1}{3}(x - 2)^5 + 1$$

and

$$y_2 = \frac{3}{5}(x + 2)^5 - 3.$$

Determine whether the graphs are increasing or decreasing. Explain.

(b) Will the graph of g always be increasing or decreasing? If so, is this behavior determined by a, h, or k? Explain.

(c) Use a graphing utility to graph the function $H(x) = x^5 - 3x^3 + 2x + 1$. Use the graph and the result of part (b) to determine whether H can be written in the form $H(x) = a(x - h)^5 + k$. Explain.

Review

In Exercises 98–101, solve the equation by factoring.

98. $2x^2 - x - 28 = 0$ 99. $3x^2 - 22x - 16 = 0$

100. $12x^2 + 11x - 5 = 0$ 101. $x^2 + 24x + 144 = 0$

In Exercises 102–105, solve the equation by completing the square.

102. $x^2 - 2x - 21 = 0$ 103. $x^2 - 8x + 2 = 0$

104. $2x^2 + 5x - 20 = 0$ 105. $3x^2 + 4x - 9 = 0$

In Exercises 106–109, factor the expression completely.

106. $5x^2 + 7x - 24$ 107. $6x^3 - 61x^2 + 10x$

108. $4x^4 - 7x^3 - 15x^2$ 109. $y^3 + 216$

110. Use the graphs of the functions f and g to answer each question.

(a) Explain why f does not have an inverse.

(b) Find $g^{-1}(1)$.

Section 2.6 ► Inverse Functions   **241**

In Exercises 69–74, use the functions $f(x) = \frac{1}{8}x - 3$ and $g(x) = x^3$ to find the indicated value or function.

69. $(f^{-1} \circ g^{-1})(1)$ **70.** $(g^{-1} \circ f^{-1})(-3)$

71. $(f^{-1} \circ f^{-1})(6)$ **72.** $(g^{-1} \circ g^{-1})(-4)$

73. $(f \circ g)^{-1}$ **74.** $g^{-1} \circ f^{-1}$

In Exercises 75–78, use the functions $f(x) = x + 4$ and $g(x) = 2x - 5$ to find the specified function.

75. $g^{-1} \circ f^{-1}$ **76.** $f^{-1} \circ g^{-1}$

77. $(f \circ g)^{-1}$ **78.** $(g \circ f)^{-1}$

79. *Hourly Wage* Your wage is $8.00 per hour plus $0.75 for each unit produced per hour. So, your hourly wage y in terms of the number of units produced is

$$y = 8 + 0.75x.$$

(a) Find the inverse of the function.

(b) What does each variable represent in the inverse function?

(c) Determine the number of units produced when your hourly wage is $22.25.

80. *Cost* Suppose you need a total of 50 pounds of two commodities costing $1.25 and $1.60 per pound, respectively.

(a) Verify that the total cost is

$$y = 1.25x + 1.60(50 - x)$$

where x is the number of pounds of the less expensive commodity.

(b) Find the inverse of the cost function. What does each variable represent in the inverse function?

(c) Use the context of the problem to determine the domain of the inverse function.

(d) Determine the number of pounds of the less expensive commodity purchased if the total cost is $73.

81. *Diesel Mechanics* The function

$$y = 0.03x^2 + 245.50, \quad 0 < x < 100$$

approximates the exhaust temperature y in degrees Fahrenheit where x is the percent load for a diesel engine.

(a) Find the inverse of the function. What does each variable represent in the inverse function?

(b) Use a graphing utility to graph the inverse function.

(c) Determine the percent load interval if the exhaust temperature of the engine must not exceed 500 degrees Fahrenheit.

82. *New Car Sales* The total value of new car sales f (in billions of dollars) in the United States from 1992 through 1997 is shown in the table. The time (in years) is given by t, with $t = 2$ corresponding to 1992. (Source: National Automobile Dealers Association)

t	2	3	4	5	6	7
$f(t)$	333.8	377.3	430.6	456.2	490.0	507.5

(a) Does f^{-1} exist?

(b) If f^{-1} exists, what does it mean in the context of the problem?

(c) If f^{-1} exists, find $f^{-1}(456.2)$.

83. If the table in Exercise 82 were extended to 1998 and if the total value of new car sales for that year was $430.6 billion, would f^{-1} exist? Explain.

84. *Cellular Phones* The average local bill (in dollars) for cellular phones in the United States from 1990 to 1997 is shown in the table. The time (in years) is given by t, with $t = 0$ corresponding to 1990. (Source: Cellular Telecommunications Industry Association)

t	0	1	2	3
$f(t)$	80.90	72.74	68.68	61.48

t	4	5	6	7
$f(t)$	56.21	51.00	47.70	42.78

(a) Find $f^{-1}(51)$.

(b) What does f^{-1} mean in the context of the problem?

(c) Use the regression feature of a graphing utility to find a linear model for the data, $y = mx + b$. Round m and b to two decimal places.

(d) Algebraically find the inverse of the linear model in part (c).

(e) Use the inverse of the linear model you found in part (d) to approximate $f^{-1}(11)$.

► Real-Life Applications

- A wide variety of real-life applications, many using current, real data are integrated throughout examples and exercises.

- The icon 🌎 indicates an example that involves a real-life application.

► Algebra of Calculus

- Special emphasis is given to the algebraic techniques used in calculus.
- Algebra of Calculus examples and exercises are integrated throughout the text.
- The symbol 🔘 indicates an example or exercise in which the Algebra of Calculus is featured.

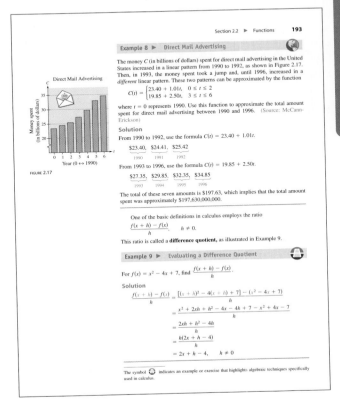

Section 2.2 ► Functions   **193**

Example 8 ► **Direct Mail Advertising**

The money C (in billions of dollars) spent for direct mail advertising in the United States increased in a linear pattern from 1990 to 1992, as shown in Figure 2.17. Then, in 1993, the money spent took a jump and, until 1996, increased in a *different* linear pattern. These two patterns can be approximated by the function

$$C(t) = \begin{cases} 23.40 + 1.01t, & 0 \le t \le 2 \\ 19.85 + 2.50t, & 3 \le t \le 6 \end{cases}$$

where $t = 0$ represents 1990. Use this function to approximate the total amount spent for direct mail advertising between 1990 and 1996. (Source: McCann-Erickson)

Solution

From 1990 to 1992, use the formula $C(t) = 23.40 + 1.01t$.

$$\underbrace{\$23.40,}_{1990} \quad \underbrace{\$24.41,}_{1991} \quad \underbrace{\$25.42}_{1992}$$

From 1993 to 1996, use the formula $C(t) = 19.85 + 2.50t$.

$$\underbrace{\$27.35,}_{1993} \quad \underbrace{\$29.85,}_{1994} \quad \underbrace{\$32.35,}_{1995} \quad \underbrace{\$34.85}_{1996}$$

The total of these seven amounts is $197.63, which implies that the total amount spent was approximately $197,630,000,000.

One of the basic definitions in calculus employs the ratio

$$\frac{f(x + h) - f(x)}{h}, \quad h \ne 0.$$

This ratio is called a **difference quotient,** as illustrated in Example 9.

Example 9 ► **Evaluating a Difference Quotient**

For $f(x) = x^2 - 4x + 7$, find $\dfrac{f(x + h) - f(x)}{h}$.

Solution

$$\frac{f(x + h) - f(x)}{h} = \frac{[(x + h)^2 - 4(x + h) + 7] - (x^2 - 4x + 7)}{h}$$

$$= \frac{x^2 + 2xh + h^2 - 4x - 4h + 7 - x^2 + 4x - 7}{h}$$

$$= \frac{2xh + h^2 - 4h}{h}$$

$$= \frac{h(2x + h - 4)}{h}$$

$$= 2x + h - 4, \quad h \ne 0$$

The symbol 🔘 indicates an example or exercise that highlights algebraic techniques specifically used in calculus.

Direct Mail Advertising

FIGURE 2.17

(Graph: Money spent (in billions of dollars) vs. Year (0 ↔ 1990))

Flexibility and Accessibility

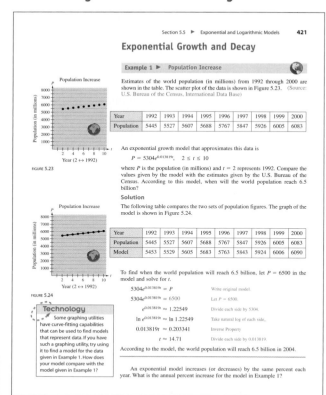

Exponential Growth and Decay

Example 1 ▶ Population Increase

Estimates of the world population (in millions) from 1992 through 2000 are shown in the table. The scatter plot of the data is shown in Figure 5.23.   (Source: U.S. Bureau of the Census, International Data Base)

Year	1992	1993	1994	1995	1996	1997	1998	1999	2000
Population	5445	5527	5607	5688	5767	5847	5926	6005	6083

An exponential growth model that approximates this data is

$$P = 5304e^{0.013819t}, \quad 2 \leq t \leq 10$$

where P is the population (in millions) and $t = 2$ represents 1992. Compare the values given by the model with the estimates given by the U.S. Bureau of the Census. According to this model, when will the world population reach 6.5 billion?

Solution

The following table compares the two sets of population figures. The graph of the model is shown in Figure 5.24.

Year	1992	1993	1994	1995	1996	1997	1998	1999	2000
Population	5445	5527	5607	5688	5767	5847	5926	6005	6083
Model	5453	5529	5605	5683	5763	5843	5924	6006	6090

To find when the world population will reach 6.5 billion, let $P = 6500$ in the model and solve for t.

$5304e^{0.013819t} = P$	Write original model.
$5304e^{0.013819t} = 6500$	Let $P = 6500$.
$e^{0.013819t} = 1.22549$	Divide each side by 5304.
$\ln e^{0.013819t} = \ln 1.22549$	Take natural log of each side.
$0.013819t = 0.203341$	Inverse Property
$t \approx 14.71$	Divide each side by 0.013819.

According to the model, the world population will reach 6.5 billion in 2004.

Technology

Some graphing utilities have curve-fitting capabilities that can be used to find models that represent data. If you have such a graphing utility, try using it to find a model for the data given in Example 1. How does your model compare with the model given in Example 1?

An exponential model increases (or decreases) by the same percent each year. What is the annual percent increase for the model in Example 1?

◀ **E x p l o r a t i o n** ▶

- Before introduction of selected topics, *Exploration* engages students in active discovery of mathematical concepts and relationships, often through the power of technology.

- *Exploration* strengthens students' critical thinking skills and helps them develop an intuitive understanding of theoretical concepts.

- *Exploration* is an optional feature and can be omitted without loss of continuity in coverage.

▶ **Additional Features**

Carefully crafted learning tools designed to create a rich learning environment can be found throughout the text. These learning tools include Study Tips, Historical Notes, Writing About Mathematics, Chapter Projects, Chapter Review Exercises, Chapter Tests, Cumulative Tests, and an extensive art program.

Technology

- Point-of-use instructions for using graphing utilities appear in the margin, encouraging the use of graphing technology as a tool for visualization of mathematical concepts, for verification of other solution methods, and for facilitation of computations.

- The use of technology is optional in this text. This feature and related exercises can easily be omitted without loss of continuity in coverage. Exercises that require the use of a graphing utility are identified by the symbol ▦ .

▶ **Examples**

- Each example was carefully chosen to illustrate a particular mathematical concept or problem-solving skill.

- Every example contains step-by-step solutions, most with side-by-side explanations that lead students through the solution process.

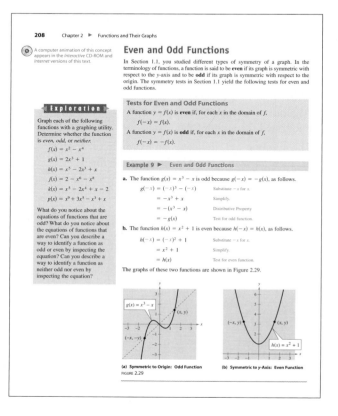

A computer animation of this concept appears in the *Interactive CD-ROM* and *Internet* versions of this text.

Even and Odd Functions

In Section 1.1, you studied different types of symmetry of a graph. In the terminology of functions, a function is said to be **even** if its graph is symmetric with respect to the y-axis and to be **odd** if its graph is symmetric with respect to the origin. The symmetry tests in Section 1.1 yield the following tests for even and odd functions.

Tests for Even and Odd Functions

A function $y = f(x)$ is **even** if, for each x in the domain of f,

$$f(-x) = f(x).$$

A function $y = f(x)$ is **odd** if, for each x in the domain of f,

$$f(-x) = -f(x).$$

◀ **E x p l o r a t i o n** ▶

Graph each of the following functions with a graphing utility. Determine whether the function is *even, odd,* or *neither.*

$f(x) = x^2 - x^4$

$g(x) = 2x^3 + 1$

$h(x) = x^5 - 2x^3 + x$

$j(x) = 2 - x^6 - x^8$

$k(x) = x^5 - 2x^4 + x - 2$

$p(x) = x^9 + 3x^5 - x^3 + x$

What do you notice about the equations of functions that are odd? What do you notice about the equations of functions that are even? Can you describe a way to identify a function as odd or even by inspecting the equation? Can you describe a way to identify a function as neither odd nor even by inspecting the equation?

Example 9 ▶ Even and Odd Functions

a. The function $g(x) = x^3 - x$ is odd because $g(-x) = -g(x)$, as follows.

$g(-x) = (-x)^3 - (-x)$	Substitute $-x$ for x.
$= -x^3 + x$	Simplify.
$= -(x^3 - x)$	Distributive Property
$= -g(x)$	Test for odd function.

b. The function $h(x) = x^2 + 1$ is even because $h(-x) = h(x)$, as follows.

$h(-x) = (-x)^2 + 1$	Substitute $-x$ for x.
$= x^2 + 1$	Simplify.
$= h(x)$	Test for even function.

The graphs of these two functions are shown in Figure 2.29.

(a) Symmetric to Origin: Odd Function
(b) Symmetric to y-Axis: Even Function
FIGURE 2.29

Supplements

Resources

Website (*college.hmco.com*)
Many additional text-specific study and interactive features for students and instructors can be found at the Houghton Mifflin website.

For the Student

Student Success Organizer

Study and Solutions Guide by Dianna L. Zook (Indiana University/Purdue University–Fort Wayne)

Graphing Technology Guide by Benjamin N. Levy and Laurel Technical Services

Instructional Videotapes by Dana Mosely

Instructional Videotapes for Graphing Calculators by Dana Mosely

For the Instructor

Instructor's Annotated Edition

Instructor Success Organizer

Complete Solutions Guide by Dianna L. Zook (Indiana University/Purdue University–Fort Wayne) and Laurel Technical Services

Test Item File

Problem Solving, Modeling, and Data Analysis Labs by Wendy Metzger (Palomar College)

Computerized Testing (Windows, Macintosh)

HMClassPrep Instructor's CD-ROM

An Introduction to Graphing Utilities

Graphing utilities such as graphing calculators and computers with graphing software are very valuable tools for visualizing mathematical principles, verifying solutions to equations, exploring mathematical ideas, and developing mathematical models. Although graphing utilities are extremely helpful in learning mathematics, their use does not mean that learning algebra is any less important. In fact, the combination of knowledge of mathematics and the use of graphing utilities allows you to explore mathematics more easily and to a greater depth. If you are using a graphing utility in this course, it is up to you to learn its capabilities and to practice using this tool to enhance your mathematical learning.

In this text there are many opportunities to use a graphing utility, some of which are described below.

Some Uses of a Graphing Utility

A graphing utility can be used to

- check or validate answers to problems obtained using algebraic methods.
- discover and explore algebraic properties, rules, and concepts.
- graph functions, and approximate solutions to equations involving functions.
- efficiently perform complicated mathematical procedures such as those found in many real-life applications.
- find mathematical models for sets of data.

In this introduction, the features of graphing utilities are discussed from a generic perspective. To learn how to use the features of a specific graphing utility, consult your user's manual or the website for this text found at *college.hmco.com*. Additionally, keystroke guides are available for most graphing utilities, and your college library may have a videotape on how to use your graphing utility.

The Equation Editor

Many graphing utilities are designed to act as "function graphers." In this course, you will study functions and their graphs in detail. You may recall from previous courses that a function can be thought of as a rule that describes the relationship between two variables. These rules are frequently written in terms of x and y. For example, the equation $y = 3x + 5$ represents y as a function of x.

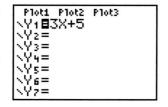

FIGURE 1

Many graphing utilities have an equation editor that requires an equation to be written in "$y =$" form in order to be entered, as shown in Figure 1. (You should note that your equation editor screen may not look like the screen shown in Figure 1.) To determine exactly how to enter an equation into your graphing utility, consult your user's manual.

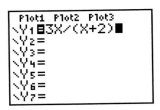

FIGURE 2

The Table Feature

Most graphing utilities are capable of displaying a table of values with x-values and one or more corresponding y-values. These tables can be used to check solutions of an equation and to generate ordered pairs to assist in graphing an equation.

To use the *table* feature, enter an equation into the equation editor in "$y =$" form. The table may have a setup screen, which allows you to select the starting x-value and the table step or x-increment. You may then have the option of automatically generating values for x and y or building your own table using the *ask* mode. In the *ask* mode, you enter a value for x and the graphing utility displays the y-value.

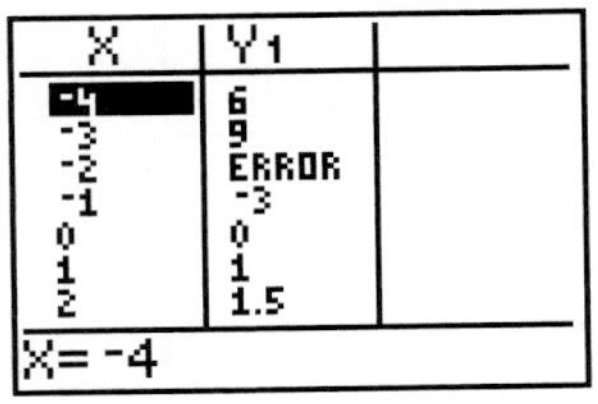

For example, enter the equation

$$y = \frac{3x}{x + 2}$$

FIGURE 3

into the equation editor, as shown in Figure 2. In the table setup screen, set the table to start at $x = -4$ and set the table step to 1. When you view the table, notice that the first x-value is -4 and each value after it increases by 1. Also notice that the Y_1 column gives the resulting y-value for each x-value, as shown in Figure 3. The table shows that the y-value when $x = -2$ is ERROR. This means that the variable x may not take on the value -2 in this equation.

With the same equation in the equation editor, set the table to *ask* mode. In this mode you do not need to set the starting x-value or the table step, because you are entering any value you choose for x. You may enter any real value for x—integers, fractions, decimals, irrational numbers, and so forth. If you enter $x = 1 + \sqrt{3}$, the graphing utility may rewrite the number as a decimal approximation, as shown in Figure 4. You can continue to build your own table by entering additional x-values in order to generate y-values.

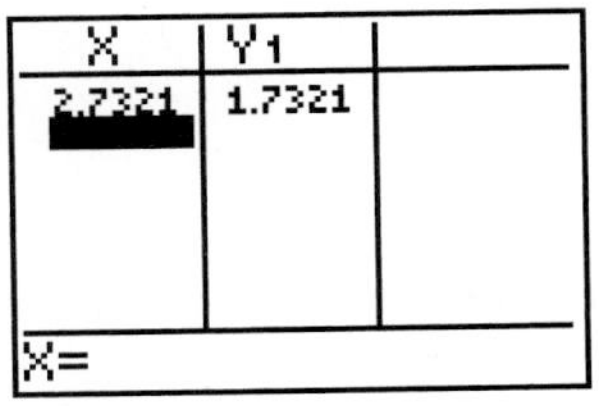

FIGURE 4

If you have several equations in the equation editor, the table may generate y-values for each equation.

Creating a Viewing Window

A **viewing window** for a graph is a rectangular portion of the coordinate plane. A viewing window is determined by the following six values.

Xmin = the smallest value of x
Xmax = the largest value of x
Xscl = the number of units per tick mark on the x-axis
Ymin = the smallest value of y
Ymax = the largest value of y
Yscl = the number of units per tick mark on the y-axis

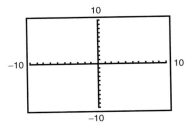

FIGURE 5

When you enter these six values into a graphing utility, you are setting the viewing window. Some graphing utilities have a standard viewing window, as shown in Figure 5.

By choosing different viewing windows for a graph, it is possible to obtain very different impressions of the graph's shape. For instance, Figure 6 shows four different viewing windows for the graph of

$$y = 0.1x^4 - x^3 + 2x^2.$$

Of these, the view shown in part (a) is the most complete.

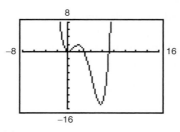

(a)

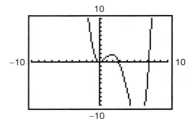

(b)

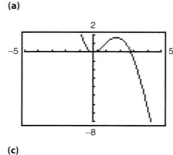

(c)

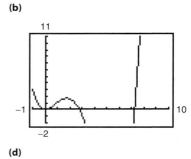

(d)

FIGURE 6

On most graphing utilities, the display screen is two-thirds as high as it is wide. On such screens, you can obtain a graph with a true geometric perspective by using a **square setting**—one in which

$$\frac{\text{Ymax} - \text{Ymin}}{\text{Xmax} - \text{Xmin}} = \frac{2}{3}.$$

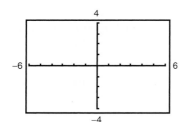

FIGURE 7

One such setting is shown in Figure 7. Notice that the x and y tick marks are equally spaced on a square setting, but not on a standard setting.

To see how the viewing window affects the geometric perspective, graph the semicircles $y_1 = \sqrt{9 - x^2}$ and $y_2 = -\sqrt{9 - x^2}$ in a standard viewing window. Then graph y_1 and y_2 in a square window. Note the difference in the shapes of the circles.

Zoom and Trace Features

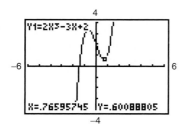

FIGURE 8

When you graph an equation, you can move from point to point along its graph using the *trace* feature. As you trace the graph, the coordinates of each point are displayed, as shown in Figure 8. The *trace* feature combined with the *zoom* feature allows you to obtain better and better approximations of desired points on a graph. For instance, you can use the *zoom* feature of a graphing utility to approximate the x-intercept(s) of a graph [the point(s) where the graph crosses the x-axis]. Suppose you want to approximate the x-intercept(s) of the graph of $y = 2x^3 - 3x + 2.$

Begin by graphing the equation, as shown in Figure 9(a). From the viewing window shown, the graph appears to have only one x-intercept. This intercept lies between -2 and -1. By zooming in on the intercept, you can improve the approximation, as shown in Figure 9(b). To three decimal places, the solution is $x \approx -1.476$.

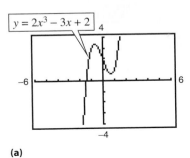

(a)

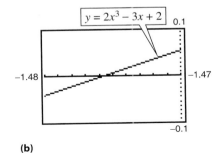

(b)

FIGURE 9

Here are some suggestions for using the *zoom* feature.

1. With each successive zoom-in, adjust the x-scale so that the viewing window shows at least one tick mark on each side of the x-intercept.

2. The error in your approximation will be less than the distance between two scale marks.

3. The *trace* feature can usually be used to add one more decimal place of accuracy without changing the viewing window.

Figure 10(a) shows the graph of $y = x^2 - 5x + 3$. Figures 10(b) and 10(c) show "zoom-in views" of the two x-intercepts. From these views, you can approximate the x-intercepts to be $x \approx 0.697$ and $x \approx 4.303$.

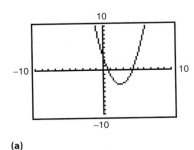

(a)

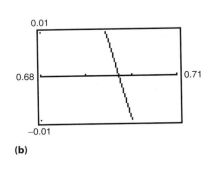

(b)

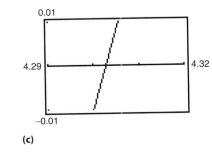

(c)

FIGURE 10

Zero or Root Feature

Using the zero or root feature, you can find the real zeros of functions of the various types studied in this text—polynomial, exponential, and logarithmic functions. To find the zeros of a function such as $f(x) = \frac{3}{4}x - 2$, first enter the function as $y_1 = \frac{3}{4}x - 2$. Then use the zero or root feature, which may require entering lower and upper bound estimates of the root, as shown in Figure 11.

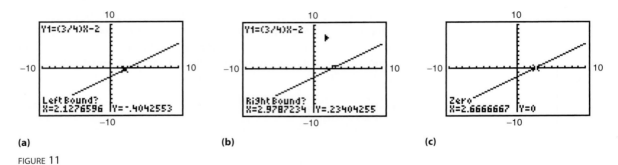

(a)   **(b)**   **(c)**

FIGURE 11

In Figure 11(c), you can see that the zero is $x = 2.6666667 \approx 2\frac{2}{3}$.

Intersect Feature

To find the points of intersection of two graphs, you can use the *intersect* feature. For instance, to find the points of intersection of the graphs of $y_1 = -x + 2$ and $y_2 = x + 4$, enter these two functions and use the *intersect* feature, as shown in Figure 12.

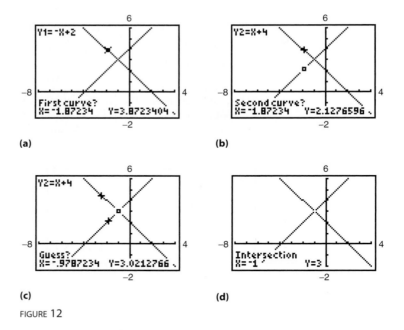

(a)   **(b)**

(c)   **(d)**

FIGURE 12

From Figure 12(d), you can see that the point of intersection is $(-1, 3)$.

Regression Capabilities

Throughout the text, you will be asked to use the regression capabilities of a graphing utility to find models for sets of data. Most graphing utilities have built-in regression programs for the following.

Regression	*Form of Model*
Linear	$y = ax + b$
Quadratic	$y = ax^2 + bx + c$
Cubic	$y = ax^3 + bx^2 + cx + d$
Quartic	$y = ax^4 + bx^3 + cx^2 + dx + e$
Logarithmic	$y = a + b\ln(x)$
Exponential	$y = ab^x$
Power	$y = ax^b$
Logistic	$y = \dfrac{c}{1 + ae^{-bx}}$
Sine	$y = a\sin(bx + c) + d$

For instance, you can find the linear regression model for the average hourly wages y (in dollars per hour) of production workers in manufacturing industries from 1987 through 1997 shown in the table. (Source: U.S. Bureau of Labor Statistics)

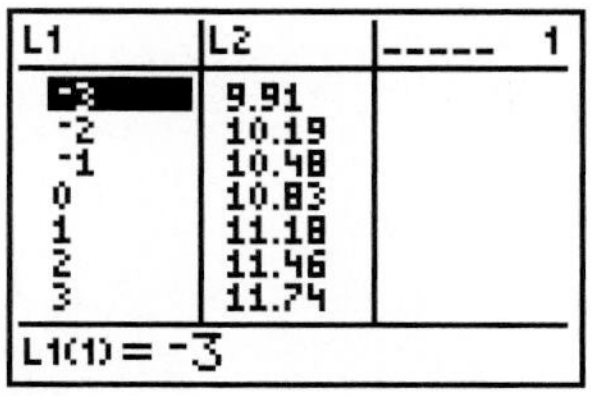

Year	1987	1988	1989	1990	1991	1992
y	9.91	10.19	10.48	10.83	11.18	11.46

Year	1993	1994	1995	1996	1997
y	11.74	12.06	12.37	12.78	13.17

First, let $x = 0$ correspond to 1990 and enter the data into the list editor, as shown in Figure 13. Note that the list in the first column contains the years and the list in the second column contains the hourly wages that correspond to the years. Run your graphing utility's built-in linear regression program to obtain the coefficients a and b for the model $y = ax + b$, as shown in Figure 14. So, a linear model for the data is

$$y \approx 0.321x + 10.83.$$

When you run some regression programs, you may obtain an "r-value," which gives a measure of how well the model fits the data. The closer the value of $|r|$ is to 1, the better the fit. For the data in the table above, $r \approx 0.999$, which implies that the model is a very good fit.

FIGURE 13

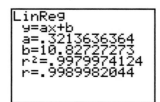

FIGURE 14

Andy Sachs/Tony Stone Images

In 1997, over $24 billion worth of corrugated and solid fiber boxes were produced to create shipping containers for products such as globes as well as containers for food and beverage products. (Source: U.S. Bureau of the Census)

P Prerequisites

► How to Study This Chapter

The Big Picture

In this chapter you will learn the following skills and concepts.

► How to represent and order real numbers and use inequalities

► How to evaluate algebraic expressions using the basic rules of algebra

► How to use properties of exponents and radicals to simplify and evaluate expressions

► How to add, subtract, multiply, and factor polynomials

► How to determine the domains of algebraic expressions and simplify rational expressions

► How to avoid common algebraic errors and use algebraic techniques common in calculus

► How to plot points in the coordinate plane and use the Distance and Midpoint Formulas

Important Vocabulary

As you encounter each new vocabulary term in this chapter, add the term and its definition to your notebook glossary.

Real numbers (p. 2)
Real number line (p. 2)
Origin (p. 2)
Order (p. 3)
Inequality (p. 3)
Infinity (p. 4)
Absolute value (p. 5)
Distance between two numbers (p. 5)
Variables (p. 6)
Algebraic expressions (p. 6)
Constant (p. 6)
Coefficient (p. 6)
Evaluate (p. 6)
Additive inverse (p. 6)
Multiplicative inverse (p. 6)
Factors (p. 8)
Prime number (p. 8)
Composite number (p. 8)
Exponential form (p. 12)
Scientific notation (p. 14)

Square root (p. 15)
Cube root (p. 15)
Principal nth root (p. 15)
Index (p. 15)
Radicand (p. 15)
Simplest form (p. 17)
Conjugate (p. 18)
Rational exponent (p. 19)
Polynomial (p. 25)
Degree (p. 25)
Factoring (p. 34)
Domain (p. 42)
Equivalent (p. 42)
Rational expression (p. 42)
Complex fractions (p. 46)
Rectangular coordinate system (p. 60)
Cartesian plane (p. 60)
Ordered pair (p. 60)
Distance Formula (p. 62)
Midpoint Formula (p. 64)

Study Tools

- Learning objectives at the beginning of each section
- Chapter Summary (p. 70)
- Review Exercises (pp. 71–73)
- Chapter Test (p. 75)

Additional Resources

- Study and Solutions Guide
- Interactive College Algebra
- Videotapes for Chapter P
- College Algebra Website
- Student Success Organizer

STUDY T!P

A good rule of thumb to use when studying algebra is to study 2 to 4 hours for every hour in class. Increase your study time if you do not make the grade you want on your first major test.

P.1 Real Numbers

▶ **What you should learn**

- How to represent and classify real numbers
- How to order real numbers and use inequalities
- How to find the absolute values of real numbers and find the distance between two real numbers
- How to evaluate algebraic expressions
- How to use the basic rules and properties of algebra

▶ **Why you should learn it**

Real numbers are used to represent many real-life quantities, such as the variance of a budget (see Exercises 83–86 on page 10).

Don Smetzer/Tony Stone Images

Real Numbers

Real numbers are used in everyday life to describe quantities such as age, miles per gallon, container size, and population. Real numbers are represented by symbols such as

$$-5, 9, 0, \frac{4}{3}, 0.666\ldots, 28.21, \sqrt{2}, \pi, \text{ and } \sqrt[3]{-32}.$$

Here are some important subsets of the real numbers.

$$\{1, 2, 3, 4, \ldots\} \qquad \text{Set of natural numbers}$$

$$\{0, 1, 2, 3, 4, \ldots\} \qquad \text{Set of whole numbers}$$

$$\{\ldots, -3, -2, -1, 0, 1, 2, 3, \ldots\} \qquad \text{Set of integers}$$

A real number is **rational** if it can be written as the ratio p/q of two integers, where $q \neq 0$. For instance, the numbers

$$\frac{1}{3} = 0.3333\ldots = 0.\overline{3}, \frac{1}{8} = 0.125, \text{ and } \frac{125}{111} = 1.126126\ldots = 1.\overline{126}$$

are rational. The decimal representation of a rational number either repeats (as in $\frac{173}{55} = 3.1\overline{45}$) or terminates (as in $\frac{1}{2} = 0.5$). A real number that cannot be written as the ratio of two integers is called **irrational.** Irrational numbers have infinite nonrepeating decimal representations. For instance, the numbers

$$\sqrt{2} \approx 1.4142136 \quad \text{and} \quad \pi \approx 3.1415927$$

are irrational. (The symbol $\approx$ means "is approximately equal to.")

Real numbers are represented graphically by a **real number line.** The point 0 on the real number line is the **origin.** Numbers to the right of 0 are positive, and numbers to the left of 0 are negative, as shown in Figure P.1. The term **nonnegative** describes a number that is either positive or zero.

FIGURE P.1 *The Real Number Line*

As illustrated in Figure P.2, there is a *one-to-one correspondence* between real numbers and points on the real number line.

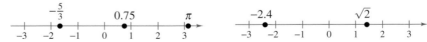

Every real number corresponds to exactly one point on the real number line.

FIGURE P.2 *One-to-One Correspondence*

Every point on the real number line corresponds to exactly one real number.

Ordering Real Numbers

One important property of real numbers is that they are *ordered*.

Definition of Order on the Real Number Line

If a and b are real numbers, a is *less than b* if $b - a$ is positive. The **order** of a and b is denoted by the **inequality**

$$a < b.$$

This relationship can also be described by saying that b is *greater than a* and writing $b > a$. The inequality $a \leq b$ means that a is *less than or equal to b*, and the inequality $b \geq a$ means that b is *greater than or equal to a*. The symbols $<$, $>$, $\leq$, and $\geq$ are *inequality symbols*.

FIGURE P.3 $a < b$ if and only if a lies to the left of b.

Geometrically, this definition implies that $a < b$ if and only if a lies to the *left* of b on the real number line, as shown in Figure P.3.

Example 1 ▶ Interpreting Inequalities

Describe the subset of real numbers represented by each inequality.

a. $x \leq 2$ **b.** $-2 \leq x < 3$

Solution

a. The inequality $x \leq 2$ denotes all real numbers less than or equal to 2, as shown in Figure P.4(a).

b. The inequality $-2 \leq x < 3$ means that $x \geq -2$ and $x < 3$. This "double inequality" denotes all real numbers between -2 and 3, including -2 but not including 3, as shown in Figure P.4(b).

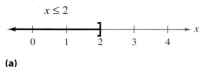

(a)

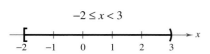

(b)

FIGURE P.4

A computer animation of this example appears in the *Interactive* CD-ROM and *Internet* versions of this text.

Inequalities can be used to describe subsets of real numbers called **intervals**. In the bounded intervals below, the real numbers a and b are the **endpoints** of each interval.

Bounded Intervals on the Real Number Line

Notation	Interval Type	Inequality	Graph
$[a, b]$	Closed	$a \leq x \leq b$	
(a, b)	Open	$a < x < b$	
$[a, b)$		$a \leq x < b$	
$(a, b]$		$a < x \leq b$	

The symbols ∞, **positive infinity,** and $-\infty$, **negative infinity,** do not represent real numbers. They are simply convenient symbols used to describe the unboundedness of an interval such as $(1, \infty)$ or $(-\infty, 3]$.

Unbounded Intervals on the Real Number Line

Notation	Interval Type	Inequality	Graph
$[a, \infty)$		$x \geq a$	
(a, ∞)	Open	$x > a$	
$(-\infty, b]$		$x \leq b$	
$(-\infty, b)$	Open	$x < b$	
$(-\infty, \infty)$	Entire real line		

Example 2 ▶ Using Inequalities to Represent Intervals

Use inequality notation to describe each of the following.

a. c is at most 2.

b. m is at least -3.

c. All x in the interval $(-3, 5]$

Solution

a. The statement "c is at most 2" can be represented by $c \leq 2$.

b. The statement "m is at least -3" can be represented by $m \geq -3$.

c. "All x in the interval $(-3, 5]$" can be represented by $-3 < x \leq 5$.

Example 3 ▶ Interpreting Intervals

Give a verbal description of each interval.

a. $(-1, 0)$ **b.** $[2, \infty)$ **c.** $(-\infty, 0)$

Solution

a. This interval consists of all real numbers that are greater than -1 and less than 0.

b. This interval consists of all real numbers that are greater than or equal to 2.

c. This interval consists of all real numbers that are less than zero (the negative real numbers).

The **Law of Trichotomy** states that for any two real numbers a and b, *precisely* one of three relationships is possible:

$$a = b, \quad a < b, \quad \text{or} \quad a > b. \qquad \text{Law of Trichotomy}$$

Absolute Value and Distance

The **absolute value** of a real number is its *magnitude*, or the distance between the origin and the point representing the real number on the real number line.

Exploration

Absolute value expressions can be evaluated on a graphing utility. When an expression such as $|3 - 8|$ is evaluated, parentheses should surround the expression, as in abs$(3 - 8)$. Evaluate each expression below. What can you conclude?

a. $|6|$ **b.** $|-1|$

c. $|5 - 2|$ **d.** $|2 - 5|$

Definition of Absolute Value

If a is a real number, then the absolute value of a is

$$|a| = \begin{cases} a, & \text{if } a \geq 0 \\ -a, & \text{if } a < 0. \end{cases}$$

Notice in this definition that the absolute value of a real number is never negative. For instance, if $a = -5$, then $|-5| = -(-5) = 5$. The absolute value of a real number is either positive or zero. Moreover, 0 is the only real number whose absolute value is 0. So, $|0| = 0$.

Example 4 ▶ Evaluating the Absolute Value of a Number

Evaluate $\dfrac{|x|}{x}$ for (a) $x > 0$ and (b) $x < 0$.

Solution

a. If $x > 0$, then $|x| = x$ and $\dfrac{|x|}{x} = \dfrac{x}{x} = 1$.

b. If $x < 0$, then $|x| = -x$ and $\dfrac{|x|}{x} = \dfrac{-x}{x} = -1$.

Properties of Absolute Values

1. $|a| \geq 0$ **2.** $|-a| = |a|$

3. $|ab| = |a||b|$ **4.** $\left|\dfrac{a}{b}\right| = \dfrac{|a|}{|b|}, \quad b \neq 0$

Distance Between Two Points on the Real Line

Let a and b be real numbers. The **distance between a and b** is

$$d(a, b) = |b - a| = |a - b|.$$

Absolute value can be used to define the distance between two points on the real number line. For instance, the distance between -3 and 4 is

$$|-3 - 4| = |-7| = 7,$$

as shown in Figure P.5.

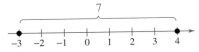

FIGURE P.5 *The distance between -3 and 4 is 7.*

Algebraic Expressions

One characteristic of algebra is the use of letters to represent numbers. The letters are **variables,** and combinations of letters and numbers are **algebraic expressions.** Here are a few examples of algebraic expressions.

$$5x, \qquad 2x - 3, \qquad \frac{4}{x^2 + 2}, \qquad 7x + y$$

> ### Definition of an Algebraic Expression
> An **algebraic expression** is a collection of letters (**variables**) and real numbers (**constants**) combined using the operations of addition, subtraction, multiplication, division, and exponentiation.

The **terms** of an algebraic expression are those parts that are separated by *addition.* For example,

$$x^2 - 5x + 8 = x^2 + (-5x) + 8$$

has three terms: x^2 and $-5x$ are the **variable terms** and 8 is the **constant term.** The numerical factor of a variable term is the **coefficient** of the variable term. For instance, the coefficient of $-5x$ is -5, and the coefficient of x^2 is 1.

To **evaluate** an algebraic expression, substitute numerical values for each of the variables in the expression. Here are two examples.

Expression	Value of Variable	Substitute	Value of Expression
$-3x + 5$	$x = 3$	$-3(3) + 5$	$-9 + 5 = -4$
$3x^2 + 2x - 1$	$x = -1$	$3(-1)^2 + 2(-1) - 1$	$3 - 2 - 1 = 0$

When an algebraic expression is evaluated, the **Substitution Principle** is used. It states that "If $a = b$, then a can be replaced by b in any expression involving a." In the first evaluation shown above, for instance, 3 is *substituted* for x in the expression $-3x + 5$.

A computer animation of this concept appears in the *Interactive* CD-ROM and *Internet* versions of this text.

Basic Rules of Algebra

There are four arithmetic operations with real numbers: *addition, multiplication, subtraction,* and *division,* denoted by the symbols $+, \times$ or $\cdot, -$, and $\div$. Of these, addition and multiplication are the two primary operations. Subtraction and division are the inverse operations of addition and multiplication, respectively.

Subtraction: Add the opposite.

$$a - b = a + (-b)$$

Division: Multiply by the reciprocal.

If $b \neq 0$, then $a/b = a\left(\dfrac{1}{b}\right) = \dfrac{a}{b}$.

In these definitions, $-b$ is the **additive inverse** (or opposite) of b, and $1/b$ is the **multiplicative inverse** (or reciprocal) of b. In the fractional form a/b, a is the **numerator** of the fraction and b is the **denominator.**

Because the properties of real numbers on page 7 are true for variables and algebraic expressions as well as for real numbers, they are often called the **Basic Rules of Algebra.**

STUDY T!P

Try to formulate a verbal description of each property. For instance, the first property states that *the order in which two real numbers are added does not affect their sum.*

Basic Rules of Algebra

Let a, b, and c be real numbers, variables, or algebraic expressions.

Property		*Example*
Commutative Property of Addition:	$a + b = b + a$	$4x + x^2 = x^2 + 4x$
Commutative Property of Multiplication:	$ab = ba$	$(4 - x)x^2 = x^2(4 - x)$
Associative Property of Addition:	$(a + b) + c = a + (b + c)$	$(x + 5) + x^2 = x + (5 + x^2)$
Associative Property of Multiplication:	$(ab)c = a(bc)$	$(2x \cdot 3y)(8) = (2x)(3y \cdot 8)$
Distributive Properties:	$a(b + c) = ab + ac$	$3x(5 + 2x) = 3x \cdot 5 + 3x \cdot 2x$
	$(a + b)c = ac + bc$	$(y + 8)y = y \cdot y + 8 \cdot y$
Additive Identity Property:	$a + 0 = a$	$5y^2 + 0 = 5y^2$
Multiplicative Identity Property:	$a \cdot 1 = a$	$(4x^2)(1) = 4x^2$
Additive Inverse Property:	$a + (-a) = 0$	$5x^3 + (-5x^3) = 0$
Multiplicative Inverse Property:	$a \cdot \dfrac{1}{a} = 1, \qquad a \neq 0$	$(x^2 + 4)\left(\dfrac{1}{x^2 + 4}\right) = 1$

Because subtraction is defined as "adding the opposite," the Distributive Properties are also true for subtraction. For instance, the "subtraction form" of $a(b + c) = ab + ac$ is

$$a(b - c) = ab - ac.$$

Properties of Negation

Let a and b be real numbers, variables, or algebraic expressions.

Property	*Example*
1. $(-1)a = -a$	$(-1)7 = -7$
2. $-(-a) = a$	$-(-6) = 6$
3. $(-a)b = -(ab) = a(-b)$	$(-5)3 = -(5 \cdot 3) = 5(-3)$
4. $(-a)(-b) = ab$	$(-2)(-x) = 2x$
5. $-(a + b) = (-a) + (-b)$	$-(x + 8) = (-x) + (-8)$
	$\qquad\qquad = -x - 8$

Properties of Equality

Let a, b, and c be real numbers, variables, or algebraic expressions.

1. If $a = b$, then $a + c = b + c$.	Add c to each side.
2. If $a = b$, then $ac = bc$.	Multiply each side by c.
3. If $a + c = b + c$, then $a = b$.	Subtract c from each side.
4. If $ac = bc$ and $c \neq 0$, then $a = b$.	Divide each side by c.

Properties of Zero

Let a and b be real numbers, variables, or algebraic expressions.

1. $a + 0 = a$ and $a - 0 = a$ 2. $a \cdot 0 = 0$

3. $\dfrac{0}{a} = 0, \qquad a \neq 0$ 4. $\dfrac{a}{0}$ is undefined.

5. **Zero-Factor Property:** If $ab = 0$, then $a = 0$ or $b = 0$.

Properties and Operations of Fractions

Let a, b, c, and d be real numbers, variables, or algebraic expressions such that $b \neq 0$ and $d \neq 0$.

1. **Equivalent Fractions:** $\dfrac{a}{b} = \dfrac{c}{d}$ if and only if $ad = bc$.

2. **Rules of Signs:** $-\dfrac{a}{b} = \dfrac{-a}{b} = \dfrac{a}{-b}$ and $\dfrac{-a}{-b} = \dfrac{a}{b}$

3. **Generate Equivalent Fractions:** $\dfrac{a}{b} = \dfrac{ac}{bc}, \qquad c \neq 0$

4. **Add or Subtract with Like Denominators:** $\dfrac{a}{b} \pm \dfrac{c}{b} = \dfrac{a \pm c}{b}$

5. **Add or Subtract with Unlike Denominators:** $\dfrac{a}{b} \pm \dfrac{c}{d} = \dfrac{ad \pm bc}{bd}$

6. **Multiply Fractions:** $\dfrac{a}{b} \cdot \dfrac{c}{d} = \dfrac{ac}{bd}$

7. **Divide Fractions:** $\dfrac{a}{b} \div \dfrac{c}{d} = \dfrac{a}{b} \cdot \dfrac{d}{c} = \dfrac{ad}{bc}, \qquad c \neq 0$

Example 5 ► Properties and Operations of Fractions

a. Equivalent fractions: $\dfrac{x}{5} = \dfrac{3 \cdot x}{3 \cdot 5} = \dfrac{3x}{15}$ **b.** Divide fractions: $\dfrac{7}{x} \div \dfrac{3}{2} = \dfrac{7}{x} \cdot \dfrac{2}{3} = \dfrac{14}{3x}$

c. Add fractions with unlike denominators: $\dfrac{x}{3} + \dfrac{2x}{5} = \dfrac{5 \cdot x + 3 \cdot 2x}{15} = \dfrac{11x}{15}$

If a, b, and c are integers such that $ab = c$, then a and b are **factors** or **divisors** of c. A **prime number** is an integer that has exactly two positive factors: itself and 1, such as 2, 3, 5, 7, and 11. The numbers 4, 6, 8, 9, and 10 are **composite** because they can be written as the product of two or more prime numbers. The number 1 is neither prime nor composite. The **Fundamental Theorem of Arithmetic** states that every positive integer greater than 1 can be written as the product of prime numbers in precisely one way (disregarding order). For instance, the *prime factorization* of 24 is $24 = 2 \cdot 2 \cdot 2 \cdot 3$.

The *Interactive* CD-ROM and *Internet* versions of this text contain step-by-step solutions to all odd-numbered Section and Review Exercises. They also provide Tutorial Exercises that link to Guided Examples for additional help.

P.1 Exercises

In Exercises 1–6, determine which numbers are (a) natural numbers, (b) integers, (c) rational numbers, and (d) irrational numbers.

1. $-9, -\frac{7}{2}, 5, \frac{2}{3}, \sqrt{2}, 0, 1, -4, 2, -11$

2. $\sqrt{5}, -7, -\frac{7}{3}, 0, 3.12, \frac{5}{4}, -3, 12, 5$

3. $2.01, 0.666\ldots, -13, 0.010110111\ldots, 1, -6$

4. $2.3030030003\ldots, 0.7575, -4.63, \sqrt{10}, -75, 15, 31$

5. $-\pi, -\frac{1}{3}, \frac{6}{3}, \frac{1}{2}\sqrt{2}, -7.5, -1, 8, -22$

6. $25, -17, -\frac{12}{5}, \sqrt{9}, 3.12, \frac{1}{2}\pi, 7, -11.1, 13$

In Exercises 7–10, use a calculator to find the decimal form of the rational number. If it is a nonterminating decimal, write the repeating pattern.

7. $\frac{5}{8}$

8. $\frac{1}{3}$

9. $\frac{41}{333}$

10. $\frac{6}{11}$

In Exercises 11–16, use a calculator with a fraction feature to write the rational number as the ratio of two integers.

11. 4.1

12. 8.5

13. $10.\overline{2}$

14. $5.\overline{45}$

15. $-2.01\overline{2}$

16. $-1.6\overline{5}$

In Exercises 17 and 18, approximate the numbers and place the correct symbol (< or >) between them.

17.

18.

In Exercises 19–24, plot the two real numbers on the real number line. Then place the appropriate inequality symbol (< or >) between them.

19. $-4, -8$

20. $-3.5, 1$

21. $\frac{3}{2}, 7$

22. $1, \frac{16}{3}$

23. $\frac{5}{6}, \frac{2}{3}$

24. $-\frac{8}{7}, -\frac{3}{7}$

In Exercises 25–34, verbally describe the subset of real numbers represented by the inequality. Then sketch the subset on the real number line. State whether the interval is bounded or unbounded.

25. $x \le 5$

26. $x \ge -2$

27. $x < 0$

28. $x > 3$

29. $x \ge 4$

30. $x < 2$

31. $-2 < x < 2$

32. $0 \le x \le 5$

33. $-1 \le x < 0$

34. $0 < x \le 6$

In Exercises 35 and 36, use a calculator to order the numbers from smallest to largest.

35. $\frac{7071}{5000}, \frac{584}{413}, \sqrt{2}, \frac{47}{33}, \frac{127}{90}$

36. $\frac{26}{15}, \sqrt{3}, 1.7320, \frac{381}{220}, \frac{2103}{1214}$

In Exercises 37–44, use inequality notation to describe the set.

37. All x in the interval $(-2, 4]$

38. All y in the interval $[-6, 0)$

39. y is nonnegative.

40. y is no more than 25.

41. t is at least 10 and at most 22.

42. k is less than 5 but no less than -3.

43. The dog's weight W is more than 65 pounds.

44. The annual rate of inflation r is expected to be at least 2.5% but no more than 5%.

In Exercises 45–48, give a verbal description of the interval.

45. $[0, 8)$

46. $[-5, 7]$

47. $(-6, \infty)$

48. $(-\infty, 4]$

In Exercises 49–58, evaluate the expression.

49. $|-10|$

50. $|0|$

51. $|3 - x|$

52. $|4 - x|$

53. $|-1| - |-2|$

54. $-3 - |-3|$

55. $\dfrac{-5}{|-5|}$

56. $-3|-3|$

57. $\dfrac{|x + 2|}{x + 2}, \quad x < -2$

58. $\dfrac{|x - 1|}{x - 1}, \quad x > 1$

In Exercises 59–64, place the correct symbol (<, >, or =) between the pair of real numbers.

59. $|-3| \quad\rule{1cm}{0.4pt}\quad -|-3|$

60. $|-4| \quad\rule{1cm}{0.4pt}\quad |4|$

61. $-5 \quad\rule{1cm}{0.4pt}\quad -|5|$

62. $-|-6| \quad\rule{1cm}{0.4pt}\quad |-6|$

63. $-|-2| \quad\rule{1cm}{0.4pt}\quad -|2|$

64. $-(-2) \quad\rule{1cm}{0.4pt}\quad -2$

In Exercises 65–72, find the distance between *a* and *b*.

65.

$a = -1$ $b = 3$

66.

$a = -4$ $b = -\frac{3}{2}$

67. $a = 126, b = 75$ **68.** $a = -126, b = -75$

69. $a = -\frac{5}{2}, b = 0$ **70.** $a = \frac{1}{4}, b = \frac{11}{4}$

71. $a = \frac{16}{5}, b = \frac{112}{75}$ **72.** $a = 9.34, b = -5.65$

In Exercises 73 and 74, use the real numbers *A*, *B*, and *C* shown on the number line. Determine the sign of each expression.

          C   *B*   *A*

73. (a) $-A$ **74.** (a) $-C$

     (b) $B - A$ (b) $A - C$

In Exercises 75–82, use absolute value notation to describe the situation.

75. While traveling, you pass milepost 7, then milepost 18. How far do you travel during that time period?

76. While traveling, you pass milepost 103, then milepost 86. How far do you travel during that time period?

77. The temperature was 60° at noon, then 23° at midnight. What was the change in temperature over the 12-hour period?

78. The temperature was 48° last night at midnight, then 82° at noon today. What was the change in temperature over the 12-hour period?

79. The distance between *x* and 5 is no more than 3.

80. The distance between *x* and −10 is at least 6.

81. *y* is at least 6 units from 0.

82. *y* is at most 2 units from *a*.

Budget Variance **In Exercises 83–86, the accounting department of a company is checking to see whether the actual expenses of a department differ from the budgeted expenses by more than $500 or by more than 5%. Fill in the missing parts of the table, and determine whether the actual expense passes the "budget variance test."**

	Budgeted Expense, *b*	Actual Expense, *a*	$|a - b|$	$0.05b$
83. Wages	$112,700	$113,356		
84. Utilities	$9,400	$9,772		
85. Taxes	$37,640	$37,335		
86. Insurance	$2,575	$2,613		

Federal Deficit **In Exercises 87–90, use the bar graph, which shows the receipts of the federal government (in billions of dollars) for selected years from 1960 through 1998. In each exercise you are given the outlay of the federal government. Find the magnitude of the surplus or deficit for the year.** (Source: U.S. Treasury Department)

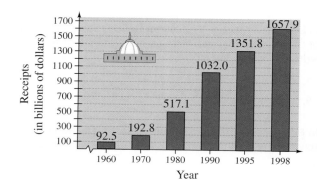

	Receipts	Outlay	$|Receipts - Outlay|$
87. 1960		$92.2 billion	
88. 1980		$590.9 billion	
89. 1990		$1253.2 billion	
90. 1998		$1667.8 billion	

In Exercises 91–96, identify the terms. Then identify the coefficients of the variable terms of the expression.

91. $7x + 4$ **92.** $6x^3 - 5x$

93. $\sqrt{3}x^2 - 8x - 11$ **94.** $3\sqrt{3}x^2 + 1$

95. $4x^3 + \dfrac{x}{2} - 5$ **96.** $3x^4 - \dfrac{x^2}{4}$

In Exercises 97–102, evaluate the expression for each value of *x*. (If not possible, state the reason.)

Expression	Values	
97. $4x - 6$	(a) $x = -1$	(b) $x = 0$
98. $9 - 7x$	(a) $x = -3$	(b) $x = 3$
99. $x^2 - 3x + 4$	(a) $x = -2$	(b) $x = 2$

Expression | *Values*

100. $-x^2 + 5x - 4$ (a) $x = -1$ (b) $x = 1$

101. $\dfrac{x + 1}{x - 1}$ (a) $x = 1$ (b) $x = -1$

102. $\dfrac{x}{x + 2}$ (a) $x = 2$ (b) $x = -2$

In Exercises 103–112, identify the rule(s) of algebra illustrated by the equation.

103. $x + 9 = 9 + x$

104. $2\left(\frac{1}{2}\right) = 1$

105. $\dfrac{1}{h + 6}(h + 6) = 1, \quad h \neq -6$

106. $(x + 3) - (x + 3) = 0$

107. $2(x + 3) = 2x + 6$

108. $(z - 2) + 0 = z - 2$

109. $1 \cdot (1 + x) = 1 + x$

110. $x + (y + 10) = (x + y) + 10$

111. $x(3y) = (x \cdot 3)y = (3x)y$

112. $\frac{1}{7}(7 \cdot 12) = \left(\frac{1}{7} \cdot 7\right)12 = 1 \cdot 12 = 12$

In Exercises 113–120, perform the operation(s). (Write fractional answers in simplest form.)

113. $\frac{3}{16} + \frac{5}{16}$ **114.** $\frac{6}{7} - \frac{4}{7}$

115. $\frac{5}{8} - \frac{5}{12} + \frac{1}{6}$ **116.** $\frac{10}{11} + \frac{6}{33} - \frac{13}{66}$

117. $12 \div \frac{1}{4}$ **118.** $-\left(6 \cdot \frac{4}{8}\right)$

119. $\frac{2x}{3} - \frac{x}{4}$ **120.** $\frac{5x}{6} \cdot \frac{2}{9}$

In Exercises 121–124, use a calculator to evaluate the expression. (Round your answer to two decimal places.)

121. $-3 + \frac{3}{7}$ **122.** $3\left(-\frac{5}{12} + \frac{3}{8}\right)$

123. $\dfrac{11.46 - 5.37}{3.91}$ **124.** $\dfrac{\frac{1}{5}(-8 - 9)}{-\frac{1}{3}}$

125. (a) Use a calculator to complete the table.

n	1	0.5	0.01	0.0001	0.000001
$5/n$					

    (b) Use the result from part (a) to make a conjecture about the value of $5/n$ as n approaches 0.

126. (a) Use a calculator to complete the table.

n	1	10	100	10,000	100,000
$5/n$					

    (b) Use the result from part (a) to make a conjecture about the value of $5/n$ as n increases without bound.

Synthesis

127. *Exploration* Consider $|u + v|$ and $|u| + |v|$.

    (a) Are the values of the expressions always equal? If not, under what conditions are they unequal?

    (b) If the two expressions are not equal for certain values of u and v, is one of the expressions always greater than the other? Explain.

128. *Think About It* Is there a difference between saying that a real number is positive and saying that a real number is nonnegative? Explain.

129. *Think About It* Because every even number is divisible by 2, is it possible that there exist any even prime numbers? Explain.

True or False? In Exercises 130 and 131, determine whether the statement is true or false. Justify your answer.

130. If $a < b$, then $\dfrac{1}{a} < \dfrac{1}{b}$, where $a \neq b \neq 0$.

131. Because $\dfrac{a + b}{c} = \dfrac{a}{c} + \dfrac{b}{c}$, then $\dfrac{c}{a + b} = \dfrac{c}{a} + \dfrac{c}{b}$.

132. *Writing* Describe the differences among the sets of natural numbers, integers, rational numbers, and irrational numbers.

133. *Writing* You may hear it said that to take the absolute value of a real number you simply remove any negative sign and make the number positive. Can it ever be true that $|a| = -a$ for a real number a? Explain.

P.2   Exponents and Radicals

▶ **What you should learn**

- How to use properties of exponents
- How to use scientific notation to represent real numbers
- How to use properties of radicals
- How to simplify and combine radicals
- How to rationalize denominators and numerators
- How to use properties of rational exponents

▶ **Why you should learn it**

Real numbers and algebraic expressions are often written with exponents and radicals. For instance, in Exercise 122 on page 23, you will use an expression involving a radical to find the size of a particle that can be carried by a stream moving at a certain velocity.

Marc Muench/Tony Stone Images

Exponents

Repeated *multiplication* can be written in **exponential form.**

Repeated Multiplication	*Exponential Form*
$a \cdot a \cdot a \cdot a \cdot a$	a^5
$(-4)(-4)(-4)$	$(-4)^3$
$(2x)(2x)(2x)(2x)$	$(2x)^4$

In general, if a is a real number and n is a positive integer, then

$$a^n = \underbrace{a \cdot a \cdot a \cdots a}_{n \text{ factors}}$$

where n is the **exponent** and a is the **base.** The expression a^n is read "a to the nth **power.**" In Property 3 below, be sure you see how to use a negative exponent.

Properties of Exponents

Let a and b be real numbers, variables, or algebraic expressions, and let m and n be integers. (All denominators and bases are nonzero.)

Property	*Example*
1. $a^m a^n = a^{m+n}$	$3^2 \cdot 3^4 = 3^{2+4} = 3^6 = 729$
2. $\dfrac{a^m}{a^n} = a^{m-n}$	$\dfrac{x^7}{x^4} = x^{7-4} = x^3$
3. $a^{-n} = \dfrac{1}{a^n} = \left(\dfrac{1}{a}\right)^n$	$y^{-4} = \dfrac{1}{y^4} = \left(\dfrac{1}{y}\right)^4$
4. $a^0 = 1, \quad a \neq 0$	$(x^2 + 1)^0 = 1$
5. $(ab)^m = a^m b^m$	$(5x)^3 = 5^3 x^3 = 125x^3$
6. $(a^m)^n = a^{mn}$	$(y^3)^{-4} = y^{3(-4)} = y^{-12} = \dfrac{1}{y^{12}}$
7. $\left(\dfrac{a}{b}\right)^m = \dfrac{a^m}{b^m}$	$\left(\dfrac{2}{x}\right)^3 = \dfrac{2^3}{x^3} = \dfrac{8}{x^3}$
8. $\lvert a^2 \rvert = \lvert a \rvert^2 = a^2$	$\lvert (-2)^2 \rvert = \lvert -2 \rvert^2 = (-2)^2 = 4$

It is important to recognize the difference between expressions such as $(-2)^4$ and -2^4. In $(-2)^4$, the parentheses indicate that the exponent applies to the negative sign as well as to the 2, but in $-2^4 = -(2^4)$, the exponent applies only to the 2. So,

$$(-2)^4 = 16 \quad \text{and} \quad -2^4 = -16.$$

STUDY T!P

Rarely in algebra is there only one way to solve a problem. Don't be concerned if the steps you use to solve a problem are not exactly the same as the steps presented in this text. The important thing is to use steps that you understand and, of course, steps that are justified by the rules of algebra. For instance, you might prefer the following steps for Example 2(d).

$$\left(\frac{3x^2}{y}\right)^{-2} = \left(\frac{y}{3x^2}\right)^2 = \frac{y^2}{9x^4}$$

Note how Property 3 is used in the first step of this solution. The fractional form of this property is

$$\left(\frac{a}{b}\right)^{-m} = \left(\frac{b}{a}\right)^m.$$

The properties of exponents listed on the preceding page apply to *all* integers m and n, not just to positive integers. For instance, by Property 2, you can write

$$\frac{3^4}{3^{-5}} = 3^{4-(-5)} = 3^{4+5} = 3^9.$$

Example 1 ▶ Using Properties of Exponents

Use the properties of exponents to simplify each expression.

a. $(-3ab^4)(4ab^{-3})$ **b.** $(2xy^2)^3$ **c.** $3a(-4a^2)^0$ **d.** $\left(\frac{5x^3}{y}\right)^2$

Solution

a. $(-3ab^4)(4ab^{-3}) = -12(a)(a)(b^4)(b^{-3}) = -12a^2b$
b. $(2xy^2)^3 = 2^3(x)^3(y^2)^3 = 8x^3y^6$
c. $3a(-4a^2)^0 = 3a(1) = 3a, \quad a \neq 0$
d. $\left(\frac{5x^3}{y}\right)^2 = \frac{5^2(x^3)^2}{y^2} = \frac{25x^6}{y^2}$

Example 2 ▶ Rewriting with Positive Exponents

Rewrite each expression with positive exponents.

a. x^{-1} **b.** $\frac{1}{3x^{-2}}$ **c.** $\frac{12a^3b^{-4}}{4a^{-2}b}$ **d.** $\left(\frac{3x^2}{y}\right)^{-2}$

Solution

a. $x^{-1} = \frac{1}{x}$ Property 3

b. $\frac{1}{3x^{-2}} = \frac{1(x^2)}{3} = \frac{x^2}{3}$ The exponent -2 does not apply to 3.

c. $\frac{12a^3b^{-4}}{4a^{-2}b} = \frac{12a^3 \cdot a^2}{4b \cdot b^4}$ Property 3

    $= \frac{3a^5}{b^5}$ Property 1

d. $\left(\frac{3x^2}{y}\right)^{-2} = \frac{3^{-2}(x^2)^{-2}}{y^{-2}}$ Properties 5 and 7

    $= \frac{3^{-2}x^{-4}}{y^{-2}}$ Property 6

    $= \frac{y^2}{3^2x^4}$ Property 3

    $= \frac{y^2}{9x^4}$ Simplify.

Technology

You can use a calculator to evaluate expressions with exponents. For instance, evaluate -3^{-2} as follows.
Scientific:

3 [+/−] [yˣ] 2 [+/−] [=]

Graphing:

[(−)] 3 [^] [(−)] 2

The display will be as follows.

−.1111111111

Historical Note
The French mathematician Nicolas Chuquet (ca. 1500) wrote *Triparty en la science des nombres*, in which a form of exponent notation was used. Our expressions $6x^3$ and $10x^2$ were written as .6.³ and .10.². Zero and negative exponents were also represented, so x^0 would be written as .1.⁰ and $3x^{-2}$ as .3.²ᵐ. Chuquet wrote that .72.¹ divided by .8.³ is .9.²ᵐ. That is, $72x \div 8x^3 = 9x^{-2}$.

Scientific Notation

Exponents provide an efficient way of writing and computing with very large (or very small) numbers. For instance, there are about 326 billion billion gallons of water on earth—that is, 326 followed by 18 zeros.

$$326,000,000,000,000,000,000$$

It is convenient to write such numbers in **scientific notation.** This notation has the form $\pm c \times 10^n$, where $1 \leq c < 10$ and n is an integer. So, the number of gallons of water on earth can be written in scientific notation as

$$3.26 \times 100,000,000,000,000,000,000 = 3.26 \times 10^{20}.$$

The *positive* exponent 20 indicates that the number is *large* (10 or more) and that the decimal point has been moved 20 places. A *negative* exponent indicates that the number is *small* (less than 1). For instance, the mass (in grams) of one electron is approximately

$$9.0 \times 10^{-28} = 0.0000000000000000000000000009.$$

28 decimal places

Example 3 ▶ Scientific Notation

a. $1.345 \times 10^2 = 134.5$

b. $0.0000782 = 7.82 \times 10^{-5}$

c. $9.36 \times 10^{-6} = 0.00000936$

d. $836,100,000 = 8.361 \times 10^8$

Most calculators switch to scientific notation when they are showing large (or small) numbers that exceed the display range. Try evaluating $86,500,000 \times 6000$. If your calculator follows standard conventions, its display should be

$$\boxed{5.19 \quad 11} \quad \text{or} \quad \boxed{5.19 \quad E \quad 11}$$

which is 5.19×10^{11}.

 The *Interactive* CD-ROM and *Internet* versions of this text offer a built-in graphing calculator, which can be used in the Examples, Explorations, Technology notes, and Exercises.

Example 4 ▶ Using Scientific Notation with a Calculator

Use a calculator to evaluate $65,000 \times 3,400,000,000$.

Solution

Because $65,000 = 6.5 \times 10^4$ and $3,400,000,000 = 3.4 \times 10^9$, you can multiply the two numbers using the following graphing calculator steps.

$$6.5 \boxed{EE} 4 \boxed{\times} 3.4 \boxed{EE} 9 \boxed{ENTER}$$

After entering these keystrokes, the calculator display should read $\boxed{2.21 \quad E \quad 14}$. Therefore, the product of the two numbers is

$$(6.5 \times 10^4)(3.4 \times 10^9) = 2.21 \times 10^{14}$$

$$= 221,000,000,000,000.$$

Radicals and Their Properties

A **square root** of a number is one of its two equal factors. For example, 5 is a square root of 25 because 5 is one of the two equal factors of 25. In a similar way, a **cube root** of a number is one of its three equal factors, as in $125 = 5^3$.

Definition of *n*th Root of a Number

Let a and b be real numbers and let $n \geq 2$ be a positive integer. If

$$a = b^n$$

then b is an ***n*th root of a.** If $n = 2$, the root is a **square root.** If $n = 3$, the root is a **cube root.**

Some numbers have more than one *n*th root. For example, both 5 and -5 are square roots of 25. The *principal square root* of 25, written as $\sqrt{25}$, is the positive root, 5. The **principal *n*th root** of a number is defined as follows.

Principal *n*th Root of a Number

Let a be a real number that has at least one *n*th root. The **principal *n*th root of a** is the *n*th root that has the same sign as a. It is denoted by a **radical symbol**

$$\sqrt[n]{a}. \qquad \text{Principal } n\text{th root}$$

The positive integer n is the **index** of the radical, and the number a is the **radicand.** If $n = 2$, we omit the index and write $\sqrt{a}$ rather than $\sqrt[2]{a}$. (The plural of index is *indices*.)

A common misunderstanding is that the square root sign implies both negative and positive roots. This is not correct. The square root sign implies only a positive root. When a negative root is needed, you must use the negative sign with the square root sign.

Incorrect: $\sqrt{4} = \pm 2$ *Correct:* $-\sqrt{4} = -2$ *and* $\sqrt{4} = 2$

Example 5 ► Evaluating Expressions Involving Radicals

a. $\sqrt{36} = 6$ because $6^2 = 36$.

b. $-\sqrt{36} = -6$ because $6^2 = 36$.

c. $\sqrt[3]{\dfrac{125}{64}} = \dfrac{5}{4}$ because $\left(\dfrac{5}{4}\right)^3 = \dfrac{5^3}{4^3} = \dfrac{125}{64}$.

d. $\sqrt[5]{-32} = -2$ because $(-2)^5 = -32$.

e. $\sqrt[4]{-81}$ is not a real number because there is no real number that can be raised to the fourth power to produce -81.

Here are some generalizations about the *n*th roots of a real number.

Generalizations About *n*th Roots of Real Numbers

Real number *a*	Integer *n*	Root(s) of *a*	Example
$a > 0$	$n > 0$, *n* is even.	$\sqrt[n]{a}, -\sqrt[n]{a}$	$\sqrt[4]{81} = 3,\ -\sqrt[4]{81} = -3$
$a > 0$ or $a < 0$	*n* is odd.	$\sqrt[n]{a}$	$\sqrt[3]{-8} = -2$
$a < 0$	*n* is even.	No real roots	$\sqrt{-4}$ is not real.
$a = 0$	*n* is even or odd.	$\sqrt[n]{0} = 0$	$\sqrt[5]{0} = 0$

Integers such as 1, 4, 9, 16, 25, and 36 are called **perfect squares** because they have integer square roots. Similarly, integers such as 1, 8, 27, 64, and 125 are called **perfect cubes** because they have integer cube roots.

Properties of Radicals

Let *a* and *b* be real numbers, variables, or algebraic expressions such that the indicated roots are real numbers, and let *m* and *n* be positive integers.

Property	*Example*
1. $\sqrt[n]{a^m} = \left(\sqrt[n]{a}\right)^m$	$\sqrt[3]{8^2} = \left(\sqrt[3]{8}\right)^2 = (2)^2 = 4$
2. $\sqrt[n]{a} \cdot \sqrt[n]{b} = \sqrt[n]{ab}$	$\sqrt{5} \cdot \sqrt{7} = \sqrt{5 \cdot 7} = \sqrt{35}$
3. $\dfrac{\sqrt[n]{a}}{\sqrt[n]{b}} = \sqrt[n]{\dfrac{a}{b}},\quad b \neq 0$	$\dfrac{\sqrt[4]{27}}{\sqrt[4]{9}} = \sqrt[4]{\dfrac{27}{9}} = \sqrt[4]{3}$
4. $\sqrt[m]{\sqrt[n]{a}} = \sqrt[mn]{a}$	$\sqrt[3]{\sqrt{10}} = \sqrt[6]{10}$
5. $\left(\sqrt[n]{a}\right)^n = a$	$\left(\sqrt{3}\right)^2 = 3$
6. For *n* even, $\sqrt[n]{a^n} = |a|$.	$\sqrt{(-12)^2} = |-12| = 12$
For *n* odd, $\sqrt[n]{a^n} = a$.	$\sqrt[3]{(-12)^3} = -12$

A common special case of Property 6 is $\sqrt{a^2} = |a|$.

Example 6 ▶ Using Properties of Radicals

Use the properties of radicals to simplify each expression.

a. $\sqrt{8} \cdot \sqrt{2}$ **b.** $\left(\sqrt[3]{5}\right)^3$ **c.** $\sqrt[3]{x^3}$ **d.** $\sqrt[6]{y^6}$

Solution

a. $\sqrt{8} \cdot \sqrt{2} = \sqrt{8 \cdot 2} = \sqrt{16} = 4$

b. $\left(\sqrt[3]{5}\right)^3 = 5$

c. $\sqrt[3]{x^3} = x$

d. $\sqrt[6]{y^6} = |y|$

Simplifying Radicals

An expression involving radicals is in **simplest form** when the following conditions are satisfied.

1. All possible factors have been removed from the radical.
2. All fractions have radical-free denominators (accomplished by a process called *rationalizing the denominator*).
3. The index of the radical is reduced.

To simplify a radical, factor the radicand into factors whose exponents are multiples of the index. The roots of these factors are written outside the radical, and the "leftover" factors make up the new radicand.

Example 7 ▶ Simplifying Even Roots

$$\overset{\substack{\text{Perfect}\\\text{4th power}}}{\qquad}\quad\overset{\substack{\text{Leftover}\\\text{factor}}}{\qquad}$$

a. $\sqrt[4]{48} = \sqrt[4]{16 \cdot 3} = \sqrt[4]{2^4 \cdot 3} = 2\sqrt[4]{3}$

$$\overset{\substack{\text{Perfect}\\\text{square}}}{\qquad}\quad\overset{\substack{\text{Leftover}\\\text{factor}}}{\qquad}$$

b. $\sqrt{75x^3} = \sqrt{25x^2 \cdot 3x}$ Find largest square factor.

$\qquad\quad = \sqrt{(5x)^2 \cdot 3x}$

$\qquad\quad = 5x\sqrt{3x}$ Find root of perfect square.

c. $\sqrt[4]{(5x)^4} = |5x| = 5|x|$

In Example 7(b), the expression $\sqrt{75x^3}$ makes sense only for nonnegative values of x.

Example 8 ▶ Simplifying Odd Roots

$$\overset{\substack{\text{Perfect}\\\text{cube}}}{\qquad}\quad\overset{\substack{\text{Leftover}\\\text{factor}}}{\qquad}$$

a. $\sqrt[3]{24} = \sqrt[3]{8 \cdot 3} = \sqrt[3]{2^3 \cdot 3} = 2\sqrt[3]{3}$

$$\overset{\substack{\text{Perfect}\\\text{cube}}}{\qquad}\quad\overset{\substack{\text{Leftover}\\\text{factor}}}{\qquad}$$

b. $\sqrt[3]{24a^4} = \sqrt[3]{8a^3 \cdot 3a}$ Find largest cube factor.

$\qquad\quad = \sqrt[3]{(2a)^3 \cdot 3a}$

$\qquad\quad = 2a\sqrt[3]{3a}$ Find root of perfect cube.

c. $\sqrt[3]{-40x^6} = \sqrt[3]{(-8x^6) \cdot 5}$ Find largest cube factor.

$\qquad\quad = \sqrt[3]{(-2x^2)^3 \cdot 5}$

$\qquad\quad = -2x^2\sqrt[3]{5}$ Find root of perfect cube.

Radical expressions can be combined (added or subtracted) if they are **like radicals**—that is, if they have the same index and radicand. For instance, $\sqrt{2}$, $3\sqrt{2}$, and $\frac{1}{2}\sqrt{2}$ are like radicals, but $\sqrt{3}$ and $\sqrt{2}$ are unlike radicals. To determine whether two radicals can be combined, you should first simplify each radical.

Example 9 ▶ Combining Radicals

a.
$$\begin{aligned}
2\sqrt{48} - 3\sqrt{27} &= 2\sqrt{16 \cdot 3} - 3\sqrt{9 \cdot 3} &&\text{Find square factors.}\\
&= 8\sqrt{3} - 9\sqrt{3} &&\text{Find square roots.}\\
&= (8 - 9)\sqrt{3} &&\text{Combine like terms.}\\
&= -\sqrt{3}
\end{aligned}$$

b.
$$\begin{aligned}
\sqrt[3]{16x} - \sqrt[3]{54x^4} &= \sqrt[3]{8 \cdot 2x} - \sqrt[3]{27 \cdot x^3 \cdot 2x} &&\text{Find cube factors.}\\
&= 2\sqrt[3]{2x} - 3x\sqrt[3]{2x} &&\text{Find cube roots.}\\
&= (2 - 3x)\sqrt[3]{2x} &&\text{Combine like terms.}
\end{aligned}$$

Rationalizing Denominators and Numerators

To rationalize a denominator or numerator of the form $a - b\sqrt{m}$ or $a + b\sqrt{m}$, multiply both numerator and denominator by a **conjugate:** $a + b\sqrt{m}$ and $a - b\sqrt{m}$ are conjugates of each other. If $a = 0$, then the rationalizing factor for $\sqrt{m}$ is itself, $\sqrt{m}$. For cube roots, choose a rationalizing factor that generates a perfect cube.

Example 10 ▶ Rationalizing Single-Term Denominators

Rationalize the denominator of each expression.

a. $\dfrac{5}{2\sqrt{3}}$ **b.** $\dfrac{2}{\sqrt[3]{5}}$

Solution

a.
$$\begin{aligned}
\frac{5}{2\sqrt{3}} &= \frac{5}{2\sqrt{3}} \cdot \frac{\sqrt{3}}{\sqrt{3}} &&\sqrt{3} \text{ is rationalizing factor.}\\
&= \frac{5\sqrt{3}}{2(3)}\\
&= \frac{5\sqrt{3}}{6}
\end{aligned}$$

b.
$$\begin{aligned}
\frac{2}{\sqrt[3]{5}} &= \frac{2}{\sqrt[3]{5}} \cdot \frac{\sqrt[3]{5^2}}{\sqrt[3]{5^2}} &&\sqrt[3]{5^2} \text{ is rationalizing factor.}\\
&= \frac{2\sqrt[3]{5^2}}{\sqrt[3]{5^3}}\\
&= \frac{2\sqrt[3]{25}}{5}
\end{aligned}$$

Example 11 ▶ Rationalizing a Denominator with Two Terms

$$\frac{2}{3 + \sqrt{7}} = \frac{2}{3 + \sqrt{7}} \cdot \frac{3 - \sqrt{7}}{3 - \sqrt{7}}$$

Multiply numerator and denominator by conjugate of denominator.

$$= \frac{2(3 - \sqrt{7})}{3(3) + 3(-\sqrt{7}) + \sqrt{7}(3) - (\sqrt{7})(\sqrt{7})}$$

Use Distributive Property.

$$= \frac{2(3 - \sqrt{7})}{(3)^2 - (\sqrt{7})^2}$$

Simplify.

$$= \frac{2(3 - \sqrt{7})}{9 - 7}$$

Square terms of denominator.

$$= \frac{2(3 - \sqrt{7})}{2} = 3 - \sqrt{7}$$

Simplify.

Sometimes it is necessary to rationalize the numerator of an expression. For instance, in Section P.5 you will use the technique shown in the next example to rationalize the numerator of an expression from calculus.

Example 12 ▶ Rationalizing a Numerator

$$\frac{\sqrt{5} - \sqrt{7}}{2} = \frac{\sqrt{5} - \sqrt{7}}{2} \cdot \frac{\sqrt{5} + \sqrt{7}}{\sqrt{5} + \sqrt{7}}$$

Multiply numerator and denominator by conjugate of numerator.

$$= \frac{(\sqrt{5})^2 - (\sqrt{7})^2}{2(\sqrt{5} + \sqrt{7})}$$

Simplify.

$$= \frac{5 - 7}{2(\sqrt{5} + \sqrt{7})}$$

Square terms of numerator.

$$= \frac{-2}{2(\sqrt{5} + \sqrt{7})} = \frac{-1}{\sqrt{5} + \sqrt{7}}$$

Simplify.

Rational Exponents

Definition of Rational Exponents

If a is a real number and n is a positive integer such that the principal nth root of a exists, then $a^{1/n}$ is defined as

$$a^{1/n} = \sqrt[n]{a}, \text{ where } 1/n \text{ is the } \textbf{rational exponent} \text{ of } a.$$

Moreover, if m is a positive integer that has no common factor with n, then

$$a^{m/n} = (a^{1/n})^m = (\sqrt[n]{a})^m \quad \text{and} \quad a^{m/n} = (a^m)^{1/n} = \sqrt[n]{a^m}.$$

The symbol ⊕ indicates an example or exercise that highlights algebraic techniques specifically used in calculus.

The *Interactive* CD-ROM and *Internet* versions of this text show every example with its solution; clicking on the *Try It!* button brings up similar problems. Guided Examples and Integrated Examples show step-by-step solutions to additional examples. Integrated Examples are related to several concepts in the section.

Technology

There are four methods of evaluating radicals on most graphing calculators. For square roots, you can use the square root key ⬚. For cube roots, you can use the cube root key ⬚. For other roots, you can first convert the radical to exponential form and then use the exponential key ⬚, or you can use the *n*th root key ⬚.

The numerator of a rational exponent denotes the *power* to which the base is raised, and the denominator denotes the *index* or the *root* to be taken.

$$b^{m/n} = \left(\sqrt[n]{b}\right)^m = \sqrt[n]{b^m}$$

When you are working with rational exponents, the properties of integer exponents still apply. For instance,

$$2^{1/2}2^{1/3} = 2^{(1/2)+(1/3)} = 2^{5/6}.$$

Example 13 ▶ Changing from Radical to Exponential Form

a. $\sqrt{3} = 3^{1/2}$

b. $\sqrt{(3xy)^5} = \sqrt[2]{(3xy)^5} = (3xy)^{(5/2)}$

c. $2x\sqrt[4]{x^3} = (2x)(x^{3/4}) = 2x^{1+(3/4)} = 2x^{7/4}$

Example 14 ▶ Changing from Exponential to Radical Form

a. $(x^2 + y^2)^{3/2} = \left(\sqrt{x^2 + y^2}\right)^3 = \sqrt{(x^2 + y^2)^3}$

b. $2y^{3/4}z^{1/4} = 2(y^3z)^{1/4} = 2\sqrt[4]{y^3z}$

c. $a^{-3/2} = \dfrac{1}{a^{3/2}} = \dfrac{1}{\sqrt{a^3}}$

d. $x^{0.2} = x^{1/5} = \sqrt[5]{x}$

Rational exponents are useful for evaluating roots of numbers on a calculator, for reducing the index of a radical, and for simplifying calculus expressions.

Example 15 ▶ Simplifying with Rational Exponents

a. $(-32)^{-4/5} = \left(\sqrt[5]{-32}\right)^{-4} = (-2)^{-4} = \dfrac{1}{(-2)^4} = \dfrac{1}{16}$

b. $(-5x^{5/3})(3x^{-3/4}) = -15x^{(5/3)-(3/4)} = -15x^{11/12}, \qquad x \neq 0$

c. $\sqrt[9]{a^3} = a^{3/9} = a^{1/3} = \sqrt[3]{a}$ Reduce index.

d. $\sqrt[3]{\sqrt{125}} = \sqrt[6]{125} = \sqrt[6]{(5)^3} = 5^{3/6} = 5^{1/2} = \sqrt{5}$

e. $(2x - 1)^{4/3}(2x - 1)^{-1/3} = (2x - 1)^{(4/3)-(1/3)}$

$$= 2x - 1, \qquad x \neq \frac{1}{2}$$

f. $\dfrac{x - 1}{(x - 1)^{-1/2}} = \dfrac{x - 1}{(x - 1)^{-1/2}} \cdot \dfrac{(x - 1)^{1/2}}{(x - 1)^{1/2}}$

$$= \frac{(x - 1)^{3/2}}{(x - 1)^0}$$

$$= (x - 1)^{3/2}, \qquad x \neq 1$$

P.2 Exercises

In Exercises 1–4, write the expression as a repeated multiplication problem.

1. 8^5

2. $(-2)^7$

3. -0.4^6

4. 11.3^4

In Exercises 5–8, write the expression using exponential notation.

5. $(4.9)(4.9)(4.9)(4.9)(4.9)(4.9)$

6. $(2\sqrt{5})(2\sqrt{5})(2\sqrt{5})(2\sqrt{5})$

7. $(-10)(-10)(-10)(-10)(-10)$

8. $-\left(\frac{3}{2} \times \frac{3}{2} \times \frac{3}{2} \times \frac{3}{2}\right)$

In Exercises 9–16, evaluate each expression.

9. (a) $3^2 \cdot 3$ (b) $3 \cdot 3^3$

10. (a) $\dfrac{5^5}{5^2}$ (b) $\dfrac{3^2}{3^4}$

11. (a) $(3^3)^2$ (b) -3^2

12. (a) $(2^3 \cdot 3^2)^2$ (b) $\left(-\frac{3}{5}\right)^3\left(\frac{5}{3}\right)^2$

13. (a) $\dfrac{3 \cdot 4^{-4}}{3^{-4} \cdot 4^{-1}}$ (b) $32(-2)^{-5}$

14. (a) $\dfrac{4 \cdot 3^{-2}}{2^{-2} \cdot 3^{-1}}$ (b) $(-2)^0$

15. (a) $2^{-1} + 3^{-1}$ (b) $(2^{-1})^{-2}$

16. (a) $3^{-1} + 2^{-2}$ (b) $(3^{-2})^2$

In Exercises 17–20, use a calculator to evaluate the expression. (If necessary, round your answer to three decimal places.)

17. $(-4)^3(5^2)$

18. $(8^{-4})(10^3)$

19. $\dfrac{3^6}{7^3}$

20. $\dfrac{4^3}{3^{-4}}$

In Exercises 21–28, evaluate the expression for the value of x.

	Expression	Value
21.	$-3x^3$	2
22.	$7x^{-2}$	4
23.	$6x^0$	10
24.	$5(-x)^3$	3
25.	$2x^3$	-3
26.	$-3x^4$	-2
27.	$4x^2$	$-\frac{1}{2}$
28.	$5(-x)^3$	$\frac{1}{3}$

In Exercises 29–34, simplify each expression.

29. (a) $(-5z)^3$ (b) $5x^4(x^2)$

30. (a) $(3x)^2$ (b) $(4x^3)^2$

31. (a) $6y^2(2y^4)^2$ (b) $\dfrac{3x^5}{x^3}$

32. (a) $(-z)^3(3z^4)$ (b) $\dfrac{25y^8}{10y^4}$

33. (a) $\dfrac{7x^2}{x^3}$ (b) $\dfrac{12(x+y)^3}{9(x+y)}$

34. (a) $\dfrac{r^4}{r^6}$ (b) $\left(\frac{4}{y}\right)^3\left(\frac{3}{y}\right)^4$

In Exercises 35–42, rewrite each expression with positive exponents and simplify.

35. (a) $(x+5)^0, \quad x \neq -5$ (b) $(2x^2)^{-2}$

36. (a) $(2x^5)^0, \quad x \neq 0$ (b) $(z+2)^{-3}(z+2)^{-1}$

37. (a) $(-2x^2)^3(4x^3)^{-1}$ (b) $\left(\frac{x}{10}\right)^{-1}$

38. (a) $(4y^{-2})(8y^4)$ (b) $\left(\frac{x^{-3}y^4}{5}\right)^{-3}$

39. (a) $(4a^{-2}b^3)^{-3}$ (b) $\left(\frac{5x^2}{y^{-2}}\right)^{-4}$

40. (a) $[(x^2y^{-2})^{-1}]^{-1}$ (b) $(5x^2z^6)^3(5x^2z^6)^{-3}$

41. (a) $3^n \cdot 3^{2n}$ (b) $\left(\frac{a^{-2}}{b^{-2}}\right)\left(\frac{b}{a}\right)^3$

42. (a) $\dfrac{x^2 \cdot x^n}{x^3 \cdot x^n}$ (b) $\left(\frac{a^{-3}}{b^{-3}}\right)\left(\frac{a}{b}\right)^3$

In Exercises 43–54, fill in the missing form of the equation.

Radical Form	Rational Exponent Form
43. $\sqrt{9} = 3$	
44. $\sqrt[3]{64} = 4$	
45.	$32^{1/5} = 2$
46.	$-(144^{1/2}) = -12$

Radical Form	*Rational Exponent Form*
47.	$196^{1/2} = 14$
48. $\sqrt[3]{614.125} = 8.5$	
49. $\sqrt[3]{-216} = -6$	
50.	$(-243)^{1/5} = -3$
51.	$27^{2/3} = 9$
52. $\left(\sqrt[4]{81}\right)^3 = 27$	
53. $\sqrt[4]{81^3} = 27$	
54.	$16^{5/4} = 32$

In Exercises 55–64, evaluate each expression without using a calculator.

55. (a) $\sqrt{9}$ (b) $\sqrt[3]{8}$

56. (a) $\sqrt{49}$ (b) $\sqrt[3]{\frac{27}{8}}$

57. (a) $-\sqrt[3]{-27}$ (b) $\dfrac{4}{\sqrt{64}}$

58. (a) $\sqrt[3]{0}$ (b) $\dfrac{\sqrt[4]{81}}{3}$

59. (a) $\left(\sqrt[3]{-125}\right)^3$ (b) $27^{1/3}$

60. (a) $\sqrt[4]{562^4}$ (b) $36^{3/2}$

61. (a) $32^{-3/5}$ (b) $\left(\frac{16}{81}\right)^{-3/4}$

62. (a) $100^{-3/2}$ (b) $\left(\frac{9}{4}\right)^{-1/2}$

63. (a) $\left(-\dfrac{1}{64}\right)^{-1/3}$ (b) $\left(\dfrac{1}{\sqrt{32}}\right)^{-2/5}$

64. (a) $\left(-\dfrac{125}{27}\right)^{-1/3}$ (b) $-\left(\dfrac{1}{125}\right)^{-4/3}$

In Exercises 65–70, use a calculator to approximate the number. (Round your answer to three decimal places.)

65. (a) $\sqrt{57}$ (b) $\sqrt[5]{-27^3}$

66. (a) $\sqrt[3]{45^2}$ (b) $\sqrt[6]{125}$

67. (a) $(1.2^{-2})\sqrt{75} + 3\sqrt{8}$ (b) $\dfrac{-3 + \sqrt{21}}{3}$

68. (a) $(15.25)^{-1.4}$ (b) $(3.4)^{2.5}$

69. (a) $(-12.4)^{-1.8}$ (b) $\left(5\sqrt{3}\right)^{-2.5}$

70. (a) $\dfrac{7 - (4.1)^{-3.2}}{2}$ (b) $\left(\dfrac{13}{3}\right)^{-3/2} - \left(-\dfrac{3}{2}\right)^{13/3}$

In Exercises 71–76, simplify by removing all possible factors from each radical.

71. (a) $\sqrt{8}$ (b) $\sqrt[3]{24}$

72. (a) $\sqrt[3]{\frac{16}{27}}$ (b) $\sqrt{\frac{75}{4}}$

73. (a) $\sqrt{72x^3}$ (b) $\sqrt{\dfrac{18^2}{z^3}}$

74. (a) $\sqrt{54xy^4}$ (b) $\sqrt{\dfrac{32a^4}{b^2}}$

75. (a) $\sqrt[3]{16x^5}$ (b) $\sqrt{75x^2y^{-4}}$

76. (a) $\sqrt[4]{(3x^2)^4}$ (b) $\sqrt[5]{96x^5}$

In Exercises 77–82, perform the operations and simplify.

77. $5^{4/3} \cdot 5^{8/3}$ **78.** $\dfrac{8^{12/5}}{8^{2/5}}$

79. $\dfrac{(2x^2)^{3/2}}{2^{1/2}x^4}$ **80.** $\dfrac{x^{4/3}y^{2/3}}{(xy)^{1/3}}$

81. $\dfrac{x^{-3} \cdot x^{1/2}}{x^{3/2} \cdot x^{-1}}$ **82.** $\dfrac{5^{-1/2} \cdot 5x^{5/2}}{(5x)^{3/2}}$

In Exercises 83–86, rationalize the denominator of each expression. Then simplify your answer.

83. (a) $\dfrac{1}{\sqrt{3}}$ (b) $\dfrac{8}{\sqrt[3]{2}}$

84. (a) $\dfrac{5}{\sqrt{10}}$ (b) $\dfrac{5}{\sqrt[3]{(5x)^2}}$

85. (a) $\dfrac{2x}{5 - \sqrt{3}}$ (b) $\dfrac{3}{\sqrt{5} + \sqrt{6}}$

86. (a) $\dfrac{5}{\sqrt{14} - 2}$ (b) $\dfrac{5}{\sqrt{10} - 5}$

⊖ **In Exercises 87–90, rationalize the numerator of each expression. Then simplify your answer.**

87. (a) $\dfrac{\sqrt{8}}{2}$ (b) $\sqrt[3]{\dfrac{9}{25}}$

88. (a) $\dfrac{\sqrt{2}}{3}$ (b) $\sqrt[4]{\dfrac{5}{4}}$

89. (a) $\dfrac{\sqrt{5} + \sqrt{3}}{3}$ (b) $\dfrac{\sqrt{7} - 3}{4}$

90. (a) $\dfrac{\sqrt{3} - \sqrt{2}}{2}$ (b) $\dfrac{2\sqrt{3} + \sqrt{3}}{3}$

In Exercises 91 and 92, reduce the index of each radical.

91. (a) $\sqrt[4]{3^2}$ (b) $\sqrt[6]{(x + 1)^4}$

92. (a) $\sqrt[6]{x^3}$ (b) $\sqrt[4]{(3x^2)^4}$

The symbol ⊖ indicates an example or exercise that highlights algebraic techniques specifically used in calculus.

In Exercises 93 and 94, write each expression as a single radical. Then simplify your answer.

93. (a) $\sqrt{\sqrt{32}}$ (b) $\sqrt{\sqrt[4]{2x}}$

94. (a) $\sqrt{\sqrt{243(x+1)}}$ (b) $\sqrt{\sqrt[3]{10a^7b}}$

In Exercises 95–100, simplify each expression.

95. (a) $2\sqrt{50} + 12\sqrt{8}$ (b) $10\sqrt{32} - 6\sqrt{18}$

96. (a) $4\sqrt{27} - \sqrt{75}$ (b) $\sqrt[3]{16} + 3\sqrt[3]{54}$

97. (a) $5\sqrt{x} - 3\sqrt{x}$ (b) $-2\sqrt{9y} + 10\sqrt{y}$

98. (a) $8\sqrt{49x} - 14\sqrt{100x}$

     (b) $-3\sqrt{48x^2} + 7\sqrt{75x^2}$

99. (a) $3\sqrt{x+1} + 10\sqrt{x+1}$

     (b) $7\sqrt{80x} - 2\sqrt{125x}$

100. (a) $-\sqrt{x^3 - 7} + 5\sqrt{x^3 - 7}$

     (b) $11\sqrt{245x^3} - 9\sqrt{45x^3}$

In Exercises 101–104, complete the statement with $<$, $=$, or $>$.

101. $\sqrt{5} + \sqrt{3}$ ▢ $\sqrt{5+3}$

102. $\sqrt{\dfrac{3}{11}}$ ▢ $\dfrac{\sqrt{3}}{\sqrt{11}}$

103. 5 ▢ $\sqrt{3^2 + 2^2}$

104. 5 ▢ $\sqrt{3^2 + 4^2}$

In Exercises 105–108, write the number in scientific notation.

105. Land area of the earth: 57,300,000 square miles

106. Light year: 9,460,000,000,000,000 kilometers

107. Relative density of hydrogen: 0.0000899 gram per cubic centimeter

108. One micron (millionth of a meter): 0.00003937 inch

In Exercises 109–112, write the number in decimal form.

109. U.S. daily Coca-Cola consumption: 6.048×10^8 servings (Source: The World of Coca-Cola Pavilion)

110. Interior temperature of the sun: 1.5×10^7 degrees Celsius

111. Charge of an electron: 1.602×10^{-19} coulomb

112. Width of a human hair: 9.0×10^{-5} meter

In Exercises 113 and 114, evaluate each expression without using a calculator.

113. (a) $\sqrt{25 \times 10^8}$ (b) $\sqrt[3]{8 \times 10^{15}}$

114. (a) $(1.2 \times 10^7)(5 \times 10^{-3})$ (b) $\dfrac{(6.0 \times 10^8)}{(3.0 \times 10^{-3})}$

In Exercises 115–118, use a calculator to evaluate each expression. (Round your answer to three decimal places.)

115. (a) $750\left(1 + \dfrac{0.11}{365}\right)^{800}$

     (b) $\dfrac{67{,}000{,}000 + 93{,}000{,}000}{0.0052}$

116. (a) $(9.3 \times 10^6)^3(6.1 \times 10^{-4})$

     (b) $\dfrac{(2.414 \times 10^4)^6}{(1.68 \times 10^5)^5}$

117. (a) $\sqrt{4.5 \times 10^9}$ (b) $\sqrt[3]{6.3 \times 10^4}$

118. (a) $(2.65 \times 10^{-4})^{1/3}$ (b) $\sqrt{9 \times 10^{-4}}$

119. *Exploration*   List all possible digits that occur in the units place of the square of a positive integer. Use that list to determine whether $\sqrt{5233}$ is an integer.

120. *Think About It*   Square the real number $2/\sqrt{5}$ and note that the radical is eliminated from the denominator. Is this equivalent to rationalizing the denominator? Why or why not?

121. *Period of a Pendulum*   The period T (in seconds) of a pendulum is

$$T = 2\pi\sqrt{\dfrac{L}{32}}$$

where L is the length of the pendulum (in feet). Find the period of a pendulum whose length is 2 feet.

122. *Erosion*   A stream of water moving at the rate of v feet per second can carry particles of size $0.03\sqrt{v}$ inches. Find the size of the largest particles that can be carried by a stream flowing at the rate of $\frac{3}{4}$ foot per second.

123. *Mathematical Modeling*   A funnel is filled with water to a height of h centimeters. The time t (in seconds) for the funnel to empty is

$$t = 0.03[12^{5/2} - (12 - h)^{5/2}], \quad 0 \le h \le 12.$$

Find t for $h = 7$ centimeters.

124. *Speed of Light*   The speed of light is 11,160,000 miles per minute. The distance from the sun to the earth is 93,000,000 miles. Find the time for light to travel from the sun to the earth.

125. *Depreciation*   Find the annual depreciation rate r from the bar graph below. To find r by the declining balances method, use the formula

$$r = 1 - \left(\frac{S}{C}\right)^{1/n}$$

where n is the useful life of the item (in years), S is the salvage value (in dollars), and C is the original cost (in dollars).

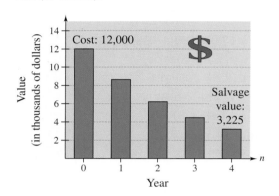

126. *Ecology*   There were 2.097×10^8 tons of municipal waste generated in the United States in 1996. Find the number of tons for each of the categories in the figure. (Source: U.S. Environmental Protection Agency)

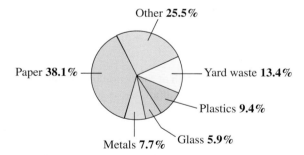

Synthesis

True or False?   **In Exercises 127 and 128, determine whether the statement is true or false. Justify your answer.**

127. $\dfrac{x^{k+1}}{x} = x^k$ **128.** $(a^n)^k = a^{n^k}$

129. Verify that $a^0 = 1$, $a \neq 0$. (*Hint*: Use the property of exponents $\dfrac{a^m}{a^n} = a^{m-n}$.)

130. Explain why each of the following pairs is not equal.

(a) $(3x)^{-1} \neq \dfrac{3}{x}$ (b) $y^3 \cdot y^2 \neq y^6$

(c) $(a^2b^3)^4 \neq a^6b^7$ (d) $(a+b)^2 \neq a^2 + b^2$

(e) $\sqrt{4x^2} \neq 2x$ (f) $\sqrt{2} + \sqrt{3} \neq \sqrt{5}$

131. Is the real number 52.7×10^5 written in scientific notation? Explain.

132. *Writing*   Johannes Kepler (1571–1630), a well-known German astronomer, discovered a relationship between the average distance of a planet from the sun and the time (or period) it takes the planet to orbit the sun. People then knew that planets that are closer to the sun take less time to complete an orbit than planets that are farther from the sun. Kepler discovered that the distance and period are related by an exact mathematical formula.

The table shows the average distance x (in astronomical units) and period y (in years) for the five planets that are closest to the sun. By completing the table, can you rediscover Kepler's relationship? Write a paragraph that summarizes your conclusions.

Planet	Mercury	Venus	Earth	Mars	Jupiter
x	0.387	0.723	1.0	1.523	5.203
$\sqrt{x}$					
y	0.241	0.615	1.0	1.881	11.861
$\sqrt[3]{y}$					

P.3   Polynomials and Special Products

▶ **What you should learn**

- How to write polynomials in standard form
- How to add, subtract, and multiply polynomials
- How to use special products to multiply polynomials
- How to use polynomials to solve real-life problems

▶ **Why you should learn it**

Polynomials can be used to model and solve real-life problems. For instance, in Exercise 106 on page 32, a polynomial is used to model the stopping distance of an automobile.

Nicholas DeVore/Tony Stone Images

Polynomials

The most common type of algebraic expression is the **polynomial.** Some examples are

$$2x + 5, \quad 3x^4 - 7x^2 + 2x + 4, \quad \text{and} \quad 5x^2y^2 - xy + 3.$$

The first two are *polynomials in x* and the third is a *polynomial in x and y.* The terms of a polynomial in x have the form ax^k, where a is the **coefficient** and k is the **degree** of the term. For instance, the polynomial

$$2x^3 - 5x^2 + 1 = 2x^3 + (-5)x^2 + (0)x + 1$$

has coefficients 2, -5, 0, and 1.

Definition of a Polynomial in x

Let $a_0, a_1, a_2, \ldots, a_n$ be real numbers and let n be a nonnegative integer. A polynomial in x is an expression of the form

$$a_n x^n + a_{n-1}x^{n-1} + \cdots + a_1 x + a_0$$

where $a_n \neq 0$. The polynomial is of **degree** n, a_n is the **leading coefficient,** and a_0 is the **constant term.**

Polynomials with one, two, and three terms are called **monomials, binomials,** and **trinomials,** respectively. In **standard form,** a polynomial is written with descending powers of x.

Example 1 ▶ Writing Polynomials in Standard Form

	Polynomial	Standard Form	Degree
a.	$4x^2 - 5x^7 - 2 + 3x$	$-5x^7 + 4x^2 + 3x - 2$	7
b.	$4 - 9x^2$	$-9x^2 + 4$	2
c.	8	$8 \; (8 = 8x^0)$	0

A polynomial that has all zero coefficients is called the zero polynomial, denoted by 0. No degree is assigned to this particular polynomial. For polynomials in more than one variable, the degree of a *term* is the sum of the exponents of the variables in the term. The degree of the *polynomial* is the highest degree of its terms. The leading coefficient of the polynomial is the coefficient of the highest degree term. Expressions such as the following are not polynomials.

$$x^3 - \sqrt{3x} = x^3 - (3x)^{1/2} \qquad \text{Exponent on } \sqrt{3x} \text{ is not an integer.}$$

$$x^2 + 5x^{-1} \qquad\qquad\qquad \text{Exponent on } 5x^{-1} \text{ is not a nonnegative integer.}$$

Operations with Polynomials

You can add and subtract polynomials in much the same way you add and subtract real numbers. Simply add or subtract the *like terms* (terms having the same variables to the same powers) by adding their coefficients. For instance, $-3xy^2$ and $5xy^2$ are like terms and their sum is

$$-3xy^2 + 5xy^2 = (-3 + 5)xy^2 = 2xy^2.$$

Example 2 ▶ **Sums and Differences of Polynomials**

a. $(5x^3 - 7x^2 - 3) + (x^3 + 2x^2 - x + 8)$

$= (5x^3 + x^3) + (2x^2 - 7x^2) - x + (8 - 3)$	Group like terms.
$= 6x^3 - 5x^2 - x + 5$	Combine like terms.

b. $(7x^4 - x^2 - 4x + 2) - (3x^4 - 4x^2 + 3x)$

$= 7x^4 - x^2 - 4x + 2 - 3x^4 + 4x^2 - 3x$	Distributive Property
$= (7x^4 - 3x^4) + (4x^2 - x^2) + (-3x - 4x) + 2$	Group like terms.
$= 4x^4 + 3x^2 - 7x + 2$	Combine like terms.

To find the product of two polynomials, use the left and right Distributive Properties. For example, if you treat $(5x + 7)$ as a single quantity, you can multiply $(3x - 2)$ by $(5x + 7)$ as follows.

$$(3x - 2)(5x + 7) = 3x(5x + 7) - 2(5x + 7)$$
$$= (3x)(5x) + (3x)(7) - (2)(5x) - (2)(7)$$
$$= 15x^2 + 21x - 10x - 14$$

Product of **First terms**	Product of **Outer terms**	Product of **Inner terms**	Product of **Last terms**

$$= 15x^2 + 11x - 14$$

Note in this **FOIL Method** that for binomials the outer (O) and inner (I) terms are alike and can be combined into one term.

Example 3 ▶ **Using the FOIL Method**

Use the FOIL Method to find the product of $(2x - 4)$ and $(x + 5)$.

Solution

$$(2x - 4)(x + 5) = \overset{F}{2x^2} + \overset{O}{10x} - \overset{I}{4x} - \overset{L}{20}$$
$$= 2x^2 + 6x - 20$$

When multiplying two polynomials, be sure to multiply *each* term of one polynomial by *each* term of the other. A vertical arrangement is helpful.

Example 4 ▶ A Vertical Arrangement for Multiplication

Multiply $(x^2 - 2x + 2)$ by $(x^2 + 2x + 2)$.

Solution

$$x^2 - 2x + 2 \qquad\qquad \text{Write in standard form.}$$
$$x^2 + 2x + 2 \qquad\qquad \text{Write in standard form.}$$
$$x^4 - 2x^3 + 2x^2 \qquad\qquad ⬅ \qquad x^2(x^2 - 2x + 2)$$
$$2x^3 - 4x^2 + 4x \qquad\qquad ⬅ \qquad 2x(x^2 - 2x + 2)$$
$$2x^2 - 4x + 4 \qquad\qquad ⬅ \qquad 2(x^2 - 2x + 2)$$
$$x^4 + 0x^3 + 0x^2 + 0x + 4 = x^4 + 4 \qquad \text{Combine like terms.}$$

So,

$$(x^2 - 2x + 2)(x^2 + 2x + 2) = x^4 + 4.$$

Special Products

Some binomial products have special forms that occur frequently in algebra.

Special Products

Let u and v be real numbers, variables, or algebraic expressions.

Special Product	*Example*
Sum and Difference of Same Terms	
$(u + v)(u - v) = u^2 - v^2$	$(x + 4)(x - 4) = x^2 - 4^2$
	$= x^2 - 16$
Square of a Binomial	
$(u + v)^2 = u^2 + 2uv + v^2$	$(x + 3)^2 = x^2 + 2(x)(3) + 3^2$
	$= x^2 + 6x + 9$
$(u - v)^2 = u^2 - 2uv + v^2$	$(3x - 2)^2 = (3x)^2 - 2(3x)(2) + 2^2$
	$= 9x^2 - 12x + 4$
Cube of a Binomial	
$(u + v)^3 = u^3 + 3u^2v + 3uv^2 + v^3$	$(x + 2)^3 = x^3 + 3x^2(2) + 3x(2^2) + 2^3$
	$= x^3 + 6x^2 + 12x + 8$
$(u - v)^3 = u^3 - 3u^2v + 3uv^2 - v^3$	$(x - 1)^3 = x^3 - 3x^2(1) + 3x(1^2) - 1^3$
	$= x^3 - 3x^2 + 3x - 1$

Example 5 ▶   Sum and Difference of Same Terms

Find the product of $(5x + 9)$ and $(5x - 9)$.

Solution

The product of a sum and a difference of the *same* two terms has no middle term and takes the form $(u + v)(u - v) = u^2 - v^2$.

$$(5x + 9)(5x - 9) = (5x)^2 - 9^2$$
$$= 25x^2 - 81$$

Example 6 ▶   Square of a Binomial

Find $(6x - 5)^2$.

Solution

The square of a binomial has the form

$$(u - v)^2 = u^2 - 2uv + v^2.$$
$$(6x - 5)^2 = (6x)^2 - 2(6x)(5) + 5^2$$
$$= 36x^2 - 60x + 25$$

Example 7 ▶   Cube of a Binomial

Find $(3x + 2)^3$.

Solution

The cube of a binomial has the form

$$(u + v)^3 = u^3 + 3u^2v + 3uv^2 + v^3.$$

Note the *decrease* of powers of u and the *increase* of powers of v.

$$(3x + 2)^3 = (3x)^3 + 3(3x)^2(2) + 3(3x)(2)^2 + 2^3$$
$$= 27x^3 + 54x^2 + 36x + 8$$

Example 8 ▶   The Product of Two Trinomials

Find the product of $(x + y - 2)$ and $(x + y + 2)$.

Solution

By grouping $x + y$ in parentheses, you can write the product of the trinomials as a special product.

$$(x + y - 2)(x + y + 2) = [(x + y) - 2][(x + y) + 2]$$
$$= (x + y)^2 - 2^2$$
$$= x^2 + 2xy + y^2 - 4$$

Application

Example 9 ▶ Volume of a Box

An open box is made by cutting squares from the corners of a piece of metal that is 16 inches by 20 inches, as shown in Figure P.6. The edge of each cut-out square is x inches. Find the volume of the box when $x = 1$, $x = 2$, and $x = 3$.

Solution

The volume of a rectangular box is equal to the product of its length, width, and height. From the figure, the length is $20 - 2x$, the width is $16 - 2x$, and the height is x. So, the volume of the box is

$$\text{Volume} = (20 - 2x)(16 - 2x)(x)$$
$$= (320 - 72x + 4x^2)(x)$$
$$= 320x - 72x^2 + 4x^3.$$

When $x = 1$ inch, the volume of the box is

$$\text{Volume} = 320(1) - 72(1^2) + 4(1^3)$$
$$= 252 \text{ cubic inches.}$$

When $x = 2$ inches, the volume of the box is

$$\text{Volume} = 320(2) - 72(2^2) + 4(2^3)$$
$$= 384 \text{ cubic inches.}$$

When $x = 3$ inches, the volume of the box is

$$\text{Volume} = 320(3) - 72(3^2) + 4(3^3)$$
$$= 420 \text{ cubic inches.}$$

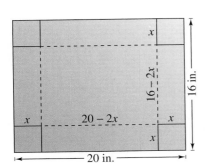

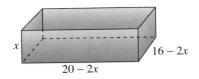

FIGURE P.6

Writing ABOUT MATHEMATICS

Mathematical Experiment In Example 9, the volume of the open metal box is given by

$$\text{Volume} = 320x - 72x^2 + 4x^3.$$

You want to create a box that has as much volume as possible. From Example 9, you know that by cutting 1-, 2-, and 3-inch squares from the corners, you can create boxes whose volumes are 252, 384, and 420 cubic inches, respectively. What are the possible values of x that make sense in the problem? Write your answer as an interval. Try several other values of x to find the size of the square that should be cut from the corners to produce a box that has maximum volume. Write a summary of your findings.

P.3 Exercises

In Exercises 1–6, match the polynomial with its description. [The polynomials are labeled (a), (b), (c), (d), (e), and (f).]

(a) $3x^2$ (b) $1 - 2x^3$

(c) $x^3 + 3x^2 + 3x + 1$ (d) 12

(e) $-3x^5 + 2x^3 + x$ (f) $\frac{2}{3}x^4 + x^2 + 10$

1. A polynomial of degree zero
2. A trinomial of degree five
3. A binomial with leading coefficient -2
4. A monomial of positive degree
5. A trinomial with leading coefficient $\frac{2}{3}$
6. A third-degree polynomial with leading coefficient 1

In Exercises 7–10, write a polynomial that fits the description. (There are many correct answers.)

7. A third-degree polynomial with leading coefficient -2
8. A fifth-degree polynomial with leading coefficient 6
9. A fourth-degree binomial with a negative leading coefficient
10. A third-degree binomial with an even leading coefficient

In Exercises 11–18, find the degree and leading coefficient of the polynomial.

11. $2x^2 - x + 1$
12. $-3x^4 + 2x^2 - 5$
13. $x^5 - 1$
14. 3
15. $1 - x + 6x^4 - 4x^5$
16. $3 + 2x$
17. $4x^3y - 3xy^2 + x^2y^3$
18. $-x^5y + 2x^2y^2 + xy^4$

In Exercises 19–24, is the expression a polynomial? If so, write the polynomial in standard form.

19. $2x - 3x^3 + 8$
20. $2x^3 + x - 3x^{-1}$
21. $\dfrac{3x + 4}{x}$
22. $\dfrac{x^2 + 2x - 3}{2}$
23. $y^2 - y^4 + y^3$
24. $\sqrt{y^2 - y^4}$

In Exercises 25–42, perform the operations and write the result in standard form.

25. $(6x + 5) - (8x + 15)$
26. $(2x^2 + 1) - (x^2 - 2x + 1)$
27. $-(x^3 - 2) + (4x^3 - 2x)$
28. $-(5x^2 - 1) - (-3x^2 + 5)$
29. $(15x^2 - 6) - (-8.3x^3 - 14.7x^2 - 17)$
30. $(15.2x^4 - 18x - 19.1) - (13.9x^4 - 9.6x + 15)$
31. $5z - [3z - (10z + 8)]$
32. $(y^3 + 1) - [(y^2 + 1) + (3y - 7)]$
33. $3x(x^2 - 2x + 1)$ 34. $y^2(4y^2 + 2y - 3)$
35. $-5z(3z - 1)$ 36. $(-3x)(5x + 2)$
37. $(1 - x^3)(4x)$ 38. $-4x(3 - x^3)$
39. $(2.5x^2 + 3)(3x)$ 40. $(2 - 3.5y)(2y^3)$
41. $-4x(\frac{1}{8}x + 3)$ 42. $2y(4 - \frac{7}{8}y)$

In Exercises 43–52, perform the operations.

43. Add $7x^3 - 2x^2 + 8$ and $-3x^3 - 4$.
44. Add $2x^5 - 3x^3 + 2x + 3$ and $4x^3 + x - 6$.
45. Subtract $x - 3$ from $5x^2 - 3x + 8$.
46. Subtract $-t^4 + 0.5t^2 - 5.6$ from $0.6t^4 - 2t^2$.
47. Multiply $-6x^2 + 15x - 4$ and $5x + 3$.
48. Multiply $4x^4 + x^3 - 6x^2 + 9$ and $x^2 + 2x + 3$.
49. $(x^2 + 9)(x^2 - x - 4)$
50. $(x - 2)(x^2 + 2x + 4)$
51. $(x^2 - x + 1)(x^2 + x + 1)$
52. $(x^2 + 3x - 2)(x^2 - 3x - 2)$

In Exercises 53–88, multiply or find the special product.

53. $(x + 3)(x + 4)$ 54. $(x - 5)(x + 10)$
55. $(3x - 5)(2x + 1)$ 56. $(7x - 2)(4x - 3)$
57. $(2x + 3)^2$ 58. $(4x + 5)^2$
59. $(2x - 5y)^2$ 60. $(5 - 8x)^2$
61. $(x + 10)(x - 10)$ 62. $(2x + 3)(2x - 3)$
63. $(x + 2y)(x - 2y)$ 64. $(2x + 3y)(2x - 3y)$
65. $[(m - 3) + n][(m - 3) - n]$
66. $[(x + y) + 1][(x + y) - 1]$
67. $[(x - 3) + y]^2$ 68. $[(x + 1) - y]^2$
69. $(2r^2 - 5)(2r^2 + 5)$

70. $(3a^3 - 4b^2)(3a^3 + 4b^2)$

71. $(x + 1)^3$

72. $(x - 2)^3$

73. $(2x - y)^3$

74. $(3x + 2y)^3$

75. $(4x^3 - 3)^2$

76. $(8x + 3)^2$

77. $\left(\frac{1}{2}x - 3\right)^2$

78. $\left(\frac{2}{3}t + 5\right)^2$

79. $\left(\frac{1}{3}x - 2\right)\left(\frac{1}{3}x + 2\right)$

80. $\left(2x + \frac{1}{5}\right)\left(2x - \frac{1}{5}\right)$

81. $(1.2x + 3)^2$

82. $(1.5y - 3)^2$

83. $(1.5x - 4)(1.5x + 4)$

84. $(2.5y + 3)(2.5y - 3)$

85. $5x(x + 1) - 3x(x + 1)$

86. $(2x - 1)(x + 3) + 3(x + 3)$

87. $(u + 2)(u - 2)(u^2 + 4)$

88. $(x + y)(x - y)(x^2 + y^2)$

In Exercises 89–92, find the product. The expressions are not polynomials, but the formulas can still be used.

89. $\left(\sqrt{x} + \sqrt{y}\right)\left(\sqrt{x} - \sqrt{y}\right)$

90. $\left(5 + \sqrt{x}\right)\left(5 - \sqrt{x}\right)$

91. $\left(x - \sqrt{5}\right)^2$

92. $\left(x + \sqrt{3}\right)^2$

93. *Pattern Recognition* Perform the multiplications.

(a) $(x - 1)(x + 1)$

(b) $(x - 1)(x^2 + x + 1)$

(c) $(x - 1)(x^3 + x^2 + x + 1)$

From the pattern formed by these products, can you predict the result of $(x - 1)(x^4 + x^3 + x^2 + x + 1)$?

94. *Think About It* When the polynomial $-x^3 + 3x^2 + 2x - 1$ is subtracted from an unknown polynomial, the difference is $5x^2 + 8$. If it is possible, find the unknown polynomial.

95. *Logical Reasoning* Verify that $(x + y)^2$ is not equal to $x^2 + y^2$ by letting $x = 3$ and $y = 4$ and evaluating both expressions. Are there any values of x or y for which $(x + y)^2 = x^2 + y^2$? Explain.

96. *Business* A manufacturer can produce and sell x radios per week. The total cost (in dollars) for producing x radios is

$$C = 73x + 25,000$$

and the total revenue (in dollars) is

$$R = 95x.$$

Find the profit P obtained by selling 5000 radios per week.

97. *Business* An artist can produce and sell x craft items per month. The total cost (in dollars) for producing x craft items is

$$C = 460 + 12x$$

and the total revenue (in dollars) is

$$R = 36x.$$

Find the profit P obtained by selling 42 craft items per month.

98. *Finance* After 2 years, an investment of $500 compounded annually at an interest rate r will yield an amount of

$$500(1 + r)^2.$$

(a) Write this polynomial in standard form.

(b) Use a calculator to evaluate the polynomial for the values of r in the table.

r	$2\frac{1}{2}\%$	3%	4%	$4\frac{1}{2}\%$	5%
$500(1 + r)^2$					

(c) What conclusion can you make from the table?

99. *Finance* After 3 years, an investment of $1200 compounded annually at an interest rate r will yield an amount of

$$1200(1 + r)^3.$$

(a) Write this polynomial in standard form.

(b) Use a calculator to evaluate the polynomial for the values of r in the table.

r	2%	3%	$3\frac{1}{2}\%$	4%	$4\frac{1}{2}\%$
$1200(1 + r)^3$					

(c) What conclusion can you make from the table?

100. *Volume of a Box* An open box is made by cutting squares from the corners of a piece of metal that is 18 centimeters by 26 centimeters (see figure). If the edge of each cut-out square is x centimeters, find the volume when $x = 1$, $x = 2$, and $x = 3$.

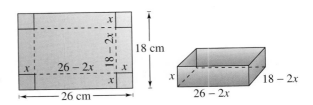

101. *Volume of a Box*   A closed box is constructed by cutting along the solid lines and folding along the broken lines on the rectangular piece of metal shown in the figure. The length and width of the rectangle are 45 centimeters and 15 centimeters, respectively. Find the volume of the box in terms of *x*. Find the volume when $x = 3$, $x = 5$, and $x = 7$.

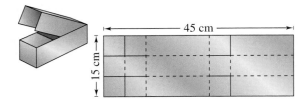

102. *Geometry*   Find the area of the shaded region in each figure. Write your result as a polynomial in standard form.

(a)   (b)

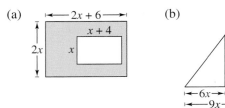

103. *Geometry*   Find the area of the shaded region in each figure. Write your result as a polynomial in standard form.

(a)   (b)

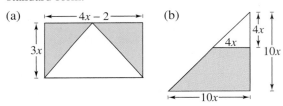

Geometry   **In Exercises 104 and 105, find a polynomial that represents the total number of square feet for the floor plan shown in the figure.**

104.

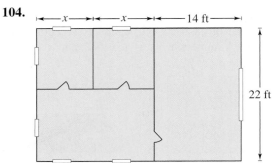

105.

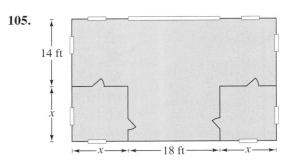

106. *Stopping Distance*   The stopping distance of an automobile is the distance traveled during the driver's reaction time plus the distance traveled after the brakes are applied. In an experiment, these distances were measured (in feet) when the automobile was traveling at a speed of *x* miles per hour, as shown in the bar graph. The distance traveled during the reaction time was

$$R = 1.1x$$

and the braking distance was

$$B = 0.14x^2 - 4.43x + 58.40.$$

(a) Determine the polynomial that represents the total stopping distance *T*.

(b) Use the result of part (a) to estimate the total stopping distance when $x = 30$, $x = 40$, and $x = 55$ miles per hour.

(c) Use the bar graph to make a statement about the total stopping distance required for increasing speeds.

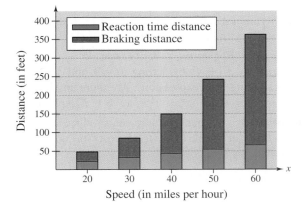

107. *Engineering*   A uniformly distributed load is placed on a 1-inch-wide steel beam. When the span of the beam is x feet and its depth is 6 inches, the safe load S is approximated by

$$S_6 = (0.06x^2 - 2.42x + 38.71)^2.$$

When the depth is 8 inches, the safe load is approximated by

$$S_8 = (0.08x^2 - 3.30x + 51.93)^2.$$

(a) Use the bar graph to estimate the difference in the safe loads for these two beams when the span is 12 feet.

(b) How does the difference in safe load change as the span increases?

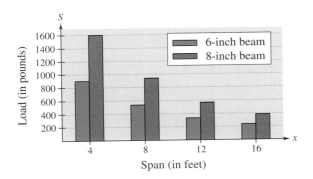

Geometry   In Exercises 108 and 109, use the area model to write two different expressions for the area. Then equate the two expressions and name the algebraic property that is illustrated.

108.

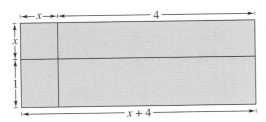

109.

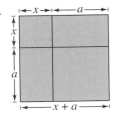

Synthesis

True or False?   **In Exercises 110 and 111, determine whether the statement is true or false. Justify your answer.**

110. The product of two binomials is always a second-degree polynomial.

111. The sum of two binomials is always a binomial.

112. Find the degree of the product of two polynomials of degrees m and n.

113. Find the degree of the sum of two polynomials of degrees m and n if $m < n$.

114. *Writing*   A student's homework paper included the following.

$$(x - 3)^2 = x^2 + 9$$

Write a paragraph fully explaining the error and give the correct method for squaring a binomial.

115. A third-degree polynomial and a fourth-degree polynomial are added.

(a) Can the sum be a fourth-degree polynomial? Explain or give an example.

(b) Can the sum be a second-degree polynomial? Explain or give an example.

(c) Can the sum be a seventh-degree polynomial? Explain or give an example.

116. *Think About It*   Must the sum of two second-degree polynomials be a second-degree polynomial? If not, give an example.

P.4 Factoring

▶ **What you should learn**

- How to remove common factors from polynomials
- How to factor special polynomial forms
- How to factor trinomials as the product of two binomials
- How to factor by grouping

▶ **Why you should learn it**

Polynomial factoring can be used to solve real-life problems. For instance, in Exercise 136 on page 41, factoring is used to develop an alternative formula for the volume of concrete used to make a cylindrical storage tank.

Arnulf Husmo/Tony Stone Images

Polynomials with Common Factors

The process of writing a polynomial as a product is called **factoring.** It is an important tool for solving equations and for simplifying rational expressions.

Unless noted otherwise, when you are asked to factor a polynomial, you can assume that you are hunting for factors with integer coefficients. If a polynomial cannot be factored using integer coefficients, then it is **prime** or **irreducible over the integers.** For instance, the polynomial $x^2 - 3$ is irreducible over the integers. Over the *real numbers*, this polynomial can be factored as

$$x^2 - 3 = (x + \sqrt{3})(x - \sqrt{3}).$$

Example 1 ▶ Recognizing Completely Factored Polynomials

a. $x^3 - x^2 + 4x - 4 = (x - 1)(x^2 + 4)$

is completely factored.

b. $x^3 - x^2 - 4x + 4 = (x - 1)(x^2 - 4)$

is not completely factored. Its complete factorization would be

$$x^3 - x^2 - 4x + 4 = (x - 1)(x + 2)(x - 2).$$

The simplest type of factoring involves a polynomial that can be written as the product of a monomial and another polynomial. The technique used here is the Distributive Property, $a(b + c) = ab + ac$, in the *reverse* direction.

$$ab + ac = a(b + c) \qquad \text{\textit{a} is a common factor.}$$

Removing (factoring out) a common factor is the first step in completely factoring a polynomial.

Example 2 ▶ Removing Common Factors

Factor each expression.

a. $6x^3 - 4x$ **b.** $-4x^2 + 12x - 16$ **c.** $(x - 2)(2x) + (x - 2)(3)$

Solution

a. $6x^3 - 4x = 2x(3x^2) - 2x(2)$ $2x$ is a common factor.

$$= 2x(3x^2 - 2)$$

b. $-4x^2 + 12x - 16 = -4(x^2) + (-4)(-3x) + (-4)4$ -4 is a common factor.

$$= -4(x^2 - 3x + 4)$$

c. $(x - 2)(2x) + (x - 2)(3) = (x - 2)(2x + 3)$ $x - 2$ is a common factor.

Factoring Special Polynomial Forms

Some polynomials have special forms that you should learn to recognize so that you can factor the polynomials easily.

Factoring Special Polynomial Forms

Factored Form	*Example*

Difference of Two Squares

$$u^2 - v^2 = (u + v)(u - v)$$

$$9x^2 - 4 = (3x)^2 - 2^2 = (3x + 2)(3x - 2)$$

Perfect Square Trinomial

$$u^2 + 2uv + v^2 = (u + v)^2$$

$$x^2 + 6x + 9 = x^2 + 2(x)(3) + 3^2 = (x + 3)^2$$

$$u^2 - 2uv + v^2 = (u - v)^2$$

$$x^2 - 6x + 9 = x^2 - 2(x)(3) + 3^2 = (x - 3)^2$$

Sum or Difference of Two Cubes

$$u^3 + v^3 = (u + v)(u^2 - uv + v^2)$$

$$x^3 + 8 = x^3 + 2^3 = (x + 2)(x^2 - 2x + 4)$$

$$u^3 - v^3 = (u - v)(u^2 + uv + v^2)$$

$$27x^3 - 1 = (3x)^3 - 1^3 = (3x - 1)(9x^2 + 3x + 1)$$

One of the easiest special polynomial forms to factor is the difference of two squares. Think of this form as follows.

$$u^2 - v^2 = (u + v)(u - v)$$

Difference   Opposite signs

To recognize perfect square terms, look for coefficients that are squares of integers and variables raised to *even powers*.

Example 3 ▶ Removing a Common Factor First

Factor $3 - 12x^2$.

Solution

$$3 - 12x^2 = 3(1 - 4x^2) \qquad \text{3 is a common factor.}$$

$$= 3[1^2 - (2x)^2]$$

$$= 3(1 + 2x)(1 - 2x) \qquad \text{Difference of two squares}$$

Example 4 ▶ Factoring the Difference of Two Squares

a. $(x + 2)^2 - y^2 = [(x + 2) + y][(x + 2) - y]$

$$= (x + 2 + y)(x + 2 - y)$$

b. $16x^4 - 81 = (4x^2)^2 - 9^2$

$$= (4x^2 + 9)(4x^2 - 9) \qquad \text{Difference of two squares}$$

$$= (4x^2 + 9)[(2x)^2 - 3^2]$$

$$= (4x^2 + 9)(2x + 3)(2x - 3) \qquad \text{Difference of two squares}$$

STUDY T!P

In Example 3, note that the first step in factoring a polynomial is to check for a common factor. Once the common factor is removed, it is often possible to recognize patterns that were not immediately obvious.

A perfect square trinomial is the square of a binomial, and it has the following form.

$$u^2 + 2uv + v^2 = (u + v)^2 \qquad \text{or} \qquad u^2 - 2uv + v^2 = (u - v)^2$$

Like signs   Like signs

Note that the first and last terms are squares and the middle term is twice the product of u and v.

Example 5 ▶ Factoring Perfect Square Trinomials

Factor each trinomial.

a. $x^2 - 10x + 25$ **b.** $16x^2 + 8x + 1$

Solution

a. $x^2 - 10x + 25 = x^2 - 2(x)(5) + 5^2$

$\qquad\qquad\qquad\quad = (x - 5)^2$

b. $16x^2 + 8x + 1 = (4x)^2 + 2(4x)(1) + 1^2$

$\qquad\qquad\qquad\quad = (4x + 1)^2$

◀ **Exploration** ▶

Rewrite $u^6 - v^6$ as the difference of two squares. Then find a formula for completely factoring $u^6 - v^6$. Use your formula to factor completely $x^6 - 1$ and $x^6 - 64$.

The next two formulas show the sums and differences of cubes. Pay special attention to the signs of the terms.

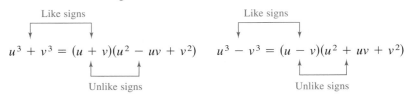

Like signs   Like signs

$$u^3 + v^3 = (u + v)(u^2 - uv + v^2) \qquad u^3 - v^3 = (u - v)(u^2 + uv + v^2)$$

Unlike signs   Unlike signs

Example 6 ▶ Factoring the Difference of Cubes

Factor $x^3 - 27$.

Solution

$x^3 - 27 = x^3 - 3^3$ Rewrite 27 as 3^3.

$\qquad\quad = (x - 3)(x^2 + 3x + 9)$ Factor.

Example 7 ▶ Factoring the Sum of Cubes

a. $y^3 + 8 = y^3 + 2^3$ Rewrite 8 as 2^3.

$\qquad\quad = (y + 2)(y^2 - 2y + 4)$ Factor.

b. $3(x^3 + 64) = 3(x^3 + 4^3)$ Rewrite 64 as 4^3.

$\qquad\qquad\quad = 3(x + 4)(x^2 - 4x + 16)$ Factor.

Trinomials with Binomial Factors

To factor a trinomial of the form $ax^2 + bx + c$, use the following pattern.

Factors of a

$$ax^2 + bx + c = (\quad x + \quad)(\quad x + \quad)$$

Factors of c

The goal is to find a combination of factors of a and c such that the outer and inner products add up to the middle term bx. For instance, in the trinomial $6x^2 + 17x + 5$, you can write

$$\begin{array}{cccc} F & O & I & L \\ \downarrow & \downarrow & \downarrow & \downarrow \end{array}$$

$$(2x + 5)(3x + 1) = 6x^2 + 2x + 15x + 5$$

$$O + I$$
$$\downarrow$$
$$= 6x^2 + 17x + 5.$$

Note that the outer (O) and inner (I) products add up to $17x$.

Example 8 ► Factoring a Trinomial: Leading Coefficient Is 1

Factor $x^2 - 7x + 12$.

Solution

The possible factorizations are

$$(x - 2)(x - 6), \quad (x - 1)(x - 12), \quad \text{and} \quad (x - 3)(x - 4).$$

Testing the middle term, you will find the correct factorization to be

$$x^2 - 7x + 12 = (x - 3)(x - 4).$$

 A computer animation of this example appears in the *Interactive* CD-ROM and *Internet* versions of this text.

Example 9 ► Factoring a Trinomial: Leading Coefficient Is Not 1

Factor $2x^2 + x - 15$.

Solution

The eight possible factorizations are as follows.

$$(2x - 1)(x + 15) \qquad (2x + 1)(x - 15)$$
$$(2x - 3)(x + 5) \qquad (2x + 3)(x - 5)$$
$$(2x - 5)(x + 3) \qquad (2x + 5)(x - 3)$$
$$(2x - 15)(x + 1) \qquad (2x + 15)(x - 1)$$

Testing the middle term, you will find the correct factorization to be

$$2x^2 + x - 15 = (2x - 5)(x + 3). \qquad O + I = 6x - 5x = x$$

Factoring by Grouping

Sometimes polynomials with more than three terms can be factored by a method called **factoring by grouping.** It is not always obvious which terms to group, and sometimes several different groupings will work.

Example 10 ▶ Factoring by Grouping

Use factoring by grouping to factor

$$x^3 - 2x^2 - 3x + 6.$$

Solution

$$
\begin{aligned}
x^3 - 2x^2 - 3x + 6 &= (x^3 - 2x^2) - (3x - 6) &&\text{Group terms.}\\
&= x^2(x - 2) - 3(x - 2) &&\text{Factor groups.}\\
&= (x - 2)(x^2 - 3) &&\text{Distributive Property}
\end{aligned}
$$

Factoring a trinomial can involve quite a bit of trial and error. Some of this trial and error can be lessened by using factoring by grouping.

Example 11 ▶ Factoring a Trinomial by Grouping

Use factoring by grouping to factor $2x^2 + 5x - 3$.

Solution

In the trinomial $2x^2 + 5x - 3$, $a = 2$ and $c = -3$ which implies that the product ac is -6. Now, -6 factors as $(6)(-1)$ and $6 - 1 = 5 = b$. So, you can rewrite the middle term as $5x = 6x - x$. This produces the following.

$$
\begin{aligned}
2x^2 + 5x - 3 &= 2x^2 + 6x - x - 3 &&\text{Rewrite middle term.}\\
&= (2x^2 + 6x) - (x + 3) &&\text{Group terms.}\\
&= 2x(x + 3) - (x + 3) &&\text{Factor groups.}\\
&= (x + 3)(2x - 1) &&\text{Distributive Property}
\end{aligned}
$$

Therefore, the trinomial factors as

$$2x^2 + 5x - 3 = (x + 3)(2x - 1).$$

Guidelines for Factoring Polynomials

1. Factor out any common factors using the Distributive Property.

2. Factor according to one of the special polynomial forms.

3. Factor as $ax^2 + bx + c = (mx + r)(nx + s)$.

4. Factor by grouping.

P.4 Exercises

In Exercises 1–4, find the greatest common factor of the expressions.

1. 90, 300

2. 36, 84, 294

3. $12x^2y^3$, $18x^2y$, $24x^3y^2$

4. $15(x + 2)^3$, $42x(x + 2)^2$

In Exercises 5–12, factor out the common factor.

5. $3x + 6$

6. $5y - 30$

7. $2x^3 - 6x$

8. $4x^3 - 6x^2 + 12x$

9. $x(x - 1) + 6(x - 1)$

10. $3x(x + 2) - 4(x + 2)$

11. $(x + 3)^2 - 4(x + 3)$

12. $(3x - 1)^2 + (3x - 1)$

In Exercises 13–18, find the greatest common factor such that the remaining factors have only integer coefficients.

13. $\frac{1}{2}x + 4$

14. $\frac{1}{3}y + 5$

15. $\frac{1}{2}x^3 + 2x^2 - 5x$

16. $\frac{1}{3}y^4 - 5y^2 + 2y$

17. $\frac{2}{3}x(x - 3) - 4(x - 3)$

18. $\frac{4}{5}y(y + 1) - 2(y + 1)$

In Exercises 19–28, factor the difference of two squares.

19. $x^2 - 36$

20. $x^2 - 49$

21. $16y^2 - 9$

22. $49 - 9y^2$

23. $16x^2 - \frac{1}{9}$

24. $\frac{4}{25}y^2 - 64$

25. $(x - 1)^2 - 4$

26. $25 - (z + 5)^2$

27. $9u^2 - 4v^2$

28. $25x^2 - 16y^2$

In Exercises 29–38, factor the perfect square trinomial.

29. $x^2 - 4x + 4$

30. $x^2 + 10x + 25$

31. $4t^2 + 4t + 1$

32. $9x^2 - 12x + 4$

33. $25y^2 - 10y + 1$

34. $36y^2 - 108y + 81$

35. $9u^2 + 24uv + 16v^2$

36. $4x^2 - 4xy + y^2$

37. $x^2 - \frac{4}{3}x + \frac{4}{9}$

38. $z^2 + z + \frac{1}{4}$

In Exercises 39–42, factor a negative real number from the polynomial and then write the polynomial factor in standard form.

39. $25 - 5x^2$

40. $5 + 3y^2 - y^3$

41. $-2t^3 + 4t + 6$

42. $-3x^5 - 3x^2 + 6x + 9$

In Exercises 43–56, factor the trinomial.

43. $x^2 + x - 2$

44. $x^2 + 5x + 6$

45. $s^2 - 5s + 6$

46. $t^2 - t - 6$

47. $20 - y - y^2$

48. $24 + 5z - z^2$

49. $x^2 - 30x + 200$

50. $x^2 - 13x + 42$

51. $3x^2 - 5x + 2$

52. $2x^2 - x - 1$

53. $5x^2 + 26x + 5$

54. $12x^2 + 7x + 1$

55. $-9z^2 + 3z + 2$

56. $-5u^2 - 13u + 6$

In Exercises 57–64, factor the sum or difference of cubes.

57. $x^3 - 8$

58. $x^3 - 27$

59. $y^3 + 64$

60. $z^3 + 125$

61. $8t^3 - 1$

62. $27x^3 + 8$

63. $u^3 + 27v^3$

64. $64x^3 - y^3$

In Exercises 65–72, factor by grouping.

65. $x^3 - x^2 + 2x - 2$

66. $x^3 + 5x^2 - 5x - 25$

67. $2x^3 - x^2 - 6x + 3$

68. $5x^3 - 10x^2 + 3x - 6$

69. $6 + 2x - 3x^3 - x^4$

70. $x^5 + 2x^3 + x^2 + 2$

71. $6x^3 - 2x + 3x^2 - 1$

72. $8x^5 - 6x^2 + 12x^3 - 9$

In Exercises 73–78, factor the trinomial by grouping.

73. $3x^2 + 10x + 8$

74. $2x^2 + 9x + 9$

75. $6x^2 + x - 2$

76. $6x^2 - x - 15$

77. $15x^2 - 11x + 2$

78. $12x^2 - 13x + 1$

In Exercises 79–113, completely factor the expression.

79. $6x^2 - 54$

80. $12x^2 - 48$

81. $x^3 - 4x^2$

82. $x^3 - 9x$

83. $x^2 - 2x + 1$

84. $16 + 6x - x^2$

85. $1 - 4x + 4x^2$

86. $-9x^2 + 6x - 1$

87. $2x^2 + 4x - 2x^3$

88. $2y^3 - 7y^2 - 15y$

89. $9x^2 + 10x + 1$

90. $13x + 6 + 5x^2$

91. $\frac{1}{81}x^2 + \frac{2}{9}x - 8$

92. $\frac{1}{8}x^2 - \frac{1}{96}x - \frac{1}{16}$

93. $3x^3 + x^2 + 15x + 5$

94. $5 - x + 5x^2 - x^3$

95. $x^4 - 4x^3 + x^2 - 4x$

96. $3u - 2u^2 + 6 - u^3$

97. $\frac{1}{4}x^3 + 3x^2 + \frac{3}{4}x + 9$

98. $25 - (z + 5)^2$

99. $(t - 1)^2 - 49$

100. $(x^2 + 1)^2 - 4x^2$

101. $(x^2 + 8)^2 - 36x^2$

102. $2t^3 - 16$

103. $5x^3 + 40$

104. $4x(2x - 1) + (2x - 1)^2$

105. $5(3 - 4x)^2 - 8(3 - 4x)(5x - 1)$

106. $2(x + 1)(x - 3)^2 - 3(x + 1)^2(x - 3)$

107. $7(3x + 2)^2(1 - x)^2 + (3x + 2)(1 - x)^3$

108. $7x(2)(x^2 + 1)(2x) - (x^2 + 1)^2(7)$

109. $3(x - 2)^2(x + 1)^4 + (x - 2)^3(4)(x + 1)^3$

110. $2x(x - 5)^4 - x^2(4)(x - 5)^3$

111. $5(x^6 + 1)^4(6x^5)(3x + 2)^3 + 3(3x + 2)^2(3)(x^6 + 1)^5$

112. $\dfrac{x^2}{2}(x^2 + 1)^4 - (x^2 + 1)^5$

113. $5w^3(9w + 1)^4(9) + (2w + 1)^5(3w^2)$

Geometric Modeling **In Exercises 114–117, match the factoring formula with the correct "geometric factoring model." [The models are labeled (a), (b), (c), and (d).] For instance, a factoring model for**

$$2x^2 + 3x + 1 = (2x + 1)(x + 1)$$

is shown in the figure.

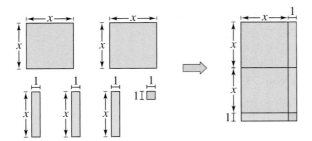

(a)

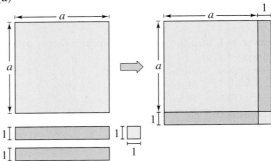

(b)

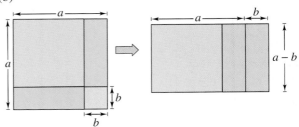

(c)

(d)

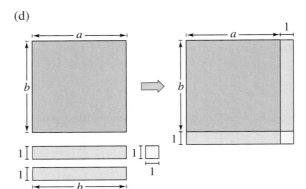

114. $a^2 - b^2 = (a + b)(a - b)$

115. $a^2 + 2ab + b^2 = (a + b)^2$

116. $a^2 + 2a + 1 = (a + 1)^2$

117. $ab + a + b + 1 = (a + 1)(b + 1)$

Geometric Modeling **In Exercises 118–121, draw a "geometric factoring model" to represent the factorization.**

118. $3x^2 + 7x + 2 = (3x + 1)(x + 2)$

119. $x^2 + 4x + 3 = (x + 3)(x + 1)$

120. $2x^2 + 7x + 3 = (2x + 1)(x + 3)$

121. $x^2 + 3x + 2 = (x + 2)(x + 1)$

Geometry **In Exercises 122–125, write an expression in factored form for the shaded portion of the figure.**

122.

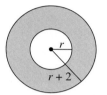

123.

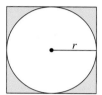

124.

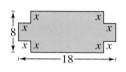

125.

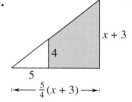

In Exercises 126–129, find all values of b for which the trinomial can be factored.

126. $x^2 + bx - 15$ **127.** $x^2 + bx + 50$

128. $x^2 + bx - 12$ **129.** $x^2 + bx + 24$

In Exercises 130–133, find two integer values of c such that the trinomial can be factored. (There are many correct answers.)

130. $2x^2 + 5x + c$ **131.** $3x^2 - 10x + c$

132. $3x^2 - x + c$ **133.** $2x^2 + 9x + c$

134. *Error Analysis*   Describe the error.

$$9x^2 - 9x - 54 = (3x + 6)(3x - 9) \quad \times$$
$$= 3(x + 2)(x - 3)$$

135. *Think About It*   Is $(3x - 6)(x + 1)$ completely factored? Explain.

136. *Geometry*   The volume of concrete used to make the cylindrical concrete storage tank shown in the figure is

$$V = \pi R^2 h - \pi r^2 h.$$

(a) Factor the expression for the volume.

(b) From the result of part (a), show that the volume of concrete is

$$2\pi(\text{average radius})(\text{thickness of the tank})h.$$

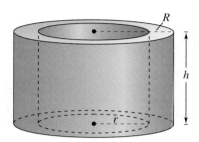

137. *Chemistry*   The rate of change of an autocatalytic chemical reaction is $kQx - kx^2$, where Q is the amount of the original substance, x is the amount of substance formed, and k is a constant of proportionality. Factor the expression.

Synthesis

138. Factor $x^{2n} - y^{2n}$ completely.

139. Factor $x^{3n} + y^{3n}$ completely.

140. Factor $x^{3n} - y^{2n}$ completely.

True or False?   **In Exercises 141 and 142, determine whether the statement is true or false. Justify your answer.**

141. The difference of two perfect squares can be factored as the product of conjugate pairs.

142. The sum of two perfect squares can be factored as the binomial sum squared.

143. Explain what is meant when it is said that a polynomial is in factored form.

P.5 Rational Expressions

▶ **What you should learn**
- How to find domains of algebraic expressions
- How to simplify rational expressions
- How to add, subtract, multiply, and divide rational expressions
- How to simplify complex fractions

▶ **Why you should learn it**

Rational expressions can be used to solve real-life problems. For instance, in Exercise 102 on page 52, a rational expression is used to model the cost per ounce of precious metals from 1992 through 1997.

Patrick Cocklin/Tony Stone Images

Domain of an Algebraic Expression

The set of real numbers for which an algebraic expression is defined is the **domain** of the expression. Two algebraic expressions are **equivalent** if they have the same domain and yield the same values for all numbers in their domain. For instance, $(x + 1) + (x + 2)$ and $2x + 3$ are equivalent because

$$(x + 1) + (x + 2) = x + 1 + x + 2$$
$$= x + x + 1 + 2$$
$$= 2x + 3.$$

Example 1 ▶ Finding the Domain of an Algebraic Expression

a. The domain of the polynomial

$$2x^3 + 3x + 4$$

is the set of all real numbers. In fact, the domain of any polynomial is the set of all real numbers, unless the domain is specifically restricted.

b. The domain of the radical expression

$$\sqrt{x - 2}$$

is the set of real numbers greater than or equal to 2, because the square root of a negative number is not a real number.

c. The domain of the expression

$$\frac{x + 2}{x - 3}$$

is the set of all real numbers except $x = 3$, which would produce an undefined division by zero.

The quotient of two algebraic expressions is a *fractional expression*. Moreover, the quotient of two *polynomials* such as

$$\frac{1}{x}, \qquad \frac{2x - 1}{x + 1}, \qquad \text{or} \qquad \frac{x^2 - 1}{x^2 + 1}$$

is a **rational expression.** Recall that a fraction is in simplest form if its numerator and denominator have no factors in common aside from ±1. To write a fraction in simplest form, divide out common factors.

$$\frac{a \cdot \cancel{c}}{b \cdot \cancel{c}} = \frac{a}{b}, \quad c \neq 0$$

The key to success in simplifying rational expressions lies in your ability to *factor* polynomials.

Simplifying Rational Expressions

When simplifying rational expressions, be sure to factor each polynomial completely before concluding that the numerator and denominator have no factors in common.

Example 2 ▶ Simplifying a Rational Expression

Write $\dfrac{x^2 + 4x - 12}{3x - 6}$ in simplest form.

Solution

$$\frac{x^2 + 4x - 12}{3x - 6} = \frac{(x + 6)(x - 2)}{3(x - 2)} \qquad \text{Factor completely.}$$

$$= \frac{x + 6}{3}, \qquad x \neq 2 \qquad \text{Divide out common factors.}$$

Note that the original expression is undefined when $x = 2$ (because division by zero is undefined). To make sure that the simplified expression is *equivalent* to the original expression, you must restrict the domain of the simplified expression by excluding the value $x = 2$.

STUDY T!P

In Example 2, do not make the mistake of trying to simplify further by dividing out terms.

$$\frac{x + 6}{3} \neq \frac{x + 6}{3}^{2} = x + 2$$

Remember that to simplify fractions, divide out common *factors*, not terms.

Sometimes it may be necessary to change the sign of a factor to simplify a rational expression, as shown in Example 3(b).

Example 3 ▶ Simplifying Rational Expressions

Write each expression in simplest form.

a. $\dfrac{x^3 - 4x}{x^2 + x - 2}$ **b.** $\dfrac{12 + x - x^2}{2x^2 - 9x + 4}$

Solution

a. $\dfrac{x^3 - 4x}{x^2 + x - 2} = \dfrac{x(x^2 - 4)}{(x + 2)(x - 1)}$

$$= \frac{x(x + 2)(x - 2)}{(x + 2)(x - 1)} \qquad \text{Factor completely.}$$

$$= \frac{x(x - 2)}{(x - 1)}, \qquad x \neq -2 \qquad \text{Divide out common factors.}$$

b. $\dfrac{12 + x - x^2}{2x^2 - 9x + 4} = \dfrac{(4 - x)(3 + x)}{(2x - 1)(x - 4)} \qquad \text{Factor completely.}$

$$= \frac{-(x - 4)(3 + x)}{(2x - 1)(x - 4)} \qquad (4 - x) = -(x - 4)$$

$$= -\frac{3 + x}{2x - 1}, \qquad x \neq 4 \qquad \text{Divide out common factors.}$$

Operations with Rational Expressions

To multiply or divide rational expressions, use the properties of fractions discussed in Section P.1. Recall that to divide fractions, you invert the divisor and multiply.

Example 4 ▶ Multiplying Rational Expressions

$$\frac{2x^2 + x - 6}{x^2 + 4x - 5} \cdot \frac{x^3 - 3x^2 + 2x}{4x^2 - 6x} = \frac{(2x - 3)(x + 2)}{(x + 5)(x - 1)} \cdot \frac{x(x - 2)(x - 1)}{2x(2x - 3)}$$

$$= \frac{(x + 2)(x - 2)}{2(x + 5)}, \qquad x \neq 0, x \neq 1, x \neq \tfrac{3}{2}$$

In Example 4, note that when multiplying rational expressions you must follow the convention of listing *by the product* all values of x that must be specifically excluded from the domain in order to make the domain of the product agree with the domains of the factors.

Example 5 ▶ Dividing Rational Expressions

$$\frac{x^3 - 8}{x^2 - 4} \div \frac{x^2 + 2x + 4}{x^3 + 8} = \frac{x^3 - 8}{x^2 - 4} \cdot \frac{x^3 + 8}{x^2 + 2x + 4} \qquad \text{Invert and multiply.}$$

$$= \frac{(x - 2)(x^2 + 2x + 4)}{(x + 2)(x - 2)} \cdot \frac{(x + 2)(x^2 - 2x + 4)}{x^2 + 2x + 4}$$

$$= x^2 - 2x + 4, \qquad x \neq \pm 2$$

To add or subtract rational expressions, you can use the LCD (least common denominator) method or the basic definition

$$\frac{a}{b} \pm \frac{c}{d} = \frac{ad \pm bc}{bd}, \qquad b \neq 0 \text{ and } d \neq 0. \qquad \text{Basic definition}$$

This definition provides an efficient way of adding or subtracting *two* fractions that have no common factors in their denominators.

Example 6 ▶ Subtracting Rational Expressions

$$\frac{x}{x - 3} - \frac{2}{3x + 4} = \frac{x(3x + 4) - 2(x - 3)}{(x - 3)(3x + 4)} \qquad \text{Basic definition}$$

$$= \frac{3x^2 + 4x - 2x + 6}{(x - 3)(3x + 4)} \qquad \text{Distributive Property}$$

$$= \frac{3x^2 + 2x + 6}{(x - 3)(3x + 4)} \qquad \text{Combine like terms.}$$

For three or more fractions, or for fractions with a repeated factor in the denominators, the LCD method works well. Recall that the least common denominator of several fractions consists of the product of all prime factors in the denominators, with each factor given the highest power of its occurrence in any denominator. Here is a numerical example.

$$\frac{1}{6} + \frac{3}{4} - \frac{2}{3} = \frac{1 \cdot 2}{6 \cdot 2} + \frac{3 \cdot 3}{4 \cdot 3} - \frac{2 \cdot 4}{3 \cdot 4} \qquad \text{The LCD is 12.}$$

$$= \frac{2}{12} + \frac{9}{12} - \frac{8}{12}$$

$$= \frac{3}{12}$$

$$= \frac{1}{4}$$

Sometimes the numerator of the answer has a factor in common with the denominator. In such cases the answer should be simplified. For instance, in the example above, $\frac{3}{12}$ was simplified to $\frac{1}{4}$.

Example 7 ▶ Combining Rational Expressions: The LCD Method

Perform the operations and simplify.

$$\frac{3}{x - 1} - \frac{2}{x} + \frac{x + 3}{x^2 - 1}$$

Solution

Using the factored denominators $(x - 1)$, x, and $(x + 1)(x - 1)$, you can see that the LCD is $x(x + 1)(x - 1)$.

$$\frac{3}{x - 1} - \frac{2}{x} + \frac{x + 3}{(x + 1)(x - 1)}$$

$$= \frac{3(x)(x + 1)}{x(x + 1)(x - 1)} - \frac{2(x + 1)(x - 1)}{x(x + 1)(x - 1)} + \frac{(x + 3)(x)}{x(x + 1)(x - 1)}$$

$$= \frac{3(x)(x + 1) - 2(x + 1)(x - 1) + (x + 3)(x)}{x(x + 1)(x - 1)}$$

$$= \frac{3x^2 + 3x - 2x^2 + 2 + x^2 + 3x}{x(x + 1)(x - 1)} \qquad \text{Distributive Property}$$

$$= \frac{3x^2 - 2x^2 + x^2 + 3x + 3x + 2}{x(x + 1)(x - 1)} \qquad \text{Group like terms.}$$

$$= \frac{2x^2 + 6x + 2}{x(x + 1)(x - 1)} \qquad \text{Combine like terms.}$$

$$= \frac{2(x^2 + 3x + 1)}{x(x + 1)(x - 1)} \qquad \text{Factor.}$$

Complex Fractions

Fractional expressions with separate fractions in the numerator, denominator, or both are called **complex fractions.** Here are two examples.

$$\frac{\left(\dfrac{1}{x}\right)}{x^2 + 1} \quad \text{and} \quad \frac{\left(\dfrac{1}{x}\right)}{\left(\dfrac{1}{x^2 + 1}\right)}$$

A complex fraction can be simplified by combining its numerator and denominator into single fractions, then inverting the denominator and multiplying.

Example 8 ▶   **Simplifying a Complex Fraction**

$$\frac{\left(\dfrac{2}{x} - 3\right)}{\left(1 - \dfrac{1}{x - 1}\right)} = \frac{\left[\dfrac{2 - 3(x)}{x}\right]}{\left[\dfrac{1(x - 1) - 1}{x - 1}\right]} \qquad \text{Combine fractions.}$$

$$= \frac{\left(\dfrac{2 - 3x}{x}\right)}{\left(\dfrac{x - 2}{x - 1}\right)} \qquad \text{Simplify.}$$

$$= \frac{2 - 3x}{x} \cdot \frac{x - 1}{x - 2} \qquad \text{Invert and multiply.}$$

$$= \frac{(2 - 3x)(x - 1)}{x(x - 2)}, \qquad x \ne 1$$

Another way to simplify a complex fraction is to multiply the numerator and denominator by the LCD of all fractions in its numerator and denominator. This method is applied to the fraction in Example 8 as follows.

$$\frac{\left(\dfrac{2}{x} - 3\right)}{\left(1 - \dfrac{1}{x - 1}\right)} = \frac{\left(\dfrac{2}{x} - 3\right)}{\left(1 - \dfrac{1}{x - 1}\right)} \cdot \frac{x(x - 1)}{x(x - 1)} \qquad \text{LCD is } x(x - 1).$$

$$= \frac{\left(\dfrac{2 - 3x}{x}\right) \cdot x(x - 1)}{\left(\dfrac{x - 2}{x - 1}\right) \cdot x(x - 1)}$$

$$= \frac{(2 - 3x)(x - 1)}{x(x - 2)}, \qquad x \ne 1$$

The next three examples illustrate some methods for simplifying rational expressions involving radicals and negative exponents. These types of expressions occur frequently in calculus.

To simplify an expression with negative exponents, one method is to begin by factoring out the common factor with the smaller exponent. Remember that when factoring, you subtract exponents. For instance, in $3x^{-5/2} + 2x^{-3/2}$ the smaller exponent is $-\frac{5}{2}$ and the common factor is $x^{-5/2}$.

$$3x^{-5/2} + 2x^{-3/2} = x^{-5/2}[3(1) + 2x^{-3/2-(-5/2)}]$$
$$= x^{-5/2}(3 + 2x^1)$$
$$= \frac{3 + 2x}{x^{5/2}}$$

Example 9 ▶ Simplifying an Expression

Simplify the following expression containing negative exponents.

$$x(1 - 2x)^{-3/2} + (1 - 2x)^{-1/2}$$

Solution

Begin by factoring out the common factor with the *smaller exponent.*

$$x(1 - 2x)^{-3/2} + (1 - 2x)^{-1/2} = (1 - 2x)^{-3/2}[x + (1 - 2x)^{(-1/2)-(-3/2)}]$$
$$= (1 - 2x)^{-3/2}[x + (1 - 2x)^1]$$
$$= \frac{1 - x}{(1 - 2x)^{3/2}}$$

A second method for simplifying an expression with negative exponents is shown in the next example.

Example 10 ▶ Simplifying a Complex Fraction

$$\frac{(4 - x^2)^{1/2} + x^2(4 - x^2)^{-1/2}}{4 - x^2}$$
$$= \frac{(4 - x^2)^{1/2} + x^2(4 - x^2)^{-1/2}}{4 - x^2} \cdot \frac{(4 - x^2)^{1/2}}{(4 - x^2)^{1/2}}$$
$$= \frac{(4 - x^2)^1 + x^2(4 - x^2)^0}{(4 - x^2)^{3/2}}$$
$$= \frac{4 - x^2 + x^2}{(4 - x^2)^{3/2}}$$
$$= \frac{4}{(4 - x^2)^{3/2}}$$

Example 11 ▶ **Rewriting a Difference Quotient**

The expression from calculus

$$\frac{\sqrt{x + h} - \sqrt{x}}{h}$$

is an example of a *difference quotient*. Rewrite this expression by rationalizing its numerator.

Solution

$$\frac{\sqrt{x + h} - \sqrt{x}}{h} = \frac{\sqrt{x + h} - \sqrt{x}}{h} \cdot \frac{\sqrt{x + h} + \sqrt{x}}{\sqrt{x + h} + \sqrt{x}}$$

$$= \frac{\left(\sqrt{x + h}\right)^2 - \left(\sqrt{x}\right)^2}{h\left(\sqrt{x + h} + \sqrt{x}\right)}$$

$$= \frac{h}{h\left(\sqrt{x + h} + \sqrt{x}\right)}$$

$$= \frac{1}{\sqrt{x + h} + \sqrt{x}}, \qquad h \neq 0$$

Notice that the original expression is undefined when $h = 0$. So, you must exclude $h = 0$ from the domain of the simplified expression so that the expressions are equivalent.

Difference quotients, such as that in Example 11, occur frequently in calculus. Often, they need to be rewritten in an equivalent form that can be evaluated when $h = 0$. Note that the equivalent form is not simpler than the original form, but it has the advantage that it is defined when $h = 0$.

Writing ABOUT MATHEMATICS

Comparing Domains of Two Expressions Complete the table by evaluating the expressions

$$\frac{x^2 - 3x + 2}{x - 2} \quad \text{and} \quad x - 1$$

for the values of x. If you have a graphing utility with a table feature, use it to help create the table. Write a short paragraph describing the equivalence or nonequivalence of the two expressions.

x	-3	-2	-1	0	1	2	3
$\dfrac{x^2 - 3x + 2}{x - 2}$							
$x - 1$							

P.5 Exercises

In Exercises 1–8, find the domain of the expression.

1. $3x^2 - 4x + 7$

2. $2x^2 + 5x - 2$

3. $4x^3 + 3, \quad x \geq 0$

4. $6x^2 - 9, \quad x > 0$

5. $\dfrac{1}{x - 2}$

6. $\dfrac{x + 1}{2x + 1}$

7. $\sqrt{x + 1}$

8. $\sqrt{6 - x}$

In Exercises 9–14, find the missing factor in the numerator such that the two fractions are equivalent.

9. $\dfrac{5}{2x} = \dfrac{5()}{6x^2}$

10. $\dfrac{3}{4} = \dfrac{3()}{4(x + 1)}$

11. $\dfrac{x + 1}{x} = \dfrac{(x + 1)()}{x(x - 2)}$

12. $\dfrac{3y - 4}{y + 1} = \dfrac{(3y - 4)()}{y^2 - 1}$

13. $\dfrac{3x}{x - 3} = \dfrac{3x()}{x^2 - 3x}$

14. $\dfrac{1 - z}{z^2} = \dfrac{(1 - z)()}{z^3 + z^2}$

In Exercises 15–32, write the rational expression in simplest form.

15. $\dfrac{15x^2}{10x}$

16. $\dfrac{18y^2}{60y^5}$

17. $\dfrac{3xy}{xy + x}$

18. $\dfrac{2x^2y}{xy - y}$

19. $\dfrac{4y - 8y^2}{10y - 5}$

20. $\dfrac{9x^2 + 9x}{2x + 2}$

21. $\dfrac{x - 5}{10 - 2x}$

22. $\dfrac{12 - 4x}{x - 3}$

23. $\dfrac{y^2 - 16}{y + 4}$

24. $\dfrac{x^2 - 25}{5 - x}$

25. $\dfrac{x^3 + 5x^2 + 6x}{x^2 - 4}$

26. $\dfrac{x^2 + 8x - 20}{x^2 + 11x + 10}$

27. $\dfrac{y^2 - 7y + 12}{y^2 + 3y - 18}$

28. $\dfrac{x^2 - 7x + 6}{x^2 + 11x + 10}$

29. $\dfrac{2 - x + 2x^2 - x^3}{x^2 - 4}$

30. $\dfrac{x^2 - 9}{x^3 + x^2 - 9x - 9}$

31. $\dfrac{z^3 - 8}{z^2 + 2z + 4}$

32. $\dfrac{y^3 - 2y^2 - 3y}{y^3 + 1}$

In Exercises 33 and 34, complete the table. What can you conclude?

33.

x	0	1	2	3	4	5	6
$\dfrac{x^2 - 2x - 3}{x - 3}$							
$x + 1$							

34.

x	0	1	2	3	4	5	6
$\dfrac{x - 3}{x^2 - x - 6}$							
$\dfrac{1}{x + 2}$							

35. *Error Analysis* Describe the error.

$$\dfrac{5x^3}{2x^3 + 4} = \dfrac{5x^3}{2x^3 + 4} = \dfrac{5}{2 + 4} = \dfrac{5}{6} \quad \times$$

36. *Error Analysis* Describe the error.

$$\dfrac{x^3 + 25x}{x^2 - 2x - 15} = \dfrac{x(x^2 + 25)}{(x - 5)(x + 3)} \quad \times$$

$$= \dfrac{x(x - 5)(x + 5)}{(x - 5)(x + 3)}$$

$$= \dfrac{x(x + 5)}{x + 3}$$

Geometry **In Exercises 37 and 38, find the ratio of the area of the shaded portion of the figure to the total area of the figure.**

37.

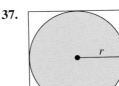

38.

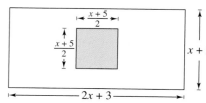

In Exercises 39–54, perform the multiplication or division and simplify.

39. $\dfrac{5}{x-1} \cdot \dfrac{x-1}{25(x-2)}$

40. $\dfrac{x+13}{x^3(3-x)} \cdot \dfrac{x(x-3)}{5}$

41. $\dfrac{(x+5)(x-3)}{x+2} \cdot \dfrac{1}{(x+5)(x+2)}$

42. $\dfrac{(x-9)(x+7)}{x+1} \cdot \dfrac{x}{9-x}$

43. $\dfrac{r}{r-1} \cdot \dfrac{r^2-1}{r^2}$

44. $\dfrac{4y-16}{5y+15} \cdot \dfrac{2y+6}{4-y}$

45. $\dfrac{t^2-t-6}{t^2+6t+9} \cdot \dfrac{t+3}{t^2-4}$

46. $\dfrac{y^3-8}{2y^3} \cdot \dfrac{4y}{y^2-5y+6}$

47. $\dfrac{x^2+xy-2y^2}{x^3+x^2y} \cdot \dfrac{x}{x^2+3xy+2y^2}$

48. $\dfrac{x^3-1}{x+1} \cdot \dfrac{x^2+1}{x^2-1}$

49. $\dfrac{3(x+y)}{4} \div \dfrac{x+y}{2}$

50. $\dfrac{x+2}{5(x-3)} \div \dfrac{x-2}{5(x-3)}$

51. $\dfrac{x^2+16}{5x+20} \div \dfrac{x+4}{5x^2-20}$

52. $\dfrac{3x+18}{x^4} \div \dfrac{x+6}{x^2}$

53. $\dfrac{x^2-36}{x} \div \dfrac{x^3-6x^2}{x^2+x}$

54. $\dfrac{x^2-14x+49}{x^2-49} \div \dfrac{3x-21}{x+7}$

In Exercises 55–66, perform the addition or subtraction and simplify.

55. $\dfrac{5}{x-1} + \dfrac{x}{x-1}$

56. $\dfrac{2x-1}{x+3} + \dfrac{1-x}{x+3}$

57. $6 - \dfrac{5}{x+3}$

58. $\dfrac{3}{x-1} - 5$

59. $\dfrac{3}{x-2} + \dfrac{5}{2-x}$

60. $\dfrac{2x}{x-5} - \dfrac{5}{5-x}$

61. $\dfrac{2}{x^2-4} - \dfrac{1}{x^2-3x+2}$

62. $\dfrac{x}{x^2+x-2} - \dfrac{1}{x+2}$

63. $\dfrac{1}{x^2-x-2} - \dfrac{x}{x^2-5x+6}$

64. $\dfrac{2}{x^2-x-2} + \dfrac{10}{x^2+2x-8}$

65. $-\dfrac{1}{x} + \dfrac{2}{x^2+1} + \dfrac{1}{x^3+x}$

66. $\dfrac{2}{x+1} + \dfrac{2}{x-1} + \dfrac{1}{x^2-1}$

🔵 In Exercises 67–74, factor the expression by removing the common factor with the smaller exponent.

67. $x^5 - 2x^{-2}$

68. $x^5 - 5x^{-3}$

69. $3x^{3/2} - 2x^{-1/2}$

70. $5x^5 - 3x^{-3/2}$

71. $x^2(x^2+1)^{-5} - (x^2+1)^{-4}$

72. $2x(x-5)^{-3} - 4x^2(x-5)^{-4}$

73. $2x^2(x-1)^{1/2} - 5(x-1)^{-1/2}$

74. $4x^3(2x-1)^{3/2} - 2x(2x-1)^{-1/2}$

Error Analysis   In Exercises 75 and 76, describe the error.

75. $\dfrac{x+4}{x+2} - \dfrac{3x-8}{x+2} = \dfrac{x+4-3x-8}{x+2}$ ✗

$$= \dfrac{-2x-4}{x+2}$$

$$= \dfrac{-2(x+2)}{x+2} = -2$$

76. $\dfrac{6-x}{x(x+2)} + \dfrac{x+2}{x^2} + \dfrac{8}{x^2(x+2)}$ ✗

$$= \dfrac{x(6-x)+(x+2)^2+8}{x^2(x+2)}$$

$$= \dfrac{6x-x^2+x^2+4+8}{x^2(x+2)}$$

$$= \dfrac{6(x+2)}{x^2(x+2)} = \dfrac{6}{x^2}$$

In Exercises 77–92, simplify the complex fraction.

77. $\dfrac{\left(\dfrac{x}{2}-1\right)}{(x-2)}$

78. $\dfrac{(x-4)}{\left(\dfrac{x}{4}-\dfrac{4}{x}\right)}$

79. $\dfrac{\left[\dfrac{x^2}{(x+1)^2}\right]}{\left[\dfrac{x}{(x+1)^3}\right]}$

80. $\dfrac{\left(\dfrac{x^2-1}{x}\right)}{\left[\dfrac{(x-1)^2}{x}\right]}$

81. $\dfrac{\left(\dfrac{1}{x}-\dfrac{1}{x+1}\right)}{\left(\dfrac{1}{x+1}\right)}$

82. $\dfrac{\left(\dfrac{5}{y}-\dfrac{6}{2y+1}\right)}{\left(\dfrac{5}{y}+4\right)}$

83. $\dfrac{\left(\dfrac{x+3}{x-3}\right)^2}{\left(\dfrac{1}{x+3}+\dfrac{1}{x-3}\right)}$

84. $\dfrac{\left(\dfrac{x+4}{x+5}-\dfrac{x}{x+1}\right)}{4}$

85. $\dfrac{\left[\dfrac{1}{(x+h)^2}-\dfrac{1}{x^2}\right]}{h}$

86. $\dfrac{\left(\dfrac{x+h}{x+h+1}-\dfrac{x}{x+1}\right)}{h}$

87. $\dfrac{\left(\sqrt{x}-\dfrac{1}{2\sqrt{x}}\right)}{\sqrt{x}}$

88. $\dfrac{\left(\dfrac{t^2}{\sqrt{t^2+1}}-\sqrt{t^2+1}\right)}{t^2}$

89. $\dfrac{3x^{1/3}-x^{-2/3}}{3x^{-2/3}}$

90. $\dfrac{-x^3(1-x^2)^{-1/2}-2x(1-x^2)^{1/2}}{x^4}$

91. $\dfrac{x(x+1)^{-3/4}-(x+1)^{1/4}}{x^2}$

92. $\dfrac{(2x+1)^{1/3}-\dfrac{4x}{3(2x+1)^{2/3}}}{(2x+1)^{2/3}}$

🔵 **In Exercises 93 and 94, rationalize the numerator of the expression.**

93. $\dfrac{\sqrt{x+2}-\sqrt{x}}{2}$

94. $\dfrac{\sqrt{z-3}-\sqrt{z}}{3}$

95. *Rate* A photocopier copies at a rate of 16 pages per minute.

(a) Find the time required to copy one page.

(b) Find the time required to copy x pages.

(c) Find the time required to copy 60 pages.

96. *Rate* After working together for t hours on a common task, two workers have done fractional parts of the job equal to $t/3$ and $t/5$, respectively. What fractional part of the task has been completed?

97. *Average* Determine the average of the two real numbers $x/3$ and $2x/5$.

98. *Partition into Equal Parts* Find three real numbers that divide the real number line between $x/3$ and $3x/4$ into four equal parts.

Finance **In Exercises 99 and 100, use the formula that gives the approximate annual interest rate r of a monthly installment loan**

$$r=\dfrac{\left[\dfrac{24(NM-P)}{N}\right]}{\left(P+\dfrac{NM}{12}\right)}$$

where N is the total number of payments, M is the monthly payment, and P is the amount financed.

99. (a) Approximate the annual interest rate for a 4-year car loan of $16,000 that has monthly payments of $400.

(b) Simplify the expression for the annual interest rate r, and then rework part (a).

100. (a) Approximate the annual interest rate for a 5-year car loan of $20,000 that has monthly payments of $400.

(b) Simplify the expression for the annual interest rate r, and then rework part (a).

101. *Refrigeration* When food (at room temperature) is placed in a refrigerator, the time required for the food to cool depends on the amount of food, the air circulation in the refrigerator, the original temperature of the food, and the temperature of the refrigerator. Consider the model that gives the temperature of food that has an original temperature of 75°F and is placed in a 40°F refrigerator

$$T=10\left(\dfrac{4t^2+16t+75}{t^2+4t+10}\right)$$

where T is the temperature (in degrees Fahrenheit) and t is the time (in hours).

(a) Complete the table.

t	0	2	4	6	8	10
T						

t	12	14	16	18	20	22
T						

(b) What value of T does the mathematical model appear to be approaching?

102. *Precious Metals* The costs per fine ounce of gold and silver for the years 1992 through 1997 are given in the table. (Source: U.S. Bureau of Mines, U.S. Geological Survey)

Year	1992	1993	1994	1995	1996	1997
Gold	$345	$361	$385	$386	$389	$333
Silver	$3.94	$4.30	$5.29	$5.15	$5.19	$4.90

Mathematical models for these data are

$$\text{Cost of gold} = \frac{-38.5t + 310.1}{0.007t^2 - 0.176t + 1}$$

and

$$\text{Cost of silver} = \frac{0.242t^2 - 1.86t + 4.02}{0.056t^2 - 0.45t + 1}$$

where $t = 2$ corresponds to the year 1992.

(a) Create a table using the models to estimate the prices of the two metals for the given years. Compare the estimates given by the models with the actual prices.

(b) Determine a model for the ratio of the price of gold to the price of silver. Use the model to find this ratio over the given years. Over this period of time, did the price of gold increase or decrease relative to the price of silver?

Probability **In Exercises 103–106, consider an experiment in which a marble is tossed into a box whose base is shown in the figure. The probability that the marble will come to rest in the shaded portion of the box is equal to the ratio of the shaded area to the total area of the figure. Find the probability.**

103.

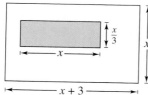

104.

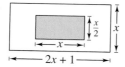

105.

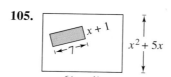

106.

Synthesis

True or False? **In Exercises 107–109, determine whether the statement is true or false. Justify your answer.**

107. $\dfrac{x^{2n} - 1^{2n}}{x^n - 1^n} = x^n + 1^n$

108. $\dfrac{x^2 - 3x + 2}{x - 1} = x - 2$ for all values of x.

109. The least common denominator of two or more fractions is always the product of all the denominators of the fractions.

110. How do you determine whether a rational expression is in simplest form?

111. *Think About It* Is the following statement true for all nonzero real numbers a and b? Explain.

$$\frac{ax - b}{b - ax} = -1$$

P.6 Errors and the Algebra of Calculus

▶ **What you should learn**
- How to avoid common algebraic errors
- How to recognize and use algebraic techniques that are common in calculus

▶ **Why you should learn it**
An efficient command of algebra is critical in the study of calculus.

Algebraic Errors to Avoid

This section contains five lists of common algebraic errors: errors involving parentheses, errors involving fractions, errors involving exponents, errors involving radicals, and errors involving dividing out. Many of these errors are made because they seem to be the *easiest* things to do.

Errors Involving Parentheses

Potential Error	Correct Form	Comment
$a - (x - b) \neq a - x - b$	$a - (x - b) = a - x + b$	Change all signs when distributing minus sign.
$(a + b)^2 \neq a^2 + b^2$	$(a + b)^2 = a^2 + 2ab + b^2$	Remember the middle term when squaring binomials.
$\left(\dfrac{1}{2}a\right)\left(\dfrac{1}{2}b\right) \neq \dfrac{1}{2}(ab)$	$\left(\dfrac{1}{2}a\right)\left(\dfrac{1}{2}b\right) = \dfrac{1}{4}(ab) = \dfrac{ab}{4}$	$\frac{1}{2}$ occurs twice as a factor.
$(3x + 6)^2 \neq 3(x + 2)^2$	$(3x + 6)^2 = [3(x + 2)]^2$ $= 3^2(x + 2)^2$	When factoring, apply exponents to all factors.

Errors Involving Fractions

Potential Error	Correct Form	Comment
$\dfrac{a}{x + b} \neq \dfrac{a}{x} + \dfrac{a}{b}$	Leave as $\dfrac{a}{x + b}$.	Do not add denominators when adding fractions.
$\dfrac{\left(\dfrac{x}{a}\right)}{b} \neq \dfrac{bx}{a}$	$\dfrac{\left(\dfrac{x}{a}\right)}{b} = \left(\dfrac{x}{a}\right)\left(\dfrac{1}{b}\right) = \dfrac{x}{ab}$	Multiply by the reciprocal when dividing fractions.
$\dfrac{1}{a} + \dfrac{1}{b} \neq \dfrac{1}{a + b}$	$\dfrac{1}{a} + \dfrac{1}{b} = \dfrac{b + a}{ab}$	Use the property for adding fractions.
$\dfrac{1}{3x} \neq \dfrac{1}{3}x$	$\dfrac{1}{3x} = \dfrac{1}{3} \cdot \dfrac{1}{x}$	Use the property for multiplying fractions.
$(1/3)x \neq \dfrac{1}{3x}$	$(1/3)x = \dfrac{1}{3} \cdot x = \dfrac{x}{3}$	Be careful when using a slash to denote division.
$(1/x) + 2 \neq \dfrac{1}{x + 2}$	$(1/x) + 2 = \dfrac{1}{x} + 2 = \dfrac{1 + 2x}{x}$	Be careful when using a slash to denote division.

Errors Involving Exponents

Potential Error	Correct Form	Comment
$(x^2)^3 \neq x^5$	$(x^2)^3 = x^{2\cdot3} = x^6$	Multiply exponents when raising a power to a power.
$x^2 \cdot x^3 \neq x^6$	$x^2 \cdot x^3 = x^{2+3} = x^5$	Add exponents when multiplying powers with like bases.
$2x^3 \neq (2x)^3$	$2x^3 = 2(x^3)$	Exponents have priority over coefficients.
$\dfrac{1}{x^2 - x^3} \neq x^{-2} - x^{-3}$	Leave as $\dfrac{1}{x^2 - x^3}$.	Do not move term-by-term from denominator to numerator.

Errors Involving Radicals

Potential Error	Correct Form	Comment
$\sqrt{5x} \neq 5\sqrt{x}$	$\sqrt{5x} = \sqrt{5}\sqrt{x}$	Radicals apply to every factor inside the radical.
$\sqrt{x^2 + a^2} \neq x + a$	Leave as $\sqrt{x^2 + a^2}$.	Do not apply radicals term-by-term.
$\sqrt{-x + a} \neq -\sqrt{x - a}$	Leave as $\sqrt{-x + a}$.	Do not factor minus signs out of square roots.

Errors Involving Dividing Out

Potential Error	Correct Form	Comment
$\dfrac{a + bx}{a} \neq 1 + bx$	$\dfrac{a + bx}{a} = \dfrac{a}{a} + \dfrac{bx}{a} = 1 + \dfrac{b}{a}x$	Divide out common factors, not common terms.
$\dfrac{a + ax}{a} \neq a + x$	$\dfrac{a + ax}{a} = \dfrac{a(1 + x)}{a} = 1 + x$	Factor before dividing out.
$1 + \dfrac{x}{2x} \neq 1 + \dfrac{1}{x}$	$1 + \dfrac{x}{2x} = 1 + \dfrac{1}{2} = \dfrac{3}{2}$	Divide out common factors.

For many people, a good way to avoid errors is to *work slowly, write neatly,* and *talk to yourself.* Each time you write a step, ask yourself why the step is algebraically legitimate. For instance, you can justify the step written below because *dividing the numerator and denominator by the same nonzero number produces an equivalent fraction.*

$$\frac{2x}{6} = \frac{2 \cdot x}{2 \cdot 3} = \frac{x}{3}$$

Example 1 ▶ Using the Property for Adding Fractions

Describe and correct the error. $\dfrac{1}{2x} + \dfrac{1}{3x} = \dfrac{1}{5x}$ ✗

Solution

When adding fractions, use the property for adding fractions: $\dfrac{1}{a} + \dfrac{1}{b} = \dfrac{b + a}{ab}$.

$$\frac{1}{2x} + \frac{1}{3x} = \frac{3x + 2x}{6x^2} = \frac{5x}{6x^2} = \frac{5}{6x}$$

Some Algebra of Calculus

In calculus it is often necessary to take a simplified algebraic expression and "unsimplify" it. See the following lists, taken from a standard calculus text.

Unusual Factoring

Expression	Useful Calculus Form	Comment
$\dfrac{5x^4}{8}$	$\dfrac{5}{8}x^4$	Write with fractional coefficient.
$\dfrac{x^2 + 3x}{-6}$	$-\dfrac{1}{6}(x^2 + 3x)$	Write with fractional coefficient.
$2x^2 - x - 3$	$2\left(x^2 - \dfrac{x}{2} - \dfrac{3}{2}\right)$	Factor out the leading coefficient.
$\dfrac{x}{2}(x + 1)^{-1/2} + (x + 1)^{1/2}$	$\dfrac{(x + 1)^{-1/2}}{2}[x + 2(x + 1)]$	Factor out factor with least power.

Writing with Negative Exponents

Expression	Useful Calculus Form	Comment
$\dfrac{9}{5x^3}$	$\dfrac{9}{5}x^{-3}$	Move the factor to the numerator and change the sign of the exponent.
$\dfrac{7}{\sqrt{2x - 3}}$	$7(2x - 3)^{-1/2}$	Move the factor to the numerator and change the sign of the exponent.

Writing a Fraction as a Sum

Expression	Useful Calculus Form	Comment
$\dfrac{x + 2x^2 + 1}{\sqrt{x}}$	$x^{1/2} + 2x^{3/2} + x^{-1/2}$	Divide each term by $x^{1/2}$.
$\dfrac{1 + x}{x^2 + 1}$	$\dfrac{1}{x^2 + 1} + \dfrac{x}{x^2 + 1}$	Rewrite the fraction as the sum of fractions.
$\dfrac{2x}{x^2 + 2x + 1}$	$\dfrac{2x + 2 - 2}{x^2 + 2x + 1}$ $= \dfrac{2x + 2}{x^2 + 2x + 1} - \dfrac{2}{(x + 1)^2}$	Add and subtract the same term. Rewrite the fraction as the difference of fractions.
$\dfrac{x^2 - 2}{x + 1}$	$x - 1 - \dfrac{1}{x + 1}$	Use long division. (See Section 3.3.)
$\dfrac{x + 7}{x^2 - x - 6}$	$\dfrac{2}{x - 3} - \dfrac{1}{x + 2}$	Use the method of partial fractions. (See Section 4.3.)

Inserting Factors and Terms

Expression	Useful Calculus Form	Comment
$(2x - 1)^3$	$\dfrac{1}{2}(2x - 1)^3(2)$	Multiply and divide by 2.
$7x^2(4x^3 - 5)^{1/2}$	$\dfrac{7}{12}(4x^3 - 5)^{1/2}(12x^2)$	Multiply and divide by 12.
$\dfrac{4x^2}{9} - 4y^2 = 1$	$\dfrac{x^2}{9/4} - \dfrac{y^2}{1/4} = 1$	Write with fractional denominators.
$\dfrac{x}{x + 1}$	$\dfrac{x + 1 - 1}{x + 1} = 1 - \dfrac{1}{x + 1}$	Add and subtract the same term.

The next five examples demonstrate many of the steps in the preceding lists.

Example 2 ▶ Factors Involving Negative Exponents

Factor $x(x + 1)^{-1/2} + (x + 1)^{1/2}$.

Solution

When multiplying factors with like bases, you add exponents. When factoring, you are undoing multiplication, and so you *subtract* exponents.

$$x(x + 1)^{-1/2} + (x + 1)^{1/2} = (x + 1)^{-1/2}[x(x + 1)^0 + (x + 1)^1]$$

$$= (x + 1)^{-1/2}[x + (x + 1)]$$

$$= (x + 1)^{-1/2}(2x + 1)$$

Here is another way to simplify the expression in Example 2.

$$x(x + 1)^{-1/2} + (x + 1)^{1/2} = x(x + 1)^{-1/2} + (x + 1)^{1/2} \cdot \frac{(x + 1)^{1/2}}{(x + 1)^{1/2}}$$

$$= \frac{x(x + 1)^0 + (x + 1)^1}{(x + 1)^{1/2}}$$

$$= \frac{2x + 1}{\sqrt{x + 1}}$$

Example 3 ▶ Inserting Factors in an Expression

Insert the required factor: $\dfrac{x + 2}{(x^2 + 4x - 3)^2} = (\quad)\dfrac{1}{(x^2 + 4x - 3)^2}(2x + 4)$.

Solution

The expression on the right side of the equation is twice the expression on the left side. To make both sides equal, insert a factor of $\frac{1}{2}$.

$$\frac{x + 2}{(x^2 + 4x - 3)^2} = \left(\frac{1}{2}\right)\frac{1}{(x^2 + 4x - 3)^2}(2x + 4)$$

Right side is multiplied and divided by 2.

Example 4 ▶ Rewriting Fractions

Explain the following.

$$\frac{4x^2}{9} - 4y^2 = \frac{x^2}{9/4} - \frac{y^2}{1/4}$$

Solution

To write the expression on the left side of the equation in the form given on the right side, multiply the numerators and denominators of both terms by $\frac{1}{4}$.

$$\frac{4x^2}{9} - 4y^2 = \frac{4x^2}{9}\left(\frac{1/4}{1/4}\right) - 4y^2\left(\frac{1/4}{1/4}\right)$$

$$= \frac{x^2}{9/4} - \frac{y^2}{1/4}$$

Example 5 ▶ Rewriting with Negative Exponents

Rewrite each expression using negative exponents.

a. $\dfrac{-4x}{(1 - 2x^2)^2}$ **b.** $\dfrac{2}{5x^3} - \dfrac{1}{\sqrt{x}} + \dfrac{3}{5(4x)^2}$

Solution

a. $\dfrac{-4x}{(1 - 2x^2)^2} = -4x(1 - 2x^2)^{-2}$

b. Begin by writing the second term in exponential form.

$$\frac{2}{5x^3} - \frac{1}{\sqrt{x}} + \frac{3}{5(4x)^2} = \frac{2}{5x^3} - \frac{1}{x^{1/2}} + \frac{3}{5(4x)^2}$$

$$= \frac{2}{5}x^{-3} - x^{-1/2} + \frac{3}{5}(4x)^{-2}$$

Example 6 ▶ Writing a Fraction as a Sum of Terms

Rewrite each fraction as the sum of three terms.

a. $\dfrac{x^2 - 4x + 8}{2x}$ **b.** $\dfrac{x + 2x^2 + 1}{\sqrt{x}}$

Solution

a. $\dfrac{x^2 - 4x + 8}{2x} = \dfrac{x^2}{2x} - \dfrac{4x}{2x} + \dfrac{8}{2x}$

$$= \frac{x}{2} - 2 + \frac{4}{x}$$

b. $\dfrac{x + 2x^2 + 1}{\sqrt{x}} = \dfrac{x}{x^{1/2}} + \dfrac{2x^2}{x^{1/2}} + \dfrac{1}{x^{1/2}}$

$$= x^{1/2} + 2x^{3/2} + x^{-1/2}$$

P.6 Exercises

In Exercises 1–22, describe and correct the error.

1. $2x - (3y + 4) = 2x - 3y + 4$ ✗

2. $5z + 3(x - 2) = 5z + 3x - 2$ ✗

3. $\dfrac{4}{16x - (2x + 1)} = \dfrac{4}{14x + 1}$ ✗

4. $\dfrac{1 - x}{(5 - x)(-x)} = \dfrac{x - 1}{x(x - 5)}$ ✗

5. $(5z)(6z) = 30z$ ✗ **6.** $x(yz) = (xy)(xz)$ ✗

7. $a\left(\dfrac{x}{y}\right) = \dfrac{ax}{ay}$ ✗ **8.** $(4x)^2 = 4x^2$ ✗

9. $\left(\dfrac{x}{y}\right)^3 = \dfrac{x^3}{y}$ ✗ **10.** $\sqrt{25 - x^2} = 5 - x$ ✗

11. $\sqrt{x + 9} = \sqrt{x} + 3$ ✗

12. $\dfrac{2x^2 + 1}{5x} = \dfrac{2x + 1}{5}$ ✗

13. $\dfrac{6x + y}{6x - y} = \dfrac{x + y}{x - y}$ ✗

14. $\dfrac{1}{a^{-1} + b^{-1}} = \left(\dfrac{1}{a + b}\right)^{-1}$ ✗

15. $\dfrac{1}{x + y^{-1}} = \dfrac{y}{x + 1}$ ✗

16. $(x^2 + 5x)^{1/2} = x(x + 5)^{1/2}$ ✗

17. $x(2x - 1)^2 = (2x^2 - x)^2$ ✗

18. $(3x^2 - 6x)^3 = 3x(x - 2)^3$ ✗

19. $\sqrt[3]{x^3 + 7x^2} = x^2\sqrt[3]{x + 7}$ ✗

20. $\dfrac{7 + 5(x + 3)}{x + 3} = 12$ ✗

21. $\dfrac{3}{x} + \dfrac{4}{y} = \dfrac{7}{x + y}$ ✗ **22.** $\dfrac{1}{2y} = (1/2)y$ ✗

27. $\frac{5}{2}z^2 - \frac{1}{4}z + 2 = (\;\rule{0.5cm}{0.15mm}\;)(10z^2 - z + 8)$

28. $x^2(x^3 - 1)^4 = (\;\rule{0.5cm}{0.15mm}\;)(x^3 - 1)^4(3x^2)$

29. $x(1 - 2x^2)^3 = (\;\rule{0.5cm}{0.15mm}\;)(1 - 2x^2)^3(-4x)$

30. $\dfrac{4x + 6}{(x^2 + 3x + 7)^3} = (\;\rule{0.5cm}{0.15mm}\;)\dfrac{1}{(x^2 + 3x + 7)^3}(2x + 3)$

31. $\dfrac{x + 1}{(x^2 + 2x - 3)^2} = (\;\rule{0.5cm}{0.15mm}\;)\dfrac{1}{(x^2 + 2x - 3)^2}(2x + 2)$

32. $\dfrac{1}{(x - 1)\sqrt{(x - 1)^4 - 4}} = \dfrac{(\;\rule{0.5cm}{0.15mm}\;)}{(x - 1)^2\sqrt{(x - 1)^4 - 4}}$

33. $\dfrac{3}{x} + \dfrac{5}{2x^2} - \dfrac{3}{2}x = (\;\rule{0.5cm}{0.15mm}\;)(6x + 5 - 3x^3)$

34. $\dfrac{(x - 1)^2}{169} + (y + 5)^2 = \dfrac{(x - 1)^3}{169(\;\rule{0.5cm}{0.15mm}\;)} + (y + 5)^2$

35. $\dfrac{9x^2}{25} + \dfrac{16y^2}{49} = \dfrac{x^2}{(\;\rule{0.5cm}{0.15mm}\;)} + \dfrac{y^2}{(\;\rule{0.5cm}{0.15mm}\;)}$

36. $\dfrac{3x^2}{4} - \dfrac{9y^2}{16} = \dfrac{x^2}{(\;\rule{0.5cm}{0.15mm}\;)} - \dfrac{y^2}{(\;\rule{0.5cm}{0.15mm}\;)}$

37. $\dfrac{x^2}{1/12} - \dfrac{y^2}{2/3} = \dfrac{12x^2}{(\;\rule{0.5cm}{0.15mm}\;)} - \dfrac{3y^2}{(\;\rule{0.5cm}{0.15mm}\;)}$

38. $\dfrac{x^2}{4/9} + \dfrac{y^2}{7/8} = \dfrac{9x^2}{(\;\rule{0.5cm}{0.15mm}\;)} + \dfrac{8y^2}{(\;\rule{0.5cm}{0.15mm}\;)}$

39. $x^{1/3} - 5x^{4/3} = x^{1/3}(\;\rule{0.5cm}{0.15mm}\;)$

40. $3(2x + 1)x^{1/2} + 4x^{3/2} = x^{1/2}(\;\rule{0.5cm}{0.15mm}\;)$

41. $(1 - 3x)^{4/3} - 4x(1 - 3x)^{1/3} = (1 - 3x)^{1/3}(\;\rule{0.5cm}{0.15mm}\;)$

42. $\dfrac{1}{2\sqrt{x}} + 5x^{3/2} - 10x^{5/2} = \dfrac{1}{2\sqrt{x}}(\;\rule{0.5cm}{0.15mm}\;)$

43. $\dfrac{1}{10}(2x + 1)^{5/2} - \dfrac{1}{6}(2x + 1)^{3/2} = \dfrac{(2x + 1)^{3/2}}{15}(\;\rule{0.5cm}{0.15mm}\;)$

44. $\dfrac{3}{7}(t + 1)^{7/3} - \dfrac{3}{4}(t + 1)^{4/3} = \dfrac{3(t + 1)^{4/3}}{28}(\;\rule{0.5cm}{0.15mm}\;)$

In Exercises 23–44, insert the required factor in the parentheses.

23. $\dfrac{3x + 2}{5} = \dfrac{1}{5}(\;\rule{0.5cm}{0.15mm}\;)$

24. $\dfrac{7x^2}{10} = \dfrac{7}{10}(\;\rule{0.5cm}{0.15mm}\;)$

25. $\frac{2}{3}x^2 + \frac{1}{3}x + 5 = \frac{1}{3}(\;\rule{0.5cm}{0.15mm}\;)$

26. $\frac{3}{4}x + \frac{1}{2} = \frac{1}{4}(\;\rule{0.5cm}{0.15mm}\;)$

In Exercises 45–50, write the fraction as a sum of two or more terms.

45. $\dfrac{16 - 5x - x^2}{x}$ **46.** $\dfrac{x^3 - 5x^2 + 4}{x^2}$

47. $\dfrac{4x^3 - 7x^2 + 1}{x^{1/3}}$ **48.** $\dfrac{2x^5 - 3x^3 + 5x - 1}{x^{3/2}}$

49. $\dfrac{3 - 5x^2 - x^4}{\sqrt{x}}$ **50.** $\dfrac{x^3 - 5x^4}{3x^2}$

In Exercises 51–62, simplify the expression.

51. $\dfrac{-2(x^2 - 3)^{-3}(2x)(x + 1)^3 - 3(x + 1)^2(x^2 - 3)^{-2}}{[(x + 1)^3]^2}$

52. $\dfrac{x^5(-3)(x^2 + 1)^{-4}(2x) - (x^2 + 1)^{-3}(5)x^4}{(x^5)^2}$

53. $\dfrac{(6x + 1)^3(27x^2 + 2) - (9x^3 + 2x)(3)(6x + 1)^2(6)}{[(6x + 1)^3]^2}$

54. $\dfrac{(4x^2 + 9)^{1/2}(2) - (2x + 3)(\frac{1}{2})(4x^2 + 9)^{-1/2}(8x)}{[(4x^2 + 9)^{1/2}]^2}$

55. $\dfrac{(x + 2)^{3/4}(x + 3)^{-2/3} - (x + 3)^{1/3}(x + 2)^{-1/4}}{[(x + 2)^{3/4}]^2}$

56. $(2x - 1)^{1/2} - (x + 2)(2x - 1)^{-1/2}$

57. $\dfrac{2(3x - 1)^{1/3} - (2x + 1)(\frac{1}{3})(3x - 1)^{-2/3}(3)}{(3x - 1)^{2/3}}$

58. $\dfrac{(x + 1)(\frac{1}{2})(2x - 3x^2)^{-1/2}(2 - 6x) - (2x - 3x^2)^{1/2}}{(x + 1)^2}$

59. $\dfrac{1}{(x^2 + 4)^{1/2}} \cdot \dfrac{1}{2}(x^2 + 4)^{-1/2}(2x)$

60. $\dfrac{1}{x^2 - 6}(2x) + \dfrac{1}{2x + 5}(2)$

61. $(x^2 + 5)^{1/2}(\frac{3}{2})(3x - 2)^{1/2}(3) +$
$\qquad (3x - 2)^{3/2}(\frac{1}{2})(x^2 + 5)^{-1/2}(2x)$

62. $(3x + 2)^{-1/2}(3)(x - 6)^{1/2}(1) +$
$\qquad (x - 6)^3(-\frac{1}{2})(3x + 2)^{-3/2}(3)$

63. (a) Verify that $y_1 = y_2$ analytically.

$y_1 = x^2(\frac{1}{3})(x^2 + 1)^{-2/3}(2x) + (x^2 + 1)^{1/3}(2x)$

$y_2 = \dfrac{2x(4x^2 + 3)}{3(x^2 + 1)^{2/3}}$

(b) Complete the table and demonstrate the equality in part (a) numerically.

x	-2	-1	$-\frac{1}{2}$	0	1	2	$\frac{5}{2}$
y_1							
y_2							

64. (a) Verify that $y_1 = y_2$ analytically.

$y_1 = -\dfrac{\sqrt{9 - x^2}}{x^2} - \dfrac{1}{\sqrt{9 - x^2}}$

$y_2 = \dfrac{-9}{x^2\sqrt{9 - x^2}}$

(b) Complete the table and demonstrate the equality in part (a) numerically.

x	-2	-1	$-\frac{1}{2}$	$\frac{1}{4}$	1	2	$\frac{5}{2}$
y_1							
y_2							

65. *Logical Reasoning* Verify that $y_1 \neq y_2$ by letting $x = 0$ and evaluating y_1 and y_2.

$y_1 = 2x\sqrt{1 - x^2} - \dfrac{x^3}{\sqrt{1 - x^2}}$

$y_2 = \dfrac{2 - 3x^2}{\sqrt{1 - x^2}}$

Change y_2 so that $y_1 = y_2$.

Synthesis

True or False? **In Exercises 66–69, determine whether the statement is true or false. Justify your answer.**

66. $x^{-1} + y^{-2} = \dfrac{y^2 + x}{xy^2}$

67. $\dfrac{1}{x^{-2} + y^{-1}} = x^2 + y$

68. $\dfrac{1}{\sqrt{x} + 4} = \dfrac{\sqrt{x} - 4}{x - 16}$

69. $\dfrac{x^2 - 9}{\sqrt{x} - 3} = \sqrt{x} + 3$

In Exercises 70–73, find and correct any errors.

70. $x^n \cdot x^{3n} = x^{3n^2}$

71. $(x^n)^{2n} + (x^{2n})^n = 2x^{2n^2}$

72. $x^{2n} + y^{2n} = (x^n + y^n)^2$

73. $\dfrac{x^{2n} \cdot x^{3n}}{x^{3n} + x^2} = \dfrac{x^{5n}}{x^{3n} + x^2}$

74. *Think About It* Suppose you are taking a course in calculus, and for one of the homework problems you obtain the following answer.

$\dfrac{1}{10}(2x - 1)^{5/2} + \dfrac{1}{6}(2x - 1)^{3/2}$

The answer in the back of the book is

$\dfrac{1}{15}(2x - 1)^{3/2}(3x + 1).$

Are these two answers equivalent? If so, show how the second answer can be obtained from the first.

P.7 Graphical Representation of Data

▶ **What you should learn**

- How to plot points in the Cartesian plane
- How to use the Distance Formula to find the distance between two points
- How to use the Midpoint Formula to find the midpoint of a line segment
- How to use a coordinate plane to model and solve real-life problems

▶ **Why you should learn it**

The Cartesian plane can be used to represent relationships between two variables. For instance, Exercise 68 on page 69 shows how to represent graphically the number of recording artists elected to the Rock and Roll Hall of Fame from 1986 to 1999.

The Cartesian Plane

Just as you can represent real numbers by points on a real number line, you can represent ordered pairs of real numbers by points in a plane called the **rectangular coordinate system,** or the **Cartesian plane,** named after the French mathematician René Descartes (1596–1650).

The Cartesian plane is formed by using two real number lines intersecting at right angles, as shown in Figure P.7. The horizontal real number line is usually called the **x-axis,** and the vertical real number line is usually called the **y-axis.** The point of intersection of these two axes is the **origin,** and the two axes divide the plane into four parts called **quadrants.**

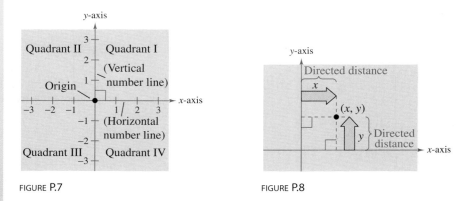

FIGURE P.7

FIGURE P.8

Each point in the plane corresponds to an **ordered pair** (x, y) of real numbers x and y, called **coordinates** of the point. The **x-coordinate** represents the directed distance from the y-axis to the point, and the **y-coordinate** represents the directed distance from the x-axis to the point, as shown in Figure P.8.

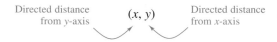

Directed distance from y-axis (x, y) Directed distance from x-axis

The notation (x, y) denotes both a point in the plane and an open interval on the real number line. The context will tell you which meaning is intended.

Example 1 ▶ Plotting Points in the Cartesian Plane

Plot the points $(-1, 2)$, $(3, 4)$, $(0, 0)$, $(3, 0)$, and $(-2, -3)$.

Solution

To plot the point $(-1, 2)$, imagine a vertical line through -1 on the x-axis and a horizontal line through 2 on the y-axis. The intersection of these two lines is the point $(-1, 2)$. The other four points can be plotted in a similar way, as shown in Figure P.9.

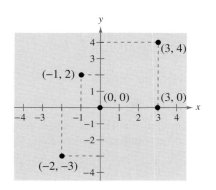

FIGURE P.9

The beauty of a rectangular coordinate system is that it allows you to see relationships between two variables. It would be difficult to overestimate the importance of Descartes's introduction of coordinates to the plane. Today, his ideas are in common use in virtually every scientific and business-related field.

A computer animation of this example appears in the *Interactive* CD-ROM and *Internet* versions of this text.

Example 2 ▶ Sketching a Scatter Plot

From 1988 through 1997, the amount *A* (in millions of dollars) spent on archery equipment in the United States is given in the table, where *t* represents the year. Sketch a scatter plot of the data.   (Source: National Sporting Goods Association)

t	1988	1989	1990	1991	1992	1993	1994	1995	1996	1997
A	235	261	265	270	334	285	306	287	272	273

Amount Spent on Archery Equipment

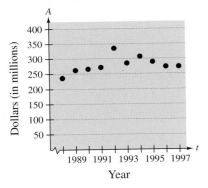

FIGURE P.10

Solution

To sketch a *scatter plot* of the data given in the table, you simply represent each pair of values by an ordered pair (*t*, *A*) and plot the resulting points, as shown in Figure P.10. For instance, the first pair of values is represented by the ordered pair (1988, 235). Note that the break in the *t*-axis indicates that the numbers between 0 and 1988 have been omitted.

In Example 2, you could have let *t* = 1 represent the year 1988. In that case, the horizontal axis would not have been broken, and the tick marks would have been labeled 1 through 10 (instead of 1988 through 1997).

Technology

The scatter plot in Example 2 is only one way to represent the data graphically. Two other techniques are shown at the right. The first is a bar graph and the second is a line graph. All three graphical representations were created with a computer. If you have access to a graphing utility, try using it to represent graphically the data given in Example 2.

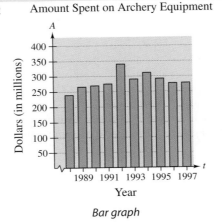

Bar graph

Line graph

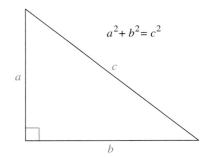

$$a^2 + b^2 = c^2$$

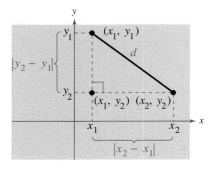

FIGURE P.12

The Distance Formula

Recall from the Pythagorean Theorem that, for a right triangle with hypotenuse of length c and sides of lengths a and b, you have

$$a^2 + b^2 = c^2 \qquad \text{Pythagorean Theorem}$$

as shown in Figure P.11. (The converse is also true. That is, if $a^2 + b^2 = c^2$, then the triangle is a right triangle.)

Suppose you want to determine the distance d between two points (x_1, y_1) and (x_2, y_2) in the plane. With these two points, a right triangle can be formed, as shown in Figure P.12. The length of the vertical side of the triangle is $|y_2 - y_1|$, and the length of the horizontal side is $|x_2 - x_1|$. By the Pythagorean Theorem, you can write

$$d^2 = |x_2 - x_1|^2 + |y_2 - y_1|^2$$
$$d = \sqrt{|x_2 - x_1|^2 + |y_2 - y_1|^2}$$
$$d = \sqrt{(x_2 - x_1)^2 + (y_2 - y_1)^2}.$$

This result is the **Distance Formula.**

The Distance Formula

The distance d between the points (x_1, y_1) and (x_2, y_2) in the plane is

$$d = \sqrt{(x_2 - x_1)^2 + (y_2 - y_1)^2}.$$

Example 3 ▶ Finding a Distance

Find the distance between the points $(-2, 1)$ and $(3, 4)$.

Solution

Let $(x_1, y_1) = (-2, 1)$ and $(x_2, y_2) = (3, 4)$. Then apply the Distance Formula.

$$d = \sqrt{(x_2 - x_1)^2 + (y_2 - y_1)^2} \qquad \text{Distance Formula}$$
$$= \sqrt{[3 - (-2)]^2 + (4 - 1)^2} \qquad \text{Substitute for } x_1, y_1, x_2, \text{ and } y_2.$$
$$= \sqrt{(5)^2 + (3)^2} \qquad \text{Simplify.}$$
$$= \sqrt{34}$$
$$\approx 5.83 \qquad \text{Use a calculator.}$$

Note in Figure P.13 that a distance of 5.83 looks about right. You can use the Pythagorean Theorem to check that the distance is correct.

$$d^2 \stackrel{?}{=} 3^2 + 5^2 \qquad \text{Pythagorean Theorem}$$
$$\left(\sqrt{34}\right)^2 \stackrel{?}{=} 3^2 + 5^2 \qquad \text{Substitute for } d.$$
$$34 = 34 \qquad \text{Distance checks. ✓}$$

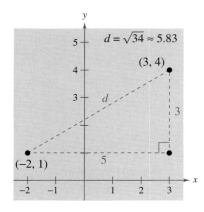

FIGURE P.13

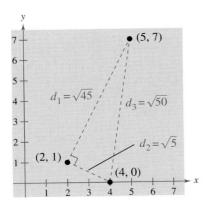

FIGURE P.14

Example 4 ▶ Verifying a Right Triangle

Show that the points $(2, 1)$, $(4, 0)$, and $(5, 7)$ are vertices of a right triangle.

Solution

The three points are plotted in Figure P.14. Using the Distance Formula, you can find the lengths of the three sides as follows.

$$d_1 = \sqrt{(5 - 2)^2 + (7 - 1)^2} = \sqrt{9 + 36} = \sqrt{45}$$

$$d_2 = \sqrt{(4 - 2)^2 + (0 - 1)^2} = \sqrt{4 + 1} = \sqrt{5}$$

$$d_3 = \sqrt{(5 - 4)^2 + (7 - 0)^2} = \sqrt{1 + 49} = \sqrt{50}$$

Because

$$d_1{}^2 + d_2{}^2 = 45 + 5 = 50 = d_3{}^2,$$

you can conclude that the triangle must be a right triangle.

The figures provided with Examples 3 and 4 were not really essential to the solution. Nevertheless, it is strongly recommended that you develop the habit of including sketches with your solutions—even if they are not required.

Example 5 ▶ Finding the Length of a Pass

A football quarterback throws a pass from the 5-yard line, 20 yards from the sideline. The pass is caught by a wide receiver on the 45-yard line, 50 yards from the same sideline, as shown in Figure P.15. How long is the pass?

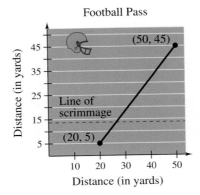

FIGURE P.15

STUDY T!P

In Example 5, the scale along the goal line does not normally appear on a football field. However, when you use coordinate geometry to solve real-life problems, you are free to place the coordinate system in any way that is convenient to the solution of the problem.

Solution

You can find the length of the pass by finding the distance between the points $(20, 5)$ and $(50, 45)$.

$$d = \sqrt{(50 - 20)^2 + (45 - 5)^2} \qquad \text{Distance Formula}$$

$$= \sqrt{900 + 1600}$$

$$= 50 \qquad \text{Simplify.}$$

So, the pass is 50 yards long.

The Midpoint Formula

To find the **midpoint** of the line segment that joins two points in a coordinate plane, you can simply find the average values of the respective coordinates of the two endpoints using the **Midpoint Formula.** (See Appendix A for a proof of the Midpoint Formula.)

> ### The Midpoint Formula
>
> The midpoint of the segment joining the points (x_1, y_1) and (x_2, y_2) is given by the Midpoint Formula
>
> $$\text{Midpoint} = \left(\frac{x_1 + x_2}{2}, \frac{y_1 + y_2}{2}\right).$$

Example 6 ▶ Finding a Line Segment's Midpoint

Find the midpoint of the line segment joining the points $(-5, -3)$ and $(9, 3)$, as shown in Figure P.16.

Solution

Let $(x_1, y_1) = (-5, -3)$ and $(x_2, y_2) = (9, 3)$.

$$\text{Midpoint} = \left(\frac{x_1 + x_2}{2}, \frac{y_1 + y_2}{2}\right) \qquad \text{Midpoint Formula}$$

$$= \left(\frac{-5 + 9}{2}, \frac{-3 + 3}{2}\right) \qquad \text{Substitute for } x_1, y_1, x_2, \text{ and } y_2.$$

$$= (2, 0) \qquad \text{Simplify.}$$

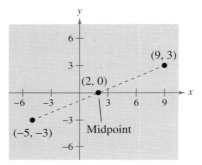

FIGURE P.16

Example 7 ▶ Estimating Annual Sales

Winn-Dixie Stores had annual sales of $1.30 billion in 1996 and $1.36 billion in 1998. Without knowing any additional information, what would you estimate the 1997 sales to have been? (Source: Winn-Dixie Stores, Inc.)

Solution

One solution to the problem is to assume that sales followed a linear pattern. With this assumption, you can estimate the 1997 sales by finding the midpoint of the segment connecting the points (1996, 1.30) and (1998, 1.36).

$$\text{Midpoint} = \left(\frac{1996 + 1998}{2}, \frac{1.30 + 1.36}{2}\right)$$

$$= (1997, 1.33)$$

So, you would estimate the 1997 sales to have been about $1.33 billion, as shown in Figure P.17. (The actual 1997 sales were $1.32 billion.)

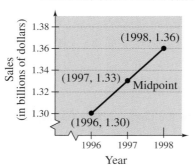

FIGURE P.17

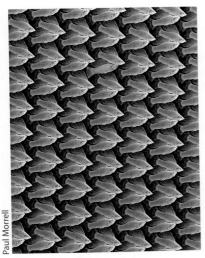

Paul Morrell

Much of computer graphics, including this computer-generated goldfish tesselation, consists of transformations of points in a coordinate plane. One type of transformation, a translation, is illustrated in Example 8. Other types include reflections, rotations, and stretches.

Application

 A computer animation of this example appears in the *Interactive* CD-ROM and *Internet* versions of this text.

Example 8 ▶ Translating Points in the Plane

The triangle in Figure P.18(a) has vertices at the points $(-1, 2)$, $(1, -4)$, and $(2, 3)$. Shift the triangle 3 units to the right and 2 units up and find the vertices of the shifted triangle, as shown in Figure P.18(b).

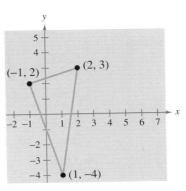

(a)

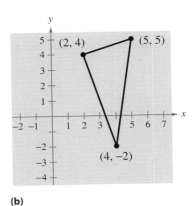

(b)

FIGURE P.18

Solution

To shift the vertices 3 units to the right, add 3 to each of the x-coordinates. To shift the vertices 2 units up, add 2 to each of the y-coordinates.

Original Point	*Translated Point*
$(-1, 2)$	$(-1 + 3, 2 + 2) = (2, 4)$
$(1, -4)$	$(1 + 3, -4 + 2) = (4, -2)$
$(2, 3)$	$(2 + 3, 3 + 2) = (5, 5)$

Writing ABOUT MATHEMATICS

Extending the Example Example 8 shows how to translate points in a coordinate plane. Write a short paragraph describing how each of the following transformed points is related to the original point.

Original Point	*Transformed Point*
(x, y)	$(-x, y)$
(x, y)	$(x, -y)$
(x, y)	$(-x, -y)$

P.7 Exercises

In Exercises 1–4, sketch the polygon with the indicated vertices.

1. Triangle: $(-1, 1)$, $(2, -1)$, $(3, 4)$
2. Triangle: $(0, 3)$, $(-1, -2)$, $(4, 8)$
3. Square: $(2, 4)$, $(5, 1)$, $(2, -2)$, $(-1, 1)$
4. Parallelogram: $(5, 2)$, $(7, 0)$, $(1, -2)$, $(-1, 0)$

In Exercises 5 and 6, approximate the coordinates of the points.

5.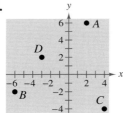
6.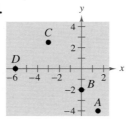

In Exercises 7–10, find the coordinates of the point.

7. The point is located 3 units to the left of the y-axis and 4 units above the x-axis.
8. The point is located 8 units below the x-axis and 4 units to the right of the y-axis.
9. The point is located 5 units below the x-axis and the coordinates of the point are equal.
10. The point is on the x-axis and 12 units to the left of the y-axis.

In Exercises 11–20, determine the quadrant(s) in which (x, y) is located so that the condition(s) is (are) satisfied.

11. $x > 0$ and $y < 0$
12. $x < 0$ and $y < 0$
13. $x = -4$ and $y > 0$
14. $x > 2$ and $y = 3$
15. $y < -5$
16. $x > 4$
17. $(x, -y)$ is in the second quadrant.
18. $(-x, y)$ is in the fourth quadrant.
19. $xy > 0$ 20. $xy < 0$

In Exercises 21–24, the polygon is shifted to a new position in the plane. Find the coordinates of the vertices of the polygon in its *new* position.

21.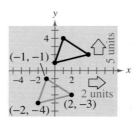
22.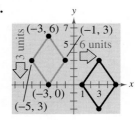

23. Original coordinates of vertices:

    $(-7, -2)$, $(-2, 2)$, $(-2, -4)$, $(-7, -4)$

    Shift: 8 units up, 4 units to the right

24. Original coordinates of vertices:

    $(5, 8)$, $(3, 6)$, $(7, 6)$, $(5, 2)$

    Shift: 6 units down, 10 units to the left

In Exercises 25 and 26, sketch a scatter plot of the data given in the table.

25. *Meteorology* The table shows the lowest temperature of record y (in degrees Fahrenheit) in Duluth, Minnesota, for each month x, where $x = 1$ represents January.   (Source: NOAA)

x	1	2	3	4	5	6
y	-39	-33	-29	-5	17	27

x	7	8	9	10	11	12
y	35	32	22	8	-23	-34

26. *Business* The table shows the number y of Wal-Mart stores for each year x from 1992 through 1999.   (Source: Wal-Mart Stores, Inc.)

x	1992	1993	1994	1995
y	2136	2440	2759	2943

x	1996	1997	1998	1999
y	3054	3406	3630	3815

Milk Prices **In Exercises 27 and 28, use the graph below,** **which shows the average retail price of one-half gallon of** **milk from 1992 to 1997.** (Source: U.S. Bureau of Labor Statistics)

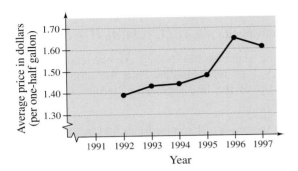

27. Approximate the highest price of one-half gallon of milk shown in the graph. When did this occur?

28. Approximate the percent change in the price of milk from the price in 1992 to the highest price shown in the graph.

Advertising **In Exercises 29 and 30, use the graph below,** **which shows the cost of a 30-second television spot (in** **thousands of dollars) during the Super Bowl from 1987 to** **1999.** (Source: USA Today Research)

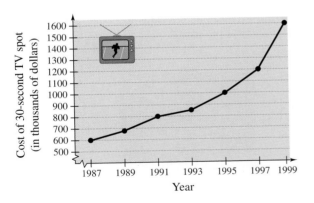

29. Approximate the percent increase in the cost of a 30-second spot from Super Bowl XXI in 1987 to Super Bowl XXXIII in 1999.

30. Estimate the increase in the cost of a 30-second spot (a) from Super Bowl XXI to Super Bowl XXVII and (b) from Super Bowl XXVII to Super Bowl XXXIII.

Labor Force **In Exercises 31 and 32, use the graph below,** **which shows the minimum wage in the United States (in** **dollars) from 1950 to 1999.** (Source: U.S. Employment Standards Administration)

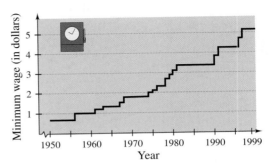

31. Which decade shows the greatest increase in minimum wage?

32. Approximate the percent increase in the minimum wage (a) from 1990 to 1995 and (b) from 1955 to 1995.

Data Analysis **In Exercises 33 and 34, use the graph** **below, which shows the mathematics entrance test scores** *x*, **and the final examination scores** *y*, **in an algebra course** **for a sample of 10 students.**

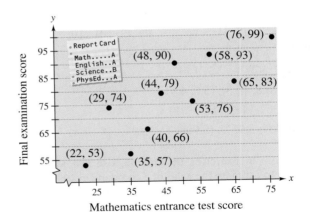

33. Find the entrance exam score of any student with a final exam score in the 80s.

34. Does a higher entrance exam score imply a higher final exam score? Explain.

In Exercises 35–38, find the distance between the points. (*Note:* **In each case the two points lie on the same horizontal or vertical line.**)

35. $(6, -3), (6, 5)$
36. $(1, 4), (8, 4)$
37. $(-3, -1), (2, -1)$
38. $(-3, -4), (-3, 6)$

In Exercises 39–42, (a) find the length of each side of the right triangle, and (b) show that these lengths satisfy the Pythagorean Theorem.

39.

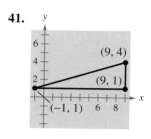

40.

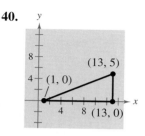

41.

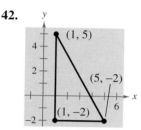

42.

In Exercises 43–54, (a) plot the points, (b) find the distance between the points, and (c) find the midpoint of the line segment joining the points.

43. $(1, 1), (9, 7)$

44. $(1, 12), (6, 0)$

45. $(-4, 10), (4, -5)$

46. $(-7, -4), (2, 8)$

47. $(-1, 2), (5, 4)$

48. $(2, 10), (10, 2)$

49. $\left(\frac{1}{2}, 1\right), \left(-\frac{5}{2}, \frac{4}{3}\right)$

50. $\left(-\frac{1}{3}, -\frac{1}{3}\right), \left(-\frac{1}{6}, -\frac{1}{2}\right)$

51. $(6.2, 5.4), (-3.7, 1.8)$

52. $(-16.8, 12.3), (5.6, 4.9)$

53. $(-36, -18), (48, -72)$

54. $(1.451, 3.051), (5.906, 11.360)$

Business In Exercises 55 and 56, use the Midpoint Formula to estimate the sales of a company in 1998, given the sales in 1996 and 2000. Assume that the sales followed a linear pattern.

55.

Year	1996	2000
Sales	$520,000	$740,000

56.

Year	1996	2000
Sales	$4,200,000	$5,650,000

In Exercises 57–60, show that the points form the vertices of the polygon.

57. Right triangle: $(4, 0), (2, 1), (-1, -5)$

58. Isosceles triangle: $(1, -3), (3, 2), (-2, 4)$

59. Parallelogram: $(2, 5), (0, 9), (-2, 0), (0, -4)$

60. Parallelogram: $(0, 1), (3, 7), (4, 4), (1, -2)$

61. A line segment has (x_1, y_1) as one endpoint and (x_m, y_m) as its midpoint. Find the other endpoint (x_2, y_2) of the line segment in terms of $x_1, y_1, x_m,$ and y_m.

62. Use the result of Exercise 61 to find the coordinates of the endpoint of a line segment if the coordinates of the other endpoint and midpoint are, respectively,
(a) $(1, -2), (4, -1)$ and (b) $(-5, 11), (2, 4)$.

63. Use the Midpoint Formula three times to find the three points that divide the line segment joining (x_1, y_1) and (x_2, y_2) into four parts.

64. Use the result of Exercise 63 to find the points that divide the line segment joining the given points into four equal parts.
(a) $(1, -2), (4, -1)$ (b) $(-2, -3), (0, 0)$

65. *Sports* In a football game, a quarterback throws a pass from the 15-yard line, 10 yards from the sideline, as shown in the figure. The pass is caught on the 40-yard line, 45 yards from the same sideline. How long is the pass?

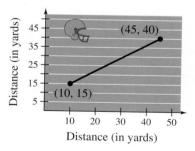

66. *Flying Distance* A plane flies in a straight line to a city that is 100 kilometers east and 150 kilometers north of the point of departure. How far does it fly?

67. *Make a Conjecture*   Plot the points $(2, 1)$, $(-3, 5)$, and $(7, -3)$ on a rectangular coordinate system. Then change the sign of the *x*-coordinate of each point and plot the three new points on the same rectangular coordinate system. Make a conjecture about the location of a point when each of the following occurs.

(a) The sign of the *x*-coordinate is changed.

(b) The sign of the *y*-coordinate is changed.

(c) The signs of both the *x*- and *y*-coordinates are changed.

68. *Rock and Roll Hall of Fame*   The graph below shows the numbers of recording artists who were elected to the Rock and Roll Hall of Fame in the years from 1986 to 1999.

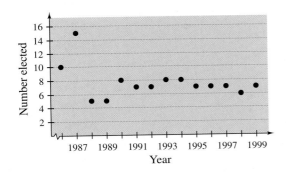

(a) Describe any trends in the data. From these trends, predict the number of artists elected in 2001.

(b) Why do you think the numbers elected in 1986 and 1987 were greater than in other years?

69. *Business*   Starbucks Corp. had annual sales of $696.5 million in 1996 and $1308.7 million in 1998. Use the Midpoint Formula to estimate the sales in 1997.   (Source: Starbucks Corp.)

70. *Business*   Lands' End, Inc. had annual sales of $1118.7 million in 1996 and $1371.4 million in 1998. Use the Midpoint Formula to estimate the sales in 1997.   (Source: Lands' End, Inc.)

Synthesis

True or False?   **In Exercises 71 and 72, determine whether the statement is true or false. Justify your answer.**

71. In order to divide a line segment into 16 equal parts, you would have to use the Midpoint Formula 16 times.

72. The points $(-8, 4)$, $(2, 11)$, and $(-5, 1)$ represent the vertices of an isosceles triangle.

73. *Think About It*   What is the *y*-coordinate of any point on the *x*-axis? What is the *x*-coordinate of any point on the *y*-axis?

74. *Think About It*   When plotting points on the rectangular coordinate system, is it true that the scales on the *x*- and *y*-axes must be the same? Explain.

75. Prove that the diagonals of the parallelogram in the figure intersect at their midpoints.

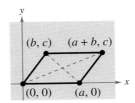

In Exercises 76–79, use the plot of the point (x_0, y_0) in the figure. Match the transformation of the point with the correct plot. [The plots are labeled (a), (b), (c), and (d).]

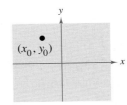

(a)

(b)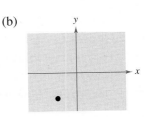

(c)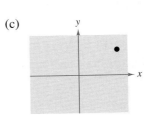

(d)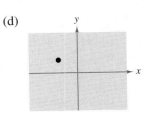

76. $(x_0, -y_0)$

77. $(-2x_0, y_0)$

78. $\left(x_0, \frac{1}{2}y_0\right)$

79. $(-x_0, -y_0)$

Chapter Summary

What did you learn?

Review Exercises

P.1 In Exercises 1 and 2, determine which numbers in the set are (a) natural numbers, (b) integers, (c) rational numbers, and (d) irrational numbers.

1. $\left\{11, -14, -\frac{8}{9}, \frac{5}{2}, \sqrt{6}, 0.4\right\}$
2. $\left\{\sqrt{15}, -22, -\frac{10}{3}, 0, 5.2, \frac{3}{7}\right\}$

In Exercises 3 and 4, use a calculator to find the decimal form of each rational number. If it is a nonterminating decimal, write the repeating pattern. Then plot the numbers on the real number line and place the appropriate inequality sign (< or >) between them.

3. (a) $\frac{5}{6}$ (b) $\frac{7}{8}$
4. (a) $\frac{9}{25}$ (b) $\frac{5}{7}$

In Exercises 5 and 6, give a verbal description of the subset of real numbers represented by the inequality, and sketch the subset on the real number line.

5. $x \le 7$
6. $x > 1$

In Exercises 7 and 8, find the distance between a and b.

7. $a = -92, b = 63$
8. $a = -112, b = -6$

In Exercises 9–12, use absolute value notation to describe the expression.

9. The distance between x and 7 is at least 4.
10. The distance between x and 25 is no more than 10.
11. The distance between y and -30 is less than 5.
12. The distance between z and -16 is greater than 8.

In Exercises 13–16, evaluate the expression for each value of x. (If not possible, state the reason.)

	Expression	*Values*	
13.	$12x - 7$	(a) $x = 0$	(b) $x = -1$
14.	$x^2 - 6x + 5$	(a) $x = -2$	(b) $x = 2$
15.	$-x^2 + x - 1$	(a) $x = 1$	(b) $x = -1$
16.	$\dfrac{x}{x - 3}$	(a) $x = -3$	(b) $x = 3$

In Exercises 17–20, identify the rule of algebra illustrated by the equation.

17. $2x + (3x - 10) = (2x + 3x) - 10$
18. $(t + 4)(2t) = (2t)(t + 4)$
19. $\dfrac{2}{y + 4} \cdot \dfrac{y + 4}{2} = 1, \quad y \ne -4$
20. $0 + (a - 5) = a - 5$

In Exercises 21–26, perform the operations without using a calculator.

21. $|-3| + 4(-2) - 6$
22. $\dfrac{|-10|}{-10}$
23. $\frac{5}{18} \div \frac{10}{3}$
24. $(16 - 8) \div 4$
25. $6[4 - 2(6 + 8)]$
26. $-4[16 - 3(7 - 10)]$

P.2 In Exercises 27 and 28, simplify each expression.

27. (a) $\dfrac{6^2 u^3 v^{-3}}{12u^{-2}v}$ (b) $\dfrac{3^{-4}m^{-1}n^{-3}}{9^{-2}mn^{-3}}$
28. (a) $(x + y^{-1})^{-1}$ (b) $\left(\dfrac{x^{-3}}{y}\right)\left(\dfrac{x}{y}\right)^{-1}$

In Exercises 29 and 30, write the number in scientific notation.

29. *Sales of K-Mart Corporation in 1998:* $33,674,000,000 (Source: K-Mart Corporation)
30. *Number of meters in 1 foot:* 0.3048

In Exercises 31 and 32, write the number in decimal form.

31. *Distance between the sun and Jupiter:* 4.836×10^8 miles
32. *Ratio of day to year:* 2.74×10^{-3}

In Exercises 33–38, simplify each expression.

33. (a) $\sqrt[3]{27^2}$ (b) $\sqrt{49^3}$
34. (a) $\sqrt[3]{\frac{64}{125}}$ (b) $\sqrt{\frac{81}{100}}$
35. (a) $\sqrt{\sqrt[3]{7}}$ (b) $\sqrt{\sqrt{19}}$
36. (a) $\left(\sqrt[3]{216}\right)^3$ (b) $\sqrt[4]{32^4}$
37. (a) $\sqrt{4x^4}$ (b) $\sqrt{\frac{18u^2}{b^3}}$
38. (a) $\sqrt[3]{\frac{2x^3}{27}}$ (b) $\sqrt[5]{64x^6}$

In Exercises 39 and 40, simplify each expression.

39. (a) $\sqrt{50} - \sqrt{18}$ (b) $2\sqrt{32} + 3\sqrt{72}$
40. (a) $\sqrt{8x^3} + \sqrt{2x}$ (b) $\sqrt{18x^5} - \sqrt{8x^3}$

41. *Writing*   Explain why $\sqrt{5u} + \sqrt{3u} \neq 2\sqrt{2u}$.

42. *Engineering*   The rectangular cross section of a wooden beam cut from a log of diameter 24 inches (see figure) will have a maximum strength if its width w and height h are

$$w = 8\sqrt{3} \quad \text{and} \quad h = \sqrt{24^2 - (8\sqrt{3})^2}.$$

Find the area of the rectangular cross section and express the answer in simplest form.

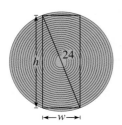

In Exercises 43 and 44, rewrite the expression by rationalizing the denominator or numerator. Simplify your answer.

43. $\dfrac{1}{2 - \sqrt{3}}$

44. $\dfrac{\sqrt{x} - 1}{2}$

In Exercises 45–48, simplify the expression.

45. $(16)^{3/2}$

46. $(64)^{-2/3}$

47. $(3x^{2/5})(2x^{1/2})$

48. $(x - 1)^{1/3}(x - 1)^{-1/4}$

In Exercises 49 and 50, fill in the missing form of the equation.

	Radical Form	*Rational Exponent Form*
49.	$\sqrt{16} = 4$	$ = 4$
50.	$ = 2$	$16^{1/4} = 2$

P.3   **In Exercises 51–54, write the polynomial in standard form.**

51. $3 - 11x^2$

52. $3x^3 - 5x^5 + x - 4$

53. $-4 - 12x^2$

54. $12x - 7x^2 + 6$

In Exercises 55–58, perform the operations and write the result in standard form.

55. $-(3x^2 + 2x) + (1 - 5x)$

56. $8y - [2y^2 - (3y - 8)]$

57. $(3x - 6)(5x + 1)$

58. $\left(x - \dfrac{1}{x}\right)(x + 2)$

In Exercises 59–62, find the product.

59. $(2x - 3)^2$

60. $(6x + 5)(6x - 5)$

61. $(3\sqrt{5} + 2)(3\sqrt{5} - 2)$

62. $(x - 4)^3$

63. *Exploration*   The surface area of a right circular cylinder is $S = 2\pi r^2 + 2\pi rh$.

(a) Draw a right circular cylinder of radius r and height h. Use the figure to explain how the surface area formula was obtained.

(b) Find the surface area when the radius is 6 inches and the height is 8 inches.

64. *Business*   The revenue from selling x units of a product at a price of p dollars per unit is $R = xp$. For a particular product the revenue is

$$R = 1600x - 0.50x^2.$$

(a) Find the revenue when the number of units sold is 3000.

(b) Find the price when the number of units sold is 2000.

P.4   **In Exercises 65–74, factor completely.**

65. $x^3 - x$

66. $x(x - 3) + 4(x - 3)$

67. $25x^2 - 49$

68. $x^2 - 12x + 36$

69. $x^3 - 64$

70. $8x^3 + 27$

71. $2x^2 + 21x + 10$

72. $3x^2 + 14x + 8$

73. $x^3 - x^2 + 2x - 2$

74. $x^3 - 4x^2 + 2x - 8$

P.5   **In Exercises 75 and 76, find the domain of the expression.**

75. $\dfrac{1}{x + 6}$

76. $\sqrt{x + 4}$

In Exercises 77 and 78, write the rational expression in simplest form.

77. $\dfrac{x^2 - 64}{5(3x + 24)}$

78. $\dfrac{x^3 + 27}{x^2 + x - 6}$

In Exercises 79–84, perform the operations and simplify.

79. $\dfrac{x^2 - 4}{x^4 - 2x^2 - 8} \cdot \dfrac{x^2 + 2}{x^2}$

80. $\dfrac{4x - 6}{(x - 1)^2} \div \dfrac{2x^2 - 3x}{x^2 + 2x - 3}$

81. $2x + \dfrac{3}{2(x-4)}$

82. $\dfrac{1}{x} - \dfrac{x-1}{x^2+1}$

83. $\dfrac{1}{x-1} + \dfrac{1-x}{x^2+x+1}$

84. $\dfrac{3x}{x+2} - \dfrac{4x^2-5}{2x^2+3x-2}$

In Exercises 85 and 86, simplify the complex fraction.

85. $\dfrac{\left[\dfrac{3a}{(a^2/x)-1}\right]}{\left(\dfrac{a}{x}-1\right)}$

86. $\dfrac{\left(\dfrac{1}{2x-3} - \dfrac{1}{2x+3}\right)}{\left(\dfrac{1}{2x} - \dfrac{1}{2x+3}\right)}$

P.6 **In Exercises 87–94, describe and correct the error.**

87. $10(4 \cdot 7) = 40 \cdot 70$ ✕

88. $\left(\tfrac{1}{3}x\right)\left(\tfrac{1}{3}y\right) = \tfrac{1}{3}xy$ ✕

89. $(2x)^4 = 2x^4$ ✕

90. $(-x)^6 = -x^6$ ✕

91. $(3^4)^4 = 3^8$ ✕

92. $\sqrt{3^2+4^2} = 3+4$ ✕

93. $(5+8)^2 = 5^2 + 8^2$ ✕

94. $(9x+12)^2 = 3(3x+4)^2$ ✕

In Exercises 95–98, insert the missing factor.

95. $\tfrac{2}{3}x^4 - \tfrac{3}{8}x^3 + \tfrac{5}{6}x^2 = \tfrac{1}{24}x^2\left(\rule{1cm}{0.3cm}\right)$

96. $\dfrac{t}{\sqrt{t+1}} - \sqrt{t+1} = \dfrac{1}{\sqrt{t+1}}\left(\rule{1cm}{0.3cm}\right)$

97. $2x(x^2-3)^{1/3} - 5(x^2-3)^{4/3} = (x^2-3)^{1/3}\left(\rule{1cm}{0.3cm}\right)$

98. $y(y-1)^{5/4} - y^2(y-1)^{1/4} = y(y-1)^{1/4}\left(\rule{1cm}{0.3cm}\right)$

In Exercises 99 and 100, factor the expression.

99. $x(x+2)^{-1/2} + (x+2)^{1/2}$

100. $\tfrac{2}{3}x(4+x)^{-1/2} - \tfrac{2}{15}(4+x)^{3/2}$

P.7 *Geometry* **In Exercises 101 and 102, plot the points and verify that the points form the polygon.**

101. Right triangle: $(2,3), (13,11), (5,22)$

102. Parallelogram: $(1,2), (8,3), (9,6), (2,5)$

In Exercises 103–106, determine the quadrant(s) in which (x, y) is located so that the condition(s) is (are) satisfied.

103. $x > 0$ and $y = -2$

104. $y > 0$

105. $(-x, y)$ is in the third quadrant.

106. $xy = 4$

In Exercises 107 and 108, (a) plot the points and (b) find the distance between the points.

107. $(-3, 8), (1, 5)$

108. $(5.6, 0), (0, 8.2)$

In Exercises 109 and 110, (a) plot the points and (b) find the midpoint of the line segment joining the points.

109. $(-2, 6), (4, -3)$

110. $(0, -1.2), (-3.6, 0)$

111. *Meteorology* The apparent temperature is a measure of relative discomfort to a person from heat and high humidity. The scatter plot shows the apparent temperatures (in degrees Fahrenheit) for a relative humidity of 75%.

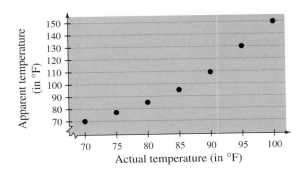

Find the change in the apparent temperature when the actual temperature changes from 70°F to 100°F.

Synthesis

True or False? **In Exercises 112–114, determine whether the statement is true or false. Justify your answer.**

112. $\dfrac{x^3-1}{x-1} = x^2 + x + 1$ for all values of x.

113. A binomial sum squared is equal to the sum of the terms squared.

114. $x^n - y^n$ factors as conjugates for all values of n.

Chapter Project ▶ Numerically Finding a Maximum Volume

Many mathematical results are discovered by calculating examples and looking for patterns. Prior to the 1950s, this mode of discovery was very time-consuming because the calculations had to be done by hand. The introduction of computer and calculator technology has removed much of the drudgery of calculation.

Using technology to conduct mathematical experiments usually involves creation of an algebraic **model** to represent the quantity under question.

Example ▶ Modeling the Volume of a Box

Consider a rectangular box with a square base and a surface area of 216 square inches. Let x represent the length (in inches) of each side of the base. Use the variable x to write a model for the volume of the box.

Solution

Begin by expressing the areas of the base, top, and sides in terms of x and h. The expression for the surface area of the box in terms of x and h is $S = x^2 + x^2 + 4xh$. Express the variable h in terms of x.

$$216 = 2x^2 + 4xh \qquad \text{Substitute 216 for } S.$$

$$\frac{54}{x} - \frac{x}{2} = h$$

Find an expression for the volume of the box in the figure above in terms of x.

$$V = x \cdot x \cdot \left(\frac{54}{x} - \frac{x}{2}\right) = 54x - \frac{1}{2}x^3, \quad 0 \le x \le \sqrt{108}$$

So, a model for the volume is $V = 54x - \frac{1}{2}x^3$.

A computer simulation to accompany this project appears in the *Interactive* CD-ROM and *Internet* versions of this text.

Base, x	Height	Surface Area	Volume
1.0	53.5	216.0	53.5
1.5	35.3	216.0	79.3
2.0	26.0	216.0	104.0
2.5	20.4	216.0	127.2
3.0	16.5	216.0	148.5
3.5	13.7	216.0	167.6
4.0	11.5	216.0	184.0
4.5	9.8	216.0	197.4
5.0	8.3	216.0	207.5
5.5	7.1	216.0	213.8
6.0	6.0	216.0	216.0
6.5	5.1	216.0	213.7
7.0	4.2	216.0	206.5
7.5	3.5	216.0	194.1
8.0	2.8	216.0	176.0
8.5	2.1	216.0	151.9
9.0	1.5	216.0	121.5
9.5	0.9	216.0	84.3
10.0	0.4	216.0	40.0

The table shows the surface areas and volumes of boxes with different bases and heights. From the table, it appears that the cube (the box whose dimensions are 6 inches by 6 inches by 6 inches) has the greatest volume.

Chapter Project Investigations

1. In the example, what happens to the height of the boxes as x gets closer and closer to 0? Of all boxes with square bases and surface areas of 216 square inches, is there a tallest? Explain your reasoning.

2. In the example, what happens to the height of the boxes as x gets closer and closer to $\sqrt{108}$? Is there a shortest box that has a square base and a surface area of 216 square inches? Explain your reasoning.

3. Complete the table. Why does it support the conclusion in the margin?

x	5.9	5.99	5.999	6.001	6.01	6.1
V	?	?	?	?	?	?

4. Of all rectangular boxes with surface areas of 216 square inches and bases of x inches by $2x$ inches, which has the maximum volume? Explain.

Chapter Test

Take this test as you would take a test in class. After you are done, check your work against the answers given in the back of the book.

1. Place $<$ or $>$ between the real numbers $-\frac{10}{3}$ and $-|-4|$.

2. Find the distance between the real numbers -5.4 and $3\frac{3}{4}$.

In Exercises 3–6, evaluate each expression without using a calculator.

3. (a) $27\left(-\frac{2}{3}\right)$ (b) $\frac{5}{18} \div \frac{15}{8}$ **4.** (a) $\left(-\frac{3}{5}\right)^3$ (b) $\left(\frac{3^2}{2}\right)^{-3}$

5. (a) $\sqrt{5} \cdot \sqrt{125}$ (b) $\frac{\sqrt{72}}{\sqrt{2}}$ **6.** (a) $\frac{5.4 \times 10^8}{3 \times 10^3}$ (b) $(3 \times 10^4)^3$

In Exercises 7 and 8, simplify each expression.

7. (a) $3z^2(2z^3)^2$ (b) $(u-2)^{-4}(u-2)^{-3}$ (c) $\left(\frac{x^{-2}y^2}{3}\right)^{-1}$

8. (a) $9z\sqrt{8z} - 3\sqrt{2z^3}$ (b) $-5\sqrt{16y} + 10\sqrt{y}$ (c) $\sqrt[3]{\frac{16}{v^5}}$

In Exercises 9–12, perform the operations and simplify.

9. $(x^2 + 3) - [3x + (8 - x^2)]$ **10.** $\left(x + \sqrt{5}\right)\left(x - \sqrt{5}\right)$

11. $\frac{8x}{x-3} + \frac{24}{3-x}$ **12.** $\frac{\left(\dfrac{2}{x} - \dfrac{2}{x+1}\right)}{\left(\dfrac{4}{x^2-1}\right)}$

13. Factor (a) $2x^4 - 3x^3 - 2x^2$ and (b) $x^3 + 2x^2 - 4x - 8$ completely.

14. Rationalize each denominator.   (a) $\frac{16}{\sqrt[3]{16}}$ (b) $\frac{6}{1 - \sqrt{3}}$

15. A hummingbird can beat its wings at a rate of 4200 times per minute. How long does it take for a hummingbird to beat its wings 10 times? one time? x times?

16. A T-shirt company can produce and sell x T-shirts per day. The total cost (in dollars) for producing x T-shirts is $C = 1480 + 6x$, and the total revenue (in dollars) is $R = 15x$. Find the profit obtained by selling 225 T-shirts per day.

17. Plot the points $(-2, 5)$ and $(6, 0)$. Find the coordinates of the midpoint of the line segment joining the points and the distance between the points.

18. Write an expression for the area of the shaded region in the figure and simplify the result.

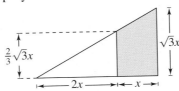

The Empire State Building on Fifth Avenue between 33rd Street and 34th Street is a famous landmark in New York City. Completed in 1931, it was the world's tallest building (1250 feet) until the 1970s.

Joe Ortner/Tony Stone Images

1 Equations and Inequalities

▶ How to Study This Chapter

The Big Picture

In this chapter you will learn the following skills and concepts.

▶ How to sketch the graph of an equation

▶ How to solve linear equations, quadratic equations, polynomial equations, radical equations, and absolute value equations

▶ How to perform operations with complex numbers

▶ How to solve linear inequalities, polynomial inequalities, rational inequalities, and inequalities involving absolute value

Important Vocabulary

As you encounter each new vocabulary term in this chapter, add the term and its definition to your notebook glossary.

Equation in two variables (p. 78)
Solution of equation in two variables (p. 78)
Graph of an equation (p. 78)
Intercepts (p. 80)
Symmetry (p. 81)
Circle (p. 83)
Equation in one variable (p. 88)
Solution of equation in one variable (p. 88)
Identity equation (p. 88)
Conditional equation (p. 88)
Linear equation in one variable (p. 88)
Equivalent equations (p. 89)

Extraneous solution (p. 91)
Quadratic equation (p. 110)
Quadratic Formula (p. 113)
Discriminant (p. 114)
Complex number (p. 124)
Imaginary number (p. 124)
Complex conjugates (p. 127)
Polynomial equation (p. 131)
Solution of an inequality (p. 143)
Graph of an inequality (p. 143)
Linear inequality in one variable (p. 145)
Double inequality (p. 146)
Critical numbers (p. 153)

Study Tools

- Learning objectives at the beginning of each section
- Chapter Summary (p. 163)
- Review Exercises (pp. 164–167)
- Chapter Test (p. 169)

Additional Resources

- Study and Solutions Guide
- Interactive College Algebra
- Videotapes for Chapter 1
- College Algebra Website
- Student Success Organizer

STUDY T!P

This text can help you best if you read the sections of the text to be covered before attending class. Then, do the homework as soon as possible after class, when the concepts are still fresh in your mind.

1.1 Graphs of Equations

▶ **What you should learn**

- How to sketch graphs of equations
- How to find *x*- and *y*-intercepts of graphs
- How to use symmetry to sketch graphs of equations
- How to find equations and sketch graphs of circles
- How to use graphs of equations in real-life problems

▶ **Why you should learn it**

The graph of an equation can help you see relationships between real-life quantities. For example, Exercise 74 on page 87 shows how a graph can be used to estimate the life expectancies of children who are born in the years 2002 and 2004.

The Graph of an Equation

In Section P.7, you used a coordinate system to represent graphically the relationship between two quantities. There, the graphical picture consisted of a collection of points in a coordinate plane.

Frequently, a relationship between two quantities is expressed as an **equation in two variables**. For instance, $y = 7 - 3x$ is an equation in x and y. An ordered pair (a, b) is a **solution** or **solution point** of an equation in x and y if the equation is true when a is substituted for x and b is substituted for y. For instance, $(1, 4)$ is a solution of $y = 7 - 3x$ because $4 = 7 - 3(1)$ is a true statement.

In this section, you will review some basic procedures for sketching the graph of an equation in two variables. The **graph of an equation** is the set of all points that are solutions of the equation.

Example 1 ▶ Sketching the Graph of an Equation

Sketch the graph of $y = 7 - 3x$.

Solution

The simplest way to sketch the graph of an equation is the *point-plotting method*. With this method, you construct a table of values that consists of several solution points of the equation. For instance, when $x = 0$,

$$y = 7 - 3(0) = 7,$$

which implies that $(0, 7)$ is a solution point of the graph.

x	0	1	2	3	4
$y = 7 - 3x$	7	4	1	-2	-5

From the table, it follows that $(0, 7)$, $(1, 4)$, $(2, 1)$, $(3, -2)$, and $(4, -5)$ are solution points of the equation. After plotting these points, you can see that they appear to lie on a line, as shown in Figure 1.1. The graph of the equation is the line that passes through the five plotted points.

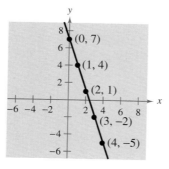

FIGURE 1.1

Example 2 ▶ Sketching the Graph of an Equation

Sketch the graph of $y = x^2 - 2$.

Solution

Begin by constructing a table of values.

x	-2	-1	0	1	2	3
$y = x^2 - 2$	2	-1	-2	-1	2	7

Next, plot the points given in the table, as shown in Figure 1.2(a). Finally, connect the points with a smooth curve, as shown in Figure 1.2(b).

STUDY TIP

One of your goals in this course is to learn to classify the basic shape of a graph from its equation. For instance, you will learn that the *linear equation* in Example 1 has the form

$$y = mx + b$$

and its graph is a straight line. Similarly, the *quadratic equation* in Example 2 has the form

$$y = ax^2 + bx + c$$

and its graph is a parabola.

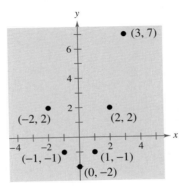

(a)

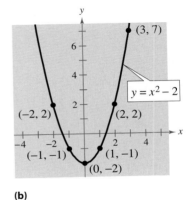

(b)

FIGURE 1.2

The point-plotting technique demonstrated in Examples 1 and 2 is easy to use, but it has some shortcomings. With too few solution points, you can badly misrepresent the graph of an equation. For instance, using only the four points

$$(-2, 2), (-1, -1), (1, -1), \text{ and } (2, 2)$$

in Figure 1.2, any one of the three graphs in Figure 1.3 would be reasonable.

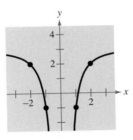
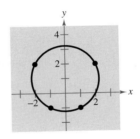
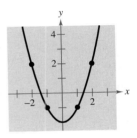

FIGURE 1.3

Intercepts of a Graph

It is often easy to determine the solution points that have zero as either the *x*-coordinate or the *y*-coordinate. These points are called **intercepts** because they are the points at which the graph intersects the *x*- or *y*-axis. It is possible for a graph to have no intercepts or several intercepts, as shown in Figure 1.4.

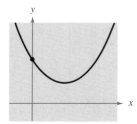

No x-intercept
One y-intercept

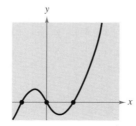

Three x-intercepts
One y-intercept

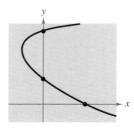

One x-intercept
Two y-intercepts

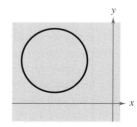

No intercepts

FIGURE 1.4

Note that an *x*-intercept is written as the ordered pair $(x, 0)$ and a *y*-intercept is written as the ordered pair $(0, y)$.

Finding Intercepts

1. To find *x*-intercepts, let *y* be zero and solve the equation for *x*.

2. To find *y*-intercepts, let *x* be zero and solve the equation for *y*.

Example 3 ▶ Finding *x*- and *y*-Intercepts

Find the *x*- and *y*-intercepts of the graph of each equation.

a. $y = x^3 - 4x$ **b.** $y^2 = x + 4$

Solution

a. Let $y = 0$. Then

$$0 = x^3 - 4x = x(x^2 - 4)$$

has solutions $x = 0$ and $x = \pm 2$.

   x-intercepts: $(0, 0), (2, 0), (-2, 0)$

Let $x = 0$. Then $y = (0)^3 - 4(0) = 0$.

   y-intercept: $(0, 0)$ (See Figure 1.5.)

b. Let $y = 0$. Then $(0)^2 = x + 4$, and $-4 = x$.

   x-intercept: $(-4, 0)$

Let $x = 0$. Then $y^2 = 0 + 4 = 4$ has solutions $y = \pm 2$.

   y-intercepts: $(0, 2), (0, -2)$ (See Figure 1.6.)

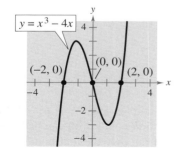

FIGURE 1.5

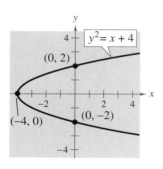

FIGURE 1.6

Symmetry

The graphs shown in Figures 1.2(b), 1.5, and 1.6 each have **symmetry** with respect to one of the coordinate axes or with respect to the origin.

Figure 1.2(b)	$y = x^2 - 2$	y-Axis symmetry
Figure 1.5	$y = x^3 - 4x$	Origin symmetry
Figure 1.6	$y^2 = x + 4$	x-Axis symmetry

Symmetry with respect to the x-axis means that if the Cartesian plane were folded along the x-axis, the portion of the graph above the x-axis would coincide with the portion below the x-axis. Symmetry with respect to the y-axis or the origin can be described in a similar manner, as shown in Figure 1.7.

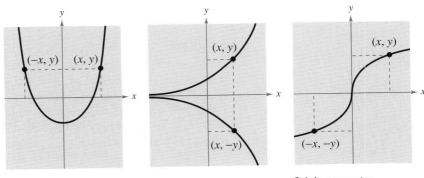

| y-Axis symmetry | x-Axis symmetry | Origin symmetry |

FIGURE 1.7

A computer animation of this concept appears in the *Interactive* CD-ROM and *Internet* versions of this text.

Knowing the symmetry of a graph *before* attempting to sketch it is helpful, because then you need only half as many solution points to sketch the graph. There are three basic types of symmetry. A graph is *symmetric with respect to the y-axis* if, whenever (x, y) is on the graph, $(-x, y)$ is also on the graph. A graph is *symmetric with respect to the x-axis* if, whenever (x, y) is on the graph, $(x, -y)$ is also on the graph. A graph is *symmetric with respect to the origin* if, whenever (x, y) is on the graph, $(-x, -y)$ is also on the graph.

Example 4 ▶ Testing for Symmetry

The graph of $y = x^2 - 2$ is symmetric with respect to the y-axis because the point $(-x, y)$ satisfies the equation.

$y = x^2 - 2$	Write original equation.
$y = (-x)^2 - 2$	Substitute $(-x, y)$ for (x, y).
$y = x^2 - 2$	Replacement yields equivalent equation.

See Figure 1.8.

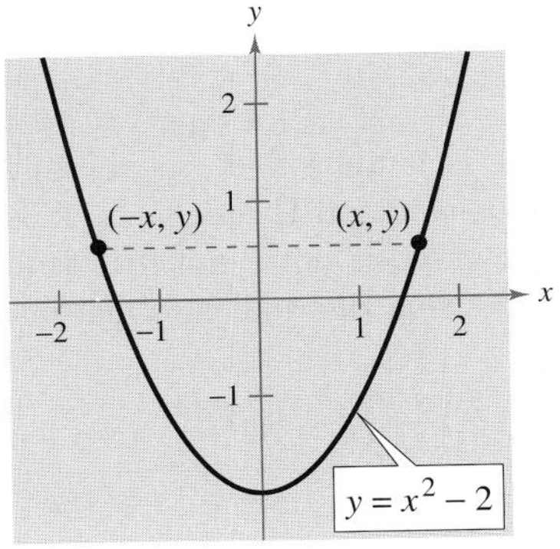

y-Axis symmetry

FIGURE 1.8

A computer animation of this example appears in the *Interactive* CD-ROM and *Internet* versions of this text.

Tests for Symmetry

1. The graph of an equation is symmetric with respect to the *y-axis* if replacing x with $-x$ yields an equivalent equation.

2. The graph of an equation is symmetric with respect to the *x-axis* if replacing y with $-y$ yields an equivalent equation.

3. The graph of an equation is symmetric with respect to the *origin* if replacing x with $-x$ *and* y with $-y$ yields an equivalent equation.

Example 5 ▶ Using Intercepts and Symmetry as Sketching Aids

Use intercepts and symmetry to sketch the graph of

$$x - y^2 = 1.$$

Solution

Letting $x = 0$, you can see that $-y^2 = 1$ or $y^2 = -1$ has no real solutions. So, there are no y-intercepts. Letting $y = 0$, you obtain $x = 1$. So, the x-intercept is $(1, 0)$. Of the three tests for symmetry, the only one that is satisfied is the test for x-axis symmetry. So, the graph is symmetric with respect to the x-axis. Using symmetry, you need only to find the solution points above the x-axis and then reflect them to obtain the graph, as shown in Figure 1.9.

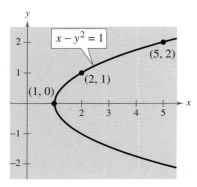

FIGURE **1.9**

y	0	1	2
$x = y^2 + 1$	1	2	5

Example 6 ▶ Sketching the Graph of an Equation

Sketch the graph of

$$y = |x - 1|.$$

Solution

Letting $x = 0$ yields $y = 1$, which means that $(0, 1)$ is the y-intercept. Letting $y = 0$ yields $x = 1$, which means that $(1, 0)$ is the x-intercept. This equation fails all three tests for symmetry and consequently its graph is not symmetric with respect to either axis or to the origin. The absolute value sign indicates that y is always nonnegative.

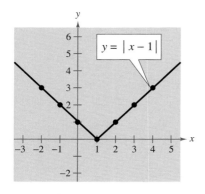

FIGURE **1.10**

x	-2	-1	0	1	2	3	4
$y = |x - 1|$	3	2	1	0	1	2	3

The graph is shown in Figure 1.10.

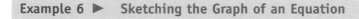

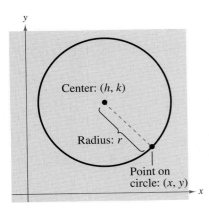

FIGURE **1.11**

Throughout this course, you will learn to recognize several types of graphs from their equations. For instance, you will learn to recognize that the graph of a second-degree equation of the form

$$y = ax^2 + bx + c$$

is a parabola (see Example 2). Another easily recognized graph is that of a **circle.**

Circles

Consider the circle shown in Figure 1.11. A point (x, y) is on the circle if and only if its distance from the center (h, k) is r. By the Distance Formula,

$$\sqrt{(x - h)^2 + (y - k)^2} = r.$$

By squaring both sides of this equation, you obtain the **standard form of the equation of a circle.**

Standard Form of the Equation of a Circle

The point (x, y) lies on the circle of radius r and center (h, k) if and only if

$$(x - h)^2 + (y - k)^2 = r^2.$$

STUDY T!P

To find the correct h and k, it may be helpful to rewrite the quantities $(x + 1)^2$ and $(y - 2)^2$.

$$(x + 1)^2 = [x - (-1)]^2,$$
$$h = -1$$
$$(y - 2)^2 = [y - (2)]^2,$$
$$k = 2$$

From this result, you can see that the standard form of the equation of a circle *with its center at the origin*, $(h, k) = (0, 0)$, is simply

$$x^2 + y^2 = r^2. \qquad \text{Circle with center at origin}$$

Example 7 ▶ Finding the Equation of a Circle

The point $(3, 4)$ lies on a circle whose center is at $(-1, 2)$, as shown in Figure 1.12. Find the standard form of the equation of this circle.

Solution

The radius of the circle is the distance between $(-1, 2)$ and $(3, 4)$.

$$r = \sqrt{(x - h)^2 + (y - k)^2} \qquad \text{Distance Formula}$$
$$r = \sqrt{[3 - (-1)]^2 + (4 - 2)^2} \qquad \text{Substitute for } x, y, h, \text{ and } k.$$
$$= \sqrt{4^2 + 2^2} \qquad \text{Simplify.}$$
$$= \sqrt{16 + 4} \qquad \text{Simplify.}$$
$$= \sqrt{20} \qquad \text{Radius}$$

Using $(h, k) = (-1, 2)$ and $r = \sqrt{20}$, the equation of the circle is

$$(x - h)^2 + (y - k)^2 = r^2$$
$$[x - (-1)]^2 + (y - 2)^2 = \left(\sqrt{20}\right)^2 \qquad \text{Substitute for } h, k, \text{ and } r.$$
$$(x + 1)^2 + (y - 2)^2 = 20. \qquad \text{Standard form}$$

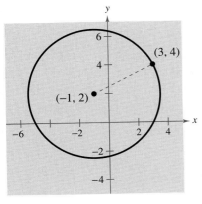

FIGURE **1.12**

Application

In this course, you will learn that there are many ways to approach a problem. Three common approaches are illustrated in Example 8.

A Numerical Approach: Construct and use a table.
A Graphical Approach: Draw and use a graph.
An Analytical Approach: Use the rules of algebra.

We strongly recommend that you develop the habit of using at least two approaches with every problem. This helps build your intuition and helps you check that your answer is reasonable.

Example 8 ▶ Recommended Weight

The median recommended weight y (in pounds) for men of medium frame who are 25 to 59 years old can be approximated by the mathematical model

$$y = 0.073x^2 - 6.986x + 288.985, \quad 62 \le x \le 76$$

where x is the man's height in inches.   (Source: Metropolitan Life Insurance Company)

a. Construct a table of values that shows the median recommended weights for men with heights of 62, 64, 66, 68, 70, 72, 74, and 76 inches.

b. Use the table of values to sketch a graph of the model. Then use the graph to estimate *graphically* the median recommended weight for a man whose height is 71 inches.

c. Use the model to confirm *analytically* the estimate you found in part (b).

Solution

a. You can use a calculator to complete the table, as shown below.

x	62	64	66	68	70	72	74	76
y	136.5	140.9	145.9	151.5	157.7	164.4	171.8	179.7

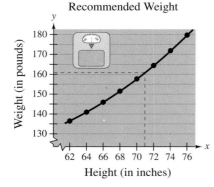

FIGURE 1.13

b. The table of values can be used to sketch the graph of the function, as shown in Figure 1.13. From the graph, you can estimate that a height of 71 inches corresponds to a weight of about 160 pounds.

c. To confirm analytically the estimate found in part (b), you can substitute 71 for x in the model.

$$y = 0.073x^2 - 6.986x + 288.985 \qquad \text{Write original model.}$$

$$= 0.073(71)^2 - 6.986(71) + 288.985 \qquad \text{Substitute 71 for } x.$$

$$\approx 160.97 \qquad \text{Use a calculator.}$$

So, the graphical estimate of 160 pounds is fairly good.

1.1 Exercises

The *Interactive* CD-ROM and *Internet* versions of this text contain step-by-step solutions to all odd-numbered Section and Review Exercises. They also provide Tutorial Exercises that link to Guided Examples for additional help.

In Exercises 1–4, determine whether each point lies on the graph of the equation.

Equation	Points	
1. $y = \sqrt{x + 4}$	(a) $(0, 2)$	(b) $(5, 3)$
2. $y = x^2 - 3x + 2$	(a) $(2, 0)$	(b) $(-2, 8)$
3. $y = 4 - |x - 2|$	(a) $(1, 5)$	(b) $(6, 0)$
4. $y = \frac{1}{3}x^3 - 2x^2$	(a) $\left(2, -\frac{16}{3}\right)$	(b) $(-3, 9)$

In Exercises 5–8, complete the table. Use the resulting solution points to sketch the graph of the equation.

5. $y = -2x + 5$

x	-1	0	1	2	$\frac{5}{2}$
y					

6. $y = \frac{3}{4}x - 1$

x	-2	0	1	$\frac{4}{3}$	2
y					

7. $y = x^2 - 3x$

x	-1	0	1	2	3
y					

8. $y = 5 - x^2$

x	-2	-1	0	1	2
y					

In Exercises 9–16, check for symmetry with respect to both axes and the origin.

9. $x^2 - y = 0$

10. $x - y^2 = 0$

11. $y = x^3$

12. $y = x^4 - x^2 + 3$

13. $y = \dfrac{x}{x^2 + 1}$

14. $y = \sqrt{9 - x^2}$

15. $xy^2 + 10 = 0$

16. $xy = 4$

In Exercises 17–20, assume that the graph has the indicated type of symmetry. Sketch the complete graph of the equation.

17.
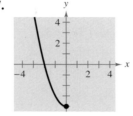
y-Axis symmetry

18.
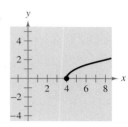
x-Axis symmetry

19.
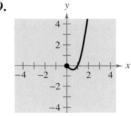
Origin symmetry

20.
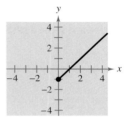
y-Axis symmetry

In Exercises 21–24, match the equation with its graph. [The graphs are labeled (a), (b), (c), and (d).]

(a)
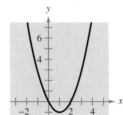

(b)
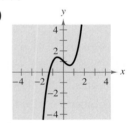

(c)
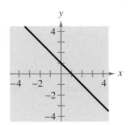

(d)
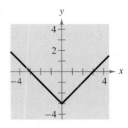

21. $y = 1 - x$

22. $y = x^2 - 2x$

23. $y = x^3 - x + 1$

24. $y = |x| - 3$

In Exercises 25–28, find the x- and y-intercepts of the graph of the equation.

25. $y = 16 - 4x^2$ **26.** $y = (x + 2)^2$

27. $y = 2x^3 - 5x^2$ **28.** $y^2 = x + 1$

In Exercises 29–40, use intercepts and symmetry to sketch the graph of the equation.

29. $y = -3x + 1$ **30.** $y = 2x - 3$

31. $y = x^2 - 2x$ **32.** $y = -x^2 - 2x$

33. $y = x^3 + 3$ **34.** $y = x^3 - 1$

35. $y = \sqrt{x - 3}$ **36.** $y = \sqrt{1 - x}$

37. $y = |x - 6|$ **38.** $y = 1 - |x|$

39. $x = y^2 - 1$ **40.** $x = y^2 - 5$

▦ In Exercises 41–52, use a graphing utility to graph the equation. Use a standard setting. Approximate any intercepts.

41. $y = 3 - \frac{1}{2}x$ **42.** $y = \frac{2}{3}x - 1$

43. $y = x^2 - 4x + 3$ **44.** $y = x^2 + x - 2$

45. $y = \dfrac{2x}{x - 1}$ **46.** $y = \dfrac{4}{x^2 + 1}$

47. $y = \sqrt[3]{x}$ **48.** $y = \sqrt[3]{x + 1}$

49. $y = x\sqrt{x + 6}$ **50.** $y = (6 - x)\sqrt{x}$

51. $y = |x + 3|$ **52.** $y = 2 - |x|$

In Exercises 53–60, find the standard form of the equation of the specified circle.

53. Center: $(0, 0)$; radius: 4

54. Center: $(0, 0)$; radius: 5

55. Center: $(2, -1)$; radius: 4

56. Center: $(-7, -4)$; radius: 7

57. Center: $(-1, 2)$; solution point: $(0, 0)$

58. Center: $(3, -2)$; solution point: $(-1, 1)$

59. Endpoints of a diameter: $(0, 0)$, $(6, 8)$

60. Endpoints of a diameter: $(-4, -1)$, $(4, 1)$

In Exercises 61–66, find the center and radius, and sketch the graph of the equation.

61. $x^2 + y^2 = 25$ **62.** $x^2 + y^2 = 16$

63. $(x - 1)^2 + (y + 3)^2 = 9$

64. $x^2 + (y - 1)^2 = 1$

65. $\left(x - \frac{1}{2}\right)^2 + \left(y - \frac{1}{2}\right)^2 = \frac{9}{4}$

66. $(x - 2)^2 + (y + 1)^2 = 3$

▦ In Exercises 67 and 68, use a graphing utility to graph y_1 and y_2. Use a square setting. Identify the graph.

67. $y_1 = 4 + \sqrt{25 - x^2}$

$y_2 = 4 - \sqrt{25 - x^2}$

68. $y_1 = 2 + \sqrt{16 - (x - 1)^2}$

$y_2 = 2 - \sqrt{16 - (x - 1)^2}$

69. *Business*   A manufacturing plant purchases a new molding machine for $225,000. The depreciated value y after t years is

$$y = 225,000 - 20,000t, \qquad 0 \le t \le 8.$$

Sketch the graph of the equation.

70. *Consumerism*   You purchase a jet ski for $8100. The depreciated value y after t years is

$$y = 8100 - 929t, \qquad 0 \le t \le 6.$$

Sketch the graph of the equation.

71. *Geometry*   A rectangle of length x and width w has a perimeter of 12 meters.

(a) Draw a rectangle that gives a visual representation of the problem. Use the specified variables to label the sides of the rectangle.

(b) Show that the width of the rectangle is $w = 6 - x$ and its area is $A = x(6 - x)$.

▦ (c) Use a graphing utility to graph the area equation.

▦ (d) From the graph in part (c), estimate the dimensions of the rectangle that yields a maximum area.

72. *Geometry*   A rectangle of length x and width w has a perimeter of 22 yards.

(a) Draw a rectangle that gives a visual representation of the problem. Use the specified variables to label the sides of the rectangle.

(b) Show that the width of the rectangle is $w = 11 - x$ and its area is $A = x(11 - x)$.

▦ (c) Use a graphing utility to graph the area equation. Be sure to adjust your window settings.

▦ (d) From the graph in part (c), estimate the dimensions of the rectangle that yields a maximum area.

The symbol ▦ indicates an exercise or parts of an exercise in which you are instructed to use a graphing utility.

Data Analysis **In Exercises 73 and 74, (a) sketch a scatter plot of the data, (b) graph the model for the data and compare the scatter plot and the graph, and (c) use the model to estimate the values of *y* for the years 2002 and 2004.**

73. *Federal Debt* The table shows the per capita federal debt of the United States for several years. (Source: U.S. Treasury Department, U.S. Bureau of the Census)

Year	1950	1960	1970	1980
Per capita debt	$1688	$1572	$1807	$3981

Year	1990	1994	1997	1998
Per capita debt	$12,848	$15,750	$20,063	$20,513

A model for the per capita debt during this period is

$y = 0.223t^3 - 0.733t^2 - 78.255t + 1837.433$

where *y* represents the per capita debt and *t* is the time in years, with $t = 0$ corresponding to 1950.

74. *Population Statistics* The table shows the life expectancy of a child (at birth) in the United States for selected years from 1920 to 2000. (Source: U.S. National Center for Health Statistics, U.S. Bureau of the Census)

Year	1920	1930	1940	1950
Life expectancy	54.1	59.7	62.9	68.2

Year	1960	1970	1980	1990	2000
Life expectancy	69.7	70.8	73.7	75.4	76.4

A model for the life expectancy during this period is

$$y = \frac{66.93 + t}{1 + 0.01t}$$

where *y* represents the life expectancy and *t* is the time in years, with $t = 0$ corresponding to 1950.

75. *Electronics* The resistance *y* in ohms of 1000 feet of solid copper wire at 77 degrees Fahrenheit can be approximated by the model

$$y = \frac{10{,}770}{x^2} - 0.37, \qquad 5 \le x \le 100$$

where *x* is the diameter of the wire in mils (0.001 in.). Use the model to estimate the resistance when $x = 50$. (Source: American Wire Gage)

Synthesis

True or False? **In Exercises 76 and 77, determine whether the statement is true or false. Justify your answer.**

76. In order to find the *y*-intercepts of the graph of an equation, let $y = 0$ and solve the equation for *x*.

77. The graph of a linear equation of the form $y = mx + b$ has one *y*-intercept.

78. *Think About It* Suppose you correctly enter an expression for the variable *y* on a graphing utility. However, no graph appears on the display when you graph the equation. Give a possible explanation and the steps you could take to remedy the problem. Illustrate your explanation with an example.

79. *Think About It* Find *a* and *b* if the graph of $y = ax^2 + bx^3$ is symmetric with respect to (a) the *y*-axis and (b) the origin. (There are many correct answers.)

80. In your own words, explain how the display of a graphing utility changes if the maximum setting for *x* is changed from 10 to 20.

Review

True or False? **In Exercises 81 and 82, determine whether the statement is true or false. Justify your answer.**

81. $\dfrac{1}{3 \cdot 4^{-1}} = 3 \cdot 4$ **82.** $(3 + 4)^2 = 3^2 + 4^2$

83. Identify the terms: $9x^5 + 4x^3 - 7$.

84. Write the expression using exponential notation: $-(7 \times 7 \times 7 \times 7)$

In Exercises 85–90, simplify the expression.

85. $\sqrt{18x} - \sqrt{2x}$ **86.** $\sqrt[4]{x^5}$

87. $\dfrac{70}{\sqrt{7x}}$ **88.** $\dfrac{55}{\sqrt{20} - 3}$

89. $\sqrt[6]{t^2}$ **90.** $\sqrt[3]{\sqrt{y}}$

1.2   Linear Equations in One Variable

▶ **What you should learn**

- How to identify different types of equations
- How to solve linear equations in one variable
- How to solve equations involving fractional expressions
- How to use linear equations to model and solve real-life problems

▶ **Why you should learn it**

Linear equations are used in many real-life applications. For example, Exercises 95 and 96 on page 96 show how linear equations can model the relationship between the length of a thigh bone and the height of a person, helping researchers learn about ancient cultures.

M. Greenlar/The Image Works

Equations and Solutions of Equations

An **equation** in x is a statement that two algebraic expressions are equal. For example,

$$3x - 5 = 7, \quad x^2 - x - 6 = 0, \quad \text{and} \quad \sqrt{2x} = 4$$

are equations. To **solve** an equation in x means to find all values of x for which the equation is true. Such values are **solutions.** For instance, $x = 4$ is a solution of the equation

$$3x - 5 = 7,$$

because $3(4) - 5 = 7$ is a true statement.

The solutions of an equation depend on the kinds of numbers being considered. For instance, in the set of rational numbers, $x^2 = 10$ has no solution because there is no rational number whose square is 10. However, in the set of real numbers, the equation has the two solutions $\sqrt{10}$ and $-\sqrt{10}$.

An equation that is true for *every* real number in the domain of the variable is called an **identity.** For example,

$$x^2 - 9 = (x + 3)(x - 3) \qquad \text{Identity}$$

is an identity because it is a true statement for any real value of x, and

$$\frac{x}{3x^2} = \frac{1}{3x} \qquad \text{Identity}$$

where $x \neq 0$, is an identity because it is true for any nonzero real value of x.

An equation that is true for just *some* (or even none) of the real numbers in the domain of the variable is called a **conditional equation.** For example, the equation

$$x^2 - 9 = 0 \qquad \text{Conditional equation}$$

is conditional because $x = 3$ and $x = -3$ are the only values in the domain that satisfy the equation. The equation $2x - 4 = 2x + 1$ is conditional because there are no real values of x for which the equation is true. Learning to solve conditional equations is the primary focus of this chapter.

Linear Equations in One Variable

Definition of a Linear Equation

A **linear equation in one variable** x is an equation that can be written in the standard form

$$ax + b = 0$$

where a and b are real numbers with $a \neq 0$.

A linear equation has exactly one solution. To see this, consider the following steps. (Remember that $a \neq 0$.)

$$ax + b = 0 \qquad \text{Write original equation.}$$

$$ax = -b \qquad \text{Subtract } b \text{ from each side.}$$

$$x = -\frac{b}{a} \qquad \text{Divide each side by } a.$$

To solve a conditional equation in x, isolate x on one side of the equation by a sequence of **equivalent** (and usually simpler) **equations,** each having the same solution(s) as the original equation. The operations that yield equivalent equations come from the Substitution Principle and the simplification techniques studied in Chapter P.

Generating Equivalent Equations

An equation can be transformed into an *equivalent equation* by one or more of the following steps.

	Given Equation	*Equivalent Equation*
1. Remove symbols of grouping, combine like terms, or simplify fractions on one or both sides of the equation.	$2x - x = 4$	$x = 4$
2. Add (or subtract) the same quantity to (from) *each* side of the equation.	$x + 1 = 6$	$x = 5$
3. Multiply (or divide) *each* side of the equation by the same *nonzero* quantity.	$2x = 6$	$x = 3$
4. Interchange the two sides of the equation.	$2 = x$	$x = 2$

Exploration

Use a graphing utility to graph the equation $y = 3x - 6$. Use the result to estimate the x-intercept of the graph. Explain how the x-intercept is related to the solution of the equation $3x - 6 = 0$, as shown in Example 1.

Example 1 ► Solving a Linear Equation

$$3x - 6 = 0 \qquad \text{Write original equation.}$$

$$3x = 6 \qquad \text{Add 6 to each side.}$$

$$x = 2 \qquad \text{Divide each side by 3.}$$

Check

After solving an equation, you should check each solution in the original equation.

$$3x - 6 = 0 \qquad \text{Write original equation.}$$

$$3(2) - 6 \overset{?}{=} 0 \qquad \text{Substitute 2 for } x.$$

$$0 = 0 \qquad \text{Solution checks. } \checkmark$$

So, the solution is 2.

Some linear equations have no solutions because all the x-terms sum to zero and a contradictory (false) statement such as $0 = 5$ or $12 = 7$ is obtained. For instance, the linear equation

$$x = x + 1$$

has no solution. Watch for this type of linear equation in the exercises.

Example 2 ▶   Solving a Linear Equation

Solve $6(x - 1) + 4 = 3(7x + 1)$.

Solution

$6(x - 1) + 4 = 3(7x + 1)$	Write original equation.
$6x - 6 + 4 = 21x + 3$	Distributive Property
$6x - 2 = 21x + 3$	Simplify.
$6x = 21x + 5$	Add 2 to each side.
$-15x = 5$	Subtract $21x$ from each side.
$x = -\dfrac{1}{3}$	Divide each side by -15.

Check

Check this solution by substitution in the original equation.

$6(x - 1) + 4 = 3(7x + 1)$	Write original equation.
$6\left(-\frac{1}{3} - 1\right) + 4 \overset{?}{=} 3\left[7\left(-\frac{1}{3}\right) + 1\right]$	Substitute $-\frac{1}{3}$ for x.
$6\left(-\frac{4}{3}\right) + 4 \overset{?}{=} 3\left[-\frac{7}{3} + 1\right]$	Simplify.
$-\frac{24}{3} + 4 \overset{?}{=} -\frac{21}{3} + 3$	Multiply.
$-8 + 4 \overset{?}{=} -7 + 3$	Simplify.
$-4 = -4$	Solution checks. ✓

So, the solution is $-\frac{1}{3}$.

Technology

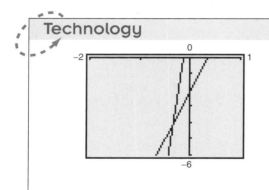

You can use a graphing utility to check that a solution is reasonable. One way to do this is to graph the left side of the equation, then graph the right side of the equation, and determine the point of intersection. For instance, in Example 2, if you graph the equations

$$y = 6(x - 1) + 4 \qquad \text{The left side}$$

$$y = 3(7x + 1) \qquad \text{The right side}$$

on the same screen, they should intersect when $x = -\frac{1}{3}$, as shown in the graph at the left.

Equations Involving Fractional Expressions

To solve an equation involving fractional expressions, find the least common denominator of all terms and multiply every term by this LCD.

Example 3 ▶ An Equation Involving Fractional Expressions

Solve $\dfrac{x}{3} + \dfrac{3x}{4} = 2$.

Solution

$$\dfrac{x}{3} + \dfrac{3x}{4} = 2 \qquad \text{Write original equation.}$$

$$(12)\dfrac{x}{3} + (12)\dfrac{3x}{4} = (12)2 \qquad \text{Multiply each term by the LCD of 12.}$$

$$4x + 9x = 24 \qquad \text{Divide out and multiply.}$$

$$13x = 24 \qquad \text{Combine like terms.}$$

$$x = \dfrac{24}{13} \qquad \text{Divide each side by 13.}$$

The solution is $\frac{24}{13}$. Check this in the original equation.

When multiplying or dividing an equation by a *variable* quantity, it is possible to introduce an extraneous solution. An **extraneous solution** is one that does not satisfy the original equation.

Example 4 ▶ An Equation with an Extraneous Solution

Solve $\dfrac{1}{x-2} = \dfrac{3}{x+2} - \dfrac{6x}{x^2 - 4}$.

Solution

The LCD is $x^2 - 4$, or $(x+2)(x-2)$. Multiply each term by this LCD.

$$\dfrac{1}{x-2}(x+2)(x-2) = \dfrac{3}{x+2}(x+2)(x-2) - \dfrac{6x}{x^2-4}(x+2)(x-2)$$

$$x + 2 = 3(x-2) - 6x, \qquad x \neq \pm 2$$

$$x + 2 = 3x - 6 - 6x$$

$$x + 2 = -3x - 6$$

$$4x = -8$$

$$x = -2$$

In the original equation, $x = -2$ yields a denominator of zero. Therefore, $x = -2$ is an extraneous solution, and the original equation has *no solution*.

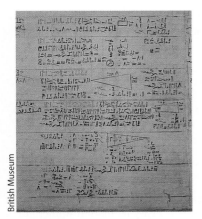

British Museum

Historical Note

This ancient Egyptian papyrus, discovered in 1858, contains one of the earliest examples of mathematical writing in existence. The papyrus itself dates back to around 1650 B.C., but it is actually a copy of writings from two centuries earlier. The algebraic equations on the papyrus were written in words. Diophantus, a Greek who lived around A.D. 250, is often called the Father of Algebra. He was the first to use abbreviated word forms in equations.

An equation with a *single fraction* on each side can be cleared of denominators by **cross multiplying,** which is equivalent to multiplying by the LCD and then dividing out. For instance, in the equation

$$\frac{2}{x - 3} = \frac{3}{x + 1}$$

the LCD is $(x - 3)(x + 1)$. Multiply both sides of the equation by this LCD.

$$\frac{2}{x - 3}(x - 3)(x + 1) = \frac{3}{x + 1}(x - 3)(x + 1)$$

$$2(x + 1) = 3(x - 3), \qquad x \neq -1, x \neq 3$$

By comparing this equation with the original, you can see that the original numerators and denominators have been "cross multiplied." That is, the left numerator was multiplied by the right denominator and the right numerator was multiplied by the left denominator.

Applications

Example 5 ▶ Graphical Estimation

The number A of amusement parks in the United States from 1990 through 1996 can be approximated by the linear model

$$A = 301t + 4806, \qquad 0 \leq t \leq 6$$

where $t = 0$ represents 1990. Use the graph given in Figure 1.14 to estimate graphically (a) the year in which there were about 5400 amusement parks and (b) the year in which there were about 6300 amusement parks. (Source: Current Business Reports, Service Annual Survey: 1996)

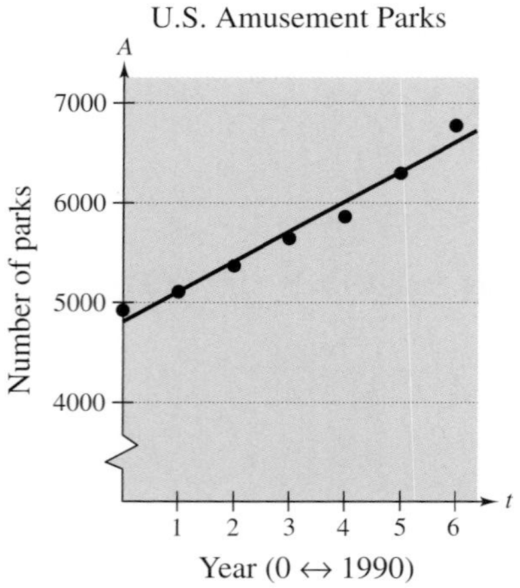

Solution

a. From the graph, you can estimate that $A = 5400$ corresponds to a t-value of about 2. This implies that there were about 5400 amusement parks in about 1992.

b. From the graph, you can estimate that $A = 6300$ corresponds to a t-value of about 5. This implies that there were about 6300 amusement parks in about 1995.

U.S. Amusement Parks

FIGURE 1.14

Example 5 uses a graphical approach. You can also answer the question analytically. For instance, to find the year in which there were 6300 amusement parks, you can solve the equation $6300 = 301t + 4806$ as follows.

$$6300 = 301t + 4806 \qquad \text{Write original equation.}$$

$$1494 = 301t \qquad \text{Subtract 4806 from each side.}$$

$$\frac{1494}{301} = t \qquad \text{Divide each side by 301.}$$

The solution is $t = \frac{1494}{301} \approx 4.96$, which agrees with the estimate found in Example 5(b).

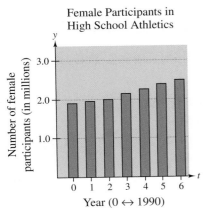

Female Participants in
High School Athletics

Number of female
participants (in millions)

FIGURE **1.15**

Example 6 ▶ Female Participants in Athletic Programs

The number of female participants in high school athletic programs (in millions) in the United States from 1990 to 1996 can be approximated by the linear model

$$y = 0.101t + 1.844, \qquad 0 \le t \le 6$$

where $t = 0$ represents 1990. From Figure 1.15, you can see that the number of female participants appears to increase in a linear pattern. Assuming that this pattern continues, find the year when there will be 3.25 million female participants. (Source: National Federation of State High School Associations)

Solution

Let $y = 3.25$ and solve the equation $3.25 = 0.101t + 1.844$ for t.

$3.25 = 0.101t + 1.844$	Write original equation.
$1.406 = 0.101t$	Subtract 1.844 from each side.
$t = \dfrac{1.406}{0.101}$	Divide each side by 0.101.
$t \approx 14$	Simplify.

Because $t = 0$ represents 1990, $t = 14$ must represent 2004. So, from this model, there will be 3.25 million female participants in 2004.

Writing ABOUT MATHEMATICS

Solving Equations Choose one of the equations below and write a step-by-step explanation of how to solve the equation, without using another equation in the explanation. Exchange your explanation with that of another student—see if he or she can correctly solve the equation just by following your instructions. Write a short paragraph describing any improvements that could be made in communicating the solution process.

a. $x - 2 + \dfrac{3x - 1}{8} = \dfrac{x + 4}{4}$

b. $t - \{7 - [t - (7 + t)]\} = 27$

c. $\dfrac{y + 6}{21} = \dfrac{2y}{3} + \dfrac{5y + 1}{7}$

d. $-4 + 3(x - 1) = -(x - 1) + 3$

1.2 Exercises

In Exercises 1–10, determine whether each value of x is a solution of the equation.

	Equation		*Values*	

1. $5x - 3 = 3x + 5$ (a) $x = 0$ (b) $x = -5$
                           (c) $x = 4$ (d) $x = 10$

2. $7 - 3x = 5x - 17$ (a) $x = -3$ (b) $x = 0$
                           (c) $x = 8$ (d) $x = 3$

3. $3x^2 + 2x - 5$ (a) $x = -3$ (b) $x = 1$
     $= 2x^2 - 2$ (c) $x = 4$ (d) $x = -5$

4. $5x^3 + 2x - 3$ (a) $x = 2$ (b) $x = -2$
     $= 4x^3 + 2x - 11$ (c) $x = 0$ (d) $x = 10$

5. $\dfrac{5}{2x} - \dfrac{4}{x} = 3$ (a) $x = -\frac{1}{2}$ (b) $x = 4$
                     (c) $x = 0$ (d) $x = \frac{1}{4}$

6. $3 + \dfrac{1}{x + 2} = 4$ (a) $x = -1$ (b) $x = -2$
                     (c) $x = 0$ (d) $x = 5$

7. $\sqrt{3x - 2} = 4$ (a) $x = 3$ (b) $x = 2$
                   (c) $x = 9$ (d) $x = -6$

8. $\sqrt[3]{x - 8} = 3$ (a) $x = 2$ (b) $x = -5$
                   (c) $x = 35$ (d) $x = 8$

9. $6x^2 - 11x - 35 = 0$ (a) $x = -\frac{5}{3}$ (b) $x = -\frac{2}{7}$
                         (c) $x = \frac{7}{2}$ (d) $x = \frac{5}{3}$

10. $10x^2 + 21x - 10 = 0$ (a) $x = \frac{2}{5}$ (b) $x = -\frac{5}{2}$
                         (c) $x = -\frac{1}{3}$ (d) $x = -2$

In Exercises 11–20, determine whether the equation is an identity or a conditional equation.

11. $2(x - 1) = 2x - 2$

12. $3(x + 2) = 5x + 4$

13. $-6(x - 3) + 5 = -2x + 10$

14. $3(x + 2) - 5 = 3x + 1$

15. $4(x + 1) - 2x = 2(x + 2)$

16. $-7(x - 3) + 4x = 3(7 - x)$

17. $x^2 - 8x - 5 = (x - 4)^2 - 11$

18. $x^2 + 2(3x - 2) = x^2 + 6x - 4$

19. $3 + \dfrac{1}{x + 1} = \dfrac{4x}{x + 1}$ **20.** $\dfrac{5}{x} + \dfrac{3}{x} = 24$

In Exercises 21 and 22, justify each step of the solution.

21.
$$4x + 32 = 83$$
$$4x + 32 - 32 = 83 - 32$$
$$4x = 51$$
$$\frac{4x}{4} = \frac{51}{4}$$
$$x = \frac{51}{4}$$

22. $3(x - 4) + 10 = 7$
$$3x - 12 + 10 = 7$$
$$3x - 2 = 7$$
$$3x - 2 + 2 = 7 + 2$$
$$3x = 9$$
$$\frac{3x}{3} = \frac{9}{3}$$
$$x = 3$$

In Exercises 23–26, solve the equation mentally.

23. $3x = 18$ **24.** $\frac{1}{2}t = 9$

25. $s + 12 = 19$ **26.** $u - 3 = 27$

In Exercises 27–42, solve the equation and check your solution.

27. $x + 11 = 15$ **28.** $7 - x = 19$

29. $7 - 2x = 25$ **30.** $7x + 2 = 23$

31. $8x - 5 = 3x + 20$ **32.** $7x + 3 = 3x - 17$

33. $2(x + 5) - 7 = 3(x - 2)$

34. $3(x + 3) = 5(1 - x) - 1$

35. $x - 3(2x + 3) = 8 - 5x$

36. $9x - 10 = 5x + 2(2x - 5)$

37. $\dfrac{5x}{4} + \dfrac{1}{2} = x - \dfrac{1}{2}$ **38.** $\dfrac{x}{5} - \dfrac{x}{2} = 3 + \dfrac{3x}{10}$

39. $\frac{3}{2}(z + 5) - \frac{1}{4}(z + 24) = 0$

40. $\dfrac{3x}{2} + \dfrac{1}{4}(x - 2) = 10$

41. $0.25x + 0.75(10 - x) = 3$

42. $0.60x + 0.40(100 - x) = 50$

In Exercises 43–46, solve the equation in two ways. Then explain which way is easier.

43. $3(x - 1) = 4$

44. $4(x + 3) = 15$

45. $\frac{1}{3}(x + 2) = 5$

46. $\frac{3}{4}(z - 4) = 6$

▦ *Graphical Analysis* In Exercises 47–52, use a graphing utility to graph the equation and approximate any *x*-intercepts. Set $y = 0$ and solve the resulting equation. Compare the results with the graph's *x*-intercepts.

47. $y = 2(x - 1) - 4$

48. $y = \frac{4}{3}x + 2$

49. $y = 20 - (3x - 10)$

50. $y = 10 + 2(x - 2)$

51. $y = -38 + 5(9 - x)$

52. $y = 6x - 6\left(\frac{16}{11} + x\right)$

In Exercises 53–74, solve the equation and check your solution. (If not possible, explain why.)

53. $x + 8 = 2(x - 2) - x$

54. $8(x + 2) - 3(2x + 1) = 2(x + 5)$

55. $\dfrac{100 - 4x}{3} = \dfrac{5x + 6}{4} + 6$

56. $\dfrac{17 + y}{y} + \dfrac{32 + y}{y} = 100$

57. $\dfrac{5x - 4}{5x + 4} = \dfrac{2}{3}$

58. $\dfrac{10x + 3}{5x + 6} = \dfrac{1}{2}$

59. $10 - \dfrac{13}{x} = 4 + \dfrac{5}{x}$

60. $\dfrac{15}{x} - 4 = \dfrac{6}{x} + 3$

61. $\dfrac{x}{x + 4} + \dfrac{4}{x + 4} + 2 = 0$

62. $3 = 2 + \dfrac{2}{z + 2}$

63. $\dfrac{1}{x} + \dfrac{2}{x - 5} = 0$

64. $\dfrac{7}{2x + 1} - \dfrac{8x}{2x - 1} = -4$

65. $\dfrac{2}{(x - 4)(x - 2)} = \dfrac{1}{x - 4} + \dfrac{2}{x - 2}$

66. $\dfrac{4}{x - 1} + \dfrac{6}{3x + 1} = \dfrac{15}{3x + 1}$

67. $\dfrac{1}{x - 3} + \dfrac{1}{x + 3} = \dfrac{10}{x^2 - 9}$

68. $\dfrac{1}{x - 2} + \dfrac{3}{x + 3} = \dfrac{4}{x^2 + x - 6}$

69. $\dfrac{3}{x^2 - 3x} + \dfrac{4}{x} = \dfrac{1}{x - 3}$

70. $\dfrac{6}{x} - \dfrac{2}{x + 3} = \dfrac{3(x + 5)}{x^2 + 3x}$

71. $(x + 2)^2 + 5 = (x + 3)^2$

72. $(x + 1)^2 + 2(x - 2) = (x + 1)(x - 2)$

73. $(x + 2)^2 - x^2 = 4(x + 1)$

74. $(2x + 1)^2 = 4(x^2 + x + 1)$

In Exercises 75–82, solve for *x*.

75. $4(x + 1) - ax = x + 5$

76. $4 - 2(x - 2b) = ax + 3$

77. $6x + ax = 2x + 5$

78. $5 + ax = 12 - bx$

79. $19x + \frac{1}{2}ax = x + 9$

80. $-5(3x - 6b) + 12 = 8 + 3ax$

81. $-2ax + 6(x + 3) = -4x + 1$

82. $\frac{4}{5}x - ax = 2\left(\frac{2}{5}x - 1\right) + 10$

In Exercises 83–88, solve the equation for *x*. (Round your solution to three decimal places.)

83. $0.275x + 0.725(500 - x) = 300$

84. $2.763 - 4.5(2.1x - 5.1432) = 6.32x + 5$

85. $\dfrac{x}{0.6321} + \dfrac{x}{0.0692} = 1000$

86. $\dfrac{x}{2.625} + \dfrac{x}{4.875} = 1$

87. $\dfrac{2}{7.398} - \dfrac{4.405}{x} = \dfrac{1}{x}$

88. $\dfrac{3}{6.350} - \dfrac{6}{x} = 18$

In Exercises 89–92, evaluate each expression in two ways. (a) Calculate entirely on your calculator by storing intermediate results and then rounding the final answer to two decimal places. (b) Round both the numerator and denominator to two decimal places before dividing, and then round the final answer to two decimal places. Does the method in part (b) introduce an additional roundoff error?

89. $\dfrac{1 + 0.73205}{1 - 0.73205}$

90. $\dfrac{1 + 0.86603}{1 - 0.86603}$

91. $\dfrac{3.33 + (1.98/0.74)}{4 + (6.25/3.15)}$

92. $\dfrac{1.73205 - 1.19195}{3 - (1.73205)(1.19195)}$

93. *Exploration*

(a) Complete the table.

x	−1	0	1	2	3	4
$3.2x - 5.8$						

(b) Use the table in part (a) to determine the interval in which the solution to the equation $3.2x - 5.8 = 0$ is located. Explain your reasoning.

(c) Complete the table.

x	1.5	1.6	1.7	1.8	1.9	2
$3.2x - 5.8$						

(d) Use the table in part (c) to determine the interval in which the solution to the equation $3.2x - 5.8 = 0$ is located. Explain how this process can be used to approximate the solution to any desired degree of accuracy.

94. *Exploration* Use the procedure in Exercise 93 to approximate the solution to the equation $0.3(x - 1.5) - 2 = 0$ accurate to two decimal places.

Anthropology **In Exercises 95 and 96, use the following information. The relationship between the length of an adult's thigh bone and the height of the adult can be approximated by the linear equations**

$y = 0.432x - 10.44$ **Female**

$y = 0.449x - 12.15$ **Male**

where y is the length of the femur (thigh bone) in inches and x is the height of the adult in inches (see figure).

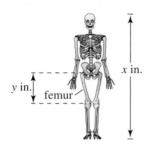

y in.   femur x in.

95. An anthropologist discovers a thigh bone belonging to an adult human female. The bone is 16 inches long. Estimate the height of the female.

96. From the foot bones of an adult human male, an anthropologist estimates that the person's height was 69 inches. A few feet away from the site where the foot bones were discovered, the anthropologist discovered a male adult thigh bone that was 19 inches long. Is it likely that both bones came from the same person?

Tax Credits **In Exercises 97–100, use the following information about a possible tax credit for a family consisting of two adults and two children (see figure).**

Earned income: E

Subsidy: $S = 10{,}000 - \frac{1}{2}E$, $0 \le E \le 20{,}000$

Total income: $T = E + S$

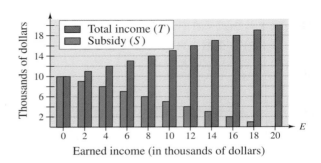

97. Express the total income T in terms of E.

98. Find the earned income E if the subsidy is $6600.

99. Find the earned income E if the total income is $13,800.

100. Find the subsidy S if the total income is $12,500.

101. *Geometry* The surface area S of the rectangular solid in the figure is

$$S = 2(24) + 2(4x) + 2(6x).$$

Find the length x of the box if the surface area is 248 square centimeters.

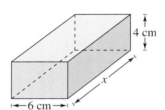

102. *Geometry* The surface area S of the regular pyramid shown in the figure is

$$S = x^2 + \frac{1}{2}(4x)(18).$$

Find the length x of the sides of the base of the pyramid if the surface area is 576 square feet.

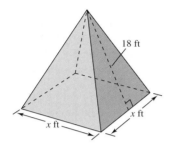

18 ft

x ft x ft

103. *Consumerism* The number of light trucks sold (in millions) in the United States from 1993 to 1997 can be approximated by the model

$$y = 0.35t + 4.31$$

where $t = 3$ represents 1993. (Source: U.S. Bureau of Economic Analysis)

(a) Use the model to create a line graph of the number of light trucks sold from 1993 to 1997.

(b) Use the graph to determine the year during which the number of light trucks sold reached 6 million.

104. *Labor Statistics* The number of married women y in the civilian work force (in millions) in the United States from 1990 to 1997 (see figure) can be approximated by the model

$$y = 0.46t + 30.81$$

where $t = 0$ represents 1990. According to this model, during which year did this number reach 33 million? Explain how to answer the question graphically and algebraically. (Source: U.S. Bureau of Labor Statistics)

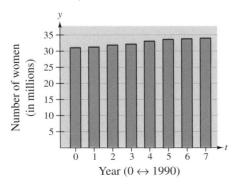

Number of women (in millions)

Year ($0 \leftrightarrow 1990$)

105. *Operating Cost* A delivery company has a fleet of vans. The annual operating cost per van is

$$C = 0.32m + 2500$$

where m is the number of miles traveled by a van in a year. What number of miles will yield an annual operating cost that is equal to $10,000?

106. *Flood Control* Suppose a river has risen 8 feet above its flood stage. The water begins to recede at a rate of 3 inches per hour. Write a mathematical model that shows the number of feet above flood stage after t hours. If the water continually recedes at this rate, when will the river be 1 foot above its flood stage?

Synthesis

True or False? In Exercises 107 and 108, determine whether the statement is true or false. Justify your answer.

107. The equation $x(3 - x) = 10$ is a linear equation.

108. The equation $x^2 + 9x - 5 = 4 - x^3$ has no real solution.

109. *Think About It* What is meant by "equivalent equations"? Give an example of two equivalent equations.

110. *Writing* In your own words, describe the steps used to transform an equation into an equivalent equation.

Review

In Exercises 111–116, simplify the expression.

111. $\dfrac{28x^4}{7x}$

112. $\dfrac{10xy^2}{xy^2 + x}$

113. $\dfrac{x^2 + 5x - 36}{2x^2 + 17x - 9}$

114. $\dfrac{x^2 - 49}{x^3 + x^2 + 3x - 21}$

115. $\dfrac{1}{x - 6} - 5$

116. $\dfrac{5}{x + 1} + \dfrac{4x}{x^2 + 4x + 3}$

In Exercises 117–122, sketch the graph of the equation.

117. $y = 3x - 5$

118. $y = -\frac{1}{2}x - \frac{9}{2}$

119. $y = -x^2 - 5x$

120. $y = \sqrt{5 - x}$

121. $y = 7 - |x|$

122. $x^2 + (y - 5)^2 = 64$

1.3 Modeling with Linear Equations

▶ **What you should learn**

- How to use a verbal model in a problem-solving plan
- How to write and use mathematical models to solve real-life problems
- How to solve mixture problems
- How to use common formulas to solve real-life problems

▶ **Why you should learn it**

You can use linear equations to determine the score you must get on a test in order to get an A for the course you are taking. See Exercise 53 on page 106.

Ulrike Welsch/PhotoEdit

Introduction to Problem Solving

In this section you will learn how algebra can be used to solve problems that occur in real-life situations. The process of translating phrases or sentences into algebraic expressions or equations is called **mathematical modeling.** A good approach to mathematical modeling is to use two stages. Begin by using the verbal description of the problem to form a *verbal model.* Then, after assigning labels to the quantities in the verbal model, form a *mathematical model* or *algebraic equation.*

| Verbal Description | | Verbal Model | | Algebraic Equation |

When you are trying to construct a verbal model, it is helpful to look for a *hidden equality*—a statement that two algebraic expressions are equal.

Example 1 ▶ Using a Verbal Model

You have accepted a job for which your annual salary will be $27,236. This salary includes a year-end bonus of $500. If you are paid twice a month, what will your gross pay be for each paycheck?

Solution

Because there are 12 months in a year and you will be paid twice a month, it follows that you will receive 24 paychecks during the year. You can construct an algebraic equation for this problem as follows. Begin with a verbal model, then assign labels, and finally form an algebraic equation.

Verbal Model: Income for year = 24 paychecks + Bonus

Labels:
Income for year = 27,236   (dollars)
Amount of each paycheck = x (dollars)
Bonus = 500   (dollars)

Equation: $27{,}236 = 24x + 500$

The algebraic equation for this problem is a *linear equation* in the variable x, which you can solve as follows.

$$27{,}236 = 24x + 500 \qquad \text{Write original equation.}$$

$$27{,}236 - 500 = 24x + 500 - 500 \qquad \text{Subtract 500 from each side.}$$

$$26{,}736 = 24x \qquad \text{Simplify.}$$

$$\frac{26{,}736}{24} = \frac{24x}{24} \qquad \text{Divide each side by 24.}$$

$$1114 = x \qquad \text{Simplify.}$$

So, your gross pay for each paycheck will be $1114.

A key step in writing a mathematical model to represent a real-life problem is translating key words and phrases into algebraic expressions and equations. The following list gives several examples.

Translating Key Words and Phrases

Key Words and Phrases	Verbal Description	Algebraic Expression or Equation
Equality:		
Equals, equal to, is, are, was, will be, represents	• The sale price S is $10 less than the list price L.	$S = L - 10$
Addition:		
Sum, plus, greater than, increased by, more than, exceeds, total of	• The sum of 5 and x • Seven more than y	$5 + x$ or $x + 5$ $7 + y$ or $y + 7$
Subtraction:		
Difference, minus, less than, decreased by, subtracted from, reduced by, the remainder	• The difference of 4 and b • Three less than z	$4 - b$ $z - 3$
Multiplication:		
Product, multiplied by, twice, times, percent of	• Two times x • Three percent of t	$2x$ $0.03t$
Division:		
Quotient, divided by, ratio, per	• The ratio of x to 8	$\dfrac{x}{8}$

STUDY T!P

Writing the units for each label in a real-life problem helps you determine the units for the answer. This is called *unit analysis*. When the same units of measure occur in the numerator and denominator of an expression, you can divide out the units. For instance, unit analysis verifies that the units for time in the formula below are hours.

$$\text{Time} = \frac{\text{distance}}{\text{rate}}$$

$$= \frac{\text{miles}}{\dfrac{\text{miles}}{\text{hour}}}$$

$$= \cancel{\text{miles}} \cdot \frac{\text{hours}}{\cancel{\text{miles}}}$$

$$= \text{hours}$$

Using Mathematical Models

Example 2 ▶ Finding the Percent of a Raise

You have accepted a job that pays $8 an hour. You are told that after a 2-month probationary period, your hourly wage will be increased to $9 an hour. What percent raise will you receive after the 2-month period?

Solution

Verbal Model: Raise = Percent · Old wage

Labels:
Old wage = 8 (dollars per hour)
New wage = 9 (dollars per hour)
Raise = 9 − 8 = 1 (dollars per hour)
Percent = r (percent in decimal form)

Equation:

$1 = r \cdot 8$ Write original equation.

$\frac{1}{8} = r$ Divide each side by 8.

$0.125 = r$ Rewrite fraction as a decimal.

You will receive a raise of 0.125 or 12.5%.

Example 3 ▶ Finding the Percent of a Benefits Package

Your annual salary is $24,000. In addition to your salary, your employer also pays for the following benefits: employer's portion of Social Security ($1836), worker's compensation ($120), unemployment compensation ($180), medical insurance ($2240), and retirement contribution ($1560). The total value of this benefits package represents what percent of your annual salary?

Solution

The total amount of your benefits package is $5936.

Verbal Model: Benefits package = Percent · Salary

Labels: Salary = 24,000   (dollars)
Benefits package = 5936   (dollars)
Percent = r (in decimal form)

Equation: $5936 = r \cdot 24{,}000$ Write original equation.

$\dfrac{5936}{24{,}000} = r$ Divide each side by 24,000.

$0.247 \approx r$ Use a calculator.

Your benefits package is approximately 0.247 or 24.7% of your salary.

Example 4 ▶ Finding the Dimensions of a Room

A rectangular family room is twice as long as it is wide, and its perimeter is 84 feet. Find the dimensions of the family room.

Solution

For this problem, it helps to sketch a picture, as shown in Figure 1.16.

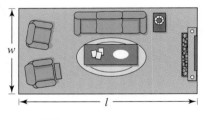

FIGURE 1.16

Verbal Model: 2 · Length + 2 · Width = Perimeter

Labels: Perimeter = 84   (feet)
Width = w (feet)
Length = $l = 2w$ (feet)

Equation: $2(2w) + 2w = 84$ Write original equation.

$6w = 84$ Group like terms.

$w = 14$ Divide each side by 6.

Because the length is twice the width, you have

$l = 2w$ Length is twice width.

$= 2(14)$ Substitute.

$= 28.$ Simplify.

So, the dimensions of the room are 14 feet by 28 feet.

Example 5 ▶ A Distance Problem

A plane is flying nonstop from New York to San Francisco, a distance of about 2700 miles, as shown in Figure 1.17. After $1\frac{1}{2}$ hours in the air, the plane flies over Chicago (a distance of 800 miles from New York). Estimate the time it will take the plane to fly from New York to San Francisco.

Solution

Verbal Model:

$$\boxed{\text{Distance}} = \boxed{\text{Rate}} \cdot \boxed{\text{Time}}$$

Labels:

Distance = 2700   (miles)

Time = t (hours)

Rate = $\dfrac{\text{distance to Chicago}}{\text{time to Chicago}} = \dfrac{800}{1.5}$ (miles per hour)

Equation:

$$2700 = \frac{800}{1.5}t$$

$$4050 = 800t$$

$$\frac{4050}{800} = t$$

$$5.06 \approx t$$

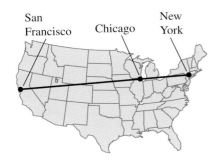

FIGURE **1.17**

The trip will take about 5.06 hours, or about 5 hours and 4 minutes.

Example 6 ▶ An Application Involving Similar Triangles

To measure the height of the twin towers of the World Trade Center, you measure the shadow cast by one of the buildings and find it to be 170.25 feet long, as shown in Figure 1.18. Then you measure the shadow cast by a 4-foot post and find it to be 6 inches long. Estimate the building's height.

Solution

To solve this problem, you use a result from geometry that states that the ratios of corresponding sides of similar triangles are equal.

Verbal Model

$$\frac{\boxed{\text{Height of building}}}{\boxed{\text{Length of building's shadow}}} = \frac{\boxed{\text{Height of post}}}{\boxed{\text{Length of post's shadow}}}$$

Labels:

Height of building = x (feet)

Length of building's shadow = 170.25   (feet)

Height of post = 4 feet = 48 inches   (inches)

Length of post's shadow = 6   (inches)

Equation:

$$\frac{x}{170.25} = \frac{48}{6}$$

$$x = 1362$$

So, the World Trade Center is about 1362 feet high.

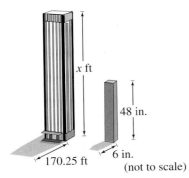

x ft

48 in.

6 in.

170.25 ft

(not to scale)

FIGURE **1.18**

Mixture Problems

Problems that involve two or more rates are called mixture problems. They are not limited to mixtures of chemical solutions, such as those in Exercises 72 to 74. For instance, Example 7 is a mixture problem involving a "mix" of investments.

Example 7 ▶ A Simple Interest Problem

You invested a total of $10,000 at $4\frac{1}{2}\%$ and $5\frac{1}{2}\%$ simple interest. During one year, the two accounts earned $508.75. How much did you invest in each?

Solution

Verbal Model: | Interest from $4\frac{1}{2}\%$ | + | Interest from $5\frac{1}{2}\%$ | = | Total interest |

Labels: Amount invested at $4\frac{1}{2}\% = x$ (dollars)
Amount invested at $5\frac{1}{2}\% = 10,000 - x$ (dollars)
Interest from $4\frac{1}{2}\% = Prt = (x)(0.045)(1)$ (dollars)
Interest from $5\frac{1}{2}\% = Prt = (10,000 - x)(0.055)(1)$ (dollars)
Total interest $= 508.75$ (dollars)

Equation: $0.045x + 0.055(10,000 - x) = 508.75$

$$-0.01x = -41.25$$

$$x = 4125, \text{ or } \$4125 \text{ at } 4\tfrac{1}{2}\%$$

$$10,000 - x = 5875, \text{ or } \$5875 \text{ at } 5\tfrac{1}{2}\%$$

Example 8 ▶ An Inventory Problem

A store has $30,000 of inventory in 12-inch and 19-inch color televisions. The profit on a 12-inch set is 22% and the profit on a 19-inch set is 40%. The profit for the entire stock is 35%. How much was invested in each type of television?

Solution

Verbal Model: | Profit from 12-inch sets | + | Profit from 19-inch sets | = | Total profit |

Labels: Inventory of 12-inch sets $= x$ (dollars)
Inventory of 19-inch sets $= 30,000 - x$ (dollars)
Profit from 12-inch sets $= 0.22x$ (dollars)
Profit from 19-inch sets $= 0.40(30,000 - x)$ (dollars)
Total profit $= 0.35(30,000) = 10,500$ (dollars)

Equation: $0.22(x) + 0.40(30,000 - x) = 10,500$

$$-0.18x = -1500$$

$$x \approx \$8333.33 \text{ invested in 12-inch sets}$$

$$30,000 - x \approx \$21,666.67 \text{ invested in 19-inch sets}$$

Common Formulas

Many common types of geometric, scientific, and investment problems use ready-made equations called **formulas.** Knowing these formulas will help you translate and solve a wide variety of real-life applications.

Common Formulas for Area *A*, Perimeter *P*, Circumference *C*, and Volume *V*

Square	*Rectangle*	*Circle*	*Triangle*
$A = s^2$	$A = lw$	$A = \pi r^2$	$A = \dfrac{1}{2}bh$
$P = 4s$	$P = 2l + 2w$	$C = 2\pi r$	$P = a + b + c$

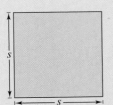 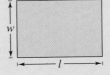 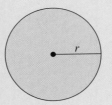 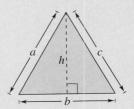

Cube	*Rectangular Solid*	*Circular Cylinder*	*Sphere*
$V = s^3$	$V = lwh$	$V = \pi r^2 h$	$V = \dfrac{4}{3}\pi r^3$

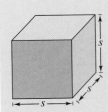 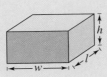  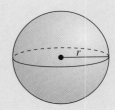

Miscellaneous Common Formulas

Temperature:

$$F = \frac{9}{5}C + 32 \qquad F = \text{degrees Fahrenheit, } C = \text{degrees Celsius}$$

Simple Interest:

$$I = Prt \qquad \begin{aligned} &I = \text{interest, } P = \text{principal,}\\ &r = \text{annual interest rate, } t = \text{time in years} \end{aligned}$$

Compound Interest:

$$A = P\left(1 + \frac{r}{n}\right)^{nt} \qquad \begin{aligned} &A = \text{balance, } P = \text{principal, } r = \text{annual interest rate,}\\ &n = \text{compoundings per year, } t = \text{time in years} \end{aligned}$$

Distance:

$$d = rt \qquad d = \text{distance traveled, } r = \text{rate, } t = \text{time}$$

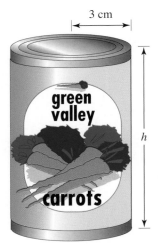

3 cm

h

FIGURE 1.19

When working with applied problems, you often need to rewrite one of the common formulas. For instance, the formula for the perimeter of a rectangle, $P = 2l + 2w$, can be rewritten or solved for w as $w = \frac{1}{2}(P - 2l)$.

Example 9 ▶ Using a Formula

A cylindrical can has a volume of 300 cubic centimeters (cm^3) and a radius of 3 centimeters (cm), as shown in Figure 1.19. Find the height of the can.

Solution

The formula for the *volume of a cylinder* is $V = \pi r^2 h$. To find the height of the can, solve for h.

$$h = \frac{V}{\pi r^2}$$

Then, using $V = 300$ and $r = 3$, find the height.

$$h = \frac{300}{\pi(3)^2} \qquad \text{Substitute 300 for } V \text{ and 3 for } r.$$

$$= \frac{300}{9\pi} \qquad \text{Simplify denominator.}$$

$$\approx 10.61 \qquad \text{Use a calculator.}$$

You can use unit analysis to check that your answer is reasonable.

$$\frac{300 \text{ cm}^3}{9\pi \text{ cm}^2} \approx 10.61 \text{ cm}$$

Writing ABOUT MATHEMATICS

Translating Algebraic Formulas Most people use algebraic formulas every day—sometimes without realizing it because they use a verbal form or think of an often-repeated calculation in steps. Translate each of the following verbal descriptions into an algebraic formula, and demonstrate the use of each formula.

a. **Designing Billboards** "The letters on a sign or billboard are designed to be readable at a certain distance. Take half the letter height in inches and multiply by 100 to find the readable distance in feet."—Thos. Hodgson, Hodgson Signs (Source: *Rules of Thumb* by Tom Parker)

b. **Percent of Calories from Fat** "To calculate percent of calories from fat, multiply grams of total fat per serving by 9, then divide by the number of calories per serving." (Source: *Good Housekeeping*)

c. **Building Stairs** "A set of steps will be comfortable to use if two times the height of one riser plus the width of one tread is equal to 26 inches." —Alice Lukens Bachelder, gardener (Source: *Rules of Thumb* by Tom Parker)

1.3 Exercises

In Exercises 1–10, write a verbal description of the algebraic expression without using the variable.

1. $x + 4$

2. $t - 10$

3. $\dfrac{u}{5}$

4. $\dfrac{2}{3}x$

5. $\dfrac{y - 4}{5}$

6. $\dfrac{z + 10}{7}$

7. $-3(b + 2)$

8. $\dfrac{-5(x - 1)}{8}$

9. $12x(x - 5)$

10. $\dfrac{(q + 4)(3 - q)}{2q}$

In Exercises 11–22, write an algebraic expression for the verbal description.

11. The sum of two consecutive natural numbers

12. The product of two consecutive natural numbers

13. The product of two consecutive odd integers, the first of which is $2n - 1$

14. The sum of the squares of two consecutive even integers, the first of which is $2n$

15. The distance traveled in t hours by a car traveling at 50 miles per hour

16. The travel time for a plane traveling at a rate of r kilometers per hour for 200 kilometers

17. The amount of acid in x liters of a 20% acid solution

18. The sale price of an item that is discounted 20% of its list price L

19. The perimeter of a rectangle with a width x and a length that is twice the width

20. The area of a triangle with base 20 inches and height h inches

21. The total cost of producing x units for which the fixed costs are $1200 and the cost per unit is $25

22. The total revenue obtained by selling x units at $3.59 per unit

In Exercises 23–26, translate the statement into an algebraic expression or equation.

23. Thirty percent of the list price L

24. The amount of water in q quarts of a product that is 35% water

25. The percent of 500 that is represented by the number N

26. The percent change in sales from one month to the next if the monthly sales are S_1 and S_2, respectively

In Exercises 27 and 28, write an expression for the area of the region in the figure.

27.

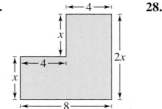

28.

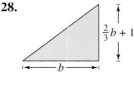

Number Problems **In Exercises 29–34, write a mathematical model for the number problem and solve.**

29. The sum of two consecutive natural numbers is 525. Find the numbers.

30. The sum of three consecutive natural numbers is 804. Find the numbers.

31. One positive number is 5 times another number. The difference between the two numbers is 148. Find the numbers.

32. One positive number is one-fifth of another number. The difference between the two numbers is 76. Find the numbers.

33. Find two consecutive integers whose product is 5 less than the square of the smaller number.

34. Find two consecutive natural numbers such that the difference of their reciprocals is one-fourth the reciprocal of the smaller number.

In Exercises 35–40, solve the percent equation.

35. What is 30% of 45?

36. What is 175% of 360?

37. 432 is what percent of 1600?

38. 459 is what percent of 340?

39. 12 is $\frac{1}{2}$% of what number?

40. 70 is 40% of what number?

41. *Finance* A family has annual loan payments equaling 58.6% of their annual income. During the year, their loan payments total $13,077.75. What is their annual income?

42. *Discount* The price of a swimming pool has been discounted 16.5%. The sale price is $1210.75. Find the original list price of the pool.

43. *Government* The circle graph shows the sources of income for the federal government in 1997. The total income was $1,579,292,000,000. Find the income for each of the categories. (Round your answers to the nearest billion dollars.) (Source: U.S. Office of Management and Budget)

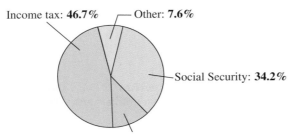

Income tax: **46.7%** — Other: **7.6%**

Social Security: **34.2%**

Corporation taxes: **11.5%**

44. *Government* The circle graph shows expenses of the federal government in 1997. The total expenses were $1,601,235,000,000. Find the expense for each of the categories. (Round your answers to the nearest billion dollars.) (Source: U.S. Office of Management and Budget)

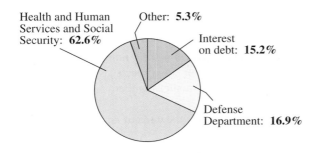

Health and Human Services and Social Security: **62.6%**

Other: **5.3%**

Interest on debt: **15.2%**

Defense Department: **16.9%**

45. *Business* The total profit for a company in February was 20% higher than it was in January. The total profit for the two months was $157,498. Find the profit for each month.

46. *Finance* Your weekly paycheck is 15% less than your coworker's. Your two paychecks total $645. Find the amount of each paycheck.

In Exercises 47–50, the values or prices of different items are given for 1980 and 1997. Find the percent change for each item. (Source: 1998 Statistical Abstract of the U.S.)

Item	1980	1997
47. Gallon of diesel fuel	$0.82	$0.64
48. Cable TV monthly basic rate	$7.69	$26.48
49. An ounce of gold	$613.00	$333.00
50. One acre of farmland	$737.00	$945.00

51. *Dimensions of a Room* A room is 1.5 times as long as it is wide, and its perimeter is 25 meters.

(a) Draw a diagram that represents the problem. Identify the length as l and the width as w.

(b) Write l in terms of w and write an equation for the perimeter in terms of w.

(c) Find the dimensions of the room.

52. *Dimensions of a Picture Frame* A picture frame has a total perimeter of 2 meters. The height of the frame is 0.62 times its width.

(a) Draw a diagram that represents the problem. Identify the width as w and the height as h.

(b) Write h in terms of w and write an equation for the perimeter in terms of w.

(c) Find the dimensions of the picture frame.

53. *Course Grade* To get an A in a course, you must have an average of at least 90 on four tests of 100 points each. The scores on your first three tests were 87, 92, and 84. What must you score on the fourth test to get an A for the course?

54. *Course Grade* Suppose you are taking a course that has four tests. The first three tests are 100 points each and the fourth test is 200 points. To get an A in the course, you must have an average of at least 90% on the four tests. Your scores on the first three tests were 87, 92, and 84. What must you score on the fourth test to get an A for the course?

55. *Travel Time* You are driving on a Canadian freeway to a town that is 300 kilometers from your home. After 30 minutes you pass a freeway exit that you know is 50 kilometers from your home. Assuming that you continue at the same constant speed, how long will it take for the entire trip?

56. *Travel Time* Two cars start at a given point and travel in the same direction at average speeds of 40 miles per hour and 55 miles per hour. How much time must elapse before the two cars are 5 miles apart?

57. *Travel Time* On the first part of a 317-mile trip, a salesman averaged 58 miles per hour. He averaged only 52 miles per hour on the last part of the trip because of an increased volume of traffic. Find the amount of time at each of the speeds if the total time was 5 hours and 45 minutes.

58. *Travel Time* Students are traveling in two cars to a football game 135 miles away. The first car leaves on time and travels at an average speed of 45 miles per hour. The second car starts $\frac{1}{2}$ hour later and travels at an average speed of 55 miles per hour. How long will it take the second car to catch up to the first car? Will the second car catch up to the first car before the first car arrives at the game?

59. *Travel Time* Two families meet at a park for a picnic. At the end of the day one family travels east at an average speed of 42 miles per hour and the other travels west at an average speed of 50 miles per hour. Both families have approximately 160 miles to travel.

    (a) Find the time it takes each family to get home.

    (b) Find the time that will have elapsed when they are 100 miles apart.

    (c) Find the distance the eastbound family has to travel after the westbound family has arrived home.

60. *Average Speed* A truck driver traveled at an average speed of 55 miles per hour on a 200-mile trip to pick up a load of freight. On the return trip (with the truck fully loaded), the average speed was 40 miles per hour. Find the average speed for the round trip.

61. *Wind Speed* An executive flew in the corporate jet to a meeting in a city 1500 kilometers away. After traveling the same amount of time on the return flight, the pilot mentioned that they still had 300 kilometers to go. If the air speed of the plane was 600 kilometers per hour, how fast was the wind blowing? (Assume that the wind direction was parallel to the flight path and constant all day.)

62. *Physics* Light travels at the speed of 3.0×10^8 meters per second. Find the time in minutes required for light to travel from the sun to the earth (a distance of 1.5×10^{11} meters).

63. *Radio Waves* Radio waves travel at the same speed as light, 3.0×10^8 meters per second. Find the time required for a radio wave to travel from Mission Control in Houston to NASA astronauts on the surface of the moon 3.86×10^8 meters away.

64. *Height of a Tree* To obtain the height of a tree (see figure), you measure the tree's shadow and find that it is 8 meters long. You also measure the shadow of a 2-meter lamppost and find that it is 75 centimeters long. How tall is the tree?

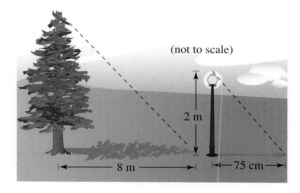

(not to scale)

2 m

|← 8 m →| |←75 cm→|

65. *Height of a Building* To obtain the height of a building (see figure), you measure the building's shadow and find that it is 80 feet long. You also measure the shadow of a 4-foot stake and find that it is $3\frac{1}{2}$ feet long. How tall is the building?

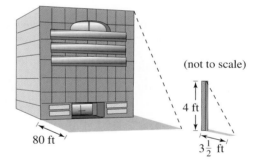

(not to scale)

4 ft

80 ft

$3\frac{1}{2}$ ft

66. *Flagpole Height* A person who is 6 feet tall walks away from a flagpole toward the tip of the shadow of the pole. When the person is 30 feet from the pole, the tips of the person's shadow and the shadow cast by the pole coincide at a point 5 feet in front of the person.

    (a) Draw a diagram that gives a visual representation of the problem. Let h represent the height of the pole.

    (b) Find the height of the pole.

67. *Shadow Length* A person who is 6 feet tall walks away from a 50-foot silo toward the tip of the silo's shadow. At a distance of 32 feet from the silo, the person's shadow begins to emerge beyond the silo's shadow. How much farther must the person walk to be completely out of the silo's shadow?

68. *Investment* You plan to invest $12,000 in two funds paying $4\frac{1}{2}\%$ and 5% simple interest. (There is more risk in the 5% fund.) Your goal is to obtain a total annual interest income of $580 from the investments. What is the smallest amount you can invest in the 5% fund in order to meet your objective?

69. *Investment* You plan to invest $25,000 in two funds paying 3% and $4\frac{1}{2}\%$ simple interest. (There is more risk in the $4\frac{1}{2}\%$ fund.) Your goal is to obtain a total annual interest income of $1000 from the investments. What is the smallest amount you can invest in the $4\frac{1}{2}\%$ fund in order to meet your objective?

70. *Business* A nursery has $20,000 of inventory in dogwood trees and red maple trees. The profit on a dogwood tree is 25% and the profit on a red maple is 17%. The profit for the entire stock is 20%. How much was invested in each type of tree?

71. *Business* An automobile dealer has $600,000 of inventory in compact cars and midsize cars. The profit on a compact car is 24% and the profit on a midsize car is 28%. The profit for the entire stock is 25%. How much was invested in each type of car?

72. *Mixture Problem* Using the values in the table, determine the amounts of solutions 1 and 2, respectively, needed to obtain the desired amount and concentration of the final mixture.

	Concentration			
	Solution 1	Solution 2	Final solution	Amount of final solution
(a)	10%	30%	25%	100 gal
(b)	25%	50%	30%	5 L
(c)	15%	45%	30%	10 qt
(d)	70%	90%	75%	25 gal

73. *Mixture Problem* A 100% concentrate is to be mixed with a mixture having a concentration of 40% to obtain 55 gallons of a mixture with a concentration of 75% (see figure). How much of the 100% concentrate will be needed?

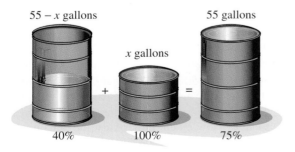

$55 - x$ gallons x gallons   55 gallons

40%   100%   75%

74. *Mixture Problem* A forester mixes gasoline and oil to make 2 gallons of mixture for his two-cycle chain-saw engine. This mixture is 32 parts gasoline and 1 part two-cycle oil. How much gasoline must be added to bring the mixture to 40 parts gasoline and 1 part oil?

75. *Mixture Problem* A grocer mixes two kinds of nuts that cost $2.49 per pound and $3.89 per pound, respectively, to make 100 pounds of a mixture that costs $3.19 per pound. How much of each kind of nut is put into the mixture?

76. *Company Costs* A company has fixed costs of $10,000 per month and variable costs of $8.50 per unit manufactured. The company has $85,000 available to cover the monthly costs. How many units can the company manufacture? (*Fixed costs* are those that occur regardless of the level of production. *Variable costs* depend on the level of production.)

77. *Company Costs* A company has fixed costs of $10,000 per month and variable costs of $9.30 per unit manufactured. The company has $85,000 available to cover the monthly costs. How many units can the company manufacture? (*Fixed costs* are those that occur regardless of the level of production. *Variable costs* depend on the level of production.)

78. *Water Depth* A trough is 12 feet long, 3 feet deep, and 3 feet wide (see figure). Find the depth of the water when the trough contains 70 gallons (1 gallon ≈ 0.13368 cubic foot).

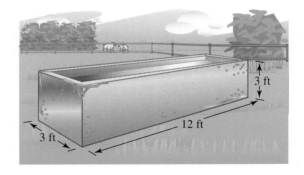

3 ft   12 ft   3 ft

Physics **In Exercises 79 and 80, suppose you have a uniform beam of length *L* with a fulcrum *x* feet from one end (see figure). If objects with weights W_1 and W_2 are placed at opposite ends of the beam, then the beam will balance if $W_1 x = W_2(L - x)$. Find *x* such that the beam will balance.**

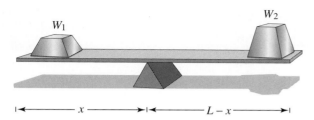

W_1 W_2

|◀— x —▶|◀— $L - x$ —▶|

FIGURE FOR 79 AND 80

79. Two children weighing 50 pounds and 75 pounds are going to play on a seesaw that is 10 feet long.

80. A person weighing 200 pounds is attempting to move a 550-pound rock with a bar that is 5 feet long.

In Exercises 81–96, solve for the indicated variable.

81. *Area of a Triangle*   Solve for h: $A = \frac{1}{2}bh$

82. *Volume of a Right Circular Cylinder*
Solve for h: $V = \pi r^2 h$

83. *Markup*   Solve for C: $S = C + RC$

84. *Investment at Simple Interest*
Solve for r: $A = P + Prt$

85. *Investment at Compound Interest*
Solve for P: $A = P\left(1 + \dfrac{r}{n}\right)^{nt}$

86. *Area of a Trapezoid*
Solve for b: $A = \frac{1}{2}(a + b)h$

87. *Area of a Sector of a Circle*
Solve for θ: $A = \dfrac{\pi r^2 \theta}{360}$

88. *Volume of a Spherical Segment*
Solve for r: $V = \frac{1}{3}\pi h^2(3r - h)$

89. *Volume of an Oblate Spheroid*
Solve for b: $V = \frac{4}{3}\pi a^2 b$

90. *Free-Falling Body*   Solve for a: $h = v_0 t + \frac{1}{2}at^2$

91. *Newton's Law of Universal Gravitation*
Solve for m_2: $F = \alpha \dfrac{m_1 m_2}{r^2}$

92. *Lensmaker's Equation*
Solve for R_1: $\dfrac{1}{f} = (n - 1)\left(\dfrac{1}{R_1} - \dfrac{1}{R_2}\right)$

93. *Capacitance in Series Circuits*
Solve for C_1: $C = \dfrac{1}{\dfrac{1}{C_1} + \dfrac{1}{C_2}}$

94. *Arithmetic Progression*
Solve for n: $L = a + (n - 1)d$

95. *Arithmetic Progression*
Solve for a: $S = \dfrac{n}{2}[2a + (n - 1)d]$

96. *Geometric Progression*   Solve for r: $S = \dfrac{rL - a}{r - 1}$

97. *Volume of a Billiard Ball*   A billiard ball has a volume of 5.96 cubic inches. Find the radius of a billiard ball.

98. *Length of a Tank*   The diameter of a cylindrical propane gas tank is 4 feet. The total volume of the tank is 603.2 cubic feet. Find the length of the tank.

Synthesis

True or False?   **In Exercises 99 and 100, determine whether the statement is true or false. Justify your answer.**

99. "8 less than z cubed divided by the difference of z squared and 9" can be written as $\dfrac{z^3 - 8}{(z - 9)^2}$.

100. The volume of a cube with a side of length 9.5 inches is greater than the volume of a sphere with a radius of 5.9 inches.

101. Consider the linear equation $ax + b = 0$.
(a) What is the sign of the solution if $ab > 0$?
(b) What is the sign of the solution if $ab < 0$?
In each case, explain your reasoning.

Review

In Exercises 102–105, simplify the expression.

102. $(5x^4)(25x^2)^{-1}$, $x \neq 0$ **103.** $\sqrt{150s^2 t^3}$

104. $\dfrac{3}{x - 5} + \dfrac{2}{5 - x}$ **105.** $\dfrac{5}{x} + \dfrac{3x}{x^2 - 9} - \dfrac{10}{x + 3}$

In Exercises 106–109, rationalize the denominator.

106. $\dfrac{10}{7\sqrt{3}}$ **107.** $\dfrac{6}{\sqrt{10} - 2}$

108. $\dfrac{5}{\sqrt{6} + \sqrt{11}}$ **109.** $\dfrac{14}{3\sqrt{10} - 1}$

1.4 Quadratic Equations

▶ **What you should learn**

- How to solve quadratic equations by factoring
- How to solve quadratic equations by extracting square roots
- How to solve quadratic equations by completing the square
- How to use the Quadratic Formula to solve quadratic equations
- How to use quadratic equations to model and solve real-life problems

▶ **Why you should learn it**

Quadratic equations can be used to model and solve real-life problems. For instance, Exercise 120 on page 121 shows how a quadratic equation can be used to model the time it takes an object to fall from the top of the CN Tower.

Chris Thomaidis/Tony Stone Images

Factoring

A **quadratic equation** in x is an equation that can be written in the general form

$$ax^2 + bx + c = 0$$

where a, b, and c are real numbers with $a \neq 0$. A quadratic equation in x is also known as a **second-degree polynomial equation** in x.

In this section, you will study four methods for solving quadratic equations: factoring, extracting square roots, completing the square, and the Quadratic Formula. The first method is based on the Zero-Factor Property given in Section P.1.

> If $ab = 0$, then $a = 0$ or $b = 0$.   Zero-Factor Property

To use this property, write the left side of the general form of a quadratic equation as the product of two linear factors. Then find the solutions of the quadratic equation by setting each linear factor equal to zero.

Example 1 ▶ Solving a Quadratic Equation by Factoring

a.

$2x^2 + 9x + 7 = 3$	Write original equation.
$2x^2 + 9x + 4 = 0$	Write in general form.
$(2x + 1)(x + 4) = 0$	Factor.
$2x + 1 = 0 \implies x = -\dfrac{1}{2}$	Set 1st factor equal to 0.
$x + 4 = 0 \implies x = -4$	Set 2nd factor equal to 0.

The solutions are $-\frac{1}{2}$ and -4. Check these in the original equation.

b.

$6x^2 - 3x = 0$	Write original equation.
$3x(2x - 1) = 0$	Factor.
$3x = 0 \implies x = 0$	Set 1st factor equal to 0.
$2x - 1 = 0 \implies x = \dfrac{1}{2}$	Set 2nd factor equal to 0.

The solutions are 0 and $\frac{1}{2}$. Check these in the original equation.

Be sure you see that the Zero-Factor Property works *only* for equations written in general form (in which the right side of the equation is zero). Therefore, all terms must be collected on one side *before* factoring. For instance, in the equation

$$(x - 5)(x + 2) = 8$$

it is *incorrect* to set each factor equal to 8. Can you solve this equation correctly?

Extracting Square Roots

There is a nice shortcut for solving quadratic equations of the form $u^2 = d$, where $d > 0$ and u is an algebraic expression. By factoring, you can see that this equation has two solutions.

$u^2 = d$	Write original equation.
$u^2 - d = 0$	Write in general form.
$(u + \sqrt{d})(u - \sqrt{d}) = 0$	Factor.
$u + \sqrt{d} = 0 \implies u = -\sqrt{d}$	Set 1st factor equal to 0.
$u - \sqrt{d} = 0 \implies u = \sqrt{d}$	Set 2nd factor equal to 0.

Because the two solutions differ only in sign, you can write the solutions together, using a "plus or minus sign," as

$$u = \pm\sqrt{d}.$$

This form of the solution is read as "u is equal to plus or minus the square root of d." Solving an equation of the form $u^2 = d$ without going through the steps of factoring is called **extracting square roots.**

Extracting Square Roots

The equation $u^2 = d$, where $d > 0$, has exactly two solutions:

$$u = \sqrt{d} \quad \text{and} \quad u = -\sqrt{d}.$$

These solutions can also be written as

$$u = \pm\sqrt{d}.$$

Example 2 ▶ Extracting Square Roots

Solve each equation by extracting square roots.

a. $4x^2 = 12$ **b.** $(x - 3)^2 = 7$

Solution

a.

$4x^2 = 12$	Write original equation.
$x^2 = 3$	Divide each side by 4.
$x = \pm\sqrt{3}$	Extract square roots.

The solutions are $\sqrt{3}$ and $-\sqrt{3}$. Check these in the original equation.

b.

$(x - 3)^2 = 7$	Write original equation.
$x - 3 = \pm\sqrt{7}$	Extract square roots.
$x = 3 \pm\sqrt{7}$	Add 3 to each side.

The solutions are $3 \pm \sqrt{7}$. Check these in the original equation.

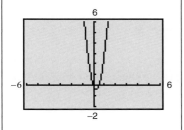

Completing the Square

The equation in Example 2(b) was given in the form $(x - 3)^2 = 7$ so that you could find the solution by extracting square roots. Suppose, however, that the equation $(x - 3)^2 = 7$ had been given in the general form

$$x^2 - 6x + 2 = 0. \qquad \text{General form}$$

This equation is equivalent to the original and so it has the same two solutions, $x = 3 \pm \sqrt{7}$. However, the left side of the equation is not factorable, and you cannot find its solutions unless you can rewrite the equation by **completing the square.**

Completing the Square

To **complete the square** for the expression

$$x^2 + bx$$

add $(b/2)^2$, which is the square of half the coefficient of x. Consequently,

$$x^2 + bx + \left(\frac{b}{2}\right)^2 = \left(x + \frac{b}{2}\right)^2.$$

When solving quadratic equations by completing the square, you must add $(b/2)^2$ to *both sides* in order to maintain equality.

Example 3 ▶   Completing the Square: Leading Coefficient Is 1

Solve $x^2 - 6x + 2 = 0$ by completing the square. Compare the solutions with those obtained in Example 2(b).

Solution

$$x^2 - 6x + 2 = 0 \qquad \text{Write original equation.}$$

$$x^2 - 6x = -2 \qquad \text{Subtract 2 from each side.}$$

$$x^2 - 6x + \left(\frac{-6}{2}\right)^2 = -2 + \left(\frac{-6}{2}\right)^2 \qquad \text{Add } \left(\frac{-6}{2}\right)^2 \text{ to each side.}$$

$$\underbrace{}_{\text{(half of } -6)^2}$$

$$x^2 - 6x + 9 = 7 \qquad \text{Simplify.}$$

$$(x - 3)^2 = 7 \qquad \text{Perfect square trinomial}$$

$$x - 3 = \pm\sqrt{7} \qquad \text{Extract square roots.}$$

$$x = 3 \pm \sqrt{7} \qquad \text{Add 3 to each side.}$$

The solutions are $x = 3 \pm \sqrt{7}$, as found in Example 2(b).

If the leading coefficient is *not* 1, you must divide both sides of the equation by the leading coefficient *before* completing the square. For instance, to complete the square for $3x^2 - 4x - 5 = 0$, first divide each term by the leading coefficient 3. Then proceed as in Example 3.

The Quadratic Formula

Often in mathematics you are taught the long way of solving a problem first. Then, the longer method is used to develop shorter techniques. The long way stresses understanding and the short way stresses efficiency.

For instance, you can think of completing the square as a "long way" of solving a quadratic equation. When you use completing the square to solve a quadratic equation, you must complete the square for *each* equation separately. In the following derivation, you complete the square *once* in a general setting to obtain the **Quadratic Formula**—a shortcut for solving a quadratic equation.

$$ax^2 + bx + c = 0 \qquad \text{Write in general form, } a \ne 0.$$

$$ax^2 + bx = -c \qquad \text{Subtract } c \text{ from each side.}$$

$$x^2 + \frac{b}{a}x = -\frac{c}{a} \qquad \text{Divide each side by } a.$$

$$x^2 + \frac{b}{a}x + \left(\frac{b}{2a}\right)^2 = -\frac{c}{a} + \left(\frac{b}{2a}\right)^2 \qquad \text{Complete the square.}$$

$$\left(\text{half of } \frac{b}{a}\right)^2$$

$$\left(x + \frac{b}{2a}\right)^2 = \frac{b^2 - 4ac}{4a^2} \qquad \text{Simplify.}$$

$$x + \frac{b}{2a} = \pm\sqrt{\frac{b^2 - 4ac}{4a^2}} \qquad \text{Extract square roots.}$$

$$x = -\frac{b}{2a} \pm \frac{\sqrt{b^2 - 4ac}}{2|a|} \qquad \text{Solutions}$$

Note that because $\pm 2|a|$ represents the same numbers as $\pm 2a$, you can omit the absolute value sign. So, the formula simplifies to

$$x = \frac{-b \pm \sqrt{b^2 - 4ac}}{2a}.$$

The Quadratic Formula

The solutions of a quadratic equation in the general form

$$ax^2 + bx + c = 0, \qquad a \ne 0$$

are given by the **Quadratic Formula**

$$x = \frac{-b \pm \sqrt{b^2 - 4ac}}{2a}.$$

The Quadratic Formula is one of the most important formulas in algebra. You should learn the verbal statement of the Quadratic Formula:

"Negative *b*, plus or minus the square root of *b* squared minus 4*ac*, *all* divided by 2*a*."

◀ **E x p l o r a t i o n** ▶

From each graph, can you tell whether the discriminant is positive, zero, or negative? Explain your reasoning. Find each discriminant to verify your answers.

a. $x^2 - 2x = 0$

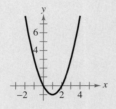

b. $x^2 - 2x + 1 = 0$

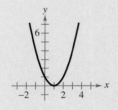

c. $x^2 - 2x + 2 = 0$

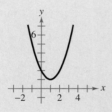

In the Quadratic Formula, the quantity under the radical sign, $b^2 - 4ac$, is the **discriminant** of the quadratic expression $ax^2 + bx + c$. It can be used to determine the nature of the solutions of a quadratic equation.

Solutions of a Quadratic Equation

The solutions of a quadratic equation $ax^2 + bx + c = 0$, $a \neq 0$, can be classified as follows. If the discriminant $b^2 - 4ac$ is

1. *positive*, then the quadratic equation has *two* distinct real solutions and its graph has *two* x-intercepts.

2. *zero*, then the quadratic equation has *one* repeated real solution and its graph has *one* x-intercept.

3. *negative*, then the quadratic equation has *no* real solutions and its graph has *no* x-intercepts.

If the discriminant of a quadratic equation is negative, as in case 3 above, then its square root is imaginary (not a real number) and the Quadratic Formula yields two complex solutions. You will study complex solutions in Section 1.5.

When using the Quadratic Formula, remember that *before* the formula can be applied, you must first write the quadratic equation in general form.

Example 4 ▶ The Quadratic Formula: Two Distinct Solutions

Use the Quadratic Formula to solve

$$x^2 + 3x = 9.$$

Solution

$$x^2 + 3x = 9 \qquad \text{Write original equation.}$$

$$x^2 + 3x - 9 = 0 \qquad \text{Write in general form.}$$

$$x = \frac{-b \pm \sqrt{b^2 - 4ac}}{2a} \qquad \text{Quadratic Formula}$$

$$x = \frac{-3 \pm \sqrt{(3)^2 - 4(1)(-9)}}{2(1)} \qquad \begin{array}{l}\text{Substitute } a = 1, b = 3, \\ \text{and } c = -9.\end{array}$$

$$x = \frac{-3 \pm \sqrt{45}}{2} \qquad \text{Simplify.}$$

$$x = \frac{-3 \pm 3\sqrt{5}}{2} \qquad \text{Simplify.}$$

The equation has two solutions:

$$x = \frac{-3 + 3\sqrt{5}}{2} \qquad \text{and} \qquad x = \frac{-3 - 3\sqrt{5}}{2}.$$

Check these in the original equation.

A computer animation of this example appears in the *Interactive* CD-ROM and *Internet* versions of this text.

Applications

Quadratic equations often occur in problems dealing with area. Here is a simple example. "A square room has an area of 144 square feet. Find the dimensions of the room." To solve this problem, let x represent the length of each side of the room. Then, by solving the equation

$$x^2 = 144,$$

you can conclude that each side of the room is 12 feet long. Note that although the equation $x^2 = 144$ has two solutions, -12 and 12, the negative solution makes no sense, so you choose the positive solution.

Example 5 ▶ Finding the Dimensions of a Room

A bedroom is 3 feet longer than it is wide (see Figure 1.20) and has an area of 154 square feet. Find the dimensions of the room.

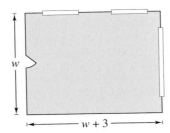

FIGURE 1.20

Solution

Verbal Model: | Width of room | · | Length of room | = | Area of room |

Labels: Width of room $= w$ (feet)
         Length of room $= w + 3$ (feet)
         Area of room $= 154$ (square feet)

Equation:

$$w(w + 3) = 154$$

$$w^2 + 3w - 154 = 0$$

$$(w - 11)(w + 14) = 0$$

$$w - 11 = 0 \implies w = 11$$

$$w + 14 = 0 \implies w = -14$$

Choosing the positive value, you find that the width is 11 feet and the length is $w + 3$, or 14 feet. You can check this solution by observing that the length is 3 feet longer than the width *and* that the product of the length and width is 154 square feet.

Another common application of quadratic equations involves an object that is falling (or projected into the air). The general equation that gives the height of such an object is called a **position equation,** and on *earth's* surface it has the form

$$s = -16t^2 + v_0 t + s_0.$$

In this equation, s represents the height of the object (in feet), v_0 represents the initial velocity of the object (in feet per second), s_0 represents the initial height of the object (in feet), and t represents the time (in seconds).

Example 6 ▶ Falling Time

A construction worker on the 24th floor of a building (see Figure 1.21) accidentally drops a wrench and yells "Look out below!" Could a person at ground level hear this warning in time to get out of the way?

Solution

Assume that each floor of the building is 10 feet high, so that the wrench is dropped from a height of 240 feet. Because sound travels at about 1100 feet per second, it follows that a person at ground level hears the warning within 1 second of the time the wrench is dropped. To set up a mathematical model for the height of the wrench, use the position equation

$$s = -16t^2 + v_0 t + s_0.$$

Because the object is dropped rather than thrown, the initial velocity is $v_0 = 0$. Moreover, because the initial height is $s_0 = 240$ feet, you have the following model.

$$s = -16t^2 + (0)t + 240 = -16t^2 + 240$$

After falling for 1 second, the height of the wrench is $-16(1)^2 + 240 = 224$. After falling for 2 seconds, the height of the wrench is $-16(2)^2 + 240 = 176$. To find the number of seconds it takes the wrench to hit the ground, let the height s be zero and solve the equation for t.

$s = -16t^2 + 240$	Write position equation.
$0 = -16t^2 + 240$	Substitute 0 for height.
$16t^2 = 240$	Add $16t^2$ to each side.
$t^2 = 15$	Divide each side by 16.
$t = \sqrt{15}$	Extract positive square root.
$t \approx 3.87$	Use a calculator.

The wrench will take about 3.87 seconds to hit the ground. If the person hears the warning 1 second after the wrench is dropped, the person still has almost 3 seconds to get out of the way.

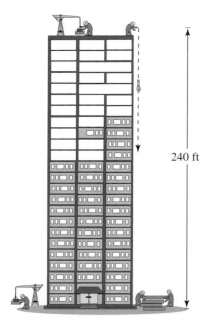

240 ft

FIGURE 1.21

The position equation used in Example 6 ignores air resistance. This implies that it is appropriate to use the position equation only to model falling objects that have little air resistance and that fall over short distances.

Example 7 ▶ Quadratic Modeling: Number of Lawyers

From 1970 to 1996, the number of lawyers practicing in the United States can be modeled by the quadratic equation

$$L = 0.139t^2 + 19.25t + 326.5$$

where L is the number of lawyers (in thousands) and t is the time, with $t = 0$ representing 1970. The number of lawyers is shown graphically in Figure 1.22. According to this model, in which year will the number of lawyers reach or surpass 1 million?   (Source: American Bar Association, American Bar Foundation)

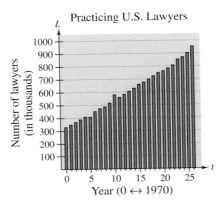

FIGURE 1.22

Solution

To find the year when the number of lawyers will reach 1 million, you need to solve the equation

$$0.139t^2 + 19.25t + 326.5 = 1000.$$

To begin, write the equation in general form.

$$0.139t^2 + 19.25t - 673.5 = 0$$

Then apply the Quadratic Formula.

$$t = \frac{-19.25 \pm \sqrt{(19.25)^2 - 4(0.139)(-673.5)}}{2(0.139)}$$

Choosing the positive solution, you find that

$$t = \frac{-19.25 + \sqrt{(19.25)^2 - 4(0.139)(-673.5)}}{2(0.139)}$$

$$\approx 28.94$$

$$\approx 29.$$

Because $t = 0$ corresponds to 1970, it follows that $t = 29$ must correspond to 1999. So, the number of lawyers should have reached 1 million by 1999.

A fourth type of application that often involves a quadratic equation is one dealing with the hypotenuse of a right triangle. These types of applications often use the Pythagorean Theorem, which states that

$$a^2 + b^2 = c^2 \qquad \text{Pythagorean Theorem}$$

where a and b are the legs of a right triangle and c is the hypotenuse.

Example 8 ▶ Cutting Across the Lawn

Your house is on a large corner lot. Several of the children in the neighborhood cut across your lawn, as shown in Figure 1.23. The distance across the lawn is 32 feet. How many feet does a person save by walking across the lawn instead of walking on the sidewalk?

Solution

In Figure 1.23, let x represent the length of the shorter part of the sidewalk. Using a ruler, you can find that the length of the longer part of the sidewalk is twice the shorter, so you can represent its length by $2x$. Now, using the Pythagorean Theorem, you have

$$x^2 + (2x)^2 = 32^2 \qquad \text{Pythagorean Theorem}$$
$$5x^2 = 1024 \qquad \text{Combine like terms.}$$
$$x^2 = 204.8 \qquad \text{Divide each side by 5.}$$
$$x = \sqrt{204.8} \qquad \text{Extract positive square root.}$$

The total distance covered by walking on the sidewalk is

$$x + 2x = 3x$$
$$= 3\sqrt{204.8}$$
$$\approx 42.9 \text{ feet.}$$

Cutting across the lawn saves a person about $42.9 - 32$, or 10.9, feet.

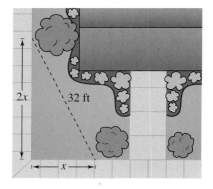

2x   32 ft   x

FIGURE 1.23

Writing ABOUT MATHEMATICS

Comparing Solution Methods In this section, you studied four algebraic methods for solving quadratic equations. Solve each of the quadratic equations below in several different ways. Write a short paragraph explaining which method(s) you prefer. Does it depend on the equation?

a. $x^2 - 4x - 5 = 0$ **b.** $x^2 - 4x = 0$

c. $x^2 - 4x - 3 = 0$ **d.** $x^2 - 4x - 6 = 0$

1.4 Exercises

In Exercises 1–6, write the quadratic equation in general form.

1. $2x^2 = 3 - 8x$

2. $x^2 = 16x$

3. $(x - 3)^2 = 3$

4. $13 - 3(x + 7)^2 = 0$

5. $\frac{1}{5}(3x^2 - 10) = 18x$

6. $x(x + 2) = 5x^2 + 1$

In Exercises 7–20, solve the quadratic equation for x by factoring.

7. $6x^2 + 3x = 0$

8. $9x^2 - 1 = 0$

9. $x^2 - 2x - 8 = 0$

10. $x^2 - 10x + 9 = 0$

11. $x^2 + 10x + 25 = 0$

12. $4x^2 + 12x + 9 = 0$

13. $3 + 5x - 2x^2 = 0$

14. $2x^2 = 19x + 33$

15. $x^2 + 4x = 12$

16. $-x^2 + 8x = 12$

17. $\frac{3}{4}x^2 + 8x + 20 = 0$

18. $\frac{1}{8}x^2 - x - 16 = 0$

19. $x^2 + 2ax + a^2 = 0$

20. $(x + a)^2 - b^2 = 0$

In Exercises 21–34, solve the equation by extracting square roots. List both the exact solution and the decimal solution rounded to two decimal places.

21. $x^2 = 49$

22. $x^2 = 169$

23. $x^2 = 11$

24. $x^2 = 32$

25. $3x^2 = 81$

26. $9x^2 = 36$

27. $(x - 12)^2 = 16$

28. $(x + 13)^2 = 25$

29. $(x + 2)^2 = 14$

30. $(x - 5)^2 = 30$

31. $(2x - 1)^2 = 18$

32. $(4x + 7)^2 = 44$

33. $(x - 7)^2 = (x + 3)^2$

34. $(x + 5)^2 = (x + 4)^2$

In Exercises 35–44, solve the quadratic equation by completing the square.

35. $x^2 - 2x = 0$

36. $x^2 + 4x = 0$

37. $x^2 + 4x - 32 = 0$

38. $x^2 - 2x - 3 = 0$

39. $x^2 + 6x + 2 = 0$

40. $x^2 + 8x + 14 = 0$

41. $9x^2 - 18x = -3$

42. $9x^2 - 12x = 14$

43. $8 + 4x - x^2 = 0$

44. $4x^2 - 4x - 99 = 0$

In Exercises 45–50, complete the square for the quadratic portion of the algebraic expression.

45. $\dfrac{1}{x^2 + 2x + 5}$

46. $\dfrac{1}{x^2 - 12x + 19}$

47. $\dfrac{4}{4x^2 + 4x - 3}$

48. $\dfrac{5}{5x^2 + 25x + 11}$

49. $\dfrac{1}{\sqrt{6x - x^2}}$

50. $\dfrac{1}{\sqrt{16 - 6x - x^2}}$

▦ *Graphical Analysis* In Exercises 51–58, use a graphing utility to graph the equation. Use the graph to approximate any x-intercepts of the graph. Set $y = 0$ and solve the resulting equation. Compare the result with the x-intercepts of the graph.

51. $y = (x + 3)^2 - 4$

52. $y = (x - 4)^2 - 1$

53. $y = 1 - (x - 2)^2$

54. $y = 9 - (x - 8)^2$

55. $y = -4x^2 + 4x + 3$

56. $y = 4x^2 - 1$

57. $y = x^2 + 3x - 4$

58. $y = x^2 - 5x - 24$

In Exercises 59–66, use the discriminant to determine the number of real solutions of the quadratic equation.

59. $2x^2 - 5x + 5 = 0$

60. $-5x^2 - 4x + 1 = 0$

61. $2x^2 - x - 1 = 0$

62. $x^2 - 4x + 4 = 0$

63. $\frac{1}{3}x^2 - 5x + 25 = 0$

64. $\frac{4}{7}x^2 - 8x + 28 = 0$

65. $0.2x^2 + 1.2x - 8 = 0$

66. $9 + 2.4x - 8.3x^2 = 0$

In Exercises 67–90, use the Quadratic Formula to solve the equation.

67. $2x^2 + x - 1 = 0$

68. $2x^2 - x - 1 = 0$

69. $16x^2 + 8x - 3 = 0$

70. $25x^2 - 20x + 3 = 0$

71. $2 + 2x - x^2 = 0$

72. $x^2 - 10x + 22 = 0$

73. $x^2 + 14x + 44 = 0$

74. $6x = 4 - x^2$

75. $x^2 + 8x - 4 = 0$

76. $4x^2 - 4x - 4 = 0$

77. $12x - 9x^2 = -3$

78. $16x^2 + 22 = 40x$

79. $9x^2 + 24x + 16 = 0$

80. $36x^2 + 24x - 7 = 0$

81. $4x^2 + 4x = 7$

82. $16x^2 - 40x + 5 = 0$

83. $28x - 49x^2 = 4$

84. $3x + x^2 - 1 = 0$

85. $8t = 5 + 2t^2$

86. $25h^2 + 80h + 61 = 0$

87. $(y - 5)^2 = 2y$

88. $(z + 6)^2 = -2z$

89. $\frac{1}{2}x^2 + \frac{3}{8}x = 2$

90. $\left(\frac{5}{7}x - 14\right)^2 = 8x$

In Exercises 91–98, use the Quadratic Formula to solve the equation. (Round your answers to three decimal places.)

91. $5.1x^2 - 1.7x - 3.2 = 0$

92. $2x^2 - 2.50x - 0.42 = 0$

93. $-0.067x^2 - 0.852x + 1.277 = 0$

94. $-0.005x^2 + 0.101x - 0.193 = 0$

95. $422x^2 - 506x - 347 = 0$

96. $1100x^2 + 326x - 715 = 0$

97. $12.67x^2 + 31.55x + 8.09 = 0$

98. $-3.22x^2 - 0.08x + 28.651 = 0$

In Exercises 99–108, solve the equation for x by any convenient method.

99. $x^2 - 2x - 1 = 0$ **100.** $11x^2 + 33x = 0$

101. $(x + 3)^2 = 81$ **102.** $x^2 - 14x + 49 = 0$

103. $x^2 - x - \frac{11}{4} = 0$ **104.** $x^2 + 3x - \frac{3}{4} = 0$

105. $(x + 1)^2 = x^2$ **106.** $a^2x^2 - b^2 = 0$

107. $3x + 4 = 2x^2 - 7$

108. $4x^2 + 2x + 4 = 2x + 8$

109. *Exploration*   Solve $3(x + 4)^2 + (x + 4) - 2 = 0$ in two ways.

(a) Let $u = x + 4$, and solve the resulting equation for u. Then solve the u-solution for x.

(b) Expand and collect like terms in the equation, and solve the resulting equation for x.

(c) Which method is easier? Explain.

Think About It   **In Exercises 110–113, write a quadratic equation that has the given solutions. (There are many correct answers.)**

110. -3 and 6 **111.** -4 and -11

112. 8 and 14 **113.** $\frac{1}{6}$ and $-\frac{2}{5}$

114. *Floor Space*   The floor of a one-story building is 14 feet longer than it is wide. The building has 1632 square feet of floor space.

(a) Draw a diagram that gives a visual representation of the floor space. Represent the width as w and show the length in terms of w.

(b) Write a quadratic equation in terms of w.

(c) Find the length and width of the floor of the building.

115. *Dimensions of a Corral*   A rancher has 100 meters of fencing to enclose two adjacent rectangular corrals (see figure). Find the dimensions such that the enclosed area will be 350 square meters.

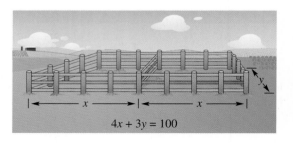

$4x + 3y = 100$

116. *Packaging*   An open box with a square base (see figure) is to be constructed from 84 square inches of material. What should be the dimensions of the base if the height of the box is to be 2 inches? (*Hint:* The surface area is $S = x^2 + 4xh$.)

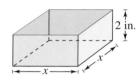

117. *Packaging*   An open box is to be made from a square piece of material by cutting 2-centimeter squares from the corners and turning up the sides (see figure). The volume of the finished box is to be 200 cubic centimeters. Find the size of the original piece of material.

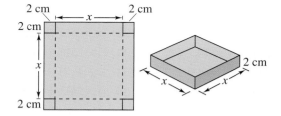

118. *Mowing the Lawn* Two people must mow a rectangular lawn 100 feet by 200 feet. Each wants to mow no more than half of the lawn. The first starts by mowing around the outside of the lawn. How wide a strip must the person mow on each of the four sides? If the mower has a 24-inch cut, approximate the required number of trips around the lawn.

119. *Seating* A rectangular classroom seats 72 students. If the seats were rearranged with three more seats in each row, the classroom would have two fewer rows. Find the original number of seats in each row.

In Exercises 120–123, use the position equation given in Example 6 as the model for the problem.

120. *CN Tower* At 1821 feet tall, the CN Tower in Toronto, Ontario is the world's tallest self-supporting structure. Suppose an object were dropped from the top of the tower.

(a) Find the position equation $s = -16t^2 + v_0t + s_0$.

(b) Complete the table.

t	0	2	4	6	8	10
s						

(c) From the table in part (b), you know that the time until the object reaches ground level is greater than how many seconds? Find the time analytically.

121. *Dropping a Ball* Suppose you drop a steel ball from the top of a building that is 1368 feet tall.

(a) Use the position equation to write a mathematical model for the height of the steel ball.

(b) Find the height of the steel ball after 4 seconds.

(c) How long will it take before the steel ball strikes the ground?

(d) Use a graphing utility with the appropriate viewing window to verify your answer to part (c).

122. *Military* A bomber flying at 32,000 feet over level terrain drops a 500-pound bomb.

(a) How long will it take until the bomb strikes the ground?

(b) If the plane is flying at 600 miles per hour, how far will the bomb travel horizontally during its descent?

123. *Baseball* You throw a baseball straight up into the air at a velocity of 45 feet per second. You release the baseball at a height of 5.5 feet and are going to catch it when it falls back to a height of 6 feet.

(a) Draw a diagram that gives a visual representation of the problem.

(b) Use the position equation to write a mathematical model for the height of the baseball.

(c) Find the height of the baseball after 0.5 second.

(d) For how many seconds will the baseball be in the air?

(e) Use a graphing utility with the appropriate viewing window to verify your answer to part (d).

124. *Geometry* The hypotenuse of an isosceles right triangle is 5 centimeters long. How long are its sides?

125. *Geometry* An equilateral triangle has a height of 10 inches. How long is one of its sides? (*Hint:* Use the height of the triangle to partition the triangle into two congruent right triangles.)

126. *Flying Speed* Two planes leave simultaneously from the same airport, one flying due north and the other due east (see figure). The northbound plane is flying 50 miles per hour faster than the eastbound plane. After 3 hours the planes are 2440 miles apart. Find the speed of each plane.

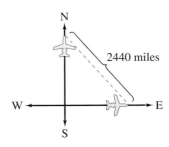

127. *Boating* A winch is used to tow a boat to a dock. The rope is attached to the boat at a point 15 feet below the level of the winch.

(a) Draw a diagram that gives a visual representation of the problem. Let l be the length of the rope and let x be the distance between the boat and the dock.

(b) Use the Pythagorean Theorem to write an equation giving the relationship between l and x. Find the distance from the boat to the dock when there is 75 feet of rope out.

128. *Economics* The demand equation for a certain product is $p = 20 - 0.0002x$, where p is the price per unit and x is the number of units sold. The total revenue for selling x units is

Revenue $= xp = x(20 - 0.0002x)$.

How many units must be sold to produce a revenue of $500,000?

129. *Economics* The demand equation for a certain product is $p = 60 - 0.0004x$, where p is the price per unit and x is the number of units sold. The total revenue for selling x units is

Revenue $= xp = x(60 - 0.0004x)$.

How many units must be sold to produce a revenue of $220,000?

Business **In Exercises 130–133, use the cost equation to find the number of units x that a manufacturer can produce for the given cost C. Round your answer to the nearest positive integer.**

130. $C = 0.125x^2 + 20x + 500$ $C = \$14,000$

131. $C = 0.5x^2 + 15x + 5000$ $C = \$11,500$

132. $C = 800 + 0.04x + 0.002x^2$ $C = \$1680$

133. $C = 800 - 10x + \dfrac{x^2}{4}$ $C = \$896$

134. *Population Statistics* The population of the United States from 1800 to 1890 can be approximated by the model

Population $= 0.6942t^2 + 6.183$

where the population is given in millions of people and t represents time, with $t = 0$ corresponding to 1800, $t = 1$ corresponding to 1810, and so on. If this model had continued to be valid up through the present time, when would the population of the United States have reached 250 million? Judging from the graph, would you say this model was a good representation of the population through 1890? (Source: U.S. Bureau of the Census)

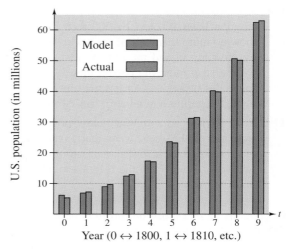

FIGURE FOR **134**

135. *Biology* The metabolic rate of ectothermic organisms increases with increasing temperature within a certain range. The graph shows experimental data for the oxygen consumption C (in microliters per gram per hour) of a beetle for certain temperatures. This data can be approximated by the model

$$C = 0.45x^2 - 1.65x + 50.75, \qquad 10 \le x \le 25$$

where x is the air temperature in degrees Celsius.

(a) Find the air temperature if the oxygen consumption is 150 microliters per gram per hour.

(b) If the temperature is increased from 10°C to 20°C, the oxygen consumption is increased by approximately what factor?

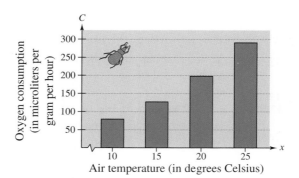

136. *Boating* The total number of dollars spent on boating in the United States from 1991 through 1996 can be approximated by the model

$$S = 0.230t^2 + 0.088t + 9.750$$

where S is the amount spent (in billions of dollars) and t is the time, with $t = 1$ corresponding to 1991. (Source: National Marine Manufacturers Association)

(a) Use a graphing utility to sketch a graph of the model over the interval $1 \leq t \leq 6$.

(b) If the model is used to forecast future sales, will sales ever exceed 25 billion dollars? If so, estimate the year.

137. *Flying Distance* A small commuter airline flies to three cities whose locations form the vertices of a right triangle (see figure). The total flight distance (from City A to City B to City C and back to City A) is 1400 kilometers. It is 600 kilometers between the two cities that are farthest apart. Approximate the other two distances between cities.

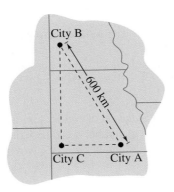

Synthesis

True or False? **In Exercises 138 and 139, determine whether the statement is true or false. Justify your answer.**

138. The quadratic equation $-3x^2 - x = 10$ has two real solutions.

139. If $(2x - 3)(x + 5) = 8$, then $2x - 3 = 8$ or $x + 5 = 8$.

140. To solve the equation

$$2x^2 + 3x = 15x$$

a student divides both sides by x and solves the equation $2x + 3 = 15$. The resulting solution ($x = 6$) satisfies the given equation. Is there an error? Explain.

141. The graphs show the solutions of equations plotted on the real number line. In each case, determine whether the solution(s) is (are) for a linear equation, a quadratic equation, both, or neither. Explain.

(a)
a b c x

(b)
a x

(c)
a b x

(d)
a b c d x

142. *Exploration* Solve the equations, given that a and b are not zero.

(a) $ax^2 + bx = 0$ (b) $ax^2 - ax = 0$

Review

In Exercises 143–146, identify the rule of algebra being demonstrated.

143. $(10x)y = 10(xy)$ **144.** $-4(x - 3) = -4x + 12$

145. $7x^4 + (-7x^4) = 0$

146. $(x + 4) + x^3 = x + (4 + x^3)$

In Exercises 147–150, simplify the expression.

147. $\left(\dfrac{6u^2}{5v^{-3}}\right)^{-1}$ **148.** $(3.5 \times 10^8)(2 \times 10^{-5})$

149. $\dfrac{x^2 - 100}{10 - x}$ **150.** $\dfrac{3}{2 - x} + \dfrac{5}{x} + \dfrac{1}{(x - 2)^2}$

In Exercises 151–154, completely factor the expression.

151. $x^5 - 27x^2$ **152.** $x^3 - 5x^2 - 14x$

153. $x^3 + 5x^2 - 2x - 10$

154. $5(x + 5)x^{1/3} + 4x^{4/3}$

In Exercises 155–160, find the product.

155. $(x + 3)(x - 6)$ **156.** $(x - 8)(x - 1)$

157. $(x + 4)(x^2 - x + 2)$

158. $(x + 9)(x^2 - 6x + 4)$

159. $(2x - 1)(3x^3 - x + 1)$

160. $(3x + 2)(2x^2 + x + 1)$

1.5 Complex Numbers

▶ **What you should learn**

• How to use the imaginary unit i to write complex numbers

• How to add, subtract, and multiply complex numbers

• How to use complex conjugates to divide complex numbers

• How to use the Quadratic Formula to find complex solutions of quadratic equations

▶ **Why you should learn it**

You can use complex numbers to model and solve real-life problems in electronics. For instance, in Exercise 84 on page 130, you will learn how to use complex numbers to find the impedance of an electrical circuit.

Chris Hamilton/The Stock Market

The Imaginary Unit i

In Section 1.4, you learned that some quadratic equations have no real solutions. For instance, the quadratic equation

$$x^2 + 1 = 0 \qquad \text{Equation with no real solution}$$

has no real solution because there is no real number x that can be squared to produce -1. To overcome this deficiency, mathematicians created an expanded system of numbers using the **imaginary unit i,** defined as

$$i = \sqrt{-1} \qquad \text{Imaginary unit}$$

where $i^2 = -1$. By adding real numbers to real multiples of this imaginary unit, the set of **complex numbers** is obtained. Each complex number can be written in the **standard form $a + bi$**. The real number a is called the **real part** of the complex number $a + bi$, and the number bi (where b is a real number) is called the **imaginary part** of the complex number.

Definition of a Complex Number

If a and b are real numbers, the number $a + bi$ is a **complex number,** and it is said to be written in **standard form.** If $b = 0$, the number $a + bi = a$ is a real number. If $b \neq 0$, the number $a + bi$ is called an **imaginary number.** A number of the form bi, where $b \neq 0$, is called a **pure imaginary number.**

The set of real numbers is a subset of the set of complex numbers, as shown in Figure 1.24. This is true because every real number a can be written as a complex number using $b = 0$. That is, for every real number a, you can write $a = a + 0i$.

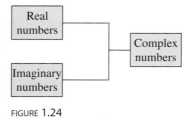

FIGURE 1.24

Equality of Complex Numbers

Two complex numbers $a + bi$ and $c + di$, written in standard form, are equal to each other

$$a + bi = c + di \qquad \text{Equality of two complex numbers}$$

if and only if $a = c$ and $b = d$.

Operations with Complex Numbers

To add (or subtract) two complex numbers, you add (or subtract) the real and imaginary parts of the numbers separately.

Addition and Subtraction of Complex Numbers

If $a + bi$ and $c + di$ are two complex numbers written in standard form, their sum and difference are defined as follows.

Sum: $(a + bi) + (c + di) = (a + c) + (b + d)i$

Difference: $(a + bi) - (c + di) = (a - c) + (b - d)i$

The **additive identity** in the complex number system is zero (the same as in the real number system). Furthermore, the **additive inverse** of the complex number $a + bi$ is

$-(a + bi) = -a - bi.$ Additive inverse

So, you have

$(a + bi) + (-a - bi) = 0 + 0i = 0.$

 The *Interactive* CD-ROM and *Internet* versions of this text show every example with its solution; clicking on the *Try It!* button brings up similar problems. Guided Examples and Integrated Examples show step-by-step solutions to additional examples. Integrated Examples are related to several concepts in the section.

Example 1 ▶ Adding and Subtracting Complex Numbers

a. $(3 - i) + (2 + 3i) = 3 - i + 2 + 3i$ Remove parentheses.

$= 3 + 2 - i + 3i$ Group like terms.

$= (3 + 2) + (-1 + 3)i$

$= 5 + 2i$ Write in standard form.

b. $2i + (-4 - 2i) = 2i - 4 - 2i$ Remove parentheses.

$= -4 + 2i - 2i$ Group like terms.

$= -4$ Write in standard form.

c. $3 - (-2 + 3i) + (-5 + i) = 3 + 2 - 3i - 5 + i$

$= 3 + 2 - 5 - 3i + i$

$= 0 - 2i$

$= -2i$

d. $(3 + 2i) + (4 - i) - (7 + i) = 3 + 2i + 4 - i - 7 - i$

$= 3 + 4 - 7 + 2i - i - i$

$= 0 + 0i$

$= 0$

Note in Example 1(b) that the sum of two complex numbers can be a real number.

◀ **Exploration** ▶

Complete the table:

$i^1 = i$ $i^7 =$ ▨

$i^2 = -1$ $i^8 =$ ▨

$i^3 = -i$ $i^9 =$ ▨

$i^4 = 1$ $i^{10} =$ ▨

$i^5 =$ ▨ $i^{11} =$ ▨

$i^6 =$ ▨ $i^{12} =$ ▨

What pattern do you see? Write a brief description of how you would find i raised to any positive integer power.

Many of the properties of real numbers are valid for complex numbers as well. Here are some examples.

Associative Properties of Addition and Multiplication

Commutative Properties of Addition and Multiplication

Distributive Property of Multiplication Over Addition

Notice below how these properties are used when two complex numbers are multiplied.

$$(a + bi)(c + di) = a(c + di) + bi(c + di) \qquad \text{Distributive Property}$$
$$= ac + (ad)i + (bc)i + (bd)i^2 \qquad \text{Distributive Property}$$
$$= ac + (ad)i + (bc)i + (bd)(-1) \qquad i^2 = -1$$
$$= ac - bd + (ad)i + (bc)i \qquad \text{Commutative Property}$$
$$= (ac - bd) + (ad + bc)i \qquad \text{Associative Property}$$

Rather than trying to memorize this multiplication rule, you should simply remember how the Distributive Property is used to multiply two complex numbers. The procedure is similar to multiplying two polynomials and combining like terms (as in the FOIL Method).

Example 2 ▶ Multiplying Complex Numbers

a. $4(-2 + 3i) = 4(-2) + 4(3i)$ Distributive Property

$\qquad\qquad = -8 + 12i$ Simplify.

b. $(i)(-3i) = -3i^2$ Multiply.

$\qquad\quad = -3(-1)$ $i^2 = -1$

$\qquad\quad = 3$ Simplify.

c. $(2 - i)(4 + 3i) = 8 + 6i - 4i - 3i^2$ Product of binomials

$\qquad\qquad\qquad = 8 + 6i - 4i - 3(-1)$ $i^2 = -1$

$\qquad\qquad\qquad = (8 + 3) + (6i - 4i)$ Group like terms.

$\qquad\qquad\qquad = 11 + 2i$ Write in standard form.

d. $(3 + 2i)(3 - 2i) = 9 - 6i + 6i - 4i^2$ Product of binomials

$\qquad\qquad\qquad = 9 - 4(-1)$ $i^2 = -1$

$\qquad\qquad\qquad = 9 + 4$ Simplify.

$\qquad\qquad\qquad = 13$ Write in standard form.

e. $(3 + 2i)^2 = 9 + 6i + 6i + 4i^2$ Product of binomials

$\qquad\qquad = 9 + 4(-1) + 12i$ $i^2 = -1$

$\qquad\qquad = 9 - 4 + 12i$ Simplify.

$\qquad\qquad = 5 + 12i$ Write in standard form.

Complex Conjugates and Division

Notice in Example 2(d) that the product of two complex numbers can be a real number. This occurs with pairs of complex numbers of the form $a + bi$ and $a - bi$, called **complex conjugates.**

$$(a + bi)(a - bi) = a^2 - abi + abi - b^2i^2$$
$$= a^2 - b^2(-1)$$
$$= a^2 + b^2$$

To find the quotient of $a + bi$ and $c + di$ where c and d are not both zero, multiply the numerator and denominator by the conjugate of the *denominator* to obtain

$$\frac{a + bi}{c + di} = \frac{a + bi}{c + di}\left(\frac{c - di}{c - di}\right)$$

$$= \frac{(ac + bd) + (bc - ad)i}{c^2 + d^2}.$$

Example 3 ▶ Dividing Complex Numbers

$$\frac{1}{1 + i} = \frac{1}{1 + i}\left(\frac{1 - i}{1 - i}\right) \qquad \text{Multiply numerator and denominator by conjugate of denominator.}$$

$$= \frac{1 - i}{1^2 - i^2} \qquad \text{Expand.}$$

$$= \frac{1 - i}{1 - (-1)} \qquad i^2 = -1$$

$$= \frac{1 - i}{2} \qquad \text{Simplify.}$$

$$= \frac{1}{2} - \frac{1}{2}i \qquad \text{Write in standard form.}$$

Example 4 ▶ Dividing Complex Numbers

$$\frac{2 + 3i}{4 - 2i} = \frac{2 + 3i}{4 - 2i}\left(\frac{4 + 2i}{4 + 2i}\right) \qquad \text{Multiply numerator and denominator by conjugate of denominator.}$$

$$= \frac{8 + 4i + 12i + 6i^2}{16 - 4i^2} \qquad \text{Expand.}$$

$$= \frac{8 - 6 + 16i}{16 + 4} \qquad i^2 = -1$$

$$= \frac{2 + 16i}{20} \qquad \text{Simplify.}$$

$$= \frac{1}{10} + \frac{4}{5}i \qquad \text{Write in standard form.}$$

Complex Solutions of Quadratic Equations

When using the Quadratic Formula to solve a quadratic equation, you often obtain a result such as $\sqrt{-3}$, which you know is not a real number. By factoring out $i = \sqrt{-1}$, you can write this number in standard form.

$$\sqrt{-3} = \sqrt{3(-1)} = \sqrt{3}\sqrt{-1} = \sqrt{3}\,i$$

The number $\sqrt{3}\,i$ is called the *principal square root* of -3.

Principal Square Root of a Negative Number

If a is a positive number, the **principal square root** of the negative number $-a$ is defined as

$$\sqrt{-a} = \sqrt{a}\,i.$$

STUDY T!P

The definition of principal square root uses the rule

$$\sqrt{ab} = \sqrt{a}\sqrt{b}$$

for $a > 0$ and $b < 0$. This rule is not valid if *both* a and b are negative. For example,

$$\sqrt{-5}\sqrt{-5} = \sqrt{5(-1)}\sqrt{5(-1)}$$
$$= \sqrt{5}\,i\sqrt{5}\,i$$
$$= \sqrt{25}\,i^2$$
$$= 5i^2 = -5$$

whereas

$$\sqrt{(-5)(-5)} = \sqrt{25} = 5.$$

To avoid problems with multiplying square roots of negative numbers, be sure to convert to standard form *before* multiplying.

Example 5 ▶ Writing Complex Numbers in Standard Form

a. $\sqrt{-3}\sqrt{-12} = \sqrt{3}\,i\sqrt{12}\,i = \sqrt{36}\,i^2 = 6(-1) = -6$

b. $\sqrt{-48} - \sqrt{-27} = \sqrt{48}\,i - \sqrt{27}\,i = 4\sqrt{3}\,i - 3\sqrt{3}\,i = \sqrt{3}\,i$

c. $\left(-1 + \sqrt{-3}\right)^2 = \left(-1 + \sqrt{3}\,i\right)^2$
$$= (-1)^2 - 2\sqrt{3}\,i + \left(\sqrt{3}\right)^2(i^2)$$
$$= 1 - 2\sqrt{3}\,i + 3(-1)$$
$$= -2 - 2\sqrt{3}\,i$$

Example 6 ▶ Complex Solutions of a Quadratic Equation

Solve (a) $x^2 + 4 = 0$ and (b) $3x^2 - 2x + 5 = 0$.

Solution

a. $x^2 + 4 = 0$ Write original equation.

$$x^2 = -4$$ Subtract 4 from each side.

$$x = \pm 2i$$ Extract square roots.

b. $3x^2 - 2x + 5 = 0$ Write original equation.

$$x = \frac{-(-2) \pm \sqrt{(-2)^2 - 4(3)(5)}}{2(3)}$$ Quadratic Formula

$$= \frac{2 \pm \sqrt{-56}}{6}$$ Simplify.

$$= \frac{2 \pm 2\sqrt{14}\,i}{6}$$ Write $\sqrt{-56}$ in standard form.

$$= \frac{1}{3} \pm \frac{\sqrt{14}}{3}\,i$$ Write in standard form.

1.5 Exercises

In Exercises 1–4, find real numbers a and b such that the equation is true.

1. $a + bi = -10 + 6i$

2. $a + bi = 13 + 4i$

3. $(a - 1) + (b + 3)i = 5 + 8i$

4. $(a + 6) + 2bi = 6 - 5i$

In Exercises 5–16, write the complex number in standard form.

5. $4 + \sqrt{-9}$

6. $3 + \sqrt{-16}$

7. $2 - \sqrt{-27}$

8. $1 + \sqrt{-8}$

9. $\sqrt{-75}$

10. $\sqrt{-4}$

11. 8

12. 45

13. $-6i + i^2$

14. $-4i^2 + 2i$

15. $\sqrt{-0.09}$

16. $\sqrt{-0.0004}$

In Exercises 17–26, perform the addition or subtraction and write the result in standard form.

17. $(5 + i) + (6 - 2i)$

18. $(13 - 2i) + (-5 + 6i)$

19. $(8 - i) - (4 - i)$

20. $(3 + 2i) - (6 + 13i)$

21. $\left(-2 + \sqrt{-8}\right) + \left(5 - \sqrt{-50}\right)$

22. $\left(8 + \sqrt{-18}\right) - \left(4 + 3\sqrt{2}i\right)$

23. $13i - (14 - 7i)$

24. $22 + (-5 + 8i) + 10i$

25. $-\left(\frac{3}{2} + \frac{5}{2}i\right) + \left(\frac{5}{3} + \frac{11}{3}i\right)$

26. $(1.6 + 3.2i) + (-5.8 + 4.3i)$

In Exercises 27–40, perform the operation and write the result in standard form.

27. $\sqrt{-6} \cdot \sqrt{-2}$

28. $\sqrt{-5} \cdot \sqrt{-10}$

29. $\left(\sqrt{-10}\right)^2$

30. $\left(\sqrt{-75}\right)^2$

31. $(1 + i)(3 - 2i)$

32. $(6 - 2i)(2 - 3i)$

33. $6i(5 - 2i)$

34. $-8i(9 + 4i)$

35. $\left(\sqrt{14} + \sqrt{10}i\right)\left(\sqrt{14} - \sqrt{10}i\right)$

36. $\left(3 + \sqrt{-5}\right)\left(7 - \sqrt{-10}\right)$

37. $(4 + 5i)^2$

38. $(2 - 3i)^2$

39. $(2 + 3i)^2 + (2 - 3i)^2$

40. $(1 - 2i)^2 - (1 + 2i)^2$

In Exercises 41–48, write the conjugate of the complex number. Then multiply the number and its conjugate.

41. $6 + 3i$

42. $7 - 12i$

43. $-1 - \sqrt{5}i$

44. $-3 + \sqrt{2}i$

45. $\sqrt{-20}$

46. $\sqrt{-15}$

47. $\sqrt{8}$

48. $1 + \sqrt{8}$

In Exercises 49–62, perform the operation and write the result in standard form.

49. $\dfrac{5}{i}$

50. $-\dfrac{14}{2i}$

51. $\dfrac{2}{4 - 5i}$

52. $\dfrac{5}{1 - i}$

53. $\dfrac{3 + i}{3 - i}$

54. $\dfrac{6 - 7i}{1 - 2i}$

55. $\dfrac{6 - 5i}{i}$

56. $\dfrac{8 + 16i}{2i}$

57. $\dfrac{3i}{(4 - 5i)^2}$

58. $\dfrac{5i}{(2 + 3i)^2}$

59. $\dfrac{2}{1 + i} - \dfrac{3}{1 - i}$

60. $\dfrac{2i}{2 + i} + \dfrac{5}{2 - i}$

61. $\dfrac{i}{3 - 2i} + \dfrac{2i}{3 + 8i}$

62. $\dfrac{1 + i}{i} - \dfrac{3}{4 - i}$

In Exercises 63–72, use the Quadratic Formula to solve the quadratic equation.

63. $x^2 - 2x + 2 = 0$

64. $x^2 + 6x + 10 = 0$

65. $4x^2 + 16x + 17 = 0$

66. $9x^2 - 6x + 37 = 0$

67. $4x^2 + 16x + 15 = 0$

68. $16t^2 - 4t + 3 = 0$

69. $\frac{3}{2}x^2 - 6x + 9 = 0$

70. $\frac{7}{8}x^2 - \frac{3}{4}x + \frac{5}{16} = 0$

71. $1.4x^2 - 2x - 10 = 0$

72. $4.5x^2 - 3x + 12 = 0$

73. Express each of the following powers of i as i, $-i$, 1, or -1.

(a) i^{40} (b) i^{25} (c) i^{50} (d) i^{67}

In Exercises 74–81, simplify the complex number and write it in standard form.

74. $-6i^3 + i^2$

75. $4i^2 - 2i^3$

76. $-5i^5$

77. $(-i)^3$

78. $\left(\sqrt{-75}\right)^3$

79. $\left(\sqrt{-2}\right)^6$

80. $\dfrac{1}{i^3}$

81. $\dfrac{1}{(2i)^3}$

82. Cube each complex number.

(a) 2   (b) $-1 + \sqrt{3}i$ (c) $-1 - \sqrt{3}i$

83. Raise each complex number to the fourth power.

(a) 2   (b) -2 (c) $2i$ (d) $-2i$

84. *Impedance* The opposition to current in an electrical circuit is called its impedance. The impedance in a parallel circuit with two pathways satisfies the equation

$$\frac{1}{z} = \frac{1}{z_1} + \frac{1}{z_2}$$

where z_1 is the impedance (in ohms) of pathway 1 and z_2 is the impedance of pathway 2. Use the table to determine the impedance of each parallel circuit. The impedance of each pathway is found by adding the impedance of each component in the pathway.

	Resistor	Inductor	Capacitor
	─⋀⋀⋀─	─◯◯◯─	─┤├─
Symbol	$a\Omega$	$b\Omega$	$c\Omega$
Impedance	a	bi	$-ci$

(a)

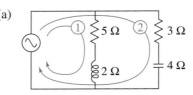

$z_1 = 5 + 2i, \quad z_2 = 3 - 4i$

(b)

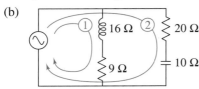

$z_1 = 9 + 16i, \quad z_2 = 20 - 10i$

Synthesis

True or False? **In Exercises 85–87, determine whether the statement is true or false. Justify your answer.**

85. There is no complex number that is equal to its conjugate.

86. $-i\sqrt{6}$ is a solution of $x^4 - x^2 + 14 = 56$.

87. $i^{44} + i^{150} - i^{74} - i^{109} + i^{61} = -1$

88. *Error Analysis* Describe the error.

$$\sqrt{-6}\sqrt{-6} = \sqrt{(-6)(-6)} = \sqrt{36} = 6 \;\; \times$$

89. Prove that the conjugate of the product of two complex numbers $a_1 + b_1i$ and $a_2 + b_2i$ is the product of their conjugates.

90. Prove that the conjugate of the sum of two complex numbers $a_1 + b_1i$ and $a_2 + b_2i$ is the sum of their conjugates.

Review

In Exercises 91–94, perform the operations and write the result in standard form.

91. $(4 + 3x) + (8 - 6x - x^2)$

92. $(x^3 - 3x^2) - (6 - 2x - 4x^2)$

93. $\left(3x - \tfrac{1}{2}\right)(x + 4)$ **94.** $(2x - 5)^2$

In Exercises 95–98, solve the equation and check your solution.

95. $-x - 12 = 19$ **96.** $8 - 3x = -34$

97. $4(5x - 6) - 3(6x + 1) = 0$

98. $5[x - (3x + 11)] = 20x - 15$

99. *Volume of an Oblate Spheroid*

Solve for a: $V = \tfrac{4}{3}\pi a^2 b$

100. *Newton's Law of Universal Gravitation*

Solve for r: $F = \alpha \dfrac{m_1 m_2}{r^2}$

101. *Mixture Problem* A 5-liter container contains a mixture with a concentration of 50%. How much of this mixture must be withdrawn and replaced by 100% concentrate to bring the mixture up to 60% concentration?

102. *Travel* A business executive drove at an average speed of 100 kilometers per hour on a 200-kilometer trip. Because of heavy traffic, the average speed on the return trip was 80 kilometers per hour. Find the average speed for the round trip.

1.6 Other Types of Equations

▶ **What you should learn**

- How to solve polynomial equations of degree three or greater
- How to solve equations involving radicals
- How to solve equations involving fractions or absolute values
- How to use polynomial equations and equations involving radicals to model and solve real-life problems

▶ **Why you should learn it**

Polynomial equations, radical equations, and absolute value equations can be used to model and solve real-life problems. For instance, Exercise 97 on page 141 shows how to solve a radical equation to find the number of airline passengers who flew based on the total monthly cost.

Jeff Greenberg/PhotoEdit

Polynomial Equations

In this section you will extend the techniques for solving equations to nonlinear and nonquadratic equations. At this point in the text, you have only four basic methods for solving nonlinear equations—*factoring, extracting roots, completing the square,* and the *Quadratic Formula.* So the main goal of this section is to learn to *rewrite* nonlinear equations in a form to which you can apply one of these methods.

Example 1 shows how to use factoring to solve a **polynomial equation,** which is an equation that can be written in the form

$$a_n x^n + a_{n-1} x^{n-1} + \cdots + a_2 x^2 + a_1 x + a_0 = 0.$$

Example 1 ▶ Solving a Polynomial Equation by Factoring

Solve $3x^4 = 48x^2$.

Solution

First write the polynomial equation in general form with zero on one side, factor the other side, and then set each factor equal to zero.

$3x^4 = 48x^2$	Write original equation.
$3x^4 - 48x^2 = 0$	Write in general form.
$3x^2(x^2 - 16) = 0$	Factor.
$3x^2(x + 4)(x - 4) = 0$	Write in factored form.
$3x^2 = 0 \implies x = 0$	Set 1st factor equal to 0.
$x + 4 = 0 \implies x = -4$	Set 2nd factor equal to 0.
$x - 4 = 0 \implies x = 4$	Set 3rd factor equal to 0.

You can check these solutions by substituting in the original equation, as follows.

Check

$3x^4 = 48x^2$	Write original equation.
$3(0)^4 = 48(0)^2$	0 checks. ✔
$3(-4)^4 = 48(-4)^2$	-4 checks. ✔
$3(4)^4 = 48(4)^2$	4 checks. ✔

After checking, you can conclude that the solutions are 0, -4, and 4.

A common mistake that is made in solving an equation such as that in Example 1 is to divide both sides of the equation by the variable factor x^2. This loses the solution $x = 0$. When solving an equation, be sure to write the equation in general form, then factor the equation and set each factor equal to zero. Don't divide both sides of an equation by a variable factor in an attempt to simplify the equation.

You can use a graphing utility to check graphically the solutions of the equation in Example 2. To do this, sketch the graph of

$$y = x^3 - 3x^2 + 3x - 9.$$

As shown below, the x-intercept of the graph occurs at the *real* zero of the function, $x = 3$, confirming the result found in Example 2.

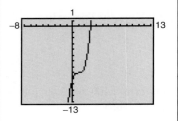

Try using a graphing utility to check the solutions found in Example 3.

For a review of factoring, see Section P.4. For instance, look at the examples of factoring special polynomial forms such as the difference of two cubes or factoring by grouping.

$$x^3 - 8 = (x - 2)(x^2 + 2x + 4) \quad \text{or} \quad x^3 - 4x^2 - 3x + 12 = (x^2 - 3)(x - 4)$$

Example 2 ▶ Solving a Polynomial Equation by Factoring

Solve $x^3 - 3x^2 + 3x - 9 = 0$.

Solution

$x^3 - 3x^2 + 3x - 9 = 0$	Write original equation.
$x^2(x - 3) + 3(x - 3) = 0$	Factor by grouping.
$(x - 3)(x^2 + 3) = 0$	Distributive Property
$x - 3 = 0 \implies x = 3$	Set 1st factor equal to 0.
$x^2 + 3 = 0 \implies x = \pm\sqrt{3}i$	Set 2nd factor equal to 0.

The solutions are 3, $\sqrt{3}i$, and $-\sqrt{3}i$.

Occasionally, mathematical models involve equations that are of quadratic type. In general, an equation is of quadratic type if it can be written in the form

$$au^2 + bu + c = 0$$

where $a \neq 0$ and u is an algebraic expression.

Example 3 ▶ Solving an Equation of Quadratic Type

Solve $x^4 - 3x^2 + 2 = 0$.

Solution

This equation is of quadratic type with $u = x^2$.

$$(x^2)^2 - 3(x^2) + 2 = 0$$

To solve this equation, you can factor the left side of the equation as the product of two second-degree polynomials.

$x^4 - 3x^2 + 2 = 0$	Write original equation.
$(x^2 - 1)(x^2 - 2) = 0$	Partially factor.
$(x + 1)(x - 1)(x^2 - 2) = 0$	Factor.
$x + 1 = 0 \implies x = -1$	Set 1st factor equal to 0.
$x - 1 = 0 \implies x = 1$	Set 2nd factor equal to 0.
$x^2 - 2 = 0 \implies x = \pm\sqrt{2}$	Set 3rd factor equal to 0.

The solutions are -1, 1, $\sqrt{2}$, and $-\sqrt{2}$. Check these in the original equation.

Equations Involving Radicals

The steps involved in solving the remaining equations in this section will often introduce *extraneous solutions*, as discussed in Section 1.2. Operations such as squaring each side of an equation, raising each side of an equation to a rational power, and multiplying each side of an equation by a variable quantity all can introduce extraneous solutions. So, when you use any of these operations, checking is crucial.

Example 4 ▶ Solving an Equation Involving a Rational Exponent

$4x^{3/2} - 8 = 0$	Write original equation.
$4x^{3/2} = 8$	Add 8 to each side.
$x^{3/2} = 2$	Divide each side by 4.
$x = 2^{2/3}$	Raise each side to the $\frac{2}{3}$ power.
$x \approx 1.587$	Round to three decimal places.

The solution is $2^{2/3}$. Check this in the original equation.

Example 5 ▶ Solving Equations Involving Radicals

STUDY T!P

The essential operations in Example 4 are isolating the factor with the rational exponent and raising each side to the *reciprocal power*. In Example 5, this is equivalent to isolating the square root and squaring each side.

a.

$\sqrt{2x + 7} - x = 2$	Write original equation.
$\sqrt{2x + 7} = x + 2$	Add x to each side.
$2x + 7 = x^2 + 4x + 4$	Square each side.
$0 = x^2 + 2x - 3$	Write in general form.
$0 = (x + 3)(x - 1)$	Factor.
$x + 3 = 0 \implies x = -3$	Set 1st factor equal to 0.
$x - 1 = 0 \implies x = 1$	Set 2nd factor equal to 0.

By checking these values, you can determine that the only solution is 1.

b.

$\sqrt{2x - 5} - \sqrt{x - 3} = 1$	Write original equation.
$\sqrt{2x - 5} = \sqrt{x - 3} + 1$	Add $\sqrt{x - 3}$ to each side.
$2x - 5 = x - 3 + 2\sqrt{x - 3} + 1$	Square each side.
$2x - 5 = x - 2 + 2\sqrt{x - 3}$	Combine like terms.
$x - 3 = 2\sqrt{x - 3}$	Add $-x + 2$ to each side.
$x^2 - 6x + 9 = 4(x - 3)$	Square each side.
$x^2 - 10x + 21 = 0$	Write in general form.
$(x - 3)(x - 7) = 0$	Factor.
$x - 3 = 0 \implies x = 3$	Set 1st factor equal to 0.
$x - 7 = 0 \implies x = 7$	Set 2nd factor equal to 0.

The solutions are 3 and 7. Check these in the original equation.

Equations with Fractions or Absolute Values

To solve an equation involving fractions, multiply each side of the equation by the least common denominator of all terms in the equation. This procedure will "clear the equation of fractions." For instance, in the equation

$$\frac{2}{x^2 + 1} + \frac{1}{x} = \frac{2}{x}$$

you can multiply each side of the equation by $x(x^2 + 1)$. Try doing this and solve the resulting equation. You should obtain one solution: $x = 1$.

Example 6 ▶ Solving an Equation Involving Fractions

Solve $\dfrac{2}{x} = \dfrac{3}{x - 2} - 1$.

Solution

For this equation, the least common denominator of the three terms is $x(x - 2)$, so you begin by multiplying each term in the equation by this expression.

$$\frac{2}{x} = \frac{3}{x - 2} - 1$$

$$x(x - 2)\frac{2}{x} = x(x - 2)\frac{3}{x - 2} - x(x - 2)(1)$$

$$2(x - 2) = 3x - x(x - 2)$$

$$2x - 4 = -x^2 + 5x$$

$$x^2 - 3x - 4 = 0$$

$$(x - 4)(x + 1) = 0$$

$$x - 4 = 0 \quad \Longrightarrow \quad x = 4$$

$$x + 1 = 0 \quad \Longrightarrow \quad x = -1$$

Check

$\dfrac{2}{x} = \dfrac{3}{x - 2} - 1$	Write original equation.
$\dfrac{2}{4} \stackrel{?}{=} \dfrac{3}{4 - 2} - 1$	Substitute 4 for x.
$\dfrac{1}{2} \stackrel{?}{=} \dfrac{3}{2} - 1$	Simplify.
$\dfrac{1}{2} = \dfrac{1}{2}$	4 checks. ✓

Similarly, substitute -1 for x to determine that -1 is also a solution. So, the solutions are 4 and -1.

To solve an equation involving an absolute value, remember that the expression inside the absolute value signs can be positive or negative. This results in *two* separate equations, each of which must be solved. For instance, the equation

$$|x - 2| = 3$$

results in the two equations

$$x - 2 = 3 \quad \text{and} \quad -(x - 2) = 3$$

which implies that the equation has two solutions: 5 and -1.

Example 7 ▶ Solving an Equation Involving Absolute Value

Solve $|x^2 - 3x| = -4x + 6$.

Solution

Because the variable expression inside the absolute value signs can be positive or negative, you must solve the following two equations.

First Equation

$x^2 - 3x = -4x + 6$	Use positive expression.
$x^2 + x - 6 = 0$	Write in general form.
$(x + 3)(x - 2) = 0$	Factor.
$x + 3 = 0 \implies x = -3$	Set 1st factor equal to 0.
$x - 2 = 0 \implies x = 2$	Set 2nd factor equal to 0.

Second Equation

$-(x^2 - 3x) = -4x + 6$	Use negative expression.
$x^2 - 7x + 6 = 0$	Write in general form.
$(x - 1)(x - 6) = 0$	Factor.
$x - 1 = 0 \implies x = 1$	Set 1st factor equal to 0.
$x - 6 = 0 \implies x = 6$	Set 2nd factor equal to 0.

Check

$$|(-3)^2 - 3(-3)| \overset{?}{=} -4(-3) + 6 \qquad \text{Substitute } -3 \text{ for } x.$$

$$18 = 18 \qquad -3 \text{ checks. } ✓$$

$$|(2)^2 - 3(2)| \overset{?}{=} -4(2) + 6 \qquad \text{Substitute } 2 \text{ for } x.$$

$$2 \ne -2 \qquad 2 \text{ does not check.}$$

$$|(1)^2 - 3(1)| \overset{?}{=} -4(1) + 6 \qquad \text{Substitute } 1 \text{ for } x.$$

$$2 = 2 \qquad 1 \text{ checks. } ✓$$

$$|(6)^2 - 3(6)| \overset{?}{=} -4(6) + 6 \qquad \text{Substitute } 6 \text{ for } x.$$

$$18 \ne -18 \qquad 6 \text{ does not check.}$$

The solutions are -3 and 1.

Applications

It would be impossible to categorize the many different types of applications that involve nonlinear and nonquadratic models. However, from the few examples and exercises that are given, you will gain some appreciation for the variety of applications that can occur.

Example 8 ▶   **Reduced Rates**

A ski club chartered a bus for a ski trip at a cost of $480. In an attempt to lower the bus fare per skier, the club invited nonmembers to go along. After five nonmembers joined the trip, the fare per skier decreased by $4.80. How many club members are going on the trip?

Solution

Begin the solution by creating a verbal model and assigning labels.

Verbal Model: $\boxed{\text{Cost per skier}} \cdot \boxed{\text{Number of skiers}} = \boxed{\text{Cost of trip}}$

Labels: Cost of trip $= 480$ (dollars)

            Number of ski club members $= x$ (people)

            Number of skiers $= x + 5$ (people)

            Original cost per member $= \dfrac{480}{x}$ (dollars per person)

            Cost per skier $= \dfrac{480}{x} - 4.80$ (dollars per person)

Equation: $\left(\dfrac{480}{x} - 4.80\right)(x + 5) = 480$ Write original equation.

                $\left(\dfrac{480 - 4.8x}{x}\right)(x + 5) = 480$ Write $\left(\dfrac{480}{x} - 4.80\right)$ as a fraction.

                $(480 - 4.8x)(x + 5) = 480x$ Multiply each side by x.

      $480x - 4.8x^2 - 24x + 2400 = 480x$ Multiply.

               $-4.8x^2 - 24x + 2400 = 0$ Subtract $480x$ from each side.

                    $x^2 + 5x - 500 = 0$ Divide each side by -4.8.

                  $(x + 25)(x - 20) = 0$ Factor.

                       $x + 25 = 0 \implies x = -25$

                       $x - 20 = 0 \implies x = 20$

Choosing the positive value of x, you can conclude that 20 ski club members are going on the trip. Check this in the original statement of the problem.

$$\left(\frac{480}{20} - 4.80\right)(20 + 5) \overset{?}{=} 480 \qquad \text{Substitute 20 for } x.$$

$$(24 - 4.80)25 \overset{?}{=} 480 \qquad \text{Simplify.}$$

$$480 = 480 \qquad \text{20 checks. } \checkmark$$

Interest in a savings account is calculated by one of three basic methods: simple interest, interest compounded n times per year, and interest compounded continuously. The next example uses the formula for interest that is compounded n times per year.

$$A = P\left(1 + \frac{r}{n}\right)^{nt}$$

In this formula, A is the balance in the account, P is the principal (or original deposit), r is the annual interest rate (in decimal form), n is the number of compoundings per year, and t is the time in years. In Chapter 5, you will study a derivation of the formula above for interest compounded continuously.

Example 9 ▶ Compound Interest

When your cousin was born, your grandparents deposited $5000 in a long-term investment in which the interest was compounded quarterly. Today, on your cousin's 25th birthday, the value of the investment is $25,062.59. What is the annual interest rate for this investment?

Solution

Formula: $\quad A = P\left(1 + \frac{r}{n}\right)^{nt}$

Labels:

Balance $= A = 25{,}062.59$		(dollars)
Principal $= P = 5000$		(dollars)
Time $= t = 25$		(years)
Compoundings per year $= n = 4$		(compoundings per year)
Annual interest rate $= r$		(percent in decimal form)

Equation: $\quad 25{,}062.59 = 5000\left(1 + \frac{r}{4}\right)^{4(25)}$ Write original equation.

$$\frac{25{,}062.59}{5000} = \left(1 + \frac{r}{4}\right)^{100}$$ Divide each side by 5000.

$$5.0125 \approx \left(1 + \frac{r}{4}\right)^{100}$$ Use a calculator.

$$(5.0125)^{1/100} = 1 + \frac{r}{4}$$ Raise each side to reciprocal power.

$$1.01625 \approx 1 + \frac{r}{4}$$ Use a calculator.

$$0.01625 = \frac{r}{4}$$ Subtract 1 from each side.

$$0.065 = r$$ Multiply each side by 4.

The annual interest rate is about 0.065, or 6.5%. Check this in the original statement of the problem.

Example 10 ▶ Market Research

The marketing department at a publishing firm is asked to determine the price of a book. The department determines that the demand for the book depends on the price of the book according to the formula

$$p = 40 - \sqrt{0.0001x + 1}, \qquad x \geq 0$$

where p is the price per book (in dollars) and x is the number of books sold at the given price. For instance, in Figure 1.25 note that if the price were $39, then (according to the model) no one would be willing to buy the book. On the other hand, if the price were $17.60, 5 million copies could be sold. If the publisher set the price at $12.95, how many copies would be sold?

Market Research

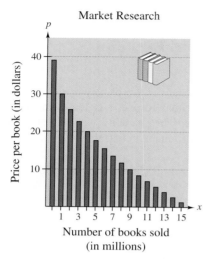

Number of books sold
(in millions)

FIGURE **1.25**

Solution

$p = 40 - \sqrt{0.0001x + 1}$	Write original model.
$12.95 = 40 - \sqrt{0.0001x + 1}$	Substitute 12.95 for p.
$-27.05 = -\sqrt{0.0001x + 1}$	Isolate the square root.
$(-27.05)^2 = \left(-\sqrt{0.0001x + 1}\right)^2$	Square each side.
$731.7025 = 0.0001x + 1$	Simplify.
$730.7025 = 0.0001x$	Subtract 1 from each side.
$7{,}307{,}025 = x$	Divide each side by 0.0001.

So, by setting the book's price at $12.95, the publisher can expect to sell about 7.3 million copies.

Writing ABOUT MATHEMATICS

In Example 10, suppose the cost of producing the book is $150,000 plus $5.50 per book. Then the cost equation is

$$C = 5.5x + 150{,}000$$

where C is measured in dollars and x represents the number of books produced. From Example 10, the total revenue is

$$R = xp$$
$$= x\left(40 - \sqrt{0.0001x + 1}\right).$$

Use the equations for R and C to find an equation for the profit P in terms of x. (*Hint:* $P = R - C$) Construct a table that gives the profits for values of x ranging from 4.0 million to 6.5 million books in increments of 0.25 million. At what point does the profit appear to be maximized? Write a paragraph discussing ways in which you could refine or verify this estimate.

1.6 Exercises

In Exercises 1–24, find all solutions of the equation. Check your solutions in the original equation.

1. $4x^4 - 18x^2 = 0$

2. $20x^3 - 125x = 0$

3. $x^4 - 81 = 0$

4. $x^6 - 64 = 0$

5. $x^3 + 216 = 0$

6. $27x^3 - 512 = 0$

7. $5x^3 + 30x^2 + 45x = 0$

8. $9x^4 - 24x^3 + 16x^2 = 0$

9. $x^3 - 3x^2 - x + 3 = 0$

10. $x^3 + 2x^2 + 3x + 6 = 0$

11. $x^4 - x^3 + x - 1 = 0$

12. $x^4 + 2x^3 - 8x - 16 = 0$

13. $x^4 - 4x^2 + 3 = 0$

14. $x^4 + 5x^2 - 36 = 0$

15. $4x^4 - 65x^2 + 16 = 0$

16. $36t^4 + 29t^2 - 7 = 0$

17. $x^6 + 7x^3 - 8 = 0$

18. $x^6 + 3x^3 + 2 = 0$

19. $\dfrac{1}{x^2} + \dfrac{8}{x} + 15 = 0$

20. $6\left(\dfrac{x}{x+1}\right)^2 + 5\left(\dfrac{x}{x+1}\right) - 6 = 0$

21. $2x + 9\sqrt{x} = 5$

22. $6x - 7\sqrt{x} - 3 = 0$

23. $3x^{1/3} + 2x^{2/3} = 5$

24. $9t^{2/3} + 24t^{1/3} + 16 = 0$

▦ *Graphical Analysis* In Exercises 25–28, (a) use a graphing utility to graph the equation; (b) use the graph to approximate any x-intercepts of the graph; (c) set $y = 0$ and solve the resulting equation; and (d) compare the result with the x-intercepts of the graph.

25. $y = x^3 - 2x^2 - 3x$

26. $y = 2x^4 - 15x^3 + 18x^2$

27. $y = x^4 - 10x^2 + 9$

28. $y = x^4 - 29x^2 + 100$

In Exercises 29–52, find all solutions of the equation. Check your solutions in the original equation.

29. $\sqrt{2x} - 10 = 0$

30. $4\sqrt{x} - 3 = 0$

31. $\sqrt{x - 10} - 4 = 0$

32. $\sqrt{5 - x} - 3 = 0$

33. $\sqrt[3]{2x + 5} + 3 = 0$

34. $\sqrt[3]{3x + 1} - 5 = 0$

35. $-\sqrt{26 - 11x} + 4 = x$

36. $x + \sqrt{31 - 9x} = 5$

37. $\sqrt{x + 1} = \sqrt{3x + 1}$

38. $\sqrt{x + 5} = \sqrt{x - 5}$

39. $\sqrt{x} - \sqrt{x - 5} = 1$

40. $\sqrt{x} + \sqrt{x - 20} = 10$

41. $\sqrt{x + 5} + \sqrt{x - 5} = 10$

42. $2\sqrt{x + 1} - \sqrt{2x + 3} = 1$

43. $\sqrt{x + 2} - \sqrt{2x - 3} = -1$

44. $4\sqrt{x - 3} - \sqrt{6x - 17} = 3$

45. $(x - 5)^{3/2} = 8$

46. $(x + 3)^{3/2} = 8$

47. $(x + 3)^{2/3} = 8$

48. $(x + 2)^{2/3} = 9$

49. $(x^2 - 5)^{3/2} = 27$

50. $(x^2 - x - 22)^{3/2} = 27$

51. $3x(x - 1)^{1/2} + 2(x - 1)^{3/2} = 0$

52. $4x^2(x - 1)^{1/3} + 6x(x - 1)^{4/3} = 0$

▦ *Graphical Analysis* In Exercises 53–56, (a) use a graphing utility to graph the equation; (b) use the graph to approximate any x-intercepts of the graph; (c) set $y = 0$ and solve the resulting equation; and (d) compare the result with the x-intercepts of the graph.

53. $y = \sqrt{11x - 30} - x$

54. $y = 2x - \sqrt{15 - 4x}$

55. $y = \sqrt{7x + 36} - \sqrt{5x + 16} - 2$

56. $y = 3\sqrt{x} - \dfrac{4}{\sqrt{x}} - 4$

In Exercises 57–70, find all solutions of the equation. Check your solutions in the original equation.

57. $\dfrac{20 - x}{x} = x$

58. $\dfrac{4}{x} - \dfrac{5}{3} = \dfrac{x}{6}$

59. $\dfrac{1}{x} - \dfrac{1}{x + 1} = 3$

60. $\dfrac{x}{x^2 - 4} + \dfrac{1}{x + 2} = 3$

61. $x = \dfrac{3}{x} + \dfrac{1}{2}$

62. $4x + 1 = \dfrac{3}{x}$

63. $\dfrac{4}{x + 1} - \dfrac{3}{x + 2} = 1$

64. $\dfrac{x + 1}{3} - \dfrac{x + 1}{x + 2} = 0$

65. $|2x - 1| = 5$

66. $|3x + 2| = 7$

67. $|x| = x^2 + x - 3$

68. $|x^2 + 6x| = 3x + 18$

69. $|x + 1| = x^2 - 5$

70. $|x - 10| = x^2 - 10x$

▨ *Graphical Analysis*   **In Exercises 71–74, (a) use a graphing utility to graph the equation; (b) use the graph to approximate any *x*-intercepts of the graph; (c) set *y* = 0 and solve the resulting equation; and (d) compare the result with the *x*-intercepts of the graph.**

71. $y = \dfrac{1}{x} - \dfrac{4}{x - 1} - 1$

72. $y = x + \dfrac{9}{x + 1} - 5$

73. $y = |x + 1| - 2$

74. $y = |x - 2| - 3$

In Exercises 75–78, find the real solutions of the equation analytically. (Round your answers to three decimal places.)

75. $3.2x^4 - 1.5x^2 - 2.1 = 0$

76. $7.08x^6 + 4.15x^3 - 9.6 = 0$

77. $1.8x - 6\sqrt{x} - 5.6 = 0$

78. $4x^{2/3} + 8x^{1/3} + 3.6 = 0$

Think About It   **In Exercises 79–86, find an equation that has the given solutions. (There are many correct answers.)**

79. $-2, 5$

80. $0, 3, 5$

81. $-\dfrac{7}{3}, \dfrac{6}{7}$

82. $-\dfrac{1}{8}, -\dfrac{4}{5}$

83. $\sqrt{3}, -\sqrt{3}, 4$

84. $2\sqrt{7}, -\sqrt{7}$

85. $-1, 1, i, -i$

86. $4i, -4i, 6, -6$

87. *Chartering a Bus*   A college charters a bus for $1700 to take a group to a museum. When six more students join the trip, the cost per student drops by $7.50. How many students were in the original group?

88. *Renting an Apartment*   Three students are planning to rent an apartment for a year and share equally in the cost. By adding a fourth person, each person could save $75 a month. How much is the monthly rent?

89. *Airspeed*   An airline runs a commuter flight between two cities that are 720 miles apart. If the average speed of the plane could be increased by 40 miles per hour, the travel time would be decreased by 12 minutes. What airspeed is required to obtain this decrease in travel time?

90. *Average Speed*   A family drove 1080 miles to their vacation lodge. Because of increased traffic density, their average speed on the return trip was decreased by 6 miles per hour and the trip took $2\frac{1}{2}$ hours longer. Determine their average speed on the way to the lodge.

91. *Mutual Funds*   A deposit of $2500 in a mutual fund reaches a balance of $3052.49 after 5 years. What annual interest rate on a certificate of deposit compounded monthly would yield an equivalent return?

92. *Mutual Funds*   A sales representative for a mutual funds company describes a "guaranteed investment fund" that the company is offering to new investors. You are told that if you deposit $10,000 in the fund you will be guaranteed a return of at least $25,000 after 20 years. (Assume the interest is compounded quarterly.)

(a) What is the annual interest rate if the investment only meets the minimum guaranteed amount?

(b) If after 20 years you receive $32,000, what is the annual interest rate?

Exploration   **In Exercises 93 and 94, find *x* such that the distance between the points is 13. Explain your results.**

93. $(1, 2), (x, -10)$

94. $(-8, 0), (x, 5)$

Exploration   **In Exercises 95 and 96, find *y* such that the distance between the points is 17. Explain your results.**

95. $(0, 0), (8, y)$

96. $(-8, 4), (7, y)$

97. *Airline Passengers* An airline offers daily flights between Chicago and Denver. The total monthly cost of these flights is

$$C = \sqrt{0.2x + 1}$$

where C is the cost (in millions of dollars) and x is the number of passengers (in thousands). The total cost of the flights for a certain month is 2.5 million dollars. How many passengers flew that month?

98. *Saturated Steam* The temperature T (in degrees Fahrenheit) of saturated steam increases as pressure increases (see figure). This relationship is approximated by the model

$$T = 75.82 - 2.11x + 43.51\sqrt{x}, \qquad 5 \le x \le 40$$

where x is the absolute pressure (in pounds per square inch).

(a) The temperature of steam at sea level ($x = 14.696$) is 212°F. Evaluate the model above at this pressure.

(b) Use the model to approximate the pressure if the temperature of the steam is 240°F.

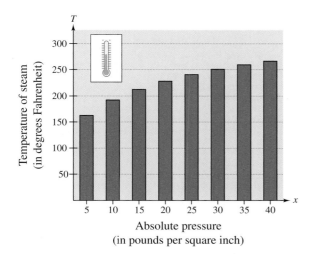

Absolute pressure
(in pounds per square inch)

99. *Economics* The demand equation for a certain product is modeled by

$$p = 40 - \sqrt{0.01x + 1}$$

where x is the number of units demanded per day and p is the price per unit. Approximate the demand if the price is \$37.55.

100. *Economics* The demand equation for a certain product is modeled by

$$p = 40 - \sqrt{0.0001x + 1}$$

where x is the number of units demanded per day and p is the price per unit. Approximate the demand if the price is \$34.70.

101. *Power Line* A power station is on one side of a river that is $\frac{3}{4}$ mile wide. A factory is 8 miles downstream on the other side of the river. It costs \$24 per foot to run power lines overland and \$30 per foot to run them underwater, as shown in the figure. The total cost of the project is \$1,098,662.40. Find the length x as labeled in the figure.

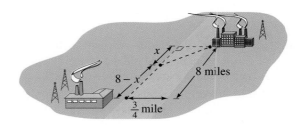

102. *Baseball* A baseball diamond has the shape of a square where the distance from home plate to second base is approximately $127\frac{1}{2}$ feet. Approximate the distance between the bases.

103. *Meteorology* A meteorologist is positioned 100 feet from the point where a weather balloon is launched. When the balloon is at height h, the distance between the meteorologist and the balloon is $d = \sqrt{100^2 + h^2}$, where d is the distance (in feet).

(a) Use a graphing utility to graph the equation. Use the trace feature to approximate the value of h when $d = 200$.

(b) Complete the table. Use the table to approximate the value of h when $d = 200$.

h	160	165	170	175	180	185
d						

(c) Find h algebraically when $d = 200$.

(d) Compare the results of each method. In each case, what information did you gain that wasn't apparent in another solution method?

104. *Geometry*   You construct a cone with a base radius of 8 inches. The surface area of the cone can be represented by the equation

$$S = 8\pi\sqrt{64 + h^2}$$

where h is the height of the cone.

(a) Use a graphing utility to graph the equation. Use the trace feature to approximate the value of h when $S = 350$ square inches.

(b) Complete the table. Use the table to approximate the value of h when $S = 350$.

h	8	9	10	11	12	13
S						

(c) Find h algebraically when $S = 350$.

(d) Compare the results of each method. In each case, what information did you gain that wasn't apparent in another solution method?

105. *Labor*   Working together, two people can complete a task in 8 hours. Working alone, how long would it take each to do the task if one person takes 2 hours longer than the other?

106. *Labor*   Working together, two people can complete a task in 12 hours. Working alone, how long would it take each to do the task if one person takes 3 hours longer than the other?

In Exercises 107 and 108, solve for the indicated variable.

107. *A Person's Tangential Speed in a Rotor*

Solve for g: $v = \sqrt{\dfrac{gR}{\mu s}}$

108. *Inductance*

Solve for Q: $i = \pm\sqrt{\dfrac{1}{LC}}\sqrt{Q^2 - q}$

Synthesis

True or False?   **In Exercises 109 and 110, determine whether the statement is true or false. Justify your answer.**

109. An equation can never have more than one extraneous solution.

110. When solving an absolute value equation, you will always have to check more than one solution.

In Exercises 111 and 112, consider an equation of the form $x + |x - a| = b$, **where a and b are constants.**

111. *Exploration*   Find a and b if the solution to the equation is $x = 9$. (There are many correct answers.)

112. *Writing*   Write a short paragraph listing the steps required to solve this equation involving absolute values.

In Exercises 113 and 114, consider an equation of the form $x + \sqrt{x - a} = b$, **where a and b are constants.**

113. *Exploration*   Find a and b if the solution to the equation is $x = 20$. (There are many correct answers.)

114. *Writing*   Write a short paragraph listing the steps required to solve this equation involving radicals.

Review

In Exercises 115–120, perform the operations and simplify.

115. $\dfrac{8}{3x} + \dfrac{3}{2x}$

116. $\dfrac{2}{x^2 - 4} - \dfrac{1}{x^2 - 3x + 2}$

117. $\dfrac{2}{z + 2} - \left(3 - \dfrac{2}{z}\right)$

118. $25y^2 \div \dfrac{xy}{5}$

119. $\dfrac{\left[\dfrac{24 - 18x}{(2 - x)^2}\right]}{\left(\dfrac{60 - 45x}{x^2 - 4x - 4}\right)}$

120. $x^2 \cdot \dfrac{x + 1}{x^2 - x} \cdot \dfrac{(5x - 5)^2}{x^2 + 6x + 5}$

In Exercises 121–124, find all real solution(s) of the equation.

121. $x^2 - 22x + 121 = 0$

122. $x(x - 20) + 3(x - 20) = 0$

123. $(x + 20)^2 = 625$

124. $5x^2 + x = 0$

1.7 Linear Inequalities

▶ **What you should learn**

- How to recognize solutions of linear inequalities
- How to use properties of inequalities to solve linear inequalities
- How to solve inequalities involving absolute values
- How to use inequalities to model and solve real-life problems

▶ **Why you should learn it**

Inequalities can be used to model and solve real-life problems. For instance, Exercise 97 on page 151 shows how to use a linear inequality to analyze data about the maximum weight a weightlifter can bench press.

Introduction

Simple inequalities were reviewed in Section P.1. There, you used the inequality symbols $<$, $\leq$, $>$, and $\geq$ to compare two numbers and to denote subsets of real numbers. For instance, the simple inequality

$$x \geq 3$$

denotes all real numbers x that are greater than or equal to 3.

In this section you will expand your work with inequalities to include more involved statements such as

$$5x - 7 < 3x + 9$$

and

$$-3 \leq 6x - 1 < 3.$$

As with an equation, you **solve an inequality** in the variable x by finding all values of x for which the inequality is true. Such values are **solutions** and are said to **satisfy** the inequality. The set of all real numbers that are solutions of an inequality is the **solution set** of the inequality. For instance, the solution set of

$$x + 1 < 4$$

is all real numbers that are less than 3.

The set of all points on the real number line that represent the solution set is the **graph of the inequality.** Graphs of many types of inequalities consist of intervals on the real number line. You can review the nine basic types of intervals on the real number line by turning to pages 3 and 4 in Section P.1. On those pages, note that each type of interval can be classified as *bounded* or *unbounded*.

Example 1 ▶ Intervals and Inequalities

Write an inequality to represent each interval and state whether the interval is bounded or unbounded.

a. $(-3, 5]$

b. $(-3, \infty)$

c. $[0, 2]$

d. $(-\infty, \infty)$

Solution

a. $(-3, 5]$ corresponds to $-3 < x \leq 5$.   Bounded

b. $(-3, \infty)$ corresponds to $-3 < x$.   Unbounded

c. $[0, 2]$ corresponds to $0 \leq x \leq 2$.   Bounded

d. $(-\infty, \infty)$ corresponds to $-\infty < x < \infty$.   Unbounded

Properties of Inequalities

The procedures for solving linear inequalities in one variable are much like those for solving linear equations. To isolate the variable, you can make use of the **properties of inequalities.** These properties are similar to the properties of equality, but there are two important exceptions. When each side of an inequality is multiplied or divided by a negative number, the direction of the inequality symbol must be reversed. Here is an example.

$$-2 < 5 \qquad \text{Write original inequality.}$$
$$(-3)(-2) > (-3)(5) \qquad \text{Multiply each side by } -3 \text{ and reverse inequality.}$$
$$6 > -15 \qquad \text{Simplify.}$$

Two inequalities that have the same solution set are **equivalent.** For instance, the inequalities

$$x + 2 < 5$$

and

$$x < 3$$

are equivalent. To obtain the second inequality from the first, you can subtract 2 from each side of the inequality. The following list describes the operations that can be used to create equivalent inequalities.

Properties of Inequalities

Let a, b, c, and d be real numbers.

1. Transitive Property

$$a < b \text{ and } b < c \implies a < c$$

2. Addition of Inequalities

$$a < b \text{ and } c < d \implies a + c < b + d$$

3. Addition of a Constant

$$a < b \implies a + c < b + c$$

4. Multiplication by a Constant

$$\text{For } c > 0, a < b \implies ac < bc$$
$$\text{For } c < 0, a < b \implies ac > bc$$

Each of the properties above is true if the symbol $<$ is replaced by $\leq$ and $>$ is replaced by $\geq$. For instance, another form of the multiplication property would be as follows.

$$\text{For } c > 0, a \leq b \implies ac \leq bc$$
$$\text{For } c < 0, a \leq b \implies ac \geq bc$$

Solving a Linear Inequality

The simplest type of inequality is a **linear inequality** in a single variable. For instance, $2x + 3 > 4$ is a linear inequality in x.

In the following examples, pay special attention to the steps in which the inequality symbol is reversed. Remember that when you multiply or divide by a negative number, you must reverse the inequality symbol.

> **STUDY T!P**
>
> Checking the solution set of an inequality is not as simple as checking the solutions of an equation. You can, however, get an indication of the validity of a solution set by substituting a few convenient values of x.

Example 2 ▶ Solving Linear Inequalities

Solve each inequality.

a. $5x - 7 > 3x + 9$ **b.** $1 - \frac{3x}{2} \geq x - 4$

Solution

a.

$5x - 7 > 3x + 9$	Write original inequality.
$5x > 3x + 16$	Add 7 to each side.
$5x - 3x > 16$	Subtract $3x$ from each side.
$2x > 16$	Combine like terms.
$x > 8$	Divide each side by 2.

The solution set is all real numbers that are greater than 8, which is denoted by $(8, \infty)$. The graph of this solution set is shown in Figure 1.26.

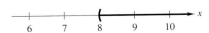

Solution interval: $(8, \infty)$
FIGURE 1.26

b.

$1 - \frac{3x}{2} \geq x - 4$	Write original inequality.
$2 - 3x \geq 2x - 8$	Multiply each side by 2.
$-3x \geq 2x - 10$	Subtract 2 from each side.
$-5x \geq -10$	Subtract $2x$ from each side.
$x \leq 2$	Divide each side by -5 and reverse the inequality.

The solution set is all real numbers that are less than or equal to 2, which is denoted by $(-\infty, 2]$. The graph of this solution set is shown in Figure 1.27.

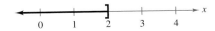

Solution interval: $(-\infty, 2]$
FIGURE 1.27

Sometimes it is possible to write two inequalities as a **double inequality.** For instance, you can write the two inequalities $-4 \le 5x - 2$ and $5x - 2 < 7$ more simply as

$$-4 \le 5x - 2 < 7.$$

This form allows you to solve the two inequalities together, as demonstrated in Example 3.

Example 3 ▶ Solving a Double Inequality

To solve a double inequality, you can isolate x as the middle term.

$$-3 \le 6x - 1 < 3 \qquad \text{Write original inequality.}$$

$$-3 + 1 \le 6x - 1 + 1 < 3 + 1 \qquad \text{Add 1 to each part.}$$

$$-2 \le 6x < 4 \qquad \text{Simplify.}$$

$$\frac{-2}{6} \le \frac{6x}{6} < \frac{4}{6} \qquad \text{Divide each part by 6.}$$

$$-\frac{1}{3} \le x < \frac{2}{3} \qquad \text{Simplify.}$$

The solution set is all real numbers that are greater than or equal to $-\frac{1}{3}$ and less than $\frac{2}{3}$, which is denoted by $\left[-\frac{1}{3}, \frac{2}{3}\right)$. The graph of this solution set is shown in Figure 1.28.

Solution interval: $\left[-\frac{1}{3}, \frac{2}{3}\right)$
FIGURE **1.28**

The double inequality in Example 3 could have been solved in two parts as follows.

$$-3 \le 6x - 1 \qquad \text{and} \qquad 6x - 1 < 3$$

$$-2 \le 6x \qquad\qquad\qquad 6x < 4$$

$$-\frac{1}{3} \le x \qquad\qquad\qquad x < \frac{2}{3}$$

The solution set consists of all real numbers that satisfy *both* inequalities. In other words, the solution set is the set of all values of x for which

$$-\frac{1}{3} \le x < \frac{2}{3}.$$

When combining two inequalities to form a double inequality, be sure that the inequalities satisfy the Transitive Property. For instance, it is *incorrect* to combine the inequalities $3 < x$ and $x \le -1$ as $3 < x \le -1$. This "inequality" is obviously wrong because 3 is not less than -1.

Technology

A graphing utility can be used to identify the solution set of the graph of an inequality. For instance, to find the solution set of $|x - 5| < 2$ (see Example 4), enter

Y1 = abs (X − 5) < 2

and press the graph key. The graph should look like the one shown below.

The solution set is indicated by the line segment above the x-axis.

Inequalities Involving Absolute Value

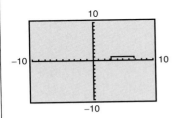

Solving an Absolute Value Inequality

Let x be a variable or an algebraic expression and let a be a real number such that $a \geq 0$.

1. The solutions of $|x| < a$ are all values of x that lie between $-a$ and a.

$$|x| < a \quad \text{if and only if} \quad -a < x < a.$$

2. The solutions of $|x| > a$ are all values of x that are less than $-a$ or greater than a.

$$|x| > a \quad \text{if and only if} \quad x < -a \quad \text{or} \quad x > a.$$

These rules are also valid if $<$ is replaced by $\leq$ and $>$ is replaced by $\geq$.

Example 4 ▶ Solving an Absolute Value Inequality

Solve each inequality.

a. $|x - 5| < 2$ **b.** $|x + 3| \geq 7$

Solution

a.

$\lvert x - 5\rvert < 2$	Write original inequality.
$-2 < x - 5 < 2$	Write equivalent inequalities.
$-2 + 5 < x - 5 + 5 < 2 + 5$	Add 5 to each part.
$3 < x < 7$	Simplify.

The solution set is all real numbers that are greater than 3 and less than 7, which is denoted by $(3, 7)$. The graph of this solution set is shown in Figure 1.29.

b.

$\lvert x + 3\rvert \geq 7$		Write original inequality.
$x + 3 \leq -7$ or	$x + 3 \geq 7$	Write equivalent inequalities.
$x + 3 - 3 \leq -7 - 3$	$x + 3 - 3 \geq 7 - 3$	Subtract 3 from each side.
$x \leq -10$	$x \geq 4$	Simplify.

The solution set is all real numbers that are less than or equal to -10 *or* greater than or equal to 4. The interval notation for this solution set is $(-\infty, -10] \cup [4, \infty)$. The symbol $\cup$ is called a *union* symbol and is used to denote the combining of two sets. The graph of this solution set is shown in Figure 1.30.

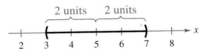

$|x - 5| < 2$: *Solutions lie inside* $(3, 7)$

FIGURE 1.29

$|x + 3| \geq 7$: *Solutions lie outside* $(-10, 4)$

FIGURE 1.30

STUDY TIP

Note that the graph of the inequality $|x - 5| < 2$ can be described as all real numbers within 2 units of 5, as shown in Figure 1.29.

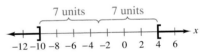

The smallest registered street-legal car in the United States was built by Arlis Sluder and is now owned by Jeff Gibson. Its overall length is $88\frac{3}{4}$ inches and its width is $40\frac{1}{2}$ inches.

Applications

The problem-solving plan described in Section 1.3 can be used to model and solve real-life problems that involve inequalities, as in Example 5.

Example 5 ▶ Comparative Shopping

A subcompact car can be rented from Company A for $180 per week with no extra charge for mileage. A similar car can be rented from Company B for $100 per week plus 20 cents for each mile driven. How many miles must you drive in a week in order for the rental fee for Company B to be more than that for Company A?

Solution

Verbal Model: Weekly cost for Company B $>$ Weekly cost for Company A

Labels:
Miles driven in one week $= m$ (miles)
Weekly cost for Company A $= 180$ (dollars)
Weekly cost for Company B $= 100 + 0.20m$ (dollars)

Inequality: $100 + 0.2m > 180$

$$0.2m > 80$$

$$m > 400 \text{ miles}$$

If you drive more than 400 miles in a week, Company B costs more.

Example 6 ▶ Accuracy of a Measurement

You go to a candy store to buy chocolates that cost $9.89 per pound. The scale that is used in the store has a state seal of approval that indicates the scale is accurate to within half an ounce. According to the scale, your purchase weighs one-half pound and costs $4.95. How much might you have been undercharged or overcharged due to an error in the scale?

Solution

Let x represent the *true* weight of the candy. Because the scale is accurate to within half an ounce (or $\frac{1}{32}$ of a pound), the difference between the exact weight (x) and the scale weight $\left(\frac{1}{2}\right)$ is less than or equal to $\frac{1}{32}$ of a pound. That is,

$$\left| x - \tfrac{1}{2} \right| \le \tfrac{1}{32}.$$

You can solve this inequality as follows.

$$-\tfrac{1}{32} \le x - \tfrac{1}{2} \le \tfrac{1}{32}$$

$$\tfrac{15}{32} \le x \le \tfrac{17}{32}$$

$$0.46875 \le x \le 0.53125$$

In other words, your "one-half pound" of candy could have weighed as little as 0.46875 pound (which would have cost $4.64) or as much as 0.53125 pound (which would have cost $5.25). So, you could have been overcharged by as much as $0.31 or undercharged by as much as $0.30.

1.7 Exercises

In Exercises 1–6, write an inequality to represent the interval, and state whether the interval is bounded or unbounded.

1. $[-1, 5]$

2. $(2, 10]$

3. $(11, \infty)$

4. $[-5, \infty)$

5. $(-\infty, -2)$

6. $(-\infty, 7]$

In Exercises 7–12, match the inequality with its graph. [The graphs are labeled (a), (b), (c), (d), (e), and (f).]

(a)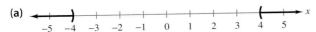

(b)

(c)

(d)

(e)

(f)

7. $x < 3$

8. $x \geq 5$

9. $-3 < x \leq 4$

10. $0 \leq x \leq \frac{9}{2}$

11. $|x| < 3$

12. $|x| > 4$

In Exercises 13–18, determine whether each value of *x* is a solution of the inequality.

Inequality	*Values*	
13. $5x - 12 > 0$	(a) $x = 3$	(b) $x = -3$
	(c) $x = \frac{5}{2}$	(d) $x = \frac{3}{2}$
14. $2x + 1 < -3$	(a) $x = 0$	(b) $x = -\frac{1}{4}$
	(c) $x = -4$	(d) $x = -\frac{3}{2}$
15. $0 < \dfrac{x - 2}{4} < 2$	(a) $x = 4$	(b) $x = 10$
	(c) $x = 0$	(d) $x = \frac{7}{2}$
16. $-1 < \dfrac{3 - x}{2} \leq 1$	(a) $x = 0$	(b) $x = -5$
	(c) $x = 1$	(d) $x = 5$
17. $|x - 10| \geq 3$	(a) $x = 13$	(b) $x = -1$
	(c) $x = 14$	(d) $x = 9$
18. $|2x - 3| < 15$	(a) $x = -6$	(b) $x = 0$
	(c) $x = 12$	(d) $x = 7$

In Exercises 19–44, solve the inequality and sketch the solution on the real number line.

19. $4x < 12$

20. $-10x < 40$

21. $2x > 3$

22. $-6x > 15$

23. $x - 5 \geq 7$

24. $x + 7 \leq 12$

25. $2x + 7 < 3 + 4x$

26. $3x + 1 \geq 2 + x$

27. $2x - 1 \geq 1 - 5x$

28. $6x - 4 \leq 2 + 8x$

29. $4 - 2x < 3(3 - x)$

30. $4(x + 1) < 2x + 3$

31. $\frac{3}{4}x - 6 \leq x - 7$

32. $3 + \frac{2}{7}x > x - 2$

33. $\frac{1}{2}(8x + 1) \geq 3x + \frac{5}{2}$

34. $9x - 1 < \frac{3}{4}(16x - 2)$

35. $3.6x + 11 \geq -3.4$

36. $15.6 - 1.3x < -5.2$

37. $1 < 2x + 3 < 9$

38. $-8 \leq -(3x + 5) < 13$

39. $-4 < \dfrac{2x - 3}{3} < 4$

40. $0 \leq \dfrac{x + 3}{2} < 5$

41. $\frac{3}{4} > x + 1 > \frac{1}{4}$

42. $-1 < 2 - \dfrac{x}{3} < 1$

43. $3.2 \leq 0.4x - 1 \leq 4.4$

44. $4.5 > \dfrac{1.5x + 6}{2} > 10.5$

In Exercises 45–60, solve the inequality and sketch the solution on the real number line. (Some equations have no solution.)

45. $|x| < 6$

46. $|x| > 4$

47. $\left|\dfrac{x}{2}\right| > 5$

48. $\left|\dfrac{x}{5}\right| > 3$

49. $|x - 5| < -1$

50. $|x - 5| \geq 0$

51. $|x - 20| \leq 6$

52. $|x - 7| < -5$

53. $|3 - 4x| \geq 9$

54. $|1 - 2x| < 5$

55. $\left|\dfrac{x - 3}{2}\right| \geq 5$

56. $\left|1 - \dfrac{2x}{3}\right| < 1$

57. $|9 - 2x| - 2 < -1$

58. $|x + 14| + 3 > 17$

59. $2|x + 10| \geq 9$

60. $3|4 - 5x| \leq 9$

▦ *Graphical Analysis* **In Exercises 61–68, use a graphing utility to graph the inequality and identify the solution set.**

61. $6x > 12$

62. $3x - 1 \le 5$

63. $5 - 2x \ge 1$

64. $3(x + 1) < x + 7$

65. $|x - 8| \le 14$

66. $|2x + 9| > 13$

67. $2|x + 7| \ge 13$

68. $\frac{1}{2}|x + 1| \le 3$

▦ *Graphical Analysis* **In Exercises 69–74, use a graphing utility to graph the equation. Use the graph to approximate the values of *x* that satisfy each inequality.**

Equation	Inequalities	
69. $y = 2x - 3$	(a) $y \ge 1$	(b) $y \le 0$
70. $y = \frac{2}{3}x + 1$	(a) $y \le 5$	(b) $y \ge 0$
71. $y = -\frac{1}{2}x + 2$	(a) $0 \le y \le 3$	(b) $y \ge 0$
72. $y = -3x + 8$	(a) $-1 \le y \le 3$	(b) $y \le 0$
73. $y = |x - 3|$	(a) $y \le 2$	(b) $y \ge 4$
74. $y = \left|\frac{1}{2}x + 1\right|$	(a) $y \le 4$	(b) $y \ge 1$

In Exercises 75–80, find the interval(s) on the real number line for which the radicand is nonnegative (greater than or equal to zero).

75. $\sqrt{x - 5}$

76. $\sqrt{x - 10}$

77. $\sqrt{x + 3}$

78. $\sqrt{3 - x}$

79. $\sqrt[4]{7 - 2x}$

80. $\sqrt[4]{6x + 15}$

81. *Think About It* The graph of $|x - 5| < 3$ can be described as all real numbers within 3 units of 5. Give a similar description of $|x - 10| < 8$.

82. *Think About It* The graph of $|x - 2| > 5$ can be described as all real numbers more than 5 units from 2. Give a similar description of $|x - 8| > 4$.

In Exercises 83–90, use absolute value notation to define the interval (or pair of intervals) on the real number line.

83.

84.

85.

86.

87. All real numbers within 10 units of 12

88. All real numbers at least 5 units from 8

89. All real numbers more than 5 units from -3

90. All real numbers no more than 7 units from -6

91. *Car Rental* You can rent a midsize car from Company A for $250 per week with unlimited mileage. A similar car can be rented from Company B for $150 per week plus 25 cents for each mile driven. How many miles must you drive in a week in order for the rental fee for Company B to be greater than that for Company A?

92. *Copying Costs* Your department sends its copying to the photocopy center of your company. The center bills your department $0.10 per page. You have investigated the possibility of buying a departmental copier for $3000. With your own copier, the cost per page would be $0.03. The expected life of the copier is 4 years. How many copies must you make in the 4-year period to justify buying the copier?

93. *Investment* In order for an investment of $1000 to grow to more than $1062.50 in 2 years, what must the annual interest rate be? $[A = P(1 + rt)]$

94. *Investment* In order for an investment of $750 to grow to more than $825 in 2 years, what must the annual interest rate be? $[A = P(1 + rt)]$

95. *Business* The revenue for selling x units of a product is $R = 115.95x$. The cost of producing x units is $C = 95x + 750$. To obtain a profit, the revenue must be greater than the cost. For what values of x will this product return a profit?

▦ **96.** *Data Analysis* The admissions office of a college wants to determine whether there is a relationship between IQ scores x and grade-point averages y after the first year of school. The table shows a sample of data from 12 students.

x	118	131	125	123	133	136
y	2.2	2.4	3.2	2.4	3.5	3.0

x	128	124	116	120	134	131
y	3.0	2.8	2.2	1.8	3.4	3.6

(a) Use a graphing utility to plot the data.

(b) A model for this data is

$$y = 0.067x - 5.638.$$

Use a graphing utility to graph the equation on the same display used in part (a).

(c) Use the graph to estimate the values of x that predict a grade-point average of at least 3.0.

(d) Use the graph to write a statement about the accuracy of the model. If you think the graph indicates that IQ scores are not particularly good predictors of grade-point average, list other factors that might influence college performance.

97. *Data Analysis*   You want to determine whether there is a relationship between an athlete's weight x (in pounds) and the athlete's maximum bench-press weight y (in pounds). The table shows a sample of data from 12 athletes.

x	165	184	150	210	196	240
y	170	185	200	255	205	295

x	202	170	185	190	230	160
y	190	175	195	185	250	155

(a) Use a graphing utility to plot the data.

(b) A model for this data is

$$y = 1.266x - 35.766.$$

Use a graphing utility to graph the equation on the same display used in part (a).

(c) Use the graph to estimate the values of x that predict a maximum bench-press weight of at least 200 pounds.

(d) Use the graph to write a statement about the accuracy of the model. If you think the graph indicates that an athlete's weight is not a particularly good indicator of the athlete's maximum bench-press weight, list other factors that might influence an individual's maximum bench-press weight.

98. *Teachers' Salaries*   The average salary S (in thousands of dollars) for elementary and secondary teachers in the United States from 1987 to 1997 is approximated by the model

$$S = 30.944 + 1.187t, \qquad -3 \le t \le 7$$

where $t = -3$ represents 1987. According to this model, when will the average salary exceed $42,000? (Source: National Education Association)

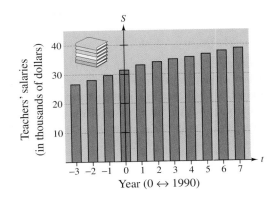

Year (0 ↔ 1990)

99. *Egg Production*   The number of dozens of eggs (in millions) produced in the United States from 1990 to 1997 can be modeled by

$$S = 107.3t + 5725.8$$

where $t = 0$ represents 1990. According to the model, when will the number of dozens of eggs produced exceed 7 billion? (Source: U.S. Department of Agriculture)

100. *Geometry*   The side of a square is measured as 10.4 inches with a possible error of $\frac{1}{16}$ inch. Using these measurements, determine the interval containing the possible areas of the square.

101. *Finance*   You buy a bag of oranges for $0.95 per pound. The weight that is listed on the bag is 4.65 pounds. The scale that weighed the bag is accurate to within 1 ounce. How much might you have been undercharged or overcharged?

102. *Height*   The heights h of two-thirds of the members of a certain population satisfy the inequality

$$\left| \frac{h - 68.5}{2.7} \right| \le 1$$

where h is measured in inches. Determine the interval on the real number line in which these heights lie.

103. *Meteorology*   A certain electronic device is to be operated in an environment with relative humidity h in the interval defined by

$$|h - 50| \le 30.$$

What are the minimum and maximum relative humidities for the operation of this device?

104. *Music*   Michael Kasha of Florida State University used physics and mathematics to design a new classical guitar. He used the model for the frequency of the vibrations on a circular plate

$$v = \frac{2.6t}{d^2} \sqrt{\frac{E}{\rho}}$$

where v is the frequency (in vibrations per second), t is the plate thickness (in millimeters), d is the diameter of the plate, E is the elasticity of the plate material, and ρ is the density of the plate material. For fixed values of d, E, and ρ, the graph of the equation is a line (see figure).

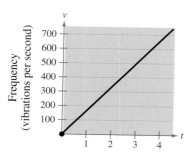

Plate thickness (millimeters)

(a) Estimate the frequency if the plate thickness is 2 millimeters.

(b) Estimate the plate thickness if the frequency is 600 vibrations per second.

(c) Approximate the interval for the plate thickness if the frequency is between 200 and 400 vibrations per second.

(d) Approximate the interval for the frequency if the plate thickness is less than 3 millimeters.

Synthesis

True or False?   **In Exercises 105 and 106, determine whether the statement is true or false. Justify your answer.**

105. If a, b, and c are real numbers, and $a \le b$, then $ac \le bc$.

106. If $-10 \le x \le 8$, then $-10 \ge -x$ and $-x \ge -8$.

107. Identify the graph of the inequality $|x - a| \ge 2$.

(a) [number line from $a-2$ to a to $a+2$]

(b) [number line from $a-2$ to a to $a+2$]

(c) [number line from $2-a$ to 2 to $2+a$]

(d) [number line from $2-a$ to 2 to $2+a$]

108. *Exploration*   Find sets of values of a, b, and c such that $0 \le x \le 10$ is a solution of the inequality $|ax - b| \le c$.

Review

In Exercises 109–112, find the distance between each pair of points. Then find the midpoint of the line segment joining the points.

109. $(-4, 2), (1, 12)$

110. $(1, -2), (10, 3)$

111. $(3, 6), (-5, -8)$

112. $(0, -3), (-6, 9)$

In Exercises 113–120, solve the equation.

113. $3(x - 1) = 30$

114. $8x - 5(x + 4) = -19$

115. $-6(2 - x) - 12 = 36$

116. $4(x + 7) - 9 = -6(-x - 1)$

117. $2x^2 - 19x - 10 = 0$

118. $3x^2 - x - 10 = 0$

119. $14x^2 + 5x - 1 = 0$

120. $x^3 + 5x^2 - 4x - 20 = 0$

121. Find the coordinates of the point located 3 units to the left of the y-axis and 10 units above the x-axis.

122. Determine the quadrants in which the point (x, y) could be located if $y > 0$.

123. Find the distance between the points $(6, 5)$ and $(1, -7)$.

1.8 Other Types of Inequalities

▶ What you should learn
- How to solve polynomial inequalities
- How to solve rational inequalities
- How to use inequalities to model and solve real-life problems

▶ Why you should learn it
Inequalities can be used to model and solve real-life problems. For instance, Exercise 71 on page 161 shows how to use a quadratic inequality to model the percent of the American population that has completed 4 years of college or more.

Peter Hvizdak/The Image Works

Polynomial Inequalities

To solve a polynomial inequality such as

$$x^2 - 2x - 3 < 0$$

you can use the fact that a polynomial can change signs only at its zeros (the x-values that make the polynomial equal to zero). Between two consecutive zeros a polynomial must be entirely positive or entirely negative. This means that when the real zeros of a polynomial are put in order, they divide the real number line into intervals in which the polynomial has no sign changes. These zeros are the **critical numbers** of the inequality, and the resulting intervals are the **test intervals** for the inequality. For instance, the polynomial above factors as

$$x^2 - 2x - 3 = (x + 1)(x - 3)$$

and has two zeros, $x = -1$ and $x = 3$. These zeros divide the real number line into three test intervals:

$$(-\infty, -1), \quad (-1, 3), \quad \text{and} \quad (3, \infty). \qquad \text{(See Figure 1.31.)}$$

So, to solve the inequality $x^2 - 2x - 3 < 0$, you need only test one value from each of these test intervals to determine whether the value satisfies the original inequality. If so, you can conclude that the interval is a solution of the inequality.

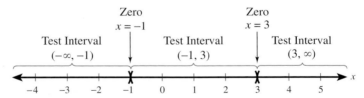

FIGURE 1.31   *Three test intervals for $x^2 - 2x - 3$*

You can use the same basic approach to determine the test intervals for any polynomial.

Finding Test Intervals for a Polynomial

To determine the intervals on which the values of a polynomial are entirely negative or entirely positive, use the following steps.

1. Find all real zeros of the polynomial, and arrange the zeros in increasing order (from smallest to largest). These zeros are the critical numbers of the polynomial.

2. Use the critical numbers of the polynomial to determine its test intervals.

3. Choose one representative x-value in each test interval and evaluate the polynomial at that value. If the value of the polynomial is negative, the polynomial will have negative values for every x-value in the interval. If the value of the polynomial is positive, the polynomial will have positive values for every x-value in the interval.

Example 1 ▶ Solving a Polynomial Inequality

Solve $x^2 - x - 6 < 0$.

Solution

By factoring the quadratic as

$$x^2 - x - 6 = (x + 2)(x - 3)$$

you can see that the critical numbers are $x = -2$ and $x = 3$. So, the polynomial's test intervals are

$$(-\infty, -2), \quad (-2, 3), \quad \text{and} \quad (3, \infty). \qquad \text{Test intervals}$$

In each test interval, choose a representative x-value and evaluate the polynomial.

Interval	x-Value	Polynomial Value	Conclusion
$(-\infty, -2)$	$x = -3$	$(-3)^2 - (-3) - 6 = 6$	Positive
$(-2, 3)$	$x = 0$	$(0)^2 - (0) - 6 = -6$	Negative
$(3, \infty)$	$x = 4$	$(4)^2 - (4) - 6 = 6$	Positive

From this you can conclude that the polynomial is positive for all x-values in $(-\infty, -2)$ and $(3, \infty)$ and is negative only for all x-values in $(-2, 3)$. This implies that the solution of the inequality $x^2 - x - 6 < 0$ is the interval $(-2, 3)$, as shown in Figure 1.32.

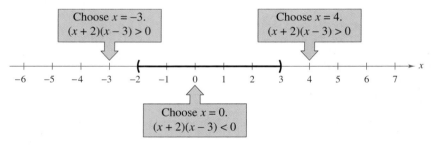

FIGURE 1.32

As with linear inequalities, you can check the reasonableness of a solution by substituting x-values into the original inequality. For instance, to check the solution found in Example 1, try substituting several x-values from the interval $(-2, 3)$ into the inequality

$$x^2 - x - 6 < 0.$$

Regardless of which x-values you choose, the inequality should be satisfied.

You can also use a graph to check the result of Example 1. Sketch the graph of $y = x^2 - x - 6$, as shown in Figure 1.33. Notice that the graph is below the x-axis on the interval $(-2, 3)$.

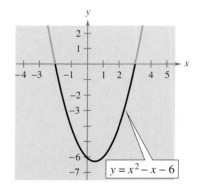

FIGURE 1.33

In Example 1, the polynomial inequality was given in general form. Whenever this is not the case, you should begin the solution process by writing the inequality in general form—with the polynomial on one side and zero on the other.

Example 2 ▶ Solving a Polynomial Inequality

Solve $2x^3 - 3x^2 - 32x > -48$

Solution

Begin by writing the inequality in general form.

$$2x^3 - 3x^2 - 32x > -48 \qquad \text{Write original inequality.}$$

$$2x^3 - 3x^2 - 32x + 48 > 0 \qquad \text{Write in general form.}$$

$$(x - 4)(x + 4)(2x - 3) > 0 \qquad \text{Factor.}$$

The critical numbers are $x = -4$, $x = \frac{3}{2}$, and $x = 4$, and the test intervals are $(-\infty, -4)$, $\left(-4, \frac{3}{2}\right)$, $\left(\frac{3}{2}, 4\right)$, and $(4, \infty)$.

Interval	x-Value	Polynomial Value	Conclusion
$(-\infty, -4)$	$x = -5$	$2(-5)^3 - 3(-5)^2 - 32(-5) + 48$	Negative
$\left(-4, \frac{3}{2}\right)$	$x = 0$	$2(0)^3 - 3(0)^2 - 32(0) + 48$	Positive
$\left(\frac{3}{2}, 4\right)$	$x = 2$	$2(2)^3 - 3(2)^2 - 32(2) + 48$	Negative
$(4, \infty)$	$x = 5$	$2(5)^3 - 3(5)^2 - 32(5) + 48$	Positive

From this you can conclude that the polynomial $2x^3 - 3x^2 - 32x + 48$ is positive on the open intervals $\left(-4, \frac{3}{2}\right)$ and $(4, \infty)$. Therefore, the solution set consists of all real numbers in the intervals $\left(-4, \frac{3}{2}\right)$ and $(4, \infty)$, as shown in Figure 1.34.

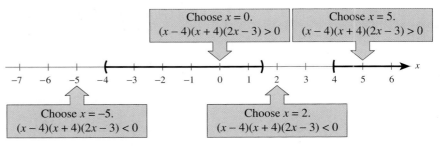

FIGURE 1.34

When solving a quadratic inequality, be sure you have accounted for the particular type of inequality symbol given in the inequality. For instance, in Example 2, note that the original inequality contained a "greater than" symbol and the solution consisted of two open intervals. If the original inequality had been

$$2x^3 - 3x^2 - 32x \geq -48$$

the solution would have consisted of the closed interval $\left[-4, \frac{3}{2}\right]$ and the interval $[4, \infty)$.

Each of the polynomial inequalities in Examples 1 and 2 has a solution set that consists of a single interval or the union of two intervals. When solving the exercises for this section, watch for unusual solution sets, as illustrated in Example 3.

Example 3 ▶ Unusual Solution Sets

a. The solution set of the following inequality consists of the entire set of real numbers, $(-\infty, \infty)$.

$$x^2 + 2x + 4 > 0$$

b. The solution set of the following inequality consists of the single real number $\{-1\}$.

$$x^2 + 2x + 1 \leq 0$$

c. The solution set of the following inequality is empty.

$$x^2 + 3x + 5 < 0$$

d. The solution set of the following inequality consists of all real numbers except the number 2.

$$x^2 - 4x + 4 > 0$$

Exploration

You can use a graphing utility to verify the results given in Example 3. For instance, the graph of

$$y = x^2 + 2x + 4$$

is shown below. Notice that the y-values are greater than 0 for all values of x, as stated in Example 3(a). Use the graphing utility to graph the following:

$$y = x^2 + 2x + 1 \qquad y = x^2 + 3x + 5 \qquad y = x^2 - 4x + 4$$

Explain how you can use the graphs to verify the results of parts (b), (c), and (d) of Example 3.

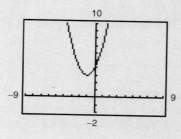

Rational Inequalities

The concepts of critical numbers and test intervals can be extended to rational inequalities. To do this, use the fact that the value of a rational expression can change sign only at its *zeros* (the *x*-values for which its numerator is zero) and its *undefined values* (the *x*-values for which its denominator is zero). These two types of numbers make up the *critical numbers* of a rational inequality.

Example 4 ▶ Solving a Rational Inequality

Solve $\dfrac{2x - 7}{x - 5} \le 3$.

Solution

Begin by writing the rational inequality in general form.

$$\frac{2x - 7}{x - 5} \le 3 \qquad \text{Write original inequality.}$$

$$\frac{2x - 7}{x - 5} - 3 \le 0 \qquad \text{Write in general form.}$$

$$\frac{2x - 7 - 3x + 15}{x - 5} \le 0 \qquad \text{Add fractions.}$$

$$\frac{-x + 8}{x - 5} \le 0 \qquad \text{Simplify.}$$

Critical numbers: $x = 5, x = 8$ Zeros and undefined values of rational expression

Test intervals: $(-\infty, 5), (5, 8), (8, \infty)$

Test: Is $\dfrac{-x + 8}{x - 5} \le 0$?

After testing these intervals, as shown in Figure 1.35, you can see that the rational expression $(-x + 8)/(x - 5)$ is negative in the open intervals $(-\infty, 5)$ and $(8, \infty)$. Moreover, because $(-x + 8)/(x - 5) = 0$ when $x = 8$, you can conclude that the solution set consists of all real numbers in the intervals $(-\infty, 5) \cup [8, \infty)$. (Be sure to use a closed interval to indicate that *x* can equal 8.)

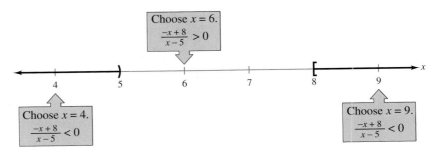

FIGURE 1.35

Applications

One common application of inequalities comes from business and involves profit, revenue, and cost. The formula that relates these three quantities is

$$\boxed{\text{Profit}} = \boxed{\text{Revenue}} - \boxed{\text{Cost}}$$

$$P = R - C.$$

Example 5 ▶ Increasing the Profit for a Product

The marketing department of a calculator manufacturer has determined that the demand for a new model of calculator is

$$p = 100 - 0.00001x, \qquad 0 \le x \le 10{,}000{,}000 \qquad \text{Demand equation}$$

where p is the price per calculator (in dollars) and x represents the number of calculators sold. (If this model is accurate, no one would be willing to pay $100 for the calculator. At the other extreme, the company couldn't *give* away more than 10 million calculators.) The revenue for selling x calculators is

$$R = xp$$
$$= x(100 - 0.00001x) \qquad \text{Revenue equation}$$

Calculators

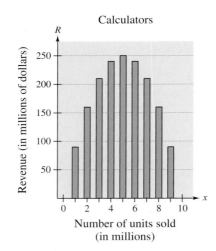

Number of units sold
(in millions)

FIGURE 1.36

as shown in Figure 1.36. The total cost of producing x calculators is $10 per calculator plus a development cost of $2,500,000. So, the total cost is

$$C = 10x + 2{,}500{,}000. \qquad \text{Cost equation}$$

What price should the company charge per calculator to obtain a profit of at least $190,000,000?

Solution

Verbal Model: $\boxed{\text{Profit}} = \boxed{\text{Revenue}} - \boxed{\text{Cost}}$

Equation: $P = R - C$

$$P = 100x - 0.00001x^2 - (10x + 2{,}500{,}000)$$

$$P = -0.00001x^2 + 90x - 2{,}500{,}000$$

To answer the question, solve the inequality

$$P \ge 190{,}000{,}000$$

$$-0.00001x^2 + 90x - 2{,}500{,}000 \ge 190{,}000{,}000.$$

When you write the inequality in general form, find the critical numbers and the test intervals, and then test a value in each test interval, you can find the solution to be

$$3{,}500{,}000 \le x \le 5{,}500{,}000$$

Calculators

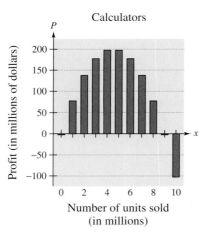

Number of units sold
(in millions)

FIGURE 1.37

as shown in Figure 1.37. Substituting the x-values in the original price equation shows that prices of

$$\$45.00 \le p \le \$65.00$$

will yield a profit of at least $190,000,000.

Another common application of inequalities is finding the domain of an expression that involves a square root, as shown in Example 6.

Example 6 ► Finding the Domain of an Expression

Find the domain of $\sqrt{64 - 4x^2}$.

Solution

Remember that the domain of an expression is the set of all x-values for which the expression is defined. Because $\sqrt{64 - 4x^2}$ is defined (has real values) only if $64 - 4x^2$ is nonnegative, the domain is given by $64 - 4x^2 \geq 0$.

$$64 - 4x^2 \geq 0 \qquad \text{Write in general form.}$$

$$16 - x^2 \geq 0 \qquad \text{Divide each side by 4.}$$

$$(4 - x)(4 + x) \geq 0 \qquad \text{Write in factored form.}$$

So, the inequality has two critical numbers: -4 and 4. You can use these two numbers to test the inequality as follows.

Critical numbers: $\quad x = -4, x = 4$

Test intervals: $\qquad (-\infty, -4), (-4, 4), (4, \infty)$

Test: $\qquad\qquad$ Is $(4 - x)(4 + x) \geq 0$?

A test shows that $64 - 4x^2$ is greater than or equal to 0 in the *closed interval* $[-4, 4]$. So, the domain of the expression $\sqrt{64 - 4x^2}$ is the interval $[-4, 4]$, as shown in Figure 1.38.

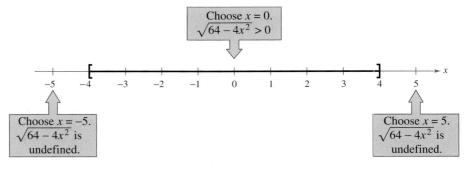

FIGURE 1.38

Writing ABOUT MATHEMATICS

Profit Analysis Consider the relationship $P = R - C$ described on page 158. Write a paragraph discussing why it might be beneficial to solve $P < 0$ if you owned a business. Use the situation described in Example 5 to illustrate your reasoning.

1.8   Exercises

In Exercises 1–4, determine whether each value of x is a solution of the inequality.

Inequality		*Values*

1. $x^2 - 3 < 0$ (a) $x = 3$ (b) $x = 0$
                      (c) $x = \frac{3}{2}$ (d) $x = -5$

2. $x^2 - x - 12 \geq 0$ (a) $x = 5$ (b) $x = 0$
                      (c) $x = -4$ (d) $x = -3$

3. $\dfrac{x + 2}{x - 4} \geq 3$ (a) $x = 5$ (b) $x = 4$
                      (c) $x = -\frac{9}{2}$ (d) $x = \frac{9}{2}$

4. $\dfrac{3x^2}{x^2 + 4} < 1$ (a) $x = -2$ (b) $x = -1$
                      (c) $x = 0$ (d) $x = 3$

In Exercises 5–8, find the critical numbers.

5. $2x^2 - x - 6$ **6.** $9x^3 - 25x^2$

7. $2 + \dfrac{3}{x - 5}$ **8.** $\dfrac{x}{x + 2} - \dfrac{2}{x - 1}$

In Exercises 9–24, solve the inequality and graph the solution on the real number line.

9. $x^2 \leq 9$ **10.** $x^2 < 5$

11. $(x + 2)^2 < 25$ **12.** $(x - 3)^2 \geq 1$

13. $x^2 + 4x + 4 \geq 9$ **14.** $x^2 - 6x + 9 < 16$

15. $x^2 + x < 6$ **16.** $x^2 + 2x > 3$

17. $x^2 + 2x - 3 < 0$ **18.** $x^2 - 4x - 1 > 0$

19. $x^2 + 8x - 5 \geq 0$

20. $-2x^2 + 6x + 15 \leq 0$

21. $x^3 - 3x^2 - x + 3 > 0$

22. $x^3 + 2x^2 - 4x - 8 \leq 0$

23. $x^3 - 2x^2 - 9x - 2 \geq -20$

24. $2x^3 + 13x^2 - 8x - 46 \geq 6$

In Exercises 25–30, solve the inequality and write the solution set in interval notation.

25. $4x^3 - 6x^2 < 0$ **26.** $4x^3 - 12x^2 > 0$

27. $x^3 - 4x \geq 0$ **28.** $2x^3 - x^4 \leq 0$

29. $(x - 1)^2(x + 2)^3 \geq 0$ **30.** $x^4(x - 3) \leq 0$

Graphical Analysis   **In Exercises 31–34, use a graphing utility to graph the equation. Use the graph to approximate the values of x that satisfy each inequality.**

Equation	*Inequalities*

31. $y = -x^2 + 2x + 3$ (a) $y \leq 0$ (b) $y \geq 3$

32. $y = \frac{1}{2}x^2 - 2x + 1$ (a) $y \leq 0$ (b) $y \geq 7$

33. $y = \frac{1}{8}x^3 - \frac{1}{2}x$ (a) $y \geq 0$ (b) $y \leq 6$

34. $y = x^3 - x^2 - 16x + 16$ (a) $y \leq 0$ (b) $y \geq 36$

In Exercises 35–48, solve the inequality and graph the solution on the real number line.

35. $\dfrac{1}{x} - x > 0$ **36.** $\dfrac{1}{x} - 4 < 0$

37. $\dfrac{x + 6}{x + 1} - 2 < 0$ **38.** $\dfrac{x + 12}{x + 2} - 3 \geq 0$

39. $\dfrac{3x - 5}{x - 5} > 4$ **40.** $\dfrac{5 + 7x}{1 + 2x} < 4$

41. $\dfrac{4}{x + 5} > \dfrac{1}{2x + 3}$ **42.** $\dfrac{5}{x - 6} > \dfrac{3}{x + 2}$

43. $\dfrac{1}{x - 3} \leq \dfrac{9}{4x + 3}$ **44.** $\dfrac{1}{x} \geq \dfrac{1}{x + 3}$

45. $\dfrac{x^2 + 2x}{x^2 - 9} \leq 0$ **46.** $\dfrac{x^2 + x - 6}{x} \geq 0$

47. $\dfrac{5}{x - 1} - \dfrac{2x}{x + 1} < 1$ **48.** $\dfrac{3x}{x - 1} \leq \dfrac{x}{x + 4} + 3$

Graphical Analysis   **In Exercises 49–52, use a graphing utility to graph the equation. Use the graph to approximate the values of x that satisfy each inequality.**

Equation	*Inequalities*

49. $y = \dfrac{3x}{x - 2}$ (a) $y \leq 0$ (b) $y \geq 6$

50. $y = \dfrac{2(x - 2)}{x + 1}$ (a) $y \leq 0$ (b) $y \geq 8$

51. $y = \dfrac{2x^2}{x^2 + 4}$ (a) $y \geq 1$ (b) $y \leq 2$

52. $y = \dfrac{5x}{x^2 + 4}$ (a) $y \geq 1$ (b) $y \leq 0$

In Exercises 53–58, find the domain of x in the expression.

53. $\sqrt{4 - x^2}$

54. $\sqrt{x^2 - 4}$

55. $\sqrt{x^2 - 7x + 12}$

56. $\sqrt{144 - 9x^2}$

57. $\sqrt{\dfrac{x}{x^2 - 2x - 35}}$

58. $\sqrt{\dfrac{x}{x^2 - 9}}$

In Exercises 59–64, solve the inequality. (Round your answers to two decimal places.)

59. $0.4x^2 + 5.26 < 10.2$

60. $-1.3x^2 + 3.78 > 2.12$

61. $-0.5x^2 + 12.5x + 1.6 > 0$

62. $1.2x^2 + 4.8x + 3.1 < 5.3$

63. $\dfrac{1}{2.3x - 5.2} > 3.4$

64. $\dfrac{2}{3.1x - 3.7} > 5.8$

65. *Physics*   A projectile is fired straight upward from ground level with an initial velocity of 160 feet per second.

(a) At what instant will it be back at ground level?

(b) When will the height exceed 384 feet?

66. *Physics*   A projectile is fired straight upward from ground level with an initial velocity of 128 feet per second.

(a) At what instant will it be back at ground level?

(b) When will the height be less than 128 feet?

67. *Geometry*   A rectangular playing field with a perimeter of 100 meters is to have an area of at least 500 square meters. Within what bounds must the length of the rectangle lie?

68. *Geometry*   A rectangular parking lot with a perimeter of 440 feet is to have an area of at least 8000 square feet. Within what bounds must the length of the rectangle lie?

69. *Investment* P dollars, invested at interest rate r compounded annually, increases to an amount

$A = P(1 + r)^2$

in 2 years. If an investment of \$1000 is to increase to an amount greater than \$1100 in 2 years, then the interest rate must be greater than what percent?

70. *Economics*   The revenue and cost equations for a product are

$R = x(50 - 0.0002x)$

$C = 12x + 150{,}000$

where R and C are measured in dollars and x represents the number of units sold. How many units must be sold to obtain a profit of at least \$1,650,000?

71. *Education*   The percent P of the U.S. population that had completed 4 years of college or more from 1960 to 1997 is approximated by the model

$P = 7.17 + 0.46t - 0.0001t^2$

where t is the time, with $t = 0$ corresponding to 1960. According to this model, during approximately what year will more than 25% of the population be college graduates?   (Source: U.S. Bureau of the Census)

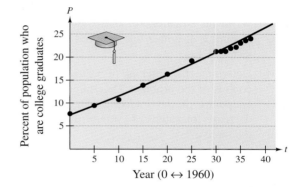

72. *Entertainment*   The receipts from Broadway shows (in millions of dollars) from 1990 to 1996 can be modeled by

$R = 4.7t^2 + 277$

where t is the time, with $t = 0$ corresponding to 1990. According to this model, when will the receipts from Broadway shows exceed \$1 billion?   (Source: The League of American Theaters and Producers, Inc.)

73. *Resistors*   When two resistors of resistance R_1 and R_2 are connected in parallel (see figure), the total resistance R satisfies the equation

$\dfrac{1}{R} = \dfrac{1}{R_1} + \dfrac{1}{R_2}.$

Find R_1 for a parallel circuit in which $R_2 = 2$ ohms and R must be at least 1 ohm.

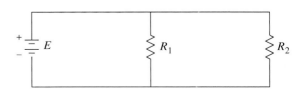

74. *Safe Load*   The maximum safe load uniformly distributed over a 1-foot section of a 2-inch-wide wooden beam is approximated by the model

$$\text{Load} = 168.5d^2 - 472.1$$

where d is the depth of the beam.

(a) Evaluate the model for $d = 4$, $d = 6$, $d = 8$, $d = 10$, and $d = 12$. Use the results to create a bar graph.

(b) Determine the minimum depth of the beam that will safely support a load of 2000 pounds.

75. *Power Supply*   Two factories are located at the coordinates $(-4, 0)$ and $(4, 0)$ with their power supply located at the point $(0, 6)$, as shown in the figure.

(a) Write an equation, in terms of y, giving the amount L of power line required to supply both factories.

(b) Determine the interval of values for y in the context of the problem. Determine L for the two endpoints of the interval. Will L increase or decrease for values of y not at the endpoints of the interval?

(c) Use a graphing utility to graph the equation in part (a) and use the graph to verify your answers to part (b).

(d) Find the values of y such that $L < 13$.

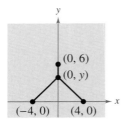

Synthesis

True or False?   **In Exercises 76 and 77, determine whether the statement is true or false. Justify your answer.**

76. The zeros of the polynomial $x^3 - 2x^2 - 11x + 12 \geq 0$ divide the real number line into four test intervals.

77. The solution set of the inequality $\frac{3}{2}x^2 + 3x + 6 \geq 0$ is the set of real numbers.

Exploration   **In Exercises 78–81, find the interval for b such that the equation has at least one real solution.**

78. $x^2 + bx + 4 = 0$ **79.** $x^2 + bx - 4 = 0$

80. $3x^2 + bx + 10 = 0$ **81.** $2x^2 + bx + 5 = 0$

82. *Conjecture*   Write a conjecture about the interval for b in Exercises 78–81. Explain your reasoning.

83. *Think About It*   What is the center of the interval for b in Exercises 78–81?

84. Consider the polynomial $(x - a)(x - b)$ and the real number line shown below.

(a) Identify the points on the line at which the polynomial is zero.

(b) In each of the three subintervals of the line, write the sign of each factor and the sign of the product.

(c) For what x-values does a polynomial change signs?

Review

In Exercises 85–88, factor the expression completely.

85. $4x^2 + 20x + 25$ **86.** $(x + 3)^2 - 16$

87. $x^2(x + 3) - 4(x + 3)$ **88.** $2x^4 - 54x$

In Exercises 89–92, write an expression for the area of the region.

89. 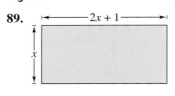   **90.**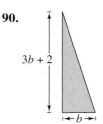

91.   **92.**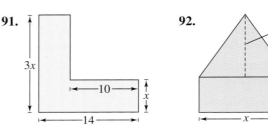

Chapter Summary

What did you learn?

Review Exercises

In Exercises 1–4, complete a table of values. Use the solution points to sketch the graph of the equation.

1. $y = 3x - 5$ **2.** $y = -\frac{1}{2}x + 2$

3. $y = x^2 - 3x$ **4.** $y = 2x^2 - x - 9$

In Exercises 5–12, sketch the graph by hand.

5. $y - 2x - 3 = 0$ **6.** $3x + 2y + 6 = 0$

7. $y = 8 - |x|$ **8.** $y = |x - 8|$

9. $y = \sqrt{5 - x}$ **10.** $y = \sqrt{x + 2}$

11. $y + 2x^2 = 0$ **12.** $y = x^2 - 4x$

In Exercises 13–22, find the x- and y-intercepts of the graph of the equation.

13. $y = 2x - 9$ **14.** $y = -x + 11$

15. $y = (x + 1)^2$ **16.** $y = 4 - (x - 4)^2$

17. $y = \frac{1}{4}x^4 - 2x^2$ **18.** $y = \frac{1}{4}x^3 - 3x$

19. $y = x\sqrt{9 - x^2}$ **20.** $y = x\sqrt{x + 3}$

21. $y = |x - 4| - 4$ **22.** $y = |x + 2| + |3 - x|$

In Exercises 23–32, use intercepts and symmetry to sketch the graph of the equation.

23. $y = -4x + 1$ **24.** $y = 5x - 6$

25. $y = 5 - x^2$ **26.** $y = x^2 - 10$

27. $y = x^3 + 3$ **28.** $y = -6 - x^3$

29. $y = \sqrt{x + 5}$ **30.** $y = -\sqrt{x - 5}$

31. $y = 1 - |x|$ **32.** $y = |x| + 9$

In Exercises 33–38, find the center and radius of the circle and sketch its graph.

33. $x^2 + y^2 = 25$ **34.** $x^2 + y^2 = 4$

35. $(x + 2)^2 + y^2 = 16$

36. $x^2 + (y - 8)^2 = 81$

37. $\left(x - \frac{1}{2}\right)^2 + (y + 1)^2 = 36$

38. $(x + 4)^2 + \left(y - \frac{3}{2}\right)^2 = 100$

39. Find the standard form of the equation of the circle for which the endpoints of a diameter are $(0, 0)$ and $(4, -6)$.

40. Find the standard form of the equation of the circle for which the endpoints of a diameter are $(-2, -3)$ and $(4, -10)$.

41. *Physics* The force F (in pounds) required to stretch a spring x inches from its natural length (see figure) is

$$F = \frac{5}{4}x, \quad 0 \le x \le 20.$$

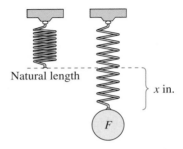

(a) Use the model to complete the table.

x	0	4	8	12	16	20
F						

(b) Sketch a graph of the model.

(c) Use the graph to estimate the force necessary to stretch the spring 10 inches.

42. *Business* The dividends declared per share of the Clorox Company from 1990 to 1998 can be approximated by the model

$$y = 0.073t + 0.644$$

where y is the dividend (in dollars) and t is the time (in years), with $t = 0$ corresponding to 1990. Sketch a graph of this equation. Use the graph to estimate the year in which the dividend per share will be $1.50. (Source: Clorox Company)

In Exercises 43–46, determine whether the equation is an identity or a conditional equation.

43. $6 - (x - 2)^2 = 2 + 4x - x^2$

44. $3(x - 2) + 2x = 2(x + 3)$

45. $-x^3 + x(7 - x) + 3 = x(-x^2 - x) + 7(x + 1) - 4$

46. $3(x^2 - 4x + 8) = -10(x + 2) - 3x^2 + 6$

In Exercises 47 and 48, determine whether each value of x is a solution of the equation.

Equation	Values
47. $3x^2 + 7x = x^2 + 4$	(a) $x = 0$ (b) $x = -4$
	(c) $x = \frac{1}{2}$ (d) $x = -1$
48. $6 + \dfrac{3}{x-4} = 5$	(a) $x = 5$ (b) $x = 0$
	(c) $x = -2$ (d) $x = 7$

In Exercises 49–56, solve the equation (if possible) and check your solution.

49. $3x - 2(x + 5) = 10$ **50.** $4x + 2(7 - x) = 5$

51. $4(x + 3) - 3 = 2(4 - 3x) - 4$

52. $\frac{1}{2}(x - 3) - 2(x + 1) = 5$

53. $\dfrac{x}{5} - 3 = \dfrac{2x}{2} + 1$ **54.** $\dfrac{4x - 3}{6} + \dfrac{x}{4} = x - 2$

55. $\dfrac{18}{x} = \dfrac{10}{x - 4}$ **56.** $\dfrac{5}{x - 2} = \dfrac{13}{2x - 3}$

57. *Geometry* The surface area of the cylinder shown in the figure is approximated by

$$S = 2(3.14)(3)^2 + 2(3.14)(3)h.$$

Find the height h of the cylinder if the surface area is 244.92 square inches.

3 in.

h

58. *Temperature* The Fahrenheit and Celsius temperature scales are related by the equation

$$C = \frac{5}{9}F - \frac{160}{9}.$$

Find the Fahrenheit temperature that corresponds to 100° Celsius.

1.3 **59.** *Business* In October, a company's total profit was 12% more than it was in September. The total profit for the two months was $689,000. Write a verbal model, assign labels, and write an algebraic equation to find the profit for each month.

60. *Consumerism* The price of a television set has been discounted $85. The sale price is $340. Write a verbal model, assign labels, and write an algebraic equation to find the percent discount.

61. *Geometry* A fitness center has two running tracks around a rectangular playing floor (see figure). The tracks are 1 meter wide and form semicircles at the narrow ends of the floor. How much longer is the running distance on the outer track than on the inner track?

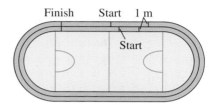

Finish Start 1 m
Start

62. *Finance* A group agrees to share equally in the cost of a $48,000 piece of machinery. If it can find two more group members, each member's share will decrease by $4000. How many are presently in the group?

63. *Business Venture* You are planning to start a small business that will require an investment of $90,000. You have found some people who are willing to share equally in the venture. If you can find three more people, each person's share will decrease by $2500. How many people have you found so far?

64. *Average Speed* You commute 56 miles one way to work. The trip to work takes 10 minutes longer than the trip home. Your average speed on the trip home is 8 miles per hour faster. What is your average speed on the trip home?

65. *Mixture Problem* A car radiator contains 10 liters of a 30% antifreeze solution. How many liters will have to be replaced with pure antifreeze if the resulting solution is to be 50% antifreeze?

66. *Investment* You invested $6000 at $4\frac{1}{2}\%$ and $5\frac{1}{2}\%$ simple interest. During one year, the two accounts earned $305. How much did you invest in each?

In Exercises 67 and 68, solve for the indicated variable.

67. *Volume of a Cone*

Solve for r: $V = \frac{1}{3}\pi r^2 h$

68. *Kinetic Energy*

Solve for m: $E = \frac{1}{2}mv^2$

69. Two cars start at a given time and travel in the same direction at average speeds of 40 miles per hour and 55 miles per hour. How much time will elapse before the two cars are 10 miles apart?

70. The volume of a circular cylinder is 81π cubic feet. The cylinder's radius is 3 feet. What is the height of the cylinder?

1.4 In Exercises 71–80, use any method to solve the equation.

71. $15 + x - 2x^2 = 0$

72. $2x^2 - x - 28 = 0$

73. $6 = 3x^2$

74. $16x^2 = 25$

75. $(x + 4)^2 = 18$

76. $(x - 8)^2 = 15$

77. $x^2 - 12x + 30 = 0$

78. $x^2 + 6x - 3 = 0$

79. $-2x^2 - 5x + 27 = 0$

80. $-20 - 3x + 3x^2 = 0$

81. *Simply Supported Beam* A simply supported 20-foot beam supports a uniformly distributed load of 1000 pounds per foot. The bending moment M (in foot-pounds) x feet from one end of the beam is given by $M = 500x(20 - x)$.

(a) Where is the bending moment zero?

(b) Use a graphing utility to graph the equation.

(c) Use the graph to determine the point on the beam where the bending moment is the greatest.

82. *Physics* A ball is thrown upward with an initial velocity of 30 feet per second from a point that is 24 feet above the ground (see figure). The height h (in feet) of the ball at time t (in seconds) after it is thrown is

$$h = -16t^2 + 30t + 24.$$

Find the time when the ball hits the ground.

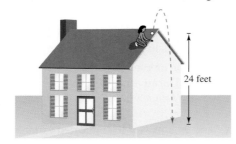

24 feet

1.5 In Exercises 83–86, write the complex number in standard form.

83. $6 + \sqrt{-4}$

84. $3 - \sqrt{-25}$

85. $i^2 + 3i$

86. $-5i + i^2$

In Exercises 87–96, perform the operations and write the result in standard form.

87. $(7 + 5i) + (-4 + 2i)$

88. $\left(\dfrac{\sqrt{2}}{2} - \dfrac{\sqrt{2}}{2}i\right) - \left(\dfrac{\sqrt{2}}{2} + \dfrac{\sqrt{2}}{2}i\right)$

89. $5i(13 - 8i)$

90. $(1 + 6i)(5 - 2i)$

91. $(10 - 8i)(2 - 3i)$

92. $i(6 + i)(3 - 2i)$

93. $\dfrac{6 + i}{4 - i}$

94. $\dfrac{3 + 2i}{5 + i}$

95. $\dfrac{4}{2 - 3i} + \dfrac{2}{1 + i}$

96. $\dfrac{1}{2 + i} - \dfrac{5}{1 + 4i}$

In Exercises 97–100, find all solutions of the equation.

97. $3x^2 + 1 = 0$

98. $2 + 8x^2 = 0$

99. $x^2 - 2x + 10 = 0$

100. $6x^2 + 3x + 27 = 0$

1.6 In Exercises 101–118, find all solutions of the equation. Check your solutions in the original equation.

101. $5x^4 - 12x^3 = 0$

102. $4x^3 - 6x^2 = 0$

103. $x^4 - 5x^2 + 6 = 0$

104. $9x^4 + 27x^3 - 4x^2 - 12x = 0$

105. $\sqrt{x + 4} = 3$

106. $\sqrt{x - 2} - 8 = 0$

107. $2\sqrt{x} - 5 = x$

108. $\sqrt{3x - 2} = 4 - x$

109. $\sqrt{2x + 3} + \sqrt{x - 2} = 2$

110. $5\sqrt{x} - \sqrt{x - 1} = 6$

111. $(x - 1)^{2/3} - 25 = 0$

112. $(x + 2)^{3/4} = 27$

113. $(x + 4)^{1/2} + 5x(x + 4)^{3/2} = 0$

114. $8x^2(x^2 - 4)^{1/3} + (x^2 - 4)^{4/3} = 0$

115. $|x - 5| = 10$

116. $|2x + 3| = 7$

117. $|x^2 - 3| = 2x$

118. $|x^2 - 6| = x$

119. *Economics* The demand equation for a product is

$$p = 42 - \sqrt{0.001x + 2}$$

where x is the number of units demanded per day and p is the price per unit. Find the demand if the price is set at $29.95.

120. *Data Analysis* The total sales S (in billions of dollars) of sporting goods in the United States from 1990 to 1997 can be approximated by the model

$$S = 48.66 - 4.11t + 1.96t^2 - 0.15t^3$$

where $t = 0$ represents 1990. The actual sales are given in the table. (Source: National Sporting Goods Association)

Year	1990	1991	1992	1993
Sales	48.3	47.1	47.1	49.1

Year	1994	1995	1996	1997
Sales	53.5	58.4	61.3	62.9

(a) Use a graphing utility to compare the data with the model.

(b) Use the model to estimate total sales in 2002.

1.7 In Exercises 121–124, write an inequality to represent the interval and state whether the interval is bounded or unbounded.

121. $(-7, 2]$ **122.** $(4, \infty)$

123. $(-\infty, -10]$ **124.** $[-2, 2]$

In Exercises 125–134, solve the inequality.

125. $9x - 8 \le 7x + 16$

126. $\frac{15}{2}x + 4 > 3x - 5$

127. $4(5 - 2x) \le \frac{1}{2}(8 - x)$

128. $\frac{1}{2}(3 - x) > \frac{1}{3}(2 - 3x)$

129. $-19 < 3x - 17 \le 34$

130. $-3 \le \dfrac{2x - 5}{3} < 5$

131. $|x| \le 4$ **132.** $|x - 2| < 1$

133. $|x - 3| > 4$ **134.** $\left|x - \frac{3}{2}\right| \ge \frac{3}{2}$

135. *Business* The revenue for selling x units of a product is $R = 125.33x$. The cost of producing x units is $C = 92x + 1200$. To obtain a profit, the revenue must be greater than the cost. Determine the smallest value of x for which this product returns a profit.

136. *Geometry* The side of a square is measured as 19.3 centimeters with a possible error of 0.5 centimeter. Using these measurements, determine the interval containing the area of the square.

1.8 In Exercises 137–144, solve the inequality.

137. $x^2 - 6x - 27 < 0$ **138.** $x^2 - 2x \ge 3$

139. $6x^2 + 5x < 4$ **140.** $2x^2 + x \ge 15$

141. $\dfrac{2}{x + 1} \le \dfrac{3}{x - 1}$ **142.** $\dfrac{x - 5}{3 - x} < 0$

143. $\dfrac{x^2 + 7x + 12}{x} \ge 0$ **144.** $\dfrac{1}{x - 2} > \dfrac{1}{x}$

145. *Investment* P dollars invested at interest rate r compounded annually increases to an amount

$$A = P(1 + r)^2$$

in 2 years. If an investment of $5000 is to increase to an amount greater than $5500 in 2 years, then the interest rate must be greater than what percent?

146. *Biology* A biologist introduces 200 ladybugs into a crop field. The population P of the ladybugs is approximated by the model

$$P = \frac{1000(1 + 3t)}{5 + t}$$

where t is the time in days. Find the time required for the population to increase to at least 2000 ladybugs.

Synthesis

True or False? In Exercises 147 and 148, determine whether the statement is true or false. Justify your answer.

147. $\sqrt{-18}\sqrt{-2} = \sqrt{(-18)(-2)}$

148. The equation $325x^2 - 717x + 398 = 0$ has no solution.

149. Explain why it is important to check your solutions to certain types of equations.

150. Create a solution that solves a real-life problem using a linear model.

151. Write quadratic equations that have (a) two distinct real solutions, (b) two complex solutions, and (c) no real solution.

152. *Error Analysis* What is wrong with the following solution?

$$|11x - 4| \ge -26$$

$11x - 4 \le -26$ or $11x - 4 \ge -26$

$11x \le -22$ $11x \ge -22$

$x \le -2$ $x \ge -2$

Chapter Project ▶ Finding Falling Times Numerically

A computer simulation to accompany this project appears in the *Interactive* CD-ROM and *Internet* versions of this text.

For many problems, you can gain insight by using a *numerical* approach instead of, or in addition to, using an *algebraic* approach or a *graphical* approach.

Example ▶ Finding the Falling Time for an Object

At 9:55 A.M. on Saturday, July 28, 1945, a terrible airplane accident occurred. A B-25 bomber crashed into the 78th and 79th floors of the Empire State Building in New York City. Hundreds of pieces of debris fell 975 feet to the streets below. How much time after hearing the crash did the people on the street have to get out of the way?

Solution

Assume that the debris *dropped* from a height of 975 feet. Using an initial velocity of $v_0 = 0$ and an initial height of $s_0 = 975$, you can model the height (in feet) of the falling debris as $s = -16t^2 + 975$, where t is the falling time in seconds. When you solve this equation for the time t that corresponds to a height of $s = 0$, you obtain $t \approx 7.8$ seconds.

This conclusion is obtained numerically using the table shown at the left. From the table, you can see that the debris took about 8 seconds to hit the ground. Because it would have taken about 1 second for the sound of the crash to reach the ground, you can conclude that the people had about 7 seconds to get out of the way.

Time, t	Height, s
0	975
1	959
2	911
3	831
4	719
5	575
6	399
7	191
7.8	1.56

Chapter Project Investigations

1. Use the model and table in the example.
   (a) Find the average velocity of the debris during its first second of fall.
   (b) Was the debris falling faster during its next second of fall? Explain.
   (c) Find the height of the debris after 7.7 seconds.
   (d) Using part (c), approximate the terminal velocity of the debris.

2. In the example, because the accident was accompanied by an explosion, some of the debris was most likely propelled downward. Assuming that the debris was propelled straight downward with an initial velocity of 100 feet per second, a model for the height (in feet) of the falling debris is $s = -16t^2 - 100t + 975$. Use this model to create a table to find how much time the people on the street had to get out of the way.

3. In Question 2, how long would the people on the street have had to get out of the way if some of the debris had been propelled downward with an initial velocity of 200 feet per second?

▶ Chapter Test

The *Interactive* CD-ROM and *Internet* versions of this text provide answers to the Chapter Tests and Cumulative Tests. They also offer Chapter Pre-Tests (which test key skills and concepts covered in previous chapters) and Chapter Post-Tests, both of which have randomly generated exercises with diagnostic capabilities.

Take this test as you would take a test in class. After you are done, check your work against the answers given in the back of the book.

In Exercises 1–6, use intercepts and symmetry to sketch the complete graph of the equation.

1. $y = 4 - \frac{3}{4}x$

2. $y = 4 - \frac{3}{4}|x|$

3. $y = 4 - (x - 2)^2$

4. $y = x - x^3$

5. $y = \sqrt{3 - x}$

6. $(x - 3)^2 + y^2 = 9$

In Exercises 7–12, solve the equation (if possible).

7. $\frac{2}{3}(x - 1) + \frac{1}{4}x = 10$

8. $(x - 3)(x + 2) = 14$

9. $\frac{x - 2}{x + 2} + \frac{4}{x + 2} + 4 = 0$

10. $x^4 + x^2 - 6 = 0$

11. $2\sqrt{x} - \sqrt{2x + 1} = 1$

12. $|3x - 1| = 7$

In Exercises 13–16, solve the inequality. Sketch the solution on the real number line.

13. $-3 \le 2(x + 4) < 14$

14. $\frac{2}{x} > \frac{5}{x + 6}$

15. $2x^2 + 5x > 12$

16. $|x - 15| \ge 5$

17. Perform the operations and write the result in standard form.

(a) $10i - (3 + \sqrt{-25})$

(b) $(2 + \sqrt{3}i)(2 - \sqrt{3}i)$

(c) $\frac{5}{2 + i}$

18. The dividends declared per share of Procter & Gamble Company from 1992 to 1998 can be approximated by the model

$$y = 0.281 + 0.091t, \quad 2 \le t \le 8$$

where y is the dividend and t is the time (in years), with $t = 2$ corresponding to 1992. Sketch a graph of this equation. (Source: Procter & Gamble Company)

19. On the first part of a 350-kilometer trip, a salesperson travels 2 hours and 15 minutes at an average speed of 100 kilometers per hour. Find the average speed required for the remainder of the trip if the salesperson needs to arrive at the destination in another hour and 20 minutes.

20. The area of the ellipse in the figure at the left below is $A = \pi ab$. If a and b satisfy the constraint $a + b = 100$, find a and b such that the area of the ellipse equals the area of the circle.

21. In order for an investment of $2000 to grow to more than $2200 in 2 years, what must the annual interest rate be? Use the formula $A = P(1 + rt)$.

Mark Joseph/Tony Stone Images

The average cost of a new domestic car increased from $18,064 in 1996 to $18,580 in 1997 even though sales (demand) and production (supply) of new domestic cars declined. (Source: U.S. Bureau of Economic Analysis)

2 Functions and Their Graphs

▶ How to Study This Chapter

The Big Picture

In this chapter you will learn the following skills and concepts.

▶ How to find and use the slopes of lines to write and graph linear equations in two variables

▶ How to evaluate functions and find their domains

▶ How to analyze graphs of functions, including linear, step, and piecewise-defined functions

▶ How to identify and graph shifts, reflections, and nonrigid transformations of functions

▶ How to find arithmetic combinations and compositions of functions

▶ How to find inverses of functions graphically and algebraically

Important Vocabulary

As you encounter each new vocabulary term in this chapter, add the term and its definition to your notebook glossary.

Linear equation in two variables (p. 172)
Slope (p. 172)
Slope-intercept form (p. 172)
Ratio (p. 174)
Rate of change (p. 174)
Point-slope form (p. 177)
Two-point form (p. 177)
General form (p. 178)
Parallel (p. 179)
Perpendicular (p. 179)
Function (p. 187)
Domain (p. 187)
Range (p. 187)
Independent variable (p. 188)
Dependent variable (p. 188)
Function notation (p. 189)
Piecewise-defined function (p. 190)
Implied domain (p. 191)
Graph of a function (p. 201)
Vertical Line Test (p. 202)

Zeros of a function (p. 203)
Increasing function (p. 204)
Decreasing function (p. 204)
Constant function (p. 204)
Relative minimum (p. 205)
Relative maximum (p. 205)
Linear function (p. 206)
Greatest integer function (p. 207)
Step function (p. 207)
Even function (p. 208)
Odd function (p. 208)
Vertical and horizontal shifts (p. 215)
Reflection (p. 217)
Vertical stretch (p. 219)
Vertical shrink (p. 219)
Arithmetic combination of functions (p. 225)
Composition of functions (p. 227)
Inverse (p. 233)
Horizontal Line Test (p. 236)

Study Tools

- Learning objectives at the beginning of each section
- Chapter Summary (p. 243)
- Review Exercises (pp. 244–247)
- Chapter Test (p. 249)
- Cumulative Test for Chapters P–2 (pp. 250–251)

Additional Resources

- Study and Solutions Guide
- Interactive College Algebra
- Videotapes for Chapter 2
- College Algebra Website
- Student Success Organizer

STUDY T!P

During class, take notes on definitions, examples, concepts, and rules—whatever it is you identify as important to the instructor. Then, as soon after class as possible, review your notes.

2.1 Linear Equations in Two Variables

▶ **What you should learn**

- How to use slope to graph linear equations in two variables
- How to find slopes of lines
- How to write linear equations in two variables
- How to use slope to identify parallel and perpendicular lines
- How to use linear equations in two variables to model and solve real-life problems

▶ **Why you should learn it**

Linear equations in two variables can be used to model and solve real-life problems. For instance, Exercise 112 on page 185 shows how to use a linear equation to model the average annual salaries of major league baseball players from 1988 to 1998.

Walter Schmid/Tony Stone Images

Using Slope

The simplest mathematical model for relating two variables is the **linear equation in two variables** $y = mx + b$. The equation is called *linear* because its graph is a line. (In mathematics, the term *line* means *straight line*.) By letting $x = 0$, you can see that the line crosses the y-axis at $y = b$, as shown in Figure 2.1. In other words, the y-intercept is $(0, b)$. The steepness or slope of the line is m.

$$y = mx + b$$

Slope ⟶   ⟵ y-Intercept

The **slope** of a nonvertical line is the number of units the line rises (or falls) vertically for each unit of horizontal change from left to right, as shown in Figure 2.1.

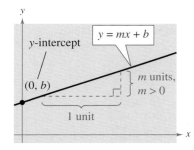
Positive slope, line rises.

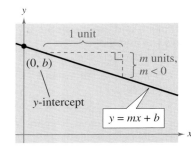
Negative slope, line falls.

FIGURE 2.1

A linear equation that is written in the form $y = mx + b$ is said to be written in **slope-intercept form.**

The Slope-Intercept Form of the Equation of a Line

The graph of the equation

$$y = mx + b$$

is a line whose slope is m and whose y-intercept is $(0, b)$.

◀ Exploration ▶

Use a graphing utility to compare the slopes of the lines $y = mx$ where $m = 0.5, 1, 2,$ and 4. Which line rises most quickly? Now, let $m = -0.5,$ $-1, -2,$ and -4. Which line falls most quickly? Use a square setting to obtain a true geometric perspective. What can you conclude about the slope and the "rate" at which the line rises or falls?

 A computer simulation of this concept appears in the *Interactive* CD-ROM and *Internet* versions of this text.

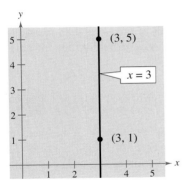

FIGURE 2.2 *Slope is undefined.*

Once you have determined the slope and the *y*-intercept of a line, it is a relatively simple matter to sketch its graph. In the following example, note that none of the lines is vertical. A vertical line has an equation of the form

$$x = a. \qquad \text{Vertical line}$$

The equation of a vertical line cannot be written in the form $y = mx + b$ because the slope of a vertical line is undefined, as indicated in Figure 2.2.

Example 1 ▶ Graphing a Linear Equation

Sketch the graph of each linear equation.

a. $y = 2x + 1$

b. $y = 2$

c. $x + y = 2$

Solution

a. Because $b = 1$, the *y*-intercept is $(0, 1)$. Moreover, because the slope is $m = 2$, the line *rises* 2 units for each unit the line moves to the right, as shown in Figure 2.3(a).

b. By writing this equation in the form $y = (0)x + 2$, you can see that the *y*-intercept is $(0, 2)$ and the slope is zero. A zero slope implies that the line is horizontal—that is, it doesn't rise *or* fall, as shown in Figure 2.3(b).

c. By writing this equation in slope-intercept form

$$\begin{aligned} x + y &= 2 && \text{Write original equation.} \\ y &= -x + 2 && \text{Subtract } x \text{ from each side.} \\ y &= (-1)x + 2 && \text{Write in slope-intercept form.} \end{aligned}$$

you can see that the *y*-intercept is $(0, 2)$. Moreover, because the slope is $m = -1$, the line *falls* 1 unit for each unit the line moves to the right, as shown in Figure 2.3(c).

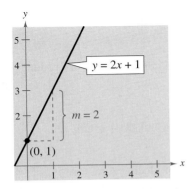

(a) When *m* is positive, the line rises.

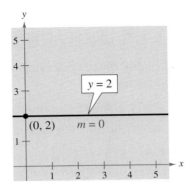

(b) When *m* is 0, the line is horizontal.

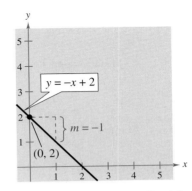

(c) When *m* is negative, the line falls.

FIGURE 2.3

In real-life problems, the slope of a line can be interpreted as either a *ratio* or a *rate*. If the *x*-axis and *y*-axis have the same unit of measure, then the slope has no units and is a **ratio**. If the *x*-axis and *y*-axis have different units of measure, then the slope is a **rate** or **rate of change.**

Example 2 ▶ Using Slope as a Ratio

The maximum recommended slope of a wheelchair ramp is $\frac{1}{12}$. A business is installing a wheelchair ramp that rises 22 inches over a horizontal length of 24 feet. Is the ramp steeper than recommended? (Source: Americans with Disabilities Act Handbook)

Solution

The horizontal length of the ramp is 24 feet or $12(24) = 288$ inches, as shown in Figure 2.4. So, the slope of the ramp is

$$\text{Slope} = \frac{\text{vertical change}}{\text{horizontal change}}$$

$$= \frac{22 \text{ in.}}{288 \text{ in.}}$$

$$\approx 0.076.$$

Because $\frac{1}{12} \approx 0.083$, the slope of the ramp is not steeper than recommended.

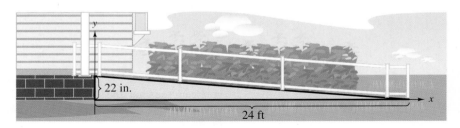

FIGURE 2.4

Example 3 ▶ Using Slope as a Rate of Change

A manufacturing company determines that the total cost in dollars of producing *x* units of a product is

$$C = 25x + 3500. \qquad \text{Cost equation}$$

Describe the practical significance of the *y*-intercept and slope of this line.

Solution

The *y*-intercept $(0, 3500)$ tells you that the cost of producing zero units is $3500. This is the *fixed cost* of production—it includes costs that must be paid regardless of the number of units produced. The slope of $m = 25$ tells you that the cost of producing each unit is $25, as shown in Figure 2.5. Economists call the cost per unit the *marginal cost*. If the production increases by 1 unit, then the "margin," or extra amount of cost, is $25. So, the cost increases at a rate of $25 per unit.

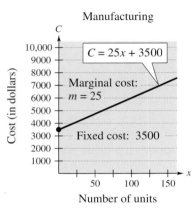

FIGURE 2.5 *Production Cost*

Finding the Slope of a Line

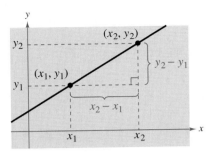

FIGURE **2.6**

Given an equation of a line, you can find its slope by writing the equation in slope-intercept form. If you are not given an equation, you can still find the slope of a line. For instance, suppose you want to find the slope of the line passing through the points (x_1, y_1) and (x_2, y_2), as shown in Figure 2.6. As you move from left to right along this line, a change of $(y_2 - y_1)$ units in the vertical direction corresponds to a change of $(x_2 - x_1)$ units in the horizontal direction.

$$y_2 - y_1 = \text{the change in } y = \text{rise}$$

and

$$x_2 - x_1 = \text{the change in } x = \text{run}$$

The ratio of $(y_2 - y_1)$ to $(x_2 - x_1)$ represents the slope of the line that passes through the points (x_1, y_1) and (x_2, y_2).

$$\text{Slope} = \frac{\text{change in } y}{\text{change in } x}$$

$$= \frac{\text{rise}}{\text{run}}$$

$$= \frac{y_2 - y_1}{x_2 - x_1}$$

The Slope of a Line Passing Through Two Points

The **slope** m of the nonvertical line through (x_1, y_1) and (x_2, y_2) is

$$m = \frac{y_2 - y_1}{x_2 - x_1}$$

where $x_1 \neq x_2$.

Historical Note

René Descartes (1596–1650) was largely responsible for the development of analytic geometry. His use of the symbol m for slope came from the French verb *monter*, meaning to mount, to climb, or to rise.

When this formula is used for slope, the *order of subtraction* is important. Given two points on a line, you are free to label either one of them as (x_1, y_1) and the other as (x_2, y_2). However, once you have done this, you must form the numerator and denominator using the same order of subtraction.

$$m = \frac{y_2 - y_1}{x_2 - x_1}$$

Correct

$$m = \frac{y_1 - y_2}{x_1 - x_2}$$

Correct

$$m = \frac{y_2 - y_1}{x_1 - x_2}$$

Incorrect

For instance, the slope of the line passing through the points $(3, 4)$ and $(5, 7)$ can be calculated as

$$m = \frac{7 - 4}{5 - 3} = \frac{3}{2}$$

or

$$m = \frac{4 - 7}{3 - 5} = \frac{-3}{-2} = \frac{3}{2}.$$

The *Interactive* CD-ROM and *Internet* versions of this text show every example with its solution; clicking on the *Try It!* button brings up similar problems. Guided Examples and Integrated Examples show step-by-step solutions to additional examples. Integrated Examples are related to several concepts in the section.

Example 4　▶　Finding the Slope of a Line Through Two Points

Find the slope of the line passing through each pair of points. (See Figure 2.7.)

a. $(-2, 0)$ and $(3, 1)$　　**b.** $(-1, 2)$ and $(2, 2)$
c. $(0, 4)$ and $(1, -1)$　　**d.** $(3, 4)$ and $(3, 1)$

Solution

a. Letting $(x_1, y_1) = (-2, 0)$ and $(x_2, y_2) = (3, 1)$, you obtain a slope of

$$m = \frac{y_2 - y_1}{x_2 - x_1} = \frac{1 - 0}{3 - (-2)} = \frac{1}{5}.$$

b. The slope of the line passing through $(-1, 2)$ and $(2, 2)$ is

$$m = \frac{2 - 2}{2 - (-1)} = \frac{0}{3} = 0.$$

c. The slope of the line passing through $(0, 4)$ and $(1, -1)$ is

$$m = \frac{-1 - 4}{1 - 0} = \frac{-5}{1} = -5.$$

d. The slope of the vertical line passing through $(3, 4)$ and $(3, 1)$ is

$$m = \frac{1 - 4}{3 - 3} = \frac{-3}{0}.$$

Because division by 0 is undefined, the slope is undefined.

STUDY T!P

In Figure 2.7, note the relationships between slope and the description of the line.

a. Positive slope; line rises from left to right

b. Zero slope; line is horizontal

c. Negative slope; line falls from left to right

d. Undefined slope; line is vertical

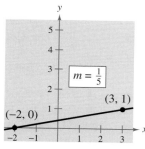

(a)

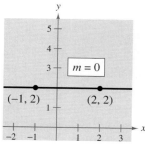

(b)

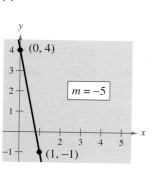

(c)

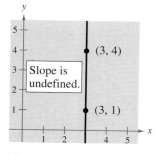

(d)

FIGURE 2.7

Writing Linear Equations in Two Variables

If (x_1, y_1) is a point on a line of slope m and (x, y) is *any other* point on the line, then

$$\frac{y - y_1}{x - x_1} = m.$$

This equation, involving the variables x and y, can be rewritten in the form

$$y - y_1 = m(x - x_1),$$

which is the **point-slope form** of the equation of a line.

Point-Slope Form of the Equation of a Line

The equation of the line with slope m passing through the point (x_1, y_1) is

$$y - y_1 = m(x - x_1).$$

The point-slope form is most useful for *finding* the equation of a line. You should remember this formula.

Example 5 ▶ Using the Point-Slope Form

Find the slope-intercept form of the equation of the line that has a slope of 3 and passes through the point $(1, -2)$, as shown in Figure 2.8.

Solution

Use the point-slope form with $m = 3$ and $(x_1, y_1) = (1, -2)$.

$y - y_1 = m(x - x_1)$	Point-slope form
$y - (-2) = 3(x - 1)$	Substitute for m, x_1, and y_1.
$y + 2 = 3x - 3$	Simplify.
$y = 3x - 5$	Write in slope-intercept form.

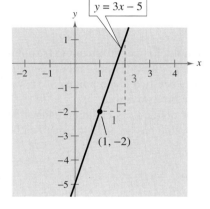

FIGURE **2.8**

The point-slope form can be used to find an equation of the line passing through points (x_1, y_1) and (x_2, y_2). To do this, first find the slope of the line

$$m = \frac{y_2 - y_1}{x_2 - x_1}, \qquad x_1 \neq x_2$$

and then use the point-slope form to obtain the equation

$$y - y_1 = \frac{y_2 - y_1}{x_2 - x_1}(x - x_1). \qquad \text{Two-point form}$$

This is sometimes called the **two-point form** of the equation of a line.

Yahoo! Inc.

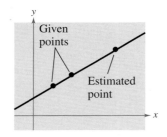

Cash flow per share (in dollars)

$y = 0.27t + 0.03$

$(2, 0.57)$

$(1, 0.30)$

$(0, 0.03)$

Year $(0 \leftrightarrow 1997)$

FIGURE **2.9**

Example 6 ▶ Predicting Cash Flow Per Share

The cash flow per share for Yahoo! Inc. was $0.03 in 1997 and $0.30 in 1998. Using only this information, write a linear equation that gives the cash flow per share in terms of the year. Then predict the cash flow for 1999.   (Source: Yahoo! Inc.)

Solution

Let $t = 0$ represent 1997. Then the two given values are represented by the data points $(0, 0.03)$ and $(1, 0.30)$. The slope of the line through these points is

$$m = \frac{0.30 - 0.03}{1 - 0}$$

$$= 0.27.$$

Using the point-slope form, you can find the equation that relates the cash flow y and the year t to be

$$y = 0.27t + 0.03.$$

According to this equation, the cash flow in 1999 was $0.57, as shown in Figure 2.9. (In this case, the prediction is quite good—the actual cash flow in 1999 was $0.55.)

The prediction method illustrated in Example 6 is called **linear extrapolation.** Note in Figure 2.10(a) that an extrapolated point does not lie between the given points. When the estimated point lies between two given points, as shown in Figure 2.10(b), the procedure is called **linear interpolation.**

Because the slope of a vertical line is not defined, its equation cannot be written in slope-intercept form. However, every line has an equation that can be written in the **general form**

$$Ax + By + C = 0 \qquad \text{General form}$$

where A and B are not both zero. For instance, the vertical line given by $x = a$ can be represented by the general form

$$x - a = 0.$$

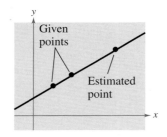

(a) Linear extrapolation

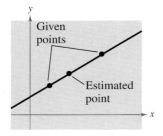

(b) Linear interpolation

FIGURE **2.10**

Equations of Lines

1. General form: $Ax + By + C = 0$

2. Vertical line: $x = a$

3. Horizontal line: $y = b$

4. Slope-intercept form: $y = mx + b$

5. Point-slope form: $y - y_1 = m(x - x_1)$

6. Two-point form: $y - y_1 = \dfrac{y_2 - y_1}{x_2 - x_1}(x - x_1)$

Parallel and Perpendicular Lines

Slope can be used to decide whether two nonvertical lines in a plane are parallel, perpendicular, or neither.

Parallel and Perpendicular Lines

1. Two distinct nonvertical lines are **parallel** if and only if their slopes are equal. That is, $m_1 = m_2$.

2. Two nonvertical lines are **perpendicular** if and only if their slopes are negative reciprocals of each other. That is, $m_1 = -1/m_2$.

Example 7 ▶ Finding Parallel and Perpendicular Lines

Find the slope-intercept form of the equation of the line that passes through the point $(2, -1)$ and is (a) parallel to and (b) perpendicular to the line $2x - 3y = 5$.

Solution

By writing the equation of the given line in slope-intercept form

$2x - 3y = 5$	Write original equation.
$-3y = -2x + 5$	Subtract $2x$ from each side.
$y = \frac{2}{3}x - \frac{5}{3}$	Write in slope-intercept form.

you can see that it has a slope of $m = \frac{2}{3}$, as shown in Figure 2.11.

a. Any line parallel to the given line must also have a slope of $\frac{2}{3}$. So, the line through $(2, -1)$ that is parallel to the given line has the following equation.

$y - (-1) = \frac{2}{3}(x - 2)$	Write in point-slope form.
$3(y + 1) = 2(x - 2)$	Multiply each side by 3.
$3y + 3 = 2x - 4$	Distributive Property
$2x - 3y - 7 = 0$	Write in general form.
$y = \frac{2}{3}x - \frac{7}{3}$	Write in slope-intercept form.

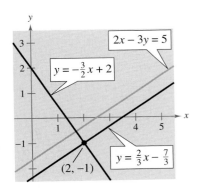

FIGURE 2.11

b. Any line perpendicular to the given line must have a slope of $-1/(2/3)$ or $-3/2$. So, the line through $(2, -1)$ that is perpendicular to the given line has the following equation.

$y - (-1) = -\frac{3}{2}(x - 2)$	Write in point-slope form.
$2(y + 1) = -3(x - 2)$	Multiply each side by 2.
$2y + 2 = -3x + 6$	Distributive Property
$3x + 2y - 4 = 0$	Write in general form.
$y = -\frac{3}{2}x + 2$	Write in slope-intercept form.

Application

Most business expenses can be deducted in the same year they occur. One exception is the cost of property that has a useful life of more than 1 year. Such costs must be *depreciated* over the useful life of the property. If the *same amount* is depreciated each year, the procedure is called *linear* or *straight-line depreciation.* The *book value* is the difference between the original value and the total amount of depreciation accumulated to date.

Example 8 ▶ Straight-Line Depreciation

Your company has purchased a \$12,000 machine that has a useful life of 8 years. The salvage value at the end of 8 years is \$2000. Write a linear equation that describes the book value of the machine each year.

Solution

Let V represent the value of the machine at the end of year t. You can represent the initial value of the machine by the data point (0, 12,000) and the salvage value of the machine by the data point (8, 2000). The slope of the line is

$$m = \frac{2000 - 12{,}000}{8 - 0} = -\$1250$$

which represents the annual depreciation in *dollars per year.* Using the point-slope form, you can write the equation of the line as follows.

$$V - 12{,}000 = -1250(t - 0) \qquad \text{Write in point-slope form.}$$

$$V = -1250t + 12{,}000 \qquad \text{Write in slope-intercept form.}$$

The table shows the book value at the end of each year, and the graph of the equation is shown in Figure 2.12.

t	0	1	2	3	4	5	6	7	8
V	12,000	10,750	9500	8250	7000	5750	4500	3250	2000

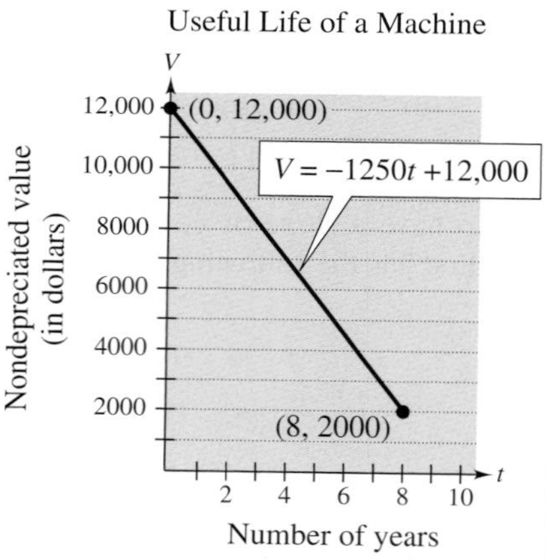

FIGURE 2.12 *Straight-Line Depreciation*

Writing ABOUT MATHEMATICS

x	y
0	80.90
1	72.74
2	68.68
3	61.48

x	y
4	56.21
5	51.00
6	47.70
7	42.78

Modeling Linear Data The table at the left shows the average monthly cellular phone bill y (in dollars) for subscribers in the United States from 1990 through 1997, where x is the time in years, with $x = 0$ corresponding to 1990. (Source: Cellular Telecommunications Industry Association)

Sketch a scatter plot of the data, and use a straightedge to sketch the best-fitting line through the points. Find the equation of the line. Explain the procedure you used. Write a short paragraph explaining the meaning of the slope and the y-intercept of the line in terms of the data. Compare the values obtained using your model with the actual values in the table. Use your model to estimate the average monthly bill in the year 2000.

2.1 Exercises

In Exercises 1 and 2, identify the line that has each slope.

1. (a) $m = \frac{2}{3}$

   (b) m is undefined.

   (c) $m = -2$

2. (a) $m = 0$

   (b) $m = -\frac{3}{4}$

   (c) $m = 1$

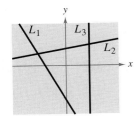
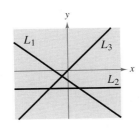

In Exercises 3 and 4, sketch the graph of the line through the point with each indicated slope on the same set of coordinate axes.

Point	Slopes
3. $(2, 3)$	(a) 0 (b) 1 (c) 2 (d) -3
4. $(-4, 1)$	(a) 3 (b) -3 (c) $\frac{1}{2}$ (d) Undefined

In Exercises 5–10, estimate the slope of the line.

5.

6.
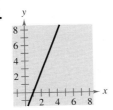

7.
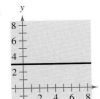

8.
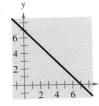

9.
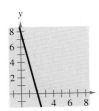

10.
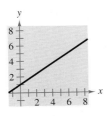

In Exercises 11–18, plot the points and find the slope of the line passing through the pair of points.

11. $(-3, -2), (1, 6)$ **12.** $(2, 4), (4, -4)$

13. $(-6, -1), (-6, 4)$ **14.** $(0, -10), (-4, 0)$

15. $\left(\frac{11}{2}, -\frac{4}{3}\right), \left(-\frac{3}{2}, -\frac{1}{3}\right)$ **16.** $\left(\frac{7}{8}, \frac{3}{4}\right), \left(\frac{5}{4}, -\frac{1}{4}\right)$

17. $(4.8, 3.1), (-5.2, 1.6)$

18. $(-1.75, -8.3), (2.25, -2.6)$

In Exercises 19–28, use the point on the line and the slope of the line to find three additional points through which the line passes. (There are many correct answers.)

Point	Slope
19. $(2, 1)$	$m = 0$
20. $(-4, 1)$	m is undefined.
21. $(5, -6)$	$m = 1$
22. $(10, -6)$	$m = -1$
23. $(-8, 1)$	m is undefined.
24. $(-3, -1)$	$m = 0$
25. $(-5, 4)$	$m = 2$
26. $(0, -9)$	$m = -2$
27. $(7, -2)$	$m = \frac{1}{2}$
28. $(-1, -6)$	$m = -\frac{1}{2}$

In Exercises 29–32, determine whether the lines L_1 and L_2 passing through the pairs of points are parallel, perpendicular, or neither.

29. L_1: $(0, -1), (5, 9)$

    L_2: $(0, 3), (4, 1)$

30. L_1: $(-2, -1), (1, 5)$

    L_2: $(1, 3), (5, -5)$

31. L_1: $(3, 6), (-6, 0)$

    L_2: $(0, -1), \left(5, \frac{7}{3}\right)$

32. L_1: $(4, 8), (-4, 2)$

    L_2: $(3, -5), \left(-1, \frac{1}{3}\right)$

33. *Business* The following are the slopes of lines representing annual sales y in terms of time x in years. Use the slopes to interpret any change in annual sales for a 1-year increase in time.

   (a) The line has a slope of $m = 135$.

   (b) The line has a slope of $m = 0$.

   (c) The line has a slope of $m = -40$.

34. *Business* The following are the slopes of lines representing daily revenues y in terms of time x in days. Use the slopes to interpret any change in daily revenues for a 1-day increase in time.

(a) The line has a slope of $m = 400$.

(b) The line has a slope of $m = 100$.

(c) The line has a slope of $m = 0$.

35. *Business* The graph shows the earnings per share of stock for the Kellogg Company for the years 1988 through 1998. (Source: Kellogg Company)

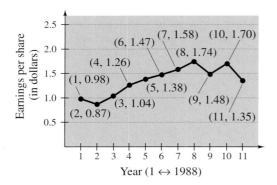

Year (1 ↔ 1988)

(a) Use the slopes to determine the years when the earnings per share showed the greatest increase and decrease.

(b) Find the slope of the line segment connecting the years 1988 and 1998.

(c) Interpret the meaning of the slope in part (b) in the context of the problem.

36. *Business* The graph shows the dividends declared per share of stock for the Colgate-Palmolive Company for the years 1988 through 1998. (Source: Colgate-Palmolive Company)

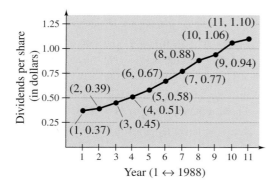

Year (1 ↔ 1988)

(a) Use the slopes to determine the years when the dividends declared per share showed the greatest increase and the smallest increase.

(b) Find the slope of the line segment connecting the years 1988 and 1998.

(c) Interpret the meaning of the slope in part (b) in the context of the problem.

37. *Road Grade* From the top of a mountain road, a surveyor takes several horizontal measurements x and several vertical measurements y, as shown in the table.

x	300	600	900	1200	1500	1800	2100
y	-25	-50	-75	-100	-125	-150	-175

(a) Sketch a scatter plot of the data.

(b) Use a straightedge to sketch the best-fitting line through the points.

(c) Find an equation for the line you sketched in part (b).

(d) Interpret the meaning of the slope of the line in part (c) in the context of the problem.

(e) The surveyor needs to put up a road sign that indicates the steepness of the road. For instance, a surveyor would put a sign that states "8% grade" on a road with a downhill grade that has a slope of $-\frac{8}{100}$. What should the sign state for the road in this problem?

38. *Road Grade* You are driving on a road that has a 6% uphill grade. This means that the slope of the road is $\frac{6}{100}$. Approximate the amount of vertical change in your position if you drive 200 feet.

39. *Height of an Attic* The "rise to run" in determining the steepness of the roof on a house is 3 to 4 (see figure). Determine the maximum height in the attic if the house is 32 feet wide.

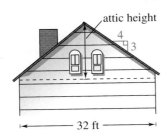

In Exercises 40–45, find the slope and *y*-intercept (if possible) of the equation of the line. Sketch a graph of the line.

40. $y = 5x + 3$

41. $y = x - 10$

42. $5x - 2 = 0$

43. $3y + 5 = 0$

44. $7x + 6y = 30$

45. $2x + 3y = 9$

In Exercises 46–57, find the slope-intercept form of the equation of the line that passes through the given point and has the indicated slope. Sketch a graph of the line.

Point	Slope
46. $(0, -2)$	$m = 3$
47. $(0, 10)$	$m = -1$
48. $(-3, 6)$	$m = -2$
49. $(0, 0)$	$m = 4$
50. $(4, 0)$	$m = -\frac{1}{3}$
51. $(-2, -5)$	$m = \frac{3}{4}$
52. $(6, -1)$	m is undefined.
53. $(-10, 4)$	$m = 0$
54. $\left(4, \frac{5}{2}\right)$	$m = \frac{4}{3}$
55. $\left(-\frac{1}{2}, \frac{3}{2}\right)$	$m = -3$
56. $(-5.1, 1.8)$	$m = 5$
57. $(2.3, -8.5)$	$m = -\frac{5}{2}$

In Exercises 58–67, find the slope-intercept form of the equation of the line passing through the points. Sketch a graph of the line.

58. $(5, -1), (-5, 5)$

59. $(4, 3), (-4, -4)$

60. $(-8, 1), (-8, 7)$

61. $(-1, 4), (6, 4)$

62. $\left(2, \frac{1}{2}\right), \left(\frac{1}{2}, \frac{5}{4}\right)$

63. $\left(1, 1\right), \left(6, -\frac{2}{3}\right)$

64. $\left(-\frac{1}{10}, -\frac{3}{5}\right), \left(\frac{9}{10}, -\frac{9}{5}\right)$

65. $\left(\frac{3}{4}, \frac{3}{2}\right), \left(-\frac{4}{3}, \frac{7}{4}\right)$

66. $(1, 0.6), (-2, -0.6)$

67. $(-8, 0.6), (2, -2.4)$

In Exercises 68–73, use the *intercept form* to find the equation of the line with the given intercepts. The intercept form of the equation of a line with intercepts $(a, 0)$ and $(0, b)$ is

$$\frac{x}{a} + \frac{y}{b} = 1, \quad a \neq 0, \ b \neq 0.$$

68. *x*-intercept: $(2, 0)$

y-intercept: $(0, 3)$

69. *x*-intercept: $(-3, 0)$

y-intercept: $(0, 4)$

70. *x*-intercept: $\left(-\frac{1}{6}, 0\right)$

y-intercept: $\left(0, -\frac{2}{3}\right)$

71. *x*-intercept: $\left(\frac{2}{3}, 0\right)$

y-intercept: $(0, -2)$

72. Point on line: $(1, 2)$

x-intercept: $(c, 0)$

y-intercept: $(0, c), \quad c \neq 0$

73. Point on line: $(-3, 4)$

x-intercept: $(d, 0)$

y-intercept: $(0, d), \quad d \neq 0$

In Exercises 74–81, write the slope-intercept forms of the equations of the lines through the given point (a) parallel to the given line and (b) perpendicular to the given line.

Point	Line
74. $(2, 1)$	$4x - 2y = 3$
75. $(-3, 2)$	$x + y = 7$
76. $\left(-\frac{2}{3}, \frac{7}{8}\right)$	$3x + 4y = 7$
77. $\left(\frac{7}{8}, \frac{3}{4}\right)$	$5x + 3y = 0$
78. $(-1, 0)$	$y = -3$
79. $(2, 5)$	$x = 4$
80. $(2.5, 6.8)$	$x - y = 4$
81. $(-3.9, -1.4)$	$6x + 2y = 9$

▦ *Graphical Interpretation* In Exercises 82–85, identify any relationships that exist among the lines, and then use a graphing utility to graph the three equations in the same viewing window. Adjust the viewing window so that the slope appears visually correct.

82. (a) $y = 2x$ (b) $y = -2x$ (c) $y = \frac{1}{2}x$

83. (a) $y = \frac{2}{3}x$ (b) $y = -\frac{3}{2}x$ (c) $y = \frac{2}{3}x + 2$

84. (a) $y = -\frac{1}{2}x$ (b) $y = -\frac{1}{2}x + 3$ (c) $y = 2x - 4$

85. (a) $y = x - 8$ (b) $y = x + 1$ (c) $y = -x + 3$

Rate of Change **In Exercises 86 and 87, you are given the dollar value of a product in 2001 and the rate at which the value of the product is expected to change during the next 5 years. Use this information to write a linear equation that gives the dollar value V of the product in terms of the year t. (Let t = 1 represent 2001.)**

2001 Value	Rate
86. $2540	$125 increase per year
87. $156	$4.50 increase per year

Graphical Interpretation **In Exercises 88–91, match the description of the situation with its graph. Also determine the slope of each graph and interpret the slope in the context of the situation. [The graphs are labeled (a), (b), (c), and (d).]**

(a)
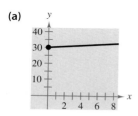

(b)
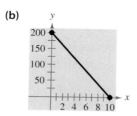

(c)
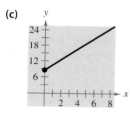

(d)
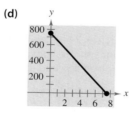

88. A person is paying $20 per week to a friend to repay a $200 loan.

89. An employee is paid $8.50 per hour plus $2 for each unit produced per hour.

90. A sales representative receives $30 per day for food plus $0.32 for each mile traveled.

91. A word processor that was purchased for $750 depreciates $100 per year.

In Exercises 92–95, find a relationship between x and y such that (x, y) is equidistant from the two points.

92. $(4, -1), (-2, 3)$

93. $(6, 5), (1, -8)$

94. $(3, \frac{5}{2}), (-7, 1)$

95. $(-\frac{1}{2}, -4), (\frac{7}{2}, \frac{5}{4})$

96. *Cash Flow per Share* The cash flow per share for America Online, Inc. was $0.26 in 1998 and $0.70 in 1999. Write a linear equation that gives the cash flow per share in terms of the year. Let $t = 0$ represent 1998. Then predict the cash flows for the years 2000 and 2001. (Source: America Online, Inc.)

97. *Number of Stores* In 1996 there were 3927 J.C. Penney stores and in 1997 there were 3981 stores. Write a linear equation that gives the number of stores in terms of the year. Let $t = 0$ represent 1996. Then predict the numbers of stores for the years 1999 and 2000. (Source: J.C. Penney Co.)

98. *Temperature* Find the equation of the line that shows the relationship between the temperature in degrees Celsius C and degrees Fahrenheit F. Remember that water freezes at 0° Celsius (32° Fahrenheit) and boils at 100° Celsius (212° Fahrenheit).

99. *Temperature* Use the result of Exercise 98 to complete the table.

C		$-10°$	$10°$			$177°$
F	$0°$			$68°$	$90°$	

100. *Annual Salary* Your salary was $28,500 in 1998 and $32,900 in 2000. If your salary follows a linear growth pattern, what will your salary be in 2003?

101. *College Enrollment* A small college had 2546 students in 1998 and 2702 students in 2000. If the enrollment follows a linear growth pattern, how many students will the college have in 2004?

102. *Business* A business purchases a piece of equipment for $875. After 5 years the equipment will be outdated and have no value. Write a linear equation giving the value V of the equipment during the 5 years it will be used.

103. *Business* A business purchases a piece of equipment for $25,000. After 10 years the equipment will have to be replaced. Its value at that time is expected to be $2000. Write a linear equation giving the value V of the equipment during the 10 years it will be used.

104. *Sales* A store is offering a 15% discount on all items. Write a linear equation giving the sale price S for an item with a list price L.

105. *Hourly Wage*   A manufacturer pays its assembly line workers $11.50 per hour. In addition, workers receive a piecework rate of $0.75 per unit produced. Write a linear equation for the hourly wage W in terms of the number of units x produced per hour.

106. *Business*   A contractor purchases a piece of equipment for $36,500. The equipment requires an average expenditure of $5.25 per hour for fuel and maintenance, and the operator is paid $11.50 per hour.

   (a) Write a linear equation giving the total cost C of operating this equipment for t hours. (Include the purchase cost of the equipment.)

   (b) Assuming that customers are charged $27 per hour of machine use, write an equation for the revenue R derived from t hours of use.

   (c) Use the formula for profit $(P = R - C)$ to write an equation for the profit derived from t hours of use.

   (d) Use the result of part (c) to find the break-even point—that is, the number of hours this equipment must be used to yield a profit of 0 dollars.

107. *Rental Demand*   A real estate office handles an apartment complex with 50 units. When the rent per unit is $580 per month, all 50 units are occupied. However, when the rent is $625 per month, the average number of occupied units drops to 47. Assume that the relationship between the monthly rent p and the demand x is linear.

   (a) Write the equation of the line giving the demand x in terms of the rent p.

   (b) Use this equation to predict the number of units occupied if the rent is $655.

   (c) Predict the number of units occupied if the rent is $595.

108. *Geometry*   The length and width of a rectangular garden are 15 meters and 10 meters, respectively. A walkway of width x surrounds the garden.

   (a) Draw a diagram that gives a visual representation of the problem.

   (b) Write the equation for the perimeter y of the walkway in terms of x.

   (c) Use a graphing utility to graph the equation for the perimeter.

   (d) Determine the slope of the graph in part (c). For each additional 1-meter increase in the width of the walkway, determine the increase in its perimeter.

109. *Monthly Salary*   A salesperson receives a monthly salary of $2500 plus a commission of 7% of sales. Write a linear equation for the salesperson's monthly wage W in terms of monthly sales S.

110. *Business Costs*   A sales representative of a company using a personal car receives $120 per day for lodging and meals plus $0.31 per mile driven. Write a linear equation giving the daily cost C to the company in terms of x, the number of miles driven.

111. *Investment*   An inheritance of $12,000 is invested in two different mutual funds. A less risky fund pays $2\frac{1}{2}\%$ simple interest and a more risky fund pays 4% simple interest.

   (a) If x dollars is invested in the fund paying $2\frac{1}{2}\%$, how much money is invested in the fund paying 4%?

   (b) Write the total annual interest y in terms of x.

   (c) Use a graphing utility to graph the function in part (b) over the interval $0 \le x \le 12,000$.

   (d) Explain why the slope of the line in part (c) is negative.

112. *Sports*   The average annual salaries of major league baseball players (in thousands of dollars) from 1988 to 1998 are shown in the scatter plot. Find the equation of the line that you think best fits these data. (Let y represent the average salary and let t represent the year, with $t = 0$ corresponding to 1988.)   (Source: Major League Baseball Player Relations Committee)

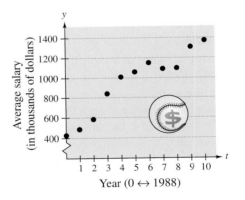

Year (0 ↔ 1988)

113. *Data Analysis* An instructor gives regular 20-point quizzes and 100-point exams in a mathematics course. Average scores for six students, given as data points (x, y) where x is the average quiz score and y is the average test score, are $(18, 87)$, $(10, 55)$, $(19, 96)$, $(16, 79)$, $(13, 76)$, and $(15, 82)$. [*Note:* There are many correct answers for parts (b)–(d).]

(a) Sketch a scatter plot of the data.

(b) Use a straightedge to sketch the best-fitting line through the points.

(c) Find an equation for the line sketched in part (b).

(d) Use the equation in part (c) to estimate the average test score for a person with an average quiz score of 17.

(e) If the instructor adds 4 points to the average test score of everyone in the class, describe the changes in the positions of the plotted points and the change in the equation of the line.

Synthesis

True or False? **In Exercises 114 and 115, determine whether the statement is true or false. Justify your answer.**

114. A line with a slope of $-\frac{5}{7}$ is steeper than a line with a slope of $-\frac{6}{7}$.

115. The line through $(-8, 2)$ and $(-1, 4)$ and the line through $(0, -4)$ and $(-7, 7)$ are parallel.

116. Explain how you could show that the points $A(2, 3)$, $B(2, 9)$, and $C(7, 2)$ are the vertices of a right triangle.

117. Explain why the slope of a vertical line is said to be undefined.

118. With the information given in the graphs, is it possible to determine the slope of each line? Is it possible that the lines could have the same slope? Explain.

(a)
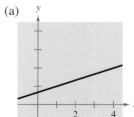

(b)
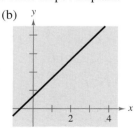

119. The slopes of two lines are -4 and $\frac{5}{2}$. Which is steeper? Explain.

120. The value V of a machine t years after it is purchased is
$$V = -4000t + 58{,}500, \quad 0 \le t \le 5.$$
Explain what the V-intercept and slope measure.

121. *Writing* Write a brief paragraph explaining whether or not any pair of points on a line can be used to calculate the slope of the line.

122. *Think About It* Is it possible for two lines with positive slopes to be perpendicular? Explain.

Review

In Exercises 123–126, match the equation with its graph. [The graphs are labeled (a), (b), (c), and (d).]

(a)
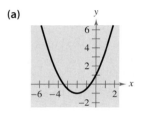

(b)
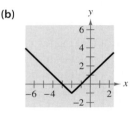

(c)
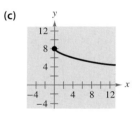

(d)
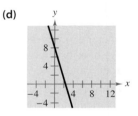

123. $y = 8 - 3x$

124. $y = 8 - \sqrt{x}$

125. $y = \frac{1}{2}x^2 + 2x + 1$

126. $y = |x + 2| - 1$

In Exercises 127–132, find all the solutions of the equation. Check your solution(s) in the original equation.

127. $-7(3 - x) = 14(x - 1)$

128. $\dfrac{8}{2x - 7} = \dfrac{4}{9 - 4x}$

129. $2x^2 - 21x + 49 = 0$

130. $x^2 - 8x + 3 = 0$

131. $\sqrt{x - 9} + 15 = 0$

132. $3x - 16\sqrt{x} + 5 = 0$

2.2 Functions

▶ **What you should learn**

- How to decide whether relations between two variables are functions
- How to use function notation and evaluate functions
- How to find the domains of functions
- How to use functions to model and solve real-life problems

▶ **Why you should learn it**

Functions can be used to model and solve real-life problems. For instance, Exercise 98 on page 199 shows how to use a function to find the force of water against the face of a dam.

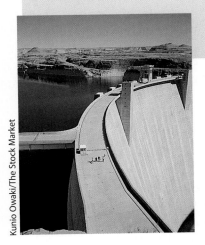

Introduction to Functions

Many everyday phenomena involve two quantities that are related to each other by some rule of correspondence. The mathematical term for such a rule of correspondence is a **relation.** Here are two examples.

1. The simple interest I earned on \$1000 for 1 year is related to the annual interest rate r by the formula $I = 1000r$.

2. The area A of a circle is related to its radius r by the formula $A = \pi r^2$.

Not all relations have simple mathematical formulas. For instance, people commonly match up NFL starting quarterbacks with touchdown passes and hours of the day with temperature. In cases 1 and 2 above, however, there is some relation that matches each item from one set with exactly one item from a different set. Such a relation is called a **function.**

Definition of a Function

A **function** f from a set A to a set B is a relation that assigns to each element x in the set A exactly one element y in the set B. The set A is the **domain** (or set of inputs) of the function f, and the set B contains the **range** (or set of outputs).

To help understand this definition, look at the function in Figure 2.13.

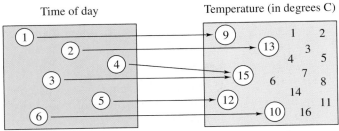

Set A is the domain.
Inputs: 1, 2, 3, 4, 5, 6

Set B contains the range.
Outputs: 9, 10, 12, 13, 15

FIGURE 2.13

This function can be represented by the following ordered pairs.

$$\{(1, 9°), (2, 13°), (3, 15°), (4, 15°), (5, 12°), (6, 10°)\}$$

In each ordered pair, the first coordinate is the input and the second coordinate is the output. In this example, note the following characteristics of a function.

1. Each element in A must be matched with an element of B.

2. Some elements in B may not be matched with any element in A.

3. Two or more elements of A may be matched with the same element of B.

The converse of the third statement is not true. That is, an element of A (the domain) cannot be matched with two different elements of B.

Functions are commonly represented in four ways.

1. *Verbally* by a sentence that describes how the input variable is related to the output variable
2. *Numerically* by a table or a list of ordered pairs that matches input values with output values
3. *Graphically* by points on a graph in a coordinate plane in which the input values are represented by the horizontal axis and the output values are represented by the vertical axis
4. *Algebraically* by an equation in two variables

In the following example, you are asked to decide whether or not the given relation is a function. To do this, you must decide whether each input value is matched with exactly one output value. If any input value is matched with two or more output values, the relation is not a function.

Example 1 ▶ Testing for Functions

Decide whether the description represents y as a function of x.

a. The input value x is the number of representatives from a state, and the output value y is the number of senators.

b.

Input x	2	2	3	4	5
Output y	11	10	8	5	1

c.

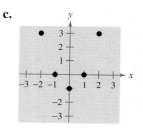

FIGURE 2.14

Solution

a. This verbal description *does* describe y as a function of x. Regardless of the value of x, the value of y is always 2. Such functions are called *constant functions*.

b. This table *does not* describe y as a function of x. The input value 2 is matched with two different y-values.

c. The graph in Figure 2.14 *does* describe y as a function of x. No input value is matched with two output values.

Representing functions by sets of ordered pairs is common in *discrete mathematics*. In algebra, however, it is more common to represent functions by equations or formulas involving two variables. For instance, the equation

$$y = x^2 \qquad \text{\small y is a function of x.}$$

represents the variable y as a function of the variable x. In this equation, x is the **independent variable** and y is the **dependent variable.** The domain of the function is the set of all values taken on by the independent variable x, and the range of the function is the set of all values taken on by the dependent variable y.

Historical Note
Leonhard Euler (1707–1783), a Swiss mathematician, is considered to have been the most prolific and productive mathematician in history. One of his greatest influences on mathematics was his use of symbols, or notation. The function notation $y = f(x)$ was introduced by Euler.

Example 2 ▶ Testing for Functions Represented Algebraically

Which of the equations represent(s) y as a function of x?

a. $x^2 + y = 1$

b. $-x + y^2 = 1$

Solution

To determine whether y is a function of x, try to solve for y in terms of x.

a. Solving for y yields

$$x^2 + y = 1$$ Write original equation.

$$y = 1 - x^2.$$ Solve for y.

To each value of x there corresponds exactly one value of y. So, y is a function of x.

b. Solving for y yields

$$-x + y^2 = 1$$ Write original equation.

$$y^2 = 1 + x$$ Add x to each side.

$$y = \pm\sqrt{1 + x}.$$ Solve for y.

The $\pm$ indicates that to a given value of x there correspond two values of y. So, y is not a function of x.

Function Notation

When an equation is used to represent a function, it is convenient to name the function so that it can be referenced easily. For example, you know that the equation $y = 1 - x^2$ describes y as a function of x. Suppose you give this function the name "f." Then you can use the following **function notation.**

Input	*Output*	*Equation*
x	$f(x)$	$f(x) = 1 - x^2$

The symbol $f(x)$ is read as *the value of f at x* or simply *f of x.* The symbol $f(x)$ corresponds to the y-value for a given x. So, you can write $y = f(x)$. Keep in mind that f is the *name* of the function, whereas $f(x)$ is the *value* of the function at x. For instance, the function

$$f(x) = 3 - 2x$$

has *function values* denoted by $f(-1)$, $f(0)$, $f(2)$, and so on. To find these values, substitute the specified input values into the given equation.

For $x = -1$, $f(-1) = 3 - 2(-1) = 3 + 2 = 5.$
For $x = 0$, $f(0) = 3 - 2(0) = 3 - 0 = 3.$
For $x = 2$, $f(2) = 3 - 2(2) = 3 - 4 = -1.$

Although f is often used as a convenient function name and x is often used as the independent variable, you can use other letters. For instance,

$$f(x) = x^2 - 4x + 7, \quad f(t) = t^2 - 4t + 7, \quad \text{and} \quad g(s) = s^2 - 4s + 7$$

all define the same function. In fact, the role of the independent variable is that of a "placeholder." Consequently, the function could be described by

$$f(\quad) = (\quad)^2 - 4(\quad) + 7.$$

Example 3 ▶ Evaluating a Function

Let $g(x) = -x^2 + 4x + 1$ and find

a. $g(2)$ **b.** $g(t)$ **c.** $g(x + 2)$.

Solution

a. Replacing x with 2 in $g(x) = -x^2 + 4x + 1$ yields the following.

$$g(2) = -(2)^2 + 4(2) + 1 = -4 + 8 + 1 = 5$$

b. Replacing x with t yields the following.

$$g(t) = -(t)^2 + 4(t) + 1 = -t^2 + 4t + 1$$

c. Replacing x with $x + 2$ yields the following.

$$g(x + 2) = -(x + 2)^2 + 4(x + 2) + 1$$
$$= -(x^2 + 4x + 4) + 4x + 8 + 1$$
$$= -x^2 - 4x - 4 + 4x + 8 + 1$$
$$= -x^2 + 5$$

> **STUDY T!P**
>
> In Example 3, note that $g(x + 2)$ is not equal to $g(x) + g(2)$. In general, $g(u + v) \neq g(u) + g(v)$.

A function defined by two or more equations over a specified domain is called a **piecewise-defined function.**

Example 4 ▶ A Piecewise-Defined Function

Evaluate the function when $x = -1, 0,$ and 1.

$$f(x) = \begin{cases} x^2 + 1, & x < 0 \\ x - 1, & x \geq 0 \end{cases}$$

Solution

Because $x = -1$ is less than 0, use $f(x) = x^2 + 1$ to obtain

$$f(-1) = (-1)^2 + 1 = 2.$$

For $x = 0$, use $f(x) = x - 1$ to obtain

$$f(0) = (0) - 1 = -1.$$

For $x = 1$, use $f(x) = x - 1$ to obtain

$$f(1) = (1) - 1 = 0.$$

The Domain of a Function

The domain of a function can be described explicitly or it can be *implied* by the expression used to define the function. The **implied domain** is the set of all real numbers for which the expression is defined. For instance, the function

$$f(x) = \frac{1}{x^2 - 4}$$ 　　Domain excludes *x*-values that result in division by zero.

has an implied domain that consists of all real x other than $x = \pm 2$. These two values are excluded from the domain because division by zero is undefined. Another common type of implied domain is that used to avoid even roots of negative numbers. For example, the function

$$f(x) = \sqrt{x}$$ 　　Domain excludes *x*-values that result in even roots of negative numbers.

is defined only for $x \geq 0$. So, its implied domain is the interval $[0, \infty)$. In general, the domain of a function *excludes* values that would cause division by zero *or* that would result in the even root of a negative number.

Example 5 ▶ Finding the Domain of a Function

Find the domain of each function.

a. f: $\{(-3, 0), (-1, 4), (0, 2), (2, 2), (4, -1)\}$ 　　**b.** $g(x) = \dfrac{1}{x + 5}$

c. Volume of a sphere: $V = \frac{4}{3}\pi r^3$ 　　**d.** $h(x) = \sqrt{4 - x^2}$

Solution

a. The domain of f consists of all first coordinates in the set of ordered pairs.

$$\text{Domain} = \{-3, -1, 0, 2, 4\}$$

b. Excluding x-values that yield zero in the denominator, the domain of g is the set of all real numbers $x \neq -5$.

c. Because this function represents the volume of a sphere, the values of the radius r must be positive. So, the domain is the set of all real numbers r such that $r > 0$.

d. This function is defined only for x-values for which

$$4 - x^2 \geq 0.$$

Using the methods described in Section 1.8, you can conclude that $-2 \leq x \leq 2$. So, the domain is the interval $[-2, 2]$.

In Example 5(c), note that the domain of a function may be implied by the physical context. For instance, from the equation $V = \frac{4}{3}\pi r^3$, you would have no reason to restrict r to positive values, but the physical context implies that a sphere cannot have a negative radius.

FIGURE **2.15**

Applications

Example 6 ▶ The Dimensions of a Container

You work in the marketing department of a soft-drink company and are experimenting with a new soft-drink can that is slightly narrower and taller than a standard can. For your experimental can, the ratio of the height to the radius is 4, as shown in Figure 2.15.

a. Express the volume of the can as a function of the radius r.

b. Express the volume of the can as a function of the height h.

Solution

a. $V(r) = \pi r^2 h = \pi r^2 (4r) = 4\pi r^3$ Write V as a function of r.

b. $V(h) = \pi \left(\dfrac{h}{4}\right)^2 h = \dfrac{\pi h^3}{16}$ Write V as a function of h.

Example 7 ▶ The Path of a Baseball

A baseball is hit at a point 3 feet above ground at a velocity of 100 feet per second and an angle of 45°. The path of the baseball is given by the function

$$f(x) = -0.0032x^2 + x + 3$$

where y and x are measured in feet, as shown in Figure 2.16. Will the baseball clear a 10-foot fence located 300 feet from home plate?

Solution

When $x = 300$, the height of the baseball is

$$f(300) = -0.0032(300)^2 + 300 + 3$$

$$= 15 \text{ feet.}$$

So, the ball will clear the fence.

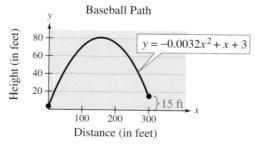

FIGURE **2.16**

In the equation in Example 7, the height of the baseball is a function of the distance from home plate.

Example 8 ▶ Direct Mail Advertising

The money C (in billions of dollars) spent for direct mail advertising in the United States increased in a linear pattern from 1990 to 1992, as shown in Figure 2.17. Then, in 1993, the money spent took a jump and, until 1996, increased in a *different* linear pattern. These two patterns can be approximated by the function

$$C(t) = \begin{cases} 23.40 + 1.01t, & 0 \le t \le 2 \\ 19.85 + 2.50t, & 3 \le t \le 6 \end{cases}$$

where $t = 0$ represents 1990. Use this function to approximate the total amount spent for direct mail advertising between 1990 and 1996. (Source: McCann-Erickson)

Solution

From 1990 to 1992, use the formula $C(t) = 23.40 + 1.01t$.

$$\underbrace{\$23.40,}_{1990} \quad \underbrace{\$24.41,}_{1991} \quad \underbrace{\$25.42}_{1992}$$

From 1993 to 1996, use the formula $C(t) = 19.85 + 2.50t$.

$$\underbrace{\$27.35,}_{1993} \quad \underbrace{\$29.85,}_{1994} \quad \underbrace{\$32.35,}_{1995} \quad \underbrace{\$34.85}_{1996}$$

The total of these seven amounts is $197.63, which implies that the total amount spent was approximately $197,630,000,000.

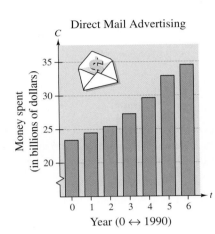

Direct Mail Advertising

Money spent (in billions of dollars) vs. Year (0 ↔ 1990)

FIGURE 2.17

One of the basic definitions in calculus employs the ratio

$$\frac{f(x + h) - f(x)}{h}, \quad h \neq 0.$$

This ratio is called a **difference quotient,** as illustrated in Example 9.

Example 9 ▶ Evaluating a Difference Quotient

For $f(x) = x^2 - 4x + 7$, find $\dfrac{f(x + h) - f(x)}{h}$.

Solution

$$\frac{f(x + h) - f(x)}{h} = \frac{[(x + h)^2 - 4(x + h) + 7] - (x^2 - 4x + 7)}{h}$$

$$= \frac{x^2 + 2xh + h^2 - 4x - 4h + 7 - x^2 + 4x - 7}{h}$$

$$= \frac{2xh + h^2 - 4h}{h}$$

$$= \frac{h(2x + h - 4)}{h}$$

$$= 2x + h - 4, \quad h \neq 0$$

The symbol ⬤ indicates an example or exercise that highlights algebraic techniques specifically used in calculus.

Summary of Function Terminology

Function: A **function** is a relationship between two variables such that to each value of the independent variable there corresponds exactly one value of the dependent variable.

Function Notation: $y = f(x)$

  f is the *name* of the function.

  y is the **dependent variable.**

  x is the **independent variable.**

  $f(x)$ is the *value of the function at x.*

Domain: The **domain** of a function is the set of all values (inputs) of the independent variable for which the function is defined. If x is in the domain of f, f is said to be *defined* at x. If x is not in the domain of f, f is said to be *undefined* at x.

Range: The **range** of a function is the set of all values (outputs) assumed by the dependent variable (that is, the set of all function values).

Implied Domain: If f is defined by an algebraic expression and the domain is not specified, the **implied domain** consists of all real numbers for which the expression is defined.

Writing ABOUT MATHEMATICS

Modeling with Piecewise-Defined Functions   The table below shows the monthly revenue y (in thousands of dollars) for one year of a landscaping business, with $x = 1$ representing January.

x	1	2	3	4	5	6
y	5.2	5.6	6.6	8.3	11.5	15.8

x	7	8	9	10	11	12
y	12.8	10.1	8.6	6.9	4.5	2.7

A mathematical model that represents these data is

$$f(x) = \begin{cases} -1.97x + 26.33 \\ 0.5x^2 - 1.47x + 6.3 \end{cases}.$$

What is the domain of each part of the piecewise-defined function? How can you tell? Explain your reasoning.

Find $f(5)$ and $f(11)$, and interpret your results in the context of the problem. How do these model values compare with the actual data values?

2.2 Exercises

In Exercises 1–4, is the relationship a function?

1. *Domain   Range*

2. *Domain   Range*

3. *Domain   Range*   **4.** *Domain   Range*

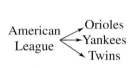   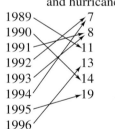

In Exercises 5–8, does the table describe a function? Explain your reasoning.

5.

Input value	-2	-1	0	1	2
Output value	-8	-1	0	1	8

6.

Input value	0	1	2	1	0
Output value	-4	-2	0	2	4

7.

Input value	10	7	4	7	10
Output value	3	6	9	12	15

8.

Input value	0	3	9	12	15
Output value	3	3	3	3	3

In Exercises 9 and 10, which sets of ordered pairs represent functions from *A* to *B*? Explain.

9. $A = \{0, 1, 2, 3\}$ and $B = \{-2, -1, 0, 1, 2\}$
  (a) $\{(0, 1), (1, -2), (2, 0), (3, 2)\}$
  (b) $\{(0, -1), (2, 2), (1, -2), (3, 0), (1, 1)\}$
  (c) $\{(0, 0), (1, 0), (2, 0), (3, 0)\}$
  (d) $\{(0, 2), (3, 0), (1, 1)\}$

10. $A = \{a, b, c\}$ and $B = \{0, 1, 2, 3\}$
  (a) $\{(a, 1), (c, 2), (c, 3), (b, 3)\}$
  (b) $\{(a, 1), (b, 2), (c, 3)\}$
  (c) $\{(1, a), (0, a), (2, c), (3, b)\}$
  (d) $\{(c, 0), (b, 0), (a, 3)\}$

Circulation of Newspapers   **In Exercises 11 and 12, use the graph, which shows the circulation (in millions) of daily newspapers in the United States.** (Source: Editor & Publisher Company)

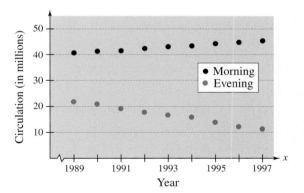

11. Is the circulation of morning newspapers a function of the year? Is the circulation of evening newspapers a function of the year? Explain.

12. Let $f(x)$ represent the circulation of evening newspapers in year x. Find $f(1994)$.

In Exercises 13–22, determine whether the equation represents *y* as a function of *x*.

13. $x^2 + y^2 = 4$

14. $x = y^2$

15. $x^2 + y = 4$

16. $x + y^2 = 4$

17. $2x + 3y = 4$

18. $(x - 2)^2 + y^2 = 4$

19. $y^2 = x^2 - 1$

20. $y = \sqrt{x + 5}$

21. $y = |4 - x|$

22. $|y| = 4 - x$

In Exercises 23 and 24, fill in the blanks using the specified function and the given values of the independent variable.

23. $f(s) = \dfrac{1}{s + 1}$

(a) $f(4) = \dfrac{1}{(\ \blacksquare\ ) + 1}$

(b) $f(0) = \dfrac{1}{(\ \blacksquare\ ) + 1}$

(c) $f(4x) = \dfrac{1}{(\ \blacksquare\ ) + 1}$

(d) $f(x + c) = \dfrac{1}{(\ \blacksquare\ ) + 1}$

24. $g(x) = x^2 - 2x$

(a) $g(2) = (\ \blacksquare\ )^2 - 2(\ \blacksquare\ )$

(b) $g(-3) = (\ \blacksquare\ )^2 - 2(\ \blacksquare\ )$

(c) $g(t + 1) = (\ \blacksquare\ )^2 - 2(\ \blacksquare\ )$

(d) $g(x + c) = (\ \blacksquare\ )^2 - 2(\ \blacksquare\ )$

In Exercises 25–36, evaluate the function at each specified value of the independent variable and simplify.

25. $f(x) = 2x - 3$

(a) $f(1)$ (b) $f(-3)$ (c) $f(x - 1)$

26. $g(y) = 7 - 3y$

(a) $g(0)$ (b) $g\left(\frac{7}{3}\right)$ (c) $g(s + 2)$

27. $V(r) = \frac{4}{3}\pi r^3$

(a) $V(3)$ (b) $V\left(\frac{3}{2}\right)$ (c) $V(2r)$

28. $h(t) = t^2 - 2t$

(a) $h(2)$ (b) $h(1.5)$ (c) $h(x + 2)$

29. $f(y) = 3 - \sqrt{y}$

(a) $f(4)$ (b) $f(0.25)$ (c) $f(4x^2)$

30. $f(x) = \sqrt{x + 8} + 2$

(a) $f(-8)$ (b) $f(1)$ (c) $f(x - 8)$

31. $q(x) = \dfrac{1}{x^2 - 9}$

(a) $q(0)$ (b) $q(3)$ (c) $q(y + 3)$

32. $q(t) = \dfrac{2t^2 + 3}{t^2}$

(a) $q(2)$ (b) $q(0)$ (c) $q(-x)$

33. $f(x) = \dfrac{|x|}{x}$

(a) $f(2)$ (b) $f(-2)$ (c) $f(x - 1)$

34. $f(x) = |x| + 4$

(a) $f(2)$ (b) $f(-2)$ (c) $f(x^2)$

35. $f(x) = \begin{cases} 2x + 1, & x < 0 \\ 2x + 2, & x \geq 0 \end{cases}$

(a) $f(-1)$ (b) $f(0)$ (c) $f(2)$

36. $f(x) = \begin{cases} x^2 + 2, & x \leq 1 \\ 2x^2 + 2, & x > 1 \end{cases}$

(a) $f(-2)$ (b) $f(1)$ (c) $f(2)$

In Exercises 37–42, complete the table.

37. $f(x) = x^2 - 3$

x	-2	-1	0	1	2
$f(x)$					

38. $g(x) = \sqrt{x - 3}$

x	3	4	5	6	7
$g(x)$					

39. $h(t) = \frac{1}{2}|t + 3|$

t	-5	-4	-3	-2	-1
$h(t)$					

40. $f(s) = \dfrac{|s - 2|}{s - 2}$

s	0	1	$\frac{3}{2}$	$\frac{5}{2}$	4
$f(s)$					

41. $f(x) = \begin{cases} -\frac{1}{2}x + 4, & x \leq 0 \\ (x - 2)^2, & x > 0 \end{cases}$

x	-2	-1	0	1	2
$f(x)$					

42. $h(x) = \begin{cases} 9 - x^2, & x < 3 \\ x - 3, & x \geq 3 \end{cases}$

x	1	2	3	4	5
$h(x)$					

In Exercises 43–50, find all real values of x such that $f(x) = 0$.

43. $f(x) = 15 - 3x$

44. $f(x) = 5x + 1$

45. $f(x) = \dfrac{3x - 4}{5}$

46. $f(x) = \dfrac{12 - x^2}{5}$

47. $f(x) = x^2 - 9$

48. $f(x) = x^2 - 8x + 15$

49. $f(x) = x^3 - x$

50. $f(x) = x^3 - x^2 - 4x + 4$

In Exercises 51–54, find the value(s) of x for which $f(x) = g(x)$.

51. $f(x) = x^2, \quad g(x) = x + 2$

52. $f(x) = x^2 + 2x + 1, \quad g(x) = 3x + 3$

53. $f(x) = \sqrt{3x} + 1, \quad g(x) = x + 1$

54. $f(x) = x^4 - 2x^2, \quad g(x) = 2x^2$

In Exercises 55–68, find the domain of the function.

55. $f(x) = 5x^2 + 2x - 1$

56. $g(x) = 1 - 2x^2$

57. $h(t) = \dfrac{4}{t}$

58. $s(y) = \dfrac{3y}{y + 5}$

59. $g(y) = \sqrt{y - 10}$

60. $f(t) = \sqrt[3]{t + 4}$

61. $f(x) = \sqrt[4]{1 - x^2}$

62. $f(x) = \sqrt[4]{x^2 + 3x}$

63. $g(x) = \dfrac{1}{x} - \dfrac{3}{x + 2}$

64. $h(x) = \dfrac{10}{x^2 - 2x}$

65. $f(s) = \dfrac{\sqrt{s - 1}}{s - 4}$

66. $f(x) = \dfrac{\sqrt{x + 6}}{6 + x}$

67. $f(x) = \dfrac{\sqrt[3]{x - 4}}{x}$

68. $f(x) = \dfrac{x - 5}{x^2 - 9}$

In Exercises 69–72, assume that the domain of f is the set $A = \{-2, -1, 0, 1, 2\}$. Determine the set of ordered pairs that represents the function f.

69. $f(x) = x^2$

70. $f(x) = \dfrac{2x}{x^2 + 1}$

71. $f(x) = \sqrt{x + 2}$

72. $f(x) = |x + 1|$

Exploration In Exercises 73–76, determine the function from

$$f(x) = cx, \ g(x) = cx^2, \ h(x) = c\sqrt{|x|}, \ \text{and} \ r(x) = \frac{c}{x}$$

and the value of the constant c that will make the function fit the data in the table.

73.

x	-4	-1	0	1	4
y	-32	-2	0	-2	-32

74.

x	-4	-1	0	1	4
y	-1	$-\frac{1}{4}$	0	$\frac{1}{4}$	1

75.

x	-4	-1	0	1	4
y	-8	-32	Undef.	32	8

76.

x	-4	-1	0	1	4
y	6	3	0	3	6

In Exercises 77–84, find the difference quotient and simplify your answer.

77. $f(x) = x^2 - x + 1, \quad \dfrac{f(2 + h) - f(2)}{h}, h \neq 0$

78. $f(x) = 5x - x^2, \quad \dfrac{f(5 + h) - f(5)}{h}, h \neq 0$

79. $f(x) = x^3, \quad \dfrac{f(x + c) - f(x)}{c}, c \neq 0$

80. $f(x) = 2x, \quad \dfrac{f(x + c) - f(x)}{c}, c \neq 0$

81. $g(x) = 3x - 1, \quad \dfrac{g(x) - g(3)}{x - 3}, x \neq 3$

82. $f(t) = \dfrac{1}{t}, \quad \dfrac{f(t) - f(1)}{t - 1}, t \neq 1$

83. $f(x) = \sqrt{5x}, \quad \dfrac{f(x) - f(5)}{x - 5}, x \neq 5$

84. $f(x) = x^{2/3} + 1, \quad \dfrac{f(x) - f(8)}{x - 8}, x \neq 8$

85. *Geometry* Express the area A of a square as a function of its perimeter P.

86. *Geometry* Express the area A of a circle as a function of its circumference C.

The symbol indicates an example or exercise that highlights algebraic techniques specifically used in calculus.

87. *Geometry*   Express the area A of an isosceles triangle with a height of 8 inches and a base of b inches as a function of the length s of one of its two equal sides.

88. *Geometry*   Express the area A of an equilateral triangle as a function of the length s of its sides.

89. *Maximum Volume*   An open box of maximum volume is to be made from a square piece of material 24 centimeters on a side by cutting equal squares from the corners and turning up the sides (see figure).

   (a) Complete six rows of a table. (The first two rows are shown.) Use the result to estimate the maximum volume.

Height x	Width	Volume V
1	$24 - 2(1)$	$1[24 - 2(1)]^2 = 484$
2	$24 - 2(2)$	$2[24 - 2(2)]^2 = 800$

   (b) Plot the points (x, V). Is V a function of x?

   (c) If V is a function of x, write the function and determine its domain.

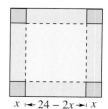
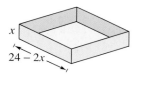

$x \leftarrow 24 - 2x \rightarrow x$

90. *Maximum Profit*   The cost per unit in the production of a certain radio model is $60. The manufacturer charges $90 per unit for orders of 100 or less. To encourage large orders, the manufacturer reduces the charge by $0.15 per radio for each unit ordered in excess of 100 (for example, there would be a charge of $87 per radio for an order size of 120).

   (a) Complete six rows of the table. Use the result to estimate the maximum profit.

Units x	Price p	Profit P
110	$90 - 10(0.15)$	$xp - 110(60)$
120	$90 - 20(0.15)$	$xp - 120(60)$

   (b) Plot the points (x, P). Is P a function of x?

   (c) If P is a function of x, write the function and determine its domain.

91. *Geometry*   A right triangle is formed in the first quadrant by the x- and y-axes and a line through the point $(2, 1)$ (see figure). Write the area A of the triangle as a function of x, and determine the domain of the function.

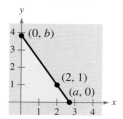

92. *Geometry*   A rectangle is bounded by the x-axis and the semicircle $y = \sqrt{36 - x^2}$ (see figure). Write the area A of the rectangle as a function of x, and determine the domain of the function.

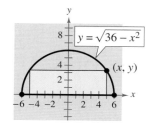

93. *Average Price*   The average price p (in thousands of dollars) of a new mobile home in the United States from 1974 to 1997 (see figure) can be approximated by the model

$$p(t) = \begin{cases} 17.27 + 1.036t, & -6 \leq t \leq 11 \\ -4.807 + 2.882t - 0.011t^2, & 12 \leq t \leq 17 \end{cases}$$

where $t = 0$ represents 1980. Use this model to find the average prices of a mobile home in 1978, 1988, 1993, and 1997.   (Source: U.S. Bureau of Census, Construction Reports)

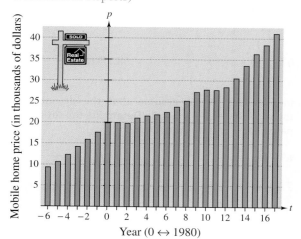

Year $(0 \leftrightarrow 1980)$

94. *Postal Regulations* A rectangular package to be sent by the U.S. Postal Service can have a maximum combined length and girth (perimeter of a cross section) of 108 inches (see figure).

(a) Write the volume V of the package as a function of x.

(b) What is the domain of the function?

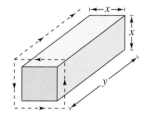

⊞ (c) Use a graphing utility to graph your function. Be sure to use the appropriate window setting.

(d) What dimensions will maximize the volume of the package? Explain your answer.

95. *Business* A company produces a product for which the variable cost is $12.30 per unit and the fixed costs are $98,000. The product sells for $17.98. Let x be the number of units produced and sold.

(a) The total cost for a business is the sum of the variable cost and the fixed costs. Write the total cost C as a function of the number of units produced.

(b) Write the revenue R as a function of the number of units sold.

(c) Write the profit P as a function of the number of units sold. (*Note:* $P = R - C$.)

96. *Total Average Cost* The inventor of a new game believes that the variable cost for producing the game is $0.95 per unit and the fixed costs are $6000. The inventor sells each game for $1.69. Let x be the number of games sold.

(a) The total cost for a business is the sum of the variable cost and the fixed costs. Write the total cost C as a function of the number of games sold.

(b) Write the average cost per unit $\overline{C} = C/x$ as a function of x.

97. *Transportation* For groups of 80 or more people, a charter bus company determines the rate per person according to the formula

$$\text{Rate} = 8 - 0.05(n - 80), \qquad n \geq 80$$

where the rate is given in dollars and n is the number of people.

(a) Express the revenue R for the bus company as a function of n.

(b) Use the function in part (a) to complete the table. What can you conclude?

n	90	100	110	120	130	140	150
$R(n)$							

98. *Physics* The force F (in tons) of water against the face of a dam is estimated by the function

$$F(y) = 149.76\sqrt{10}\,y^{5/2}$$

where y is the depth of the water in feet.

(a) Complete the table. What can you conclude from the table?

y	5	10	20	30	40
$F(y)$					

(b) Use the table to approximate the depth at which the force against the dam is 1,000,000 tons. How could you find a better estimate?

99. *Height of a Balloon* A balloon carrying a transmitter ascends vertically from a point 3000 feet from the receiving station.

(a) Draw a diagram that gives a visual representation of the problem. Let h represent the height of the balloon and let d represent the distance between the balloon and the receiving station.

(b) Express the height of the balloon as a function of d. What is the domain of the function?

100. *Lynx Population*   A study was done on the lynx population of the Yukon Territory in Canada. The graph shows the lynx population from 1988 through 1995 in a 350-square-kilometer region of the Yukon Territory. Let $f(t)$ represent the number of lynx in year t.   (Source: Kluane Boreal Forest Ecosystem Project)

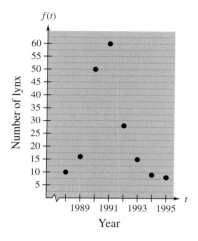

(a) Find

$$\frac{f(1994) - f(1991)}{1994 - 1991}$$

and interpret the result in the context of the problem.

(b) An approximate formula for the function $f(t)$ is

$$N(x) = \frac{434x + 4387}{45x^2 - 55x + 100}$$

where N is the number of lynx and x is time in years, with $x = 0$ corresponding to 1990. Complete the table and compare the result with the data.

x	-2	-1	0	1
N				

x	2	3	4	5
N				

101. *Path of a Ball*   The height y (in feet) of a baseball thrown by a child is

$$y = -\frac{1}{10}x^2 + 3x + 6,$$

where x is the horizontal distance (in feet) from where the ball was thrown. Will the ball fly over the head of another child 30 feet away trying to catch the ball? (Assume that the child who is trying to catch the ball holds a baseball glove at a height of 5 feet.)

Synthesis

True or False?   **In Exercises 102 and 103, determine whether the statement is true or false. Justify your answer.**

102. The domain of the function $f(x) = x^4 - 1$ is $(-\infty, \infty)$, and the range of $f(x)$ is $(0, \infty)$.

103. The set of ordered pairs $\{(-8, -2), (-6, 0), (-4, 0), (-2, 2), (0, 4), (2, -2)\}$ represents a function.

104. Does the relationship shown in the figure represent a function from set A to set B? Explain.

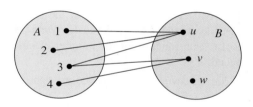

105. *Think About It*   In your own words, explain the meanings of *domain* and *range*.

106. *Think About It*   Describe an advantage of function notation.

Review

In Exercises 107–110, solve the equation.

107. $\dfrac{t}{3} + \dfrac{t}{5} = 1$

108. $\dfrac{3}{t} + \dfrac{5}{t} = 1$

109. $\dfrac{3}{x(x + 1)} - \dfrac{4}{x} = \dfrac{1}{x + 1}$

110. $\dfrac{12}{x} - 3 = \dfrac{4}{x} + 9$

In Exercises 111–114, find the equation of the line passing through the given pair of points.

111. $(-2, -5), (4, -1)$

112. $(10, 0), (1, 9)$

113. $(-6, 5), (3, -5)$

114. $\left(-\frac{1}{2}, 3\right), \left(\frac{11}{2}, -\frac{1}{3}\right)$

2.3 Analyzing Graphs of Functions

▶ **What you should learn**

- How to use the Vertical Line Test for functions
- How to find the zeros of functions
- How to determine intervals on which functions are increasing or decreasing
- How to identify and graph linear functions
- How to identify and graph step functions and other piecewise-defined functions
- How to identify even and odd functions

▶ **Why you should learn it**

Graphs of functions can help you visualize relationships between variables in real life. For instance, Exercise 103 on page 211 shows how the graph of a step function visually represents the cost of a telephone call.

The Graph of a Function

In Section 2.2, you studied functions from an algebraic point of view. In this section, you will study functions from a graphical perspective.

The **graph of a function** f is the collection of ordered pairs $(x, f(x))$ such that x is in the domain of f. As you study this section, remember that

$$x = \text{the directed distance from the } y\text{-axis}$$

$$f(x) = \text{the directed distance from the } x\text{-axis}$$

as shown in Figure 2.18.

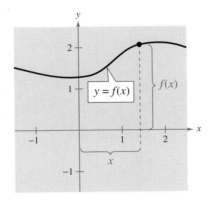

FIGURE 2.18

Example 1 ▶ Finding the Domain and Range of a Function

Use the graph of the function f, shown in Figure 2.19, to find (a) the domain of f, (b) the function values $f(-1)$ and $f(2)$, and (c) the range of f.

Solution

a. The closed dot at $(-1, -5)$ indicates that $x = -1$ is in the domain of f, whereas the open dot at $(4, 0)$ indicates $x = 4$ is not in the domain. So, the domain of f is all x in the interval $[-1, 4)$.

b. Because $(-1, -5)$ is a point on the graph of f, it follows that $f(-1) = -5$. Similarly, because $(2, 4)$ is a point on the graph of f, it follows that $f(2) = 4$.

c. Because the graph does not extend below $f(-1) = -5$ or above $f(2) = 4$, the range of f is the interval $[-5, 4]$.

FIGURE 2.19

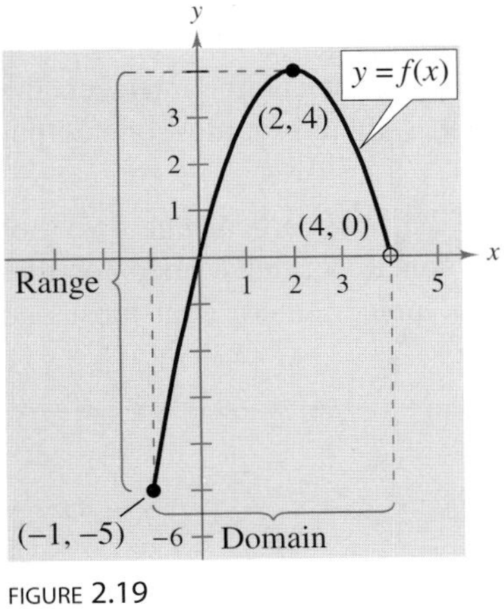

The use of dots (open or closed) at the extreme left and right points of a graph indicates that the graph does not extend beyond these points. If no such dots are shown, assume that the graph extends beyond these points.

By the definition of a function, at most one y-value corresponds to a given x-value. This means that the graph of a function cannot have two or more different points with the same x-coordinate, and no two points on the graph of a function can be vertically above and below each other. It follows, then, that a vertical line can intersect the graph of a function at most once. This observation provides a convenient visual test called the **Vertical Line Test** for functions.

Vertical Line Test for Functions

A set of points in a coordinate plane is the graph of y as a function of x if and only if no vertical line intersects the graph at more than one point.

Example 2 ▶   Vertical Line Test for Functions

Use the Vertical Line Test to decide whether the graphs in Figure 2.20 represent y as a function of x.

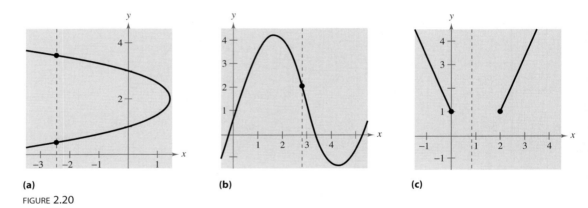

(a)   **(b)**   **(c)**

FIGURE 2.20

Solution

a. This *is not* a graph of y as a function of x because you can find a vertical line that intersects the graph twice. That is, for a particular input x, there is more than one output y.

b. This *is* a graph of y as a function of x because every vertical line intersects the graph at most once. That is, for a particular input x, there is at most one output y.

c. This *is* a graph of y as a function of x. (Note that if a vertical line does not intersect the graph, it simply means that the function is undefined for that particular value of x.) That is, for a particular input x, there is at most one output y.

Zeros of a Function

If the graph of a function of x has an x-intercept at $(a, 0)$, then a is a **zero** of the function.

Zeros of a Function

The **zeros of a function** f of x are the x-values for which $f(x) = 0$.

Example 3 ▶ Finding Zeros of a Function

Find the zeros of each function.

a. $f(x) = 3x^2 + x - 10$ **b.** $g(x) = \sqrt{10 - x^2}$ **c.** $h(t) = \dfrac{2t - 3}{t + 5}$

Solution

To find the zeros of a function, set the function equal to zero and solve for the independent variable.

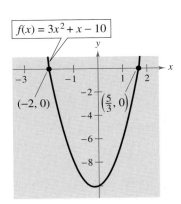

$f(x) = 3x^2 + x - 10$

(a) Zeros of f: $x = -2, x = \frac{5}{3}$

a.

$3x^2 + x - 10 = 0$	Set $f(x)$ equal to 0.
$(3x - 5)(x + 2) = 0$	Factor.
$3x - 5 = 0 \implies x = \frac{5}{3}$	Set 1st factor equal to 0.
$x + 2 = 0 \implies x = -2$	Set 2nd factor equal to 0.

The zeros of f are $\frac{5}{3}$ and -2. In Figure 2.21(a), note that the graph of f has $\left(\frac{5}{3}, 0\right)$ and $(-2, 0)$ as its x-intercepts.

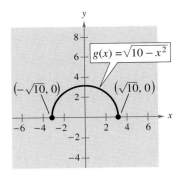

$g(x) = \sqrt{10 - x^2}$

(b) Zeros of g: $x = \pm\sqrt{10}$

b.

$\sqrt{10 - x^2} = 0$	Set $g(x)$ equal to 0.
$10 - x^2 = 0$	Square each side.
$10 = x^2$	Add x^2 to each side.
$\pm\sqrt{10} = x$	Extract square root.

The zeros of g are $-\sqrt{10}$ and $\sqrt{10}$. In Figure 2.21(b), note that the graph of g has $\left(-\sqrt{10}, 0\right)$ and $\left(\sqrt{10}, 0\right)$ as its x-intercepts.

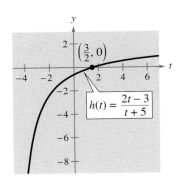

$h(t) = \dfrac{2t - 3}{t + 5}$

(c) Zero of h: $t = \frac{3}{2}$

FIGURE 2.21

c.

$\dfrac{2t - 3}{t + 5} = 0$	Set $h(t)$ equal to 0.
$2t - 3 = 0$	Set numerator equal to 0.
$2t = 3$	Add 3 to each side.
$t = \dfrac{3}{2}$	Divide each side by 2.

The zero of h is $\frac{3}{2}$. In Figure 2.21(c), note that the graph of h has $\left(\frac{3}{2}, 0\right)$ as its x-intercept.

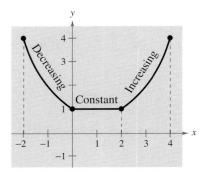

FIGURE 2.22

Increasing and Decreasing Functions

The more you know about the graph of a function, the more you know about the function itself. Consider the graph shown in Figure 2.22. As you move from *left to right,* this graph decreases, then is constant, and then increases.

Increasing, Decreasing, and Constant Functions

A function f is **increasing** on an interval if, for any x_1 and x_2 in the interval, $x_1 < x_2$ implies $f(x_1) < f(x_2)$.

A function f is **decreasing** on an interval if, for any x_1 and x_2 in the interval, $x_1 < x_2$ implies $f(x_1) > f(x_2)$.

A function f is **constant** on an interval if, for any x_1 and x_2 in the interval, $f(x_1) = f(x_2)$.

Example 4 ▶ Increasing and Decreasing Functions

In Figure 2.23, use the graphs to describe the increasing or decreasing behavior of each function.

Solution

a. This function is increasing over the entire real line.

b. This function is increasing on the interval $(-\infty, -1)$, decreasing on the interval $(-1, 1)$, and increasing on the interval $(1, \infty)$.

c. This function is increasing on the interval $(-\infty, 0)$, constant on the interval $(0, 2)$, and decreasing on the interval $(2, \infty)$.

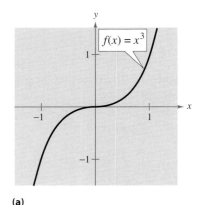

(a)

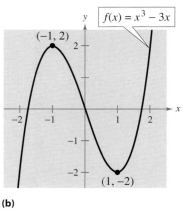

(b)

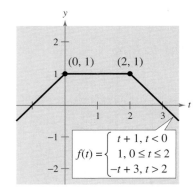

(c)

FIGURE 2.23

To decide whether a function is increasing, decreasing, or constant, you can evaluate the function for several values of x. For instance, the table below verifies that the function in Example 4(a) is increasing over the entire real line.

x	-100	-10	-1	0	1	10	100
$f(x) = x^3$	$-1{,}000{,}000$	-1000	-1	0	1	1000	$1{,}000{,}000$

The points at which a function changes its increasing, decreasing, or constant behavior are helpful in determining the **relative maximum** or **relative minimum** values of the function.

Definition of Relative Minimum and Relative Maximum

A function value $f(a)$ is called a **relative minimum** of f if there exists an interval (x_1, x_2) that contains a such that

$$x_1 < x < x_2 \quad \text{implies} \quad f(a) \le f(x).$$

A function value $f(a)$ is called a **relative maximum** of f if there exists an interval (x_1, x_2) that contains a such that

$$x_1 < x < x_2 \quad \text{implies} \quad f(a) \ge f(x).$$

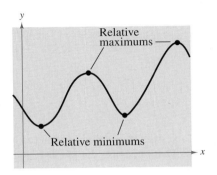

FIGURE 2.24

Figure 2.24 shows several different examples of relative minimums and relative maximums. In Section 3.1, you will study a technique for finding the *exact point* at which a second-degree polynomial function has a relative minimum or relative maximum. For the time being, however, you can use a graphing utility to find reasonable approximations of these points.

Example 5 ▶ Approximating a Relative Minimum

Use a graphing utility to approximate the relative minimum of the function $f(x) = 3x^2 - 4x - 2$.

Solution

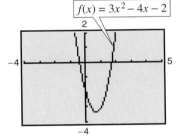

FIGURE 2.25

The graph of f is shown in Figure 2.25. By using the zoom and trace features of a graphing utility, you can estimate that the function has a relative minimum at the point

$(0.67, -3.33)$.   Relative minimum

Later, in Section 3.1, you will be able to determine that the exact point at which the relative minimum occurs is $\left(\frac{2}{3}, -\frac{10}{3}\right)$.

Note in Example 5 that you can also use the table feature of a graphing utility to approximate numerically the relative minimum of the function. Using a table that begins at 0.6 and increments the value of x by 0.01, you can approximate the minimum of $f(x) = 3x^2 - 4x - 2$ to be $(0.67, -3.33)$.

Technology

If you use a graphing utility to estimate the x- and y-values of a relative minimum or relative maximum, the automatic zoom feature will often produce graphs that are nearly flat. To overcome this problem, you can manually change the vertical setting of the viewing window. The graph will stretch vertically if the values of Y_{min} and Y_{max} are closer together.

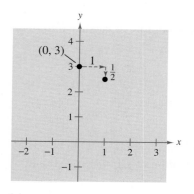

(a)

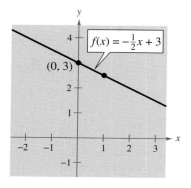

(b)

FIGURE 2.26

Linear Functions

A **linear function** of x is a function of the form

$$f(x) = mx + b. \qquad \text{Linear function}$$

In Section 2.1, you learned that the graph of such a function is a line that has a slope of m and a y-intercept of $(0, b)$.

Example 6 ▶ Graphing a Linear Function

Sketch the graph of the linear function

$$f(x) = -\frac{1}{2}x + 3.$$

Solution

The graph of this function is a line that has a slope of $m = -\frac{1}{2}$ and a y-intercept of $(0, 3)$. To sketch the line, plot the y-intercept. Then, because the slope is $-\frac{1}{2}$, move 1 unit to the right and $\frac{1}{2}$ unit *down* and plot a second point, as shown in Figure 2.26(a). Finally, draw the line that passes through these two points, as shown in Figure 2.26(b).

Example 7 ▶ Writing a Linear Function

Write the linear function f for which $f(1) = 3$ and $f(4) = 0$.

Solution

The graph of a linear function is a line. So, you need to find an equation of the line that passes through $(1, 3)$ and $(4, 0)$. The slope of the line is

$$m = \frac{y_2 - y_1}{x_2 - x_1}$$

$$= \frac{0 - 3}{4 - 1}$$

$$= -1.$$

Using the point-slope form of the equation of a line, you can write

$$y - y_1 = m(x - x_1) \qquad \text{Point-slope form}$$

$$y - 3 = -1(x - 1) \qquad \text{Substitute for } y_1, m, \text{ and } x_1.$$

$$y = -x + 4. \qquad \text{Simplify.}$$

So, the linear function is $f(x) = -x + 4$. You can check this result as follows.

$$f(1) = -(1) + 4 = 3 \checkmark$$

$$f(4) = -(4) + 4 = 0 \checkmark$$

Step Functions and Piecewise-Defined Functions

The **greatest integer function** is denoted by $[\![x]\!]$ and is defined as follows.

$$[\![x]\!] = \text{the greatest integer less than or equal to } x$$

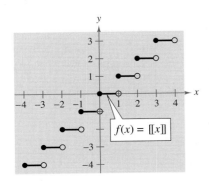

FIGURE 2.27

The graph of this function is shown in Figure 2.27. Note that the graph of the greatest integer function jumps vertically 1 unit at each integer and is constant (a horizontal line segment) between each pair of consecutive integers. The greatest integer function is an example of a category of functions called **step functions** whose graphs resemble a set of stair steps. Some values of the greatest integer function are as follows.

$$[\![-1]\!] = -1 \qquad [\![-0.5]\!] = -1 \qquad [\![0]\!] = 0$$

$$[\![0.5]\!] = 0 \qquad [\![1]\!] = 1 \qquad [\![1.5]\!] = 1$$

The range of the greatest integer function is the set of all integers. (If you use a graphing utility to sketch a step function, you should set the utility to *dot* mode rather than *connected* mode.)

In Section 2.2, you learned that a piecewise-defined function is a function that is defined by two or more equations over a specified domain. To sketch the graph of a piecewise-defined function, you need to sketch the graph of each equation on the appropriate portion of the domain.

Example 8 ▶ Graphing a Piecewise-Defined Function

Sketch the graph of $f(x) = \begin{cases} 2x + 3, & x \le 1 \\ -x + 4, & x > 1 \end{cases}$.

Solution

This piecewise-defined function is composed of two linear functions. At $x = 1$ and to the left of $x = 1$ the graph is the line $y = 2x + 3$, and to the right of $x = 1$ the graph is the line $y = -x + 4$, as shown in Figure 2.28.

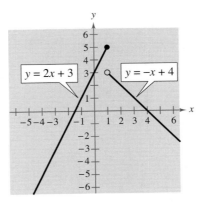

FIGURE 2.28

A computer animation of this concept appears in the *Interactive* CD-ROM and *Internet* versions of this text.

Even and Odd Functions

In Section 1.1, you studied different types of symmetry of a graph. In the terminology of functions, a function is said to be **even** if its graph is symmetric with respect to the *y*-axis and to be **odd** if its graph is symmetric with respect to the origin. The symmetry tests in Section 1.1 yield the following tests for even and odd functions.

Graph each of the following functions with a graphing utility. Determine whether the function is *even, odd,* or *neither.*

$$f(x) = x^2 - x^4$$

$$g(x) = 2x^3 + 1$$

$$h(x) = x^5 - 2x^3 + x$$

$$j(x) = 2 - x^6 - x^8$$

$$k(x) = x^5 - 2x^4 + x - 2$$

$$p(x) = x^9 + 3x^5 - x^3 + x$$

What do you notice about the equations of functions that are odd? What do you notice about the equations of functions that are even? Can you describe a way to identify a function as odd or even by inspecting the equation? Can you describe a way to identify a function as neither odd nor even by inspecting the equation?

Tests for Even and Odd Functions

A function $y = f(x)$ is **even** if, for each x in the domain of f,

$$f(-x) = f(x).$$

A function $y = f(x)$ is **odd** if, for each x in the domain of f,

$$f(-x) = -f(x).$$

Example 9 ▶ Even and Odd Functions

a. The function $g(x) = x^3 - x$ is odd because $g(-x) = -g(x)$, as follows.

$$g(-x) = (-x)^3 - (-x) \qquad \text{Substitute } -x \text{ for } x.$$

$$= -x^3 + x \qquad \text{Simplify.}$$

$$= -(x^3 - x) \qquad \text{Distributive Property}$$

$$= -g(x) \qquad \text{Test for odd function.}$$

b. The function $h(x) = x^2 + 1$ is even because $h(-x) = h(x)$, as follows.

$$h(-x) = (-x)^2 + 1 \qquad \text{Substitute } -x \text{ for } x.$$

$$= x^2 + 1 \qquad \text{Simplify.}$$

$$= h(x) \qquad \text{Test for even function.}$$

The graphs of these two functions are shown in Figure 2.29.

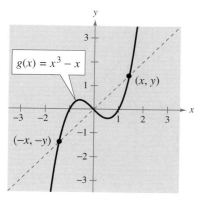

(a) **Symmetric to Origin: Odd Function**

FIGURE 2.29

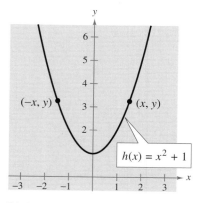

(b) **Symmetric to y-Axis: Even Function**

2.3 Exercises

In Exercises 1–8, find the domain and range of the function.

1. $f(x) = \frac{2}{3}x - 4$

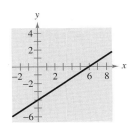

2. $f(x) = x^3 - 3x + 2$

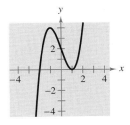

3. $f(x) = 1 - x^2$

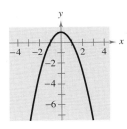

4. $f(x) = \sqrt{x - 1}$

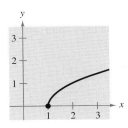

5. $f(x) = \sqrt{x^2 - 1}$

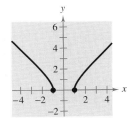

6. $f(x) = \frac{1}{2}|x - 2|$

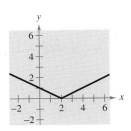

7. $h(x) = \sqrt{16 - x^2}$

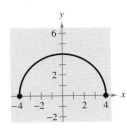

8. $g(x) = \dfrac{|x - 1|}{x - 1}$

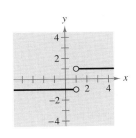

In Exercises 9–14, use the Vertical Line Test to determine whether y is a function of x.

9. $y = \frac{1}{2}x^2$

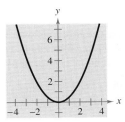

10. $y = \frac{1}{4}x^3$

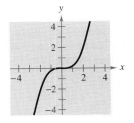

11. $x - y^2 = 1$

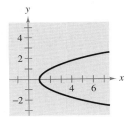

12. $x^2 + y^2 = 25$

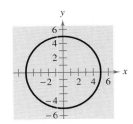

13. $x^2 = 2xy - 1$

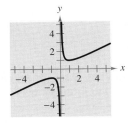

14. $x = |y + 2|$

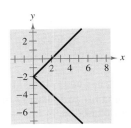

In Exercises 15–22, find the zeros of the function by factoring.

15. $f(x) = 2x^2 - 7x - 30$

16. $f(x) = 3x^2 + 22x - 16$

17. $f(x) = \dfrac{x}{9x^2 - 4}$

18. $f(x) = \dfrac{x^2 - 9x + 14}{4x}$

19. $f(x) = \frac{1}{2}x^3 - x$

20. $f(x) = x^3 - 4x^2 - 9x + 36$

21. $f(x) = 4x^3 - 24x^2 - x + 6$

22. $f(x) = 9x^4 - 25x^2$

In Exercises 23–28, find the zeros of the function algebraically. Verify your results graphically.

23. $f(x) = 3 + \dfrac{5}{x}$

24. $f(x) = x(x - 7)$

25. $f(x) = \sqrt{2x + 11}$

26. $f(x) = \sqrt{3x - 14} - 8$

27. $f(x) = \dfrac{3x - 1}{x - 6}$

28. $f(x) = \dfrac{2x^2 - 9}{3 - x}$

In Exercises 29–32, (a) determine the intervals over which the function is increasing, decreasing, or constant, and (b) determine whether the function is even, odd, or neither.

29. $f(x) = \frac{3}{2}x$

30. $f(x) = x^2 - 4x$

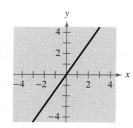 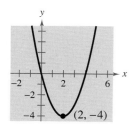

31. $f(x) = x^3 - 3x^2 + 2$

32. $f(x) = \sqrt{x^2 - 1}$

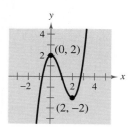 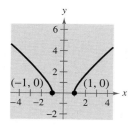

In Exercises 33–50, (a) use a graphing utility to graph the function and visually determine the intervals over which the function is increasing, decreasing, or constant, and (b) make a table of values to verify whether the function is increasing, decreasing, or constant.

33. $f(x) = 3$

34. $g(x) = x$

35. $f(x) = 5 - 3x$

36. $f(x) = -x - 10$

37. $g(s) = \dfrac{s^2}{4}$

38. $h(x) = x^2 - 4$

39. $f(t) = -t^4$

40. $f(x) = 3x^4 - 6x^2$

41. $f(x) = \sqrt{1 - x}$

42. $f(x) = x\sqrt{x + 3}$

43. $f(x) = x^{3/2}$

44. $f(x) = x^{2/3}$

45. $g(t) = \sqrt[3]{t - 1}$

46. $f(x) = \sqrt[4]{x + 5}$

47. $f(x) = |x + 2|$

48. $f(x) = |x + 1| + |x - 1|$

49. $f(x) = \begin{cases} x + 3, & x \le 0 \\ 3, & 0 < x \le 2 \\ 2x - 1, & x > 2 \end{cases}$

50. $f(x) = \begin{cases} 2x + 1, & x \le -1 \\ x^2 - 2, & x > -1 \end{cases}$

In Exercises 51–56, use a graphing utility to approximate the relative minimum/relative maximum of each function.

51. $f(x) = (x - 4)(x + 2)$

52. $f(x) = 3x^2 - 2x - 5$

53. $f(x) = x(x - 2)(x + 3)$

54. $f(x) = x^3 - 3x^2 - x + 1$

55. $f(x) = 2x^3 - 5x^2 - 4x - 1$

56. $f(x) = 8x^4 - 3x - 1$

In Exercises 57–64, sketch the graph of the linear function. Label the y-intercept.

57. $f(x) = 2x - 1$

58. $f(x) = 11 - 3x$

59. $f(x) = -x - \frac{3}{4}$

60. $f(x) = 3x - \frac{5}{2}$

61. $f(x) = -\frac{1}{6}x - \frac{5}{2}$

62. $f(x) = \frac{5}{6} - \frac{2}{3}x$

63. $f(x) = -1.8 + 2.5x$

64. $f(x) = 10.2 + 3.1x$

In Exercises 65–72, write the linear function that has the indicated function values.

65. $f(1) = 4, f(0) = 6$

66. $f(-3) = -8, f(1) = 2$

67. $f(5) = -4, f(-2) = 17$

68. $f(3) = 9, f(-1) = -11$

69. $f(-5) = -5, f(5) = -1$

70. $f(-10) = 12, f(16) = -1$

71. $f(\frac{1}{2}) = -6, f(4) = -3$

72. $f(\frac{2}{3}) = -\frac{15}{2}, f(-4) = -11$

In Exercises 73 and 74, sketch the graph of the function. Describe how it differs from the graph of $f(x) = [\![x]\!]$.

73. $f(x) = [\![x]\!] - 2$

74. $f(x) = [\![x - 1]\!]$

In Exercises 75–78, graph the function.

75. $f(x) = \begin{cases} 2x + 3, & x < 0 \\ 3 - x, & x \ge 0 \end{cases}$

76. $f(x) = \begin{cases} \sqrt{4 + x}, & x < 0 \\ \sqrt{4 - x}, & x \ge 0 \end{cases}$

77. $f(x) = \begin{cases} x^2 + 5, & x \le 1 \\ -x^2 + 4x + 3, & x > 1 \end{cases}$

78. $f(x) = \begin{cases} 1 - (x - 1)^2, & x \le 2 \\ \sqrt{x - 2}, & x > 2 \end{cases}$

In Exercises 79–90, graph the function and determine the interval(s) for which $f(x) \ge 0$.

79. $f(x) = 4 - x$

80. $f(x) = 4x + 2$

81. $f(x) = x^2 - 9$

82. $f(x) = x^2 - 4x$

83. $f(x) = 1 - x^4$

84. $f(x) = \sqrt{x + 2}$

85. $f(x) = x^2 + 1$

86. $f(x) = -(1 + |x|)$

87. $f(x) = -5$

88. $f(x) = \frac{1}{2}(2 + |x|)$

89. $f(x) = \begin{cases} 1 - 2x^2, & x \le -2 \\ -x + 8, & x > -2 \end{cases}$

90. $f(x) = \begin{cases} \sqrt{x - 5}, & x > 5 \\ x^2 + x - 1, & x \le 5 \end{cases}$

In Exercises 91 and 92, use a graphing utility to graph the function. State the domain and range of the function. Describe the pattern of the graph.

91. $s(x) = 2\left(\frac{1}{4}x - \left[\!\left[\frac{1}{4}x\right]\!\right]\right)$

92. $g(x) = 2\left(\frac{1}{4}x - \left[\!\left[\frac{1}{4}x\right]\!\right]\right)^2$

In Exercises 93–98, determine whether the function is even, odd, or neither.

93. $f(x) = x^6 - 2x^2 + 3$

94. $h(x) = x^3 - 5$

95. $g(x) = x^3 - 5x$

96. $f(x) = x\sqrt{1 - x^2}$

97. $f(t) = t^2 + 2t - 3$

98. $g(s) = 4s^{2/3}$

Think About It **In Exercises 99–102, find the coordinates of a second point on the graph of a function f if the given point is on the graph and the function is (a) even and (b) odd.**

99. $\left(-\frac{3}{2}, 4\right)$

100. $\left(-\frac{5}{3}, -7\right)$

101. $(4, 9)$

102. $(5, -1)$

103. *Communications* The cost of using a telephone calling card is \$1.05 for the first minute and \$0.38 for each additional minute or portion of a minute.

(a) A customer needs a model for the cost C of using a calling card for a call lasting t minutes. Which of the following is the appropriate model? Explain.

$C_1(t) = 1.05 + 0.38[\![t - 1]\!]$

$C_2(t) = 1.05 - 0.38[\![-(t - 1)]\!]$

(b) Graph the appropriate model. Determine the cost of a call lasting 18 minutes and 45 seconds.

104. *Delivery Charges* Suppose that the cost of sending an overnight package from New York to Atlanta is \$9.80 for a package weighing up to but not including 1 pound and \$2.50 for each additional pound or portion of a pound. Use the greatest integer function to create a model for the cost C of overnight delivery of a package weighing x pounds, $x > 0$. Sketch the graph of the function.

105. *Electronics* The number of lumens (time rate of flow of light) L from a fluorescent lamp can be approximated by the model

$$L = -0.294x^2 + 97.744x - 664.875, \quad 20 \le x \le 90$$

where x is the wattage of the lamp. Use a graphing utility to graph the function and estimate the wattage necessary to obtain 2000 lumens.

In Exercises 106–109, write the height h of the rectangle as a function of x.

106.

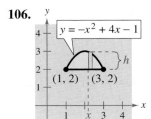

107.

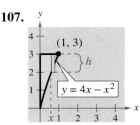

108.

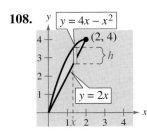

109.

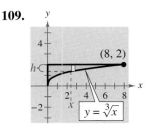

In Exercises 110–113, write the length L of the rectangle as a function of y.

110.

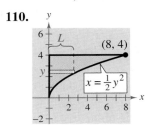

111.

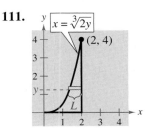

112.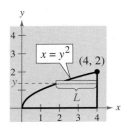

113.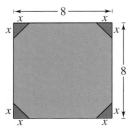

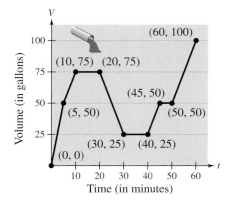

FIGURE FOR 115

114. *Data Analysis*   The table shows the amount y (in billions of dollars) of the merchandise trade balance of the United States for the years 1990 through 1997. The merchandise trade balance is the difference between the values of exports and imports. A negative merchandise trade balance indicates that imports exceeded exports. (Source: U.S. International Trade Administration)

Year	1990	1991	1992	1993
y	−101.7	−66.7	−84.5	−115.6

Year	1994	1995	1996	1997
y	−150.6	−158.7	−170.2	−181.5

   (a) Use the regression feature of a graphing utility to find a cubic model (a model of the form $y = ax^3 + bx^2 + cx + d$) for the data. Let x be the time (in years), with $x = 0$ corresponding to 1990.

   (b) What is the domain of the model?

   (c) Use a graphing utility to graph the data and the model in the same viewing window.

   (d) For which year does the model most accurately estimate the actual data? During which year is it least accurate?

115. *Geometry*   Corners of equal size are cut from a square with sides of length 8 meters (see figure).

   (a) Write the area A of the resulting figure as a function of x. Determine the domain of the function.

   (b) Use a graphing utility to graph the area function over its domain. Use the graph to find the range of the function.

   (c) Identify the resulting figure for the maximum value of x in the domain of the function. What is the length of each side of the figure?

116. *Coordinate Axis Scale*   Each function models the specified data for the years 1993 through 2000, with $t = 3$ corresponding to 1993. Estimate a reasonable scale for the vertical axis (e.g., hundreds, thousands, millions, etc.) of the graph and justify your answer. (There are many correct answers.)

   (a) $f(t)$ represents the average salary of college professors.

   (b) $f(t)$ represents the U.S. population.

   (c) $f(t)$ represents the percent of the civilian work force that is unemployed.

117. *Fluid Flow*   The intake pipe of a 100-gallon tank has a flow rate of 10 gallons per minute, and two drainpipes have flow rates of 5 gallons per minute each. The figure shows the volume V of fluid in the tank as a function of time t. Determine the combination of the input pipe and drain pipes in which the fluid is flowing in specific subintervals of the 1 hour of time shown on the graph. (There are many correct answers.)

Synthesis

True or False? **In Exercises 118 and 119, determine whether the statement is true or false. Justify your answer.**

118. A function with a square root cannot have a domain that is the set of real numbers.

119. A piecewise-defined function will always have at least one x-intercept or at least one y-intercept.

120. Prove that a function of the following form is odd.

$$y = a_{2n+1}x^{2n+1} + a_{2n-1}x^{2n-1} + \cdots$$
$$+ a_3x^3 + a_1x$$

121. Prove that a function of the following form is even.

$$y = a_{2n}x^{2n} + a_{2n-2}x^{2n-2} + \cdots$$
$$+ a_2x^2 + a_0$$

122. If f is an even function, determine whether g is even, odd, or neither. Explain.

(a) $g(x) = -f(x)$ (b) $g(x) = f(-x)$

(c) $g(x) = f(x) - 2$ (d) $g(x) = f(x - 2)$

123. *Think About It* Does the graph in Exercise 11 represent x as a function of y? Explain.

124. *Think About It* Does the graph in Exercise 12 represent x as a function of y? Explain.

125. *Writing* Use a graphing utility to graph each function. Write a paragraph describing any similarities and differences you observe among the graphs.

(a) $y = x$ (b) $y = x^2$

(c) $y = x^3$ (d) $y = x^4$

(e) $y = x^5$ (f) $y = x^6$

126. *Conjecture* Use the results of Exercise 125 to make a conjecture about the graphs of $y = x^7$ and $y = x^8$. Use a graphing utility to graph the functions and compare the results with your conjecture.

Review

In Exercises 127–130, solve the equation.

127. $x^2 - 10x = 0$ **128.** $100 - (x - 5)^2 = 0$

129. $x^3 - x = 0$ **130.** $16x^2 - 40x + 25 = 0$

In Exercises 131–134, evaluate the function at each specified value of the independent variable and simplify.

131. $f(x) = 5x - 8$

(a) $f(9)$ (b) $f(-4)$ (c) $f(x - 7)$

132. $f(x) = x^2 - 10x$

(a) $f(4)$ (b) $f(-8)$ (c) $f(x - 4)$

133. $f(x) = \sqrt{x - 12} - 9$

(a) $f(12)$ (b) $f(40)$ (c) $f(-\sqrt{36})$

134. $f(x) = x^4 - x - 5$

(a) $f(-1)$ (b) $f(\frac{1}{2})$ (c) $f(2\sqrt{3})$

In Exercises 135 and 136, find the difference quotient and simplify your answer.

135. $f(x) = x^2 - 2x + 9$, $\dfrac{f(3 + h) - f(3)}{h}$, $h \neq 0$

136. $f(x) = 5 + 6x - x^2$, $\dfrac{f(6 + h) - f(6)}{h}$, $h \neq 0$

2.4 Shifting, Reflecting, and Stretching Graphs

▶ **What you should learn**

- How to recognize graphs of common functions
- How to use vertical and horizontal shifts to sketch graphs of functions
- How to use reflections to sketch graphs of functions
- How to use nonrigid transformations to sketch graphs of functions

▶ **Why you should learn it**

Knowing the graphs of common functions and knowing how to shift, reflect, and stretch graphs of functions can help you sketch a wide variety of simple functions by hand. This skill is useful in sketching graphs of functions that model real-life data, such as in Exercise 59 on page 223, where you are asked to sketch a function that models the amount of fuel used by trucks from 1980 through 1996.

Summary of Graphs of Common Functions

One of the goals of this text is to enable you to recognize the basic shapes of the graphs of different types of functions. For instance, from your study of lines in Section 2.1, you can determine the basic shape of the graph of the linear function $f(x) = mx + b$. Specifically, you know that the graph of this function is a line whose slope is m and whose y-intercept is b.

The six graphs shown in Figure 2.30 represent the most commonly used functions in algebra. Familiarity with the basic characteristics of these simple graphs will help you analyze the shapes of more complicated graphs.

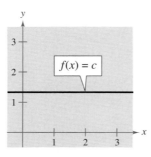

(a) Constant Function

$f(x) = c$

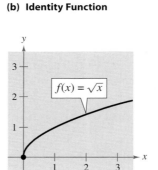

(b) Identity Function

$f(x) = x$

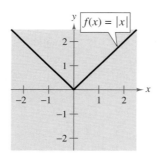

(c) Absolute Value Function

$f(x) = |x|$

(d) Square Root Function

$f(x) = \sqrt{x}$

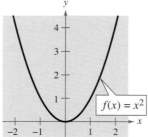

(e) Quadratic Function

$f(x) = x^2$

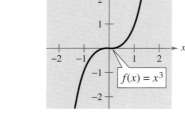

(f) Cubic Function

$f(x) = x^3$

FIGURE 2.30

Shifting Graphs

Many functions have graphs that are simple transformations of the common graphs summarized on page 214. For example, you can obtain the graph of

$$h(x) = x^2 + 2$$

by shifting the graph of $f(x) = x^2$ *up* 2 units, as shown in Figure 2.31. In function notation, h and f are related as follows.

$$h(x) = x^2 + 2$$
$$= f(x) + 2 \qquad \text{Upward shift of 2}$$

Similarly, you can obtain the graph of

$$g(x) = (x - 2)^2$$

by shifting the graph of $f(x) = x^2$ to the *right* 2 units, as shown in Figure 2.32. In this case, the functions g and f have the following relationship.

$$g(x) = (x - 2)^2$$
$$= f(x - 2) \qquad \text{Right shift of 2}$$

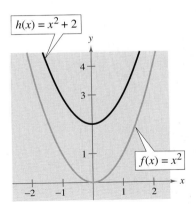

FIGURE 2.31

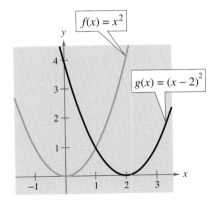

FIGURE 2.32

The following list summarizes this discussion about horizontal and vertical shifts.

Vertical and Horizontal Shifts

Let c be a positive real number. **Vertical and horizontal shifts** in the graph of $y = f(x)$ are represented as follows.

1. Vertical shift c units *upward:* $h(x) = f(x) + c$

2. Vertical shift c units *downward:* $h(x) = f(x) - c$

3. Horizontal shift c units to the *right:* $h(x) = f(x - c)$

4. Horizontal shift c units to the *left:* $h(x) = f(x + c)$

In items 3 and 4, be sure you see that $h(x) = f(x - c)$ corresponds to a *right* shift and $h(x) = f(x + c)$ corresponds to a *left* shift for $c > 0$.

Some graphs can be obtained from a combination of vertical and horizontal shifts, as demonstrated in Example 1(b). Vertical and horizontal shifts generate a *family of functions*, each with the same shape but at different locations in the plane.

Example 1 ▶ Shifts in the Graph of a Function

Use the graph of $f(x) = x^3$ to sketch the graph of each function.

a. $g(x) = x^3 - 1$

b. $h(x) = (x + 2)^3 + 1$

Solution

Relative to the graph of $f(x) = x^3$, the graph of $g(x) = x^3 - 1$ is a downward shift of 1 unit. The graph of $h(x) = (x + 2)^3 + 1$ involves a left shift of 2 units *and* an upward shift of 1 unit. The graphs of both functions are compared with the graph of $f(x) = x^3$ in Figure 2.33.

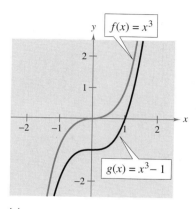

(a)

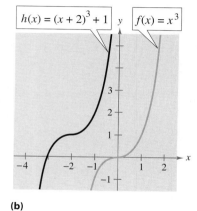

(b)

FIGURE 2.33

In part (b) of Figure 2.33, notice that the same result is obtained if the vertical shift precedes the horizontal shift *or* if the horizontal shift precedes the vertical shift.

◀ Exploration ▶

Graphing utilities are ideal tools for exploring translations of functions. Graph f, g, and h on the same screen. Before looking at the graphs, try to predict how the graphs of g and h relate to the graph of f.

a. $f(x) = x^2$, $g(x) = (x - 4)^2$, $h(x) = (x - 4)^2 + 3$

b. $f(x) = x^2$, $g(x) = (x + 1)^2$, $h(x) = (x + 1)^2 - 2$

c. $f(x) = x^2$, $g(x) = (x + 4)^2$, $h(x) = (x + 4)^2 + 2$

Reflecting Graphs

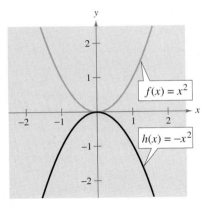

FIGURE 2.34

A computer animation of this concept appears in the *Interactive* CD-ROM and *Internet* versions of this text.

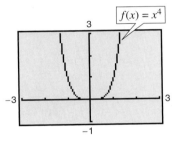

FIGURE 2.35

The second common type of transformation is a **reflection.** For instance, if you consider the x-axis to be a mirror, the graph of

$$h(x) = -x^2$$

is the mirror image (or reflection) of the graph of $f(x) = x^2$, as shown in Figure 2.34.

Reflections in the Coordinate Axes

Reflections in the coordinate axes of the graph of $y = f(x)$ are represented as follows.

1. Reflection in the x-axis: $\qquad h(x) = -f(x)$

2. Reflection in the y-axis: $\qquad h(x) = f(-x)$

Example 2 ▶ Finding Equations from Graphs

The graph of the function

$$f(x) = x^4$$

is shown in Figure 2.35. Each of the graphs in Figure 2.36 is a transformation of the graph of f. Find an equation for each of these functions.

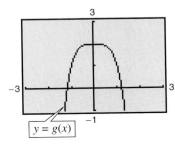

(a)

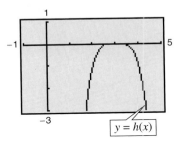

(b)

FIGURE 2.36

Solution

a. The graph of g is a reflection in the x-axis *followed by* an upward shift of 2 units of the graph of $f(x) = x^4$. So, the equation for g is

$$g(x) = -x^4 + 2.$$

b. The graph of h is a horizontal shift of 3 units to the right *followed by* a reflection in the x-axis of the graph of $f(x) = x^4$. So, the equation for h is

$$h(x) = -(x - 3)^4.$$

Example 3 ▶ Reflections and Shifts

Compare the graph of each function with the graph of $f(x) = \sqrt{x}$.

a. $g(x) = -\sqrt{x}$ **b.** $h(x) = \sqrt{-x}$ **c.** $k(x) = -\sqrt{x + 2}$

Solution

a. The graph of g is a reflection of the graph of f in the x-axis because

$$g(x) = -\sqrt{x}$$
$$= -f(x).$$

b. The graph of h is a reflection of the graph of f in the y-axis because

$$h(x) = \sqrt{-x}$$
$$= f(-x).$$

c. The graph of k is a left shift of 2 units, followed by a reflection in the x-axis because

$$k(x) = -\sqrt{x + 2}$$
$$= -f(x + 2).$$

The graphs of all three functions are shown in Figure 2.37.

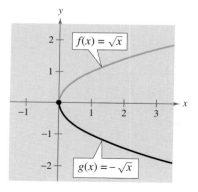

(a)

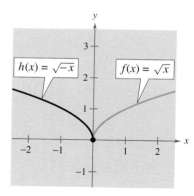

(b)

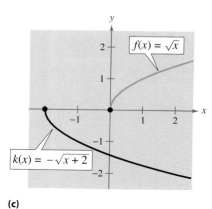

(c)

FIGURE 2.37

When sketching the graphs of functions involving square roots, remember that the domain must be restricted to exclude negative numbers inside the radical. For instance, here are the domains of the functions in Example 3.

Domain of $g(x) = -\sqrt{x}$: $x \geq 0$

Domain of $h(x) = \sqrt{-x}$: $x \leq 0$

Domain of $k(x) = -\sqrt{x + 2}$: $x \geq -2$

A computer animation of this concept appears in the *Interactive* CD-ROM and *Internet* versions of this text.

Nonrigid Transformations

Horizontal shifts, vertical shifts, and reflections are **rigid transformations** because the basic shape of the graph is unchanged. These transformations change only the *position* of the graph in the xy-plane. **Nonrigid transformations** are those that cause a *distortion*—a change in the shape of the original graph. For instance, a nonrigid transformation of the graph of $y = f(x)$ is represented by $g(x) = cf(x)$, where the transformation is a **vertical stretch** if $c > 1$ and a **vertical shrink** if $0 < c < 1$.

◀ **E x p l o r a t i o n** ▶

Sketch the graphs of $f(x) = 2x^2$ and $h(x) = x^2$ on the same set of axes. Describe the effect of multiplying x^2 by a number greater than 1. Then sketch the graphs of $g(x) = \frac{1}{2}x^2$ and $h(x) = x^2$ on the same set of axes. Describe the effect of multiplying x^2 by a number less than 1. Can you think of an easy way to remember this generalization?

Example 4 ▶ **Nonrigid Transformations**

Compare the graph of each function with the graph of $f(x) = |x|$.

a. $h(x) = 3|x|$

b. $g(x) = \frac{1}{3}|x|$

Solution

a. Relative to the graph of $f(x) = |x|$, the graph of
$$h(x) = 3|x|$$
$$= 3f(x)$$
is a vertical stretch (each y-value is multiplied by 3) of the graph of f.

b. Similarly, the function
$$g(x) = \frac{1}{3}|x|$$
$$= \frac{1}{3}f(x)$$
indicates that the graph of g is a vertical shrink $\left(\text{each } y\text{-value is multiplied by } \frac{1}{3}\right)$ of the graph of f.

The graphs of both functions are shown in Figure 2.38.

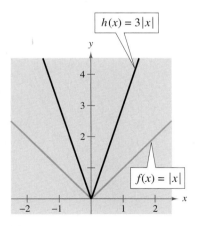

(a)

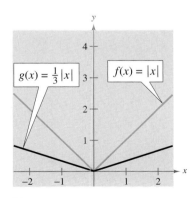

(b)

FIGURE 2.38

2.4 Exercises

1. Sketch (on the same set of coordinate axes) a graph of each function for $c = -2, 0$, and 2.

(a) $f(x) = x^3 + c$

(b) $f(x) = (x - c)^3$

2. Sketch (on the same set of coordinate axes) a graph of each function for $c = -1, 1$, and 3.

(a) $f(x) = |x| + c$

(b) $f(x) = |x - c|$

(c) $f(x) = |x + 4| + c$

3. Sketch (on the same set of coordinate axes) a graph of each function for $c = -3, -1, 1$, and 3.

(a) $f(x) = \sqrt{x} + c$

(b) $f(x) = \sqrt{x - c}$

(c) $f(x) = \sqrt{x - 3} + c$

4. Sketch (on the same set of coordinate axes) a graph of each function for $c = -3, -1, 1$, and 3.

(a) $f(x) = \begin{cases} x^2 + c, & x < 0 \\ -x^2 + c, & x \geq 0 \end{cases}$

(b) $f(x) = \begin{cases} (x + c)^2, & x < 0 \\ -(x + c)^2, & x \geq 0 \end{cases}$

5. Use the graph of f to sketch each graph.

(a) $y = f(x) + 2$

(b) $y = f(x - 2)$

(c) $y = 2f(x)$

(d) $y = -f(x)$

(e) $y = f(x + 3)$

(f) $y = f(-x)$

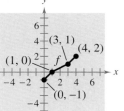

6. Use the graph of f to sketch each graph.

(a) $y = f(-x)$

(b) $y = f(x) + 4$

(c) $y = 2f(x)$

(d) $y = -f(x - 4)$

(e) $y = f(x) - 3$

(f) $y = -f(x) - 1$

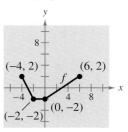

7. Use the graph of f to sketch each graph.

(a) $y = f(x) - 1$

(b) $y = f(x - 1)$

(c) $y = f(-x)$

(d) $y = f(x + 1)$

(e) $y = -f(x - 2)$

(f) $y = \frac{1}{2}f(x)$

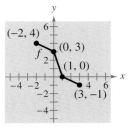

8. Use the graph of f to sketch each graph.

(a) $y = f(x - 5)$

(b) $y = -f(x) + 3$

(c) $y = \frac{1}{3}f(x)$

(d) $y = -f(x + 1)$

(e) $y = f(-x)$

(f) $y = f(x) - 10$

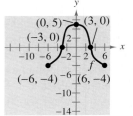

9. Use the graph of $f(x) = x^2$ to write an equation for each function whose graph is shown.

(a)

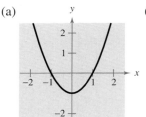

(b)

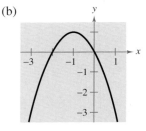

(c)

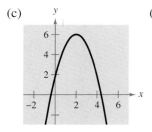

(d)

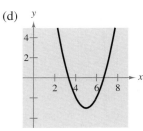

10. Use the graph of $f(x) = x^3$ to write an equation for each function whose graph is shown.

(a)
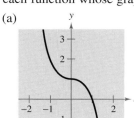

(b)
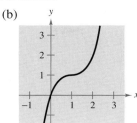

(c)
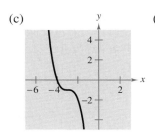

(d)
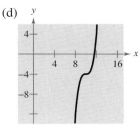

11. Use the graph of $f(x) = |x|$ to write an equation for each function whose graph is shown.

(a)
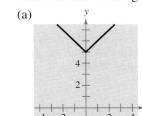

(b)
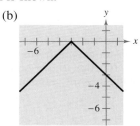

(c)
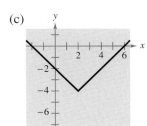

(d)
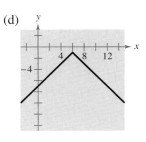

12. Use the graph of $f(x) = \sqrt{x}$ to write an equation for each function whose graph is shown.

(a)
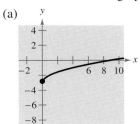

(b)
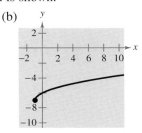

(c)
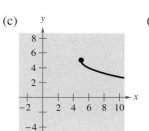

(d)
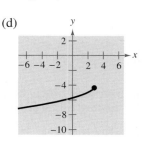

In Exercises 13–18, identify the common function and the transformation shown in the graph. Write an equation for the function shown in the graph.

13.
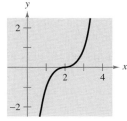

14.
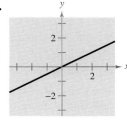

15.
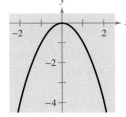

16.
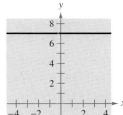

17.
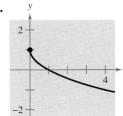

18.
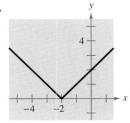

In Exercises 19–34, describe the transformation that occurs in the function. Then sketch its graph.

19. $f(x) = 12 - x^2$ **20.** $f(x) = (x - 8)^2$

21. $f(x) = x^3 + 7$ **22.** $f(x) = -x^3 - 1$

23. $f(x) = 2 - (x + 5)^2$ **24.** $f(x) = -(x + 10)^2 + 5$

25. $f(x) = (x - 1)^3 + 2$ **26.** $f(x) = (x + 3)^3 - 10$

27. $f(x) = -|x| - 2$ **28.** $f(x) = 6 - |x + 5|$

29. $f(x) = -|x + 4| + 8$ **30.** $f(x) = |-x + 3| + 9$

31. $f(x) = \sqrt{x - 9}$ **32.** $f(x) = \sqrt{x + 4} + 8$

33. $f(x) = \sqrt{7 - x} - 2$ **34.** $f(x) = -\sqrt{x + 1} - 6$

In Exercises 35–42, write an equation for the function that is described by the given characteristics.

35. The shape of $f(x) = x^2$, but moved 2 units to the right and 8 units down

36. The shape of $f(x) = x^2$, but moved 3 units to the left, 7 units up, and reflected in the x-axis

37. The shape of $f(x) = x^3$, but moved 13 units to the right

38. The shape of $f(x) = x^3$, but moved 6 units to the left, 6 units down, and reflected in the y-axis

39. The shape of $f(x) = |x|$, but moved 10 units up and reflected in the x-axis

40. The shape of $f(x) = |x|$, but moved 1 unit to the left and 7 units down

41. The shape of $f(x) = \sqrt{x}$, but moved 6 units to the left and reflected in both the x-axis and the y-axis

42. The shape of $f(x) = \sqrt{x}$, but moved 9 units down and reflected in both the x-axis and the y-axis

43. Use the graph of $f(x) = x^2$ to write an equation for each function whose graph is shown.

(a)
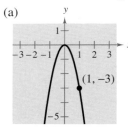

(b)
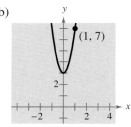

44. Use the graph of $f(x) = x^3$ to write an equation for each function whose graph is shown.

(a)
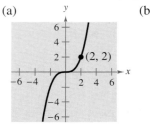

(b)
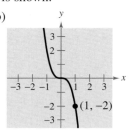

45. Use the graph of $f(x) = |x|$ to write an equation for each function whose graph is shown.

(a)
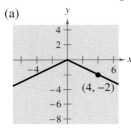

(b)
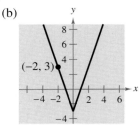

46. Use the graph of $f(x) = \sqrt{x}$ to write an equation for each function whose graph is shown.

(a)
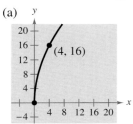

(b)
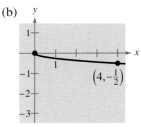

In Exercises 47–52, identify the common function and the transformation shown in the graph. Write an equation for the function shown in the graph. Then use a graphing utility to verify your answer.

47.
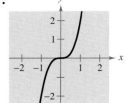

48.
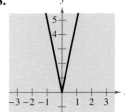

49.
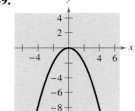

50.
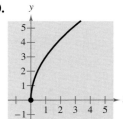

51.

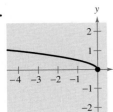

52.

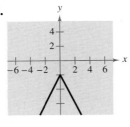

⊞ *Graphical Analysis*   In Exercises 53–56, use the viewing window shown to write a possible equation for the transformation of the common function.

53.

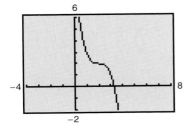

54.

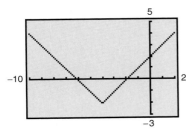

55.

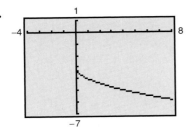

56.

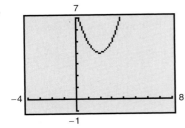

Graphical Reasoning   **In Exercises 57 and 58, use the graph of f to sketch the graph of g.**

57.

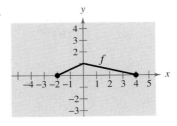

(a) $g(x) = f(x) + 2$ (b) $g(x) = f(x) - 1$

(c) $g(x) = f(-x)$ (d) $g(x) = -2f(x)$

58.

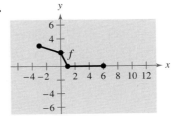

(a) $g(x) = f(x) - 5$ (b) $g(x) = f(x) + \frac{1}{2}$

(c) $g(x) = f(-x)$ (d) $g(x) = -4f(x)$

59. *Fuel Use*   The amount of fuel F (in billions of gallons) used by trucks from 1980 through 1996 can be approximated by the function

$$F = f(t) = 20.46 + 0.04t^2$$

where $t = 0$ represents 1980.   (Source: U.S. Federal Highway Administration)

(a) Describe the transformation of the common function $f(x) = x^2$. Then sketch the graph over the interval $0 \le t \le 16$.

(b) Rewrite the function so that $t = 0$ represents 1990. Explain how you got your answer.

60. *Finance*   The amount (in trillions of dollars) of mortgage debt M outstanding in the United States from 1985 through 1997 can be approximated by the function

$$M = f(t) = 1.5\sqrt{t} - 1.25$$

where $t = 5$ represents 1985.   (Source: Board of Governors of the Federal Reserve System)

(a) Describe the transformation of the common function $f(x) = \sqrt{x}$. Then sketch the graph over the interval $5 \le t \le 17$.

(b) Rewrite the function so that $t = 5$ represents 1995. Explain how you got your answer.

Synthesis

In Exercises 61 and 62, determine whether the statement is true or false. Justify your answer.

61. The graphs of $f(x) = |x| + 6$ and $f(x) = |-x| + 6$ are identical.

62. If the graph of the common function $f(x) = x^2$ is moved 6 units to the right, 3 units up, and reflected in the x-axis, then the point $(-2, 19)$ will lie on the graph of the transformation.

63. *Describing Profits* Management originally predicted that the profits from the sales of a new product would be approximated by the graph of the function f shown. The actual profits are shown by the function g along with a verbal description. Use the concepts of transformations of graphs to write g in terms of f.

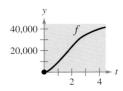

(a) The profits were only three-fourths as large as expected.

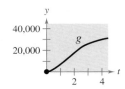

(b) The profits were consistently $10,000 greater than predicted.

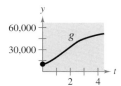

(c) There was a 2-year delay in the introduction of the product. After sales began, profits grew as expected.

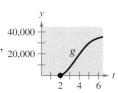

64. Explain why the graph of $y = -f(x)$ is a reflection of the graph of $y = f(x)$ about the x-axis.

65. The graph of $y = f(x)$ passes through the points $(0, 1)$, $(1, 2)$, and $(2, 3)$. Find the corresponding points on the graph of $y = f(x + 2) - 1$.

66. *Think About It* You can use two methods to graph a function: plotting points, or translating a common function as shown in this section. Which method of graphing do you prefer to use? Explain.

Review

In Exercises 67–74, perform the operations and simplify.

67. $\dfrac{4}{x} + \dfrac{4}{1 - x}$

68. $\dfrac{2}{x + 5} - \dfrac{2}{x - 5}$

69. $\dfrac{3}{x - 1} - \dfrac{2}{x(x - 1)}$

70. $\dfrac{x}{x - 5} + \dfrac{1}{2}$

71. $(x - 4)\left(\dfrac{1}{\sqrt{x^2 - 4}}\right)$

72. $\left(\dfrac{x}{x^2 - 4}\right)\left(\dfrac{x^2 - x - 2}{x^2}\right)$

73. $(x^2 - 9) \div \left(\dfrac{x + 3}{5}\right)$

74. $\left(\dfrac{x}{x^2 - 3x - 28}\right) \div \left(\dfrac{x^2 + 3x}{x^2 + 5x + 4}\right)$

In Exercises 75 and 76, evaluate the function at the specified values of the independent variable and simplify.

75. $f(x) = x^2 - 6x + 11$

  (a) $f(-3)$ (b) $f\left(-\tfrac{1}{2}\right)$ (c) $f(x - 3)$

76. $f(x) = \sqrt{x + 10} - 3$

  (a) $f(-10)$ (b) $f(26)$ (c) $f(x - 10)$

In Exercises 77–80, find the domain of the function.

77. $f(x) = \dfrac{2}{11 - x}$

78. $f(x) = \dfrac{\sqrt{x - 3}}{x - 8}$

79. $f(x) = \sqrt{81 - x^2}$

80. $f(x) = \sqrt[3]{4 - x^2}$

2.5 Combinations of Functions

▶ **What you should learn**

- How to add, subtract, multiply, and divide functions
- How to find compositions of one function with another function
- How to use combinations of functions to model and solve real-life problems

▶ **Why you should learn it**

Combinations of functions can be used to model and solve real-life problems. For instance, Exercises 33 and 34 on page 231 show how to use combinations of functions to analyze U.S. health expenditures.

Charles Gupton/Tony Stone Images

A computer animation of this concept appears in the *Interactive* CD-ROM and *Internet* versions of this text.

Arithmetic Combinations of Functions

Just as two real numbers can be combined by the operations of addition, subtraction, multiplication, and division to form other real numbers, two *functions* can be combined to create new functions. For example, the functions $f(x) = 2x - 3$ and $g(x) = x^2 - 1$ can be combined to form the sum, difference, product, and quotient of f and g.

$$f(x) + g(x) = (2x - 3) + (x^2 - 1)$$
$$= x^2 + 2x - 4 \qquad \text{Sum}$$
$$f(x) - g(x) = (2x - 3) - (x^2 - 1)$$
$$= -x^2 + 2x - 2 \qquad \text{Difference}$$
$$f(x)g(x) = (2x - 3)(x^2 - 1)$$
$$= 2x^3 - 3x^2 - 2x + 3 \qquad \text{Product}$$
$$\frac{f(x)}{g(x)} = \frac{2x - 3}{x^2 - 1}, \quad x \neq \pm 1 \qquad \text{Quotient}$$

The domain of an **arithmetic combination** of functions f and g consists of all real numbers that are common to the domains of f and g. In the case of the quotient $f(x)/g(x)$, there is the further restriction that $g(x) \neq 0$.

Sum, Difference, Product, and Quotient of Functions

Let f and g be two functions with overlapping domains. Then, for all x common to both domains, the *sum, difference, product,* and *quotient* of f and g are defined as follows.

1. *Sum:* $\qquad (f + g)(x) = f(x) + g(x)$

2. *Difference:* $\qquad (f - g)(x) = f(x) - g(x)$

3. *Product:* $\qquad (fg)(x) = f(x) \cdot g(x)$

4. *Quotient:* $\qquad \left(\dfrac{f}{g}\right)(x) = \dfrac{f(x)}{g(x)}, \qquad g(x) \neq 0$

Example 1 ▶ Finding the Sum of Two Functions

Given $f(x) = 2x + 1$ and $g(x) = x^2 + 2x - 1$, find $(f + g)(x)$. Then evaluate the sum when $x = -1$.

Solution

$$(f + g)(x) = f(x) + g(x)$$
$$= (2x + 1) + (x^2 + 2x - 1)$$
$$= x^2 + 4x$$

When $x = -1$, the value of the sum is $(f + g)(-1) = (-1)^2 + 4(-1) = -3$.

Example 2 ▶   Finding the Difference of Two Functions

Given $f(x) = 2x + 1$ and $g(x) = x^2 + 2x - 1$, find $(f - g)(x)$. Then evaluate the difference when $x = 2$.

Solution

The difference of f and g is

$$(f - g)(x) = f(x) - g(x)$$
$$= (2x + 1) - (x^2 + 2x - 1)$$
$$= -x^2 + 2.$$

When $x = 2$, the value of this difference is

$$(f - g)(2) = -(2)^2 + 2$$
$$= -2.$$

In Examples 1 and 2, both f and g have domains that consist of all real numbers. So, the domains of $(f + g)$ and $(f - g)$ are also the set of all real numbers. Remember that any restrictions on the domains of f and g must be considered when forming the sum, difference, product, or quotient of f and g.

Example 3 ▶   Finding the Domains of Quotients of Functions

Find the domains of $\left(\dfrac{f}{g}\right)(x)$ and $\left(\dfrac{g}{f}\right)(x)$ for the functions

$$f(x) = \sqrt{x} \quad \text{and} \quad g(x) = \sqrt{4 - x^2}.$$

Solution

The quotient of f and g is

$$\left(\frac{f}{g}\right)(x) = \frac{f(x)}{g(x)} = \frac{\sqrt{x}}{\sqrt{4 - x^2}}$$

and the quotient of g and f is

$$\left(\frac{g}{f}\right)(x) = \frac{g(x)}{f(x)} = \frac{\sqrt{4 - x^2}}{\sqrt{x}}.$$

The domain of f is $[0, \infty)$ and the domain of g is $[-2, 2]$. The intersection of these domains is $[0, 2]$. So, the domains of $\left(\dfrac{f}{g}\right)$ and $\left(\dfrac{g}{f}\right)$ are as follows.

$$\text{Domain of } \left(\frac{f}{g}\right): [0, 2) \qquad \text{Domain of } \left(\frac{g}{f}\right): (0, 2]$$

Can you see why these two domains differ slightly?

Composition of Functions

A computer animation of this concept appears in the *Interactive* CD-ROM and *Internet* versions of this text.

Another way of combining two functions is to form the **composition** of one with the other. For instance, if $f(x) = x^2$ and $g(x) = x + 1$, the composition of f with g is

$$f(g(x)) = f(x + 1)$$
$$= (x + 1)^2.$$

This composition is denoted as $(f \circ g)$.

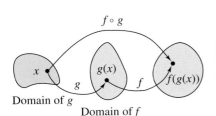

Domain of g

Domain of f

FIGURE **2.39**

Definition of Composition of Two Functions

The **composition** of the function f with the function g is

$$(f \circ g)(x) = f(g(x)).$$

The domain of $(f \circ g)$ is the set of all x in the domain of g such that $g(x)$ is in the domain of f. (See Figure 2.39.)

The following tables of values help illustrate the composition of the functions f and g given in Example 4.

x	0	1	2	3
$g(x)$	4	3	0	−5

$g(x)$	4	3	0	−5
$f(g(x))$	6	5	2	−3

x	0	1	2	3
$f(g(x))$	6	5	2	−3

Note that the first two tables can be combined (or "composed") to produce the values given in the third table.

Example 4 ▶ Composition of Functions

Given $f(x) = x + 2$ and $g(x) = 4 - x^2$, find the following.

a. $(f \circ g)(x)$ **b.** $(g \circ f)(x)$ **c.** $(g \circ f)(-2)$

Solution

a. The composition of f with g is as follows.

$$(f \circ g)(x) = f(g(x)) \qquad \text{Definition of } f \circ g$$
$$= f(4 - x^2) \qquad \text{Definition of } g(x)$$
$$= (4 - x^2) + 2 \qquad \text{Definition of } f(x)$$
$$= -x^2 + 6 \qquad \text{Simplify.}$$

b. The composition of g with f is as follows.

$$(g \circ f)(x) = g(f(x)) \qquad \text{Definition of } g \circ f$$
$$= g(x + 2) \qquad \text{Definition of } f(x)$$
$$= 4 - (x + 2)^2 \qquad \text{Definition of } g(x)$$
$$= 4 - (x^2 + 4x + 4) \qquad \text{Expand.}$$
$$= -x^2 - 4x \qquad \text{Simplify.}$$

Note that, in this case, $(f \circ g)(x) \neq (g \circ f)(x)$.

c. Using the result of part (b), you can write the following.

$$(g \circ f)(-2) = -(-2)^2 - 4(-2) \qquad \text{Substitute.}$$
$$= -4 + 8 \qquad \text{Simplify.}$$
$$= 4 \qquad \text{Simplify.}$$

Example 5 ▶   **Finding the Domain of a Composite Function**

Find the composition $(f \circ g)(x)$ for the functions

$$f(x) = x^2 - 9 \quad \text{and} \quad g(x) = \sqrt{9 - x^2}.$$

Then find the domain of $(f \circ g)$.

Solution

$$
\begin{aligned}
(f \circ g)(x) &= f(g(x)) \\
&= f\left(\sqrt{9 - x^2}\right) \\
&= \left(\sqrt{9 - x^2}\right)^2 - 9 \\
&= 9 - x^2 - 9 \\
&= -x^2
\end{aligned}
$$

From this, it might appear that the domain of the composition is the set of all real numbers. This, however, is not true because the domain of g is $-3 \le x \le 3$.

In Examples 4 and 5 you formed the composition of two given functions. In calculus, it is also important to be able to identify two functions that make up a given composite function. For instance, the function h given by

$$h(x) = (3x - 5)^3$$

is the composition of f with g, where $f(x) = x^3$ and $g(x) = 3x - 5$. That is,

$$h(x) = (3x - 5)^3 = [g(x)]^3 = f(g(x)).$$

Basically, to "decompose" a composite function, look for an "inner" and an "outer" function. In the function h above, $g(x) = 3x - 5$ is the inner function and $f(x) = x^3$ is the outer function.

Example 6 ▶   **Identifying a Composite Function**

Express the function

$$h(x) = \frac{1}{(x - 2)^2}$$

as a composition of two functions.

Solution

One way to write h as a composition of two functions is to take the inner function to be $g(x) = x - 2$ and the outer function to be

$$f(x) = \frac{1}{x^2} = x^{-2}.$$

Then you can write

$$h(x) = \frac{1}{(x - 2)^2} = (x - 2)^{-2} = f(x - 2) = f(g(x)).$$

Application

Example 7 ▶ Bacteria Count

The number N of bacteria in a refrigerated food is

$$N(T) = 20T^2 - 80T + 500, \qquad 2 \le T \le 14$$

where T is the temperature of the food. When the food is removed from refrigeration, the temperature is

$$T(t) = 4t + 2, \qquad 0 \le t \le 3$$

where t is the time in hours. (a) Find the composite $N(T(t))$ and interpret its meaning in context. (b) Find the time when the bacterial count reaches 2000.

Solution

a. $N(T(t)) = 20(4t + 2)^2 - 80(4t + 2) + 500$

$$= 20(16t^2 + 16t + 4) - 320t - 160 + 500$$

$$= 320t^2 + 320t + 80 - 320t - 160 + 500$$

$$= 320t^2 + 420$$

The composite function $N(T(t))$ represents the number of bacteria in the food as a function of time.

b. The bacterial count will reach 2000 when $320t^2 + 420 = 2000$. Solve this equation to find that the count will reach 2000 when $t \approx 2.2$ hours. When you solve this equation, note that the negative value is rejected because it is not in the domain of the composite function.

Writing **ABOUT MATHEMATICS**

Analyzing Arithmetic Combinations of Functions

a. Use the graphs of f and $(f + g)$ in Figure 2.40 to make a table showing the values of $g(x)$ when $x = 1, 2, 3, 4, 5,$ and 6. Explain your reasoning.

b. Use the graphs of f and $(f - h)$ in Figure 2.40 to make a table showing the values of $h(x)$ when $x = 1, 2, 3, 4, 5,$ and 6. Explain your reasoning.

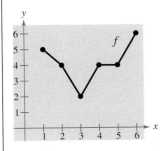
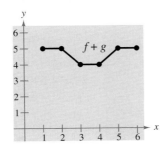
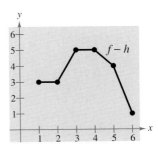

FIGURE **2.40**

2.5 Exercises

In Exercises 1–4, use the graphs of f and g to graph $h(x) = (f + g)(x)$.

1.

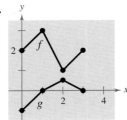

2.

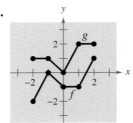

3.

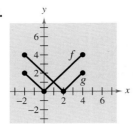

4.

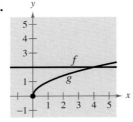

In Exercises 5–12, find (a) $(f + g)(x)$, (b) $(f - g)(x)$, (c) $(fg)(x)$, and (d) $(f/g)(x)$. What is the domain of f/g?

5. $f(x) = x + 2$, $g(x) = x - 2$

6. $f(x) = 2x - 5$, $g(x) = 2 - x$

7. $f(x) = x^2$, $g(x) = 2 - x$

8. $f(x) = 2x - 5$, $g(x) = 4$

9. $f(x) = x^2 + 6$, $g(x) = \sqrt{1 - x}$

10. $f(x) = \sqrt{x^2 - 4}$, $g(x) = \dfrac{x^2}{x^2 + 1}$

11. $f(x) = \dfrac{1}{x}$, $g(x) = \dfrac{1}{x^2}$

12. $f(x) = \dfrac{x}{x + 1}$, $g(x) = x^3$

In Exercises 13–24, evaluate the indicated function for $f(x) = x^2 + 1$ and $g(x) = x - 4$.

13. $(f + g)(2)$

14. $(f - g)(-1)$

15. $(f - g)(0)$

16. $(f + g)(1)$

17. $(f - g)(3t)$

18. $(f + g)(t - 2)$

19. $(fg)(6)$

20. $(fg)(-6)$

21. $\left(\dfrac{f}{g}\right)(5)$

22. $\left(\dfrac{f}{g}\right)(0)$

23. $\left(\dfrac{f}{g}\right)(-1) - g(3)$

24. $(2f)(5)$

In Exercises 25–28, graph the functions $f, g,$ and $f + g$ on the same set of coordinate axes.

25. $f(x) = \frac{1}{2}x$, $g(x) = x - 1$

26. $f(x) = \frac{1}{3}x$, $g(x) = -x + 4$

27. $f(x) = x^2$, $g(x) = -2x$

28. $f(x) = 4 - x^2$, $g(x) = x$

▦ *Graphical Reasoning* In Exercises 29 and 30, use a graphing utility to sketch the graphs of $f, g,$ and $f + g$ in the same viewing window. Which function contributes most to the magnitude of the sum when $0 \le x \le 2$? Which function contributes most to the magnitude of the sum when $x > 6$?

29. $f(x) = 3x$, $g(x) = -\dfrac{x^3}{10}$

30. $f(x) = \dfrac{x}{2}$, $g(x) = \sqrt{x}$

31. *Stopping Distance* While traveling in a car at x miles per hour, you are required to stop quickly to avoid an accident. The distance (in feet) the car travels during your reaction time is given by $R(x) = \frac{3}{4}x$. The distance (in feet) traveled while you are braking is

$$B(x) = \frac{1}{15}x^2.$$

Find the function that represents the total stopping distance (in feet) T. Graph the functions $R, B,$ and T on the same set of coordinate axes for $0 \le x \le 60$.

▦ **32.** *Business* You own two restaurants. From 1995 to 2000, the sales R_1 (in thousands of dollars) for one restaurant can be modeled by

$$R_1 = 480 - 8t - 0.8t^2, \qquad t = 0, 1, 2, 3, 4, 5$$

where $t = 0$ represents 1995. During the same 6-year period, the sales R_2 (in thousands of dollars) for the second restaurant can be modeled by

$$R_2 = 254 + 0.78t, \qquad t = 0, 1, 2, 3, 4, 5.$$

Write a function that represents the total sales for the two restaurants. Use a graphing utility to graph the total sales function.

Health Care Costs In Exercises 33 and 34, use the table, which gives the total amount spent (in billions of dollars) on health services and supplies in the United States (including Puerto Rico) for the years 1990 through 1996. The variables y_1, y_2, and y_3 represent out-of-pocket payments, insurance premiums, and other types of payments, respectively. (Source: U.S. Health Care Financing Administration)

Year	1990	1991	1992	1993	1994	1995	1996
y_1	144.4	151.6	159.5	163.6	164.8	166.7	171.2
y_2	238.6	259.4	282.5	303.3	315.6	326.9	337.3
y_3	21.8	24.0	25.1	27.3	29.6	31.7	32.4

33. Use a graphing utility to find a mathematical model for each of the variables. Let $t = 0$ represent 1990. Find a quadratic model for y_1 and linear models for y_2 and y_3.

34. Use a graphing utility to graph y_1, y_2, y_3, and $y_1 + y_2 + y_3$ in the same viewing window. Use the model to estimate the total amount spent on health services and supplies in the year 2000.

35. *Graphical Reasoning* An electronically controlled thermostat in a home is programmed to lower the temperature automatically during the night. The temperature in the house T (in degrees Fahrenheit) is given in terms of t, the time in hours on a 24-hour clock (see figure).

(a) Explain why T is a function of t.

(b) Approximate $T(4)$ and $T(15)$.

(c) Suppose the thermostat were reprogrammed to produce a temperature H where $H(t) = T(t - 1)$. How would this change the temperature?

(d) Suppose the thermostat were reprogrammed to produce a temperature H where $H(t) = T(t) - 1$. How would this change the temperature?

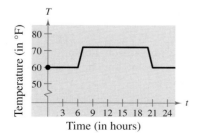

36. *Think About It* Write a piecewise-defined function that represents the graph in Exercise 35.

In Exercises 37–40, find (a) $f \circ g$, (b) $g \circ f$, and (c) $f \circ f$.

37. $f(x) = x^2$, $\qquad$ $g(x) = x - 1$

38. $f(x) = \sqrt[3]{x - 1}$, $\qquad$ $g(x) = x^3 + 1$

39. $f(x) = 3x + 5$, $\qquad$ $g(x) = 5 - x$

40. $f(x) = x^3$, $\qquad$ $g(x) = \dfrac{1}{x}$

In Exercises 41–50, find (a) $f \circ g$ and (b) $g \circ f$. Find the domain of each function and each composite function.

41. $f(x) = \sqrt{x + 4}$, $\qquad$ $g(x) = x^2$

42. $f(x) = \sqrt[3]{x - 5}$, $\qquad$ $g(x) = x^3 + 1$

43. $f(x) = \frac{1}{3}x - 3$, $\qquad$ $g(x) = 3x + 1$

44. $f(x) = x^2 + 1$, $\qquad$ $g(x) = \sqrt{x}$

45. $f(x) = x^4$, $\qquad$ $g(x) = x^4$

46. $f(x) = \sqrt{x}$, $\qquad$ $g(x) = 2x - 3$

47. $f(x) = |x|$, $\qquad$ $g(x) = x + 6$

48. $f(x) = x^{2/3}$, $\qquad$ $g(x) = x^6$

49. $f(x) = \dfrac{1}{x}$, $\qquad$ $g(x) = x + 3$

50. $f(x) = \dfrac{3}{x^2 - 1}$, $\qquad$ $g(x) = x + 1$

In Exercises 51–54, use the graphs of f and g to evaluate the functions.

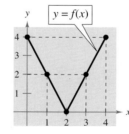

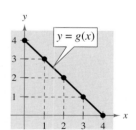

51. (a) $(f + g)(3)$ $\qquad$ (b) $\left(\dfrac{f}{g}\right)(2)$

52. (a) $(f - g)(1)$ $\qquad$ (b) $(fg)(4)$

53. (a) $(f \circ g)(2)$ $\qquad$ (b) $(g \circ f)(2)$

54. (a) $(f \circ g)(1)$ $\qquad$ (b) $(g \circ f)(3)$

In Exercises 55–62, find two functions f and g such that $(f \circ g)(x) = h(x)$. (There is more than one correct answer.)

55. $h(x) = (2x + 1)^2$

56. $h(x) = (1 - x)^3$

57. $h(x) = \sqrt[3]{x^2 - 4}$

58. $h(x) = \sqrt{9 - x}$

59. $h(x) = \dfrac{1}{x + 2}$

60. $h(x) = \dfrac{4}{(5x + 2)^2}$

61. $h(x) = \dfrac{-x^2 + 3}{4 - x^2}$

62. $h(x) = \dfrac{27x^3 + 6x}{10 - 27x^3}$

63. *Geometry* A square concrete foundation was prepared as a base for a cylindrical tank (see figure).

(a) Express the radius r of the tank as a function of the length x of the sides of the square.

(b) Express the area A of the circular base of the tank as a function of the radius r.

(c) Find and interpret $(A \circ r)(x)$.

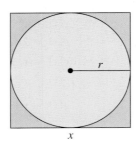

64. *Physics* A pebble is dropped into a calm pond, causing ripples in the form of concentric circles. The radius (in feet) of the outer ripple is $r(t) = 0.6t$, where t is the time in seconds after the pebble strikes the water. The area of the circle is given by the function $A(r) = \pi r^2$. Find and interpret $(A \circ r)(t)$.

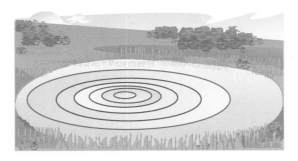

65. *Economics* The weekly cost of producing x units in a manufacturing process is given by the function

$$C(x) = 60x + 750.$$

The number of units produced in t hours is $x(t) = 50t$. Find and interpret $(C \circ x)(t)$.

Synthesis

True or False? In Exercises 66 and 67, determine whether the statement is true or false. Justify your answer.

66. If $f(x) = x + 1$ and $g(x) = 6x$, then $(f \circ g)(x) = (g \circ f)(x)$.

67. If you are given two functions $f(x)$ and $g(x)$, you can calculate $(f \circ g)(x)$ if and only if the range of g is a subset of the domain of f.

68. *Think About It* You are a sales representative for an automobile manufacturer. You are paid an annual salary, plus a bonus of 3% of your sales over \$500,000. Consider the two functions

$$f(x) = x - 500,000 \qquad \text{and} \qquad g(x) = 0.03x.$$

If x is greater than \$500,000, which of the following represents your bonus? Explain your reasoning.

(a) $f(g(x))$ (b) $g(f(x))$

69. Prove that the product of two odd functions is an even function, and that the product of two even functions is an even function.

70. *Conjecture* Use examples to hypothesize whether the product of an odd function and an even function is even or odd. Then prove your hypothesis.

Review

Average Rate of Change In Exercises 71–74, find the difference quotient

$$\frac{f(x + h) - f(x)}{h}$$

and simplify your answer.

71. $f(x) = 3x - 4$

72. $f(x) = 1 - x^2$

73. $f(x) = \dfrac{4}{x}$

74. $f(x) = \sqrt{2x + 1}$

In Exercises 75–78, find an equation of the line that passes through the given point and has the indicated slope. Sketch a graph of the line.

75. $(2, -4), m = 3$

76. $(-6, 3), m = -1$

77. $(8, -1), m = -\frac{3}{2}$

78. $(7, 0), m = \frac{5}{7}$

2.6 Inverse Functions

▶ **What you should learn**

- How to find inverse functions informally and verify that two functions are inverses of each other
- How to use graphs of functions to decide whether functions have inverses
- How to find inverse functions algebraically

▶ **Why you should learn it**

Inverse functions can be used to model and solve real-life problems. For instance, Exercise 81 on page 241 shows how the inverse of a function can be used to determine when damage to a diesel engine may occur.

David J. Sams/Stock Boston

The Inverse of a Function

Recall from Section 2.2 that a function can be represented by a set of ordered pairs. For instance, the function $f(x) = x + 4$ from the set $A = \{1, 2, 3, 4\}$ to the set $B = \{5, 6, 7, 8\}$ can be written as follows.

$$f(x) = x + 4: \ \{(1, 5), (2, 6), (3, 7), (4, 8)\}$$

In this case, by interchanging the first and second coordinates of each of these ordered pairs, you can form the **inverse function** of f, which is denoted by f^{-1}. It is a function from the set B to the set A, and can be written as follows.

$$f^{-1}(x) = x - 4: \ \{(5, 1), (6, 2), (7, 3), (8, 4)\}$$

Note that the domain of f is equal to the range of f^{-1}, and vice versa, as shown in Figure 2.41. Also note that the functions f and f^{-1} have the effect of "undoing" each other. In other words, when you form the composition of f with f^{-1} or the composition of f^{-1} with f, you obtain the identity function.

$$f(f^{-1}(x)) = f(x - 4) = (x - 4) + 4 = x$$
$$f^{-1}(f(x)) = f^{-1}(x + 4) = (x + 4) - 4 = x$$

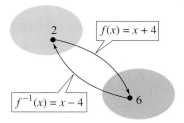

FIGURE 2.41

Example 1 ▶ Finding Inverse Functions Informally

Find the inverse of $f(x) = 4x$. Then verify that both $f(f^{-1}(x))$ and $f^{-1}(f(x))$ are equal to the identity function.

Solution

The function f *multiplies* each input by 4. To "undo" this function, you need to *divide* each input by 4. So, the inverse function of $f(x) = 4x$ is

$$f^{-1}(x) = \frac{x}{4}.$$

You can verify that both $f(f^{-1}(x))$ and $f^{-1}(f(x))$ are equal to the identity function as follows.

$$f(f^{-1}(x)) = f\left(\frac{x}{4}\right) = 4\left(\frac{x}{4}\right) = x \qquad f^{-1}(f(x)) = f^{-1}(4x) = \frac{4x}{4} = x$$

◀ **Exploration** ▶

Consider the functions

$$f(x) = x + 2$$

and

$$f^{-1}(x) = x - 2.$$

Evaluate $f(f^{-1}(x))$ and $f^{-1}(f(x))$ for the indicated values of x. What can you conclude about the functions?

x	-10	0	7	45
$f(f^{-1}(x))$				
$f^{-1}(f(x))$				

Definition of the Inverse of a Function

Let f and g be two functions such that

$$f(g(x)) = x \qquad \text{for every } x \text{ in the domain of } g$$

and

$$g(f(x)) = x \qquad \text{for every } x \text{ in the domain of } f.$$

Under these conditions, the function g is the **inverse** of the function f. The function g is denoted by f^{-1} (read "f-inverse"). So,

$$f(f^{-1}(x)) = x \qquad \text{and} \qquad f^{-1}(f(x)) = x.$$

The domain of f must be equal to the range of f^{-1}, and the range of f must be equal to the domain of f^{-1}.

Don't be confused by the use of -1 to denote the inverse function f^{-1}. In this text, whenever f^{-1} is written, it *always* refers to the inverse of the function f and *not* to the reciprocal of $f(x)$.

If the function g is the inverse of the function f, it must also be true that the function f is the inverse of the function g. For this reason, you can say that the functions f and g are *inverses of each other.*

Example 2 ▶ Verifying Inverse Functions

Which of the functions is the inverse of $f(x) = \dfrac{5}{x - 2}$?

$$g(x) = \frac{x - 2}{5} \qquad h(x) = \frac{5}{x} + 2$$

Solution

By forming the composition of f with g, you have

$$f(g(x)) = f\left(\frac{x - 2}{5}\right)$$

$$= \frac{5}{\left(\dfrac{x - 2}{5}\right) - 2} \qquad \text{Substitute } \frac{x - 2}{5} \text{ for } x.$$

$$= \frac{25}{x - 12} \qquad \text{Simplify.}$$

$$\neq x.$$

Because this composition is not equal to the identity function x, it follows that g *is not* the inverse of f. By forming the composition of f with h, you have

$$f(h(x)) = f\left(\frac{5}{x} + 2\right) = \frac{5}{\left(\dfrac{5}{x} + 2\right) - 2} = \frac{5}{\left(\dfrac{5}{x}\right)} = x.$$

So, it appears that h *is* the inverse of f. You can confirm this by showing that the composition of h with f is also equal to the identity function. (Try doing this.)

The Graph of the Inverse of a Function

The graphs of a function f and its inverse f^{-1} are related to each other in the following way. If the point (a, b) lies on the graph of f, then the point (b, a) must lie on the graph of f^{-1}, and vice versa. This means that the graph of f^{-1} is a *reflection* of the graph of f in the line $y = x$, as shown in Figure 2.42.

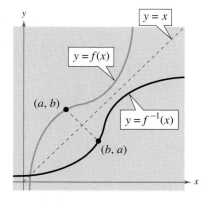

FIGURE **2.42**

Example 3 ▶ The Graphs of f and f^{-1}

Sketch the graphs of the inverse functions $f(x) = 2x - 3$ and $f^{-1}(x) = \frac{1}{2}(x + 3)$ on the same rectangular coordinate system and show that the graphs are reflections of each other in the line $y = x$.

Solution

The graphs of f and f^{-1} are shown in Figure 2.43. It appears that the graphs are reflections of each other in the line $y = x$. You can further verify this reflective property by testing a few points on each graph. Note in the following list that if the point (a, b) is on the graph of f, the point (b, a) is on the graph of f^{-1}.

Graph of $f(x) = 2x - 3$	*Graph of $f^{-1}(x) = \frac{1}{2}(x + 3)$*
$(-1, -5)$	$(-5, -1)$
$(0, -3)$	$(-3, 0)$
$(1, -1)$	$(-1, 1)$
$(2, 1)$	$(1, 2)$
$(3, 3)$	$(3, 3)$

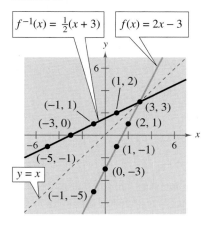

FIGURE **2.43**

Example 4 ▶ Finding Inverse Functions Graphically

Sketch the graphs of the inverse functions $f(x) = x^2, x \geq 0$ and $f^{-1}(x) = \sqrt{x}$ on the same rectangular coordinate system and show that the graphs are reflections of each other in the line $y = x$.

Solution

The graphs of f and f^{-1} are shown in Figure 2.44. It appears that the graphs are reflections of each other in the line $y = x$. You can further verify this reflective property by testing a few points on each graph. Note in the following list that if the point (a, b) is on the graph of f, the point (b, a) is on the graph of f^{-1}.

Graph of $f(x) = x^2, \quad x \geq 0$	*Graph of $f^{-1}(x) = \sqrt{x}$*
$(0, 0)$	$(0, 0)$
$(1, 1)$	$(1, 1)$
$(2, 4)$	$(4, 2)$
$(3, 9)$	$(9, 3)$

Try showing that $f(f^{-1}(x)) = x$ and $f^{-1}(f(x)) = x$.

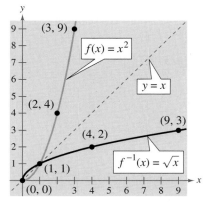

FIGURE **2.44**

The reflective property of the graphs of inverse functions gives you a nice *geometric* test for determining whether a function has an inverse. This test is called the **Horizontal Line Test** for inverse functions.

Horizontal Line Test for Inverse Functions

A function f has an inverse function if and only if no *horizontal* line intersects the graph of f at more than one point.

STUDY T!P

Not every function has an inverse function. Consider the following table of values for the function $f(x) = x^2$.

x	-2	-1	0	1	2	3
$f(x)$	4	1	0	1	4	9

The table of values made up by interchanging the rows of the table does not represent a function because the input $x = 4$ is matched with two different outputs: $y = -2$ and $y = 2$.

x	4	1	0	1	4	9
y	-2	-1	0	1	2	3

So, $f(x) = x^2$ does not have an inverse function.

Example 5 ▶ Applying the Horizontal Line Test

a. The graph of the function $f(x) = x^3 - 1$ is shown in Figure 2.45(a). Because no horizontal line intersects the graph of f at more than one point, you can conclude that f *does* have an inverse function.

b. The graph of the function $f(x) = x^2 - 1$ is shown in Figure 2.45(b). Because it is possible to find a horizontal line that intersects the graph of f at more than one point, you can conclude that f *does not* have an inverse function.

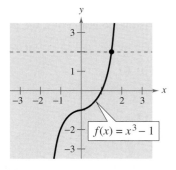

$f(x) = x^3 - 1$

(a)

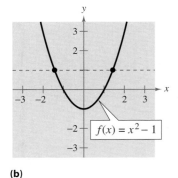

$f(x) = x^2 - 1$

(b)

FIGURE 2.45

Finding the Inverse of a Function Algebraically

For simple functions (such as the one in Example 1), you can find inverse functions by inspection. For more complicated functions, however, it is best to use the following guidelines. The key step in these guidelines is Step 3—interchanging the roles of x and y. This step corresponds to the fact that inverse functions have ordered pairs with the coordinates reversed.

Finding the Inverse of a Function

1. Use the Horizontal Line Test to decide whether f has an inverse.

2. In the equation for $f(x)$, replace $f(x)$ by y.

3. Interchange the roles of x and y, and solve for y.

4. Replace y by $f^{-1}(x)$ in the new equation.

5. Verify that f and f^{-1} are inverses of each other by showing that the domain of f is equal to the range of f^{-1}, the range of f is equal to the domain of f^{-1}, and $f(f^{-1}(x)) = x = f^{-1}(f(x))$.

Example 6 ▶ Finding the Inverse of a Function Algebraically

Find the inverse of $f(x) = \dfrac{5 - 3x}{2}$.

Solution

The graph of f is a line, as shown in Figure 2.46. This graph passes the Horizontal Line Test. So, you know that f has an inverse.

$$f(x) = \frac{5 - 3x}{2} \qquad \text{Write original function.}$$

$$y = \frac{5 - 3x}{2} \qquad \text{Replace } f(x) \text{ by } y.$$

$$x = \frac{5 - 3y}{2} \qquad \text{Interchange } x \text{ and } y.$$

$$2x = 5 - 3y \qquad \text{Multiply each side by 2.}$$

$$3y = 5 - 2x \qquad \text{Isolate the } y\text{-term.}$$

$$y = \frac{5 - 2x}{3} \qquad \text{Solve for } y.$$

$$f^{-1}(x) = \frac{5 - 2x}{3} \qquad \text{Replace } y \text{ by } f^{-1}(x).$$

Note that both f and f^{-1} have domains and ranges that consist of the entire set of real numbers. Check that $f(f^{-1}(x)) = x$ and $f^{-1}(f(x)) = x$.

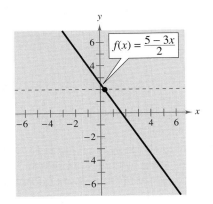

FIGURE **2.46**

Example 7 ▶ Finding the Inverse of a Function

Find the inverse of $f(x) = \sqrt[3]{x + 1}$.

Solution

The graph of f is a curve, as shown in Figure 2.47. Because this graph passes the Horizontal Line Test, you know that f has an inverse function.

$$f(x) = \sqrt[3]{x + 1} \qquad \text{Write original function.}$$

$$y = \sqrt[3]{x + 1} \qquad \text{Replace } f(x) \text{ by } y.$$

$$x = \sqrt[3]{y + 1} \qquad \text{Interchange } x \text{ and } y.$$

$$x^3 = y + 1 \qquad \text{Cube each side.}$$

$$x^3 - 1 = y \qquad \text{Solve for } y.$$

$$x^3 - 1 = f^{-1}(x) \qquad \text{Replace } y \text{ by } f^{-1}(x).$$

Both f and f^{-1} have domains and ranges that consist of the entire set of real numbers.

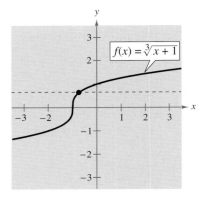

$f(x) = \sqrt[3]{x + 1}$

FIGURE 2.47

Check

$$f(f^{-1}(x)) = f(x^3 - 1)$$
$$= \sqrt[3]{(x^3 - 1) + 1}$$
$$= \sqrt[3]{x^3}$$
$$= x \checkmark$$

$$f^{-1}(f(x)) = f^{-1}\left(\sqrt[3]{x + 1}\right)$$
$$= \left(\sqrt[3]{x + 1}\right)^3 - 1$$
$$= x + 1 - 1$$
$$= x \checkmark$$

You can verify the result of Example 7 numerically as shown in the tables below.

x	-28	-9	-2	-1	0	7	26
$f(x)$	-3	-2	-1	0	1	2	3

x	-3	-2	-1	0	1	2	3
$f^{-1}(x)$	-28	-9	-2	-1	0	7	26

Writing ABOUT MATHEMATICS

The Existence of an Inverse Function Write a short paragraph describing why the following functions do or do not have inverse functions.

a. Let x represent the retail price of an item (in dollars), and let $f(x)$ represent the sales tax on the item. Assume that the sales tax is 6% of the retail price *and* that the sales tax is rounded to the nearest cent. Does this function have an inverse? (*Hint:* Can you undo this function? For instance, if you know that the sales tax is $0.12, can you determine *exactly* what the retail price is?)

b. Let x represent the temperature in degrees Celsius, and let $f(x)$ represent the temperature in degrees Fahrenheit. Does this function have an inverse? (*Hint:* The formula for converting from degrees Celsius to degrees Fahrenheit is $F = \frac{9}{5}C + 32$.)

2.6 Exercises

In Exercises 1–4, match the graph of the function with the graph of its inverse. [The graphs of the inverse functions are labeled (a), (b), (c), and (d).]

(a)
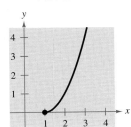

(b)
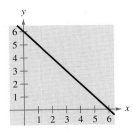

(c)
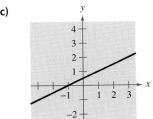

(d)
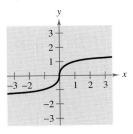

1.
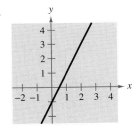

2.
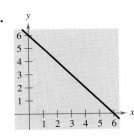

3.
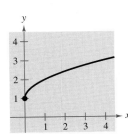

4.
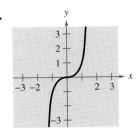

In Exercises 5–12, find the inverse of f informally. Verify that $f(f^{-1}(x)) = x$ and $f^{-1}(f(x)) = x$.

5. $f(x) = 6x$

6. $f(x) = \frac{1}{3}x$

7. $f(x) = x + 9$

8. $f(x) = x - 4$

9. $f(x) = 3x + 1$

10. $f(x) = \dfrac{x - 1}{5}$

11. $f(x) = \sqrt[3]{x}$

12. $f(x) = x^5$

In Exercises 13–24, show that f and g are inverse functions (a) algebraically and (b) graphically.

13. $f(x) = 2x,$ $\qquad\qquad g(x) = \dfrac{x}{2}$

14. $f(x) = x - 5,$ $\qquad\qquad g(x) = x + 5$

15. $f(x) = 5x + 1,$ $\qquad\qquad g(x) = \dfrac{x - 1}{5}$

16. $f(x) = 3 - 4x,$ $\qquad\qquad g(x) = \dfrac{3 - x}{4}$

17. $f(x) = x^3,$ $\qquad\qquad g(x) = \sqrt[3]{x}$

18. $f(x) = \dfrac{1}{x},$ $\qquad\qquad g(x) = \dfrac{1}{x}$

19. $f(x) = \sqrt{x - 4},$ $\qquad g(x) = x^2 + 4, \quad x \ge 0$

20. $f(x) = 1 - x^3,$ $\qquad g(x) = \sqrt[3]{1 - x}$

21. $f(x) = 9 - x^2, \quad x \ge 0, \quad g(x) = \sqrt{9 - x}, \quad x \le 9$

22. $f(x) = \dfrac{1}{1 + x}, \quad x \ge 0$

$\qquad g(x) = \dfrac{1 - x}{x}, \quad 0 < x \le 1$

23. $f(x) = \dfrac{x - 1}{x + 5},$ $\qquad\qquad g(x) = -\dfrac{5x + 1}{x - 1}$

24. $f(x) = \dfrac{x + 3}{x - 2},$ $\qquad\qquad g(x) = \dfrac{2x + 3}{x - 1}$

In Exercises 25 and 26, does the function have an inverse?

25.

x	-1	0	1	2	3	4
$f(x)$	-2	1	2	1	-2	-6

26.

x	-3	-2	-1	0	2	3
$f(x)$	10	6	4	1	-3	-10

In Exercises 27 and 28, use the table of values for $y = f(x)$ to complete a table for $y = f^{-1}(x)$.

27.

x	-2	-1	0	1	2	3
$f(x)$	-2	0	2	4	6	8

28.

x	-3	-2	-1	0	1	2
$f(x)$	-10	-7	-4	-1	2	5

In Exercises 29–32, does the function have an inverse?

29.
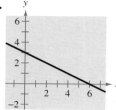

30.
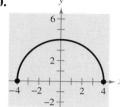

31.
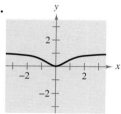

32.
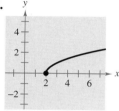

In Exercises 33–38, use a graphing utility to graph the function and use the Horizontal Line Test to determine whether the function has an inverse.

33. $g(x) = \dfrac{4 - x}{6}$

34. $f(x) = 10$

35. $h(x) = |x + 4| - |x - 4|$

36. $g(x) = (x + 5)^3$

37. $f(x) = -2x\sqrt{16 - x^2}$

38. $f(x) = \frac{1}{8}(x + 2)^2 - 1$

In Exercises 39–54, find the inverse of the function f. Then graph both f and f^{-1} on the same set of coordinate axes.

39. $f(x) = 2x - 3$

40. $f(x) = 3x + 1$

41. $f(x) = x^5 - 2$

42. $f(x) = x^3 + 1$

43. $f(x) = \sqrt{x}$

44. $f(x) = x^2, \quad x \geq 0$

45. $f(x) = \sqrt{4 - x^2}, \quad 0 \leq x \leq 2$

46. $f(x) = x^2 - 2, \quad x \leq 0$

47. $f(x) = \dfrac{4}{x}$

48. $f(x) = -\dfrac{2}{x}$

49. $f(x) = \dfrac{x + 1}{x - 2}$

50. $f(x) = \dfrac{x - 3}{x + 2}$

51. $f(x) = \sqrt[3]{x - 1}$

52. $f(x) = x^{3/5}$

53. $f(x) = \dfrac{6x + 4}{4x + 5}$

54. $f(x) = \dfrac{8x - 4}{2x + 6}$

In Exercises 55–68, determine whether the function has an inverse. If it does, find the inverse.

55. $f(x) = x^4$

56. $f(x) = \dfrac{1}{x^2}$

57. $g(x) = \dfrac{x}{8}$

58. $f(x) = 3x + 5$

59. $p(x) = -4$

60. $f(x) = \dfrac{3x + 4}{5}$

61. $f(x) = (x + 3)^2, \quad x \geq -3$

62. $q(x) = (x - 5)^2$

63. $f(x) = \begin{cases} x + 3, & x < 0 \\ 6 - x, & x \geq 0 \end{cases}$

64. $f(x) = \begin{cases} -x, & x \leq 0 \\ x^2 - 3x, & x > 0 \end{cases}$

65. $h(x) = \dfrac{1}{x}$

66. $f(x) = |x - 2|, \quad x \leq 2$

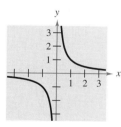

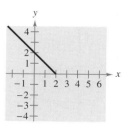

67. $f(x) = \sqrt{2x + 3}$

68. $f(x) = \sqrt{x - 2}$

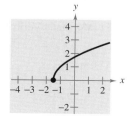

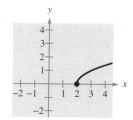

In Exercises 69–74, use the functions $f(x) = \frac{1}{8}x - 3$ and $g(x) = x^3$ to find the indicated value or function.

69. $(f^{-1} \circ g^{-1})(1)$

70. $(g^{-1} \circ f^{-1})(-3)$

71. $(f^{-1} \circ f^{-1})(6)$

72. $(g^{-1} \circ g^{-1})(-4)$

73. $(f \circ g)^{-1}$

74. $g^{-1} \circ f^{-1}$

In Exercises 75–78, use the functions $f(x) = x + 4$ and $g(x) = 2x - 5$ to find the specified function.

75. $g^{-1} \circ f^{-1}$

76. $f^{-1} \circ g^{-1}$

77. $(f \circ g)^{-1}$

78. $(g \circ f)^{-1}$

79. *Hourly Wage*   Your wage is $8.00 per hour plus $0.75 for each unit produced per hour. So, your hourly wage y in terms of the number of units produced is

$$y = 8 + 0.75x.$$

(a) Find the inverse of the function.

(b) What does each variable represent in the inverse function?

(c) Determine the number of units produced when your hourly wage is $22.25.

80. *Cost*   Suppose you need a total of 50 pounds of two commodities costing $1.25 and $1.60 per pound, respectively.

(a) Verify that the total cost is

$$y = 1.25x + 1.60(50 - x)$$

where x is the number of pounds of the less expensive commodity.

(b) Find the inverse of the cost function. What does each variable represent in the inverse function?

(c) Use the context of the problem to determine the domain of the inverse function.

(d) Determine the number of pounds of the less expensive commodity purchased if the total cost is $73.

81. *Diesel Mechanics*   The function

$$y = 0.03x^2 + 245.50, \qquad 0 < x < 100$$

approximates the exhaust temperature y in degrees Fahrenheit where x is the percent load for a diesel engine.

(a) Find the inverse of the function. What does each variable represent in the inverse function?

(b) Use a graphing utility to graph the inverse function.

(c) Determine the percent load interval if the exhaust temperature of the engine must not exceed 500 degrees Fahrenheit.

82. *New Car Sales*   The total value of new car sales f (in billions of dollars) in the United States from 1992 through 1997 is shown in the table. The time (in years) is given by t, with $t = 2$ corresponding to 1992. (Source: National Automobile Dealers Association)

t	2	3	4	5	6	7
$f(t)$	333.8	377.3	430.6	456.2	490.0	507.5

(a) Does f^{-1} exist?

(b) If f^{-1} exists, what does it mean in the context of the problem?

(c) If f^{-1} exists, find $f^{-1}(456.2)$.

83. If the table in Exercise 82 were extended to 1998 and if the total value of new car sales for that year was $430.6 billion, would f^{-1} exist? Explain.

84. *Cellular Phones*   The average local bill (in dollars) for cellular phones in the United States from 1990 to 1997 is shown in the table. The time (in years) is given by t, with $t = 0$ corresponding to 1990. (Source: Cellular Telecommunications Industry Association)

t	0	1	2	3
$f(t)$	80.90	72.74	68.68	61.48

t	4	5	6	7
$f(t)$	56.21	51.00	47.70	42.78

(a) Find $f^{-1}(51)$.

(b) What does f^{-1} mean in the context of the problem?

(c) Use the regression feature of a graphing utility to find a linear model for the data, $y = mx + b$. Round m and b to two decimal places.

(d) Algebraically find the inverse of the linear model in part (c).

(e) Use the inverse of the linear model you found in part (d) to approximate $f^{-1}(11)$.

85. *Soft Drink Consumption* The per capita consumption of regular soft drinks f (in gallons) in the United States from 1991 through 1996 is shown in the table. The time (in years) is given by t, with $t = 1$ corresponding to 1991. (Source: U.S. Department of Agriculture)

t	1	2	3	4	5	6
$f(t)$	36.3	36.9	38.4	39.5	39.8	40.2

(a) Does f^{-1} exist? If so, what does it represent in the context of the problem?

(b) If f^{-1} exists, what is $f^{-1}(39.8)$?

Synthesis

True or False? In Exercises 86–89, determine whether the statement is true or false. Justify your answer.

86. If f is an even function, f^{-1} exists.

87. If the inverse of f exists and the y-intercept of the graph of f exists, the y-intercept of f is an x-intercept of f^{-1}.

88. If $f(x) = x^n$ where n is odd, f^{-1} exists.

89. There exists no function f such that $f = f^{-1}$.

In Exercises 90–93, use the graph of the function f to create a table of values for the given points. Then create a second table that can be used to find f^{-1} and sketch the graph of f^{-1} if possible.

90.

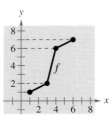

91.

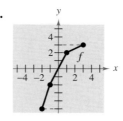

92.

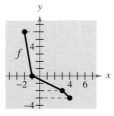

93.

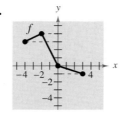

94. *Think About It* The function
$$f(x) = k(2 - x - x^3)$$
has an inverse, and $f^{-1}(3) = -2$. Find k.

Review

In Exercises 95–102, solve the equation by any convenient method.

95. $x^2 = 64$

96. $(x - 5)^2 = 8$

97. $4x^2 - 12x + 9 = 0$

98. $9x^2 + 12x + 3 = 0$

99. $x^2 - 6x + 4 = 0$

100. $2x^2 - 4x - 6 = 0$

101. $50 + 5x = 3x^2$

102. $2x^2 + 4x - 9 = 2(x - 1)^2$

In Exercises 103–106, find the domain of the function.

103. $f(x) = \sqrt[3]{x + 4}$

104. $f(x) = \sqrt{x + 6}$

105. $g(x) = \dfrac{2}{x^2 - 4x}$

106. $h(x) = \dfrac{x}{5x + 7}$

107. Find two consecutive positive even integers whose product is 288.

108. *Landscaping* Two people must mow a rectangular lawn measuring 100 feet by 200 feet. The second person agrees to mow three-fourths of the lawn and starts by mowing around the outside. How wide a strip must the person mow on each of the four sides? If the mower has a 24-inch cut, approximate the required number of trips around the lawn.

109. *Geometry* A triangular sign has a height that is equal to its base. The area of the sign is 10 square feet. Find the base and height of the sign.

110. *Geometry* A triangular sign has a height that is twice its base. The area of the sign is 10 square feet. Find the base and height of the sign.

Chapter Summary

What did you learn?

Section 2.1	Review Exercises
☐ How to use slope to graph linear equations in two variables	1–14
☐ How to find slopes of lines	15–18
☐ How to write linear equations in two variables	19–26
☐ How to use slope to identify parallel and perpendicular lines	27, 28
☐ How to use linear equations in two variables to model and solve real-life problems	29–32

Section 2.2	
☐ How to decide whether relations between two variables are functions	33–38
☐ How to use function notation and evaluate functions	39, 40
☐ How to find the domains of functions	41–46
☐ How to use functions to model and solve real-life problems	47–50

Section 2.3	
☐ How to use the Vertical Line Test for functions	51–54
☐ How to find the zeros of functions	55–58
☐ How to determine intervals on which functions are increasing or decreasing	59–62
☐ How to identify and graph linear functions	63–66
☐ How to identify and graph step functions and other piecewise-defined functions	67, 68
☐ How to identify even and odd functions	69–72

Section 2.4	
☐ How to recognize graphs of common functions	73–76
☐ How to use vertical and horizontal shifts to sketch graphs of functions	77–80
☐ How to use reflections to sketch graphs of functions	81–84
☐ How to use nonrigid transformations to sketch graphs of functions	85–88

Section 2.5	
☐ How to add, subtract, multiply, and divide functions	89–92
☐ How to find compositions of one function with another function	93–96
☐ How to use combinations of functions to model and solve real-life problems	97, 98

Section 2.6	
☐ How to find inverse functions informally and verify that two functions are inverses of each other	99–102
☐ How to use graphs of functions to decide whether functions have inverses	103–108
☐ How to find inverse functions algebraically	109–114

Review Exercises

2.1 **In Exercises 1 and 2, identify the line that has each slope.**

1. (a) $m = \frac{3}{2}$
   (b) $m = 0$
   (c) $m = -3$
   (d) $m = -\frac{1}{5}$

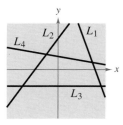

2. (a) m is undefined.
   (b) $m = -1$
   (c) $m = \frac{5}{2}$
   (d) $m = \frac{1}{2}$

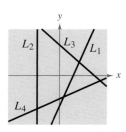

In Exercises 3–10, sketch the graph of the linear equation.

3. $y = -2x - 7$

4. $y = 4x - 3$

5. $y = 6$

6. $x = -3$

7. $y = 3x + 13$

8. $y = -10x + 9$

9. $y = -\frac{5}{2}x - 1$

10. $y = \frac{5}{6}x + 5$

In Exercises 11 and 12, use the concept of slope to find t such that the three points are on the same line.

11. $(-2, 5), (0, t), (1, 1)$ 12. $(-6, 1), (1, t), (10, 5)$

In Exercises 13 and 14, use the point on the line and the slope of the line to find three additional points through which the line passes. (There are many correct answers.)

Point	Slope
13. $(2, -1)$	$m = \frac{1}{4}$
14. $(-3, 5)$	$m = -\frac{3}{2}$

In Exercises 15–18, plot the points and find the slope of the line passing through the pair of points.

15. $(3, -4), (-7, 1)$ 16. $(-1, 8), (6, 5)$

17. $(-4.5, 6), (2.1, 3)$ 18. $(-3, 2), (8, 2)$

In Exercises 19–22, find an equation of the line that passes through the points.

19. $(0, 0), (0, 10)$

20. $(2, 5), (-2, -1)$

21. $(-1, 4), (2, 0)$

22. $(11, -2), (6, -1)$

In Exercises 23–26, find an equation of the line that passes through the given point and has the specified slope. Sketch the graph of the line.

Point	Slope
23. $(0, -5)$	$m = \frac{3}{2}$
24. $(-2, 6)$	$m = 0$
25. $(10, -3)$	$m = -\frac{1}{2}$
26. $(-8, 5)$	Undefined

In Exercises 27 and 28, write an equation of the line through the point (a) parallel to the given line and (b) perpendicular to the given line.

Point	Line
27. $(3, -2)$	$5x - 4y = 8$
28. $(-8, 3)$	$2x + 3y = 5$

Rate of Change   **In Exercises 29 and 30, you are given the dollar value of a product in the year 2000 *and* the rate at which the value of the item is expected to change during the next 5 years. Write a linear equation that gives the dollar value V of the product in terms of the year t. (Let $t = 0$ represent 2000.)**

2000 Value	Rate
29. $12,500	$850 increase per year
30. $72.95	$5.15 increase per year

31. *Business*   During the second and third quarters of the year, a business had sales of $160,000 and $185,000, respectively. If the growth of sales follows a linear pattern, estimate sales during the fourth quarter.

32. *Inflation*   The dollar value of a product in 1999 is $85, and the product is expected to increase in value at a rate of $3.75 per year.

(a) Write a linear equation that gives the dollar value V of the product in terms of the year t. (Let $t = 9$ represent 1999.)

(b) Use a graphing utility to graph the sales equation.

(c) Move the cursor along the graph of the sales model to estimate the dollar value of the product in 2005.

2.2   In Exercises 33 and 34, determine which of the sets of ordered pairs represents a function from *A* to *B*. Give reasons for your answers.

33. $A = \{10, 20, 30, 40\}$ and $B = \{0, 2, 4, 6\}$

(a) $\{(20, 4), (40, 0), (20, 6), (30, 2)\}$

(b) $\{(10, 4), (20, 4), (30, 4), (40, 4)\}$

(c) $\{(40, 0), (30, 2), (20, 4), (10, 6)\}$

(d) $\{(20, 2), (10, 0), (40, 4)\}$

34. $A = \{u, v, w\}$ and $B = \{-2, -1, 0, 1, 2\}$

(a) $\{(v, -1), (u, 2), (w, 0), (u, -2)\}$

(b) $\{(u, -2), (v, 2), (w, 1)\}$

(c) $\{(u, 2), (v, 2), (w, 1), (w, 1)\}$

(d) $\{(w, -2), (v, 0), (w, 2)\}$

In Exercises 35–38, determine whether the equation represents *y* as a function of *x*.

35. $16x - y^4 = 0$

36. $2x - y - 3 = 0$

37. $y = \sqrt{1 - x}$

38. $|y| = x + 2$

In Exercises 39 and 40, evaluate the function as indicated. Simplify your answers.

39. $f(x) = x^2 + 1$

(a) $f(2)$ (b) $f(-4)$

(c) $f(t^2)$ (d) $-f(x)$

40. $g(x) = x^{4/3}$

(a) $g(8)$ (b) $g(t + 1)$

(c) $\dfrac{g(8) - g(1)}{8 - 1}$ (d) $g(-x)$

In Exercises 41–46, determine the domain of the function. Verify your result with a graph.

41. $f(x) = \sqrt{25 - x^2}$

42. $f(x) = 3x + 4$

43. $g(s) = \dfrac{5}{3s - 9}$

44. $f(x) = \sqrt{x^2 + 8x}$

45. $h(x) = \dfrac{x}{x^2 - x - 6}$

46. $h(t) = |t + 1|$

47. *Physics*   The velocity of a ball thrown vertically upward from ground level is $v(t) = -32t + 48$, where t is the time in seconds and v is the velocity in feet per second.

(a) Find the velocity when $t = 1$.

(b) Find the time when the ball reaches its maximum height. [*Hint:* Find the time when $v(t) = 0$.]

(c) Find the velocity when $t = 2$.

48. *Total Cost*   A company produces a product for which the variable cost is $5.35 per unit and the fixed costs are $16,000. The company sells the product for $8.20 and can sell all that it produces.

(a) Find the total cost as a function of x, the number of units produced.

(b) Find the profit as a function of x.

49. *Geometry*   A wire 24 inches long is to be cut into four pieces to form a rectangle with one side of length x.

(a) Express the area A of the rectangle as a function of x.

(b) Determine the domain of the function.

50. *Mixture Problem*   From a full 50-liter container of a 40% concentration of acid, x liters is removed and replaced with 100% acid.

(a) Write the amount of acid in the final mixture as a function of x.

(b) Determine the domain and range of the function.

(c) Determine x if the final mixture is 50% acid.

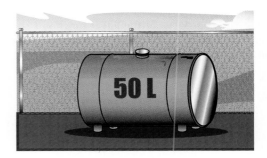

2.3 In Exercises 51–54, use the Vertical Line Test to determine whether y is a function of x.

51. $y = (x - 3)^2$

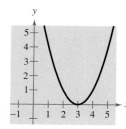

52. $y = -\frac{3}{5}x^3 - 2x + 1$

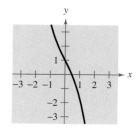

53. $x - 4 = y^2$

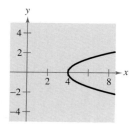

54. $x = -|4 - y|$

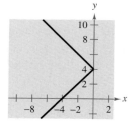

In Exercises 55–58, find the zeros of the function.

55. $f(x) = 3x^2 - 16x + 21$

56. $f(x) = 5x^2 + 4x - 1$

57. $f(x) = \dfrac{8x + 3}{11 - x}$

58. $f(x) = x^3 - x^2 - 25x + 25$

Graphical Analysis In Exercises 59–62, use a graphing utility to graph the function to approximate the intervals in which the function is increasing, decreasing, or constant.

59. $g(x) = |x + 2| - |x - 2|$

60. $f(x) = (x^2 - 4)^2$

61. $h(x) = 4x^3 - x^4$

62. $g(x) = \sqrt[3]{x(x + 3)^2}$

In Exercises 63–66, write the linear function f so that the following are true. Then use a graphing utility to graph the function.

63. $f(2) = -6$, $f(-1) = 3$

64. $f(0) = -5$, $f(4) = -8$

65. $f\left(-\frac{4}{5}\right) = 2$, $f\left(\frac{11}{5}\right) = 7$

66. $f(3.3) = 5.6$, $f(-4.7) = -1.4$

In Exercises 67 and 68, graph the function.

67. $f(x) = \begin{cases} 5x - 3, & x \geq -1 \\ -4x + 5, & x < -1 \end{cases}$

68. $f(x) = \begin{cases} x^2 - 2, & x < -2 \\ 5, & -2 \leq x \leq 0 \\ 8x - 5, & x > 0 \end{cases}$

In Exercises 69–72, determine whether the function is even, odd, or neither.

69. $f(x) = x^5 + 4x - 7$

70. $f(x) = x^4 - 20x^2$

71. $f(x) = 2x\sqrt{x^2 + 3}$

72. $f(x) = \sqrt[5]{6x^2}$

2.4 In Exercises 73–76, identify the common function and describe the transformation shown in the graph.

73.

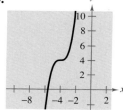

74.

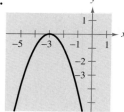

75.

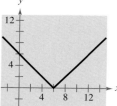

76.

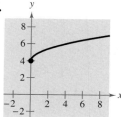

In Exercises 77–88, identify the transformation of the graph of f and sketch the graph of h.

77. $f(x) = x^2$, $\quad h(x) = x^2 - 9$

78. $f(x) = x^3$, $\quad h(x) = (x - 2)^3 + 2$

79. $f(x) = \sqrt{x}$, $\quad h(x) = \sqrt{x - 7}$

80. $f(x) = |x|$, $\quad h(x) = |x + 3| - 5$

81. $f(x) = x^2$, $\quad h(x) = -(x + 3)^2 + 1$

82. $f(x) = x^3,$ $h(x) = -(x-5)^3 - 5$

83. $f(x) = \sqrt{x},$ $h(x) = -\sqrt{x+1} + 9$

84. $f(x) = |x|,$ $h(x) = -|-x+4| + 6$

85. $f(x) = x^2,$ $h(x) = -2(x+1)^2 - 3$

86. $f(x) = x^3,$ $h(x) = -\frac{1}{3}x^3$

87. $f(x) = \sqrt{x},$ $h(x) = -2\sqrt{x-4}$

88. $f(x) = |x|,$ $h(x) = \frac{1}{2}|x| - 1$

2.5 In Exercises 89–96, let $f(x) = 3 - 2x$, $g(x) = \sqrt{x}$, and $h(x) = 3x^2 + 2$. Find the indicated value.

89. $(f - g)(4)$

90. $(f + g)(4)$

91. $(fh)(1)$

92. $\left(\dfrac{f}{h}\right)(0)$

93. $(h \circ g)(7)$

94. $(f \circ h)(3)$

95. $(g \circ f)(-2)$

96. $(g \circ h)(-1)$

Data Analysis In Exercises 97 and 98, use the table, which shows the total values (in billions of dollars) of U.S. imports from Mexico and Canada for the years 1992 through 1997. The variables y_1 and y_2 represent the total values of imports from Mexico and Canada, respectively. (Source: U.S. Bureau of the Census)

Year	1992	1993	1994	1995	1996	1997
y_1	35.2	39.9	49.5	62.1	74.3	85.9
y_2	98.6	111.2	128.4	144.4	155.9	168.2

97. Use a graphing utility to find quadratic models for each of the variables. Let $t = 2$ represent 1992.

98. Use a graphing utility to graph y_1, y_2, and $y_1 + y_2$ in the same viewing window. Use the model to estimate the total value of U.S. imports from Canada and Mexico in 2002.

2.6 In Exercises 99–102, find the inverse of f informally. Verify that $f(f^{-1}(x)) = x = f^{-1}(f(x))$.

99. $f(x) = 6x$

100. $f(x) = \frac{1}{12}x$

101. $f(x) = x - 7$

102. $f(x) = x + 5$

In Exercises 103–108, use a graphing utility to graph each function and determine whether the function has an inverse.

103. $f(x) = 3x^3 - 5$

104. $f(x) = -\frac{1}{4}x^2 - 3$

105. $f(x) = -\sqrt{4-x}$

106. $f(x) = -x\sqrt{25 - 3x^2}$

107. $f(x) = -|x+2| + |7 - x|$

108. $f(x) = \dfrac{x^2 + 5}{x - 3}$

In Exercises 109–112, (a) find f^{-1}, (b) sketch the graphs of f and f^{-1} on the same coordinate system, and (c) verify that $f^{-1}(f(x)) = x = f(f^{-1}(x))$.

109. $f(x) = \frac{1}{2}x - 3$

110. $f(x) = 5x - 7$

111. $f(x) = \sqrt{x+1}$

112. $f(x) = x^3 + 2$

In Exercises 113 and 114, restrict the domain of the function f to an interval over which the function is increasing and determine f^{-1} over that interval.

113. $f(x) = 2(x - 4)^2$

114. $f(x) = |x - 2|$

Synthesis

True or False? In Exercises 115 and 116, determine whether the statement is true or false. Justify your answer.

115. Relative to the graph of $f(x) = \sqrt{x}$, the function $h(x) = -\sqrt{x + 9} - 13$ is shifted 9 units to the left and 13 units down, then reflected in the x-axis.

116. If f and g are two inverse functions, then the domain of g is equal to the range of f.

117. Explain how to tell whether a relation between two variables is a function.

118. Explain the difference between the Vertical Line Test and the Horizontal Line Test.

Chapter Project ▶ A Graphical Approach to Maximization

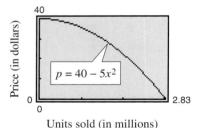

Price (in dollars)

$p = 40 - 5x^2$

Units sold (in millions)

In business, a **demand function** gives the price per unit p in terms of the number of units sold x. The demand function whose graph is shown at the left is

$$p = 40 - 5x^2, \qquad 0 \le x \le \sqrt{8} \qquad \text{Demand function}$$

where x is measured in millions of units. Note that as the price decreases, the number of units sold increases. The **revenue** R (in millions of dollars) is determined by multiplying the number of units sold by the price per unit. So,

$$R = xp = x(40 - 5x^2), \qquad 0 \le x \le \sqrt{8}. \qquad \text{Revenue function}$$

Example ▶ Finding the Maximum Revenue

Use a graphing utility to sketch the graph of the revenue function $R = 40x - 5x^3$ for $0 \le x \le \sqrt{8}$. How many units should be sold to obtain maximum revenue? What price per unit should be charged to obtain maximum revenue?

Solution

To begin, you need to determine a viewing window that will display the part of the graph that is important to this problem. The domain is given, so you can set the x-boundaries of the graph between 0 and $\sqrt{8}$. To determine the y-boundaries, however, you need to experiment a little. After calculating several values of R, you could decide to use y-boundaries between 0 and 50, as shown in the graph at the left. Next, you can use the trace key to find that the maximum revenue of about \$43.5 million occurs when x is approximately 1.64 million units. To find the price per unit that corresponds to this maximum revenue, you can substitute $x = 1.64$ into the demand function to obtain

$$p = 40 - 5(1.64)^2 \approx \$26.55.$$

Revenue (in millions of dollars)

Units sold (in millions)

Chapter Project Investigations

1. For the demand function $p = 40 - 5x^2$, match each of the points $(0, 40)$ and $\left(\sqrt{8}, 0\right)$ with statement (a) or (b). Explain your reasoning.

   (a) No one will buy the product at this price.

   (b) You can't give more than this number away.

2. Use a graphing utility to zoom in on the maximum point of the revenue function in the example. (Use a setting of $1.62 \le x \le 1.65$ and $43.5 \le y \le 43.6$.) Use the trace feature to improve the accuracy of the approximation obtained in the example. Do you think this improved accuracy is appropriate in the context of this particular problem? Does it change the price?

3. For the revenue function discussed in the example, the cost of producing each unit is \$15, so the total cost of producing x million units is $C = 15x$. Use a graphing utility to graph the profit function $P = R - C$ to determine how many units should be sold to obtain maximum profit. What price per unit should be charged to obtain maximum profit?

▶ Chapter Test

 The *Interactive* CD-ROM and *Internet* versions of this text provide answers to the Chapter Tests and Cumulative Tests. They also offer Chapter Pre-Tests (which test key skills and concepts covered in previous chapters) and Chapter Post-Tests, both of which have randomly generated exercises with diagnostic capabilities.

Take this test as you would take a test in class. After you are done, check your work against the answers given in the back of the book.

In Exercises 1 and 2, find an equation of the line passing through the given points. Then sketch a graph of the line.

1. $(2, -3), (-4, 9)$

2. $(3, 0.8), (7, -6)$

3. Find an equation of the line that passes through the point $(3, 8)$ and is (a) parallel to and (b) perpendicular to the line $-4x + 7y = -5$.

In Exercises 4 and 5, evaluate the function at each specified value.

4. $f(x) = |x + 2| - 15$

    (a) $f(-8)$ (b) $f(14)$ (c) $f(x - 6)$

5. $f(x) = \dfrac{\sqrt{x + 9}}{x^2 - 81}$

    (a) $f(7)$ (b) $f(-5)$ (c) $f(x - 9)$

In Exercises 6 and 7, determine the domain of the function.

6. $f(x) = \sqrt{100 - x^2}$

7. $f(x) = |-x + 6| + 2$

In Exercises 8–10, (a) use a graphing utility to graph the function, (b) approximate the intervals over which the function is increasing, decreasing, or constant, and (c) determine whether the function is even, odd, or neither.

8. $f(x) = 2x^6 + 5x^4 - x^2$

9. $f(x) = 4x\sqrt{3 - x}$

10. $f(x) = |x + 5|$

11. Sketch the graph of $f(x) = \begin{cases} 3x + 7, & x \le -3 \\ 4x^2 - 1, & x > -3 \end{cases}$.

In Exercises 12–14, identify the common function in the transformation. Then sketch a graph of the function.

12. $h(x) = -x^3 - 7$

13. $h(x) = -\sqrt{x + 5} + 8$

14. $h(x) = \frac{1}{4}|x + 1| - 3$

In Exercises 15–18, let $f(x) = 3x^2 - 7$ and $g(x) = -x^2 - 4x + 5$. Find the indicated value.

15. $(f + g)(2)$

16. $(f - g)(-3)$

17. $(fg)(0)$

18. $(g \circ f)(-1)$

In Exercises 19–21, determine whether the function has an inverse, and if so, find the inverse function.

19. $f(x) = x^3 + 8$

20. $f(x) = |x^2 - 3| + 6$

21. $f(x) = \dfrac{3x\sqrt{x}}{8}$

22. It costs a company $58 to produce 6 units of a product and $78 to produce 10 units. How much does it cost to produce 25 units, assuming that the cost function is linear?

Cumulative Test for Chapters P–2

Take this test to review the material from earlier chapters. After you are done, check your work against the answers given in the back of the book.

In Exercises 1 and 2, simplify the expression.

1. $\dfrac{8x^2y^{-3}}{30x^{-1}y^2}$

2. $\sqrt{24x^4y^3}$

In Exercises 3–5, perform the operations and simplify the result.

3. $4x - [2x + 3(2 - x)]$

4. $(x - 2)(x^2 + x - 3)$

5. $\dfrac{2}{s + 3} - \dfrac{1}{s + 1}$

In Exercises 6–8, factor the expression completely.

6. $25 - (x - 2)^2$

7. $x - 5x^2 - 6x^3$

8. $54 - 16x^3$

In Exercises 9–11, graph the equation without using a graphing utility.

9. $x - 3y + 12 = 0$

10. $y = x^2 - 9$

11. $y = \sqrt{4 - x}$

In Exercises 12 and 13, write an expression for the area of the region.

12.

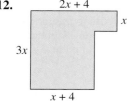

13.

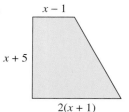

In Exercises 14–19, solve the equation by any convenient method. State the method you used.

14. $x^2 - 4x + 3 = 0$

15. $-2x^2 + 8x + 12 = 0$

16. $\frac{3}{4}x^2 = 12$

17. $3x^2 + 5x - 6 = 0$

18. $3x^2 + 9x + 1 = 0$

19. $\frac{1}{2}x^2 - 7 = 25$

In Exercises 20–25, solve the equation (if possible).

20. $x^4 + 12x^3 + 4x^2 + 48x = 0$

21. $8x^3 - 48x^2 + 72x = 0$

22. $x^{2/3} + 13 = 17$

23. $\sqrt{x + 10} = x - 2$

24. $|4(x - 2)| = 28$

25. $|x - 12| = -2$

In Exercises 26–28, determine whether each value of x is a solution of the inequality.

26. $4x + 2 > 7$

    (a) $x = -1$ (b) $x = \frac{1}{2}$

    (c) $x = \frac{3}{2}$ (d) $x = 2$

27. $3 - \frac{1}{2}x \le -2$

    (a) $x = -10$ (b) $x = 9$

    (c) $x = 10$ (d) $x = 12$

28. $|5x - 1| < 4$

    (a) $x = -1$ (b) $x = -\frac{1}{2}$

    (c) $x = 1$ (d) $x = 2$

In Exercises 29–32, solve the inequality and sketch the solution on the real number line.

29. $|x + 1| \le 6$ **30.** $|7 + 8x| > 5$

31. $5x^2 + 12x + 7 \ge 0$ **32.** $-x^2 + x + 4 < 0$

33. Find an equation for the line passing through $\left(-\frac{1}{2}, 1\right)$ and $(3, 8)$.

34. Explain why the graph at the left does not represent y as a function of x.

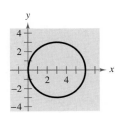

FIGURE FOR **34**

35. Evaluate (if possible) the function $f(x) = \dfrac{x}{x - 2}$ for each value.

    (a) $f(6)$ (b) $f(2)$ (c) $f(s + 2)$

36. Describe how the graph of each function would differ from the graph of $y = \sqrt[3]{x}$. (*Note:* It is not necessary to sketch the graphs.)

    (a) $r(x) = \frac{1}{2}\sqrt[3]{x}$ (b) $h(x) = \sqrt[3]{x} + 2$ (c) $g(x) = \sqrt[3]{x + 2}$

In Exercises 37 and 38, find (a) $(f + g)(x)$, (b) $(f - g)(x)$, (c) $(fg)(x)$, and (d) $(f/g)(x)$. What is the domain of f/g?

37. $f(x) = x - 3, \quad g(x) = 4x + 1$ **38.** $f(x) = \sqrt{x - 1}, \quad g(x) = x^2 + 1$

In Exercises 39 and 40, find (a) $f \circ g$ and (b) $g \circ f$.

39. $f(x) = 2x^2, \quad g(x) = \sqrt{x + 6}$ **40.** $f(x) = x - 2, \quad g(x) = |x|$

41. Determine whether $h(x) = 5x - 2$ has an inverse. If so, find it.

42. A group of n people decide to buy a \$36,000 minibus. Each person will pay an equal share of the cost. If three additional people join the group, the cost per person will decrease by \$1000. Find n.

43. For groups of 80 or more, a charter bus company determines the rate per person according to the formula

    Rate $= \$8.00 - \$0.05(n - 80), \quad n \ge 80$.

    (a) Determine the revenue R as a function of n.

    (b) Use a graphing utility to graph the revenue function. Move the cursor along the function to estimate the number of passengers that will maximize the revenue.

Greg Probst/Tony Stone Images

Production of wheat in the United States increased from 2183 million bushels in 1995 to 2527 million bushels in 1997. During that time the price per bushel dropped from $4.55 to $3.45. (Source: U.S. Department of Agriculture)

3 Polynomial Functions

▶ How to Study This Chapter

The Big Picture

In this chapter you will learn the following skills and concepts.

▶ How to sketch and analyze graphs of functions

▶ How to sketch and analyze graphs of polynomial functions

▶ How to use long division and synthetic division to divide polynomials by other polynomials

▶ How to determine the number of rational and real zeros of polynomial functions, and find the zeros

▶ How to write mathematical models for direct, inverse, and joint variation

▶ How to use the least squares regression feature of a graphing utility to find mathematical models of actual data

Important Vocabulary

As you encounter each new vocabulary term in this chapter, add the term and its definition to your notebook glossary.

Polynomial function (p. 254)
Constant function (p. 254)
Linear function (p. 254)
Quadratic function (p. 254)
Parabola (p. 254)
Axis (p. 255)
Vertex (p. 255)
Standard form of a quadratic function (p. 257)
Continuous (p. 265)
Leading Coefficient Test (p. 267)
Repeated zero (p. 269)
Multiplicity (p. 269)
Intermediate Value Theorem (p. 272)
Long division (p. 278)
Division Algorithm (p. 279)
Improper (p. 279)
Proper (p. 279)
Synthetic division (p. 281)
Remainder Theorem (p. 282)
Factor Theorem (p. 282)

Fundamental Theorem of Algebra (p. 288)
Linear Factorization Theorem (p. 288)
Rational Zero Test (p. 289)
Conjugates (p. 292)
Irreducible over the reals (p. 293)
Descartes's Rule of Signs (p. 295)
Variation in sign (p. 295)
Upper bound (p. 296)
Lower bound (p. 296)
Vary directly (p. 304)
Directly proportional (p. 304)
Constant of variation (p. 304)
Vary inversely (p. 306)
Inversely proportional (p. 306)
Vary jointly (p. 307)
Jointly proportional (p. 307)
Sum of square differences (p. 308)
Least squares regression line (p. 308)

Study Tools

- Learning objectives at the beginning of each section
- Chapter Summary (p. 315)
- Review Exercises (pp. 316–319)
- Chapter Test (p. 321)

Additional Resources

- Study and Solutions Guide
- Interactive College Algebra
- Videotapes for Chapter 3
- College Algebra Website
- Student Success Organizer

STUDY T!P

The best time to do homework is right after class, when concepts are still fresh in your mind. This increases your chances of retaining the information in long-term memory.

3.1 Quadratic Functions

▶ **What you should learn**

- How to analyze graphs of quadratic functions
- How to write quadratic functions in standard form and use the results to sketch graphs of functions
- How to use quadratic functions to model and solve real-life problems

▶ **Why you should learn it**

Quadratic functions can be used to model data to analyze consumer behavior. For instance, Exercise 91 on page 264 shows how a quadratic function can model VCR usage in the United States.

Tony Freeman/PhotoEdit

The Graph of a Quadratic Function

In this and the next section, you will study the graphs of polynomial functions.

Definition of Polynomial Function

Let n be a nonnegative integer and let $a_n, a_{n-1}, \ldots, a_2, a_1, a_0$ be real numbers with $a_n \neq 0$. The function

$$f(x) = a_n x^n + a_{n-1} x^{n-1} + \cdots + a_2 x^2 + a_1 x + a_0$$

is called a **polynomial function of x with degree n.**

Polynomial functions are classified by degree. For instance, the polynomial function

$$f(x) = a, \qquad a \neq 0 \qquad \text{Constant function}$$

has degree 0 and is called a **constant function.** In Chapter 2, you learned that the graph of this type of function is a horizontal line. The polynomial function

$$f(x) = ax + b, \qquad a \neq 0 \qquad \text{Linear function}$$

has degree 1 and is called a **linear function.** In Chapter 2, you learned that the graph of the linear function $f(x) = ax + b$ is a line whose slope is a and whose y-intercept is $(0, b)$. In this section you will study second-degree polynomial functions, which are called **quadratic functions.**

For instance, each of the following functions is a quadratic function.

$$f(x) = x^2 + 6x + 2$$

$$g(x) = 2(x + 1)^2 - 3$$

$$h(x) = 9 + \tfrac{1}{4}x^2$$

$$k(x) = -3x^2 + 4$$

$$m(x) = (x - 2)(x + 1)$$

Definition of Quadratic Function

Let a, b, and c be real numbers with $a \neq 0$. The function

$$f(x) = ax^2 + bx + c \qquad \text{Quadratic function}$$

is called a **quadratic function.**

The graph of a quadratic function is a special type of "U"-shaped curve that is called a **parabola.** Parabolas occur in many real-life applications—especially those involving reflective properties of satellite dishes and flashlight reflectors. You will study these properties in Section 4.4.

All parabolas are symmetric with respect to a line called the **axis of symmetry,** or simply the **axis** of the parabola. The point where the axis intersects the parabola is the **vertex** of the parabola, as shown in Figure 3.1. If the leading coefficient is positive, the graph of $f(x) = ax^2 + bx + c$ is a parabola that opens upward. If the leading coefficient is negative, the graph of $f(x) = ax^2 + bx + c$ is a parabola that opens downward.

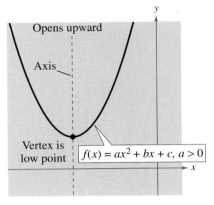

Leading coefficient is positive.

FIGURE 3.1

Leading coefficient is negative.

The simplest type of quadratic function is

$$f(x) = ax^2.$$

Its graph is a parabola whose vertex is $(0, 0)$. If $a > 0$, the vertex is the point with the *minimum* y-value on the graph, and if $a < 0$, the vertex is the point with the *maximum* y-value on the graph, as shown in Figure 3.2.

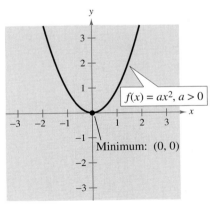

Leading coefficient is positive.

FIGURE 3.2

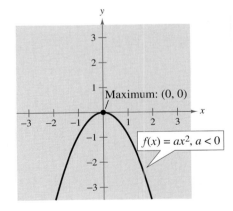

Leading coefficient is negative.

When sketching the graph of $f(x) = ax^2$, it is helpful to use the graph of $y = x^2$ as a reference, as discussed in Section 2.4.

◀ **Exploration** ▶

Graph $y = ax^2$ for $a = -2, -1,$ $-0.5, 0.5, 1,$ and 2. How does changing the value of a affect the graph?

Graph $y = (x - h)^2$ for $h = -4,$ $-2, 2,$ and 4. How does changing the value of h affect the graph?

Graph $y = x^2 + k$ for $k = -4,$ $-2, 2,$ and 4. How does changing the value of k affect the graph?

Example 1 ▶ Sketching Graphs of Quadratic Functions

a. Compare the graphs of $y = x^2$ and $f(x) = \frac{1}{3}x^2$.
b. Compare the graphs of $y = x^2$ and $g(x) = 2x^2$.

Solution

a. Compared with $y = x^2$, each output of $f(x) = \frac{1}{3}x^2$ "shrinks" by a factor of $\frac{1}{3}$, creating the broader parabola shown in Figure 3.3(a).

b. Compared with $y = x^2$, each output of $g(x) = 2x^2$ "stretches" by a factor of 2, creating the narrower parabola shown in Figure 3.3(b).

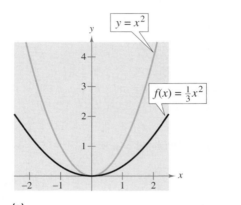

(a)

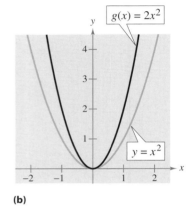

(b)

FIGURE **3.3**

In Example 1, note that the coefficient a determines how widely the parabola given by $f(x) = ax^2$ opens. If $|a|$ is small, the parabola opens more widely than if $|a|$ is large.

Recall from Section 2.4 that the graphs of $y = f(x \pm c)$, $y = f(x) \pm c$, $y = f(-x)$, and $y = -f(x)$ are rigid transformations of the graph of $y = f(x)$. For instance, in Figure 3.4, notice how the graph of $y = x^2$ can be transformed to produce the graphs of $f(x) = -x^2 + 1$ and $g(x) = (x + 2)^2 - 3$.

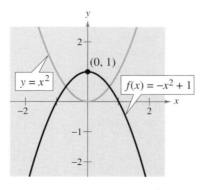

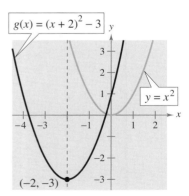

FIGURE **3.4**

The Standard Form of a Quadratic Function

The **standard form** of a quadratic function is

$$f(x) = a(x - h)^2 + k.$$

This form is especially convenient for sketching a parabola because it identifies the vertex of the parabola.

> ### Standard Form of a Quadratic Function
>
> The quadratic function
>
> $$f(x) = a(x - h)^2 + k, \qquad a \neq 0$$
>
> is in **standard form.** The graph of f is a parabola whose axis is the vertical line $x = h$ and whose vertex is the point (h, k). If $a > 0$, the parabola opens upward, and if $a < 0$, the parabola opens downward.

To write a quadratic function in standard form, you can use the process of *completing the square,* as illustrated in Example 2.

Example 2 ▶ Graphing a Parabola in Standard Form

Sketch the graph of

$$f(x) = 2x^2 + 8x + 7$$

and identify the vertex and the axis of the parabola.

Solution

Begin by writing the quadratic function in standard form. Notice that the first step in completing the square is to factor out any coefficient of x^2 that is not 1.

$$
\begin{aligned}
f(x) &= 2x^2 + 8x + 7 & &\text{Write original function.}\\
&= 2(x^2 + 4x) + 7 & &\text{Factor 2 out of } x\text{-terms.}\\
&= 2(x^2 + 4x + 4 - 4) + 7 & &\text{Add and subtract 4 within parentheses.}\\
& \qquad\qquad\quad 2^2 \\
&= 2(x^2 + 4x + 4) - 2(4) + 7 & &\text{Regroup terms.}\\
&= 2(x^2 + 4x + 4) - 8 + 7 & &\text{Simplify.}\\
&= 2(x + 2)^2 - 1 & &\text{Write in standard form.}
\end{aligned}
$$

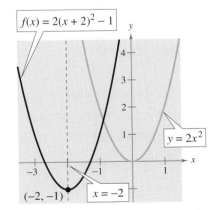

$f(x) = 2(x + 2)^2 - 1$

$y = 2x^2$

$(-2, -1)$ $x = -2$

FIGURE **3.5**

From this form, you can see that the graph of f is a parabola that opens upward and has its vertex at $(-2, -1)$. This corresponds to a left shift of 2 units and a downward shift of 1 unit relative to the graph of $y = 2x^2$, as shown in Figure 3.5. In the figure, you can see that the axis of the parabola is the vertical line through the vertex, $x = -2$.

To find the x-intercepts of the graph of $f(x) = ax^2 + bx + c$, you must solve the equation $ax^2 + bx + c = 0$. If $ax^2 + bx + c$ does not factor, you can use the Quadratic Formula to find the x-intercepts. Remember, however, that a parabola may have no x-intercepts.

Example 3 ▶ Finding the Vertex and x-Intercepts of a Parabola

Sketch the graph of $f(x) = -x^2 + 6x - 8$ and identify the vertex and x-intercepts.

Solution

As in Example 2, begin by writing the quadratic function in standard form.

$$f(x) = -x^2 + 6x - 8 \qquad \text{Write original function.}$$
$$= -(x^2 - 6x) - 8 \qquad \text{Factor } -1 \text{ out of } x\text{-terms.}$$
$$= -(x^2 - 6x + 9 - 9) - 8 \qquad \text{Add and subtract 9 within parentheses.}$$

$$(-3)^2$$

$$= -(x^2 - 6x + 9) - (-9) - 8 \qquad \text{Regroup terms.}$$
$$= -(x - 3)^2 + 1 \qquad \text{Write in standard form.}$$

From this form, you can see that the vertex is $(3, 1)$. To find the x-intercepts of the graph, solve the equation $-x^2 + 6x - 8 = 0$.

$$-x^2 + 6x - 8 = 0 \qquad \text{Write original equation.}$$
$$-(x^2 - 6x + 8) = 0 \qquad \text{Factor out } -1.$$
$$-(x - 2)(x - 4) = 0 \qquad \text{Factor.}$$
$$x - 2 = 0 \quad \Longrightarrow \quad x = 2 \qquad \text{Set 1st factor equal to 0.}$$
$$x - 4 = 0 \quad \Longrightarrow \quad x = 4 \qquad \text{Set 2nd factor equal to 0.}$$

The x-intercepts are $(2, 0)$ and $(4, 0)$. So, the graph of f is a parabola that opens downward, as shown in Figure 3.6.

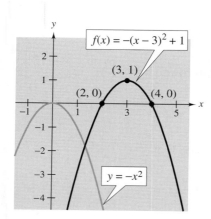

FIGURE 3.6

Example 4 ▶ Finding the Equation of a Parabola

Find the standard form of the equation of the parabola whose vertex is $(1, 2)$ and that passes through the point $(0, 0)$, as shown in Figure 3.7.

Solution

Because the vertex of the parabola is at $(h, k) = (1, 2)$, the equation has the form

$$f(x) = a(x - 1)^2 + 2. \qquad \text{Substitute for } h \text{ and } k \text{ in standard form.}$$

Because the parabola passes through the point $(0, 0)$, it follows that $f(0) = 0$. So,

$$0 = a(0 - 1)^2 + 2 \quad \Longrightarrow \quad a = -2 \qquad \text{Substitute 0 for } x; \text{ solve for } a.$$

which implies that the equation is

$$f(x) = -2(x - 1)^2 + 2. \qquad \text{Substitute for } a \text{ in standard form.}$$

So, the equation of this parabola is $y = -2(x - 1)^2 + 2$.

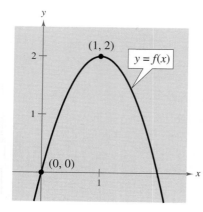

FIGURE 3.7

Applications

Many applications involve finding the maximum or minimum value of a quadratic function. Some quadratic functions are not easily written in standard form. For such functions, it is useful to have an alternative method for finding the vertex. For a quadratic function in the form $f(x) = ax^2 + bx + c$, the vertex occurs when $x = -b/2a$.

Vertex of a Parabola

The vertex of the graph of $f(x) = ax^2 + bx + c$ is $\left(-\dfrac{b}{2a}, f\left(-\dfrac{b}{2a}\right)\right)$.

A computer simulation of this example appears in the *Interactive* CD-ROM and *Internet* versions of this text.

Example 5 ▶ The Maximum Height of a Baseball

A baseball is hit at a point 3 feet above the ground at a velocity of 100 feet per second and at an angle of 45° with respect to the ground. The path of the baseball is given by the function

$$f(x) = -0.0032x^2 + x + 3$$

where $f(x)$ is the height of the baseball (in feet) and x is the horizontal distance from home plate (in feet). What is the maximum height reached by the baseball?

Solution

For this quadratic function, you have

$$f(x) = ax^2 + bx + c$$
$$= -0.0032x^2 + x + 3.$$

So, $a = -0.0032$ and $b = 1$. Because the function has a maximum at $x = -b/2a$, you can conclude that the baseball reaches its maximum height when it is x feet from home plate, where x is

$$x = -\frac{b}{2a}$$

$$= -\frac{1}{2(-0.0032)} \qquad \text{Substitute for } a \text{ and } b.$$

$$= 156.25 \text{ feet.}$$

To find the maximum height, you must determine the value of the function when $x = 156.25$.

$$f(156.25) = -0.0032(156.25)^2 + 156.25 + 3$$

$$= 81.125 \text{ feet.}$$

The path of the baseball is shown in Figure 3.8. You can estimate from the graph in Figure 3.8 that the ball hits the ground at a distance of about 320 feet from home plate. The actual distance is the x-intercept of the graph of f, which you can find by solving the equation $-0.0032x^2 + x + 3 = 0$ and taking the positive solution, $x \approx 315.5$.

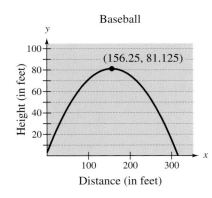

FIGURE **3.8**

3.1 Exercises

The *Interactive* CD-ROM and *Internet* versions of this text contain step-by-step solutions to all odd-numbered Section and Review Exercises. They also provide Tutorial Exercises that link to Guided Examples for additional help.

In Exercises 1–8, match the quadratic function with its graph. [The graphs are labeled (a), (b), (c), (d), (e), (f), (g), and (h).]

(a)

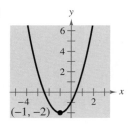

(b)

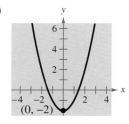

(c)

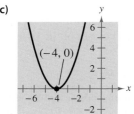

(d)

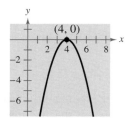

(e)

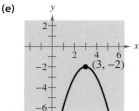

(f)

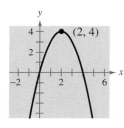

(g)

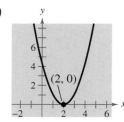

(h)

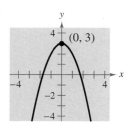

1. $f(x) = (x - 2)^2$
2. $f(x) = (x + 4)^2$
3. $f(x) = x^2 - 2$
4. $f(x) = 3 - x^2$
5. $f(x) = 4 - (x - 2)^2$
6. $f(x) = (x + 1)^2 - 2$
7. $f(x) = -(x - 3)^2 - 2$
8. $f(x) = -(x - 4)^2$

Exploration In Exercises 9–12, graph each equation. Compare the graph of each function with the graph of $y = x^2$.

9. (a) $f(x) = \frac{1}{2}x^2$ (b) $g(x) = -\frac{1}{8}x^2$
   (c) $h(x) = \frac{3}{2}x^2$ (d) $k(x) = -3x^2$

10. (a) $f(x) = x^2 + 1$ (b) $g(x) = x^2 - 1$
    (c) $h(x) = x^2 + 3$ (d) $k(x) = x^2 - 3$

11. (a) $f(x) = (x - 1)^2$ (b) $g(x) = (x + 1)^2$
    (c) $h(x) = (x - 3)^2$ (d) $k(x) = (x + 3)^2$

12. (a) $f(x) = -\frac{1}{2}(x - 2)^2 + 1$
    (b) $g(x) = \frac{1}{2}(x - 2)^2 + 1$
    (c) $h(x) = -\frac{1}{2}(x + 2)^2 - 1$
    (d) $k(x) = \frac{1}{2}(x + 2)^2 - 1$

In Exercises 13–28, sketch the graph of the quadratic function without using a graphing utility. Identify the vertex and x-intercepts.

13. $f(x) = x^2 - 5$
14. $h(x) = 25 - x^2$
15. $f(x) = \frac{1}{2}x^2 - 4$
16. $f(x) = 16 - \frac{1}{4}x^2$
17. $f(x) = (x + 5)^2 - 6$
18. $f(x) = (x - 6)^2 + 3$
19. $h(x) = x^2 - 8x + 16$
20. $g(x) = x^2 + 2x + 1$
21. $f(x) = x^2 - x + \frac{5}{4}$
22. $f(x) = x^2 + 3x + \frac{1}{4}$
23. $f(x) = -x^2 + 2x + 5$
24. $f(x) = -x^2 - 4x + 1$
25. $h(x) = 4x^2 - 4x + 21$
26. $f(x) = 2x^2 - x + 1$
27. $f(x) = \frac{1}{4}x^2 - 2x - 12$
28. $f(x) = -\frac{1}{3}x^2 + 3x - 6$

In Exercises 29–36, use a graphing utility to graph the quadratic function. Identify the vertex and x-intercepts. Then check your results algebraically by completing the square.

29. $f(x) = -(x^2 + 2x - 3)$
30. $f(x) = -(x^2 + x - 30)$
31. $g(x) = x^2 + 8x + 11$
32. $f(x) = x^2 + 10x + 14$
33. $f(x) = 2x^2 - 16x + 31$
34. $f(x) = -4x^2 + 24x - 41$
35. $g(x) = \frac{1}{2}(x^2 + 4x - 2)$
36. $f(x) = \frac{3}{5}(x^2 + 6x - 5)$

In Exercises 37–42, find the standard form of the equation of the parabola.

37.

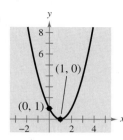

38.

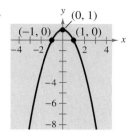

39.

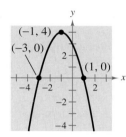

40.

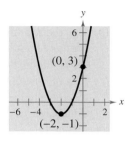

41.

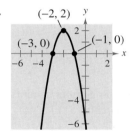

42.

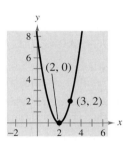

In Exercises 43–52, find the quadratic function that has the indicated vertex and whose graph passes through the given point.

43. Vertex: $(-2, 5)$; Point: $(0, 9)$

44. Vertex: $(4, -1)$; Point: $(2, 3)$

45. Vertex: $(3, 4)$; Point: $(1, 2)$

46. Vertex: $(2, 3)$; Point: $(0, 2)$

47. Vertex: $(5, 12)$; Point: $(7, 15)$

48. Vertex: $(-2, -2)$; Point: $(-1, 0)$

49. Vertex: $\left(-\frac{1}{4}, \frac{3}{2}\right)$; Point: $(-2, 0)$

50. Vertex: $\left(\frac{5}{2}, -\frac{3}{4}\right)$; Point: $(-2, 4)$

51. Vertex: $\left(-\frac{5}{2}, 0\right)$; Point: $\left(-\frac{7}{2}, -\frac{16}{3}\right)$

52. Vertex: $(6, 6)$; Point: $\left(\frac{61}{10}, \frac{3}{2}\right)$

Graphical Reasoning In Exercises 53–56, determine the x-intercepts of the graph visually. Explain how the x-intercepts relate to the solutions of the quadratic equation when $y = 0$. Then find the x-intercepts algebraically to confirm your results.

53. $y = x^2 - 16$

54. $y = x^2 - 6x + 9$

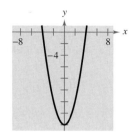

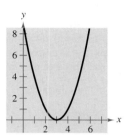

55. $y = x^2 - 4x - 5$

56. $y = 2x^2 + 5x - 3$

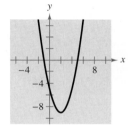

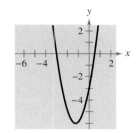

In Exercises 57–64, use a graphing utility to graph the quadratic function. Find the x-intercepts of the graph and compare them with the solutions of the corresponding quadratic equation when $y = 0$.

57. $f(x) = x^2 - 4x$

58. $f(x) = -2x^2 + 10x$

59. $f(x) = x^2 - 9x + 18$

60. $f(x) = x^2 - 8x - 20$

61. $f(x) = 2x^2 - 7x - 30$

62. $f(x) = 4x^2 + 25x - 21$

63. $f(x) = -\frac{1}{2}(x^2 - 6x - 7)$

64. $f(x) = \frac{7}{10}(x^2 + 12x - 45)$

In Exercises 65–70, find two quadratic functions, one that opens upward and one that opens downward, whose graphs have the given x-intercepts. (There are many correct answers.)

65. $(-1, 0), (3, 0)$

66. $(-5, 0), (5, 0)$

67. $(0, 0), (10, 0)$

68. $(4, 0), (8, 0)$

69. $(-3, 0), \left(-\frac{1}{2}, 0\right)$

70. $\left(-\frac{5}{2}, 0\right), (2, 0)$

In Exercises 71–74, find two positive real numbers whose product is a maximum.

71. The sum is 110.   **72.** The sum is S.

73. The sum of the first and twice the second is 24.

74. The sum of the first and three times the second is 42.

Geometry   **In Exercises 75 and 76, consider a rectangle of length x and perimeter P. (a) Express the area A as a function of x and determine the domain of the function. (b) Graph the area function. (c) Find the length and width of the rectangle of maximum area.**

75. $P = 100$ feet   **76.** $P = 36$ meters

77. *Numerical, Graphical, and Analytical Analysis*   A rancher has 200 feet of fencing to enclose two adjacent rectangular corrals (see figure).

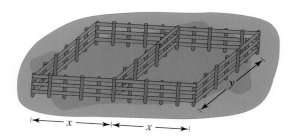

(a) Complete six rows of a table such as the one below, showing possible values for $x, y,$ and the area of the corral.

x	y	Area
2	$\frac{1}{3}[200 - 4(2)]$	$2xy = 256$
4	$\frac{1}{3}[200 - 4(4)]$	$2xy \approx 491$

(b) Use a graphing utility to generate additional rows of the table. Use the table to estimate the dimensions that will enclose the maximum area.

(c) Write the area A as a function of x.

(d) Use a graphing utility to graph the area function. Use the graph to approximate the dimensions that will produce the maximum enclosed area.

(e) Write the area function in standard form to find analytically the dimensions that will produce the maximum area.

78. *Geometry*   An indoor physical fitness room consists of a rectangular region with a semicircle on each end (see figure). The perimeter of the room is to be a 200-meter single-lane running track.

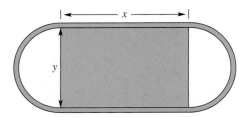

(a) Determine the radius of the semicircular ends of the room. Determine the distance, in terms of y, around the inside edge of the two semicircular parts of the track.

(b) Use the result of part (a) to write an equation, in terms of x and y, for the distance traveled in one lap around the track. Solve for y.

(c) Use the result of part (b) to write the area A of the rectangular region as a function of x. What dimensions will produce a maximum area of the rectangle?

79. *Maximum Revenue*   Find the number of units sold that produces a maximum revenue

$$R = 900x - 0.1x^2$$

where R is the total revenue (in dollars) and x is the number of units sold.

80. *Maximum Revenue*   Find the number of units sold that produces a maximum revenue

$$R = 100x - 0.0002x^2$$

where R is the total revenue (in dollars) and x is the number of units sold.

81. *Minimum Cost*   A manufacturer of lighting fixtures has daily production costs of

$$C = 800 - 10x + 0.25x^2$$

where C is the total cost (in dollars) and x is the number of units produced. How many fixtures should be produced each day to yield a minimum cost?

82. *Minimum Cost*   A textile manufacturer has daily production costs of

$$C = 100{,}000 - 110x + 0.045x^2$$

where C is the total cost (in dollars) and x is the number of units produced. How many units should be produced each day to yield a minimum cost?

83. *Maximum Profit*   The profit for a company is

$$P = -0.0002x^2 + 140x - 250,000$$

where x is the number of units sold. What sales level will yield a maximum profit?

84. *Maximum Profit*   The profit P (in hundreds of dollars) that a company makes depends on the amount x (in hundreds of dollars) the company spends on advertising according to the model

$$P = 230 + 20x - 0.5x^2.$$

What expenditure for advertising will yield a maximum profit?

85. *Physics*   The height y (in feet) of a ball thrown by a child is

$$y = -\frac{1}{12}x^2 + 2x + 4$$

where x is the horizontal distance (in feet) from the point at which the ball is thrown.

(a) How high is the ball when it leaves the child's hand? (*Hint:* Find y when $x = 0$.)

(b) What is the maximum height of the ball?

(c) How far from the child does the ball strike the ground?

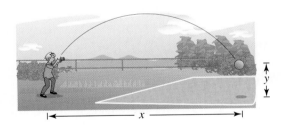

86. *Forestry*   The number of board feet in a 16-foot log is approximated by the model

$$V = 0.77x^2 - 1.32x - 9.31, \qquad 5 \le x \le 40$$

where V is the number of board feet and x is the diameter (in inches) of the log at the small end. (One board foot is a measure of volume equivalent to a board that is 12 inches wide, 12 inches long, and 1 inch thick.)

(a) Sketch a graph of the function.

(b) Estimate the number of board feet in a 16-foot log with a diameter of 16 inches.

(c) Estimate the diameter of a 16-foot log that produced 500 board feet.

87. *Physics*   The path of a diver is

$$y = -\frac{4}{9}x^2 + \frac{24}{9}x + 12$$

where y is the height (in feet) and x is the horizontal distance from the end of the diving board (in feet). What is the maximum height of the diver?

88. *Physics*   The number of horsepower y required to overcome wind drag on a certain automobile is approximated by

$$y = 0.002s^2 + 0.005s - 0.029, \qquad 0 \le s \le 100$$

where s is the speed of the car (in miles per hour).

(a) Use a graphing utility to graph the function.

(b) Graphically estimate the maximum speed of the car if the power required to overcome wind drag is not to exceed 10 horsepower. Verify analytically.

89. *Graphical Analysis*   From 1950 to 1990, the average annual consumption C of cigarettes by Americans (18 and older) for selected years can be modeled by

$$C = 3248.89 + 108.64t - 2.97t^2, \qquad 0 \le t \le 40$$

where t is the year, with $t = 0$ corresponding to 1950. (Source: U.S. Department of Agriculture)

(a) Use a graphing utility to graph the model.

(b) Use the graph of the model to approximate the maximum average annual consumption. Beginning in 1966, all cigarette packages were required by law to carry a health warning. Do you think the warning had any effect? Explain.

(c) In 1960, the U.S. population (18 and over) was 116,530,000. Of those, about 48,500,000 were smokers. What was the average annual cigarette consumption *per smoker* in 1960? What was the average daily cigarette consumption *per smoker*?

90. *Maximum Fuel Economy* A study was done to compare the speed x (in miles per hour) with the mileage y (in miles per gallon) of an automobile. The results are shown in the table. (Source: Federal Highway Administration)

Speed x	15	20	25	30	35	40	45
Mileage y	22.3	25.5	27.5	29.0	28.8	30.0	29.9

Speed x	50	55	60	65	70	75
Mileage y	30.2	30.4	28.8	27.4	25.3	23.3

(a) Use a graphing utility to plot the data and find the quadratic model that best fits the data. Graph the model in the same viewing window as the data.

(b) Estimate the speed for which the miles per gallon is greatest.

91. *Data Analysis* The numbers y (in millions) of VCRs in use in the United States for the years 1987 through 1996 are shown in the table. The variable t represents time (in years), with $t = 7$ corresponding to 1987. (Source: Television Bureau of Advertising, Inc.)

t	7	8	9	10	11	12	13	14	15	16
y	43	51	58	63	67	69	72	74	77	79

(a) Use a graphing utility to sketch a scatter plot of the data.

(b) Use the regression feature of a graphing utility to find a quadratic model for the data.

(c) Use a graphing utility to graph the model in the same viewing window as the scatter plot.

(d) Do you think the model can be used to predict VCR use in 2005? Explain.

Synthesis

True or False? **In Exercises 92 and 93, determine whether the statement is true or false. Justify your answer.**

92. The function $f(x) = -12x^2 - 1$ has no x-intercepts.

93. The graphs of $f(x) = -4x^2 - 10x + 7$ and $g(x) = 12x^2 + 30x + 1$ have the same axis of symmetry.

94. Write the quadratic equation $f(x) = ax^2 + bx + c$ in standard form to verify that the vertex occurs at
$$\left(-\frac{b}{2a}, f\left(-\frac{b}{2a}\right)\right).$$

95. *Business* The profit P (in millions of dollars) for a company is modeled by a quadratic function of the form
$$P = at^2 + bt + c$$
where t represents the year. If you were president of the company, which of the models below would you prefer? Explain your reasoning.

(a) a is positive and $-b/(2a) \leq t$.

(b) a is positive and $t \leq -b/(2a)$.

(c) a is negative and $-b/(2a) \leq t$.

(d) a is negative and $t \leq -b/(2a)$.

96. Is it possible for a quadratic equation to have only one x-intercept? Explain.

97. Assume that the function $f(x) = ax^2 + bx + c$ $(a \neq 0)$ has two real zeros. Show that the x-coordinate of the vertex of the graph is the average of the zeros of f. (*Hint:* Use the Quadratic Formula.)

98. Use a graphing utility to demonstrate the result of Exercise 97 for each of the following functions.

(a) $f(x) = \frac{1}{2}(x - 3)^2 - 2$

(b) $f(x) = 6 - \frac{2}{3}(x + 1)^2$

Review

In Exercises 99–102, find the equation of the line in slope-intercept form that has the given characteristics.

99. Contains the points $(-4, 3)$ and $(2, 1)$

100. Contains the point $(\frac{7}{2}, 2)$ and has a slope of $\frac{3}{2}$

101. Contains the point $(0, 3)$ and is perpendicular to the line $4x + 5y = 10$

102. Contains the point $(-8, 4)$ and is parallel to the line $y = -3x + 2$

In Exercises 103–108, let $f(x) = 14x - 3$ and let $g(x) = 8x^2$. Find the indicated value.

103. $(f + g)(-3)$

104. $(g - f)(2)$

105. $(fg)\left(-\frac{4}{7}\right)$

106. $\left(\frac{f}{g}\right)(-1.5)$

107. $(f \circ g)(-1)$

108. $(g \circ f)(0)$

3.2 Polynomial Functions of Higher Degree

Graphs of Polynomial Functions

In this section, you will study basic features of the graphs of polynomial functions. The first feature is that the graph of a polynomial function is **continuous.** Essentially, this means that the graph of a polynomial function has no breaks, holes, or gaps, as shown in Figure 3.9(a).

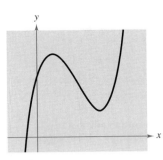

(a) Polynomial functions have continuous graphs.

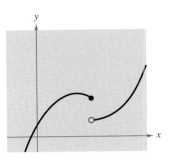

(b) Functions with graphs that are not continuous are not polynomial functions.

FIGURE 3.9

The second feature is that the graph of a polynomial function has only smooth, rounded turns, as shown in Figure 3.10(a). A polynomial function cannot have a sharp turn. For instance, the function $f(x) = |x|$, which has a sharp turn at the point $(0, 0)$, as shown in Figure 3.10(b), is not a polynomial function.

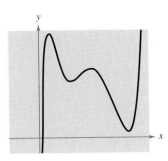

(a) Polynomial functions have graphs with rounded turns.

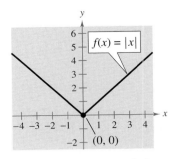

(b) Functions whose graphs have sharp turns are not polynomial functions.

FIGURE 3.10

The graphs of polynomial functions of degree greater than 2 are more difficult to analyze than the graphs of polynomials of degree 0, 1, or 2. However, using the features presented in this section, together with point plotting, intercepts, and symmetry, you should be able to make reasonably accurate sketches *by hand.*

The polynomial functions that have the simplest graphs are monomials of the form $f(x) = x^n$, where n is an integer greater than zero. From Figure 3.11, you can see that when n is *even* the graph is similar to the graph of $f(x) = x^2$ and when n is *odd* the graph is similar to the graph of $f(x) = x^3$. Moreover, the greater the value of n, the flatter the graph near the origin.

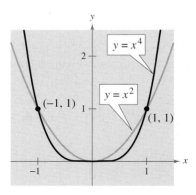

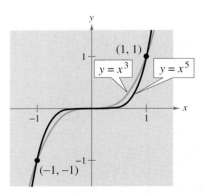

(a) **If n is even, the graph of $y = x^n$ touches the axis at the x-intercept.**

(b) **If n is odd, the graph of $y = x^n$ crosses the axis at the x-intercept.**

FIGURE 3.11

Example 1 ▶   Sketching Transformations of Monomial Functions

Sketch the graph of each function.

a. $f(x) = -x^5$ **b.** $h(x) = (x + 1)^4$

Solution

a. Because the degree of $f(x) = -x^5$ is odd, its graph is similar to the graph of $y = x^3$. In Figure 3.12(a), note that the negative coefficient has the effect of reflecting the graph in the x-axis.

b. The graph of $h(x) = (x + 1)^4$ as shown in Figure 3.12(b), is a left shift by 1 unit of the graph of $y = x^4$.

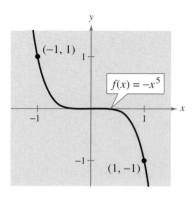

(a)

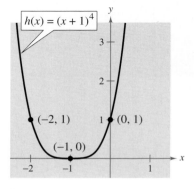

(b)

FIGURE 3.12

For each function, identify the degree of the function and whether the degree of the function is even or odd. Identify the leading coefficient and whether the leading coefficient is greater than 0 or less than 0. Use a graphing utility to graph each function. Describe the relationship between the degree and leading coefficient of the function and the right- and left-hand behavior of the graph of the function.

a. $f(x) = x^3 - 2x^2 - x + 1$

b. $f(x) = 2x^5 + 2x^2 - 5x + 1$

c. $f(x) = -2x^5 - x^2 + 5x + 3$

d. $f(x) = -x^3 + 5x - 2$

e. $f(x) = 2x^2 + 3x - 4$

f. $f(x) = x^4 - 3x^2 + 2x - 1$

g. $f(x) = x^2 + 3x + 2$

STUDY T!P

The notation "$f(x) \rightarrow -\infty$ as $x \rightarrow -\infty$" indicates that the graph falls to the left. The notation "$f(x) \rightarrow \infty$ as $x \rightarrow \infty$" indicates that the graph rises to the right.

The Leading Coefficient Test

In Example 1, note that both graphs eventually rise or fall without bound as x moves to the right. Whether the graph of a polynomial eventually rises or falls can be determined by the function's degree (even or odd) and by its leading coefficient, as indicated in the **Leading Coefficient Test.**

Leading Coefficient Test

As x moves without bound to the left or to the right, the graph of the polynomial function $f(x) = a_n x^n + \cdots + a_1 x + a_0$ eventually rises or falls in the following manner.

1. When n is *odd:*

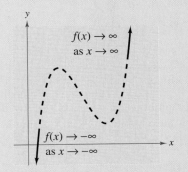

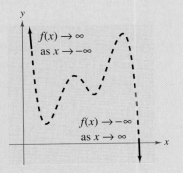

If the leading coefficient is positive $(a_n > 0)$, the graph falls to the left and rises to the right.

If the leading coefficient is negative $(a_n < 0)$, the graph rises to the left and falls to the right.

2. When n is *even:*

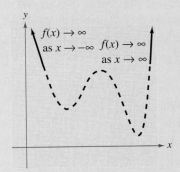

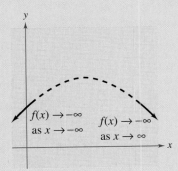

If the leading coefficient is positive $(a_n > 0)$, the graph rises to the left and right.

If the leading coefficient is negative $(a_n < 0)$, the graph falls to the left and right.

The dashed portions of the graphs indicate that the test determines *only* the right-hand and left-hand behavior of the graph.

Example 2 ▶   **Applying the Leading Coefficient Test**

Describe the right-hand and left-hand behavior of the graph of $f(x) = -x^3 + 4x$.

Solution

Because the degree is odd and the leading coefficient is negative, the graph rises to the left and falls to the right, as shown in Figure 3.13.

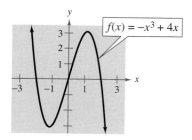

FIGURE 3.13

In Example 2, note that the Leading Coefficient Test only tells you whether the graph *eventually* rises or falls to the right or left. Other characteristics of the graph, such as intercepts and minimum and maximum points, must be determined by other tests.

Example 3 ▶   **Applying the Leading Coefficient Test**

Describe the right-hand and left-hand behavior of the graph of each function.

a. $f(x) = x^4 - 5x^2 + 4$ **b.** $f(x) = x^5 - x$

Solution

a. Because the degree is even and the leading coefficient is positive, the graph rises to the left and right, as shown in Figure 3.14(a).

b. Because the degree is odd and the leading coefficient is positive, the graph falls to the left and rises to the right, as shown in Figure 3.14(b).

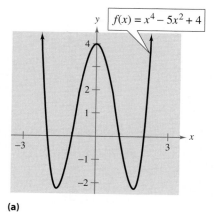

(a)

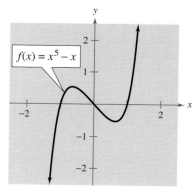

(b)

FIGURE 3.14

Zeros of Polynomial Functions

It can be shown that for a polynomial function f of degree n, the following statements are true. (Remember that the *zeros* of a function of x are the x-values for which the function is zero.)

1. The graph of f has, at most, $n - 1$ turning points. (Turning points are points at which the graph changes from increasing to decreasing or vice versa.)

2. The function f has, at most, n real zeros. (You will study this result in detail in Section 3.4 on the Fundamental Theorem of Algebra.)

   Finding the zeros of polynomial functions is one of the most important problems in algebra. There is a strong interplay between graphical and algebraic approaches to this problem. Sometimes you can use information about the graph of a function to help find its zeros, and in other cases you can use information about the zeros of a function to help sketch its graph.

Real Zeros of Polynomial Functions

If f is a polynomial function and a is a real number, the following statements are equivalent.

1. $x = a$ is a *zero* of the function f.

2. $x = a$ is a *solution* of the polynomial equation $f(x) = 0$.

3. $(x - a)$ is a *factor* of the polynomial $f(x)$.

4. $(a, 0)$ is an *x-intercept* of the graph of f.

Example 4 ▶ Finding the Zeros of a Polynomial Function

Find all real zeros of $f(x) = -2x^4 + 2x^2$. Use the graph in Figure 3.15 to determine the number of turning points of the graph of the function.

Solution

In this case, the polynomial factors as follows.

$$f(x) = -2x^4 + 2x^2 \qquad \text{Write original function.}$$

$$= -2x^2(x^2 - 1) \qquad \text{Remove common monomial factor.}$$

$$= -2x^2(x - 1)(x + 1) \qquad \text{Factor completely.}$$

So, the real zeros are $x = 0$, $x = 1$, and $x = -1$, and the corresponding x-intercepts are $(0, 0)$, $(1, 0)$, and $(-1, 0)$, as shown in Figure 3.15. Note in the figure that the graph has three turning points. This is consistent with the fact that a fourth-degree polynomial can have *at most* three turning points.

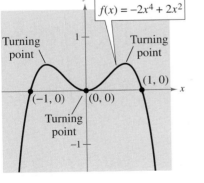

FIGURE 3.15

In Example 4, the real zero arising from $-2x^2 = 0$ is called a **repeated zero**. In general, a factor $(x - a)^k$, $k > 1$, yields a repeated zero $x = a$ of **multiplicity** k. If k is odd, the graph *crosses* the x-axis at $x = a$. If k is even, the graph *touches* the x-axis (but does not cross the x-axis) at $x = a$. In Example 4, the factor $-2x^2$ yields the repeated zero $x = 0$. Because k is even, the graph touches the x-axis at $x = 0$, as shown in Figure 3.15.

Technology

Example 5 uses an "algebraic approach" to describe the graph of the function. A graphing utility is a complement to this approach. Remember that an important aspect of using a graphing utility is to find a viewing window that shows all significant features of the graph. For instance, which of the graphs below shows all of the significant features of the function in Example 5?

a.

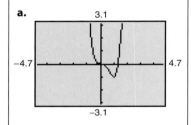

b.

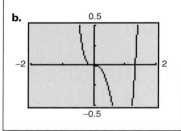

Example 5 ▶ Sketching the Graph of a Polynomial Function

Sketch the graph of $f(x) = 3x^4 - 4x^3$.

Solution

1. *Apply the Leading Coefficient Test.* Because the leading coefficient is positive and the degree is even, you know that the graph eventually rises to the left and to the right (see Figure 3.16a).

2. *Find the Zeros of the Polynomial.* By factoring

$$f(x) = 3x^4 - 4x^3 = x^3(3x - 4)$$ Remove common factor.

you can see that the zeros of f are $x = 0$ and $x = \frac{4}{3}$ (both of odd multiplicity). So, the x-intercepts occur at $(0, 0)$ and $\left(\frac{4}{3}, 0\right)$. Add these points to your graph, as shown in Figure 3.16(a).

3. *Plot a Few Additional Points.* To sketch the graph by hand, find a few additional points, as shown in the table. Then plot the points (see Figure 3.16b).

x	-1	0.5	1	1.5
$f(x)$	7	-0.3125	-1	1.6875

4. *Draw the Graph.* Draw a continuous curve through the points, as shown in Figure 3.16(b). Because both zeros are of odd multiplicity, you know that the graph should cross the x-axis at $x = 0$ and $x = \frac{4}{3}$. If you are unsure of the shape of that portion of the graph, plot some additional points.

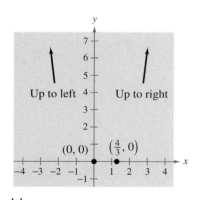

(a)

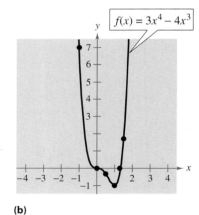

(b)

FIGURE **3.16**

The *Interactive* CD-ROM and *Internet* versions of this text offer a built-in graphing calculator, which can be used in the Examples, Explorations, Technology notes, and Exercises.

A polynomial function is written in **standard form** if its terms are written in descending order of exponents from left to right. Before applying the Leading Coefficient Test to a polynomial function, it is a good idea to check that the polynomial function is written in standard form. For instance, if the function in Example 5 had been given as $f(x) = -4x^3 + 3x^4$, it might have appeared that the leading coefficient was negative.

Example 6 ▶ Sketching the Graph of a Polynomial Function

Sketch the graph of $f(x) = -2x^3 + 6x^2 - \frac{9}{2}x$.

Solution

1. *Apply the Leading Coefficient Test.* Because the leading coefficient is negative and the degree is odd, you know that the graph eventually rises to the left and falls to the right (see Figure 3.17a).

2. *Find the Zeros of the Polynomial.* By factoring

$$f(x) = -2x^3 + 6x^2 - \frac{9}{2}x$$

$$= -\frac{1}{2}x(4x^2 - 12x + 9) \qquad \text{Remove common factor.}$$

$$= -\frac{1}{2}x(2x - 3)^2 \qquad \text{Factor completely.}$$

you can see that the zeros of f are $x = 0$ (odd multiplicity) and $x = \frac{3}{2}$ (even multiplicity). So, the x-intercepts occur at $(0, 0)$ and $\left(\frac{3}{2}, 0\right)$. Add these points to your graph, as shown in Figure 3.17(a).

3. *Plot a Few Additional Points.* To sketch the graph by hand, find a few additional points, as shown in the table. Then plot the points (see Figure 3.17b).

x	-0.5	0.5	1	2
$f(x)$	4	-1	-0.5	-1

4. *Draw the Graph.* Draw a continuous curve through the points, as shown in Figure 3.17(b). As indicated by the multiplicities of the zeros, the graph crosses the x-axis at $(0, 0)$ but does not cross the x-axis at $\left(\frac{3}{2}, 0\right)$.

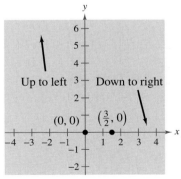

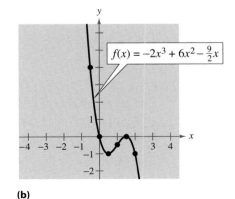

(a)

(b)

FIGURE **3.17**

The Intermediate Value Theorem

The next theorem, called the **Intermediate Value Theorem,** tells you of the existence of real zeros of polynomial functions. This theorem implies that if $(a, f(a))$ and $(b, f(b))$ are two points on the graph of a polynomial function such that $f(a) \neq f(b)$, then for any number d between $f(a)$ and $f(b)$ there must be a number c between a and b such that $f(c) = d$. (See Figure 3.18.)

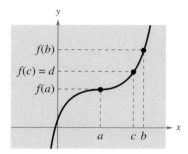

FIGURE 3.18

Intermediate Value Theorem

Let a and b be real numbers such that $a < b$. If f is a polynomial function such that $f(a) \neq f(b)$, then, in the interval $[a, b]$, f takes on every value between $f(a)$ and $f(b)$.

The Intermediate Value Theorem helps you locate the real zeros of a polynomial function in the following way. If you can find a value $x = a$ at which a polynomial function is positive, and another value $x = b$ at which it is negative, you can conclude that the function has at least one real zero between these two values. For example, the function

$$f(x) = x^3 + x^2 + 1$$

is negative when $x = -2$ and positive when $x = -1$. Therefore, it follows from the Intermediate Value Theorem that f must have a real zero somewhere between -2 and -1, as shown in Figure 3.19.

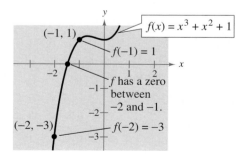

FIGURE 3.19

By continuing this line of reasoning, you can approximate any real zeros of a polynomial function to any desired accuracy. This concept is further demonstrated in Example 7.

Example 7 ▶ Approximating a Zero of a Polynomial Function

Use the Intermediate Value Theorem to approximate the real zero of

$$f(x) = x^3 - x^2 + 1.$$

Solution

Begin by computing a few function values, as follows.

x	-2	-1	0	1
$f(x)$	-11	-1	1	1

Because $f(-1)$ is negative and $f(0)$ is positive, you can apply the Intermediate Value Theorem to conclude that the function has a zero between -1 and 0. To pinpoint this zero more closely, divide the interval $[-1, 0]$ into tenths and evaluate the function at each point. When you do this, you will find that

$$f(-0.8) = -0.152$$

and

$$f(-0.7) = 0.167.$$

So, f must have a zero between -0.8 and -0.7, as shown in Figure 3.20. For a more accurate approximation, compute function values between $f(-0.8)$ and $f(-0.7)$ and apply the Intermediate Value Theorem again. By continuing this process you can approximate this zero to any desired accuracy.

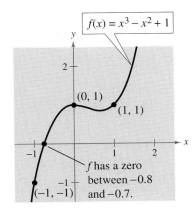

$f(x) = x^3 - x^2 + 1$

$(0, 1)$

$(1, 1)$

f has a zero between -0.8 and -0.7.

$(-1, -1)$

FIGURE 3.20

Writing ABOUT MATHEMATICS

Creating Polynomial Functions Suppose you are a math instructor and are writing a quiz for your algebra class. You want to make up several polynomial functions for your students to investigate. Write a paragraph explaining how you could find polynomial functions that have reasonably simple zeros. Justify your reasoning. Then use the methods you described to find polynomial functions that have the following zeros. For (a) and (c), find two different polynomial functions having the specified zeros.

a. $-5, \frac{1}{2}$ **b.** $-\frac{1}{6}, 7$

c. $2, 4, -3$ **d.** $3, -\frac{2}{3}, \frac{3}{4}$

e. $-3, -3, -3, -3$ **f.** $1, -2, 3, -4$

3.2 Exercises

In Exercises 1–8, match the polynomial function with its graph. [The graphs are labeled (a) through (h).]

(a)
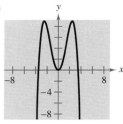

(b)
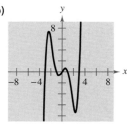

(c)
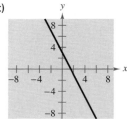

(d)
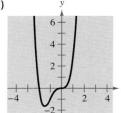

(e)
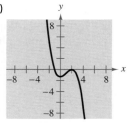

(f)
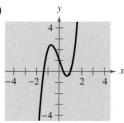

(g)
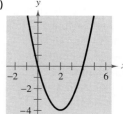

(h)
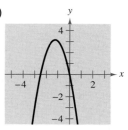

1. $f(x) = -2x + 3$
2. $f(x) = x^2 - 4x$
3. $f(x) = -2x^2 - 5x$
4. $f(x) = 2x^3 - 3x + 1$
5. $f(x) = -\frac{1}{4}x^4 + 3x^2$
6. $f(x) = -\frac{1}{3}x^3 + x^2 - \frac{4}{3}$
7. $f(x) = x^4 + 2x^3$
8. $f(x) = \frac{1}{5}x^5 - 2x^3 + \frac{9}{5}x$

In Exercises 9–12, sketch the graph of $y = x^n$ and each transformation.

9. $y = x^3$
   (a) $f(x) = (x - 2)^3$
   (b) $f(x) = x^3 - 2$
   (c) $f(x) = -\frac{1}{2}x^3$
   (d) $f(x) = (x - 2)^3 - 2$

10. $y = x^5$
   (a) $f(x) = (x + 1)^5$
   (b) $f(x) = x^5 + 1$
   (c) $f(x) = 1 - \frac{1}{2}x^5$
   (d) $f(x) = -\frac{1}{2}(x + 1)^5$

11. $y = x^4$
   (a) $f(x) = (x + 3)^4$
   (b) $f(x) = x^4 - 3$
   (c) $f(x) = 4 - x^4$
   (d) $f(x) = \frac{1}{2}(x - 1)^4$

12. $y = x^6$
   (a) $f(x) = -\frac{1}{8}x^6$
   (b) $f(x) = (x + 2)^6 - 4$
   (c) $f(x) = x^6 - 4$
   (d) $f(x) = -\frac{1}{4}x^6 + 1$

In Exercises 13–22, determine the right-hand and left-hand behavior of the graph of the polynomial function.

13. $f(x) = \frac{1}{3}x^3 + 5x$
14. $f(x) = 2x^2 - 3x + 1$
15. $g(x) = 5 - \frac{7}{2}x - 3x^2$
16. $h(x) = 1 - x^6$
17. $f(x) = -2.1x^5 + 4x^3 - 2$
18. $f(x) = 2x^5 - 5x + 7.5$
19. $f(x) = 6 - 2x + 4x^2 - 5x^3$
20. $f(x) = \dfrac{3x^4 - 2x + 5}{4}$
21. $h(t) = -\frac{2}{3}(t^2 - 5t + 3)$
22. $f(s) = -\frac{7}{8}(s^3 + 5s^2 - 7s + 1)$

▦ *Graphical Analysis*   In Exercises 23–26, use a graphing utility to graph the functions f and g in the same viewing window. Zoom out sufficiently far to show that the right-hand and left-hand behaviors of f and g appear identical.

23. $f(x) = 3x^3 - 9x + 1$, $g(x) = 3x^3$
24. $f(x) = -\frac{1}{3}(x^3 - 3x + 2)$, $g(x) = -\frac{1}{3}x^3$
25. $f(x) = -(x^4 - 4x^3 + 16x)$, $g(x) = -x^4$
26. $f(x) = 3x^4 - 6x^2$, $g(x) = 3x^4$

In Exercises 27–42, find all the real zeros of the polynomial function.

27. $f(x) = x^2 - 25$

28. $f(x) = 49 - x^2$

29. $h(t) = t^2 - 6t + 9$

30. $f(x) = x^2 + 10x + 25$

31. $f(x) = \frac{1}{3}x^2 + \frac{1}{3}x - \frac{2}{3}$

32. $f(x) = \frac{1}{2}x^2 + \frac{5}{2}x - \frac{3}{2}$

33. $f(x) = 3x^2 - 12x + 3$

34. $g(x) = 5(x^2 - 2x - 1)$

35. $f(t) = t^3 - 4t^2 + 4t$

36. $f(x) = x^4 - x^3 - 20x^2$

37. $g(t) = \frac{1}{2}t^4 - \frac{1}{2}$

38. $f(x) = x^5 + x^3 - 6x$

39. $g(t) = t^5 - 6t^3 + 9t$

40. $f(x) = 2x^4 - 2x^2 - 40$

41. $f(x) = 5x^4 + 15x^2 + 10$

42. $f(x) = x^3 - 4x^2 - 25x + 100$

Graphical Analysis **In Exercises 43–46, use a graphing utility to graph the function. Use the graph to approximate any x-intercepts of the graph. Set $y = 0$ and solve the resulting equation. Compare the result with any x-intercepts of the graph.**

43. $y = 4x^3 - 20x^2 + 25x$

44. $y = 4x^3 + 4x^2 - 7x + 2$

45. $y = x^5 - 5x^3 + 4x$

46. $y = \frac{1}{4}x^3(x^2 - 9)$

In Exercises 47–56, find a polynomial function that has the given zeros. (There are many correct answers.)

47. $0, 10$

48. $0, -3$

49. $2, -6$

50. $-4, 5$

51. $0, -2, -3$

52. $0, 2, 5$

53. $4, -3, 3, 0$

54. $-2, -1, 0, 1, 2$

55. $1 + \sqrt{3}, 1 - \sqrt{3}$

56. $2, 4 + \sqrt{5}, 4 - \sqrt{5}$

In Exercises 57–66, find a polynomial of degree n that has the given zeros. (There are many correct answers.)

57. Zero: $x = -2$ Degree: $n = 2$

58. Zeros: $x = -8, -4$ Degree: $n = 2$

59. Zeros: $x = -3, 0, 1$ Degree: $n = 3$

60. Zeros: $x = -2, 4, 7$ Degree: $n = 3$

61. Zeros: $x = 0, \sqrt{3}, -\sqrt{3}$ Degree: $n = 3$

62. Zero: $x = 9$ Degree: $n = 3$

63. Zeros: $x = -5, 1, 2$ Degree: $n = 4$

64. Zeros: $x = -4, -1, 3, 6$ Degree: $n = 4$

65. Zeros: $x = 0, -4$ Degree: $n = 5$

66. Zeros: $x = -3, 1, 5, 6$ Degree: $n = 5$

In Exercises 67–80, sketch the graph of the function by (a) applying the Leading Coefficient Test, (b) finding the zeros of the polynomial, (c) plotting sufficient solution points, and (d) drawing a continuous curve through the points.

67. $f(x) = x^3 - 9x$

68. $g(x) = x^4 - 4x^2$

69. $f(t) = \frac{1}{4}(t^2 - 2t + 15)$

70. $g(x) = -x^2 + 10x - 16$

71. $f(x) = x^3 - 3x^2$

72. $f(x) = 1 - x^3$

73. $f(x) = 3x^3 - 15x^2 + 18x$

74. $f(x) = -4x^3 + 4x^2 + 15x$

75. $f(x) = -5x^2 - x^3$

76. $f(x) = -48x^2 + 3x^4$

77. $f(x) = x^2(x - 4)$

78. $h(x) = \frac{1}{3}x^3(x - 4)^2$

79. $g(t) = -\frac{1}{4}(t - 2)^2(t + 2)^2$

80. $g(x) = \frac{1}{10}(x + 1)^2(x - 3)^3$

In Exercises 81–84, use a graphing utility to graph the function. Use the *zero* or *root* feature to approximate zeros of the function. Determine the multiplicity of each zero.

81. $f(x) = x^3 - 4x$

82. $f(x) = \frac{1}{4}x^4 - 2x^2$

83. $g(x) = \frac{1}{5}(x + 1)^2(x - 3)(2x - 9)$

84. $h(x) = \frac{1}{5}(x + 2)^2(3x - 5)^2$

In Exercises 85–88, use the Intermediate Value Theorem and a graphing utility to find intervals 1 unit in length in which the polynomial function is guaranteed to have a zero. (See Example 7.)

85. $f(x) = x^3 - 3x^2 + 3$

86. $f(x) = 0.11x^3 - 2.07x^2 + 9.81x - 6.88$

87. $g(x) = 3x^4 + 4x^3 - 3$

88. $h(x) = x^4 - 10x^2 + 3$

89. *Numerical and Graphical Analysis* An open box is to be made from a square piece of material, 36 inches on a side, by cutting equal squares of length x from the corners and turning up the sides (see figure).

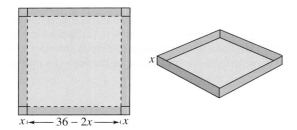

(a) Complete four rows of a table such as the one below.

Box height	Box width	Box volume
1	$36 - 2(1)$	$1[36 - 2(1)]^2 = 1156$
2	$36 - 2(2)$	$2[36 - 2(2)]^2 = 2048$

(b) Use a graphing utility to generate additional rows of the table. Use the table to estimate the dimensions that will produce a maximum volume.

(c) Verify that the volume of the box is given by the function

$$V(x) = x(36 - 2x)^2.$$

Determine the domain of the function.

(d) Use a graphing utility to graph V and use the graph to estimate the value of x for which $V(x)$ is maximum. Compare your result with that of part (b).

90. *Maximum Volume* An open box with locking tabs is to be made from a square piece of material 24 inches on a side. This is to be done by cutting equal squares from the corners and folding along the dashed lines shown in the figure.

(a) Verify that the volume of the box is given by the function

$$V(x) = 8x(6 - x)(12 - x).$$

(b) Determine the domain of the function V.

(c) Sketch the graph of the function and estimate the value of x for which $V(x)$ is maximum.

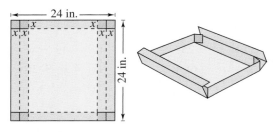

FIGURE FOR **90**

91. *Business* The total revenue R (in millions of dollars) for a company is related to its advertising expense by the function

$$R = \frac{1}{100,000}(-x^3 + 600x^2), \qquad 0 \le x \le 400$$

where x is the amount spent on advertising (in tens of thousands of dollars). Use the graph of this function, shown in the figure, to estimate the point on the graph at which the function is increasing most rapidly. This point is called the *point of diminishing returns* because any expense above this amount will yield less return per dollar invested in advertising.

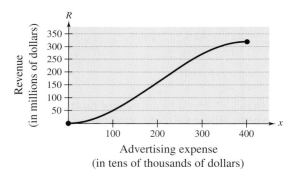

92. *Environment* The growth of a red oak tree is approximated by the function

$$G = -0.003t^3 + 0.137t^2 + 0.458t - 0.839$$

where G is the height of the tree (in feet) and t ($2 \le t \le 34$) is its age (in years). Use a graphing utility to graph the function and estimate the age of the tree when it is growing most rapidly. This point is called the *point of diminishing returns* because the increase in size will be less with each additional year. (*Hint:* Use a viewing window in which $-10 \le x \le 45$ and $-5 \le y \le 60$.)

Synthesis

True or False? **In Exercises 93 and 94, determine whether the statement is true or false. Justify your answer.**

93. A fifth-degree polynomial can have five turning points in its graph.

94. It is possible for a sixth-degree polynomial to have only one solution.

95. *Graphical Analysis* Describe a polynomial function that could represent the graph. (Indicate the degree of the function and the sign of its leading coefficient.)

(a)

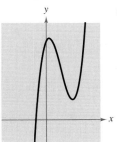

(b)

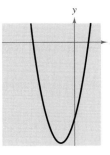

(c)

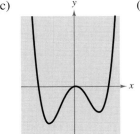

(d)

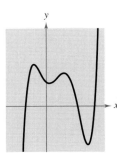

96. *Graphical Reasoning* Sketch a graph of the function $f(x) = x^4$. Explain how the graph of g differs (if it does) from the graph of f. Determine whether g is odd, even, or neither.

(a) $g(x) = f(x) + 2$ (b) $g(x) = f(x + 2)$

(c) $g(x) = f(-x)$ (d) $g(x) = -f(x)$

(e) $g(x) = f\left(\frac{1}{2}x\right)$ (f) $g(x) = \frac{1}{2}f(x)$

(g) $g(x) = f\left(x^{3/4}\right)$ (h) $g(x) = (f \circ f)(x)$

97. *Exploration* Explore the transformations of the form $g(x) = a(x - h)^5 + k$.

(a) Use a graphing utility to graph the functions

$$y_1 = -\frac{1}{3}(x - 2)^5 + 1$$

and

$$y_2 = \frac{3}{5}(x + 2)^5 - 3.$$

Determine whether the graphs are increasing or decreasing. Explain.

(b) Will the graph of g always be increasing or decreasing? If so, is this behavior determined by a, h, or k? Explain.

(c) Use a graphing utility to graph the function

$$H(x) = x^5 - 3x^3 + 2x + 1.$$

Use the graph and the result of part (b) to determine whether H can be written in the form $H(x) = a(x - h)^5 + k$. Explain.

Review

In Exercises 98–101, solve the equation by factoring.

98. $2x^2 - x - 28 = 0$ **99.** $3x^2 - 22x - 16 = 0$

100. $12x^2 + 11x - 5 = 0$ **101.** $x^2 + 24x + 144 = 0$

In Exercises 102–105, solve the equation by completing the square.

102. $x^2 - 2x - 21 = 0$ **103.** $x^2 - 8x + 2 = 0$

104. $2x^2 + 5x - 20 = 0$ **105.** $3x^2 + 4x - 9 = 0$

In Exercises 106–109, factor the expression completely.

106. $5x^2 + 7x - 24$ **107.** $6x^3 - 61x^2 + 10x$

108. $4x^4 - 7x^3 - 15x^2$ **109.** $y^3 + 216$

110. Use the graphs of the functions f and g to answer each question.

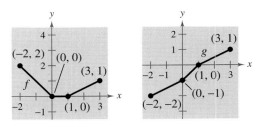

(a) Explain why f does not have an inverse.

(b) Find $g^{-1}(1)$.

3.3 Polynomial and Synthetic Division

▶ **Why you should learn it**

Polynomial division can help you rewrite polynomials that are used to model real-life problems. Rewriting a model can sometimes make the important properties of the model more apparent. For instance, Example 7 on page 284 shows how polynomial division can be used to rewrite the model for a person's take-home pay.

Long Division of Polynomials

In this section, you will study two procedures for *dividing* polynomials. These procedures are especially valuable in factoring and finding the zeros of polynomial functions. To begin, suppose you are given the graph of

$$f(x) = 6x^3 - 19x^2 + 16x - 4.$$

Notice that a zero of f occurs at $x = 2$, as shown in Figure 3.21. Because $x = 2$ is a zero of f, you know that $(x - 2)$ is a factor of $f(x)$. This means that there exists a second-degree polynomial $q(x)$ such that

$$f(x) = (x - 2) \cdot q(x).$$

To find $q(x)$, you can use **long division,** as illustrated in Example 1.

Example 1 ▶ Long Division of Polynomials

Divide $6x^3 - 19x^2 + 16x - 4$ by $x - 2$, and use the result to factor the polynomial completely.

Solution

Think $\dfrac{6x^3}{x} = 6x^2$.

Think $\dfrac{-7x^2}{x} = -7x$.

Think $\dfrac{2x}{x} = 2$.

$$
\begin{array}{r}
6x^2 - 7x + 2 \\
x - 2 \overline{)\, 6x^3 - 19x^2 + 16x - 4} \\
\underline{6x^3 - 12x^2} \\
-7x^2 + 16x \\
\underline{-7x^2 + 14x} \\
2x - 4 \\
\underline{2x - 4} \\
0
\end{array}
$$

Multiply: $6x^2(x - 2)$.

Subtract.

Multiply: $-7x(x - 2)$.

Subtract.

Multiply: $2(x - 2)$.

Subtract.

From this division, you can conclude that

$$6x^3 - 19x^2 + 16x - 4 = (x - 2)(6x^2 - 7x + 2)$$

and by factoring the quadratic $6x^2 - 7x + 2$, you have

$$6x^3 - 19x^2 + 16x - 4 = (x - 2)(2x - 1)(3x - 2).$$

Note that this factorization agrees with the graph shown in Figure 3.21 in that the three x-intercepts occur at $x = 2$, $x = \frac{1}{2}$, and $x = \frac{2}{3}$.

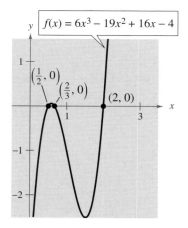

FIGURE **3.21**

In Example 1, $x - 2$ is a factor of the polynomial $6x^3 - 19x^2 + 16x - 4$, and the long division process produces a remainder of zero. Often, long division will produce a nonzero remainder. For instance, if you divide $x^2 + 3x + 5$ by $x + 1$, you obtain the following.

$$
\begin{array}{r}
x + 2 \quad \longleftarrow \text{ Quotient} \\
\text{Divisor} \longrightarrow x + 1 \overline{) x^2 + 3x + 5} \quad \longleftarrow \text{ Dividend} \\
\underline{x^2 + x} \\
2x + 5 \\
\underline{2x + 2} \\
3 \quad \longleftarrow \text{ Remainder}
\end{array}
$$

In fractional form, you can write this result as follows.

$$
\underbrace{\frac{x^2 + 3x + 5}{x + 1}}_{\text{Divisor}} = \underbrace{x + 2}_{\text{Quotient}} + \underbrace{\frac{3}{x + 1}}_{\text{Divisor}}
$$

Dividend, Quotient, Remainder

This implies that

$$
x^2 + 3x + 5 = (x + 1)(x + 2) + 3 \qquad \text{Multiply each side by } (x + 1).
$$

which illustrates the following theorem, called the **Division Algorithm.**

The Division Algorithm

If $f(x)$ and $d(x)$ are polynomials such that $d(x) \neq 0$, and the degree of $d(x)$ is less than or equal to the degree of $f(x)$, there exist unique polynomials $q(x)$ and $r(x)$ such that

$$
f(x) = d(x)q(x) + r(x)
$$

Dividend   Quotient

Divisor   Remainder

where $r(x) = 0$ *or* the degree of $r(x)$ is less than the degree of $d(x)$. If the remainder $r(x)$ is zero, $d(x)$ *divides evenly* into $f(x)$.

The Division Algorithm can also be written as

$$
\frac{f(x)}{d(x)} = q(x) + \frac{r(x)}{d(x)}.
$$

In the Division Algorithm, the rational expression $f(x)/d(x)$ is **improper** because the degree of $f(x)$ is greater than or equal to the degree of $d(x)$. On the other hand, the rational expression $r(x)/d(x)$ is **proper** because the degree of $r(x)$ is less than the degree of $d(x)$.

Example 2 ▶   **Long Division of Polynomials**

Divide $x^3 - 1$ by $x - 1$.

Solution

Because there is no x^2-term or x-term in the dividend, you need to line up the subtraction by using zero coefficients (or leaving spaces) for the missing terms.

$$
\begin{array}{r}
x^2 + x + 1 \\
x - 1 \overline{)\, x^3 + 0x^2 + 0x - 1} \\
\underline{x^3 - x^2} \\
x^2 + 0x \\
\underline{x^2 - x} \\
x - 1 \\
\underline{x - 1} \\
0
\end{array}
$$

So, $x - 1$ divides evenly into $x^3 - 1$ and you can write

$$\frac{x^3 - 1}{x - 1} = x^2 + x + 1, \quad x \neq 1.$$

You can check the result of a division problem by multiplying. For instance, in Example 2, try checking that

$$(x - 1)(x^2 + x + 1) = x^3 - 1.$$

A computer animation of this example appears in the *Interactive* CD-ROM and *Internet* versions of this text.

Example 3 ▶   **Long Division of Polynomials**

Divide $2x^4 + 4x^3 - 5x^2 + 3x - 2$ by $x^2 + 2x - 3$.

Solution

$$
\begin{array}{r}
2x^2 + 1 \\
x^2 + 2x - 3 \overline{)\, 2x^4 + 4x^3 - 5x^2 + 3x - 2} \\
\underline{2x^4 + 4x^3 - 6x^2} \\
x^2 + 3x - 2 \\
\underline{x^2 + 2x - 3} \\
x + 1
\end{array}
$$

Note that the first subtraction eliminated two terms from the dividend. When this happens, the quotient skips a term. You can write the result as

$$\frac{2x^4 + 4x^3 - 5x^2 + 3x - 2}{x^2 + 2x - 3} = 2x^2 + 1 + \frac{x + 1}{x^2 + 2x - 3}.$$

Synthetic Division

There is a nice shortcut for long division of polynomials when dividing by divisors of the form $x - k$. This shortcut is called **synthetic division.** The pattern for synthetic division of a cubic polynomial is summarized as follows. (The pattern for higher-degree polynomials is similar.)

Synthetic Division (for a Cubic Polynomial)

To divide $ax^3 + bx^2 + cx + d$ by $x - k$, use the following pattern.

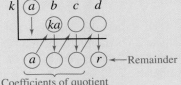

Vertical pattern: Add terms.
Diagonal pattern: Multiply by k.

Synthetic division works *only* for divisors of the form $x - k$. [Remember that $x + k = x - (-k)$.] You cannot use synthetic division to divide a polynomial by a quadratic such as $x^2 - 3$.

A computer animation of this example appears in the *Interactive* CD-ROM and *Internet* versions of this text.

Example 4 ▶ Using Synthetic Division

Use synthetic division to divide $x^4 - 10x^2 - 2x + 4$ by $x + 3$.

Solution

You should set up the array as follows. Note that a zero is included for each missing term in the dividend.

$$-3 \mid \quad 1 \quad 0 \quad -10 \quad -2 \quad 4$$

Then, use the synthetic division pattern by adding terms in columns and multiplying the results by -3.

Divisor: $x + 3$ Dividend: $x^4 - 10x^2 - 2x + 4$

$$
\begin{array}{r|rrrrr}
-3 & 1 & 0 & -10 & -2 & 4 \\
   &   & -3 & 9 & 3 & -3 \\
\hline
   & 1 & -3 & -1 & 1 & \boxed{1}
\end{array}
$$

← Remainder: 1

Quotient: $x^3 - 3x^2 - x + 1$

So, you have

$$\frac{x^4 - 10x^2 - 2x + 4}{x + 3} = x^3 - 3x^2 - x + 1 + \frac{1}{x + 3}.$$

The Remainder and Factor Theorems

The remainder obtained in the synthetic division process has an important interpretation, as described in the **Remainder Theorem.** (A proof is given in Appendix A.)

The Remainder Theorem

If a polynomial $f(x)$ is divided by $x - k$, then the remainder is

$$r = f(k).$$

The Remainder Theorem tells you that synthetic division can be used to evaluate a polynomial function. That is, to evaluate a polynomial function $f(x)$ when $x = k$, divide $f(x)$ by $x - k$. The remainder will be $f(k)$, as illustrated in Example 5.

Example 5 ▶ Using the Remainder Theorem

Use the Remainder Theorem to evaluate the following function at $x = -2$.

$$f(x) = 3x^3 + 8x^2 + 5x - 7$$

Solution

Using synthetic division, you obtain the following.

$$
\begin{array}{r|rrrr}
-2 & 3 & 8 & 5 & -7 \\
   &   & -6 & -4 & -2 \\
\hline
   & 3 & 2 & 1 & -9
\end{array}
$$

Because the remainder is $r = -9$, you can conclude that

$$f(-2) = -9.$$

This means that $(-2, -9)$ is a point on the graph of f. You can check this by substituting $x = -2$ in the original function.

Check

$$
\begin{aligned}
f(-2) &= 3(-2)^3 + 8(-2)^2 + 5(-2) - 7 \\
&= 3(-8) + 8(4) - 10 - 7 \\
&= -9
\end{aligned}
$$

Another important theorem is the **Factor Theorem,** which is stated below. This theorem states that you can test to see whether a polynomial has $(x - k)$ as a factor by evaluating the polynomial at $x = k$. If the result is 0, $(x - k)$ is a factor. (A proof is given in Appendix A.)

The Factor Theorem

A polynomial $f(x)$ has a factor $(x - k)$ if and only if $f(k) = 0$.

Example 6 ▶   **Factoring a Polynomial: Repeated Division**

Show that $(x - 2)$ and $(x + 3)$ are factors of

$$f(x) = 2x^4 + 7x^3 - 4x^2 - 27x - 18.$$

Then find the remaining factors of $f(x)$.

Solution

Using synthetic division with the factor $(x - 2)$, you obtain the following.

$$
\begin{array}{r|rrrrr}
2 & 2 & 7 & -4 & -27 & -18 \\
  &   & 4 & 22 & 36 & 18 \\
\hline
  & 2 & 11 & 18 & 9 & 0
\end{array}
$$
⟶ 0 remainder, so $f(2) = 0$ and $(x - 2)$ is a factor.

Use the result of this division to perform synthetic division again with the factor $(x + 3)$.

$$
\begin{array}{r|rrrr}
-3 & 2 & 11 & 18 & 9 \\
   &   & -6 & -15 & -9 \\
\hline
   & 2 & 5 & 3 & 0
\end{array}
$$
⟶ 0 remainder, so $f(-3) = 0$ and $(x + 3)$ is a factor.

Because the resulting quadratic expression factors as

$$2x^2 + 5x + 3 = (2x + 3)(x + 1)$$

the complete factorization of $f(x)$ is

$$f(x) = (x - 2)(x + 3)(2x + 3)(x + 1).$$

Note that this factorization implies that f has four real zeros:

$$2, -3, -\tfrac{3}{2}, \text{ and } -1.$$

This is confirmed by the graph of f, which is shown in Figure 3.22.

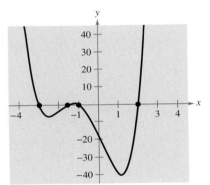

FIGURE **3.22**

Using the Remainder in Synthetic Division

The remainder r, obtained in the synthetic division of $f(x)$ by $x - k$, provides the following information.

1. The remainder r gives the value of f at $x = k$. That is, $r = f(k)$.

2. If $r = 0$, $(x - k)$ is a factor of $f(x)$.

3. If $r = 0$, $(k, 0)$ is an x-intercept of the graph of f.

Throughout this text, we have emphasized the importance of developing several problem-solving strategies. In the exercises for this section, try using more than one strategy to solve several of the exercises. For instance, if you find that $x - k$ divides evenly into $f(x)$ (with no remainder), try sketching the graph of f. You should find that $(k, 0)$ is an x-intercept of the graph.

Application

Example 7 ▶ Take-Home Pay

The 1999 monthly take-home pay for an employee who is single and claimed one deduction is given by the function

$$y = -0.00002436x^2 + 0.79337x + 42.096, \qquad 500 \le x \le 5000$$

where y represents the take-home pay and x represents the gross monthly salary. Find a function that gives the take-home pay as a *percent* of the gross monthly salary.   (Source: UA Corporate Accounting Software, based on a state and local income tax rate of 3.8%)

Solution

Because the gross monthly salary is given by x and the take-home pay is given by y, the percent of gross monthly salary that the person takes home is

$$P = \frac{y}{x}$$

$$= \frac{-0.00002436x^2 + 0.79337x + 42.096}{x}$$

$$= -0.00002436x + 0.79337 + \frac{42.096}{x}.$$

The graphs of these functions are shown in Figure 3.23(a) and (b). Note in Figure 3.23(b) that as a person's gross monthly salary increases, the *percent* that he or she takes home decreases.

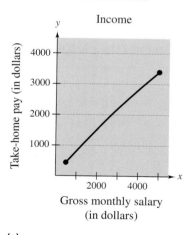

(a)

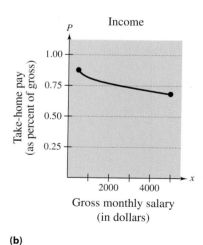

(b)

FIGURE 3.23

In Example 7, the difference between gross pay and take-home pay results primarily from federal deductions for income tax, Social Security tax, and Medicare tax. Because the federal income tax is graduated, the percent that is deducted depends on a person's income. People who earn more do not simply pay more tax—they pay *much* more tax because the tax rate increases.

3.3 Exercises

Analytical Analysis **In Exercises 1–4, use long division to verify that $y_1 = y_2$.**

1. $y_1 = \dfrac{4x}{x - 1}$, $y_2 = 4 + \dfrac{4}{x - 1}$

2. $y_1 = \dfrac{3x - 5}{x - 3}$, $y_2 = 3 + \dfrac{4}{x - 3}$

3. $y_1 = \dfrac{x^2}{x + 2}$, $y_2 = x - 2 + \dfrac{4}{x + 2}$

4. $y_1 = \dfrac{x^4 - 3x^2 - 1}{x^2 + 5}$, $y_2 = x^2 - 8 + \dfrac{39}{x^2 + 5}$

⊞ *Graphical Analysis* **In Exercises 5 and 6, use a graphing utility to graph the two equations in the same viewing window. Use the graphs to verify that the expressions are equivalent. Use long division to verify the results algebraically.**

5. $y_1 = \dfrac{x^5 - 3x^3}{x^2 + 1}$, $y_2 = x^3 - 4x + \dfrac{4x}{x^2 + 1}$

6. $y_1 = \dfrac{x^3 - 2x^2 + 5}{x^2 + x + 1}$, $y_2 = x - 3 + \dfrac{2(x + 4)}{x^2 + x + 1}$

In Exercises 7–20, use long division to divide.

7. $(2x^2 + 10x + 12) \div (x + 3)$

8. $(5x^2 - 17x - 12) \div (x - 4)$

9. $(4x^3 - 7x^2 - 11x + 5) \div (4x + 5)$

10. $(6x^3 - 16x^2 + 17x - 6) \div (3x - 2)$

11. $(x^4 + 5x^3 + 6x^2 - x - 2) \div (x + 2)$

12. $(x^3 + 4x^2 - 3x - 12) \div (x - 3)$

13. $(7x + 3) \div (x + 2)$ **14.** $(8x - 5) \div (2x + 1)$

15. $(6x^3 + 10x^2 + x + 8) \div (2x^2 + 1)$

16. $(x^3 - 9) \div (x^2 + 1)$

17. $\dfrac{x^4 + 3x^2 + 1}{x^2 - 2x + 3}$ **18.** $\dfrac{x^5 + 7}{x^3 - 1}$

19. $\dfrac{x^4}{(x - 1)^3}$ **20.** $\dfrac{2x^3 - 4x^2 - 15x + 5}{(x - 1)^2}$

In Exercises 21–38, use synthetic division to divide.

21. $(3x^3 - 17x^2 + 15x - 25) \div (x - 5)$

22. $(5x^3 + 18x^2 + 7x - 6) \div (x + 3)$

23. $(4x^3 - 9x + 8x^2 - 18) \div (x + 2)$

24. $(9x^3 - 16x - 18x^2 + 32) \div (x - 2)$

25. $(-x^3 + 75x - 250) \div (x + 10)$

26. $(3x^3 - 16x^2 - 72) \div (x - 6)$

27. $(5x^3 - 6x^2 + 8) \div (x - 4)$

28. $(5x^3 + 6x + 8) \div (x + 2)$

29. $\dfrac{10x^4 - 50x^3 - 800}{x - 6}$

30. $\dfrac{x^5 - 13x^4 - 120x + 80}{x + 3}$

31. $\dfrac{x^3 + 512}{x + 8}$ **32.** $\dfrac{x^3 - 729}{x - 9}$

33. $\dfrac{-3x^4}{x - 2}$ **34.** $\dfrac{-3x^4}{x + 2}$

35. $\dfrac{180x - x^4}{x - 6}$ **36.** $\dfrac{5 - 3x + 2x^2 - x^3}{x + 1}$

37. $\dfrac{4x^3 + 16x^2 - 23x - 15}{x + \frac{1}{2}}$ **38.** $\dfrac{3x^3 - 4x^2 + 5}{x - \frac{3}{2}}$

In Exercises 39–46, express the function in the form $f(x) = (x - k)q(x) + r$ for the given value of k, and demonstrate that $f(k) = r$.

Function	Value of k
39. $f(x) = x^3 - x^2 - 14x + 11x$	$k = 4$
40. $f(x) = x^3 - 5x^2 - 11x + 8$	$k = -2$
41. $f(x) = 15x^4 + 10x^3 - 6x^2 + 14$	$k = -\frac{2}{3}$
42. $f(x) = 10x^3 - 22x^2 - 3x + 4$	$k = \frac{1}{5}$
43. $f(x) = x^3 + 3x^2 - 2x - 14$	$k = \sqrt{2}$
44. $f(x) = x^3 + 2x^2 - 5x - 4$	$k = -\sqrt{5}$
45. $f(x) = -4x^3 + 6x^2 + 12x + 4$	$k = 1 - \sqrt{3}$
46. $f(x) = -3x^3 + 8x^2 + 10x - 8$	$k = 2 + \sqrt{2}$

In Exercises 47–50, use synthetic division to find each function value. Verify using another method.

47. $f(x) = 4x^3 - 13x + 10$

(a) $f(1)$ (b) $f(-2)$

(c) $f\left(\frac{1}{2}\right)$ (d) $f(8)$

48. $g(x) = x^6 - 4x^4 + 3x^2 + 2$

(a) $g(2)$ (b) $g(-4)$

(c) $g(3)$ (d) $g(-1)$

49. $h(x) = 3x^3 + 5x^2 - 10x + 1$

    (a) $h(3)$ (b) $h\left(\frac{1}{3}\right)$

    (c) $h(-2)$ (d) $h(-5)$

50. $f(x) = 0.4x^4 - 1.6x^3 + 0.7x^2 - 2$

    (a) $f(1)$ (b) $f(-2)$

    (c) $f(5)$ (d) $f(-10)$

In Exercises 51–58, use synthetic division to show that x is a solution of the third-degree polynomial equation, and use the result to factor the polynomial completely. List all real zeros of the function.

Polynomial Equation	*Value of x*
51. $x^3 - 7x + 6 = 0$	$x = 2$
52. $x^3 - 28x - 48 = 0$	$x = -4$
53. $2x^3 - 15x^2 + 27x - 10 = 0$	$x = \frac{1}{2}$
54. $48x^3 - 80x^2 + 41x - 6 = 0$	$x = \frac{2}{3}$
55. $x^3 + 2x^2 - 3x - 6 = 0$	$x = \sqrt{3}$
56. $x^3 + 2x^2 - 2x - 4 = 0$	$x = \sqrt{2}$
57. $x^3 - 3x^2 + 2 = 0$	$x = 1 + \sqrt{3}$
58. $x^3 - x^2 - 13x - 3 = 0$	$x = 2 - \sqrt{5}$

In Exercises 59–66, (a) verify the given factors of the function f, (b) find the remaining factors of f, (c) use your results to write the complete factorization of f, (d) list all real zeros of f, and ▦ (e) confirm your results by using a graphing utility to graph the function.

Function	*Factors*
59. $f(x) = 2x^3 + x^2 - 5x + 2$	$(x + 2), (x - 1)$
60. $f(x) = 3x^3 + 2x^2 - 19x + 6$	$(x + 3), (x - 2)$
61. $f(x) = x^4 - 4x^3 - 15x^2$ $+ 58x - 40$	$(x - 5), (x + 4)$
62. $f(x) = 8x^4 - 14x^3 - 71x^2$ $- 10x + 24$	$(x + 2), (x - 4)$
63. $f(x) = 6x^3 + 41x^2 - 9x - 14$	$(2x + 1), (3x - 2)$
64. $f(x) = 10x^3 - 11x^2 - 72x + 45$	$(2x + 5), (5x - 3)$
65. $f(x) = 2x^3 - x^2 - 10x + 5$	$(2x - 1), \left(x + \sqrt{5}\right)$
66. $f(x) = x^3 + 3x^2 - 48x - 144$	$\left(x + 4\sqrt{3}\right), (x + 3)$

▦ *Graphical Analysis* **In Exercises 67–70, (a) use the root-finding capabilities of a graphing utility to approximate the zeros of the function accurate to three decimal places. (b) Determine one of the exact zeros and use synthetic division to verify your result. Then factor the polynomial completely.**

67. $f(x) = x^3 - 2x^2 - 5x + 10$

68. $g(x) = x^3 - 4x^2 - 2x + 8$

69. $h(t) = t^3 - 2t^2 - 7t + 2$

70. $f(s) = s^3 - 12s^2 + 40s - 24$

In Exercises 71–76, simplify the rational expression.

71. $\dfrac{4x^3 - 8x^2 + x + 3}{2x - 3}$

72. $\dfrac{x^3 + x^2 - 64x - 64}{x + 8}$

73. $\dfrac{x^3 + 3x^2 - x - 3}{x + 1}$

74. $\dfrac{2x^3 + 3x^2 - 3x - 2}{x - 1}$

75. $\dfrac{x^4 + 6x^3 + 11x^2 + 6x}{x^2 + 3x + 2}$

76. $\dfrac{x^4 + 9x^3 - 5x^2 - 36x + 4}{x^2 - 4}$

77. *Data Analysis* The average monthly basic rates R for cable television in the United States (in dollars) for the years 1988 through 1997 are shown in the table, where t represents the time (in years), with $t = 0$ corresponding to 1990. (Source: Paul Kagan Associates, Inc.)

t	-2	-1	0	1	2
R	13.86	15.21	16.78	18.10	19.08

t	3	4	5	6	7
R	19.39	21.62	23.07	24.41	26.48

▦ (a) Use a graphing utility to sketch a scatter plot of the data.

▦ (b) Use the regression feature of a graphing utility to find a cubic model for the data. Then graph the model in the same viewing window as the scatter plot. Compare the model with the data.

(c) Use the model to create a table of estimated values of R. Compare the estimated values with the actual data.

(d) Use synthetic division to evaluate the model for the year 2002. Even though the model is relatively accurate for estimating the given data, do you think it is accurate for predicting future cable rates? Explain.

78. *Data Analysis* The numbers of United States military personnel M (in thousands) on active duty for the years 1989 through 1996 are shown in the table, where t represents the time (in years), with $t = 0$ corresponding to 1990. (Source: U.S. Department of Defense)

t	-1	0	1	2
M	2130	2044	1986	1807

t	3	4	5	6
M	1705	1611	1518	1472

(a) Use a graphing utility to sketch a scatter plot of the data.

(b) Use the regression feature of a graphing utility to find a cubic model for the data. Then graph the model in the same viewing window as the scatter plot. Compare the model with the data.

(c) Use the model to create a table of estimated values of M. Compare the estimated values with the actual data.

(d) Use synthetic division to evaluate the model for the year 2001. Even though the model is relatively accurate for estimating the given data, would you use this model to predict the number of military personnel in the future? Explain.

Synthesis

True or False? **In Exercises 79 and 80, determine whether the statement is true or false. Justify your answer.**

79. If $(7x + 4)$ is a factor of some polynomial function f, then $\frac{4}{7}$ is a zero of f.

80. $(2x - 1)$ is a factor of the polynomial
$6x^6 + x^5 - 92x^4 + 45x^3 + 184x^2 + 4x - 48$.

Think About It **In Exercises 81 and 82, perform the division by assuming that n is a positive integer.**

81. $\dfrac{x^{3n} + 9x^{2n} + 27x^n + 27}{x^n + 3}$

82. $\dfrac{x^{3n} - 3x^{2n} + 5x^n - 6}{x^n - 2}$

83. *Writing* Briefly explain what it means for a divisor to divide evenly into a dividend.

84. *Writing* Briefly explain how to check polynomial division, and justify your reasoning. Give an example.

Exploration **In Exercises 85 and 86, find the constant c such that the denominator will divide evenly into the numerator.**

85. $\dfrac{x^3 + 4x^2 - 3x + c}{x - 5}$

86. $\dfrac{x^5 - 2x^2 + x + c}{x + 2}$

Think About It **In Exercises 87 and 88, answer the questions about the division**

$$\frac{f(x)}{x - k}$$

where $f(x) = (x + 3)^2(x - 3)(x + 1)^3$.

87. What is the remainder when $k = -3$? Explain.

88. If it is necessary to find $f(2)$, is it easier to evaluate the function directly or to use synthetic division? Explain.

89. *Exploration* Use the form

$$f(x) = (x - k)q(x) + r$$

to create a cubic function that (a) passes through the point $(2, 5)$ and rises to the right, and (b) passes through the point $(-3, 1)$ and falls to the right. (There are many correct answers.)

Review

In Exercises 90–95, use any method to solve the quadratic equation.

90. $9x^2 - 25 = 0$

91. $16x^2 - 21 = 0$

92. $5x^2 - 3x - 14 = 0$

93. $8x^2 - 22x + 15 = 0$

94. $2x^2 + 6x + 3 = 0$

95. $x^2 + 3x - 3 = 0$

In Exercises 96–99, find a polynomial function that has the given zeros.

96. $0, 3, 4$

97. $-6, 1$

98. $-3, 1 + \sqrt{2}, 1 - \sqrt{2}$

99. $1, -2, 2 + \sqrt{3}, 2 - \sqrt{3}$

3.4 Zeros of Polynomial Functions

▶ **Why you should learn it**

Finding zeros of polynomial functions is an important part of solving real-life problems. For instance, Exercise 109 on page 301 shows how the zeros of a polynomial function can help you analyze the profit function for a business.

Don Smetzer/Tony Stone Images

The Fundamental Theorem of Algebra

You know that an nth-degree polynomial can have at most n real zeros. In the complex number system, this statement can be improved. That is, in the complex number system, every nth-degree polynomial function has *precisely* n zeros. This important result is derived from the **Fundamental Theorem of Algebra,** first proved by the German mathematician Carl Friedrich Gauss (1777–1855).

The Fundamental Theorem of Algebra

If $f(x)$ is a polynomial of degree n, where $n > 0$, then f has at least one zero in the complex number system.

Using the Fundamental Theorem of Algebra and the equivalence of zeros and factors, you obtain the **Linear Factorization Theorem.** (A proof is given in Appendix A.)

Linear Factorization Theorem

If $f(x)$ is a polynomial of degree n, where $n > 0$, then f has precisely n linear factors

$$f(x) = a_n(x - c_1)(x - c_2) \cdots (x - c_n)$$

where $c_1, c_2, \ldots, c_n$ are complex numbers.

Note that the Fundamental Theorem of Algebra and the Linear Factorization Theorem tell you only that the zeros or factors of a polynomial exist, not how to find them. Such theorems are called *existence theorems*. To find the zeros of a polynomial function, you still must rely on other techniques.

Example 1 ▶ Zeros of Polynomial Functions

a. The first-degree polynomial $f(x) = x - 2$ has exactly *one* zero: $x = 2$.

b. Counting multiplicity, the second-degree polynomial function

$$f(x) = x^2 - 6x + 9 = (x - 3)(x - 3)$$

has exactly *two* zeros: $x = 3$ and $x = 3$. (This is called a *repeated zero*.)

c. The third-degree polynomial function

$$f(x) = x^3 + 4x = x(x^2 + 4) = x(x - 2i)(x + 2i)$$

has exactly *three* zeros: $x = 0$, $x = 2i$, and $x = -2i$.

d. The fourth-degree polynomial function

$$f(x) = x^4 - 1 = (x - 1)(x + 1)(x - i)(x + i)$$

has exactly *four* zeros: $x = 1$, $x = -1$, $x = i$, and $x = -i$.

The Rational Zero Test

The **Rational Zero Test** relates the possible rational zeros of a polynomial (having integer coefficients) to the leading coefficient and to the constant term of the polynomial.

The Rational Zero Test

If the polynomial $f(x) = a_nx^n + a_{n-1}x^{n-1} + \cdots + a_2x^2 + a_1x + a_0$ has *integer* coefficients, every rational zero of f has the form

$$\text{Rational zero} = \frac{p}{q}$$

where p and q have no common factors other than 1, and

p = a factor of the constant term a_0

q = a factor of the leading coefficient a_n.

To use the Rational Zero Test, you should first list all rational numbers whose numerators are factors of the constant term and whose denominators are factors of the leading coefficient.

$$\text{Possible rational zeros} = \frac{\text{factors of constant term}}{\text{factors of leading coefficient}}$$

Having formed this list of *possible rational zeros*, use a trial-and-error method to determine which, if any, are actual zeros of the polynomial. Note that when the leading coefficient is 1, the possible rational zeros are simply the factors of the constant term.

Example 2 ▶ Rational Zero Test with Leading Coefficient of 1

Find the rational zeros of

$$f(x) = x^3 + x + 1.$$

Solution

Because the leading coefficient is 1, the possible rational zeros are ± 1, the factors of the constant term. By testing these possible zeros, you can see that neither works.

$$f(1) = (1)^3 + 1 + 1 = 3$$

$$f(-1) = (-1)^3 + (-1) + 1 = -1$$

So, you can conclude that the given polynomial has *no* rational zeros. Note from the graph of f in Figure 3.24 that f does have one real zero between -1 and 0. However, by the Rational Zero Test, you know that this real zero is *not* a rational number.

Historical Note

Jean Le Rond d'Alembert (1717–1783) worked independently of Carl Gauss in trying to prove the Fundamental Theorem of Algebra. His efforts were such that, in France, the Fundamental Theorem of Algebra is frequently known as the Theorem of d'Alembert.

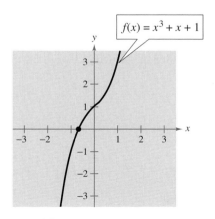

$f(x) = x^3 + x + 1$

FIGURE 3.24

Example 3 ▶   Rational Zero Test with Leading Coefficient of 1

Find the rational zeros of $f(x) = x^4 - x^3 + x^2 - 3x - 6$.

Solution

Because the leading coefficient is 1, the possible rational zeros are the factors of the constant term.

Possible rational zeros: $\pm 1, \pm 2, \pm 3, \pm 6$

A test of these possible zeros shows that $x = -1$ and $x = 2$ are the only two that work. Check the others to be sure.

If the leading coefficient of a polynomial is not 1, the list of possible rational zeros can increase dramatically. In such cases the search can be shortened in several ways: (1) a programmable calculator can be used to speed up the calculations; (2) a graph, drawn either by hand or with a graphing utility, can give a good estimate of the locations of the zeros; and (3) synthetic division can be used to test the possible rational zeros.

To see how to use synthetic division to test the possible rational zeros, take another look at the function $f(x) = x^4 - x^3 + x^2 - 3x - 6$ given in Example 3. To test that $x = -1$ and $x = 2$ are zeros of f, you can apply synthetic division successively, as follows.

$$
\begin{array}{r|rrrrr}
-1 & 1 & -1 & 1 & -3 & -6 \\
   &   & -1 & 2 & -3 & 6 \\
\hline
   & 1 & -2 & 3 & -6 & 0 \\
\end{array}
$$

$$
\begin{array}{r|rrrr}
2 & 1 & -2 & 3 & -6 \\
  &   & 2 & 0 & 6 \\
\hline
  & 1 & 0 & 3 & 0 \\
\end{array}
$$

So, you have

$$f(x) = (x + 1)(x - 2)(x^2 + 3).$$

Because the factor $(x^2 + 3)$ produces no real zeros, you can conclude that $x = -1$ and $x = 2$ are the only *real* zeros of f, which is verified in Figure 3.25.

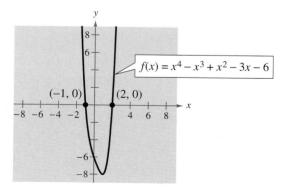

FIGURE 3.25

Finding the first zero is often the hardest part. After that, the search is simplified by using the lower-degree polynomial obtained in synthetic division.

Example 4 ▶ Using the Rational Zero Test

Find the rational zeros of $f(x) = 2x^3 + 3x^2 - 8x + 3$.

Solution

The leading coefficient is 2 and the constant term is 3.

$$\textit{Possible rational zeros:} \frac{\text{Factors of } 3}{\text{Factors of } 2} = \frac{\pm 1, \pm 3}{\pm 1, \pm 2} = \pm 1, \pm 3, \pm \frac{1}{2}, \pm \frac{3}{2}$$

By synthetic division, you can determine that $x = 1$ is a zero.

$$
\begin{array}{r|rrrr}
1 & 2 & 3 & -8 & 3 \\
  &   & 2 & 5 & -3 \\
\hline
  & 2 & 5 & -3 & 0 \\
\end{array}
$$

So, $f(x)$ factors as

$$
\begin{aligned}
f(x) &= (x - 1)(2x^2 + 5x - 3) \\
     &= (x - 1)(2x - 1)(x + 3)
\end{aligned}
$$

and you can conclude that the rational zeros of f are $x = 1$, $x = \frac{1}{2}$, and $x = -3$.

Example 5 ▶ Using the Rational Zero Test

Find all the real zeros of $f(x) = -10x^3 + 15x^2 + 16x - 12$.

Solution

The leading coefficient is -10 and the constant term is -12.

$$\textit{Possible rational zeros:} \frac{\text{Factors of } -12}{\text{Factors of } -10} = \frac{\pm 1, \pm 2, \pm 3, \pm 4, \pm 6, \pm 12}{\pm 1, \pm 2, \pm 5, \pm 10}$$

With so many possibilities (32, in fact), it is worth your time to stop and sketch a graph. From Figure 3.26, it looks like three reasonable choices would be $x = -\frac{6}{5}$, $x = \frac{1}{2}$, and $x = 2$. Testing these by synthetic division shows that only $x = 2$ works. So, you have

$$f(x) = (x - 2)(-10x^2 - 5x + 6).$$

Using the Quadratic Formula, you find that the two additional zeros are irrational numbers.

$$x = \frac{-(-5) + \sqrt{265}}{-20} \approx -1.0639$$

and

$$x = \frac{-(-5) - \sqrt{265}}{-20} \approx 0.5639$$

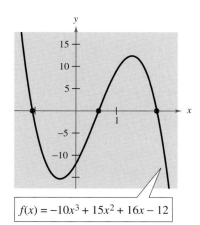

$f(x) = -10x^3 + 15x^2 + 16x - 12$

FIGURE **3.26**

Conjugate Pairs

In Example 1(c) and (d), note that the pairs of complex zeros are **conjugates.** That is, they are of the form $a + bi$ and $a - bi$.

Complex Zeros Occur in Conjugate Pairs

Let $f(x)$ be a polynomial function that has *real coefficients*. If $a + bi$, where $b \neq 0$, is a zero of the function, the conjugate $a - bi$ is also a zero of the function.

Be sure you see that this result is true only if the polynomial function has *real coefficients*. For instance, the result applies to the function $f(x) = x^2 + 1$ but not to the function $g(x) = x - i$.

Example 6 ▶ Finding a Polynomial with Given Zeros

Find a fourth-degree polynomial function with real coefficients that has -1, -1, and $3i$ as zeros.

Solution

Because $3i$ is a zero *and* the polynomial is stated to have real coefficients, you know that the conjugate $-3i$ must also be a zero. So, from the Linear Factorization Theorem, $f(x)$ can be written as

$$f(x) = a(x + 1)(x + 1)(x - 3i)(x + 3i).$$

For simplicity, let $a = 1$ to obtain

$$f(x) = (x^2 + 2x + 1)(x^2 + 9)$$

$$= x^4 + 2x^3 + 10x^2 + 18x + 9.$$

Factoring a Polynomial

The Linear Factorization Theorem shows that you can write any nth-degree polynomial as the product of n linear factors.

$$f(x) = a_n(x - c_1)(x - c_2)(x - c_3) \cdots (x - c_n)$$

However, this result includes the possibility that some of the values of c_i are complex. The following theorem says that even if you do not want to get involved with "complex factors," you can still write $f(x)$ as the product of linear and/or quadratic factors. A proof of this theorem is given in Appendix A.

Factors of a Polynomial

Every polynomial of degree $n > 0$ with real coefficients can be written as the product of linear and quadratic factors with real coefficients, where the quadratic factors have no real zeros.

A quadratic factor with no real zeros is said to be *prime* or **irreducible over the reals.** Be sure you see that this is not the same as being *irreducible over the rationals.* For example, the quadratic

$$x^2 + 1 = (x - i)(x + i)$$

is irreducible over the reals (and therefore over the rationals). On the other hand, the quadratic

$$x^2 - 2 = \left(x - \sqrt{2}\right)\left(x + \sqrt{2}\right)$$

is irreducible over the rationals but *reducible* over the reals.

Example 7 ▶ Finding the Zeros of a Polynomial Function

Find all the zeros of

$$f(x) = x^4 - 3x^3 + 6x^2 + 2x - 60$$

given that $1 + 3i$ is a zero of f.

Solution

Because complex zeros occur in conjugate pairs, you know that $1 - 3i$ is also a zero of f. This means that both

$$[x - (1 + 3i)] \quad \text{and} \quad [x - (1 - 3i)]$$

are factors of f. Multiplying these two factors produces

$$[x - (1 + 3i)][x - (1 - 3i)] = [(x - 1) - 3i][(x - 1) + 3i]$$
$$= (x - 1)^2 - 9i^2$$
$$= x^2 - 2x + 1 - 9(-1)$$
$$= x^2 - 2x + 10.$$

Using long division, you can divide $x^2 - 2x + 10$ into f to obtain the following.

$$
\require{enclose}
\begin{array}{r}
x^2 - x - 6 \\[-3pt]
x^2 - 2x + 10 \enclose{longdiv}{x^4 - 3x^3 + 6x^2 + 2x - 60} \\[-3pt]
\underline{x^4 - 2x^3 + 10x^2} \\[-3pt]
-x^3 - 4x^2 + 2x \\[-3pt]
\underline{-x^3 + 2x^2 - 10x} \\[-3pt]
-6x^2 + 12x - 60 \\[-3pt]
\underline{-6x^2 + 12x - 60} \\[-3pt]
0
\end{array}
$$

Therefore, you have

$$f(x) = (x^2 - 2x + 10)(x^2 - x - 6)$$
$$= (x^2 - 2x + 10)(x - 3)(x + 2)$$

and you can conclude that the zeros of f are $1 + 3i$, $1 - 3i$, 3, and -2.

STUDY T!P

In Example 7, if you had not been told that $1 + 3i$ is a zero of f, you could still find all zeros of the function by using synthetic division to find the real zeros -2 and 3. Then you could factor the polynomial as $(x + 2)(x - 3)(x^2 - 2x + 10)$. Finally, by using the Quadratic Formula, you could determine that the zeros are -2, 3, $1 + 3i$, and $1 - 3i$.

Example 8 shows how to find all the zeros of a polynomial function, including complex zeros.

Example 8 ▶ Finding the Zeros of a Polynomial Function

Write

$$f(x) = x^5 + x^3 + 2x^2 - 12x + 8$$

as the product of linear factors, and list all of its zeros.

Solution

The possible rational zeros are $\pm 1, \pm 2, \pm 4,$ and ± 8. Synthetic division produces the following.

$$
\begin{array}{r|rrrrrr}
1 & 1 & 0 & 1 & 2 & -12 & 8 \\
  &   & 1 & 1 & 2 & 4 & -8 \\
\hline
  & 1 & 1 & 2 & 4 & -8 & 0 \\
\end{array}
\longrightarrow \text{1 is a zero.}
$$

$$
\begin{array}{r|rrrrr}
1 & 1 & 1 & 2 & 4 & -8 \\
  &   & 1 & 2 & 4 & 8 \\
\hline
  & 1 & 2 & 4 & 8 & 0 \\
\end{array}
\longrightarrow \text{1 is a repeated zero.}
$$

$$
\begin{array}{r|rrrr}
-2 & 1 & 2 & 4 & 8 \\
   &   & -2 & 0 & -8 \\
\hline
   & 1 & 0 & 4 & 0 \\
\end{array}
\longrightarrow \text{-2 is a zero.}
$$

So, you have

$$f(x) = x^5 + x^3 + 2x^2 - 12x + 8$$
$$= (x - 1)(x - 1)(x + 2)(x^2 + 4).$$

By factoring $x^2 + 4$ as

$$x^2 - (-4) = \left(x - \sqrt{-4}\right)\left(x + \sqrt{-4}\right)$$
$$= (x - 2i)(x + 2i)$$

you obtain

$$f(x) = (x - 1)(x - 1)(x + 2)(x - 2i)(x + 2i)$$

which gives the following five zeros of f.

$$1, \quad 1, \quad -2, \quad 2i, \quad \text{and} \quad -2i$$

Note from the graph of f shown in Figure 3.27 that the *real* zeros are the only ones that appear as x-intercepts.

In Example 8, the fifth-degree polynomial function has three real zeros. In such cases, you can use the zoom and trace features or the zero or root feature of a graphing utility to approximate the real zeros. You can then use these real zeros to determine the complex zeros algebraically.

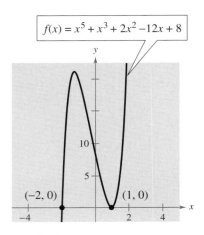

$f(x) = x^5 + x^3 + 2x^2 - 12x + 8$

$(-2, 0)$ $(1, 0)$

FIGURE 3.27

Other Tests for Zeros of Polynomials

You know that an nth-degree polynomial function can have *at most* n real zeros. Of course, many nth-degree polynomials do not have that many real zeros. For instance, $f(x) = x^2 + 1$ has no real zeros, and $f(x) = x^3 + 1$ has only one real zero. The following theorem, called **Descartes's Rule of Signs,** sheds more light on the number of real zeros of a polynomial.

Descartes's Rule of Signs

Let $f(x) = a_n x^n + a_{n-1} x^{n-1} + \cdots + a_2 x^2 + a_1 x + a_0$ be a polynomial with real coefficients and $a_0 \neq 0$.

1. The number of *positive real zeros* of f is either equal to the number of variations in sign of $f(x)$ *or* less than that number by an even integer.

2. The number of *negative real zeros* of f is either equal to the number of variations in sign of $f(-x)$ *or* less than that number by an even integer.

A **variation in sign** means that two consecutive coefficients have opposite signs.

When using Descartes's Rule of Signs, a zero of multiplicity m should be counted as m zeros. For instance, the polynomial $x^3 - 3x + 2$ has two variations in sign, and so has either two positive or no positive real zeros. Because

$$x^3 - 3x + 2 = (x - 1)(x - 1)(x + 2)$$

you can see that the two positive real zeros are $x = 1$ of multiplicity 2.

Example 9 ▶ Using Descartes's Rule of Signs

Describe the possible real zeros of $f(x) = 3x^3 - 5x^2 + 6x - 4$.

Solution

The original polynomial has *three* variations in sign.

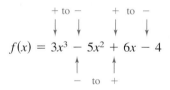

The polynomial

$$f(-x) = 3(-x)^3 - 5(-x)^2 + 6(-x) - 4$$
$$= -3x^3 - 5x^2 - 6x - 4$$

has no variations in sign. So, from Descartes's Rule of Signs, the polynomial $f(x) = 3x^3 - 5x^2 + 6x - 4$ has either three positive real zeros or one positive real zero, and has no negative real zeros. From the graph in Figure 3.28, you can see that the function has only one real zero (it is a positive number, near 1).

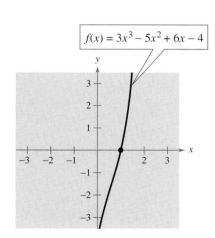

FIGURE **3.28**

Another test for zeros of a polynomial function is related to the sign pattern in the last row of the synthetic division array. This test can give you an upper or lower bound of the real zeros of f. A real number b is an **upper bound** for the real zeros of f if no zeros are greater than b. Similarly, b is a **lower bound** if no real zeros of f are less than b.

Upper and Lower Bound Rules

Let $f(x)$ be a polynomial with real coefficients and a positive leading coefficient. Suppose $f(x)$ is divided by $x - c$, using synthetic division.

1. If $c > 0$ and each number in the last row is either positive or zero, c is an *upper bound* for the real zeros of f.

2. If $c < 0$ and the numbers in the last row are alternately positive and negative (zero entries count as positive or negative), c is a *lower bound* for the real zeros of f.

Example 10 ▶ Finding the Zeros of a Polynomial Function

Find the real zeros of

$$f(x) = 6x^3 - 4x^2 + 3x - 2.$$

Solution

The possible real zeros are as follows.

$$\frac{\text{Factors of 2}}{\text{Factors of 6}} = \frac{\pm 1, \pm 2}{\pm 1, \pm 2, \pm 3, \pm 6}$$

$$= \pm 1, \pm \frac{1}{2}, \pm \frac{1}{3}, \pm \frac{1}{6}, \pm \frac{2}{3}, \pm 2$$

Because $f(x)$ has three variations in sign and $f(-x)$ has none, you can apply Descartes's Rule of Signs to conclude that there are three positive real zeros or one positive real zero, and no negative zeros. Trying $x = 1$ produces the following.

$$
\begin{array}{r|rrrr}
1 & 6 & -4 & 3 & -2 \\
  &   & 6 & 2 & 5 \\
\hline
  & 6 & 2 & 5 & 3
\end{array}
$$

So, $x = 1$ is not a zero, but because the last row has all positive entries, you know that $x = 1$ is an upper bound for the real zeros. So, you can restrict the search to zeros between 0 and 1. By trial and error, you can determine that $x = \frac{2}{3}$ is a zero. So,

$$f(x) = \left(x - \frac{2}{3}\right)(6x^2 + 3).$$

Because $6x^2 + 3$ has no real zeros, it follows that $x = \frac{2}{3}$ is the only real zero.

Before concluding this section, here are two additional hints that can help you find the real zeros of a polynomial.

1. If the terms of $f(x)$ have a common monomial factor, it should be factored out before applying the tests in this section. For instance, by writing

$$f(x) = x^4 - 5x^3 + 3x^2 + x$$
$$= x(x^3 - 5x^2 + 3x + 1)$$

you can see that $x = 0$ is a zero of f and that the remaining zeros can be obtained by analyzing the cubic factor.

2. If you are able to find all but two zeros of $f(x)$, you can always use the Quadratic Formula on the remaining quadratic factor. For instance, if you succeeded in writing

$$f(x) = x^4 - 5x^3 + 3x^2 + x$$
$$= x(x - 1)(x^2 - 4x - 1)$$

you can apply the Quadratic Formula to $x^2 - 4x - 1$ to conclude that the two remaining zeros are

$$x = 2 + \sqrt{5} \qquad \text{and} \qquad x = 2 - \sqrt{5}.$$

A computer animation of this concept appears in the *Interactive* CD-ROM and *Internet* versions of this text.

Example 11 ▶ Using a Polynomial Model

You are designing candle-making kits. Each kit will contain 25 cubic inches of candle wax and a mold for making a pyramid-shaped candle. You want the height of the candle to be 2 inches less than the length of each side of the candle's square base. What should the dimensions of your candle mold be?

Solution

The volume of a pyramid is $V = \frac{1}{3}Bh$, where B is the area of the base and h is the height. The area of the base is x^2 and the height is $(x - 2)$. So, the volume of the pyramid is

$$V = \frac{1}{3}Bh = \frac{1}{3}x^2(x - 2).$$

Substituting 25 for the volume yields the following.

$$25 = \frac{1}{3}x^2(x - 2) \qquad \text{Substitute 25 for } V.$$

$$75 = x^3 - 2x^2 \qquad \text{Multiply each side by 3.}$$

$$0 = x^3 - 2x^2 - 75 \qquad \text{Write in general form.}$$

The possible rational zeros are

$$x = \frac{\pm 1, \pm 3, \pm 5, \pm 15, \pm 25, \pm 75}{\pm 1}.$$

Using synthetic division, you can determine that $x = 5$ is a solution. The other two solutions, which satisfy $x^2 + 3x + 15 = 0$, are imaginary and can be discarded.

The base of the candle mold should be 5 inches by 5 inches. The height of the mold should be $5 - 2 = 3$ inches.

3.4 Exercises

In Exercises 1–6, find all the zeros of the function.

1. $f(x) = x(x - 6)^2$ **2.** $f(x) = x^2(x + 3)(x^2 - 1)$

3. $g(x) = (x - 2)(x + 4)^3$

4. $f(x) = (x + 5)(x - 8)^2$

5. $f(x) = (x + 6)(x + i)(x - i)$

6. $h(t) = (t - 3)(t - 2)(t - 3i)(t + 3i)$

In Exercises 7–10, use the Rational Zero Test to list all possible rational zeros of f. Verify that the zeros of f shown on the graph are contained in the list.

7. $f(x) = x^3 + 3x^2 - x - 3$

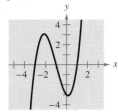

8. $f(x) = x^3 - 4x^2 - 4x + 16$

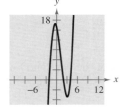

9. $f(x) = 2x^4 - 17x^3 + 35x^2 + 9x - 45$

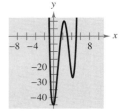

10. $f(x) = 4x^5 - 8x^4 - 5x^3 + 10x^2 + x - 2$

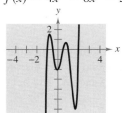

In Exercises 11–20, find all the real zeros of the function.

11. $f(x) = x^3 - 6x^2 + 11x - 6$

12. $f(x) = x^3 - 7x - 6$

13. $g(x) = x^3 - 4x^2 - x + 4$

14. $h(x) = x^3 - 9x^2 + 20x - 12$

15. $h(t) = t^3 + 12t^2 + 21t + 10$

16. $p(x) = x^3 - 9x^2 + 27x - 27$

17. $C(x) = 2x^3 + 3x^2 - 1$

18. $f(x) = 3x^3 - 19x^2 + 33x - 9$

19. $f(x) = 9x^4 - 9x^3 - 58x^2 + 4x + 24$

20. $f(x) = 2x^4 - 15x^3 + 23x^2 + 15x - 25$

In Exercises 21–24, find all real solutions of the polynomial equation.

21. $z^4 - z^3 - 2z - 4 = 0$

22. $x^4 - 13x^2 - 12x = 0$

23. $2y^4 + 7y^3 - 26y^2 + 23y - 6 = 0$

24. $x^5 - x^4 - 3x^3 + 5x^2 - 2x = 0$

In Exercises 25–28, (a) list the possible rational zeros of f, (b) sketch the graph of f so that some of the possible zeros in part (a) can be disregarded, and then (c) determine all real zeros of f.

25. $f(x) = x^3 + x^2 - 4x - 4$

26. $f(x) = -3x^3 + 20x^2 - 36x + 16$

27. $f(x) = -4x^3 + 15x^2 - 8x - 3$

28. $f(x) = 4x^3 - 12x^2 - x + 15$

In Exercises 29–32, (a) list the possible rational zeros of f, (b) use a graphing utility to graph f so that some of the possible zeros in part (a) can be disregarded, and then (c) determine all real zeros of f.

29. $f(x) = -2x^4 + 13x^3 - 21x^2 + 2x + 8$

30. $f(x) = 4x^4 - 17x^2 + 4$

31. $f(x) = 32x^3 - 52x^2 + 17x + 3$

32. $f(x) = 4x^3 + 7x^2 - 11x - 18$

Graphical Analysis In Exercises 33–36, (a) use the root-finding capabilities of a graphing utility to approximate the zeros of the function accurate to three decimal places. (b) Determine one of the exact zeros and use synthetic division to verify your result, and then factor the polynomial completely.

33. $f(x) = x^4 - 3x^2 + 2$

34. $P(t) = t^4 - 7t^2 + 12$

35. $h(x) = x^5 - 7x^4 + 10x^3 + 14x^2 - 24x$

36. $g(x) = 6x^4 - 11x^3 - 51x^2 + 99x - 27$

In Exercises 37–42, find a polynomial function with integer coefficients that has the given zeros. (There are many correct answers.)

37. $1, 5i, -5i$

38. $4, 3i, -3i$

39. $6, -5 + 2i, -5 - 2i$

40. $2, 4 + i, 4 - i$

41. $\frac{2}{3}, -1, 3 + \sqrt{2}i$

42. $-5, -5, 1 + \sqrt{3}i$

In Exercises 43–46, write the polynomial (a) as the product of factors that are irreducible over the *rationals*, (b) as the product of linear and quadratic factors that are irreducible over the *reals*, and (c) in completely factored form.

43. $f(x) = x^4 + 6x^2 - 27$

44. $f(x) = x^4 - 2x^3 - 3x^2 + 12x - 18$
   (*Hint:* One factor is $x^2 - 6$.)

45. $f(x) = x^4 - 4x^3 + 5x^2 - 2x - 6$
   (*Hint:* One factor is $x^2 - 2x - 2$.)

46. $f(x) = x^4 - 3x^3 - x^2 - 12x - 20$
   (*Hint:* One factor is $x^2 + 4$.)

In Exercises 47–54, use the given zero to find all the zeros of the function.

Function	Zero
47. $f(x) = 2x^3 + 3x^2 + 50x + 75$	$5i$
48. $f(x) = x^3 + x^2 + 9x + 9$	$3i$
49. $f(x) = 2x^4 - x^3 + 7x^2 - 4x - 4$	$2i$
50. $g(x) = x^3 - 7x^2 - x + 87$	$5 + 2i$
51. $g(x) = 4x^3 + 23x^2 + 34x - 10$	$-3 + i$
52. $h(x) = 3x^3 - 4x^2 + 8x + 8$	$1 - \sqrt{3}i$
53. $f(x) = x^4 + 3x^3 - 5x^2 - 21x + 22$	$-3 + \sqrt{2}i$
54. $f(x) = x^3 + 4x^2 + 14x + 20$	$-1 - 3i$

In Exercises 55–72, find all the zeros of the function and write the polynomial as a product of linear factors.

55. $f(x) = x^2 + 25$

56. $f(x) = x^2 - x + 56$

57. $h(x) = x^2 - 4x + 1$

58. $g(x) = x^2 + 10x + 23$

59. $f(x) = x^4 - 81$

60. $f(y) = y^4 - 625$

61. $f(z) = z^2 - 2z + 2$

62. $h(x) = x^3 - 3x^2 + 4x - 2$

63. $g(x) = x^3 - 6x^2 + 13x - 10$

64. $f(x) = x^3 - 2x^2 - 11x + 52$

65. $h(x) = x^3 - x + 6$

66. $h(x) = x^3 + 9x^2 + 27x + 35$

67. $f(x) = 5x^3 - 9x^2 + 28x + 6$

68. $g(x) = 3x^3 - 4x^2 + 8x + 8$

69. $g(x) = x^4 - 4x^3 + 8x^2 - 16x + 16$

70. $h(x) = x^4 + 6x^3 + 10x^2 + 6x + 9$

71. $f(x) = x^4 + 10x^2 + 9$

72. $f(x) = x^4 + 29x^2 + 100$

In Exercises 73–78, find all the zeros of the function. When there is an extended list of possible rational zeros, use a graphing utility to graph the function in order to discard any rational zeros that are obviously not zeros of the function.

73. $f(x) = x^3 + 24x^2 + 214x + 740$

74. $f(s) = 2s^3 - 5s^2 + 12s - 5$

75. $f(x) = 16x^3 - 20x^2 - 4x + 15$

76. $f(x) = 9x^3 - 15x^2 + 11x - 5$

77. $f(x) = 2x^4 + 5x^3 + 4x^2 + 5x + 2$

78. $g(x) = x^5 - 8x^4 + 28x^3 - 56x^2 + 64x - 32$

In Exercises 79–86, use Descartes's Rule of Signs to determine the possible number of positive and negative zeros of the function.

79. $g(x) = 5x^5 + 10x$

80. $h(x) = 4x^2 - 8x + 3$

81. $h(x) = 3x^4 + 2x^2 + 1$

82. $h(x) = 2x^4 - 3x + 2$

83. $g(x) = 2x^3 - 3x^2 - 3$

84. $f(x) = 4x^3 - 3x^2 + 2x - 1$

85. $f(x) = -5x^3 + x^2 - x + 5$

86. $f(x) = 3x^3 + 2x^2 + x + 3$

In Exercises 87–90, use synthetic division to verify the upper and lower bounds of the real zeros of f.

87. $f(x) = x^4 - 4x^3 + 15$

   (a) Upper: $x = 4$ (b) Lower: $x = -1$

88. $f(x) = 2x^3 - 3x^2 - 12x + 8$

   (a) Upper: $x = 4$ (b) Lower: $x = -3$

89. $f(x) = x^4 - 4x^3 + 16x - 16$

   (a) Upper: $x = 5$ (b) Lower: $x = -3$

90. $f(x) = 2x^4 - 8x + 3$

   (a) Upper: $x = 3$ (b) Lower: $x = -4$

In Exercises 91–94, find all the real zeros of the function.

91. $f(x) = 4x^3 - 3x - 1$

92. $f(z) = 12z^3 - 4z^2 - 27z + 9$

93. $f(y) = 4y^3 + 3y^2 + 8y + 6$

94. $g(x) = 3x^3 - 2x^2 + 15x - 10$

In Exercises 95–98, find all the rational zeros of the polynomial function.

95. $P(x) = x^4 - \frac{25}{4}x^2 + 9 = \frac{1}{4}(4x^4 - 25x^2 + 36)$

96. $f(x) = x^3 - \frac{3}{2}x^2 - \frac{23}{2}x + 6 = \frac{1}{2}(2x^3 - 3x^2 - 23x + 12)$

97. $f(x) = x^3 - \frac{1}{4}x^2 - x + \frac{1}{4} = \frac{1}{4}(4x^3 - x^2 - 4x + 1)$

98. $f(z) = z^3 + \frac{11}{6}z^2 - \frac{1}{2}z - \frac{1}{3} = \frac{1}{6}(6z^3 + 11z^2 - 3z - 2)$

In Exercises 99–102, match the cubic function with the numbers of rational and irrational zeros.

(a) Rational zeros: 0;   Irrational zeros: 1

(b) Rational zeros: 3;   Irrational zeros: 0

(c) Rational zeros: 1;   Irrational zeros: 2

(d) Rational zeros: 1;   Irrational zeros: 0

99. $f(x) = x^3 - 1$

100. $f(x) = x^3 - 2$

101. $f(x) = x^3 - x$

102. $f(x) = x^3 - 2x$

103. *Think About It*   A third-degree polynomial function f has real zeros -2, $\frac{1}{2}$, and 3, and its leading coefficient is negative. Write an equation for f. Sketch the graph of f. How many different polynomial functions are possible for f?

104. *Think About It*   Sketch the graph of a fifth-degree polynomial function, whose leading coefficient is positive, that has one root at $x = 3$ of multiplicity 2.

105. Use the information in the table.

Interval	Value of $f(x)$
$(-\infty, -2)$	Positive
$(-2, 1)$	Negative
$(1, 4)$	Negative
$(4, \infty)$	Positive

(a) What are the three zeros of the polynomial function f?

(b) What can be said about the behavior of the graph of f at $x = 1$?

(c) What is the least possible degree of f? Explain. Can the degree of f ever be odd? Explain.

(d) Is the leading coefficient of f positive or negative? Explain.

(e) Write an equation for f.

(f) Sketch a graph of the function you wrote in part (e).

106. Use the information in the table.

Interval	Value of $f(x)$
$(-\infty, -2)$	Negative
$(-2, 0)$	Positive
$(0, 2)$	Positive
$(2, \infty)$	Negative

(a) What are the three zeros of the polynomial function f?

(b) What can be said about the behavior of the graph of f at $x = 0$?

(c) What is the least possible degree of f? Explain. Can the degree of f ever be odd? Explain.

(d) Is the leading coefficient of f positive or negative? Explain.

(e) Write an equation for f.

(f) Sketch a graph of the function you wrote in part (e).

107. *Geometry* An open box is to be made from a rectangular piece of material, 15 centimeters by 9 centimeters, by cutting equal squares from the corners and turning up the sides.

(a) Let x represent the length of the sides of the squares removed. Draw a diagram showing the squares removed from the original piece of material and the resulting dimensions of the open box.

(b) Use the diagram to write the volume V of the box as a function of x. Determine the domain of the function.

(c) Sketch the graph of the function and approximate the dimensions of the box that yield a maximum volume.

(d) Find values of x such that $V = 56$. Which of these values is a physical impossibility in the construction of the box? Explain.

108. *Geometry* A rectangular package to be sent by a delivery service (see figure) can have a maximum combined length and girth (perimeter of a cross section) of 120 inches.

(a) Show that the volume of the package is
$$V(x) = 4x^2(30 - x).$$

(b) Use a graphing utility to graph the function and approximate the dimensions of the package that yield a maximum volume.

(c) Find values of x such that $V = 13,500$. Which of these values is a physical impossibility in the construction of the package? Explain.

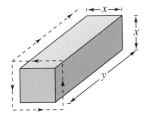

109. *Advertising Cost* A company that produces portable cassette players estimates that the profit for selling a particular model is
$$P = -76x^3 + 4830x^2 - 320,000, \quad 0 \le x \le 60$$
where P is the profit (in dollars) and x is the advertising expense (in tens of thousands of dollars). Using this model, find the smaller of two advertising amounts that yield a profit of $2,500,000.

110. *Advertising Cost* A company that manufactures bicycles estimates that the profit for selling a particular model is
$$P = -45x^3 + 2500x^2 - 275,000, \quad 0 \le x \le 50$$
where P is the profit (in dollars) and x is the advertising expense (in tens of thousands of dollars). Using this model, find the smaller of two advertising amounts that yield a profit of $800,000.

111. *Business* The ordering and transportation cost for the components used in manufacturing a certain product is
$$C = 100\left(\frac{200}{x^2} + \frac{x}{x + 30}\right), \quad 1 \le x$$
where C is the cost (in thousands of dollars) and x is the order size (in hundreds). In calculus, it can be shown that the cost is a minimum when
$$3x^3 - 40x^2 - 2400x - 36,000 = 0.$$
Use a calculator to approximate the optimal order size to the nearest hundred units.

112. *Foreign Trade* The values (in billions of dollars) of goods imported into the United States for the years 1988 through 1997 are shown in the table. (Source: U.S. International Trade Administration)

Year	1988	1989	1990	1991	1992
Imports	441.0	473.2	495.3	488.5	532.7

Year	1993	1994	1995	1996	1997
Imports	580.7	663.3	743.4	795.3	870.7

(a) A model for this data is
$$I = -0.222t^3 + 6.432t^2 + 23.328t + 473.991$$
where I is the annual value of goods imported (in billions of dollars) and t is the time (in years), with $t = 0$ corresponding to 1990. Use a graphing utility to sketch a scatter plot of the data and the model in the same viewing window. How do they compare?

(b) According to this model, when did the annual value of imports reach 750 billion dollars?

(c) According to the right-hand behavior of the model, will the value of imports continue to increase? Explain.

113. *Physics*   A baseball is thrown upward from ground level with an initial velocity of 48 feet per second, and its height h (in feet) is

$$h = -16t^2 + 48t, \quad 0 \le t \le 3$$

where t is the time (in seconds). Suppose you are told the ball reaches a height of 64 feet. Is this possible?

114. *Economics*   The demand equation for a certain product is $p = 140 - 0.0001x$, where p is the unit price (in dollars) of the product and x is the number of units produced and sold. The cost equation for the product is $C = 80x + 150,000$, where C is the total cost (in dollars) and x is the number of units produced. The total profit obtained by producing and selling x units is

$$P = R - C = xp - C.$$

You are working in the marketing department of the company that produces this product, and you are asked to determine a price p that will yield a profit of 9 million dollars. Is this possible? Explain.

Synthesis

True or False?   **In Exercises 115 and 116, decide whether the statement is true or false. Justify your answer.**

115. It is possible for a third-degree polynomial function with integer coefficients to have no real zeros.

116. If $x = -i$ is a zero of the function $f(x) = x^3 + ix^2 + ix - 1$, then $x = i$ must also be a zero of f.

Think About It   **In Exercises 117–122, determine (if possible) the zeros of the function g if the function f has zeros at $x = r_1$, $x = r_2$, and $x = r_3$.**

117. $g(x) = -f(x)$

118. $g(x) = 3f(x)$

119. $g(x) = f(x - 5)$

120. $g(x) = f(2x)$

121. $g(x) = 3 + f(x)$

122. $g(x) = f(-x)$

123. *Exploration*   Use a graphing utility to graph the function $f(x) = x^4 - 4x^2 + k$ for different values of k. Find values of k such that the zeros of f satisfy the specified characteristics. (Some parts do not have unique answers.)

(a) Four real zeros

(b) Two real zeros, each of multiplicity 2

(c) Two real zeros and two complex roots

(d) Four complex zeros

124. *Think About It*   Will the answers to Exercise 123 change for the function g?

(a) $g(x) = f(x - 2)$

(b) $g(x) = f(2x)$

125. (a) Find a quadratic function f (with integer coefficients) that has $\pm \sqrt{b}i$ as zeros. Assume that b is a positive integer.

(b) Find a quadratic function f (with integer coefficients) that has $a \pm bi$ as zeros. Assume that b is a positive integer.

126. *Graphical Reasoning*   The graph of one of the following functions is shown below. Identify the function shown in the graph. Explain why each of the others is not the correct function. Use a graphing utility to verify your result.

(a) $f(x) = x^2(x + 2)(x - 3.5)$

(b) $g(x) = (x + 2)(x - 3.5)$

(c) $h(x) = (x + 2)(x - 3.5)(x^2 + 1)$

(d) $k(x) = (x + 1)(x + 2)(x - 3.5)$

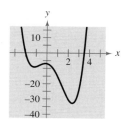

Review

In Exercises 127–132, perform the operation and simplify.

127. $(-3 + 6i) - (8 - 3i)$

128. $(12 - 5i) + 16i$

129. $(6 - 2i)(1 + 7i)$

130. $(9 - 5i)(9 + 5i)$

131. $\dfrac{1 + i}{1 - i}$

132. $(3 + i)^3$

In Exercises 133–138, use the graph of f to graph the function g.

133. $g(x) = f(x - 2)$

134. $g(x) = f(x) - 2$

135. $g(x) = 2f(x)$

136. $g(x) = f(-x)$

137. $g(x) = f(2x)$

138. $g(x) = f\left(\tfrac{1}{2}x\right)$

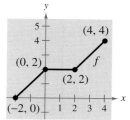

3.5 Mathematical Modeling

▶ **Why you should learn it**

You can use functions as models to represent a wide variety of real-life data sets. For instance, Exercise 69 on page 313 shows how a linear function can be used to model the salaries of professional hockey players from 1990 through 1996.

Introduction

You have already studied some techniques for fitting models to data. For instance, in Section 2.1, you learned how to find the equation of a line that passes through two points. In this section, you will study other techniques for fitting models to data: *direct and inverse variation* and *least squares regression*. The resulting models are either polynomial functions or rational functions. (Rational functions will be studied in Chapter 4.)

Example 1 ▶ A Mathematical Model

The numbers of insured commercial banks y (in thousands) in the United States for the years 1987 to 1996 are shown in the table.   (Source: Federal Deposit Insurance Corporation)

Year	1987	1988	1989	1990	1991	1992	1993	1994	1995	1996
y	13.70	13.12	12.71	12.34	11.92	11.46	10.96	10.45	9.94	9.53

A linear model that approximates this data is

$$y = -0.459t + 16.89, \quad 7 \le t \le 16$$

where $t = 7$ corresponds to 1987. Plot the actual data *and* the model on the same graph. How closely does the model represent the data?

Solution

The actual data is plotted in Figure 3.29, along with the graph of the linear model. From the graph, it appears that the model is a "good fit" for the actual data. You can see how well the model fits by comparing the actual values of y with the values of y given by the model. The values given by the model are labeled $y*$ in the table below.

t	7	8	9	10	11	12	13	14	15	16
y	13.70	13.12	12.71	12.34	11.92	11.46	10.96	10.45	9.94	9.53
$y*$	13.68	13.22	12.76	12.30	11.84	11.38	10.92	10.46	10.01	9.55

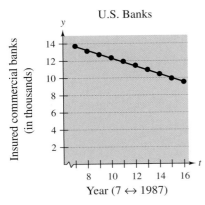

U.S. Banks

Insured commercial banks (in thousands)

Year (7 ↔ 1987)

FIGURE **3.29**

Note in Example 1 that you could have chosen any two points to find a line that fits the data. However, the linear model above was found using the regression feature of a graphing utility and is the line that *best* fits the data. This concept of a "best-fitting" line is discussed later in this section.

Direct Variation

There are two basic types of linear models. The more general model has a y-intercept that is nonzero.

$$y = mx + b, \quad b \neq 0$$

The simpler model

$$y = kx$$

has a y-intercept that is zero. In the simpler model, y is said to **vary directly** as x, or to be **directly proportional** to x.

Direct Variation

The following statements are equivalent.

1. y **varies directly** as x.

2. y is **directly proportional** to x.

3. $y = kx$ for some nonzero constant k.

k is the **constant of variation** or the **constant of proportionality.**

Example 2 ▶ Direct Variation

In Pennsylvania, the state income tax is directly proportional to *gross income*. Suppose you were working in Pennsylvania and your state income tax deduction was \$42 for a gross monthly income of \$1500. Find a mathematical model that gives the Pennsylvania state income tax in terms of gross income.

Solution

Verbal Model:	State income tax	$= k \cdot$	Gross income

Labels:
State income tax $= y$ (dollars)
Gross income $= x$ (dollars)
Income tax rate $= k$ (percent in decimal form)

Equation: $y = kx$

To solve for k, substitute the given information into the equation $y = kx$, and then solve for k.

$$y = kx \qquad \text{Write direct variation model.}$$

$$42 = k(1500) \qquad \text{Substitute } y = 42 \text{ and } x = 1500.$$

$$0.028 = k \qquad \text{Simplify.}$$

So, the equation (or model) for state income tax in Pennsylvania is

$$y = 0.028x.$$

In other words, Pennsylvania has a state income tax rate of 2.8% of gross income. The graph of this equation is shown in Figure 3.30.

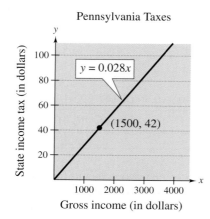

Pennsylvania Taxes

$y = 0.028x$

$(1500, 42)$

State income tax (in dollars)

Gross income (in dollars)

FIGURE **3.30**

Direct Variation as *n*th Power

Another type of direct variation relates one variable to a *power* of another variable. For example, in the formula for the area of a circle

$$A = \pi r^2$$

the area A is directly proportional to the square of the radius r. Note that for this formula, π is the constant of proportionality.

STUDY T!P

Note that the direct variation model $y = kx$ is a special case of $y = kx^n$ with $n = 1$.

Direct Variation as *n*th Power

The following statements are equivalent.

1. y **varies directly as the *n*th power** of x.

2. y is **directly proportional to the *n*th power** of x.

3. $y = kx^n$ for some constant k.

Example 3 ▶ Direct Variation as *n*th Power

The distance a ball rolls down an inclined plane is directly proportional to the square of the time it rolls. During the first second the ball rolls 8 feet. (See Figure 3.31.)

a. Write an equation relating the distance traveled to the time.

b. How far will the ball roll during the first 3 seconds?

Solution

a. Letting d be the distance (in feet) the ball rolls and letting t be the time (in seconds), you have

$$d = kt^2.$$

Now, because $d = 8$ when $t = 1$, you can see that $k = 8$, as follows.

$$d = kt^2$$
$$8 = k(1)^2$$
$$8 = k$$

So, the equation relating distance to time is

$$d = 8t^2.$$

b. When $t = 3$, the distance traveled is $d = 8(3^2) = 8(9) = 72$ feet.

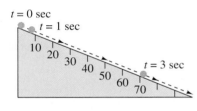

$t = 0$ sec

$t = 1$ sec

10 20 30 40 50 60 70

$t = 3$ sec

FIGURE **3.31**

In Examples 2 and 3, the direct variations are such that an *increase* in one variable corresponds to an *increase* in the other variable. This is also true in the model $d = \frac{1}{5}F$, $F > 0$, where an increase in F results in an increase in d. You should not, however, assume that this always occurs with direct variation. For example, in the model $y = -3x$, an increase in x results in a *decrease* in y, and yet y is said to vary directly as x.

Inverse Variation

Inverse Variation

The following statements are equivalent.

1. y **varies inversely** as x.

2. y is **inversely proportional** to x.

3. $y = \dfrac{k}{x}$ for some constant k.

If x and y are related by an equation of the form $y = k/x^n$, then y varies inversely as the nth power of x (or y is inversely proportional to the nth power of x).

Example 4 ▶ Inverse Variation

A gas law states that the volume of an enclosed gas varies directly as the temperature *and* inversely as the pressure, as shown in Figure 3.32. The pressure of a gas is 0.75 kilogram per square centimeter when the temperature is 294 K and the volume is 8000 cubic centimeters.

a. Write an equation relating pressure, temperature, and volume.

b. Find the pressure when the temperature is 300 K and the volume is 7000 cubic centimeters.

Solution

a. Let V be volume (in cubic centimeters), let P be pressure (in kilograms per square centimeter), and let T be temperature (in Kelvin). Because V varies directly as T and inversely as P,

$$V = \frac{kT}{P}.$$

Now, because $P = 0.75$ when $T = 294$ and $V = 8000$,

$$8000 = \frac{k(294)}{0.75}$$

$$\frac{8000(0.75)}{294} = k$$

$$k = \frac{6000}{294} = \frac{1000}{49}.$$

So, the equation relating pressure, temperature, and volume is

$$V = \frac{1000}{49}\left(\frac{T}{P}\right).$$

b. When $T = 300$ and $V = 7000$, the pressure is

$$P = \frac{1000}{49}\left(\frac{300}{7000}\right) = \frac{300}{343} \approx 0.87 \text{ kilogram per square centimeter.}$$

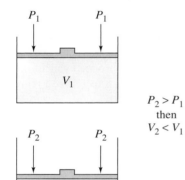

FIGURE 3.32 *If the temperature is held constant and pressure increases, volume decreases.*

Joint Variation

In Example 4, note that when a direct variation and an inverse variation occur in the same statement, they are coupled with the word "and." To describe two different *direct* variations in the same statement, the word **jointly** is used.

Joint Variation

The following statements are equivalent.

1. z **varies jointly** as x and y.

2. z is **jointly proportional** to x and y.

3. $z = kxy$ for some constant k.

If x, y, and z are related by an equation of the form

$$z = kx^n y^m$$

then z varies jointly as the nth power of x and the mth power of y.

Example 5 ▶ Joint Variation

The *simple* interest for a certain savings account is jointly proportional to the time and the principal. After one quarter (3 months), the interest on a principal of $5000 is $43.75.

a. Write an equation relating the interest, principal, and time.

b. Find the interest after three quarters.

Solution

a. Let $I =$ interest (in dollars), $P =$ principal (in dollars), and $t =$ time (in years). Because I is jointly proportional to P and t,

$$I = kPt.$$

For $I = 43.75$, $P = 5000$, and $t = \frac{1}{4}$,

$$43.75 = k(5000)\left(\frac{1}{4}\right)$$

which implies that $k = 4(43.75)/5000 = 0.035$. So, the equation relating interest, principal, and time is

$$I = 0.035Pt$$

which is the familiar equation for simple interest where the constant of proportionality, 0.035, represents an annual interest rate of 3.5%.

b. When $P = \$5000$ and $t = \frac{3}{4}$, the interest is

$$I = (0.035)(5000)\left(\frac{3}{4}\right) = \$131.25.$$

Technology

Most graphing utilities have a built-in linear regression program. When you run such a program, the "r-value" or **correlation coefficient,** gives a measure of how well the model fits the data. The closer the value of $|r|$ is to 1, the better the fit. For the data in Example 6, $r \approx 0.955$, which implies that the model is a good fit. This is confirmed in the table at the right.

A computer simulation of this example appears in the *Interactive* CD-ROM and *Internet* versions of this text.

Least Squares Regression

So far in this text, you have worked with many different types of mathematical models that approximate real-life data. For instance, in Example 1 on page 303 you analyzed a model for data on the number of insured commercial banks in the United States.

To find such a model, statisticians use a measure called the **sum of square differences,** which is the sum of the squares of the differences between actual data values and model values. The "best-fitting" linear model is the one with the least sum of square differences. This best-fitting linear model is called the **least squares regression line.** You can approximate this line visually by plotting the data points and drawing the line that appears to fit best—or you can enter the data points into a calculator or computer and use the calculator's or computer's linear regression program.

Example 6 ▶ Finding a Least Squares Regression Line

The amounts of total annual prize money p (in millions of dollars) awarded at the Indianapolis 500 race from 1989 to 1997 are shown in the table. Construct a scatter plot that represents the data and find a linear model that approximates the data. (Source: Indianapolis Motor Speedway Hall of Fame)

Year	1989	1990	1991	1992	1993	1994	1995	1996	1997
p	5.71	6.33	7.01	7.53	7.68	7.86	8.06	8.11	8.61

Solution

Let $t = 9$ represent 1989. The scatter plot for the points is shown in Figure 3.33. Using the least squares regression feature of a graphing utility, you can determine that the equation of the line is

$$p = 0.323t + 3.24.$$

To check this model, compare the actual p-values with the p-values given by the model, which are labeled $p*$ in the table below.

t	9	10	11	12	13	14	15	16	17
p	5.71	6.33	7.01	7.53	7.68	7.86	8.06	8.11	8.61
$p*$	6.15	6.47	6.79	7.12	7.44	7.76	8.09	8.41	8.73

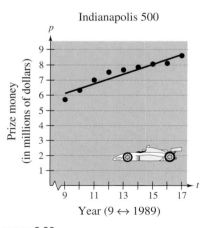

Indianapolis 500

FIGURE 3.33

3.5 Exercises

1. *Employment*   The total numbers of employees (in thousands) in the United States from 1990 to 1997 are given by the following ordered pairs.

(1990, 125,840)   (1994, 131,056)
(1991, 126,346)   (1995, 132,304)
(1992, 128,105)   (1996, 133,943)
(1993, 129,200)   (1997, 136,297)

A linear model that approximates this data is

$y = 125{,}151.5 + 1495.68t, \quad 0 \le t \le 7$

where y represents the number of employees (in thousands) and $t = 0$ represents 1990. Plot the actual data and the model on the same graph. How closely does the model represent the data?   (Source: U.S. Bureau of Labor Statistics)

2. *Sports*   The winning times (in minutes) in the women's 400-meter freestyle swimming event in the Olympics from 1936 to 1996 are given by the following ordered pairs.

(1936, 5.44)   (1972, 4.32)
(1948, 5.30)   (1976, 4.16)
(1952, 5.20)   (1980, 4.15)
(1956, 4.91)   (1984, 4.12)
(1960, 4.84)   (1988, 4.06)
(1964, 4.72)   (1992, 4.12)
(1968, 4.53)   (1996, 4.12)

A linear model that approximates this data is

$y = 5.35 - 0.027t, \quad -4 \le t \le 56$

where y represents the winning time in minutes and $t = 0$ represents 1940. Plot the actual data and the model on the same graph. How closely does the model represent the data?   (Source: ESPN)

Think About It   **In Exercises 3 and 4, use the graph to determine whether y varies directly as some power of x or inversely as some power of x. Explain.**

3.

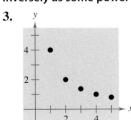

4.

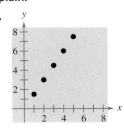

In Exercises 5–8, use the given value of k to complete the table for the direct variation model $y = kx^2$. Plot the points on a rectangular coordinate system.

x	2	4	6	8	10
$y = kx^2$					

5. $k = 1$ **6.** $k = 2$
7. $k = \frac{1}{2}$ **8.** $k = \frac{1}{4}$

In Exercises 9–12, use the given value of k to complete the table for the inverse variation model

$y = \dfrac{k}{x^2}.$

Plot the points on a rectangular coordinate system.

x	2	4	6	8	10
$y = \dfrac{k}{x^2}$					

9. $k = 2$ **10.** $k = 5$
11. $k = 10$ **12.** $k = 20$

In Exercises 13–16, determine whether the variation model is of the form

$y = kx \quad \text{or} \quad y = \dfrac{k}{x}$

and find k.

13.

x	5	10	15	20	25
y	1	$\frac{1}{2}$	$\frac{1}{3}$	$\frac{1}{4}$	$\frac{1}{5}$

14.

x	5	10	15	20	25
y	2	4	6	8	10

15.

x	5	10	15	20	25
y	-3.5	-7	-10.5	-14	-17.5

16.

x	5	10	15	20	25
y	24	12	8	6	$\frac{24}{5}$

Direct Variation   **In Exercises 17–20, assume that** *y* **is directly proportional to** *x***. Use the given** *x***-value and** *y***-value to find a linear model that relates** *y* **and** *x***.**

x-Value	y-Value		x-Value	y-Value
17. $x = 5$	$y = 12$		**18.** $x = 2$	$y = 14$
19. $x = 10$	$y = 2050$		**20.** $x = 6$	$y = 580$

21. *Simple Interest*   The simple interest on an investment is directly proportional to the amount of the investment. By investing $2500 in a certain bond issue, you obtained an interest payment of $87.50 after 1 year. Find a mathematical model that gives the interest *I* for this bond issue after 1 year in terms of the amount invested *P*.

22. *Simple Interest*   The simple interest on an investment is directly proportional to the amount of the investment. By investing $5000 in a municipal bond, you obtained an interest payment of $187.50 after 1 year. Find a mathematical model that gives the interest *I* for this municipal bond after 1 year in terms of the amount invested *P*.

23. *Measurement*   On a yardstick with scales in inches and centimeters, you notice that 13 inches is approximately the same length as 33 centimeters. Use this information to find a mathematical model that relates centimeters to inches. Then use the model to complete the table.

Inches	5	10	20	25	30
Centimeters					

24. *Measurement*   When buying gasoline, you notice that 14 gallons of gasoline is approximately the same amount of gasoline as 53 liters. Use this information to find a linear model that relates gallons to liters. Use the model to complete the table.

Gallons	5	10	20	25	30
Liters					

25. *Taxes*   Property tax is based on the assessed value of the property. A house that has an assessed value of $150,000 has a property tax of $5520. Find a mathematical model that gives the amount of property tax *y* in terms of the assessed value *x* of the property. Use the model to find the property tax on a house that has an assessed value of $200,000.

26. *Taxes*   State sales tax is based on retail price. An item that sells for $145.99 has a sales tax of $10.22. Find a mathematical model that gives the amount of sales tax *y* in terms of the retail price *x*. Use the model to find the sales tax on a $540.50 purchase.

Physics   **In Exercises 27–30, use Hooke's Law for springs, which states that the distance a spring is stretched (or compressed) varies directly as the force on the spring.**

27. A force of 265 newtons stretches a spring 0.15 meter (see figure).

(a) How far will a force of 90 newtons stretch the spring?

(b) What force is required to stretch the spring 0.1 meter?

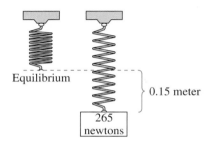

28. A force of 220 newtons stretches a spring 0.12 meter. What force is required to stretch the spring 0.16 meter?

29. The coiled spring of a toy supports the weight of a child. The spring is compressed a distance of 1.9 inches by the weight of a 25-pound child. The toy will not work properly if its spring is compressed more than 3 inches. What is the weight of the heaviest child who should be allowed to use the toy?

30. An overhead garage door has two springs, one on each side of the door (see figure). A force of 15 pounds is required to stretch each spring 1 foot. Because of a pulley system, the springs stretch only one-half the distance the door travels. The door moves a total of 8 feet, and the springs are at their natural length when the door is open. Find the combined lifting force applied to the door by the springs when the door is closed.

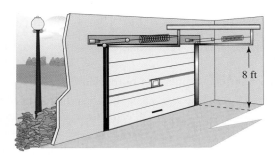

FIGURE FOR 30

In Exercises 31–40, find a mathematical model for the verbal statement.

31. *A* varies directly as the square of *r*.

32. *V* varies directly as the cube of *e*.

33. *y* varies inversely as the square of *x*.

34. *h* varies inversely as the square root of *s*.

35. *F* varies directly as *g* and inversely as r^2.

36. *z* is jointly proportional to the square of *x* and y^3.

37. Boyle's Law: For a constant temperature, the pressure *P* of a gas is inversely proportional to the volume *V* of the gas.

38. Newton's Law of Cooling: The rate of change *R* of the temperature of an object is proportional to the difference between the temperature *T* of the object and the temperature T_e of the environment in which the object is placed.

39. Newton's Law of Universal Gravitation: The gravitational attraction *F* between two objects of masses m_1 and m_2 is proportional to the product of the masses and inversely proportional to the square of the distance *r* between the objects.

40. Logistic growth: The rate of growth *R* of a population is jointly proportional to the size *S* of the population and the difference between *S* and the maximum population size *L* that the environment can support.

In Exercises 41–46, write a sentence using the variation terminology of this section to describe the formula.

41. Area of a triangle: $A = \frac{1}{2}bh$

42. Surface area of a sphere: $S = 4\pi r^2$

43. Volume of a sphere: $V = \frac{4}{3}\pi r^3$

44. Volume of a right circular cylinder: $V = \pi r^2 h$

45. Average speed: $r = \dfrac{d}{t}$

46. Free vibrations: $\omega = \sqrt{\dfrac{kg}{W}}$

In Exercises 47–54, find a mathematical model representing the statement. (In each case, determine the constant of proportionality.)

47. *A* varies directly as r^2. ($A = 9\pi$ when $r = 3$.)

48. *y* varies inversely as *x*. ($y = 3$ when $x = 25$.)

49. *y* is inversely proportional to *x*. ($y = 7$ when $x = 4$.)

50. *z* varies jointly as *x* and *y*. ($z = 64$ when $x = 4$ and $y = 8$.)

51. *F* is jointly proportional to *r* and the third power of *s*. ($F = 4158$ when $r = 11$ and $s = 3$.)

52. *P* varies directly as *x* and inversely as the square of *y*. $\left(P = \frac{28}{3} \text{ when } x = 42 \text{ and } y = 9.\right)$

53. *z* varies directly as the square of *x* and inversely as *y*. ($z = 6$ when $x = 6$ and $y = 4$.)

54. *v* varies jointly as *p* and *q* and inversely as the square of *s*. ($v = 1.5$ when $p = 4.1$, $q = 6.3$, and $s = 1.2$.)

Ecology **In Exercises 55 and 56, use the fact that the diameter of the largest particle that can be moved by a stream varies approximately directly as the square of the velocity of the stream.**

55. A stream with a velocity of $\frac{1}{4}$ mile per hour can move coarse sand particles about 0.02 inch in diameter. Approximate the velocity required to carry particles 0.12 inch in diameter.

56. A stream of velocity *v* can move particles of diameter *d* or less. By what factor does *d* increase when the velocity is doubled?

Resistance **In Exercises 57 and 58, use the fact that the resistance of a wire carrying an electrical current is directly proportional to its length and inversely proportional to its cross-sectional area.**

57. If #28 copper wire (which has a diameter of 0.0126 inch) has a resistance of 66.17 ohms per thousand feet, what length of #28 copper wire will produce a resistance of 33.5 ohms?

58. A 14-foot piece of copper wire produces a resistance of 0.05 ohm. Use the constant of proportionality from Exercise 57 to find the diameter of the wire.

59. *Free Fall* Neglecting air resistance, the distance *s* an object falls varies directly as the square of the duration *t* of the fall. An object falls a distance of 144 feet in 3 seconds. How far will it fall in 5 seconds?

60. *Stopping Distance* The stopping distance d of an automobile is directly proportional to the square of its speed s. A car required 75 feet to stop when its speed was 30 miles per hour. Estimate the stopping distance if the brakes are applied when the car is traveling at 50 miles per hour.

61. *Spending* The prices of three sizes of pizza at a pizza shop are as follows.

9-inch: $8.78
12-inch: $11.78
15-inch: $14.18

You would expect that the price of a certain size of pizza would be directly proportional to its surface area. Is that the case for this pizza shop? If not, which size of pizza is the best buy?

62. *Economics* A company has found that the demand for its product varies inversely as the price of the product. When the price is $3.75, the demand is 500 units. Approximate the demand when the price is $4.25.

63. *Fluid Flow* The velocity v of a fluid flowing in a conduit is inversely proportional to the cross-sectional area of the conduit. (Assume that the volume of the flow per unit of time is held constant.)

(a) Determine the change in the velocity of water flowing from a hose when a person places a finger over the end of the hose to decrease its cross-sectional area by 25%.

(b) Use the fluid velocity model in part (a) to determine the effect on the velocity of a stream when it is dredged to increase its cross-sectional area by one-third.

64. *Beam Load* The maximum load that can be safely supported by a horizontal beam varies jointly as the width of the beam and the square of its depth, and inversely as the length of the beam. Determine the change in the maximum safe load under the following conditions.

(a) The width and length of the beam are doubled.

(b) The width and depth of the beam are doubled.

(c) All three of the dimensions are doubled.

(d) The depth of the beam is halved.

65. *Data Analysis* An experiment in a physics lab requires a student to measure the compressed length x (in centimeters) of a spring when a force of F pounds is applied. The data is shown in the table.

F	0	2	4	6	8	10	12
x	0	1.15	2.3	3.45	4.6	5.75	6.9

(a) Sketch a scatter plot of the data.

(b) Does it appear that the data can be modeled by Hooke's Law? If so, estimate k. (See Exercises 27–30.)

(c) Use the model in part (b) to approximate the force required to compress the spring 9 centimeters.

66. *Data Analysis* An oceanographer took readings of the water temperature C (in degrees Celsius) at depth d (in meters). The data collected is shown in the table.

d	1000	2000	3000	4000	5000
C	4.2°	1.9°	1.4°	1.2°	0.9°

(a) Sketch a scatter plot of the data.

(b) Does it appear that the data can be modeled by the inverse proportion model $C = k/d$? If so, estimate k.

(c) Use a graphing utility to plot the data points and the inverse model in part (b).

(d) Use the model to approximate the depth at which the water temperature is 3°C.

67. *Data Analysis* A light probe is located x centimeters from a light source, and the intensity y (in microwatts per square centimeter) of the light is measured. The results are shown in the table.

x	30	34	38	42	46	50
y	0.1881	0.1543	0.1172	0.0998	0.0775	0.0645

A model for the data is $y = 262.76/x^{2.12}$.

(a) Use a graphing utility to plot the data points and the model in the same viewing window.

(b) Use the model to approximate the light intensity 25 centimeters from the light source.

68. *Illumination* The illumination from a light source varies inversely as the square of the distance from the light source. When the distance from a light source is doubled, how does the illumination change? Discuss this model in terms of the data given in Exercise 67. Give a possible explanation of the difference.

69. *Hockey Salaries* The average annual salaries of professional hockey players (in thousands of dollars) from 1990 to 1996 are shown in the table. (Source: The News and Observer Publishing Company)

Year	1990	1991	1992	1993	1994	1995	1996
Salary	253	351	434	560	733	892	982

(a) Use the regression feature of a graphing utility to find the least squares regression line that fits this data. [Let y represent the average salary (in thousands of dollars) and let $t = 0$ represent 1990.]

(b) Sketch a scatter plot of the data and graph the linear model you found in part (a) on the same set of axes.

(c) Use the model to estimate the average salaries in 1997, 1998, and 1999.

(d) Use your school's library or some other reference source to analyze the accuracy of the salary estimates in part (c).

70. *Sports* The lengths (in feet) of the winning men's discus throws in the Olympics from 1904 to 1996 are listed below. (Source: ESPN)

1904	128.9	1936	165.6	1972	211.3
1908	134.2	1948	173.2	1976	221.4
1912	148.3	1952	180.5	1980	218.7
1920	146.6	1956	184.9	1984	218.5
1924	151.3	1960	194.2	1988	225.8
1928	155.3	1964	200.1	1992	213.7
1932	162.3	1968	212.5	1996	227.7

(a) Use the regression feature of a graphing utility to find the least squares regression line that fits this data. [Let y represent the length of the winning discus throw (in feet) and let $t = 4$ represent 1904.]

(b) Sketch a scatter plot of the data and graph the linear model you found in part (a) on the same set of axes.

(c) Use the model to estimate the winning men's discus throw in the year 2000.

(d) Use your school's library or some other reference source to analyze the accuracy of the estimate in part (c).

71. *Business* The total assets (in millions of dollars) for First Virginia Banks, Inc. from 1990 to 1998 are listed below. (Source: First Virginia Banks, Inc.)

1990	5384.2	1993	7036.9	1996	8236.1
1991	6119.3	1994	7865.4	1997	9011.6
1992	6840.6	1995	8221.5	1998	9564.7

(a) Use the regression feature of a graphing utility to find the least squares regression line that fits this data. [Let y represent the total assets (in millions of dollars) and let $t = 0$ represent 1990.]

(b) Use a graphing utility to sketch a scatter plot of the data and the graph of the model in the same viewing window.

(c) Use the model to estimate the assets of First Virginia Banks, Inc. in 1999.

(d) Use your school's library or some other reference source to analyze the accuracy of the estimate in part (c).

72. *Energy* The table gives the oil production x (in thousands of barrels per day) in Canada and the oil production y (in thousands of barrels per day) in the United States for the years 1991 through 1996. (Source: U.S. Energy Information Administration)

x	1548	1605	1679	1746	1805	1837
y	7417	7171	6847	6662	6560	6465

(a) Use the regression feature of a graphing utility to find the least squares regression line that fits this data.

(b) Sketch a scatter plot of the data and graph the linear model on the same set of axes.

(c) Use the model to estimate oil production in the United States if oil production in Canada is 2000 thousand barrels per day.

(d) Interpret the meaning of the slope of the linear model in the context of the problem.

🖩 **73.** *Sales* The table gives the amounts x (in millions of dollars) of home computer sales by factories and the amounts y (in millions of dollars) of personal word processor sales by factories for the years 1991 through 1996 in the United States. (Source: Electronic Industries Association)

x	4287	6825	8190	10,088	12,600	15,040
y	600	555	558	504	451	404

(a) Use the regression feature of a graphing utility to find the least squares regression line that fits this data.

(b) Sketch a scatter plot of the data and graph the linear model on the same set of axes.

(c) Use the model to estimate the amount of personal word processor sales if the amount of home computer sales is $18,000 million.

Synthesis

True or False? In Exercises 74 and 75, decide whether the statement is true or false. Justify your answer.

74. If y varies directly as x, then if x increases, y will increase as well.

75. In the equation for kinetic energy, $E = \frac{1}{2}mv^2$, the amount of kinetic energy E is directly proportional to the mass m of an object and the square of its velocity v.

In Exercises 76–79, discuss how well the data shown in the scatter plot can be approximated by a linear model.

76.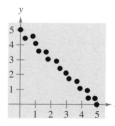

77.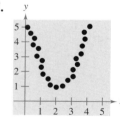

78.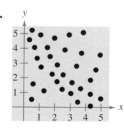

79.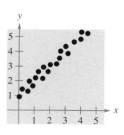

In Exercises 80–83, sketch the line that you think best approximates the data in the scatter plot. Then find an equation of the line.

80.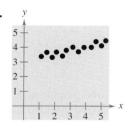

81.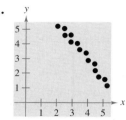

82.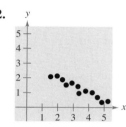

83.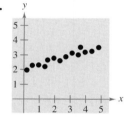

84. *Writing* A linear mathematical model for predicting prize winnings at a race is based on data for 3 years. Write a paragraph discussing the potential accuracy or inaccuracy of such a model.

Review

In Exercises 85–88, solve the inequality and graph the solution on the real number line.

85. $(x - 5)^2 \geq 1$

86. $3(x + 1)(x - 3) < 0$

87. $6x^3 - 30x^2 > 0$

88. $x^4(x - 8) \geq 0$

In Exercises 89 and 90, evaluate the function at each value of the independent variable and simplify.

89. $f(x) = \dfrac{x^2 + 5}{x - 3}$

    (a) $f(0)$ (b) $f(-3)$ (c) $f(4)$

90. $f(x) = \begin{cases} -x^2 + 10, & x \geq -2 \\ 6x^2 - 1, & x < -2 \end{cases}$

    (a) $f(-2)$ (b) $f(1)$ (c) $f(-8)$

In Exercises 91–94, find the domain of the function.

91. $f(x) = -10x^2 - x - 1$ **92.** $f(x) = \sqrt[3]{x - 2}$

93. $f(x) = \dfrac{x - 1}{x + 7}$ **94.** $f(x) = \dfrac{\sqrt{x - 3}}{x - 6}$

Chapter Summary

What did you learn?

Section 3.1	Review Exercises
☐ How to analyze graphs of quadratic functions	1–6
☐ How to write quadratic functions in standard form and use the results to sketch graphs of functions	7–18
☐ How to use quadratic functions to model and solve real-life problems	19–23

Section 3.2	
☐ How to use transformations to sketch graphs of polynomial functions	24–29
☐ How to use the Leading Coefficient Test to determine the end behavior of graphs of polynomial functions	30–37
☐ How to use zeros of polynomial functions as sketching aids	38–43
☐ How to use the Intermediate Value Theorem to help locate zeros of polynomial functions	44–47

Section 3.3	
☐ How to use long division to divide polynomials by other polynomials	48–53
☐ How to use synthetic division to divide polynomials by binomials of the form $(x - k)$	54–61
☐ How to use the Remainder Theorem and the Factor Theorem	62–65
☐ How to use polynomial division to answer questions about real-life problems	66

Section 3.4	
☐ How to use the Fundamental Theorem of Algebra to determine the number of zeros of polynomial functions	67–72
☐ How to find rational zeros of polynomial functions	73–80
☐ How to find conjugate pairs of complex zeros	81, 82
☐ How to find zeros of polynomials by factoring	83–86
☐ How to use additional properties of zeros of polynomial functions	87–92

Section 3.5	
☐ How to use mathematical models to approximate sets of data points	93
☐ How to write mathematical models for direct variation	94
☐ How to write mathematical models for direct variation as an nth power	95, 96
☐ How to write mathematical models for inverse variation	97
☐ How to write mathematical models for joint variation	98
☐ How to use the least squares regression feature of a graphing utility to find mathematical models	99

Review Exercises

3.1 In Exercises 1–4, find the quadratic function that has the indicated vertex and whose graph passes through the given point.

1.

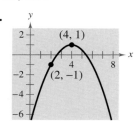

2.

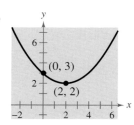

3. Vertex: $(1, -4)$; Point: $(2, -3)$

4. Vertex: $(2, 3)$; Point: $(-1, 6)$

Graphical Reasoning In Exercises 5 and 6, use a graphing utility to graph each equation in the same viewing window with the graph of $y = x^2$. Describe how each graph differs from the graph of $y = x^2$.

5. (a) $y = 2x^2$ (b) $y = -2x^2$
    (c) $y = x^2 + 2$ (d) $y = (x + 2)^2$

6. (a) $y = x^2 - 4$ (b) $y = 4 - x^2$
    (c) $y = (x - 3)^2$ (d) $y = \frac{1}{2}x^2 - 1$

In Exercises 7–18, write the quadratic function in standard form and sketch its graph.

7. $g(x) = x^2 - 2x$ **8.** $f(x) = 6x - x^2$

9. $f(x) = x^2 + 8x + 10$ **10.** $h(x) = 3 + 4x - x^2$

11. $f(t) = -2t^2 + 4t + 1$ **12.** $f(x) = x^2 - 8x + 12$

13. $h(x) = 4x^2 + 4x + 13$ **14.** $f(x) = x^2 - 6x + 1$

15. $h(x) = x^2 + 5x - 4$ **16.** $f(x) = 4x^2 + 4x + 5$

17. $f(x) = \frac{1}{3}(x^2 + 5x - 4)$

18. $f(x) = \frac{1}{2}(6x^2 - 24x + 22)$

19. *Numerical, Graphical, and Analytical Analysis* A rectangle is inscribed in the region bounded by the x-axis, the y-axis, and the graph of $x + 2y - 8 = 0$, as shown in the figure.

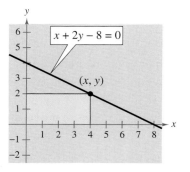

(a) Complete six rows of a table such as the one below.

x	y	Area
1	$4 - \frac{1}{2}(1)$	$(1)\left[4 - \frac{1}{2}(1)\right] = \frac{7}{2}$
2	$4 - \frac{1}{2}(2)$	$(2)\left[4 - \frac{1}{2}(2)\right] = 6$

(b) Use a graphing utility to generate additional rows of the table. Use the table to estimate the dimensions that will produce the maximum area.

(c) Write the area A as a function of x. Determine the domain of the function in the context of the problem.

(d) Use a graphing utility to graph the area function. Use the graph to approximate the dimensions that will produce the maximum area.

(e) Write the area function in standard form to find analytically the dimensions that will produce the maximum area.

20. *Geometry* The perimeter of a rectangle is 200 meters.

(a) Draw a rectangle that gives a visual representation of the problem. Label the length and width in terms of x and y, respectively.

(b) Write y as a function of x. Use the result to write the area as a function of x.

(c) Of all possible rectangles with perimeters of 200 meters, find the dimensions of the one with the maximum area.

21. *Maximum Profit* A real estate office handles 50 apartment units. When the rent is $540 per month, all units are occupied. However, for each $30 increase in rent, one unit becomes vacant. Each occupied unit requires an average of $18 per month for service and repairs. What rent should be charged to obtain the maximum profit?

22. *Minimum Cost* A manufacturer has daily production costs of

$$C = 70,000 - 120x + 0.055x^2$$

where C is the total cost (in dollars) and x is the number of units produced. How many units should be produced each day to yield a minimum cost?

23. *Sociology* The average age of the groom at a wedding for a given age of the bride can be approximated by the model $y = -0.00428x^2 + 1.442x - 3.136$, $20 \le x \le 55$, where y is the age of the groom and x is the age of the bride. For what age of the bride is the average age of the groom 30? (Source: U.S. National Center for Health Statistics)

3.2 In Exercises 24–29, sketch the graphs of $y = x^n$ and the transformation.

24. $y = x^3$, $f(x) = -(x-4)^3$
25. $y = x^3$, $f(x) = -4x^3$
26. $y = x^4$, $f(x) = 2 - x^4$
27. $y = x^4$, $f(x) = 2(x-2)^4$
28. $y = x^5$, $f(x) = (x-3)^5$
29. $y = x^5$, $f(x) = \frac{1}{2}x^5 + 3$

Graphical Analysis In Exercises 30–33, use a graphing utility to graph the functions f and g in the same viewing window. Zoom out sufficiently far to show that the right-hand and left-hand behaviors of f and g appear identical. Describe the viewing window.

30. $f(x) = \frac{1}{2}x^3 - 2x + 1$, $g(x) = \frac{1}{2}x^3$
31. $f(x) = -\frac{1}{5}(x^3 - 6x^2 + 15)$, $g(x) = -\frac{1}{5}x^3$
32. $f(x) = -x^4 + 2x^3$, $g(x) = -x^4$
33. $f(x) = x^5 - 5x$, $g(x) = x^5$

In Exercises 34–37, determine the right-hand and left-hand behavior of the graph of the polynomial function.

34. $f(x) = -x^2 + 6x + 9$
35. $f(x) = \frac{1}{2}x^3 + 2x$
36. $g(x) = \frac{3}{4}(x^4 + 3x^2 + 2)$
37. $h(x) = -x^5 - 7x^2 + 10x$

In Exercises 38–43, find the zeros of the function and sketch its graph.

38. $f(x) = 2x^2 + 11x - 21$
39. $f(x) = x(x+3)^2$
40. $f(t) = t^3 - 3t$
41. $f(x) = x^3 - 8x^2$
42. $f(x) = -12x^3 + 20x^2$
43. $g(x) = x^4 - x^3 - 2x^2$

In Exercises 44–47, use the Intermediate Value Theorem and a graphing utility to find intervals 1 unit in length in which the polynomial function is guaranteed to have a zero.

44. $f(x) = 3x^3 - x^2 + 3$
45. $f(x) = 0.25x^3 - 3.65x + 6.12$
46. $f(x) = x^4 - 5x - 1$
47. $f(x) = 7x^4 + 3x^3 - 8x^2 + 2$

3.3 In Exercises 48–53, use long division to divide.

48. $\dfrac{24x^2 - x - 8}{3x - 2}$

49. $\dfrac{4x + 7}{3x - 2}$

50. $\dfrac{5x^3 - 13x^2 - x + 2}{x^2 - 3x + 1}$

51. $\dfrac{3x^4}{x^2 - 1}$

52. $\dfrac{x^4 - 3x^3 + 4x^2 - 6x + 3}{x^2 + 2}$

53. $\dfrac{6x^4 + 10x^3 + 13x^2 - 5x + 2}{2x^2 - 1}$

In Exercises 54–57, use synthetic division to divide.

54. $\dfrac{6x^4 - 4x^3 - 27x^2 + 18x}{x - \frac{2}{3}}$

55. $\dfrac{0.1x^3 + 0.3x^2 - 0.5}{x - 5}$

56. $\dfrac{2x^3 - 19x^2 + 38x + 24}{x - 4}$

57. $\dfrac{3x^3 + 20x^2 + 29x - 12}{x + 3}$

In Exercises 58 and 59, use synthetic division to determine whether the given values of x are zeros of the function.

58. $f(x) = 20x^4 + 9x^3 - 14x^2 - 3x$
(a) $x = -1$ (b) $x = \frac{3}{4}$
(c) $x = 0$ (d) $x = 1$

59. $f(x) = 3x^3 - 8x^2 - 20x + 16$
(a) $x = 4$ (b) $x = -4$
(c) $x = \frac{2}{3}$ (d) $x = -1$

In Exercises 60 and 61, use synthetic division to find the specified value of the function.

60. $f(x) = x^4 + 10x^3 - 24x^2 + 20x + 44$

    (a) $f(-3)$ (b) $f(-1)$

61. $g(t) = 2t^5 - 5t^4 - 8t + 20$

    (a) $g(-4)$ (b) $g(\sqrt{2})$

In Exercises 62–65, (a) verify the given factor(s) of the function f, (b) find the remaining factors of f, (c) use your results to write the complete factorization of f, (d) list all real zeros of f, and ⊞ (e) confirm your results by using a graphing utility to graph the function.

	Function	*Factors*
62.	$f(x) = x^3 + 4x^2 - 25x - 28$	$(x - 4)$
63.	$f(x) = 2x^3 + 11x^2 - 21x - 90$	$(x + 6)$
64.	$f(x) = x^4 - 4x^3 - 7x^2 + 22x + 24$	$(x + 2)(x - 3)$
65.	$f(x) = x^4 - 11x^3 + 41x^2 - 61x + 30$	$(x - 2)(x - 5)$

⊞ **66.** *Data Analysis*   The values V (in billions of dollars) of farm real estate in the United States for the years 1990 through 1997 are shown in the table. The variable t represents the year, with $t = 0$ corresponding to 1990.   (Source: U.S. Department of Agriculture)

t	0	1	2	3
V	671.4	688	695.5	717.1

t	4	5	6	7
V	759.2	807	860.9	912.3

(a) Use a graphing utility to sketch a scatter plot of the data.

(b) Use the regression feature of a graphing utility to find a cubic model for the given data. Then graph the model in the same viewing window as the scatter plot. Compare the model with the data.

(c) Use the model to create a table of estimated values of V. Compare the estimated values with the actual data.

(d) Use synthetic division to evaluate the model for the year 2001. Do you think the model is accurate in predicting the future value of farm real estate? Explain.

3.4 **In Exercises 67–72, find all the zeros of the function.**

67. $f(x) = 3x(x - 2)^2$ **68.** $f(x) = (x - 4)(x + 9)^2$

69. $f(x) = x^2 - 9x + 8$

70. $f(x) = x^3 + 6x$

71. $f(x) = (x + 4)(x - 6)(x - 2i)(x + 2i)$

72. $f(x) = (x - 8)(x - 5)^2(x - 3 + i)(x - 3 - i)$

In Exercises 73 and 74, use the Rational Zero Test to list all possible rational zeros of f.

73. $f(x) = -4x^3 + 8x^2 - 3x + 15$

74. $f(x) = 3x^4 + 4x^3 - 5x^2 - 8$

In Exercises 75–80, find all the real zeros of the function.

75. $f(x) = x^3 - 2x^2 - 21x - 18$

76. $f(x) = 3x^3 - 20x^2 + 7x + 30$

77. $f(x) = x^3 - 10x^2 + 17x - 8$

78. $f(x) = x^3 + 9x^2 + 24x + 20$

79. $f(x) = x^4 + x^3 - 11x^2 + x - 12$

80. $f(x) = 25x^4 + 25x^3 - 154x^2 - 4x + 24$

In Exercises 81 and 82, find a polynomial with real coefficients that has the given zeros.

81. $\frac{2}{3}, 4, \sqrt{3}i$

82. $2, -3, 1 - 2i$

In Exercises 83–86, find all the zeros of the function.

83. $f(x) = 4x^3 - 11x^2 + 10x - 3$

84. $f(x) = x^3 - 1.3x^2 - 1.7x + 0.6$

85. $f(x) = 6x^4 - 25x^3 + 14x^2 + 27x - 18$

86. $f(x) = 5x^4 + 126x^2 + 25$

⊞ **In Exercises 87–90, use a graphing utility to (a) graph the function, (b) determine the number of real zeros of the function, and (c) approximate the real zeros of the function to the nearest hundredth.**

87. $f(x) = x^4 + 2x + 1$

88. $g(x) = x^3 - 3x^2 + 3x + 2$

89. $h(x) = x^3 - 6x^2 + 12x - 10$

90. $f(x) = x^5 + 2x^3 - 3x - 20$

In Exercises 91 and 92, use Descartes's Rule of Signs to determine the possible numbers of positive and negative zeros of the function.

91. $g(x) = 5x^3 + 3x^2 - 6x + 9$

92. $h(x) = -2x^5 + 4x^3 - 2x^2 + 5$

3.5 **93.** *Data Analysis*   The sales S (in billions of dollars) of recreational vehicles in the United States for the years 1988 through 1997 are shown in the table. (Source: National Sporting Goods Association)

Year	8	9	10	11	12
S	4.8	4.5	4.1	3.6	4.4

Year	13	14	15	16	17
S	4.8	5.7	5.9	6.3	6.5

A model for this data is

$S = 6.7 + 2.60t - 0.742t^2 + 0.0611t^3 - 0.00156t^4$

where t is the time in years, with $t = 8$ corresponding to 1988.

(a) Use a graphing utility to sketch a scatter plot of the data and the model in the same viewing window. How do they compare?

(b) The table shows that sales were down from 1989 through 1991. Give a possible explanation. Does the model show the downturn in sales?

(c) Use a graphing utility to approximate the magnitude of the decrease in sales during the slump described in part (b). Was the actual decrease more or less than indicated by the model?

(d) Use the model to estimate sales in 2001. Is this model accurate in predicting future sales? Explain.

94. *Measurement*   You notice a billboard indicating that it is 2.5 miles or 4 kilometers to the next restaurant of a national fast-food chain. Use this information to find a linear model that relates miles to kilometers. Use the model to complete the table.

Miles	2	5	10	12
Kilometers				

95. *Energy*   The power P produced by a wind turbine is proportional to the cube of the wind speed S. A wind speed of 27 miles per hour produces a power output of 750 kilowatts. Find the output for a wind speed of 40 miles per hour.

96. *Frictional Force*   The frictional force F between the tires and the road required to keep a car on a curved section of a highway is directly proportional to the square of the speed s of the car. If the speed of the car is doubled, the force will change by what factor?

In Exercises 97 and 98, find a mathematical model representing the statement. (In each case, determine the constant of proportionality.)

97. y is inversely proportional to x. ($y = 9$ when $x = 5.5$.)

98. F is jointly proportional to x and the square root of y. ($F = 6$ when $x = 9$ and $y = 4$.)

99. *Employment*   The table shows the average hourly wages (y_1) for workers in the mining industry and the average hourly wages (y_2) for workers in the construction industry in the United States for the years 1994 through 1997, where t is the time in years, with $t = 4$ corresponding to 1994.   (Source: U.S. Bureau of Labor Statistics)

t	4	5	6	7
y_1	$14.89	$15.30	$15.60	$16.17
y_2	$14.69	$15.08	$15.43	$16.03

(a) Use the regression feature of a graphing utility to find the least squares regression lines for mining wages versus time and for construction wages versus time.

(b) Use a graphing utility to sketch a scatter plot of the data. Graph the linear models you found in part (a) on the same set of axes.

(c) Interpret the slope of each model in the context of the problem.

(d) Use the models to estimate the wages in each industry for the year 2002.

Synthesis

True or False?   **In Exercises 100 and 101, determine whether the statement is true or false. Justify your answer.**

100. A fourth-degree polynomial can have -5, $-8i$, $4i$, and 5 as its zeros.

101. If y is directly proportional to x, then x is directly proportional to y.

Chapter Project ▶ Finding Points of Intersection Graphically

Example ▶ Approximating Points of Intersection

Approximate the points of intersection of the circle and parabola given by

$$x^2 + y^2 - 3x + 5y - 11 = 0 \quad \text{and} \quad y = x^2 - 4x + 5$$

using the zoom and trace features of a graphing utility.

Solution

Begin by writing the circle as the union of two functions.

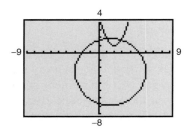

FIGURE 3.34

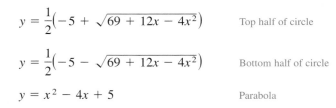

$$y = \frac{1}{2}\left(-5 + \sqrt{69 + 12x - 4x^2}\right) \qquad \text{Top half of circle}$$

$$y = \frac{1}{2}\left(-5 - \sqrt{69 + 12x - 4x^2}\right) \qquad \text{Bottom half of circle}$$

$$y = x^2 - 4x + 5 \qquad \text{Parabola}$$

Next, graph all three functions in the same viewing window, as shown in Figure 3.34. Use the zoom and trace features of the graphing utility to estimate that the points of intersection are roughly $(1, 1.9)$ and $(2.8, 1.7)$.

Chapter Project Investigations

1. Using a setting of $1.05 \le x \le 1.06$ and $1.89 \le y \le 1.90$, graph the top half of the circle and the parabola in the example on the same screen. Then use the trace feature to approximate (accurate to three decimal places) the y-coordinate of the point of intersection that is shown on the screen.

2. Another method for finding the points of intersection is to substitute $x^2 - 4x + 5$ for y in the equation of the circle to get a fourth-degree polynomial equation. Graph this polynomial function.

   (a) Find a setting that allows you to approximate the solution $x \approx 1.055$ of the polynomial equation to two more decimal places.

   (b) Find a setting that allows you to approximate the solution $x \approx 2.841$ to two more decimal places.

3. Use a graphing utility to find the points of intersection of the circle and the parabola given by $x^2 + y^2 - 5x + 4y - 13 = 0$ and $y = x^2 - 3x + 2$.

4. The *market equilibrium* of a commodity is the quantity (and corresponding price) at which the supply of the commodity and the demand for the commodity are equal. The supply and demand curves for a business dealing with wheat are $p = 1.45 + 0.00014x^2$ and $p = (2.388 - 0.007x)^2$, respectively, where p is the price (in dollars per bushel) and x is the quantity (in bushels per day). Use a graphing utility to graph the supply and demand equations and find the market equilibrium. (*Hint:* The market equilibrium is the point of intersection of the graphs for $x > 0$.)

Chapter Test

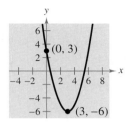

(0, 3)

(3, −6)

FIGURE FOR **3**

The *Interactive* CD-ROM and *Internet* versions of this text provide answers to the Chapter Tests and Cumulative Tests. They also offer Chapter Pre-Tests (which test key skills and concepts covered in previous chapters) and Chapter Post-Tests, both of which have randomly generated exercises with diagnostic capabilities.

Take this test as you would take a test in class. After you are done, check your work against the answers given in the back of the book.

1. Describe how the graph of g differs from the graph of $f(x) = x^2$.

   (a) $g(x) = 2 - x^2$ (b) $g(x) = \left(x - \frac{3}{2}\right)^2$

2. Identify the vertex and intercepts of the graph of $y = x^2 + 4x + 3$.

3. Find an equation of the parabola shown in the figure at the left.

4. The path of a ball is given by $y = -\frac{1}{20}x^2 + 3x + 5$, where y is the height (in feet) of the ball and x is the horizontal distance (in feet) from where the ball was thrown.

   (a) Find the maximum height of the ball.

   (b) Which constant determines the height at which the ball was thrown? Does changing this constant change the coordinates of the maximum height of the ball? Explain.

5. Determine the right-hand and left-hand behavior of the graph of the function $h(t) = -\frac{3}{4}t^5 + 2t^2$. Then sketch its graph.

6. Divide by long division.

$$\frac{3x^3 + 4x - 1}{x^2 + 1}$$

7. Divide by synthetic division.

$$\frac{2x^4 - 5x^2 - 3}{x - 2}$$

8. Use synthetic division to show that $x = \sqrt{3}$ is a solution of the equation $4x^3 - x^2 - 12x + 3 = 0$. Use the result to factor the polynomial completely and list all the real solutions of the equation.

In Exercises 9 and 10, list all the possible rational zeros of the function. Use a graphing utility to graph the function and find all the rational zeros.

9. $g(t) = 2t^4 - 3t^3 + 16t - 24$ **10.** $h(x) = 3x^5 + 2x^4 - 3x - 2$

In Exercises 11 and 12, use the root-finding capabilities of a graphing utility to approximate the real zeros of the function, accurate to three decimal places.

11. $f(x) = x^4 - x^3 - 1$ **12.** $f(x) = 3x^5 + 2x^4 - 12x - 8$

In Exercises 13 and 14, find a polynomial function with integer coefficients that has the given zeros.

13. $0, 3, 3 + i, 3 - i$ **14.** $1 + \sqrt{3}i, 1 - \sqrt{3}i, 2, 2$

In Exercises 15–17, find a mathematical model representing the statement. (In each case, determine the constant of proportionality.)

15. v varies directly as the square root of s. ($v = 24$ when $s = 16$.)

16. A varies jointly as x and y. ($A = 500$ when $x = 15$ and $y = 8$.)

17. b varies inversely as a. ($b = 32$ when $a = 1.5$.)

Garry Hunter/Tony Stone Images

Compact discs (CDs) can be used to store data, sound, photographs, and video. The information stored is digitally encoded as a series of microscopic pits on a polished surface. The information is then read from the CD using a laser beam.

4 Rational Functions and Conics

▶ How to Study This Chapter

The Big Picture

In this chapter you will learn the following skills and concepts.

▶ How to determine the domains of rational functions and find asymptotes of rational functions.

▶ How to sketch the graphs of rational functions.

▶ How to recognize and find partial fraction decompositions of rational expressions.

▶ How to recognize, graph, and write equations of circles, ellipses, parabolas, and hyperbolas (vertex or center at origin).

▶ How to recognize, graph, and write equations of conics that have been shifted vertically or horizontally in the plane.

Important Vocabulary

As you encounter each new vocabulary term in this chapter, add the term and its definition to your notebook glossary.

Rational function (p. 324)
Vertical asymptote (p. 325)
Horizontal asymptote (p. 325)
Slant (or oblique) asymptote (p. 336)
Partial fraction (p. 342)
Partial fraction decomposition (p. 342)
Basic equation (p. 343)
Conic section or conic (p. 350)
Degenerate conic (p. 350)
Parabola (p. 351)
Directrix (p. 351)
Focus (p. 351)
Standard form of the equation of a parabola (p. 351)

Ellipse (p. 353)
Foci (p. 353)
Vertices (p. 353)
Major axis (p. 353)
Center (p. 353)
Minor axis (p. 353)
Standard form of the equation of an ellipse (p. 353)
Hyperbola (p. 355)
Branches (p. 355)
Transverse axis (p. 355)
Standard form of the equation of a hyperbola (p. 355)
Conjugate axis (p. 356)
Asymptotes of a hyperbola (p. 357)

Study Tools

- Learning objectives at the beginning of each section
- Chapter Summary (p. 371)
- Review Exercises (pp. 372–375)
- Chapter Test (p. 377)

Additional Resources

- Study and Solutions Guide
- Interactive College Algebra
- Videotapes for Chapter 4
- College Algebra Website
- Student Success Organizer

STUDY T!P

Don't let yourself fall behind in the course. If you are having trouble, seek help immediately—from your instructor, a math tutor, a study partner, or additional study aids such as videotapes and tutorial software.

4.1   Rational Functions and Asymptotes

▶ **What you should learn**

• How to find the domains of rational functions
• How to find the horizontal and vertical asymptotes of graphs of rational functions
• How to use rational functions to model and solve real-life problems

▶ **Why you should learn it**

Rational functions can be used to model and solve real-life problems relating to the environment. For instance, Exercise 35 on page 331 shows how a rational function can be used to model the cost of removing pollutants from a river.

David Woodfall/Tony Stone Images

Introduction

A **rational function** can be written in the form

$$f(x) = \frac{N(x)}{D(x)}$$

where $N(x)$ and $D(x)$ are polynomials and $D(x)$ is not the zero polynomial. In this section it is assumed that $N(x)$ and $D(x)$ have no common factors.

In general, the *domain* of a rational function of x includes all real numbers except x-values that make the denominator zero. Much of the discussion of rational functions will focus on their graphical behavior near these x-values.

Example 1 ▶   Finding the Domain of a Rational Function

Find the domain of $f(x) = 1/x$ and discuss the behavior of f near any excluded x-values.

Solution

Because the denominator is zero when $x = 0$, the domain of f is all real numbers except $x = 0$. To determine the behavior of f near this excluded value, evaluate $f(x)$ to the left and right of $x = 0$, as indicated in the following tables.

x	-1	-0.5	-0.1	-0.01	-0.001	⟶ 0
$f(x)$	-1	-2	-10	-100	-1000	⟶ $-\infty$

x	0 ⟵		0.001	0.01	0.1	0.5	1
$f(x)$	∞ ⟵		1000	100	10	2	1

Note that as x approaches 0 *from the left*, $f(x)$ decreases without bound. In contrast, as x approaches 0 *from the right*, $f(x)$ increases without bound. The graph of f is shown in Figure 4.1.

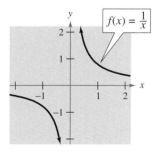

FIGURE **4.1**

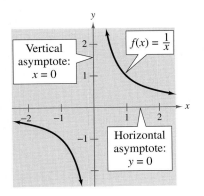

FIGURE **4.2**

Horizontal and Vertical Asymptotes

In Example 1, the behavior of f near $x = 0$ is denoted as follows.

$$f(x) \longrightarrow -\infty \text{ as } x \longrightarrow 0^- \qquad f(x) \longrightarrow \infty \text{ as } x \longrightarrow 0^+$$

$f(x)$ decreases without bound as x approaches 0 from the left. $f(x)$ increases without bound as x approaches 0 from the right.

The line $x = 0$ is a **vertical asymptote** of the graph of f, as shown in Figure 4.2. From this figure, you can see that the graph of f also has a **horizontal asymptote**— the line $y = 0$. This means that the values of $f(x) = 1/x$ approach zero as x increases or decreases without bound.

$$f(x) \longrightarrow 0 \text{ as } x \longrightarrow -\infty \qquad f(x) \longrightarrow 0 \text{ as } x \longrightarrow \infty$$

$f(x)$ approaches 0 as x decreases without bound. $f(x)$ approaches 0 as x increases without bound.

Definition of Vertical and Horizontal Asymptotes

1. The line $x = a$ is a **vertical asymptote** of the graph of f if

$$f(x) \longrightarrow \infty \quad \text{or} \quad f(x) \longrightarrow -\infty$$

as $x \longrightarrow a$, either from the right or from the left.

2. The line $y = b$ is a **horizontal asymptote** of the graph of f if

$$f(x) \longrightarrow b$$

as $x \longrightarrow \infty$ or $x \longrightarrow -\infty$.

Eventually (as $x \longrightarrow \infty$ or $x \longrightarrow -\infty$), the distance between the horizontal asymptote and the points on the graph must approach zero. Figure 4.3 shows the horizontal and vertical asymptotes of the graphs of three rational functions.

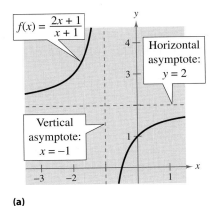

(a)

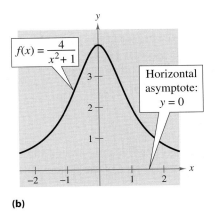

(b)

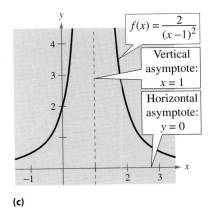

(c)

FIGURE **4.3**

The graphs of $f(x) = 1/x$ in Figure 4.2 and $f(x) = (2x + 1)/(x + 1)$ in Figure 4.3(a) are **hyperbolas.** You will study hyperbolas in Sections 4.4 and 4.5.

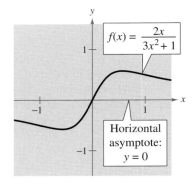

(a)

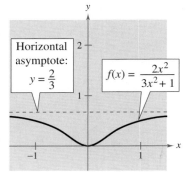

(b)

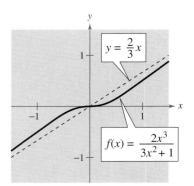

(c)

FIGURE **4.4**

Asymptotes of a Rational Function

Let f be the rational function given by

$$f(x) = \frac{N(x)}{D(x)} = \frac{a_n x^n + a_{n-1} x^{n-1} + \cdots + a_1 x + a_0}{b_m x^m + b_{m-1} x^{m-1} + \cdots + b_1 x + b_0}$$

where $N(x)$ and $D(x)$ have no common factors.

1. The graph of f has *vertical* asymptotes at the zeros of $D(x)$.

2. The graph of f has one or no *horizontal* asymptote determined by comparing the degrees of $N(x)$ and $D(x)$.

   a. If $n < m$, the graph of f has the line $y = 0$ (the x-axis) as a horizontal asymptote.

   b. If $n = m$, the graph of f has the line $y = a_n/b_m$ as a horizontal asymptote.

   c. If $n > m$, the graph of f has no horizontal asymptote.

Example 2 ▶ Finding Horizontal Asymptotes

a. The graph of

$$f(x) = \frac{2x}{3x^2 + 1}$$

has the line $y = 0$ (the x-axis) as a horizontal asymptote, as shown in Figure 4.4(a). Note that the degree of the numerator is *less than* the degree of the denominator.

b. The graph of

$$f(x) = \frac{2x^2}{3x^2 + 1}$$

has the line $y = \frac{2}{3}$ as a horizontal asymptote, as shown in Figure 4.4(b). Note that the degree of the numerator is *equal to* the degree of the denominator, and the horizontal asymptote is given by the ratio of the leading coefficients of the numerator and denominator.

c. The graph of

$$f(x) = \frac{2x^3}{3x^2 + 1}$$

has no horizontal asymptote because the degree of the numerator is greater than the degree of the denominator. See Figure 4.4(c).

Although the graph of the function in Example 2(c) does not have a horizontal asymptote, it does have a *slant asymptote*—the line $y = \frac{2}{3}x$. You will study slant asymptotes in the next section.

Applications

There are many examples of asymptotic behavior in real life. For instance, Example 3 shows how a vertical asymptote can be used to analyze the cost of removing pollutants from smokestack emissions.

Example 3 ▶ Cost-Benefit Model

A utility company burns coal to generate electricity. The cost of removing a certain *percent* of the pollutants from smokestack emissions is typically not a linear function. That is, if it costs C dollars to remove 25% of the pollutants, it would cost more than $2C$ dollars to remove 50% of the pollutants. As the percent of removed pollutants approaches 100%, the cost tends to increase without bound, becoming prohibitive. Suppose that the cost C (in dollars) of removing $p\%$ of the smokestack pollutants is $C = 80{,}000p/(100 - p)$ for $0 \le p < 100$. Sketch the graph of this function. Suppose you are a member of a state legislature considering a law that would require utility companies to remove 90% of the pollutants from their smokestack emissions. If the current law requires 85% removal, how much additional cost would the utility company incur as a result of the new law?

Solution

The graph of this function is shown in Figure 4.5. Note that the graph has a vertical asymptote at $p = 100$. Because the current law requires 85% removal, the current cost to the utility company is

$$C = \frac{80{,}000(85)}{100 - 85} \approx \$453{,}333. \qquad \text{Evaluate } C \text{ when } p = 85.$$

If the new law increases the percent removal to 90%, the cost to the utility company will be

$$C = \frac{80{,}000(90)}{100 - 90} = \$720{,}000. \qquad \text{Evaluate } C \text{ when } p = 90.$$

So, the new law would require the utility company to spend an additional

$$720{,}000 - 453{,}333 = \$266{,}667. \qquad \begin{array}{l}\text{Subtract 85\% removal cost} \\ \text{from 90\% removal cost.}\end{array}$$

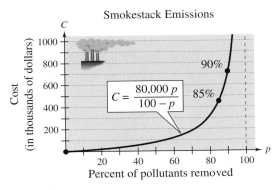

FIGURE **4.5**

Average Cost of a Product

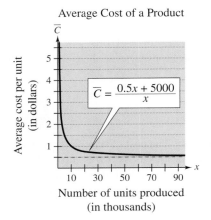

FIGURE 4.6

Example 4 ▶   **Average Cost of Producing a Product**

A business has a cost function of $C = 0.5x + 5000$, where C is measured in dollars and x is the number of units produced. The *average cost per unit* is

$$\overline{C} = \frac{C}{x} = \frac{0.5x + 5000}{x}.$$

Find the average cost per unit when $x = 1000$, 5000, $10{,}000$, and $100{,}000$. What is the horizontal asymptote for this function, and what does it represent?

Solution

When $x = 1000$, $\overline{C} = \dfrac{0.5(1000) + 5000}{1000} = \$5.50.$

When $x = 5000$, $\overline{C} = \dfrac{0.5(5000) + 5000}{5000} = \$1.50.$

When $x = 10{,}000$, $\overline{C} = \dfrac{0.5(10{,}000) + 5000}{10{,}000} = \$1.00.$

When $x = 100{,}000$, $\overline{C} = \dfrac{0.5(100{,}000) + 5000}{100{,}000} = \$0.55.$

As shown in Figure 4.6, the horizontal asymptote is the line $\overline{C} = 0.50$. This line represents the least possible unit cost for the product. This example points out one of the major problems of a small business. That is, it is difficult to have competitively low prices when the production level is low.

Writing ABOUT MATHEMATICS

Common Factors in the Numerator and Denominator When sketching the graph of a rational function, be sure that the rational function has no factor that is common to its numerator and denominator. To see why, consider the function

$$f(x) = \frac{2x^2 + x - 1}{x + 1} = \frac{(x + 1)(2x - 1)}{x + 1}$$

which has a common factor of $x + 1$ in the numerator and denominator. Sketch the graph of this function. Does it have a vertical asymptote at $x = -1$?

Decide whether each function below has a vertical asymptote. Write a short paragraph to explain your reasoning. Include a graph of each function in your explanation.

a. $f(x) = \dfrac{x^2 - 4}{x + 2}$ **b.** $f(x) = \dfrac{x^2 - 4}{x}$ **c.** $f(x) = \dfrac{x^2 - 4}{2 - x}$

4.1 Exercises

In Exercises 1–6, (a) complete each table, (b) determine the vertical and horizontal asymptotes of the function, and (c) find the domain of the function.

x	$f(x)$
0.5	
0.9	
0.99	
0.999	

x	$f(x)$
1.5	
1.1	
1.01	
1.001	

x	$f(x)$
5	
10	
100	
1000	

1. $f(x) = \dfrac{1}{x - 1}$

2. $f(x) = \dfrac{5x}{x - 1}$

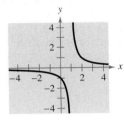

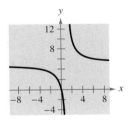

3. $f(x) = \dfrac{4x}{|x - 1|}$

4. $f(x) = \dfrac{2}{|x - 1|}$

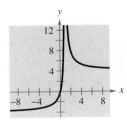

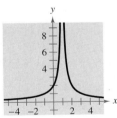

5. $f(x) = \dfrac{3x^2}{x^2 - 1}$

6. $f(x) = \dfrac{4x}{x^2 - 1}$

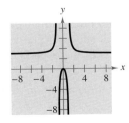

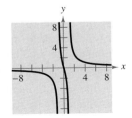

In Exercises 7–14, find the domain of the function and identify any horizontal and vertical asymptotes.

7. $f(x) = \dfrac{1}{x^2}$

8. $f(x) = \dfrac{4}{(x - 2)^3}$

9. $f(x) = \dfrac{2 + x}{2 - x}$

10. $f(x) = \dfrac{1 - 5x}{1 + 2x}$

11. $f(x) = \dfrac{x^3}{x^2 - 1}$

12. $f(x) = \dfrac{2x^2}{x + 1}$

13. $f(x) = \dfrac{3x^2 + 1}{x^2 + x + 9}$

14. $f(x) = \dfrac{3x^2 + x - 5}{x^2 + 1}$

In Exercises 15–20, match the rational function with its graph. [The graphs are labeled (a) through (f).]

(a)
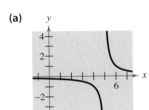

(b)
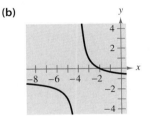

(c)
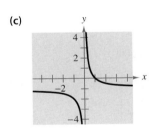

(d)
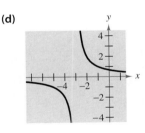

(e)
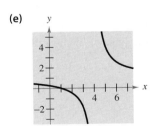

(f)
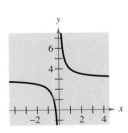

15. $f(x) = \dfrac{2}{x + 3}$

16. $f(x) = \dfrac{1}{x - 5}$

17. $f(x) = \dfrac{3x + 1}{x}$

18. $f(x) = \dfrac{1 - x}{x}$

19. $f(x) = \dfrac{x - 1}{x - 4}$

20. $f(x) = -\dfrac{x + 2}{x + 4}$

Analytical and Numerical Analysis **In Exercises 21–24, (a) determine the domain of f and g, (b) find any vertical asymptotes of f, (c) complete the table, and (d) explain how the two functions differ.**

21. $f(x) = \dfrac{x^2 - 4}{x + 2}$, $g(x) = x - 2$

x	-4	-3	-2.5	-2	-1.5	-1	0
$f(x)$							
$g(x)$							

22. $f(x) = \dfrac{x^2(x - 2)}{x^2 - 2x}$, $g(x) = x$

x	-1	-0.5	0	1	2	3	4
$f(x)$							
$g(x)$							

23. $f(x) = \dfrac{x - 2}{x^2 - 2x}$, $g(x) = \dfrac{1}{x}$

x	-1	-0.5	0	0.5	2	3	4
$f(x)$							
$g(x)$							

24. $f(x) = \dfrac{2x - 8}{x^2 - 9x + 20}$, $g(x) = \dfrac{2}{x - 5}$

x	0	1	2	3	4	5	6
$f(x)$							
$g(x)$							

In Exercises 25–28, find the zeros (if any) of the rational function.

25. $g(x) = \dfrac{x^2 - 1}{x + 1}$

26. $h(x) = 2 + \dfrac{5}{x^2 + 2}$

27. $f(x) = 1 - \dfrac{3}{x - 3}$

28. $g(x) = \dfrac{x^3 - 8}{x^2 + 1}$

Exploration **In Exercises 29–32, (a) determine the value that the function f approaches as the magnitude of x increases. Is f(x) greater than or less than this functional value when (b) x is positive and large in magnitude and (c) x is negative and large in magnitude?**

29. $f(x) = 4 - \dfrac{1}{x}$

30. $f(x) = 2 + \dfrac{1}{x - 3}$

31. $f(x) = \dfrac{2x - 1}{x - 3}$

32. $f(x) = \dfrac{2x - 1}{x^2 + 1}$

Data Analysis **In Exercises 33 and 34, consider a physics laboratory experiment designed to determine an unknown mass. A flexible metal meter stick is clamped to a table with 50 centimeters overhanging the edge (see figure). Known masses M ranging from 200 grams to 2000 grams are attached to the end of the meter stick. For each mass the meter stick is displaced vertically and then allowed to oscillate. The average duration t of one oscillation (in seconds) for each mass is recorded in the table.**

M	200	400	600	800	1000
t	0.450	0.597	0.721	0.831	0.906

M	1200	1400	1600	1800	2000
t	1.003	1.088	1.168	1.218	1.338

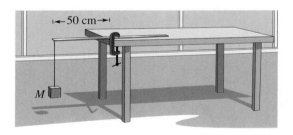

33. A model for the data that can be used to predict the time of one oscillation is

$$t = \frac{38M + 16{,}965}{10(M + 5000)}.$$

Use this model to create a table showing the predicted time for each of the masses shown in the table. Compare the predicted times with the experimental times. What can you conclude?

34. Use the model to approximate the mass of an object for which $t = 1.056$ seconds.

35. *Pollution* The cost (in millions of dollars) of removing $p\%$ of the industrial and municipal pollutants discharged into a river is

$$C = \frac{255p}{100 - p}, \qquad 0 \le p < 100.$$

(a) Find the cost of removing 10% of the pollutants.

(b) Find the cost of removing 40% of the pollutants.

(c) Find the cost of removing 75% of the pollutants.

(d) According to this model, would it be possible to remove 100% of the pollutants? Explain.

36. *Recycling* In a pilot project, a rural township is given recycling bins for separating and storing recyclable products. The cost (in dollars) for supplying bins to $p\%$ of the population is

$$C = \frac{25,000p}{100 - p}, \qquad 0 \le p < 100.$$

(a) Find the cost if 15% of the population gets bins.

(b) Find the cost if 50% of the population gets bins.

(c) Find the cost if 90% of the population gets bins.

(d) According to this model, would it be possible to supply bins to 100% of the residents? Explain.

37. *Population Growth* The game commission introduces 100 deer into newly acquired state game lands. The population N of the herd is

$$N = \frac{20(5 + 3t)}{1 + 0.04t}, \qquad t \ge 0$$

where t is the time in years.

(a) Find the population when t is 5, 10, and 25.

(b) What is the limiting size of the herd as time increases?

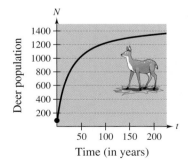

Time (in years)

38. *Food Consumption* A biology class performs an experiment comparing the quantity of food consumed by a certain kind of moth with the quantity supplied. The model for the experimental data is

$$y = \frac{1.568x - 0.001}{6.360x + 1}, \qquad x > 0$$

where x is the quantity (in milligrams) of food supplied and y is the quantity (in milligrams) eaten. At what level of consumption will the moth become satiated?

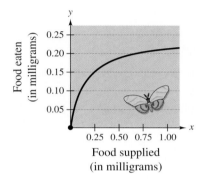

Food supplied
(in milligrams)

39. *Human Memory Model* Psychologists have developed mathematical models to predict performance as a function of the number of trials n for a certain task. Consider the learning curve

$$P = \frac{0.5 + 0.9(n - 1)}{1 + 0.9(n - 1)}, \qquad n > 0$$

where P is the fraction of correct responses after n trials.

(a) Complete the table for this model. What does it suggest?

n	1	2	3	4	5	6	7	8	9	10
P										

(b) According to this model, what is the limiting percent of correct responses as n increases?

40. *Human Memory Model* How would the limiting percent of correct responses change if the human memory model in Exercise 39 were changed to

$$P = \frac{0.5 + 0.6(n - 1)}{1 + 0.8(n - 1)}, \qquad n > 0?$$

41. *Military* The numbers of United States military reserve personnel M (in thousands) for the years 1990 through 1997 are shown in the table. (Source: U.S. Department of Defense)

Year	1990	1991	1992	1993
M	1671	1786	1883	1867

Year	1994	1995	1996	1997
M	1805	1659	1550	1461

A model for this data is

$$M = \frac{1671.92 + 130.23t}{1 - 0.02t + 0.02t^2}$$

where t is time (in years), with $t = 0$ corresponding to 1990.

(a) Use a graphing utility to plot the data and graph the model in the same viewing window.

(b) Use the model to estimate the number of military reserve personnel in 2003.

(c) Would this model be useful for estimating the number of military reserve personnel after 2003? Explain.

42. *Business Partnerships* The numbers of business partnerships P (in thousands) in the United States for the years 1988 through 1995 are shown in the table. (Source: U.S. Internal Revenue Service)

Year	1988	1989	1990	1991
P	1654	1635	1554	1515

Year	1992	1993	1994	1995
P	1485	1468	1494	1581

A model for this data is

$$P = \frac{1568.84 - 406.82t + 53.70t^2}{1 - 0.23t + 0.03t^2}$$

where t is the time (in years), with $t = 0$ corresponding to 1990.

(a) Use a graphing utility to plot the data and graph the model in the same viewing window.

(b) Use the model to estimate the number of partnerships in 2002.

(c) Would this model be useful for estimating the number of partnerships after 2002? Explain.

Synthesis

True or False? **In Exercises 43 and 44, determine whether the statement is true or false. Justify your answer.**

43. A polynomial can have infinitely many vertical asymptotes.

44. $f(x) = x^3 - 2x^2 - 5x + 6$ is a rational function.

Think About It **In Exercises 45–48, write a rational function f that has the specified characteristics.**

45. Vertical asymptotes: $x = -2, x = 1$

46. Vertical asymptote: None
Horizontal asymptote: $y = 0$

47. Vertical asymptote: None
Horizontal asymptote: $y = 2$

48. Vertical asymptotes: $x = 0, x = \frac{5}{2}$
Horizontal asymptote: $y = -3$

49. Give an example of a rational function whose domain is the set of all real numbers. Give an example of a rational function whose domain is the set of all real numbers except $x = 20$.

50. Describe what is meant by an asymptote of a graph.

Review

In Exercises 51–54, find the inverse of the function f. Then graph both f and f^{-1} in the same coordinate plane.

51. $f(x) = 8x - 7$

52. $f(x) = \frac{1}{9}x$

53. $f(x) = \frac{7 - 2x}{5}$

54. $f(x) = \frac{x - 3}{6}$

In Exercises 55–58, divide using long division.

55. $(x^2 + 5x + 6) \div (x - 4)$

56. $(x^2 - 10x + 15) \div (x - 3)$

57. $(2x^2 + x - 11) \div (x + 5)$

58. $(4x^2 + 3x - 10) \div (x + 6)$

4.2 Graphs of Rational Functions

▶ What you should learn

- How to analyze and sketch graphs of rational functions
- How to sketch graphs of rational functions that have slant asymptotes
- How to use graphs of rational functions to model and solve real-life problems

▶ Why you should learn it

You can use rational functions to model average speed over a distance. For example, see Exercise 75 on page 341.

Baron Wolman/Tony Stone Images

Analyzing Graphs of Rational Functions

Guidelines for Analyzing Graphs of Rational Functions

Let $f(x) = N(x)/D(x)$, where $N(x)$ and $D(x)$ are polynomials with no common factors.

1. Find and plot the y-intercept (if any) by evaluating $f(0)$.

2. Find the zeros of the numerator (if any) by solving the equation $N(x) = 0$. Then plot the corresponding x-intercepts.

3. Find the zeros of the denominator (if any) by solving the equation $D(x) = 0$. Then sketch the corresponding vertical asymptotes.

4. Find and sketch the horizontal asymptote (if any) by using the rule for finding the horizontal asymptote of a rational function.

5. Plot at least one point *between* and one point *beyond* each x-intercept and vertical asymptote.

6. Use smooth curves to complete the graph between and beyond the vertical asymptotes.

Testing for symmetry can be useful, especially for simple rational functions. For example, the graph of $f(x) = 1/x$ is symmetric with respect to the origin, and the graph of $g(x) = 1/x^2$ is symmetric with respect to the y-axis.

Technology

Some graphing utilities have difficulty sketching graphs of rational functions that have vertical asymptotes. Often, the utility will connect parts of the graph that are not supposed to be connected. For instance, the screen below on the left shows the graph of $f(x) = 1/(x - 2)$.

Notice that the graph should consist of two *separated* portions—one to the left of $x = 2$ and the other to the right of $x = 2$. To eliminate this problem, you can try changing the *mode* of the graphing utility to *dot mode*. The problem with this is that the graph is then represented as a collection of dots (as shown in the screen below on the right) rather than as a smooth curve.

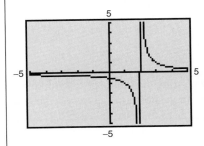
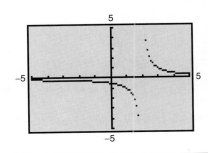

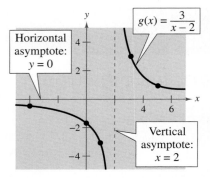

Horizontal asymptote: $y = 0$

$g(x) = \dfrac{3}{x-2}$

Vertical asymptote: $x = 2$

FIGURE **4.7**

Example 1 ▶   Sketching the Graph of a Rational Function

Sketch the graph of the function and state its domain.

$$g(x) = \dfrac{3}{x-2}$$

Solution

y-Intercept: $\left(0, -\frac{3}{2}\right)$, from $g(0) = -\frac{3}{2}$

x-Intercept: None, because $3 \neq 0$

Vertical asymptote: $x = 2$, zero of denominator

Horizontal asymptote: $y = 0$, because degree of $N(x) <$ degree of $D(x)$

Additional points:

x	-4	1	3	5
$g(x)$	-0.5	-3	3	1

By plotting the intercepts, asymptotes, and a few additional points, you can obtain the graph shown in Figure 4.7. The domain of g is all real numbers except $x = 2$.

The graph of g in Example 1 is a vertical stretch and a right shift of the graph of $f(x) = 1/x$ because

$$g(x) = \dfrac{3}{x-2} = 3\left(\dfrac{1}{x-2}\right) = 3f(x-2).$$

Example 2 ▶   Sketching the Graph of a Rational Function

Sketch the graph of the function and state its domain.

$$f(x) = \dfrac{2x-1}{x}$$

Solution

y-Intercept: None, because $x = 0$ is not in the domain

x-Intercept: $\left(\frac{1}{2}, 0\right)$, from $2x - 1 = 0$

Vertical asymptote: $x = 0$, zero of denominator

Horizontal asymptote: $y = 2$, because degree of $N(x) =$ degree of $D(x)$

Additional points:

x	-4	-1	$\frac{1}{4}$	4
$f(x)$	2.25	3	-2	1.75

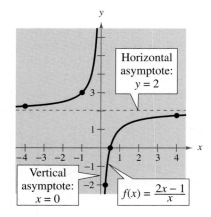

Horizontal asymptote: $y = 2$

Vertical asymptote: $x = 0$

$f(x) = \dfrac{2x-1}{x}$

FIGURE **4.8**

By plotting the intercepts, asymptotes, and a few additional points, you can obtain the graph shown in Figure 4.8. The domain of f is all real numbers except $x = 0$.

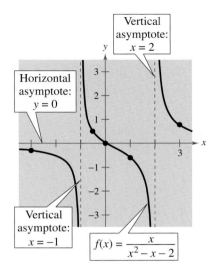

Horizontal asymptote: $y = 0$

Vertical asymptote: $x = 2$

Vertical asymptote: $x = -1$

$f(x) = \dfrac{x}{x^2 - x - 2}$

FIGURE 4.9

A computer animation of this example appears in the *Interactive* CD-ROM and *Internet* versions of this text.

Example 3 ▶ Sketching the Graph of a Rational Function

Sketch the graph of

$$f(x) = \frac{x}{x^2 - x - 2}.$$

Solution

Factor the denominator to determine more easily the zeros of the denominator.

$$f(x) = \frac{x}{x^2 - x - 2} = \frac{x}{(x + 1)(x - 2)}$$

y-Intercept: $(0, 0)$, because $f(0) = 0$

x-Intercept: $(0, 0)$

Vertical asymptotes: $x = -1$, $x = 2$, zeros of denominator

Horizontal asymptote: $y = 0$, because degree of $N(x) <$ degree of $D(x)$

Additional points:

x	-3	-0.5	1	3
$f(x)$	-0.3	0.4	-0.5	0.75

The graph is shown in Figure 4.9.

Example 4 ▶ Sketching the Graph of a Rational Function

Sketch the graph of

$$f(x) = \frac{2(x^2 - 9)}{x^2 - 4}.$$

Solution

By factoring the numerator and denominator, you have

$$f(x) = \frac{2(x^2 - 9)}{x^2 - 4} = \frac{2(x - 3)(x + 3)}{(x - 2)(x + 2)}.$$

y-Intercept: $\left(0, \frac{9}{2}\right)$, because $f(0) = \frac{9}{2}$

x-Intercepts: $(-3, 0)$ and $(3, 0)$

Vertical asymptotes: $x = -2$, $x = 2$, zeros of denominator

Horizontal asymptote: $y = 2$, because degree of $N(x) =$ degree of $D(x)$

Symmetry: With respect to y-axis, because $f(-x) = f(x)$

Additional points:

x	0.5	2.5	6
$f(x)$	4.67	-2.44	1.69

The graph is shown in Figure 4.10.

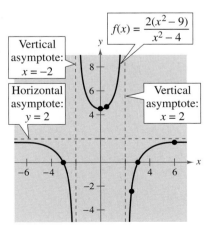

Vertical asymptote: $x = -2$

Horizontal asymptote: $y = 2$

Vertical asymptote: $x = 2$

$f(x) = \dfrac{2(x^2 - 9)}{x^2 - 4}$

FIGURE 4.10

Slant Asymptotes

If the degree of the numerator of a rational function is exactly *one more* than the degree of the denominator, the graph of the function has a **slant** (or **oblique**) **asymptote.** For example, the graph of

$$f(x) = \frac{x^2 - x}{x + 1}$$

has a slant asymptote, as shown in Figure 4.11. To find the equation of a slant asymptote, use long division. For instance, by dividing $x + 1$ into $x^2 - x$, you obtain

$$f(x) = \frac{x^2 - x}{x + 1}$$

$$= \underbrace{x - 2}_{\substack{\text{Slant asymptote} \\ (y = x - 2)}} + \frac{2}{x + 1}.$$

In Figure 4.11, notice that the graph of f approaches the line $y = x - 2$ as x moves to the right or left.

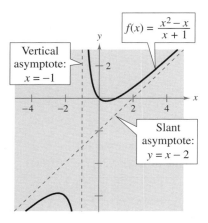

Vertical asymptote: $x = -1$

$f(x) = \dfrac{x^2 - x}{x + 1}$

Slant asymptote: $y = x - 2$

FIGURE **4.11**

Example 5 ▶ A Rational Function with a Slant Asymptote

Sketch the graph of

$$f(x) = \frac{x^2 - x - 2}{x - 1}.$$

Solution

First write $f(x)$ in two different ways. Factoring the numerator

$$f(x) = \frac{x^2 - x - 2}{x - 1} = \frac{(x - 2)(x + 1)}{x - 1}$$

allows you to recognize the x-intercepts. Long division

$$f(x) = \frac{x^2 - x - 2}{x - 1} = x - \frac{2}{x - 1}$$

allows you to recognize that the line $y = x$ is a slant asymptote of the graph.

y-Intercept: $(0, 2)$, because $f(0) = 2$

x-Intercepts: $(-1, 0)$ and $(2, 0)$

Vertical asymptote: $x = 1$, zero of denominator

Slant asymptote: $y = x$

Additional points:

x	-2	0.5	1.5	3
$f(x)$	-1.33	4.5	-2.5	2

The graph is shown in Figure 4.12.

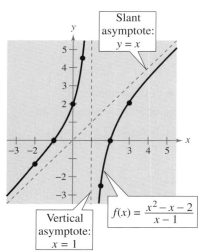

Slant asymptote: $y = x$

$f(x) = \dfrac{x^2 - x - 2}{x - 1}$

Vertical asymptote: $x = 1$

FIGURE **4.12**

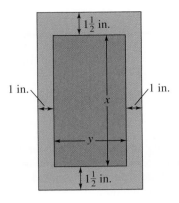

FIGURE 4.13

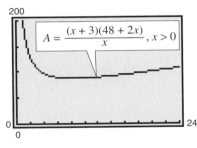

FIGURE 4.14

The *Interactive* CD-ROM and *Internet* versions of this text offer a built-in graphing calculator, which can be used in the Examples, Explorations, Technology notes, and Exercises.

Application

Example 6 ▶ Finding a Minimum Area

A rectangular page is designed to contain 48 square inches of print. The margins at the top and bottom of the page are each $1\frac{1}{2}$ inches deep. The margins on each side are 1 inch wide. What should the dimensions of the page be so that the least amount of paper is used?

Solution

Let A be the area to be minimized. From Figure 4.13, you can write

$$A = (x + 3)(y + 2).$$

The printed area inside the margins is modeled by $48 = xy$ or $y = 48/x$. To find the minimum area, rewrite the equation for A in terms of just one variable by substituting $48/x$ for y.

$$A = (x + 3)\left(\frac{48}{x} + 2\right)$$

$$= \frac{(x + 3)(48 + 2x)}{x}, \qquad x > 0$$

The graph of this rational function is shown in Figure 4.14. Because x represents the height of the printed area, you need consider only the portion of the graph for which x is positive. Using a graphing utility, you can approximate the minimum value of A to occur when $x \approx 8.5$ inches. The corresponding value of y is $48/8.5 \approx 5.6$ inches. So, the dimensions should be

$$x + 3 \approx 11.5 \text{ inches} \quad \text{by} \quad y + 2 \approx 7.6 \text{ inches.}$$

If you go on to take a course in calculus, you will learn an analytic technique for finding the exact value of x that produces a minimum area. In this case, that value is $x = 6\sqrt{2} \approx 8.485$.

Writing **ABOUT MATHEMATICS**

Asymptotes of Graphs of Rational Functions Do you think it is possible for the graph of a rational function to cross its horizontal asymptote or its slant asymptote? Use the graphs of the following functions to investigate these questions. Write a summary of your conclusions. Explain your reasoning.

a. $f(x) = \dfrac{x}{x^2 + 1}$

b. $g(x) = \dfrac{x^3}{x^2 + 1}$

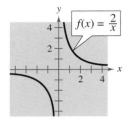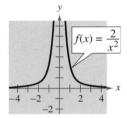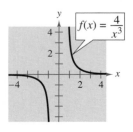

4.2 Exercises

In Exercises 1–4, use the graph of $f(x) = 2/x$ to sketch the graph of g.

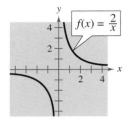

1. $g(x) = \dfrac{2}{x} + 3$

2. $g(x) = \dfrac{2}{x - 3}$

3. $g(x) = -\dfrac{2}{x}$

4. $g(x) = \dfrac{1}{x + 2}$

In Exercises 5–8, use the graph of $f(x) = 2/x^2$ to sketch the graph of g.

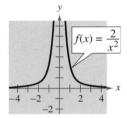

5. $g(x) = \dfrac{2}{x^2} - 1$

6. $g(x) = -\dfrac{2}{x^2}$

7. $g(x) = \dfrac{2}{(x - 1)^2}$

8. $g(x) = \dfrac{1}{2x^2}$

In Exercises 9–12, use the graph of $f(x) = 4/x^3$ to sketch the graph of g.

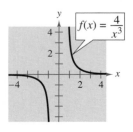

9. $g(x) = \dfrac{4}{(x + 3)^3}$

10. $g(x) = \dfrac{4}{x^3} + 3$

11. $g(x) = -\dfrac{4}{x^3}$

12. $g(x) = \dfrac{1}{x^3}$

In Exercises 13–34, (a) identify all intercepts, (b) find any vertical and horizontal asymptotes, (c) check for symmetry, and (d) plot additional solution points as needed and sketch the graph of the rational function.

13. $f(x) = \dfrac{1}{x + 2}$

14. $f(x) = \dfrac{1}{x - 3}$

15. $h(x) = \dfrac{-1}{x + 2}$

16. $g(x) = \dfrac{1}{3 - x}$

17. $C(x) = \dfrac{5 + 2x}{1 + x}$

18. $P(x) = \dfrac{1 - 3x}{1 - x}$

19. $g(x) = \dfrac{1}{x + 2} + 2$

20. $f(x) = 2 - \dfrac{3}{x^2}$

21. $f(x) = \dfrac{x^2}{x^2 + 9}$

22. $f(t) = \dfrac{1 - 2t}{t}$

23. $h(x) = \dfrac{x^2}{x^2 - 9}$

24. $g(x) = \dfrac{x}{x^2 - 9}$

25. $g(s) = \dfrac{s}{s^2 + 1}$

26. $f(x) = -\dfrac{1}{(x - 2)^2}$

27. $g(x) = \dfrac{4(x + 1)}{x(x - 4)}$

28. $h(x) = \dfrac{2}{x^2(x - 2)}$

29. $f(x) = \dfrac{3x}{x^2 - x - 2}$

30. $f(x) = \dfrac{2x}{x^2 + x - 2}$

31. $f(x) = \dfrac{6x}{x^2 - 5x - 14}$

32. $f(x) = \dfrac{3(x^2 + 1)}{x^2 + 2x - 15}$

33. $f(x) = \dfrac{2x^2 - x - 1}{x^3 - 2x^2 - x + 2}$

34. $f(x) = \dfrac{x^2 - x - 2}{x^3 - 2x^2 - 5x + 6}$

Analytical, Numerical, and Graphical Analysis **In** Exercises 35–38, do the following.

(a) Determine the domains of f and g.

(b) Find any vertical asymptotes of f.

(c) Compare the functions by completing the table.

(d) Use a graphing utility to graph f and g in the same viewing window.

(e) Explain why the graphing utility may not show the difference in the domains of f and g.

35. $f(x) = \dfrac{x^2 - 1}{x + 1}$, $g(x) = x - 1$

x	−3	−2	−1.5	−1	−0.5	0	1
f(x)							
g(x)							

36. $f(x) = \dfrac{x^2(x - 2)}{x^2 - 2x}$, $g(x) = x$

x	−1	0	1	1.5	2	2.5	3
f(x)							
g(x)							

37. $f(x) = \dfrac{x - 2}{x^2 - 2x}$, $g(x) = \dfrac{1}{x}$

x	−0.5	0	0.5	1	1.5	2	3
f(x)							
g(x)							

38. $f(x) = \dfrac{2x - 6}{x^2 - 7x + 12}$, $g(x) = \dfrac{2}{x - 4}$

x	0	1	2	3	4	5	6
f(x)							
g(x)							

In Exercises 39–44, sketch the graph of the function. State the domain of the function and identify any vertical or horizontal asymptotes.

39. $h(t) = \dfrac{4}{t^2 + 1}$

40. $g(x) = -\dfrac{x}{(x - 2)^2}$

41. $f(t) = \dfrac{2t^2}{t^2 - 4}$

42. $f(x) = \dfrac{x + 4}{x^2 + x - 6}$

43. $f(x) = \dfrac{20x}{x^2 + 1} - \dfrac{1}{x}$

44. $f(x) = 5\left(\dfrac{1}{x - 4} - \dfrac{1}{x + 2}\right)$

In Exercises 45–52, (a) identify all intercepts, (b) find any vertical and slant asymptotes, (c) check for symmetry, (d) and plot additional solution points as needed and sketch the graph of the rational function.

45. $f(x) = \dfrac{2x^2 + 1}{x}$

46. $f(x) = \dfrac{1 - x^2}{x}$

47. $g(x) = \dfrac{x^2 + 1}{x}$

48. $h(x) = \dfrac{x^2}{x - 1}$

49. $f(x) = \dfrac{x^3}{x^2 - 1}$

50. $g(x) = \dfrac{x^3}{2x^2 - 8}$

51. $f(x) = \dfrac{x^2 - x + 1}{x - 1}$

52. $f(x) = \dfrac{2x^2 - 5x + 5}{x - 2}$

In Exercises 53–56, use a graphing utility to graph the rational function. Give the domain of the function and identify any asymptotes. Then zoom out sufficiently far so that the graph appears as a line. Identify the line.

53. $f(x) = \dfrac{x^2 + 5x + 8}{x + 3}$

54. $f(x) = \dfrac{2x^2 + x}{x + 1}$

55. $g(x) = \dfrac{1 + 3x^2 - x^3}{x^2}$

56. $h(x) = \dfrac{12 - 2x - x^2}{2(4 + x)}$

Graphical Reasoning **In** Exercises 57–60, (a) use the graph to determine any x-intercepts of the rational function and (b) set y = 0 and solve the resulting equation to confirm your result in part (a).

57. $y = \dfrac{x + 1}{x - 3}$

58. $y = \dfrac{2x}{x - 3}$

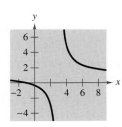

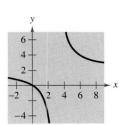

59. $y = \dfrac{1}{x} - x$

60. $y = x - 3 + \dfrac{2}{x}$

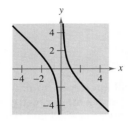

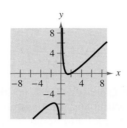

⊞ *Graphical Reasoning* **In Exercises 61–64, (a) use a graphing utility to graph the function and determine any x-intercepts and (b) set y = 0 and solve the resulting equation to confirm your result in part (a).**

61. $y = \dfrac{1}{x + 5} + \dfrac{4}{x}$

62. $y = 20\left(\dfrac{2}{x + 1} - \dfrac{3}{x}\right)$

63. $y = x - \dfrac{6}{x - 1}$

64. $y = x - \dfrac{9}{x}$

65. *Concentration of a Mixture* A 1000-liter tank contains 50 liters of a 25% brine solution. You add x liters of a 75% brine solution to the tank.

(a) Show that the concentration C, the proportion of brine to total solution, in the final mixture is

$$C = \dfrac{3x + 50}{4(x + 50)}.$$

(b) Determine the domain of the function based on the physical constraints of the problem.

(c) Graph the concentration function. As the tank is filled, what happens to the rate at which the concentration of brine is increasing? What percent does the concentration of brine appear to approach?

66. *Geometry* A rectangular region of length x and width y has an area of 500 square meters.

(a) Express the width y as a function of x.

(b) Determine the domain of the function based on the physical constraints of the problem.

(c) Sketch a graph of the function and determine the width of the rectangle when x = 30 meters.

67. *Geometry* A page that is x inches wide and y inches high contains 30 square inches of print. The top and bottom margins are 1 inch deep and the margins on each side are 2 inches wide (see figure).

(a) Show that the total area A on the page is

$$A = \dfrac{2x(x + 11)}{x - 4}.$$

(b) Determine the domain of the function based on the physical constraints of the problem.

⊞ (c) Use a graphing utility to obtain a graph of the area function and approximate the page size for which the least amount of paper will be used.

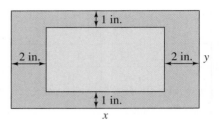

68. *Geometry* A right triangle is formed in the first quadrant by the x-axis, the y-axis, and a line segment through the point (3, 2) (see figure).

(a) Show that an equation of the line segment is

$$y = \dfrac{2(a - x)}{a - 3}, \qquad 0 \le x \le a.$$

(b) Show that the area of the triangle is

$$A = \dfrac{a^2}{a - 3}.$$

(c) Graph the area function. From the graph, estimate the value of a that yields a minimum area. Estimate the minimum area.

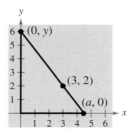

⊞ **In Exercises 69 and 70, use a graphing utility to obtain the graph of the function and locate any relative maximum or minimum points on the graph.**

69. $f(x) = \dfrac{3(x + 1)}{x^2 + x + 1}$

70. $C(x) = x + \dfrac{32}{x}$

71. *Minimum Cost* The ordering and transportation cost C for the components used in manufacturing a certain product is

$$C = 100\left(\frac{200}{x^2} + \frac{x}{x + 30}\right), \qquad x \geq 1$$

where C is the cost (in thousands of dollars) and x is the order size (in hundreds). Use a graphing utility to obtain the graph of the cost function. From the graph, estimate the order size that minimizes cost.

72. *Minimum Cost* The cost of producing x units is $C = 0.2x^2 + 10x + 5$, and the average cost per unit is

$$\overline{C} = \frac{C}{x} = \frac{0.2x^2 + 10x + 5}{x}, \qquad x > 0.$$

Sketch the graph of the average cost function and estimate the number of units that should be produced to minimize the average cost per unit.

73. *Minimum Area* A rectangular page is designed to contain 64 square inches of print. The margins at the top and bottom of the page are each 1 inch deep. The margins on each side are $1\frac{1}{2}$ inches wide. What should the dimensions of the page be so that the least amount of paper is used?

74. *Medicine* The concentration of a certain chemical in the bloodstream t hours after injection into muscle tissue is

$$C = \frac{3t^2 + t}{t^3 + 50}, \qquad t > 0.$$

(a) Determine the horizontal asymptote of the function and interpret its meaning in the context of the problem.

(b) Use a graphing utility to graph the function and approximate the time when the bloodstream concentration is greatest.

75. *Average Speed* A driver averaged 50 miles per hour on the round trip between home and a city 100 miles away. The average speeds for going and returning were x and y miles per hour, respectively.

(a) Show that $y = \dfrac{25x}{x - 25}$.

(b) Determine the vertical and horizontal asymptotes of the function.

(c) Complete the table.

x	30	35	40	45	50	55	60
y							

(d) Are the results in the table unexpected? Explain.

(e) Is it possible to average 20 miles per hour in one direction and still average 50 miles per hour on the round trip? Explain.

Synthesis

True or False? **In Exercises 76 and 77, determine whether the statement is true or false. Justify your answer.**

76. If the graph of a rational function f has a vertical asymptote at $x = 5$, it is possible to sketch the graph without lifting your pencil from the paper.

77. A rational function can never cross one of its asymptotes.

78. *Think About It* Write a rational function satisfying the following criteria.

Vertical asymptote: $x = 2$

Slant asymptote: $y = x + 1$

Zero of the function: $x = -2$

Think About It **In Exercises 79 and 80, use a graphing utility to obtain the graph of the function. Explain why there is no vertical asymptote when a superficial examination of the function may indicate that there should be one.**

79. $h(x) = \dfrac{6 - 2x}{3 - x}$ **80.** $g(x) = \dfrac{x^2 + x - 2}{x - 1}$

81. *Writing* Write a paragraph discussing whether every rational function has a vertical asymptote.

Review

In Exercises 82–85, completely factor the expression.

82. $x^2 - 15x + 56$ **83.** $3x^2 + 23x - 36$

84. $x^3 - 5x^2 + 4x - 20$ **85.** $x^3 + 6x^2 - 2x - 12$

In Exercises 86–89, solve the inequality and show the solution on the real number line.

86. $10 - 3x \leq 0$ **87.** $5 - 2x > 5(x + 1)$

88. $|4(x - 2)| < 20$ **89.** $\frac{1}{2}|2x + 3| \geq 5$

4.3 Partial Fractions

▶ **What you should learn**

- How to recognize partial fraction decompositions of rational expressions
- How to find partial fraction decompositions of rational expressions

▶ **Why you should learn it**

Partial fractions can help you analyze the behavior of a rational function. For instance, Exercise 57 on page 349 shows how partial fractions can help you analyze the exhaust temperatures of a diesel engine.

Introduction

In this section, you will learn to write a rational expression as the sum of two or more simpler rational expressions. For example, the rational expression

$$\frac{x + 7}{x^2 - x - 6}$$

can be written as the sum of two fractions with first-degree denominators. That is,

Partial fraction decomposition
of $\dfrac{x + 7}{x^2 - x - 6}$

$$\frac{x + 7}{x^2 - x - 6} = \underbrace{\frac{2}{x - 3}}_{\substack{\text{Partial} \\ \text{fraction}}} + \underbrace{\frac{-1}{x + 2}}_{\substack{\text{Partial} \\ \text{fraction}}}.$$

Each fraction on the right side of the equation is a **partial fraction,** and together they make up the **partial fraction decomposition** of the left side.

Decomposition of $N(x)/D(x)$ into Partial Fractions

1. *Divide if improper:* If $N(x)/D(x)$ is an improper fraction [degree of $N(x) \geq$ degree of $D(x)$], divide the denominator into the numerator to obtain

$$\frac{N(x)}{D(x)} = (\text{polynomial}) + \frac{N_1(x)}{D(x)}$$

and apply Steps 2, 3, and 4 (below) to the proper rational expression $N_1(x)/D(x)$. Note that $N_1(x)$ is the remainder from the division of $N(x)$ by $D(x)$.

2. *Factor the denominator:* Completely factor the denominator into factors of the form

$$(px + q)^m \quad \text{and} \quad (ax^2 + bx + c)^n$$

where $(ax^2 + bx + c)$ is irreducible.

3. *Linear factors:* For *each* factor of the form $(px + q)^m$, the partial fraction decomposition must include the following sum of m fractions.

$$\frac{A_1}{(px + q)} + \frac{A_2}{(px + q)^2} + \cdots + \frac{A_m}{(px + q)^m}$$

4. *Quadratic factors:* For *each* factor of the form $(ax^2 + bx + c)^n$, the partial fraction decomposition must include the following sum of n fractions.

$$\frac{B_1x + C_1}{ax^2 + bx + c} + \frac{B_2x + C_2}{(ax^2 + bx + c)^2} + \cdots + \frac{B_nx + C_n}{(ax^2 + bx + c)^n}$$

Michael Rosenfeld/Tony Stone Images

Partial Fraction Decomposition

Algebraic techniques for determining the constants in the numerators of partial fractions are demonstrated in the examples that follow. Note that the techniques vary slightly, depending on the type of factors of the denominator: linear or quadratic, distinct or repeated.

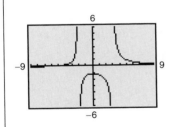

Example 1 ▶ **Distinct Linear Factors**

Write the partial fraction decomposition for $\dfrac{x + 7}{x^2 - x - 6}$.

Solution

The expression is not improper, so factor the denominator. Because $x^2 - x - 6 = (x - 3)(x + 2)$, you should include one partial fraction with a constant numerator for each linear factor of the denominator. Write the form of the decomposition as follows.

$$\frac{x + 7}{x^2 - x - 6} = \frac{A}{x - 3} + \frac{B}{x + 2} \qquad \text{Write form of decomposition.}$$

Multiplying both sides of this equation by the least common denominator, $(x - 3)(x + 2)$, leads to the **basic equation**

$$x + 7 = A(x + 2) + B(x - 3). \qquad \text{Basic equation}$$

Because this equation is true for all x, you can substitute any *convenient* values of x that will help determine the constants A and B. Values of x that are especially convenient are ones that make the factors $(x + 2)$ and $(x - 3)$ equal to zero. For instance, let $x = -2$. Then

$$-2 + 7 = A(-2 + 2) + B(-2 - 3) \qquad \text{Substitute } -2 \text{ for } x.$$
$$5 = A(0) + B(-5)$$
$$5 = -5B$$
$$-1 = B.$$

To solve for A, let $x = 3$ and obtain

$$3 + 7 = A(3 + 2) + B(3 - 3) \qquad \text{Substitute 3 for } x.$$
$$10 = A(5) + B(0)$$
$$10 = 5A$$
$$2 = A.$$

So, the decomposition is

$$\frac{x + 7}{x^2 - x - 6} = \frac{2}{x - 3} + \frac{-1}{x + 2}$$

as indicated at the beginning of this section. Check this result by combining the two partial fractions on the right side of the equation, or by using your graphing utility.

The next example shows how to find the partial fraction decomposition for a rational expression whose denominator has a repeated linear factor.

Example 2 ▶ Repeated Linear Factors

Write the partial fraction decomposition for $\dfrac{x^4 + 2x^3 + 6x^2 + 20x + 6}{x^3 + 2x^2 + x}$.

Solution

This rational expression is improper, so you should begin by dividing the numerator by the denominator to obtain

$$x + \frac{5x^2 + 20x + 6}{x^3 + 2x^2 + x}.$$

Because the denominator of the remainder factors as

$$x^3 + 2x^2 + x = x(x^2 + 2x + 1) = x(x + 1)^2$$

you should include one partial fraction with a constant numerator for each power of x and $(x + 1)$ and write the form of the decomposition as follows.

$$\frac{5x^2 + 20x + 6}{x(x + 1)^2} = \frac{A}{x} + \frac{B}{x + 1} + \frac{C}{(x + 1)^2}$$

Multiplying by the LCD, $x(x + 1)^2$, leads to the basic equation

$$5x^2 + 20x + 6 = A(x + 1)^2 + Bx(x + 1) + Cx. \qquad \text{Basic equation}$$

Letting $x = -1$ eliminates the A- and B-terms and yields

$$5(-1)^2 + 20(-1) + 6 = A(-1 + 1)^2 + B(-1)(-1 + 1) + C(-1)$$

$$5 - 20 + 6 = 0 + 0 - C$$

$$C = 9.$$

Letting $x = 0$ eliminates the B- and C-terms and yields

$$5(0)^2 + 20(0) + 6 = A(0 + 1)^2 + B(0)(0 + 1) + C(0)$$

$$6 = A(1) + 0 + 0$$

$$6 = A.$$

At this point, you have exhausted the most convenient choices for x, so to find the value of B, use *any other value* for x along with the known values of A and C. So, using $x = 1$, $A = 6$, and $C = 9$,

$$5(1)^2 + 20(1) + 6 = 6(1 + 1)^2 + B(1)(1 + 1) + 9(1)$$

$$31 = 6(4) + 2B + 9$$

$$-2 = 2B$$

$$-1 = B.$$

Therefore, the partial fraction decomposition is

$$\frac{x^4 + 2x^3 + 6x^2 + 20x + 6}{x^3 + 2x^2 + x} = x + \frac{6}{x} + \frac{-1}{x + 1} + \frac{9}{(x + 1)^2}.$$

The procedure used to solve for the constants in Examples 1 and 2 works well when the factors of the denominator are linear. However, when the denominator contains irreducible quadratic factors, you should use a different procedure, which involves writing the right side of the basic equation in polynomial form and *equating the coefficients* of like terms.

Example 3 ▶ Distinct Linear and Quadratic Factors

Write the partial fraction decomposition for

$$\frac{3x^2 + 4x + 4}{x^3 + 4x}.$$

Solution

Because the denominator factors as

$$x^3 + 4x = x(x^2 + 4)$$

you should include one partial fraction with a constant numerator and one partial fraction with a linear numerator and write the form of the decomposition as follows.

$$\frac{3x^2 + 4x + 4}{x^3 + 4x} = \frac{A}{x} + \frac{Bx + C}{x^2 + 4}$$

Multiplying by the LCD, $x(x^2 + 4)$, yields the basic equation

$$3x^2 + 4x + 4 = A(x^2 + 4) + (Bx + C)x. \qquad \text{Basic equation}$$

Expanding this basic equation and collecting like terms produces

$$3x^2 + 4x + 4 = Ax^2 + 4A + Bx^2 + Cx$$

$$= (A + B)x^2 + Cx + 4A. \qquad \text{Polynomial form}$$

Finally, because two polynomials are equal if and only if the coefficients of like terms are equal,

$$3x^2 + 4x + 4 = (A + B)x^2 + Cx + 4A \qquad \text{Equate coefficients of like terms.}$$

you obtain the following equations.

$$3 = A + B, \qquad 4 = C, \qquad \text{and} \qquad 4 = 4A$$

So, $A = 1$ and $C = 4$. Moreover, substituting $A = 1$ in the equation $3 = A + B$ yields

$$3 = 1 + B$$

$$2 = B.$$

Therefore, the partial fraction decomposition is

$$\frac{3x^2 + 4x + 4}{x^3 + 4x} = \frac{1}{x} + \frac{2x + 4}{x^2 + 4}.$$

JEAN BERNOUILLI
Professeur de Mathématique
Né à Bâle en Suisse l'an 1667.

Historical Note
John Bernoulli (1667–1748), a Swiss mathematician, introduced the method of partial fractions and was instrumental in the early development of calculus. Bernoulli was a professor at the University of Basel and taught many outstanding students, the most famous of whom was Leonhard Euler.

The next example shows how to find the partial fraction decomposition for a rational function whose denominator has a repeated quadratic factor.

Example 4 ▶ Repeated Quadratic Factors

Write the partial fraction decomposition for

$$\frac{8x^3 + 13x}{(x^2 + 2)^2}.$$

Solution

You need to include one partial fraction with a linear numerator for each power of $(x^2 + 2)$.

$$\frac{8x^3 + 13x}{(x^2 + 2)^2} = \frac{Ax + B}{x^2 + 2} + \frac{Cx + D}{(x^2 + 2)^2}$$

Multiplying by the LCD, $(x^2 + 2)^2$, yields the basic equation

$$8x^3 + 13x = (Ax + B)(x^2 + 2) + Cx + D \qquad \text{Basic equation}$$

$$= Ax^3 + 2Ax + Bx^2 + 2B + Cx + D$$

$$= Ax^3 + Bx^2 + (2A + C)x + (2B + D). \qquad \text{Polynomial form}$$

Equating coefficients of like terms

$$8x^3 + 0x^2 + 13x + 0 = Ax^3 + Bx^2 + (2A + C)x + (2B + D)$$

produces

$$8 = A,\ 0 = B,\ 13 = 2A + C,\ \text{and}\ 0 = 2B + D. \qquad \text{Equate coefficients.}$$

Finally, use the values $A = 8$ and $B = 0$ to obtain the following.

$$13 = 2A + C$$

$$= 2(8) + C$$

$$-3 = C$$

$$0 = 2B + D$$

$$= 2(0) + D$$

$$0 = D$$

Therefore,

$$\frac{8x^3 + 13x}{(x^2 + 2)^2} = \frac{8x}{x^2 + 2} + \frac{-3x}{(x^2 + 2)^2}.$$

By equating coefficients of like terms in Examples 3 and 4, you obtained several equations involving A, B, C, and D, which were solved by substitution. In a later chapter you will study a more general method for solving systems of equations.

Guidelines for Solving the Basic Equation

Linear Factors

1. Substitute the *zeros* of the distinct linear factors into the basic equation.

2. For repeated linear factors, use the coefficients determined above to rewrite the basic equation. Then substitute *other* convenient values of x and solve for the remaining coefficients.

Quadratic Factors

1. Expand the basic equation.

2. Collect terms according to powers of x.

3. Equate the coefficients of like terms to obtain equations involving A, B, C, and so on.

4. Use substitution to solve for A, B, C, ...

Keep in mind that for *improper* rational expressions such as

$$\frac{N(x)}{D(x)} = \frac{2x^3 + x^2 - 7x + 7}{x^2 + x - 2}$$

you must first divide before applying partial fraction decomposition.

Writing ABOUT MATHEMATICS

Error Analysis Suppose you are tutoring a student in algebra. In trying to find a partial fraction decomposition, your student writes the following.

$$\frac{x^2 + 1}{x(x - 1)} = \frac{A}{x} + \frac{B}{x - 1}$$

$$\frac{x^2 + 1}{x(x - 1)} = \frac{A(x - 1)}{x(x - 1)} + \frac{Bx}{x(x - 1)}$$

$$x^2 + 1 = A(x - 1) + Bx \qquad \text{Basic equation}$$

By substituting $x = 0$ and $x = 1$ into the basic equation, your student concludes that $A = -1$ and $B = 2$. However, in checking this solution, your student obtains the following. What has gone wrong?

$$\frac{-1}{x} + \frac{2}{x - 1} = \frac{(-1)(x - 1) + 2(x)}{x(x - 1)}$$

$$= \frac{x + 1}{x(x - 1)}$$

$$\neq \frac{x^2 + 1}{x(x - 1)}.$$

4.3 Exercises

In Exercises 1–4, match the rational expression with the form of its decomposition. [The decompositions are labeled (a), (b), (c), and (d).]

(a) $\dfrac{A}{x} + \dfrac{B}{x+2} + \dfrac{C}{x-2}$

(b) $\dfrac{A}{x} + \dfrac{B}{x-4}$

(c) $\dfrac{A}{x} + \dfrac{B}{x^2} + \dfrac{C}{x-4}$

(d) $\dfrac{A}{x} + \dfrac{Bx+C}{x^2+4}$

1. $\dfrac{3x-1}{x(x-4)}$

2. $\dfrac{3x-1}{x^2(x-4)}$

3. $\dfrac{3x-1}{x(x^2+4)}$

4. $\dfrac{3x-1}{x(x^2-4)}$

In Exercises 5–14, write the form of the partial fraction decomposition of the rational expression. Do not solve for the constants.

5. $\dfrac{7}{x^2-14x}$

6. $\dfrac{x-2}{x^2+4x+3}$

7. $\dfrac{12}{x^3-10x^2}$

8. $\dfrac{x^2-3x+2}{4x^3+11x^2}$

9. $\dfrac{4x^2+3}{(x-5)^3}$

10. $\dfrac{6x+5}{(x+2)^4}$

11. $\dfrac{2x-3}{x^3+10x}$

12. $\dfrac{x-6}{2x^3+8x}$

13. $\dfrac{x-1}{x(x^2+1)^2}$

14. $\dfrac{x+4}{x^2(3x-1)^2}$

In Exercises 15–38, write the partial fraction decomposition for the rational expression. Check your result algebraically.

15. $\dfrac{1}{x^2-1}$

16. $\dfrac{1}{4x^2-9}$

17. $\dfrac{1}{x^2+x}$

18. $\dfrac{3}{x^2-3x}$

19. $\dfrac{1}{2x^2+x}$

20. $\dfrac{5}{x^2+x-6}$

21. $\dfrac{3}{x^2+x-2}$

22. $\dfrac{x+1}{x^2+4x+3}$

23. $\dfrac{x^2+12x+12}{x^3-4x}$

24. $\dfrac{x+2}{x(x-4)}$

25. $\dfrac{4x^2+2x-1}{x^2(x+1)}$

26. $\dfrac{2x-3}{(x-1)^2}$

27. $\dfrac{3x}{(x-3)^2}$

28. $\dfrac{6x^2+1}{x^2(x-1)^2}$

29. $\dfrac{x^2-1}{x(x^2+1)}$

30. $\dfrac{x}{(x-1)(x^2+x+1)}$

31. $\dfrac{x}{x^3-x^2-2x+2}$

32. $\dfrac{x+6}{x^3-3x^2-4x+12}$

33. $\dfrac{x^2}{x^4-2x^2-8}$

34. $\dfrac{2x^2+x+8}{(x^2+4)^2}$

35. $\dfrac{x}{16x^4-1}$

36. $\dfrac{x+1}{x^3+x}$

37. $\dfrac{x^2+5}{(x+1)(x^2-2x+3)}$

38. $\dfrac{x^2-4x+7}{(x+1)(x^2-2x+3)}$

In Exercises 39–44, write the partial fraction decomposition for the improper rational expression.

39. $\dfrac{x^2-x}{x^2+x+1}$

40. $\dfrac{x^2-4x}{x^2+x+6}$

41. $\dfrac{2x^3-x^2+x+5}{x^2+3x+2}$

42. $\dfrac{x^3+2x^2-x+1}{x^2+3x-4}$

43. $\dfrac{x^4}{(x-1)^3}$

44. $\dfrac{16x^4}{(2x-1)^3}$

In Exercises 45–52, write the partial fraction decomposition for the rational expression. Use a graphing utility to check your result graphically.

45. $\dfrac{5-x}{2x^2+x-1}$

46. $\dfrac{3x^2-7x-2}{x^3-x}$

47. $\dfrac{x-1}{x^3+x^2}$

48. $\dfrac{4x^2-1}{2x(x+1)^2}$

49. $\dfrac{x^2+x+2}{(x^2+2)^2}$

50. $\dfrac{x^3}{(x+2)^2(x-2)^2}$

51. $\dfrac{2x^3-4x^2-15x+5}{x^2-2x-8}$

52. $\dfrac{x^3-x+3}{x^2+x-2}$

Graphical Analysis   **In Exercises 53–56, write the partial fraction decomposition for the rational function. Identify the graph of the rational function and the graph of each term of its decomposition. State any relationship between the vertical asymptotes of the rational function and the vertical asymptotes of the terms of the decomposition.**

53. $y = \dfrac{x - 12}{x(x - 4)}$

54. $y = \dfrac{2(x + 1)^2}{x(x^2 + 1)}$

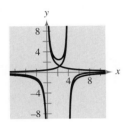
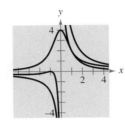

55. $y = \dfrac{2(4x - 3)}{x^2 - 9}$

56. $y = \dfrac{2(4x^2 - 15x + 39)}{x^2(x^2 - 10x + 26)}$

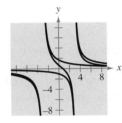
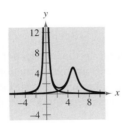

57. *Thermodynamics*   The magnitude of the range of exhaust temperatures in degrees Fahrenheit in an experimental diesel engine is approximated by

$$R = \frac{2000(4 - 3x)}{(11 - 7x)(7 - 4x)}, \qquad 0 < x \leq 1$$

where x is the relative load.

(a) Write the partial fraction decomposition of the expression.

(b) The decomposition in part (a) is the difference of two fractions. The absolute values of the terms give the expected maximum and minimum temperatures of the exhaust gases for different loads.

Ymax = |1st term|

Ymin = |2nd term|

Write the equations for Ymax and Ymin. Then use a graphing utility to graph each equation in the same viewing window.

Synthesis

True or False?   **In Exercises 58 and 59, determine whether the statement is true or false. Justify your answer.**

58. For the rational expression

$$\frac{x}{(x + 10)(x - 10)^2}$$

the partial fraction decomposition is of the form

$$\frac{A}{x + 10} + \frac{B}{(x - 10)^2}.$$

59. When writing the partial fraction decomposition for the expression

$$\frac{x^3 + x - 2}{x^2 - 5x + 14}$$

the first step is to factor the denominator.

In Exercises 60–63, write the partial fraction decomposition for the rational expression. Check your result algebraically. Then assign a value to the constant a to check the result graphically.

60. $\dfrac{1}{a^2 - x^2}$

61. $\dfrac{1}{x(x + a)}$

62. $\dfrac{1}{y(a - y)}$

63. $\dfrac{1}{(x + 1)(a - x)}$

Review

In Exercises 64–71, sketch the graph of the function.

64. $f(x) = -3x + 7$

65. $f(x) = 6 - x$

66. $f(x) = -2x^2$

67. $f(x) = \frac{1}{4}x^2 + 1$

68. $f(x) = x^2 - 9x + 18$

69. $f(x) = 2x^2 - 9x - 5$

70. $f(x) = -x^2(x - 3)$

71. $f(x) = \frac{1}{2}x^3 - 1$

In Exercises 72–77, sketch the graph of the rational function.

72. $f(x) = \dfrac{6}{x - 1}$

73. $f(x) = \dfrac{1 - 4x}{x}$

74. $f(x) = \dfrac{x^2 + x - 6}{x + 5}$

75. $f(x) = \dfrac{3x - 1}{x^2 + 4x - 12}$

76. $f(x) = \dfrac{3(x^2 - 1)}{x^2 + 4x - 5}$

77. $f(x) = \dfrac{2x - 3}{x^2 - 16}$

4.4 Conics

▶ What you should learn

- How to recognize the four basic conics: circles, ellipses, parabolas, and hyperbolas
- How to recognize, graph, and write equations of parabolas (vertex at origin)
- How to recognize, graph, and write equations of ellipses (center at origin)
- How to recognize, graph, and write equations of hyperbolas (center at origin)

▶ Why you should learn it

Conics have been used for hundreds of years to model and solve engineering problems. For instance, Exercise 32 on page 360 shows how a parabola can be used to model the cables of a suspension bridge.

Introduction

Conic sections were discovered during the classical Greek period, 600 to 300 B.C. This early Greek study was largely concerned with the geometric properties of conics. It was not until the early 17th century that the broad applicability of conics became apparent and played a prominent role in the early development of calculus.

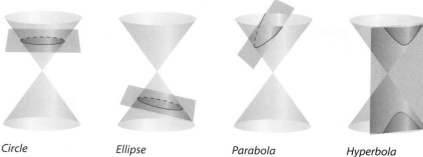

Circle Ellipse Parabola Hyperbola

FIGURE 4.15 *Basic Conics*

A **conic section** (or simply **conic**) is the intersection of a plane and a double-napped cone. Notice in Figure 4.15 that in the formation of the four basic conics, the intersecting plane does not pass through the vertex of the cone. When the plane does pass through the vertex, the resulting figure is a **degenerate conic,** as shown in Figure 4.16.

Point Line Two Intersecting Lines

FIGURE 4.16 *Degenerate Conics*

There are several ways to approach the study of conics. You could begin by defining conics in terms of the intersections of planes and cones, as the Greeks did, or you could define them algebraically, in terms of the general second-degree equation

$$Ax^2 + Bxy + Cy^2 + Dx + Ey + F = 0.$$

However, you will study a third approach, in which each of the conics is defined as a *locus* (collection) of points satisfying a certain geometric property. For example, in Section 1.1 you saw how the definition of a circle as *the collection of all points (x, y) that are equidistant from a fixed point (h, k)* led easily to the standard equation of a circle

$$(x - h)^2 + (y - k)^2 = r^2.$$ Equation of a circle

A computer animation of this concept appears in the *Interactive* CD-ROM and *Internet* versions of this text.

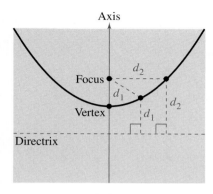

FIGURE 4.17 *Parabola*

Parabolas

In Section 3.1, you learned that the graph of the quadratic function

$$f(x) = ax^2 + bx + c$$

is a parabola that opens upward or downward. The following definition of a parabola is more general in the sense that it is independent of the orientation of the parabola.

Definition of a Parabola

A **parabola** is the set of all points (x, y) in a plane that are equidistant from a fixed line, the **directrix,** and a fixed point, the **focus,** not on the line. (See Figure 4.17.) The **vertex** is the midpoint between the focus and the directrix. The **axis** of the parabola is the line passing through the focus and the vertex.

A proof of the following standard form of the equation of a parabola is given in Appendix A.

Standard Equation of a Parabola (Vertex at Origin)

The **standard form of the equation of a parabola** with vertex at $(0, 0)$ and directrix $y = -p$ is

$$x^2 = 4py, \qquad p \neq 0. \qquad \text{Vertical axis}$$

For directrix $x = -p$, the equation is

$$y^2 = 4px, \qquad p \neq 0. \qquad \text{Horizontal axis}$$

The focus is on the axis p units (directed distance) from the vertex.

Notice that a parabola can have a vertical or a horizontal axis. Examples of each are shown in Figure 4.18.

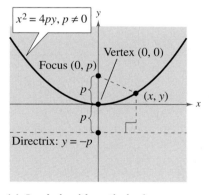

(a) Parabola with vertical axis

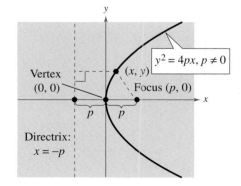

(b) Parabola with horizontal axis

FIGURE 4.18

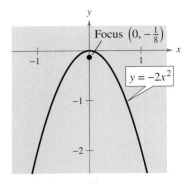

FIGURE **4.19**

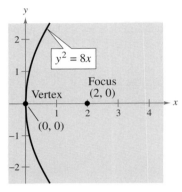

FIGURE **4.20**

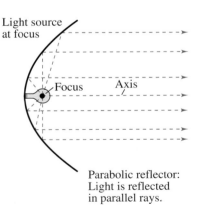

Parabolic reflector:
Light is reflected
in parallel rays.

FIGURE **4.21**

Example 1 ▶ Finding the Focus of a Parabola

Find the focus of the parabola whose equation is $y = -2x^2$.

Solution

Because the squared term in the equation involves x, you know that the axis is vertical, and the equation is of the form

$$x^2 = 4py.$$

You can write the given equation in this form as follows.

$$x^2 = -\frac{1}{2}y$$

$$x^2 = 4\left(-\frac{1}{8}\right)y \qquad \text{Write in standard form.}$$

So, $p = -\frac{1}{8}$. Because p is negative, the parabola opens downward (see Figure 4.19), and the focus of the parabola is

$$(0, p) = \left(0, -\frac{1}{8}\right). \qquad \text{Focus}$$

Example 2 ▶ A Parabola with a Horizontal Axis

Write the standard form of the equation of the parabola with vertex at the origin and focus at $(2, 0)$.

Solution

The axis of the parabola is horizontal, passing through $(0, 0)$ and $(2, 0)$, as shown in Figure 4.20. So, the standard form is

$$y^2 = 4px.$$

Because the focus is $p = 2$ units from the vertex, the equation is

$$y^2 = 4(2)x$$

$$y^2 = 8x.$$

Parabolas occur in a wide variety of applications. For instance, a parabolic reflector can be formed by revolving a parabola about its axis. The resulting surface has the property that all incoming rays parallel to the axis are reflected through the focus of the parabola. This is the principle behind the construction of the parabolic mirrors used in reflecting telescopes. Conversely, the light rays emanating from the focus of a parabolic reflector used in a flashlight are all parallel to one another, as shown in Figure 4.21.

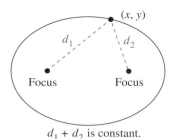

$d_1 + d_2$ is constant.

FIGURE **4.22**

A computer animation of this concept appears in the *Interactive* CD-ROM and *Internet* versions of this text.

FIGURE **4.23**

◀ **E x p l o r a t i o n** ▶

An ellipse can be drawn using two thumbtacks placed at the foci of the ellipse, a string of fixed length (greater than the distance between the tacks), and a pencil, as shown in Figure 4.23. Try doing this. Vary the length of the string and the distance between the thumbtacks. Explain how to obtain ellipses that are almost circular. Explain how to obtain ellipses that are long and narrow.

Ellipses

Definition of an Ellipse

An **ellipse** is the set of all points (x, y) in a plane the sum of whose distances from two distinct fixed points **(foci)** is constant. See Figure 4.22.

The line through the foci intersects the ellipse at two points **(vertices).** The chord joining the vertices is the **major axis,** and its midpoint is the **center** of the ellipse. The chord perpendicular to the major axis at the center is the **minor axis.**

   You can visualize the definition of an ellipse by imagining two thumbtacks placed at the foci, as shown in Figure 4.23. If the ends of a fixed length of string are fastened to the thumbtacks and the string is *drawn taut* with a pencil, the path traced by the pencil will be an ellipse.

   The standard form of the equation of an ellipse takes one of two forms, depending on whether the major axis is horizontal or vertical.

Standard Equation of an Ellipse (Center at Origin)

The **standard form of the equation of an ellipse** centered at the origin with major and minor axes of lengths $2a$ and $2b$ (where $0 < b < a$) is

$$\frac{x^2}{a^2} + \frac{y^2}{b^2} = 1 \qquad \text{or} \qquad \frac{x^2}{b^2} + \frac{y^2}{a^2} = 1.$$

The vertices and foci lie on the major axis, a and c units, respectively, from the center, as shown in Figure 4.24. Moreover, a, b, and c are related by the equation $c^2 = a^2 - b^2$.

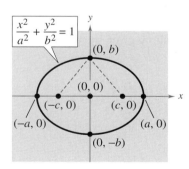

(a) Major axis is horizontal; minor axis is vertical.

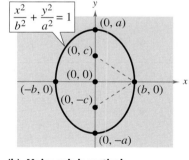

(b) Major axis is vertical; minor axis is horizontal.

FIGURE **4.24**

   In Figure 4.24(a), note that because the sum of the distances from a point on the ellipse to the two foci is constant,

Sum of distances from $(0, b)$ to foci $=$ sum of distances from $(a, 0)$ to foci

$$2\sqrt{b^2 + c^2} = (a + c) + (a - c)$$
$$\sqrt{b^2 + c^2} = a$$
$$c^2 = a^2 - b^2.$$

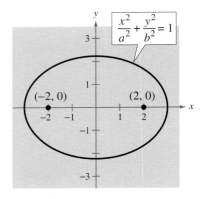

FIGURE **4.25**

Technology

Conics can be graphed using a graphing calculator by first solving for y. For example,

$$\frac{(x-2)^2}{9} + \frac{(y-1)^2}{1} = 1$$

can be graphed by graphing both

$$y = 1 - \sqrt{1 - \frac{(x-2)^2}{9}}$$

and

$$y = 1 + \sqrt{1 - \frac{(x-2)^2}{9}}.$$

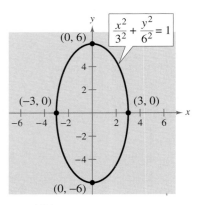

FIGURE **4.26**

Example 3 ▶ **Finding the Standard Equation of an Ellipse**

Find the standard form of the equation of the ellipse that has a major axis of length 6 and foci at $(-2, 0)$ and $(2, 0)$, as shown in Figure 4.25.

Solution

Because the foci occur at $(-2, 0)$ and $(2, 0)$, the center of the ellipse is $(0, 0)$ and the major axis is horizontal. So, the ellipse has an equation of the form

$$\frac{x^2}{a^2} + \frac{y^2}{b^2} = 1. \qquad \text{Write in standard form.}$$

Because the length of the major axis is 6,

$$2a = 6. \qquad \text{Length of major axis}$$

This implies that $a = 3$. Moreover, the distance from the center to either focus is $c = 2$. Finally,

$$b^2 = a^2 - c^2$$
$$= 3^2 - 2^2$$
$$= 9 - 4$$
$$= 5.$$

Substituting $a^2 = 3^2$ and $b^2 = \left(\sqrt{5}\right)^2$ yields the equation

$$\frac{x^2}{3^2} + \frac{y^2}{\left(\sqrt{5}\right)^2} = 1.$$

Example 4 ▶ **Sketching an Ellipse**

Sketch the ellipse given by $4x^2 + y^2 = 36$, and identify the vertices.

Solution

$$4x^2 + y^2 = 36 \qquad \text{Write original equation.}$$

$$\frac{4x^2}{36} + \frac{y^2}{36} = \frac{36}{36} \qquad \text{Divide each side by 36.}$$

$$\frac{x^2}{3^2} + \frac{y^2}{6^2} = 1 \qquad \text{Write in standard form.}$$

Because the denominator of the y^2-term is larger than the denominator of the x^2-term, you can conclude that the major axis is vertical. Moreover, because $a = 6$, the vertices are $(0, -6)$ and $(0, 6)$. Finally, because $b = 3$, the endpoints of the minor axis (or co-vertices) are $(-3, 0)$ and $(3, 0)$, as shown in Figure 4.26. Note that you can sketch the ellipse by locating the endpoints of the two axes. Because 3^2 is the denominator of the x^2-term, move 3 units to the *right and left* of the center to locate the endpoints of the horizontal axis. Similarly, because 6^2 is the denominator of the y^2-term, move 6 units *up and down* from the center to locate the endpoints of the vertical axis.

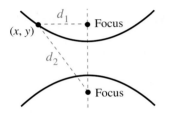

$d_2 - d_1$ is a positive constant.

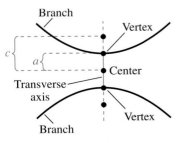

FIGURE 4.27

Hyperbolas

The definition of a **hyperbola** is similar to that of an ellipse. The difference is that for an ellipse the *sum* of the distances between the foci and a point on the ellipse is constant, whereas for a hyperbola the *difference* of the distances between the foci and a point on the hyperbola is constant.

Definition of a Hyperbola

A **hyperbola** is the set of all points (x, y) in a plane the difference of whose distances from two distinct fixed points (**foci**) is a positive constant. See Figure 4.27.

The graph of a hyperbola has two disconnected parts (**branches**). The line through the two foci intersects the hyperbola at two points (**vertices**). The line segment connecting the vertices is the **transverse axis,** and the midpoint of the transverse axis is the **center** of the hyperbola.

Standard Equation of a Hyperbola (Center at Origin)

The **standard form of the equation of a hyperbola** with center at the origin (where $a \neq 0$ and $b \neq 0$) is

$$\frac{x^2}{a^2} - \frac{y^2}{b^2} = 1 \qquad \text{Transverse axis is horizontal.}$$

or

$$\frac{y^2}{a^2} - \frac{x^2}{b^2} = 1. \qquad \text{Transverse axis is vertical.}$$

The vertices and foci are, respectively, a and c units from the center. Moreover, a, b, and c are related by the equation $b^2 = c^2 - a^2$. See Figure 4.28.

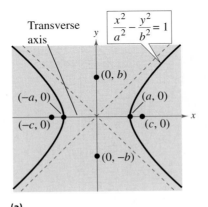

(a)

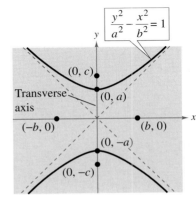

(b)

FIGURE 4.28

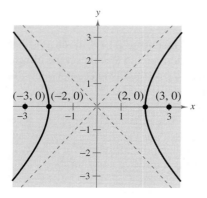

FIGURE 4.29

Example 5 ▶ Finding the Standard Equation of a Hyperbola

Find the standard form of the equation of the hyperbola with foci at $(-3, 0)$ and $(3, 0)$ and vertices at $(-2, 0)$ and $(2, 0)$, as shown in Figure 4.29.

Solution

From the graph, you can determine that $c = 3$, because the foci are 3 units from the center. Moreover, $a = 2$ because the vertices are 2 units from the center. So, it follows that

$$b^2 = c^2 - a^2$$
$$= 3^2 - 2^2$$
$$= 9 - 4$$
$$= 5.$$

Because the transverse axis is horizontal, the standard form of the equation is

$$\frac{x^2}{a^2} - \frac{y^2}{b^2} = 1.$$

Finally, substitute $a^2 = 2^2$ and $b^2 = \left(\sqrt{5}\right)^2$ to obtain

$$\frac{x^2}{2^2} - \frac{y^2}{\left(\sqrt{5}\right)^2} = 1. \qquad \text{Write in standard form.}$$

An important aid in sketching the graph of a hyperbola is the determination of its *asymptotes*, as shown in Figure 4.30. Each hyperbola has two asymptotes that intersect at the center of the hyperbola. Furthermore, the asymptotes pass through the corners of a rectangle of dimensions $2a$ by $2b$. The line segment of length $2b$ joining $(0, b)$ and $(0, -b)$ [or $(-b, 0)$ and $(b, 0)$] is the **conjugate axis** of the hyperbola.

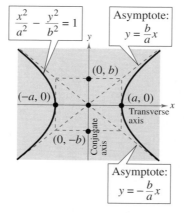

(a) **Transverse axis is horizontal; conjugate axis is vertical.**

(b) **Transverse axis is vertical; conjugate axis is horizontal.**

FIGURE 4.30

Asymptotes of a Hyperbola (Center at Origin)

The **asymptotes of a hyperbola** with center at $(0, 0)$ are

$$y = \frac{b}{a}x \quad \text{and} \quad y = -\frac{b}{a}x \qquad \text{Transverse axis is horizontal.}$$

or

$$y = \frac{a}{b}x \quad \text{and} \quad y = -\frac{a}{b}x. \qquad \text{Transverse axis is vertical.}$$

◀ **Exploration** ▶

Use a graphing utility to graph the hyperbola in Example 6. Does your graph look like that shown in Figure 4.31(b)? If not, what must you do to obtain both the upper and lower portions of the hyperbola? Explain your reasoning.

Example 6 ▶ Sketching the Graph of a Hyperbola

Sketch the graph of the hyperbola whose equation is

$$4x^2 - y^2 = 16.$$

Solution

$$4x^2 - y^2 = 16 \qquad \text{Write original equation.}$$

$$\frac{4x^2}{16} - \frac{y^2}{16} = \frac{16}{16} \qquad \text{Divide each side by 16.}$$

$$\frac{x^2}{2^2} - \frac{y^2}{4^2} = 1 \qquad \text{Write in standard form.}$$

Because the x^2-term is positive, you can conclude that the transverse axis is horizontal and the vertices occur at $(-2, 0)$ and $(2, 0)$. Moreover, the endpoints of the conjugate axis occur at $(0, -4)$ and $(0, 4)$, and you can sketch the rectangle shown in Figure 4.31(a). Finally, by drawing the asymptotes through the corners of this rectangle, you can complete the sketch shown in Figure 4.31(b).

The *Interactive* CD-ROM and *Internet* versions of this text show every example with its solution; clicking on the *Try It!* button brings up similar problems. Guided Examples and Integrated Examples show step-by-step solutions to additional examples. Integrated Examples are related to several concepts in the section.

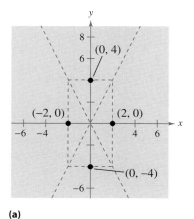

(a)
FIGURE 4.31

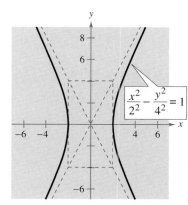

(b)

Example 7 ▶   Finding the Standard Equation of a Hyperbola

Find the standard form of the equation of the hyperbola that has vertices at $(0, -3)$ and $(0, 3)$ and asymptotes $y = -2x$ and $y = 2x$, as shown in Figure 4.32.

Solution

Because the transverse axis is vertical, the asymptotes are of the forms

$$y = \frac{a}{b}x \quad \text{and} \quad y = -\frac{a}{b}x.$$

Using the fact that $y = 2x$ and $y = -2x$, you can determine that

$$\frac{a}{b} = 2.$$

Because $a = 3$, you can determine that $b = \frac{3}{2}$. Finally, you can conclude that the hyperbola has the following equation.

$$\frac{y^2}{3^2} - \frac{x^2}{(3/2)^2} = 1 \qquad \text{Write in standard form.}$$

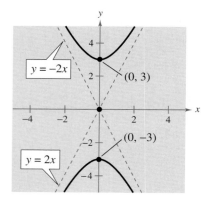

FIGURE 4.32

Writing ABOUT MATHEMATICS

Hyperbolas in Application At the beginning of this section, you learned that each type of conic section can be formed by the intersection of a plane and a double-napped cone. The figure below shows three examples of how such intersections can occur in physical situations.

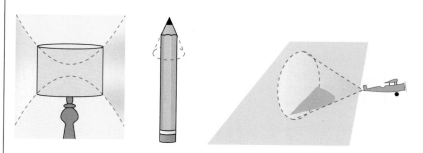

Identify the cone and hyperbola (or portion of a hyperbola) in each of the three situations. Write a short paragraph describing other examples of physical situations in which hyperbolas are formed.

4.4 Exercises

In Exercises 1–10, match the equation with its graph. If the graph of an equation is not shown, write "not shown." [The graphs are labeled (a), (b), (c), (d), (e), (f), (g), and (h).]

(a)
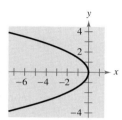

(b)
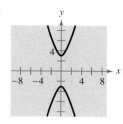

(c)
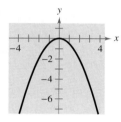

(d)
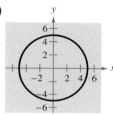

(e)
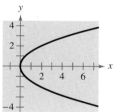

(f)
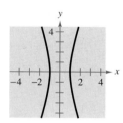

(g)
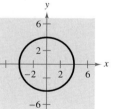

(h)
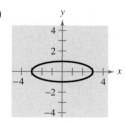

1. $x^2 = 2y$

2. $x^2 = -2y$

3. $y^2 = 2x$

4. $y^2 = -2x$

5. $9x^2 + y^2 = 9$

6. $x^2 + 9y^2 = 9$

7. $9x^2 - y^2 = 9$

8. $y^2 - 9x^2 = 9$

9. $x^2 + y^2 = 49$

10. $x^2 + y^2 = 16$

In Exercises 11–16, find the vertex and focus of the parabola and sketch its graph.

11. $y = \frac{1}{2}x^2$

12. $y = 2x^2$

13. $y^2 = -6x$

14. $y^2 = 3x$

15. $x^2 + 8y = 0$

16. $x + y^2 = 0$

In Exercises 17–26, find an equation of the parabola with vertex at the origin.

17. Focus: $(2, 0)$

18. Focus: $(-2, 0)$

19. Focus: $\left(0, -\frac{3}{2}\right)$

20. Focus: $(0, -2)$

21. Directrix: $y = -1$

22. Directrix: $y = 2$

23. Directrix: $x = 3$

24. Directrix: $x = -2$

25. Passes through the point $(4, 6)$; horizontal axis

26. Passes through the point $(-2, -2)$; vertical axis

In Exercises 27–30, find the equation in standard form of the parabola and determine the coordinates of the focus.

27.
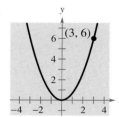

28.
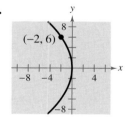

29.
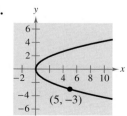

30.
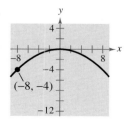

31. *Satellite Antenna* Write an equation for a cross section of the parabolic television dish antenna shown in the figure.

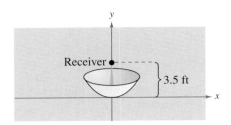

32. *Suspension Bridge* Each cable of a suspension bridge is suspended (in the shape of a parabola) between two towers that are 140 meters apart. The top of each tower is 30 meters above the roadway. The cables touch the roadway at the midpoint between the towers.

(a) Draw a sketch of the bridge. Locate the origin of a rectangular coordinate system at the center of the roadway. Label the coordinates of the known points.

(b) Write an equation that models the cables.

(c) Complete the table by finding the height y of the suspension cables over the roadway at a distance of x meters from the center of the bridge.

x	0	20	40	60	70
y					

33. *Beam Deflection* A simply supported beam (see figure) is 64 feet long and has a load at the center. The deflection of the beam at its center is 1 inch. The shape of the deflected beam is parabolic.

(a) Find an equation of the parabola. (Assume that the origin is at the center of the beam.)

(b) How far from the center of the beam is the deflection $\frac{1}{2}$ inch?

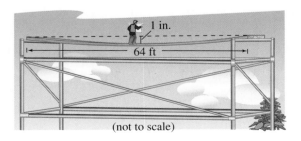

(not to scale)

In Exercises 34–41, find the center and vertices of the ellipse and sketch its graph.

34. $\dfrac{x^2}{25} + \dfrac{y^2}{16} = 1$ **35.** $\dfrac{x^2}{144} + \dfrac{y^2}{169} = 1$

36. $\dfrac{x^2}{25/9} + \dfrac{y^2}{16/9} = 1$ **37.** $\dfrac{x^2}{4} + \dfrac{y^2}{1/4} = 1$

38. $\dfrac{x^2}{9} + \dfrac{y^2}{5} = 1$ **39.** $\dfrac{x^2}{28} + \dfrac{y^2}{64} = 1$

40. $4x^2 + y^2 = 1$ **41.** $4x^2 + 9y^2 = 36$

In Exercises 42–45, use a graphing utility to graph the ellipse. (*Hint:* Use two equations.)

42. $5x^2 + 3y^2 = 15$ **43.** $x^2 + 4y^2 = 4$

44. $9x^2 + y^2 = 81$ **45.** $4x^2 + 25y^2 = 100$

In Exercises 46–55, find an equation of the ellipse with center at the origin.

46.

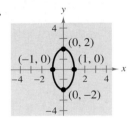

47.

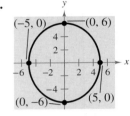

48.

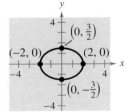

49.

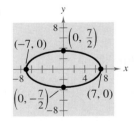

50. Vertices: $(\pm 5, 0)$; Foci: $(\pm 2, 0)$

51. Vertices: $(0, \pm 8)$; Foci: $(0, \pm 4)$

52. Foci: $(\pm 5, 0)$; Major axis of length 12

53. Foci: $(\pm 2, 0)$; Major axis of length 8

54. Vertices: $(0, \pm 5)$; Passes through the point $(4, 2)$

55. Major axis vertical; Passes through the points $(0, 4)$ and $(2, 0)$

56. *Architecture* A fireplace arch is to be constructed in the shape of a semiellipse (see figure). The opening is to have a height of 2 feet at the center and a width of 6 feet along the base. The contractor draws the outline of the ellipse on the wall by the method shown in Figure 4.23. Give the required positions of the tacks and the length of the string.

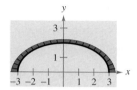

57. *Architecture* A semielliptical arch over a tunnel for a road through a mountain has a major axis of 100 feet and a height at the center of 30 feet.

(a) Make a sketch that illustrates the problem. Draw a rectangular coordinate system on the tunnel with the center of the road entering the tunnel at the origin. Identify the coordinates of the known points.

(b) Find an equation of the semielliptical arch over the tunnel.

(c) Determine the height of the arch 5 feet from each edge of the tunnel.

58. *Geometry* A line segment through a focus of an ellipse with endpoints on the ellipse and perpendicular to the major axis (see figure) is called a **latus rectum** of the ellipse. Therefore, an ellipse has two latera recta. Knowing the length of the latera recta is helpful in sketching an ellipse because it yields other points on the curve. Show that the length of each latus rectum is $2b^2/a$.

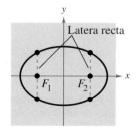

In Exercises 59–62, sketch the graph of the ellipse, using the latera recta (see Exercise 58).

59. $\dfrac{x^2}{4} + \dfrac{y^2}{1} = 1$

60. $\dfrac{x^2}{9} + \dfrac{y^2}{16} = 1$

61. $9x^2 + 4y^2 = 36$

62. $5x^2 + 3y^2 = 15$

In Exercises 63–70, find the center and vertices of the hyperbola and sketch its graph, using asymptotes as sketching aids.

63. $x^2 - y^2 = 1$

64. $\dfrac{x^2}{9} - \dfrac{y^2}{16} = 1$

65. $\dfrac{y^2}{1} - \dfrac{x^2}{4} = 1$

66. $\dfrac{y^2}{9} - \dfrac{x^2}{1} = 1$

67. $\dfrac{y^2}{25} - \dfrac{x^2}{144} = 1$

68. $\dfrac{x^2}{36} - \dfrac{y^2}{4} = 1$

69. $4y^2 - x^2 = 1$

70. $4y^2 - 9x^2 = 36$

In Exercises 71–78, find an equation of the specified hyperbola with center at the origin.

71. Vertices: $(0, \pm 2)$; Foci: $(0, \pm 4)$

72. Vertices: $(\pm 3, 0)$; Foci: $(\pm 5, 0)$

73. Vertices: $(\pm 1, 0)$; Asymptotes: $y = \pm 3x$

74. Vertices: $(0, \pm 3)$; Asymptotes: $y = \pm 3x$

75. Foci: $(0, \pm 8)$; Asymptotes: $y = \pm 4x$

76. Foci: $(\pm 10, 0)$; Asymptotes: $y = \pm\frac{3}{4}x$

77.
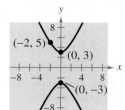

78.
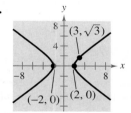

79. *Optics* A hyperbolic mirror (used in some telescopes) has the property that a light ray directed at the focus will be reflected to the other focus. The focus of a hyperbolic mirror (see figure) has coordinates $(24, 0)$. Find the vertex of the mirror if its mount at the top edge of the mirror has coordinates $(24, 24)$.

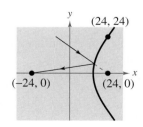

80. *Navigation* Long-distance radio navigation for aircraft and ships uses synchronized pulses transmitted by widely separated transmitting stations. These pulses travel at the speed of light (186,000 miles per second). The difference in the times of arrival of these pulses at an aircraft or ship is constant on a hyperbola having the transmitting stations as foci.

Assume that two stations 300 miles apart are positioned on a rectangular coordinate system at points with coordinates $(-150, 0)$ and $(150, 0)$ and that a ship is traveling on a path with coordinates $(x, 75)$, as shown in the figure. Find the x-coordinate of the position of the ship if the time difference between the pulses from the transmitting stations is 1000 microseconds (0.001 second).

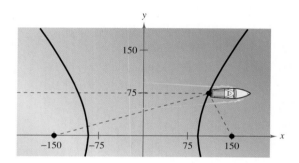

Synthesis

True or False? **In Exercises 81–83, determine whether the statement is true or false. Justify your answer.**

81. The equation $x^2 - y^2 = 144$ represents a circle.

82. The major axis of the ellipse $y^2 + 16x^2 = 64$ is vertical.

83. It is possible for a parabola to intersect its directrix.

84. *Exploration* Consider the ellipse

$$\frac{x^2}{a^2} + \frac{y^2}{b^2} = 1, \qquad a + b = 20.$$

(a) The area of the ellipse is $A = \pi a b$. Write the area of the ellipse as a function of a.

(b) Find the equation of an ellipse with an area of 264 square centimeters.

(c) Complete the table using your equation from part (a), and make a conjecture about the shape of the ellipse with maximum area.

a	8	9	10	11	12	13
A						

(d) Use a graphing utility to graph the area function and use the graph to make a conjecture about the shape of the ellipse that yields a maximum area.

85. *Think About It* Is the graph of $x^2 + 4y^4 = 4$ an ellipse? Explain.

86. *Think About It* The graph of $x^2 - y^2 = 0$ is a degenerate conic. Sketch this graph.

87. *Writing* Write a paragraph discussing the changes in the shape and orientation of the graph of the ellipse

$$\frac{x^2}{a^2} + \frac{y^2}{4^2} = 1$$

as a increases from 1 to 8.

88. Use the definition of an ellipse to derive the standard form of the equation of an ellipse.

89. Use the definition of a hyperbola to derive the standard form of the equation of a hyperbola.

Review

In Exercises 90–93, sketch the graph of the quadratic function without using a graphing utility. Identify the vertex.

90. $f(x) = x^2 - 8$

91. $f(x) = 25 - x^2$

92. $f(x) = x^2 + 8x + 12$

93. $f(x) = x^2 - 4x - 21$

94. Find a polynomial with integer coefficients that has the zeros 3, $2 + i$, and $2 - i$.

95. Find all the zeros of $f(x) = 2x^3 - 3x^2 + 50x - 75$ if one of the zeros is $x = \frac{3}{2}$.

96. List the possible rational zeros of the function $g(x) = 6x^4 + 7x^3 - 29x^2 - 28x + 20$.

In Exercises 97–100, write the partial fraction decomposition for the rational expression. Check your result algebraically.

97. $\dfrac{x + 8}{x^2 + 3x - 18}$

98. $\dfrac{x - 2}{x^2 - 6x - 7}$

99. $\dfrac{x - 6}{x^2 - 8x + 15}$

100. $\dfrac{x + 1}{x^2 - 14x - 15}$

4.5 Translations of Conics

Nordisk Pressfoto Copenhagen
Hulton Getty Picture Library

Vertical and Horizontal Shifts of Conics

In Section 4.4 you looked at conic sections whose graphs were in *standard position*. In this section you will study the equations of conic sections that have been shifted vertically or horizontally in the plane.

Standard Forms of Equations of Conics

Circle: Center $= (h, k)$; Radius $= r$

$$(x - h)^2 + (y - k)^2 = r^2$$

Ellipse: Center $= (h, k)$

Major axis length $= 2a$; Minor axis length $= 2b$

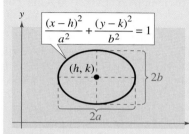

$$\frac{(x-h)^2}{a^2} + \frac{(y-k)^2}{b^2} = 1$$

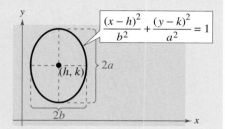

$$\frac{(x-h)^2}{b^2} + \frac{(y-k)^2}{a^2} = 1$$

Hyperbola: Center $= (h, k)$

Transverse axis length $= 2a$; Conjugate axis length $= 2b$

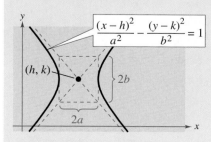

$$\frac{(x-h)^2}{a^2} - \frac{(y-k)^2}{b^2} = 1$$

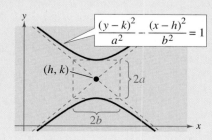

$$\frac{(y-k)^2}{a^2} - \frac{(x-h)^2}{b^2} = 1$$

Parabola: Vertex $= (h, k)$

Directed distance from vertex to focus $= p$

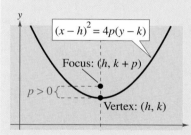

$$(x - h)^2 = 4p(y - k)$$

Focus: $(h, k + p)$

Vertex: (h, k)

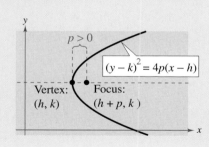

$$(y - k)^2 = 4p(x - h)$$

Vertex: (h, k)

Focus: $(h + p, k)$

Example 1 ▶ Equations of Conic Sections

a. The graph of $(x - 1)^2 + (y + 2)^2 = 3^2$ is a circle whose center is the point $(1, -2)$ and whose radius is 3, as shown in Figure 4.33(a).

b. The graph of

$$\frac{(x - 2)^2}{3^2} + \frac{(y - 1)^2}{2^2} = 1$$

is an ellipse whose center is the point $(2, 1)$. The major axis of the ellipse is horizontal and of length $2(3) = 6$, and the minor axis of the ellipse is vertical and of length $2(2) = 4$, as shown in Figure 4.33(b).

c. The graph of

$$\frac{(x - 3)^2}{1^2} - \frac{(y - 2)^2}{3^2} = 1$$

is a hyperbola whose center is the point $(3, 2)$. The transverse axis is horizontal and of length $2(1) = 2$, and the conjugate axis is vertical and of length $2(3) = 6$, as shown in Figure 4.33(c).

d. The graph of $(x - 2)^2 = 4(-1)(y - 3)$ is a parabola whose vertex is the point $(2, 3)$. The axis of the parabola is vertical. The focus is 1 unit above or below the vertex. Moreover, because $p = -1$, it follows that the focus lies *below* the vertex. [See Figure 4.33(d)].

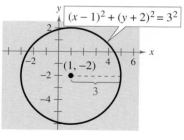

(a)

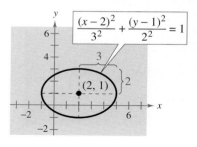

(b)

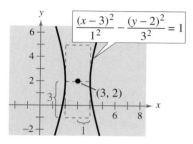

(c)

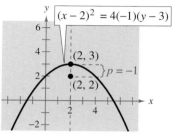

(d)

FIGURE 4.33

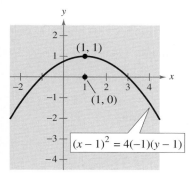

FIGURE **4.34**

Equations of Conics in Standard Form

Example 2 ▶ Finding the Standard Form of a Parabola

Find the vertex and focus of the parabola

$$x^2 - 2x + 4y - 3 = 0.$$

Solution

Complete the square to write the equation in standard form.

$x^2 - 2x + 4y - 3 = 0$	Write original equation.
$x^2 - 2x = -4y + 3$	Group terms.
$x^2 - 2x + 1 = -4y + 3 + 1$	Add 1 to each side.
$(x - 1)^2 = -4y + 4$	Write in completed square form.
$(x - 1)^2 = 4(-1)(y - 1)$	Write in standard form, $(x - h)^2 = 4p(y - k)$.

From this standard form, it follows that $h = 1$, $k = 1$, and $p = -1$. Because the axis is vertical and p is negative, the parabola opens downward. The vertex is $(h, k) = (1, 1)$ and the focus is $(h, k + p) = (1, 0)$. (See Figure 4.34.)

Example 3 ▶ Sketching an Ellipse

Sketch the graph of the ellipse

$$x^2 + 4y^2 + 6x - 8y + 9 = 0.$$

Solution

Complete the square to write the equation in standard form.

$x^2 + 4y^2 + 6x - 8y + 9 = 0$	Write original equation.
$(x^2 + 6x + \ \ ) + (4y^2 - 8y + \ \ ) = -9$	Group terms.
$(x^2 + 6x + \ \ ) + 4(y^2 - 2y + \ \ ) = -9$	Factor 4 out of y-terms.
$(x^2 + 6x + 9) + 4(y^2 - 2y + 1) = -9 + 9 + 4(1)$	Add 9 and $4(1) = 4$ to each side.
$(x + 3)^2 + 4(y - 1)^2 = 4$	Write in completed square form.
$\dfrac{(x + 3)^2}{4} + \dfrac{(y - 1)^2}{1} = 1$	$\dfrac{(x - h)^2}{a^2} + \dfrac{(y - k)^2}{b^2} = 1$

From this standard form, it follows that the center is $(h, k) = (-3, 1)$. Because the denominator of the x-term is $4 = a^2 = 2^2$, the endpoints of the major axis lie 2 units to the right and left of the center. Similarly, because the denominator of the y-term is $1 = b^2 = 1^2$, the endpoints of the minor axis lie 1 unit up and down from the center. The ellipse is shown in Figure 4.35.

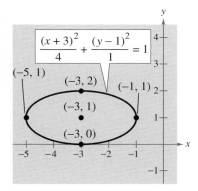

FIGURE **4.35**

Example 4 ▶ Sketching a Hyperbola

Sketch the graph of the hyperbola given by the equation

$$y^2 - 4x^2 + 4y + 24x - 41 = 0.$$

Solution

Complete the square to write the equation in standard form.

$y^2 - 4x^2 + 4y + 24x - 41 = 0$	Write original equation.
$(y^2 + 4y + \ ) - (4x^2 - 24x + \ ) = 41$	Group terms.
$(y^2 + 4y + \ ) - 4(x^2 - 6x + \ ) = 41$	Factor 4 out of x-terms.
$(y^2 + 4y + 4) - 4(x^2 - 6x + 9) = 41 + 4 - 4(9)$	Add 4 and subtract $4(9) = 36$.
$(y + 2)^2 - 4(x - 3)^2 = 9$	Write in completed square form.
$\dfrac{(y + 2)^2}{9} - \dfrac{4(x - 3)^2}{9} = 1$	Divide each side by 9.
$\dfrac{(y + 2)^2}{9} - \dfrac{(x - 3)^2}{\dfrac{9}{4}} = 1$	Change 4 to $\dfrac{1}{\frac{1}{4}}$.
$\dfrac{(y + 2)^2}{3^2} - \dfrac{(x - 3)^2}{\left(\dfrac{3}{2}\right)^2} = 1$	$\dfrac{(y - k)^2}{a^2} - \dfrac{(x - h)^2}{b^2} = 1$

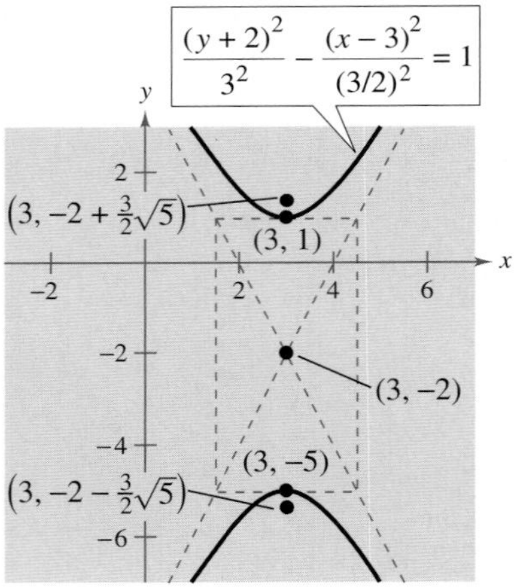

From this standard form, it follows that the transverse axis is vertical and the center lies at $(h, k) = (3, -2)$. Because the denominator of the y-term is $a^2 = 3^2$, you know that the vertices occur 3 units above and below the center.

$$(3, -5) \qquad \text{and} \qquad (3, 1) \qquad \text{Vertices}$$

To sketch the hyperbola, draw a rectangle whose top and bottom pass through the vertices. Because the denominator of the x-term is $b^2 = \left(\frac{3}{2}\right)^2$, locate the sides of the rectangle $\frac{3}{2}$ units to the right and left of the center, as shown in Figure 4.36. Finally, sketch the asymptotes by drawing lines through the opposite corners of the rectangle. Using these asymptotes, you can complete the graph of the hyperbola, as shown in Figure 4.36.

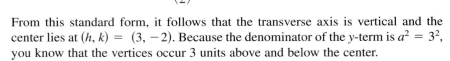

$$\frac{(y+2)^2}{3^2} - \frac{(x-3)^2}{(3/2)^2} = 1$$

FIGURE 4.36

To find the foci in Example 4, first find c.

$$c^2 = a^2 + b^2$$

$$= 9 + \frac{9}{4}$$

$$= \frac{45}{4} \quad \Longrightarrow \quad c = \frac{3\sqrt{5}}{2}$$

Because the transverse axis is vertical, the foci lie c units above and below the center.

$$\left(3, -2 + \tfrac{3}{2}\sqrt{5}\right) \qquad \text{and} \qquad \left(3, -2 - \tfrac{3}{2}\sqrt{5}\right) \qquad \text{Foci}$$

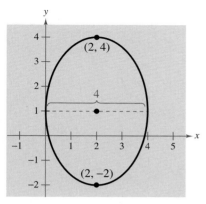

FIGURE **4.37**

Example 5 ▶ Writing the Equation of an Ellipse

Write the standard form of the equation of the ellipse whose vertices are $(2, -2)$ and $(2, 4)$. The length of the minor axis of the ellipse is 4, as shown in Figure 4.37.

Solution

The center of the ellipse lies at the midpoint of its vertices. So, the center is

$$(h, k) = (2, 1). \qquad \text{Center}$$

Because the vertices lie on a vertical line and are 6 units apart, it follows that the major axis is vertical and has a length of $2a = 6$. So, $a = 3$. Moreover, because the minor axis has a length of 4, it follows that $2b = 4$, which implies that $b = 2$. Therefore, the standard form of the ellipse is as follows.

$$\frac{(x - h)^2}{b^2} + \frac{(y - k)^2}{a^2} = 1 \qquad \text{Major axis is vertical.}$$

$$\frac{(x - 2)^2}{2^2} + \frac{(y - 1)^2}{3^2} = 1 \qquad \text{Write in standard form.}$$

An interesting application of conic sections involves the orbits of comets in our solar system. Of the 610 comets identified prior to 1970, 245 have elliptical orbits, 295 have parabolic orbits, and 70 have hyperbolic orbits. For example, Halley's comet has an elliptical orbit, and reappearance of this comet can be predicted every 76 years. The center of the sun is a focus of each of these orbits, and each orbit has a vertex at the point where the comet is closest to the sun, as shown in Figure 4.38.

If p is the distance between the vertex and the focus, and v is the speed of the comet at the vertex, then the orbit is:

an *ellipse* if $v < \sqrt{\dfrac{2GM}{p}}$

a *parabola* if $v = \sqrt{\dfrac{2GM}{p}}$

a *hyperbola* if $v > \sqrt{\dfrac{2GM}{p}}$

where M is the mass of the sun and G is the universal gravitational constant, which is approximately equal to 6.67×10^{-8} cm³/(kg · sec²).

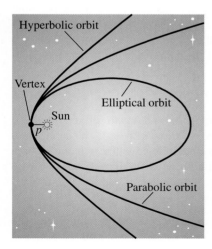

FIGURE **4.38**

Writing ABOUT MATHEMATICS

Identifying Equations of Conics Use the Internet to research information about the orbits of comets in our solar system. What can you find about the orbits of comets that have been identified since 1970? Write a summary of your results. Identify your source. Does it seem reliable?

4.5 Exercises

In Exercises 1–6, identify the center and radius of each circle.

1. $x^2 + y^2 = 49$ **2.** $x^2 + y^2 = 1$

3. $(x + 3)^2 + (y - 8)^2 = 16$

4. $(x + 9)^2 + (y + 1)^2 = 36$

5. $(x - 1)^2 + y^2 = 10$ **6.** $x^2 + (y + 12)^2 = 24$

In Exercises 7–10, write the equation of the circle in standard form, and then identify its center and radius.

7. $x^2 + y^2 - 2x + 6y + 9 = 0$

8. $x^2 + y^2 - 10x - 6y + 25 = 0$

9. $4x^2 + 4y^2 + 12x - 24y + 41 = 0$

10. $9x^2 + 9y^2 + 54x - 36y + 17 = 0$

In Exercises 11–18, find the vertex, focus, and directrix of the parabola, and sketch its graph.

11. $(x - 1)^2 + 8(y + 2) = 0$

12. $(x + 3) + (y - 2)^2 = 0$

13. $\left(y + \frac{1}{2}\right)^2 = 2(x - 5)$ **14.** $\left(x + \frac{1}{2}\right)^2 = 4(y - 3)$

15. $y = \frac{1}{4}(x^2 - 2x + 5)$ **16.** $4x - y^2 - 2y - 33 = 0$

17. $y^2 + 6y + 8x + 25 = 0$

18. $y^2 - 4y - 4x = 0$

 In Exercises 19–22, find the vertex, focus, and directrix of the parabola, and use a graphing utility to graph the parabola.

19. $y = -\frac{1}{6}(x^2 + 4x - 2)$

20. $x^2 - 2x + 8y + 9 = 0$

21. $y^2 + x + y = 0$ **22.** $y^2 - 4x - 4 = 0$

In Exercises 23–30, find an equation of the parabola.

23.

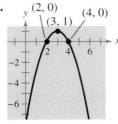

24.

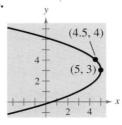

25. Vertex: $(3, 2)$; Focus: $(1, 2)$

26. Vertex: $(-1, 2)$; Focus: $(-1, 0)$

27. Vertex: $(0, 4)$; Directrix: $y = 2$

28. Vertex: $(-2, 1)$; Directrix: $x = 1$

29. Focus: $(2, 2)$; Directrix: $x = -2$

30. Focus: $(0, 0)$; Directrix: $y = 4$

31. *Satellite Orbit* An earth satellite in a 100-mile-high circular orbit around the earth has a velocity of approximately 17,500 miles per hour (see figure). If this velocity is multiplied by $\sqrt{2}$, the satellite will have the minimum velocity necessary to escape the earth's gravity and it will follow a parabolic path with the center of the earth as the focus.

(a) Find the escape velocity of the satellite.

(b) Find an equation of its path (assume that the radius of the earth is 4000 miles).

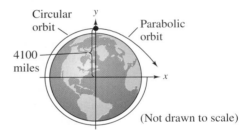

(Not drawn to scale)

32. *Projectile Motion* A bomber is flying at an altitude of 30,000 feet and a speed of 540 miles per hour (792 feet per second). How many feet will a bomb dropped from the plane travel horizontally before it hits the target if the path of the bomb is modeled by

$$y = 30,000 - \frac{x^2}{39,204}?$$

33. *Path of a Projectile* The path of a softball is modeled by $y = -0.08x^2 + x + 4$. The coordinates x and y are measured in feet, with $x = 0$ corresponding to the position from which the ball was thrown.

(a) Use a graphing utility to graph the trajectory of the softball.

(b) Use the trace feature of the graphing utility to approximate the highest point and the range of the trajectory.

34. *Maximum Revenue* The revenue R generated by the sale of x units of a product is $R = 375x - \frac{3}{2}x^2$.

(a) Use a graphing utility to graph the function.

(b) Use the trace feature of the graphing utility to approximate *graphically* the sales that will maximize the revenue.

(c) Use a table to approximate *numerically* the sales that will maximize the revenue.

(d) Find the coordinates of the vertex to determine *analytically* the sales that will maximize the revenue.

(e) Compare the results of parts (b)–(d). What did you learn by using all three approaches?

In Exercises 35–42, find the center, foci, and vertices of the ellipse, and sketch its graph.

35. $\dfrac{(x-1)^2}{9} + \dfrac{(y-5)^2}{25} = 1$

36. $\dfrac{(x-6)^2}{4} + \dfrac{(y+7)^2}{16} = 1$

37. $(x+2)^2 + \dfrac{(y+4)^2}{1/4} = 1$

38. $\dfrac{(x-3)^2}{25/9} + (y-8)^2 = 1$

39. $9x^2 + 4y^2 + 36x - 24y + 36 = 0$

40. $9x^2 + 4y^2 - 36x + 8y + 31 = 0$

41. $16x^2 + 25y^2 - 32x + 50y + 16 = 0$

42. $9x^2 + 25y^2 - 36x - 50y + 61 = 0$

In Exercises 43–52, find an equation for the ellipse.

43.

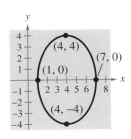

44.

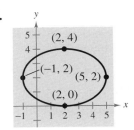

45. Vertices: $(0, 2)$, $(4, 2)$; Minor axis of length 2

46. Foci: $(0, 0)$, $(4, 0)$; Major axis of length 8

47. Foci: $(0, 0)$, $(0, 8)$; Major axis of length 16

48. Center: $(2, -1)$; Vertex: $\left(2, \frac{1}{2}\right)$;
Minor axis of length 2

49. Vertices: $(3, 1)$, $(3, 9)$; Minor axis of length 6

50. Center: $(3, 2)$; $a = 3c$; Foci: $(1, 2)$, $(5, 2)$

51. Center: $(0, 4)$; $a = 2c$; Vertices: $(-4, 4)$, $(4, 4)$

52. Vertices: $(5, 0)$, $(5, 12)$;
Endpoints of the minor axis: $(0, 6)$, $(10, 6)$

In Exercises 53–57, use or determine the eccentricity of the ellipse, which is defined by $e = c/a$. The eccentricity measures the flatness of the ellipse.

53. Find an equation of the ellipse with vertices $(\pm 5, 0)$ and eccentricity $e = \frac{3}{5}$.

54. Find an equation of the ellipse with vertices $(0, \pm 8)$ and eccentricity $e = \frac{1}{2}$.

55. *Distance from the Sun* The planet Pluto moves in an elliptical orbit with the sun at one of the foci, as shown in the figure. The length of half of the major axis, a, is 3.666×10^9 miles, and the eccentricity is 0.248. Find the smallest distance (*perihelion*) and the greatest distance (*aphelion*) of Pluto from the center of the sun.

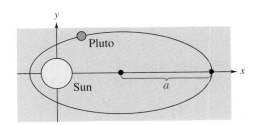

56. *Eccentricity of an Orbit* The first artificial satellite to orbit the earth was Sputnik I (launched by the former Soviet Union in 1957). As shown in the figure, its highest point above the earth's surface was 938 kilometers, and its lowest point was 212 kilometers. The center of the earth is the focus of the elliptical orbit, and the radius of the earth is 6378 kilometers. Find the eccentricity of the orbit.

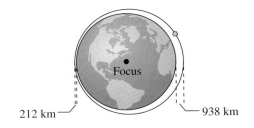

57. *Eccentricity of Orbit* Saturn moves in an elliptical orbit with the sun at one of the foci. The smallest distance and the greatest distance of the planet from the sun are 1.3495×10^9 and 1.5045×10^9 kilometers, respectively. Find the eccentricity of the orbit.

58. *Australian Football* In Australia, football by *Australian Rules* (or rugby) is played on elliptical fields. The field can be a maximum of 170 yards wide and a maximum of 200 yards long. Let the center of a field of maximum size be represented by the point $(0, 85)$. Write an equation of the ellipse that represents this field. (Source: *Oxford Companion to World Sports and Games*)

In Exercises 59–68, find the center, foci, and vertices of the hyperbola, and sketch its graph. Sketch the asymptotes as an aid in obtaining the graph of the hyperbola.

59. $\dfrac{(x-1)^2}{4} - \dfrac{(y+2)^2}{1} = 1$

60. $\dfrac{(x+1)^2}{144} - \dfrac{(y-4)^2}{25} = 1$

61. $(y+6)^2 - (x-2)^2 = 1$

62. $\dfrac{(y-1)^2}{1/4} - \dfrac{(x+3)^2}{1/9} = 1$

63. $9x^2 - y^2 - 36x - 6y + 18 = 0$

64. $x^2 - 9y^2 + 36y - 72 = 0$

65. $x^2 - 9y^2 + 2x - 54y - 80 = 0$

66. $16y^2 - x^2 + 2x + 64y + 63 = 0$

67. $9y^2 - 4x^2 + 8x + 18y + 41 = 0$

68. $11y^2 - 3x^2 + 12x + 44y + 48 = 0$

In Exercises 69–78, find an equation for the hyperbola.

69.

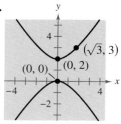

70.

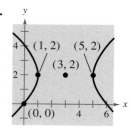

71. Vertices: $(2, 0)$, $(6, 0)$; Foci: $(0, 0)$, $(8, 0)$

72. Vertices: $(2, 3)$, $(2, -3)$; Foci: $(2, 5)$, $(2, -5)$

73. Vertices: $(4, 1)$, $(4, 9)$; Foci: $(4, 0)$, $(4, 10)$

74. Vertices: $(-2, 1)$, $(2, 1)$; Foci: $(-3, 1)$, $(3, 1)$

75. Vertices: $(2, 3)$, $(2, -3)$;
Passes through the point $(0, 5)$

76. Vertices: $(-2, 1)$, $(2, 1)$;
Passes through the point $(4, 3)$

77. Vertices: $(0, 2)$, $(6, 2)$;
Asymptotes: $y = \frac{2}{3}x$, $y = 4 - \frac{2}{3}x$

78. Vertices: $(3, 0)$, $(3, 4)$;
Asymptotes: $y = \frac{2}{3}x$, $y = 4 - \frac{2}{3}x$

In Exercises 79–86, classify the graph of the equation as a circle, a parabola, an ellipse, or a hyperbola.

79. $x^2 + y^2 - 6x + 4y + 9 = 0$

80. $x^2 + 4y^2 - 6x + 16y + 21 = 0$

81. $4x^2 - y^2 - 4x - 3 = 0$

82. $y^2 - 4y - 4x = 0$

83. $4x^2 + 3y^2 + 8x - 24y + 51 = 0$

84. $4y^2 - 2x^2 - 4y - 8x - 15 = 0$

85. $25x^2 - 10x - 200y - 119 = 0$

86. $4x^2 + 4y^2 - 16y + 15 = 0$

Synthesis

True or False? **In Exercises 87 and 88, determine whether the statement is true or false. Justify your answer.**

87. The conic represented by the equation $3x^2 + 2y^2 - 18x - 16y + 58 = 0$ is an ellipse.

88. The graphs of $x^2 + 10y - 10x + 5 = 0$ and $x^2 + 16y^2 + 10x - 32y - 23 = 0$ do not intersect.

Review

In Exercises 89–92, identify the rule of algebra that is illustrated.

89. $(a + 4) - (a + 4) = 0$

90. $u + (v + 10) = (u + v) + 10$

91. $(x + 3)(a + b) = x(a + b) + 3(a + b)$

92. $\dfrac{1}{z - 3}(z - 3) = 1, \quad z \neq 3$

In Exercises 93 and 94, determine whether f and g are inverse functions. If g is not the inverse of f, find the correct inverse.

93. $f(x) = 10 - 7x, \quad g(x) = \dfrac{x - 10}{7}$

94. $f(x) = \sqrt{x + 8}, \quad g(x) = x^2 + 8$

Chapter Summary

What did you learn?

Review Exercises

4.1 In Exercises 1–4, find the domain of the rational function.

1. $f(x) = \dfrac{5x}{x + 12}$

2. $f(x) = \dfrac{3x^2}{1 + 3x}$

3. $f(x) = \dfrac{8}{x^2 - 10x + 24}$

4. $f(x) = \dfrac{x^2 + x - 2}{x^2 + 4}$

In Exercises 5–8, identify any horizontal or vertical asymptotes.

5. $f(x) = \dfrac{4}{x + 3}$

6. $f(x) = \dfrac{2x^2 + 5x - 3}{x^2 + 2}$

7. $g(x) = \dfrac{x^2}{x^2 - 4}$

8. $g(x) = \dfrac{1}{(x - 3)^2}$

9. *Average Cost* A business has a cost of $C = 0.5x + 500$ for producing x units of a product. The average cost per unit is

$$\overline{C} = \frac{C}{x} = \frac{0.5x + 500}{x}, \qquad x > 0.$$

Determine the average cost per unit as x increases without bound. (Find the horizontal asymptote.)

10. *Seizure of Illegal Drugs* The cost (in millions of dollars) for the federal government to seize $p\%$ of a certain illegal drug as it enters the country is

$$C = \frac{528p}{100 - p}, \qquad 0 \le p \le 100.$$

(a) Find the cost of seizing 25% of the drug.

(b) Find the cost of seizing 50% of the drug.

(c) Find the cost of seizing 75% of the drug.

(d) According to this model, would it be possible to seize 100% of the drug?

11. *Physics* The rise of distilled water in a tube of diameter x inches is approximated by the model

$$y = \left(\frac{0.80 - 0.54x}{1 + 2.72x}\right)^2, \qquad x > 0$$

where y is measured in inches. Approximate the diameter of the tube that will cause the water to rise 0.1 inch.

12. *Population Growth* The Parks and Wildlife Commission introduces 80,000 fish into a large human-made lake. The population of the fish (in thousands) is

$$N = \frac{20(4 + 3t)}{1 + 0.05t}, \qquad t \ge 0$$

where t is time in years.

(a) Find the population when t is 5, 10, and 25.

(b) What is the limiting number of fish in the lake as time increases?

4.2 In Exercises 13–24, identify intercepts, check for symmetry, identify any vertical or horizontal asymptotes, and sketch the graph of the rational function.

13. $f(x) = \dfrac{-5}{x^2}$

14. $f(x) = \dfrac{4}{x}$

15. $g(x) = \dfrac{2 + x}{1 - x}$

16. $h(x) = \dfrac{x - 3}{x - 2}$

17. $p(x) = \dfrac{x^2}{x^2 + 1}$

18. $f(x) = \dfrac{2x}{x^2 + 4}$

19. $f(x) = \dfrac{x}{x^2 + 1}$

20. $h(x) = \dfrac{4}{(x - 1)^2}$

21. $f(x) = \dfrac{-6x^2}{x^2 + 1}$

22. $y = \dfrac{2x^2}{x^2 - 4}$

23. $y = \dfrac{x}{x^2 - 1}$

24. $g(x) = \dfrac{-2}{(x + 3)^2}$

In Exercises 25–28, find the equation of the slant asymptote and sketch the graph of the rational function.

25. $f(x) = \dfrac{2x^3}{x^2 + 1}$

26. $f(x) = \dfrac{x^2 + 1}{x + 1}$

27. $f(x) = \dfrac{x^2 + 3x - 10}{x + 2}$

28. $f(x) = \dfrac{x^3}{x^2 - 4}$

29. *Average Cost* The cost of producing x units of a product is C, and the average cost per unit is

$$\overline{C} = \frac{C}{x} = \frac{100,000 + 0.9x}{x}, \qquad x > 0.$$

(a) Graph the average cost function.

(b) Find the average cost of producing $x = 1000$, 10,000, and 100,000 units.

(c) By increasing the level of production, what is the smallest average cost per unit you can obtain? Explain your reasoning.

30. *Numerical and Graphical Analysis* A right triangle is formed in the first quadrant by the x- and y-axes and a line through the point $(2, 3)$.

  (a) Draw a diagram that illustrates the problem. Label the known and unknown quantities.

  (b) Verify that the area of the triangle is

$$A = \frac{3x^2}{2(x - 2)}, \qquad x > 2.$$

  (c) Create a table that gives values of area for various values of x. Start the table with $x = 2.5$ and increment x in steps of 0.5. Continue until you can approximate the dimensions of the triangle of minimum area.

  (d) Use a graphing utility to graph the area function. Use the graph to approximate the dimensions of the triangle of minimum area.

  (e) Determine the slant asymptote of the area function. Explain its meaning.

31. *Photosynthesis* The amount y of CO_2 uptake in milligrams per square decimeter per hour at optimal temperatures and with the natural supply of CO_2 is approximated by the model

$$y = \frac{18.47x - 2.96}{0.23x + 1}, \qquad x > 0$$

where x is the light intensity (in watts per square meter). Use a graphing utility to graph the function and determine the limiting amount of CO_2 uptake.

4.3 In Exercises 32–35, write the form of the partial fraction decomposition for the rational expression. Do not solve for the constants.

32. $\dfrac{3}{x^2 + 20x}$

33. $\dfrac{x - 8}{x^2 - 3x - 28}$

34. $\dfrac{3x - 4}{x^3 - 5x^2}$

35. $\dfrac{x - 2}{x(x^2 + 2)^2}$

In Exercises 36–43, write the partial fraction decomposition for the rational expression.

36. $\dfrac{4 - x}{x^2 + 6x + 8}$

37. $\dfrac{-x}{x^2 + 3x + 2}$

38. $\dfrac{x^2}{x^2 + 2x - 15}$

39. $\dfrac{9}{x^2 - 9}$

40. $\dfrac{x^2 + 2x}{x^3 - x^2 + x - 1}$

41. $\dfrac{4x - 2}{3(x - 1)^2}$

42. $\dfrac{3x^3 + 4x}{(x^2 + 1)^2}$

43. $\dfrac{4x^2}{(x - 1)(x^2 + 1)}$

4.4 In Exercises 44–51, identify the conic.

44. $y^2 = -12x$

45. $16x^2 + y^2 = 16$

46. $\dfrac{x^2}{9} - \dfrac{y^2}{1} = 1$

47. $\dfrac{x^2}{1} + \dfrac{y^2}{9} = 1$

48. $x^2 + 20y = 0$

49. $x^2 + y^2 = 100$

50. $\dfrac{y^2}{49} - \dfrac{x^2}{144} = 1$

51. $\dfrac{x^2}{49} + \dfrac{y^2}{144} = 1$

In Exercises 52–55, write the equation of the specified parabola.

52.

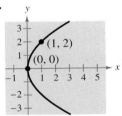

53.

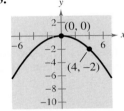

54. Vertex: $(0, 0)$; Focus: $(-6, 0)$

55. Vertex: $(0, 0)$; Focus: $(0, 3)$

56. *Satellite Antenna* A cross section of a large parabolic antenna (see figure) is

$$y = \frac{x^2}{200}, \qquad -100 \le x \le 100.$$

The receiving and transmitting equipment is positioned at the focus. Find the coordinates of the focus.

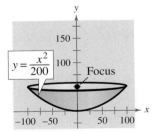

57. *Business* The sales S (in millions of dollars) for Owens Corning for the years 1994 through 1998 are shown in the table. (Source: Owens Corning)

Year	1994	1995	1996	1997	1998
S	3351	3612	3832	4373	5009

A model for this data is

$$S = 76.5t^2 - 510.3t + 4190.2$$

where t is the time (in years), with $t = 4$ corresponding to 1994.

(a) Use a graphing utility to plot the data in the table and graph the model in the same viewing window.

(b) Use the model to estimate the earnings for the year 2000.

In Exercises 58–61, write the equation of the specified ellipse whose center is the origin.

58.

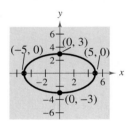

59.

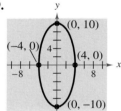

60. Vertices: $(0, \pm 6)$; Passes through the point $(2, 2)$

61. Vertices: $(\pm 7, 0)$; Foci: $(\pm 6, 0)$

62. *Architecture* A semielliptical archway is to be formed over the entrance to an estate (see figure). The arch is to be set on pillars that are 10 feet apart and is to have a height (atop the pillars) of 4 feet. Where should the foci be placed in order to sketch the arch?

In Exercises 63–66, write the equation of the specified hyperbola whose center is the origin.

63.

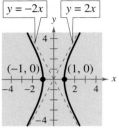

64.

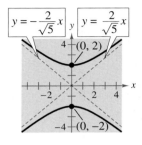

65. Vertices: $(0, \pm 1)$; Foci: $(0, \pm 3)$

66. Vertices: $(\pm 4, 0)$; Foci: $(\pm 6, 0)$

4.5 **In Exercises 67–74, identify the conic by writing its equation in standard form. Then sketch its graph.**

67. $x^2 - 6x + 2y + 9 = 0$

68. $y^2 - 12y - 8x + 20 = 0$

69. $x^2 + y^2 - 2x - 4y + 5 = 0$

70. $16x^2 + 16y^2 - 16x + 24y - 3 = 0$

71. $x^2 + 9y^2 + 10x - 18y + 25 = 0$

72. $4x^2 + y^2 - 16x + 15 = 0$

73. $9x^2 - y^2 - 72x + 8y + 119 = 0$

74. $x^2 - 9y^2 + 10x + 18y + 7 = 0$

In Exercises 75–80, find an equation of the specified parabola.

75.

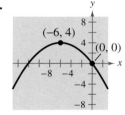

76.

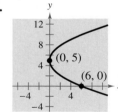

77. Vertex: $(4, 2)$; Focus: $(4, 0)$

78. Vertex: $(2, 0)$; Focus: $(0, 0)$

79. Vertex: $(0, 2)$; Horizontal axis; Passes through the point $(-1, 0)$

80. Vertex: $(2, 2)$; Directrix: $y = 0$

In Exercises 81–86, find an equation of the specified ellipse.

81.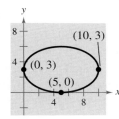

82.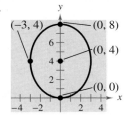

83. Vertices: $(-3, 0)$, $(7, 0)$; Foci: $(0, 0)$, $(4, 0)$

84. Vertices: $(2, 0)$, $(2, 4)$; Foci: $(2, 1)$, $(2, 3)$

85. Vertices: $(0, 1)$, $(4, 1)$; Co-vertices: $(2, 0)$, $(2, 2)$

86. Vertices: $(-4, -1)$, $(-4, 11)$;
Co-vertices: $(-6, 5)$, $(-2, 5)$

In Exercises 87–92, find an equation of the specified hyperbola.

87.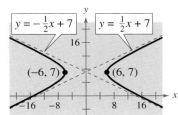

88.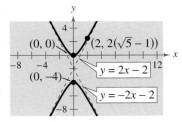

89. Vertices: $(-10, 3)$, $(6, 3)$; Foci: $(-12, 3)$, $(8, 3)$

90. Vertices: $(2, 2)$, $(-2, 2)$; Foci: $(4, 2)$, $(-4, 2)$

91. Foci: $(0, 0)$, $(8, 0)$; Asymptotes: $y = \pm 2(x - 4)$

92. Foci: $(3, \pm 2)$; Asymptotes: $y = \pm 2(x - 3)$

93. *Architecture* A parabolic archway (see figure) is 12 meters high at the vertex. At a height of 10 meters, the width of the archway is 8 meters. How wide is the archway at ground level?

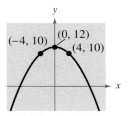

FIGURE FOR 93

94. *Architecture* A church window (see figure) is bounded above by a parabola and below by the arc of a circle.

(a) Find equations for the parabola and the circle.

(b) Complete the table by filling in the vertical distance d between the circle and parabola for each given value of x.

x	0	1	2	3	4
d					

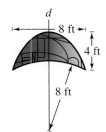

95. *Heating and Plumbing* Find the diameter d of the largest water pipe that can be placed in the corner behind the ventilation duct in the figure.

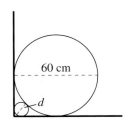

Synthesis

True or False? **In Exercises 96 and 97, determine whether the statement is true or false. Justify your answer.**

96. The domain of a rational function can never be the set of all real numbers.

97. The graph of the equation $Ax^2 + Bxy + Cy^2 + Dx + Ey + F = 0$ can be a single point.

Chapter Project ▶ Graphical Approach to Finding Average Cost

A graphing utility can be used to investigate asymptotic behavior. For instance, the example below shows how to use a graphing utility to visualize how the average cost of producing a product changes as the number of units produced increases.

Example ▶ Finding the Average Cost of a Product

You have started a small business that presses and packages compact discs. Your initial investment is $250,000, and the cost of producing and packaging each disc is $0.32. Describe the *average cost per unit* of producing each disc as a function of x (the number of units produced).

Solution

The total cost C of producing x units is $C = 250{,}000 + 0.32x$. To obtain the average cost per unit $\overline{C}$ of producing x units, divide the total cost by x.

$$\overline{C} = \frac{C}{x} = \frac{250{,}000 + 0.32x}{x} = \frac{250{,}000}{x} + 0.32$$

The graph of the average cost function is shown at the left, together with the graph of the horizontal line $y = 0.32$. Notice that as the number of units produced increases, the average cost per unit gets closer and closer to the unit cost of $0.32.

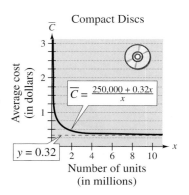

Compact Discs

Average cost (in dollars)

$$\overline{C} = \frac{250{,}000 + 0.32x}{x}$$

$y = 0.32$

Number of units (in millions)

Chapter Project Investigations

1. In the example, would it be possible to produce enough units so that the average cost was exactly $0.32? Would it be possible to produce enough units so that the average cost was arbitrarily close to $0.32?

2. In the example, how many units should be produced to obtain an average cost per unit of $0.33? To answer this question, did you use a numerical approach, an analytical approach, or a graphical approach? Explain why you chose the approach you used.

3. As the production level in the example increases and your business grows, suppose you must invest additional capital to buy more equipment, advertise, build new offices, and so on. You derive the model $C = 250{,}000 + 0.32x + 0.000001x^2$ for the total cost of producing x units. Describe the average cost per unit as a function of the number of units produced.

4. Use a graphing utility to graph the average cost function in Question 3. This graph has a slant asymptote. Find the equation of the slant asymptote, sketch its graph, and interpret its meaning in the context of the problem.

5. Use the zoom feature of a graphing utility to approximate the *minimum* average cost of the function you graphed in Question 4. At what production level does this minimum average cost occur?

Chapter Test

The *Interactive* CD-ROM and *Internet* versions of this text provide answers to the Chapter Tests and Cumulative Tests. They also offer Chapter Pre-Tests (which test key skills and concepts covered in previous chapters) and Chapter Post-Tests, both of which have randomly generated exercises with diagnostic capabilities.

Take this test as you would take a test in class. After you are done, check your work against the answers given in the back of the book.

In Exercises 1–3, find the domain of the function and identify any asymptotes.

1. $y = \dfrac{2}{4 - x}$
2. $f(x) = \dfrac{3 - x^2}{3 + x^2}$
3. $g(x) = \dfrac{x^2 + 2x - 3}{x - 2}$

In Exercises 4 and 5, graph the function. Identify any intercepts and asymptotes.

4. $h(x) = \dfrac{4}{x^2} - 1$
5. $g(x) = \dfrac{x^2 + 2}{x - 1}$

6. Find a rational function that has vertical asymptotes at $x = \pm 2$ and a horizontal asymptote at $y = 3$.

7. A triangle is formed by the coordinate axes and a line through the point $(2, 1)$, as shown in the figure.

(a) Verify that $y = 1 + \dfrac{2}{x - 2}$.

(b) Write the area A of the triangle as a function of x. Determine the domain of the function in the context of the problem.

(c) Graph the area function. Estimate the minimum area of the triangle from the graph.

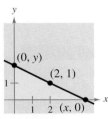

FIGURE FOR **7**

In Exercises 8–11, write the partial fraction decomposition for the rational expression.

8. $\dfrac{2x + 5}{x^2 - x - 2}$
9. $\dfrac{3x^2 - 2x + 4}{x^2(2 - x)}$
10. $\dfrac{x^2 + 5}{x^3 - x}$
11. $\dfrac{x^2 - 4}{x^3 + 2x}$

In Exercises 12–15, graph the conic and identify any vertices and foci.

12. $y^2 - 8x = 0$
13. $y^2 - 4x + 4 = 0$

14. $x^2 - \dfrac{y^2}{4} = 1$
15. $x^2 - 4y^2 - 4x = 0$

16. Find an equation of the ellipse with vertices $(0, 2)$ and $(8, 2)$ and minor axis of length 4.

17. Find an equation of the hyperbola with vertices $(0, \pm 3)$ and asymptotes $y = \pm \frac{3}{2}x$.

18. Use a graphing utility to graph the conics $x^2 + y^2 = 36$ and $x^2 - (y^2/4) = 1$ in the same viewing window and approximate the coordinates of any points of intersection.

19. The moon orbits the earth in an elliptical path with the center of the earth at one focus, as shown in the figure. The major and minor axes of the orbit have lengths of 768,806 kilometers and 767,746 kilometers. Find the smallest distance (perigee) and the greatest distance (apogee) from the center of the moon to the center of the earth.

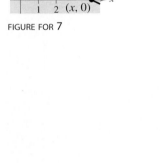

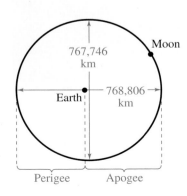

FIGURE FOR **19**

In the United States in 1997, personal savings as a percent of disposable income was 3.9% and the disposable per capita income was $21,969. So, the per capita personal savings in 1997 was $856.79. (Source: U.S. Bureau of Economic Analysis)

Bill Aron/PhotoEdit

5 Exponential and Logarithmic Functions

▶ How to Study This Chapter

The Big Picture

In this chapter you will learn the following skills and concepts.

▶ How to recognize and evaluate exponential and logarithmic functions

▶ How to graph exponential and logarithmic functions

▶ How to rewrite logarithmic functions with a different base

▶ How to use properties of logarithms to evaluate, rewrite, expand, or condense logarithmic expressions

▶ How to solve exponential and logarithmic equations

▶ How to use exponential growth models, exponential decay models, Gaussian models, logistic growth models, and logarithmic models to solve real-life problems

Important Vocabulary

As you encounter each new vocabulary term in this chapter, add the term and its definition to your notebook glossary.

Algebraic functions (p. 380)
Transcendental functions (p. 380)
Exponential function *f* with base *a* (p. 380)
Natural base *e* (p. 384)
Natural exponential function (p. 384)
Continuous compounding (p. 385)
Logarithmic function with base *a* (p. 392)
Common logarithmic function (p. 393)

Natural logarithmic function (p. 396)
Exponential growth model (p. 420)
Exponential decay model (p. 420)
Gaussian model (p. 420)
Logistic growth model (p. 420)
Logarithmic models (p. 420)
Bell-shaped curve (p. 424)
Logistic curve (p. 425)
Sigmoidal curve (p. 425)

Study Tools

• Learning objectives at the beginning of each section
• Chapter Summary (p. 433)
• Review Exercises (pp. 434–437)
• Chapter Test (p. 439)
• Cumulative Test for Chapters 3–5 (pp. 440, 441)

Additional Resources

• Study and Solutions Guide
• Interactive College Algebra
• Videotapes for Chapter 5
• College Algebra Website
• Student Success Organizer

STUDY T!P

Attend every class. Arrive on time with your text, a pen or pencil and paper for notes, and your calculator. If you must miss class, get the notes from another student, go to your tutor for help, or view the appropriate mathematics videotape.

5.1 Exponential Functions and Their Graphs

▶ **What you should learn**

- How to recognize and evaluate exponential functions with base a
- How to graph exponential functions
- How to recognize and evaluate exponential functions with base e
- How to use exponential functions to model and solve real-life applications

▶ **Why you should learn it**

Exponential functions can be used to model and solve real-life problems. For instance, Exercise 74 on page 390 shows how to use an exponential function to model the amount of defoliation caused by the gypsy moth.

Exponential Functions

So far, this book has dealt only with **algebraic functions,** which include polynomial functions and rational functions. In this chapter you will study two types of nonalgebraic functions—*exponential* functions and *logarithmic* functions. These functions are examples of **transcendental functions.**

Definition of Exponential Function

The **exponential function** f **with base** a is denoted by

$$f(x) = a^x$$

where $a > 0$, $a \neq 1$, and x is any real number.

The base $a = 1$ is excluded because it yields $f(x) = 1^x = 1$. This is a constant function, not an exponential function.

You already know how to evaluate a^x for integer and rational values of x. For example, you know that $4^3 = 64$ and $4^{1/2} = 2$. However, to evaluate 4^x for any real number x, you need to interpret forms with *irrational* exponents. For the purposes of this book, it is sufficient to think of

$$a^{\sqrt{2}} \quad (\text{where } \sqrt{2} \approx 1.41421356)$$

as the number that has the successively closer approximations

$$a^{1.4},\ a^{1.41},\ a^{1.414},\ a^{1.4142},\ a^{1.41421},\ \ldots .$$

Example 1 shows how to use a calculator to evaluate exponential expressions.

Example 1 ▶ Evaluating Exponential Expressions

Use a calculator to evaluate each expression.

a. $2^{-3.1}$

b. $2^{-\pi}$

c. $12^{5/7}$

d. $(0.6)^{3/2}$

Solution

Number	Graphing Calculator Keystrokes	Display
a. $2^{-3.1}$	2 $\boxed{\wedge}$ $\boxed{(-)}$ 3.1 $\boxed{\text{ENTER}}$	0.1166291
b. $2^{-\pi}$	2 $\boxed{\wedge}$ $\boxed{(-)}$ π $\boxed{\text{ENTER}}$	0.1133147
c. $12^{5/7}$	12 $\boxed{\wedge}$ $\boxed{(}$ 5 $\boxed{\div}$ 7 $\boxed{)}$ $\boxed{\text{ENTER}}$	5.8998877
d. $(0.6)^{3/2}$	.6 $\boxed{\wedge}$ $\boxed{(}$ 3 $\boxed{\div}$ 2 $\boxed{)}$ $\boxed{\text{ENTER}}$	0.4647580

Graphs of Exponential Functions

The graphs of all exponential functions have similar characteristics, as shown in Examples 2, 3, and 4.

Example 2 ▶ Graphs of $y = a^x$

In the same coordinate plane, sketch the graph of each function.

a. $f(x) = 2^x$ **b.** $g(x) = 4^x$

Solution

The table below lists some values for each function, and Figure 5.1 shows their graphs. Note that both graphs are increasing. Moreover, the graph of $g(x) = 4^x$ is increasing more rapidly than the graph of $f(x) = 2^x$.

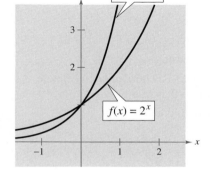

FIGURE **5.1**

x	-2	-1	0	1	2	3
2^x	$\frac{1}{4}$	$\frac{1}{2}$	1	2	4	8
4^x	$\frac{1}{16}$	$\frac{1}{4}$	1	4	16	64

The table in Example 2 was evaluated by hand. You could, of course, use a graphing utility to construct tables with even more values.

Example 3 ▶ Graphs of $y = a^{-x}$

In the same coordinate plane, sketch the graph of each function.

a. $F(x) = 2^{-x}$ **b.** $G(x) = 4^{-x}$

Solution

The table below lists some values for each function, and Figure 5.2 shows their graphs. Note that both graphs are decreasing. Moreover, the graph of $G(x) = 4^{-x}$ is decreasing more rapidly than the graph of $F(x) = 2^{-x}$.

FIGURE **5.2**

x	-3	-2	-1	0	1	2
2^{-x}	8	4	2	1	$\frac{1}{2}$	$\frac{1}{4}$
4^{-x}	64	16	4	1	$\frac{1}{4}$	$\frac{1}{16}$

In Example 3, note that the functions $F(x) = 2^{-x}$ and $G(x) = 4^{-x}$ can be rewritten with positive exponents.

$$F(x) = 2^{-x} = \left(\frac{1}{2}\right)^x \quad \text{and} \quad G(x) = 4^{-x} = \left(\frac{1}{4}\right)^x$$

Comparing the functions in Examples 2 and 3, observe that

$$F(x) = 2^{-x} = f(-x) \qquad \text{and} \qquad G(x) = 4^{-x} = g(-x).$$

Consequently, the graph of F is a reflection (in the y-axis) of the graph of f. The graphs of G and g have the same relationship. The graphs in Figures 5.1 and 5.2 are typical of the exponential functions a^x and a^{-x}. They have one y-intercept and one horizontal asymptote (the x-axis), and they are continuous. The basic characteristics of these exponential functions are summarized in Figures 5.3 and 5.4.

STUDY TIP

Notice that the range of an exponential function is $(0, \infty)$, which means that $a^x > 0$ for all values of x.

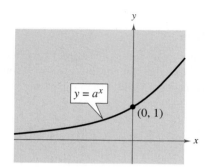

Graph of $y = a^x, a > 1$
- Domain: $(-\infty, \infty)$
- Range: $(0, \infty)$
- Intercept: $(0, 1)$
- Increasing
- x-Axis is a horizontal asymptote $(a^x \to 0 \text{ as } x \to -\infty)$.
- Continuous

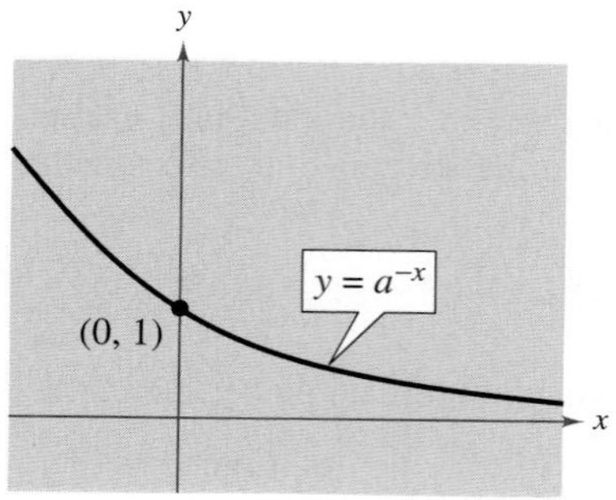

FIGURE 5.3

Graph of $y = a^{-x}, a > 1$
- Domain: $(-\infty, \infty)$
- Range: $(0, \infty)$
- Intercept: $(0, 1)$
- Decreasing
- x-Axis is a horizontal asymptote $(a^{-x} \to 0 \text{ as } x \to \infty)$.
- Continuous

FIGURE 5.4

The *Interactive* CD-ROM and *Internet* versions of this text offer a built-in graphing calculator, which can be used in the Examples, Explorations, Technology notes, and Exercises.

◀ **Exploration** ▶

Use a graphing utility to graph

$$y = a^x$$

for $a = 3$, 5, and 7 in the same viewing window. (Use a viewing window in which $-2 \le x \le 1$ and $0 \le y \le 2$.) For instance, the graph of

$$y = 3^x$$

is shown in Figure 5.5. How do the graphs compare with each other? Which graph is on the top in the interval $(-\infty, 0)$? Which is on the bottom? Which graph is on the top in the interval $(0, \infty)$? Which is on the bottom?

Repeat this experiment with the graphs of $y = b^x$ for $b = \frac{1}{3}, \frac{1}{5}$, and $\frac{1}{7}$. (Use a viewing window in which $-1 \le x \le 2$ and $0 \le y \le 2$.) What can you conclude about the shape of the graph of $y = b^x$ and the value of b?

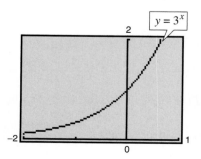

FIGURE 5.5

In the following example, notice how the graph of $y = a^x$ can be used to sketch the graphs of functions of the form $f(x) = b \pm a^{x+c}$.

Example 4 ▶ Transformations of Graphs of Exponential Functions

Each of the following graphs is a transformation of the graph of $f(x) = 3^x$, as shown in Figure 5.6.

a. Because $g(x) = 3^{x+1} = f(x + 1)$, the graph of g can be obtained by shifting the graph of f 1 unit to the left.

b. Because $h(x) = 3^x - 2 = f(x) - 2$, the graph of h can be obtained by shifting the graph of f down 2 units.

c. Because $k(x) = -3^x = -f(x)$, the graph of k can be obtained by reflecting the graph of f in the x-axis.

d. Because $j(x) = 3^{-x} = f(-x)$, the graph of j can be obtained by reflecting the graph of f in the y-axis.

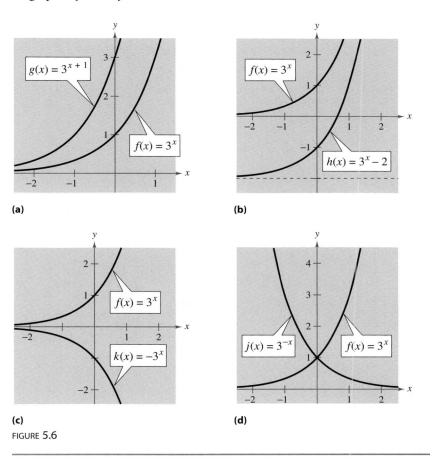

(a)

(b)

(c)

(d)

FIGURE 5.6

In Figure 5.6, notice that the transformations in parts (a), (c), and (d) keep the x-axis as a horizontal asymptote, but the transformation in part (b) yields a new horizontal asymptote of $y = -2$. Also, be sure to note how the y-intercept is affected by each transformation.

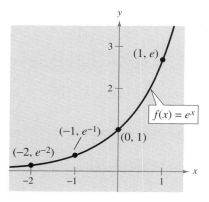

FIGURE **5.7**

The Natural Base e

In many applications, the most convenient choice for a base is the irrational number

$$e \approx 2.718281828 \ldots .$$

This number is called the **natural base.** The function $f(x) = e^x$ is called the **natural exponential function.** Its graph is shown in Figure 5.7. Be sure you see that for the exponential function $f(x) = e^x$, e is the constant $2.71828183 \ldots$, whereas x is the variable.

Example 5 ▶ Evaluating the Natural Exponential Function

Use a calculator to evaluate each expression.

a. e^{-2}

b. e^{-1}

c. e^{1}

d. e^{2}

Solution

Number	Graphing Calculator Keystrokes	Display
a. e^{-2}	e^x $(-)$ 2 ENTER	0.1353353
b. e^{-1}	e^x $(-)$ 1 ENTER	0.3678794
c. e^{1}	e^x 1 ENTER	2.7182818
d. e^{2}	e^x 2 ENTER	7.3890561

Example 6 ▶ Graphing Natural Exponential Functions

Sketch the graph of each natural exponential function.

a. $f(x) = 2e^{0.24x}$

b. $g(x) = \frac{1}{2}e^{-0.58x}$

Solution

To sketch these two graphs, you can use a graphing utility to construct a table of values, as shown below. After constructing the table, plot the points and connect them with smooth curves, as shown in Figure 5.8. Note that the graph in part (a) is increasing whereas the graph in part (b) is decreasing.

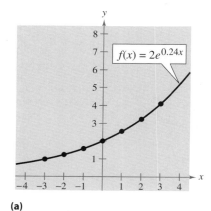

(a)

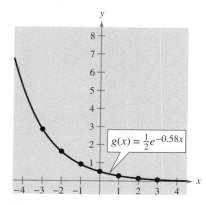

(b)

FIGURE **5.8**

x	-3	-2	-1	0	1	2	3
$f(x)$	0.974	1.238	1.573	2.000	2.542	3.232	4.109
$g(x)$	2.849	1.595	0.893	0.500	0.280	0.157	0.088

Applications

One of the most familiar examples of exponential growth is that of an investment earning *continuously compounded interest*. On page 137 in Section 1.6, you were introduced to the formula for the balance in an account that is compounded *n* times per year. Using exponential functions, you can now *develop* that formula and show how it leads to continuous compounding.

Suppose a principal *P* is invested at an annual interest rate *r*, compounded once a year. If the interest is added to the principal at the end of the year, the new balance P_1 is

$$P_1 = P + Pr$$
$$= P(1 + r).$$

This pattern of multiplying the previous principal by $1 + r$ is then repeated each successive year, as shown below.

Year	Balance After Each Compounding
0	$P = P$
1	$P_1 = P(1 + r)$
2	$P_2 = P_1(1 + r) = P(1 + r)(1 + r) = P(1 + r)^2$
3	$P_3 = P_2(1 + r) = P(1 + r)^2(1 + r) = P(1 + r)^3$
⋮	
t	$P_t = P(1 + r)^t$

To accommodate more frequent (quarterly, monthly, or daily) compounding of interest, let *n* be the number of compoundings per year and let *t* be the number of years. Then the rate per compounding is r/n and the account balance after *t* years is

$$A = P\left(1 + \frac{r}{n}\right)^{nt}. \qquad \text{Amount (balance) with } n \text{ compoundings per year}$$

If you let the number of compoundings *n* increase without bound, the process approaches what is called **continuous compounding.** In the formula for *n* compoundings per year, let $m = n/r$. This produces

$$A = P\left(1 + \frac{r}{n}\right)^{nt} \qquad \text{Amount with } n \text{ compoundings per year}$$

$$= P\left(1 + \frac{r}{mr}\right)^{mrt} \qquad \text{Substitute } mr \text{ for } n.$$

$$= P\left(1 + \frac{1}{m}\right)^{mrt} \qquad \text{Simplify.}$$

$$= P\left[\left(1 + \frac{1}{m}\right)^m\right]^{rt}. \qquad \text{Property of exponents}$$

As *m* increases without bound, it can be shown that $[1 + (1/m)]^m$ approaches *e*. (Try the values $m = 10$, 10,000, and 10,000,000.) From this, you can conclude that the formula for continuous compounding is

$$A = Pe^{rt}. \qquad \text{Substitute } e \text{ for } (1 + 1/m)^m.$$

Formulas for Compound Interest

After t years, the balance A in an account with principal P and annual interest rate r (in decimal form) is given by the following formulas.

1. For n compoundings per year: $A = P\left(1 + \dfrac{r}{n}\right)^{nt}$

2. For continuous compounding: $A = Pe^{rt}$

Example 7 ▶ Compound Interest

A total of $12,000 is invested at an annual interest rate of 9%. Find the balance after 5 years if it is compounded

a. quarterly.

b. monthly.

c. continuously.

Solution

a. For quarterly compoundings, you have $n = 4$. So, in 5 years at 9%, the balance is

$$A = P\left(1 + \frac{r}{n}\right)^{nt}$$ Formula for compound interest

$$= 12{,}000\left(1 + \frac{0.09}{4}\right)^{4(5)}$$ Substitute for P, r, n, and t.

$$\approx \$18{,}726.11.$$ Use a calculator.

b. For monthly compoundings, you have $n = 12$. So, in 5 years at 9%, the balance is

$$A = P\left(1 + \frac{r}{n}\right)^{nt}$$ Formula for compound interest

$$= 12{,}000\left(1 + \frac{0.09}{12}\right)^{12(5)}$$ Substitute for P, r, n, and t.

$$\approx \$18{,}788.17.$$ Use a calculator.

c. For continuous compounding, the balance is

$$A = Pe^{rt}$$ Formula for continuous compounding

$$= 12{,}000e^{0.09(5)}$$ Substitute for P, r, n, and t.

$$\approx \$18{,}819.75.$$ Use a calculator.

In Example 7, note that continuous compounding yields more than quarterly or monthly compounding. This is typical of the two types of compounding. That is, for a given principal, interest rate, and time, continuous compounding will always yield a larger balance than compounding n times a year.

Example 8 ▶ Radioactive Decay

In 1986, a nuclear reactor accident occurred in Chernobyl in what was then the Soviet Union. The explosion spread highly toxic radioactive chemicals, such as plutonium, over hundreds of square miles, and the government evacuated the city and the surrounding area. To see why the city is now uninhabited, consider the model

$$P = 10e^{-0.00002845t}.$$

This model represents the amount of plutonium that remains (from an initial amount of 10 pounds) after t years. Sketch the graph of this function over the interval from $t = 0$ to $t = 100,000$. How much of the 10 pounds will remain in the year 2002? How much of the 10 pounds will remain after 100,000 years?

Solution

The graph of this function is shown in Figure 5.9. Note from this graph that plutonium has a *half-life* of about 24,360 years. That is, after 24,360 years, *half* of the original amount will remain. After another 24,360 years, one-quarter of the original amount will remain, and so on. In the year 2002 ($t = 16$), there will still be

$$P = 10e^{-0.00002845(16)}$$

$$= 10e^{-0.0004552}$$

$$\approx 9.995 \text{ pounds}$$

of plutonium remaining. After 100,000 years, there will still be

$$P = 10e^{-0.00002845(100,000)}$$

$$= 10e^{-2.845}$$

$$\approx 0.58 \text{ pound}$$

of plutonium remaining.

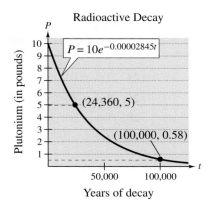

FIGURE 5.9

Writing ABOUT MATHEMATICS

Identifying Exponential Functions Which of the following functions generated the two tables below? Discuss how you were able to decide. What do these functions have in common? Are any the same? If so, explain why.

a. $f_1(x) = 2^{(x+3)}$ **b.** $f_2(x) = 8\left(\frac{1}{2}\right)^x$ **c.** $f_3(x) = \left(\frac{1}{2}\right)^{(x-3)}$

d. $f_4(x) = \left(\frac{1}{2}\right)^x + 7$ **e.** $f_5(x) = 7 + 2^x$ **f.** $f_6(x) = (8)2^x$

x	-1	0	1	2	3
$g(x)$	7.5	8	9	11	15

x	-2	-1	0	1	2
$h(x)$	32	16	8	4	2

Create two different exponential functions of the forms $y = a(b)^x$ and $y = c^x + d$ with y-intercepts of $(0, -3)$.

5.1 Exercises

 The *Interactive* CD-ROM and *Internet* versions of this text contain step-by-step solutions to all odd-numbered Section and Review Exercises. They also provide Tutorial Exercises that link to Guided Examples for additional help.

In Exercises 1–10, evaluate the expression. Round your result to three decimal places.

1. $(3.4)^{5.6}$

2. $5000(2^{-1.5})$

3. $(1.005)^{400}$

4. $8^{2\pi}$

5. $5^{-\pi}$

6. $\sqrt[3]{4395}$

7. $100^{\sqrt{2}}$

8. $e^{1/2}$

9. $e^{-3/4}$

10. $e^{3.2}$

In Exercises 11–14, match the exponential function with its graph. [The graphs are labeled (a), (b), (c), and (d).]

(a)

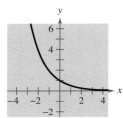

(b)

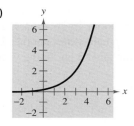

(c)

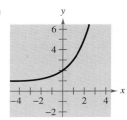

(d)

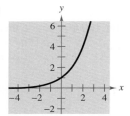

11. $f(x) = 2^x$

12. $f(x) = 2^x + 1$

13. $f(x) = 2^{-x}$

14. $f(x) = 2^{x-2}$

In Exercises 15–22, use the graph of f to describe the transformation that yields the graph of g.

15. $f(x) = 3^x,\quad g(x) = 3^{x-4}$

16. $f(x) = 4^x,\quad g(x) = 4^x + 1$

17. $f(x) = -2^x,\quad g(x) = 5 - 2^x$

18. $f(x) = 10^x,\quad g(x) = 10^{-x+3}$

19. $f(x) = \left(\frac{3}{5}\right)^x,\quad g(x) = -\left(\frac{3}{5}\right)^{x+4}$

20. $f(x) = \left(\frac{7}{2}\right)^x,\quad g(x) = -\left(\frac{7}{2}\right)^{-x+6}$

21. $f(x) = 0.3^x,\quad g(x) = -0.3^x + 5$

22. $f(x) = 3.6^x,\quad g(x) = -3.6^{-x} + 8$

 In Exercises 23–36, use a graphing utility to construct a table of values for each function. Then sketch the graph of the function.

23. $f(x) = \left(\frac{1}{2}\right)^x$

24. $f(x) = 6^x$

25. $f(x) = \left(\frac{1}{2}\right)^{-x}$

26. $f(x) = 6^{-x}$

27. $f(x) = 2^{x-1}$

28. $f(x) = 3^{x+2}$

29. $f(x) = e^x$

30. $f(x) = e^{-x}$

31. $f(x) = 3e^{x+4}$

32. $f(x) = 2e^{-0.5x}$

33. $f(x) = 2e^{x-2} + 4$

34. $f(x) = 2 + e^{x-5}$

35. $f(x) = 4^{x-3} + 3$

36. $f(x) = -4^{x-3} - 3$

In Exercises 37–54, use a graphing utility to graph the exponential function.

37. $g(x) = 5^x$

38. $f(x) = \left(\frac{3}{2}\right)^x$

39. $f(x) = \left(\frac{1}{5}\right)^x = 5^{-x}$

40. $h(x) = \left(\frac{3}{2}\right)^{-x}$

41. $h(x) = 5^{x-2}$

42. $g(x) = \left(\frac{3}{2}\right)^{x+2}$

43. $g(x) = 5^{-x} - 3$

44. $f(x) = \left(\frac{3}{2}\right)^{-x} + 2$

45. $y = 2^{-x^2}$

46. $y = 3^{-|x|}$

47. $y = 3^{x-2} + 1$

48. $y = 4^{x+1} - 2$

49. $y = 1.08^{-5x}$

50. $y = 1.08^{5x}$

51. $s(t) = 2e^{0.12t}$

52. $s(t) = 3e^{-0.2t}$

53. $g(x) = 1 + e^{-x}$

54. $h(x) = e^{x-2}$

Finance In Exercises 55–58, complete the table to determine the balance A for P dollars invested at rate r for t years and compounded n times per year.

n	1	2	4	12	365	Continuous
A						

55. $P = \$2500,\ r = 8\%,\ t = 10$ years

56. $P = \$1000,\ r = 6\%,\ t = 10$ years

57. $P = \$2500,\ r = 8\%,\ t = 20$ years

58. $P = \$1000,\ r = 6\%,\ t = 40$ years

Finance In Exercises 59–62, complete the table to determine the balance A for $12,000 invested at rate r for t years, compounded continuously.

t	1	10	20	30	40	50
A						

59. $r = 8\%$

60. $r = 6\%$

61. $r = 6.5\%$

62. $r = 7.5\%$

63. *Finance* On the day of a child's birth, a deposit of $25,000 is made in a trust fund that pays 8.75% interest, compounded continuously. Determine the balance in this account on the child's 25th birthday.

64. *Finance* A deposit of $5000 is made in a trust fund that pays 7.5% interest, compounded continuously. It is specified that the balance will be given to the college from which the donor graduated after the money has earned interest for 50 years. How much will the college receive?

65. *Graphical Reasoning* There are two options for investing $500. The first earns 7% compounded annually and the second earns 7% simple interest. The figure shows the growth of each investment over a 30-year period.

(a) Identify which graph represents each type of investment in the figure. Explain your reasoning.

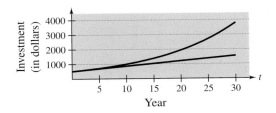

(b) Verify your answer in part (a) by finding the equations that model the investment growth and graphing the models.

66. *Depreciation* After t years, the value of a car that cost $20,000 is modeled by

$$V(t) = 20{,}000\left(\frac{3}{4}\right)^t.$$

Graph the function and determine the value of the car 2 years after it was purchased.

67. *Inflation* If the annual rate of inflation averages 4% over the next 10 years, the approximate cost C of goods or services during any year in that decade will be modeled by

$$C(t) = P(1.04)^t$$

where t is the time in years and P is the present cost. If the price of an oil change for your car is presently $23.95, estimate the price 10 years from now.

68. *Economics* The demand equation for a certain product is

$$p = 5000\left(1 - \frac{4}{4 + e^{-0.002x}}\right).$$

(a) Use a graphing utility to graph the demand function for $x > 0$ and $p > 0$.

(b) Find the price p for a demand of $x = 500$ units.

(c) Use the graph in part (a) to approximate the greatest price that will still yield a demand of at least 600 units.

69. *Population Growth* A certain type of bacterium increases according to the model $P(t) = 100e^{0.2197t}$, where t is the time in hours. Find (a) $P(0)$, (b) $P(5)$, and (c) $P(10)$.

70. *Population Growth* The population of a town increases according to the model $P(t) = 2500e^{0.0293t}$, where t is the time in years, with $t = 0$ corresponding to 1990. Use the model to estimate the population in (a) 2000 and (b) 2010.

71. *Radioactive Decay* Let Q represent a mass of radioactive radium (^{226}Ra) (in grams), whose half-life is 1620 years. The quantity of radium present after t years is

$$Q = 25\left(\frac{1}{2}\right)^{t/1620}.$$

(a) Determine the initial quantity (when $t = 0$).

(b) Determine the quantity present after 1000 years.

(c) Use a graphing utility to graph the function over the interval $t = 0$ to $t = 5000$.

72. *Radioactive Decay* Let Q represent a mass of carbon 14(^{14}C) (in grams), whose half-life is 5730 years. The quantity of carbon 14 present after t years is

$$Q = 10\left(\frac{1}{2}\right)^{t/5730}.$$

(a) Determine the initial quantity (when $t = 0$).

(b) Determine the quantity present after 2000 years.

(c) Sketch the graph of this function over the interval $t = 0$ to $t = 10{,}000$.

73. *Data Analysis* A meteorologist measures the atmospheric pressure P (in pascals) at altitude h (in kilometers). The data is shown in the table.

h	0	5	10	15	20
P	101,293	54,735	23,294	12,157	5069

A model for the data is given by

$$P = 102{,}303e^{-0.137h}.$$

(a) Sketch a scatter plot of the data and graph the model on the same set of axes.

(b) Create a table that compares the model with the sample data.

(c) Estimate the atmospheric pressure at a height of 8 kilometers.

(d) Use the graph in part (a) to estimate the altitude at which the atmospheric pressure is 21,000 pascals.

74. *Data Analysis* To estimate the amount of defoliation caused by the gypsy moth during a given year, a forester counts the number x of egg masses on $\frac{1}{40}$ of an acre (circle of radius 18.6 feet) in the fall. The percent of defoliation y the next spring is shown in the table.   (Source: USDA, Forest Service)

x	0	25	50	75	100
y	12	44	81	96	99

(a) A model for the data is

$$y = \frac{300}{3 + 17e^{-0.065x}}.$$

Use a graphing utility to create a scatter plot of the data and graph the model in the same viewing window.

(b) Create a table that compares the model with the sample data.

(c) Estimate the percent of defoliation if 36 egg masses are counted on $\frac{1}{40}$ acre.

(d) Use the graph in part (a) to estimate the number of egg masses per $\frac{1}{40}$ acre if you observe that $\frac{2}{3}$ of a forest is defoliated the following spring.

Synthesis

True or False? **In Exercises 75 and 76, determine whether the statement is true or false. Justify your answer.**

75. The x-axis is an asymptote for the graph of $f(x) = 10^x$.

76. $e = \dfrac{271{,}801}{99{,}990}$.

Think About It **In Exercises 77–80, use properties of exponents to determine which functions (if any) are the same.**

77. $f(x) = 3^{x-2}$
$g(x) = 3^x - 9$
$h(x) = \frac{1}{9}(3^x)$

78. $f(x) = 4^x + 12$
$g(x) = 2^{2x+6}$
$h(x) = 64(4^x)$

79. $f(x) = 16(4^{-x})$
$g(x) = \left(\frac{1}{4}\right)^{x-2}$
$h(x) = 16(2^{-2x})$

80. $f(x) = 5^{-x} + 3$
$g(x) = 5^{3-x}$
$h(x) = -5^{x-3}$

81. Graph the functions $y = 3^x$ and $y = 4^x$ and use the graphs to solve the inequalities.

(a) $4^x < 3^x$

(b) $4^x > 3^x$

82. Graph the functions $y = \left(\frac{1}{2}\right)^x$ and $y = \left(\frac{1}{4}\right)^x$ and use the graphs to solve the inequalities.

(a) $\left(\frac{1}{4}\right)^x < \left(\frac{1}{2}\right)^x$

(b) $\left(\frac{1}{4}\right)^x > \left(\frac{1}{2}\right)^x$

83. Use a graphing utility to graph each function. Use the graph to find any asymptotes of the function.

(a) $f(x) = \dfrac{8}{1 + e^{-0.5x}}$

(b) $g(x) = \dfrac{8}{1 + e^{-0.5/x}}$

84. Use a graphing utility to graph each function. Use the graph to find where the function is increasing and decreasing, and approximate any relative maximum or minimum values.

(a) $f(x) = x^2 e^{-x}$

(b) $g(x) = x2^{3-x}$

85. Use a graphing utility to graph $y_1 = e^x$ and each of the functions $y_2 = x^2$, $y_3 = x^3$, $y_4 = \sqrt{x}$, and $y_5 = |x|$. Which function increases at the fastest rate as x approaches $+\infty$?

86. *Conjecture*　Use the result of Exercise 85 to make a conjecture about the rate of growth of $y_1 = e^x$ and $y = x^n$, where n is a natural number and x approaches $+\infty$.

87. *Writing*　Use the results of Exercises 85 and 86 to describe what is implied when it is stated that a quantity is growing exponentially.

⊞ **88.** *Graphical Analysis*　Use a graphing utility to graph

$$f(x) = \left(1 + \frac{0.5}{x}\right)^x \quad \text{and} \quad g(x) = e^{0.5}$$

in the same viewing window. What is the relationship between f and g as x increases without bound?

89. *Conjecture*　Use the result of Exercise 88 to make a conjecture about the value of $[1 + (r/x)]^x$ as x increases without bound. Create a table that illustrates your conjecture for $r = 1$.

90. *Think About It*　Which functions are exponential?

　(a) $3x$　　　(b) $3x^2$

　(c) 3^x　　　(d) 2^{-x}

91. *Think About It*　Without using a calculator, why do you know that $2^{\sqrt{2}}$ is greater than 2, but less than 4?

⊞ **92.** *Exploration*　Use a graphing utility to compare the graph of the function $y = e^x$ with the graph of each given function. [$n!$ (read "n factorial") is defined as $n! = 1 \cdot 2 \cdot 3 \cdots (n-1) \cdot n$.]

　(a) $y_1 = 1 + \dfrac{x}{1!}$

　(b) $y_2 = 1 + \dfrac{x}{1!} + \dfrac{x^2}{2!}$

　(c) $y_3 = 1 + \dfrac{x}{1!} + \dfrac{x^2}{2!} + \dfrac{x^3}{3!}$

93. *Pattern Recognition*　Identify the pattern of successive polynomials given in Exercise 92. Extend the pattern one more term and compare the graph of the resulting polynomial function with the graph of $y = e^x$. What do you think this pattern implies?

94. Given the exponential function

$$f(x) = a^x$$

　show that

　(a) $f(u + v) = f(u) \cdot f(v)$.

　(b) $f(2x) = [f(x)]^2$.

Review

In Exercises 95–98, solve for y.

95. $2x - 7y + 14 = 0$

96. $x^2 + 3y = 4$

97. $x^2 + y^2 = 25$

98. $x - |y| = 2$

In Exercises 99–102, sketch the graph of the rational function.

99. $f(x) = \dfrac{2}{9 + x}$

100. $f(x) = \dfrac{4x - 3}{x}$

101. $f(x) = \dfrac{6}{x^2 + 5x - 24}$

102. $f(x) = \dfrac{x^2 - 7x + 12}{x + 2}$

In Exercises 103–106, identify the conic and sketch its graph.

103. $16x^2 + 4y^2 = 64$

104. $x^2 + y^2 = 1$

105. $\dfrac{(x + 3)^2}{9} - \dfrac{(y - 2)^2}{36} = 1$

106. $\dfrac{(x - 1)^2}{25} + \dfrac{(y - 4)^2}{49} = 1$

5.2 Logarithmic Functions and Their Graphs

Mary Kate Kenny

Logarithmic Functions

In Section 2.6, you studied the concept of the inverse of a function. There, you learned that if a function has the property that no horizontal line intersects the graph of the function more than once, the function must have an inverse. By looking back at the graphs of the exponential functions introduced in Section 5.1, you will see that every function of the form

$$f(x) = a^x$$

passes the Horizontal Line Test and therefore must have an inverse. This inverse function is called the **logarithmic function with base a.**

Definition of Logarithmic Function with Base a

For $x > 0$ and $0 < a \neq 1$,

$$y = \log_a x \text{ if and only if } x = a^y.$$

The function given by

$$f(x) = \log_a x$$

is called the **logarithmic function with base a.**

The equations

$$y = \log_a x \qquad \text{and} \qquad x = a^y$$

are equivalent. The first equation is in logarithmic form and the second is in exponential form.

When evaluating logarithms, remember that *a logarithm is an exponent.* This means that $\log_a x$ is the exponent to which a must be raised to obtain x. For instance, $\log_2 8 = 3$ because 2 must be raised to the third power to get 8.

Example 1 ▶ Evaluating Logarithms

Use the definition of logarithmic function to evaluate the logarithms.

a. $\log_2 32$ **b.** $\log_3 27$ **c.** $\log_4 2$

d. $\log_{10} \frac{1}{100}$ **e.** $\log_3 1$ **f.** $\log_2 2$

Solution

a. $\log_2 32 = 5$ because $2^5 = 32$.

b. $\log_3 27 = 3$ because $3^3 = 27$.

c. $\log_4 2 = \frac{1}{2}$ because $4^{1/2} = \sqrt{4} = 2$.

d. $\log_{10} \frac{1}{100} = -2$ because $10^{-2} = \frac{1}{10^2} = \frac{1}{100}$.

e. $\log_3 1 = 0$ because $3^0 = 1$.

f. $\log_2 2 = 1$ because $2^1 = 2$.

Exploration

Complete the table for
$f(x) = 10^x$.

x	-2	-1	0	1	2
$f(x)$					

Complete the table for
$f(x) = \log_{10} x$.

x	$\frac{1}{100}$	$\frac{1}{10}$	1	10	100
$f(x)$					

Compare the two tables. What is the relationship between
$f(x) = 10^x$ and $f(x) = \log_{10} x$?

The logarithmic function with base 10 is called the **common logarithmic function**. On most calculators, this function is denoted by $\boxed{\text{LOG}}$. Because $\log_a x$ is the inverse function of a^x, it follows that the domain of $\log_a x$ is the range of a^x, $(0, \infty)$. In other words, $\log_a x$ is defined only if x is positive.

Example 2 ▶ Evaluating Logarithms on a Calculator

Use a calculator to evaluate each expression.

a. $\log_{10} 10$

b. $2\log_{10} 2.5$

c. $\log_{10}(-2)$

Solution

Number	Graphing Calculator Keystrokes	Display
a. $\log_{10} 10$	$\boxed{\text{LOG}}$ 10 $\boxed{\text{ENTER}}$	1
b. $2\log_{10} 2.5$	2 $\boxed{\times}$ $\boxed{\text{LOG}}$ 2.5 $\boxed{\text{ENTER}}$	0.7958800
c. $\log_{10}(-2)$	$\boxed{\text{LOG}}$ $\boxed{(-)}$ 2 $\boxed{\text{ENTER}}$	ERROR

Note that the calculator displays an error message (or a complex number) when you try to evaluate $\log_{10}(-2)$. The reason for this is that the domain of every logarithmic function is the set of *positive real numbers*.

The following properties follow directly from the definition of the logarithmic function with base a.

Properties of Logarithms

1. $\log_a 1 = 0$ because $a^0 = 1$.

2. $\log_a a = 1$ because $a^1 = a$.

3. $\log_a a^x = x$ and $a^{\log_a x} = x$ Inverse Properties

4. If $\log_a x = \log_a y$, then $x = y$.   One-to-One Property

Example 3 ▶ Using Properties of Logarithms

Solve each equation for x.

a. $\log_2 x = \log_2 3$

b. $\log_4 4 = x$

Solution

a. Using the One-to-One Property (Property 4), you can conclude that $x = 3$.

b. Using Property 2, you can conclude that $x = 1$.

Graphs of Logarithmic Functions

To sketch the graph of

$$y = \log_a x$$

you can use the fact that the graphs of inverse functions are reflections of each other in the line $y = x$.

A computer animation of this example appears in the *Interactive* CD-ROM and *Internet* versions of this text.

Example 4 ▶ Graphs of Exponential and Logarithmic Functions

In the same coordinate plane, sketch the graph of each function.

a. $f(x) = 2^x$ **b.** $g(x) = \log_2 x$

Solution

a. For $f(x) = 2^x$, construct a table of values.

x	-2	-1	0	1	2	3
$f(x) = 2^x$	$\frac{1}{4}$	$\frac{1}{2}$	1	2	4	8

By plotting these points and connecting them with a smooth curve, you obtain the graph shown in Figure 5.10.

b. Because $g(x) = \log_2 x$ is the inverse of $f(x) = 2^x$, the graph of g is obtained by plotting the points $(f(x), x)$ and connecting them with a smooth curve. The graph of g is a reflection of the graph of f in the line $y = x$, as shown in Figure 5.10.

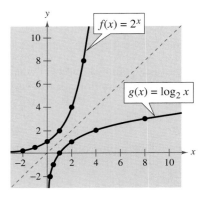

FIGURE **5.10**

Example 5 ▶ Sketching the Graph of a Logarithmic Function

Sketch the graph of the common logarithmic function

$$f(x) = \log_{10} x.$$

Identify the x-intercept and the vertical asymptote.

Solution

Begin by constructing a table of values. Note that some of the values can be obtained without a calculator by using the Inverse Property of Logarithms. Others require a calculator. Next, plot the points and connect them with a smooth curve, as shown in Figure 5.11. The x-intercept of the graph is $(1, 0)$ and the vertical asymptote is $x = 0$ (y-axis).

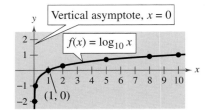

FIGURE **5.11**

	Without calculator				With calculator		
x	$\frac{1}{100}$	$\frac{1}{10}$	1	10	2	5	8
$\log_{10} x$	-2	-1	0	1	0.301	0.699	0.903

The nature of the graph in Figure 5.11 is typical of functions of the form $f(x) = \log_a x, a > 1$. They have one x-intercept and one vertical asymptote. Notice how slowly the graph rises for $x > 1$. In Figure 5.11 you would need to move out to $x = 1000$ before the graph rose to $y = 3$. The basic characteristics of logarithmic graphs are summarized in Figure 5.12.

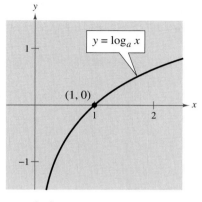

Graph of $y = \log_a x, a > 1$

- Domain: $(0, \infty)$
- Range: $(-\infty, \infty)$
- Intercept: $(1, 0)$
- Increasing
- y-axis is a vertical asymptote
  $(\log_a x \to -\infty$ as $x \to 0^+)$
- Continuous
- Reflection of graph of $y = a^x$
  about the line $y = x$

FIGURE 5.12

In the next example, the graph of $\log_a x$ is used to sketch the graphs of functions of the form $y = b \pm \log_a(x + c)$. Notice how a horizontal shift of the graph results in a horizontal shift of the vertical asymptote.

Example 6 ▶ Shifting Graphs of Logarithmic Functions

The graph of each of the following functions is similar to the graph of $f(x) = \log_{10} x$, as shown in Figure 5.13.

a. Because

$$g(x) = \log_{10}(x - 1) = f(x - 1),$$

the graph of g can be obtained by shifting the graph of f 1 unit to the right.

b. Because

$$h(x) = 2 + \log_{10} x = 2 + f(x),$$

the graph of h can be obtained by shifting the graph of f 2 units up.

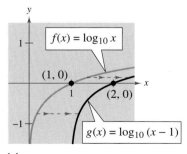

(a)

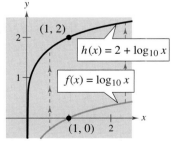

(b)

FIGURE 5.13

The Natural Logarithmic Function

As with exponential functions, the most widely used base for logarithmic functions is the number e, where

$$e \approx 2.718281828 \ldots .$$

The logarithmic function with base e is the **natural logarithmic function** and is denoted by the special symbol $\ln x$, read as "the natural log of x" or "el en of x."

The Natural Logarithmic Function

The function defined by

$$f(x) = \log_e x = \ln x, \quad x > 0$$

is called the **natural logarithmic function.**

The four properties of logarithms listed on page 393 are also valid for natural logarithms.

Properties of Natural Logarithms

1. $\ln 1 = 0$ because $e^0 = 1$.

2. $\ln e = 1$ because $e^1 = e$.

3. $\ln e^x = x$ and $e^{\ln x} = x$ Inverse Properties

4. If $\ln x = \ln y$, then $x = y$.   One-to-One Property

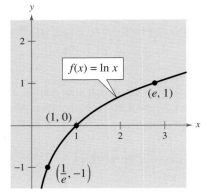

FIGURE 5.14

The graph of the natural logarithmic function is shown in Figure 5.14. Try using a graphing utility to confirm this graph. What is the domain of the natural logarithmic function?

Example 7 ▶ Using Properties of Natural Logarithms

Use the properties of natural logarithms to simplify the expression.

a. $\ln \dfrac{1}{e}$ **b.** $e^{\ln 5}$

c. $\dfrac{\ln 1}{3}$ **d.** $2 \ln e$

Solution

a. $\ln \dfrac{1}{e} = \ln e^{-1} = -1$ Inverse Property

b. $e^{\ln 5} = 5$ Inverse Property

c. $\dfrac{\ln 1}{3} = \dfrac{0}{3} = 0$ Property 1

d. $2 \ln e = 2(1) = 2$ Property 2

On most calculators, the natural logarithm is denoted by ⌊LN⌋, as illustrated in Example 8.

Example 8 ▶ Evaluating the Natural Logarithmic Function

Use a calculator to evaluate each expression.

a. $\ln 2$ **b.** $\ln 0.3$ **c.** $\ln e^2$ **d.** $\ln(-1)$ **e.** $\ln\left(1 + \sqrt{2}\right)$

Solution

	Number	Graphing Calculator Keystrokes	Display
a.	$\ln 2$	⌊LN⌋ 2 ⌊ENTER⌋	0.6931472
b.	$\ln 0.3$	⌊LN⌋ .3 ⌊ENTER⌋	−1.2039728
c.	$\ln e^2$	⌊LN⌋ ⌊eˣ⌋ 2 ⌊ENTER⌋	2
d.	$\ln(-1)$	⌊LN⌋ ⌊(−)⌋ 1 ⌊ENTER⌋	ERROR
e.	$\ln\left(1 + \sqrt{2}\right)$	⌊LN⌋ ⌊(⌋ 1 ⌊+⌋ ⌊√⌋ 2 ⌊)⌋ ⌊ENTER⌋	0.8813736

In Example 8, be sure you see that $\ln(-1)$ gives an error message on most calculators. This occurs because the domain of $\ln x$ is the set of positive real numbers (see Figure 5.14). So, $\ln(-1)$ is undefined.

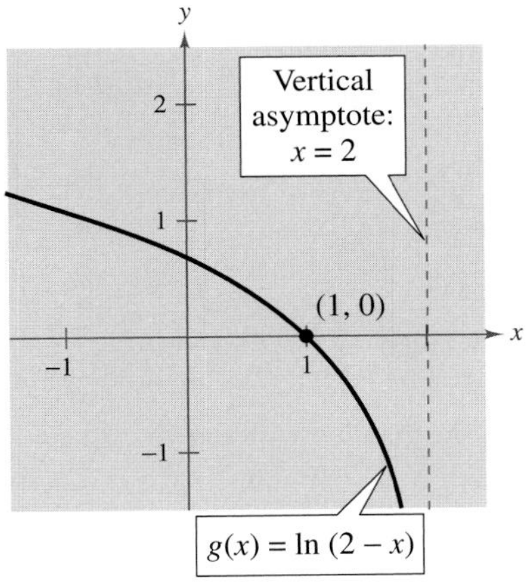

Example 9 ▶ Finding the Domains of Logarithmic Functions

Find the domain of each function.

a. $f(x) = \ln(x - 2)$ **b.** $g(x) = \ln(2 - x)$ **c.** $h(x) = \ln x^2$

Solution

a. Because $\ln(x - 2)$ is defined only if

$$x - 2 > 0,$$

it follows that the domain of f is $(2, \infty)$.

b. Because $\ln(2 - x)$ is defined only if

$$2 - x > 0,$$

it follows that the domain of g is $(-\infty, 2)$. The graph of g is shown in Figure 5.15.

c. Because $\ln x^2$ is defined only if

$$x^2 > 0,$$

it follows that the domain of h is all real numbers except $x = 0$.

Vertical asymptote: $x = 2$

$(1, 0)$

$g(x) = \ln(2 - x)$

FIGURE **5.15**

In Example 9, suppose you had been asked to analyze the function given by $h(x) = \ln|x - 2|$. How would the domain of this function compare with the domains of the functions given in parts (a) and (b) of the example?

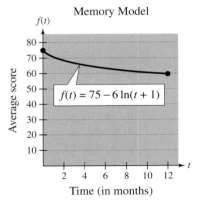

FIGURE 5.16

Application

Example 10 ▶ Human Memory Model

Students participating in a psychological experiment attended several lectures on a subject and were given an exam. Every month for a year after the exam, the students were retested to see how much of the material they remembered. The average scores for the group are given by the *human memory model*

$$f(t) = 75 - 6 \ln(t + 1), \quad 0 \le t \le 12$$

where t is the time in months. The graph of f is shown in Figure 5.16.

a. What was the average score on the original $(t = 0)$ exam?

b. What was the average score at the end of $t = 2$ months?

c. What was the average score at the end of $t = 6$ months?

Solution

a. The original average score was

$$f(0) = 75 - 6 \ln(0 + 1) \qquad \text{Substitute 0 for } t.$$
$$= 75 - 6 \ln 1 \qquad \text{Simplify.}$$
$$= 75 - 6(0) \qquad \text{Property of natural logarithms}$$
$$= 75. \qquad \text{Solution}$$

b. After 2 months, the average score was

$$f(2) = 75 - 6 \ln(2 + 1)$$
$$= 75 - 6 \ln 3$$
$$\approx 75 - 6(1.0986)$$
$$\approx 68.4.$$

c. After 6 months, the average score was

$$f(6) = 75 - 6 \ln(6 + 1)$$
$$= 75 - 6 \ln 7$$
$$\approx 75 - 6(1.9459)$$
$$\approx 63.3.$$

Writing ABOUT MATHEMATICS

Analyzing a Human Memory Model Use a graphing utility to determine the time in months when the average score in Example 10 was 60. Explain your method of solving the problem. Describe another way that you can use a graphing utility to determine the answer.

5.2 Exercises

In Exercises 1–8, write the logarithmic equation in exponential form. For example, the exponential form of $\log_5 25 = 2$ is $5^2 = 25$.

1. $\log_4 64 = 3$

2. $\log_3 81 = 4$

3. $\log_7 \frac{1}{49} = -2$

4. $\log_{10} \frac{1}{1000} = -3$

5. $\log_{32} 4 = \frac{2}{5}$

6. $\log_{16} 8 = \frac{3}{4}$

7. $\ln 1 = 0$

8. $\ln 4 = 1.386 \ldots$

In Exercises 9–18, write the exponential equation in logarithmic form. For example, the logarithmic form of $2^3 = 8$ is $\log_2 8 = 3$.

9. $5^3 = 125$

10. $8^2 = 64$

11. $81^{1/4} = 3$

12. $9^{3/2} = 27$

13. $6^{-2} = \frac{1}{36}$

14. $10^{-3} = 0.001$

15. $e^3 = 20.0855 \ldots$

16. $e^x = 4$

17. $e^0 = 1$

18. $u^v = w$

In Exercises 19–32, evaluate the expression without using a calculator.

19. $\log_2 16$

20. $\log_2 \frac{1}{8}$

21. $\log_{16} 4$

22. $\log_{27} 9$

23. $\log_7 1$

24. $\log_{10} 1000$

25. $\log_{10} 0.01$

26. $\log_{10} 10$

27. $\log_8 32$

28. $\log_9 243$

29. $\ln e^3$

30. $\ln e^{-2}$

31. $\log_a a^2$

32. $\log_b b^{-3}$

In Exercises 33–44, use a calculator to evaluate the logarithm. Round to three decimal places.

33. $\log_{10} 345$

34. $\log_{10} 145$

35. $\log_{10} \frac{4}{5}$

36. $\log_{10} 12.5$

37. $\ln 18.42$

38. $\ln \sqrt{42}$

39. $3 \ln 0.32$

40. $2 \ln 0.75$

41. $\ln\left(1 + \sqrt{3}\right)$

42. $\ln\left(\sqrt{5} - 2\right)$

43. $\ln \frac{2}{3}$

44. $\ln \frac{1}{2}$

In Exercises 45–50, use the graph of $y = \log_3 x$ to match the given function with its graph. [The graphs are labeled (a), (b), (c), (d), (e), and (f).]

(a)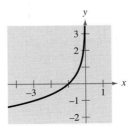

(b)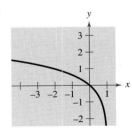

(c)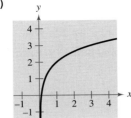

(d)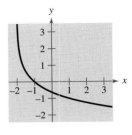

(e)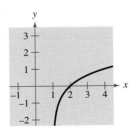

(f)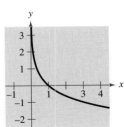

45. $f(x) = \log_3 x + 2$

46. $f(x) = -\log_3 x$

47. $f(x) = -\log_3(x + 2)$

48. $f(x) = \log_3(x - 1)$

49. $f(x) = \log_3(1 - x)$

50. $f(x) = -\log_3(-x)$

In Exercises 51–62, find the domain, x-intercept, and vertical asymptote of the logarithmic function and sketch its graph.

51. $f(x) = \log_4 x$

52. $g(x) = \log_6 x$

53. $y = -\log_3 x + 2$

54. $h(x) = \log_4(x - 3)$

55. $f(x) = -\log_6(x + 2)$

56. $y = \log_5(x - 1) + 4$

57. $y = \log_{10}\left(\dfrac{x}{5}\right)$

58. $y = \log_{10}(-x)$

59. $f(x) = \ln(x - 2)$

60. $h(x) = \ln(x + 1)$

61. $g(x) = \ln(-x)$

62. $f(x) = \ln(3 - x)$

In Exercises 63–68, use a graphing utility to graph the function. Be sure to use an appropriate viewing window.

63. $f(x) = \log_{10}(x + 1)$ **64.** $f(x) = \log_{10}(x - 1)$

65. $f(x) = \ln(x - 1)$ **66.** $f(x) = \ln(x + 2)$

67. $f(x) = \ln x + 2$ **68.** $f(x) = 3 \ln x - 1$

69. *Human Memory Model*   Students in a mathematics class were given an exam and then retested monthly with an equivalent exam. The average scores for the class are given by the human memory model

$$f(t) = 80 - 17 \log_{10}(t + 1), \qquad 0 \le t \le 12$$

where t is the time in months.

(a) What was the average score on the original exam $(t = 0)$?

(b) What was the average score after 4 months?

(c) What was the average score after 10 months?

70. *Population Growth*   The population of a town will double in

$$t = \frac{10 \ln 2}{\ln 67 - \ln 50} \text{ years.}$$

Find t.

71. *Population*   The time t in years for the world population to double if it is increasing at a continuous rate of r is given by

$$t = \frac{\ln 2}{r}.$$

(a) Complete the table.

r	0.005	0.01	0.015	0.02	0.025	0.03
t						

(b) Use a reference source to decide which value of r best approximates the actual rate of growth for the world population.

72. *Finance*   A principal P, invested at $9\frac{1}{2}\%$ and compounded continuously, increases to an amount K times the original principal after t years, where t is given by

$$t = \frac{\ln K}{0.095}.$$

(a) Complete the table and interpret your results.

K	1	2	4	6	8	10	12
t							

(b) Sketch a graph of the function.

Ventilation   In Exercises 73 and 74, use the model

$$y = 80.4 - 11 \ln x, \qquad 100 \le x \le 1500$$

which approximates the minimum required ventilation rate in terms of the air space per child in a public school classroom. In the model, x is the air space per child in cubic feet and y is the ventilation rate in cubic feet per minute.

73. Use a graphing utility to graph the function and approximate the required ventilation rate if there is 300 cubic feet of air space per child.

74. A classroom is designed for 30 students. The air conditioning system in the room has the capacity of moving 450 cubic feet of air per minute.

(a) Determine the ventilation rate per child, assuming that the room is filled to capacity.

(b) Use the graph in Exercise 73 to estimate the air space required per child.

(c) Determine the minimum number of square feet of floor space required for the room if the ceiling height is 30 feet.

75. *Work*   The work (in foot-pounds) done in compressing a volume of 9 cubic feet at a pressure of 15 pounds per square inch to a volume of 3 cubic feet is

$$W = 19,440(\ln 9 - \ln 3).$$

Find W.

76. *Sound Intensity*   The relationship between the number of decibels β and the intensity of a sound I in watts per square meter is

$$\beta = 10 \log_{10}\left(\frac{I}{10^{-12}}\right).$$

(a) Determine the number of decibels of a sound with an intensity of 1 watt per square meter.

(b) Determine the number of decibels of a sound with an intensity of 10^{-2} watt per square meter.

(c) The intensity of the sound in part (a) is 100 times as great as that in part (b). Is the number of decibels 100 times as great? Explain.

Monthly Payment   **In Exercises 77–80, use the model**

$$t = 12.542 \ln\left(\frac{x}{x - 1000}\right), \qquad x > 1000$$

which approximates the length of a home mortgage of $150,000 at 8% in terms of the monthly payment. In the model, t is the length of the mortgage in years and x is the monthly payment in dollars (see figure).

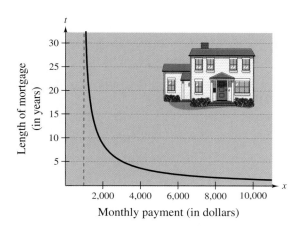

77. Use the model to approximate the length of a $150,000 mortgage at 8% if the monthly payment is $1100.65.

78. Use the model to approximate the length of a $150,000 mortgage at 8% if the monthly payment is $1254.68.

79. Approximate the total amount paid over the term of the mortgage in Exercise 77 with a monthly payment of $1100.65. What is the total interest charge?

80. Approximate the total amount paid over the term of the mortgage in Exercise 78 with a monthly payment of $1254.68. What is the total interest charge?

81. (a) Complete the table for the function

$$f(x) = \frac{\ln x}{x}.$$

x	1	5	10	10^2	10^4	10^6
$f(x)$						

   (b) Use the table in part (a) to determine what value $f(x)$ approaches as x increases without bound.

   (c) Use a graphing utility to confirm the result of part (b).

Synthesis

True or False?   **In Exercises 82 and 83, determine whether the statement is true or false. Justify your answer.**

82. You can determine the graph of $f(x) = \log_6 x$ by graphing $g(x) = 6^x$ and reflecting it about the x-axis.

83. The graph of $f(x) = \log_3 x$ contains the point $(27, 3)$.

In Exercises 84–87, describe the relationship between the graphs of f and g. What is the relationship between the functions f and g?

84. $f(x) = 3^x$
    $g(x) = \log_3 x$

85. $f(x) = 5^x$
    $g(x) = \log_5 x$

86. $f(x) = e^x$
    $g(x) = \ln x$

87. $f(x) = 10^x$
    $g(x) = \log_{10} x$

88. *Graphical Analysis*   Use a graphing utility to graph f and g in the same viewing window and determine which is increasing at the greater rate as x approaches $+\infty$. What can you conclude about the rate of growth of the natural logarithmic function?

   (a) $f(x) = \ln x,$ $g(x) = \sqrt{x}$

   (b) $f(x) = \ln x,$ $g(x) = \sqrt[4]{x}$

89. *Exploration*   The table of values was obtained by evaluating a function. Determine which of the statements may be true and which must be false.

x	1	2	8
y	0	1	3

   (a) y is an exponential function of x.

   (b) y is a logarithmic function of x.

   (c) x is an exponential function of y.

   (d) y is a linear function of x.

90. *Exploration*   Use a graphing utility to compare the graph of the function $y = \ln x$ with the graph of each given function.

   (a) $y_1 = x - 1$

   (b) $y_2 = (x - 1) - \frac{1}{2}(x - 1)^2$

   (c) $y_3 = (x - 1) - \frac{1}{2}(x - 1)^2 + \frac{1}{3}(x - 1)^3$

91. *Pattern Recognition* Identify the pattern of successive polynomials given in Exercise 90. Extend the pattern one more term and compare the graph of the resulting polynomial function with the graph of $y = \ln x$. What do you think the pattern implies?

92. *Exploration* Answer the following questions for the function $f(x) = \log_{10} x$. Do not use a calculator.

(a) What is the domain of f?

(b) What is f^{-1}?

(c) If x is a real number between 1000 and 10,000, in which interval will $f(x)$ be found?

(d) In which interval will x be found if $f(x)$ is negative?

(e) If $f(x)$ is increased by 1 unit, x must have been increased by what factor?

(f) If $f(x_1) = 3n$ and $f(x_2) = n$, what is the ratio of x_1 to x_2?

In Exercises 93–96, (a) use a graphing utility to graph the function, (b) use the graph to determine the intervals in which the function is increasing and decreasing, and (c) approximate any relative maximum or minimum values of the function.

93. $f(x) = |\ln x|$

94. $h(x) = \ln(x^2 + 1)$

95. $f(x) = \dfrac{x}{2} - \ln\dfrac{x}{4}$

96. $g(x) = \dfrac{12 \ln x}{x}$

Review

In Exercises 97–100, translate the statement into an algebraic expression.

97. The product of 8 and n is decreased by 3.

98. The total hourly wage for an employee is $9.25 per hour plus 75 cents for each of q units produced per hour.

99. The total cost for auto repairs if the cost of parts was $83.95 and there were t hours of labor at $37.50 per hour

100. The area of a rectangle if the length is 10 units more than the width w

In Exercises 101–104, find the asymptotes of the rational function.

101. $f(x) = \dfrac{4}{-8 - x}$

102. $f(x) = \dfrac{2x^3 - 3}{x^2}$

103. $f(x) = \dfrac{x + 5}{2x^2 + x - 15}$

104. $f(x) = \dfrac{2x^2(x - 5)}{x - 7}$

In Exercises 105–108, evaluate the expression. Round your result to three decimal places.

105. e^6

106. $e^{3/2}$

107. e^{-4}

108. $4e^{-6}$

▶ **What you should learn**

- How to rewrite logarithmic functions with a different base
- How to use properties of logarithms to evaluate or rewrite logarithmic expressions
- How to use properties of logarithms to expand or condense logarithmic expressions
- How to use logarithmic functions to model and solve real-life applications

▶ **Why you should learn it**

Logarithmic functions are often used to model scientific observations. For instance, Exercise 85 on page 408 shows how to use a logarithmic function to model human memory.

Gary Conner/PhotoEdit

Change of Base

Most calculators have only two types of log keys, one for common logarithms (base 10) and one for natural logarithms (base e). Although common logs and natural logs are the most frequently used, you may occasionally need to evaluate logarithms to other bases. To do this, you can use the following *change-of-base formula*.

Change-of-Base Formula

Let a, b, and x be positive real numbers such that $a \neq 1$ and $b \neq 1$. Then $\log_a x$ can be converted to a different base as follows.

Base b	Base 10	Base e
$\log_a x = \dfrac{\log_b x}{\log_b a}$	$\log_a x = \dfrac{\log_{10} x}{\log_{10} a}$	$\log_a x = \dfrac{\ln x}{\ln a}$

One way to look at the change-of-base formula is that logarithms to base a are simply *constant multiples* of logarithms to base b. The constant multiplier is $1/(\log_b a)$.

Example 1 ▶ Changing Bases Using Common Logarithms

a. $\log_4 30 = \dfrac{\log_{10} 30}{\log_{10} 4}$ $\log_a b = \dfrac{\log_{10} b}{\log_{10} a}$

       $\approx \dfrac{1.47712}{0.60206}$ Use a calculator.

         ≈ 2.4534 Use a calculator.

b. $\log_2 14 = \dfrac{\log_{10} 14}{\log_{10} 2} \approx \dfrac{1.14613}{0.30103} \approx 3.8074$

Example 2 ▶ Changing Bases Using Natural Logarithms

a. $\log_4 30 = \dfrac{\ln 30}{\ln 4}$ $\log_a b = \dfrac{\ln b}{\ln a}$

       $\approx \dfrac{3.40120}{1.38629}$ Use a calculator.

         ≈ 2.4534 Use a calculator.

b. $\log_2 14 = \dfrac{\ln 14}{\ln 2} \approx \dfrac{2.63906}{0.693147} \approx 3.8074$

Properties of Logarithms

You know from the preceding section that the logarithmic function with base a is the *inverse* of the exponential function with base a. So, it makes sense that the properties of exponents should have corresponding properties involving logarithms. For instance, the exponential property $a^0 = 1$ has the corresponding logarithmic property $\log_a 1 = 0$.

Properties of Logarithms

Let a be a positive number such that $a \neq 1$, and let n be a real number. If u and v are positive real numbers, the following properties are true.

1. $\log_a(uv) = \log_a u + \log_a v$ 1. $\ln(uv) = \ln u + \ln v$

2. $\log_a \dfrac{u}{v} = \log_a u - \log_a v$ 2. $\ln \dfrac{u}{v} = \ln u - \ln v$

3. $\log_a u^n = n \log_a u$ 3. $\ln u^n = n \ln u$

A proof of the first property listed above is given in Appendix A.

Example 3 ► Using Properties of Logarithms

Write the logarithm in terms of $\ln 2$ and $\ln 3$.

a. $\ln 6$ **b.** $\ln \dfrac{2}{27}$

Solution

a. $\ln 6 = \ln(2 \cdot 3)$ Rewrite 6 as $2 \cdot 3$.

$\qquad = \ln 2 + \ln 3$ Property 1

b. $\ln \dfrac{2}{27} = \ln 2 - \ln 27$ Property 2

$\qquad\qquad = \ln 2 - \ln 3^3$ Rewrite 27 as 3^3.

$\qquad\qquad = \ln 2 - 3 \ln 3$ Property 3

Example 4 ► Using Properties of Logarithms

Use the properties of logarithms to verify that $-\log_{10} \frac{1}{100} = \log_{10} 100$.

Solution

$-\log_{10} \frac{1}{100} = -\log_{10}(100^{-1})$ Rewrite $\frac{1}{100}$ as 100^{-1}.

$\qquad\qquad = -(-1)\log_{10} 100$ Property 3

$\qquad\qquad = \log_{10} 100$ Simplify.

Try checking this result on your calculator.

The Granger Collection

Historical Note
John Napier, a Scottish mathematician, developed logarithms as a way to simplify some of the tedious calculations of his day. Beginning in 1594, Napier worked about 20 years on the invention of logarithms. Napier was only partially successful in his quest to simplify tedious calculations. Nonetheless, the development of logarithms was a step forward and received immediate recognition.

Rewriting Logarithmic Expressions

The properties of logarithms are useful for rewriting logarithmic expressions in forms that simplify the operations of algebra. This is true because these properties convert complicated products, quotients, and exponential forms into simpler sums, differences, and products, respectively.

Example 5 ▶ Expanding Logarithmic Expressions

Expand the logarithmic expressions.

a. $\log_{10} 5x^3y$ **b.** $\ln \dfrac{\sqrt{3x - 5}}{7}$

Solution

a. $\log_{10} 5x^3y = \log_{10} 5 + \log_{10} x^3y$ Property 1

$\qquad\qquad\quad = \log_{10} 5 + \log_{10} x^3 + \log_{10} y$ Property 1

$\qquad\qquad\quad = \log_{10} 5 + 3 \log_{10} x + \log_{10} y$ Property 3

b. $\ln \dfrac{\sqrt{3x - 5}}{7} = \ln \dfrac{(3x - 5)^{1/2}}{7}$ Rewrite using rational exponent.

$\qquad\qquad\quad = \ln(3x - 5)^{1/2} - \ln 7$ Property 2

$\qquad\qquad\quad = \dfrac{1}{2} \ln(3x - 5) - \ln 7$ Property 3

In Example 5, the properties of logarithms were used to *expand* logarithmic expressions. In Example 6, this procedure is reversed and the properties of logarithms are used to *condense* logarithmic expressions.

Example 6 ▶ Condensing Logarithmic Expressions

Condense the logarithmic expressions.

a. $\frac{1}{2} \log_{10} x + 3 \log_{10}(x + 1)$

b. $2 \ln(x + 2) - \ln x$

Solution

a. $\frac{1}{2} \log_{10} x + 3 \log_{10}(x + 1) = \log_{10} x^{1/2} + \log_{10}(x + 1)^3$ Property 3

$\qquad\qquad\qquad\qquad\qquad = \log_{10}\left[\sqrt{x} \cdot (x + 1)^3 \right]$ Property 1

b. $2 \ln(x + 2) - \ln x = \ln(x + 2)^2 - \ln x$ Property 3

$\qquad\qquad\qquad\qquad = \ln \dfrac{(x + 2)^2}{x}$ Property 2

◀ **Exploration** ▶

Use a graphing utility to graph the functions

$$y = \ln x - \ln(x - 3)$$

and

$$y = \ln \frac{x}{x - 3}$$

in the same viewing window. Does the graphing utility show the functions with the same domain? If so, should it? Explain your reasoning.

Application

One method of determining how the x- and y-values for a set of nonlinear data are related begins by taking the natural log of each of the x- and y-values. If the points are graphed and fall on a straight line, then you can determine that the x- and y-values are related by the equation

$$\ln y = m \ln x$$

where m is the slope of the straight line.

Example 7 ▶ Finding a Mathematical Model

The table gives the mean distance x and the period y of the six planets that are closest to the sun. In the table, the mean distance is given in terms of astronomical units (where the earth's mean distance is defined as 1.0), and the period is given in terms of years. Find an equation that expresses y as a function of x.

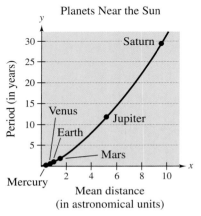

Planets Near the Sun

Period (in years)

Mean distance
(in astronomical units)

FIGURE 5.17

Planet	Mercury	Venus	Earth	Mars	Jupiter	Saturn
Period, y	0.241	0.615	1.0	1.881	11.862	29.458
Mean distance, x	0.387	0.723	1.0	1.524	5.203	9.539

Solution

The points in the table are plotted in Figure 5.17. From this figure it is not clear how to find an equation that relates y and x. To solve this problem, take the natural log of each of the x- and y-values given in the table. This produces the following results.

Planet	Mercury	Venus	Earth	Mars	Jupiter	Saturn
$\ln y$	−1.423	−0.486	0.0	0.632	2.473	3.383
$\ln x$	−0.949	−0.324	0.0	0.421	1.649	2.255

Now, by plotting the points in the second table, you can see that all six of the points appear to lie in a line (see Figure 5.18). You can use a graphical approach or the algebraic approach discussed in Section 3.5 to find that the slope of this line is $\frac{3}{2}$. You can therefore conclude that

$$\ln y = \frac{3}{2} \ln x.$$

Try to convert this to $y = f(x)$ form. You will get a function of the form $y = ax^b$, which is a power model (the variable x is raised to a power, b).

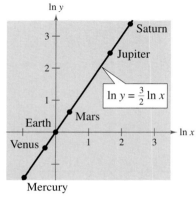

FIGURE 5.18

5.3 Exercises

In Exercises 1–8, evaluate the logarithm using the change-of-base formula. Round your result to three decimal places.

1. $\log_3 7$

2. $\log_7 4$

3. $\log_{1/2} 4$

4. $\log_{1/4} 5$

5. $\log_9 0.4$

6. $\log_{20} 0.125$

7. $\log_{15} 1250$

8. $\log_3 0.015$

In Exercises 9–16, rewrite the logarithm as a ratio of (a) common logarithms and (b) natural logarithms.

9. $\log_5 x$

10. $\log_3 x$

11. $\log_{1/5} x$

12. $\log_{1/3} x$

13. $\log_x \frac{3}{10}$

14. $\log_x \frac{3}{4}$

15. $\log_{2.6} x$

16. $\log_{7.1} x$

In Exercises 17–22, use the change-of-base formula to rewrite the logarithm as a ratio of logarithms. Then use a graphing utility to sketch the graph.

17. $f(x) = \log_2 x$

18. $f(x) = \log_4 x$

19. $f(x) = \log_{1/2} x$

20. $f(x) = \log_{1/4} x$

21. $f(x) = \log_{11.8} x$

22. $f(x) = \log_{12.4} x$

In Exercises 23–42, use the properties of logarithms to expand the expression as a sum, difference, and/or constant multiple of logarithms. (Assume all variables are positive.)

23. $\log_{10} 5x$

24. $\log_{10} 10z$

25. $\log_{10} \frac{5}{x}$

26. $\log_{10} \frac{y}{2}$

27. $\log_8 x^4$

28. $\log_6 z^{-3}$

29. $\ln \sqrt{z}$

30. $\ln \sqrt[3]{t}$

31. $\ln xyz$

32. $\ln \frac{xy}{z}$

33. $\ln \sqrt{a-1}, \ a > 1$

34. $\ln\left(\frac{x^2-1}{x^3}\right), \ x > 1$

35. $\ln z(z-1)^2, \ z > 1$

36. $\ln \frac{x}{\sqrt{x^2+1}}$

37. $\ln \sqrt[3]{\frac{x}{y}}$

38. $\ln \sqrt{\frac{x^2}{y^3}}$

39. $\ln \frac{x^4\sqrt{y}}{z^5}$

40. $\ln \sqrt{x^2(x+2)}$

41. $\log_b \frac{x^2}{y^2z^3}$

42. $\log_b \frac{\sqrt{x}\,y^4}{z^4}$

In Exercises 43–62, condense the expression to the logarithm of a single quantity.

43. $\ln x + \ln 3$

44. $\ln y + \ln t$

45. $\log_4 z - \log_4 y$

46. $\log_5 8 - \log_5 t$

47. $2 \log_2(x+4)$

48. $-4 \log_6 2x$

49. $\frac{1}{4} \log_3 5x$

50. $\frac{2}{3} \log_7(z-2)$

51. $\ln x - 3 \ln(x+1)$

52. $2 \ln 8 + 5 \ln z$

53. $\ln(x-2) - \ln(x+2)$

54. $3 \ln x + 4 \ln y - 4 \ln z$

55. $\ln x - 4[\ln(x+2) + \ln(x-2)]$

56. $4[\ln z + \ln(z+5)] - 2 \ln(z-5)$

57. $\frac{1}{3}[2 \ln(x+3) + \ln x - \ln(x^2-1)]$

58. $2[\ln x - \ln(x+1) - \ln(x-1)]$

59. $\frac{1}{3}[\ln y + 2 \ln(y+4)] - \ln(y-1)$

60. $\frac{1}{2}[\ln(x+1) + 2 \ln(x-1)] + 6 \ln x$

61. $2 \ln 3 - \frac{1}{2} \ln(x^2+1)$

62. $\frac{3}{2} \ln 5t^6 - \frac{3}{4} \ln t^4$

In Exercises 63 and 64, compare the logarithmic quantities. If two are equal, explain why.

63. $\dfrac{\log_2 32}{\log_2 4}, \quad \log_2 \dfrac{32}{4}, \quad \log_2 32 - \log_2 4$

64. $\log_7 \sqrt{70}, \quad \log_7 35, \quad \frac{1}{2} + \log_7 \sqrt{10}$

In Exercises 65–78, find the exact value of the logarithm without using a calculator. (If this is not possible, state the reason.)

65. $\log_3 9$

66. $\log_6 \sqrt[3]{6}$

67. $\log_4 16^{1.2}$

68. $\log_5 \frac{1}{125}$

69. $\log_3(-9)$

70. $\log_2(-16)$

71. $\log_5 75 - \log_5 3$

72. $\log_4 2 + \log_4 32$

73. $\ln e^2 - \ln e^5$

74. $3 \ln e^4$

75. $\log_{10} 0$

76. $\ln 1$

77. $\ln e^{4.5}$

78. $\ln \sqrt[4]{e^3}$

In Exercises 79–84, use the properties of logarithms to rewrite and simplify the logarithmic expression.

79. $\log_4 8$

80. $\log_2(4^2 \cdot 3^4)$

81. $\log_5 \frac{1}{250}$

82. $\log_{10} \frac{9}{300}$

83. $\ln(5e^6)$

84. $\ln \frac{6}{e^2}$

85. *Human Memory Model* Students participating in a psychological experiment attended several lectures and were given an exam. Every month for a year after the exam, the students were retested to see how much of the material they remembered. The average scores for the group can be modeled by the memory model

$$f(t) = 90 - 15 \log_{10}(t + 1), \quad 0 \le t \le 12$$

where t is the time in months.

(a) What was the average score on the original exam ($t = 0$)?

(b) What was the average score after 6 months?

(c) What was the average score after 12 months?

(d) When will the average score decrease to 75?

(e) Use the properties of logarithms to write the function in another form.

(f) Sketch the graph of the function over the specified domain.

86. *Sound Intensity* The relationship between the number of decibels β and the intensity of a sound I in watts per square meter is

$$\beta = 10 \log_{10}\left(\frac{I}{10^{-12}}\right).$$

Use the properties of logarithms to write the formula in simpler form, and determine the number of decibels of a sound with an intensity of 10^{-6} watt per square meter.

Synthesis

True or False? **In Exercises 87–92, determine whether the statement is true or false given that $f(x) = \ln x$. Justify your answer.**

87. $f(0) = 0$

88. $f(ax) = f(a) + f(x), \quad a > 0, x > 0$

89. $f(x - 2) = f(x) - f(2), \quad x > 2$

90. $\sqrt{f(x)} = \frac{1}{2}f(x)$

91. If $f(u) = 2f(v)$, then $v = u^2$.

92. If $f(x) < 0$, then $0 < x < 1$.

93. Prove that $\log_b \frac{u}{v} = \log_b u - \log_b v$.

94. Prove that $\log_b u^n = n \log_b u$.

In Exercises 95 and 96, use a graphing utility to graph the two functions in the same viewing window. Use the graphs to verify that the expressions are equivalent.

95. $f(x) = \log_{10} x$

$g(x) = \frac{\ln x}{\ln 10}$

96. $f(x) = \ln x$

$g(x) = \frac{\log_{10} x}{\log_{10} e}$

97. *Think About It* Sketch the graphs of

$$f(x) = \ln \frac{x}{2}, \quad g(x) = \frac{\ln x}{\ln 2}, \quad h(x) = \ln x - \ln 2$$

on the same set of axes. Which two functions have identical graphs? Explain why.

98. *Exploration* Approximate the natural logarithms of as many integers as possible between 1 and 20 given that $\ln 2 \approx 0.6931$, $\ln 3 \approx 1.0986$, and $\ln 5 \approx 1.6094$. (Do not use a calculator.)

Review

In Exercises 99–102, simplify the expression.

99. $\frac{24xy^{-2}}{16x^{-3}y}$

100. $\left(\frac{2x^2}{3y}\right)^{-3}$

101. $(18x^3y^4)^{-3}(18x^3y^4)^3$

102. $xy(x^{-1} + y^{-1})^{-1}$

In Exercises 103–108, use a calculator to evaluate the expression. Round your result to three decimal places.

103. $(2.8)^{7.6}$

104. $40(6^{-3.2})$

105. $7^{-\pi}$

106. $1.4^{3\pi}$

107. $\sqrt[4]{350}$

108. $96^{\sqrt{3}}$

In Exercises 109–112, use a calculator to evaluate the logarithm. Round your result to three decimal places.

109. $\log_{10} 26$

110. $\log_{10} \frac{5}{8}$

111. $\ln 10.6$

112. $\ln(\sqrt{7} + 1)$

5.4 Exponential and Logarithmic Equations

▶ **What you should learn**

- How to solve simple exponential and logarithmic equations
- How to solve more complicated exponential equations
- How to solve more complicated logarithmic equations
- How to use exponential and logarithmic equations to model and solve real-life applications

▶ **Why you should learn it**

Applications of exponential and logarithmic equations are found in consumer safety testing. For instance, Exercise 122 on page 419 shows how to use a logarithmic function to model crumple zones for automobile crash tests.

Introduction

So far in this chapter, you have studied the definitions, graphs, and properties of exponential and logarithmic functions. In this section, you will study procedures for *solving equations* involving these exponential and logarithmic functions.

There are two basic strategies for solving exponential or logarithmic equations. The first is based on the One-to-One Properties and the second is based on the Inverse Properties. For $a > 0$ and $a \neq 1$, the following properties are true for all x and y for which $\log_a x$ and $\log_a y$ are defined.

One-to-One Properties

$a^x = a^y$ if and only if $x = y$.

$\log_a x = \log_a y$ if and only if $x = y$.

Inverse Properties

$a^{\log_a x} = x$

$\log_a a^x = x$

Example 1 ▶ Solving Simple Equations

	Original Equation	*Rewritten Equation*	*Solution*	*Property*
a.	$2^x = 32$	$2^x = 2^5$	$x = 5$	One-to-One
b.	$\ln x - \ln 3 = 0$	$\ln x = \ln 3$	$x = 3$	One-to-One
c.	$\left(\frac{1}{3}\right)^x = 9$	$3^{-x} = 3^2$	$x = -2$	One-to-One
d.	$e^x = 7$	$\ln e^x = \ln 7$	$x = \ln 7$	Inverse
e.	$\ln x = -3$	$e^{\ln x} = e^{-3}$	$x = e^{-3}$	Inverse
f.	$\log_{10} x = -1$	$10^{\log_{10} x} = 10^{-1}$	$x = 10^{-1} = \frac{1}{10}$	Inverse

---

The strategies used in Example 1 are summarized as follows.

Strategies for Solving Exponential and Logarithmic Equations

1. Rewrite the given equation in a form that allows the use of the One-to-One Properties of exponential or logarithmic functions.

2. Rewrite an *exponential* equation in logarithmic form and apply the Inverse Property of logarithmic functions.

3. Rewrite a *logarithmic* equation in exponential form and apply the Inverse Property of exponential functions.

Solving Exponential Equations

Example 2 ▶ Solving Exponential Equations

Solve each equation and approximate the result to three decimal places.

a. $e^x = 72$ **b.** $3(2^x) = 42$

Solution

a.
$e^x = 72$	Write original equation.
$\ln e^x = \ln 72$	Take natural log of each side.
$x = \ln 72$	Inverse Property
$x \approx 4.277$	Use a calculator.

The solution is $\ln 72 \approx 4.277$. Check this in the original equation.

b.
$3(2^x) = 42$	Write original equation.
$2^x = 14$	Divide each side by 3.
$\log_2 2^x = \log_2 14$	Take log (base 2) of each side.
$x = \log_2 14$	Inverse Property
$x = \dfrac{\ln 14}{\ln 2}$	Change-of-base formula
$x \approx 3.807$	Use a calculator.

The solution is $\log_2 14 \approx 3.807$. Check this in the original equation.

In Example 2(a), the exact solution is $x = \ln 72$ and the approximate solution is $x \approx 4.277$. An exact answer is preferred when the solution is an intermediate step in a larger problem. For a final answer, an approximate solution is easier to comprehend.

Example 3 ▶ Solving an Exponential Equation

Solve $e^x + 5 = 60$ and approximate the result to three decimal places.

Solution

$e^x + 5 = 60$	Write original equation.
$e^x = 55$	Subtract 5 from each side.
$\ln e^x = \ln 55$	Take natural log of each side.
$x = \ln 55$	Inverse Property
$x \approx 4.007$	Use a calculator.

The solution is $\ln 55 \approx 4.007$. Check this in the original equation.

Technology

When solving an exponential or logarithmic equation, remember that you can check your solution graphically by "graphing the left and right sides separately" and using the intersect feature of your graphing utility to determine the point of intersection. For instance, to check the solution of the equation in Example 2(a), you can graph

$y = e^x$ and $y = 72$

in the same viewing window, as shown below. Using the intersect feature of your graphing utility, you can determine that the graphs intersect when $x \approx 4.277$, which confirms the solution found in Example 2(a).

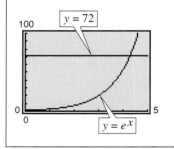

Example 4 ▶ Solving an Exponential Equation

Solve $2(3^{2t-5}) - 4 = 11$ and approximate the result to three decimal places.

Solution

$$2(3^{2t-5}) - 4 = 11 \qquad \text{Write original equation.}$$

$$2(3^{2t-5}) = 15 \qquad \text{Add 4 to each side.}$$

$$3^{2t-5} = \frac{15}{2} \qquad \text{Divide each side by 2.}$$

$$\log_3 3^{2t-5} = \log_3 \frac{15}{2} \qquad \text{Take log (base 3) of each side.}$$

$$2t - 5 = \log_3 \frac{15}{2} \qquad \text{Inverse Property}$$

$$2t = 5 + \log_3 7.5 \qquad \text{Add 5 to each side.}$$

$$t = \frac{5}{2} + \frac{1}{2}\log_3 7.5 \qquad \text{Divide each side by 2.}$$

$$t \approx 3.417 \qquad \text{Use a calculator.}$$

The solution is $\frac{5}{2} + \frac{1}{2}\log_3 7.5 \approx 3.417$. Check this in the original equation.

When an equation involves two or more exponential expressions, you can still use a procedure similar to that demonstrated in Examples 2, 3, and 4. However, the algebra is a bit more complicated.

Example 5 ▶ Solving an Exponential Equation of Quadratic Type

Solve $e^{2x} - 3e^x + 2 = 0$.

Solution

$$e^{2x} - 3e^x + 2 = 0 \qquad \text{Write original equation.}$$

$$(e^x)^2 - 3e^x + 2 = 0 \qquad \text{Write in quadratic form.}$$

$$(e^x - 2)(e^x - 1) = 0 \qquad \text{Factor.}$$

$$e^x - 2 = 0 \qquad \text{Set 1st factor equal to 0.}$$

$$x = \ln 2 \qquad \text{Solution}$$

$$e^x - 1 = 0 \qquad \text{Set 2nd factor equal to 0.}$$

$$x = 0 \qquad \text{Solution}$$

The solutions are $\ln 2$ and 0. Check these in the original equation.

In Example 5, use a graphing utility to graph $y = e^{2x} - 3e^x + 2$. The graph should have two x-intercepts: one at $x = \ln 2 \approx 0.693$ and one at $x = 0$.

The *Interactive* CD-ROM and *Internet* versions of this text show every example with its solution; clicking on the *Try It!* button brings up similar problems. Guided Examples and Integrated Examples show step-by-step solutions to additional examples. Integrated Examples are related to several concepts in the section.

Solving Logarithmic Equations

To solve a logarithmic equation such as

$$\ln x = 3 \qquad \qquad \text{Logarithmic form}$$

write the equation in exponential form as follows.

$$e^{\ln x} = e^3 \qquad \qquad \text{Exponentiate each side.}$$

$$x = e^3 \qquad \qquad \text{Exponential form}$$

This procedure is called *exponentiating* both sides of an equation.

Example 6 ▶ Solving a Logarithmic Equation

a. Solve $\ln x = 2$.   **b.** Solve $\log_3(5x - 1) = \log_3(x + 7)$.

Solution

a. $\ln x = 2$ 　　　　　　　　　Write original equation.

$$e^{\ln x} = e^2 \qquad \qquad \text{Exponentiate each side.}$$

$$x = e^2 \qquad \qquad \text{Inverse Property}$$

The solution is e^2. Check this in the original equation.

b. $\log_3(5x - 1) = \log_3(x + 7)$ 　　Write original equation.

$$5x - 1 = x + 7 \qquad \qquad \text{One-to-One Property}$$

$$4x = 8 \qquad \qquad \text{Add } -x \text{ and 1 to each side.}$$

$$x = 2 \qquad \qquad \text{Divide each side by 4.}$$

The solution is 2. Check this in the original equation.

Example 7 ▶ Solving a Logarithmic Equation

Solve $5 + 2 \ln x = 4$ and approximate the result to three decimal places.

Solution

$$5 + 2 \ln x = 4 \qquad \qquad \text{Write original equation.}$$

$$2 \ln x = -1 \qquad \qquad \text{Subtract 5 from each side.}$$

$$\ln x = -\frac{1}{2} \qquad \qquad \text{Divide each side by 2.}$$

$$e^{\ln x} = e^{-1/2} \qquad \qquad \text{Exponentiate each side.}$$

$$x = e^{-1/2} \qquad \qquad \text{Inverse Property}$$

$$x \approx 0.607 \qquad \qquad \text{Use a calculator.}$$

The solution is $e^{-1/2} \approx 0.607$. To check this result graphically, you can use a graphing utility to graph $y = 5 + 2 \ln x$ and $y = 4$ in the same viewing window, as shown in Figure 5.19.

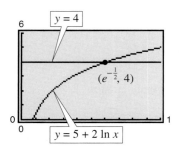

FIGURE 5.19

Example 8 ▶ Solving a Logarithmic Equation

Solve $2 \log_5 3x = 4$.

Solution

$2 \log_5 3x = 4$	Write original equation.
$\log_5 3x = 2$	Divide each side by 2.
$5^{\log_5 3x} = 5^2$	Exponentiate each side (base 5).
$3x = 25$	Inverse Property
$x = \dfrac{25}{3}$	Divide each side by 3.

The solution is $\frac{25}{3}$. Check this in the original equation. Or try performing a graphical check by graphing the functions

$$y = 2 \log_5 3x \qquad \text{and} \qquad y = 4$$

on the same screen. The two graphs should intersect when $x = \frac{25}{3}$ and $y = 4$.

Because the domain of a logarithmic function generally does not include all real numbers, you should be sure to check for extraneous solutions of logarithmic equations.

STUDY TIP

In Example 9 the domain of $\log_{10} 5x$ is $x > 0$ and the domain of $\log_{10}(x - 1)$ is $x > 1$, so the domain of the original equation is $x > 1$. Because the domain is all real numbers greater than 1, the solution $x = -4$ is extraneous.

Example 9 ▶ Checking for Extraneous Solutions

Solve $\log_{10} 5x + \log_{10}(x - 1) = 2$.

Solution

$\log_{10} 5x + \log_{10}(x - 1) = 2$	Write original equation.
$\log_{10}[5x(x - 1)] = 2$	Product Property of Logarithms
$10^{\log_{10}(5x^2 - 5x)} = 10^2$	Exponentiate each side (base 10).
$5x^2 - 5x = 100$	Inverse Property
$x^2 - x - 20 = 0$	Write in general form.
$(x - 5)(x + 4) = 0$	Factor.
$x - 5 = 0$	Set 1st factor equal to 0.
$x = 5$	Solution
$x + 4 = 0$	Set 2nd factor equal to 0.
$x = -4$	Solution

The solutions appear to be 5 and -4. However, when you check these in the original equation or use a graphical check, you can see that $x = 5$ is the only solution.

Applications

Example 10 ▶ Doubling an Investment

You have deposited $500 in an account that pays 6.75% interest, compounded continuously. How long will it take your money to double?

Solution

Using the formula for continuous compounding, you can find that the balance in the account is

$$A = Pe^{rt}$$

$$A = 500e^{0.0675t}.$$

To find the time required for the balance to double, let $A = 1000$ and solve the resulting equation for t.

$500e^{0.0675t} = 1000$	Let $A = 1000$.
$e^{0.0675t} = 2$	Divide each side by 500.
$\ln e^{0.0675t} = \ln 2$	Take natural log of each side.
$0.0675t = \ln 2$	Inverse Property
$t = \dfrac{\ln 2}{0.0675}$	Divide each side by 0.0675.
$t \approx 10.27$	Use a calculator.

The balance in the account will double after approximately 10.27 years. This result is demonstrated graphically in Figure 5.20.

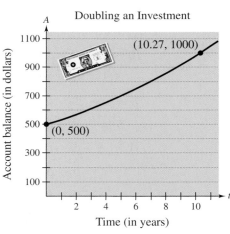

FIGURE 5.20

In Example 10, an approximate answer of 10.27 years is given. The exact solution, $(\ln 2)/0.0675$ years, does not make sense as an answer.

Example 11 ▶ Consumer Price Index for Sugar

From 1970 to 1997, the Consumer Price Index (CPI) value y for a fixed amount of sugar for the year t can be modeled by the equation

$$y = -171.8 + 87.1 \ln t$$

where $t = 10$ represents 1970 (see Figure 5.21). During which year did the price of sugar reach 4.5 times its 1970 price of 30.5 on the CPI? (Source: U.S. Bureau of Labor Statistics)

Solution

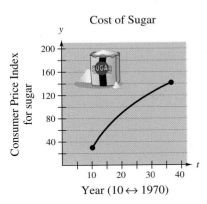

FIGURE **5.21**

$-171.8 + 87.1 \ln t = y$	Write original equation.
$-171.8 + 87.1 \ln t = 137.25$	Let $y = (4.5)(30.5) = 137.25$.
$87.1 \ln t = 309.05$	Add 171.8 to each side.
$\ln t \approx 3.548$	Divide each side by 87.1.
$e^{\ln t} \approx e^{3.548}$	Exponentiate each side.
$t \approx e^{3.548}$	Inverse Property
$t \approx 35$	Use a calculator.

The solution is $t \approx 35$ years. Because $t = 10$ represents 1970, it follows that the price of sugar reached 4.5 times its 1970 price in 1995.

Writing ABOUT MATHEMATICS

Comparing Mathematical Models The table gives the numbers y (in millions) of single compact discs (CDs) shipped annually by manufacturers from 1993 through 1997, where $x = 3$ represents 1993. (Source: Recording Industry Association of America)

x	3	4	5	6	7
y	7.8	9.3	21.5	43.2	66.7

a. Create a scatter plot of the data. Find a linear model for the data, and add its graph to your scatter plot. According to this model, when will the annual shipment of single CDs reach 100 million?

b. Create a new table giving values for $\ln x$ and $\ln y$ and create a scatter plot of this transformed data. Use the method illustrated in Example 7 in Section 5.3 to find a model for the transformed data, and add its graph to your scatter plot. According to this model, when will the annual shipment of single CDs reach 100 million?

c. Solve the model in part (b) for y, and add its graph to your scatter plot in part (a). Which model better fits the original data? Which model will better predict future shipments? Explain.

5.4 Exercises

In Exercises 1–6, determine whether the x-values are solutions or approximate solutions of the equation.

1. $4^{2x-7} = 64$

   (a) $x = 5$

   (b) $x = 2$

2. $2^{3x+1} = 32$

   (a) $x = -1$

   (b) $x = 2$

3. $3e^{x+2} = 75$

   (a) $x = -2 + e^{25}$

   (b) $x = -2 + \ln 25$

   (c) $x \approx 1.219$

4. $5^{2x+3} = 812$

   (a) $x = -1.5 + \log_5 \sqrt{812}$

   (b) $x \approx 0.581$

   (c) $x = \dfrac{1}{2}\left(-3 + \dfrac{\ln 812}{\ln 5}\right)$

5. $\log_4(3x) = 3$

   (a) $x \approx 20.356$

   (b) $x = -4$

   (c) $x = \frac{64}{3}$

6. $\ln(x - 1) = 3.8$

   (a) $x = 1 + e^{3.8}$

   (b) $x \approx 45.701$

   (c) $x = 1 + \ln 3.8$

In Exercises 7–30, solve for x.

7. $4^x = 16$

8. $3^x = 243$

9. $5^x = 625$

10. $3^x = 729$

11. $7^x = \frac{1}{49}$

12. $8^x = 4$

13. $\left(\frac{1}{2}\right)^x = 32$

14. $\left(\frac{1}{4}\right)^x = 64$

15. $\left(\frac{3}{4}\right)^x = \frac{27}{64}$

16. $\left(\frac{2}{3}\right)^x = \frac{4}{9}$

17. $3^{x-1} = 27$

18. $2^{x-3} = 32$

19. $\ln x - \ln 2 = 0$

20. $\ln x - \ln 5 = 0$

21. $e^x = 2$

22. $e^x = 4$

23. $\ln x = -1$

24. $\ln x = -7$

25. $\log_4 x = 3$

26. $\log_x 625 = 4$

27. $\log_{10} x - 2 = 0$

28. $\log_{10} x + 3 = 0$

29. $\log_{10} x = -1$

30. $\ln(2x - 1) = 0$

In Exercises 31–34, approximate the point of intersection of the graphs of f and g. Then solve the equation $f(x) = g(x)$ algebraically.

31. $f(x) = 2^x$

   $g(x) = 8$

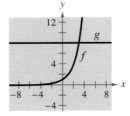

32. $f(x) = 27^x$

   $g(x) = 9$

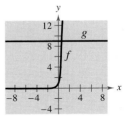

33. $f(x) = \log_3 x$

   $g(x) = 2$

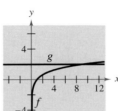

34. $f(x) = \ln(x - 4)$

   $g(x) = 0$

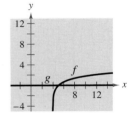

In Exercises 35–44, apply the inverse properties of ln x and e^x to simplify the expression.

35. $\log_{10} 10^{x^2}$

36. $\log_6 6^{2x-1}$

37. $8^{\log_8(x-2)}$

38. $4^{\log_4 x^3}$

39. $\ln e^{7x+2}$

40. $\ln e^{x^4}$

41. $e^{\ln(5x+2)}$

42. $e^{\ln x^2}$

43. $-1 + \ln e^{2x}$

44. $-8 + e^{\ln x^3}$

In Exercises 45–56, solve the exponential equation with base e algebraically. Approximate the result to three decimal places.

45. $e^x = 10$

46. $4e^x = 91$

47. $7 - 2e^x = 5$

48. $-14 + 3e^x = 11$

49. $e^{3x} = 12$

50. $e^{2x} = 50$

51. $500e^{-x} = 300$

52. $1000e^{-4x} = 75$

53. $e^{2x} - 4e^x - 5 = 0$

54. $e^{2x} - 5e^x + 6 = 0$

55. $20(100 - e^{x/2}) = 500$

56. $\dfrac{400}{1 + e^{-x}} = 350$

In Exercises 57–64, solve the exponential equation with base a algebraically. Approximate the result to three decimal places.

57. $10^x = 42$ **58.** $10^x = 570$

59. $3^{2x} = 80$ **60.** $6^{5x} = 3000$

61. $5^{-t/2} = 0.20$ **62.** $4^{-3t} = 0.10$

63. $2^{3-x} = 565$ **64.** $8^{-2-x} = 431$

 In Exercises 65–72, use a graphing utility to graph the function. Approximate its zero to three decimal places.

65. $g(x) = 6e^{1-x} - 25$ **66.** $f(x) = -4e^{-x-1} + 15$

67. $f(x) = 3e^{3x/2} - 962$ **68.** $g(x) = 8e^{-2x/3} - 11$

69. $g(t) = e^{0.09t} - 3$ **70.** $f(x) = -e^{1.8x} + 7$

71. $h(t) = e^{0.125t} - 8$ **72.** $f(x) = e^{2.724x} - 29$

In Exercises 73–82, solve the exponential equation. Approximate the result to three decimal places.

73. $8(10^{3x}) = 12$ **74.** $5(10^{x-6}) = 7$

75. $3(5^{x-1}) = 21$ **76.** $8(3^{6-x}) = 40$

77. $\left(1 + \dfrac{0.065}{365}\right)^{365t} = 4$ **78.** $\left(4 - \dfrac{2.471}{40}\right)^{9t} = 21$

79. $\left(1 + \dfrac{0.10}{12}\right)^{12t} = 2$ **80.** $\left(16 - \dfrac{0.878}{26}\right)^{3t} = 30$

81. $\dfrac{3000}{2 + e^{2x}} = 2$ **82.** $\dfrac{119}{e^{6x} - 14} = 7$

In Exercises 83–96, solve the natural logarithmic equation algebraically. Approximate the result to three decimal places.

83. $\ln x = -3$

84. $\ln x = 2$

85. $\ln 2x = 2.4$

86. $\ln 4x = 1$

87. $3 \ln 5x = 10$

88. $2 \ln x = 7$

89. $\ln \sqrt{x + 2} = 1$

90. $\ln \sqrt{x - 8} = 5$

91. $\ln(x + 1)^2 = 2$

92. $\ln x + \ln(x + 1) = 1$

93. $\ln x + \ln(x - 2) = 1$

94. $\ln x + \ln(x + 3) = 1$

95. $\ln(x + 5) = \ln(x - 1) - \ln(x + 1)$

96. $\ln(x + 1) - \ln(x - 2) = \ln x^2$

In Exercises 97–106, solve the logarithmic equation algebraically. Approximate the result to three decimal places.

97. $\log_{10}(z - 3) = 2$

98. $\log_{10} x^2 = 6$

99. $6 \log_3(0.5x) = 11$

100. $5 \log_{10}(x - 2) = 11$

101. $\log_{10}(x + 4) - \log_{10} x = \log_{10}(x + 2)$

102. $\log_2 x + \log_2(x + 2) = \log_2(x + 6)$

103. $\log_4 x - \log_4(x - 1) = \frac{1}{2}$

104. $\log_3 x + \log_3(x - 8) = 2$

105. $\log_{10} 8x - \log_{10}(1 + \sqrt{x}) = 2$

106. $\log_{10} 4x - \log_{10}(12 + \sqrt{x}) = 2$

In Exercises 107–110, use a graphing utility to approximate the point of intersection of the graphs. Approximate the result to three decimal places.

107. $y_1 = 7$ **108.** $y_1 = 500$
$y_2 = 2^x$ $y_2 = 1500e^{-x/2}$

109. $y_1 = 3$ **110.** $y_1 = 10$
$y_2 = \ln x$ $y_2 = 4 \ln(x - 2)$

Finance In Exercises 111 and 112, find the time required for a $1000 investment to double at interest rate r, compounded continuously.

111. $r = 0.085$ **112.** $r = 0.12$

Finance In Exercises 113 and 114, find the time required for a $1000 investment to triple at interest rate r, compounded continuously.

113. $r = 0.085$ **114.** $r = 0.12$

115. *Economics*   The demand equation for a certain product is
$$p = 500 - 0.5(e^{0.004x}).$$
Find the demand x for a price of (a) $p = \$350$ and (b) $p = \$300$.

116. *Economics*   The demand equation for a certain product is
$$p = 5000\left(1 - \dfrac{4}{4 + e^{-0.002x}}\right).$$
Find the demand x for a price of (a) $p = \$600$ and (b) $p = \$400$.

117. *Forest Yield*   The yield V (in millions of cubic feet per acre) for a forest at age t years is

$$V = 6.7e^{-48.1/t}.$$

⊞ (a) Use a graphing utility to graph the function.

(b) Determine the horizontal asymptote of the function. Interpret its meaning in the context of the problem.

(c) Find the time necessary to obtain a yield of 1.3 million cubic feet.

118. *Trees per Acre*   The number of trees per acre N of a certain species is approximated by the model

$$N = 68(10^{-0.04x}), \qquad 5 \le x \le 40$$

where x is the average diameter of the trees 3 feet above the ground. Use the model to approximate the average diameter of the trees in a test plot when $N = 21$.

119. *Average Heights*   The percent of American males between the ages of 18 and 24 who are no more than x inches tall is

$$m(x) = \frac{100}{1 + e^{-0.6114(x - 69.71)}}.$$

The percent of American females between the ages of 18 and 24 who are no more than x inches tall is

$$f(x) = \frac{100}{1 + e^{-0.66607(x - 64.51)}}$$

where m and f are the percents and x is the height in inches. (Source: U.S. National Center for Health Statistics)

(a) Use the graph to determine any horizontal asymptotes of the functions. What do they mean?

(b) What is the average height of each sex?

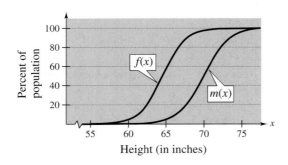

120. *Human Learning Model*   In a group project in learning theory, a mathematical model for the proportion P of correct responses after n trials was found to be

$$P = \frac{0.83}{1 + e^{-0.2n}}.$$

⊞ (a) Use a graphing utility to graph the function.

(b) Use the graph to determine any horizontal asymptotes of the function. Interpret the meaning of the upper asymptote in the context of this problem.

(c) After how many trials will 60% of the responses be correct?

121. *Data Analysis*   An object at a temperature of 160°C was removed from a furnace and placed in a room at 20°C. The temperature T of the object was measured each hour h and recorded in the table.

h	0	1	2	3	4	5
T	160°	90°	56°	38°	29°	24°

A model for this data is

$$T = 20[1 + 7(2^{-h})].$$

(a) The graph of this model is shown in the figure. Use the graph to identify the horizontal asymptote of the model and interpret the asymptote in the context of the problem.

(b) Use the model to approximate the time when the temperature of the object was 100°C.

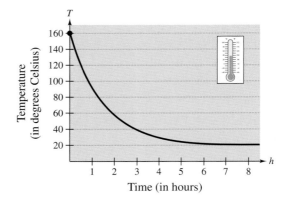

122. *Automobiles* Automobiles are designed with crumple zones that help protect their occupants in crashes. The crumple zones allow the occupants to move short distances when the automobiles come to abrupt stops. The greater the distance moved, the fewer g's the crash victims experience. (One g is equal to the acceleration due to gravity. For very short periods of time, humans have withstood as much as 40 g's.) In crash tests with vehicles moving at 90 kilometers per hour, analysts measured the numbers y of g's experienced during deceleration by crash dummies that were permitted to move x meters during impact. The data is shown in the table.

x	0.2	0.4	0.6	0.8	1
y	158	80	53	40	32

A model for this data is

$$y = -3.00 + 11.88 \ln x + \frac{36.94}{x}.$$

(a) Use a graphing utility to graph the data points and the model in the same viewing window. How do they compare?

(b) Use the model to estimate the distance traveled during impact if the passenger deceleration must not exceed 30 g's.

(c) Do you think it is practical to lower the number of g's experienced during impact to fewer than 23? Explain your reasoning.

Synthesis

True or False? **In Exercises 123–126, rewrite each verbal statement as an equation. Then decide whether the statement is true or false. Justify your answer.**

123. The logarithm of the product of two numbers is equal to the sum of the logarithms of the numbers.

124. The logarithm of the sum of two numbers is equal to the product of the logarithms of the numbers.

125. The logarithm of the difference of two numbers is equal to the difference of the logarithms of the numbers.

126. The logarithm of the quotient of two numbers is equal to the difference of the logarithms of the numbers.

127. *Finance* You are investing P dollars at an annual interest rate of r, compounded continuously, for t years. Which of the following would result in the highest value of the investment? Explain your reasoning.

(a) Double the amount you invest.

(b) Double your interest rate.

(c) Double the number of years.

128. *Think About It* Are the times required for the investments in Exercises 111 and 112 to quadruple twice as long as the times for them to double? Give a reason for your answer and verify your answer algebraically.

129. *Writing* Write a paragraph explaining whether or not the time required for an investment to double is dependent on the size of the investment.

Review

In Exercises 130–133, simplify the expression.

130. $\sqrt{48x^2y^5}$

131. $\sqrt{32} - 2\sqrt{25}$

132. $\sqrt[3]{25} \cdot \sqrt[3]{15}$

133. $\dfrac{3}{\sqrt{10} - 2}$

In Exercises 134–138, find a mathematical model for the verbal statement.

134. M varies directly as the cube of p.

135. t varies inversely as the cube of s.

136. d varies jointly as a and b.

137. x is inversely proportional to $b - 3$.

138. c varies inversely as the square root of w.

In Exercises 139–142, evaluate the logarithm using the change-of-base formula. Approximate your result to three decimal places.

139. $\log_6 9$

140. $\log_3 4$

141. $\log_{3/4} 5$

142. $\log_8 22$

5.5 Exponential and Logarithmic Models

Amy C. Etra/PhotoEdit

▶ **What you should learn**

- How to recognize the five most common types of models involving exponential and logarithmic functions
- How to use exponential growth and decay functions to model and solve real-life problems
- How to use Gaussian functions to model and solve real-life problems
- How to use logistic growth functions to model and solve real-life problems
- How to use logarithmic functions to model and solve real-life problems

▶ **Why you should learn it**

Real-life applications of mathematics often involve deciding which mathematical model to use for a given situation. For instance, Exercise 45 on page 429 compares an exponential decay model and a linear model for the depreciation of a computer over 3 years.

Introduction

The five most common types of mathematical models involving exponential functions and logarithmic functions are as follows.

1. **Exponential growth model:** $y = ae^{bx}, \quad b > 0$
2. **Exponential decay model:** $y = ae^{-bx}, \quad b > 0$
3. **Gaussian model:** $y = ae^{-(x-b)^2/c}$
4. **Logistic growth model:** $y = \dfrac{a}{1 + be^{-rx}}$
5. **Logarithmic models:** $y = a + b \ln x, \quad y = a + b \log_{10} x$

The graphs of the basic forms of these functions are shown in Figure 5.22.

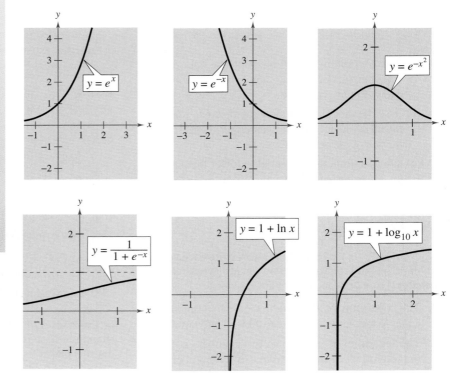

FIGURE 5.22

You can often gain quite a bit of insight into a situation modeled by an exponential or logarithmic function by identifying and interpreting the function's asymptotes. Use the graphs in Figure 5.22 to identify the asymptotes of each function.

Exponential Growth and Decay

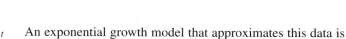

Example 1 ▶ Population Increase

Estimates of the world population (in millions) from 1992 through 2000 are shown in the table. The scatter plot of the data is shown in Figure 5.23. (Source: U.S. Bureau of the Census, International Data Base)

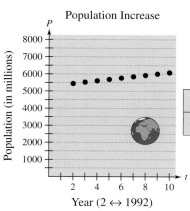

Population Increase

FIGURE 5.23

Year	1992	1993	1994	1995	1996	1997	1998	1999	2000
Population	5445	5527	5607	5688	5767	5847	5926	6005	6083

An exponential growth model that approximates this data is

$$P = 5304e^{0.013819t}, \quad 2 \le t \le 10$$

where P is the population (in millions) and $t = 2$ represents 1992. Compare the values given by the model with the estimates given by the U.S. Bureau of the Census. According to this model, when will the world population reach 6.5 billion?

Solution

The following table compares the two sets of population figures. The graph of the model is shown in Figure 5.24.

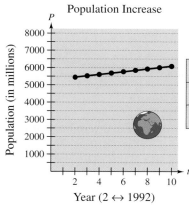

Population Increase

FIGURE 5.24

Year	1992	1993	1994	1995	1996	1997	1998	1999	2000
Population	5445	5527	5607	5688	5767	5847	5926	6005	6083
Model	5453	5529	5605	5683	5763	5843	5924	6006	6090

To find when the world population will reach 6.5 billion, let $P = 6500$ in the model and solve for t.

$$5304e^{0.013819t} = P \qquad \text{Write original model.}$$

$$5304e^{0.013819t} = 6500 \qquad \text{Let } P = 6500.$$

$$e^{0.013819t} \approx 1.22549 \qquad \text{Divide each side by 5304.}$$

$$\ln e^{0.013819t} \approx \ln 1.22549 \qquad \text{Take natural log of each side.}$$

$$0.013819t \approx 0.203341 \qquad \text{Inverse Property}$$

$$t \approx 14.71 \qquad \text{Divide each side by 0.013819.}$$

According to the model, the world population will reach 6.5 billion in 2004.

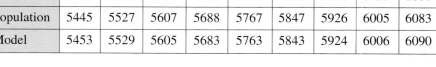

Technology

Some graphing utilities have curve-fitting capabilities that can be used to find models that represent data. If you have such a graphing utility, try using it to find a model for the data given in Example 1. How does your model compare with the model given in Example 1?

An exponential model increases (or decreases) by the same percent each year. What is the annual percent increase for the model in Example 1?

In Example 1, you were given the exponential growth model. But suppose this model were not given; how could you find such a model? One technique for doing this is demonstrated in Example 2.

Example 2 ▶ Modeling Population Growth

In a research experiment, a population of fruit flies is increasing according to the law of exponential growth. After 2 days there are 100 flies, and after 4 days there are 300 flies. How many flies will there be after 5 days?

Solution

Let y be the number of flies at time t. From the given information, you know that $y = 100$ when $t = 2$ and $y = 300$ when $t = 4$. Substituting this information into the model $y = ae^{bt}$ produces

$$100 = ae^{2b} \qquad \text{and} \qquad 300 = ae^{4b}.$$

To solve for b, solve for a in the first equation.

$$100 = ae^{2b} \qquad \Longrightarrow \qquad a = \frac{100}{e^{2b}} \qquad\qquad \text{Solve for } a \text{ in the first equation.}$$

Then substitute the result into the second equation.

$$300 = ae^{4b} \qquad\qquad\qquad \text{Write second equation.}$$

$$300 = \left(\frac{100}{e^{2b}}\right)e^{4b} \qquad\qquad \text{Substitute } 100/e^{2b} \text{ for } a.$$

$$\frac{300}{100} = e^{2b} \qquad\qquad\qquad \text{Divide each side by 100.}$$

$$\ln 3 = 2b \qquad\qquad\qquad \text{Take natural log of each side.}$$

$$\frac{1}{2}\ln 3 = b \qquad\qquad\qquad \text{Solve for } b.$$

Using $b = \frac{1}{2}\ln 3$ and the equation you found for a, you can determine that

$$a = \frac{100}{e^{2[(1/2)\ln 3]}} \qquad\qquad \text{Substitute } (1/2)\ln 3 \text{ for } b.$$

$$= \frac{100}{e^{\ln 3}} \qquad\qquad\qquad \text{Simplify.}$$

$$= \frac{100}{3} \qquad\qquad\qquad \text{Inverse Property}$$

$$\approx 33. \qquad\qquad\qquad \text{Simplify.}$$

So, with $a \approx 33$ and $b = \frac{1}{2}\ln 3 \approx 0.5493$, the exponential growth model is

$$y = 33e^{0.5493t}$$

as shown in Figure 5.25. This implies that, after 5 days, the population is

$$y = 33e^{0.5493(5)} \approx 514 \text{ flies.}$$

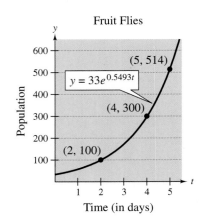

Fruit Flies

$y = 33e^{0.5493t}$

(5, 514)

(4, 300)

(2, 100)

Population

Time (in days)

FIGURE **5.25**

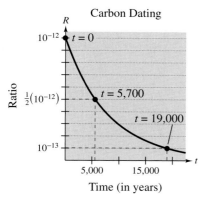

Carbon Dating

FIGURE 5.26

In living organic material, the ratio of the number of radioactive carbon isotopes (carbon 14) to the number of nonradioactive carbon isotopes (carbon 12) is about 1 to 10^{12}. When organic material dies, its carbon 12 content remains fixed, whereas its radioactive carbon 14 begins to decay with a half-life of about 5700 years. To estimate the age of dead organic material, scientists use the following formula, which denotes the ratio of carbon 14 to carbon 12 present at any time t (in years).

$$R = \frac{1}{10^{12}}e^{-t/8223} \qquad \text{Carbon dating model}$$

The graph of R is shown in Figure 5.26. Note that R decreases as t increases.

Example 3 ▶ Carbon Dating

The ratio of carbon 14 to carbon 12 in a newly discovered fossil is

$$R = \frac{1}{10^{13}}.$$

Estimate the age of the fossil.

Solution

In the carbon dating model, substitute the given value of R to obtain the following.

$$\frac{1}{10^{12}}e^{-t/8223} = R \qquad \text{Write original model.}$$

$$\frac{e^{-t/8223}}{10^{12}} = \frac{1}{10^{13}} \qquad \text{Let } R = \frac{1}{10^{13}}.$$

$$e^{-t/8223} = \frac{1}{10} \qquad \text{Multiply each side by } 10^{12}.$$

$$\ln e^{-t/8223} = \ln \frac{1}{10} \qquad \text{Take natural log of each side.}$$

$$-\frac{t}{8223} \approx -2.3026 \qquad \text{Inverse Property}$$

$$t \approx 18{,}934 \qquad \text{Multiply each side by } -8223.$$

So, to the nearest thousand years, you can estimate the age of the fossil to be 19,000 years.

The carbon dating model in Example 3 assumed that the carbon 14/carbon 12 ratio was one part in 10,000,000,000,000. Suppose an error in measurement occurred and the actual ratio was only one part in 8,000,000,000,000. The fossil age corresponding to the actual ratio would then be approximately 17,000 years. Try checking this result.

Gaussian Models

As mentioned at the beginning of this section, Gaussian models are of the form

$$y = ae^{-(x-b)^2/c}.$$

This type of model is commonly used in probability and statistics to represent populations that are **normally distributed.** One model for this situation takes the form

$$y = \frac{1}{\sigma\sqrt{2\pi}}e^{-x^2/(2\sigma^2)}$$

where σ is the standard deviation (σ is the lowercase Greek letter sigma). The graph of a Gaussian model is called a **bell-shaped curve.** Try assigning a value to σ and sketching a normal distribution curve with a graphing utility. Can you see why it is called a bell-shaped curve?

The average value for a population can be found from the bell-shaped curve by observing where the maximum y-value of the function occurs. The x-value corresponding to the maximum y-value of the function represents the average value of the independent variable—in this case, x.

Example 4 ▶ SAT Scores

In 1997, the Scholastic Aptitude Test (SAT) math scores for college-bound seniors roughly followed a normal distribution

$$y = 0.0036e^{-(x-511)^2/25,088}, \quad 200 \le x \le 800$$

where x is the SAT score for mathematics. Sketch the graph of this function. From the graph, estimate the average SAT score. (Source: College Board)

Solution

The graph of the function is given in Figure 5.27. From the graph, you can see that the average mathematics score for college-bound seniors in 1997 was 511.

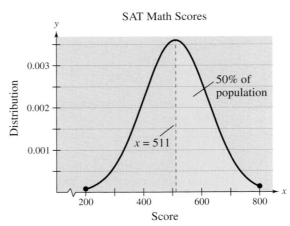

FIGURE **5.27**

Logistic Growth Models

Some populations initially have rapid growth, followed by a declining rate of growth, as indicated by the graph in Figure 5.28. One model for describing this type of growth pattern is the **logistic curve** given by the function

$$y = \frac{a}{1 + be^{-rx}}$$

where y is the population size and x is the time. An example is a bacteria culture that is initially allowed to grow under ideal conditions, and then under less favorable conditions that inhibit growth. A logistic growth curve is also called a **sigmoidal curve.**

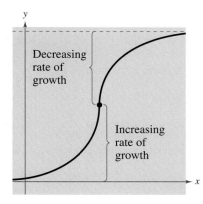

FIGURE 5.28

Example 5 ▶ Spread of a Virus

On a college campus of 5000 students, one student returns from vacation with a contagious and long-lasting flu virus. The spread of the virus is modeled by

$$y = \frac{5000}{1 + 4999e^{-0.8t}}, \quad 0 \le t$$

where y is the total number of students infected after t days. The college will cancel classes when 40% or more of the students are infected.

a. How many students are infected after 5 days?

b. After how many days will the college cancel classes?

Solution

a. After 5 days, the number of students infected is

$$y = \frac{5000}{1 + 4999e^{-0.8(5)}} = \frac{5000}{1 + 4999e^{-4}} \approx 54.$$

b. Classes are canceled when the number infected is $(0.40)(5000) = 2000.$

$$2000 = \frac{5000}{1 + 4999e^{-0.8t}}$$

$$1 + 4999e^{-0.8t} = 2.5$$

$$e^{-0.8t} \approx \frac{1.5}{4999}$$

$$\ln e^{-0.8t} \approx \ln \frac{1.5}{4999}$$

$$-0.8t \approx \ln \frac{1.5}{4999}$$

$$t = -\frac{1}{0.8} \ln \frac{1.5}{4999}$$

$$t \approx 10.1$$

So, after 10 days, at least 40% of the students will be infected, and classes will be canceled. The graph of the function is shown in Figure 5.29.

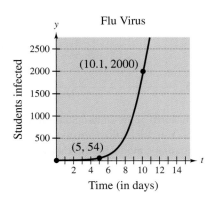

FIGURE 5.29

Twenty seconds of a 7.2 magnitude earthquake in Kobe, Japan on January 17, 1995 left damage approaching $60 billion.

Logarithmic Models

Example 6 ▶   Magnitude of Earthquakes

On the Richter scale, the magnitude R of an earthquake of intensity I is

$$R = \log_{10} \frac{I}{I_0}$$

where $I_0 = 1$ is the minimum intensity used for comparison. Find the intensities per unit of area for the following earthquakes. (Intensity is a measure of the wave energy of an earthquake.)

a. Tokyo and Yokohama, Japan in 1923: $R = 8.3$.

b. Kobe, Japan in 1995: $R = 7.2$.

Solution

a. Because $I_0 = 1$ and $R = 8.3$, you have

$$8.3 = \log_{10} \frac{I}{1} \qquad \text{Substitute 1 for } I_0 \text{ and 8.3 for } R.$$

$$10^{8.3} = 10^{\log_{10} I} \qquad \text{Exponentiate each side.}$$

$$I = 10^{8.3} \approx 199{,}526{,}000. \qquad \text{Inverse property of exponents and logs}$$

b. For $R = 7.2$, you have

$$7.2 = \log_{10} \frac{I}{1} \qquad \text{Substitute 1 for } I_0 \text{ and 7.2 for } R.$$

$$10^{7.2} = 10^{\log_{10} I} \qquad \text{Exponentiate each side.}$$

$$I = 10^{7.2} \approx 15{,}849{,}000. \qquad \text{Inverse property of exponents and logs}$$

Note that an increase of 1.1 units on the Richter scale (from 7.2 to 8.3) represents an increase in intensity by a factor of

$$\frac{199{,}526{,}000}{15{,}849{,}000} \approx 13.$$

In other words, the earthquake in 1923 had an intensity about 13 times greater than that of the 1995 quake.

t	Year	Population
1	1810	7.23
3	1830	12.87
5	1850	23.19
7	1870	39.82
9	1890	62.95
11	1910	91.97
13	1930	122.78
15	1950	151.33
17	1970	203.30
19	1990	250.00

Writing ABOUT MATHEMATICS

Comparing Population Models The population (in millions) of the United States from 1810 to 1990 is given in the table.   (Source: U.S. Bureau of the Census)
Least squares regression analysis gives the best quadratic model for this data as $P = 0.6569t^2 + 0.305t + 6.12$ and the best exponential model for this data as $P = 8.325e^{0.195t}$. Which model better fits the data? Describe the method you used to reach your conclusion.

5.5 Exercises

In Exercises 1–6, match the function with its graph. [The graphs are labeled (a) through (f).]

(a)
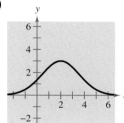

(b)
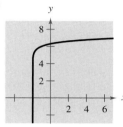

(c)
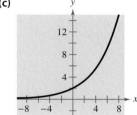

(d)
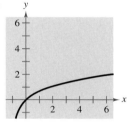

(e)
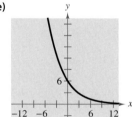

(f)
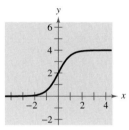

1. $y = 2e^{x/4}$

2. $y = 6e^{-x/4}$

3. $y = 6 + \log_{10}(x + 2)$

4. $y = 3e^{-(x-2)^2/5}$

5. $y = \ln(x + 1)$

6. $y = \dfrac{4}{1 + e^{-2x}}$

Finance In Exercises 7–14, complete the table for a savings account in which interest is compounded continuously.

	Initial Investment	Annual % Rate	Time to Double	Amount After 10 Years
7.	$1000	12%		
8.	$20,000	$10\frac{1}{2}\%$		
9.	$750		$7\frac{3}{4}$ yr	
10.	$10,000		12 yr	
11.	$500			$1505.00
12.	$600			$19,205.00
13.		4.5%		$10,000.00
14.		8%		$20,000.00

Finance In Exercises 15 and 16, determine the principal P that must be invested at rate r, compounded monthly, so that $500,000 will be available for retirement in t years.

15. $r = 7\frac{1}{2}\%, t = 20$

16. $r = 12\%, t = 40$

Finance In Exercises 17 and 18, determine the time necessary for $1000 to double if it is invested at interest rate r compounded (a) annually, (b) monthly, (c) daily, and (d) continuously.

17. $r = 11\%$

18. $r = 10\frac{1}{2}\%$

19. *Finance* Complete the table for the time t necessary for P dollars to triple if interest is compounded continuously at rate r.

r	2%	4%	6%	8%	10%	12%
t						

20. *Modeling Data* Draw a scatter plot of the data in Exercise 19. Use the curve-fitting capabilities of a graphing utility to find a model for the data.

21. *Finance* Complete the table for the time t necessary for P dollars to triple if interest is compounded annually at rate r.

r	2%	4%	6%	8%	10%	12%
t						

22. *Modeling Data* Draw a scatter plot of the data in Exercise 21. Use the curve-fitting capabilities of a graphing utility to find a model for the data.

23. *Finance* If $1 is invested in an account over a 10-year period, the amount in the account, where t represents the time in years, is

$$A = 1 + 0.075[\![t]\!] \quad \text{or} \quad A = e^{0.07t}$$

depending on whether the account pays simple interest at $7\frac{1}{2}\%$ or continuous compound interest at 7%. Graph each function on the same set of axes. Which grows at the faster rate? (Remember that $[\![t]\!]$ is the greatest integer function discussed in Section 2.3.)

▦ **24.** *Finance* If $1 is invested in an account over a 10-year period, the amount in the account, where t represents the time in years, is

$$A = 1 + 0.06[\![t]\!] \quad \text{or} \quad A = \left(1 + \frac{0.055}{365}\right)^{[\![365t]\!]}$$

depending on whether the account pays simple interest at 6% or compound interest at $5\frac{1}{2}\%$ compounded daily. Use a graphing utility to graph each function in the same viewing window. Which grows at the faster rate?

In Exercises 25–30, complete the table for the radioactive isotope.

Isotope	Half-life (years)	Initial Quantity	Amount After 1000 Years
25. ^{226}Ra	1620	10 g	
26. ^{226}Ra	1620		1.5 g
27. ^{14}C	5730		2 g
28. ^{14}C	5730	3 g	
29. ^{239}Pu	24,360		2.1 g
30. ^{239}Pu	24,360		0.4 g

In Exercises 31–34, find the exponential model $y = ae^{bx}$ that fits the points in the graph or table.

31.

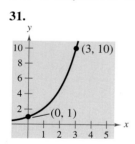

32.

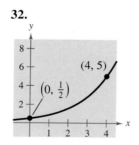

33.

x	0	4
y	5	1

34.

x	0	3
y	1	$\frac{1}{4}$

35. *Population* The population P of a city is

$$P = 105,300e^{0.015t}$$

where $t = 0$ represents the year 2000. According to this model, when will the population reach 150,000?

36. *Population* The population P of a city is

$$P = 240,360e^{0.012t}$$

where $t = 0$ represents the year 2000. According to this model, when will the population reach 275,000?

37. *Population* The population P of a city is

$$P = 2500e^{kt}$$

where $t = 0$ represents the year 2000. In 1945, the population was 1350. Find the value of k, and use this result to predict the population in the year 2010.

38. *Population* The population P of a city is

$$P = 140,500e^{kt}$$

where $t = 0$ represents the year 2000. In 1960, the population was 100,250. Find the value of k, and use this result to predict the population in the year 2020.

39. *Population* The table gives the population (in millions) of a country in 1997 and the projected population (in millions) for the year 2020. (Source: U.S. Bureau of the Census, International Data Base)

Country	1997	2020
Croatia	5.0	4.8
Mali	9.9	20.4
Singapore	3.5	4.3
Sweden	8.9	9.5

(a) Find the exponential growth model $y = ae^{bt}$ for the population in each country by letting $t = 0$ correspond to 1997. Use the model to predict the population of each country in 2030.

(b) You can see that the populations of Mali and Sweden are growing at different rates. What constant in the equation $y = ae^{bt}$ is determined by these different growth rates? Discuss the relationship between the different growth rates and the magnitude of the constant.

(c) You can see that the population of Singapore is increasing while the population of Croatia is decreasing. What constant in the equation $y = ae^{bt}$ reflects this difference? Explain.

40. *Bacteria Growth* The number of bacteria N in a culture is modeled by

$$N = 100e^{kt}$$

where t is the time in hours. If $N = 300$ when $t = 5$, estimate the time required for the population to double in size.

41. *Bacteria Growth* The number of bacteria N in a culture is modeled by

$$N = 250e^{kt}$$

where t is the time in hours. If $N = 280$ when $t = 10$, estimate the time required for the population to double in size.

42. *Radioactive Decay* The half-life of radioactive radium (^{226}Ra) is 1620 years. What percent of a present amount of radioactive radium will remain after 100 years?

43. *Radioactive Decay* Carbon 14 dating assumes that the carbon dioxide on earth today has the same radioactive content as it did centuries ago. If this is true, the amount of ^{14}C absorbed by a tree that grew several centuries ago should be the same as the amount of ^{14}C absorbed by a tree growing today. A piece of ancient charcoal contains only 15% as much radioactive carbon as a piece of modern charcoal. How long ago was the tree burned to make the ancient charcoal if the half-life of ^{14}C is 5730 years?

44. *Depreciation* A car that cost $22,000 new has a book value of $13,000 after 2 years.

(a) Find the straight-line model $V = mt + b$.

(b) Find the exponential model $V = ae^{kt}$.

▦ (c) Use a graphing utility to graph the two models in the same viewing window. Which model depreciates faster in the first 2 years?

(d) Find the book values of the car after 1 year and after 3 years using each model.

(e) Interpret the slope of the straight-line model.

45. *Depreciation* A computer that costs $2000 new has a book value of $500 after 2 years.

(a) Find the straight-line model $V = mt + b$.

(b) Find the exponential model $V = ae^{kt}$.

▦ (c) Use a graphing utility to graph the two models in the same viewing window. Which model depreciates faster in the first 2 years?

(d) Find the book values of the computer after 1 year and after 3 years using each model.

(e) Interpret the slope of the straight-line model.

46. *Business* The sales S (in thousands of units) of a new product after it has been on the market t years are modeled by

$$S(t) = 100(1 - e^{kt}).$$

Fifteen thousand units of the new product were sold the first year.

(a) Complete the model by solving for k.

(b) Sketch the graph of the model.

(c) Use the model to estimate the number of units sold after 5 years.

47. *Business* After discontinuing all advertising for a certain product in 1998, the manufacturer noted that sales began to drop according to the model

$$S = \frac{500,000}{1 + 0.6e^{kt}}$$

where S represents the number of units sold and $t = 0$ represents 1998. In 2000, the company sold 300,000 units.

(a) Complete the model by solving for k.

(b) Estimate sales in 2003.

48. *Business* The sales S (in thousands of units) of a product after x hundred dollars is spent on advertising are modeled by

$$S = 10(1 - e^{kx}).$$

When $500 is spent on advertising, 2500 units are sold.

(a) Complete the model by solving for k.

(b) Estimate the number of units that will be sold if advertising expenditures are raised to $700.

49. *Business* Because of a slump in the economy, a company finds that its annual profits have dropped from $742,000 in 1998 to $632,000 in 2000. If the profit follows an exponential pattern of decline, what is the expected profit for 2001? (Let $t = 0$ represent 1998.)

50. *Learning Curve* The management at a factory has found that the maximum number of units a worker can produce in a day is 30. The learning curve for the number of units N produced per day after a new employee has worked t days is

$$N = 30(1 - e^{kt}).$$

After 20 days on the job, a new employee produces 19 units.

(a) Find the learning curve for this employee (first, find the value of k).

(b) How many days should pass before this employee is producing 25 units per day?

(c) Is the employee's production increasing at a linear rate? Explain your reasoning.

51. *Population Growth* A conservation organization releases 100 animals of an endangered species into a game preserve. The organization believes that the preserve has a carrying capacity of 1000 animals and that the growth of the herd will be modeled by the logistic curve

$$p(t) = \frac{1000}{1 + 9e^{-0.1656t}}$$

where t is measured in months (see figure).

(a) Use a graphing utility to graph the function. Use the graph to determine the horizontal asymptotes, and interpret the meaning of the larger p-value in the context of the problem.

(b) Estimate the population after 5 months.

(c) After how many months will the population be 500?

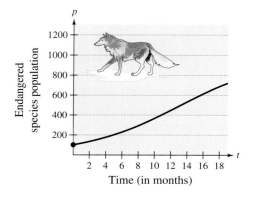

Time (in months)

Geology **In Exercises 52 and 53, use the Richter scale for measuring the magnitudes of earthquakes.**

52. Find the magnitude R of an earthquake of intensity I (let $I_0 = 1$).

(a) $I = 80,500,000$

(b) $I = 48,275,000$

(c) $I = 251,200$

53. Find the intensity I of an earthquake measuring R on the Richter scale (let $I_0 = 1$).

(a) Chile in 1906, $R = 8.6$

(b) Los Angeles in 1971, $R = 6.7$

(c) Taiwan in 1999, $R = 7.7$

Physics **In Exercises 54–57, use the following information for determining sound intensity. The level of sound β, in decibels, with an intensity of I is**

$$\beta = 10 \log_{10} \frac{I}{I_0}$$

where I_0 is an intensity of 10^{-12} watt per square meter, corresponding roughly to the faintest sound that can be heard by the human ear. In Exercises 54 and 55, find the level of sound, β.

54. (a) $I = 10^{-10}$ watt per m^2 (faint whisper)

(b) $I = 10^{-5}$ watt per m^2 (busy street corner)

(c) $I = 10^{-2.5}$ watt per m^2 (air hammer)

(d) $I = 10^0$ watt per m^2 (threshold of pain)

55. (a) $I = 10^{-9}$ watt per m^2 (whisper)

(b) $I = 10^{-3.5}$ watt per m^2 (jet 4 miles from takeoff)

(c) $I = 10^{-3}$ watt per m^2 (diesel truck at 25 feet)

(d) $I = 10^{-0.5}$ watt per m^2 (auto horn at 3 feet)

56. Due to the installation of noise suppression materials, the noise level in an auditorium was reduced from 93 to 80 decibels. Find the percent decrease in the intensity level of the noise as a result of the installation of these materials.

57. Due to the installation of a muffler, the noise level of an engine was reduced from 88 to 72 decibels. Find the percent decrease in the intensity level of the noise as a result of the installation of the muffler.

Chemistry **In Exercises 58–63, use the acidity model given by pH $= -\log_{10}[H^+]$, where acidity (pH) is a measure of the hydrogen ion concentration $[H^+]$ (measured in moles of hydrogen per liter) of a solution.**

58. Find the pH if $[H^+] = 2.3 \times 10^{-5}$.

59. Find the pH if $[H^+] = 11.3 \times 10^{-6}$.

60. Compute $[H^+]$ for a solution in which pH $= 5.8$.

61. Compute $[H^+]$ for a solution in which pH $= 3.2$.

62. A certain fruit has a pH of 2.5 and an antacid tablet has a pH of 9.5. The hydrogen ion concentration of the fruit is how many times the concentration of the tablet?

63. If the pH of a solution is decreased by 1 unit, the hydrogen ion concentration is increased by what factor?

64. Finance A $120,000 home mortgage for 35 years at $7\frac{1}{2}\%$ has a monthly payment of $809.39. Part of the monthly payment goes for the interest charge on the unpaid balance, and the remainder of the payment is used to reduce the principal. The amount that goes for interest is

$$u = M - \left(M - \frac{Pr}{12}\right)\left(1 + \frac{r}{12}\right)^{12t}$$

and the amount that goes toward reduction of the principal is

$$v = \left(M - \frac{Pr}{12}\right)\left(1 + \frac{r}{12}\right)^{12t}.$$

In these formulas, P is the size of the mortgage, r is the interest rate, M is the monthly payment, and t is the time in years.

(a) Use a graphing utility to graph each function in the same viewing window. (The viewing window should show all 35 years of mortgage payments.)

(b) In the early years of the mortgage, the larger part of the monthly payment goes for what purpose? Approximate the time when the monthly payment is evenly divided between interest and principal reduction.

(c) Repeat parts (a) and (b) for a repayment period of 20 years ($M = \$966.71$). What can you conclude?

65. Finance The total interest u paid on a home mortgage of P dollars at interest rate r for t years is

$$u = P\left[\frac{rt}{1 - \left(\dfrac{1}{1 + r/12}\right)^{12t}} - 1\right].$$

Consider a $120,000 home mortgage at $7\frac{1}{2}\%$.

(a) Use a graphing utility to graph the total interest function.

(b) Approximate the length of the mortgage for which the total interest paid is the same as the size of the mortgage. Is it possible that some people are paying twice as much in interest charges as the size of the mortgage?

66. Data Analysis The table shows the time t (in seconds) required to attain a speed of s miles per hour from a standing start for a particular car.

s	30	40	50	60	70	80	90
t	3.4	5.0	7.0	9.3	12.0	15.8	20.0

Two models for this data are as follows.

$$t_1 = 40.757 + 0.556s - 15.817 \ln s$$

$$t_2 = 1.2259 + 0.0023s^2$$

(a) Use a graphing utility to fit a linear model t_3 and an exponential model t_4 to the data.

(b) Use a graphing utility to graph the data points and each model.

(c) Create a table comparing the data with estimates obtained from each model.

(d) Use the results of part (c) to find the sum of the absolute values of the differences between the data and estimated values given by each model. Based on the four sums, which model do you think better fits the data? Explain.

67. Forensics At 8:30 A.M., a coroner was called to the home of a person who had died during the night. In order to estimate the time of death, the coroner took the person's temperature twice. At 9:00 A.M. the temperature was 85.7°F, and at 9:30 A.M. the temperature was 82.8°F. From these two temperatures the coroner was able to determine that the time elapsed since death and the body temperature were related by the formula

$$t = -2.5 \ln \frac{T - 70}{98.6 - 70}$$

where t is the time in hours elapsed since the person died and T is the temperature (in degrees Fahrenheit) of the person's body. Assume that the person had a normal body temperature of 98.6°F at death, and that the room temperature was a constant 70°F. (This formula is derived from a general cooling principle called Newton's Law of Cooling.) Use the formula to estimate the time of death of the person.

Synthesis

True or False? In Exercises 68–70, determine whether the statement is true or false. Justify your answer.

68. The domain of a logistic growth function cannot be the set of real numbers.

69. A logistic growth function will always have an x-intercept.

70. The graph of $f(x) = \dfrac{4}{1 + 6e^{-2x}} + 5$ is the graph of $g(x) = \dfrac{4}{1 + 6e^{-2x}}$ shifted to the right 5 units.

71. Identify each model as linear, logarithmic, exponential, logistic, or none of the above. Explain your reasoning.

(a)
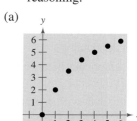

(b)
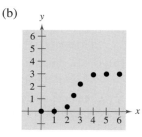

(c)
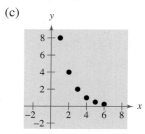

(d)
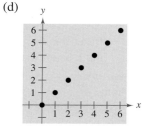

(e)
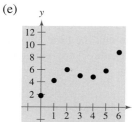

(f)
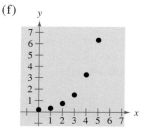

72. *Writing*   Use your school's library or some other reference source to write a paper describing John Napier's work with logarithms.

73. *Writing*   Before the development of electronic calculators and graphing utilities, some computations were done on slide rules. Use your school's library or some other reference source to write a paper describing the use of logarithmic scales on a slide rule.

Review

In Exercises 74–77, divide using synthetic division.

74. $\dfrac{4x^3 + 4x^2 - 39x + 36}{x + 4}$

75. $\dfrac{8x^3 - 36x^2 + 54x - 27}{x - \frac{3}{2}}$

76. $(2x^3 - 8x^2 + 3x - 9) \div (x - 4)$

77. $(x^4 - 3x + 1) \div (x + 5)$

In Exercises 78–87, sketch the graph of the equation.

78. $y = 10 - 3x$

79. $y = -4x - 1$

80. $y = -2x^2 - 3$

81. $y = 2x^2 - 7x - 30$

82. $3x^2 - 4y = 0$

83. $-x^2 - 8y = 0$

84. $y = \dfrac{4}{1 - 3x}$

85. $y = \dfrac{x^2}{-x - 2}$

86. $x^2 + (y - 8)^2 = 25$

87. $(x - 4)^2 + (y + 7)^2 = 4$

In Exercises 88–91, graph the exponential function.

88. $f(x) = 2^{x-1} + 5$

89. $f(x) = -2^{-x-1} - 1$

90. $f(x) = 3^x - 4$

91. $f(x) = -3^x + 4$

Chapter Summary

What did you learn?

Review Exercises

5.1 In Exercises 1–6, evaluate the expression. Approximate your result to three decimal places.

1. $(6.1)^{2.4}$

2. $-14(5^{-0.8})$

3. $2^{-0.5\pi}$

4. $\sqrt[5]{1278}$

5. $60^{\sqrt{3}}$

6. $7^{-\sqrt{11}}$

In Exercises 7–10, match the function with its graph. [The graphs are labeled (a) through (d).]

(a)

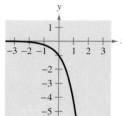

(b)

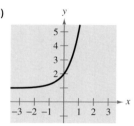

(c)

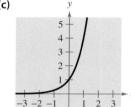

(d)

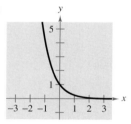

7. $f(x) = 4^x$

8. $f(x) = 4^{-x}$

9. $f(x) = -4^x$

10. $f(x) = 4^x + 1$

In Exercises 11–18, use a graphing utility to construct a table of values. Then sketch the graph of the function.

11. $f(x) = 4^{-x} + 4$

12. $f(x) = -4^x - 3$

13. $f(x) = -2.65^{x+1}$

14. $f(x) = 2.65^{x-1}$

15. $f(x) = 5^{x-2} + 4$

16. $f(x) = 2^{x-6} - 5$

17. $f(x) = \left(\frac{1}{2}\right)^{-x} + 3$

18. $f(x) = \left(\frac{1}{8}\right)^{x+2} - 5$

In Exercises 19–22, evaluate the expression. Approximate your result to three decimal places.

19. e^8

20. $e^{5/8}$

21. $e^{-1.7}$

22. $e^{0.278}$

In Exercises 23–26, use a graphing utility to construct a table of values. Then sketch the graph of the function.

23. $h(x) = e^{-x/2}$

24. $h(x) = 2 - e^{-x/2}$

25. $f(x) = e^{x+2}$

26. $s(t) = 4e^{-2/t}, \quad t > 0$

In Exercises 27 and 28, complete the table to determine the balance A for P dollars invested at rate r for t years and compounded n times per year.

n	1	2	4	12	365	Continuous
A						

27. $P = \$3500$, $r = 6.5\%$, $t = 10$ years

28. $P = \$2000$, $r = 5\%$, $t = 30$ years

In Exercises 29 and 30, complete the table to determine the amount P that should be invested at rate r to produce a balance of \$200,000 in t years.

t	1	10	20	30	40	50
P						

29. $r = 8\%$, compounded continuously

30. $r = 6\%$, compounded monthly

31. *Waiting Times*   The average time between incoming calls at a switchboard is 3 minutes. The probability of waiting less than t minutes until the next incoming call is approximated by the model

$$F(t) = 1 - e^{-t/3}.$$

If a call has just come in, find the probability that the next call will be within

(a) $\frac{1}{2}$ minute.   (b) 2 minutes.   (c) 5 minutes.

32. *Depreciation*   After t years, the value of a car that cost \$14,000 is

$$V(t) = 14{,}000\left(\frac{3}{4}\right)^t.$$

(a) Use a graphing utility to graph the function.

(b) Find the value of the car 2 years after it was purchased.

(c) According to the model, when does the car depreciate most rapidly? Is this realistic? Explain.

33. *Finance*   On the day a person was born, a deposit of \$50,000 was made in a trust fund that pays 8.75% interest, compounded continuously.

(a) Find the balance on the person's 35th birthday.

(b) How much longer would the person have to wait to get twice as much?

34. *Fuel Efficiency* A certain automobile gets 28 miles per gallon of gasoline for speeds up to 50 miles per hour. Over 50 miles per hour, the number of miles per gallon drops at a rate of 12% for each additional 10 miles per hour. If s is the speed and y is the number of miles per gallon, then

$$y = 28e^{0.6 - 0.012s}, \qquad s \geq 50.$$

Use this model to complete the table.

s	50	55	60	65	70
y					

5.2 In Exercises 35 and 36, write the exponential equation in logarithmic form.

35. $4^3 = 64$ **36.** $25^{3/2} = 125$

In Exercises 37–40, evaluate the expression by hand.

37. $\log_{10} 1000$ **38.** $\log_9 3$

39. $\log_2 \dfrac{1}{8}$ **40.** $\log_a \dfrac{1}{a}$

In Exercises 41–46, sketch the graph of the function. Identify any asymptotes.

41. $g(x) = \log_7 x$

42. $g(x) = \log_5 x$

43. $f(x) = \log_{10}\left(\dfrac{x}{3}\right)$

44. $f(x) = 6 + \log_{10} x$

45. $f(x) = 4 - \log_{10}(x + 5)$

46. $f(x) = \log_{10}(x - 3) + 1$

In Exercises 47–52, use your calculator to evaluate each expression. Approximate your result to three decimal places if necessary.

47. $\ln 22.6$ **48.** $\ln 0.98$

49. $\ln e^{-12}$ **50.** $\ln e^7$

51. $\ln\left(\sqrt{7} + 5\right)$ **52.** $\ln\left(\dfrac{\sqrt{3}}{8}\right)$

In Exercises 53–56, sketch the graph of the function. Identify any asymptotes.

53. $f(x) = \ln x + 3$ **54.** $f(x) = \ln(x - 3)$

55. $h(x) = \ln(x^2)$ **56.** $f(x) = \frac{1}{4} \ln x$

57. *Snow Removal* The number of miles s of roads cleared of snow is approximated by the model

$$s = 25 - \frac{13 \ln(h/12)}{\ln 3}, \qquad 2 \leq h \leq 15$$

where h is the depth of the snow in inches. Use this model to find s when $h = 10$ inches.

5.3 In Exercises 58–61, evaluate the logarithm using the change-of-base formula. Do each problem twice, once with common logarithms and once with natural logarithms. Approximate the results to three decimal places.

58. $\log_4 9$ **59.** $\log_{12} 200$

60. $\log_{1/2} 5$ **61.** $\log_3 0.28$

In Exercises 62–66, verify each statement using the properties of logarithms.

62. $\ln 8 + \ln 5 = \ln 40$ **63.** $-\ln\left(\frac{1}{12}\right) = \ln 12$

64. $\ln \sqrt[4]{\dfrac{x}{y}} = \dfrac{1}{4} \ln x - \dfrac{1}{4} \ln y$

65. $\log_8\left(\dfrac{\sqrt{x}}{y^3}\right) = \dfrac{1}{2} \log_8 x - 3 \log_8 y$

66. $\log_{10}\left(\dfrac{p^2 q^3}{r}\right) = 2 \log_{10} p + 3 \log_{10} q - \log_{10} r$

In Exercises 67–70, use the properties of logarithms to write the expression as a sum, difference, and/or multiple of logarithms.

67. $\log_5 5x^2$ **68.** $\log_7 \dfrac{\sqrt{x}}{4}$

69. $\log_{10} \dfrac{5\sqrt{y}}{x^2}$ **70.** $\ln \left|\dfrac{x - 1}{x + 1}\right|$

In Exercises 71–74, write the expression as the logarithm of a single quantity.

71. $\log_2 5 + \log_2 x$

72. $\log_6 y - 2 \log_6 z$

73. $\frac{1}{2} \ln|2x - 1| - 2 \ln|x + 1|$

74. $5 \ln|x - 2| - \ln|x + 2| - 3 \ln|x|$

75. *Climb Rate* The time t, in minutes, for a small plane to climb to an altitude of h feet is modeled by

$$t = 50 \log_{10} \frac{18{,}000}{18{,}000 - h}$$

where 18,000 feet is the plane's absolute ceiling.

(a) Determine the domain of the function appropriate for the context of the problem.

🔲 (b) Use a graphing utility to graph the time function and identify any asymptotes.

(c) As the plane approaches its absolute ceiling, what can be said about the time required to increase its altitude further?

(d) Find the time for the plane to climb to an altitude of 4000 feet.

5.4 **In Exercises 76–81, solve for x.**

76. $8^x = 512$ **77.** $3^x = 729$

78. $6^x = \frac{1}{216}$ **79.** $6^{x-2} = 1296$

80. $\log_7 x = 4$ **81.** $\log_x 243 = 5$

In Exercises 82–91, solve the exponential equation. Approximate your result to three decimal places.

82. $e^x = 12$ **83.** $e^{3x} = 25$

84. $3e^{-5x} = 132$ **85.** $14e^{3x+2} = 560$

86. $e^x + 13 = 35$ **87.** $e^x - 28 = -8$

88. $-4(5^x) = -68$ **89.** $2(12^x) = 190$

90. $e^{2x} - 7e^x + 10 = 0$ **91.** $e^{2x} - 6e^x + 8 = 0$

🔲 **In Exercises 92–95, use a graphing utility to solve the equation. Approximate the result to two decimal places.**

92. $2^{0.6x} - 3x = 0$ **93.** $4^{-0.2x} + x = 0$

94. $25e^{-0.3x} = 12$

95. $4e^{1.2x} = 9$

In Exercises 96–107, solve the logarithmic equation. Approximate the result to three decimal places.

96. $\ln 3x = 8.2$ **97.** $\ln 5x = 7.2$

98. $2 \ln 4x = 15$ **99.** $4 \ln 3x = 15$

100. $\ln x - \ln 3 = 2$ **101.** $\ln \sqrt{x + 8} = 3$

102. $\ln \sqrt{x + 1} = 2$ **103.** $\ln x - \ln 5 = 4$

104. $\log_{10}(x - 1) = \log_{10}(x - 2) - \log_{10}(x + 2)$

105. $\log_{10}(x + 2) - \log_{10} x = \log_{10}(x + 5)$

106. $\log_{10}(1 - x) = -1$

107. $\log_{10}(-x - 4) = 2$

🔲 **In Exercises 108–111, use a graphing utility to solve the equation. Approximate the result to two decimal places.**

108. $2 \ln(x + 3) + 3x = 8$

109. $6 \log_{10}(x^2 + 1) - x = 0$

110. $4 \ln(x + 5) - x = 10$

111. $x - 2 \log_{10}(x + 4) = 0$

112. *Finance* $7550 is deposited in an account that pays 7.25% interest, compounded continuously. How long will it take the money to triple?

113. *Finance* $2440 is deposited in an account that pays 6.5% interest, compounded continuously. How long will it take the money to quadruple?

114. *Economics*   The demand equation for a certain product is modeled by

$$p = 500 - 0.5e^{0.004x}.$$

Find the demand x for a price of (a) $p = $450 and (b) $p = $400.

5.5 **In Exercises 115–120, match the function with its graph. [The graphs are labeled (a) through (f).]**

(a)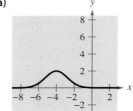

(b)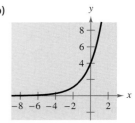

(c)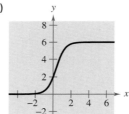

(d)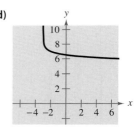

(e)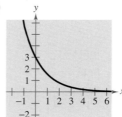

(f)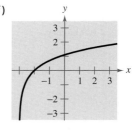

115. $y = 3e^{-2x/3}$ **116.** $y = 4e^{2x/3}$

117. $y = \ln(x + 3)$ **118.** $y = 7 - \log_{10}(x + 3)$

119. $y = 2e^{-(x+4)^2/3}$ **120.** $y = \dfrac{6}{1 + 2e^{-2x}}$

121. *Population*   The population of a town is modeled by

$$P = 12,620e^{0.0118t}$$

where $t = 0$ represents the year 2000. According to this model, when will the population reach 17,000?

122. *Radioactive Decay*   The half-life of radioactive uranium II (^{234}U) is 250,000 years. What percent of a present amount of radioactive uranium II will remain after 5000 years?

123. *Finance*   A deposit of $10,000 is made in a savings account for which the interest is compounded continuously. The balance will double in 5 years.

(a) What is the annual interest rate for this account?

(b) Find the balance after 1 year.

In Exercises 124 and 125, find the exponential function $y = ae^{bx}$ that passes through the points.

124. $(0, 2), (4, 3)$

125. $\left(0, \frac{1}{2}\right), (5, 5)$

126. *Test Scores*   The test scores for a biology test follow a normal distribution modeled by

$$y = 0.0499e^{-(x-71)^2/128}, \quad 40 \le x \le 100$$

where x is the test score.

▦ (a) Use a graphing utility to sketch the graph of the equation.

(b) From the graph, estimate the average test score.

127. *Typing Speed*   In a typing class, the average number of words per minute typed after t weeks of lessons was found to be

$$N = \frac{157}{1 + 5.4e^{-0.12t}}.$$

Find the time necessary to type (a) 50 words per minute and (b) 75 words per minute.

128. *Physics*   The relationship between the number of decibels β and the intensity of a sound I in watts per square centimeter is

$$\beta = 10 \log_{10}\left(\frac{I}{10^{-16}}\right).$$

Determine the intensity of a sound in watts per square centimeter if the decibel level is 125.

129. *Geology*   On the Richter scale, the magnitude R of an earthquake of intensity I is

$$R = \log_{10}\frac{I}{I_0}$$

where $I_0 = 1$ is the minimum intensity used for comparison. Find the intensity per unit of area for the following values of R.

(a) $R = 8.4$

(b) $R = 6.85$

(c) $R = 9.1$

Synthesis

True or False?   **In Exercises 130–135, determine whether the equation or statement is true or false. Justify your answer.**

130. $\log_b b^{2x} = 2x$

131. $e^{x-1} = \dfrac{e^x}{e}$

132. $\ln(x + y) = \ln x + \ln y$

133. $\ln(x + y) = \ln(x \cdot y)$

134. $\log_{10}\left(\dfrac{10}{x}\right) = 1 - \log_{10} x$

135. The domain of the function $f(x) = \ln x$ is the set of all real numbers.

136. The graphs of $y = e^{kt}$ are shown for $k = a, b, c,$ and d. Use the graphs to order $a, b, c,$ and d. Which of the four values are negative? Which are positive?

(a)

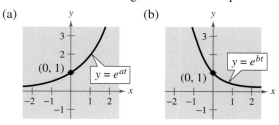

(b)

(c)

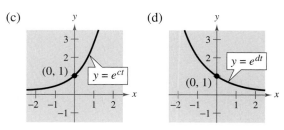

(d)

Chapter Project ▶ Graphical Approach to Compound Interest

A graphing utility can be used to investigate the rates of growth of different types of compound interest.

Example ▶ Comparing Balances

You are depositing $1000 in a savings account. Which of the following will produce the largest balance?

a. 6% annual interest rate, compounded annually

b. 6% annual interest rate, compounded continuously

c. 6.25% annual interest rate, compounded quarterly

Solution

One way to compare all three options is to sketch their graphs in the same viewing window.

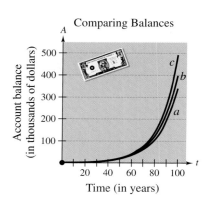

Comparing Balances

Option (a)

$$A = 1000(1 + 0.06)^t$$

Option (b)

$$A = 1000e^{0.06t}$$

Option (c)

$$A = 1000\left(1 + \frac{0.0625}{4}\right)^{4t}$$

The graphs are shown at the left. From the graphs, you can conclude that option (c) is better than option (b), and option (b) is better than option (a). Note that for the first 50 years, there is little difference in the graphs. Between 50 and 100 years, however, the balances obtained begin to differ significantly. At the end of 100 years, the balances are (a) $339,302, (b) $403,429, and (c) $493,575.

Chapter Project Investigations

1. Which would produce a larger balance: an annual interest rate of 8.05% compounded monthly or an annual interest rate of 8% compounded continuously? Explain.

2. You deposit $1000 in each of two savings accounts. The interest for the accounts is paid according to the two options described in Question 1. How long would it take for the balance in one of the accounts to exceed the balance in the other account by $100? By $100,000?

3. No income tax is due on the interest earned in some types of investments. You deposit $25,000 in an account. Which of the following plans is better? Explain.

   (a) *Tax-free* The account pays 5% compounded annually. There is no income tax due on the earned interest.

   (b) *Tax-deferred* The account pays 7% compounded annually. At maturity, the earned interest is taxable at a rate of 40%.

Chapter Test

The *Interactive* CD-ROM and *Internet* versions of this text provide answers to the Chapter Tests and Cumulative Tests. They also offer Chapter Pre-Tests (which test key skills and concepts covered in previous chapters) and Chapter Post-Tests, both of which have randomly generated exercises with diagnostic capabilities.

Take this test as you would take a test in class. After you are done, check your work against the answers given in the back of the book.

In Exercises 1–4, evaluate the expression. Approximate your result to three decimal places.

1. $12.4^{2.79}$ **2.** $4^{3\pi/2}$ **3.** $e^{-7/10}$ **4.** $e^{3.1}$

In Exercises 5–7, construct a table of values. Then sketch the graph of the function.

5. $f(x) = 10^{-x}$ **6.** $f(x) = -6^{x-2}$ **7.** $f(x) = 1 - e^{2x}$

8. Evaluate (a) $\log_7 7^{-0.89}$ and   (b) $4.6 \ln e^2$.

In Exercises 9–11, construct a table of values. Then sketch the graph of the function. Identify any asymptotes.

9. $f(x) = -\log_{10} x - 6$ **10.** $f(x) = \ln(x - 4)$ **11.** $f(x) = 1 + \ln(x + 6)$

In Exercises 12–14, evaluate the expression. Approximate your result to three decimal places.

12. $\log_7 44$ **13.** $\log_{2/5} 0.9$ **14.** $\log_{24} 68$

In Exercises 15 and 16, use the properties of logarithms to write the expression as a sum, difference, and/or multiple of logarithms.

15. $\log_2 3a^4$ **16.** $\ln \dfrac{5\sqrt{x}}{6}$

In Exercises 17 and 18, rewrite the expression as the logarithm of a single quantity.

17. $\log_3 13 + \log_3 y$ **18.** $4 \ln x - 4 \ln y$

In Exercises 19 and 20, solve the equation algebraically. Approximate your result to three decimal places.

19. $\dfrac{1025}{8 + e^{4x}} = 5$ **20.** $\log_{10} x - \log_{10}(8 - 5x) = 2$

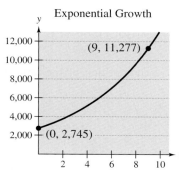

Exponential Growth

(9, 11,277)

(0, 2,745)

FIGURE FOR **21**

21. Find an exponential growth model for the graph in the figure.

22. The half-life of radioactive actinium (^{227}Ac) is 22 years. What percent of a present amount of radioactive actinium will remain after 19 years?

23. A model that can be used for predicting the height H (in centimeters) of a child based on his or her age is

$$H = 70.228 + 5.104x + 9.222 \ln x, \qquad \tfrac{1}{4} \le x \le 6$$

where x is the age of the child in years.   (Source: Snapshots of Applications in Mathematics)

(a) Construct a table of values. Then sketch the graph of the model.

(b) Use the graph from part (a) to estimate the height of a 4-year-old child. Then calculate the actual height using the model.

Cumulative Test for Chapters 3–5

Take this test to review the material from earlier chapters. After you are done, check your work against the answers given in the back of the book.

1. Find the quadratic function whose graph has a vertex at $(-8, 5)$ and passes through the point $(-4, -7)$.

In Exercises 2–4, sketch the graph of the function without the aid of a graphing utility.

2. $h(x) = -(x^2 + 4x)$ **3.** $f(t) = \frac{1}{4}t(t - 2)^2$ **4.** $g(s) = s^2 + 4s + 10$

In Exercises 5 and 6, find all the zeros of the function.

5. $f(x) = x^3 + 2x^2 + 4x + 8$ **6.** $f(x) = x^4 + 4x^3 - 21x^2$

7. Divide: $\dfrac{6x^3 - 4x^2}{2x^2 + 1}$.

8. Use synthetic division to divide $2x^4 + 3x^3 - 6x + 5$ by $x + 2$.

9. Use a graphing utility to approximate the real zero of the function $g(x) = x^3 + 3x^2 - 6$ to the nearest hundredth.

10. Find a polynomial with integer coefficients that has -5, -2, and $2 + \sqrt{3}i$ as its zeros.

In Exercises 11–13, sketch the graph of the rational function by hand. Be sure to identify all intercepts and asymptotes.

11. $f(x) = \dfrac{2x}{x - 3}$ **12.** $f(x) = \dfrac{4x^2}{x - 5}$ **13.** $f(x) = \dfrac{2x}{x^2 - 9}$

In Exercises 14 and 15, write the partial fraction decomposition of the rational expression. Check your result algebraically.

14. $\dfrac{8}{x^2 - 4x - 21}$ **15.** $\dfrac{5x}{(x - 4)^2}$

In Exercises 16 and 17, sketch a graph of the conic.

16. $6x - y^2 = 0$ **17.** $\dfrac{(x - 2)^2}{4} + \dfrac{(y + 1)^2}{9} = 1$

18. Find an equation of the parabola in the figure.

19. Find an equation of the hyperbola with foci $(0, 0)$ and $(0, 4)$ and asymptotes $y = \pm\frac{1}{2}x + 2$.

In Exercises 20 and 21, use the graph of f to describe the transformation that yields the graph of g. Use a graphing utility to graph both equations in the same viewing window.

20. $f(x) = \left(\frac{2}{5}\right)^x, \quad g(x) = -\left(\frac{2}{5}\right)^{-x+3}$ **21.** $f(x) = 2.2^x, \quad g(x) = -2.2^x + 4$

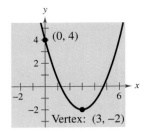

FIGURE FOR **18**

In Exercises 22–25, use a calculator to evaluate each expression. Approximate your result to three decimal places.

22. $\log_{10} 98$ **23.** $\log_{10}\left(\frac{6}{7}\right)$ **24.** $\ln\sqrt{31}$ **25.** $\ln\left(\sqrt{40} - 5\right)$

In Exercises 26–28, evaluate the logarithm using the change-of-base formula. Approximate your answer to three decimal places.

26. $\log_7 1.8$ **27.** $\log_3 0.149$ **28.** $\log_{1/2} 17$

29. Use the logarithmic properties to expand $\ln\left(\dfrac{x^2 - 16}{x^4}\right)$, where $x > 4$.

30. Write $2 \ln x - \frac{1}{2}\ln(x + 5)$ as a logarithm of a single quantity.

In Exercises 31 and 32, solve the equation.

31. $6e^{2x} = 72$ **32.** $\log_2 x + \log_2 5 = 6$

33. Use a graphing utility to graph $f(x) = \dfrac{1000}{1 + 4e^{-0.2x}}$ and determine the horizontal asymptotes.

34. Let x be the amount (in hundreds of dollars) that a company spends on advertising, and let P be the profit (in thousands of dollars), where

$$P = 230 + 20x - \frac{1}{2}x^2.$$

What amount of advertising will yield a maximum profit?

35. The numbers of used cars C (in millions) sold in the United States from 1990 through 1997 are shown in the table, where t represents the time in years, with $t = 0$ corresponding to 1990. (Source: National Automobile Dealers Association)

t	0	1	2	3	4	5	6	7
C	14.18	14.27	15.14	16.30	17.76	18.48	19.17	19.19

    (a) Use a graphing utility to sketch a scatter plot of the data.

    (b) A model for this data is given by $C = -0.025x^2 + 1.018x + 13.679$. Use a graphing utility to graph the model in the same viewing window used for the scatter plot.

    (c) Do you think this model could be used to predict the numbers of used cars sold in the future? Explain.

36. On the day a grandchild is born, a grandparent deposits $2500 in a fund earning 7.5%, compounded continuously. Determine the balance in the account at the time of the grandchild's 25th birthday.

37. The number of bacteria N in a culture is given by the model

$$N = 175e^{kt}$$

where t is the time in hours. If $N = 420$ when $t = 8$, estimate the time required for the population to double in size.

Lee Snider/The Image Works

In 1996, 57 million newspapers were printed daily in the United States. With a population of over 265 million, there were about 215 newspapers per 1000 people. (Source: U.S. Bureau of the Census and Editor & Publisher, Co.)

6 Systems of Equations and Inequalities

▶ **How to Study This Chapter**

The Big Picture

In this chapter you will learn the following skills and concepts.

▶ How to solve systems of equations by substitution, by elimination, by Gaussian elimination, and by graphing

▶ How to recognize linear systems in row-echelon form and to use back-substitution to solve the systems

▶ How to solve nonsquare systems of equations

▶ How to sketch the graphs of inequalities in two variables and to solve systems of inequalities

▶ How to solve linear programming problems

▶ How to use systems of equations and inequalities to model and solve real-life problems

Important Vocabulary

As you encounter each new vocabulary term in this chapter, add the term and its definition to your notebook glossary.

System of equations (p. 444)
Solution of a system of equations (p. 444)
Solving a system of equations (p. 444)
Method of substitution (p. 444)
Graphical method (p. 448)
Points of intersection (p. 448)
Break-even point (p. 449)
Method of elimination (p. 455)
Equivalent systems (p. 456)
Consistent system (p. 458)
Inconsistent system (p. 458)
Row-echelon form (p. 467)
Ordered triple (p. 467)
Row operations (p. 468)

Gaussian elimination (p. 468)
Nonsquare system of equations (p. 472)
Position equation (p. 473)
Solution of an inequality (p. 480)
Graph of an inequality (p. 480)
Linear inequalities (p. 481)
Solution of a system of inequalities (p. 482)
Consumer surplus (p. 485)
Producer surplus (p. 485)
Optimization (p. 491)
Linear programming (p. 491)
Objective function (p. 491)
Constraints (p. 491)
Feasible solutions (p. 491)

Study Tools

- Learning objectives at the beginning of each section
- Chapter Summary (p. 501)
- Review Exercises (pp. 502–505)
- Chapter Test (p. 507)

Additional Resources

- Study and Solutions Guide
- Interactive College Algebra
- Videotapes for Chapter 6
- College Algebra Website
- Student Success Organizer

STUDY T!P

One way to check your work is to plug your answer into the equation or inequality, then solve to see if the numbers on each side are equal. Working on your "checking skills" should improve your test scores.

6.1 Solving Systems of Equations

▶ **What you should learn**

• How to use the method of substitution to solve systems of equations in two variables

• How to use a graphical approach to solve systems of equations in two variables

• How to use systems of equations to model and solve real-life problems

▶ **Why you should learn it**

Systems of equations help you solve real-life problems. For instance, Exercise 71 on page 453 shows how you can use a system of equations to compare the compensation plans of two different job offers.

Steve Smith/FPG International

The Method of Substitution

Up to this point in the book, most problems have involved either a function of one variable or a single equation in two variables. However, many problems in science, business, and engineering involve two or more equations in two or more variables. To solve such problems, you need to find solutions of a **system of equations.** Here is an example of a system of two equations in two unknowns.

$$\begin{cases} 2x + y = 5 & \text{Equation 1} \\ 3x - 2y = 4 & \text{Equation 2} \end{cases}$$

A **solution** of this system is an ordered pair that satisfies each equation in the system. Finding the set of all solutions is called **solving the system of equations.** For instance, the ordered pair $(2, 1)$ is a solution of this system. To check this, you can substitute 2 for x and 1 for y in *each* equation.

Check $(2, 1)$ in Equation 1:

$$2x + y = 5 \qquad \text{Write Equation 1.}$$
$$2(2) + 1 \stackrel{?}{=} 5 \qquad \text{Substitute 2 for } x \text{ and 1 for } y.$$
$$4 + 1 = 5 \qquad \text{Solution checks in Equation 1. } \checkmark$$

Check $(2, 1)$ in Equation 2:

$$3x - 2y = 4 \qquad \text{Write Equation 2.}$$
$$3(2) - 2(1) \stackrel{?}{=} 4 \qquad \text{Substitute 2 for } x \text{ and 1 for } y.$$
$$6 - 2 = 4 \qquad \text{Solution checks in Equation 2. } \checkmark$$

In this chapter you will study four ways to solve equations, beginning with the **method of substitution.**

Method	Section	Type of System
1. Substitution	6.1	Linear or nonlinear, two variables
2. Graphical method	6.1	Linear or nonlinear, two variables
3. Elimination	6.2	Linear, two variables
4. Gaussian elimination	6.3	Linear, three or more variables

Method of Substitution

1. *Solve* one of the equations for one variable in terms of the other.

2. *Substitute* the expression found in Step 1 into the other equation to obtain an equation in one variable.

3. *Solve* the equation obtained in Step 2.

4. *Back-substitute* the value obtained in Step 3 into the expression obtained in Step 1 to find the value of the other variable.

5. *Check* that the solution satisfies *each* of the original equations.

Use a graphing utility to graph $y = 4 - x$ and $y = x - 2$ in the same viewing window. Use the trace feature to find the coordinates of the point of intersection. Are the coordinates the same as the solution found in Example 1? Explain.

The *Interactive* CD-ROM and *Internet* versions of this text offer a built-in graphing calculator, which can be used in the Examples, Explorations, Technology notes, and Exercises.

STUDY T!P

Because many steps are required to solve a system of equations, it is very easy to make errors in arithmetic. So, we strongly suggest that you always check your solution by substituting it into each equation in the original system.

Example 1 ▶ Solving a System of Equations by Substitution

Solve the system of equations.

$$\begin{cases} x + y = 4 & \text{Equation 1} \\ x - y = 2 & \text{Equation 2} \end{cases}$$

Solution

Begin by solving for y in Equation 1.

$$y = 4 - x \qquad \text{Solve for } y \text{ in Equation 1.}$$

Next, substitute this expression for y into Equation 2 and solve the resulting single-variable equation for x.

$$x - y = 2 \qquad \text{Write Equation 2.}$$
$$x - (4 - x) = 2 \qquad \text{Substitute } 4 - x \text{ for } y.$$
$$x - 4 + x = 2 \qquad \text{Simplify.}$$
$$2x = 6 \qquad \text{Combine like terms.}$$
$$x = 3 \qquad \text{Divide each side by 2.}$$

Finally, you can solve for y by *back-substituting* $x = 3$ into the equation $y = 4 - x$, to obtain

$$y = 4 - x \qquad \text{Write revised Equation 1.}$$
$$y = 4 - 3 \qquad \text{Substitute 3 for } x.$$
$$y = 1. \qquad \text{Solve for } y.$$

The solution is the ordered pair $(3, 1)$. You can check this as follows.

Check

Substitute $(3, 1)$ into Equation 1:

$$x + y = 4 \qquad \text{Write Equation 1.}$$
$$3 + 1 \overset{?}{=} 4 \qquad \text{Substitute for } x \text{ and } y.$$
$$4 = 4 \qquad \text{Solution checks in Equation 1. } ✓$$

Substitute $(3, 1)$ into Equation 2:

$$x - y = 2 \qquad \text{Write Equation 2.}$$
$$3 - 1 \overset{?}{=} 2 \qquad \text{Substitute for } x \text{ and } y.$$
$$2 = 2 \qquad \text{Solution checks in Equation 2. } ✓$$

Because $(3, 1)$ satisfies both equations in the system, it is a solution of the system of equations.

The term *back-substitution* implies that you work *backwards*. First you solve for one of the variables, and then you substitute that value *back* into one of the equations in the system to find the value of the other variable.

Example 2 ▶   Solving a System by Substitution

A total of $12,000 is invested in two funds paying 9% and 11% simple interest. The yearly interest is $1180. How much is invested at each rate?

Solution

Verbal Model:

$$\boxed{\begin{array}{c}9\% \\ \text{fund}\end{array}} + \boxed{\begin{array}{c}11\% \\ \text{fund}\end{array}} = \boxed{\begin{array}{c}\text{Total} \\ \text{investment}\end{array}}$$

$$\boxed{\begin{array}{c}9\% \\ \text{interest}\end{array}} + \boxed{\begin{array}{c}11\% \\ \text{interest}\end{array}} = \boxed{\begin{array}{c}\text{Total} \\ \text{interest}\end{array}}$$

Labels: Amount in 9% fund $= x$ (dollars)
Interest for 9% fund $= 0.09x$ (dollars)
Amount in 11% fund $= y$ (dollars)
Interest for 11% fund $= 0.11y$ (dollars)
Total investment $= \$12,000$ (dollars)
Total interest $= \$1180$ (dollars)

System:
$$\begin{cases} x + y = 12{,}000 & \text{Equation 1} \\ 0.09x + 0.11y = 1{,}180 & \text{Equation 2} \end{cases}$$

To begin, it is convenient to multiply both sides of Equation 2 by 100. This eliminates the need to work with decimals.

$$100(0.09x + 0.11y) = 100(1180) \qquad \text{Multiply each side by 100.}$$

$$9x + 11y = 118{,}000 \qquad \text{Revised Equation 2}$$

To solve this system, you can solve for x in Equation 1.

$$x = 12{,}000 - y \qquad \text{Revised Equation 1}$$

Then, substitute this expression for x into revised Equation 2 and solve the resulting equation for y.

$$9x + 11y = 118{,}000 \qquad \text{Write revised Equation 2.}$$

$$9(12{,}000 - y) + 11y = 118{,}000 \qquad \text{Substitute } 12{,}000 - y \text{ for } x.$$

$$108{,}000 - 9y + 11y = 118{,}000 \qquad \text{Distributive Property}$$

$$2y = 10{,}000 \qquad \text{Combine like terms.}$$

$$y = 5000 \qquad \text{Divide each side by 2.}$$

Next, back-substitute the value $y = 5000$ to solve for x.

$$x = 12{,}000 - y \qquad \text{Write revised Equation 1.}$$

$$x = 12{,}000 - 5000 \qquad \text{Substitute 5000 for } y.$$

$$x = 7000 \qquad \text{Simplify.}$$

The solution is (7000, 5000). So, $7000 is invested at 9% and $5000 is invested at 11%. Check this in the original problem.

Technology

One way to check the answers you obtain in this section is to use a graphing utility. For instance, enter the two equations in Example 2

$$y_1 = 12{,}000 - x$$

$$y_2 = \frac{1180 - 0.09x}{0.11}$$

and find an appropriate viewing window that shows where the lines intersect. Then use the zoom and trace features to find their point of intersection. Does this point agree with the solution obtained at the right?

The equations in Examples 1 and 2 are linear. Substitution can also be used to solve systems in which one or both of the equations are nonlinear.

A computer animation of this concept appears in the *Interactive* CD-ROM and *Internet* versions of this text.

◀ Exploration ▶

Use a graphing utility to graph the two equations in Example 3

$$y_1 = x^2 + 4x - 7$$

$$y_2 = 2x + 1$$

in the same viewing window. How many solutions do you think this system has? Repeat this experiment for the equations in Example 4. How many solutions does this system have? Explain your reasoning.

Example 3 ▶ Substitution: Two-Solution Case

Solve the system of equations.

$$\begin{cases} x^2 + 4x - y = 7 & \text{Equation 1} \\ 2x - y = -1 & \text{Equation 2} \end{cases}$$

Solution

Begin by solving for y in Equation 2 to obtain

$$y = 2x + 1. \qquad \text{Solve for } y \text{ in Equation 2.}$$

Next, substitute this expression for y into Equation 1 and solve for x.

$x^2 + 4x - y = 7$	Write Equation 1.
$x^2 + 4x - (2x + 1) = 7$	Substitute $2x + 1$ for y.
$x^2 + 2x - 1 = 7$	Simplify.
$x^2 + 2x - 8 = 0$	General form
$(x + 4)(x - 2) = 0$	Factor.
$x = -4, 2$	Solve for x.

Back-substituting these values of x to solve for the corresponding values of y produces the solutions $(-4, -7)$ and $(2, 5)$. Check these in the original system.

Example 4 ▶ Substitution: No-Real-Solution Case

Solve the system of equations.

$$\begin{cases} -x + y = 4 & \text{Equation 1} \\ x^2 + y = 3 & \text{Equation 2} \end{cases}$$

Solution

Begin by solving for y in Equation 1 to obtain

$$y = x + 4. \qquad \text{Solve for } y \text{ in Equation 1.}$$

Next, substitute this expression for y into Equation 2 and solve for x.

$x^2 + y = 3$	Write Equation 2.
$x^2 + (x + 4) = 3$	Substitute $x + 4$ for y.
$x^2 + x + 1 = 0$	Simplify.
$x = \dfrac{-1 \pm \sqrt{1^2 - 4(1)(1)}}{2}$	Quadratic Formula
$x = \dfrac{-1 \pm \sqrt{-3}}{2}$	Simplify.

Because the discriminant is negative, the equation $x^2 + x + 1 = 0$ has no (real) solution. So, this system has no (real) solution.

Graphical Approach to Finding Solutions

Technology

Most graphing calculators have built-in programs that approximate the point(s) of intersection of two graphs. Typically, you must enter the equations of the graphs and visually locate a point of intersection before running the intersection program.

Use this feature to find the points of intersection for the graphs in Figure 6.1. Be sure to adjust your viewing window so that you see all the points of intersection.

From Examples 2, 3, and 4, you can see that a system of two equations in two unknowns can have exactly one solution, more than one solution, or no solution. By using a **graphical method,** you can gain insight about the number of solutions and the location(s) of the solution(s) of a system of equations by graphing each of the equations in the same coordinate plane. The solutions of the system correspond to the **points of intersection** of the graphs. For instance, the two equations in Figure 6.1(a) graph as two lines with *a single point* of intersection; the two equations in Figure 6.1(b) graph as a parabola and a line with *two points* of intersection; and the two equations in Figure 6.1(c) graph as a line and a parabola that have *no points* of intersection.

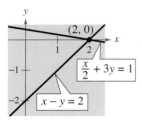
(a) One intersection point

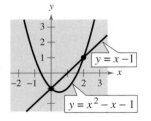
(b) Two intersection points

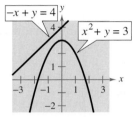
(c) No intersection points

FIGURE 6.1

 The *Interactive* CD-ROM and *Internet* versions of this text show every example with its solution; clicking on the *Try It!* button brings up similar problems. Guided Examples and Integrated Examples show step-by-step solutions to additional examples. Integrated Examples are related to several concepts in the section.

Example 5 ▶ Solving a System of Equations Graphically

Solve the system of equations.

$$\begin{cases} y = \ln x & \text{Equation 1} \\ x + y = 1 & \text{Equation 2} \end{cases}$$

Solution

Sketch the graphs of the two equations, as shown in Figure 6.2. From the graphs, it is clear that there is only one point of intersection and that $(1, 0)$ is the solution point. You can confirm this by substituting 1 for x and 0 for y in *both* equations.

Check $(1, 0)$ in Equation 1:

$y = \ln x$ Write Equation 1.

$0 = \ln 1$ Equation 1 checks. ✓

Check $(1, 0)$ in Equation 2:

$x + y = 1$ Write Equation 2.

$1 + 0 = 1$ Equation 2 checks. ✓

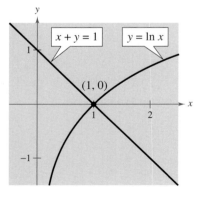

FIGURE 6.2

Example 5 shows the value of a graphical approach to solving systems of equations in two variables. Notice what would happen if you tried only the substitution method in Example 5. You would obtain the equation $x + \ln x = 1$. It would be difficult to solve this equation for x using standard algebraic techniques.

Applications

The total cost C of producing x units of a product typically has two components—the initial cost and the cost per unit. When enough units have been sold so that the total revenue R equals the total cost C, the sales are said to have reached the **break-even point.** You will find that the break-even point corresponds to the point of intersection of the cost and revenue curves.

Example 6 ▶ Break-Even Analysis

A small business invests $10,000 in equipment to produce a product. Each unit of the product costs $0.65 to produce and is sold for $1.20. How many items must be sold before the business breaks even?

Solution

The total cost of producing x units is

$$\begin{array}{c}\text{Total} \\ \text{cost}\end{array} = \begin{array}{c}\text{Cost per} \\ \text{unit}\end{array} \cdot \begin{array}{c}\text{Number} \\ \text{of units}\end{array} + \begin{array}{c}\text{Initial} \\ \text{cost}\end{array}$$

$$C = 0.65x + 10,000. \qquad \text{Equation 1}$$

The revenue obtained by selling x units is

$$\begin{array}{c}\text{Total} \\ \text{revenue}\end{array} = \begin{array}{c}\text{Price per} \\ \text{unit}\end{array} \cdot \begin{array}{c}\text{Number} \\ \text{of units}\end{array}$$

$$R = 1.20x. \qquad \text{Equation 2}$$

Because the break-even point occurs when $R = C$, you have $C = 1.20x$, and the system of equations to solve is

$$\begin{cases} C = 0.65x + 10,000 \\ C = 1.20x \end{cases}.$$

Now you can solve by substitution.

$1.20x = 0.65x + 10,000$	Substitute 1.2x for C in Equation 1.
$0.55x = 10,000$	Subtract 0.65x from each side.
$x = \dfrac{10,000}{0.55}$	Divide each side by 0.55.
$x \approx 18,182$ units	Use a calculator.

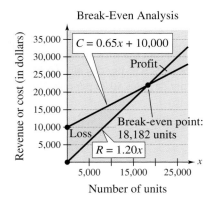

Break-Even Analysis

$C = 0.65x + 10,000$

Profit

Break-even point: 18,182 units

Loss

$R = 1.20x$

Number of units

Revenue or cost (in dollars)

FIGURE **6.3**

Note in Figure 6.3 that sales less than the break-even point correspond to an overall loss, whereas sales greater than the break-even point correspond to a profit.

Another way to view the solution in Example 6 is to consider the profit function

$$P = R - C.$$

The break-even point occurs when the profit is 0, which is the same as saying that $R = C$.

Example 7 ▶ State Population

From 1990 to 1997, the population of Arizona was increasing at a faster rate than the population of Alabama. Models that approximate the two populations P (in thousands) are

$$\begin{cases} P = 3631.2 + 131.7t & \text{Arizona} \\ P = 4055.6 + 39.9t & \text{Alabama} \end{cases}$$

where $t = 0$ represents 1990 (see Figure 6.4). According to these two models, when would you expect the population of Arizona to have exceeded the population of Alabama?   (Source: U.S. Bureau of the Census)

Solution

Because the first equation has already been solved for P in terms of t, substitute this value into the second equation and solve for t, as follows.

$$3631.2 + 131.7t = 4055.6 + 39.9t \qquad \text{Substitute for } P \text{ in Equation 2.}$$

$$131.7t - 39.9t = 4055.6 - 3631.2 \qquad \text{Subtract } 39.9t \text{ and } 3631.2 \text{ from each side.}$$

$$91.8t = 424.4 \qquad \text{Combine like terms.}$$

$$t \approx 4.6 \qquad \text{Divide each side by } 91.8.$$

So, from the given models, you would expect that the population of Arizona exceeded the population of Alabama after $t \approx 4.6$ years, which was sometime during 1994.

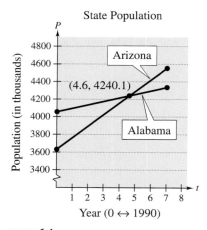

FIGURE **6.4**

Writing ABOUT MATHEMATICS

Interpreting Points of Intersection You plan to rent a 14-foot truck for a 2-day local move. At truck rental agency A, you can rent a truck for $29.95 per day plus $0.49 per mile. At agency B, you can rent a truck for $50 per day plus $0.25 per mile. The total cost y (in dollars) for the truck from agency A is

$$y = (\$29.95 \text{ per day})(2 \text{ days}) + 0.49x$$

$$= 59.90 + 0.49x$$

where x is the total number of miles the truck is driven.

a. Write a total cost equation in terms of x and y for the total cost of the truck from agency B.

b. Use a graphing utility to graph the two equations and find the point of intersection. Interpret the meaning of the point of intersection in the context of the problem.

c. Which agency should you choose if you plan to travel a total of 100 miles over the 2-day move? Why?

d. How does the situation change if you plan to drive 200 miles over the 2-day move?

6.1 Exercises

In Exercises 1–4, determine which ordered pairs are solutions of the system of equations.

1. $\begin{cases} 4x - y = 1 \\ 6x + y = -6 \end{cases}$ (a) $(0, -3)$ (b) $(-1, -4)$
  (c) $\left(-\frac{3}{2}, -2\right)$ (d) $\left(-\frac{1}{2}, -3\right)$

2. $\begin{cases} 4x^2 + y = 3 \\ -x - y = 11 \end{cases}$ (a) $(2, -13)$ (b) $(2, -9)$
  (c) $\left(-\frac{3}{2}, -\frac{31}{3}\right)$ (d) $\left(-\frac{7}{4}, -\frac{37}{4}\right)$

3. $\begin{cases} y = -2e^x \\ 3x - y = 2 \end{cases}$ (a) $(-2, 0)$ (b) $(0, -2)$
  (c) $(0, -3)$ (d) $(-1, 2)$

4. $\begin{cases} -\log x + 3 = y \\ \frac{1}{9}x + y = \frac{28}{9} \end{cases}$ (a) $\left(9, \frac{37}{9}\right)$ (b) $(10, 2)$
  (c) $(1, 3)$ (d) $(2, 4)$

In Exercises 5–14, solve the system by the method of substitution. Check your solution graphically.

5. $\begin{cases} 2x + y = 6 \\ -x + y = 0 \end{cases}$

6. $\begin{cases} x - y = -4 \\ x + 2y = 5 \end{cases}$

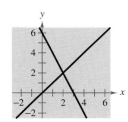

7. $\begin{cases} x - y = -4 \\ x^2 - y = -2 \end{cases}$

8. $\begin{cases} 3x + y = 2 \\ x^3 - 2 + y = 0 \end{cases}$

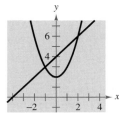
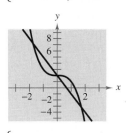

9. $\begin{cases} -2x + y = -5 \\ x^2 + y^2 = 25 \end{cases}$

10. $\begin{cases} x + y = 0 \\ x^3 - 5x - y = 0 \end{cases}$

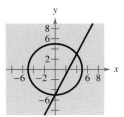
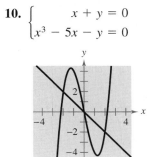

11. $\begin{cases} x^2 + y = 0 \\ x^2 - 4x - y = 0 \end{cases}$

12. $\begin{cases} y = -2x^2 + 2 \\ y = 2(x^4 - 2x^2 + 1) \end{cases}$

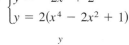
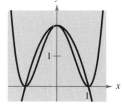

13. $\begin{cases} y = x^3 - 3x^2 + 1 \\ y = x^2 - 3x + 1 \end{cases}$

14. $\begin{cases} y = x^3 - 3x^2 + 4 \\ y = -2x + 4 \end{cases}$

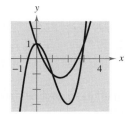
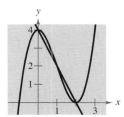

In Exercises 15–28, solve the system by the method of substitution.

15. $\begin{cases} x - y = 0 \\ 5x - 3y = 10 \end{cases}$

16. $\begin{cases} x + 2y = 1 \\ 5x - 4y = -23 \end{cases}$

17. $\begin{cases} 2x - y + 2 = 0 \\ 4x + y - 5 = 0 \end{cases}$

18. $\begin{cases} 6x - 3y - 4 = 0 \\ x + 2y - 4 = 0 \end{cases}$

19. $\begin{cases} 1.5x + 0.8y = 2.3 \\ 0.3x - 0.2y = 0.1 \end{cases}$

20. $\begin{cases} 0.5x + 3.2y = 9.0 \\ 0.2x - 1.6y = -3.6 \end{cases}$

21. $\begin{cases} \frac{1}{5}x + \frac{1}{2}y = 8 \\ x + y = 20 \end{cases}$

22. $\begin{cases} \frac{1}{2}x + \frac{3}{4}y = 10 \\ \frac{3}{4}x - y = 4 \end{cases}$

23. $\begin{cases} 6x + 5y = -3 \\ -x - \frac{5}{6}y = -7 \end{cases}$

24. $\begin{cases} -\frac{2}{3}x + y = 2 \\ 2x - 3y = 6 \end{cases}$

25. $\begin{cases} x^2 - y = 0 \\ 2x + y = 0 \end{cases}$

26. $\begin{cases} x - 2y = 0 \\ 3x - y^2 = 0 \end{cases}$

27. $\begin{cases} x^3 - y = 0 \\ x - y = 0 \end{cases}$

28. $\begin{cases} y = -x \\ y = x^3 + 3x^2 + 2x \end{cases}$

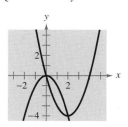

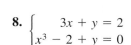
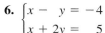

In Exercises 29–42, solve the system graphically.

29. $\begin{cases} -x + 2y = 2 \\ 3x + y = 15 \end{cases}$

30. $\begin{cases} x + y = 0 \\ 3x - 2y = 10 \end{cases}$

31. $\begin{cases} x - 3y = -2 \\ 5x + 3y = 17 \end{cases}$

32. $\begin{cases} -x + 2y = 1 \\ x - y = 2 \end{cases}$

33. $\begin{cases} x + y = 4 \\ x^2 + y^2 - 4x = 0 \end{cases}$

34. $\begin{cases} -x + y = 3 \\ x^2 - 6x - 27 + y^2 = 0 \end{cases}$

35. $\begin{cases} x - y + 3 = 0 \\ x^2 - 4x + 7 = y \end{cases}$

36. $\begin{cases} y^2 - 4x + 11 = 0 \\ -\frac{1}{2}x + y = -\frac{1}{2} \end{cases}$

37. $\begin{cases} 7x + 8y = 24 \\ x - 8y = 8 \end{cases}$

38. $\begin{cases} x - y = 0 \\ 5x - 2y = 6 \end{cases}$

39. $\begin{cases} 3x - 2y = 0 \\ x^2 - y^2 = 4 \end{cases}$

40. $\begin{cases} 2x - y + 3 = 0 \\ x^2 + y^2 - 4x = 0 \end{cases}$

41. $\begin{cases} x^2 + y^2 = 25 \\ 3x^2 - 16y = 0 \end{cases}$

42. $\begin{cases} x^2 + y^2 = 25 \\ (x - 8)^2 + y^2 = 41 \end{cases}$

In Exercises 43–50, use a graphing utility to approximate all points of intersection of the graphs.

43. $\begin{cases} y = e^x \\ x - y + 1 = 0 \end{cases}$

44. $\begin{cases} y = -4e^{-x} \\ y + 3x + 8 = 0 \end{cases}$

45. $\begin{cases} x + 2y = 8 \\ y = \log_2 x \end{cases}$

46. $\begin{cases} y = -2 + \ln(x - 1) \\ 3y + 2x = 9 \end{cases}$

47. $\begin{cases} y = \sqrt{x} \\ y = x \end{cases}$

48. $\begin{cases} x - y = 3 \\ x - y^2 = 1 \end{cases}$

49. $\begin{cases} x^2 + y^2 = 169 \\ x^2 - 8y = 104 \end{cases}$

50. $\begin{cases} x^2 + y^2 = 4 \\ 2x^2 - y = 2 \end{cases}$

In Exercises 51–62, solve the system graphically or algebraically. Explain your choice of method.

51. $\begin{cases} y = 2x \\ y = x^2 + 1 \end{cases}$

52. $\begin{cases} x + y = 4 \\ x^2 + y = 2 \end{cases}$

53. $\begin{cases} 3x - 7y + 6 = 0 \\ x^2 - y^2 = 4 \end{cases}$

54. $\begin{cases} x^2 + y^2 = 25 \\ 2x + y = 10 \end{cases}$

55. $\begin{cases} x - 2y = 4 \\ x^2 - y = 0 \end{cases}$

56. $\begin{cases} y = (x + 1)^3 \\ y = \sqrt{x - 1} \end{cases}$

57. $\begin{cases} y - e^{-x} = 1 \\ y - \ln x = 3 \end{cases}$

58. $\begin{cases} x^2 + y = 4 \\ e^x - y = 0 \end{cases}$

59. $\begin{cases} y = x^4 - 2x^2 + 1 \\ y = 1 - x^2 \end{cases}$

60. $\begin{cases} y = x^3 - 2x^2 + x - 1 \\ y = -x^2 + 3x - 1 \end{cases}$

61. $\begin{cases} xy - 1 = 0 \\ 2x - 4y + 7 = 0 \end{cases}$

62. $\begin{cases} x - 2y = 1 \\ y = \sqrt{x - 1} \end{cases}$

Break-Even Analysis **In Exercises 63–66, find the sales necessary to break even ($R = C$) for the cost C of x units and the revenue R obtained by selling x units. (Round to the nearest whole unit.)**

63. $C = 8650x + 250{,}000, \quad R = 9950x$

64. $C = 2.65x + 350{,}000, \quad R = 4.15x$

65. $C = 5.5\sqrt{x} + 10{,}000, \quad R = 3.29x$

66. $C = 7.8\sqrt{x} + 18{,}500, \quad R = 12.84x$

67. *Break-Even Point* A small business invests $16,000 to produce an item that will sell for $5.95. Each unit can be produced for $3.45.

 (a) How many units must be sold to break even?

 (b) How many units must be sold to make a profit of $6000?

68. *Break-Even Point* A small business invests $5000 to produce an item that will sell for $34.10. Each unit can be produced for $21.60.

 (a) How many units must be sold to break even?

 (b) How many units must be sold to make a profit of $8500?

69. *Finance* A total of $25,000 is invested in two funds paying 6% and 8.5% simple interest. The 6% investment has a lower risk. The investor wants a yearly interest income of $2000 from the investment.

 (a) Write a system of equations in which one equation represents the total amount invested and the other equation represents the $2000 required in interest. Let x and y represent the amounts invested at 6% and 8.5%, respectively.

 (b) Use a graphing utility to graph the two equations. As the amount invested at 6% increases, how does the amount invested at 8.5% change? How does the amount of interest income change? Explain.

 (c) What amount should be invested at 6% to meet the requirement of $2000 per year in interest?

70. *Finance*   A total of $20,000 is invested in two funds paying 6.5% and 8.5% simple interest. The 6.5% investment has a lower risk. The investor wants a yearly interest check of $1600 from the investments.

(a) Write a system of equations in which one equation represents the total amount invested and the other equation represents the $1600 required in interest. Let x and y represent the amounts invested at 6.5% and 8.5%, respectively.

(b) Use a graphing utility to graph the two equations. As the amount invested at 6.5% increases, how does the amount invested at 8.5% change? How does the amount of interest change? Explain.

(c) What amount should be invested at 6.5% to meet the requirement of $1600 per year in interest?

71. *Choice of Two Jobs*   You are offered two jobs selling dental supplies. One company offers a straight commission of 6% of sales. The other company offers a salary of $350 per week plus 3% of sales. How much would you have to sell in a week in order to make the straight commission offer better?

72. *Choice of Two Jobs*   You are offered two different jobs selling college textbooks. One company offers an annual salary of $25,000 plus a year-end bonus of 2% of your total sales. The other company offers an annual salary of $20,000 plus a year-end bonus of 3% of your total sales. Determine the annual sales required to make the second offer better.

73. *Log Volume*   You are offered two different rules for estimating the number of board feet in a 16-foot log. (A board foot is a unit of measure for lumber equal to a board 1 foot square and 1 inch thick.) The first rule is the *Doyle Log Rule* and is modeled by

$$V = (D - 4)^2, \qquad 5 \le D \le 40$$

and the other is the *Scribner Log Rule* and is modeled by

$$V = 0.79D^2 - 2D - 4, \qquad 5 \le D \le 40$$

where D is the diameter (in inches) of the log and V is its volume in board feet.

(a) Use a graphing utility to graph the two log rules in the same viewing window.

(b) For what diameter do the two scales agree?

(c) If you were selling large logs by the board foot, which scale would you use?

74. *Economics*   The supply and demand curves for a business dealing with wheat are

Supply: $p = 1.45 + 0.00014x^2$

Demand: $p = (2.388 - 0.007x)^2$

where p is the price in dollars per bushel and x is the quantity in bushels per day. Use a graphing utility to graph the supply and demand equations and find the market equilibrium. (The *market equilibrium* is the point of intersection of the graphs for $x > 0$.)

Geometry   **In Exercises 75–78, find the dimensions of the rectangle meeting the specified conditions.**

75. The perimeter is 30 meters and the length is 3 meters greater than the width.

76. The perimeter is 280 centimeters and the width is 20 centimeters less than the length.

77. The perimeter is 42 inches and the width is three-fourths the length.

78. The perimeter is 210 feet and the length is $1\frac{1}{2}$ times the width.

79. *Geometry*   What are the dimensions of a rectangular tract of land if its perimeter is 40 kilometers and its area is 96 square kilometers?

80. *Geometry*   What are the dimensions of an isosceles right triangle with a 2-inch hypotenuse and an area of 1 square inch?

81. *Data Analysis*   The table gives the amount y, in millions of short tons, of paperboard produced in the years 1993 through 1996 in the United States. (Source: American Forest and Paper Association)

Year	1993	1994	1995	1996
y	43.1	45.7	46.6	47.9

(a) Use the regression features of a graphing utility to find a linear model f and a quadratic model g that represent the data in the interval from 1993 through 1996. Let $t = 3$ represent 1993.

(b) Use a graphing utility to graph the data and the two models in the same viewing window.

(c) Approximate the points of intersection of the graphs of the models.

Synthesis

True or False? **In Exercises 82 and 83, determine whether the statement is true or false. Justify your answer.**

82. In order to solve a system of equations by substitution, you must always solve for y in one of the two equations and then back-substitute.

83. If a system consists of a parabola and a circle, then the system can have at most two solutions.

84. What is meant by a solution of a system of equations in two variables?

85. *Think About It* When solving a system of equations by substitution, how do you recognize that the system has no solution?

86. *Writing* Write a brief paragraph describing any advantages of substitution over the graphical method of solving a system of equations.

87. *Exploration* Find an equation of a line whose graph intersects the graph of the parabola $y = x^2$ at (a) two points, (b) one point, and (c) no points. (There is more than one correct answer.)

88. *Conjecture* Consider the system of equations

$$\begin{cases} y = b^x \\ y = x^b \end{cases}.$$

(a) Use a graphing utility to graph the system for $b = 1, 2, 3,$ and 4.

(b) For a fixed even value of $b > 1$, make a conjecture about the number of points of intersection of the graphs in part (a).

Review

In Exercises 89–94, find the general form of the equation of the line through the two points.

89. $(-2, 7), (5, 5)$
90. $(3.5, 4), (10, 6)$
91. $(6, 3), (10, 3)$
92. $(4, -2), (4, 5)$
93. $\left(\frac{3}{5}, 0\right), (4, 6)$
94. $\left(-\frac{7}{3}, 8\right), \left(\frac{5}{2}, \frac{1}{2}\right)$

In Exercises 95–98, find the domain of the function and identify any horizontal or vertical asymptotes.

95. $f(x) = \dfrac{5}{x - 6}$
96. $f(x) = \dfrac{2x - 7}{3x + 2}$
97. $f(x) = \dfrac{x^2 + 2}{x^2 - 16}$
98. $f(x) = 3 - \dfrac{2}{x^2}$

In Exercises 99–102, sketch the graph of the equation.

99. $y = -2^{0.5x}$
100. $y = 2^{-0.5x}$
101. $y = 3e^{-x-4}$
102. $y = \dfrac{2 + e^{-x}}{5}$

6.2 Two-Variable Linear Systems

▶ **What you should learn**

- How to use the method of elimination to solve systems of linear equations in two variables
- How to interpret graphically the numbers of solutions of systems of linear equations in two variables
- How to use systems of equations in two variables to model and solve real-life problems

▶ **Why you should learn it**

You can use systems of equations in two variables to model and solve real-life problems. For instance, Exercise 64 on page 465 shows how to use a system of equations to analyze an investment portfolio.

Jon Riley/Tony Stone Images

The Method of Elimination

In Section 6.1, you studied two methods for solving a system of equations: substitution and graphing. Now you will study the **method of elimination.** The key step in this method is to obtain, for one of the variables, coefficients that differ only in sign so that *adding* the equations eliminates the variable.

$$3x + 5y = 7 \qquad \text{Equation 1}$$
$$\underline{-3x - 2y = -1} \qquad \text{Equation 2}$$
$$3y = 6 \qquad \text{Add equations.}$$

Note that by adding the two equations, you eliminate the x-terms and obtain a single equation in y. Solving this equation for y produces $y = 2$, which you can then back-substitute into one of the original equations to solve for x.

Example 1 ▶ Solving a System of Equations by Elimination

Solve the system of linear equations.

$$\begin{cases} 3x + 2y = 4 & \text{Equation 1} \\ 5x - 2y = 8 & \text{Equation 2} \end{cases}$$

Solution

Because the coefficients of y differ only in sign, you can eliminate the y-terms by adding the two equations.

$$3x + 2y = 4 \qquad \text{Write Equation 1.}$$
$$\underline{5x - 2y = 8} \qquad \text{Write Equation 2.}$$
$$8x = 12 \qquad \text{Add equations.}$$

Therefore, $x = \frac{3}{2}$. By back-substituting this value into Equation 1, you can solve for y.

$$3x + 2y = 4 \qquad \text{Write Equation 1.}$$
$$3\left(\frac{3}{2}\right) + 2y = 4 \qquad \text{Substitute } \tfrac{3}{2} \text{ for } x.$$
$$\frac{9}{2} + 2y = 4 \qquad \text{Simplify.}$$
$$y = -\frac{1}{4} \qquad \text{Solve for } y.$$

The solution is $\left(\frac{3}{2}, -\frac{1}{4}\right)$. Check this in the original system.

Try using the method of substitution to solve the system given in Example 1. Which method do you think is easier? Many people find that the method of elimination is more efficient.

| **Example 2** ▶ | **Solving a System of Equations by Elimination** |

Solve the system of linear equations.

$$\begin{cases} 2x - 3y = -7 \\ 3x + y = -5 \end{cases}$$ Equation 1
                                                                   Equation 2

Solution

For this system, you can obtain coefficients that differ only in sign by multiplying Equation 2 by 3.

$2x - 3y = -7$	⟹	$2x - 3y = -7$	Write Equation 1.
$3x + y = -5$	⟹	$9x + 3y = -15$	Multiply Equation 2 by 3.
		$11x = -22$	Add equations.

So, you can see that $x = -2$. By back-substituting this value of x into Equation 1, you can solve for y.

$2x - 3y = -7$	Write Equation 1.
$2(-2) - 3y = -7$	Substitute -2 for x.
$-3y = -3$	Collect like terms.
$y = 1$	Solve for y.

The solution is $(-2, 1)$. Check this in the original system.

Check

$2(-2) - 3(1) \overset{?}{=} -7$	Substitute into Equation 1.
$-4 - 3 = -7$	Equation 1 checks. ✓
$3(-2) + 1 \overset{?}{=} -5$	Substitute into Equation 2.
$-6 + 1 = -5$	Equation 2 checks. ✓

In Example 2, the two systems of linear equations

$$\begin{cases} 2x - 3y = -7 \\ 3x + y = -5 \end{cases}$$

and

$$\begin{cases} 2x - 3y = -7 \\ 9x + 3y = -15 \end{cases}$$

are called **equivalent systems** because they have precisely the same solution set. The operations that can be performed on a system of linear equations to produce an equivalent system are (1) interchanging any two equations, (2) multiplying an equation by a nonzero constant, and (3) adding a multiple of one equation to any other equation in the system.

STUDY TIP

To obtain coefficients (for one of the variables) that differ only in sign, you often need to multiply one or both of the equations by suitably chosen constants.

Method of Elimination

1. *Obtain coefficients* for x (or y) that differ only in sign by multiplying all terms of one or both equations by suitably chosen constants.

2. *Add* the equations to eliminate one variable and solve the resulting equation.

3. *Back-substitute* the value obtained in Step 2 into either of the original equations and solve for the other variable.

4. *Check* your solution in both of the original equations.

◀ **Exploration** ▶

Sketch the graph of each of the following systems of equations.

a. $\begin{cases} y = 5x + 1 \\ y - x = -5 \end{cases}$

b. $\begin{cases} 3y = 4x - 1 \\ -8x + 2 = -6y \end{cases}$

c. $\begin{cases} 2y = -x + 3 \\ -4 = y + \frac{1}{2}x \end{cases}$

Determine the number of solutions each system has. Explain your reasoning.

Example 3 ▶ Solving a System of Equations by Elimination

Solve the system of linear equations.

$$\begin{cases} 5x + 3y = 9 & \text{Equation 1} \\ 2x - 4y = 14 & \text{Equation 2} \end{cases}$$

Solution

You can obtain coefficients that differ only in sign by multiplying Equation 1 by 4 and multiplying Equation 2 by 3.

$$5x + 3y = 9 \quad \Longrightarrow \quad 20x + 12y = 36 \qquad \text{Multiply Equation 1 by 4.}$$

$$\underline{2x - 4y = 14} \quad \Longrightarrow \quad \underline{6x - 12y = 42} \qquad \text{Multiply Equation 2 by 3.}$$

$$26x \qquad\quad = 78 \qquad \text{Add equations.}$$

From this equation, you can see that $x = 3$. By back-substituting this value of x into Equation 2, you can solve for y.

$$2x - 4y = 14 \qquad \text{Write Equation 2.}$$

$$2(3) - 4y = 14 \qquad \text{Substitute 3 for } x.$$

$$-4y = 8 \qquad \text{Collect like terms.}$$

$$y = -2 \qquad \text{Solve for } y.$$

The solution is $(3, -2)$. Check this in the original system.

Remember that you can check the solution of a system of equations graphically. For instance, to check the solution found in Example 3, graph both equations in the same viewing window, as shown in Figure 6.5. Notice that the two lines intersect at $(3, -2)$.

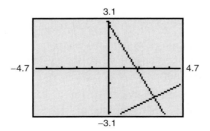

FIGURE **6.5**

Graphical Interpretation of Solutions

It is possible for a *general* system of equations to have exactly one solution, two or more solutions, or no solution. If a system of *linear* equations has two different solutions, it must have an *infinite* number of solutions. To see why this is true, consider the following graphical interpretations of a system of two linear equations in two variables.

Graphical Interpretations of Solutions

For a system of two linear equations in two variables, the number of solutions is one of the following.

Number of Solutions	*Graphical Interpretation*
1. Exactly one solution	The two lines intersect at one point.
2. Infinitely many solutions	The two lines are identical.
3. No solution	The two lines are parallel.

A system of linear equations is **consistent** if it has at least one solution. It is **inconsistent** if it has no solution.

Example 4 ▶ Recognizing Graphs of Linear Systems

Match the system of linear equations with its graph in Figure 6.6. State whether the system is consistent or inconsistent and describe the number of solutions.

a. $\begin{cases} 2x - 3y = 3 \\ -4x + 6y = 6 \end{cases}$ **b.** $\begin{cases} 2x - 3y = 3 \\ x + 2y = 5 \end{cases}$ **c.** $\begin{cases} 2x - 3y = 3 \\ -4x + 6y = -6 \end{cases}$

i. 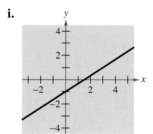   **ii.** 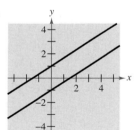   **iii.**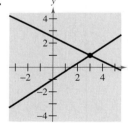

FIGURE 6.6

Solution

a. The graph of system (a) is a pair of parallel lines (ii). The lines have no point of intersection, so the system has no solution. The system is inconsistent.

b. The graph of system (b) is a pair of intersecting lines (iii). The lines have one point of intersection, so the system has exactly one solution. The system is consistent.

c. The graph of system (c) is a pair of lines that coincide (i). The lines have infinitely many points of intersection, so the system has infinitely many solutions. The system is consistent.

In Examples 5 and 6, note how you can use the method of elimination to determine that a system of linear equations has no solution or infinitely many solutions.

Example 5 ▶ No-Solution Case: Method of Elimination

Solve the system of linear equations.

$$\begin{cases} x - 2y = 3 & \text{Equation 1} \\ -2x + 4y = 1 & \text{Equation 2} \end{cases}$$

Solution

To obtain coefficients that differ only in sign, multiply Equation 1 by 2.

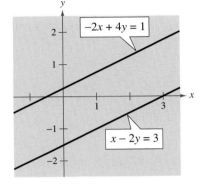

FIGURE 6.7

$x - 2y = 3$	⟹	$2x - 4y = 6$	Multiply Equation 1 by 2.
$-2x + 4y = 1$	⟹	$-2x + 4y = 1$	Write Equation 2.
		$0 = 7$	False statement

Because there are no values of x and y for which $0 = 7$, you can conclude that the system is inconsistent and has no solution. The lines corresponding to the two equations in this system are shown in Figure 6.7. Note that the two lines are parallel and therefore have no point of intersection.

In Example 5, note that the occurrence of a false statement, such as $0 = 7$, indicates that the system has no solution. In the next example, note that the occurrence of a statement that is true for all values of the variables, such as $0 = 0$, indicates that the system has infinitely many solutions.

Example 6 ▶ Many-Solutions Case: Method of Elimination

Solve the system of linear equations.

$$\begin{cases} 2x - y = 1 & \text{Equation 1} \\ 4x - 2y = 2 & \text{Equation 2} \end{cases}$$

Solution

To obtain coefficients that differ only in sign, multiply Equation 2 by $-\frac{1}{2}$.

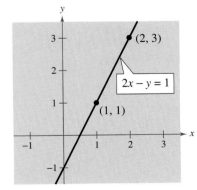

FIGURE 6.8

$2x - y = 1$	⟹	$2x - y = 1$	Write Equation 1.
$4x - 2y = 2$	⟹	$-2x + y = -1$	Multiply Equation 2 by $-\frac{1}{2}$.
		$0 = 0$	Add equations.

Because the two equations turn out to be equivalent (have the same solution set), you can conclude that the system has infinitely many solutions. The solution set consists of all points (x, y) lying on the line $2x - y = 1$ as shown in Figure 6.8. Letting $x = a$, where a is any real number, you can see that the solutions to the system are $(a, 2a + 1)$.

Technology

The general solution of the linear system

$$\begin{cases} ax + by = c \\ dx + ey = f \end{cases}$$

is $x = (ce - bf)/(ae - db)$ and $y = (af - cd)/(ae - db)$. If $ae - db = 0$, the system does not have a unique solution. Graphing utility programs for solving such a system can be found at our website *college/hmco.com.* Try using the program for your graphing utility to solve the system in Example 7.

Example 7 illustrates a strategy for solving a system of linear equations that has decimal coefficients.

Example 7 ▶ A Linear System Having Decimal Coefficients

Solve the system of linear equations.

$$\begin{cases} 0.02x - 0.05y = -0.38 & \text{Equation 1} \\ 0.03x + 0.04y = 1.04 & \text{Equation 2} \end{cases}$$

Solution

Because the coefficients in this system have two decimal places, you can begin by multiplying each equation by 100. (This produces a system in which the coefficients are all integers.)

$$\begin{cases} 2x - 5y = -38 & \text{Revised Equation 1} \\ 3x + 4y = 104 & \text{Revised Equation 2} \end{cases}$$

Now, to obtain coefficients that differ only in sign, multiply Equation 1 by 3 and multiply Equation 2 by -2.

$$2x - 5y = -38 \implies 6x - 15y = -114 \quad \text{Multiply Equation 1 by 3.}$$

$$\underline{3x + 4y = 104} \implies \underline{-6x - 8y = -208} \quad \text{Multiply Equation 2 by } -2.$$

$$-23y = -322 \quad \text{Add equations.}$$

So, you can conclude that

$$y = \frac{-322}{-23} = 14.$$

Back-substituting this value into Equation 2 produces the following.

$$3x + 4y = 104 \qquad \text{Write revised Equation 2.}$$

$$3x + 4(14) = 104 \qquad \text{Substitute 14 for } y.$$

$$3x = 48 \qquad \text{Collect like terms.}$$

$$x = 16 \qquad \text{Solve for } x.$$

The solution is $(16, 14)$. Check this in the original system.

Check

$$0.02x - 0.05y = -0.38 \qquad \text{Write original Equation 1.}$$

$$0.02(16) - 0.05(14) \stackrel{?}{=} -0.38 \qquad \text{Substitute into Equation 1.}$$

$$0.32 - 0.70 = -0.38 \qquad \text{Equation 1 checks. ✓}$$

$$0.03x + 0.04y = 1.04 \qquad \text{Write original Equation 2.}$$

$$0.03(16) + 0.04(14) \stackrel{?}{=} 1.04 \qquad \text{Substitute into Equation 2.}$$

$$0.48 + 0.56 = 1.04 \qquad \text{Equation 2 checks. ✓}$$

Applications

At this point, you may be asking the question "How can I tell which application problems can be solved using a system of linear equations?" The answer comes from the following considerations.

1. Does the problem involve more than one unknown quantity?

2. Are there two (or more) equations or conditions to be satisfied?

If one or both of these conditions occur, the appropriate mathematical model for the problem may be a system of linear equations. Example 8 shows how to construct such a model.

> **Example 8 ▶ An Application of a Linear System**

An airplane flying into a headwind travels the 2000-mile flying distance between two cities in 4 hours and 24 minutes. On the return flight, the same distance is traveled in 4 hours. Find the air speed of the plane and the speed of the wind, assuming that both remain constant.

Solution

The two unknown quantities are the speeds of the wind and the plane. If r_1 is the speed of the plane and r_2 is the speed of the wind, then

$$r_1 - r_2 = \text{speed of the plane against the wind}$$

$$r_1 + r_2 = \text{speed of the plane with the wind}$$

as shown in Figure 6.9. Using the formula

$$\text{Distance} = (\text{rate})(\text{time})$$

for these two speeds, you obtain the following equations.

$$2000 = (r_1 - r_2)\left(4 + \frac{24}{60}\right)$$

$$2000 = (r_1 + r_2)(4)$$

These two equations simplify as follows.

$$\begin{cases} 5000 = 11r_1 - 11r_2 & \text{Equation 1} \\ 500 = r_1 + r_2 & \text{Equation 2} \end{cases}$$

By elimination, the solution is

$$r_1 = \frac{5250}{11} \approx 477.27 \text{ miles per hour} \qquad \text{Speed of plane}$$

$$r_2 = \frac{250}{11} \approx 22.73 \text{ miles per hour.} \qquad \text{Speed of wind}$$

Check this solution in the original statement of the problem.

Original flight

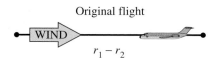

$r_1 - r_2$

Return flight

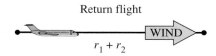

$r_1 + r_2$

FIGURE **6.9**

In a free market, the demands for many products are related to the prices of the products. As the prices decrease, the demands by consumers increase and the amounts that producers are able or willing to supply decrease.

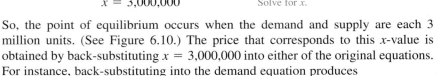

Example 9 ▶ Finding the Point of Equilibrium

The demand and supply functions for a certain type of calculator are

$$\begin{cases} p = 150 - 0.00001x & \text{Demand equation} \\ p = 60 + 0.00002x & \text{Supply equation} \end{cases}$$

where p is the price in dollars and x represents the number of units. Find the point of equilibrium for this market. The point of equilibrium is the price p and number of units x that satisfy both the demand and supply equations.

Solution

Begin by substituting the value of p given in the supply equation into the demand equation.

$$p = 150 - 0.00001x \qquad \text{Write demand equation.}$$

$$60 + 0.00002x = 150 - 0.00001x \qquad \text{Substitute } 60 + 0.00002x \text{ for } p.$$

$$0.00003x = 90 \qquad \text{Collect like terms.}$$

$$x = 3{,}000{,}000 \qquad \text{Solve for } x.$$

So, the point of equilibrium occurs when the demand and supply are each 3 million units. (See Figure 6.10.) The price that corresponds to this x-value is obtained by back-substituting $x = 3{,}000{,}000$ into either of the original equations. For instance, back-substituting into the demand equation produces

$$p = 150 - 0.00001(3{,}000{,}000)$$

$$= 150 - 30$$

$$= \$120.$$

The solution is $(3{,}000{,}000, \ 120)$. You can check this as follows.

Check

Substitute $(3{,}000{,}000, \ 120)$ into the demand equation.

$$p = 150 - 0.00001x \qquad \text{Write demand equation.}$$

$$120 \overset{?}{=} 150 - 0.00001(3{,}000{,}000) \qquad \text{Substitute 120 for } p \text{ and 3,000,000 for } x.$$

$$120 = 120 \qquad \text{Solution checks in demand equation. ✓}$$

Substitute $(3{,}000{,}000, \ 120)$ into the supply equation.

$$p = 60 + 0.00002x \qquad \text{Write supply equation.}$$

$$120 \overset{?}{=} 60 + 0.00002(3{,}000{,}000) \qquad \text{Substitute 120 for } p \text{ and 3,000,000 for } x.$$

$$120 = 120 \qquad \text{Solution checks in supply equation. ✓}$$

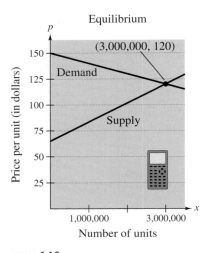

Equilibrium

$(3{,}000{,}000, \ 120)$

Demand

Supply

FIGURE 6.10

6.2 Exercises

In Exercises 1–10, solve by elimination. Match each line with its equation.

1. $\begin{cases} 2x + y = 5 \\ x - y = 1 \end{cases}$

2. $\begin{cases} x + 3y = 1 \\ -x + 2y = 4 \end{cases}$

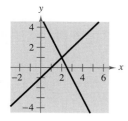
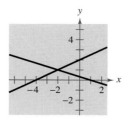

3. $\begin{cases} x + y = 0 \\ 3x + 2y = 1 \end{cases}$

4. $\begin{cases} 2x - y = 3 \\ 4x + 3y = 21 \end{cases}$

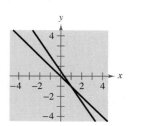
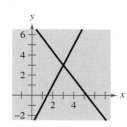

5. $\begin{cases} x - y = 2 \\ -2x + 2y = 5 \end{cases}$

6. $\begin{cases} 3x + 2y = 3 \\ 6x + 4y = 14 \end{cases}$

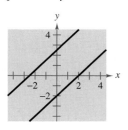
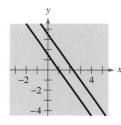

7. $\begin{cases} 3x - 2y = 5 \\ -6x + 4y = -10 \end{cases}$

8. $\begin{cases} 9x - 3y = -15 \\ -3x + y = 5 \end{cases}$

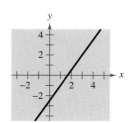
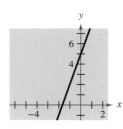

9. $\begin{cases} 9x + 3y = 1 \\ 3x - 6y = 5 \end{cases}$

10. $\begin{cases} 5x + 3y = -18 \\ 2x - 6y = 1 \end{cases}$

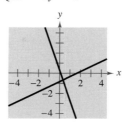
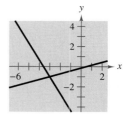

In Exercises 11–30, solve the system by elimination and check any solutions algebraically.

11. $\begin{cases} x + 2y = 4 \\ x - 2y = 1 \end{cases}$

12. $\begin{cases} 3x - 5y = 2 \\ 2x + 5y = 13 \end{cases}$

13. $\begin{cases} 2x + 3y = 18 \\ 5x - y = 11 \end{cases}$

14. $\begin{cases} x + 7y = 12 \\ 3x - 5y = 10 \end{cases}$

15. $\begin{cases} 3x + 2y = 10 \\ 2x + 5y = 3 \end{cases}$

16. $\begin{cases} 2r + 4s = 5 \\ 16r + 50s = 55 \end{cases}$

17. $\begin{cases} 5u + 6v = 24 \\ 3u + 5v = 18 \end{cases}$

18. $\begin{cases} 3x + 11y = 4 \\ -2x - 5y = 9 \end{cases}$

19. $\begin{cases} 1.8x + 1.2y = 4 \\ 9x + 6y = 3 \end{cases}$

20. $\begin{cases} 3.1x - 2.9y = -10.2 \\ 15.5x - 14.5y = 21 \end{cases}$

21. $\begin{cases} \dfrac{x}{4} + \dfrac{y}{6} = 1 \\ x - y = 3 \end{cases}$

22. $\begin{cases} \dfrac{2}{3}x + \dfrac{1}{6}y = \dfrac{2}{3} \\ 4x + y = 4 \end{cases}$

23. $\begin{cases} 2.5x - 3y = 1.5 \\ 2x - 2.4y = 1.2 \end{cases}$

24. $\begin{cases} 6.3x + 7.2y = 5.4 \\ 5.6x + 6.4y = 4.8 \end{cases}$

25. $\begin{cases} 0.05x - 0.03y = 0.21 \\ 0.07x + 0.02y = 0.16 \end{cases}$

26. $\begin{cases} 0.2x - 0.5y = -27.8 \\ 0.3x + 0.4y = 68.7 \end{cases}$

27. $\begin{cases} 4b + 3m = 3 \\ 3b + 11m = 13 \end{cases}$

28. $\begin{cases} 2x + 5y = 8 \\ 5x + 8y = 10 \end{cases}$

29. $\begin{cases} \dfrac{x + 3}{4} + \dfrac{y - 1}{3} = 1 \\ 2x - y = 12 \end{cases}$

30. $\begin{cases} \dfrac{x - 1}{2} + \dfrac{y + 2}{3} = 4 \\ x - 2y = 5 \end{cases}$

In Exercises 31–38, use a graphing utility to graph the lines in the system. Use the graphs to determine if the system is consistent or inconsistent. If the system is consistent, determine the number of solutions.

31. $\begin{cases} 2x - 5y = 0 \\ x - y = 3 \end{cases}$

32. $\begin{cases} 2x + y = 5 \\ x - 2y = -1 \end{cases}$

33. $\begin{cases} \frac{3}{5}x - y = 3 \\ -3x + 5y = 9 \end{cases}$

34. $\begin{cases} 4x - 6y = 9 \\ \frac{16}{3}x - 8y = 12 \end{cases}$

35. $\begin{cases} x + 7y = 2 \\ 4x - y = 9 \end{cases}$

36. $\begin{cases} 8x - 14y = 5 \\ 2x - 3.5y = 1.25 \end{cases}$

37. $\begin{cases} -x + 7y = 3 \\ -\frac{1}{7}x + y = 5 \end{cases}$

38. $\begin{cases} -7x + 6y = -4 \\ y + \frac{7}{6}x = -1 \end{cases}$

In Exercises 39–46, use a graphing utility to graph the two equations. Use the graphs to approximate the solutions of the system.

39. $\begin{cases} 8x + 9y = 42 \\ 6x - y = 16 \end{cases}$

40. $\begin{cases} 4y = -8 \\ 7x - 2y = 25 \end{cases}$

41. $\begin{cases} \frac{3}{2}x - \frac{1}{5}y = 8 \\ -2x + 3y = 3 \end{cases}$

42. $\begin{cases} \frac{3}{4}x - \frac{5}{2}y = -9 \\ -x + 6y = 28 \end{cases}$

43. $\begin{cases} 0.5x + 2.2y = 9 \\ 6x + 0.4y = -22 \end{cases}$

44. $\begin{cases} 2.4x + 3.8y = -17.6 \\ 4x - 0.2y = -3.2 \end{cases}$

45. $\begin{cases} 7x - 2y = 24 \\ 5x + 6y = -20 \end{cases}$

46. $\begin{cases} 10x - 13y = -20 \\ 8x + 11y = -16 \end{cases}$

In Exercises 47–54, use any method to solve the system.

47. $\begin{cases} 3x - 5y = 7 \\ 2x + y = 9 \end{cases}$

48. $\begin{cases} -x + 3y = 17 \\ 4x + 3y = 7 \end{cases}$

49. $\begin{cases} y = 2x - 5 \\ y = 5x - 11 \end{cases}$

50. $\begin{cases} 7x + 3y = 16 \\ y = x + 2 \end{cases}$

51. $\begin{cases} x - 5y = 21 \\ 6x + 5y = 21 \end{cases}$

52. $\begin{cases} y = -3x - 8 \\ y = 15 - 2x \end{cases}$

53. $\begin{cases} -2x + 8y = 19 \\ y = x - 3 \end{cases}$

54. $\begin{cases} 4x - 3y = 6 \\ -5x + 7y = -1 \end{cases}$

Supply and Demand   In Exercises 55–58, find the point of equilibrium of the given demand and supply equations.

Demand	Supply
55. $p = 50 - 0.5x$	$p = 0.125x$
56. $p = 100 - 0.05x$	$p = 25 + 0.1x$
57. $p = 140 - 0.00002x$	$p = 80 + 0.00001x$
58. $p = 400 - 0.0002x$	$p = 225 + 0.0005x$

59. *Airplane Speed*   An airplane flying into a headwind travels the 1800-mile flying distance between two cities in 3 hours and 36 minutes. On the return flight, the distance is traveled in 3 hours. Find the air speed of the plane and the speed of the wind, assuming that both remain constant.

60. *Airplane Speed*   Two planes start from the same airport and fly in opposite directions. The second plane starts $\frac{1}{2}$ hour after the first plane, but its speed is 80 kilometers per hour faster. Find the air speed of each plane if 2 hours after the first plane departs the planes are 3200 kilometers apart.

61. *Acid Mixture*   Ten liters of a 30% acid solution is obtained by mixing a 20% solution with a 50% solution.

(a) Write a system of equations in which one equation represents the amount of final mixture required and the other represents the amount of acid in the final mixture. Let x and y represent the amounts of the 20% and 50% solutions, respectively.

(b) Use a graphing utility to graph the two equations in part (a). As the amount of the 20% solution increases, how does the amount of the 50% solution change?

(c) How much of each solution is required to obtain the specified concentration of the final mixture?

62. *Fuel Mixture*   Five hundred gallons of 89 octane gasoline is obtained by mixing 87 octane gasoline with 92 octane gasoline.

(a) Write a system of equations in which one equation represents the amount of final mixture required and the other represents the amounts of 87 and 92 octane gasolines in the final mixture. Let x and y represent the numbers of gallons of 87 octane and 92 octane gasolines, respectively.

(b) Use a graphing utility to graph the two equations in part (a). As the amount of 87 octane gasoline increases, how does the amount of 92 octane gasoline change?

(c) How much of each type of gasoline is required to obtain the 500 gallons of 89 octane gasoline?

63. *Finance*   A total of $12,000 is invested in two corporate bonds that pay 7.5% and 9% simple interest. The investor wants an annual interest income of $990 from the investments. What amount should be invested in the 7.5% bond?

64. *Finance* A total of $32,000 is invested in two municipal bonds that pay 5.75% and 6.25% simple interest. The investor wants an annual interest income of $1900 from the investments. What amount should be invested in the 5.75% bond?

65. *Ticket Sales* At a local championship basketball game, 1435 tickets were sold. A student admission ticket cost $1.50 and an adult admission ticket cost $5.00. The total ticket receipts for the ball game were $3552.50. How many of each ticket were sold?

66. *Clearance Sale* A department store held a sale to sell all of the 214 winter jackets that remained after the season ended. Until noon, each jacket in the store was priced at $31.95. At noon the price of the jackets was further reduced to $18.95. After the last jacket was sold, total receipts for the clearance sale were $5108.30. How many jackets were sold before noon and how many were sold after noon?

67. *Production* A plastics factory uses two different machines working continuously to produce deodorant containers. One machine produces the containers 1.8 times faster than the second machine. If 1764 containers are produced, how many are produced by each machine?

68. *Balloons* A child and his father blow up balloons together for a party. The child inflates two balloons for every three done by his father. How many balloons are inflated by each person to total 80?

Fitting a Line to Data **In Exercises 69–74, find the least squares regression line $y = ax + b$ for the points**

$(x_1, y_1), (x_2, y_2), \ldots, (x_n, y_n)$

by solving the system for a and b. Then use the linear regression capabilities of a graphing utility to confirm the result. (For an explanation of how the coefficients of a and b in the system are obtained, see Appendix B.)

69. $\begin{cases} 5b + 10a = 20.2 \\ 10b + 30a = 50.1 \end{cases}$ **70.** $\begin{cases} 5b + 10a = 11.7 \\ 10b + 30a = 25.6 \end{cases}$

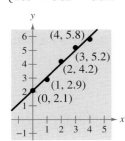

71. $\begin{cases} 7b + 21a = 35.1 \\ 21b + 91a = 114.2 \end{cases}$ **72.** $\begin{cases} 6b + 15a = 23.6 \\ 15b + 55a = 48.8 \end{cases}$

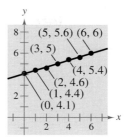
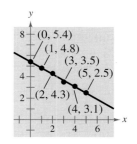

73. $\begin{cases} 4b + 4a = 8 \\ 4b + 6a = 4 \end{cases}$ **74.** $\begin{cases} 8b + 28a = 8 \\ 28b + 116a = 37 \end{cases}$

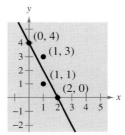
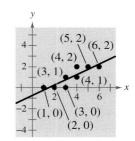

75. *Data Analysis* A store manager wants to know the demand for a certain product as a function of the price. The daily sales for the different prices of the product are given in the table.

Price (x)	$1.00	$1.25	$1.50
Demand (y)	450	375	330

(a) Find the least squares regression line $y = ax + b$ for the data by solving the system for a and b.

$$\begin{cases} 3.00b + 3.7500a = 1155.00 \\ 3.75b + 4.8125a = 1413.75 \end{cases}$$

(b) Use the linear regression capabilities of a graphing utility to confirm the result.

(c) Plot the data and the linear regression equation.

(d) Use the line to predict the demand when the price is $1.40.

76. *Data Analysis* A farmer used four test plots to determine the relationship between wheat yield in bushels per acre and the amount of fertilizer in hundreds of pounds per acre. The results are given in the table.

Fertilizer (x)	1.0	1.5	2.0	2.5
Yield (y)	32	41	48	53

(a) Find the least squares regression line $y = ax + b$ for the data by solving the system for a and b.

$$\begin{cases} 4b + 7.0a = 174 \\ 7b + 13.5a = 322 \end{cases}$$

(b) Use the linear regression capabilities of a graphing utility to confirm the result.

(c) Plot the data and the linear regression equation.

(d) Use the line to predict the yield for a fertilizer application of 160 pounds per acre.

Synthesis

True or False? **In Exercises 77–79, determine whether the statement is true or false. Justify your answer.**

77. If two lines do not have exactly one point of intersection, then they must be parallel.

78. Solving a system of equations graphically will always give an exact solution.

79. If a system of linear equations has no solution, then the lines must be parallel.

Exploration **In Exercises 80–83, find a system of linear equations that has the given solution. (There is more than one correct answer.)**

80. $(6, 3)$

81. $(8, -2)$

82. $\left(3, \frac{5}{2}\right)$

83. $\left(-\frac{2}{3}, -10\right)$

Think About It **In Exercises 84 and 85, the graphs of the two equations appear to be parallel. Yet, when the system is solved algebraically, you find that the system does have a solution. Find the solution and explain why it does not appear on the portion of the graph that is shown.**

84. $\begin{cases} 100y - x = 200 \\ 99y - x = -198 \end{cases}$

85. $\begin{cases} 21x - 20y = 0 \\ 13x - 12y = 120 \end{cases}$

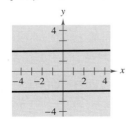

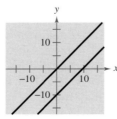

86. *Writing* Briefly explain whether or not it is possible for a consistent system of linear equations to have exactly two solutions.

87. *Think About It* Give examples of (a) a system of linear equations that has no solution and (b) a system that has an infinite number of solutions.

Exploration **In Exercises 88 and 89, find the value of k such that the system of linear equations is inconsistent.**

88. $\begin{cases} 4x - 8y = -3 \\ 2x + ky = 16 \end{cases}$

89. $\begin{cases} 15x + 3y = 6 \\ -10x + ky = 9 \end{cases}$

Review

In Exercises 90–97, solve the inequality and graph the solution on a real number line.

90. $-11 - 6x \geq 33$

91. $2(x - 3) > -5x + 1$

92. $8x - 15 \leq -4(2x - 1)$

93. $-6 \leq 3x - 10 < 6$

94. $|x - 8| < 10$

95. $|x + 10| \geq -3$

96. $2x^2 + 3x - 35 < 0$

97. $3x^2 + 12x > 0$

In Exercises 98 and 99, write the partial fraction decomposition for the rational expression.

98. $\dfrac{x - 1}{x^2 + 11x + 30}$

99. $\dfrac{3}{x(x^2 - 1)}$

In Exercises 100–103, write the expression as the logarithm of a single quantity.

100. $\ln x + \ln 6$

101. $\ln x - 5 \ln(x + 3)$

102. $\log_9 12 - \log_9 x$

103. $\frac{1}{4} \log_6 3x$

In Exercises 104 and 105, solve the system by the method of substitution.

104. $\begin{cases} 2x - y = 4 \\ -4x + 2y = -12 \end{cases}$

105. $\begin{cases} 30x - 40y - 33 = 0 \\ 10x + 20y - 21 = 0 \end{cases}$

6.3 | Multivariable Linear Systems

▶ **Why you should learn it**

Systems of linear equations in three or more variables can be used to model and solve real-life problems. For instance, Exercise 72 on page 478 shows how to use a system of linear equations to analyze the reproduction rates of deer in a wildlife preserve.

Jeanne Drake/Tony Stone Images

Row-Echelon Form and Back-Substitution

The method of elimination can be applied to a system of linear equations in more than two variables. In fact, this method easily adapts to computer use for solving linear systems with dozens of variables.

When elimination is used to solve a system of linear equations, the goal is to rewrite the system in a form to which back-substitution can be applied. To see how this works, consider the following two systems of linear equations.

System of Three Linear Equations in Three Variables: (See Example 3.)

$$\begin{cases} x - 2y + 3z = 9 \\ -x + 3y \quad\quad = -4 \\ 2x - 5y + 5z = 17 \end{cases}$$

Equivalent System in Row-Echelon Form: (See Example 1.)

$$\begin{cases} x - 2y + 3z = 9 \\ \quad\quad y + 3z = 5 \\ \quad\quad\quad\quad z = 2 \end{cases}$$

The second system is said to be in **row-echelon form,** which means that it has a "stair-step" pattern with leading coefficients of 1. After comparing the two systems, it should be clear that it is easier to solve the system in row-echelon form.

Example 1 ▶ Using Back-Substitution in Row-Echelon Form

Solve the system of linear equations.

$$\begin{cases} x - 2y + 3z = 9 & \text{Equation 1} \\ \quad\quad y + 3z = 5 & \text{Equation 2} \\ \quad\quad\quad\quad z = 2 & \text{Equation 3} \end{cases}$$

Solution

From Equation 3, you know the value of z. To solve for y, substitute $z = 2$ into Equation 2 to obtain

$$y + 3(2) = 5 \quad\quad \text{Substitute 2 for } z.$$
$$y = -1. \quad\quad \text{Solve for } y.$$

Finally, substitute $y = -1$ and $z = 2$ into Equation 1 to obtain

$$x - 2(-1) + 3(2) = 9 \quad\quad \text{Substitute } -1 \text{ for } y \text{ and 2 for } z.$$
$$x = 1. \quad\quad \text{Solve for } x.$$

The solution is $x = 1$, $y = -1$, and $z = 2$, which can be written as the **ordered triple** $(1, -1, 2)$. Check this in the original system of equations.

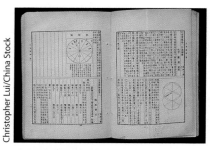

Christopher Lui/China Stock

Historical Note

One of the most influential Chinese mathematics books was the *Chui-chang suan-shu* or *Nine Chapters on the Mathematical Art* (written in approximately 250 B.C.). Chapter Eight of the *Nine Chapters* contained solutions of systems of linear equations using positive and negative numbers. One such system was

$$\begin{cases} 3x + 2y + \ z = 39 \\ 2x + 3y + \ z = 34. \\ \ x + 2y + 3z = 26 \end{cases}$$

This system was solved using column operations on a matrix. Matrices (plural for matrix) will be discussed in the next chapter.

STUDY T!P

As demonstrated in the first step in the solution of Example 2, interchanging rows is the easiest way of obtaining a leading coefficient of 1.

Gaussian Elimination

Two systems of equations are *equivalent* if they have the same solution set. To solve a system that is not in row-echelon form, first convert it to an *equivalent* system that is in row-echelon form by using the following operations.

Operations That Produce Equivalent Systems

Each of the following **row operations** on a system of linear equations produces an *equivalent* system of linear equations.

1. Interchange two equations.

2. Multiply one of the equations by a nonzero constant.

3. Add a multiple of one of the equations to another equation to replace the latter equation.

To see how this is done, let's take another look at the method of elimination, as applied to a system of two linear equations.

Example 2 ▶ Using Gaussian Elimination to Solve a System

Solve the system of linear equations.

$$\begin{cases} 3x - 2y = -1 \\ \ x - \ y = \ \ 0 \end{cases}$$

Solution

There are two strategies that seem reasonable: eliminate the variable x or eliminate the variable y. The following steps show how to use the first strategy.

$$\begin{cases} \ x - \ y = \ \ 0 \\ 3x - 2y = -1 \end{cases}$$ Interchange two equations in the system.

$$-3x + 3y = \ \ 0$$ Multiply the first equation by -3.

$$\begin{array}{r} -3x + 3y = \ \ 0 \\ \underline{3x - 2y = -1} \\ y = -1 \end{array}$$ Add the multiple of the first equation to the second equation to obtain a new equation.

$$\begin{cases} \ x - \ y = \ \ 0 \\ \qquad \ y = -1 \end{cases}$$ New system in row-echelon form

Now, using back-substitution, you can determine that the solution is $y = -1$ and $x = -1$, which can be written as the ordered pair $(-1, -1)$. Check this in the original system of equations.

As shown in Example 2, rewriting a system of linear equations in row-echelon form usually involves a chain of equivalent systems, each of which is obtained by using one of the three basic row operations listed above. This process is called **Gaussian elimination,** after the German mathematician Carl Friedrich Gauss (1777–1855).

Example 3 ▶ Using Gaussian Elimination to Solve a System

Solve the system of linear equations.

$$\begin{cases} x - 2y + 3z = 9 & \text{Equation 1} \\ -x + 3y \quad\quad = -4 & \text{Equation 2} \\ 2x - 5y + 5z = 17 & \text{Equation 3} \end{cases}$$

Solution

Because the leading coefficient of the first equation is 1, you can begin by saving the x at the upper left and eliminating the other x-terms from the first column.

$$\begin{array}{ll} x - 2y + 3z = 9 & \text{Write Equation 1.} \\ \underline{-x + 3y \quad\quad = -4} & \text{Write Equation 2.} \\ \quad\quad y + 3z = 5 & \text{Add Equation 1 to Equation 2.} \end{array}$$

$$\begin{cases} x - 2y + 3z = 9 \\ \quad\quad y + 3z = 5 \\ 2x - 5y + 5z = 17 \end{cases}$$

Adding the first equation to the second equation produces a new second equation.

$$\begin{array}{ll} -2x + 4y - 6z = -18 & \text{Multiply Equation 1 by } -2. \\ \underline{2x - 5y + 5z = \quad 17} & \text{Write Equation 3.} \\ \quad\quad -y - z = -1 & \text{Add revised Equation 1 to Equation 3.} \end{array}$$

$$\begin{cases} x - 2y + 3z = 9 \\ \quad\quad y + 3z = 5 \\ \quad\quad -y - z = -1 \end{cases}$$

Adding -2 times the first equation to the third equation produces a new third equation.

Now that all but the first x have been eliminated from the first column, go to work on the second column. (You need to eliminate y from the third equation.)

$$\begin{cases} x - 2y + 3z = 9 \\ \quad\quad y + 3z = 5 \\ \quad\quad 2z = 4 \end{cases}$$

Adding the second equation to the third equation produces a new third equation.

Finally, you need a coefficient of 1 for z in the third equation.

$$\begin{cases} x - 2y + 3z = 9 \\ \quad\quad y + 3z = 5 \\ \quad\quad z = 2 \end{cases}$$

Multiplying the third equation by $\frac{1}{2}$ produces a new third equation.

This is the same system that was solved in Example 1, and, as in that example, you can conclude that the solution is

$$x = 1, \quad y = -1, \quad \text{and} \quad z = 2.$$

In Example 3, you can check the solution by substituting $x = 1$, $y = -1$, and $z = 2$ into each original equation, as follows.

Equation 1: $\quad 1 - 2(-1) + 3(2) = \quad 9$ ✓

Equation 2: $-1 + 3(-1) \quad\quad = -4$ ✓

Equation 3: $2(1) - 5(-1) + 5(2) = \quad 17$ ✓

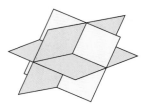

(a) Solution: one point

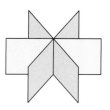

(b) Solution: one line

(c) Solution: one plane

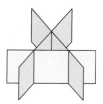

(d) Solution: none

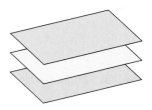

(e) Solution: none

FIGURE 6.11

The next example involves an inconsistent system—one that has no solution. The key to recognizing an inconsistent system is that at some stage in the elimination process you obtain a false statement such as $0 = -2$.

Example 4 ▶ An Inconsistent System

Solve the system of linear equations.

$$\begin{cases} x - 3y + z = 1 & \text{Equation 1} \\ 2x - y - 2z = 2 & \text{Equation 2} \\ x + 2y - 3z = -1 & \text{Equation 3} \end{cases}$$

Solution

$$\begin{cases} x - 3y + z = 1 \\ 5y - 4z = 0 \\ x + 2y - 3z = -1 \end{cases}$$

> Adding -2 times the first equation to the second equation produces a new second equation.

$$\begin{cases} x - 3y + z = 1 \\ 5y - 4z = 0 \\ 5y - 4z = -2 \end{cases}$$

> Adding -1 times the first equation to the third equation produces a new third equation.

$$\begin{cases} x - 3y + z = 1 \\ 5y - 4z = 0 \\ 0 = -2 \end{cases}$$

> Adding -1 times the second equation to the third equation produces a new third equation.

Because the third "equation" is impossible, you can conclude that this system is inconsistent and therefore has no solution. Moreover, because this system is equivalent to the original system, you can conclude that the original system also has no solution.

As with a system of linear equations in two variables, the solution(s) of a system of linear equations in more than two variables must fall into one of three categories.

The Number of Solutions of a Linear System

For a system of linear equations, exactly one of the following is true.

1. There is exactly one solution.

2. There are infinitely many solutions.

3. There is no solution.

In Section 6.2, you learned that a system of two linear equations in two variables can be represented graphically as a pair of lines that are intersecting, coincident, or parallel. A system of three linear equations in three variables has a similar graphical representation—it can be represented as three planes in space that intersect in one point [see Figure 6.11(a)], intersect in a line or a plane [see Figures 6.11(b) and 6.11(c)], or have no points common to all three planes [see Figures 6.11(d) and 6.11(e)].

Example 5 ▶ A System with Infinitely Many Solutions

Solve the system of linear equations.

$$\begin{cases} x + y - 3z = -1 & \text{Equation 1} \\ y - z = 0 & \text{Equation 2} \\ -x + 2y \phantom{{}- 3z} = 1 & \text{Equation 3} \end{cases}$$

Solution

$$\begin{cases} x + y - 3z = -1 \\ y - z = 0 \\ 3y - 3z = 0 \end{cases}$$

> Adding the first equation to the third equation produces a new third equation.

$$\begin{cases} x + y - 3z = -1 \\ y - z = 0 \\ 0 = 0 \end{cases}$$

> Adding -3 times the second equation to the third equation produces a new third equation.

This means that Equation 3 depends on Equations 1 and 2 in the sense that it gives us no additional information about the variables. So, the original system is equivalent to the system

$$\begin{cases} x + y - 3z = -1 \\ y - z = 0. \end{cases}$$

In this last equation, solve for y in terms of z to obtain $y = z$. Back-substituting for y in the previous equation produces $x = 2z - 1$. Finally, letting $z = a$, where a is a real number, you can see that the solutions to the given system are all of the form

$$x = 2a - 1, \qquad y = a, \qquad \text{and} \qquad z = a.$$

So, every ordered triple of the form

$$(2a - 1, a, a), \qquad a \text{ is a real number}$$

is a solution of the system.

STUDY TIP

In Example 5, x and y are solved in terms of the third variable z. To write a solution to the system that does not use any of the three variables of the system, let a represent any real number and let $z = a$. Then solve for x and y. The solution can then be written in terms of a, which is not one of the variables of the system.

In Example 5, there are other ways to write the same infinite set of solutions. For instance, the solutions could have been written as

$$\left(b, \tfrac{1}{2}(b + 1), \tfrac{1}{2}(b + 1)\right), \qquad b \text{ is a real number.}$$

To convince yourself that this description produces the same set of solutions, consider the following.

STUDY TIP

When comparing descriptions of an infinite solution set, keep in mind that there is more than one way to describe the set.

Substitution	Solution	
$a = 0$	$(2(0) - 1, 0, 0) = (-1, 0, 0)$	Same solution
$b = -1$	$\left(-1, \tfrac{1}{2}(-1 + 1), \tfrac{1}{2}(-1 + 1)\right) = (-1, 0, 0)$	
$a = 1$	$(2(1) - 1, 1, 1) = (1, 1, 1)$	Same solution
$b = 1$	$\left(1, \tfrac{1}{2}(1 + 1), \tfrac{1}{2}(1 + 1)\right) = (1, 1, 1)$	
$a = 2$	$(2(2) - 1, 2, 2) = (3, 2, 2)$	Same solution
$b = 3$	$\left(3, \tfrac{1}{2}(3 + 1), \tfrac{1}{2}(3 + 1)\right) = (3, 2, 2)$	

Nonsquare Systems

So far, each system of linear equations you have looked at has been *square*, which means that the number of equations is equal to the number of variables. In a **nonsquare** system, the number of equations differs from the number of variables. A system of linear equations cannot have a unique solution unless there are at least as many equations as there are variables in the system.

Example 6 ▶ A System with Fewer Equations Than Variables

Solve the system of linear equations.

$$\begin{cases} x - 2y + z = 2 & \text{Equation 1} \\ 2x - y - z = 1 & \text{Equation 2} \end{cases}$$

Solution

Begin by rewriting the system in row-echelon form.

$$\begin{cases} x - 2y + z = 2 \\ 3y - 3z = -3 \end{cases}$$

> Adding -2 times the first equation to the second equation produces a new second equation.

$$\begin{cases} x - 2y + z = 2 \\ y - z = -1 \end{cases}$$

> Multiplying the second equation by $\frac{1}{3}$ produces a new second equation.

Solving for y in terms of z, you get

$$y = z - 1$$

and back-substitution into Equation 1 yields

$$x - 2(z - 1) + z = 2 \qquad \text{Substitute for } y \text{ in Equation 1.}$$
$$x - 2z + 2 + z = 2 \qquad \text{Distributive Property}$$
$$x = z. \qquad \text{Solve for } x.$$

Finally, by letting $z = a$, where a is a real number, you have the solution

$$x = a, \qquad y = a - 1, \qquad \text{and} \qquad z = a.$$

So, every ordered triple of the form

$$(a, a - 1, a), \qquad a \text{ is a real number}$$

is a solution of the system. Because there were originally three variables and only two equations, the system cannot have a unique solution.

In Example 6, try choosing some values of a to obtain different solutions of the system, such as $(1, 0, 1)$, $(2, 1, 2)$, and $(3, 2, 3)$. Then check each of the solutions in the original system.

Applications

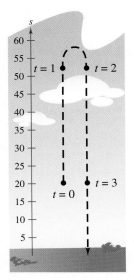

FIGURE **6.12**

Example 7 ▶ Vertical Motion

The height at time t of an object that is moving in a (vertical) line with constant acceleration a is given by the **position equation**

$$s = \tfrac{1}{2}at^2 + v_0 t + s_0.$$

The height s is measured in feet, t is measured in seconds, v_0 is the initial velocity (at $t = 0$), and s_0 is the initial height. Find the values of a, v_0, and s_0 if $s = 52$ at $t = 1$, $s = 52$ at $t = 2$, and $s = 20$ at $t = 3$. (See Figure 6.12.)

Solution

By substituting t and s into the position equation you can obtain three linear equations in a, v_0, and s_0.

When $t = 1$: $\tfrac{1}{2}a(1)^2 + v_0(1) + s_0 = 52$ ⟹ $a + 2v_0 + 2s_0 = 104$

When $t = 2$: $\tfrac{1}{2}a(2)^2 + v_0(2) + s_0 = 52$ ⟹ $2a + 2v_0 + s_0 = 52$

When $t = 3$: $\tfrac{1}{2}a(3)^2 + v_0(3) + s_0 = 20$ ⟹ $9a + 6v_0 + 2s_0 = 40$

Solving this system yields $a = -32$, $v_0 = 48$, and $s_0 = 20$. This solution results in a position equation of $s = -16t^2 + 48t + 20$ and implies that the object was thrown upward at a velocity of 48 feet per second from a height of 20 feet.

Example 8 ▶ Partial Fractions

Write the partial fraction decomposition for $\dfrac{3x + 4}{x^3 - 2x - 4}$.

Solution

Because $x^3 - 2x - 4 = (x - 2)(x^2 + 2x + 2)$, you can write

$$\frac{3x + 4}{x^3 - 2x - 4} = \frac{A}{x - 2} + \frac{Bx + C}{x^2 + 2x + 2}$$

$$3x + 4 = A(x^2 + 2x + 2) + (Bx + C)(x - 2)$$

$$3x + 4 = (A + B)x^2 + (2A - 2B + C)x + (2A - 2C).$$

By equating coefficients of like powers on both sides of the expanded equation, you obtain the following system in A, B, and C.

$$\begin{cases} A + B & = 0 \\ 2A - 2B + C & = 3 \\ 2A \quad\quad - 2C & = 4 \end{cases}$$

You can solve this system to find that $A = 1$, $B = -1$, and $C = -1$. So, the partial fraction decomposition is

$$\frac{3x + 4}{x^3 - 2x - 4} = \frac{1}{x - 2} + \frac{-x - 1}{x^2 + 2x + 2} = \frac{1}{x - 2} - \frac{x + 1}{x^2 + 2x + 2}.$$

Example 9 ▶ Data Analysis: Curve-Fitting

Find a quadratic equation, $y = ax^2 + bx + c$, whose graph passes through the points $(-1, 3)$, $(1, 1)$, and $(2, 6)$.

Solution

Because the graph of $y = ax^2 + bx + c$ passes through the points $(-1, 3)$, $(1, 1)$, and $(2, 6)$, you can write the following.

When $x = -1, y = 3$: $\quad a(-1)^2 + b(-1) + c = 3$

When $x = 1, y = 1$: $\quad a(1)^2 + b(1) + c = 1$

When $x = 2, y = 6$: $\quad a(2)^2 + b(2) + c = 6$

This produces the following system of linear equations.

$$\begin{cases} a - b + c = 3 & \text{Equation 1} \\ a + b + c = 1 & \text{Equation 2} \\ 4a + 2b + c = 6 & \text{Equation 3} \end{cases}$$

The solution of this system is $a = 2$, $b = -1$, and $c = 0$. So, the equation of the parabola is $y = 2x^2 - x$, as shown in Figure 6.13.

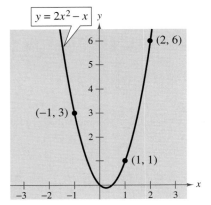

FIGURE 6.13

Example 10 ▶ Investment Analysis

An inheritance of \$12,000 was invested among three funds: a money-market fund that paid 5% annually, municipal bonds that paid 6% annually, and mutual funds that paid 12% annually. The amount invested in mutual funds was \$4000 more than the amount invested in municipal bonds. The total interest earned during the first year was \$1120. How much was invested in each type of fund?

Solution

Let x, y, and z represent the amounts invested in the money-market fund, municipal bonds, and mutual funds, respectively. From the given information, you can write the following equations.

$$\begin{cases} x + y + z = 12{,}000 & \text{Equation 1} \\ z = y + 4000 & \text{Equation 2} \\ 0.05x + 0.06y + 0.12z = 1120 & \text{Equation 3} \end{cases}$$

Rewriting this system in standard form without decimals produces the following.

$$\begin{cases} x + y + z = 12{,}000 & \text{Equation 1} \\ -y + z = 4{,}000 & \text{Equation 2} \\ 5x + 6y + 12z = 112{,}000 & \text{Equation 3} \end{cases}$$

Using Gaussian elimination to solve this system yields $x = 2000$, $y = 3000$, and $z = 7000$. So, \$2000 was invested in the money-market fund, \$3000 was invested in municipal bonds, and \$7000 was invested in mutual funds.

6.3 Exercises

In Exercises 1–4, determine which ordered triples are solutions of the system of equations.

1. $\begin{cases} 3x - y + z = 1 \\ 2x - 3z = -14 \\ 5y + 2z = 8 \end{cases}$

(a) $(2, 0, -3)$ (b) $(-2, 0, 8)$

(c) $(0, -1, 3)$ (d) $(-1, 0, 4)$

2. $\begin{cases} 3x + 4y - z = 17 \\ 5x - y + 2z = -2 \\ 2x - 3y + 7z = -21 \end{cases}$

(a) $(3, -1, 2)$ (b) $(1, 3, -2)$

(c) $(4, 1, -3)$ (d) $(1, -2, 2)$

3. $\begin{cases} 4x + y - z = 0 \\ -8x - 6y + z = -\frac{7}{4} \\ 3x - y = -\frac{9}{4} \end{cases}$

(a) $\left(\frac{1}{2}, -\frac{3}{4}, -\frac{7}{4}\right)$ (b) $\left(-\frac{3}{2}, \frac{5}{4}, -\frac{5}{4}\right)$

(c) $\left(-\frac{1}{2}, \frac{3}{4}, -\frac{5}{4}\right)$ (d) $\left(-\frac{1}{2}, \frac{1}{6}, -\frac{3}{4}\right)$

4. $\begin{cases} -4x - y - 8z = -6 \\ y + z = 0 \\ 4x - 7y = 6 \end{cases}$

(a) $(-2, -2, 2)$ (b) $\left(-\frac{33}{2}, -10, 10\right)$

(c) $\left(\frac{1}{8}, -\frac{1}{2}, \frac{1}{2}\right)$ (d) $\left(-\frac{11}{2}, -4, 4\right)$

In Exercises 5–10, use back-substitution to solve the system of linear equations.

5. $\begin{cases} 2x - y + 5z = 24 \\ y + 2z = 6 \\ z = 4 \end{cases}$

6. $\begin{cases} 4x - 3y - 2z = 21 \\ 6y - 5z = -8 \\ z = -2 \end{cases}$

7. $\begin{cases} 2x + y - 3z = 10 \\ y = 2 \\ y - z = 4 \end{cases}$

8. $\begin{cases} x = 8 \\ 2x + 3y = 10 \\ x - y + 2z = 22 \end{cases}$

9. $\begin{cases} 4x - 2y + z = 8 \\ 2z = 4 \\ -y + z = 4 \end{cases}$

10. $\begin{cases} 5x - 8z = 22 \\ 3y - 5z = 10 \\ z = -4 \end{cases}$

In Exercises 11 and 12, perform the row operation and write the equivalent system.

11. Add Equation 1 to Equation 2.

$\begin{cases} x - 2y + 3z = 5 & \text{Equation 1} \\ -x + 3y - 5z = 4 & \text{Equation 2} \\ 2x - 3z = 0 & \text{Equation 3} \end{cases}$

What did this operation accomplish?

12. Add -2 times Equation 1 to Equation 3.

$\begin{cases} x - 2y + 3z = 5 & \text{Equation 1} \\ -x + 3y - 5z = 4 & \text{Equation 2} \\ 2x - 3z = 0 & \text{Equation 3} \end{cases}$

What did this operation accomplish?

In Exercises 13–38, solve the system of linear equations and check any solution algebraically.

13. $\begin{cases} x + y + z = 6 \\ 2x - y + z = 3 \\ 3x - z = 0 \end{cases}$

14. $\begin{cases} x + y + z = 3 \\ x - 2y + 4z = 5 \\ 3y + 4z = 5 \end{cases}$

15. $\begin{cases} 2x + 2z = 2 \\ 5x + 3y = 4 \\ 3y - 4z = 4 \end{cases}$

16. $\begin{cases} 2x + 4y + z = 1 \\ x - 2y - 3z = 2 \\ x + y - z = -1 \end{cases}$

17. $\begin{cases} 6y + 4z = -12 \\ 3x + 3y = 9 \\ 2x - 3z = 10 \end{cases}$

18. $\begin{cases} 2x + 4y - z = 7 \\ 2x - 4y + 2z = -6 \\ x + 4y + z = 0 \end{cases}$

19. $\begin{cases} 2x + y - z = 7 \\ x - 2y + 2z = -9 \\ 3x - y + z = 5 \end{cases}$

20. $\begin{cases} 5x - 3y + 2z = 3 \\ 2x + 4y - z = 7 \\ x - 11y + 4z = 3 \end{cases}$

21. $\begin{cases} 3x - 5y + 5z = 1 \\ 5x - 2y + 3z = 0 \\ 7x - y + 3z = 0 \end{cases}$

22. $\begin{cases} 2x + y + 3z = 1 \\ 2x + 6y + 8z = 3 \\ 6x + 8y + 18z = 5 \end{cases}$

23. $\begin{cases} x + 2y - 7z = -4 \\ 2x + y + z = 13 \\ 3x + 9y - 36z = -33 \end{cases}$

24. $\begin{cases} 2x + y - 3z = 4 \\ 4x + 2z = 10 \\ -2x + 3y - 13z = -8 \end{cases}$

25. $\begin{cases} 3x - 3y + 6z = 6 \\ x + 2y - z = 5 \\ 5x - 8y + 13z = 7 \end{cases}$

26. $\begin{cases} x \quad\quad + 4z = \quad 13 \\ 4x - 2y + \quad z = \quad 7 \\ 2x - 2y - 7z = -19 \end{cases}$

27. $\begin{cases} x - 2y + 5z = 2 \\ 4x \quad\quad - z = 0 \end{cases}$

28. $\begin{cases} x - 3y + 2z = 18 \\ 5x - 13y + 12z = 80 \end{cases}$

29. $\begin{cases} 2x - 3y + z = -2 \\ -4x + 9y \quad = 7 \end{cases}$

30. $\begin{cases} 2x + 3y + 3z = 7 \\ 4x + 18y + 15z = 44 \end{cases}$

31. $\begin{cases} x \quad\quad\quad + 3w = 4 \\ \quad 2y - z - w = 0 \\ \quad 3y \quad - 2w = 1 \\ 2x - y + 4z \quad = 5 \end{cases}$

32. $\begin{cases} x + y + z + w = 6 \\ 2x + 3y \quad - w = 0 \\ -3x + 4y + z + 2w = 4 \\ x + 2y - z + w = 0 \end{cases}$

33. $\begin{cases} x \quad\quad + 4z = 1 \\ x + y + 10z = 10 \\ 2x - y + 2z = -5 \end{cases}$

34. $\begin{cases} 2x - 2y - 6z = -4 \\ -3x + 2y + 6z = 1 \\ x - y - 5z = -3 \end{cases}$

35. $\begin{cases} 2x + 3y \quad = 0 \\ 4x + 3y - z = 0 \\ 8x + 3y + 3z = 0 \end{cases}$

36. $\begin{cases} 4x + 3y + 17z = 0 \\ 5x + 4y + 22z = 0 \\ 4x + 2y + 19z = 0 \end{cases}$

37. $\begin{cases} 12x + 5y + z = 0 \\ 23x + 4y - z = 0 \end{cases}$

38. $\begin{cases} 2x - y - z = 0 \\ -2x + 6y + 4z = 2 \end{cases}$

In Exercises 39–42, find the equation of the parabola

$y = ax^2 + bx + c$

that passes through the given points. To verify your result, use a graphing utility to plot the points and graph the parabola.

39. $(0, 0), (2, -2), (4, 0)$ **40.** $(0, 3), (1, 4), (2, 3)$

41. $(2, 0), (3, -1), (4, 0)$ **42.** $(1, 3), (2, 2), (3, -3)$

In Exercises 43–46, find the equation of the circle

$x^2 + y^2 + Dx + Ey + F = 0$

that passes through the given points. To verify your result, use a graphing utility to plot the points and graph the circle.

43. $(0, 0), (2, 2), (4, 0)$ **44.** $(0, 0), (0, 6), (3, 3)$

45. $(-3, -1), (2, 4), (-6, 8)$ **46.** $(0, 0), (0, -2), (3, 0)$

Vertical Motion   **In Exercises 47–50, an object moving vertically is at the given heights at specified times. Find the position equation $s = \frac{1}{2}at^2 + v_0t + s_0$ for the object.**

47. At $t = 1$ second, $s = 128$ feet

At $t = 2$ seconds, $s = 80$ feet

At $t = 3$ seconds, $s = 0$ feet

48. At $t = 1$ second, $s = 48$ feet

At $t = 2$ seconds, $s = 64$ feet

At $t = 3$ seconds, $s = 48$ feet

49. At $t = 1$ second, $s = 452$ feet

At $t = 2$ seconds, $s = 372$ feet

At $t = 3$ seconds, $s = 260$ feet

50. At $t = 1$ second, $s = 132$ feet

At $t = 2$ seconds, $s = 100$ feet

At $t = 3$ seconds, $s = 36$ feet

51. *Football*   Two teams playing in a football game scored a total of 72 points. The points came from a total of 20 different scoring plays, which were a combination of touchdowns, extra-point kicks, and field goals, worth 6 points, 1 point, and 3 points, respectively. The same number of extra points were scored as field goals were kicked. How many touchdowns, extra-point kicks, and field goals were scored?

52. *Basketball*   The Aeros scored a total of 104 points in a basketball game. The scoring resulted from a combination of 3-point baskets, 2-point baskets, and 1-point free-throws. There were twice as many 2-point baskets as free-throws scored and twice as many free-throws as 3-point baskets. What combination of scoring accounted for the Aeros' 104 points?

53. *Finance*   A small corporation borrowed $775,000 to expand its product line. Some of the money was borrowed at 8%, some at 9%, and some at 10%. How much was borrowed at each rate if the annual interest owed was $67,500 and the amount borrowed at 8% was four times the amount borrowed at 10%?

54. *Finance*   A small corporation borrowed $800,000 to expand its product line. Some of the money was borrowed at 8%, some at 9%, and some at 10%. How much was borrowed at each rate if the annual interest owed was $67,000 and the amount borrowed at 8% was five times the amount borrowed at 10%?

Finance In Exercises 55 and 56, consider an investor with a portfolio totaling $500,000 that is invested in certificates of deposit, municipal bonds, blue-chip stocks, and growth or speculative stocks. How much is invested in each type of investment?

55. The certificates of deposit pay 10% annually, and the municipal bonds pay 8% annually. Over a 5-year period, the investor expects the blue-chip stocks to return 12% annually and the growth stocks to return 13% annually. The investor wants a combined annual return of 10% and also wants to have only one-fourth of the portfolio invested in stocks.

56. The certificates of deposit pay 9% annually, and the municipal bonds pay 5% annually. Over a 5-year period, the investor expects the blue-chip stocks to return 12% annually and the growth stocks to return 14% annually. The investor wants a combined annual return of 10% and also wants to have only one-fourth of the portfolio invested in stocks.

57. *Agriculture* A mixture of 12 liters of chemical A, 16 liters of chemical B, and 26 liters of chemical C is required to kill a certain destructive crop insect. Commercial spray X contains 1, 2, and 2 parts, respectively, of these chemicals. Commercial spray Y contains only chemical C. Commercial spray Z contains only chemicals A and B in equal amounts. How much of each type of commercial spray is needed to get the desired mixture?

58. *Chemistry* A chemist needs 10 liters of a 25% acid solution. The solution is to be mixed from three solutions whose concentrations are 10%, 20%, and 50%. How many liters of each solution should the chemist use to satisfy the following?

    (a) Use as little as possible of the 50% solution.

    (b) Use as much as possible of the 50% solution.

    (c) Use 2 liters of the 50% solution.

59. *Truck Scheduling* A small company that manufactures products A and B has an order for 15 units of product A and 16 units of product B. The company has trucks of three different sizes that can haul the products, as shown in the table.

Truck	Large	Medium	Small
Product A	6	4	0
Product B	3	4	3

How many trucks of each size are needed to deliver the order? Give two possible solutions.

60. *Electrical Network* Applying Kirchhoff's Laws to the electrical network in the figure, the currents I_1, I_2, and I_3 are the solution of the system

$$\begin{cases} I_1 - I_2 + I_3 = 0 \\ 3I_1 + 2I_2 \quad\quad = 7 \\ \quad\quad 2I_2 + 4I_3 = 8. \end{cases}$$

Find the currents.

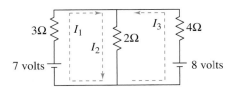

61. *Pulley System* A system of pulleys is loaded with 128-pound and 32-pound weights (see figure). The tensions t_1 and t_2 in the ropes and the acceleration a of the 32-pound weight are found by solving the system of equations

$$\begin{cases} t_1 - 2t_2 \quad\quad = 0 \\ t_1 \quad\quad - 2a = 128 \\ \quad\quad t_2 + a = 32 \end{cases}$$

where t_1 and t_2 are measured in pounds and a is measured in feet per second squared. Solve this system.

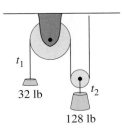

62. *Pulley System* If the 32-pound weight in the pulley system in Exercise 61 is replaced by a 64-pound weight, the new pulley system will be modeled by the following system of equations.

$$\begin{cases} t_1 - 2t_2 \quad\quad = 0 \\ t_1 \quad\quad - 2a = 128 \\ \quad\quad t_2 + 2a = 64 \end{cases}$$

Solve this system and use your answer for the acceleration to describe what (if anything) is happening in the pulley system.

Partial Fraction Decomposition   **In Exercises 63–66, write the partial fraction decomposition for the rational expression.**

63. $\dfrac{1}{x^3 - x} = \dfrac{A}{x} + \dfrac{B}{x - 1} + \dfrac{C}{x + 1}$

64. $\dfrac{3}{x^2 + x - 2} = \dfrac{A}{x - 1} + \dfrac{B}{x + 2}$

65. $\dfrac{x^2 - 3x - 3}{x(x - 2)(x + 3)} = \dfrac{A}{x} + \dfrac{B}{x - 2} + \dfrac{C}{x + 3}$

66. $\dfrac{12}{x(x - 2)(x + 3)} = \dfrac{A}{x} + \dfrac{B}{x - 2} + \dfrac{C}{x + 3}$

Fitting a Parabola   **In Exercises 67–70, find the least squares regression parabola $y = ax^2 + bx + c$ for the points $(x_1, y_1), (x_2, y_2), \ldots, (x_n, y_n)$ by solving the following system of linear equations for a, b, and c. Then use the least squares regression capabilities of a graphing utility to confirm the result. (For an explanation of how the coefficients of a, b, and c in the system are obtained, see Appendix B.)**

67. $\begin{cases} 4c \quad\quad + 40a = \quad 19 \\ \quad\quad 40b \quad\quad = -12 \\ 40c \quad\quad + 544a = \quad 160 \end{cases}$

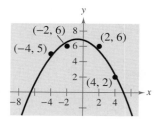

68. $\begin{cases} 5c \quad\quad + 10a = \quad 8 \\ \quad\quad 10b \quad\quad = 12 \\ 10c \quad\quad + 34a = 22 \end{cases}$

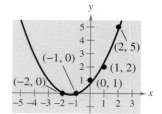

69. $\begin{cases} 4c + \quad 9b + \quad 29a = \quad 20 \\ 9c + 29b + \quad 99a = \quad 70 \\ 29c + 99b + 353a = 254 \end{cases}$

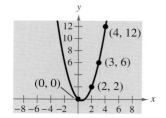

70. $\begin{cases} 4c + \quad 6b + 14a = 25 \\ 6c + 14b + 36a = 21 \\ 14c + 36b + 98a = 33 \end{cases}$

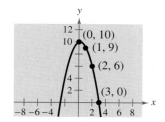

71. *Data Analysis*   In testing a new braking system on an automobile, the speed in miles per hour and the stopping distance in feet were recorded in the table.

Speed (x)	30	40	50
Stopping distance (y)	55	105	188

(a) Find a quadratic equation that models the data.

(b) Graph the parabola and the data on the same set of axes.

(c) Use the model to estimate the stopping distance if the speed is 70 miles per hour.

72. *Data Analysis*   A wildlife management team studied the reproduction rates of deer in three tracts of a wildlife preserve. Each tract contained 5 acres. In each tract the number of females and the percent of females that had offspring the following year were recorded. The results are given in the table.

Number (x)	100	120	140
Percent (y)	75	68	55

(a) Find a quadratic equation that models the data.

(b) Use a graphing utility to graph the parabola and the data in the same viewing window.

(c) Use the model to estimate the percent of females that had offspring if $x = 170$.

⬤ *Advanced Applications* **In Exercises 73–76, find x, y, and λ satisfying the system. These systems arise in certain optimization problems in calculus, and λ is called a Lagrange multiplier.**

73. $\begin{cases} y + \lambda = 0 \\ x + \lambda = 0 \\ x + y - 10 = 0 \end{cases}$

74. $\begin{cases} 2x + \lambda = 0 \\ 2y + \lambda = 0 \\ x + y - 4 = 0 \end{cases}$

75. $\begin{cases} 2x - 2x\lambda = 0 \\ -2y + \lambda = 0 \\ y - x^2 = 0 \end{cases}$

76. $\begin{cases} 2 + 2y + 2\lambda = 0 \\ 2x + 1 + \lambda = 0 \\ 2x + y - 100 = 0 \end{cases}$

Synthesis

True or False? **In Exercises 77 and 78, determine whether the statement is true or false. Justify your answer.**

77. The system $\begin{cases} x + 3y - 6z = -16 \\ 2y - z = -1 \\ z = 3 \end{cases}$ is in row-echelon form.

78. If a system of three linear equations is inconsistent, then its graph has no points common to all three equations.

79. *Think About It* Are the following two systems of equations equivalent? Give reasons for your answer.

$\begin{cases} x + 3y - z = 6 \\ 2x - y + 2z = 1 \\ 3x + 2y - z = 2 \end{cases}$ $\begin{cases} x + 3y - z = 6 \\ -7y + 4z = 1 \\ -7y - 4z = -16 \end{cases}$

80. *Think About It* One of the following systems is inconsistent and the other has one solution. How can you identify each by observation?

$\begin{cases} 3x - 5y = 3 \\ -12x + 20y = 8 \end{cases}$ $\begin{cases} 3x - 5y = 3 \\ 9x - 20y = 6 \end{cases}$

81. *Think About It* When using Gaussian elimination to solve a system of linear equations, how can you recognize that the system has no solution? Give an example that illustrates your answer.

Exploration **In Exercises 82–85, find a system of linear equations that has the ordered triple as a solution. (The answer is not unique.)**

82. $(4, -1, 2)$

83. $(-5, -2, 1)$

84. $\left(3, -\frac{1}{2}, \frac{7}{4}\right)$

85. $\left(-\frac{3}{2}, 4, -7\right)$

Review

In Exercises 86–89, solve the percent problem.

86. What is $7\frac{1}{2}\%$ of 85?

87. 225 is what percent of 150?

88. 0.5% of what number is 400?

89. 48% of what number is 132?

In Exercises 90–93, (a) determine the real zeros of f and (b) sketch the graph of f.

90. $f(x) = x^3 + x^2 - 12x$

91. $f(x) = -8x^4 + 32x^2$

92. $f(x) = 2x^3 + 5x^2 - 21x - 36$

93. $f(x) = 6x^3 - 29x^2 - 6x + 5$

In Exercises 94–97, use a graphing utility to construct a table of values. Then sketch the graph of the equation by hand.

94. $y = 4^{x-4} - 5$

95. $y = \left(\frac{5}{2}\right)^{-x+1} - 4$

96. $y = 1.9^{-0.8x} + 3$

97. $y = 3.5^{-x+2} + 6$

In Exercises 98 and 99, solve the system by elimination.

98. $\begin{cases} 2x + y = 120 \\ x + 2y = 120 \end{cases}$

99. $\begin{cases} 6x - 5y = 3 \\ 10x - 12y = 5 \end{cases}$

6.4   Systems of Inequalities

▶ **What you should learn**

- How to sketch the graphs of inequalities in two variables
- How to solve systems of inequalities
- How to use systems of inequalities in two variables to model and solve real-life problems

▶ **Why you should learn it**

You can use systems of inequalities in two variables to model and solve real-life problems. For instance, Exercise 72 on page 489 shows how to use a system of inequalities to analyze the compositions of dietary supplements.

Frank Siteman/PhotoEdit

The Graph of an Inequality

The statements $3x - 2y < 6$ and $2x^2 + 3y^2 \geq 6$ are inequalities in two variables. An ordered pair (a, b) is a **solution of an inequality** in x and y if the inequality is true when a and b are substituted for x and y, respectively. The **graph of an inequality** is the collection of all solutions of the inequality. To sketch the graph of an inequality, begin by sketching the graph of the *corresponding equation*. The graph of the equation will normally separate the plane into two or more regions. In each such region, one of the following must be true.

1. *All* points in the region are solutions of the inequality.
2. *No* point in the region is a solution of the inequality.

So, you can determine whether the points in an entire region satisfy the inequality by simply testing *one* point in the region.

Sketching the Graph of an Inequality in Two Variables

1. Replace the inequality sign by an equal sign, and sketch the graph of the resulting equation. (Use a dashed line for < or > and a solid line for ≤ or ≥.)

2. Test one point in each of the regions formed by the graph in Step 1. If the point satisfies the inequality, shade the entire region to denote that every point in the region satisfies the inequality.

Example 1 ▶ Sketching the Graph of an Inequality

To sketch the graph of $y \geq x^2 - 1$, begin by graphing the corresponding *equation* $y = x^2 - 1$, which is a parabola, as shown in Figure 6.14. By testing a point *above* the parabola $(0, 0)$ and a point *below* the parabola $(0, -2)$, you can see that the points that satisfy the inequality are those lying above (or on) the parabola.

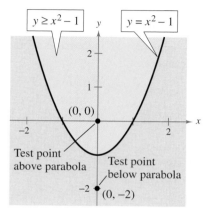

FIGURE 6.14

The inequality given in Example 1 is a nonlinear inequality in two variables. Most of the following examples involve **linear inequalities** such as $ax + by < c$. The graph of a linear inequality is a half-plane lying on one side of the line $ax + by = c$.

Example 2 ▶ Sketching the Graph of a Linear Inequality

Sketch the graph of each linear inequality.

a. $x > -2$ **b.** $y \leq 3$

Solution

a. The graph of the corresponding equation $x = -2$ is a vertical line. The points that satisfy the inequality $x > -2$ are those lying to the right of this line, as shown in Figure 6.15.

b. The graph of the corresponding equation $y = 3$ is a horizontal line. The points that satisfy the inequality $y \leq 3$ are those lying below (or on) this line, as shown in Figure 6.16.

Technology

A graphing utility can be used to graph an inequality or a system of inequalities. Consult the user's guide for your graphing utility for keystrokes. The graph of the inequality from Example 3 is shown below.

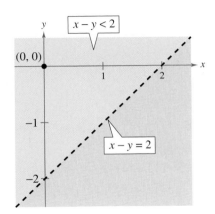

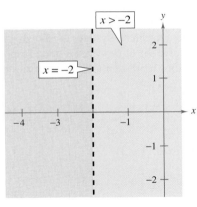

FIGURE 6.15

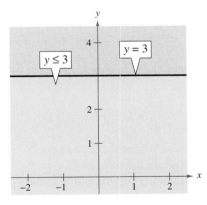

FIGURE 6.16

Example 3 ▶ Sketching the Graph of a Linear Inequality

Sketch the graph of $x - y < 2$.

Solution

The graph of the corresponding equation $x - y = 2$ is a line, as shown in Figure 6.17. Because the origin $(0, 0)$ satisfies the inequality, the graph consists of the half-plane lying above the line. (Try checking a point below the line. Regardless of which point you choose, you will see that it does not satisfy the inequality.)

To graph a linear inequality, it can help to write the inequality in slope-intercept form. For instance, by writing $x - y < 2$ in the form

$$y > x - 2$$

you can see that the solution points lie *above* the line $x - y = 2$ (or $y = x - 2$), as shown in Figure 6.17.

FIGURE 6.17

Systems of Inequalities

Many practical problems in business, science, and engineering involve systems of linear inequalities. A **solution** of a system of inequalities in x and y is a point (x, y) that satisfies each inequality in the system.

To sketch the graph of a system of inequalities in two variables, first sketch the graph of each individual inequality (on the same coordinate system) and then find the region that is *common* to every graph in the system. For systems of *linear* inequalities, it is helpful to find the vertices of the solution region.

Example 4 ▶ Solving a System of Inequalities

Sketch the graph (and label the vertices) of the solution set of the system.

$$\begin{cases} x - y < 2 & \text{Inequality 1} \\ x > -2 & \text{Inequality 2} \\ y \le 3 & \text{Inequality 3} \end{cases}$$

Solution

The graphs of these inequalities are shown in Figures 6.15 to 6.17. The triangular region common to all three graphs can be found by superimposing the graphs on the same coordinate system, as shown in Figure 6.18. To find the vertices of the region, solve the three systems of corresponding equations obtained by taking *pairs* of equations representing the boundaries of the individual regions.

Vertex A: $(-2, -4)$

$$\begin{cases} x - y = 2 & \text{Boundary of Inequality 1} \\ x = -2 & \text{Boundary of Inequality 2} \end{cases}$$

Vertex B: $(5, 3)$

$$\begin{cases} x - y = 2 & \text{Boundary of Inequality 1} \\ y = 3 & \text{Boundary of Inequality 3} \end{cases}$$

Vertex C: $(-2, 3)$

$$\begin{cases} x = -2 & \text{Boundary of Inequality 2} \\ y = 3 & \text{Boundary of Inequality 3} \end{cases}$$

STUDY T!P

Using different colored pencils to shade the solution of each inequality in a system will make identifying the solution of the system of inequalities easier.

 A computer animation of this example appears in the *Interactive* CD-ROM and *Internet* versions of this text.

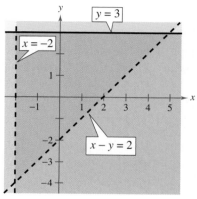

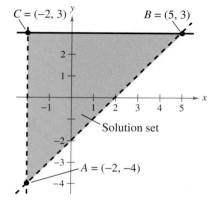

FIGURE **6.18**

For the triangular region shown in Figure 6.18, each point of intersection of a pair of boundary lines corresponds to a vertex. With more complicated regions, two border lines can sometimes intersect at a point that is not a vertex of the region, as shown in Figure 6.19. To keep track of which points of intersection are actually vertices of the region, you should sketch the region and refer to your sketch as you find each point of intersection.

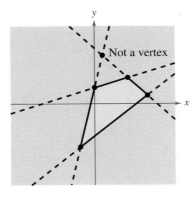

FIGURE 6.19

Example 5 ▶ Solving a System of Inequalities

Sketch the region containing all points that satisfy the system.

$$\begin{cases} x^2 - y \leq 1 & \text{Inequality 1} \\ -x + y \leq 1 & \text{Inequality 2} \end{cases}$$

Solution

As shown in Figure 6.20, the points that satisfy the inequality

$$x^2 - y \leq 1 \qquad \text{Inequality 1}$$

are the points lying above (or on) the parabola given by

$$y = x^2 - 1. \qquad \text{Parabola}$$

The points satisfying the inequality

$$-x + y \leq 1 \qquad \text{Inequality 2}$$

are the points lying below (or on) the line given by

$$y = x + 1. \qquad \text{Line}$$

To find the points of intersection of the parabola and the line, solve the system of corresponding equations.

$$\begin{cases} x^2 - y = 1 \\ -x + y = 1 \end{cases}$$

Using the method of substitution, you can find the solutions to be $(-1, 0)$ and $(2, 3)$, as shown in Figure 6.20.

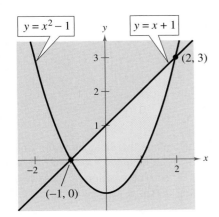

FIGURE 6.20

When solving a system of inequalities, you should be aware that the system might have no solution *or* it might be represented by an unbounded region in the plane. These two possibilities are shown in Examples 6 and 7.

Example 6 ▶   A System with No Solution

Sketch the solution set of the system.

$$\begin{cases} x + y > 3 & \text{Inequality 1} \\ x + y < -1 & \text{Inequality 2} \end{cases}$$

Solution

From the way the system is written, it is clear that the system has no solution, because the quantity $(x + y)$ cannot be both less than -1 and greater than 3. Graphically, the inequality $x + y > 3$ is represented by the half-plane lying above the line $x + y = 3$, and the inequality $x + y < -1$ is represented by the half-plane lying below the line $x + y = -1$, as shown in Figure 6.21. These two half-planes have no points in common. So the system of inequalities has no solution.

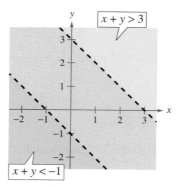

FIGURE 6.21

Example 7 ▶   An Unbounded Solution Set

Sketch the solution set of the system.

$$\begin{cases} x + y < 3 & \text{Inequality 1} \\ x + 2y > 3 & \text{Inequality 2} \end{cases}$$

Solution

The graph of the inequality $x + y < 3$ is the half-plane that lies below the line $x + y = 3$, as shown in Figure 6.22. The graph of the inequality $x + 2y > 3$ is the half-plane that lies above the line $x + 2y = 3$. The intersection of these two half-planes is an *infinite wedge* that has a vertex at $(3, 0)$.

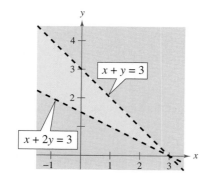

FIGURE 6.22

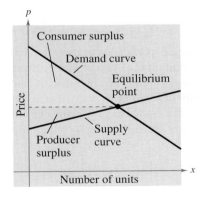

FIGURE **6.23**

Applications

Example 9 in Section 6.2 discussed the *point of equilibrium* for a system of demand and supply functions. The next example discusses two related concepts that economists call **consumer surplus** and **producer surplus**. As shown in Figure 6.23, the consumer surplus is defined as the area of the region that lies *below* the demand curve, *above* the horizontal line passing through the equilibrium point, and to the right of the *p*-axis. Similarly, the producer surplus is defined as the area of the region that lies *above* the supply curve, *below* the horizontal line passing through the equilibrium point, and to the right of the *p*-axis. The consumer surplus is a measure of the amount that consumers would have been willing to pay *above what they actually paid,* whereas the producer surplus is a measure of the amount that producers would have been willing to receive *below what they actually received.*

Example 8 ▶ Consumer Surplus and Producer Surplus

The demand and supply functions for a certain type of calculator are given by

$$\begin{cases} p = 150 - 0.00001x & \text{Demand equation} \\ p = 60 + 0.00002x & \text{Supply equation} \end{cases}$$

where p is the price in dollars and x represents the number of units. Find the consumer surplus and producer surplus for these two equations.

Solution

Begin by finding the point of equilibrium (when supply and demand are equal) by solving the equation

$$60 + 0.00002x = 150 - 0.00001x.$$

In Example 9 in Section 6.2, you saw that the solution is $x = 3,000,000$, which corresponds to an equilibrium price of $p = \$120$. So, the consumer surplus and producer surplus are the areas of the following triangular regions.

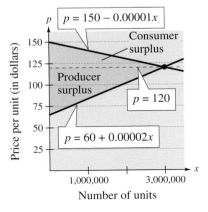

FIGURE **6.24**

Consumer Surplus	*Producer Surplus*
$\begin{cases} p \leq 150 - 0.00001x \\ p \geq 120 \\ x \geq 0 \end{cases}$	$\begin{cases} p \geq 60 + 0.00002x \\ p \leq 120 \\ x \geq 0 \end{cases}$

In Figure 6.24, you can see that the consumer and producer surpluses are defined as the areas of the shaded triangles.

$$\begin{aligned} \text{Consumer surplus} &= \frac{1}{2}(\text{base})(\text{height}) \\ &= \frac{1}{2}(30)(3,000,000) = \$45,000,000 \end{aligned}$$

$$\begin{aligned} \text{Producer surplus} &= \frac{1}{2}(\text{base})(\text{height}) \\ &= \frac{1}{2}(60)(3,000,000) = \$90,000,000 \end{aligned}$$

Example 9 ▶   Nutrition

The minimum daily requirements from the liquid portion of a diet are 300 calories, 36 units of vitamin A, and 90 units of vitamin C. A cup of dietary drink X provides 60 calories, 12 units of vitamin A, and 10 units of vitamin C. A cup of dietary drink Y provides 60 calories, 6 units of vitamin A, and 30 units of vitamin C. Set up a system of linear inequalities that describes how many cups of each drink should be consumed each day to meet the minimum daily requirements for calories and vitamins.

Solution

Begin by letting x and y represent the following.

$$x = \text{number of cups of dietary drink X}$$

$$y = \text{number of cups of dietary drink Y}$$

To meet the minimum daily requirements, the following inequalities must be satisfied.

$$\begin{cases} 60x + 60y \geq 300 & \text{Calories} \\ 12x + 6y \geq 36 & \text{Vitamin A} \\ 10x + 30y \geq 90 & \text{Vitamin C} \\ x \geq 0 \\ y \geq 0 \end{cases}$$

The last two inequalities are included because x and y cannot be negative. The graph of this system of inequalities is shown in Figure 6.25. (More is said about this application in Example 6 in Section 6.5.)

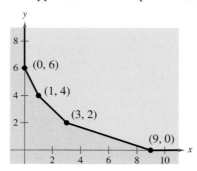

FIGURE 6.25

Writing ABOUT MATHEMATICS

Creating a System of Inequalities Plot the points (0, 0), (4, 0), (3, 2), and (0, 2) in a coordinate plane. Draw the quadrilateral that has these four points as its vertices. Write a system of linear inequalities that has the quadrilateral as its solution. Explain how you found the system of inequalities.

6.4 Exercises

In Exercises 1–16, sketch the graph of the inequality.

1. $x \geq 2$ **2.** $x \leq 4$

3. $y \geq -1$ **4.** $y \leq 3$

5. $y < 2 - x$ **6.** $y > 2x - 4$

7. $2y - x \geq 4$ **8.** $5x + 3y \geq -15$

9. $(x + 1)^2 + (y - 2)^2 < 9$

10. $y^2 - x < 0$

11. $y \leq \dfrac{1}{1 + x^2}$ **12.** $y > \dfrac{-15}{x^2 + x + 4}$

13. $y < \ln x$ **14.** $y \geq 6 - \ln(x + 5)$

15. $y < 3^{-x-4}$ **16.** $y \leq 2^{2x-0.5} - 7$

In Exercises 17–24, use a graphing utility to graph the inequality. Shade the region representing the solution.

17. $y \geq \frac{2}{3}x - 1$ **18.** $y \leq 6 - \frac{3}{2}x$

19. $y < -3.8x + 1.1$ **20.** $y \geq -20.74 + 2.66x$

21. $x^2 + 5y - 10 \leq 0$ **22.** $2x^2 - y - 3 > 0$

23. $\frac{5}{2}y - 3x^2 - 6 \geq 0$ **24.** $-\frac{1}{10}x^2 - \frac{3}{8}y < -\frac{1}{4}$

In Exercises 25–28, write an inequality for the shaded region shown in the figure.

25.

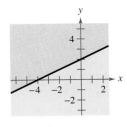

26.

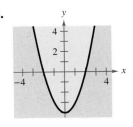

27.

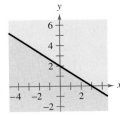

28.

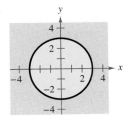

In Exercises 29–32, determine which ordered pairs are solutions of the system of linear inequalities.

29. $\begin{cases} x \geq -4 \\ y > -3 \\ y \leq -8x - 3 \end{cases}$

(a) $(0, 0)$ (b) $(-1, -3)$

(c) $(-4, 0)$ (d) $(-3, 11)$

30. $\begin{cases} -2x + 5y \geq 3 \\ y < 4 \\ -4x + 2y < 7 \end{cases}$

(a) $(0, 2)$ (b) $(-6, 4)$

(c) $(-8, -2)$ (d) $(-3, 2)$

31. $\begin{cases} 3x + y > 1 \\ -y - \frac{1}{2}x^2 \leq -4 \\ -15x + 4y > 0 \end{cases}$

(a) $(0, 10)$ (b) $(0, -1)$

(c) $(2, 9)$ (d) $(-1, 6)$

32. $\begin{cases} x^2 + y^2 \geq 36 \\ -3x + y \leq 10 \\ \frac{2}{3}x - y \geq 5 \end{cases}$

(a) $(-1, 7)$ (b) $(-5, 1)$

(c) $(6, 0)$ (d) $(4, -8)$

In Exercises 33–46, sketch the graph and label the vertices of the solution of the system of inequalities.

33. $\begin{cases} x + y \leq 1 \\ -x + y \leq 1 \\ y \geq 0 \end{cases}$ **34.** $\begin{cases} 3x + 2y < 6 \\ x > 0 \\ y > 0 \end{cases}$

35. $\begin{cases} x^2 + y \leq 5 \\ x \geq -1 \\ y \geq 0 \end{cases}$ **36.** $\begin{cases} 2x^2 + y \geq 2 \\ x \leq 2 \\ y \leq 1 \end{cases}$

37. $\begin{cases} -3x + 2y < 6 \\ x - 4y > -2 \\ 2x + y < 3 \end{cases}$ **38.** $\begin{cases} x - 7y > -36 \\ 5x + 2y > 5 \\ 6x - 5y > 6 \end{cases}$

39. $\begin{cases} 2x + y > 2 \\ 6x + 3y < 2 \end{cases}$ **40.** $\begin{cases} x - 2y < -6 \\ 5x - 3y > -9 \end{cases}$

41. $\begin{cases} x > y^2 \\ x < y + 2 \end{cases}$ **42.** $\begin{cases} x - y^2 > 0 \\ x - y > 2 \end{cases}$

43. $\begin{cases} x^2 + y^2 \leq 9 \\ x^2 + y^2 \geq 1 \end{cases}$ **44.** $\begin{cases} x^2 + y^2 \leq 25 \\ 4x - 3y \leq 0 \end{cases}$

45. $\begin{cases} 3x + 4 \geq y^2 \\ x - y < 0 \end{cases}$ **46.** $\begin{cases} x < 2y - y^2 \\ 0 < x + y \end{cases}$

▦ In Exercises 47–52, use a graphing utility to graph the inequalities. Shade the region representing the solution of the system.

47. $\begin{cases} y \le \sqrt{3x} + 1 \\ y \ge x^2 + 1 \end{cases}$ **48.** $\begin{cases} y < -x^2 + 2x + 3 \\ y > x^2 - 4x + 3 \end{cases}$

49. $\begin{cases} y < x^3 - 2x + 1 \\ y > -2x \\ x \le 1 \end{cases}$ **50.** $\begin{cases} y \ge x^4 - 2x^2 + 1 \\ y \le 1 - x^2 \end{cases}$

51. $\begin{cases} x^2 y \ge 1 \\ 0 < x \le 4 \\ y \le 4 \end{cases}$

52. $\begin{cases} y \le e^{-x^2/2} \\ y \ge 0 \\ -2 \le x \le 2 \end{cases}$

In Exercises 53–62, derive a set of inequalities to describe the region.

53.
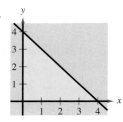

54.
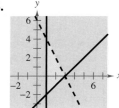

55.
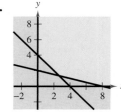

56.
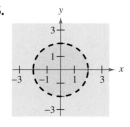

57.
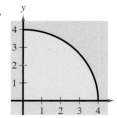

58.
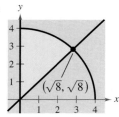

59. Rectangle: Vertices at $(2, 1)$, $(5, 1)$, $(5, 7)$, $(2, 7)$

60. Parallelogram: Vertices at $(0, 0)$, $(4, 0)$, $(1, 4)$, $(5, 4)$

61. Triangle: Vertices at $(0, 0)$, $(5, 0)$, $(2, 3)$

62. Triangle: Vertices at $(-1, 0)$, $(1, 0)$, $(0, 1)$

Economics In Exercises 63–66, find the consumer surplus and producer surplus for the demand and supply equations.

	Demand	*Supply*
63.	$p = 50 - 0.5x$	$p = 0.125x$
64.	$p = 100 - 0.05x$	$p = 25 + 0.1x$
65.	$p = 140 - 0.00002x$	$p = 80 + 0.00001x$
66.	$p = 400 - 0.0002x$	$p = 225 + 0.0005x$

67. *Business* A furniture company can sell all the tables and chairs it produces. Each table requires 1 hour in the assembly center and $1\frac{1}{3}$ hours in the finishing center. Each chair requires $1\frac{1}{2}$ hours in the assembly center and $1\frac{1}{2}$ hours in the finishing center. The company's assembly center is available 12 hours per day, and its finishing center is available 15 hours per day. Find and graph a system of inequalities describing all possible production levels.

68. *Business* A store sells two models of computers. Because of the demand, the store stocks at least twice as many units of model A as of model B. The costs to the store for the two models are $800 and $1200, respectively. The management does not want more than $20,000 in computer inventory at any one time, and it wants at least four model A computers and two model B computers in inventory at all times. Devise a system of inequalities describing all possible inventory levels, and graph the system.

69. *Finance* A person plans to invest up to $20,000 in two different interest-bearing accounts. Each account is to contain at least $5000. Moreover, the amount in one account should be at least twice the amount in the other account. Find a system of inequalities to describe the various amounts that can be deposited in each account, and graph the system.

70. *Entertainment* For a concert event, there are $30 reserved seat tickets and $20 general admission tickets. There are 2000 reserved seats available, and fire regulations limit the number of paid ticket holders to 3000. The promoter must take in at least $75,000 in ticket sales. Find a system of inequalities describing the different numbers of tickets that can be sold. Graph the system.

71. *Shipping*   A warehouse supervisor is told to ship at least 50 packages of gravel that weigh 55 pounds each and at least 40 bags of stone that weigh 70 pounds each. The maximum weight capacity in the truck he is loading is 7500 pounds. Find a system of inequalities describing the numbers of bags of stone and gravel that he can send. Graph the system.

72. *Nutrition*   A dietitian is asked to design a special dietary supplement using two different foods. Each ounce of food X contains 20 units of calcium, 15 units of iron, and 10 units of vitamin B. Each ounce of food Y contains 10 units of calcium, 10 units of iron, and 20 units of vitamin B. The minimum daily requirements of the diet are 300 units of calcium, 150 units of iron, and 200 units of vitamin B. Find and graph a system of inequalities describing the different amounts of food X and food Y that can be used.

73. *Physical Fitness Facility*   An indoor running track is to be constructed with a space for body-building equipment inside the track (see figure). The track must be at least 125 meters long, and the body-building space must have an area of at least 500 square meters. Find a system of inequalities describing the requirements of the facility. Graph the system.

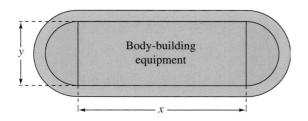

Synthesis

True or False?   **In Exercises 74 and 75, determine whether the statement is true or false. Justify your answer.**

74. The area of the figure defined by the system

$$\begin{cases} x \geq -3 \\ x \leq 6 \\ y \leq 5 \\ y \geq -6 \end{cases}$$

is 99 square units.

75. The graph below shows the solution of the system

$$\begin{cases} y \leq 6 \\ -4x - 9y > 6. \\ 3x + y^2 \geq 2 \end{cases}$$

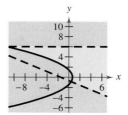

In Exercises 76–79, match the system of inequalities with the graph of its solution. [The graphs are labeled (a), (b), (c), and (d).]

(a)
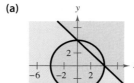

(b)
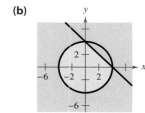

(c)
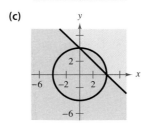

(d)
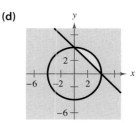

76. $\begin{cases} x^2 + y^2 \leq 16 \\ x + y \geq 4 \end{cases}$

77. $\begin{cases} x^2 + y^2 \leq 16 \\ x + y \leq 4 \end{cases}$

78. $\begin{cases} x^2 + y^2 \geq 16 \\ x + y \geq 4 \end{cases}$

79. $\begin{cases} x^2 + y^2 \geq 16 \\ x + y \leq 4 \end{cases}$

80. The graph of the solution of the inequality $x + 2y < 6$ is shown in the figure. Describe how the solution set would change for each of the following.

(a) $x + 2y \leq 6$ (b) $x + 2y > 6$

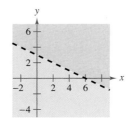

81. *Graphical Reasoning* Two concentric circles have radii x and y, where $y > x$. The area between the circles must be at least 10 square units.

(a) Find a system of inequalities describing the constraints on the circles.

(b) Use a graphing utility to graph the inequality in part (a). Graph the line $y = x$ in the same viewing window.

(c) Identify the graph of the line in relation to the boundary of the inequality. Explain its meaning in the context of the problem.

82. *Think About It* After graphing the boundary of an inequality in x and y, how do you decide on which side of the boundary the solution set of the inequality lies?

83. *Writing* Explain the difference between the graph of the inequality $x \leq 4$ on the real number line and on the rectangular coordinate system.

Review

In Exercises 84–89, find the equation of the line passing through the two points.

84. $(-2, 6), (4, -4)$

85. $(-8, 0), (3, -1)$

86. $\left(\frac{3}{4}, -2\right), \left(-\frac{7}{2}, 5\right)$

87. $\left(-\frac{1}{2}, 0\right), \left(\frac{11}{2}, 12\right)$

88. $(3.4, -5.2), (-2.6, 0.8)$

89. $(-4.1, -3.8), (2.9, 8.2)$

90. *Mortgage Loans* The table shows the numbers of outstanding mortgage loans M (in millions) in the United States for the years 1992 to 1997. (Source: Mortgage Bankers Association of America)

Year	1992	1993	1994	1995	1996	1997
M	42.6	45.3	47.6	49.2	50.1	51.2

Use your graphing utility to find a linear model and a quadratic model that represent the data. Let $t = 0$ represent 1990. Use your graphing utility to plot the actual data and the models in the same viewing window. How closely do the models represent the data?

In Exercises 91–96, evaluate the expression. Round your result to three decimal places.

91. $(2.7)^{3.99}$

92. $150(4^{-2.6})$

93. $1.5^{-3\pi}$

94. $(3.815)^6$

95. $e^{-11/4}$

96. $e^{-\sqrt{13}}$

In Exercises 97 and 98, solve the system of linear equations and check any solution algebraically.

97. $\begin{cases} -x - 2y + 3z = -23 \\ 2x + 6y - z = 17 \\ 5y + z = 8 \end{cases}$

98. $\begin{cases} 7x - 3y + 5z = -28 \\ 4x + 4z = -16 \\ 7x + 2y - z = 0 \end{cases}$

6.5 Linear Programming

What you should learn

- How to solve linear programming problems
- How to use linear programming to model and solve real-life problems

Why you should learn it

Linear programming is often useful in making real-life economic decisions. For example, Exercise 35 on page 498 shows how a merchant could use linear programming to analyze the profitability of two models of compact disc players.

Linear Programming: A Graphical Approach

Many applications in business and economics involve a process called **optimization,** in which you are asked to find the minimum or maximum of a quantity. In this section you will study an optimization strategy called **linear programming.**

A two-dimensional linear programming problem consists of a linear **objective function** and a system of linear inequalities called **constraints.** The objective function gives the quantity that is to be maximized (or minimized), and the constraints determine the set of **feasible solutions.** For example, suppose you are asked to maximize the value of

$$z = ax + by \qquad \text{Objective function}$$

subject to a set of constraints that determines the shaded region in Figure 6.26.

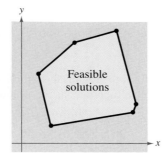

FIGURE 6.26

Because every point in the shaded region satisfies each constraint, it is not clear how you should find the point that yields a maximum value of z. Fortunately, it can be shown that if there is an optimal solution, it must occur at one of the vertices. This means that *you can find the maximum value of z by testing z at each of the vertices.*

Optimal Solution of a Linear Programming Problem

If a linear programming problem has a solution, it must occur at a vertex of the set of feasible solutions. If there is more than one solution, at least one of them must occur at such a vertex. In either case, the value of the objective function is unique.

Some guidelines for solving a linear programming problem in two variables are listed at the top of the next page.

Solving a Linear Programming Problem

1. Sketch the region corresponding to the system of constraints. (The points inside or on the boundary of the region are *feasible solutions*.)

2. Find the vertices of the region.

3. Test the objective function at each of the vertices and select the values of the variables that optimize the objective function. For a bounded region, both a minimum and a maximum value will exist. (For an unbounded region, *if* an optimal solution exists, it will occur at a vertex.)

A computer animation of this example appears in the *Interactive* CD-ROM and *Internet* versions of this text.

Example 1 ▶ Solving a Linear Programming Problem

Find the maximum value of

$$z = 3x + 2y \qquad \text{Objective function}$$

subject to the following constraints.

$$\left.\begin{array}{r} x \geq 0 \\ y \geq 0 \\ x + 2y \leq 4 \\ x - y \leq 1 \end{array}\right\} \qquad \text{Constraints}$$

Solution

The constraints form the region shown in Figure 6.27. At the four vertices of this region, the objective function has the following values.

At $(0, 0)$: $z = 3(0) + 2(0) = 0$
At $(1, 0)$: $z = 3(1) + 2(0) = 3$
At $(2, 1)$: $z = 3(2) + 2(1) = 8$ Maximum value of z
At $(0, 2)$: $z = 3(0) + 2(2) = 4$

So, the maximum value of z is 8, and this occurs when $x = 2$ and $y = 1$.

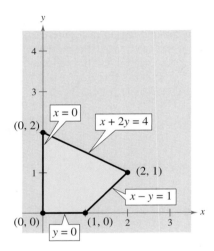

FIGURE **6.27**

In Example 1, try testing some of the *interior* points in the region. You will see that the corresponding values of z are less than 8. Here are some examples.

At $(1, 1)$: $z = 3(1) + 2(1) = 5$

At $\left(\dfrac{1}{2}, \dfrac{3}{2}\right)$: $z = 3\left(\dfrac{1}{2}\right) + 2\left(\dfrac{3}{2}\right) = \dfrac{9}{2}$

To see why the maximum value of the objective function in Example 1 must occur at a vertex, consider writing the objective function in slope-intercept form

$$y = -\frac{3}{2}x + \frac{z}{2} \qquad \text{Family of lines}$$

where $z/2$ is the y-intercept of the objective function. This equation represents a family of lines, each of slope $-\frac{3}{2}$. Of these infinitely many lines, you want the one that has the largest z-value while still intersecting the region determined by the constraints. In other words, of all the lines whose slope is $-\frac{3}{2}$, you want the one that has the largest y-intercept *and* intersects the given region, as shown in Figure 6.28. From the graph you can see that such a line will pass through one (or more) of the vertices of the region.

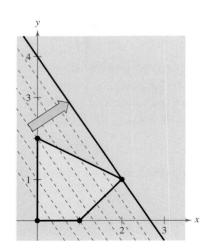

FIGURE **6.28**

The next example shows that the same basic procedure can be used to solve a problem in which the objective function is to be *minimized*.

Example 2 ▶ Minimizing an Objective Function

Find the minimum value of

$$z = 5x + 7y \qquad \text{Objective function}$$

where $x \geq 0$ and $y \geq 0$, subject to the following constraints.

$$\left.\begin{array}{r} 2x + 3y \geq 6 \\ 3x - y \leq 15 \\ -x + y \leq 4 \\ 2x + 5y \leq 27 \end{array}\right\} \qquad \text{Constraints}$$

Solution

The region bounded by the constraints is shown in Figure 6.29. By testing the objective function at each vertex, you obtain the following.

At $(0, 2)$: $z = 5(0) + 7(2) = 14$ Minimum value of z
At $(0, 4)$: $z = 5(0) + 7(4) = 28$
At $(1, 5)$: $z = 5(1) + 7(5) = 40$
At $(6, 3)$: $z = 5(6) + 7(3) = 51$
At $(5, 0)$: $z = 5(5) + 7(0) = 25$
At $(3, 0)$: $z = 5(3) + 7(0) = 15$

So, the minimum value of z is 14, and this occurs when $x = 0$ and $y = 2$.

Example 3 ▶ Maximizing an Objective Function

Find the maximum value of

$$z = 5x + 7y \qquad \text{Objective function}$$

where $x \geq 0$ and $y \geq 0$, subject to the following constraints.

$$\left.\begin{array}{r} 2x + 3y \geq 6 \\ 3x - y \leq 15 \\ -x + y \leq 4 \\ 2x + 5y \leq 27 \end{array}\right\} \qquad \text{Constraints}$$

Solution

This linear programming problem is identical to that given in Example 2 above, *except* that the objective function is maximized instead of minimized. Using the values of z at the vertices shown above, you can conclude that the maximum value of

$$z = 5(6) + 7(3) = 51$$

occurs when $x = 6$ and $y = 3$.

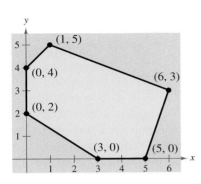

FIGURE 6.29

Edward W. Souza/News Service/Stanford University

Historical Note
George Dantzig (1914–) was the first to propose the simplex method, or linear programming, in 1947. This technique defined the steps needed to find the optimal solution to a complex multivariable problem.

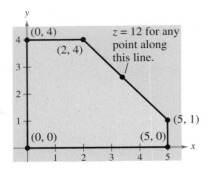

FIGURE **6.30**

It is possible for the maximum (or minimum) value in a linear programming problem to occur at *two* different vertices. For instance, at the vertices of the region shown in Figure 6.30, the objective function

$$z = 2x + 2y \qquad \text{Objective function}$$

has the following values.

At $(0, 0)$: $z = 2(0) + 2(0) = 0$
At $(0, 4)$: $z = 2(0) + 2(4) = 8$
At $(2, 4)$: $z = 2(2) + 2(4) = 12$ Maximum value of z
At $(5, 1)$: $z = 2(5) + 2(1) = 12$ Maximum value of z
At $(5, 0)$: $z = 2(5) + 2(0) = 10$

In this case, you can conclude that the objective function has a maximum value not only at the vertices $(2, 4)$ and $(5, 1)$; it also has a maximum value (of 12) at *any point on the line segment connecting these two vertices*. Note that the objective function in slope-intercept form

$$y = -x + \frac{1}{2}z$$

has the same slope as the line through the vertices $(2, 4)$ and $(5, 1)$.

Some linear programming problems have no optimal solutions. This can occur if the region determined by the constraints is *unbounded*. Example 4 illustrates such a problem.

Example 4 ▶ An Unbounded Region

Find the maximum value of

$$z = 4x + 2y \qquad \text{Objective function}$$

where $x \geq 0$ and $y \geq 0$, subject to the following constraints.

$$\left.\begin{array}{r} x + 2y \geq 4 \\ 3x + y \geq 7 \\ -x + 2y \leq 7 \end{array}\right\} \qquad \text{Constraints}$$

Solution

The region determined by the constraints is shown in Figure 6.31. For this unbounded region, there is no maximum value of z. To see this, note that the point $(x, 0)$ lies in the region for all values of $x \geq 4$. Substituting this point into the objective function, you get

$$z = 4(x) + 2(0) = 4x.$$

By choosing x to be large, you can obtain values of z that are as large as you want. So, there is no maximum value of z. For this problem, there *is* a minimum value of z.

At $(1, 4)$: $z = 4(1) + 2(4) = 12$
At $(2, 1)$: $z = 4(2) + 2(1) = 10$ Minimum value of z
At $(4, 0)$: $z = 4(4) + 2(0) = 16$

So, the minimum value of z is 10, and this occurs when $x = 2$ and $y = 1$.

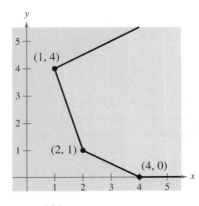

FIGURE **6.31**

Applications

Example 5 shows how linear programming can be used to find the maximum profit in a business application.

Example 5 ▶ Maximum Profit

A manufacturer wants to maximize the profit for two products. Product I yields a profit of \$1.50 per unit, and product II yields a profit of \$2.00 per unit. Market tests and available resources have indicated the following constraints.

1. The combined production level should not exceed 1200 units per month.
2. The demand for product II is no more than half the demand for product I.
3. The production level of product I should be less than or equal to 600 units plus three times the production level of product II.

Solution

If you let x be the number of units of product I and y be the number of units of product II, the objective function (for the combined profit) is given by

$$P = 1.5x + 2y. \qquad \text{Objective function}$$

The three constraints translate into the following linear inequalities.

1. $x + y \le 1200$ ⟹ $x + y \le 1200$
2. $y \le \frac{1}{2}x$ ⟹ $-x + 2y \le 0$
3. $x \le 600 + 3y$ ⟹ $x - 3y \le 600$

Because neither x nor y can be negative, you also have the two additional constraints of $x \ge 0$ and $y \ge 0$. Figure 6.32 shows the region determined by the constraints. To find the maximum profit, test the values of P at the vertices of the region.

$$
\begin{aligned}
\text{At } (0, 0): \quad & P = 1.5(0) & + 2(0) & = 0 \\
\text{At } (800, 400): \quad & P = 1.5(800) & + 2(400) & = 2000 \quad \text{Maximum profit}\\
\text{At } (1050, 150): \quad & P = 1.5(1050) & + 2(150) & = 1875 \\
\text{At } (600, 0): \quad & P = 1.5(600) & + 2(0) & = 900
\end{aligned}
$$

So, the maximum profit is \$2000, and it occurs when the monthly production consists of 800 units of product I and 400 units of product II.

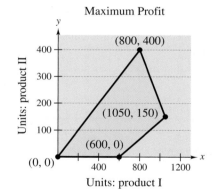

Maximum Profit

Units: product I

FIGURE 6.32

In Example 5, the manufacturer improved the production of product I so that it yielded a profit of \$2.50 per unit. The maximum profit can then be found using the objective function

$$P = 2.5x + 2y. \qquad \text{Objective function}$$

By testing the values of P at the vertices of the region, you find the maximum profit is \$2925 and that it now occurs when $x = 1050$ and $y = 150$.

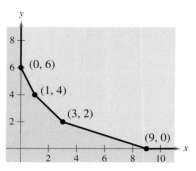

FIGURE **6.33**

Example 6 ▶ Minimum Cost

The minimum daily requirements from the liquid portion of a diet are 300 calories, 36 units of vitamin A, and 90 units of vitamin C. A cup of dietary drink X costs $0.12 and provides 60 calories, 12 units of vitamin A, and 10 units of vitamin C. A cup of dietary drink Y costs $0.15 and provides 60 calories, 6 units of vitamin A, and 30 units of vitamin C. How many cups of each drink should be consumed each day to minimize the cost and still meet the daily requirements?

Solution

As in Example 9 on page 486, let x be the number of cups of dietary drink X and let y be the number of cups of dietary drink Y.

$$\left.\begin{array}{lrr}
\text{For calories:} & 60x + 60y \geq & 300 \\
\text{For vitamin A:} & 12x + 6y \geq & 36 \\
\text{For vitamin C:} & 10x + 30y \geq & 90 \\
& x \geq & 0 \\
& y \geq & 0
\end{array}\right\} \quad \text{Constraints}$$

The cost C is given by $C = 0.12x + 0.15y$.   Objective function

The graph of the region corresponding to the constraints is shown in Figure 6.33. To determine the minimum cost, test C at each vertex of the region.

$$\begin{array}{lll}
\text{At } (0, 6): & C = 0.12(0) + 0.15(6) = 0.90 & \\
\text{At } (1, 4): & C = 0.12(1) + 0.15(4) = 0.72 & \\
\text{At } (3, 2): & C = 0.12(3) + 0.15(2) = 0.66 & \text{Minimum value of } C \\
\text{At } (9, 0): & C = 0.12(9) + 0.15(0) = 1.08 &
\end{array}$$

So, the minimum cost is $0.66 per day, and this occurs when three cups of drink X and two cups of drink Y are consumed each day.

Writing ABOUT MATHEMATICS

Creating a Linear Programming Problem   Sketch the region determined by the following constraints.

$$\left.\begin{array}{r}
x + 2y \leq 8 \\
x + y \leq 5 \\
x \geq 0 \\
y \geq 0
\end{array}\right\} \quad \text{Constraints}$$

Find, if possible, an objective function of the form $z = ax + by$ that has a maximum at the indicated vertex of the region.

a. $(0, 4)$ **b.** $(2, 3)$

c. $(5, 0)$ **d.** $(0, 0)$

Explain how you found each objective function.

6.5 Exercises

In Exercises 1–12, find the minimum and maximum values of the objective function and where they occur, subject to the indicated constraints. (For each exercise, the graph of the region determined by the constraints is provided.)

1. Objective function:

$z = 4x + 3y$

Constraints:

$x \geq 0$

$y \geq 0$

$x + y \leq 5$

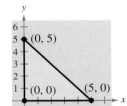

2. Objective function:

$z = 2x + 8y$

Constraints:

$x \geq 0$

$y \geq 0$

$2x + y \leq 4$

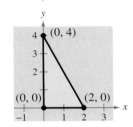

3. Objective function:

$z = 3x + 8y$

Constraints:
(See Exercise 1.)

4. Objective function:

$z = 7x + 3y$

Constraints:
(See Exercise 2.)

5. Objective function:

$z = 3x + 2y$

Constraints:

$x \geq 0$

$y \geq 0$

$x + 3y \leq 15$

$4x + y \leq 16$

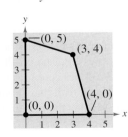

6. Objective function:

$z = 4x + 5y$

Constraints:

$x \geq 0$

$2x + 3y \geq 6$

$3x - y \leq 9$

$x + 4y \leq 16$

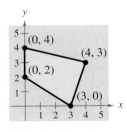

7. Objective function:

$z = 5x + 0.5y$

Constraints:
(See Exercise 5.)

8. Objective function:

$z = 2x + y$

Constraints:
(See Exercise 6.)

9. Objective function:

$z = 10x + 7y$

Constraints:

$0 \leq x \leq 60$

$0 \leq y \leq 45$

$5x + 6y \leq 420$

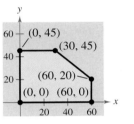

10. Objective function:

$z = 25x + 35y$

Constraints:

$x \geq 0$

$y \geq 0$

$8x + 9y \leq 7200$

$8x + 9y \geq 3600$

11. Objective function:

$z = 25x + 30y$

Constraints:
(See Exercise 9.)

12. Objective function:

$z = 15x + 20y$

Constraints:
(See Exercise 10.)

In Exercises 13–20, sketch the region determined by the constraints. Then find the minimum and maximum values of the objective function and where they occur, subject to the constraints.

13. Objective function:

$z = 6x + 10y$

Constraints:

$x \geq 0$

$y \geq 0$

$2x + 5y \leq 10$

14. Objective function:

$z = 7x + 8y$

Constraints:

$x \geq 0$

$y \geq 0$

$x + \frac{1}{2}y \leq 4$

15. Objective function:

$z = 9x + 24y$

Constraints:
(See Exercise 13.)

16. Objective function:

$z = 7x + 2y$

Constraints:
(See Exercise 14.)

17. Objective function:

$z = 4x + 5y$

Constraints:

$x \geq 0$

$y \geq 0$

$x + y \geq 8$

$3x + 5y \geq 30$

18. Objective function:

$z = 4x + 5y$

Constraints:

$x \geq 0$

$y \geq 0$

$2x + 2y \leq 10$

$x + 2y \leq 6$

19. Objective function:

$z = 2x + 7y$

Constraints:
(See Exercise 17.)

20. Objective function:

$z = 2x - y$

Constraints:
(See Exercise 18.)

In Exercises 21–24, use a graphing utility to sketch the region determined by the constraints. Then find the minimum and maximum values of the objective function and where they occur, subject to the constraints.

21. Objective function:

$z = 4x + y$

Constraints:

$$x \geq 0$$
$$y \geq 0$$
$$x + 2y \leq 40$$
$$2x + 3y \geq 72$$

22. Objective function:

$z = x$

Constraints:

$$x \geq 0$$
$$y \geq 0$$
$$2x + 3y \leq 60$$
$$2x + y \leq 28$$
$$4x + y \leq 48$$

23. Objective function:

$z = x + 4y$

Constraints:
(See Exercise 21.)

24. Objective function:

$z = y$

Constraints:
(See Exercise 22.)

In Exercises 25–28, find the maximum value of the objective function and where it occurs, subject to the constraints $x \geq 0, y \geq 0, 3x + y \leq 15,$ and $4x + 3y \leq 30.$

25. $z = 2x + y$

26. $z = 5x + y$

27. $z = x + y$

28. $z = 3x + y$

In Exercises 29–32, find the maximum value of the objective function and where it occurs, subject to the constraints $x \geq 0, y \geq 0, x + 4y \leq 20, x + y \leq 18,$ and $2x + 2y \leq 21.$

29. $z = x + 5y$

30. $z = 2x + 4y$

31. $z = 4x + 5y$

32. $z = 4x + y$

33. *Maximum Profit* A manufacturer produces two models of bicycles. The times (in hours) required for assembling, painting, and packaging each model are listed in the table.

Process	Model A	Model B
Assembling	2	2.5
Painting	4	1
Packaging	1	0.75

The total times available for assembling, painting, and packaging are 4000 hours, 4800 hours, and 1500 hours, respectively. The profits per unit are $45 for model A and $50 for model B. How many of each type should be produced to maximize profit? What is the maximum profit?

34. *Maximum Profit* A manufacturer produces two models of bicycles. The times (in hours) required for assembling, painting, and packaging each model are listed in the table.

Process	Model A	Model B
Assembling	2.5	3
Painting	2	1
Packaging	0.75	1.25

The total times available for assembling, painting, and packaging are 4000 hours, 2500 hours, and 1500 hours, respectively. The profits per unit are $50 for model A and $52 for model B. How many of each type should be produced to maximize profit? What is the maximum profit?

35. *Maximum Profit* A merchant plans to sell two models of compact disc players at costs of $150 and $200. The $150 model yields a profit of $25 per unit and the $200 model yields a profit of $40 per unit. The merchant estimates that the total monthly demand will not exceed 250 units. The merchant does not want to invest more than $40,000 in inventory for these products. Find the number of units of each model that should be stocked in order to maximize profit. What is the maximum profit?

36. *Maximum Profit* A fruit grower has 150 acres of land available to raise two crops, A and B. It takes 1 day to trim an acre of crop A and 2 days to trim an acre of crop B, and there are 240 days per year available for trimming. It takes 0.3 day to pick an acre of crop A and 0.1 day to pick an acre of crop B, and there are 30 days available for picking. The profit is $140 per acre for crop A and $235 per acre for crop B. Find the number of acres of each fruit that should be planted to maximize profit. What is the maximum profit?

37. *Minimum Cost* A farming cooperative mixes two brands of cattle feed. Brand X costs $25 per bag and contains 2 units of nutritional element A, 2 units of element B, and 2 units of element C. Brand Y costs $20 per bag and contains 1 unit of nutritional element A, 9 units of element B, and 3 units of element C. The minimum requirements of nutrients A, B, and C are 12 units, 36 units, and 24 units, respectively. Find the number of bags of each brand that should be mixed to produce a mixture having a minimum cost. What is the minimum cost?

38. *Minimum Cost* Two gasolines, type A and type B, have octane ratings of 80 and 92, respectively. Type A costs $1.13 per gallon and type B costs $1.28 per gallon. Determine the blend of minimum cost with an octane rating of at least 90. What is the minimum cost? (*Hint:* Let x be the fraction of each gallon that is type A and let y be the fraction that is type B.)

39. *Maximum Revenue* An accounting firm has 800 hours of staff time and 96 hours of reviewing time available each week. The firm charges $2000 for an audit and $300 for a tax return. Each audit requires 100 hours of staff time and 8 hours of review time. Each tax return requires 12.5 hours of staff time and 2 hours of review time. What numbers of audits and tax returns will yield the maximum revenue? What is the maximum revenue?

40. *Maximum Revenue* The accounting firm in Exercise 39 lowers its charge for an audit to $1000. What numbers of audits and tax returns will yield the maximum revenue? What is the maximum revenue?

In Exercises 41–46, the linear programming problem has an unusual characteristic. Sketch a graph of the solution region for the problem and describe the unusual characteristic. Find the maximum value of the objective function and where it occurs.

41. Objective function:

$z = 2.5x + y$

Constraints:

$x \geq 0$

$y \geq 0$

$3x + 5y \leq 15$

$5x + 2y \leq 10$

42. Objective function:

$z = x + y$

Constraints:

$x \geq 0$

$y \geq 0$

$-x + y \leq 1$

$-x + 2y \leq 4$

43. Objective function:

$z = -x + 2y$

Constraints:

$x \geq 0$

$y \geq 0$

$x \leq 10$

$x + y \leq 7$

44. Objective function:

$z = x + y$

Constraints:

$x \geq 0$

$y \geq 0$

$-x + y \leq 0$

$-3x + y \geq 3$

45. Objective function:

$z = 3x + 4y$

Constraints:

$x \geq 0$

$y \geq 0$

$x + y \leq 1$

$2x + y \leq 4$

46. Objective function:

$z = x + 2y$

Constraints:

$x \geq 0$

$y \geq 0$

$x + 2y \leq 4$

$2x + y \leq 4$

Synthesis

True or False? In Exercises 47 and 48, determine whether the statement is true or false. Justify your answer.

47. If an objective function has a maximum value at the vertices $(4, 7)$ and $(8, 3)$, you can conclude that it also has a maximum value at the points $(4.5, 6.5)$ and $(7.8, 3.2)$.

48. When solving a linear programming problem, if the objective function has a maximum value at more than one vertex, you can assume that there are an infinite number of points that will produce the maximum value.

In Exercises 49 and 50, determine values of t such that the objective function has a maximum value at the indicated vertex.

49. Objective function:

$z = 3x + ty$

Constraints:

$x \geq 0$

$y \geq 0$

$x + 3y \leq 15$

$4x + y \leq 16$

(a) $(0, 5)$

(b) $(3, 4)$

50. Objective function:

$z = 3x + ty$

Constraints:

$x \geq 0$

$y \geq 0$

$x + 2y \geq 4$

$x - y \leq 1$

(a) $(2, 1)$

(b) $(0, 2)$

Think About It In Exercises 51–54, find an objective function that has a maximum or minimum value at the indicated vertex of the constraint region shown below. (There are many correct answers.)

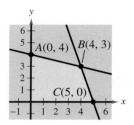

51. The maximum occurs at vertex A.

52. The maximum occurs at vertex B.

53. The maximum occurs at vertex C.

54. The minimum occurs at vertex C.

Review

In Exercises 55–58, simplify the compound fraction.

55. $\dfrac{\dfrac{9}{x}}{\dfrac{6}{x} + 2}$

56. $\dfrac{1 + \dfrac{2}{x}}{x - \dfrac{4}{x}}$

57. $\dfrac{\dfrac{4}{x^2 - 9} + \dfrac{2}{x - 2}}{\dfrac{1}{x + 3} + \dfrac{1}{x - 3}}$

58. $\dfrac{\dfrac{1}{x + 1} + \dfrac{1}{2}}{\dfrac{3}{2x^2 + 4x + 2}}$

In Exercises 59–64, sketch the graph of the conic.

59. $y^2 = 6x$

60. $3x^2 - 4 = y$

61. $\dfrac{x^2}{9} + \dfrac{y^2}{49} = 1$

62. $\dfrac{y^2}{16} - \dfrac{x^2}{169} = 1$

63. $\dfrac{(x + 5)^2}{4} - \dfrac{y^2}{36} = 1$

64. $\dfrac{(x - 3)^2}{25} + \dfrac{(y - 6)^2}{36} = 1$

In Exercises 65–70, solve the equation algebraically. Round the result to three decimal places.

65. $e^{2x} + 2e^x - 15 = 0$

66. $e^{2x} - 10e^x + 24 = 0$

67. $8(62 - e^{x/4}) = 192$

68. $\dfrac{150}{e^{-x} - 4} = 75$

69. $7 \ln 3x = 12$

70. $\ln(x + 9)^2 = 2$

Chapter Summary

What did you learn?

Section 6.1	Review Exercises
☐ How to use the method of substitution to solve systems of equations in two variables	1–4
☐ How to use a graphical approach to solve systems of equations in two variables	5–10
☐ How to use systems of equations to model and solve real-life problems	11–13

Section 6.2	
☐ How to use the method of elimination to solve systems of linear equations in two variables	14–21
☐ How to interpret graphically the numbers of solutions of systems of equations in two variables	22–25
☐ How to use systems of equations in two variables to model and solve real-life problems	26–30

Section 6.3	
☐ How to recognize linear systems in row-echelon form and use back-substitution to solve the systems	31, 32
☐ How to use Gaussian elimination to solve systems of linear equations	33–42
☐ How to solve nonsquare systems of linear equations	43, 44
☐ How to use systems of linear equations in three or more variables to model and solve application problems	45–47

Section 6.4	
☐ How to sketch the graphs of inequalities in two variables	48–51
☐ How to solve systems of inequalities	52–59
☐ How to use systems of inequalities in two variables to model and solve real-life problems	60–63

Section 6.5	
☐ How to solve linear programming problems	64–67
☐ How to use linear programming to model and solve real-life problems	68–71

Review Exercises

6.1 **In Exercises 1–4, solve the system by the method of substitution.**

1. $\begin{cases} x^2 - y^2 = 9 \\ x - y = 1 \end{cases}$

2. $\begin{cases} x^2 + y^2 = 169 \\ 3x + 2y = 39 \end{cases}$

3. $\begin{cases} y = 2x^2 \\ y = x^4 - 2x^2 \end{cases}$

4. $\begin{cases} x = y + 3 \\ x = y^2 + 1 \end{cases}$

In Exercises 5–8, solve the system graphically.

5. $\begin{cases} 2x - y = 10 \\ x + 5y = -6 \end{cases}$

6. $\begin{cases} y^2 - 2y + x = 0 \\ x + y = 0 \end{cases}$

7. $\begin{cases} y = -2e^{-x} \\ 2e^x + y = 0 \end{cases}$

8. $\begin{cases} y = 2(6 - x) \\ y = 2^{x-2} \end{cases}$

In Exercises 9 and 10, use a graphing utility to solve the system of equations. Find the solution accurate to two decimal places.

9. $\begin{cases} y = 2x^2 - 4x + 1 \\ y = x^2 - 4x + 3 \end{cases}$

10. $\begin{cases} y = \ln(x - 1) - 3 \\ y = 4 - \frac{1}{2}x \end{cases}$

11. *Break-Even Point* You set up a business and make an initial investment of $50,000. The unit cost of the product is $2.15 and the selling price is $6.95. How many units must you sell to break even?

12. *Choice of Two Jobs* You are offered two sales jobs. One company offers an annual salary of $22,500 plus a year-end bonus of 1.5% of your total sales. The other company offers an annual salary of $20,000 plus a year-end bonus of 2% of your total sales. What amount of sales will make the second offer better? Explain.

13. *Geometry* The perimeter of a rectangle is 480 meters and its length is 150% of its width. Find the dimensions of the rectangle.

6.2 **In Exercises 14–21, solve the system by elimination.**

14. $\begin{cases} 2x - y = 2 \\ 6x + 8y = 39 \end{cases}$

15. $\begin{cases} 40x + 30y = 24 \\ 20x - 50y = -14 \end{cases}$

16. $\begin{cases} 0.2x + 0.3y = 0.14 \\ 0.4x + 0.5y = 0.20 \end{cases}$

17. $\begin{cases} 12x + 42y = -17 \\ 30x - 18y = 19 \end{cases}$

18. $\begin{cases} 3x - 2y = 0 \\ 3x + 2(y + 5) = 10 \end{cases}$

19. $\begin{cases} 7x + 12y = 63 \\ 2x + 3(y + 2) = 21 \end{cases}$

20. $\begin{cases} 1.25x - 2y = 3.5 \\ 5x - 8y = 14 \end{cases}$

21. $\begin{cases} 1.5x + 2.5y = 8.5 \\ 6x + 10y = 24 \end{cases}$

In Exercises 22–25, use a graphing utility to graph the lines in the system. Use the graph to determine if the system is consistent or inconsistent. If the system is consistent, determine the number of solutions.

22. $\begin{cases} -3x - 5y = -1 \\ 6x + y = 4 \end{cases}$

23. $\begin{cases} \frac{1}{5}x = -4 + y \\ 5y = x \end{cases}$

24. $\begin{cases} 6x - 14.4y = 1.8 \\ 1.2x - 2.88y = 0.36 \end{cases}$

25. $\begin{cases} \frac{8}{5}x - y = 3 \\ -5y + 8x = -2 \end{cases}$

26. *Acid Mixture* Two hundred liters of a 75% acid solution is obtained by mixing a 90% solution with a 50% solution. How many liters of each must be used to obtain the desired mixture?

27. *Compact Disc Sales* Suppose you are the manager of a music store. At the end of one week you are going over receipts for the previous week's sales. Six hundred and fifty compact discs were sold. One type of compact disc sold for $9.95 and another sold for $14.95. The total compact disc receipts were $7717.50. The cash register that was supposed to record the number of each type of compact disc sold malfunctioned. Can you recover the information? If so, how many of each type of compact disc were sold?

28. *Flying Speeds* Two planes leave Pittsburgh and Philadelphia at the same time, each going to the other city. One plane flies 25 miles per hour faster than the other. Find the air speed of each plane if the cities are 275 miles apart and the planes pass one another after 40 minutes of flying time.

Economics **In Exercises 29 and 30, find the point of equilibrium.**

Demand Function	Supply Function
29. $p = 37 - 0.0002x$	$p = 22 + 0.00001x$
30. $p = 120 - 0.0001x$	$p = 45 + 0.0002x$

6.3 **In Exercises 31 and 32, use back-substitution to solve the system.**

31. $\begin{cases} x - 4y + 3z = 3 \\ -y + z = -1 \\ z = -5 \end{cases}$

32. $\begin{cases} x - 7y + 8z = 85 \\ y - 9z = -35 \\ z = 3 \end{cases}$

In Exercises 33–36, use Gaussian elimination to solve the system of equations.

33. $\begin{cases} x + 2y + 6z = 4 \\ -3x + 2y - z = -4 \\ 4x + 2z = 16 \end{cases}$

34. $\begin{cases} x + 3y - z = 13 \\ 2x - 5z = 23 \\ 4x - y - 2z = 14 \end{cases}$ 35. $\begin{cases} x - 2y + z = -6 \\ 2x - 3y = -7 \\ -x + 3y - 3z = 11 \end{cases}$

36. $\begin{cases} 2x + 6z = -9 \\ 3x - 2y + 11z = -16 \\ 3x - y + 7z = -11 \end{cases}$

In Exercises 37 and 38, find the equation of the parabola $y = ax^2 + bx + c$ that passes through the points. Use a graphing utility to verify your result.

37.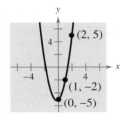

38.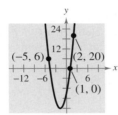

In Exercises 39 and 40, find the equation of the circle $x^2 + y^2 + Dx + Ey + F = 0$ that passes through the points. Use a graphing utility to verify your result.

39.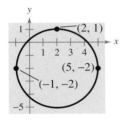

40.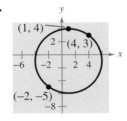

41. *Fitting a Line to Data* Solve the system of equations to find the least squares regression line $y = ax + b$ for the points in the figure. (For an explanation of how the coefficients of a and b in the system are obtained, see Appendix B.)

$\begin{cases} 5b + 10a = 17.8 \\ 10b + 30a = 45.7 \end{cases}$

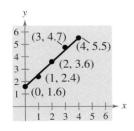

42. *Data Analysis* Let x and y represent the median ages at first marriage for women and men, respectively. In a sample of 6 years, these median ages are given by the following ordered pairs. (Source: U.S. Center for Health Statistics)

(23.0, 24.8),   (23.3, 25.1),   (23.6, 25.3),

(23.7, 25.5),   (23.9, 25.9),   (24.0, 25.9)

(a) Find the least squares regression line $y = ax + b$ for the data by solving the following system of linear equations.

$\begin{cases} 6b + 141.5a = 152.5 \\ 141.5b + 3337.75a = 3597.27 \end{cases}$

(b) Use a graphing utility to plot the data and graph the regression line in the same viewing window.

(c) Use the graph to determine whether the line is a good model for the data. Explain.

(d) What information is given by the slope of the regression line? Explain.

In Exercises 43 and 44, solve the nonsquare system of equations.

43. $\begin{cases} 5x - 12y + 7z = 16 \\ 3x - 7y + 4z = 9 \end{cases}$ 44. $\begin{cases} 2x + 5y - 19z = 34 \\ 3x + 8y - 31z = 54 \end{cases}$

45. *Agriculture* A mixture of 6 gallons of chemical A, 8 gallons of chemical B, and 13 gallons of chemical C is required to kill a certain destructive crop insect. Commercial spray X contains 1, 2, and 2 parts, respectively, of these chemicals. Commercial spray Y contains only chemical C. Commercial spray Z contains chemicals A, B, and C in equal amounts. How much of each type of commercial spray is needed to get the desired mixture?

46. *Finance* An inheritance of $40,000 was divided among three investments yielding $3500 in interest per year. The interest rates for the three investments were 7%, 9%, and 11%. Find the amount placed in each investment if the second and third were $3000 and $5000 less than the first, respectively.

47. *Fitting a Parabola to Data* Solve the system of equations to find the least squares regression parabola $y = ax^2 + bx + c$ for the points in the figure. (For an explanation of how the coefficients of a, b, and c in the system are obtained, see Appendix B.)

$$\begin{cases} 5c & + 10a = 9.1 \\ 10b & = 8.0 \\ 10c & + 34a = 19.8 \end{cases}$$

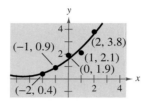

6.4 In Exercises 48–51, sketch the graph of the inequality.

48. $y \leq 5 - \frac{1}{2}x$

49. $3y - x \geq 7$

50. $y - 4x^2 > -1$

51. $y \leq 2 \ln x - 6$

In Exercises 52–59, sketch a graph and label the vertices of the solution set of the system of inequalities.

52. $\begin{cases} x + 2y \leq 160 \\ 3x + y \leq 180 \\ x \geq 0 \\ y \geq 0 \end{cases}$

53. $\begin{cases} 2x + 3y \leq 24 \\ 2x + y \leq 16 \\ x \geq 0 \\ y \geq 0 \end{cases}$

54. $\begin{cases} 3x + 2y \geq 24 \\ x + 2y \geq 12 \\ 2 \leq x \leq 15 \\ y \leq 15 \end{cases}$

55. $\begin{cases} 2x + y \geq 16 \\ x + 3y \geq 18 \\ 0 \leq x \leq 25 \\ 0 \leq y \leq 25 \end{cases}$

56. $\begin{cases} y < x + 1 \\ y > x^2 - 1 \end{cases}$

57. $\begin{cases} y \leq 6 - 2x - x^2 \\ y \geq x + 6 \end{cases}$

58. $\begin{cases} 2x - 3y \geq 0 \\ 2x - y \leq 8 \\ y \geq 0 \end{cases}$

59. $\begin{cases} x^2 + y^2 \leq 9 \\ (x - 3)^2 + y^2 \leq 9 \end{cases}$

In Exercises 60 and 61, determine a system of inequalities that models the description. Use a graphing utility to graph and shade the solution of the system.

60. *Fruit Distribution* A Pennsylvania fruit grower has 1500 bushels of apples that are to be divided between markets in Harrisburg and Philadelphia. These two markets need at least 400 bushels and 600 bushels, respectively.

61. *Inventory Costs* A warehouse operator has 24,000 square feet of floor space in which to store two products. Each unit of product I requires 20 square feet of floor space and costs \$12 per day to store. Each unit of product II requires 30 square feet of floor space and costs \$8 per day to store. The total storage cost per day cannot exceed \$12,400.

In Exercises 62 and 63, find the consumer surplus and producer surplus for the demand and supply equations. Sketch the graph of the equations and shade the regions representing the consumer surplus and producer surplus.

	Demand	*Supply*
62.	$p = 160 - 0.0001x$	$p = 70 + 0.0002x$
63.	$p = 130 - 0.0002x$	$p = 30 + 0.0003x$

6.5 In Exercises 64–67, find the required optimum value of the objective function and where it occurs, subject to the indicated constraints.

64. Maximize:

$z = 3x + 4y$

Constraints:

$$\begin{align} x &\geq 0 \\ y &\geq 0 \\ 2x + 5y &\leq 50 \\ 4x + y &\leq 28 \end{align}$$

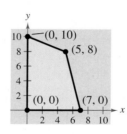

65. Minimize:

$z = 10x + 7y$

Constraints:

$$\begin{align} x &\geq 0 \\ y &\geq 0 \\ 2x + y &\geq 100 \\ x + y &\geq 75 \end{align}$$

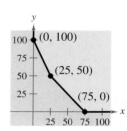

66. Minimize:

$z = 1.75x + 2.25y$

Constraints:

$$\begin{align} x &\geq 0 \\ y &\geq 0 \\ 2x + y &\geq 25 \\ 3x + 2y &\geq 45 \end{align}$$

67. Maximize:

$z = 50x + 70y$

Constraints:

$$\begin{align} x &\geq 0 \\ y &\geq 0 \\ x + 2y &\leq 1500 \\ 5x + 2y &\leq 3500 \end{align}$$

68. *Maximum Revenue* A student is working part time as a hairdresser to pay college expenses. The student may work no more than 24 hours per week. Haircuts cost $25 and require an average of 20 minutes, and permanents cost $70 and require an average of 1 hour and 10 minutes. What combination of haircuts and/or permanents will yield a maximum revenue? What is the maximum revenue?

69. *Maximum Profit* A manufacturer produces products A and B yielding profits of $18 and $24, respectively. Each product must go through three processes with the required times per unit shown in the table.

Process	Hours for Product A	Hours for Product B	Hours Available per Day
I	4	2	24
II	1	2	9
III	1	1	8

Find the daily production level for each unit to maximize the profit. What is the maximum profit?

70. *Minimum Cost* A pet supply company mixes two brands of dry dog food. Brand X costs $15 per bag and contains 8 units of nutritional element A, 1 unit of nutritional element B, and 2 units of nutritional element C. Brand Y costs $30 per bag and contains 2 units of nutritional element A, 1 unit of nutritional element B, and 7 units of nutritional element C. Each bag of mixed dog food must contain at least 16 units, 5 units, and 20 units of nutritional elements A, B, and C, respectively. Find the numbers of bags of brands X and Y that should be mixed to produce a mixture meeting the minimum nutritional requirements and having a minimum cost. What is the minimum cost?

71. *Minimum Cost* Two gasolines, type A and type B, have octane ratings of 80 and 92, respectively. Type A costs $1.25 per gallon and type B costs $1.55 per gallon. Determine the blend of minimum cost with an octane rating of at least 88. What is the minimum cost? (*Hint:* Let x be the fraction of each gallon that is type A and let y be the fraction that is type B.)

Synthesis

True or False? **In Exercises 72 and 73, determine whether the statement is true or false. Justify your answer.**

72. The system

$$\begin{cases} y \le 5 \\ y \ge -2 \\ y \ge \frac{7}{2}x - 9 \\ y \ge -\frac{7}{2}x + 26 \end{cases}$$

represents the region covered by an isosceles trapezoid.

73. It is possible for an objective function of a linear programming problem to have exactly 10 maximum value points.

Exploration **In Exercises 74–77, find a system of linear equations having the ordered pair as a solution. (There is more than one correct answer.)**

74. $(-6, 8)$ **75.** $(5, -4)$

76. $\left(\frac{4}{3}, 3\right)$ **77.** $\left(-1, \frac{9}{4}\right)$

Exploration **In Exercises 78–81, find a system of linear equations having the ordered triple as a solution. (There is more than one correct answer.)**

78. $(4, -1, 3)$ **79.** $(-3, 5, 6)$

80. $\left(5, \frac{3}{2}, 2\right)$ **81.** $\left(\frac{3}{4}, -2, 8\right)$

82. *Exploration* Find k_1 and k_2 such that the system of equations has an infinite number of solutions.

$$\begin{cases} 3x - 5y = 8 \\ 2x + k_1 y = k_2 \end{cases}$$

83. When solving a system of equations by substitution, how do you recognize that the system has no solution?

84. When solving a system of equations by elimination, how do you recognize that the system has no solution?

85. A system of two equations in two unknowns is solved and has a finite number of solutions. Determine the maximum number of solutions of the system satisfying each of the following.

(a) Both equations are linear.

(b) One equation is linear and the other is quadratic.

(c) Both equations are quadratic.

86. How can you tell graphically that a system of linear equations in two variables has no solution? Give an example.

Chapter Project ▶ Fitting Models to Data

Many of the models in this book were created with a statistical method called *least squares regression analysis.* This procedure can be performed easily with a computer or graphing calculator.

Example ▶ Fitting a Line to Data

The numbers of morning and evening newspapers published in the United States from 1990 through 1997 are shown in the table. Use the data to project the numbers of morning and evening newspapers that will be published in 2003. In the table, $t = 0$ represents 1990. (Source: Editor and Publisher Company)

Year, t	0	1	2	3	4	5	6	7
Morning	559	571	596	623	635	656	686	705
Evening	1084	1042	996	954	935	891	846	816

Solution

Begin by finding a computer or graphing calculator that will perform linear regression analysis. After entering the data and running the program, you should obtain the following models.

$$y = 554.3 + 21.3t \qquad \text{Morning newspapers}$$
$$y = 1078.4 - 38.0t \qquad \text{Evening newspapers}$$

With these models, you can project the numbers of newspapers in 2003. If the trend continues to follow the pattern from 1990 through 1997, the numbers of newspapers in 2003 should be about

$$y = 554.3 + 21.3(13) \approx 831 \qquad \text{Morning newspapers}$$
$$y = 1078.4 - 38.0(13) \approx 584. \qquad \text{Evening newspapers}$$

The graphs of the data and the models are shown at the left.

Newspapers Published

Newspapers

Evening
newspapers

Morning
newspapers

Year (0 ↔ 1990)

Year, t	Morning
0	41.3
1	41.5
2	42.4
3	43.1
4	43.4
5	44.3
6	44.8
7	45.4

Chapter Project Investigations

1. Use the models from the example to predict the year when the number of morning papers published will be equal to the number of evening papers published.
2. The total numbers (in millions) of *all* morning newspapers *sold* each day in the United States from 1990 to 1997 are shown in the table at the left. Find a linear model that represents this data. Use your model to project the number of morning papers that will be sold each day in 2003.
3. From 1990 through 1997, both the number of morning newspapers published and the total number of papers sold increased. Did the average circulation per morning paper (number sold ÷ number published) increase or decrease? Explain.

▶ Chapter Test

Take this test as you would take a test in class. After you are done, check your work against the answers given in the back of the book.

In Exercises 1–3, solve the system by the method of substitution.

1. $\begin{cases} x - y = -7 \\ 4x + 5y = 8 \end{cases}$

2. $\begin{cases} y = x - 1 \\ y = (x - 1)^3 \end{cases}$

3. $\begin{cases} 2x - y^2 = 0 \\ x - y = 4 \end{cases}$

In Exercises 4–6, solve the system graphically.

4. $\begin{cases} 2x - 3y = 0 \\ 2x + 3y = 12 \end{cases}$

5. $\begin{cases} y = 9 - x^2 \\ y = x + 3 \end{cases}$

6. $\begin{cases} y - \ln x = 12 \\ 7x - 2y + 11 = -6 \end{cases}$

In Exercises 7 and 8, solve the linear system by elimination.

7. $\begin{cases} 2x + 3y = 17 \\ 5x - 4y = -15 \end{cases}$

8. $\begin{cases} x - 2y + 3z = 11 \\ 2x - z = 3 \\ 3y + z = -8 \end{cases}$

9. Find a system of linear equations that has the solution $\left(\frac{4}{3}, -5\right)$.

10. Find the equation of the parabola $y = ax^2 + bx + c$ passing through the points $(0, 6)$, $(-2, 2)$, and $\left(3, \frac{9}{2}\right)$.

11. Find a system of linear equations that has the solution $\left(-\frac{1}{2}, 6, -\frac{5}{4}\right)$.

In Exercises 12–14, sketch the graph and label the vertices of the solution of the system of inequalities.

12. $\begin{cases} 2x + y \le 4 \\ 2x - y \ge 0 \\ x \ge 0 \end{cases}$

13. $\begin{cases} y < -x^2 + x + 4 \\ y > 4x \end{cases}$

14. $\begin{cases} x^2 + y^2 \le 16 \\ x \ge 1 \\ y \ge -3 \end{cases}$

15. Find the maximum value of the objective function $z = 20x + 12y$ and where it occurs, subject to the constraints $x \ge 0$, $y \ge 0$, $x + 4y \le 32$, and $3x + 2y \le 36$.

16. A merchant plans to sell two models of compact disc players at costs of $275 and $400. The $275 model yields a profit of $55 and the $400 model yields a profit of $75. The merchant estimates that the total monthly demand will not exceed 300 units. The merchant does not want to invest more than $100,000 in inventory for these products. Find the number of units of each model that should be stocked in order to maximize profit. What is the maximum profit?

17. A manufacturer produces two types of television stands. The amounts of time for assembling, staining, and packaging the two models are shown in the table at the left. The total amounts of time available for assembling, staining, and packaging are 4000, 8950, and 2650 hours, respectively. The profits per unit are $30 (model I) and $40 (model II). How many of each model should be produced to maximize the profit? What is the maximum profit?

	Model I	Model II
Assembling	0.5	0.75
Staining	2.0	1.5
Packaging	0.5	0.5

Peter Turnley/CORBIS

In 1998, retail sales of jewelry and precious metals reached $22.3 billion. (Source: U.S. Bureau of the Census)

7 Matrices and Determinants

▶ **How to Study This Chapter**

The Big Picture

In this chapter you will learn the following skills and concepts.

▶ How to use matrices, Gaussian elimination, and Gauss-Jordan elimination to solve systems of linear equations

▶ How to add and subtract matrices, multiply matrices by real numbers, and multiply two matrices

▶ How to find the inverses of matrices and use inverse matrices to solve systems of linear equations

▶ How to find minors, cofactors, and determinants of square matrices

▶ How to use Cramer's Rule to solve systems of linear equations

▶ How to use determinants and matrices to model and solve problems

Important Vocabulary

As you encounter each new vocabulary term in this chapter, add the term and its definition to your notebook glossary.

Matrix (p. 510)
Entry of a matrix (p. 510)
Order of a matrix (p. 510)
Square matrix (p. 510)
Main diagonal (p. 510)
Row matrix (p. 510)
Column matrix (p. 510)
Augmented matrix (p. 511)
Coefficient matrix (p. 511)
Elementary row operations (p. 512)
Row-equivalent matrices (p. 512)
Row-echelon form (p. 514)

Reduced row-echelon form (p. 514)
Gauss-Jordan elimination (p. 517)
Scalar (p. 526)
Scalar multiple (p. 526)
Zero matrix (p. 529)
Matrix multiplication (p. 530)
Identity matrix of order n (p. 532)
Inverse of a matrix (p. 539)
Determinant (pp. 543, 548)
Minors (p. 550)
Cofactors (p. 550)
Cramer's Rule (p. 556)

Study Tools

- Learning objectives at the beginning of each section
- Chapter Summary (p. 568)
- Review Exercises (pp. 569–573)
- Chapter Test (p. 575)

Additional Resources

- Study and Solutions Guide
- Interactive College Algebra
- Videotapes for Chapter 7
- College Algebra Website
- Student Success Organizer

STUDY T!P

Read the text before class. Write down any questions you may have about the material. Ask your instructor these questions during class. That way, you will understand the material better.

7.1 Matrices and Systems of Equations

▶ **What you should learn**

- How to write a matrix and identify its order
- How to perform elementary row operations on matrices
- How to use matrices and Gaussian elimination to solve systems of linear equations
- How to use matrices and Gauss-Jordan elimination to solve systems of linear equations

▶ **Why you should learn it**

You can use matrices to solve systems of linear equations in two or more variables. For instance, Exercise 88 on page 523 shows how a matrix can be used to help find a model for the parabolic path of a baseball.

Amwell/Tony Stone Images

Matrices

In this section you will study a streamlined technique for solving systems of linear equations. This technique involves the use of a rectangular array of real numbers called a **matrix.** The plural of matrix is *matrices.*

Definition of Matrix

If m and n are positive integers, an $m \times n$ matrix (read "m by n") is a rectangular array

$$\begin{bmatrix} a_{11} & a_{12} & a_{13} & \cdots & a_{1n} \\ a_{21} & a_{22} & a_{23} & \cdots & a_{2n} \\ a_{31} & a_{32} & a_{33} & \cdots & a_{3n} \\ \vdots & \vdots & \vdots & & \vdots \\ a_{m1} & a_{m2} & a_{m3} & \cdots & a_{mn} \end{bmatrix} \Bigg\} \ m \text{ rows}$$

$$\underbrace{\phantom{a_{11} \quad a_{12} \quad a_{13} \quad \cdots \quad a_{1n}}}_{n \text{ columns}}$$

in which each **entry,** a_{ij}, of the matrix is a number. An $m \times n$ matrix has m rows (horizontal lines) and n columns (vertical lines).

The entry in the ith row and jth column is denoted by the *double subscript* notation a_{ij}. A matrix having m rows and n columns is said to be of **order** $m \times n$. If $m = n$, the matrix is **square** of order n. For a square matrix, the entries $a_{11}, a_{22}, a_{33}, \ldots$ are the **main diagonal** entries.

Example 1 ▶ Order of Matrices

Determine the order of each matrix.

a. $[2]$

b. $\begin{bmatrix} 1 & -3 & 0 & \frac{1}{2} \end{bmatrix}$

c. $\begin{bmatrix} 0 & 0 \\ 0 & 0 \end{bmatrix}$

d. $\begin{bmatrix} 5 & 0 \\ 2 & -2 \\ -7 & 4 \end{bmatrix}$

Solution

a. This matrix has *one* row and *one* column. The order of the matrix is 1×1.

b. This matrix has *one* row and *four* columns. The order of the matrix is 1×4.

c. This matrix has *two* rows and *two* columns. The order of the matrix is 2×2.

d. This matrix has *three* rows and *two* columns. The order of the matrix is 3×2.

A matrix that has only one row is called a **row matrix,** and a matrix that has only one column is called a **column matrix.**

A matrix derived from a system of linear equations (each written in standard form with the constant term on the right) is the **augmented matrix** of the system. Moreover, the matrix derived from the coefficients of the system (but not including the constant terms) is the **coefficient matrix** of the system.

System
$$\begin{cases} x - 4y + 3z = 5 \\ -x + 3y - z = -3 \\ 2x \quad\quad - 4z = 6 \end{cases}$$

Augmented Matrix
$$\left[\begin{array}{ccc:c} 1 & -4 & 3 & 5 \\ -1 & 3 & -1 & -3 \\ 2 & 0 & -4 & 6 \end{array} \right]$$

Coefficient Matrix
$$\left[\begin{array}{ccc} 1 & -4 & 3 \\ -1 & 3 & -1 \\ 2 & 0 & -4 \end{array} \right]$$

Note the use of 0 for the missing *y*-variable in the third equation, and also note the fourth column of constant terms in the augmented matrix.

When forming either the coefficient matrix or the augmented matrix of a system, you should begin by vertically aligning the variables in the equations and using zeros for the missing variables.

Example 2 ▶ Writing an Augmented Matrix

Write the augmented matrix for the system of linear equations.

$$\begin{cases} x + 3y - w = 9 \\ -y + 4z + 2w = -2 \\ x - 5z - 6w = 0 \\ 2x + 4y - 3z = 4 \end{cases}$$

What is the order of the augmented matrix?

Solution

Begin by rewriting the linear system and aligning the variables.

$$\begin{cases} x + 3y \quad\quad - w = 9 \\ -y + 4z + 2w = -2 \\ x \quad\quad - 5z - 6w = 0 \\ 2x + 4y - 3z \quad\quad = 4 \end{cases}$$

Next, use the coefficients as the matrix entries. Include zeros for the missing coefficients.

$$\begin{array}{c} R_1 \\ R_2 \\ R_3 \\ R_4 \end{array} \left[\begin{array}{cccc:c} 1 & 3 & 0 & -1 & 9 \\ 0 & -1 & 4 & 2 & -2 \\ 1 & 0 & -5 & -6 & 0 \\ 2 & 4 & -3 & 0 & 4 \end{array} \right]$$

The augmented matrix has 4 rows and 5 columns, so it is a 4-by-5 matrix. The notation R_n is used to designate each row in the matrix. For example, Row 1 is represented by R_1.

Elementary Row Operations

In Section 6.3, you studied three operations that can be used on a system of linear equations to produce an equivalent system.

1. Interchange two equations.

2. Multiply an equation by a nonzero constant.

3. Add a multiple of an equation to another equation.

In matrix terminology these three operations correspond to **elementary row operations.** An elementary row operation on an augmented matrix of a given system of linear equations produces a new augmented matrix corresponding to a new (but equivalent) system of linear equations. Two matrices are **row-equivalent** if one can be obtained from the other by a sequence of elementary row operations.

Elementary Row Operations

1. Interchange two rows.

2. Multiply a row by a nonzero constant.

3. Add a multiple of a row to another row.

Although elementary row operations are simple to perform, they involve a lot of arithmetic. Because it is easy to make a mistake, you should get in the habit of noting the elementary row operations performed in each step so that you can go back and check your work.

Example 3 ▶ Elementary Row Operations

a. Interchange the first and second rows.

Original Matrix

$$\begin{bmatrix} 0 & 1 & 3 & 4 \\ -1 & 2 & 0 & 3 \\ 2 & -3 & 4 & 1 \end{bmatrix}$$

New Row-Equivalent Matrix

$$\begin{array}{c} R_2 \\ R_1 \end{array} \begin{bmatrix} -1 & 2 & 0 & 3 \\ 0 & 1 & 3 & 4 \\ 2 & -3 & 4 & 1 \end{bmatrix}$$

b. Multiply the first row by $\frac{1}{2}$.

Original Matrix

$$\begin{bmatrix} 2 & -4 & 6 & -2 \\ 1 & 3 & -3 & 0 \\ 5 & -2 & 1 & 2 \end{bmatrix}$$

New Row-Equivalent Matrix

$$\frac{1}{2}R_1 \rightarrow \begin{bmatrix} 1 & -2 & 3 & -1 \\ 1 & 3 & -3 & 0 \\ 5 & -2 & 1 & 2 \end{bmatrix}$$

c. Add -2 times the first row to the third row.

Original Matrix

$$\begin{bmatrix} 1 & 2 & -4 & 3 \\ 0 & 3 & -2 & -1 \\ 2 & 1 & 5 & -2 \end{bmatrix}$$

New Row-Equivalent Matrix

$$-2R_1 + R_3 \rightarrow \begin{bmatrix} 1 & 2 & -4 & 3 \\ 0 & 3 & -2 & -1 \\ 0 & -3 & 13 & -8 \end{bmatrix}$$

Note that the elementary row operation is written beside the row that is *changed.*

In Example 3 in Section 6.3, you used Gaussian elimination with back-substitution to solve a system of linear equations. The next example demonstrates the matrix version of Gaussian elimination. The two methods are essentially the same. The basic difference is that with matrices you do not need to keep writing the variables.

Example 4 ▶ Comparing Linear Systems and Matrix Operations

The *Interactive* CD-ROM and *Internet* versions of this text offer a built-in graphing calculator, which can be used in the Examples, Explorations, Technology notes, and Exercises.

<table>
<tr><td>*Linear System*</td><td>*Associated Augmented Matrix*</td></tr>
</table>

$$\begin{cases} x - 2y + 3z = 9 \\ -x + 3y \quad\quad = -4 \\ 2x - 5y + 5z = 17 \end{cases}$$

$$\begin{bmatrix} 1 & -2 & 3 & \vdots & 9 \\ -1 & 3 & 0 & \vdots & -4 \\ 2 & -5 & 5 & \vdots & 17 \end{bmatrix}$$

Add the first equation to the second equation.

Add the first row to the second row.

$$\begin{cases} x - 2y + 3z = 9 \\ y + 3z = 5 \\ 2x - 5y + 5z = 17 \end{cases}$$

$$R_1 + R_2 \rightarrow \begin{bmatrix} 1 & -2 & 3 & \vdots & 9 \\ 0 & 1 & 3 & \vdots & 5 \\ 2 & -5 & 5 & \vdots & 17 \end{bmatrix}$$

Add -2 times the first equation to the third equation.

Add -2 times the first row to the third row.

$$\begin{cases} x - 2y + 3z = 9 \\ y + 3z = 5 \\ -y - z = -1 \end{cases}$$

$$-2R_1 + R_3 \rightarrow \begin{bmatrix} 1 & -2 & 3 & \vdots & 9 \\ 0 & 1 & 3 & \vdots & 5 \\ 0 & -1 & -1 & \vdots & -1 \end{bmatrix}$$

Add the second equation to the third equation.

Add the second row to the third row.

$$\begin{cases} x - 2y + 3z = 9 \\ y + 3z = 5 \\ 2z = 4 \end{cases}$$

$$R_2 + R_3 \rightarrow \begin{bmatrix} 1 & -2 & 3 & \vdots & 9 \\ 0 & 1 & 3 & \vdots & 5 \\ 0 & 0 & 2 & \vdots & 4 \end{bmatrix}$$

Multiply the third equation by $\frac{1}{2}$.

Multiply the third row by $\frac{1}{2}$.

$$\begin{cases} x - 2y + 3z = 9 \\ y + 3z = 5 \\ z = 2 \end{cases}$$

$$\tfrac{1}{2}R_3 \rightarrow \begin{bmatrix} 1 & -2 & 3 & \vdots & 9 \\ 0 & 1 & 3 & \vdots & 5 \\ 0 & 0 & 1 & \vdots & 2 \end{bmatrix}$$

At this point, you can use back-substitution to find x and y.

$$y + 3(2) = 5 \qquad \text{Substitute 2 for } z.$$

$$y = -1 \qquad \text{Solve for } y.$$

$$x - 2(-1) + 3(2) = 9 \qquad \text{Substitute } -1 \text{ for } y \text{ and 2 for } z.$$

$$x = 1 \qquad \text{Solve for } x.$$

The solution is $x = 1$, $y = -1$, and $z = 2$.

Remember that you can check a solution by substituting the values of x, y, and z into each equation in the original system.

The last matrix in Example 4 is said to be in **row-echelon form.** The term *echelon* refers to the stair-step pattern formed by the nonzero elements of the matrix. To be in this form, a matrix must have the following properties.

Row-Echelon Form and Reduced Row-Echelon Form

A matrix in **row-echelon form** has the following properties.

1. All rows consisting entirely of zeros occur at the bottom of the matrix.

2. For each row that does not consist entirely of zeros, the first nonzero entry is 1 (called a leading 1).

3. For two successive (nonzero) rows, the leading 1 in the higher row is farther to the left than the leading 1 in the lower row.

A matrix in *row-echelon form* is in **reduced row-echelon form** if every column that has a leading 1 has zeros in every position above and below its leading 1.

Example 5 ▶ Row-Echelon Form

The following matrices are in row-echelon form.

a. $\begin{bmatrix} 1 & 2 & -1 & 4 \\ 0 & 1 & 0 & 3 \\ 0 & 0 & 1 & -2 \end{bmatrix}$
b. $\begin{bmatrix} 0 & 1 & 0 & 5 \\ 0 & 0 & 1 & 3 \\ 0 & 0 & 0 & 0 \end{bmatrix}$

c. $\begin{bmatrix} 1 & -5 & 2 & -1 & 3 \\ 0 & 0 & 1 & 3 & -2 \\ 0 & 0 & 0 & 1 & 4 \\ 0 & 0 & 0 & 0 & 1 \end{bmatrix}$
d. $\begin{bmatrix} 1 & 0 & 0 & -1 \\ 0 & 1 & 0 & 2 \\ 0 & 0 & 1 & 3 \\ 0 & 0 & 0 & 0 \end{bmatrix}$

The matrices in (b) and (d) also happen to be in *reduced* row-echelon form. The following matrices are not in row-echelon form.

e. $\begin{bmatrix} 1 & 2 & -3 & 4 \\ 0 & 2 & 1 & -1 \\ 0 & 0 & 1 & -3 \end{bmatrix}$ First nonzero row entry in Row 2 is not 1.

f. $\begin{bmatrix} 1 & 2 & -1 & 2 \\ 0 & 0 & 0 & 0 \\ 0 & 1 & 2 & -4 \end{bmatrix}$ Row consisting entirely of zeros is not at the bottom of the matrix.

Every matrix is row-equivalent to a matrix in row-echelon form. For instance, in Example 5, you can change the matrix in part (e) to row-echelon form by multiplying its second row by $\frac{1}{2}$.

$$\frac{1}{2}R_2 \rightarrow \begin{bmatrix} 1 & 2 & -3 & 4 \\ 0 & 1 & \frac{1}{2} & -\frac{1}{2} \\ 0 & 0 & 1 & -3 \end{bmatrix}$$

What elementary row operation could you perform on the matrix in part (f) so that it would be in row-echelon form?

Gaussian Elimination with Back-Substitution

Gaussian elimination with back-substitution works well for solving systems of linear equations by hand or with a computer. For this algorithm, the order in which the elementary row operations are performed is important. We suggest operating from left to right by columns, using elementary row operations to obtain zeros in all entries directly below the leading 1's.

A computer animation of this example appears in the *Interactive* CD-ROM and *Internet* versions of this text.

Example 6 ▶ **Gaussian Elimination with Back-Substitution**

Solve the system
$$\begin{cases} y + z - 2w = -3 \\ x + 2y - z = 2 \\ 2x + 4y + z - 3w = -2 \\ x - 4y - 7z - w = -19 \end{cases}.$$

Solution

$$\begin{array}{c} \overset{R_2}{\underset{R_1}{\rightarrow}} \end{array} \left[\begin{array}{cccc:c} 1 & 2 & -1 & 0 & 2 \\ 0 & 1 & 1 & -2 & -3 \\ 2 & 4 & 1 & -3 & -2 \\ 1 & -4 & -7 & -1 & -19 \end{array}\right]$$

Interchange R_1 and R_2 so first column has leading 1 in upper left corner.

$$\begin{array}{c} \\ \\ -2R_1 + R_3 \rightarrow \\ -R_1 + R_4 \rightarrow \end{array} \left[\begin{array}{cccc:c} 1 & 2 & -1 & 0 & 2 \\ 0 & 1 & 1 & -2 & -3 \\ 0 & 0 & 3 & -3 & -6 \\ 0 & -6 & -6 & -1 & -21 \end{array}\right]$$

Perform operations on R_3 and R_4 so first column has zeros below its leading 1.

$$\begin{array}{c} \\ \\ \\ 6R_2 + R_4 \rightarrow \end{array} \left[\begin{array}{cccc:c} 1 & 2 & -1 & 0 & 2 \\ 0 & 1 & 1 & -2 & -3 \\ 0 & 0 & 3 & -3 & -6 \\ 0 & 0 & 0 & -13 & -39 \end{array}\right]$$

Perform operations on R_4 so second column has zeros below its leading 1.

$$\begin{array}{c} \\ \\ \tfrac{1}{3}R_3 \rightarrow \\ \\ \end{array} \left[\begin{array}{cccc:c} 1 & 2 & -1 & 0 & 2 \\ 0 & 1 & 1 & -2 & -3 \\ 0 & 0 & 1 & -1 & -2 \\ 0 & 0 & 0 & -13 & -39 \end{array}\right]$$

Perform operations on R_3 so third column has a leading 1.

$$\begin{array}{c} \\ \\ \\ -\tfrac{1}{13}R_4 \rightarrow \end{array} \left[\begin{array}{cccc:c} 1 & 2 & -1 & 0 & 2 \\ 0 & 1 & 1 & -2 & -3 \\ 0 & 0 & 1 & -1 & -2 \\ 0 & 0 & 0 & 1 & 3 \end{array}\right]$$

Perform operations on R_4 so fourth column has a leading 1.

The matrix is now in row-echelon form, and the corresponding system is

$$\begin{cases} x + 2y - z = 2 \\ y + z - 2w = -3 \\ z - w = -2 \\ w = 3. \end{cases}$$

Using back-substitution, the solution is $x = -1$, $y = 2$, $z = 1$, and $w = 3$.

> ## Gaussian Elimination with Back-Substitution
>
> 1. Write the augmented matrix of the system of linear equations.
>
> 2. Use elementary row operations to rewrite the augmented matrix in row-echelon form.
>
> 3. Write the system of linear equations corresponding to the matrix in row-echelon form, and use back-substitution to find the solution.

When solving a system of linear equations, remember that it is possible for the system to have no solution. If, in the elimination process, you obtain a row with zeros except for the last entry, it is unnecessary to continue the elimination process. You can simply conclude that the system has no solution, or is *inconsistent*.

Example 7 ▶ A System with No Solution

Solve the system.

$$\begin{cases} x - y + 2z = 4 \\ x \quad\quad + z = 6 \\ 2x - 3y + 5z = 4 \\ 3x + 2y - z = 1 \end{cases}$$

Solution

$$\begin{bmatrix} 1 & -1 & 2 & \vdots & 4 \\ 1 & 0 & 1 & \vdots & 6 \\ 2 & -3 & 5 & \vdots & 4 \\ 3 & 2 & -1 & \vdots & 1 \end{bmatrix}$$

Write augmented matrix.

$$\begin{matrix} \\ -R_1 + R_2 \to \\ -2R_1 + R_3 \to \\ -3R_1 + R_4 \to \end{matrix} \begin{bmatrix} 1 & -1 & 2 & \vdots & 4 \\ 0 & 1 & -1 & \vdots & 2 \\ 0 & -1 & 1 & \vdots & -4 \\ 0 & 5 & -7 & \vdots & -11 \end{bmatrix}$$

Perform row operations.

$$\begin{matrix} \\ \\ R_2 + R_3 \to \\ \end{matrix} \begin{bmatrix} 1 & -1 & 2 & \vdots & 4 \\ 0 & 1 & -1 & \vdots & 2 \\ 0 & 0 & 0 & \vdots & -2 \\ 0 & 5 & -7 & \vdots & -11 \end{bmatrix}$$

Perform row operations.

Note that the third row of this matrix consists of zeros except for the last entry. This means that the original system of linear equations is inconsistent. You can see why this is true by converting back to a system of linear equations.

$$\begin{cases} x - y + 2z = 4 \\ y - z = 2 \\ 0 = -2 \\ 5y - 7z = -11 \end{cases}$$

Because the third equation is not possible, the system has no solution.

Gauss-Jordan Elimination

With Gaussian elimination, elementary row operations are applied to a matrix to obtain a (row-equivalent) row-echelon form of the matrix. A second method of elimination, called **Gauss-Jordan elimination,** after Carl Friedrich Gauss and Wilhelm Jordan (1842–1899), continues the reduction process until a *reduced* row-echelon form is obtained. This procedure is demonstrated in Example 8.

Example 8 ▶ Gauss-Jordan Elimination

Technology

For a demonstration of a graphical approach to Gauss-Jordan elimination on a 2×3 matrix, see the graphing calculator program available for several models of graphing calculators at our website *college.hmco.com.*

Use Gauss-Jordan elimination to solve the system $\begin{cases} x - 2y + 3z = 9 \\ -x + 3y = -4. \\ 2x - 5y + 5z = 17 \end{cases}$

Solution

In Example 4, Gaussian elimination was used to obtain the row-echelon form of the linear system above.

$$\begin{bmatrix} 1 & -2 & 3 & \vdots & 9 \\ 0 & 1 & 3 & \vdots & 5 \\ 0 & 0 & 1 & \vdots & 2 \end{bmatrix}$$

Now, apply elementary row operations until you obtain zeros above each of the leading 1's, as follows.

$$\begin{matrix} 2R_2 + R_1 \rightarrow \end{matrix} \begin{bmatrix} 1 & 0 & 9 & \vdots & 19 \\ 0 & 1 & 3 & \vdots & 5 \\ 0 & 0 & 1 & \vdots & 2 \end{bmatrix}$$

Perform operations on R_1 so second column has zeros above its leading 1.

$$\begin{matrix} -9R_3 + R_1 \rightarrow \\ -3R_3 + R_2 \rightarrow \end{matrix} \begin{bmatrix} 1 & 0 & 0 & \vdots & 1 \\ 0 & 1 & 0 & \vdots & -1 \\ 0 & 0 & 1 & \vdots & 2 \end{bmatrix}$$

Perform operations on R_1 and R_2 so third column has zeros above its leading 1.

The matrix is now in reduced row-echelon form. Now, converting back to a system of linear equations, you have

$$\begin{cases} x = 1 \\ y = -1. \\ z = 2 \end{cases}$$

Now you can simply read the solution.

The elimination procedures described in this section sometimes result in fractional coefficients. For instance, for the system

$$\begin{cases} 2x - 5y + 5z = 17 \\ 3x - 2y + 3z = 11 \\ -3x + 3y = -6 \end{cases}$$

the procedure would have required multiplication of the first row by $\frac{1}{2}$. You can sometimes avoid fractions by judiciously choosing the order in which you apply elementary row operations.

Example 9 ▶ **A System with an Infinite Number of Solutions**

Solve the system.

$$\begin{cases} 2x + 4y - 2z = 0 \\ 3x + 5y = 1 \end{cases}$$

Solution

$$\begin{bmatrix} 2 & 4 & -2 & \vdots & 0 \\ 3 & 5 & 0 & \vdots & 1 \end{bmatrix}$$

$$\frac{1}{2}R_1 \rightarrow \begin{bmatrix} 1 & 2 & -1 & \vdots & 0 \\ 3 & 5 & 0 & \vdots & 1 \end{bmatrix}$$

$$-3R_1 + R_2 \rightarrow \begin{bmatrix} 1 & 2 & -1 & \vdots & 0 \\ 0 & -1 & 3 & \vdots & 1 \end{bmatrix}$$

$$-R_2 \rightarrow \begin{bmatrix} 1 & 2 & -1 & \vdots & 0 \\ 0 & 1 & -3 & \vdots & -1 \end{bmatrix}$$

$$-2R_2 + R_1 \rightarrow \begin{bmatrix} 1 & 0 & 5 & \vdots & 2 \\ 0 & 1 & -3 & \vdots & -1 \end{bmatrix}$$

The corresponding system of equations is

$$\begin{cases} x + 5z = 2 \\ y - 3z = -1 \end{cases}.$$

Solving for x and y in terms of z, you have

$$x = -5z + 2$$

and

$$y = 3z - 1.$$

To write a solution to the system that does not use any of the three variables of the system, let a represent any real number and let

$$z = a.$$

Now substitute a for z in the equations for x and y.

$$x = -5z + 2 = -5a + 2$$

$$y = 3z - 1 = 3a - 1$$

So, the solution set has the form

$$(-5a + 2, 3a - 1, a)$$

where a is any real number. Try substituting values for a to obtain a few solutions. Then check each solution in the original equation.

It is worth noting that the row-echelon form of a matrix is not unique. That is, two different sequences of elementary row operations may yield different row-echelon forms. This is demonstrated in Example 10.

STUDY T!P

In Example 9, x and y are solved for in terms of the third variable z. To write a solution to the system that does not use any of the three variables of the system, let a represent any real number and let $z = a$. Then solve for x and y. The solution can then be written in terms of a, which is not one of the variables of the system.

Example 10 ▶ Comparing Row-Echelon Forms

Compare the following row-echelon form with the one found in Example 4. Is it the same? Does it yield the same solution?

$$\begin{cases} x - 2y + 3z = 9 \\ -x + 3y = -4 \\ 2x - 5y + 5z = 17 \end{cases}$$

$$\begin{bmatrix} 1 & -2 & 3 & \vdots & 9 \\ -1 & 3 & 0 & \vdots & -4 \\ 2 & -5 & 5 & \vdots & 17 \end{bmatrix}$$

$$\begin{matrix} R_2 \\ R_1 \end{matrix} \begin{bmatrix} -1 & 3 & 0 & \vdots & -4 \\ 1 & -2 & 3 & \vdots & 9 \\ 2 & -5 & 5 & \vdots & 17 \end{bmatrix}$$

$$-R_1 \rightarrow \begin{bmatrix} 1 & -3 & 0 & \vdots & 4 \\ 1 & -2 & 3 & \vdots & 9 \\ 2 & -5 & 5 & \vdots & 17 \end{bmatrix}$$

$$\begin{matrix} \\ -R_1 + R_2 \rightarrow \\ -2R_1 + R_3 \rightarrow \end{matrix} \begin{bmatrix} 1 & -3 & 0 & \vdots & 4 \\ 0 & 1 & 3 & \vdots & 5 \\ 0 & 1 & 5 & \vdots & 9 \end{bmatrix}$$

$$-R_2 + R_3 \rightarrow \begin{bmatrix} 1 & -3 & 0 & \vdots & 4 \\ 0 & 1 & 3 & \vdots & 5 \\ 0 & 0 & 2 & \vdots & 4 \end{bmatrix}$$

$$\tfrac{1}{2}R_3 \rightarrow \begin{bmatrix} 1 & -3 & 0 & \vdots & 4 \\ 0 & 1 & 3 & \vdots & 5 \\ 0 & 0 & 1 & \vdots & 2 \end{bmatrix}$$

Solution

This row-echelon form is different from that obtained in Example 4. The corresponding system of linear equations for this row-echelon matrix is

$$\begin{cases} x - 3y = 4 \\ y + 3z = 5. \\ z = 2 \end{cases}$$

Using back-substitution on this system, you obtain the solution $x = 1$, $y = -1$, and $z = 2$, which is the same solution that was obtained in Example 4.

You have seen that the row-echelon form of a given matrix *is not* unique; however, the *reduced* row-echelon form of a given matrix *is* unique. Try applying Gauss-Jordan elimination to the row-echelon matrix in Example 10 to see that you obtain the same reduced row-echelon form as in Example 8.

7.1 Exercises

In Exercises 1–6, determine the order of the matrix.

1. $\begin{bmatrix} 7 & 0 \end{bmatrix}$

2. $\begin{bmatrix} 5 & -3 & 8 & 7 \end{bmatrix}$

3. $\begin{bmatrix} 2 \\ 36 \\ 3 \end{bmatrix}$

4. $\begin{bmatrix} -3 & 7 & 15 & 0 \\ 0 & 0 & 3 & 3 \\ 1 & 1 & 6 & 7 \end{bmatrix}$

5. $\begin{bmatrix} 33 & 45 \\ -9 & 20 \end{bmatrix}$

6. $\begin{bmatrix} -7 & 6 & 4 \\ 0 & -5 & 1 \end{bmatrix}$

In Exercises 7–12, write the augmented matrix for the system of linear equations.

7. $\begin{cases} 4x - 3y = -5 \\ -x + 3y = 12 \end{cases}$

8. $\begin{cases} 7x + 4y = 22 \\ 5x - 9y = 15 \end{cases}$

9. $\begin{cases} x + 10y - 2z = 2 \\ 5x - 3y + 4z = 0 \\ 2x + y = 6 \end{cases}$

10. $\begin{cases} -x - 8y + 5z = 8 \\ -7x - 15z = -38 \\ 3x - y + 8z = 20 \end{cases}$

11. $\begin{cases} 7x - 5y + z = 13 \\ 19x - 8z = 10 \end{cases}$

12. $\begin{cases} 9x + 2y - 3z = 20 \\ -25y + 11z = -5 \end{cases}$

In Exercises 13–18, write the system of linear equations represented by the augmented matrix. (Use variables x, y, z, and w.)

13. $\begin{bmatrix} 1 & 2 & \vdots & 7 \\ 2 & -3 & \vdots & 4 \end{bmatrix}$

14. $\begin{bmatrix} 7 & -5 & \vdots & 0 \\ 8 & 3 & \vdots & -2 \end{bmatrix}$

15. $\begin{bmatrix} 2 & 0 & 5 & \vdots & -12 \\ 0 & 1 & -2 & \vdots & 7 \\ 6 & 3 & 0 & \vdots & 2 \end{bmatrix}$

16. $\begin{bmatrix} 4 & -5 & -1 & \vdots & 18 \\ -11 & 0 & 6 & \vdots & 25 \\ 3 & 8 & 0 & \vdots & -29 \end{bmatrix}$

17. $\begin{bmatrix} 9 & 12 & 3 & 0 & \vdots & 0 \\ -2 & 18 & 5 & 2 & \vdots & 10 \\ 1 & 7 & -8 & 0 & \vdots & -4 \\ 3 & 0 & 2 & 0 & \vdots & -10 \end{bmatrix}$

18. $\begin{bmatrix} 6 & 2 & -1 & -5 & \vdots & -25 \\ -1 & 0 & 7 & 3 & \vdots & 7 \\ 4 & -1 & -10 & 6 & \vdots & 23 \\ 0 & 8 & 1 & -11 & \vdots & -21 \end{bmatrix}$

In Exercises 19–22, determine whether the matrix is in row-echelon form. If it is, determine if it is also in reduced row-echelon form.

19. $\begin{bmatrix} 1 & 0 & 0 & 0 \\ 0 & 1 & 1 & 5 \\ 0 & 0 & 0 & 0 \end{bmatrix}$

20. $\begin{bmatrix} 1 & 3 & 0 & 0 \\ 0 & 0 & 1 & 8 \\ 0 & 0 & 0 & 0 \end{bmatrix}$

21. $\begin{bmatrix} 2 & 0 & 4 & 0 \\ 0 & -1 & 3 & 6 \\ 0 & 0 & 1 & 5 \end{bmatrix}$

22. $\begin{bmatrix} 1 & 0 & 2 & 1 \\ 0 & 1 & -3 & 10 \\ 0 & 0 & 1 & 0 \end{bmatrix}$

In Exercises 23–26, fill in the blank(s) using elementary row operations to form a row-equivalent matrix.

23. $\begin{bmatrix} 1 & 4 & 3 \\ 2 & 10 & 5 \end{bmatrix}$

$\begin{bmatrix} 1 & 4 & 3 \\ 0 & \blacksquare & -1 \end{bmatrix}$

24. $\begin{bmatrix} 3 & 6 & 8 \\ 4 & -3 & 6 \end{bmatrix}$

$\begin{bmatrix} 1 & \blacksquare & \frac{8}{3} \\ 4 & -3 & 6 \end{bmatrix}$

25. $\begin{bmatrix} 1 & 1 & 4 & -1 \\ 3 & 8 & 10 & 3 \\ -2 & 1 & 12 & 6 \end{bmatrix}$

$\begin{bmatrix} 1 & 1 & 4 & -1 \\ 0 & 5 & \blacksquare & \blacksquare \\ 0 & 3 & \blacksquare & \blacksquare \end{bmatrix}$

$\begin{bmatrix} 1 & 1 & 4 & -1 \\ 0 & 1 & -\frac{2}{5} & \frac{6}{5} \\ 0 & 3 & \blacksquare & \blacksquare \end{bmatrix}$

26. $\begin{bmatrix} 2 & 4 & 8 & 3 \\ 1 & -1 & -3 & 2 \\ 2 & 6 & 4 & 9 \end{bmatrix}$

$\begin{bmatrix} 1 & \blacksquare & \blacksquare & \blacksquare \\ 1 & -1 & -3 & 2 \\ 2 & 6 & 4 & 9 \end{bmatrix}$

$\begin{bmatrix} 1 & 2 & 4 & \frac{3}{2} \\ 0 & \blacksquare & -7 & \frac{1}{2} \\ 0 & 2 & \blacksquare & \blacksquare \end{bmatrix}$

In Exercises 27-30, identify the elementary row operation(s) being performed to obtain the new row-equivalent matrix.

Original Matrix	New Row-Equivalent Matrix

27. $\begin{bmatrix} -2 & 5 & 1 \\ 3 & -1 & -8 \end{bmatrix}$ $\qquad$ $\begin{bmatrix} 13 & 0 & -39 \\ 3 & -1 & -8 \end{bmatrix}$

Original Matrix	New Row-Equivalent Matrix

28. $\begin{bmatrix} 3 & -1 & -4 \\ -4 & 3 & 7 \end{bmatrix}$ $\qquad$ $\begin{bmatrix} 3 & -1 & -4 \\ 5 & 0 & -5 \end{bmatrix}$

Original Matrix	New Row-Equivalent Matrix

29. $\begin{bmatrix} 0 & -1 & -5 & 5 \\ -1 & 3 & -7 & 6 \\ 4 & -5 & 1 & 3 \end{bmatrix}$ $\begin{bmatrix} -1 & 3 & -7 & 6 \\ 0 & -1 & -5 & 5 \\ 0 & 7 & -27 & 27 \end{bmatrix}$

Original Matrix	New Row-Equivalent Matrix

30. $\begin{bmatrix} -1 & -2 & 3 & -2 \\ 2 & -5 & 1 & -7 \\ 5 & 4 & -7 & 6 \end{bmatrix}$ $\begin{bmatrix} -1 & -2 & 3 & -2 \\ 0 & -9 & 7 & -11 \\ 0 & -6 & 8 & -4 \end{bmatrix}$

31. Perform the sequence of row operations on the matrix. What did the operations accomplish?

$$\begin{bmatrix} 1 & 2 & 3 \\ 2 & -1 & -4 \\ 3 & 1 & -1 \end{bmatrix}$$

(a) Add -2 times R_1 to R_2.

(b) Add -3 times R_1 to R_3.

(c) Add -1 times R_2 to R_3.

(d) Multiply R_2 by $-\frac{1}{5}$.

(e) Add -2 times R_2 to R_1.

32. Perform the sequence of row operations on the matrix. What did the operations accomplish?

$$\begin{bmatrix} 7 & 1 \\ 0 & 2 \\ -3 & 4 \\ 4 & 1 \end{bmatrix}$$

(a) Add R_3 to R_4.

(b) Interchange R_1 and R_4.

(c) Add 3 times R_1 to R_3.

(d) Add -7 times R_1 to R_4.

(e) Multiply R_2 by $\frac{1}{2}$.

(f) Add the appropriate multiples of R_2 to R_1, R_3, and R_4.

In Exercises 33–36, write the matrix in row-echelon form. Remember that the row-echelon form for a matrix is not unique.

33. $\begin{bmatrix} 1 & 1 & 0 & 5 \\ -2 & -1 & 2 & -10 \\ 3 & 6 & 7 & 14 \end{bmatrix}$

34. $\begin{bmatrix} 1 & 2 & -1 & 3 \\ 3 & 7 & -5 & 14 \\ -2 & -1 & -3 & 8 \end{bmatrix}$

35. $\begin{bmatrix} 1 & -1 & -1 & 1 \\ 5 & -4 & 1 & 8 \\ -6 & 8 & 18 & 0 \end{bmatrix}$

36. $\begin{bmatrix} 1 & -3 & 0 & -7 \\ -3 & 10 & 1 & 23 \\ 4 & -10 & 2 & -24 \end{bmatrix}$

In Exercises 37–42, use the matrix capabilities of a graphing utility to write the matrix in *reduced* row-echelon form.

37. $\begin{bmatrix} 3 & 3 & 3 \\ -1 & 0 & -4 \\ 2 & 4 & -2 \end{bmatrix}$

38. $\begin{bmatrix} 1 & 3 & 2 \\ 5 & 15 & 9 \\ 2 & 6 & 10 \end{bmatrix}$

39. $\begin{bmatrix} 1 & 2 & 3 & -5 \\ 1 & 2 & 4 & -9 \\ -2 & -4 & -4 & 3 \\ 4 & 8 & 11 & -14 \end{bmatrix}$

40. $\begin{bmatrix} -2 & 3 & -1 & -2 \\ 4 & -2 & 5 & 8 \\ 1 & 5 & -2 & 0 \\ 3 & 8 & -10 & -30 \end{bmatrix}$

41. $\begin{bmatrix} -3 & 5 & 1 & 12 \\ 1 & -1 & 1 & 4 \end{bmatrix}$

42. $\begin{bmatrix} 5 & 1 & 2 & 4 \\ -1 & 5 & 10 & -32 \end{bmatrix}$

In Exercises 43–46, write the system of linear equations represented by the augmented matrix. Then use back-substitution to solve. (Use variables x, y, and z.)

43. $\begin{bmatrix} 1 & -2 & \vdots & 4 \\ 0 & 1 & \vdots & -3 \end{bmatrix}$

44. $\begin{bmatrix} 1 & 5 & \vdots & 0 \\ 0 & 1 & \vdots & -1 \end{bmatrix}$

45. $\begin{bmatrix} 1 & -1 & 2 & \vdots & 4 \\ 0 & 1 & -1 & \vdots & 2 \\ 0 & 0 & 1 & \vdots & -2 \end{bmatrix}$

46. $\begin{bmatrix} 1 & 2 & -2 & \vdots & -1 \\ 0 & 1 & 1 & \vdots & 9 \\ 0 & 0 & 1 & \vdots & -3 \end{bmatrix}$

In Exercises 47–50, an augmented matrix that represents a system of linear equations (in variables x, y, and z) has been reduced using Gauss-Jordan elimination. Write the solution represented by the augmented matrix.

47. $\begin{bmatrix} 1 & 0 & \vdots & 3 \\ 0 & 1 & \vdots & -4 \end{bmatrix}$

48. $\begin{bmatrix} 1 & 0 & \vdots & -6 \\ 0 & 1 & \vdots & 10 \end{bmatrix}$

49. $\begin{bmatrix} 1 & 0 & 0 & \vdots & -4 \\ 0 & 1 & 0 & \vdots & -10 \\ 0 & 0 & 1 & \vdots & 4 \end{bmatrix}$

50. $\begin{bmatrix} 1 & 0 & 0 & \vdots & 5 \\ 0 & 1 & 0 & \vdots & -3 \\ 0 & 0 & 1 & \vdots & 0 \end{bmatrix}$

In Exercises 51–70, solve the system of equations if possible. Use Gaussian elimination with back-substitution or Gauss-Jordan elimination.

51. $\begin{cases} x + 2y = 7 \\ 2x + y = 8 \end{cases}$

52. $\begin{cases} 2x + 6y = 16 \\ 2x + 3y = 7 \end{cases}$

53. $\begin{cases} 3x - 2y = -27 \\ x + 3y = 13 \end{cases}$

54. $\begin{cases} -x + y = 4 \\ 2x - 4y = -34 \end{cases}$

55. $\begin{cases} -2x + 6y = -22 \\ x + 2y = -9 \end{cases}$

56. $\begin{cases} 5x - 5y = -5 \\ -2x - 3y = 7 \end{cases}$

57. $\begin{cases} -x + 2y = 1.5 \\ 2x - 4y = 3 \end{cases}$

58. $\begin{cases} x - 3y = 5 \\ -2x + 6y = -10 \end{cases}$

59. $\begin{cases} x - 3z = -2 \\ 3x + y - 2z = 5 \\ 2x + 2y + z = 4 \end{cases}$

60. $\begin{cases} 2x - y + 3z = 24 \\ 2y - z = 14 \\ 7x - 5y = 6 \end{cases}$

61. $\begin{cases} -x + y - z = -14 \\ 2x - y + z = 21 \\ 3x + 2y + z = 19 \end{cases}$

62. $\begin{cases} 2x + 2y - z = 2 \\ x - 3y + z = -28 \\ -x + y = 14 \end{cases}$

63. $\begin{cases} x + 2y - 3z = -28 \\ 4y + 2z = 0 \\ -x + y - z = -5 \end{cases}$

64. $\begin{cases} 3x - 2y + z = 15 \\ -x + y + 2z = -10 \\ x - y - 4z = 14 \end{cases}$

65. $\begin{cases} x + y - 5z = 3 \\ x - 2z = 1 \\ 2x - y - z = 0 \end{cases}$

66. $\begin{cases} 2x + 3z = 3 \\ 4x - 3y + 7z = 5 \\ 8x - 9y + 15z = 9 \end{cases}$

67. $\begin{cases} x + 2y + z + 2w = 8 \\ 3x + 7y + 6z + 9w = 26 \end{cases}$

68. $\begin{cases} 4x + 12y - 7z - 20w = 22 \\ 3x + 9y - 5z - 28w = 30 \end{cases}$

69. $\begin{cases} -x + y = -22 \\ 3x + 4y = 4 \\ 4x - 8y = 32 \end{cases}$

70. $\begin{cases} x + 2y = 0 \\ x + y = 6 \\ 3x - 2y = 8 \end{cases}$

In Exercises 71–76, use the matrix capabilities of a graphing utility to reduce the augmented matrix corresponding to the system of equations, and solve the system.

71. $\begin{cases} 3x + 3y + 12z = 6 \\ x + y + 4z = 2 \\ 2x + 5y + 20z = 10 \\ -x + 2y + 8z = 4 \end{cases}$

72. $\begin{cases} 2x + 10y + 2z = 6 \\ x + 5y + 2z = 6 \\ x + 5y + z = 3 \\ -3x - 15y - 3z = -9 \end{cases}$

73. $\begin{cases} 2x + y - z + 2w = -6 \\ 3x + 4y + w = 1 \\ x + 5y + 2z + 6w = -3 \\ 5x + 2y - z - w = 3 \end{cases}$

74. $\begin{cases} x + 2y + 2z + 4w = 11 \\ 3x + 6y + 5z + 12w = 30 \\ x + 3y - 3z + 2w = -5 \\ 6x - y - z + w = -9 \end{cases}$

75. $\begin{cases} x + y + z + w = 0 \\ 2x + 3y + z - 2w = 0 \\ 3x + 5y + z = 0 \end{cases}$

76. $\begin{cases} x + 2y + z + 3w = 0 \\ x - y + w = 0 \\ y - z + 2w = 0 \end{cases}$

In Exercises 77–80, determine whether the two systems of linear equations yield the same solutions. If so, find the solutions.

77. (a) $\begin{cases} x - 2y + z = -6 \\ y - 5z = 16 \\ z = -3 \end{cases}$ (b) $\begin{cases} x + y - 2z = 6 \\ y + 3z = -8 \\ z = -3 \end{cases}$

78. (a) $\begin{cases} x - 3y + 4z = -11 \\ y - z = -4 \\ z = 2 \end{cases}$

(b) $\begin{cases} x + 4y = -11 \\ y + 3z = 4 \\ z = 2 \end{cases}$

79. (a) $\begin{cases} x - 4y + 5z = 27 \\ y - 7z = -54 \\ z = 8 \end{cases}$

(b) $\begin{cases} x - 6y + z = 15 \\ y + 5z = 42 \\ z = 8 \end{cases}$

80. (a) $\begin{cases} x + 3y - z = 19 \\ y + 6z = -18 \\ z = -4 \end{cases}$

(b) $\begin{cases} x - y + 3z = -15 \\ y - 2z = 14 \\ z = -4 \end{cases}$

81. Use the system $\begin{cases} x + 3y + z = 3 \\ x + 5y + 5z = 1 \\ 2x + 6y + 3z = 8 \end{cases}$ to write two different matrices in row-echelon form that yield the same solutions.

82. *Electrical Network* The currents in an electrical network are given by the solution of the system

$$\begin{cases} I_1 - I_2 + I_3 = 0 \\ 3I_1 + 4I_2 = 18 \\ I_2 + 3I_3 = 6 \end{cases}$$

where I_1, I_2, and I_3 are measured in amperes. Solve the system of equations.

83. *Partial Fractions* Write the partial fraction decomposition for the rational expression

$$\frac{4x^2}{(x + 1)^2(x - 1)} = \frac{A}{x - 1} + \frac{B}{x + 1} + \frac{C}{(x + 1)^2}.$$

84. *Finance* A small corporation borrowed $1,500,000 to expand its product line. Some of the money was borrowed at 7%, some at 8%, and some at 10%. How much was borrowed at each rate if the annual interest was $130,500 and the amount borrowed at 10% was 4 times the amount borrowed at 7%?

85. *Finance* A small corporation borrowed $500,000 to expand its product line. Some of the money was borrowed at 9%, some at 10%, and some at 12%. How much was borrowed at each rate if the annual interest was $52,000 and the amount borrowed at 10% was $2\frac{1}{2}$ times the amount borrowed at 9%?

In Exercises 86 and 87, find the specified equation that passes through the points. Use a graphing utility to verify your results.

86. Parabola:

$y = ax^2 + bx + c$

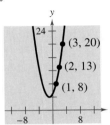

87. Parabola:

$y = ax^2 + bx + c$

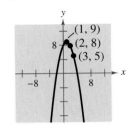

88. *Mathematical Modeling* A videotape of the path of a ball thrown by a baseball player was analyzed with a grid covering the TV screen (see figure). The tape was paused three times, and the position of the ball was measured each time. The coordinates were approximately $(0, 5.0)$, $(15, 9.6)$, and $(30, 12.4)$. (The x-coordinate measures the horizontal distance from the player in feet, and the y-coordinate is the height of the ball in feet.)

(a) Find the equation of the parabola $y = ax^2 + bx + c$ that passes through the three points.

(b) Use a graphing utility to graph the parabola. Approximate the maximum height of the ball and the point at which the ball struck the ground.

(c) Find analytically the maximum height of the ball and the point at which it struck the ground.

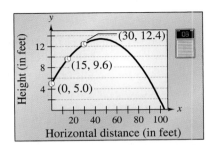

89. *Data Analysis* The bar graph gives the value y, in millions of dollars, for new orders of civil jet transport aircraft built by U.S. companies for the years 1995 through 1997. (Source: Aerospace Industries Association of America)

(a) Find the equation of the parabola that passes through the points. Let $t = 5$ represent 1995.

(b) Use a graphing utility to graph the parabola.

(c) Use the equation in part (a) to estimate y in the year 2000. Is the estimate reasonable? Explain.

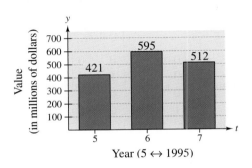

Network Analysis **In Exercises 90 and 91, answer the questions about the specified network. (In a network it is assumed that the total flow into each junction is equal to the total flow out of each junction.)**

90. Water flowing through a network of pipes (in thousands of cubic meters per hour) is shown in the figure.

    (a) Solve this system for the water flow represented by x_i, $i = 1, 2, \ldots, 7$.

    (b) Find the network flow pattern when $x_6 = x_7 = 0$.

    (c) Find the network flow pattern when $x_5 = 1000$ and $x_6 = 0$.

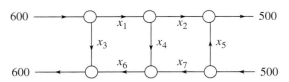

91. The flow of traffic (in vehicles per hour) through a network of streets is shown in the figure.

    (a) Solve this system for the traffic flow represented by x_i, $i = 1, 2, \ldots, 5$.

    (b) Find the traffic flow when $x_2 = 200$ and $x_3 = 50$.

    (c) Find the traffic flow when $x_2 = 150$ and $x_3 = 0$.

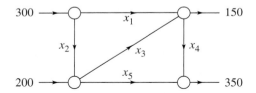

Synthesis

True or False? **In Exercises 92–94, determine whether the statement is true or false. Justify your answer.**

92. $\begin{bmatrix} 5 & 0 & -2 & 7 \\ -1 & 3 & -6 & 0 \end{bmatrix}$ is a 4×2 matrix.

93. The matrix $\begin{bmatrix} 0 & 0 & 0 & 0 \\ 0 & 0 & 1 & -4 \\ 0 & 1 & 0 & 2 \\ 1 & 0 & 0 & 5 \end{bmatrix}$ is in reduced row-echelon form.

94. Gaussian elimination reduces a matrix until a reduced row-echelon form is obtained.

95. *Think About It* The augmented matrix represents a system of linear equations (in variables x, y, and z) that has been reduced using Gauss-Jordan elimination. Write a system of equations with nonzero coefficients that is represented by the reduced matrix. (The answer is not unique.)

$$\begin{bmatrix} 1 & 0 & 3 & \vdots & -2 \\ 0 & 1 & 4 & \vdots & 1 \\ 0 & 0 & 0 & \vdots & 0 \end{bmatrix}$$

96. *Think About It*
    (a) Describe the row-echelon form of an augmented matrix that corresponds to a system of linear equations that is inconsistent.

    (b) Describe the row-echelon form of an augmented matrix that corresponds to a system of linear equations that has an infinite number of solutions.

97. Describe the three elementary row operations that can be performed on an augmented matrix.

98. What is the relationship between the three elementary row operations performed on an augmented matrix and the operations that lead to equivalent systems of equations?

99. *Writing* In your own words, describe the difference between a matrix in row-echelon form and a matrix in reduced row-echelon form.

Review

In Exercises 100 and 101, sketch the graph of the function. Identify any asymptotes.

100. $f(x) = \dfrac{7}{-x - 1}$

101. $f(x) = \dfrac{4x}{5x^2 + 2}$

In Exercises 102–107, identify the equation as a(n) line, circle, parabola, ellipse, or hyperbola.

102. $x^2 + y^2 = 100$

103. $y = -x^2 + 8$

104. $5x - 2y = 8$

105. $\dfrac{x^2}{9} - \dfrac{(y - 3)^2}{36} = 1$

106. $\dfrac{(x + 4)^2}{9} + \dfrac{(y - 6)^2}{16} = 1$

107. $(x + 5)^2 + (y - 7)^2 = 15$

In Exercises 108–111, sketch the graph of the function. Do not use a graphing utility.

108. $f(x) = 2^{x-1}$

109. $g(x) = 3^{-x+2}$

110. $h(x) = \ln(x - 1)$

111. $f(x) = 3 + \ln x$

7.2 Operations with Matrices

Jim Brown/The Stock Market

Equality of Matrices

In Section 7.1, you used matrices to solve systems of linear equations. Matrices, however, can do much more than this. There is a rich mathematical theory of matrices, and its applications are numerous. This section and the next two introduce some fundamentals of matrix theory. It is standard mathematical convention to represent matrices in any of the following three ways.

1. A matrix can be denoted by an uppercase letter such as A, B, or C.
2. A matrix can be denoted by a representative element enclosed in brackets, such as $[a_{ij}]$, $[b_{ij}]$, or $[c_{ij}]$.
3. A matrix can be denoted by a rectangular array of numbers such as

$$A = [a_{ij}] = \begin{bmatrix} a_{11} & a_{12} & a_{13} & \cdots & a_{1n} \\ a_{21} & a_{22} & a_{23} & \cdots & a_{2n} \\ a_{31} & a_{32} & a_{33} & \cdots & a_{3n} \\ \vdots & \vdots & \vdots & & \vdots \\ a_{m1} & a_{m2} & a_{m3} & \cdots & a_{mn} \end{bmatrix}.$$

Two matrices $A = [a_{ij}]$ and $B = [b_{ij}]$ are equal if they have the same order ($m \times n$) and $a_{ij} = b_{ij}$ for $1 \leq i \leq m$ and $1 \leq j \leq n$. In other words, two matrices are equal if their corresponding entries are equal.

Example 1 ▶ Equality of Matrices

Solve for a_{11}, a_{12}, a_{21}, and a_{22} in the following matrix equation.

$$\begin{bmatrix} a_{11} & a_{12} \\ a_{21} & a_{22} \end{bmatrix} = \begin{bmatrix} 2 & -1 \\ -3 & 0 \end{bmatrix}$$

Solution

Because two matrices are equal only if their corresponding entries are equal, you can conclude that

$$a_{11} = 2, \quad a_{12} = -1, \quad a_{21} = -3, \quad \text{and} \quad a_{22} = 0.$$

Be sure you see that for two matrices to be equal, they must have the same order *and* their corresponding entries must be equal. For instance,

$$\begin{bmatrix} 2 & -1 \\ \sqrt{4} & \frac{1}{2} \end{bmatrix} = \begin{bmatrix} 2 & -1 \\ 2 & 0.5 \end{bmatrix}$$

but

$$\begin{bmatrix} 2 & -1 \\ 3 & 4 \\ 0 & 0 \end{bmatrix} \neq \begin{bmatrix} 2 & -1 \\ 3 & 4 \end{bmatrix}.$$

The Granger Collection

Matrix Addition and Scalar Multiplication

In this section, three basic matrix operations will be covered. The first two are matrix addition and scalar multiplication. With matrix addition, you can add two matrices (of the same order) by adding their corresponding entries.

Definition of Matrix Addition

If $A = [a_{ij}]$ and $B = [b_{ij}]$ are matrices of order $m \times n$, their sum is the $m \times n$ matrix given by

$$A + B = [a_{ij} + b_{ij}].$$

The sum of two matrices of different orders is undefined.

Example 2 ▶ Addition of Matrices

a.
$$\begin{bmatrix} -1 & 2 \\ 0 & 1 \end{bmatrix} + \begin{bmatrix} 1 & 3 \\ -1 & 2 \end{bmatrix} = \begin{bmatrix} -1+1 & 2+3 \\ 0-1 & 1+2 \end{bmatrix}$$
$$= \begin{bmatrix} 0 & 5 \\ -1 & 3 \end{bmatrix}$$

b.
$$\begin{bmatrix} 0 & 1 & -2 \\ 1 & 2 & 3 \end{bmatrix} + \begin{bmatrix} 0 & 0 & 0 \\ 0 & 0 & 0 \end{bmatrix} = \begin{bmatrix} 0 & 1 & -2 \\ 1 & 2 & 3 \end{bmatrix}$$

c.
$$\begin{bmatrix} 1 \\ -3 \\ -2 \end{bmatrix} + \begin{bmatrix} -1 \\ 3 \\ 2 \end{bmatrix} = \begin{bmatrix} 0 \\ 0 \\ 0 \end{bmatrix}$$

d. The sum of

$$A = \begin{bmatrix} 2 & 1 & 0 \\ 4 & 0 & -1 \\ 3 & -2 & 2 \end{bmatrix} \qquad \text{and}$$

$$B = \begin{bmatrix} 0 & 1 \\ -1 & 3 \\ 2 & 4 \end{bmatrix}$$

is undefined because A and B have different orders.

Historical Note

Arthur Cayley (1821–1895), a British mathematician, invented matrices around 1858. Cayley was a Cambridge University graduate and a lawyer by profession. His ground-breaking work on matrices was begun as he studied the theory of transformations. Cayley also was instrumental in the development of determinants. Cayley and two American mathematicians, Benjamin Peirce (1809–1880) and his son Charles S. Peirce (1839–1914), are credited with developing "matrix algebra."

In operations with matrices, numbers are usually referred to as **scalars.** In this text, scalars will always be real numbers. You can multiply a matrix A by a scalar c by multiplying each entry in A by c.

Definition of Scalar Multiplication

If $A = [a_{ij}]$ is an $m \times n$ matrix and c is a scalar, the **scalar multiple** of A by c is the $m \times n$ matrix given by

$$cA = [ca_{ij}].$$

The symbol $-A$ represents the negation of A, or the scalar product $(-1)A$. Moreover, if A and B are of the same order, then $A - B$ represents the sum of A and $(-1)B$. That is,

$$A - B = A + (-1)B. \qquad \text{Subtraction of matrices}$$

Example 3 ▶ Scalar Multiplication and Matrix Subtraction

For the following matrices, find (a) $3A$, (b) $-B$, and (c) $3A - B$.

$$A = \begin{bmatrix} 2 & 2 & 4 \\ -3 & 0 & -1 \\ 2 & 1 & 2 \end{bmatrix} \quad \text{and} \quad B = \begin{bmatrix} 2 & 0 & 0 \\ 1 & -4 & 3 \\ -1 & 3 & 2 \end{bmatrix}$$

Solution

STUDY T!P

The order of operations for matrix expressions is similar to that for real numbers. In particular, you perform scalar multiplication before matrix addition and subtraction, as shown in Example 3(c).

a. $3A = 3\begin{bmatrix} 2 & 2 & 4 \\ -3 & 0 & -1 \\ 2 & 1 & 2 \end{bmatrix}$ 　　Scalar multiplication

$= \begin{bmatrix} 3(2) & 3(2) & 3(4) \\ 3(-3) & 3(0) & 3(-1) \\ 3(2) & 3(1) & 3(2) \end{bmatrix}$ 　　Multiply each entry by 3.

$= \begin{bmatrix} 6 & 6 & 12 \\ -9 & 0 & -3 \\ 6 & 3 & 6 \end{bmatrix}$ 　　Simplify.

b. $-B = (-1)\begin{bmatrix} 2 & 0 & 0 \\ 1 & -4 & 3 \\ -1 & 3 & 2 \end{bmatrix}$ 　　Definition of negation

$= \begin{bmatrix} -2 & 0 & 0 \\ -1 & 4 & -3 \\ 1 & -3 & -2 \end{bmatrix}$ 　　Multiply each entry by -1.

c. $3A - B = \begin{bmatrix} 6 & 6 & 12 \\ -9 & 0 & -3 \\ 6 & 3 & 6 \end{bmatrix} - \begin{bmatrix} 2 & 0 & 0 \\ 1 & -4 & 3 \\ -1 & 3 & 2 \end{bmatrix}$ 　　Matrix subtraction

$= \begin{bmatrix} 4 & 6 & 12 \\ -10 & 4 & -6 \\ 7 & 0 & 4 \end{bmatrix}$ 　　Subtract corresponding entries.

It is often convenient to rewrite the scalar multiple cA by factoring c out of every entry in the matrix. For instance, in the following example, the scalar $\frac{1}{2}$ has been factored out of the matrix.

$$\begin{bmatrix} \frac{1}{2} & -\frac{3}{2} \\ \frac{5}{2} & \frac{1}{2} \end{bmatrix} = \begin{bmatrix} \frac{1}{2}(1) & \frac{1}{2}(-3) \\ \frac{1}{2}(5) & \frac{1}{2}(1) \end{bmatrix}$$

$$= \frac{1}{2}\begin{bmatrix} 1 & -3 \\ 5 & 1 \end{bmatrix}$$

The properties of matrix addition and scalar multiplication are similar to those of addition and multiplication of real numbers.

Properties of Matrix Addition and Scalar Multiplication

Let A, B, and C be $m \times n$ matrices and let c and d be scalars.

1. $A + B = B + A$	Commutative Property of Matrix Addition
2. $A + (B + C) = (A + B) + C$	Associative Property of Matrix Addition
3. $(cd)A = c(dA)$	Associative Property of Scalar Multiplication
4. $1A = A$	Scalar Identity
5. $c(A + B) = cA + cB$	Distributive Property
6. $(c + d)A = cA + dA$	Distributive Property

Note that the Associative Property of Matrix Addition allows you to write expressions such as $A + B + C$ without ambiguity because the same sum occurs no matter how the matrices are grouped. In other words, you obtain the same sum whether you group $A + B + C$ as $(A + B) + C$ or as $A + (B + C)$. This same reasoning applies to sums of four or more matrices.

Example 4 ▶ Addition of More than Two Matrices

By adding corresponding entries, you obtain the following sum of four matrices.

$$\begin{bmatrix} 1 \\ 2 \\ -3 \end{bmatrix} + \begin{bmatrix} -1 \\ -1 \\ 2 \end{bmatrix} + \begin{bmatrix} 0 \\ 1 \\ 4 \end{bmatrix} + \begin{bmatrix} 2 \\ -3 \\ -2 \end{bmatrix} = \begin{bmatrix} 2 \\ -1 \\ 1 \end{bmatrix}$$

Example 5 ▶ Using the Distributive Property

$$3\left(\begin{bmatrix} -2 & 0 \\ 4 & 1 \end{bmatrix} + \begin{bmatrix} 4 & -2 \\ 3 & 7 \end{bmatrix} \right) = 3\begin{bmatrix} -2 & 0 \\ 4 & 1 \end{bmatrix} + 3\begin{bmatrix} 4 & -2 \\ 3 & 7 \end{bmatrix}$$

$$= \begin{bmatrix} -6 & 0 \\ 12 & 3 \end{bmatrix} + \begin{bmatrix} 12 & -6 \\ 9 & 21 \end{bmatrix}$$

$$= \begin{bmatrix} 6 & -6 \\ 21 & 24 \end{bmatrix}$$

STUDY T!P

In Example 5, you could add the two matrices first and then multiply the matrix by 3. The result would be the same.

Technology

Most graphing utilities can add and subtract matrices and multiply matrices by scalars. Try using a graphing utility to find the sum of the matrices

$$A = \begin{bmatrix} 2 & -3 \\ -1 & 0 \end{bmatrix} \quad \text{and} \quad B = \begin{bmatrix} -1 & 4 \\ 2 & -5 \end{bmatrix}.$$

One important property of addition of real numbers is that the number 0 is the additive identity. That is, $c + 0 = c$ for any real number c. For matrices, a similar property holds. That is, if A is an $m \times n$ matrix and O is the $m \times n$ **zero matrix** consisting entirely of zeros, then $A + O = A$.

In other words, O is the **additive identity** for the set of all $m \times n$ matrices. For example, the following matrices are the additive identities for the set of all 2×3 and 2×2 matrices.

$$O = \begin{bmatrix} 0 & 0 & 0 \\ 0 & 0 & 0 \end{bmatrix} \quad \text{and} \quad O = \begin{bmatrix} 0 & 0 \\ 0 & 0 \end{bmatrix}$$

$$\underbrace{\hspace{3cm}}_{\text{Zero } 2 \times 3 \text{ matrix}} \qquad \underbrace{\hspace{2.5cm}}_{\text{Zero } 2 \times 2 \text{ matrix}}$$

The algebra of real numbers and the algebra of matrices have many similarities. For example, compare the following solutions.

Real Numbers	*m × n Matrices*
(Solve for x.)	*(Solve for X.)*
$x + a = b$	$X + A = B$
$x + a + (-a) = b + (-a)$	$X + A + (-A) = B + (-A)$
$x + 0 = b - a$	$X + O = B - A$
$x = b - a$	$X = B - A$

The algebra of real numbers and the algebra of matrices also have important differences, which will be discussed later.

Example 6 ▶ Solving a Matrix Equation

Solve for X in the equation $3X + A = B$, where

$$A = \begin{bmatrix} 1 & -2 \\ 0 & 3 \end{bmatrix} \quad \text{and} \quad B = \begin{bmatrix} -3 & 4 \\ 2 & 1 \end{bmatrix}.$$

Solution

Begin by solving the equation for X to obtain

$$3X = B - A$$

$$X = \frac{1}{3}(B - A).$$

Now, using the matrices A and B, you have

$$X = \frac{1}{3}\left(\begin{bmatrix} -3 & 4 \\ 2 & 1 \end{bmatrix} - \begin{bmatrix} 1 & -2 \\ 0 & 3 \end{bmatrix} \right) \qquad \text{Substitute the matrices.}$$

$$= \frac{1}{3}\begin{bmatrix} -4 & 6 \\ 2 & -2 \end{bmatrix} \qquad \text{Subtract matrix } B \text{ from matrix } A.$$

$$= \begin{bmatrix} -\frac{4}{3} & 2 \\ \frac{2}{3} & -\frac{2}{3} \end{bmatrix}. \qquad \text{Multiply the matrix by } \frac{1}{3}.$$

Matrix Multiplication

The third basic matrix operation is **matrix multiplication.** At first glance the definition may seem unusual. You will see later, however, that this definition of the product of two matrices has many practical applications.

Definition of Matrix Multiplication

If $A = [a_{ij}]$ is an $m \times n$ matrix and $B = [b_{ij}]$ is an $n \times p$ matrix, the product AB is an $m \times p$ matrix

$$AB = [c_{ij}]$$

where $c_{ij} = a_{i1}b_{1j} + a_{i2}b_{2j} + a_{i3}b_{3j} + \cdots + a_{in}b_{nj}.$

The definition of matrix multiplication indicates a *row-by-column* multiplication, where the entry in the ith row and jth column of the product AB is obtained by multiplying the entries in the ith row of A by the corresponding entries in the jth column of B and then adding the results. Example 7 illustrates this process.

A computer animation of this example appears in the *Interactive* CD-ROM and *Internet* versions of this text.

Example 7 ▶ Finding the Product of Two Matrices

Find the product AB where

$$A = \begin{bmatrix} -1 & 3 \\ 4 & -2 \\ 5 & 0 \end{bmatrix} \quad \text{and} \quad B = \begin{bmatrix} -3 & 2 \\ -4 & 1 \end{bmatrix}.$$

Solution

First, note that the product AB is defined because the number of columns of A is equal to the number of rows of B. Moreover, the product AB has order 3×2, and is of the form

$$\begin{bmatrix} -1 & 3 \\ 4 & -2 \\ 5 & 0 \end{bmatrix} \begin{bmatrix} -3 & 2 \\ -4 & 1 \end{bmatrix} = \begin{bmatrix} c_{11} & c_{12} \\ c_{21} & c_{22} \\ c_{31} & c_{32} \end{bmatrix}.$$

To find the entries of the product, multiply each row of A by each column of B, as follows. Use a graphing utility to check this result.

$$AB = \begin{bmatrix} -1 & 3 \\ 4 & -2 \\ 5 & 0 \end{bmatrix} \begin{bmatrix} -3 & 2 \\ -4 & 1 \end{bmatrix}$$

$$= \begin{bmatrix} (-1)(-3) + (3)(-4) & (-1)(2) + (3)(1) \\ (4)(-3) + (-2)(-4) & (4)(2) + (-2)(1) \\ (5)(-3) + (0)(-4) & (5)(2) + (0)(1) \end{bmatrix}$$

$$= \begin{bmatrix} -9 & 1 \\ -4 & 6 \\ -15 & 10 \end{bmatrix}$$

Technology

Some graphing utilities are able to add, subtract, and multiply matrices. If you have such a graphing utility, enter the matrices

$$A = \begin{bmatrix} 1 & 2 & 3 \\ 2 & -5 & 1 \end{bmatrix} \quad \text{and}$$

$$B = \begin{bmatrix} -3 & 2 & 1 \\ 4 & -2 & 0 \\ 1 & 2 & 3 \end{bmatrix}$$

and find their product AB. You should get:

$$\begin{bmatrix} 8 & 4 & 10 \\ -25 & 16 & 5 \end{bmatrix}.$$

Be sure you understand that for the product of two matrices to be defined, the number of *columns* of the first matrix must equal the number of *rows* of the second matrix. That is, the middle two indices must be the same. The outside two indices give the order of the product, as shown below.

$$\underset{m \times n}{A} \quad \times \quad \underset{n \times p}{B} \quad = \quad \underset{m \times p}{AB}$$

Equal

Order of AB

Example 8 ▶ Patterns in Matrix Multiplication

a. $\underset{2 \times 3}{\begin{bmatrix} 1 & 0 & 3 \\ 2 & -1 & -2 \end{bmatrix}} \underset{3 \times 3}{\begin{bmatrix} -2 & 4 & 2 \\ 1 & 0 & 0 \\ -1 & 1 & -1 \end{bmatrix}} = \underset{2 \times 3}{\begin{bmatrix} -5 & 7 & -1 \\ -3 & 6 & 6 \end{bmatrix}}$

b. $\underset{2 \times 2}{\begin{bmatrix} 3 & 4 \\ -2 & 5 \end{bmatrix}} \underset{2 \times 2}{\begin{bmatrix} 1 & 0 \\ 0 & 1 \end{bmatrix}} = \underset{2 \times 2}{\begin{bmatrix} 3 & 4 \\ -2 & 5 \end{bmatrix}}$

c. $\underset{2 \times 2}{\begin{bmatrix} 1 & 2 \\ 1 & 1 \end{bmatrix}} \underset{2 \times 2}{\begin{bmatrix} -1 & 2 \\ 1 & -1 \end{bmatrix}} = \underset{2 \times 2}{\begin{bmatrix} 1 & 0 \\ 0 & 1 \end{bmatrix}}$

d. $\underset{3 \times 3}{\begin{bmatrix} 6 & 2 & 0 \\ 3 & -1 & 2 \\ 1 & 4 & 6 \end{bmatrix}} \underset{3 \times 1}{\begin{bmatrix} 1 \\ 2 \\ -3 \end{bmatrix}} = \underset{3 \times 1}{\begin{bmatrix} 10 \\ -5 \\ -9 \end{bmatrix}}$

e. $\underset{1 \times 3}{\begin{bmatrix} 1 & -2 & -3 \end{bmatrix}} \underset{3 \times 1}{\begin{bmatrix} 2 \\ -1 \\ 1 \end{bmatrix}} = \underset{1 \times 1}{\begin{bmatrix} 1 \end{bmatrix}}$

f. $\underset{3 \times 1}{\begin{bmatrix} 2 \\ -1 \\ 1 \end{bmatrix}} \underset{1 \times 3}{\begin{bmatrix} 1 & -2 & -3 \end{bmatrix}} = \underset{3 \times 3}{\begin{bmatrix} 2 & -4 & -6 \\ -1 & 2 & 3 \\ 1 & -2 & -3 \end{bmatrix}}$

g. The product AB for the following matrices is not defined.

$$A = \underset{3 \times 2}{\begin{bmatrix} -2 & 1 \\ 1 & -3 \\ 1 & 4 \end{bmatrix}} \quad \text{and} \quad B = \underset{3 \times 4}{\begin{bmatrix} -2 & 3 & 1 & 4 \\ 0 & 1 & -1 & 2 \\ 2 & -1 & 0 & 1 \end{bmatrix}}$$

In parts (e) and (f) of Example 8, note that the two products are different. Matrix multiplication is not, in general, commutative. That is, for most matrices, $AB \neq BA$.

The general pattern for matrix multiplication is as follows. To obtain the entry in the ith row and the jth column of the product AB, use the ith row of A and the jth column of B.

$$\begin{bmatrix} a_{11} & a_{12} & a_{13} & \cdots & a_{1n} \\ a_{21} & a_{22} & a_{23} & \cdots & a_{2n} \\ a_{31} & a_{32} & a_{33} & \cdots & a_{3n} \\ \vdots & \vdots & \vdots & & \vdots \\ a_{i1} & a_{i2} & a_{i3} & \cdots & a_{in} \\ \vdots & \vdots & \vdots & & \vdots \\ a_{m1} & a_{m2} & a_{m3} & \cdots & a_{mn} \end{bmatrix} \begin{bmatrix} b_{11} & b_{12} & \cdots & b_{1j} & \cdots & b_{1p} \\ b_{21} & b_{22} & \cdots & b_{2j} & \cdots & b_{2p} \\ b_{31} & b_{32} & \cdots & b_{3j} & \cdots & b_{3p} \\ \vdots & \vdots & & \vdots & & \vdots \\ b_{n1} & b_{n2} & \cdots & b_{nj} & \cdots & b_{np} \end{bmatrix} = \begin{bmatrix} c_{11} & c_{12} & \cdots & c_{1j} & \cdots & c_{1p} \\ c_{21} & c_{22} & \cdots & c_{2j} & \cdots & c_{2p} \\ \vdots & \vdots & & \vdots & & \vdots \\ c_{i1} & c_{i2} & \cdots & c_{ij} & \cdots & c_{ip} \\ \vdots & \vdots & & \vdots & & \vdots \\ c_{m1} & c_{m2} & \cdots & c_{mj} & \cdots & c_{mp} \end{bmatrix}$$

$$a_{i1}b_{1j} + a_{i2}b_{2j} + a_{i3}b_{3j} + \cdots + a_{in}b_{nj} = c_{ij}$$

Properties of Matrix Multiplication

Let A, B, and C be matrices and let c be a scalar.

1. $A(BC) = (AB)C$ — Associative Property of Multiplication
2. $A(B + C) = AB + AC$ — Distributive Property
3. $(A + B)C = AC + BC$ — Distributive Property
4. $c(AB) = (cA)B = A(cB)$

Definition of Identity Matrix

The $n \times n$ matrix that consists of 1's on its main diagonal and 0's elsewhere is called the **identity matrix of order n** and is denoted by

$$I_n = \begin{bmatrix} 1 & 0 & 0 & \cdots & 0 \\ 0 & 1 & 0 & \cdots & 0 \\ 0 & 0 & 1 & \cdots & 0 \\ \vdots & \vdots & \vdots & & \vdots \\ 0 & 0 & 0 & \cdots & 1 \end{bmatrix}.$$ Identity matrix

Note that an identity matrix must be *square*. When the order is understood to be n, you can denote I_n simply by I.

If A is an $n \times n$ matrix, the identity matrix has the property that $AI_n = A$ and $I_nA = A$. For example,

$$\begin{bmatrix} 3 & -2 & 5 \\ 1 & 0 & 4 \\ -1 & 2 & -3 \end{bmatrix} \begin{bmatrix} 1 & 0 & 0 \\ 0 & 1 & 0 \\ 0 & 0 & 1 \end{bmatrix} = \begin{bmatrix} 3 & -2 & 5 \\ 1 & 0 & 4 \\ -1 & 2 & -3 \end{bmatrix}$$

and

$$\begin{bmatrix} 1 & 0 & 0 \\ 0 & 1 & 0 \\ 0 & 0 & 1 \end{bmatrix} \begin{bmatrix} 3 & -2 & 5 \\ 1 & 0 & 4 \\ -1 & 2 & -3 \end{bmatrix} = \begin{bmatrix} 3 & -2 & 5 \\ 1 & 0 & 4 \\ -1 & 2 & -3 \end{bmatrix}.$$

Applications

Matrix multiplication can be used to represent a system of linear equations. Note how the system

$$\begin{cases} a_{11}x_1 + a_{12}x_2 + a_{13}x_3 = b_1 \\ a_{21}x_1 + a_{22}x_2 + a_{23}x_3 = b_2 \\ a_{31}x_1 + a_{32}x_2 + a_{33}x_3 = b_3 \end{cases}$$

can be written as the matrix equation $AX = B$, where A is the *coefficient matrix* of the system, and X and B are column matrices.

$$\begin{bmatrix} a_{11} & a_{12} & a_{13} \\ a_{21} & a_{22} & a_{23} \\ a_{31} & a_{32} & a_{33} \end{bmatrix} \quad \begin{bmatrix} x_1 \\ x_2 \\ x_3 \end{bmatrix} = \begin{bmatrix} b_1 \\ b_2 \\ b_3 \end{bmatrix}$$
$$\quad A \qquad\qquad \times \quad X \ = \ B$$

 A computer animation of this example appears in the *Interactive* CD-ROM and *Internet* versions of this text.

Example 9 ▶ Solving a System of Linear Equations

Consider the following system of linear equations.

$$\begin{cases} x_1 - 2x_2 + x_3 = -4 \\ x_2 + 2x_3 = 4 \\ 2x_1 + 3x_2 - 2x_3 = 2 \end{cases}$$

a. Write this system as a matrix equation, $AX = B$.

b. Use Gauss-Jordan elimination on the augmented matrix $[A \vdots B]$ to solve for the matrix X.

Solution

a. In matrix form, $AX = B$, the system can be written as follows.

$$\begin{bmatrix} 1 & -2 & 1 \\ 0 & 1 & 2 \\ 2 & 3 & -2 \end{bmatrix} \begin{bmatrix} x_1 \\ x_2 \\ x_3 \end{bmatrix} = \begin{bmatrix} -4 \\ 4 \\ 2 \end{bmatrix}$$

b. The augmented matrix is formed by adjoining matrix B to matrix A.

$$[A \vdots B] = \begin{bmatrix} 1 & -2 & 1 & \vdots & -4 \\ 0 & 1 & 2 & \vdots & 4 \\ 2 & 3 & -2 & \vdots & 2 \end{bmatrix}$$

Using Gauss-Jordan elimination, you can rewrite this equation as

$$[I \vdots X] = \begin{bmatrix} 1 & 0 & 0 & \vdots & -1 \\ 0 & 1 & 0 & \vdots & 2 \\ 0 & 0 & 1 & \vdots & 1 \end{bmatrix}.$$

So, the solution of the system of linear equations is $x_1 = -1$, $x_2 = 2$, and $x_3 = 1$, and the solution of the matrix equation is

$$X = \begin{bmatrix} x_1 \\ x_2 \\ x_3 \end{bmatrix} = \begin{bmatrix} -1 \\ 2 \\ 1 \end{bmatrix}.$$

Check this solution in the original system of equations.

Example 10 ▶ **Softball Team Expenses**

Two softball teams submit equipment lists to their sponsors.

	Women's Team	Men's Team
Bats	12	15
Balls	45	38
Gloves	15	17

Each bat costs $48, each ball costs $4, and each glove costs $42. Use matrices to find the total cost of equipment for each team.

Solution

The equipment lists and the costs per item can be written in matrix form as

$$E = \begin{bmatrix} 12 & 15 \\ 45 & 38 \\ 15 & 17 \end{bmatrix}$$

and

$$C = \begin{bmatrix} 48 & 4 & 42 \end{bmatrix}.$$

The total cost of equipment for each team is given by the product

$$CE = \begin{bmatrix} 48 & 4 & 42 \end{bmatrix} \begin{bmatrix} 12 & 15 \\ 45 & 38 \\ 15 & 17 \end{bmatrix}$$

$$= \begin{bmatrix} 48(12) + 4(45) + 42(15) & 48(15) + 4(38) + 42(17) \end{bmatrix}$$

$$= \begin{bmatrix} 1386 & 1586 \end{bmatrix}.$$

So, the total cost of equipment for the women's team is $1386 and the total cost of equipment for the men's team is $1586.

Writing ABOUT MATHEMATICS

Problem Posing   Write a matrix multiplication application problem that uses the matrix

$$A = \begin{bmatrix} 20 & 42 & 33 \\ 17 & 30 & 50 \end{bmatrix}.$$

Exchange problems with another student in your class. Form the matrices that represent the problem, and solve the problem. Interpret your solution in the context of the problem. Check with the creator of the problem to see if you are correct. Discuss other ways to represent and/or approach the problem.

7.2 Exercises

In Exercises 1–4, find *x* and *y*.

1. $\begin{bmatrix} x & -2 \\ 7 & y \end{bmatrix} = \begin{bmatrix} -4 & -2 \\ 7 & 22 \end{bmatrix}$

2. $\begin{bmatrix} -5 & x \\ y & 8 \end{bmatrix} = \begin{bmatrix} -5 & 13 \\ 12 & 8 \end{bmatrix}$

3. $\begin{bmatrix} 16 & 4 & 5 & 4 \\ -3 & 13 & 15 & 6 \\ 0 & 2 & 4 & 0 \end{bmatrix} = \begin{bmatrix} 16 & 4 & 2x+1 & 4 \\ -3 & 13 & 15 & 3x \\ 0 & 2 & 3y-5 & 0 \end{bmatrix}$

4. $\begin{bmatrix} x+2 & 8 & -3 \\ 1 & 2y & 2x \\ 7 & -2 & y+2 \end{bmatrix} = \begin{bmatrix} 2x+6 & 8 & -3 \\ 1 & 18 & -8 \\ 7 & -2 & 11 \end{bmatrix}$

In Exercises 5–12, if possible, find (a) *A* + *B*, (b) *A* − *B*, (c) 3*A*, and (d) 3*A* − 2*B*.

5. $A = \begin{bmatrix} 1 & -1 \\ 2 & -1 \end{bmatrix}$, $B = \begin{bmatrix} 2 & -1 \\ -1 & 8 \end{bmatrix}$

6. $A = \begin{bmatrix} 1 & 2 \\ 2 & 1 \end{bmatrix}$, $B = \begin{bmatrix} -3 & -2 \\ 4 & 2 \end{bmatrix}$

7. $A = \begin{bmatrix} 6 & -1 \\ 2 & 4 \\ -3 & 5 \end{bmatrix}$, $B = \begin{bmatrix} 1 & 4 \\ -1 & 5 \\ 1 & 10 \end{bmatrix}$

8. $A = \begin{bmatrix} 2 & 1 & 1 \\ -1 & -1 & 4 \end{bmatrix}$, $B = \begin{bmatrix} 2 & -3 & 4 \\ -3 & 1 & -2 \end{bmatrix}$

9. $A = \begin{bmatrix} 2 & 2 & -1 & 0 & 1 \\ 1 & 1 & -2 & 0 & -1 \end{bmatrix}$,

$B = \begin{bmatrix} 1 & 1 & -1 & 1 & 0 \\ -3 & 4 & 9 & -6 & -7 \end{bmatrix}$

10. $A = \begin{bmatrix} -1 & 4 & 0 \\ 3 & -2 & 2 \\ 5 & 4 & -1 \\ 0 & 8 & -6 \\ -4 & -1 & 0 \end{bmatrix}$, $B = \begin{bmatrix} -3 & 5 & 1 \\ 2 & -4 & -7 \\ 10 & -9 & -1 \\ 3 & 2 & -4 \\ 0 & 1 & -2 \end{bmatrix}$

11. $A = \begin{bmatrix} 6 & 0 & 3 \\ -1 & -4 & 0 \end{bmatrix}$, $B = \begin{bmatrix} 8 & -1 \\ 4 & -3 \end{bmatrix}$

12. $A = \begin{bmatrix} 3 \\ 2 \\ -1 \end{bmatrix}$, $B = \begin{bmatrix} -4 & 6 & 2 \end{bmatrix}$

In Exercises 13–18, evaluate the expression.

13. $\begin{bmatrix} -5 & 0 \\ 3 & -6 \end{bmatrix} + \begin{bmatrix} 7 & 1 \\ -2 & -1 \end{bmatrix} + \begin{bmatrix} -10 & -8 \\ 14 & 6 \end{bmatrix}$

14. $\begin{bmatrix} 6 & 8 \\ -1 & 0 \end{bmatrix} + \begin{bmatrix} 0 & 5 \\ -3 & -1 \end{bmatrix} + \begin{bmatrix} -11 & -7 \\ 2 & -1 \end{bmatrix}$

15. $4\left(\begin{bmatrix} -4 & 0 & 1 \\ 0 & 2 & 3 \end{bmatrix} - \begin{bmatrix} 2 & 1 & -2 \\ 3 & -6 & 0 \end{bmatrix} \right)$

16. $\frac{1}{2}([5 \quad -2 \quad 4 \quad 0] + [14 \quad 6 \quad -18 \quad 9])$

17. $-3\left(\begin{bmatrix} 0 & -3 \\ 7 & 2 \end{bmatrix} + \begin{bmatrix} -6 & 3 \\ 8 & 1 \end{bmatrix} \right) - 2\begin{bmatrix} 4 & -4 \\ 7 & -9 \end{bmatrix}$

18. $-1\begin{bmatrix} 4 & 11 \\ -2 & -1 \\ 9 & 3 \end{bmatrix} + \frac{1}{6}\left(\begin{bmatrix} -5 & -1 \\ 3 & 4 \\ 0 & 13 \end{bmatrix} + \begin{bmatrix} 7 & 5 \\ -9 & -1 \\ 6 & -1 \end{bmatrix} \right)$

In Exercises 19–22, use the matrix capabilities of a graphing utility to evaluate each expression. Round your results to three decimal places, if necessary.

19. $\frac{3}{7}\begin{bmatrix} 2 & 5 \\ -1 & -4 \end{bmatrix} + 6\begin{bmatrix} -3 & 0 \\ 2 & 2 \end{bmatrix}$

20. $55\left(\begin{bmatrix} 14 & -11 \\ -22 & 19 \end{bmatrix} + \begin{bmatrix} -22 & 20 \\ 13 & 6 \end{bmatrix} \right)$

21. $-\begin{bmatrix} 3.211 & 6.829 \\ -1.004 & 4.914 \\ 0.055 & -3.889 \end{bmatrix} - \begin{bmatrix} -1.630 & -3.090 \\ 5.256 & 8.335 \\ -9.768 & 4.251 \end{bmatrix}$

22. $-12\left(\begin{bmatrix} 6 & 20 \\ 1 & -9 \\ -2 & 5 \end{bmatrix} + \begin{bmatrix} 14 & -15 \\ -8 & -6 \\ 7 & 0 \end{bmatrix} + \begin{bmatrix} -31 & -19 \\ 16 & 10 \\ 24 & -10 \end{bmatrix} \right)$

In Exercises 23–26, solve for *X* when

$A = \begin{bmatrix} -2 & -1 \\ 1 & 0 \\ 3 & -4 \end{bmatrix}$ and $B = \begin{bmatrix} 0 & 3 \\ 2 & 0 \\ -4 & -1 \end{bmatrix}$.

23. $X = 3A - 2B$

24. $2X = 2A - B$

25. $2X + 3A = B$

26. $2A + 4B = -2X$

In Exercises 27–32, find (a) *AB*, (b) *BA*, and, if possible, (c) *A*². (*Note: A*² = *AA*.)

27. $A = \begin{bmatrix} 1 & 2 \\ 4 & 2 \end{bmatrix}$, $B = \begin{bmatrix} 2 & -1 \\ -1 & 8 \end{bmatrix}$

28. $A = \begin{bmatrix} 2 & -1 \\ 1 & 4 \end{bmatrix}$, $B = \begin{bmatrix} 0 & 0 \\ 3 & -3 \end{bmatrix}$

29. $A = \begin{bmatrix} 3 & -1 \\ 1 & 3 \end{bmatrix}$, $B = \begin{bmatrix} 1 & -3 \\ 3 & 1 \end{bmatrix}$

30. $A = \begin{bmatrix} 1 & -1 \\ 1 & 1 \end{bmatrix}$, $B = \begin{bmatrix} 1 & 3 \\ -3 & 1 \end{bmatrix}$

31. $A = \begin{bmatrix} 7 \\ 8 \\ -1 \end{bmatrix}$, $B = \begin{bmatrix} 1 & 1 & 2 \end{bmatrix}$

32. $A = \begin{bmatrix} 3 & 2 & 1 \end{bmatrix}$, $B = \begin{bmatrix} 2 \\ 3 \\ 0 \end{bmatrix}$

In Exercises 33–40, find AB, if possible.

33. $A = \begin{bmatrix} 2 & 1 \\ -3 & 4 \\ 1 & 6 \end{bmatrix}$, $B = \begin{bmatrix} 0 & -1 & 0 \\ 4 & 0 & 2 \\ 8 & -1 & 7 \end{bmatrix}$

34. $A = \begin{bmatrix} 1 & 0 & 3 & -2 \\ 6 & 13 & 8 & -17 \end{bmatrix}$, $B = \begin{bmatrix} 1 & 6 \\ 4 & 2 \end{bmatrix}$

35. $A = \begin{bmatrix} 0 & -1 & 0 \\ 4 & 0 & 2 \\ 8 & -1 & 7 \end{bmatrix}$, $B = \begin{bmatrix} 2 & 1 \\ -3 & 4 \\ 1 & 6 \end{bmatrix}$

36. $A = \begin{bmatrix} -1 & 3 \\ 4 & -5 \\ 0 & 2 \end{bmatrix}$, $B = \begin{bmatrix} 1 & 2 \\ 0 & 7 \end{bmatrix}$

37. $A = \begin{bmatrix} 1 & 0 & 0 \\ 0 & 4 & 0 \\ 0 & 0 & -2 \end{bmatrix}$, $B = \begin{bmatrix} 3 & 0 & 0 \\ 0 & -1 & 0 \\ 0 & 0 & 5 \end{bmatrix}$

38. $A = \begin{bmatrix} 5 & 0 & 0 \\ 0 & -8 & 0 \\ 0 & 0 & 7 \end{bmatrix}$, $B = \begin{bmatrix} \frac{1}{5} & 0 & 0 \\ 0 & -\frac{1}{8} & 0 \\ 0 & 0 & \frac{1}{2} \end{bmatrix}$

39. $A = \begin{bmatrix} 0 & 0 & 5 \\ 0 & 0 & -3 \\ 0 & 0 & 4 \end{bmatrix}$, $B = \begin{bmatrix} 6 & -11 & 4 \\ 8 & 16 & 4 \\ 0 & 0 & 0 \end{bmatrix}$

40. $A = \begin{bmatrix} 10 \\ 12 \end{bmatrix}$, $B = \begin{bmatrix} 6 & -2 & 1 & 6 \end{bmatrix}$

In Exercises 41–46, use the matrix capabilities of a graphing utility to find AB.

41. $A = \begin{bmatrix} 5 & 6 & -3 \\ -2 & 5 & 1 \\ 10 & -5 & 5 \end{bmatrix}$, $B = \begin{bmatrix} 1 & -1 & 2 \\ 8 & 1 & 4 \\ 4 & -2 & 9 \end{bmatrix}$

42. $A = \begin{bmatrix} 11 & -12 & 4 \\ 14 & 10 & 12 \\ 6 & -2 & 9 \end{bmatrix}$, $B = \begin{bmatrix} 12 & 10 \\ -5 & 12 \\ 15 & 16 \end{bmatrix}$

43. $A = \begin{bmatrix} -3 & 8 & -6 & 8 \\ -12 & 15 & 9 & 6 \\ 5 & -1 & 1 & 5 \end{bmatrix}$, $B = \begin{bmatrix} 3 & 1 & 6 \\ 24 & 15 & 14 \\ 16 & 10 & 21 \\ 8 & -4 & 10 \end{bmatrix}$

44. $A = \begin{bmatrix} -2 & 4 & 8 \\ 21 & 5 & 6 \\ 13 & 2 & 6 \end{bmatrix}$, $B = \begin{bmatrix} 2 & 0 \\ -7 & 15 \\ 32 & 14 \\ 0.5 & 1.6 \end{bmatrix}$

45. $A = \begin{bmatrix} 9 & 10 & -38 & 18 \\ 100 & -50 & 250 & 75 \end{bmatrix}$,

$B = \begin{bmatrix} 52 & -85 & 27 & 45 \\ 40 & -35 & 60 & 82 \end{bmatrix}$

46. $A = \begin{bmatrix} 15 & -18 \\ -4 & 12 \\ -8 & 22 \end{bmatrix}$, $B = \begin{bmatrix} -7 & 22 & 1 \\ 8 & 16 & 24 \end{bmatrix}$

In Exercises 47–50, use the matrix capabilities of a graphing utility to evaluate each expression.

47. $\begin{bmatrix} 3 & 1 \\ 0 & -2 \end{bmatrix}\begin{bmatrix} 1 & 0 \\ -2 & 2 \end{bmatrix}\begin{bmatrix} 1 & 0 \\ 2 & 4 \end{bmatrix}$

48. $-3\left(\begin{bmatrix} 6 & 5 & -1 \\ 1 & -2 & 0 \end{bmatrix}\begin{bmatrix} 0 & 3 \\ -1 & -3 \\ 4 & 1 \end{bmatrix} \right)$

49. $\begin{bmatrix} 0 & 2 & -2 \\ 4 & 1 & 2 \end{bmatrix}\left(\begin{bmatrix} 4 & 0 \\ 0 & -1 \\ -1 & 2 \end{bmatrix} + \begin{bmatrix} -2 & 3 \\ -3 & 5 \\ 0 & -3 \end{bmatrix} \right)$

50. $\begin{bmatrix} 3 \\ -1 \\ 5 \\ 7 \end{bmatrix}\left(\begin{bmatrix} 5 & -6 \end{bmatrix} + \begin{bmatrix} 7 & -1 \end{bmatrix} + \begin{bmatrix} -8 & 9 \end{bmatrix} \right)$

In Exercises 51–58, (a) write each system of linear equations as a matrix equation, $AX = B$, and (b) use Gauss-Jordan elimination on the augmented matrix $[A \vdots B]$ to solve for the matrix X.

51. $\begin{cases} -x_1 + x_2 = 4 \\ -2x_1 + x_2 = 0 \end{cases}$

52. $\begin{cases} 2x_1 + 3x_2 = 5 \\ x_1 + 4x_2 = 10 \end{cases}$

53. $\begin{cases} -2x_1 - 3x_2 = -4 \\ 6x_1 + x_2 = -36 \end{cases}$

54. $\begin{cases} -4x_1 + 9x_2 = -13 \\ x_1 - 3x_2 = 12 \end{cases}$

55. $\begin{cases} x_1 - 2x_2 + 3x_3 = 9 \\ -x_1 + 3x_2 - x_3 = -6 \\ 2x_1 - 5x_2 + 5x_3 = 17 \end{cases}$

56. $\begin{cases} x_1 + x_2 - 3x_3 = 9 \\ -x_1 + 2x_2 = 6 \\ x_1 - x_2 + x_3 = -5 \end{cases}$

57. $\begin{cases} x_1 - 5x_2 + 2x_3 = -20 \\ -3x_1 + x_2 - x_3 = 8 \\ -2x_2 + 5x_3 = -16 \end{cases}$

58. $\begin{cases} x_1 - x_2 + 4x_3 = 17 \\ x_1 + 3x_2 = -11 \\ -6x_2 + 5x_3 = 40 \end{cases}$

Think About It In Exercises 59–68, let matrices A, B, C, and D be of orders 2 × 3, 2 × 3, 3 × 2, and 2 × 2, respectively. Determine whether the matrices are of proper order to perform the operation(s). If so, give the order of the answer.

59. $A + 2C$

60. $B - 3C$

61. AB

62. BC

63. $BC - D$

64. $CB - D$

65. $(CA)D$

66. $(BC)D$

67. $D(A - 3B)$

68. $(BC - D)A$

69. *Manufacturing* A certain corporation has three factories, each of which manufactures two products. The number of units of product i produced at factory j in one day is represented by a_{ij} in the matrix

$$A = \begin{bmatrix} 70 & 50 & 25 \\ 35 & 100 & 70 \end{bmatrix}.$$ Find the production levels if production is increased by 20%. (*Hint*: Because an increase of 20% corresponds to 100% + 20%, multiply the given matrix by 1.2.)

70. *Manufacturing* A certain corporation has four factories, each of which manufactures two products. The number of units of product i produced at factory j in one day is represented by a_{ij} in the matrix

$$A = \begin{bmatrix} 100 & 90 & 70 & 30 \\ 40 & 20 & 60 & 60 \end{bmatrix}.$$ Find the production levels if production is increased by 10%.

71. *Agriculture* A fruit grower raises two crops, which are shipped to three outlets. The number of units of product i that are shipped to outlet j is represented by a_{ij} in the matrix

$$A = \begin{bmatrix} 125 & 100 & 75 \\ 100 & 175 & 125 \end{bmatrix}.$$

The profit per unit is represented by the matrix

$$B = [\$3.50 \quad \$6.00].$$

Find the product BA, and state what each entry of the product represents.

72. *Revenue* A manufacturer produces three models of a product, which are shipped to two warehouses. The number of units of model i that are shipped to warehouse j is represented by a_{ij} in the matrix

$$A = \begin{bmatrix} 5{,}000 & 4{,}000 \\ 6{,}000 & 10{,}000 \\ 8{,}000 & 5{,}000 \end{bmatrix}.$$

The price per unit is represented by the matrix

$$B = [\$20.50 \quad \$26.50 \quad \$29.50].$$

Compute BA and interpret the result.

73. *Business* A company sells five models of computers through three retail outlets. The inventories are represented by S.

$$\begin{array}{c} \text{Model} \\ \begin{array}{ccccc} A & B & C & D & E \end{array} \\ S = \begin{bmatrix} 3 & 2 & 2 & 3 & 0 \\ 0 & 2 & 3 & 4 & 3 \\ 4 & 2 & 1 & 3 & 2 \end{bmatrix} \begin{array}{c} 1 \\ 2 \\ 3 \end{array} \end{array} \quad \text{Outlet}$$

The wholesale and retail prices are represented by T.

$$\begin{array}{c} \text{Price} \\ \begin{array}{cc} \text{Wholesale} & \text{Retail} \end{array} \\ T = \begin{bmatrix} \$840 & \$1100 \\ \$1200 & \$1350 \\ \$1450 & \$1650 \\ \$2650 & \$3000 \\ \$3050 & \$3200 \end{bmatrix} \begin{array}{c} A \\ B \\ C \\ D \\ E \end{array} \end{array} \quad \text{Model}$$

Compute ST and interpret the result.

74. *Voting Preferences* The matrix

$$\begin{array}{c} \text{From} \\ \begin{array}{ccc} R & D & I \end{array} \\ P = \begin{bmatrix} 0.6 & 0.1 & 0.1 \\ 0.2 & 0.7 & 0.1 \\ 0.2 & 0.2 & 0.8 \end{bmatrix} \begin{array}{c} R \\ D \\ I \end{array} \end{array} \quad \text{To}$$

is called a stochastic matrix. Each entry $p_{ij} (i \neq j)$ represents the proportion of the voting population that changes from party i to party j, and p_{ii} represents the proportion that remains loyal to the party from one election to the next. Compute and interpret P^2.

75. *Labor/Wage Requirements* A company that manufactures boats has the following labor-hour and wage requirements.

Labor per boat

Department

$$S = \begin{bmatrix} 1.0 \text{ hr} & 0.5 \text{ hr} & 0.2 \text{ hr} \\ 1.6 \text{ hr} & 1.0 \text{ hr} & 0.2 \text{ hr} \\ 2.5 \text{ hr} & 2.0 \text{ hr} & 0.4 \text{ hr} \end{bmatrix} \begin{matrix} \text{Small} \\ \text{Medium} \\ \text{Large} \end{matrix} \text{Boat size}$$

with columns Cutting, Assembly, Packaging.

Wages per hour

Plant

$$T = \begin{bmatrix} \$12 & \$10 \\ \$9 & \$8 \\ \$8 & \$7 \end{bmatrix} \begin{matrix} \text{Cutting} \\ \text{Assembly} \\ \text{Packaging} \end{matrix} \text{Department}$$

with columns A, B.

Compute ST and interpret the result.

76. *Voting Preference* Use a graphing utility to find P^3, P^4, P^5, P^6, P^7, and P^8 for the matrix given in Exercise 74. Can you detect a pattern as P is raised to higher powers?

Synthesis

True or False? In Exercises 77–79, determine whether the statement is true or false. Justify your answer.

77. Two matrices can be added only if they have the same order.

78. $\begin{bmatrix} -6 & -2 \\ 2 & -6 \end{bmatrix}\begin{bmatrix} 4 & 0 \\ 0 & -1 \end{bmatrix} = \begin{bmatrix} 4 & 0 \\ 0 & -1 \end{bmatrix}\begin{bmatrix} -6 & -2 \\ 2 & -6 \end{bmatrix}$

79. $\begin{bmatrix} -2 & 4 \\ -3 & 0 \\ 6 & 1 \end{bmatrix}\begin{bmatrix} 1 & 1 \\ 1 & 1 \end{bmatrix} = \begin{bmatrix} -2 & 4 \\ -3 & 0 \\ 6 & 1 \end{bmatrix}$

80. *Think About It* If a, b, and c are real numbers such that $c \neq 0$ and $ac = bc$, then $a = b$. However, if A, B, and C are nonzero matrices such that $AC = BC$, then A is not necessarily equal to B. Illustrate this using the following matrices.

$$A = \begin{bmatrix} 0 & 1 \\ 0 & 1 \end{bmatrix}, \quad B = \begin{bmatrix} 1 & 0 \\ 1 & 0 \end{bmatrix}, \quad C = \begin{bmatrix} 2 & 3 \\ 2 & 3 \end{bmatrix}$$

81. *Think About It* If a and b are real numbers such that $ab = 0$, then $a = 0$ or $b = 0$. However, if A and B are matrices such that $AB = O$, it is *not* necessarily true that $A = O$ or $B = O$. Illustrate this using the following matrices.

$$A = \begin{bmatrix} 3 & 3 \\ 4 & 4 \end{bmatrix}, \quad B = \begin{bmatrix} 1 & -1 \\ -1 & 1 \end{bmatrix}$$

Exploration In Exercises 82 and 83, let $i = \sqrt{-1}$.

82. Consider the matrix

$$A = \begin{bmatrix} i & 0 \\ 0 & i \end{bmatrix}.$$

Find A^2, A^3, and A^4. Identify any similarities with i^2, i^3, and i^4.

83. Consider the matrix

$$A = \begin{bmatrix} 0 & -i \\ i & 0 \end{bmatrix}.$$

Find and identify A^2.

84. *Exploration* Let A and B be unequal diagonal matrices of the same order. (A diagonal matrix is a square matrix in which each entry not on the main diagonal is zero.) Determine the products AB for several pairs of such matrices. Make a conjecture about a quick rule for such products.

Review

In Exercises 85–90, solve the equation.

85. $3x^2 + 20x - 32 = 0$

86. $8x^2 - 10x - 3 = 0$

87. $4x^3 + 10x^2 - 3x = 0$

88. $3x^3 + 22x^2 - 45x = 0$

89. $3x^3 - 12x^2 + 5x - 20 = 0$

90. $2x^3 - 5x^2 - 12x + 30 = 0$

In Exercises 91–94, solve the system of linear equations both graphically and algebraically.

91. $\begin{cases} -x + 4y = -9 \\ 5x - 8y = 39 \end{cases}$

92. $\begin{cases} 8x - 3y = -17 \\ -6x + 7y = 27 \end{cases}$

93. $\begin{cases} -x + 2y = -5 \\ -3x - y = -8 \end{cases}$

94. $\begin{cases} 6x - 13y = 11 \\ 9x + 5y = 41 \end{cases}$

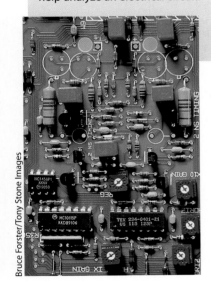

Bruce Forster/Tony Stone Images

▶ **What you should learn**

- How to verify that two matrices are inverses of each other
- How to use Gauss-Jordan elimination to find the inverses of matrices
- How to use a formula to find the inverses of 2 × 2 matrices
- How to use inverse matrices to solve systems of linear equations

▶ **Why you should learn it**

You can use inverse matrices to model and solve real-life problems. For instance, Exercises 73 and 74 on page 547 show how to use an inverse matrix to help analyze an electrical circuit.

The Inverse of a Matrix

This section further develops the algebra of matrices. To begin, consider the real number equation $ax = b$. To solve this equation for x, multiply each side of the equation by a^{-1} (provided that $a \neq 0$).

$$ax = b$$
$$(a^{-1}a)x = a^{-1}b$$
$$(1)x = a^{-1}b$$
$$x = a^{-1}b$$

The number a^{-1} is called the *multiplicative inverse of a* because $a^{-1}a = 1$. The definition of the multiplicative **inverse of a matrix** is similar.

Definition of the Inverse of a Square Matrix

Let A be an $n \times n$ matrix and let I_n be the $n \times n$ identity matrix. If there exists matrix A^{-1} such that

$$AA^{-1} = I_n = A^{-1}A$$

then A^{-1} is called the **inverse** of A. The symbol A^{-1} is read "A inverse."

Example 1 ▶ The Inverse of a Matrix

Show that B is the inverse of A, where

$$A = \begin{bmatrix} -1 & 2 \\ -1 & 1 \end{bmatrix}$$

and

$$B = \begin{bmatrix} 1 & -2 \\ 1 & -1 \end{bmatrix}.$$

Solution

To show that B is the inverse of A, show that $AB = I = BA$, as follows.

$$AB = \begin{bmatrix} -1 & 2 \\ -1 & 1 \end{bmatrix}\begin{bmatrix} 1 & -2 \\ 1 & -1 \end{bmatrix} = \begin{bmatrix} -1+2 & 2-2 \\ -1+1 & 2-1 \end{bmatrix} = \begin{bmatrix} 1 & 0 \\ 0 & 1 \end{bmatrix}$$

$$BA = \begin{bmatrix} 1 & -2 \\ 1 & -1 \end{bmatrix}\begin{bmatrix} -1 & 2 \\ -1 & 1 \end{bmatrix} = \begin{bmatrix} -1+2 & 2-2 \\ -1+1 & 2-1 \end{bmatrix} = \begin{bmatrix} 1 & 0 \\ 0 & 1 \end{bmatrix}$$

Recall that it is not always true that $AB = BA$, even if both products are defined. However, if A and B are both square matrices and $AB = I_n$, it can be shown that $BA = I_n$. So, in Example 1, you need only to check that $AB = I_2$.

Finding Inverse Matrices

If a matrix A has an inverse, A is called *invertible* (or *nonsingular*); otherwise, A is called *singular*. A nonsquare matrix cannot have an inverse. To see this, note that if A is of order $m \times n$ and B is of order $n \times m$ (where $m \neq n$), the products AB and BA are of different orders and so cannot be equal to each other. Not all square matrices have inverses (see the matrix at the bottom of page 542). If, however, a matrix does have an inverse, that inverse is unique. Example 2 shows how to use a system of equations to find the inverse of a matrix.

Example 2 ▶ Finding the Inverse of a Matrix

Find the inverse of

$$A = \begin{bmatrix} 1 & 4 \\ -1 & -3 \end{bmatrix}.$$

Solution

To find the inverse of A, try to solve the matrix equation $AX = I$ for X.

$$\overset{A}{\begin{bmatrix} 1 & 4 \\ -1 & -3 \end{bmatrix}} \overset{X}{\begin{bmatrix} x_{11} & x_{12} \\ x_{21} & x_{22} \end{bmatrix}} = \overset{I}{\begin{bmatrix} 1 & 0 \\ 0 & 1 \end{bmatrix}}$$

$$\begin{bmatrix} x_{11} + 4x_{21} & x_{12} + 4x_{22} \\ -x_{11} - 3x_{21} & -x_{12} - 3x_{22} \end{bmatrix} = \begin{bmatrix} 1 & 0 \\ 0 & 1 \end{bmatrix}$$

Equating corresponding entries, you obtain two systems of linear equations.

$$\begin{cases} x_{11} + 4x_{21} = 1 \\ -x_{11} - 3x_{21} = 0 \end{cases}$$ Linear system with two variables, x_{11} and x_{21}.

$$\begin{cases} x_{12} + 4x_{22} = 0 \\ -x_{12} - 3x_{22} = 1 \end{cases}$$ Linear system with two variables, x_{12} and x_{22}.

From the first system you can determine that $x_{11} = -3$ and $x_{21} = 1$, and from the second system you can determine that $x_{12} = -4$ and $x_{22} = 1$. Therefore, the inverse of A is

$$X = A^{-1}$$

$$= \begin{bmatrix} -3 & -4 \\ 1 & 1 \end{bmatrix}.$$

You can use matrix multiplication to check this result.

Check

$$AA^{-1} = \begin{bmatrix} 1 & 4 \\ -1 & -3 \end{bmatrix} \begin{bmatrix} -3 & -4 \\ 1 & 1 \end{bmatrix} = \begin{bmatrix} 1 & 0 \\ 0 & 1 \end{bmatrix} \checkmark$$

$$A^{-1}A = \begin{bmatrix} -3 & -4 \\ 1 & 1 \end{bmatrix} \begin{bmatrix} 1 & 4 \\ -1 & -3 \end{bmatrix} = \begin{bmatrix} 1 & 0 \\ 0 & 1 \end{bmatrix} \checkmark$$

In Example 2, note that the two systems of linear equations have the *same coefficient matrix A.* Rather than solve the two systems represented by

$$\begin{bmatrix} 1 & 4 & \vdots & 1 \\ -1 & -3 & \vdots & 0 \end{bmatrix}$$

and

$$\begin{bmatrix} 1 & 4 & \vdots & 0 \\ -1 & -3 & \vdots & 1 \end{bmatrix}$$

separately, you can solve them *simultaneously* by *adjoining* the identity matrix to the coefficient matrix to obtain

$$\begin{matrix} A & & I \\ \begin{bmatrix} 1 & 4 & \vdots & 1 & 0 \\ -1 & -3 & \vdots & 0 & 1 \end{bmatrix} \end{matrix}.$$

This "doubly augmented" matrix can be represented as $[A \vdots I]$. By applying Gauss-Jordan elimination to this matrix, you can solve *both* systems with a single elimination process.

$$\begin{bmatrix} 1 & 4 & \vdots & 1 & 0 \\ -1 & -3 & \vdots & 0 & 1 \end{bmatrix}$$

$$R_1 + R_2 \rightarrow \begin{bmatrix} 1 & 4 & \vdots & 1 & 0 \\ 0 & 1 & \vdots & 1 & 1 \end{bmatrix}$$

$$-4R_2 + R_1 \rightarrow \begin{bmatrix} 1 & 0 & \vdots & -3 & -4 \\ 0 & 1 & \vdots & 1 & 1 \end{bmatrix}$$

So, from the "doubly augmented" matrix $[A \vdots I]$, you obtained the matrix $[I \vdots A^{-1}]$.

$$\begin{matrix} A & I & & I & A^{-1} \\ \begin{bmatrix} 1 & 4 & \vdots & 1 & 0 \\ -1 & -3 & \vdots & 0 & 1 \end{bmatrix} & \Longrightarrow & \begin{bmatrix} 1 & 0 & \vdots & -3 & -4 \\ 0 & 1 & \vdots & 1 & 1 \end{bmatrix} \end{matrix}$$

This procedure (or algorithm) works for an arbitrary square matrix that has an inverse.

Technology

You can find the inverse of a matrix with a graphing utility. Enter the matrix in the graphing utility, press the inverse key $[x^{-1}]$, and press [ENTER]. The inverse of the matrix will be displayed on the screen.

Finding an Inverse Matrix

Let A be a square matrix of order n.

1. Write the $n \times 2n$ matrix that consists of the given matrix A on the left and the $n \times n$ identity matrix I on the right to obtain $[A \vdots I]$.

2. If possible, row reduce A to I using elementary row operations on the *entire* matrix $[A \vdots I]$. The result will be the matrix $[I \vdots A^{-1}]$. If this is not possible, A is not invertible.

3. Check your work by multiplying to see that $AA^{-1} = I = A^{-1}A$.

Example 3 ▶ **Finding the Inverse of a Matrix**

Find the inverse of

$$A = \begin{bmatrix} 1 & -1 & 0 \\ 1 & 0 & -1 \\ 6 & -2 & -3 \end{bmatrix}.$$

Solution

Begin by adjoining the identity matrix to A to form the matrix

$$[A \ \vdots \ I] = \begin{bmatrix} 1 & -1 & 0 & \vdots & 1 & 0 & 0 \\ 1 & 0 & -1 & \vdots & 0 & 1 & 0 \\ 6 & -2 & -3 & \vdots & 0 & 0 & 1 \end{bmatrix}.$$

Use elementary row operations to obtain the form $[I \ \vdots \ A^{-1}]$, as follows.

$$\begin{matrix} \\ -R_1 + R_2 \rightarrow \\ -6R_1 + R_3 \rightarrow \end{matrix} \begin{bmatrix} 1 & -1 & 0 & \vdots & 1 & 0 & 0 \\ 0 & 1 & -1 & \vdots & -1 & 1 & 0 \\ 0 & 4 & -3 & \vdots & -6 & 0 & 1 \end{bmatrix}$$

$$\begin{matrix} R_2 + R_1 \rightarrow \\ \\ -4R_2 + R_3 \rightarrow \end{matrix} \begin{bmatrix} 1 & 0 & -1 & \vdots & 0 & 1 & 0 \\ 0 & 1 & -1 & \vdots & -1 & 1 & 0 \\ 0 & 0 & 1 & \vdots & -2 & -4 & 1 \end{bmatrix}$$

$$\begin{matrix} R_3 + R_1 \rightarrow \\ R_3 + R_2 \rightarrow \\ \end{matrix} \begin{bmatrix} 1 & 0 & 0 & \vdots & -2 & -3 & 1 \\ 0 & 1 & 0 & \vdots & -3 & -3 & 1 \\ 0 & 0 & 1 & \vdots & -2 & -4 & 1 \end{bmatrix}$$

Therefore, the matrix A is invertible and its inverse is

$$A^{-1} = \begin{bmatrix} -2 & -3 & 1 \\ -3 & -3 & 1 \\ -2 & -4 & 1 \end{bmatrix}.$$

Try confirming this result by multiplying A and A^{-1} to obtain I.

Check

$$AA^{-1} = \begin{bmatrix} 1 & -1 & 0 \\ 1 & 0 & -1 \\ 6 & -2 & -3 \end{bmatrix} \begin{bmatrix} -2 & -3 & 1 \\ -3 & -3 & 1 \\ -2 & -4 & 1 \end{bmatrix} = \begin{bmatrix} 1 & 0 & 0 \\ 0 & 1 & 0 \\ 0 & 0 & 1 \end{bmatrix} = I$$

The process shown in Example 3 applies to any $n \times n$ matrix A. If A has an inverse, this process will find it. On the other hand, if A does not have an inverse (if A is *singular*), the process will tell you so. That is, matrix A will not reduce to the identity matrix. For instance, the following matrix has no inverse.

$$A = \begin{bmatrix} 1 & 2 & 0 \\ 3 & -1 & 2 \\ -2 & 3 & -2 \end{bmatrix}$$

Explain how the elimination process shows that this matrix is singular.

The Inverse of a 2 × 2 Matrix

Using Gauss-Jordan elimination to find the inverse of a matrix works well (even as a computer technique) for matrices of order 3×3 or greater. For 2×2 matrices, however, many people prefer to use a formula for the inverse rather than Gauss-Jordan elimination. This simple formula, which works *only* for 2×2 matrices, is explained as follows. If A is a 2×2 matrix given by

$$A = \begin{bmatrix} a & b \\ c & d \end{bmatrix}$$

then A is invertible if and only if $ad - bc \neq 0$. Moreover, if $ad - bc \neq 0$, the inverse is given by

$$A^{-1} = \frac{1}{ad - bc} \begin{bmatrix} d & -b \\ -c & a \end{bmatrix}. \qquad \text{Formula for inverse of matrix } A$$

Try verifying this inverse by multiplication. The denominator $ad - bc$ is called the **determinant** of the 2×2 matrix A. You will study determinants in the next section.

Example 4 ▶ Finding the Inverse of a 2 × 2 Matrix

If possible, find the inverse of the matrix.

a. $A = \begin{bmatrix} 3 & -1 \\ -2 & 2 \end{bmatrix}$ **b.** $B = \begin{bmatrix} 3 & -1 \\ -6 & 2 \end{bmatrix}$

Solution

a. For the matrix A, apply the formula for the inverse of a 2×2 matrix to obtain

$$ad - bc = (3)(2) - (-1)(-2)$$
$$= 4.$$

Because this quantity is not zero, the inverse is formed by interchanging the entries on the main diagonal, changing the signs of the other two entries, and multiplying by the scalar $\frac{1}{4}$, as follows.

$$A^{-1} = \frac{1}{4} \begin{bmatrix} 2 & 1 \\ 2 & 3 \end{bmatrix} \qquad \text{Substitute for } a, b, c, d, \text{ and the determinant.}$$

$$= \begin{bmatrix} \frac{1}{2} & \frac{1}{4} \\ \frac{1}{2} & \frac{3}{4} \end{bmatrix} \qquad \text{Multiply by the scalar } \frac{1}{4}.$$

b. For the matrix B, you have

$$ad - bc = (3)(2) - (-1)(-6)$$
$$= 0$$

which means that B is not invertible.

Systems of Linear Equations

You know that a system of linear equations can have exactly one solution, infinitely many solutions, or no solution. If the coefficient matrix A of a *square* system (a system that has the same number of equations as variables) is invertible, the system has a unique solution, which is defined as follows.

A System of Equations with a Unique Solution

If A is an invertible matrix, the system of linear equations represented by $AX = B$ has a unique solution given by

$$X = A^{-1}B.$$

Example 5 ► Solving a System Using an Inverse

You are going to invest \$10,000 in AAA-rated bonds, AA-rated bonds, and B-rated bonds and want an annual return of \$730. The average yields are 6% on AAA bonds, 7.5% on AA bonds, and 9.5% on B bonds. You will invest twice as much in AAA bonds as in B bonds. Your investment can be represented as

$$\begin{cases} x + y + z = 10{,}000 \\ 0.06x + 0.075y + 0.095z = 730 \\ x - 2z = 0 \end{cases}$$

where x, y, and z represent the amounts invested in AAA, AA, and B bonds, respectively. Use an inverse matrix to solve the system.

Solution

Begin by writing the system in the matrix form $AX = B$.

$$\begin{bmatrix} 1 & 1 & 1 \\ 0.06 & 0.075 & 0.095 \\ 1 & 0 & -2 \end{bmatrix} \begin{bmatrix} x \\ y \\ z \end{bmatrix} = \begin{bmatrix} 10{,}000 \\ 730 \\ 0 \end{bmatrix}$$

Then, use Gauss-Jordan elimination to find A^{-1}.

$$A^{-1} = \begin{bmatrix} 15 & -200 & -2 \\ -21.5 & 300 & 3.5 \\ 7.5 & -100 & -1.5 \end{bmatrix}$$

Finally, multiply B by A^{-1} on the left to obtain the solution.

$$X = A^{-1}B$$

$$= \begin{bmatrix} 15 & -200 & -2 \\ -21.5 & 300 & 3.5 \\ 7.5 & -100 & -1.5 \end{bmatrix} \begin{bmatrix} 10{,}000 \\ 730 \\ 0 \end{bmatrix} = \begin{bmatrix} 4000 \\ 4000 \\ 2000 \end{bmatrix}$$

The solution to the system is $x = 4000$, $y = 4000$, and $z = 2000$. So, you will invest \$4000 in AAA bonds, \$4000 in AA bonds, and \$2000 in B bonds.

7.3 Exercises

In Exercises 1–10, show that B is the inverse of A.

1. $A = \begin{bmatrix} 2 & 1 \\ 5 & 3 \end{bmatrix}$, $B = \begin{bmatrix} 3 & -1 \\ -5 & 2 \end{bmatrix}$

2. $A = \begin{bmatrix} 1 & -1 \\ -1 & 2 \end{bmatrix}$, $B = \begin{bmatrix} 2 & 1 \\ 1 & 1 \end{bmatrix}$

3. $A = \begin{bmatrix} 1 & 2 \\ 3 & 4 \end{bmatrix}$, $B = \begin{bmatrix} -2 & 1 \\ \frac{3}{2} & -\frac{1}{2} \end{bmatrix}$

4. $A = \begin{bmatrix} 1 & -1 \\ 2 & 3 \end{bmatrix}$, $B = \begin{bmatrix} \frac{3}{5} & \frac{1}{5} \\ -\frac{2}{5} & \frac{1}{5} \end{bmatrix}$

5. $A = \begin{bmatrix} 2 & -17 & 11 \\ -1 & 11 & -7 \\ 0 & 3 & -2 \end{bmatrix}$,

$B = \begin{bmatrix} 1 & 1 & 2 \\ 2 & 4 & -3 \\ 3 & 6 & -5 \end{bmatrix}$

6. $A = \begin{bmatrix} -4 & 1 & 5 \\ -1 & 2 & 4 \\ 0 & -1 & -1 \end{bmatrix}$,

$B = \begin{bmatrix} -\frac{1}{2} & 1 & \frac{3}{2} \\ \frac{1}{4} & -1 & -\frac{11}{4} \\ -\frac{1}{4} & 1 & \frac{7}{4} \end{bmatrix}$

7. $A = \begin{bmatrix} 2 & 0 & 1 & 1 \\ 3 & 0 & 0 & 1 \\ -1 & 1 & -2 & 1 \\ 4 & -1 & 1 & 0 \end{bmatrix}$,

$B = \begin{bmatrix} -1 & 2 & -1 & -1 \\ -4 & 9 & -5 & -6 \\ 0 & 1 & -1 & -1 \\ 3 & -5 & 3 & 3 \end{bmatrix}$

8. $A = \begin{bmatrix} -2 & 0 & 1 & 0 \\ 1 & -1 & -3 & 0 \\ -2 & -1 & 0 & -2 \\ 0 & 1 & 3 & -1 \end{bmatrix}$,

$B = \begin{bmatrix} -3 & -3 & 1 & -2 \\ 12 & 14 & -5 & 10 \\ -5 & -6 & 2 & -4 \\ -3 & -4 & 1 & -3 \end{bmatrix}$

9. $A = \begin{bmatrix} -2 & 2 & 3 \\ 1 & -1 & 0 \\ 0 & 1 & 4 \end{bmatrix}$, $B = \frac{1}{3}\begin{bmatrix} -4 & -5 & 3 \\ -4 & -8 & 3 \\ 1 & 2 & 0 \end{bmatrix}$

10. $A = \begin{bmatrix} -1 & 1 & 0 & -1 \\ 1 & -1 & 1 & 0 \\ -1 & 1 & 2 & 0 \\ 0 & -1 & 1 & 1 \end{bmatrix}$,

$B = \frac{1}{3}\begin{bmatrix} -3 & 1 & 1 & -3 \\ -3 & -1 & 2 & -3 \\ 0 & 1 & 1 & 0 \\ -3 & -2 & 1 & 0 \end{bmatrix}$

In Exercises 11–26, find the inverse of the matrix (if it exists).

11. $\begin{bmatrix} 2 & 0 \\ 0 & 3 \end{bmatrix}$

12. $\begin{bmatrix} 1 & 2 \\ 3 & 7 \end{bmatrix}$

13. $\begin{bmatrix} 1 & -2 \\ 2 & -3 \end{bmatrix}$

14. $\begin{bmatrix} -7 & 33 \\ 4 & -19 \end{bmatrix}$

15. $\begin{bmatrix} -1 & 1 \\ -2 & 1 \end{bmatrix}$

16. $\begin{bmatrix} 11 & 1 \\ -1 & 0 \end{bmatrix}$

17. $\begin{bmatrix} 2 & 4 \\ 4 & 8 \end{bmatrix}$

18. $\begin{bmatrix} 2 & 3 \\ 1 & 4 \end{bmatrix}$

19. $\begin{bmatrix} 2 & 7 & 1 \\ -3 & -9 & 2 \end{bmatrix}$

20. $\begin{bmatrix} -2 & 5 \\ 6 & -15 \\ 0 & 1 \end{bmatrix}$

21. $\begin{bmatrix} 1 & 1 & 1 \\ 3 & 5 & 4 \\ 3 & 6 & 5 \end{bmatrix}$

22. $\begin{bmatrix} 1 & 2 & 2 \\ 3 & 7 & 9 \\ -1 & -4 & -7 \end{bmatrix}$

23. $\begin{bmatrix} 1 & 0 & 0 \\ 3 & 4 & 0 \\ 2 & 5 & 5 \end{bmatrix}$

24. $\begin{bmatrix} 1 & 0 & 0 \\ 3 & 0 & 0 \\ 2 & 5 & 5 \end{bmatrix}$

25. $\begin{bmatrix} -8 & 0 & 0 & 0 \\ 0 & 1 & 0 & 0 \\ 0 & 0 & 4 & 0 \\ 0 & 0 & 0 & -5 \end{bmatrix}$

26. $\begin{bmatrix} 1 & 3 & -2 & 0 \\ 0 & 2 & 4 & 6 \\ 0 & 0 & -2 & 1 \\ 0 & 0 & 0 & 5 \end{bmatrix}$

In Exercises 27–38, use the matrix capabilities of a graphing utility to find the inverse of the matrix (if it exists).

27. $\begin{bmatrix} 1 & 2 & -1 \\ 3 & 7 & -10 \\ -5 & -7 & -15 \end{bmatrix}$

28. $\begin{bmatrix} 10 & 5 & -7 \\ -5 & 1 & 4 \\ 3 & 2 & -2 \end{bmatrix}$

29. $\begin{bmatrix} 1 & 1 & 2 \\ 3 & 1 & 0 \\ -2 & 0 & 3 \end{bmatrix}$

30. $\begin{bmatrix} 3 & 2 & 2 \\ 2 & 2 & 2 \\ -4 & 4 & 3 \end{bmatrix}$

31. $\begin{bmatrix} -\frac{1}{2} & \frac{3}{4} & \frac{1}{4} \\ 1 & 0 & -\frac{3}{2} \\ 0 & -1 & \frac{1}{2} \end{bmatrix}$

32. $\begin{bmatrix} -\frac{5}{6} & \frac{1}{3} & \frac{11}{6} \\ 0 & \frac{2}{3} & 2 \\ 1 & -\frac{1}{2} & -\frac{5}{2} \end{bmatrix}$

33. $\begin{bmatrix} 0.1 & 0.2 & 0.3 \\ -0.3 & 0.2 & 0.2 \\ 0.5 & 0.4 & 0.4 \end{bmatrix}$ **34.** $\begin{bmatrix} 0.6 & 0 & -0.3 \\ 0.7 & -1 & 0.2 \\ 1 & 0 & -0.9 \end{bmatrix}$

35. $\begin{bmatrix} 1 & 0 & 3 & 0 \\ 0 & 2 & 0 & 4 \\ 1 & 0 & 3 & 0 \\ 0 & 2 & 0 & 4 \end{bmatrix}$ **36.** $\begin{bmatrix} 4 & 8 & -7 & 14 \\ 2 & 5 & -4 & 6 \\ 0 & 2 & 1 & -7 \\ 3 & 6 & -5 & 10 \end{bmatrix}$

37. $\begin{bmatrix} -1 & 0 & 1 & 0 \\ 0 & 2 & 0 & -1 \\ 2 & 0 & -1 & 0 \\ 0 & -1 & 0 & 1 \end{bmatrix}$

38. $\begin{bmatrix} 1 & -2 & -1 & -2 \\ 3 & -5 & -2 & -3 \\ 2 & -5 & -2 & -5 \\ -1 & 4 & 4 & 11 \end{bmatrix}$

In Exercises 39–44, use the formula on page 543 to find the inverse of the matrix.

39. $\begin{bmatrix} 5 & -2 \\ 2 & 3 \end{bmatrix}$ **40.** $\begin{bmatrix} 7 & 12 \\ -8 & -5 \end{bmatrix}$

41. $\begin{bmatrix} -4 & -6 \\ 2 & 3 \end{bmatrix}$ **42.** $\begin{bmatrix} -12 & 3 \\ 5 & -2 \end{bmatrix}$

43. $\begin{bmatrix} \frac{7}{2} & -\frac{3}{4} \\ \frac{1}{5} & \frac{4}{5} \end{bmatrix}$ **44.** $\begin{bmatrix} -\frac{1}{4} & \frac{9}{4} \\ \frac{5}{3} & \frac{8}{9} \end{bmatrix}$

In Exercises 45–48, use an inverse matrix to solve the system of linear equations. (Use the inverse matrix found in Exercise 13.)

45. $\begin{cases} x - 2y = 5 \\ 2x - 3y = 10 \end{cases}$ **46.** $\begin{cases} x - 2y = 0 \\ 2x - 3y = 3 \end{cases}$

47. $\begin{cases} x - 2y = 4 \\ 2x - 3y = 2 \end{cases}$ **48.** $\begin{cases} x - 2y = 1 \\ 2x - 3y = -2 \end{cases}$

In Exercises 49 and 50, use an inverse matrix to solve the system of linear equations. (Use the inverse matrix found in Exercise 21.)

49. $\begin{cases} x + y + z = 0 \\ 3x + 5y + 4z = 5 \\ 3x + 6y + 5z = 2 \end{cases}$ **50.** $\begin{cases} x + y + z = -1 \\ 3x + 5y + 4z = 2 \\ 3x + 6y + 5z = 0 \end{cases}$

In Exercises 51 and 52, use an inverse matrix to solve the system of linear equations. (Use the inverse matrix found in Exercise 38.)

51. $\begin{cases} x_1 - 2x_2 - x_3 - 2x_4 = 0 \\ 3x_1 - 5x_2 - 2x_3 - 3x_4 = 1 \\ 2x_1 - 5x_2 - 2x_3 - 5x_4 = -1 \\ -x_1 + 4x_2 + 4x_3 + 11x_4 = 2 \end{cases}$

52. $\begin{cases} x_1 - 2x_2 - x_3 - 2x_4 = 1 \\ 3x_1 - 5x_2 - 2x_3 - 3x_4 = -2 \\ 2x_1 - 5x_2 - 2x_3 - 5x_4 = 0 \\ -x_1 + 4x_2 + 4x_3 + 11x_4 = -3 \end{cases}$

In Exercises 53–62, use an inverse matrix to solve (if possible) the system of linear equations.

53. $\begin{cases} 3x + 4y = -2 \\ 5x + 3y = 4 \end{cases}$ **54.** $\begin{cases} 18x + 12y = 13 \\ 30x + 24y = 23 \end{cases}$

55. $\begin{cases} -0.4x + 0.8y = 1.6 \\ 2x - 4y = 5 \end{cases}$ **56.** $\begin{cases} 0.2x - 0.6y = 2.4 \\ -x + 1.4y = -8.8 \end{cases}$

57. $\begin{cases} 3x + 6y = 6 \\ 6x + 14y = 11 \end{cases}$ **58.** $\begin{cases} 3x + 2y = 1 \\ 2x + 10y = 6 \end{cases}$

59. $\begin{cases} -\frac{1}{4}x + \frac{3}{8}y = -2 \\ \frac{3}{2}x + \frac{3}{4}y = -12 \end{cases}$ **60.** $\begin{cases} \frac{5}{6}x - y = -20 \\ \frac{4}{3}x - \frac{7}{2}y = -51 \end{cases}$

61. $\begin{cases} 4x - y + z = -5 \\ 2x + 2y + 3z = 10 \\ 5x - 2y + 6z = 1 \end{cases}$ **62.** $\begin{cases} 4x - 2y + 3z = -2 \\ 2x + 2y + 5z = 16 \\ 8x - 5y - 2z = 4 \end{cases}$

In Exercises 63–68, use the matrix capabilities of a graphing utility to solve (if possible) the system of linear equations.

63. $\begin{cases} 5x - 3y + 2z = 2 \\ 2x + 2y - 3z = 3 \\ x - 7y + 8z = -4 \end{cases}$

64. $\begin{cases} 3x - 2y + z = -29 \\ -4x + y - 3z = 37 \\ x - 5y + z = -24 \end{cases}$

65. $\begin{cases} 2x + 3y + 5z = 4 \\ 3x + 5y + 9z = 7 \\ 5x + 9y + 17z = 13 \end{cases}$

66. $\begin{cases} -8x + 7y - 10z = -151 \\ 12x + 3y - 5z = 86 \\ 15x - 9y + 2z = 187 \end{cases}$

67. $\begin{cases} 7x - 3y + 2w = 41 \\ -2x + y - w = -13 \\ 4x + z - 2w = 12 \\ -x + y - w = -8 \end{cases}$

68. $\begin{cases} 2x + 5y + w = 11 \\ x + 4y + 2z - 2w = -7 \\ 2x - 2y + 5z + w = 3 \\ x - 3w = -1 \end{cases}$

Finance In Exercises 69–72, consider a person who invests in AAA-rated bonds, A-rated bonds, and B-rated bonds. The average yields are 6.5% on AAA bonds, 7% on A bonds, and 9% on B bonds. The person invests twice as much in B bonds as in A bonds. Let x, y, and z represent the amounts invested in AAA, A, and B bonds, respectively.

$$\begin{cases} x + y + z = \text{(total investment)} \\ 0.065x + 0.07y + 0.09z = \text{(annual return)} \\ 2y - z = 0 \end{cases}$$

Use the inverse of the coefficient matrix of this system to find the amount invested in each type of bond.

69. Total investment = $10,000
    Annual return = $705

70. Total investment = $10,000
    Annual return = $760

71. Total investment = $12,000
    Annual return = $835

72. Total investment = $500,000
    Annual return = $38,000

Circuit Analysis In Exercises 73 and 74, consider the circuit in the figure. The currents I_1, I_2, and I_3, in amperes, are the solution of the system of linear equations

$$\begin{cases} 2I_1 + 4I_3 = E_1 \\ I_2 + 4I_3 = E_2 \\ I_1 + I_2 - I_3 = 0 \end{cases}$$

where E_1 and E_2 are voltages. Use the inverse of the coefficient matrix of this system to find the unknown currents for the voltages.

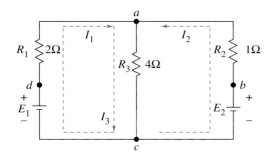

73. $E_1 = 14$ volts, $E_2 = 28$ volts
74. $E_1 = 24$ volts, $E_2 = 23$ volts

Synthesis

True or False? In Exercises 75–77, determine whether the statement is true or false. Justify your answer.

75. Multiplication of an invertible matrix and its inverse is commutative.

76. If you multiply two square matrices and obtain the identity matrix, you can assume the matrices are inverses of one another.

77. All nonsquare matrices do not have inverses.

78. If A is a 2×2 matrix

$$A = \begin{bmatrix} a & b \\ c & d \end{bmatrix}$$

then A is invertible if and only if $ad - bc \neq 0$. If $ad - bc \neq 0$, verify that the inverse is

$$A^{-1} = \frac{1}{ad - bc} \begin{bmatrix} d & -b \\ -c & a \end{bmatrix}.$$

79. *Writing* Write a brief paragraph explaining the advantage of using inverse matrices to solve the systems of linear equations in Exercises 45–52.

80. *Exploration* Consider matrices of the form

$$A = \begin{bmatrix} a_{11} & 0 & 0 & 0 & \cdots & 0 \\ 0 & a_{22} & 0 & 0 & \cdots & 0 \\ 0 & 0 & a_{33} & 0 & \cdots & 0 \\ \vdots & \vdots & \vdots & \vdots & & \vdots \\ 0 & 0 & 0 & 0 & \cdots & a_{nn} \end{bmatrix}.$$

(a) Write a 2×2 matrix and a 3×3 matrix in the form of A. Find the inverse of each.

(b) Use the result of part (a) to make a conjecture about the inverses of matrices in the form of A.

Review

In Exercises 81–84, solve the equation.

81. $3^{x/2} = 315$

82. $2000e^{-x/5} = 400$

83. $\log_2 x - 2 = 4.5$

84. $\ln x + \ln(x - 1) = 0$

In Exercises 85–88, perform the matrix operation.

85. $-3 \begin{bmatrix} -4 & 6 \\ 2 & -8 \\ 1 & 12 \end{bmatrix}$

86. $\frac{1}{8} \begin{bmatrix} -6 & 2 & 0 \\ -4 & -24 & 18 \end{bmatrix}$

87. $\begin{bmatrix} 2 & 7 \\ -3 & -1 \end{bmatrix} - 4 \begin{bmatrix} -1 & 2 \\ 6 & -5 \end{bmatrix}$

88. $8 \begin{bmatrix} 2 & -3 \\ 1 & 0 \end{bmatrix} + \begin{bmatrix} 12 & 17 \\ -7 & 9 \end{bmatrix}$

7.4 The Determinant of a Square Matrix

▶ **What you should learn**

- How to find the determinants of 2 × 2 matrices
- How to find minors and cofactors of square matrices
- How to find the determinants of square matrices

▶ **Why you should learn it**

Determinants are often used in other branches of mathematics. For instance, Exercises 79–84 on page 555 show some types of determinants that are useful when changes in variables are made in calculus.

The Determinant of a 2 × 2 Matrix

Every *square* matrix can be associated with a real number called its **determinant.** Determinants have many uses, and several will be discussed in this and the next section. Historically, the use of determinants arose from special number patterns that occur when systems of linear equations are solved. For instance, the system

$$\begin{cases} a_1x + b_1y = c_1 \\ a_2x + b_2y = c_2 \end{cases}$$

has a solution

$$x = \frac{c_1b_2 - c_2b_1}{a_1b_2 - a_2b_1} \quad \text{and} \quad y = \frac{a_1c_2 - a_2c_1}{a_1b_2 - a_2b_1}$$

provided that $a_1b_2 - a_2b_1 \neq 0$. Note that the denominators of the two fractions are the same. This denominator is called the *determinant* of the coefficient matrix of the system.

Coefficient Matrix   *Determinant*

$$A = \begin{bmatrix} a_1 & b_1 \\ a_2 & b_2 \end{bmatrix} \qquad \det(A) = a_1b_2 - a_2b_1$$

The determinant of the matrix A can also be denoted by vertical bars on both sides of the matrix, as indicated in the following definition.

Definition of the Determinant of a 2 × 2 Matrix

The **determinant** of the matrix

$$A = \begin{bmatrix} a_1 & b_1 \\ a_2 & b_2 \end{bmatrix}$$

is given by

$$\det(A) = |A| = \begin{vmatrix} a_1 & b_1 \\ a_2 & b_2 \end{vmatrix} = a_1b_2 - a_2b_1.$$

In this book, $\det(A)$ and $|A|$ are used interchangeably to represent the determinant of A. Although vertical bars are also used to denote the absolute value of a real number, the context will show which use is intended.

A convenient method for remembering the formula for the determinant of a 2 × 2 matrix is shown in the diagram.

$$\det(A) = \begin{vmatrix} a_1 & b_1 \\ a_2 & b_2 \end{vmatrix} = a_1b_2 - a_2b_1$$

Note that the determinant is the difference of the products of the two diagonals of the matrix.

Example 1 ► The Determinant of a 2 × 2 Matrix

Find the determinant of each matrix.

a. $A = \begin{bmatrix} 2 & -3 \\ 1 & 2 \end{bmatrix}$

b. $B = \begin{bmatrix} 2 & 1 \\ 4 & 2 \end{bmatrix}$

c. $C = \begin{bmatrix} 0 & \frac{3}{2} \\ 2 & 4 \end{bmatrix}$

Solution

a. $\det(A) = \begin{vmatrix} 2 & -3 \\ 1 & 2 \end{vmatrix}$

$= 2(2) - 1(-3)$

$= 4 + 3 = 7$

b. $\det(B) = \begin{vmatrix} 2 & 1 \\ 4 & 2 \end{vmatrix}$

$= 2(2) - 4(1)$

$= 4 - 4 = 0$

c. $\det(C) = \begin{vmatrix} 0 & \frac{3}{2} \\ 2 & 4 \end{vmatrix}$

$= 0(4) - 2\left(\frac{3}{2}\right)$

$= 0 = -3$

◄ **E x p l o r a t i o n** ►

Use a graphing utility to find the determinant of the following matrix.

$$A = \begin{bmatrix} 1 & 2 \\ -1 & 0 \\ 3 & -2 \end{bmatrix}$$

What message appears on the screen? Why does the graphing utility display this message?

Notice in Example 1 that the determinant of a matrix can be positive, zero, or negative.

The determinant of a matrix of order 1 × 1 is defined simply as the entry of the matrix. For instance, if $A = [-2]$, then $\det(A) = -2$.

Technology

Most graphing utilities can evaluate the determinant of a matrix. For instance, you can evaluate the determinant of

$$A = \begin{bmatrix} 2 & -3 \\ 1 & 2 \end{bmatrix}$$

by entering the matrix as $[A]$ and then choosing the determinant feature. The result should be 7, as in Example 1(a). Try evaluating determinants of other matrices.

Minors and Cofactors

To define the determinant of a square matrix of order 3×3 or higher, it is convenient to introduce the concepts of **minors** and **cofactors**.

Sign Pattern for Cofactors

$$\begin{bmatrix} + & - & + \\ - & + & - \\ + & - & + \end{bmatrix}$$

3×3 matrix

$$\begin{bmatrix} + & - & + & - \\ - & + & - & + \\ + & - & + & - \\ - & + & - & + \end{bmatrix}$$

4×4 matrix

$$\begin{bmatrix} + & - & + & - & + & \cdots \\ - & + & - & + & - & \cdots \\ + & - & + & - & + & \cdots \\ - & + & - & + & - & \cdots \\ + & - & + & - & + & \cdots \\ \vdots & \vdots & \vdots & \vdots & \vdots & \end{bmatrix}$$

$n \times n$ matrix

Minors and Cofactors of a Square Matrix

If A is a square matrix, the **minor** M_{ij} of the entry a_{ij} is the determinant of the matrix obtained by deleting the ith row and jth column of A. The **cofactor** C_{ij} of the entry a_{ij} is

$$C_{ij} = (-1)^{i+j} M_{ij}.$$

In the sign pattern for cofactors at the left, notice that *odd* positions (where $i + j$ is odd) have negative signs and *even* positions (where $i + j$ is even) have positive signs.

Example 2 ▶ Finding the Minors and Cofactors of a Matrix

Find all the minors and cofactors of

$$A = \begin{bmatrix} 0 & 2 & 1 \\ 3 & -1 & 2 \\ 4 & 0 & 1 \end{bmatrix}.$$

Solution

To find the minor M_{11}, delete the first row and first column of A and evaluate the determinant of the resulting matrix.

$$\begin{bmatrix} \cancel{0} & \cancel{2} & \cancel{1} \\ \cancel{3} & -1 & 2 \\ \cancel{4} & 0 & 1 \end{bmatrix}, \quad M_{11} = \begin{vmatrix} -1 & 2 \\ 0 & 1 \end{vmatrix} = -1(1) - 0(2) = -1$$

Similarly, to find M_{12}, delete the first row and second column.

$$\begin{bmatrix} \cancel{0} & \cancel{2} & \cancel{1} \\ 3 & \cancel{-1} & 2 \\ 4 & \cancel{0} & 1 \end{bmatrix}, \quad M_{12} = \begin{vmatrix} 3 & 2 \\ 4 & 1 \end{vmatrix} = 3(1) - 4(2) = -5$$

Continuing this pattern, you obtain the minors.

$$M_{11} = -1 \qquad M_{12} = -5 \qquad M_{13} = 4$$
$$M_{21} = 2 \qquad M_{22} = -4 \qquad M_{23} = -8$$
$$M_{31} = 5 \qquad M_{32} = -3 \qquad M_{33} = -6$$

Now, to find the cofactors, combine the checkerboard pattern of signs for a 3×3 matrix (at left above) with these minors.

$$C_{11} = -1 \qquad C_{12} = 5 \qquad C_{13} = 4$$
$$C_{21} = -2 \qquad C_{22} = -4 \qquad C_{23} = 8$$
$$C_{31} = 5 \qquad C_{32} = 3 \qquad C_{33} = -6$$

The Determinant of a Square Matrix

The definition given below is called *inductive* because it uses determinants of matrices of order $n - 1$ to define determinants of matrices of order n.

Determinant of a Square Matrix

If A is a square matrix (of order 2×2 or greater), the determinant of A is the sum of the entries in any row (or column) of A multiplied by their respective cofactors. For instance, expanding along the first row yields

$$|A| = a_{11}C_{11} + a_{12}C_{12} + \cdots + a_{1n}C_{1n}.$$

Applying this definition to find a determinant is called *expanding by cofactors*.

Try checking that for a 2×2 matrix

$$A = \begin{bmatrix} a_1 & b_1 \\ a_2 & b_2 \end{bmatrix}$$

this definition of the determinant yields $|A| = a_1 b_2 - a_2 b_1$, as previously defined.

Example 3 ▶ The Determinant of a Matrix of Order 3 × 3

Find the determinant of

$$A = \begin{bmatrix} 0 & 2 & 1 \\ 3 & -1 & 2 \\ 4 & 0 & 1 \end{bmatrix}.$$

Solution

Note that this is the same matrix that was in Example 2. There you found the cofactors of the entries in the first row to be

$$C_{11} = -1, \quad C_{12} = 5, \quad \text{and} \quad C_{13} = 4.$$

Therefore, by the definition of a determinant, you have

$$
\begin{aligned}
|A| &= a_{11}C_{11} + a_{12}C_{12} + a_{13}C_{13} &&\text{First-row expansion} \\
&= 0(-1) + 2(5) + 1(4) \\
&= 14.
\end{aligned}
$$

In Example 3 the determinant was found by expanding by the cofactors in the first row. You could have used any row or column. For instance, you could have expanded along the second row to obtain

$$
\begin{aligned}
|A| &= a_{21}C_{21} + a_{22}C_{22} + a_{23}C_{23} &&\text{Second-row expansion} \\
&= 3(-2) + (-1)(-4) + 2(8) \\
&= 14.
\end{aligned}
$$

When expanding by cofactors, you do not need to find cofactors of zero entries, because zero times its cofactor is zero.

$$a_{ij}C_{ij} = (0)C_{ij} = 0$$

So, the row (or column) containing the most zeros is usually the best choice for expansion by cofactors. This is demonstrated in the next example.

Example 4 ▶   The Determinant of a Matrix of Order 4 × 4

Find the determinant of

$$A = \begin{bmatrix} 1 & -2 & 3 & 0 \\ -1 & 1 & 0 & 2 \\ 0 & 2 & 0 & 3 \\ 3 & 4 & 0 & 2 \end{bmatrix}.$$

Solution

After inspecting this matrix, you can see that three of the entries in the third column are zeros. So, you can eliminate some of the work in the expansion by using the third column.

$$|A| = 3(C_{13}) + 0(C_{23}) + 0(C_{33}) + 0(C_{43})$$

Because C_{23}, C_{33}, and C_{43} have zero coefficients, you need only find the cofactor C_{13}. To do this, delete the first row and third column of A and evaluate the determinant of the resulting matrix.

$$C_{13} = (-1)^{1+3} \begin{vmatrix} -1 & 1 & 2 \\ 0 & 2 & 3 \\ 3 & 4 & 2 \end{vmatrix} \qquad \text{Delete 1st row and 3rd column.}$$

$$= \begin{vmatrix} -1 & 1 & 2 \\ 0 & 2 & 3 \\ 3 & 4 & 2 \end{vmatrix} \qquad \text{Simplify.}$$

Expanding by cofactors in the second row yields

$$C_{13} = 0(-1)^3 \begin{vmatrix} 1 & 2 \\ 4 & 2 \end{vmatrix} + 2(-1)^4 \begin{vmatrix} -1 & 2 \\ 3 & 2 \end{vmatrix} + 3(-1)^5 \begin{vmatrix} -1 & 1 \\ 3 & 4 \end{vmatrix}$$

$$= 0 + 2(1)(-8) + 3(-1)(-7)$$

$$= 5.$$

So, you obtain

$$|A| = 3C_{13}$$

$$= 3(5)$$

$$= 15.$$

Try using a graphing utility to confirm the result of Example 4.

The *Interactive* CD-ROM and *Internet* versions of this text contain step-by-step solutions to all odd-numbered Section and Review Exercises. They also provide Tutorial Exercises that link to Guided Examples for additional help.

7.4 Exercises

In Exercises 1–16, find the determinant of the matrix.

1. $[5]$

2. $[-8]$

3. $\begin{bmatrix} 2 & 1 \\ 3 & 4 \end{bmatrix}$

4. $\begin{bmatrix} -3 & 1 \\ 5 & 2 \end{bmatrix}$

5. $\begin{bmatrix} 5 & 2 \\ -6 & 3 \end{bmatrix}$

6. $\begin{bmatrix} 2 & -2 \\ 4 & 3 \end{bmatrix}$

7. $\begin{bmatrix} -7 & 0 \\ 3 & 0 \end{bmatrix}$

8. $\begin{bmatrix} 4 & -3 \\ 0 & 0 \end{bmatrix}$

9. $\begin{bmatrix} 2 & 6 \\ 0 & 3 \end{bmatrix}$

10. $\begin{bmatrix} 2 & -3 \\ -6 & 9 \end{bmatrix}$

11. $\begin{bmatrix} -3 & -2 \\ -6 & -1 \end{bmatrix}$

12. $\begin{bmatrix} 4 & 7 \\ -2 & 5 \end{bmatrix}$

13. $\begin{bmatrix} 9 & 0 \\ 7 & 8 \end{bmatrix}$

14. $\begin{bmatrix} 0 & 6 \\ -3 & 2 \end{bmatrix}$

15. $\begin{bmatrix} -\frac{1}{2} & \frac{1}{3} \\ -6 & \frac{1}{3} \end{bmatrix}$

16. $\begin{bmatrix} \frac{2}{3} & \frac{4}{3} \\ -1 & -\frac{1}{3} \end{bmatrix}$

In Exercises 17–22, use the matrix capabilities of a graphing utility to find the determinant of the matrix.

17. $\begin{bmatrix} 0.3 & 0.2 & 0.2 \\ 0.2 & 0.2 & 0.2 \\ -0.4 & 0.4 & 0.3 \end{bmatrix}$

18. $\begin{bmatrix} 0.1 & 0.2 & 0.3 \\ -0.3 & 0.2 & 0.2 \\ 0.5 & 0.4 & 0.4 \end{bmatrix}$

19. $\begin{bmatrix} 0.9 & 0.7 & 0 \\ -0.1 & 0.3 & 1.3 \\ -2.2 & 4.2 & 6.1 \end{bmatrix}$

20. $\begin{bmatrix} 0.1 & 0.1 & -4.3 \\ 7.5 & 6.2 & 0.7 \\ 0.3 & 0.6 & -1.2 \end{bmatrix}$

21. $\begin{bmatrix} 1 & 4 & -2 \\ 3 & 6 & -6 \\ -2 & 1 & 4 \end{bmatrix}$

22. $\begin{bmatrix} 2 & 3 & 1 \\ 0 & 5 & -2 \\ 0 & 0 & -2 \end{bmatrix}$

In Exercises 23–30, find all (a) minors and (b) cofactors of the matrix.

23. $\begin{bmatrix} 3 & 4 \\ 2 & -5 \end{bmatrix}$

24. $\begin{bmatrix} 11 & 0 \\ -3 & 2 \end{bmatrix}$

25. $\begin{bmatrix} 3 & 1 \\ -2 & -4 \end{bmatrix}$

26. $\begin{bmatrix} -6 & 5 \\ 7 & -2 \end{bmatrix}$

27. $\begin{bmatrix} 4 & 0 & 2 \\ -3 & 2 & 1 \\ 1 & -1 & 1 \end{bmatrix}$

28. $\begin{bmatrix} 1 & -1 & 0 \\ 3 & 2 & 5 \\ 4 & -6 & 4 \end{bmatrix}$

29. $\begin{bmatrix} 3 & -2 & 8 \\ 3 & 2 & -6 \\ -1 & 3 & 6 \end{bmatrix}$

30. $\begin{bmatrix} -2 & 9 & 4 \\ 7 & -6 & 0 \\ 6 & 7 & -6 \end{bmatrix}$

In Exercises 31–36, find the determinant of the matrix by the method of expansion by cofactors. Expand using the indicated row or column.

31. $\begin{bmatrix} -3 & 2 & 1 \\ 4 & 5 & 6 \\ 2 & -3 & 1 \end{bmatrix}$

  (a) Row 1

  (b) Column 2

32. $\begin{bmatrix} -3 & 4 & 2 \\ 6 & 3 & 1 \\ 4 & -7 & -8 \end{bmatrix}$

  (a) Row 2

  (b) Column 3

33. $\begin{bmatrix} 5 & 0 & -3 \\ 0 & 12 & 4 \\ 1 & 6 & 3 \end{bmatrix}$

  (a) Row 2

  (b) Column 2

34. $\begin{bmatrix} 10 & -5 & 5 \\ 30 & 0 & 10 \\ 0 & 10 & 1 \end{bmatrix}$

  (a) Row 3

  (b) Column 1

35. $\begin{bmatrix} 6 & 0 & -3 & 5 \\ 4 & 13 & 6 & -8 \\ -1 & 0 & 7 & 4 \\ 8 & 6 & 0 & 2 \end{bmatrix}$

  (a) Row 2

  (b) Column 2

36. $\begin{bmatrix} 10 & 8 & 3 & -7 \\ 4 & 0 & 5 & -6 \\ 0 & 3 & 2 & 7 \\ 1 & 0 & -3 & 2 \end{bmatrix}$

  (a) Row 3

  (b) Column 1

In Exercises 37–52, find the determinant of the matrix. Expand by cofactors on the row or column that appears to make the computations easiest.

37. $\begin{bmatrix} 2 & -1 & 0 \\ 4 & 2 & 1 \\ 4 & 2 & 1 \end{bmatrix}$

38. $\begin{bmatrix} -2 & 2 & 3 \\ 1 & -1 & 0 \\ 0 & 1 & 4 \end{bmatrix}$

39. $\begin{bmatrix} 6 & 3 & -7 \\ 0 & 0 & 0 \\ 4 & -6 & 3 \end{bmatrix}$

40. $\begin{bmatrix} 1 & 1 & 2 \\ 3 & 1 & 0 \\ -2 & 0 & 3 \end{bmatrix}$

41. $\begin{bmatrix} -1 & 2 & -5 \\ 0 & 3 & 4 \\ 0 & 0 & 3 \end{bmatrix}$

42. $\begin{bmatrix} 1 & 0 & 0 \\ -4 & -1 & 0 \\ 5 & 1 & 5 \end{bmatrix}$

43. $\begin{bmatrix} 1 & 4 & -2 \\ 3 & 2 & 0 \\ -1 & 4 & 3 \end{bmatrix}$

44. $\begin{bmatrix} 2 & -1 & 3 \\ 1 & 4 & 4 \\ 1 & 0 & 2 \end{bmatrix}$

45. $\begin{bmatrix} 2 & 4 & 6 \\ 0 & 3 & 1 \\ 0 & 0 & -5 \end{bmatrix}$

46. $\begin{bmatrix} -3 & 0 & 0 \\ 7 & 11 & 0 \\ 1 & 2 & 2 \end{bmatrix}$

47. $\begin{bmatrix} 2 & 6 & 6 & 2 \\ 2 & 7 & 3 & 6 \\ 1 & 5 & 0 & 1 \\ 3 & 7 & 0 & 7 \end{bmatrix}$ **48.** $\begin{bmatrix} 3 & 6 & -5 & 4 \\ -2 & 0 & 6 & 0 \\ 1 & 1 & 2 & 2 \\ 0 & 3 & -1 & -1 \end{bmatrix}$

49. $\begin{bmatrix} 5 & 3 & 0 & 6 \\ 4 & 6 & 4 & 12 \\ 0 & 2 & -3 & 4 \\ 0 & 1 & -2 & 2 \end{bmatrix}$ **50.** $\begin{bmatrix} 1 & 4 & 3 & 2 \\ -5 & 6 & 2 & 1 \\ 0 & 0 & 0 & 0 \\ 3 & -2 & 1 & 5 \end{bmatrix}$

51. $\begin{bmatrix} 3 & 2 & 4 & -1 & 5 \\ -2 & 0 & 1 & 3 & 2 \\ 1 & 0 & 0 & 4 & 0 \\ 6 & 0 & 2 & -1 & 0 \\ 3 & 0 & 5 & 1 & 0 \end{bmatrix}$

52. $\begin{bmatrix} 5 & 2 & 0 & 0 & -2 \\ 0 & 1 & 4 & 3 & 2 \\ 0 & 0 & 2 & 6 & 3 \\ 0 & 0 & 3 & 4 & 1 \\ 0 & 0 & 0 & 0 & 2 \end{bmatrix}$

▦ **In Exercises 53–60, use the matrix capabilities of a graphing utility to evaluate the determinant.**

53. $\begin{vmatrix} 3 & 8 & -7 \\ 0 & -5 & 4 \\ 8 & 1 & 6 \end{vmatrix}$ **54.** $\begin{vmatrix} 5 & -8 & 0 \\ 9 & 7 & 4 \\ -8 & 7 & 1 \end{vmatrix}$

55. $\begin{vmatrix} 7 & 0 & -14 \\ -2 & 5 & 4 \\ -6 & 2 & 12 \end{vmatrix}$ **56.** $\begin{vmatrix} 3 & 0 & 0 \\ -2 & 5 & 0 \\ 12 & 5 & 7 \end{vmatrix}$

57. $\begin{vmatrix} 1 & -1 & 8 & 4 \\ 2 & 6 & 0 & -4 \\ 2 & 0 & 2 & 6 \\ 0 & 2 & 8 & 0 \end{vmatrix}$ **58.** $\begin{vmatrix} 0 & -3 & 8 & 2 \\ 8 & 1 & -1 & 6 \\ -4 & 6 & 0 & 9 \\ -7 & 0 & 0 & 14 \end{vmatrix}$

59. $\begin{vmatrix} 3 & -2 & 4 & 3 & 1 \\ -1 & 0 & 2 & 1 & 0 \\ 5 & -1 & 0 & 3 & 2 \\ 4 & 7 & -8 & 0 & 0 \\ 1 & 2 & 3 & 0 & 2 \end{vmatrix}$

60. $\begin{vmatrix} -2 & 0 & 0 & 0 & 0 \\ 0 & 3 & 0 & 0 & 0 \\ 0 & 0 & -1 & 0 & 0 \\ 0 & 0 & 0 & 2 & 0 \\ 0 & 0 & 0 & 0 & -4 \end{vmatrix}$

In Exercises 61–68, find (a) $|A|$, (b) $|B|$, (c) AB, and (d) $|AB|$.

61. $A = \begin{bmatrix} -1 & 0 \\ 0 & 3 \end{bmatrix}$, $B = \begin{bmatrix} 2 & 0 \\ 0 & -1 \end{bmatrix}$

62. $A = \begin{bmatrix} -2 & 1 \\ 4 & -2 \end{bmatrix}$, $B = \begin{bmatrix} 1 & 2 \\ 0 & -1 \end{bmatrix}$

63. $A = \begin{bmatrix} 4 & 0 \\ 3 & -2 \end{bmatrix}$, $B = \begin{bmatrix} -1 & 1 \\ -2 & 2 \end{bmatrix}$

64. $A = \begin{bmatrix} 5 & 4 \\ 3 & -1 \end{bmatrix}$, $B = \begin{bmatrix} 0 & 6 \\ 1 & -2 \end{bmatrix}$

65. $A = \begin{bmatrix} 0 & 1 & 2 \\ -3 & -2 & 1 \\ 0 & 4 & 1 \end{bmatrix}$, $B = \begin{bmatrix} 3 & -2 & 0 \\ 1 & -1 & 2 \\ 3 & 1 & 1 \end{bmatrix}$

66. $A = \begin{bmatrix} 3 & 2 & 0 \\ -1 & -3 & 4 \\ -2 & 0 & 1 \end{bmatrix}$, $B = \begin{bmatrix} -3 & 0 & 1 \\ 0 & 2 & -1 \\ -2 & -1 & 1 \end{bmatrix}$

67. $A = \begin{bmatrix} -1 & 2 & 1 \\ 1 & 0 & 1 \\ 0 & 1 & 0 \end{bmatrix}$, $B = \begin{bmatrix} -1 & 0 & 0 \\ 0 & 2 & 0 \\ 0 & 0 & 3 \end{bmatrix}$

68. $A = \begin{bmatrix} 2 & 0 & 1 \\ 1 & -1 & 2 \\ 3 & 1 & 0 \end{bmatrix}$, $B = \begin{bmatrix} 2 & -1 & 4 \\ 0 & 1 & 3 \\ 3 & -2 & 1 \end{bmatrix}$

In Exercises 69–74, evaluate the determinant(s) to verify the equation.

69. $\begin{vmatrix} w & x \\ y & z \end{vmatrix} = -\begin{vmatrix} y & z \\ w & x \end{vmatrix}$

70. $\begin{vmatrix} w & cx \\ y & cz \end{vmatrix} = c\begin{vmatrix} w & x \\ y & z \end{vmatrix}$

71. $\begin{vmatrix} w & x \\ y & z \end{vmatrix} = \begin{vmatrix} w & x + cw \\ y & z + cy \end{vmatrix}$

72. $\begin{vmatrix} w & x \\ cw & cx \end{vmatrix} = 0$

73. $\begin{vmatrix} 1 & x & x^2 \\ 1 & y & y^2 \\ 1 & z & z^2 \end{vmatrix} = (y - x)(z - x)(z - y)$

74. $\begin{vmatrix} a + b & a & a \\ a & a + b & a \\ a & a & a + b \end{vmatrix} = b^2(3a + b)$

In Exercises 75–78, solve for x.

75. $\begin{vmatrix} x - 1 & 2 \\ 3 & x - 2 \end{vmatrix} = 0$

76. $\begin{vmatrix} x - 2 & -1 \\ -3 & x \end{vmatrix} = 0$

77. $\begin{vmatrix} x + 3 & 2 \\ 1 & x + 2 \end{vmatrix} = 0$

78. $\begin{vmatrix} x + 4 & -2 \\ 7 & x - 5 \end{vmatrix} = 0$

🔵 In Exercises 79–84, evaluate the determinant in which the entries are functions. Determinants of this type occur when changes in variables are made in calculus.

79. $\begin{vmatrix} 4u & -1 \\ -1 & 2v \end{vmatrix}$

80. $\begin{vmatrix} 3x^2 & -3y^2 \\ 1 & 1 \end{vmatrix}$

81. $\begin{vmatrix} e^{2x} & e^{3x} \\ 2e^{2x} & 3e^{3x} \end{vmatrix}$

82. $\begin{vmatrix} e^{-x} & xe^{-x} \\ -e^{-x} & (1-x)e^{-x} \end{vmatrix}$

83. $\begin{vmatrix} x & \ln x \\ 1 & 1/x \end{vmatrix}$

84. $\begin{vmatrix} x & x\ln x \\ 1 & 1+\ln x \end{vmatrix}$

Synthesis

True or False? In Exercises 85 and 86, determine whether the statement is true or false. Justify your answer.

85. If a square matrix has an entire row of zeros, the determinant will always be zero.

86. If two columns of a square matrix are the same, then the determinant of the matrix will be zero.

87. *Exploration* Find square matrices A and B to demonstrate that

$$|A + B| \neq |A| + |B|.$$

88. *Exploration* Consider square matrices in which the entries are consecutive integers. An example of such a matrix is

$$\begin{bmatrix} 4 & 5 & 6 \\ 7 & 8 & 9 \\ 10 & 11 & 12 \end{bmatrix}.$$

(a) Use a graphing utility to evaluate the determinants of four matrices of this type. Make a conjecture based on the results.

(b) Verify your conjecture.

89. *Writing* Write a brief paragraph explaining the difference between a square matrix and its determinant.

90. *Think About It* If A is a matrix of order 3×3 such that $|A| = 5$, is it possible to find $|2A|$? Explain.

Review

In Exercises 91–94, find the equation of the conic satisfying the conditions.

91. Parabola: Vertex: $(0, 3)$; Focus: $(2, 3)$

92. Ellipse: Vertices: $(0, \pm 4)$; Foci: $(0, \pm 3)$

93. Ellipse: Vertices: $(\pm 8, 0)$; Foci: $(\pm 6, 0)$

94. Hyperbola: Vertices: $(\pm 5, 0)$; Foci: $(\pm 6, 0)$

In Exercises 95 and 96, sketch the graph of the system of inequalities.

95. $\begin{cases} x + y \leq 8 \\ x \geq -3 \\ 2x - y < 5 \end{cases}$

96. $\begin{cases} -x - y > 4 \\ y \leq 1 \\ 7x + 4y \leq -10 \end{cases}$

In Exercises 97–100, find the inverse of the matrix (if it exists).

97. $\begin{bmatrix} -4 & 1 \\ 8 & -1 \end{bmatrix}$

98. $\begin{bmatrix} -5 & -8 \\ 3 & 6 \end{bmatrix}$

99. $\begin{bmatrix} -7 & 2 & 9 \\ 2 & -4 & -6 \\ 3 & 5 & 2 \end{bmatrix}$

100. $\begin{bmatrix} -6 & 2 & 0 \\ 1 & 3 & -2 \\ -2 & 0 & 1 \end{bmatrix}$

7.5 Applications of Matrices and Determinants

▶ **What you should learn**

- How to use Cramer's Rule to solve systems of linear equations
- How to use determinants to find the areas of triangles
- How to use a determinant to find an equation of a line passing through two points
- How to use matrices to code and decode messages

▶ **Why you should learn it**

You can use determinants and matrices to model and solve real-life problems. For instance, Exercise 27 on page 565 shows how matrices can be used to estimate the area of a region of land.

Layne Kennedy/CORBIS

Cramer's Rule

So far, you have studied three methods for solving a system of linear equations: substitution, elimination with equations, and elimination with matrices. In this section, you will study one more method, **Cramer's Rule,** named after Gabriel Cramer (1704–1752). This rule uses determinants to write the solution of a system of linear equations. To see how Cramer's Rule works, take another look at the solution described at the beginning of Section 7.4. There, it was pointed out that the system

$$\begin{cases} a_1x + b_1y = c_1 \\ a_2x + b_2y = c_2 \end{cases}$$

has a solution

$$x = \frac{c_1b_2 - c_2b_1}{a_1b_2 - a_2b_1}$$

and

$$y = \frac{a_1c_2 - a_2c_1}{a_1b_2 - a_2b_1}$$

provided that $a_1b_2 - a_2b_1 \neq 0$. Each numerator and denominator in this solution can be expressed as a determinant, as follows.

$$x = \frac{c_1b_2 - c_2b_1}{a_1b_2 - a_2b_1} = \frac{\begin{vmatrix} c_1 & b_1 \\ c_2 & b_2 \end{vmatrix}}{\begin{vmatrix} a_1 & b_1 \\ a_2 & b_2 \end{vmatrix}}$$

$$y = \frac{a_1c_2 - a_2c_1}{a_1b_2 - a_2b_1} = \frac{\begin{vmatrix} a_1 & c_1 \\ a_2 & c_2 \end{vmatrix}}{\begin{vmatrix} a_1 & b_1 \\ a_2 & b_2 \end{vmatrix}}$$

Relative to the original system, the denominator for x and y is simply the determinant of the *coefficient* matrix of the system. This determinant is denoted by D. The numerators for x and y are denoted by D_x and D_y, respectively. They are formed by using the column of constants as replacements for the coefficients of x and y, as follows.

Coefficient Matrix	D	D_x	D_y
$\begin{bmatrix} a_1 & b_1 \\ a_2 & b_2 \end{bmatrix}$	$\begin{vmatrix} a_1 & b_1 \\ a_2 & b_2 \end{vmatrix}$	$\begin{vmatrix} c_1 & b_1 \\ c_2 & b_2 \end{vmatrix}$	$\begin{vmatrix} a_1 & c_1 \\ a_2 & c_2 \end{vmatrix}$

Example 1 ▶ **Using Cramer's Rule for a 2 × 2 System**

Use Cramer's Rule to solve the system of linear equations.

$$\begin{cases} 4x - 2y = 10 \\ 3x - 5y = 11 \end{cases}$$

Solution

To begin, find the determinant of the coefficient matrix.

$$D = \begin{vmatrix} 4 & -2 \\ 3 & -5 \end{vmatrix} = -20 - (-6) = -14$$

Because this determinant is not zero, you can apply Cramer's Rule to find the solution, as follows.

$$x = \frac{D_x}{D} = \frac{\begin{vmatrix} 10 & -2 \\ 11 & -5 \end{vmatrix}}{-14} \qquad y = \frac{D_y}{D} = \frac{\begin{vmatrix} 4 & 10 \\ 3 & 11 \end{vmatrix}}{-14}$$

$$= \frac{(-50) - (-22)}{-14} \qquad\qquad = \frac{44 - 30}{-14}$$

$$= \frac{-28}{-14} \qquad\qquad\qquad = \frac{14}{-14}$$

$$= 2 \qquad\qquad\qquad\qquad = -1$$

Therefore, the solution is $x = 2$ and $y = -1$. Check this in the original system.

Cramer's Rule generalizes easily to systems of n equations in n variables. The value of each variable is given as the quotient of two determinants. The denominator is the determinant of the coefficient matrix, and the numerator is the determinant of the matrix formed by replacing the column corresponding to the variable (being solved for) with the column representing the constants. For instance, the solution for x_3 in the system

$$a_{11}x_1 + a_{12}x_2 + a_{13}x_3 = b_1$$
$$a_{21}x_1 + a_{22}x_2 + a_{23}x_3 = b_2$$
$$a_{31}x_1 + a_{32}x_2 + a_{33}x_3 = b_3$$

is given by

$$x_3 = \frac{|A_3|}{|A|} = \frac{\begin{vmatrix} a_{11} & a_{12} & b_1 \\ a_{21} & a_{22} & b_2 \\ a_{31} & a_{32} & b_3 \end{vmatrix}}{\begin{vmatrix} a_{11} & a_{12} & a_{13} \\ a_{21} & a_{22} & a_{23} \\ a_{31} & a_{32} & a_{33} \end{vmatrix}}.$$

STUDY T!P

When using Cramer's Rule, remember that the method does not apply if the determinant of the coefficient matrix is zero.

Cramer's Rule

If a system of n linear equations in n variables has a coefficient matrix A with a nonzero determinant $|A|$, the solution of the system is

$$x_1 = \frac{|A_1|}{|A|}, \quad x_2 = \frac{|A_2|}{|A|}, \quad \ldots, \quad x_n = \frac{|A_n|}{|A|}$$

where the ith column of A_i is the column of constants in the system of equations. If the determinant of the coefficient matrix is zero, the system has either no solution or infinitely many solutions.

Example 2 ▶ Using Cramer's Rule for a 3 × 3 System

Use Cramer's Rule to solve the system of linear equations.

$$\begin{cases} -x + 2y - 3z = 1 \\ 2x \qquad + z = 0 \\ 3x - 4y + 4z = 2 \end{cases}$$

Solution

The coefficient matrix

$$\begin{bmatrix} -1 & 2 & -3 \\ 2 & 0 & 1 \\ 3 & -4 & 4 \end{bmatrix}$$

can be expanded along the second row, as follows.

$$D = 2(-1)^3 \begin{vmatrix} 2 & -3 \\ -4 & 4 \end{vmatrix} + 0(-1)^4 \begin{vmatrix} -1 & -3 \\ 3 & 4 \end{vmatrix} + 1(-1)^5 \begin{vmatrix} -1 & 2 \\ 3 & -4 \end{vmatrix}$$

$$= -2(-4) + 0 - 1(-2)$$

$$= 10$$

Because this determinant is not zero, you can apply Cramer's Rule to find the solution, as follows.

$$x = \frac{D_x}{D} = \frac{\begin{vmatrix} 1 & 2 & -3 \\ 0 & 0 & 1 \\ 2 & -4 & 4 \end{vmatrix}}{10} = \frac{8}{10} = \frac{4}{5}$$

$$y = \frac{D_y}{D} = \frac{\begin{vmatrix} -1 & 1 & -3 \\ 2 & 0 & 1 \\ 3 & 2 & 4 \end{vmatrix}}{10} = \frac{-15}{10} = -\frac{3}{2}$$

$$z = \frac{D_z}{D} = \frac{\begin{vmatrix} -1 & 2 & 1 \\ 2 & 0 & 0 \\ 3 & -4 & 2 \end{vmatrix}}{10} = \frac{-16}{10} = -\frac{8}{5}$$

The solution is $\left(\frac{4}{5}, -\frac{3}{2}, -\frac{8}{5}\right)$. Check this in the original system.

Area of a Triangle

Another application of matrices and determinants is finding the area of a triangle whose vertices are given as points in a coordinate plane.

Area of a Triangle

The area of a triangle with vertices (x_1, y_1), (x_2, y_2), and (x_3, y_3) is

$$\text{Area} = \pm\frac{1}{2}\begin{vmatrix} x_1 & y_1 & 1 \\ x_2 & y_2 & 1 \\ x_3 & y_3 & 1 \end{vmatrix}$$

where the symbol $\pm$ indicates that the appropriate sign should be chosen to yield a positive area.

Example 3 ▶ Finding the Area of a Triangle

Find the area of a triangle whose vertices are $(1, 0)$, $(2, 2)$, and $(4, 3)$, as shown in Figure 7.1.

Solution

Let $(x_1, y_1) = (1, 0)$, $(x_2, y_2) = (2, 2)$, and $(x_3, y_3) = (4, 3)$. Then, to find the area of the triangle, evaluate the determinant.

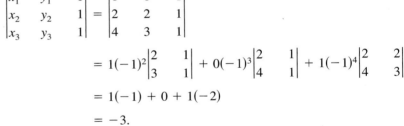

$$\begin{vmatrix} x_1 & y_1 & 1 \\ x_2 & y_2 & 1 \\ x_3 & y_3 & 1 \end{vmatrix} = \begin{vmatrix} 1 & 0 & 1 \\ 2 & 2 & 1 \\ 4 & 3 & 1 \end{vmatrix}$$

$$= 1(-1)^2\begin{vmatrix} 2 & 1 \\ 3 & 1 \end{vmatrix} + 0(-1)^3\begin{vmatrix} 2 & 1 \\ 4 & 1 \end{vmatrix} + 1(-1)^4\begin{vmatrix} 2 & 2 \\ 4 & 3 \end{vmatrix}$$

$$= 1(-1) + 0 + 1(-2)$$

$$= -3.$$

Using this value, you can conclude that the area of the triangle is

$$\text{Area} = -\frac{1}{2}\begin{vmatrix} 1 & 0 & 1 \\ 2 & 2 & 1 \\ 4 & 3 & 1 \end{vmatrix} \qquad \text{Choose } (-) \text{ so that the area is positive.}$$

$$= -\frac{1}{2}(-3)$$

$$= \frac{3}{2}.$$

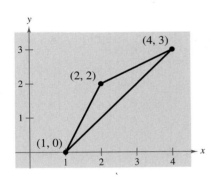

y

3 ─ (4, 3)

2 ─ (2, 2)

1 ─

(1, 0)

      1   2   3   4 → x

FIGURE **7.1**

Try using determinants to find the area of a triangle with vertices $(3, -1)$, $(7, -1)$, and $(7, 5)$. Confirm your answer by plotting the points in a coordinate plane and using the formula

$$\text{Area} = \frac{1}{2}(\text{base})(\text{height}).$$

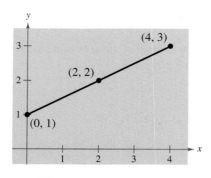

FIGURE **7.2**

Lines in a Plane

What if the three points in Example 3 had been on the same line? What would have happened had the area formula been applied to three such points? The answer is that the determinant would have been zero. Consider, for instance, the three collinear points $(0, 1)$, $(2, 2)$, and $(4, 3)$, as shown in Figure 7.2. The area of the "triangle" that has these three points as vertices is

$$\frac{1}{2}\begin{vmatrix} 0 & 1 & 1 \\ 2 & 2 & 1 \\ 4 & 3 & 1 \end{vmatrix} = \frac{1}{2}\left[0(-1)^2\begin{vmatrix} 2 & 1 \\ 3 & 1 \end{vmatrix} + 1(-1)^3\begin{vmatrix} 2 & 1 \\ 4 & 1 \end{vmatrix} + 1(-1)^4\begin{vmatrix} 2 & 2 \\ 4 & 3 \end{vmatrix}\right]$$

$$= \frac{1}{2}[0(-1) + 1(2) + 1(-2)]$$

$$= 0.$$

The result is generalized as follows.

Test for Collinear Points

Three points (x_1, y_1), (x_2, y_2), and (x_3, y_3) are collinear (lie on the same line) if and only if

$$\begin{vmatrix} x_1 & y_1 & 1 \\ x_2 & y_2 & 1 \\ x_3 & y_3 & 1 \end{vmatrix} = 0.$$

Example 4 ▶ Testing for Collinear Points

Determine whether the points $(-2, -2)$, $(1, 1)$, and $(7, 5)$ lie on the same line. (See Figure 7.3.)

Solution

Letting $(x_1, y_1) = (-2, -2)$, $(x_2, y_2) = (1, 1)$, and $(x_3, y_3) = (7, 5)$, you have

$$\begin{vmatrix} x_1 & y_1 & 1 \\ x_2 & y_2 & 1 \\ x_3 & y_3 & 1 \end{vmatrix} = \begin{vmatrix} -2 & -2 & 1 \\ 1 & 1 & 1 \\ 7 & 5 & 1 \end{vmatrix}$$

$$= -2(-1)^2\begin{vmatrix} 1 & 1 \\ 5 & 1 \end{vmatrix} + (-2)(-1)^3\begin{vmatrix} 1 & 1 \\ 7 & 1 \end{vmatrix} + 1(-1)^4\begin{vmatrix} 1 & 1 \\ 7 & 5 \end{vmatrix}$$

$$= -2(-4) + (-2)(6) + 1(-2)$$

$$= -6.$$

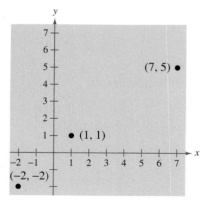

FIGURE **7.3**

Because the value of this determinant is *not* zero, you can conclude that the three points do not lie on the same line.

The test for collinear points can be adapted to another use. That is, if you are given two points on a rectangular coordinate system, you can find an equation of the line passing through the two points, as follows.

Two-Point Form of the Equation of a Line

An equation of the line passing through the distinct points (x_1, y_1) and (x_2, y_2) is given by

$$\begin{vmatrix} x & y & 1 \\ x_1 & y_1 & 1 \\ x_2 & y_2 & 1 \end{vmatrix} = 0.$$

Example 5 ▶ Finding an Equation of a Line

Find an equation of the line passing through the two points $(2, 4)$ and $(-1, 3)$, as shown in Figure 7.4.

Solution

Applying the determinant formula for the equation of a line produces

$$\begin{vmatrix} x & y & 1 \\ 2 & 4 & 1 \\ -1 & 3 & 1 \end{vmatrix} = 0.$$

To evaluate this determinant, you can expand by cofactors along the first row to obtain the following.

$$x(-1)^2\begin{vmatrix} 4 & 1 \\ 3 & 1 \end{vmatrix} + y(-1)^3\begin{vmatrix} 2 & 1 \\ -1 & 1 \end{vmatrix} + 1(-1)^4\begin{vmatrix} 2 & 4 \\ -1 & 3 \end{vmatrix} = 0$$

$$x(1)(1) + y(-1)(3) + (1)(1)(10) = 0$$

$$x - 3y + 10 = 0$$

Therefore, an equation of the line is

$$x - 3y + 10 = 0.$$

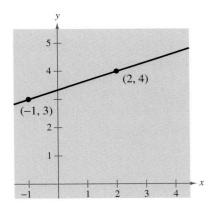

FIGURE **7.4**

Note that this method of finding the equation of a line works for all lines, including horizontal and vertical lines. For instance, the equation of the vertical line through $(2, 0)$ and $(2, 2)$ is

$$\begin{vmatrix} x & y & 1 \\ 2 & 0 & 1 \\ 2 & 2 & 1 \end{vmatrix} = 0$$

$$4 - 2x = 0$$

$$x = 2.$$

Cryptography

A **cryptogram** is a message written according to a secret code. (The Greek word *kryptos* means "hidden.") Matrix multiplication can be used to encode and decode messages. To begin, you need to assign a number to each letter in the alphabet (with 0 assigned to a blank space), as follows.

0 = _	9 = I	18 = R
1 = A	10 = J	19 = S
2 = B	11 = K	20 = T
3 = C	12 = L	21 = U
4 = D	13 = M	22 = V
5 = E	14 = N	23 = W
6 = F	15 = O	24 = X
7 = G	16 = P	25 = Y
8 = H	17 = Q	26 = Z

Then the message is converted to numbers and partitioned into **uncoded row matrices,** each having n entries, as demonstrated in Example 6.

Example 6 ▶ Forming Uncoded Row Matrices

Write the uncoded row matrices of order 1×3 for the message

MEET ME MONDAY.

Solution

Partitioning the message (including blank spaces, but ignoring punctuation) into groups of three produces the following uncoded row matrices.

$$\underset{\text{M\ \ \ E\ \ \ E}}{[13 \quad 5 \quad 5]} \quad \underset{\text{T\ \ \ \ \ \ M}}{[20 \quad 0 \quad 13]} \quad \underset{\text{E\ \ \ \ \ \ M}}{[5 \quad 0 \quad 13]} \quad \underset{\text{O\ \ N\ \ \ D}}{[15 \quad 14 \quad 4]} \quad \underset{\text{A\ \ \ Y}}{[1 \quad 25 \quad 0]}$$

Note that a blank space is used to fill out the last uncoded row matrix.

To encode a message, choose an $n \times n$ invertible matrix such as

$$A = \begin{bmatrix} 1 & -2 & 2 \\ -1 & 1 & 3 \\ 1 & -1 & -4 \end{bmatrix}$$

and multiply the uncoded row matrices by A (on the right) to obtain **coded row matrices.** Here is an example.

Uncoded Matrix *Encoding Matrix A* *Coded Matrix*

$$[13 \quad 5 \quad 5] \begin{bmatrix} 1 & -2 & 2 \\ -1 & 1 & 3 \\ 1 & -1 & -4 \end{bmatrix} = [13 \quad -26 \quad 21]$$

This technique is further illustrated in Example 7.

Example 7 ► Encoding a Message

Use the following matrix to encode the message MEET ME MONDAY.

$$A = \begin{bmatrix} 1 & -2 & 2 \\ -1 & 1 & 3 \\ 1 & -1 & -4 \end{bmatrix}$$

Solution

The coded row matrices are obtained by multiplying each of the uncoded row matrices found in Example 6 by the matrix A, as follows.

Uncoded Matrix Encoding Matrix A Coded Matrix

$$\begin{bmatrix} 13 & 5 & 5 \end{bmatrix} \begin{bmatrix} 1 & -2 & 2 \\ -1 & 1 & 3 \\ 1 & -1 & -4 \end{bmatrix} = \begin{bmatrix} 13 & -26 & 21 \end{bmatrix}$$

$$\begin{bmatrix} 20 & 0 & 13 \end{bmatrix} \begin{bmatrix} 1 & -2 & 2 \\ -1 & 1 & 3 \\ 1 & -1 & -4 \end{bmatrix} = \begin{bmatrix} 33 & -53 & -12 \end{bmatrix}$$

$$\begin{bmatrix} 5 & 0 & 13 \end{bmatrix} \begin{bmatrix} 1 & -2 & 2 \\ -1 & 1 & 3 \\ 1 & -1 & -4 \end{bmatrix} = \begin{bmatrix} 18 & -23 & -42 \end{bmatrix}$$

$$\begin{bmatrix} 15 & 14 & 4 \end{bmatrix} \begin{bmatrix} 1 & -2 & 2 \\ -1 & 1 & 3 \\ 1 & -1 & -4 \end{bmatrix} = \begin{bmatrix} 5 & -20 & 56 \end{bmatrix}$$

$$\begin{bmatrix} 1 & 25 & 0 \end{bmatrix} \begin{bmatrix} 1 & -2 & 2 \\ -1 & 1 & 3 \\ 1 & -1 & -4 \end{bmatrix} = \begin{bmatrix} -24 & 23 & 77 \end{bmatrix}$$

So, the sequence of coded row matrices is

$$\begin{bmatrix} 13 & -26 & 21 \end{bmatrix} \begin{bmatrix} 33 & -53 & -12 \end{bmatrix} \begin{bmatrix} 18 & -23 & -42 \end{bmatrix} \begin{bmatrix} 5 & -20 & 56 \end{bmatrix} \begin{bmatrix} -24 & 23 & 77 \end{bmatrix}.$$

Finally, removing the matrix notation produces the following cryptogram.

$$13 \ -26 \ 21 \ 33 \ -53 \ -12 \ 18 \ -23 \ -42 \ 5 \ -20 \ 56 \ -24 \ 23 \ 77$$

For those who do not know the matrix A, decoding the cryptogram found in Example 7 is difficult. But for an authorized receiver who knows the matrix A, decoding is simple. The receiver need only multiply the coded row matrices by A^{-1} (on the right) to retrieve the uncoded row matrices. Here is an example.

$$\underbrace{\begin{bmatrix} 13 & -26 & 21 \end{bmatrix}}_{\text{Coded}} A^{-1} = \underbrace{\begin{bmatrix} 13 & 5 & 5 \end{bmatrix}}_{\text{Uncoded}}$$

Example 8 ▶ Decoding a Message

Use the inverse of the matrix

$$A = \begin{bmatrix} 1 & -2 & 2 \\ -1 & 1 & 3 \\ 1 & -1 & -4 \end{bmatrix}$$

to decode the cryptogram

$$13 \;-26\; 21\; 33 \;-53 \;-12\; 18 \;-23 \;-42\; 5 \;-20\; 56 \;-24\; 23\; 77.$$

Solution

Partition the message into groups of three to form the coded row matrices. Then, multiply each coded row matrix by A^{-1} (on the right).

Coded Matrix   *Decoding Matrix A^{-1}*   *Decoded Matrix*

$$\begin{bmatrix} 13 & -26 & 21 \end{bmatrix} \begin{bmatrix} -1 & -10 & -8 \\ -1 & -6 & -5 \\ 0 & -1 & -1 \end{bmatrix} = \begin{bmatrix} 13 & 5 & 5 \end{bmatrix}$$

$$\begin{bmatrix} 33 & -53 & -12 \end{bmatrix} \begin{bmatrix} -1 & -10 & -8 \\ -1 & -6 & -5 \\ 0 & -1 & -1 \end{bmatrix} = \begin{bmatrix} 20 & 0 & 13 \end{bmatrix}$$

$$\begin{bmatrix} 18 & -23 & -42 \end{bmatrix} \begin{bmatrix} -1 & -10 & -8 \\ -1 & -6 & -5 \\ 0 & -1 & -1 \end{bmatrix} = \begin{bmatrix} 5 & 0 & 13 \end{bmatrix}$$

$$\begin{bmatrix} 5 & -20 & 56 \end{bmatrix} \begin{bmatrix} -1 & -10 & -8 \\ -1 & -6 & -5 \\ 0 & -1 & -1 \end{bmatrix} = \begin{bmatrix} 15 & 14 & 4 \end{bmatrix}$$

$$\begin{bmatrix} -24 & 23 & 77 \end{bmatrix} \begin{bmatrix} -1 & -10 & -8 \\ -1 & -6 & -5 \\ 0 & -1 & -1 \end{bmatrix} = \begin{bmatrix} 1 & 25 & 0 \end{bmatrix}$$

So, the message is as follows.

$$\begin{bmatrix} 13 & 5 & 5 \end{bmatrix} \begin{bmatrix} 20 & 0 & 13 \end{bmatrix} \begin{bmatrix} 5 & 0 & 13 \end{bmatrix} \begin{bmatrix} 15 & 14 & 4 \end{bmatrix} \begin{bmatrix} 1 & 25 & 0 \end{bmatrix}$$
$$\text{M} \quad \text{E} \quad \text{E} \qquad \text{T} \qquad \text{M} \quad \text{E} \qquad \text{M} \quad \text{O} \quad \text{N} \qquad \text{D} \quad \text{A} \quad \text{Y}$$

Writing ABOUT MATHEMATICS

Cryptography Use your school's library or some other reference source to research information about another type of cryptography. Write a short paragraph describing how mathematics is used to code and decode messages.

7.5 Exercises

In Exercises 1–8, use Cramer's Rule to solve (if possible) the system of equations.

1. $\begin{cases} 3x + 4y = -2 \\ 5x + 3y = 4 \end{cases}$

2. $\begin{cases} -4x - 7y = 47 \\ -x + 6y = -27 \end{cases}$

3. $\begin{cases} -0.4x + 0.8y = 1.6 \\ 0.2x + 0.3y = 2.2 \end{cases}$

4. $\begin{cases} 2.4x - 1.3y = 14.63 \\ -4.6x + 0.5y = -11.51 \end{cases}$

5. $\begin{cases} 4x - y + z = -5 \\ 2x + 2y + 3z = 10 \\ 5x - 2y + 6z = 1 \end{cases}$

6. $\begin{cases} 4x - 2y + 3z = -2 \\ 2x + 2y + 5z = 16 \\ 8x - 5y - 2z = 4 \end{cases}$

7. $\begin{cases} x + 2y + 3z = -3 \\ -2x + y - z = 6 \\ 3x - 3y + 2z = -11 \end{cases}$

8. $\begin{cases} 5x - 4y + z = -14 \\ -x + 2y - 2z = 10 \\ 3x + y + z = 1 \end{cases}$

In Exercises 9–12, use a graphing utility and Cramer's Rule to solve (if possible) the system of equations.

9. $\begin{cases} 3x + 3y + 5z = 1 \\ 3x + 5y + 9z = 2 \\ 5x + 9y + 17z = 4 \end{cases}$

10. $\begin{cases} x + 2y - z = -7 \\ 2x - 2y - 2z = -8 \\ -x + 3y + 4z = 8 \end{cases}$

11. $\begin{cases} 2x + y + 2z = 6 \\ -x + 2y - 3z = 0 \\ 3x + 2y - z = 6 \end{cases}$

12. $\begin{cases} 2x + 3y + 5z = 4 \\ 3x + 5y + 9z = 7 \\ 5x + 9y + 17z = 13 \end{cases}$

In Exercises 13–22, use a determinant and the given vertices of a triangle to find the area of the triangle.

13.

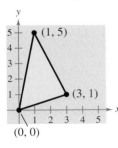

14.

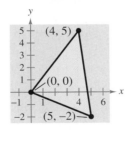

15.

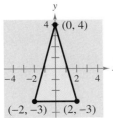

16.

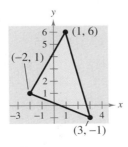

17.

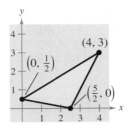

18.

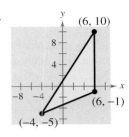

19. $(-2, 4), (2, 3), (-1, 5)$

20. $(0, -2), (-1, 4), (3, 5)$

21. $(-3, 5), (2, 6), (3, -5)$

22. $(-2, 4), (1, 5), (3, -2)$

In Exercises 23 and 24, find a value of x such that the triangle with the given vertices has an area of 4.

23. $(-5, 1), (0, 2), (-2, x)$

24. $(-4, 2), (-3, 5), (-1, x)$

In Exercises 25 and 26, find a value of x such that the triangle with the given vertices has an area of 6.

25. $(-2, -3), (1, -1), (-8, x)$

26. $(1, 0), (5, -3), (-3, x)$

27. *Area of a Region* A large region of forest has been infested with gypsy moths. The region is roughly triangular, as shown in the figure. From the northernmost vertex A of the region, the distances to the other vertices are 25 miles south and 10 miles east (for vertex B), and 20 miles south and 28 miles east (for vertex C). Use a graphing utility to approximate the number of square miles in this region.

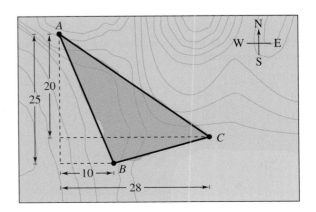

28. *Area of a Region* You own a triangular tract of land, as shown in the figure. To estimate the number of square feet in the tract, you start at one vertex, walk 65 feet east and 50 feet north to the second vertex, and then walk 85 feet west and 30 feet north to the third vertex. Use a graphing utility to determine how many square feet there are in the tract of land.

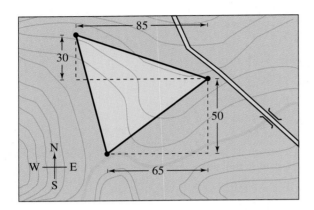

In Exercises 29–34, use a determinant to determine whether the points are collinear.

29. $(3, -1), (0, -3), (12, 5)$

30. $(-3, -5), (6, 1), (10, 2)$

31. $(2, -\frac{1}{2}), (-4, 4), (6, -3)$

32. $(0, 1), (4, -2), (-2, \frac{5}{2})$

33. $(0, 2), (1, 2.4), (-1, 1.6)$

34. $(2, 3), (3, 3.5), (-1, 2)$

In Exercises 35 and 36, find x such that the points are collinear.

35. $(2, -5), (4, x), (5, -2)$

36. $(-6, 2), (-5, x), (-3, 5)$

In Exercises 37–42, use a determinant to find an equation of the line through the points.

37. $(0, 0), (5, 3)$

38. $(0, 0), (-2, 2)$

39. $(-4, 3), (2, 1)$

40. $(10, 7), (-2, -7)$

41. $(-\frac{1}{2}, 3), (\frac{5}{2}, 1)$

42. $(\frac{2}{3}, 4), (6, 12)$

In Exercises 43 and 44, find the uncoded 1 × 3 row matrices for the message. Then encode the message using the matrix.

Message *Matrix*

43. TROUBLE IN RIVER CITY
$\begin{bmatrix} 1 & -1 & 0 \\ 1 & 0 & -1 \\ -6 & 2 & 3 \end{bmatrix}$

44. PLEASE SEND MONEY
$\begin{bmatrix} 4 & 2 & 1 \\ -3 & -3 & -1 \\ 3 & 2 & 1 \end{bmatrix}$

In Exercises 45–48, write a cryptogram for the given message using the matrix *A*.

$A = \begin{bmatrix} 1 & 2 & 2 \\ 3 & 7 & 9 \\ -1 & -4 & -7 \end{bmatrix}.$

45. CALL AT NOON

46. ICEBERG DEAD AHEAD

47. HAPPY BIRTHDAY

48. OPERATION OVERLOAD

In Exercises 49–52, use A^{-1} to decode the cryptogram.

49. $A = \begin{bmatrix} 1 & 2 \\ 3 & 5 \end{bmatrix}$

11 21 64 112 25 50 29 53 23 46
40 75 55 92

50. $A = \begin{bmatrix} -5 & 2 \\ -7 & 3 \end{bmatrix}$

-136 58 -173 72 -120 51 -95 38
-178 73 -70 28 -242 101 -115 47
-90 36 -115 49 -199 82

51. $A = \begin{bmatrix} 1 & -1 & 0 \\ 1 & 0 & -1 \\ -6 & 2 & 3 \end{bmatrix}$

9 -1 -9 38 -19 -19 28 -9 -19
-80 25 41 -64 21 31 9 -5 -4

52. $A = \begin{bmatrix} 3 & -4 & 2 \\ 0 & 2 & 1 \\ 4 & -5 & 3 \end{bmatrix}$

112 -140 83 19 -25 13 72 -76 61
95 -118 71 20 21 38 35 -23 36 42
-48 32

In Exercises 53 and 54, decode the cryptogram by using the inverse of the matrix A.

$$A = \begin{bmatrix} 1 & 2 & 2 \\ 3 & 7 & 9 \\ -1 & -4 & -7 \end{bmatrix}$$

53. 20 17 -15 -12 -56 -104 1 -25 -65
62 143 181

54. 13 -9 -59 61 112 106 -17 -73 -131
11 24 29 65 144 172

55. The following cryptogram was encoded with a 2×2 matrix.

8 21 -15 -10 -13 -13 5 10 5 25 5
19 -1 6 20 40 -18 -18 1 16

The last word of the message is _RON. What is the message?

56. The following cryptogram was encoded with a 2×2 matrix.

5 2 25 11 -2 -7 -15 -15 32 14
-8 -13 38 19 -19 -19 37 16

The last word of the message is _SUE. What is the message?

Synthesis

True or False? **In Exercises 57 and 58, determine whether the statement is true or false. Justify your answer.**

57. You cannot use Cramer's Rule when solving a system of linear equations if the determinant of the coefficient matrix is zero.

58. In a system of linear equations, if the determinant of the coefficient matrix is zero, the system has no solution.

59. *Writing*   At this point in the book, you have learned several methods for solving a system of linear equations. Briefly describe which method(s) you find easiest to use and which method(s) you find most difficult to use.

Review

In Exercises 60–63, use any method to solve the system of equations.

60. $\begin{cases} -x - 7y = -22 \\ 5x + y = -26 \end{cases}$

61. $\begin{cases} 3x + 8y = 11 \\ -2x + 12y = -16 \end{cases}$

62. $\begin{cases} -x - 3y + 5z = -14 \\ 4x + 2y - z = -1 \\ 5x - 3y + 2z = -11 \end{cases}$

63. $\begin{cases} 5x - y - z = 7 \\ -2x + 3y + z = -5 \\ 4x + 10y - 5z = -37 \end{cases}$

In Exercises 64 and 65, sketch the constraint region. Then find the minimum and maximum values of the objective function and where they occur, subject to the constraints.

64. Objective function:

$z = 6x + 4y$

Constraints:

$x \geq 0$

$y \geq 0$

$x + 6y \leq 30$

$6x + y \leq 40$

65. Objective function:

$z = 6x + 7y$

Constraints:

$x \geq 0$

$y \geq 0$

$4x + 3y \geq 24$

$x + 3y \geq 15$

In Exercises 66–69, find the determinant of the matrix.

66. $\begin{bmatrix} -8 & 5 \\ 7 & -9 \end{bmatrix}$

67. $\begin{bmatrix} 2.4 & -4.7 \\ -1.4 & -3 \end{bmatrix}$

68. $\begin{bmatrix} -6 & -1 & 5 \\ -3 & 0 & 4 \\ -2 & 4 & -5 \end{bmatrix}$

69. $\begin{bmatrix} 1 & 4 & -3 \\ 7 & -1 & 2 \\ 6 & 0 & 5 \end{bmatrix}$

Chapter Summary

What did you learn?

	Review Exercises
Section 7.1	
☐ How to write a matrix and identify its order	1–8
☐ How to perform elementary row operations on matrices	9, 10
☐ How to use matrices and Gaussian elimination to solve systems of linear equations	11–26
☐ How to use matrices and Gauss-Jordan elimination to solve systems of linear equations	27–32
Section 7.2	
☐ How to decide whether two matrices are equal	33–36
☐ How to add and subtract matrices and multiply matrices by real numbers	37–50
☐ How to multiply two matrices	51–66
☐ How to use matrix operations to model and solve real-life problems	67
Section 7.3	
☐ How to verify that two matrices are inverses of each other	68–71
☐ How to use Gauss-Jordan elimination to find the inverses of matrices	72–79
☐ How to use a formula to find the inverses of 2×2 matrices	80–83
☐ How to use inverse matrices to solve systems of linear equations	84–95
Section 7.4	
☐ How to find the determinants of 2×2 matrices	96–99
☐ How to find minors and cofactors of square matrices	100–103
☐ How to find the determinants of square matrices	104–107
Section 7.5	
☐ How to use Cramer's Rule to solve systems of linear equations	108–115
☐ How to use determinants to find the areas of triangles	116–119
☐ How to use a determinant to find an equation of a line passing through two points	120–123
☐ How to use matrices to code and decode messages	124–127

Review Exercises

7.1 In Exercises 1–4, determine the order of the matrix.

1. $\begin{bmatrix} -4 \\ 0 \\ 5 \end{bmatrix}$

2. $\begin{bmatrix} 3 & -1 & 0 & 6 \\ -2 & 7 & 1 & 4 \end{bmatrix}$

3. $\begin{bmatrix} 3 \end{bmatrix}$

4. $\begin{bmatrix} 6 & 2 & -5 & 8 & 0 \end{bmatrix}$

In Exercises 5 and 6, form the augmented matrix for the system of linear equations.

5. $\begin{cases} 3x - 10y = 15 \\ 5x + 4y = 22 \end{cases}$

6. $\begin{cases} 8x - 7y + 4z = 12 \\ 3x - 5y + 2z = 20 \\ 5x + 3y - 3z = 26 \end{cases}$

In Exercises 7 and 8, write the system of linear equations represented by the augmented matrix. (Use variables x, y, z, and w.)

7. $\begin{bmatrix} 5 & 1 & 7 & \vdots & -9 \\ 4 & 2 & 0 & \vdots & 10 \\ 9 & 4 & 2 & \vdots & 3 \end{bmatrix}$

8. $\begin{bmatrix} 13 & 16 & 7 & 3 & \vdots & 2 \\ 1 & 21 & 8 & 5 & \vdots & 12 \\ 4 & 10 & -4 & 3 & \vdots & -1 \end{bmatrix}$

In Exercises 9 and 10, write the matrix in reduced row-echelon form.

9. $\begin{bmatrix} 0 & 1 & 1 \\ 1 & 2 & 3 \\ 2 & 2 & 2 \end{bmatrix}$

10. $\begin{bmatrix} 1 & 1 & 1 & 0 \\ 1 & 1 & 0 & 1 \\ 1 & 0 & 1 & 1 \\ 0 & 1 & 1 & 1 \end{bmatrix}$

In Exercises 11–14, the row-echelon form of an augmented matrix that corresponds to a system of linear equations is given. Use the matrix to determine whether the system is consistent or inconsistent, and if consistent, determine the number of solutions.

11. $\begin{bmatrix} 1 & 2 & 3 & \vdots & 9 \\ 0 & 1 & -2 & \vdots & 2 \\ 0 & 0 & 0 & \vdots & 0 \end{bmatrix}$

12. $\begin{bmatrix} 1 & 2 & 3 & \vdots & 9 \\ 0 & 1 & -2 & \vdots & 2 \\ 0 & 0 & 0 & \vdots & 8 \end{bmatrix}$

13. $\begin{bmatrix} 1 & 2 & 3 & \vdots & 9 \\ 0 & 1 & -2 & \vdots & 2 \\ 0 & 0 & 1 & \vdots & -3 \end{bmatrix}$

14. $\begin{bmatrix} 1 & 2 & 3 & 10 & 6 & \vdots & 0 \\ 0 & 1 & -5 & -2 & 0 & \vdots & 5 \\ 0 & 0 & 1 & 12 & 0 & \vdots & -2 \\ 0 & 0 & 0 & 1 & 1 & \vdots & 0 \end{bmatrix}$

In Exercises 15–24, use Gaussian elimination with back-substitution to solve the system of equations.

15. $\begin{cases} 5x + 4y = 2 \\ -x + y = -22 \end{cases}$

16. $\begin{cases} 2x - 5y = 2 \\ 3x - 7y = 1 \end{cases}$

17. $\begin{cases} 0.3x - 0.1y = -0.13 \\ 0.2x - 0.3y = -0.25 \end{cases}$

18. $\begin{cases} 0.2x - 0.1y = 0.07 \\ 0.4x - 0.5y = -0.01 \end{cases}$

19. $\begin{cases} 2x + 3y + z = 10 \\ 2x - 3y - 3z = 22 \\ 4x - 2y + 3z = -2 \end{cases}$

20. $\begin{cases} 2x + 3y + 3z = 3 \\ 6x + 6y + 12z = 13 \\ 12x + 9y - z = 2 \end{cases}$

21. $\begin{cases} 2x + y + 2z = 4 \\ 2x + 2y = 5 \\ 2x - y + 6z = 2 \end{cases}$

22. $\begin{cases} x + 2y + 6z = 1 \\ 2x + 5y + 15z = 4 \\ 3x + y + 3z = -6 \end{cases}$

23. $\begin{cases} 2x + y + z = 6 \\ -2y + 3z - w = 9 \\ 3x + 3y - 2z - 2w = -11 \\ x + z + 3w = 14 \end{cases}$

24. $\begin{cases} x + 2y + w = 3 \\ -3y + 3z = 0 \\ 4x + 4y + z + 2w = 0 \\ 2x + z = 3 \end{cases}$

Partial Fractions In Exercises 25 and 26, write the partial fraction decomposition for the rational expression.

25. $\dfrac{x + 9}{(x + 1)(x + 2)^2} = \dfrac{A}{x + 1} + \dfrac{B}{x + 2} + \dfrac{C}{(x + 2)^2}$

26. $\dfrac{3x - 2}{(x - 4)(x - 3)^2} = \dfrac{A}{x - 4} + \dfrac{B}{x - 3} + \dfrac{C}{(x - 3)^2}$

In Exercises 27–30, use Gauss-Jordan elimination to solve the system of equations.

27. $\begin{cases} -x + y + 2z = 1 \\ 2x + 3y + z = -2 \\ 5x + 4y + 2z = 4 \end{cases}$
28. $\begin{cases} 4x + 4y + 4z = 5 \\ 4x - 2y - 8z = 1 \\ 5x + 3y + 8z = 6 \end{cases}$

29. $\begin{cases} 2x - y + 9z = -8 \\ -x - 3y + 4z = -15 \\ 5x + 2y - z = 17 \end{cases}$

30. $\begin{cases} -3x + y + 7z = -20 \\ 5x - 2y - z = 34 \\ -x + y + 4z = -8 \end{cases}$

In Exercises 31 and 32, use the matrix capabilities of a graphing utility to reduce the augmented matrix corresponding to the system of equations, and solve the system.

31. $\begin{cases} 3x - y + 5z - 2w = -44 \\ x + 6y + 4z - w = 1 \\ 5x - y + z + 3w = -15 \\ 4y - z - 8w = 58 \end{cases}$

32. $\begin{cases} 4x + 12y + 2z = 20 \\ x + 6y + 4z = 12 \\ x + 6y + z = 8 \\ -2x - 10y - 2z = -10 \end{cases}$

7.2 In Exercises 33–36, find x and y.

33. $\begin{bmatrix} -1 & x \\ y & 9 \end{bmatrix} = \begin{bmatrix} -1 & 12 \\ -7 & 9 \end{bmatrix}$

34. $\begin{bmatrix} -1 & 0 \\ x & 5 \\ -4 & y \end{bmatrix} = \begin{bmatrix} -1 & 0 \\ 8 & 5 \\ -4 & 0 \end{bmatrix}$

35. $\begin{bmatrix} x+3 & 4 & -4y \\ 0 & -3 & 2 \\ -2 & y+5 & 6x \end{bmatrix} = \begin{bmatrix} 5x-1 & 4 & -44 \\ 0 & -3 & 2 \\ -2 & 16 & 6 \end{bmatrix}$

36. $\begin{bmatrix} -9 & 4 & 2 & -5 \\ 0 & -3 & 7 & -4 \\ 6 & -1 & 1 & 0 \end{bmatrix} = \begin{bmatrix} -9 & 4 & x-10 & -5 \\ 0 & -3 & 7 & 2y \\ \frac{1}{2}x & -1 & 1 & 0 \end{bmatrix}$

In Exercises 37–40, determine if the matrix operation $A + 3B$ can be performed. If not, state why.

37. $A = \begin{bmatrix} 2 & -2 \\ 3 & 5 \end{bmatrix}$, $B = \begin{bmatrix} -3 & 10 \\ 12 & 8 \end{bmatrix}$

38. $A = \begin{bmatrix} 5 & 4 \\ -7 & 2 \\ 11 & 2 \end{bmatrix}$, $B = \begin{bmatrix} 4 & 12 \\ 20 & 40 \\ 15 & 30 \end{bmatrix}$

39. $A = \begin{bmatrix} 5 & 4 \\ -7 & 2 \\ 11 & 2 \end{bmatrix}$, $B = \begin{bmatrix} 4 & 12 \\ 20 & 40 \end{bmatrix}$

40. $A = \begin{bmatrix} 6 & -5 & 7 \end{bmatrix}$, $B = \begin{bmatrix} -1 \\ 4 \\ 8 \end{bmatrix}$

In Exercises 41–44, perform the matrix operations. If it is not possible, explain why.

41. $\begin{bmatrix} 7 & 3 \\ -1 & 5 \end{bmatrix} + \begin{bmatrix} 10 & -20 \\ 14 & -3 \end{bmatrix}$

42. $\begin{bmatrix} -11 & 16 & 19 \\ -7 & -2 & 1 \end{bmatrix} - \begin{bmatrix} 6 & 0 \\ 8 & -4 \\ -2 & 10 \end{bmatrix}$

43. $-2\begin{bmatrix} 1 & 2 \\ 5 & -4 \\ 6 & 0 \end{bmatrix} + 8\begin{bmatrix} 7 & 1 \\ 1 & 2 \\ 1 & 4 \end{bmatrix}$

44. $-\begin{bmatrix} 8 & -1 & 8 \\ -2 & 4 & 12 \\ 0 & -6 & 0 \end{bmatrix} - 5\begin{bmatrix} -2 & 0 & -4 \\ 3 & -1 & 1 \\ 6 & 12 & -8 \end{bmatrix}$

In Exercises 45 and 46, use a graphing utility to perform the matrix operations.

45. $3\begin{bmatrix} 8 & -2 & 5 \\ 1 & 3 & -1 \end{bmatrix} + 6\begin{bmatrix} 4 & -2 & -3 \\ 2 & 7 & 6 \end{bmatrix}$

46. $-5\begin{bmatrix} 2 & 0 \\ 7 & -2 \\ 8 & 2 \end{bmatrix} + 4\begin{bmatrix} 4 & -2 \\ 6 & 11 \\ -1 & 3 \end{bmatrix}$

In Exercises 47–50, solve for X when

$$A = \begin{bmatrix} -4 & 0 \\ 1 & -5 \\ -3 & 2 \end{bmatrix} \quad \text{and} \quad B = \begin{bmatrix} 1 & 2 \\ -2 & 1 \\ 4 & 4 \end{bmatrix}.$$

47. $X = 3A - 2B$
48. $6X = 4A + 3B$
49. $3X + 2A = B$
50. $2A - 5B = 3X$

In Exercises 51–54, determine if the matrix operation AB can be performed. If not, state why.

51. $A = \begin{bmatrix} 2 & -2 \\ 3 & 5 \end{bmatrix}$, $B = \begin{bmatrix} -3 & 10 \\ 12 & 8 \end{bmatrix}$

52. $A = \begin{bmatrix} 5 & 4 \\ -7 & 2 \\ 11 & 2 \end{bmatrix}$, $B = \begin{bmatrix} 4 & 12 \\ 20 & 40 \\ 15 & 30 \end{bmatrix}$

53. $A = \begin{bmatrix} 5 & 4 \\ -7 & 2 \\ 11 & 2 \end{bmatrix}$, $B = \begin{bmatrix} 4 & 12 \\ 20 & 40 \end{bmatrix}$

54. $A = \begin{bmatrix} 6 & -5 & 7 \end{bmatrix}$, $B = \begin{bmatrix} -1 \\ 4 \\ 8 \end{bmatrix}$

In Exercises 55–62, perform the matrix operations. If it is not possible, explain why.

55. $\begin{bmatrix} 1 & 2 \\ 5 & -4 \\ 6 & 0 \end{bmatrix} \begin{bmatrix} 6 & -2 & 8 \\ 4 & 0 & 0 \end{bmatrix}$

56. $\begin{bmatrix} 1 & 5 & 6 \\ 2 & -4 & 0 \end{bmatrix} \begin{bmatrix} 6 & -2 & 8 \\ 4 & 0 & 0 \end{bmatrix}$

57. $\begin{bmatrix} 1 & 5 & 6 \\ 2 & -4 & 0 \end{bmatrix} \begin{bmatrix} 6 & 4 \\ -2 & 0 \\ 8 & 0 \end{bmatrix}$

58. $\begin{bmatrix} 1 & 3 & 2 \\ 0 & 2 & -4 \\ 0 & 0 & 3 \end{bmatrix} \begin{bmatrix} 4 & -3 & 2 \\ 0 & 3 & -1 \\ 0 & 0 & 2 \end{bmatrix}$

59. $\begin{bmatrix} 4 \\ 6 \end{bmatrix} \begin{bmatrix} 6 & -2 \end{bmatrix}$

60. $\begin{bmatrix} 4 & -2 & 6 \end{bmatrix} \begin{bmatrix} -2 & 1 \\ 0 & -3 \\ 2 & 0 \end{bmatrix}$

61. $\begin{bmatrix} 2 & 1 \\ 6 & 0 \end{bmatrix} \left(\begin{bmatrix} 4 & 2 \\ -3 & 1 \end{bmatrix} + \begin{bmatrix} -2 & 4 \\ 0 & 4 \end{bmatrix} \right)$

62. $-3 \begin{bmatrix} 1 & -1 \\ 4 & 2 \end{bmatrix} \left(\begin{bmatrix} 0 & 3 \\ 1 & 2 \end{bmatrix} \begin{bmatrix} 1 & 0 \\ 5 & -3 \end{bmatrix} \right)$

In Exercises 63 and 64, use a graphing utility to perform the matrix operations.

63. $\begin{bmatrix} 4 & 1 \\ 11 & -7 \\ 12 & 3 \end{bmatrix} \begin{bmatrix} 3 & -5 & 6 \\ 2 & -2 & -2 \end{bmatrix}$

64. $\begin{bmatrix} -2 & 3 & 10 \\ 4 & -2 & 2 \end{bmatrix} \begin{bmatrix} 1 & 1 \\ -5 & 2 \\ 3 & 2 \end{bmatrix}$

65. Write the system of linear equations represented by the matrix equation

$$\begin{bmatrix} 5 & 4 \\ -1 & 1 \end{bmatrix} \begin{bmatrix} x \\ y \end{bmatrix} = \begin{bmatrix} 2 \\ -22 \end{bmatrix}.$$

66. Write the matrix equation $AX = B$ for the system of linear equations.

$$\begin{cases} 2x + 3y + z = 10 \\ 2x - 3y - 3z = 22 \\ 4x - 2y + 3z = -2 \end{cases}$$

67. *Manufacturing*   A manufacturing company produces three models of a product that are shipped to two warehouses. The number of units of model i that are shipped to warehouse j is represented by a_{ij} in the matrix

$$A = \begin{bmatrix} 8200 & 7400 \\ 6500 & 9800 \\ 5400 & 4800 \end{bmatrix}.$$

The price per unit is represented by the matrix $B = \begin{bmatrix} \$10.25 & \$14.50 & \$17.75 \end{bmatrix}$.

(a) Use a graphing utility to compute BA and interpret the result.

(b) Suppose the numbers of units of each model shipped to the warehouses are increased by 25% for the following shipment. Use your graphing utility to find a new matrix A_n representing the number of units being shipped to the warehouses. Then calculate BA_n.

7.3   **In Exercises 68–71, show that B is the inverse of A.**

68. $A = \begin{bmatrix} -4 & -1 \\ 7 & 2 \end{bmatrix}$, $B = \begin{bmatrix} -2 & -1 \\ 7 & 4 \end{bmatrix}$

69. $A = \begin{bmatrix} 5 & -1 \\ 11 & -2 \end{bmatrix}$, $B = \begin{bmatrix} -2 & 1 \\ -11 & 5 \end{bmatrix}$

70. $A = \begin{bmatrix} 1 & 1 & 0 \\ 1 & 0 & 1 \\ 6 & 2 & 3 \end{bmatrix}$, $B = \begin{bmatrix} -2 & -3 & 1 \\ 3 & 3 & -1 \\ 2 & 4 & -1 \end{bmatrix}$

71. $A = \begin{bmatrix} 1 & -1 & 0 \\ -1 & 0 & -1 \\ 8 & -4 & 2 \end{bmatrix}$, $B = \begin{bmatrix} -2 & 1 & \frac{1}{2} \\ -3 & 1 & \frac{1}{2} \\ 2 & -2 & -\frac{1}{2} \end{bmatrix}$

In Exercises 72–75, use Gauss-Jordan elimination to find the inverse of the matrix (if it exists).

72. $\begin{bmatrix} -6 & 5 \\ -5 & 4 \end{bmatrix}$

73. $\begin{bmatrix} -3 & -5 \\ 2 & 3 \end{bmatrix}$

74. $\begin{bmatrix} -1 & -2 & -2 \\ 3 & 7 & 9 \\ 1 & 4 & 7 \end{bmatrix}$

75. $\begin{bmatrix} 0 & -2 & 1 \\ -5 & -2 & -3 \\ 7 & 3 & 4 \end{bmatrix}$

In Exercises 76–79, use a graphing utility to find the inverse of the matrix (if it exists).

76. $\begin{bmatrix} 2 & 6 \\ 3 & -6 \end{bmatrix}$

77. $\begin{bmatrix} 3 & -10 \\ 4 & 2 \end{bmatrix}$

78. $\begin{bmatrix} 2 & 0 & 3 \\ -1 & 1 & 1 \\ 2 & -2 & 1 \end{bmatrix}$

79. $\begin{bmatrix} 1 & 4 & 6 \\ 2 & -3 & 1 \\ -1 & 18 & 16 \end{bmatrix}$

In Exercises 80–83, let

$$A = \begin{bmatrix} a & b \\ c & d \end{bmatrix}.$$

Find the inverse of the above matrix

$$A^{-1} = \frac{1}{ad - bc} \begin{bmatrix} d & -b \\ -c & a \end{bmatrix}$$

(if it exists).

80. $\begin{bmatrix} -7 & 2 \\ -8 & 2 \end{bmatrix}$ **81.** $\begin{bmatrix} 10 & 4 \\ 7 & 3 \end{bmatrix}$

82. $\begin{bmatrix} -\frac{1}{2} & 20 \\ \frac{3}{10} & -6 \end{bmatrix}$ **83.** $\begin{bmatrix} -\frac{3}{4} & \frac{5}{2} \\ -\frac{4}{5} & -\frac{8}{3} \end{bmatrix}$

In Exercises 84–91, use an inverse matrix to solve (if possible) the system of linear equations.

84. $\begin{cases} -x + 4y = 8 \\ 2x - 7y = -5 \end{cases}$ **85.** $\begin{cases} 5x - y = 13 \\ -9x + 2y = -24 \end{cases}$

86. $\begin{cases} -3x + 10y = 8 \\ 5x - 17y = -13 \end{cases}$ **87.** $\begin{cases} 4x - 2y = -10 \\ -19x + 9y = 47 \end{cases}$

88. $\begin{cases} 3x + 2y - z = 6 \\ x - y + 2z = -1 \\ 5x + y + z = 7 \end{cases}$

89. $\begin{cases} -x + 4y - 2z = 12 \\ 2x - 9y + 5z = -25 \\ -x + 5y - 4z = 10 \end{cases}$

90. $\begin{cases} -2x + y + 2z = -13 \\ -x - 4y + z = -11 \\ -y - z = 0 \end{cases}$

91. $\begin{cases} 3x - y + 5z = -14 \\ -x + y + 6z = 8 \\ -8x + 4y - z = 44 \end{cases}$

In Exercises 92–95, use a graphing utility to solve (if possible) the system of linear equations using the inverse of the coefficient matrix.

92. $\begin{cases} x + 2y = -1 \\ 3x + 4y = -5 \end{cases}$ **93.** $\begin{cases} x + 3y = 23 \\ -6x + 2y = -18 \end{cases}$

94. $\begin{cases} -3x - 3y - 4z = 2 \\ y + z = -1 \\ 4x + 3y + 4z = -1 \end{cases}$

95. $\begin{cases} x - 3y - 2z = 8 \\ -2x + 7y + 3z = -19 \\ x - y - 3z = 3 \end{cases}$

7.4 In Exercises 96–99, find the determinant of the matrix.

96. $\begin{bmatrix} 8 & 5 \\ 2 & -4 \end{bmatrix}$ **97.** $\begin{bmatrix} -9 & 11 \\ 7 & -4 \end{bmatrix}$

98. $\begin{bmatrix} 50 & -30 \\ 10 & 5 \end{bmatrix}$ **99.** $\begin{bmatrix} 14 & -24 \\ 12 & -15 \end{bmatrix}$

In Exercises 100–103, find all (a) minors and (b) cofactors of the matrix.

100. $\begin{bmatrix} 2 & -1 \\ 7 & 4 \end{bmatrix}$ **101.** $\begin{bmatrix} 3 & 6 \\ 5 & -4 \end{bmatrix}$

102. $\begin{bmatrix} 3 & 2 & -1 \\ -2 & 5 & 0 \\ 1 & 8 & 6 \end{bmatrix}$ **103.** $\begin{bmatrix} 8 & 3 & 4 \\ 6 & 5 & -9 \\ -4 & 1 & 2 \end{bmatrix}$

In Exercises 104–107, find the determinant of the matrix. Expand by cofactors on the row or column that appears to make the computations easiest.

104. $\begin{bmatrix} -2 & 4 & 1 \\ -6 & 0 & 2 \\ 5 & 3 & 4 \end{bmatrix}$ **105.** $\begin{bmatrix} 4 & 7 & -1 \\ 2 & -3 & 4 \\ -5 & 1 & -1 \end{bmatrix}$

106. $\begin{bmatrix} 3 & 0 & -4 & 0 \\ 0 & 8 & 1 & 2 \\ 6 & 1 & 8 & 2 \\ 0 & 3 & -4 & 1 \end{bmatrix}$ **107.** $\begin{bmatrix} -5 & 6 & 0 & 0 \\ 0 & 1 & -1 & 2 \\ -3 & 4 & -5 & 1 \\ 1 & 6 & 0 & 3 \end{bmatrix}$

7.5 In Exercises 108–111, use Cramer's Rule to solve (if possible) the system of equations.

108. $\begin{cases} 5x - 2y = 6 \\ -11x + 3y = -23 \end{cases}$

109. $\begin{cases} 3x + 8y = -7 \\ 9x - 5y = 37 \end{cases}$

110. $\begin{cases} -2x + 3y - 5z = -11 \\ 4x - y + z = -3 \\ -x - 4y + 6z = 15 \end{cases}$

111. $\begin{cases} 5x - 2y + z = 15 \\ 3x - 3y - z = -7 \\ 2x - y - 7z = -3 \end{cases}$

In Exercises 112–115, use Cramer's Rule to solve.

112. *Mixture Problem* A florist wants to arrange a dozen flowers consisting of two varieties: carnations and roses. Carnations cost $1.50 each and roses cost $3.50 each. How many of each should the florist use so that the arrangement will cost $30.00?

113. *Mixture Problem* One hundred liters of a 60% acid solution is obtained by mixing a 75% solution with a 50% solution. How many liters of each must be used to obtain the desired mixture?

114. *Fitting a Parabola to Three Points* Find an equation of the parabola $y = ax^2 + bx + c$ that passes through the points $(-1, 2)$, $(0, 3)$, and $(1, 6)$.

115. *Break-Even Point* A small business invests $25,000 in equipment to produce a product. Each unit of the product costs $3.75 to produce and is sold for $5.25. How many items must be sold before the business breaks even?

In Exercises 116–119, use a determinant to find the area of the triangle with the given vertices.

116. $(1, 0)$, $(5, 0)$, $(5, 8)$ **117.** $(-4, 0)$, $(4, 0)$, $(0, 6)$

118. $\left(\frac{1}{2}, 1\right)$, $\left(2, -\frac{5}{2}\right)$, $\left(\frac{3}{2}, 1\right)$ **119.** $\left(\frac{3}{2}, 1\right)$, $\left(4, -\frac{1}{2}\right)$, $(4, 2)$

In Exercises 120–123, use a determinant to find an equation of the line through the points.

120. $(-4, 0)$, $(4, 4)$ **121.** $(2, 5)$, $(6, -1)$

122. $\left(-\frac{5}{2}, 3\right)$, $\left(\frac{7}{2}, 1\right)$ **123.** $(-0.8, 0.2)$, $(0.7, 3.2)$

In Exercises 124 and 125, find the uncoded 1×3 row matrices for the message. Then encode the message using the matrix.

Message	Matrix
124. LOOK OUT BELOW	$\begin{bmatrix} 2 & -2 & 0 \\ 3 & 0 & -3 \\ -6 & 2 & 3 \end{bmatrix}$
125. RETURN TO BASE	$\begin{bmatrix} 2 & 1 & 0 \\ -6 & -6 & -2 \\ 3 & 2 & 1 \end{bmatrix}$

In Exercises 126 and 127, decode the cryptogram by using the inverse of the matrix

$$A = \begin{bmatrix} -5 & 4 & -3 \\ 10 & -7 & 6 \\ 8 & -6 & 5 \end{bmatrix}.$$

126. -5 11 -2 370 -265 225 -57 48 -33 32 -15 20 245 -171 147

127. 145 -105 92 264 -188 160 23 -16 15 129 -84 78 -9 8 -5 159 -118 100 219 -152 133 370 -265 225 -105 84 -63

Synthesis

True or False? **In Exercises 128 and 129, determine whether the statement is true or false. Justify your answer.**

128. It is possible to find the determinant of a 4×5 matrix.

129. $\begin{vmatrix} a_{11} & a_{12} & a_{13} \\ a_{21} & a_{22} & a_{23} \\ a_{31} + c_1 & a_{32} + c_2 & a_{33} + c_3 \end{vmatrix} =$

$\begin{vmatrix} a_{11} & a_{12} & a_{13} \\ a_{21} & a_{22} & a_{23} \\ a_{31} & a_{32} & a_{33} \end{vmatrix} + \begin{vmatrix} a_{11} & a_{12} & a_{13} \\ a_{21} & a_{22} & a_{23} \\ c_1 & c_2 & c_3 \end{vmatrix}$

130. Under what conditions does a matrix have an inverse?

131. What is meant by the cofactor of an entry of a matrix? How are cofactors used to find the determinant of the matrix?

132. Three people were asked to solve a system of equations using an augmented matrix. Each person reduced the matrix to row-echelon form. The reduced matrices were

$$\begin{bmatrix} 1 & 2 & \vdots & 3 \\ 0 & 1 & \vdots & 1 \end{bmatrix}$$

$$\begin{bmatrix} 1 & 0 & \vdots & 1 \\ 0 & 1 & \vdots & 1 \end{bmatrix}$$

and

$$\begin{bmatrix} 1 & 2 & \vdots & 3 \\ 0 & 0 & \vdots & 0 \end{bmatrix}.$$

Can all three be right? Explain.

133. *Think About It* Describe the row-echelon form of an augmented matrix that corresponds to a system of linear equations that has a unique solution.

134. Solve the equation $\begin{vmatrix} 2 - \lambda & 5 \\ 3 & -8 - \lambda \end{vmatrix} = 0$.

Chapter Project ▶ Solving Systems of Equations

Matrices have always been a powerful mathematical tool—especially for dealing with problems that involve a lot of data. With technology available to perform the calculations, matrices have also become a practical mathematical tool.

Example ▶ Solving a Metallurgy Problem

	Alloy X	Alloy Y	Alloy Z
Carbon	1%	1%	4%
Chromium	0%	15%	3%
Iron	99%	84%	93%

Three iron alloys contain different percents of carbon, chromium, and iron. Alloy X is a type of wrought iron, alloy Y is a type of stainless steel, and alloy Z is a type of cast iron. How much of each of the three alloys can you make with 15 tons of carbon, 39 tons of chromium, and 546 tons of iron?

Solution

Let x, y, and z represent the amounts of the three iron alloys. You can model the situation with the linear system

$$\begin{cases} 0.01x + 0.01y + 0.04z = 15 & \text{Carbon} \\ 0.15y + 0.03z = 39. & \text{Chromium} \\ 0.99x + 0.84y + 0.93z = 546 & \text{Iron} \end{cases}$$

The matrix equation $AX = B$ that represents this system is

$$\begin{bmatrix} 0.01 & 0.01 & 0.04 \\ 0 & 0.15 & 0.03 \\ 0.99 & 0.84 & 0.93 \end{bmatrix} \begin{bmatrix} x \\ y \\ z \end{bmatrix} = \begin{bmatrix} 15 \\ 39 \\ 546 \end{bmatrix}.$$

With a graphing utility or computer, you can solve the equation as follows.

$$X = A^{-1}B = \begin{bmatrix} -25.4 & -5.4 & 1.267 \\ -6.6 & 6.733 & 0.067 \\ 33 & -0.333 & -0.333 \end{bmatrix} \begin{bmatrix} 15 \\ 39 \\ 546 \end{bmatrix} = \begin{bmatrix} 100 \\ 200 \\ 300 \end{bmatrix}$$

So, you can make 100 tons of alloy X, 200 tons of alloy Y, and 300 tons of alloy Z.

Chapter Project Investigation

Three different gold alloys contain the percents of gold, copper, and silver shown in the matrix. You have 20,144 grams of gold, 766 grams of copper, and 1990 grams of silver. How much of each alloy can you make?

Percent by Weight

	Alloy X	Alloy Y	Alloy Z
Gold	94%	92%	80%
Copper	4%	2%	4%
Silver	2%	6%	16%

Chapter Test

The *Interactive* CD-ROM and *Internet* versions of this text provide answers to the Chapter Tests and Cumulative Tests. They also offer Chapter Pre-Tests (which test key skills and concepts covered in previous chapters) and Chapter Post-Tests, both of which have randomly generated exercises with diagnostic capabilities.

Take this test as you would take a test in class. After you are done, check your work against the answers given in the back of the book.

In Exercises 1 and 2, write the matrix in reduced row-echelon form.

1. $\begin{bmatrix} 1 & -1 & 5 \\ 6 & 2 & 3 \\ 5 & 3 & -3 \end{bmatrix}$

2. $\begin{bmatrix} 1 & 0 & -1 & 2 \\ -1 & 1 & 1 & -3 \\ 1 & 1 & -1 & 1 \\ 3 & 2 & -3 & 4 \end{bmatrix}$

3. Write the augmented matrix corresponding to the system of equations, and solve the system $\begin{cases} 4x + 3y - 2z = 14 \\ -x - y + 2z = -5. \\ 3x + y - 4z = 8 \end{cases}$

4. Find the equation of the parabola $y = ax^2 + bx + c$ that passes through the points in the figure.

5. Find (a) $A - B$, (b) $3A$, (c) $3A - 2B$, and (d) AB (if possible).

$$A = \begin{bmatrix} 5 & 4 \\ -4 & -4 \end{bmatrix}, \quad B = \begin{bmatrix} 4 & -1 \\ -4 & 0 \end{bmatrix}$$

In Exercises 6 and 7, find the inverse of the matrix (if it exists).

6. $\begin{bmatrix} -6 & 4 \\ 10 & -5 \end{bmatrix}$

7. $\begin{bmatrix} -2 & 4 & -6 \\ 2 & 1 & 0 \\ 4 & -2 & 5 \end{bmatrix}$

8. Use the result of Exercise 6 to solve the system.

$$\begin{cases} -6x + 4y = 10 \\ 10x - 5y = 20 \end{cases}$$

In Exercises 9 and 10, evaluate the determinant of the matrix.

9. $\begin{bmatrix} -9 & 4 \\ 13 & 16 \end{bmatrix}$

10. $\begin{bmatrix} \frac{5}{2} & \frac{13}{4} \\ -8 & \frac{6}{5} \end{bmatrix}$

In Exercises 11 and 12, use Cramer's Rule to solve (if possible) the system of equations.

11. $\begin{cases} 7x + 6y = 9 \\ -2x - 11y = -49 \end{cases}$

12. $\begin{cases} 6x - y + 2z = -4 \\ -2x + 3y - z = 10 \\ 4x - 4y + z = -18 \end{cases}$

13. Use a determinant to find the area of the triangle in the figure.

14. Find the uncoded 1×3 row matrices for the message KNOCK ON WOOD. Then encode the message using the matrix A at the left.

15. One hundred liters of a 50% solution is obtained by mixing a 60% solution with a 20% solution. How many liters of each must be used to obtain the desired mixture?

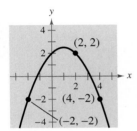

FIGURE FOR 4

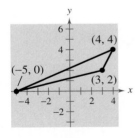

FIGURE FOR 13

$$A = \begin{bmatrix} 1 & -1 & 0 \\ 1 & 0 & -1 \\ 6 & -2 & -3 \end{bmatrix}$$

MATRIX FOR 14

Siegfried Layda/Tony Stone Images

In 1997, 831.8 million trips were taken in the United States for pleasure or vacation. The average distance traveled per trip was 901 miles. (Source: Travel Industry Association of America)

8 Sequences, Series, and Probability

▶ How to Study This Chapter

The Big Picture

In this chapter you will learn the following skills and concepts.

▶ How to use sequence, factorial, and summation notation to write the terms and sum of a sequence

▶ How to recognize, write, and manipulate arithmetic sequences and geometric sequences

▶ How to use mathematical induction to prove a statement involving a positive integer n

▶ How to use the Binomial Theorem and Pascal's Triangle to calculate binomial coefficients and binomial expansions

▶ How to solve counting problems using the Fundamental Counting Principle, permutations, and combinations

▶ How to find the probabilities of events and their complements

Important Vocabulary

As you encounter each new vocabulary term in this chapter, add the term and its definition to your notebook glossary.

Infinite sequence (p. 578)
Terms of a sequence (p. 578)
Finite sequence (p. 578)
Recursive (p. 580)
Factorial (p. 580)
Summation or sigma notation (p. 582)
Infinite series (p. 583)
Finite series (p. 583)
nth partial sum (p. 583)
Arithmetic sequence (p. 589)
Common difference (p. 589)
Geometric sequence (p. 598)
Common ratio (p. 598)
Infinite geometric series (p. 602)
Geometric series (p. 602)
Mathematical induction (p. 608)
First differences (p. 615)

Second differences (p. 615)
Binomial coefficients (p. 618)
Binomial Theorem (p. 618)
Pascal's Triangle (p. 620)
Fundamental Counting Principle (p. 627)
Permutation (p. 628)
Distinguishable permutations (p. 630)
Combination (p. 631)
Experiment (p. 636)
Outcomes (p. 636)
Sample space (p. 636)
Event (p. 636)
Probability (p. 637)
Mutually exclusive (p. 640)
Independent events (p. 642)
Complement of an event (p. 643)

Study Tools

- Learning objectives at the beginning of each section
- Chapter Summary (p. 649)
- Review Exercises (pp. 650–653)
- Chapter Test (p. 655)
- Cumulative Test for Chapters 6–8 (pp. 656–657)

Additional Resources

- Study and Solutions Guide
- Interactive College Algebra
- Videotapes for Chapter 8
- College Algebra Website
- Student Success Organizer

STUDY TIP

Try working with a partner on assignments. You may find that teaching others is an excellent way to learn.

8.1 Sequences and Series

▶ **Why you should learn it**

Sequences and series can be used to model real-life problems. For instance, Exercise 109 on page 587 uses a sequence to model the net income of Wal-Mart from 1990 through 1998.

John Mailer, Jr./The Image Works

Sequences

In mathematics, the word *sequence* is used in much the same way as in ordinary English. Saying that a collection is listed *in sequence* means that it is ordered so that it has a first member, a second member, a third member, and so on.

Mathematically, you can think of a sequence as a *function* whose domain is the set of positive integers.

$$f(1) = a_1, \ f(2) = a_2, \ f(3) = a_3, \ f(4) = a_4, \ \ldots, f(n) = a_n, \ \ldots$$

Rather than using function notation, however, sequences are usually written using subscript notation, as indicated in the following definition.

Definition of a Sequence

An **infinite sequence** is a function whose domain is the set of positive integers. The function values

$$a_1, a_2, a_3, a_4, \ \ldots, a_n, \ \ldots$$

are the **terms** of the sequence. If the domain of the function consists of the first n positive integers only, the sequence is a **finite sequence.**

On occasion it is convenient to begin subscripting a sequence with 0 instead of 1 so that the terms of the sequence become

$$a_0, a_1, a_2, a_3, \ \ldots \ .$$

Example 1 ▶ Finding Terms of a Sequence

a. The first four terms of the sequence given by $a_n = 3n - 2$ are

$$a_1 = 3(1) - 2 = 1 \qquad \text{1st term}$$
$$a_2 = 3(2) - 2 = 4 \qquad \text{2nd term}$$
$$a_3 = 3(3) - 2 = 7 \qquad \text{3rd term}$$
$$a_4 = 3(4) - 2 = 10. \qquad \text{4th term}$$

b. The first four terms of the sequence given by $a_n = 3 + (-1)^n$ are

$$a_1 = 3 + (-1)^1 = 3 - 1 = 2 \qquad \text{1st term}$$
$$a_2 = 3 + (-1)^2 = 3 + 1 = 4 \qquad \text{2nd term}$$
$$a_3 = 3 + (-1)^3 = 3 - 1 = 2 \qquad \text{3rd term}$$
$$a_4 = 3 + (-1)^4 = 3 + 1 = 4. \qquad \text{4th term}$$

Example 2 ▶ **Finding Terms of a Sequence**

The first five terms of the sequence given by

$$a_n = \frac{(-1)^n}{2n - 1}$$

are as follows.

$$a_1 = \frac{(-1)^1}{2(1) - 1} = \frac{-1}{2 - 1} = -1 \qquad \text{1st term}$$

$$a_2 = \frac{(-1)^2}{2(2) - 1} = \frac{1}{4 - 1} = \frac{1}{3} \qquad \text{2nd term}$$

$$a_3 = \frac{(-1)^3}{2(3) - 1} = \frac{-1}{6 - 1} = -\frac{1}{5} \qquad \text{3rd term}$$

$$a_4 = \frac{(-1)^4}{2(4) - 1} = \frac{1}{8 - 1} = \frac{1}{7} \qquad \text{4th term}$$

$$a_5 = \frac{(-1)^5}{2(5) - 1} = \frac{-1}{10 - 1} = -\frac{1}{9} \qquad \text{5th term}$$

Simply listing the first few terms is not sufficient to define a unique sequence—the nth term *must be given*. To see this, consider the following sequences, both of which have the same first three terms.

$$\frac{1}{2}, \frac{1}{4}, \frac{1}{8}, \frac{1}{16}, \cdots, \frac{1}{2^n}, \cdots$$

$$\frac{1}{2}, \frac{1}{4}, \frac{1}{8}, \frac{1}{15}, \cdots, \frac{6}{(n + 1)(n^2 - n + 6)}, \cdots$$

Technology

To graph a sequence using a graphing utility, set the mode to *seq* and *dot* and enter the sequence. The graph of the sequence in Example 3(a) is shown below.

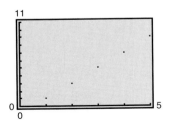

You can use the trace key to identify its terms.

Example 3 ▶ **Finding the nth Term of a Sequence**

Write an expression for the apparent nth term (a_n) of each sequence.

a. $1, 3, 5, 7, \ldots$ **b.** $2, 5, 10, 17, \ldots$

Solution

a. n: 1   2   3   4   . . . n

     Terms: 1   3   5   7   . . . a_n

     Apparent pattern: Each term is 1 less than twice n, which implies that

$$a_n = 2n - 1.$$

b. n: 1   2   3   4   . . . n

     Terms: 2   5   10   17   . . . a_n

     Apparent pattern: Each term is 1 more than the square of n, which implies that

$$a_n = n^2 + 1.$$

Some sequences are defined **recursively.** To define a sequence recursively, you need to be given one or more of the first few terms. All other terms of the sequence are then defined using previous terms. A well-known example is the Fibonacci sequence shown in Example 4.

Example 4 ▶ The Fibonacci Sequence: A Recursive Sequence

The Fibonacci sequence is defined recursively, as follows.

$$a_0 = 1, \ a_1 = 1, \ a_k = a_{k-2} + a_{k-1}, \text{ where } k \geq 2$$

Write the first six terms of this sequence.

Solution

$a_0 = 1$	0th term is given.
$a_1 = 1$	1st term is given.
$a_{2-2} + a_{2-1} = a_0 + a_1 = 1 + 1 = 2$	Use recursive formula.
$a_{3-2} + a_{3-1} = a_1 + a_2 = 1 + 2 = 3$	Use recursive formula.
$a_{4-2} + a_{4-1} = a_2 + a_3 = 2 + 3 = 5$	Use recursive formula.
$a_{5-2} + a_{5-1} = a_3 + a_4 = 3 + 5 = 8$	Use recursive formula.

Factorial Notation

Some very important sequences in mathematics involve terms that are defined with special types of products called **factorials.**

Definition of Factorial

If n is a positive integer, n **factorial** is defined by

$$n! = 1 \cdot 2 \cdot 3 \cdot 4 \cdots (n-1) \cdot n.$$

As a special case, zero factorial is defined as $0! = 1$.

Here are some values of $n!$ for the first several nonnegative integers. Notice that $0!$ is 1 by definition.

$$0! = 1$$

$$1! = 1$$

$$2! = 1 \cdot 2 = 2$$

$$3! = 1 \cdot 2 \cdot 3 = 6$$

$$4! = 1 \cdot 2 \cdot 3 \cdot 4 = 24$$

$$5! = 1 \cdot 2 \cdot 3 \cdot 4 \cdot 5 = 120$$

The value of n does not have to be very large before the value of $n!$ becomes huge. For instance, $10! = 3,628,800$.

Factorials follow the same conventions for order of operations as do exponents. For instance,

$$2n! = 2(n!)$$
$$= 2(1 \cdot 2 \cdot 3 \cdot 4 \cdots n)$$

whereas $(2n)! = 1 \cdot 2 \cdot 3 \cdot 4 \cdots 2n$.

Example 5 ▶ Finding Terms of a Sequence Involving Factorials

List the first five terms of the sequence given by

$$a_n = \frac{2^n}{n!}.$$

Begin with $n = 0$. Then plot the points on a set of coordinate axes.

Solution

$$a_0 = \frac{2^0}{0!} = \frac{1}{1} = 1 \qquad \text{0th term}$$

$$a_1 = \frac{2^1}{1!} = \frac{2}{1} = 2 \qquad \text{1st term}$$

$$a_2 = \frac{2^2}{2!} = \frac{4}{2} = 2 \qquad \text{2nd term}$$

$$a_3 = \frac{2^3}{3!} = \frac{8}{6} = \frac{4}{3} \qquad \text{3rd term}$$

$$a_4 = \frac{2^4}{4!} = \frac{16}{24} = \frac{2}{3} \qquad \text{4th term}$$

Figure 8.1 shows the first five terms of the sequence.

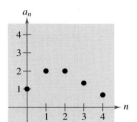

FIGURE **8.1**

When working with fractions involving factorials, you will often find that the fractions can be reduced.

Example 6 ▶ Evaluating Factorial Expressions

Evaluate each factorial expression.

a. $\dfrac{8!}{2! \cdot 6!}$ **b.** $\dfrac{2! \cdot 6!}{3! \cdot 5!}$ **c.** $\dfrac{n!}{(n-1)!}$

Solution

a. $\dfrac{8!}{2! \cdot 6!} = \dfrac{1 \cdot 2 \cdot 3 \cdot 4 \cdot 5 \cdot 6 \cdot 7 \cdot 8}{1 \cdot 2 \cdot 1 \cdot 2 \cdot 3 \cdot 4 \cdot 5 \cdot 6} = \dfrac{7 \cdot 8}{2} = 28$

b. $\dfrac{2! \cdot 6!}{3! \cdot 5!} = \dfrac{1 \cdot 2 \cdot 1 \cdot 2 \cdot 3 \cdot 4 \cdot 5 \cdot 6}{1 \cdot 2 \cdot 3 \cdot 1 \cdot 2 \cdot 3 \cdot 4 \cdot 5} = \dfrac{6}{3} = 2$

c. $\dfrac{n!}{(n-1)!} = \dfrac{1 \cdot 2 \cdot 3 \cdots (n-1) \cdot n}{1 \cdot 2 \cdot 3 \cdots (n-1)} = n$

Summation Notation

There is a convenient notation for the sum of the terms of a finite sequence. It is called **summation notation** or **sigma notation** because it involves the use of the uppercase Greek letter sigma, written as Σ.

Definition of Summation Notation

The sum of the first n terms of a sequence is represented by

$$\sum_{i=1}^{n} a_i = a_1 + a_2 + a_3 + a_4 + \cdots + a_n$$

where i is called the **index of summation,** n is the **upper limit of summation,** and 1 is the **lower limit of summation.**

Example 7 ▶ Summation Notation for Sums

Find each sum.

a. $\sum_{i=1}^{5} 3i$ **b.** $\sum_{k=3}^{6}(1+k^2)$ **c.** $\sum_{i=0}^{8}\frac{1}{i!}$

Solution

a. $\sum_{i=1}^{5} 3i = 3(1) + 3(2) + 3(3) + 3(4) + 3(5)$

$= 3(1 + 2 + 3 + 4 + 5)$

$= 3(15)$

$= 45$

b. $\sum_{k=3}^{6}(1+k^2) = (1+3^2) + (1+4^2) + (1+5^2) + (1+6^2)$

$= 10 + 17 + 26 + 37$

$= 90$

c. $\sum_{i=0}^{8}\frac{1}{i!} = \frac{1}{0!} + \frac{1}{1!} + \frac{1}{2!} + \frac{1}{3!} + \frac{1}{4!} + \frac{1}{5!} + \frac{1}{6!} + \frac{1}{7!} + \frac{1}{8!}$

$= 1 + 1 + \frac{1}{2} + \frac{1}{6} + \frac{1}{24} + \frac{1}{120} + \frac{1}{720} + \frac{1}{5040} + \frac{1}{40,320}$

≈ 2.71828

For this summation, note that the sum is very close to the irrational number $e \approx 2.718281828$. It can be shown that as more terms of the sequence whose nth term is $1/n!$ are added, the sum becomes closer and closer to e.

In Example 7, note that the lower limit of a summation does not have to be 1. Also note that the index of summation does not have to be the letter i. For instance, in part (b), the letter k is the index of summation.

Variations in the upper and lower limits of summation can produce quite different-looking summation notations for *the same sum*. For example, the following two sums have the same terms.

$$\sum_{i=1}^{3} 3(2^i) = 3(2^1 + 2^2 + 2^3)$$

$$\sum_{i=0}^{2} 3(2^{i+1}) = 3(2^1 + 2^2 + 2^3)$$

Properties of Sums

1. $\sum_{i=1}^{n} ca_i = c \sum_{i=1}^{n} a_i,$ c is any constant.

2. $\sum_{i=1}^{n} (a_i + b_i) = \sum_{i=1}^{n} a_i + \sum_{i=1}^{n} b_i$

3. $\sum_{i=1}^{n} (a_i - b_i) = \sum_{i=1}^{n} a_i - \sum_{i=1}^{n} b_i$

A proof of Property 1 is given in Appendix A.

Series

Many applications involve the sum of the terms of a finite or infinite sequence. Such a sum is called a **series.**

Definition of a Series

Consider the infinite sequence $a_1, a_2, a_3, \ldots, a_i, \ldots$

1. The sum of all terms of the infinite sequence is called an **infinite series** and is denoted by

$$a_1 + a_2 + a_3 + \cdots + a_i + \cdots = \sum_{i=1}^{\infty} a_i.$$

2. The sum of the first n terms of the sequence is called a **finite series** or the **nth partial sum** of the sequence and is denoted by

$$a_1 + a_2 + a_3 + \cdots + a_n = \sum_{i=1}^{n} a_i.$$

Example 8 ► Finding the Sum of a Series

For the series $\sum_{i=1}^{\infty} \frac{3}{10^i}$, find (a) the third partial sum and (b) the sum.

Solution

a. The third partial sum is

$$\sum_{i=1}^{3} \frac{3}{10^i} = \frac{3}{10^1} + \frac{3}{10^2} + \frac{3}{10^3} = 0.3 + 0.03 + 0.003 = 0.333.$$

b. The sum of the series is

$$\sum_{i=1}^{\infty} \frac{3}{10^i} = \frac{3}{10^1} + \frac{3}{10^2} + \frac{3}{10^3} + \frac{3}{10^4} + \frac{3}{10^5} + \cdots$$

$$= 0.3 + 0.03 + 0.003 + 0.0003 + 0.00003 + \cdots$$

$$= 0.33333\ldots = \frac{1}{3}.$$

Application

Sequences have many applications in business and science. One is illustrated in Example 9.

Example 9 ▶ Population of the United States

For the years 1960 to 1997, the resident population of the United States can be approximated by the model

$$a_n = \sqrt{33{,}282 + 801.3n + 6.12n^2} \qquad n = 0, 1, \ldots, 37$$

where a_n is the population in millions and n represents the calendar year, with $n = 0$ corresponding to 1960. Find the last five terms of this finite sequence, which represent the U.S. population for the years 1993 to 1997. (Source: U.S. Bureau of the Census)

Solution

The last five terms of this finite sequence are as follows.

$$a_{33} = \sqrt{33{,}282 + 801.3(33) + 6.12(33)^2} \approx 257.7 \qquad \text{1993 population}$$

$$a_{34} = \sqrt{33{,}282 + 801.3(34) + 6.12(34)^2} \approx 260.0 \qquad \text{1994 population}$$

$$a_{35} = \sqrt{33{,}282 + 801.3(35) + 6.12(35)^2} \approx 262.3 \qquad \text{1995 population}$$

$$a_{36} = \sqrt{33{,}282 + 801.3(36) + 6.12(36)^2} \approx 264.7 \qquad \text{1996 population}$$

$$a_{37} = \sqrt{33{,}282 + 801.3(37) + 6.12(37)^2} \approx 267.0 \qquad \text{1997 population}$$

The bar graph in Figure 8.2 graphically represents the population given by this sequence for the entire 38-year period from 1960 to 1997.

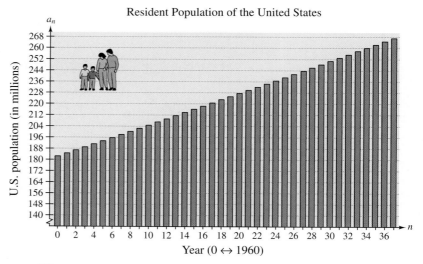

FIGURE 8.2

The *Interactive* CD-ROM and *Internet* versions of this text contain step-by-step solutions to all odd-numbered Section and Review Exercises. They also provide Tutorial Exercises that link to Guided Examples for additional help.

8.1 Exercises

In Exercises 1–24, write the first five terms of the sequence. (Assume that n begins with 1.)

1. $a_n = 3n + 1$ **2.** $a_n = 5n - 3$

3. $a_n = 2^n$ **4.** $a_n = \left(\frac{1}{2}\right)^n$

5. $a_n = (-2)^n$ **6.** $a_n = \left(-\frac{1}{2}\right)^n$

7. $a_n = \dfrac{n + 2}{n}$ **8.** $a_n = \dfrac{n}{n + 2}$

9. $a_n = \dfrac{6n}{3n^2 - 1}$ **10.** $a_n = \dfrac{3n^2 - n + 4}{2n^2 + 1}$

11. $a_n = \dfrac{1 + (-1)^n}{n}$ **12.** $a_n = 1 + (-1)^n$

13. $a_n = 2 - \dfrac{1}{3^n}$ **14.** $a_n = \dfrac{2^n}{3^n}$

15. $a_n = \dfrac{1}{n^{3/2}}$ **16.** $a_n = \dfrac{10}{n^{2/3}}$

17. $a_n = \dfrac{3^n}{n!}$ **18.** $a_n = \dfrac{n!}{n}$

19. $a_n = \dfrac{(-1)^n}{n^2}$ **20.** $a_n = (-1)^n\left(\dfrac{n}{n + 1}\right)$

21. $a_n = \frac{2}{3}$ **22.** $a_n = 0.3$

23. $a_n = n(n - 1)(n - 2)$ **24.** $a_n = n(n^2 - 6)$

In Exercises 25–30, find the indicated term of the sequence.

25. $a_n = (-1)^n(3n - 2)$

$a_{25} = $

26. $a_n = (-1)^{n-1}[n(n - 1)]$

$a_{16} = $

27. $a_n = \dfrac{2^n}{n!}$ **28.** $a_n = \dfrac{n!}{2n}$

$a_{10} = $ $a_8 = $

29. $a_n = \dfrac{4n}{2n^2 - 3}$ **30.** $a_n = \dfrac{4n^2 - n + 3}{n(n - 1)(n + 2)}$

$a_{11} = $ $a_{13} = $

In Exercises 31–36, use a graphing utility to graph the first 10 terms of the sequence.

31. $a_n = \dfrac{3}{4}n$ **32.** $a_n = 2 - \dfrac{4}{n}$

33. $a_n = 16(-0.5)^{n-1}$ **34.** $a_n = 8(0.75)^{n-1}$

35. $a_n = \dfrac{2n}{n + 1}$ **36.** $a_n = \dfrac{n^2}{n^2 + 2}$

In Exercises 37–40, match the sequence with the graph of its first 10 terms. [The graphs are labeled (a), (b), (c), and (d).]

(a)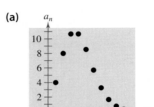

(b)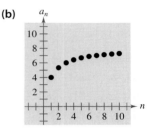

(c)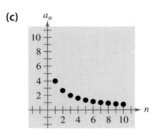

(d)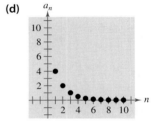

37. $a_n = \dfrac{8}{n + 1}$ **38.** $a_n = \dfrac{8n}{n + 1}$

39. $a_n = 4(0.5)^{n-1}$ **40.** $a_n = \dfrac{4^n}{n!}$

In Exercises 41–54, write an expression for the *most apparent* nth term of the sequence. (Assume that n begins with 1.)

41. $1, 4, 7, 10, 13, \ldots$ **42.** $3, 7, 11, 15, 19, \ldots$

43. $0, 3, 8, 15, 24, \ldots$ **44.** $2, -4, 6, -8, 10, \ldots$

45. $\dfrac{-2}{3}, \dfrac{3}{4}, \dfrac{-4}{5}, \dfrac{5}{6}, \dfrac{-6}{7}, \ldots$

46. $\dfrac{1}{2}, \dfrac{-1}{4}, \dfrac{1}{8}, \dfrac{-1}{16}, \ldots$ **47.** $\dfrac{2}{1}, \dfrac{3}{3}, \dfrac{4}{5}, \dfrac{5}{7}, \dfrac{6}{9}, \ldots$

48. $\dfrac{1}{3}, \dfrac{2}{9}, \dfrac{4}{27}, \dfrac{8}{81}, \ldots$ **49.** $1, \dfrac{1}{4}, \dfrac{1}{9}, \dfrac{1}{16}, \dfrac{1}{25}, \ldots$

50. $1, \dfrac{1}{2}, \dfrac{1}{6}, \dfrac{1}{24}, \dfrac{1}{120}, \ldots$ **51.** $1, -1, 1, -1, 1, \ldots$

52. $1, 2, \dfrac{2^2}{2}, \dfrac{2^3}{6}, \dfrac{2^4}{24}, \dfrac{2^5}{120}, \ldots$

53. $1 + \frac{1}{1}, 1 + \frac{1}{2}, 1 + \frac{1}{3}, 1 + \frac{1}{4}, 1 + \frac{1}{5}, \ldots$

54. $1 + \frac{1}{2}, 1 + \frac{3}{4}, 1 + \frac{7}{8}, 1 + \frac{15}{16}, 1 + \frac{31}{32}, \ldots$

In Exercises 55–58, write the first five terms of the sequence defined recursively.

55. $a_1 = 28$, $a_{k+1} = a_k - 4$

56. $a_1 = 15$, $a_{k+1} = a_k + 3$

57. $a_1 = 3$, $a_{k+1} = 2(a_k - 1)$

58. $a_1 = 32$, $a_{k+1} = \frac{1}{2}a_k$

In Exercises 59–62, write the first five terms of the sequence defined recursively. Use the pattern to write the nth term of the sequence as a function of n. (Assume that n begins with 1.)

59. $a_1 = 6$, $a_{k+1} = a_k + 2$

60. $a_1 = 25$, $a_{k+1} = a_k - 5$

61. $a_1 = 81$, $a_{k+1} = \frac{1}{3}a_k$

62. $a_1 = 14$, $a_{k+1} = (-2)a_k$

In Exercises 63–70, simplify the ratio of factorials.

63. $\dfrac{4!}{6!}$ **64.** $\dfrac{5!}{8!}$ **65.** $\dfrac{10!}{8!}$ **66.** $\dfrac{25!}{23!}$

67. $\dfrac{(n+1)!}{n!}$ **68.** $\dfrac{(n+2)!}{n!}$

69. $\dfrac{(2n-1)!}{(2n+1)!}$ **70.** $\dfrac{(3n+1)!}{(3n)!}$

In Exercises 71–82, find the sum.

71. $\displaystyle\sum_{i=1}^{5}(2i+1)$ **72.** $\displaystyle\sum_{i=1}^{6}(3i-1)$

73. $\displaystyle\sum_{k=1}^{4}10$ **74.** $\displaystyle\sum_{k=1}^{5}5$

75. $\displaystyle\sum_{i=0}^{4}i^2$ **76.** $\displaystyle\sum_{i=0}^{5}2i^2$

77. $\displaystyle\sum_{k=0}^{3}\frac{1}{k^2+1}$ **78.** $\displaystyle\sum_{j=3}^{5}\frac{1}{j^2-3}$

79. $\displaystyle\sum_{k=2}^{5}(k+1)^2(k-3)$ **80.** $\displaystyle\sum_{i=1}^{4}[(i-1)^2+(i+1)^3]$

81. $\displaystyle\sum_{i=1}^{4}2^i$ **82.** $\displaystyle\sum_{j=0}^{4}(-2)^j$

In Exercises 83–86, use a calculator to find the sum.

83. $\displaystyle\sum_{j=1}^{6}(24-3j)$ **84.** $\displaystyle\sum_{j=1}^{10}\frac{3}{j+1}$

85. $\displaystyle\sum_{k=0}^{4}\frac{(-1)^k}{k+1}$ **86.** $\displaystyle\sum_{k=0}^{4}\frac{(-1)^k}{k!}$

In Exercises 87–96, use sigma notation to write the sum.

87. $\dfrac{1}{3(1)} + \dfrac{1}{3(2)} + \dfrac{1}{3(3)} + \cdots + \dfrac{1}{3(9)}$

88. $\dfrac{5}{1+1} + \dfrac{5}{1+2} + \dfrac{5}{1+3} + \cdots + \dfrac{5}{1+15}$

89. $\left[2\left(\frac{1}{8}\right) + 3\right] + \left[2\left(\frac{2}{8}\right) + 3\right] + \cdots + \left[2\left(\frac{8}{8}\right) + 3\right]$

90. $\left[1 - \left(\frac{1}{6}\right)^2\right] + \left[1 - \left(\frac{2}{6}\right)^2\right] + \cdots + \left[1 - \left(\frac{6}{6}\right)^2\right]$

91. $3 - 9 + 27 - 81 + 243 - 729$

92. $1 - \frac{1}{2} + \frac{1}{4} - \frac{1}{8} + \cdots - \frac{1}{128}$

93. $\dfrac{1}{1^2} - \dfrac{1}{2^2} + \dfrac{1}{3^2} - \dfrac{1}{4^2} + \cdots - \dfrac{1}{20^2}$

94. $\dfrac{1}{1 \cdot 3} + \dfrac{1}{2 \cdot 4} + \dfrac{1}{3 \cdot 5} + \cdots + \dfrac{1}{10 \cdot 12}$

95. $\frac{1}{4} + \frac{3}{8} + \frac{7}{16} + \frac{15}{32} + \frac{31}{64}$

96. $\frac{1}{2} + \frac{2}{4} + \frac{6}{8} + \frac{24}{16} + \frac{120}{32} + \frac{720}{64}$

In Exercises 97–100, find the indicated partial sum of the series.

97. $\displaystyle\sum_{i=1}^{\infty}5\left(\frac{1}{2}\right)^i$,

Fourth partial sum

98. $\displaystyle\sum_{i=1}^{\infty}2\left(\frac{1}{3}\right)^i$,

Fifth partial sum

99. $\displaystyle\sum_{n=1}^{\infty}4\left(-\frac{1}{2}\right)^n$,

Third partial sum

100. $\displaystyle\sum_{n=1}^{\infty}8\left(-\frac{1}{4}\right)^n$,

Fourth partial sum

In Exercises 101–104, find the sum of the infinite series.

101. $\displaystyle\sum_{i=1}^{\infty}6\left(\frac{1}{10}\right)^i$ **102.** $\displaystyle\sum_{k=1}^{\infty}\left(\frac{1}{10}\right)^k$

103. $\displaystyle\sum_{k=1}^{\infty}7\left(\frac{1}{10}\right)^k$ **104.** $\displaystyle\sum_{i=1}^{\infty}2\left(\frac{1}{10}\right)^i$

105. *Compound Interest*   A deposit of $5000 is made in an account that earns 8% interest compounded quarterly. The balance in the account after n quarters is

$$A_n = 5000\left(1 + \frac{0.08}{4}\right)^n, \quad n = 1, 2, 3, \ldots.$$

(a) Compute the first eight terms of this sequence.

(b) Find the balance in this account after 10 years by computing the 40th term of the sequence.

106. *Compound Interest* A deposit of $100 is made *each month* in an account that earns 12% interest compounded monthly. The balance in the account after n months is

$$A_n = 100(101)[(1.01)^n - 1], \quad n = 1, 2, 3, \ldots.$$

(a) Compute the first six terms of this sequence.

(b) Find the balance in this account after 5 years by computing the 60th term of the sequence.

(c) Find the balance in this account after 20 years by computing the 240th term of the sequence.

107. *Per-Patient Hospital Care* The average cost to community hospitals per patient per day from 1989 to 1996 can be approximated by the model

$$a_n = 696 + 66.4n - 2.37n^2, \quad n = -1, 0, 1, \ldots, 6$$

where a_n is the per-patient cost (in dollars) and n is the year, with $n = 0$ corresponding to 1990. Find the terms of this finite sequence. Use a graphing utility to construct a bar graph that represents the sequence. (Source: American Hospital Association)

108. *Federal Debt* From 1989 to 1996, the federal debt of the United States rose from just under $3 trillion to over $5 trillion. The federal debt during the years from 1989 to 1996 is approximated by the model

$$a_n = \sqrt{10.9 + 2.8n}, \quad n = -1, 0, 1, \ldots, 6$$

where a_n is the debt in trillions of dollars and n is the year, with $n = 0$ corresponding to 1990. Find the terms of this finite sequence. Use a graphing utility to construct a bar graph that represents the sequence. (Source: Treasury Department, U.S. Bureau of the Census)

109. *Corporate Income* The net incomes a_n (in millions of dollars) of Wal-Mart for the years 1990 through 1998 are shown in the figure. These incomes can be approximated by the model

$$a_n = 1215 + 608.2n - 114.83n^2 + 11.00n^3,$$

$$n = 0, \ldots, 8$$

where $n = 0$ represents 1990. Use this model to approximate the total net income from 1990 through 1998. Compare this sum with the result of adding the incomes shown in the figure. (Source: Wal-Mart)

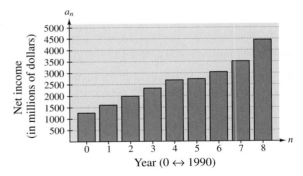

FIGURE FOR **109**

110. *Corporate Dividends* The dividends a_n (in dollars) declared per share of common stock of Procter & Gamble Company for the years 1990 through 1998 are shown in the figure. These dividends can be approximated by the model

$$a_n = 4.27 + 0.294n - 2.934 \ln n,$$

$$n = 10, \ldots, 18$$

where $n = 10$ represents 1990. Use this model to approximate the total dividends per share of common stock from 1990 through 1998. Compare this sum with the result of adding the dividends shown in the figure. (Source: Procter & Gamble Company)

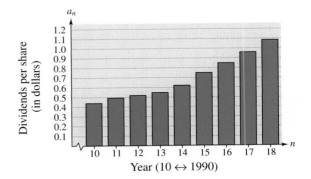

Synthesis

Fibonacci Sequence **In Exercises 111 and 112, use the Fibonacci sequence. (See Example 4.)**

111. Write the first 12 terms of the Fibonacci sequence a_n and the first 10 terms of the sequence given by

$$b_n = \frac{a_{n+1}}{a_n}, \quad n \geq 1.$$

112. Using the definition for b_n in Exercise 111, show that b_n can be defined recursively by

$$b_n = 1 + \frac{1}{b_{n-1}}.$$

Arithmetic Mean **In Exercises 113–116, use the following definition of the arithmetic mean $\bar{x}$ of a set of n measurements $x_1, x_2, x_3, \ldots, x_n$.**

$$\bar{x} = \frac{1}{n}\sum_{i=1}^{n} x_i$$

113. Find the arithmetic mean of the six checking account balances $327.15, $785.69, $433.04, $265.38, $604.12, and $590.30. Use the statistical capabilities of a graphing utility to verify your result.

114. Find the arithmetic mean of the following prices per gallon for regular unleaded gasoline at five gasoline stations in a city: $1.279, $1.259, $1.289, $1.329, and $1.349. Use the statistical capabilities of a graphing utility to verify your result.

115. Prove that $\displaystyle\sum_{i=1}^{n}(x_i - \bar{x}) = 0$.

116. Prove that $\displaystyle\sum_{i=1}^{n}(x_i - \bar{x})^2 = \sum_{i=1}^{n}x_i^2 - \frac{1}{n}\left(\sum_{i=1}^{n}x_i\right)^2$.

True or False? **In Exercises 117 and 118, determine whether the statement is true or false. Justify your answer.**

117. $\displaystyle\sum_{i=1}^{4}(i^2 + 2i) = \sum_{i=1}^{4}i^2 + 2\sum_{i=1}^{4}i$

118. $\displaystyle\sum_{j=1}^{4}2^j = \sum_{j=3}^{6}2^{j-2}$

Review

In Exercises 119–122, find (a) $A - B$, (b) $4B - 3A$, (c) AB, and (d) BA.

119. $A = \begin{bmatrix} 6 & 5 \\ 3 & 4 \end{bmatrix}$, $B = \begin{bmatrix} -2 & 4 \\ 6 & -3 \end{bmatrix}$

120. $A = \begin{bmatrix} 10 & 7 \\ -4 & 6 \end{bmatrix}$, $B = \begin{bmatrix} 0 & -12 \\ 8 & 11 \end{bmatrix}$

121. $A = \begin{bmatrix} -2 & -3 & 6 \\ 4 & 5 & 7 \\ 1 & 7 & 4 \end{bmatrix}$, $B = \begin{bmatrix} 1 & 4 & 2 \\ 0 & 1 & 6 \\ 0 & 3 & 1 \end{bmatrix}$

122. $A = \begin{bmatrix} -1 & 4 & 0 \\ 5 & 1 & 2 \\ 0 & -1 & 3 \end{bmatrix}$, $B = \begin{bmatrix} 0 & 4 & 0 \\ 3 & 1 & -2 \\ -1 & 0 & 2 \end{bmatrix}$

In Exercises 123–126, find the determinant of the matrix.

123. $A = \begin{bmatrix} 3 & 5 \\ -1 & 7 \end{bmatrix}$

124. $A = \begin{bmatrix} -2 & 8 \\ 12 & 15 \end{bmatrix}$

125. $A = \begin{bmatrix} 3 & 4 & 5 \\ 0 & 7 & 3 \\ 4 & 9 & -1 \end{bmatrix}$

126. $A = \begin{bmatrix} 16 & 11 & 10 & 2 \\ 9 & 8 & 3 & 7 \\ -2 & -1 & 12 & 3 \\ -4 & 6 & 2 & 1 \end{bmatrix}$

▶ **What you should learn**

- How to recognize and write arithmetic sequences
- How to find an *n*th partial sum of an arithmetic sequence
- How to use arithmetic sequences to model and solve real-life problems

▶ **Why you should learn it**

Arithmetic sequences have practical real-life applications. For instance, Exercise 79 on page 596 uses an arithmetic sequence to find the number of bricks needed to lay a brick patio.

Frank Pedrick/The Image Works

Arithmetic Sequences

A sequence whose consecutive terms have a common difference is called an **arithmetic sequence.**

Definition of an Arithmetic Sequence

A sequence is arithmetic if the differences between consecutive terms are the same. So, the sequence

$$a_1, a_2, a_3, a_4, \ldots, a_n, \ldots$$

is arithmetic if there is a number d such that

$$a_2 - a_1 = a_3 - a_2 = a_4 - a_3 = \cdots = d$$

and so on. The number d is the **common difference** of the arithmetic sequence.

Example 1 ▶ Examples of Arithmetic Sequences

a. The sequence whose *n*th term is $4n + 3$ is arithmetic. For this sequence, the common difference between consecutive terms is 4.

$$7, 11, 15, 19, \ldots, 4n + 3, \ldots$$

$$11 - 7 = 4$$

b. The sequence whose *n*th term is $7 - 5n$ is arithmetic. For this sequence, the common difference between consecutive terms is -5.

$$2, -3, -8, -13, \ldots, 7 - 5n, \ldots$$

$$-3 - 2 = -5$$

c. The sequence whose *n*th term is $\frac{1}{4}(n + 3)$ is arithmetic. For this sequence, the common difference between consecutive terms is $\frac{1}{4}$.

$$1, \frac{5}{4}, \frac{3}{2}, \frac{7}{4}, \ldots, \frac{n + 3}{4}, \ldots$$

$$\frac{5}{4} - 1 = \frac{1}{4}$$

The sequence $1, 4, 9, 16, \ldots$, whose *n*th term is n^2 is *not* arithmetic. The difference between the first two terms is

$$a_2 - a_1 = 4 - 1 = 3$$

but the difference between the second and third terms is

$$a_3 - a_2 = 9 - 4 = 5.$$

In Example 1, notice that each of the arithmetic sequences has an nth term that is of the form $dn + c$, where the common difference of the sequence is d. An arithmetic sequence may be thought of as a linear function whose domain is the set of natural numbers.

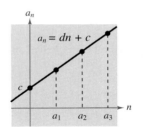

FIGURE **8.3**

The nth Term of an Arithmetic Sequence

The nth term of an arithmetic sequence has the form

$$a_n = dn + c$$

where d is the common difference between consecutive terms of the sequence and $c = a_1 - d$. A graphical representation of this definition is shown in Figure 8.3.

Example 2 ▶ Finding the nth Term of an Arithmetic Sequence

Find a formula for the nth term of the arithmetic sequence whose common difference is 3 and whose first term is 2.

Solution

Because the sequence is arithmetic, you know that the formula for the nth term is of the form $a_n = dn + c$. Moreover, because the common difference is $d = 3$, the formula must have the form

$$a_n = 3n + c. \qquad \text{Substitute 3 for } d.$$

Because $a_1 = 2$, it follows that

$$c = a_1 - d$$
$$= 2 - 3 \qquad \text{Substitute 2 for } a_1 \text{ and 3 for } d.$$
$$= -1.$$

So, the formula for the nth term is

$$a_n = 3n - 1.$$

The sequence therefore has the following form.

$$2, 5, 8, 11, 14, \ldots, 3n - 1, \ldots$$

Another way to find a formula for the nth term of the sequence in Example 2 is to begin by writing the terms of the sequence.

a_1	a_2	a_3	a_4	a_5	a_6	a_7	$\cdots$
2	$2 + 3$	$5 + 3$	$8 + 3$	$11 + 3$	$14 + 3$	$17 + 3$	$\cdots$
2	5	8	11	14	17	20	$\cdots$

From these terms, you can reason that the nth term is of the form

$$a_n = dn + c = 3n - 1.$$

Example 3 ▶ Writing the Terms of an Arithmetic Sequence

The fourth term of an arithmetic sequence is 20, and the 13th term is 65. Write the first several terms of this sequence.

Solution

The fourth and 13th terms of the sequence are related by

$$a_{13} = a_4 + 9d.$$

Using $a_4 = 20$ and $a_{13} = 65$, you can conclude that $d = 5$, which implies that the sequence is as follows.

a_1	a_2	a_3	a_4	a_5	a_6	a_7	a_8	a_9	a_{10}	a_{11} ...
5	10	15	20	25	30	35	40	45	50	55 ...

If you know the nth term of an arithmetic sequence *and* you know the common difference of the sequence, you can find the $(n + 1)$th term by using the *recursive formula*

$$a_{n+1} = a_n + d. \qquad \text{Recursive formula}$$

With this formula, you can find any term of an arithmetic sequence, *provided* that you know the preceding term. For instance, if you know the first term, you can find the second term. Then, knowing the second term, you can find the third term, and so on.

If you substitute $a_1 - d$ for c in the formula $a_n = dn + c$, the nth term of an arithmetic sequence has the alternative recursive formula

$$a_n = a_1 + (n - 1)d. \qquad \text{Alternative recursive formula}$$

Use this formula to solve Example 4. You should get the same answer.

Example 4 ▶ Using a Recursive Formula

Find the ninth term of the arithmetic sequence that begins with 2 and 9.

Solution

For this sequence, the common difference is $d = 9 - 2 = 7$. There are two ways to find the ninth term. One way is simply to write out the first nine terms (by repeatedly adding 7).

2, 9, 16, 23, 30, 37, 44, 51, 58

Another way to find the ninth term is first to find a formula for the nth term. Because the first term is 2, it follows that

$$c = a_1 - d = 2 - 7 = -5.$$

Therefore, a formula for the nth term is

$$a_n = 7n - 5,$$

which implies that the ninth term is

$$a_9 = 7(9) - 5 = 58.$$

The Sum of a Finite Arithmetic Sequence

There is a simple formula for the *sum* of a finite arithmetic sequence. A proof is given in Appendix A.

The Sum of a Finite Arithmetic Sequence

The sum of a finite arithmetic sequence with n terms is

$$S_n = \frac{n}{2}(a_1 + a_n).$$

Example 5 ▶ Finding the Sum of a Finite Arithmetic Sequence

Find the sum: $1 + 3 + 5 + 7 + 9 + 11 + 13 + 15 + 17 + 19$.

Solution

To begin, notice that the sequence is arithmetic (with a common difference of 2). Moreover, the sequence has 10 terms. So, the sum of the sequence is

$$S_n = 1 + 3 + 5 + 7 + 9 + 11 + 13 + 15 + 17 + 19$$

$$= \frac{n}{2}(a_1 + a_n)$$

$$= \frac{10}{2}(1 + 19) \qquad \text{Substitute 10 for } n, \text{ 1 for } a_1, \text{ 19 for } a_n.$$

$$= 5(20) = 100.$$

Example 6 ▶ Finding the Sum of a Finite Arithmetic Sequence

Find the sum of the integers (a) from 1 to 100 and (b) from 1 to N.

Solution

The integers from 1 to 100 form an arithmetic sequence that has 100 terms. So, you can use the formula for the sum of an arithmetic sequence, as follows.

a. $S_n = 1 + 2 + 3 + 4 + 5 + 6 + \cdots + 99 + 100$

$$= \frac{n}{2}(a_1 + a_n)$$

$$= \frac{100}{2}(1 + 100) \qquad \text{Substitute 100 for } n, \text{ 1 for } a_1, \text{ 100 for } a_n.$$

$$= 50(101) = 5050$$

b. $S_n = 1 + 2 + 3 + 4 + \cdots + N$

$$= \frac{n}{2}(a_1 + a_n)$$

$$= \frac{N}{2}(1 + N) \qquad \text{Substitute } N \text{ for } n, \text{ 1 for } a_1, \text{ } N \text{ for } a_n.$$

Corbis-Bettmann

Historical Note
A teacher of Carl Friedrich Gauss (1777–1855) asked him to add all the integers from 1 to 100. When Gauss returned with the correct answer after only a few moments, the teacher could only look at him in astounded silence. This is what Gauss did:

$$1 + 2 + 3 + \cdots + 100$$

$$\underline{100 + 99 + 98 + \cdots + 1}$$

$$101 + 101 + 101 + \cdots + 101$$

$$\frac{100 \times 101}{2} = 5050$$

The sum of the first n terms of an infinite sequence is the nth partial sum.

Example 7 ▶ Finding a Partial Sum of an Arithmetic Sequence

Find the 150th partial sum of the arithmetic sequence

$$5, 16, 27, 38, 49, \ldots .$$

Solution

For this arithmetic sequence, $a_1 = 5$ and $d = 16 - 5 = 11$. So,

$$c = a_1 - d = 5 - 11 = -6$$

and the nth term is $a_n = 11n - 6$. Therefore, $a_{150} = 11(150) - 6 = 1644$, and the sum of the first 150 terms is

$$S_n = \frac{n}{2}(a_1 + a_n)$$

$$= \frac{150}{2}(5 + 1644)$$

$$= 75(1649)$$

$$= 123{,}675.$$

Applications

Example 8 ▶ Seating Capacity

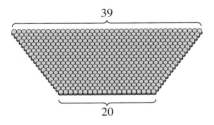

FIGURE **8.4**

An auditorium has 20 rows of seats. There are 20 seats in the first row, 21 seats in the second row, 22 seats in the third row, and so on (see Figure 8.4). How many seats are there in all 20 rows?

Solution

The numbers of seats in the 20 rows form an arithmetic sequence in which the common difference is $d = 1$. Because

$$c = a_1 - d = 20 - 1 = 19$$

you can determine that the formula for the nth term of the sequence is $a_n = n + 19$. Therefore, the 20th term in the sequence is $a_{20} = 20 + 19 = 39$, and the total number of seats is

$$S_n = 20 + 21 + 22 + \cdots + 39$$

$$= \frac{n}{2}(a_1 + a_{20})$$

$$= \frac{20}{2}(20 + 39)$$

$$= 10(59)$$

$$= 590.$$

Example 9 ▶ Total Sales

A small business sells $10,000 worth of products during its first year. The owner of the business has set a goal of increasing annual sales by $7500 each year for 9 years. Assuming that this goal is met, find the total sales during the first 10 years this business is in operation.

Solution

The annual sales form an arithmetic sequence in which $a_1 = 10{,}000$ and $d = 7500$. So,

$$c = a_1 - d$$

$$= 10{,}000 - 7500$$

$$= 2500$$

and the nth term of the sequence is

$$a_n = 7500n + 2500. \qquad \text{See Figure 8.5.}$$

This implies that the 10th term of the sequence is

$$a_{10} = 77{,}500.$$

The sum of the first 10 terms of the sequence is

$$S_n = \frac{n}{2}(a_1 + a_{10})$$

$$= \frac{10}{2}(10{,}000 + 77{,}500)$$

$$= 5(87{,}500)$$

$$= 437{,}500.$$

So, the total sales for the first 10 years are $437,500.

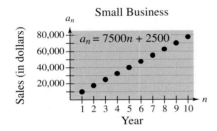

FIGURE **8.5**

Writing ABOUT MATHEMATICS

Numerical Relationships Decide whether it is possible to fill in the blanks in each of the sequences such that the resulting sequence is arithmetic. If so, find a recursive formula for the sequence.

a. -7, ▢, ▢, ▢, ▢, ▢, 11

b. 17, ▢, ▢, ▢, ▢, ▢, ▢, ▢, 71

c. $2, 6$, ▢, ▢, 162

d. $4, 7.5$, ▢, ▢, ▢, ▢, ▢, ▢, ▢, 39

e. $8, 12$, ▢, ▢, ▢, 60.75

8.2 Exercises

In Exercises 1–10, determine whether the sequence is arithmetic. If it is, find the common difference.

1. $10, 8, 6, 4, 2, \ldots$

2. $4, 7, 10, 13, 16, \ldots$

3. $1, 2, 4, 8, 16, \ldots$

4. $80, 40, 20, 10, 5, \ldots$

5. $\frac{9}{4}, 2, \frac{7}{4}, \frac{3}{2}, \frac{5}{4}, \ldots$

6. $3, \frac{5}{2}, 2, \frac{3}{2}, 1, \ldots$

7. $\frac{1}{3}, \frac{2}{3}, 1, \frac{4}{3}, \frac{5}{6}, \ldots$

8. $5.3, 5.7, 6.1, 6.5, 6.9, \ldots$

9. $\ln 1, \ln 2, \ln 3, \ln 4, \ln 5, \ldots$

10. $1^2, 2^2, 3^2, 4^2, 5^2, \ldots$

In Exercises 11–18, write the first five terms of the sequence. Determine whether the sequence is arithmetic, and if it is, find the common difference.

11. $a_n = 5 + 3n$

12. $a_n = 100 - 3n$

13. $a_n = 3 - 4(n - 2)$

14. $a_n = 1 + (n - 1)4$

15. $a_n = (-1)^n$

16. $a_n = 2^{n-1}$

17. $a_n = \dfrac{(-1)^n 3}{n}$

18. $a_n = (2^n)n$

In Exercises 19–24, write the first five terms of the arithmetic sequence. Find the common difference and write the nth term of the sequence as a function of n.

19. $a_1 = 15, \quad a_{k+1} = a_k + 4$

20. $a_1 = 6, \quad a_{k+1} = a_k + 5$

21. $a_1 = 200, \quad a_{k+1} = a_k - 10$

22. $a_1 = 72, \quad a_{k+1} = a_k - 6$

23. $a_1 = \frac{5}{8}, \quad a_{k+1} = a_k - \frac{1}{8}$

24. $a_1 = 0.375, \quad a_{k+1} = a_k + 0.25$

In Exercises 25–32, write the first five terms of the arithmetic sequence.

25. $a_1 = 5, d = 6$

26. $a_1 = 5, d = -\frac{3}{4}$

27. $a_1 = -2.6, d = -0.4$

28. $a_1 = 16.5, d = 0.25$

29. $a_1 = 2, a_{12} = 46$

30. $a_4 = 16, a_{10} = 46$

31. $a_8 = 26, a_{12} = 42$

32. $a_3 = 19, a_{15} = -1.7$

In Exercises 33–44, find a formula for a_n for the arithmetic sequence.

33. $a_1 = 1, d = 3$

34. $a_1 = 15, d = 4$

35. $a_1 = 100, d = -8$

36. $a_1 = 0, d = -\frac{2}{3}$

37. $a_1 = x, d = 2x$

38. $a_1 = -y, d = 5y$

39. $4, \frac{3}{2}, -1, -\frac{7}{2}, \ldots$

40. $10, 5, 0, -5, -10, \ldots$

41. $a_1 = 5, a_4 = 15$

42. $a_1 = -4, a_5 = 16$

43. $a_3 = 94, a_6 = 85$

44. $a_5 = 190, a_{10} = 115$

In Exercises 45–48, match the sequence with its graph. [The graphs are labeled (a), (b), (c), and (d).]

(a)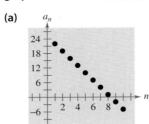

(b)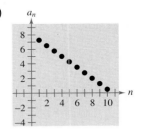

(c)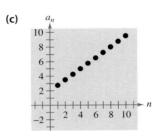

(d)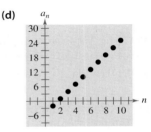

45. $a_n = -\frac{3}{4}n + 8$

46. $a_n = 3n - 5$

47. $a_n = 2 + \frac{3}{4}n$

48. $a_n = 25 - 3n$

In Exercises 49–52, use a graphing utility to graph the first 10 terms of the sequence.

49. $a_n = 15 - \frac{3}{2}n$

50. $a_n = -5 + 2n$

51. $a_n = 0.2n + 3$

52. $a_n = -0.3n + 8$

In Exercises 53–60, find the indicated nth partial sum of the arithmetic sequence.

53. $8, 20, 32, 44, \ldots, \quad n = 10$

54. $2, 8, 14, 20, \ldots, \quad n = 25$

55. $4.2, 3.7, 3.2, 2.7, \ldots, \quad n = 12$

56. $0.5, 0.9, 1.3, 1.7, \ldots, \quad n = 10$

57. $40, 37, 34, 31, \ldots, \quad n = 10$

58. $75, 70, 65, 60, \ldots, \quad n = 25$

59. $a_1 = 100, a_{25} = 220, \quad n = 25$

60. $a_1 = 15, a_{100} = 307, \quad n = 100$

In Exercises 61–68, find the partial sum.

61. $\displaystyle\sum_{n=1}^{50} n$

62. $\displaystyle\sum_{n=1}^{100} 2n$

63. $\displaystyle\sum_{n=10}^{100} 6n$

64. $\displaystyle\sum_{n=51}^{100} 7n$

65. $\displaystyle\sum_{n=11}^{30} n - \sum_{n=1}^{10} n$

66. $\displaystyle\sum_{n=51}^{100} n - \sum_{n=1}^{50} n$

67. $\displaystyle\sum_{n=1}^{400} (2n - 1)$

68. $\displaystyle\sum_{n=1}^{250} (1000 - n)$

In Exercises 69–74, use a calculator to find the partial sum.

69. $\displaystyle\sum_{n=1}^{20} (2n + 5)$

70. $\displaystyle\sum_{n=0}^{50} (1000 - 5n)$

71. $\displaystyle\sum_{n=1}^{100} \frac{n + 4}{2}$

72. $\displaystyle\sum_{n=0}^{100} \frac{8 - 3n}{16}$

73. $\displaystyle\sum_{i=1}^{60} \left(250 - \tfrac{8}{3}i\right)$

74. $\displaystyle\sum_{j=1}^{200} (4.5 + 0.025j)$

Job Offer   **In Exercises 75 and 76, consider a job offer with the given starting salary and the given annual raise.**

(a) Determine the salary during the sixth year of employment.

(b) Determine the total compensation from the company through six full years of employment.

	Starting Salary	*Annual Raise*
75.	$32,500	$1500
76.	$36,800	$1750

77. *Seating Capacity*   Determine the seating capacity of an auditorium with 30 rows of seats if there are 20 seats in the first row, 24 seats in the second row, 28 seats in the third row, and so on.

78. *Seating Capacity*   Determine the seating capacity of an auditorium with 36 rows of seats if there are 15 seats in the first row, 18 seats in the second row, 21 seats in the third row, and so on.

79. *Brick Pattern*   A brick patio has the approximate shape of a trapezoid (see figure). The patio has 18 rows of bricks. The first row has 14 bricks and the 18th row has 31 bricks. How many bricks are in the patio?

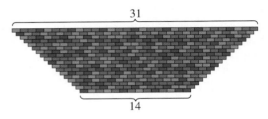

31

14

FIGURE FOR 79

80. *Brick Pattern*   A triangular brick wall is made by cutting some bricks in half to use in the first column of every other row. The wall has 28 rows. The top row is one-half brick wide and the bottom row is 14 bricks wide. How many bricks are used in the finished wall?

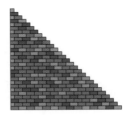

81. *Falling Object*   An object with negligible air resistance is dropped from a plane. During the first second of fall, the object falls 4.9 meters; during the second second, it falls 14.7 meters; during the third second, it falls 24.5 meters; during the fourth second, it falls 34.3 meters. If this arithmetic pattern continues, how many meters will the object fall in 10 seconds?

82. *Falling Object*   An object with negligible air resistance is dropped from a height of 1500 meters. During the first second of fall, the object falls 4.9 meters; during the second second, it falls 14.7 meters; during the third second, it falls 24.5 meters; during the fourth second, it falls 34.3 meters. If this arithmetic pattern continues, how many meters will the object fall in 17 seconds?

Synthesis

True or False?   **In Exercises 83 and 84, determine whether the statement is true or false. Justify your answer.**

83. Given an arithmetic sequence for which only the first and second terms are known, it is possible to find the *n*th term.

84. If the only known information about a finite arithmetic sequence is its first term and its last term, then it is possible to find the sum of the sequence.

85. *Writing* Describe the geometric pattern of the graph of an arithmetic sequence. Explain.

86. *Writing* Explain how to use the first two terms of an arithmetic sequence to find the nth term.

87. Find the sum of the first 100 odd integers.

88. Find the sum of the integers from -10 to 50.

89. *Pattern Recognition*

(a) Compute the following sums of positive odd integers.

$$1 + 3 = \boxed{}$$

$$1 + 3 + 5 = \boxed{}$$

$$1 + 3 + 5 + 7 = \boxed{}$$

$$1 + 3 + 5 + 7 + 9 = \boxed{}$$

$$1 + 3 + 5 + 7 + 9 + 11 = \boxed{}$$

(b) Use the sums in part (a) to make a conjecture about the sums of positive odd integers. Check your conjecture for the sum

$$1 + 3 + 5 + 7 + 9 + 11 + 13 = \boxed{}.$$

(c) Verify your conjecture analytically.

90. *Think About It* The following operations are performed on each term of an arithmetic sequence. Determine if the resulting sequence is arithmetic, and if so, state the common difference.

(a) A constant C is added to each term.

(b) Each term is multiplied by a nonzero constant C.

(c) Each term is squared.

91. *Think About It* The sum of the first 20 terms of an arithmetic sequence with a common difference of 3 is 650. Find the first term.

92. *Think About It* The sum of the first n terms of an arithmetic sequence with first term a_1 and common difference d is S_n. Determine the sum if each term is increased by 5. Explain.

Review

In Exercises 93–96, use Gaussian elimination to solve the system of equations.

93. $\begin{cases} 8x + 2y - 3z = 0 \\ 4x - 2y - 6z = 0 \\ -2x - 3y - 3z = 0 \end{cases}$

94. $\begin{cases} 2x + y - 3z = 0 \\ -x - y + 2z = 0 \\ 4x - 3y - z = 0 \end{cases}$

95. $\begin{cases} 9x + 2z = 0 \\ x - 3y + 2z = 0 \\ 7y - z = 0 \end{cases}$

96. $\begin{cases} 5x - 3y + 7z = 0 \\ -x + 6y + 4z = 0 \\ 6x + 3y + 4z = 0 \end{cases}$

In Exercises 97–100, use matrices to solve the system of equations.

97. $\begin{cases} 2x - 6y = 2 \\ y = 1 \end{cases}$

98. $\begin{cases} 7x + 3y = 0 \\ x + y = 4 \end{cases}$

99. $\begin{cases} 3x + y + z = 4 \\ x + 7y + 5z = 2 \\ 9x - 7y - 5z = 8 \end{cases}$

100. $\begin{cases} 3x + y = 0 \\ x - 2z = 0 \\ 2x - 3y + z = 0 \end{cases}$

8.3 Geometric Sequences and Series

▶ **What you should learn**

- How to recognize and write geometric sequences
- How to find the nth partial sum of a geometric sequence
- How to find the sum of an infinite geometric series
- How to use geometric sequences to model and solve real-life problems

▶ **Why you should learn it**

Geometric sequences can be used to model and solve real-life problems. For instance, Exercises 95 and 96 on page 606 uses a geometric sequence in retirement planning.

Ferguson/PhotoEdit

Geometric Sequences

In Section 8.2, you learned that a sequence whose consecutive terms have a common *difference* is an arithmetic sequence. In this section, you will study another important type of sequence called a **geometric sequence.** Consecutive terms of a geometric sequence have a common *ratio.*

Definition of a Geometric Sequence

A sequence is **geometric** if the ratios of consecutive terms are the same.

$$\frac{a_2}{a_1} = r, \quad \frac{a_3}{a_2} = r, \quad \frac{a_4}{a_3} = r, \ldots, \qquad r \neq 0$$

The number r is the **common ratio** of the sequence.

Example 1 ▶ Examples of Geometric Sequences

a. The sequence whose nth term is 2^n is geometric. For this sequence, the common ratio of consecutive terms is 2.

$$2, 4, 8, 16, \ldots, 2^n, \ldots$$

$$\frac{4}{2} = 2$$

b. The sequence whose nth term is $4(3^n)$ is geometric. For this sequence, the common ratio of consecutive terms is 3.

$$12, 36, 108, 324, \ldots, 4(3^n), \ldots$$

$$\frac{36}{12} = 3$$

c. The sequence whose nth term is $\left(-\frac{1}{3}\right)^n$ is geometric. For this sequence, the common ratio of consecutive terms is $-\frac{1}{3}$.

$$-\frac{1}{3}, \frac{1}{9}, -\frac{1}{27}, \frac{1}{81}, \ldots, \left(-\frac{1}{3}\right)^n, \ldots$$

$$\frac{1/9}{-1/3} = -\frac{1}{3}$$

The sequence $1, 4, 9, 16, \ldots$, whose nth term is n^2 is *not* geometric. The ratio of the second term to the first term is

$$\frac{a_2}{a_1} = \frac{4}{1} = 4$$

but the ratio of the third term to the second term is

$$\frac{a_3}{a_2} = \frac{9}{4}.$$

In Example 1, notice that each of the geometric sequences has an nth term that is of the form ar^n, where the common ratio of the sequence is r. A geometric sequence may be thought of as an exponential function whose domain is the set of natural numbers.

The nth Term of a Geometric Sequence

The nth term of a geometric sequence has the form

$$a_n = a_1 r^{n-1}$$

where r is the common ratio of consecutive terms of the sequence. So, every geometric sequence can be written in the following form.

$$a_1, \quad a_2, \quad a_3, \quad a_4, \quad a_5, \ldots \ldots \quad a_n, \ldots \ldots$$
$$\downarrow \quad \downarrow \quad \downarrow \quad \downarrow \quad \downarrow \qquad\qquad \downarrow$$
$$a_1, a_1 r, a_1 r^2, a_1 r^3, a_1 r^4, \ldots, a_1 r^{n-1}, \ldots$$

If you know the nth term of a geometric sequence, you can find the $(n+1)$th term by multiplying by r. That is, $a_{n+1} = ra_n$.

Example 2 ▶ Finding the Terms of a Geometric Sequence

Write the first five terms of the geometric sequence whose first term is $a_1 = 3$ and whose common ratio is $r = 2$. Then plot the points on a set of coordinate axes.

Solution

Starting with 3, repeatedly multiply by 2 to obtain the following.

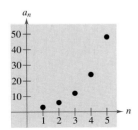

FIGURE 8.6

$a_1 = 3$	1st term
$a_2 = 3(2^1) = 6$	2nd term
$a_3 = 3(2^2) = 12$	3rd term
$a_4 = 3(2^3) = 24$	4th term
$a_5 = 3(2^4) = 48$	5th term

Figure 8.6 shows the first five terms of the geometric sequence.

Example 3 ▶ Finding a Term of a Geometric Sequence

Find the 15th term of the geometric sequence whose first term is 20 and whose common ratio is 1.05.

Solution

$$
\begin{aligned}
a_{15} &= a_1 r^{n-1} && \text{Formula for geometric sequence} \\
&= 20(1.05)^{15-1} && \text{Substitute for } a_1, r, \text{ and } n. \\
&\approx 39.599 && \text{Use a calculator.}
\end{aligned}
$$

Example 4 ▶ Finding a Term of a Geometric Sequence

Find the 12th term of the geometric sequence

$$5, 15, 45, \ldots.$$

Solution

The common ratio of this sequence is

$$r = \frac{15}{5} = 3.$$

Because the first term is $a_1 = 5$, you can determine the 12th term ($n = 12$) to be

$$
\begin{aligned}
a_{12} &= a_1 r^{n-1} && \text{Formula for geometric sequence} \\
&= 5(3)^{12-1} && \text{Substitute for } a_1, r, \text{ and } n. \\
&= 5(177{,}147) && \text{Use a calculator.} \\
&= 885{,}735. && \text{Simplify.}
\end{aligned}
$$

If you know any two terms of a geometric sequence, you can use that information to find a formula for the nth term of the sequence.

Example 5 ▶ Finding a Term of a Geometric Sequence

The fourth term of a geometric sequence is 125, and the 10th term is 125/64. Find the 14th term. (Assume that the terms of the sequence are positive.)

Solution

The 10th term is related to the fourth term by the equation

$$a_{10} = a_4 r^6. \qquad \text{Multiply 4th term by } r^{10-4}.$$

Because $a_{10} = 125/64$ and $a_4 = 125$, you can solve for r as follows.

$$
\begin{aligned}
\frac{125}{64} &= 125 r^6 && \text{Substitute } \tfrac{125}{64} \text{ for } a_{10} \text{ and } 125 \text{ for } a_4. \\
\frac{1}{64} &= r^6 && \text{Divide each side by } 125. \\
\frac{1}{2} &= r && \text{Take the sixth root of each side.}
\end{aligned}
$$

You can obtain the 14th term by multiplying the 10th term by r^4.

$$
\begin{aligned}
a_{14} &= a_{10} r^4 && \text{Multiply the 10th term by } r^{14-10}. \\
&= \frac{125}{64}\left(\frac{1}{2}\right)^4 && \text{Substitute } \tfrac{125}{64} \text{ for } a_{10} \text{ and } \tfrac{1}{2} \text{ for } r. \\
&= \frac{125}{1024} && \text{Simplify.}
\end{aligned}
$$

The Sum of a Finite Geometric Sequence

The formula for the sum of a *finite* geometric sequence is as follows. A proof is given in Appendix A.

The Sum of a Finite Geometric Sequence

The sum of the geometric sequence

$$a_1, \; a_1 r, \; a_1 r^2, \; a_1 r^3, \; a_1 r^4, \ldots, a_1 r^{n-1}$$

with common ratio $r \neq 1$ is given by

$$S_n = a_1 \left(\frac{1 - r^n}{1 - r} \right).$$

Example 6 ▶ Finding the Sum of a Finite Geometric Sequence

Find the sum $\displaystyle\sum_{n=1}^{12} 4(0.3)^n$.

Solution

By writing out a few terms, you have

$$\sum_{n=1}^{12} 4(0.3)^n = 4(0.3)^1 + 4(0.3)^2 + 4(0.3)^3 + \cdots + 4(0.3)^{12}.$$

Now, because $a_1 = 4(0.3)$, $r = 0.3$, and $n = 12$, you can apply the formula for the sum of a finite geometric sequence to obtain

$$\sum_{n=1}^{12} 4(0.3)^n = a_1 \left(\frac{1 - r^n}{1 - r} \right)$$

$$= 4(0.3) \left[\frac{1 - (0.3)^{12}}{1 - 0.3} \right]$$

$$\approx 1.714.$$

When using the formula for the sum of a geometric sequence, be careful to check that the index begins at $i = 1$. If the index begins at $i = 0$, you must adjust the formula for the nth partial sum. For instance, if the index in Example 6 had begun with $n = 0$, the sum would have been

$$\sum_{n=0}^{12} 4(0.3)^n = 4(0.3)^0 + \sum_{n=1}^{12} 4(0.3)^n$$

$$= 4 + \sum_{n=1}^{12} 4(0.3)^n$$

$$\approx 4 + 1.714$$

$$= 5.714.$$

A computer animation of this concept appears in the *Interactive* CD-ROM and *Internet* versions of this text.

Geometric Series

The summation of the terms of an infinite geometric sequence is called an **infinite geometric series** or simply a **geometric series.**

The formula for the sum of a *finite* geometric sequence can, depending on the value of r, be extended to produce a formula for the sum of an *infinite* geometric series. Specifically, if the common ratio r has the property that $|r| < 1$, it can be shown that r^n becomes arbitrarily close to zero as n increases without bound. Consequently,

$$a_1\left(\frac{1 - r^n}{1 - r}\right) \longrightarrow a_1\left(\frac{1 - 0}{1 - r}\right) \qquad \text{as} \qquad n \longrightarrow \infty.$$

This result is summarized as follows.

The Sum of an Infinite Geometric Series

If $|r| < 1$, the infinite geometric series

$$a_1, a_1r, a_1r^2, a_1r^3, \ldots, a_1r^{n-1}, \ldots$$

has the sum

$$S = \frac{a_1}{1 - r}.$$

Example 7 ▶ Finding the Sum of an Infinite Geometric Sequence

Find the following sums.

a. $\displaystyle\sum_{n=1}^{\infty} 4(0.6)^{n-1}$

b. $3 + 0.3 + 0.03 + 0.003 + \cdots$

Solution

a. $\displaystyle\sum_{n=1}^{\infty} 4(0.6)^{n-1} = 4 + 4(0.6) + 4(0.6)^2 + 4(0.6)^3 + \cdots + 4(0.6)^{n-1} + \cdots$

$$= \frac{4}{1 - (0.6)} \qquad \frac{a_1}{1 - r}$$

$$= 10$$

b. $3 + 0.3 + 0.03 + 0.003 + \cdots = 3 + 3(0.1) + 3(0.1)^2 + 3(0.1)^3 + \cdots$

$$= \frac{3}{1 - (0.1)} \qquad \frac{a_1}{1 - r}$$

$$= \frac{10}{3}$$

$$\approx 3.33$$

Application

Example 8 ▶ **Compound Interest**

A deposit of $50 is made on the first day of each month in a savings account that pays 6% compounded monthly. What is the balance of this annuity at the end of 2 years?

Solution

The first deposit will gain interest for 24 months, and its balance will be

$$A_{24} = 50\left(1 + \frac{0.06}{12}\right)^{24}$$

$$= 50(1.005)^{24}.$$

The second deposit will gain interest for 23 months, and its balance will be

$$A_{23} = 50\left(1 + \frac{0.06}{12}\right)^{23}$$

$$= 50(1.005)^{23}.$$

The last deposit will gain interest for only 1 month, and its balance will be

$$A_1 = 50\left(1 + \frac{0.06}{12}\right)^{1}$$

$$= 50(1.005).$$

The total balance in the annuity will be the sum of the balances of the 24 deposits. Using the formula for the sum of a finite geometric sequence, with $A_1 = 50(1.005)$ and $r = 1.005$, you have

$$S_{24} = 50(1.005)\left[\frac{1 - (1.005)^{24}}{1 - 1.005}\right] \qquad \text{Substitute for } A_1 \text{ and } r.$$

$$= \$1277.96.$$

Writing ABOUT MATHEMATICS

An Experiment You will need a piece of string or yarn, a pair of scissors, and a tape measure. Measure out any length of string at least 5 feet long. Double over the string and cut it in half. Take one of the resulting halves, double it over, and cut it in half. Continue this process until you are no longer able to cut a length of string in half. How many cuts were you able to make? Construct a sequence of the resulting string lengths after each cut, starting with the original length of the string. Find a formula for the nth term of this sequence. How many cuts could you theoretically make? Discuss why you were not able to make that many cuts.

8.3 Exercises

In Exercises 1–10, determine whether the sequence is geometric. If it is, find the common ratio.

1. $5, 15, 45, 135, \ldots$

2. $3, 12, 48, 192, \ldots$

3. $3, 12, 21, 30, \ldots$

4. $36, 27, 18, 9, \ldots$

5. $1, -\frac{1}{2}, \frac{1}{4}, -\frac{1}{8}, \ldots$

6. $5, 1, 0.2, 0.04, \ldots$

7. $\frac{1}{8}, \frac{1}{4}, \frac{1}{2}, 1, \ldots$

8. $9, -6, 4, -\frac{8}{3}, \ldots$

9. $1, \frac{1}{2}, \frac{1}{3}, \frac{1}{4}, \ldots$

10. $\frac{1}{5}, \frac{2}{7}, \frac{3}{9}, \frac{4}{11}, \ldots$

In Exercises 11–20, write the first five terms of the geometric sequence.

11. $a_1 = 2, r = 3$

12. $a_1 = 6, r = 2$

13. $a_1 = 1, r = \frac{1}{2}$

14. $a_1 = 1, r = \frac{1}{3}$

15. $a_1 = 5, r = -\frac{1}{10}$

16. $a_1 = 6, r = -\frac{1}{4}$

17. $a_1 = 1, r = e$

18. $a_1 = 3, r = \sqrt{5}$

19. $a_1 = 2, r = \dfrac{x}{4}$

20. $a_1 = 5, r = 2x$

In Exercises 21–26, write the first five terms of the geometric sequence. Determine the common ratio and write the nth term of the sequence as a function of n.

21. $a_1 = 64, \quad a_{k+1} = \frac{1}{2}a_k$

22. $a_1 = 81, \quad a_{k+1} = \frac{1}{3}a_k$

23. $a_1 = 7, \quad a_{k+1} = 2a_k$

24. $a_1 = 5, \quad a_{k+1} = -2a_k$

25. $a_1 = 6, \quad a_{k+1} = -\frac{3}{2}a_k$

26. $a_1 = 48, \quad a_{k+1} = -\frac{1}{2}a_k$

In Exercises 27–38, find the nth term of the geometric sequence.

27. $a_1 = 4, r = \frac{1}{2}, n = 10$

28. $a_1 = 5, r = \frac{3}{2}, n = 8$

29. $a_1 = 6, r = -\frac{1}{3}, n = 12$

30. $a_1 = 64, r = -\frac{1}{4}, n = 10$

31. $a_1 = 100, r = e^x, n = 9$

32. $a_1 = 1, r = \sqrt{3}, n = 8$

33. $a_1 = 500, r = 1.02, n = 40$

34. $a_1 = 1000, r = 1.005, n = 60$

35. $a_1 = 16, a_4 = \frac{27}{4}, n = 3$

36. $a_2 = 3, a_5 = \frac{3}{64}, n = 1$

37. $a_4 = -18, a_7 = \frac{2}{3}, n = 6$

38. $a_3 = \frac{16}{3}, a_5 = \frac{64}{27}, n = 7$

In Exercises 39–42, match the sequence with its graph. [The graphs are labeled (a), (b), (c), and (d).]

(a)

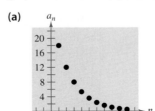

(b)

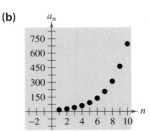

(c)

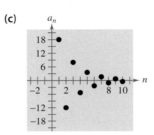

(d)

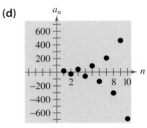

39. $a_n = 18\left(\frac{2}{3}\right)^{n-1}$

40. $a_n = 18\left(-\frac{2}{3}\right)^{n-1}$

41. $a_n = 18\left(\frac{3}{2}\right)^{n-1}$

42. $a_n = 18\left(-\frac{3}{2}\right)^{n-1}$

In Exercises 43–46, use a graphing utility to graph the first 10 terms of the sequence.

43. $a_n = 12(-0.75)^{n-1}$

44. $a_n = 12(-0.4)^{n-1}$

45. $a_n = 2(1.3)^{n-1}$

46. $a_n = 2(-1.4)^{n-1}$

In Exercises 47–56, find the sum of the finite geometric series.

47. $\displaystyle\sum_{n=1}^{9} 2^{n-1}$

48. $\displaystyle\sum_{n=1}^{9} (-2)^{n-1}$

49. $\displaystyle\sum_{i=1}^{7} 64\left(-\frac{1}{2}\right)^{i-1}$

50. $\displaystyle\sum_{i=1}^{6} 32\left(\frac{1}{4}\right)^{i-1}$

51. $\displaystyle\sum_{n=0}^{20} 3\left(\frac{3}{2}\right)^n$

52. $\displaystyle\sum_{n=0}^{15} 2\left(\frac{4}{3}\right)^n$

53. $\displaystyle\sum_{n=0}^{5} 300(1.06)^n$

54. $\displaystyle\sum_{n=0}^{6} 500(1.04)^n$

55. $\displaystyle\sum_{i=1}^{10} 8\left(-\frac{1}{4}\right)^{i-1}$

56. $\displaystyle\sum_{i=1}^{10} 5\left(-\frac{1}{3}\right)^{i-1}$

In Exercises 57–62, use summation notation to express the sum.

57. $5 + 15 + 45 + \cdots + 3645$

58. $7 + 14 + 28 + \cdots + 896$

59. $2 - \dfrac{1}{2} + \dfrac{1}{8} - \cdots + \dfrac{1}{2048}$

60. $15 - 3 + \dfrac{3}{5} - \cdots - \dfrac{3}{625}$

61. $0.1 + 0.4 + 1.6 + \cdots + 102.4$

62. $32 + 24 + 18 + \cdots + 10.125$

In Exercises 63–76, find the sum of the infinite geometric series.

63. $\displaystyle\sum_{n=0}^{\infty} \left(\frac{1}{2}\right)^n$

64. $\displaystyle\sum_{n=0}^{\infty} 2\left(\frac{2}{3}\right)^n$

65. $\displaystyle\sum_{n=0}^{\infty} \left(-\frac{1}{2}\right)^n$

66. $\displaystyle\sum_{n=0}^{\infty} 2\left(-\frac{2}{3}\right)^n$

67. $\displaystyle\sum_{n=0}^{\infty} 4\left(\frac{1}{4}\right)^n$

68. $\displaystyle\sum_{n=0}^{\infty} \left(\frac{1}{10}\right)^n$

69. $\displaystyle\sum_{n=0}^{\infty} (0.4)^n$

70. $\displaystyle\sum_{n=0}^{\infty} 4(0.2)^n$

71. $\displaystyle\sum_{n=0}^{\infty} -3(0.9)^n$

72. $\displaystyle\sum_{n=0}^{\infty} -10(0.2)^n$

73. $8 + 6 + \frac{9}{2} + \frac{27}{8} + \cdots$

74. $9 + 6 + 4 + \frac{8}{3} + \cdots$

75. $\frac{1}{9} - \frac{1}{3} + 1 - 3 + \cdots$

76. $-\frac{125}{36} + \frac{25}{6} - 5 + 6 - \cdots$

In Exercises 77–80, find the rational number representation of the repeating decimal.

77. $0.\overline{36}$

78. $0.\overline{297}$

79. $0.3\overline{18}$

80. $1.3\overline{8}$

▦ *Graphical Reasoning* **In Exercises 81 and 82, use a graphing utility to graph the function. Identify the horizontal asymptote of the graph and determine its relationship to the sum.**

81. $f(x) = 6\left[\dfrac{1 - (0.5)^x}{1 - (0.5)}\right]$, $\quad\displaystyle\sum_{n=0}^{\infty} 6\left(\frac{1}{2}\right)^n$

82. $f(x) = 2\left[\dfrac{1 - (0.8)^x}{1 - (0.8)}\right]$, $\quad\displaystyle\sum_{n=0}^{\infty} 2\left(\frac{4}{5}\right)^n$

83. *Compound Interest* A principal of \$1000 is invested at 6% interest. Find the amount after 10 years if the interest is compounded (a) annually, (b) semiannually, (c) quarterly, (d) monthly, and (e) daily.

84. *Compound Interest* A principal of \$2500 is invested at 8% interest. Find the amount after 20 years if the interest is compounded (a) annually, (b) semiannually, (c) quarterly, (d) monthly, and (e) daily.

85. *Depreciation* A company buys a machine for \$135,000 and it depreciates at a rate of 30% per year. (In other words, at the end of each year the depreciated value is 70% of what it was at the beginning of the year.) Find the depreciated value of the machine after 5 full years.

86. *Population Growth* A city of 250,000 people is growing at a rate of 1.3% per year. Estimate the population of the city 30 years from now.

87. *Annuities* A deposit of \$100 is made at the beginning of each month in an account that pays 6%, compounded monthly. The balance A in the account at the end of 5 years is

$$A = 100\left(1 + \frac{0.06}{12}\right)^1 + \cdots + 100\left(1 + \frac{0.06}{12}\right)^{60}.$$

Find A.

88. *Annuities* A deposit of \$50 is made at the beginning of each month in an account that pays 8%, compounded monthly. The balance A in the account at the end of 5 years is

$$A = 50\left(1 + \frac{0.08}{12}\right)^1 + \cdots + 50\left(1 + \frac{0.08}{12}\right)^{60}.$$

Find A.

89. *Annuities*　A deposit of P dollars is made at the beginning of each month in an account earning an annual interest rate r, compounded monthly. The balance A after t years is

$$A = P\left(1 + \frac{r}{12}\right) + P\left(1 + \frac{r}{12}\right)^2 + \cdots +$$

$$P\left(1 + \frac{r}{12}\right)^{12t}.$$

Show that the balance is

$$A = P\left[\left(1 + \frac{r}{12}\right)^{12t} - 1\right]\left(1 + \frac{12}{r}\right).$$

90. *Annuities*　A deposit of P dollars is made at the beginning of each month in an account earning an annual interest rate r, compounded continuously. The balance A after t years is

$$A = Pe^{r/12} + Pe^{2r/12} + \cdots + Pe^{12tr/12}.$$

Show that the balance is

$$A = \frac{Pe^{r/12}(e^{rt} - 1)}{e^{r/12} - 1}.$$

Annuities　**In Exercises 91–94, consider making monthly deposits of P dollars in a savings account earning an annual interest rate r. Use the results of Exercises 89 and 90 to find the balance A after t years if the interest is compounded (a) monthly and (b) continuously.**

91. $P = \$50$, $r = 7\%$, $t = 20$ years

92. $P = \$75$, $r = 9\%$, $t = 25$ years

93. $P = \$100$, $r = 10\%$, $t = 40$ years

94. $P = \$20$, $r = 6\%$, $t = 50$ years

95. *Annuities*　Consider an initial deposit of P dollars in an account earning an annual interest rate r, compounded monthly. At the end of each month a withdrawal of W dollars will occur and the account will be depleted in t years. The amount of the initial deposit required is

$$P = W\left(1 + \frac{r}{12}\right)^{-1} + W\left(1 + \frac{r}{12}\right)^{-2} + \cdots +$$

$$W\left(1 + \frac{r}{12}\right)^{-12t}.$$

Show that the initial deposit is

$$P = W\left(\frac{12}{r}\right)\left[1 - \left(1 + \frac{r}{12}\right)^{-12t}\right].$$

96. *Annuities*　Determine the amount required in a retirement account for an individual who retires at age 65 and wants an income of $2000 from the account each month for 20 years. Use the result of Exercise 95 and assume that the account earns 9% compounded monthly.

97. *Geometry*　The sides of a square are 16 inches in length. A new square is formed by connecting the midpoints of the sides of the original square, and two of the resulting triangles are shaded (see figure). If this process is repeated five more times, determine the total area of the shaded region.

98. *Corporate Revenue*　The annual revenues a_n (in billions of dollars) for Coca-Cola Enterprises for 1990 through 1996 can be approximated by the model

$$a_n = 3.978e^{0.11n}, \qquad n = 0, 1, \ldots, 6$$

where $n = 0$ represents 1990. Use this model and the formula for the sum of a finite geometric sequence to approximate the total revenue earned during this 7-year period. (Source: Coca-Cola Enterprises, Inc.)

99. *Think About It*　Suppose you work for a company that pays $0.01 the first day, $0.02 the second day, $0.04 the third day, and so on. If the daily wage keeps doubling, what would your total income be for working (a) 29 days, (b) 30 days, and (c) 31 days?

100. *Salary*　A company has a job opening with a salary of $30,000 for the first year. Suppose that during the next 39 years, there is a 5% raise each year. Find the total compensation over the 40-year period.

101. *Distance* A ball is dropped from a height of 16 feet. Each time it drops h feet, it rebounds $0.81h$ feet.

(a) Find the total distance traveled by the ball.

(b) The ball takes the following times for each fall.

$s_1 = -16t^2 + 16,$ $s_1 = 0$ if $t = 1$

$s_2 = -16t^2 + 16(0.81),$ $s_2 = 0$ if $t = 0.9$

$s_3 = -16t^2 + 16(0.81)^2,$ $s_3 = 0$ if $t = (0.9)^2$

$s_4 = -16t^2 + 16(0.81)^3,$ $s_4 = 0$ if $t = (0.9)^3$

   ⋮   ⋮

$s_n = -16t^2 + 16(0.81)^{n-1},$ $s_n = 0$ if $t = (0.9)^{n-1}$

Beginning with s_2, the ball takes the same amount of time to bounce up as it does to fall, and so the total time elapsed before it comes to rest is

$$t = 1 + 2\sum_{n=1}^{\infty}(0.9)^n.$$

Find this total.

Synthesis

True or False? **In Exercises 102 and 103, determine whether the statement is true or false. Justify your answer.**

102. A sequence is geometric if the ratios of consecutive differences of consecutive terms are the same.

103. You can find the nth term of a geometric sequence by multiplying its common ratio by the first term of the sequence raised to the $(n-1)$th power.

104. *Writing* Write a brief paragraph explaining why the terms of a geometric sequence decrease in magnitude when $-1 < r < 1$.

105. *Writing* Write a brief paragraph explaining how to use the first two terms of a geometric sequence to find the nth term.

Review

In Exercises 106–109, evaluate the function for $f(x) = 3x + 1$ and $g(x) = x^2 - 1$.

106. $g(x + 1)$ **107.** $f(x + 1)$

108. $f(g(x + 1))$ **109.** $g(f(x + 1))$

In Exercises 110–113, completely factor the expression over the rational numbers.

110. $9x^3 - 64x$ **111.** $x^2 + 4x - 63$

112. $6x^2 - 13x - 5$ **113.** $16x^2 - 4x^4$

In Exercises 114–119, perform the indicated operation and simplify.

114. $\dfrac{3}{x+3} \cdot \dfrac{x(x+3)}{x-3}$

115. $\dfrac{x-2}{x+7} \cdot \dfrac{2x(x+7)}{6x(x-2)}$

116. $\dfrac{x}{3} \div \dfrac{3x}{6x+3}$

117. $\dfrac{x-5}{x-3} \div \dfrac{10-2x}{2(3-x)}$

118. $5 + \dfrac{7}{x+2} + \dfrac{2}{x-2}$

119. $8 - \dfrac{x-1}{x+4} - \dfrac{4}{x-1} - \dfrac{x+4}{(x-1)(x+4)}$

8.4 Mathematical Induction

▶ **What you should learn**

- How to use mathematical induction to prove a statement
- How to find the sums of powers of integers
- How to recognize patter ns and write the nth term of a sequence
- How to find finite differences of a sequence

▶ **Why you should learn it**

Mathematical induction can be used to prove statements involving a positive integer n. For instance, in Exercises 41–49 on pages 616 and 617, you are asked to use mathematical induction to prove properties of real numbers.

Introduction

In this section you will study a form of mathematical proof called **mathematical induction.** It is important that you see clearly the logical need for it, so let's take a closer look at the problem discussed in Example 5 on page 592.

$$S_1 = 1 = 1^2$$

$$S_2 = 1 + 3 = 2^2$$

$$S_3 = 1 + 3 + 5 = 3^2$$

$$S_4 = 1 + 3 + 5 + 7 = 4^2$$

$$S_5 = 1 + 3 + 5 + 7 + 9 = 5^2$$

Judging from the pattern formed by these first five sums, it appears that the sum of the first n odd integers is

$$S_n = 1 + 3 + 5 + 7 + 9 + \cdots + (2n - 1) = n^2.$$

Although this particular formula *is* valid, it is important for you to see that recognizing a pattern and then simply *jumping to the conclusion* that the pattern must be true for all values of n is *not* a logically valid method of proof. There are many examples in which a pattern appears to be developing for small values of n and then at some point the pattern fails. One of the most famous cases of this was the conjecture by the French mathematician Pierre de Fermat (1601–1665), who speculated that all numbers of the form

$$F_n = 2^{2^n} + 1, \quad n = 0, 1, 2, \ldots$$

are prime. For $n = 0, 1, 2, 3,$ and 4, the conjecture is true.

$$F_0 = 3$$

$$F_1 = 5$$

$$F_2 = 17$$

$$F_3 = 257$$

$$F_4 = 65{,}537$$

The size of the next Fermat number ($F_5 = 4{,}294{,}967{,}297$) is so great that it was difficult for Fermat to determine whether it was prime or not. However, another well-known mathematician, Leonhard Euler (1707–1783), later found the factorization

$$F_5 = 4{,}294{,}967{,}297$$

$$= 641(6{,}700{,}417)$$

which proved that F_5 is not prime and therefore Fermat's conjecture was false.

Just because a rule, pattern, or formula seems to work for several values of n, you cannot simply decide that it is valid for all values of n without going through a *legitimate proof.* Mathematical induction is one method of proof.

The Principle of Mathematical Induction

Let P_n be a statement involving the positive integer n. If

1. P_1 is true, and

2. the truth of P_k implies the truth of P_{k+1} for every positive k,

then P_n must be true for all positive integers n.

To apply the Principle of Mathematical Induction, you need to be able to determine the statement P_{k+1} for a given statement P_k.

Example 1 ▶ A Preliminary Example

Find P_{k+1} for the following.

a. $P_k : S_k = \dfrac{k^2(k+1)^2}{4}$

b. $P_k : S_k = 1 + 5 + 9 + \cdots + [4(k-1) - 3] + (4k - 3)$

c. $P_k : 3^k \geq 2k + 1$

Solution

a. $P_{k+1} : S_{k+1} = \dfrac{(k+1)^2(k+1+1)^2}{4}$ Replace k by $k + 1$.

$= \dfrac{(k+1)^2(k+2)^2}{4}$.   Simplify.

b. $P_{k+1} : S_{k+1} = 1 + 5 + 9 + \cdots + \{4[(k+1) - 1] - 3\} + [4(k+1) - 3]$
$= 1 + 5 + 9 + \cdots + (4k - 3) + (4k + 1)$.

c. $P_{k+1} : 3^{k+1} \geq 2(k+1) + 1$
$3^{k+1} \geq 2k + 3$.

A well-known illustration used to explain why the Principle of Mathematical Induction works is the unending line of dominoes shown in Figure 8.7.

FIGURE **8.7**

If the line actually contains infinitely many dominoes, it is clear that you could not knock the entire line down by knocking down only *one domino* at a time. However, suppose it were true that each domino would knock down the next one as it fell. Then you could knock them all down simply by pushing the first one and starting a chain reaction. Mathematical induction works in the same way. If the truth of P_k implies the truth of P_{k+1} and if P_1 is true, the chain reaction proceeds as follows: P_1 implies P_2, P_2 implies P_3, P_3 implies P_4, and so on.

When using mathematical induction to prove a *summation* formula (such as the one in Example 2), it is helpful to think of S_{k+1} as

$$S_{k+1} = S_k + a_{k+1}$$

where a_{k+1} is the $(k+1)$th term of the original sum.

Example 2 ▶ Using Mathematical Induction

Use mathematical induction to prove the following formula.

$$S_n = 1 + 3 + 5 + 7 + \cdots + (2n - 1)$$
$$= n^2$$

Solution

Mathematical induction consists of two distinct parts. First, you must show that the formula is true when $n = 1$.

1. When $n = 1$, the formula is valid, because

$$S_1 = 1 = 1^2.$$

The second part of mathematical induction has two steps. The first step is to assume that the formula is valid for *some* integer k. The second step is to use this assumption to prove that the formula is valid for the next integer, $k + 1$.

2. Assuming that the formula

$$S_k = 1 + 3 + 5 + 7 + \cdots + (2k - 1)$$
$$= k^2$$

is true, you must show that the formula $S_{k+1} = (k+1)^2$ is true.

$$
\begin{aligned}
S_{k+1} &= 1 + 3 + 5 + 7 + \cdots + (2k - 1) + [2(k+1) - 1] \\
&= [1 + 3 + 5 + 7 + \cdots + (2k - 1)] + (2k + 2 - 1) \\
&= S_k + (2k + 1) \qquad \text{Group terms to form } S_k. \\
&= k^2 + 2k + 1 \qquad \text{Replace } S_k \text{ by } k^2. \\
&= (k + 1)^2
\end{aligned}
$$

Combining the results of parts (1) and (2), you can conclude by mathematical induction that the formula is valid for *all* positive integer values of n.

It occasionally happens that a statement involving natural numbers is not true for the first $k - 1$ positive integers but is true for all values of $n \geq k$. In these instances, you use a slight variation of the Principle of Mathematical Induction in which you verify P_k rather than P_1. This variation is called the *extended principle of mathematical induction*. To see the validity of this, note from Figure 8.7 that all but the first $k - 1$ dominoes can be knocked down by knocking over the kth domino. This suggests that you can prove a statement P_n to be true for $n \geq k$ by showing that P_k is true and that P_k implies P_{k+1}. In Exercises 41–49 of this section you are asked to apply this extension of mathematical induction.

Example 3 ▶ Using Mathematical Induction

Use mathematical induction to prove the formula

$$S_n = 1^2 + 2^2 + 3^2 + 4^2 + \cdots + n^2$$
$$= \frac{n(n+1)(2n+1)}{6}.$$

Solution

1. When $n = 1$, the formula is valid, because

$$S_1 = 1^2 = \frac{1(2)(3)}{6}.$$

2. Assuming that

$$S_k = 1^2 + 2^2 + 3^2 + 4^2 + \cdots + k^2$$
$$= \frac{k(k+1)(2k+1)}{6}$$

you must show that

$$S_{k+1} = \frac{(k+1)(k+2)(2k+3)}{6}.$$

To do this, write the following.

$$S_{k+1} = S_k + a_{k+1}$$
$$= (1^2 + 2^2 + 3^2 + 4^2 + \cdots + k^2) + (k+1)^2$$
$$= \frac{k(k+1)(2k+1)}{6} + (k+1)^2$$
$$= \frac{k(k+1)(2k+1) + 6(k+1)^2}{6}$$
$$= \frac{(k+1)[k(2k+1) + 6(k+1)]}{6}$$
$$= \frac{(k+1)(2k^2 + 7k + 6)}{6}$$
$$= \frac{(k+1)(k+2)(2k+3)}{6}$$

Combining the results of parts (1) and (2), you can conclude by mathematical induction that the formula is valid for *all* $n \geq 1$.

When proving a formula using mathematical induction, the only statement that you *need* to verify is P_1. As a check, however, it is good to try verifying other statements. For instance, in Example 3, try verifying P_2 and P_3.

Sums of Powers of Integers

The formula in Example 3 is one of a collection of useful summation formulas. This and other formulas dealing with the sums of various powers of the first n positive integers are as follows.

Sums of Powers of Integers

1. $1 + 2 + 3 + 4 + \cdots + n = \dfrac{n(n + 1)}{2}$

2. $1^2 + 2^2 + 3^2 + 4^2 + \cdots + n^2 = \dfrac{n(n + 1)(2n + 1)}{6}$

3. $1^3 + 2^3 + 3^3 + 4^3 + \cdots + n^3 = \dfrac{n^2(n + 1)^2}{4}$

4. $1^4 + 2^4 + 3^4 + 4^4 + \cdots + n^4 = \dfrac{n(n + 1)(2n + 1)(3n^2 + 3n - 1)}{30}$

5. $1^5 + 2^5 + 3^5 + 4^5 + \cdots + n^5 = \dfrac{n^2(n + 1)^2(2n^2 + 2n - 1)}{12}$

Example 4 ▶ Finding a Sum of Powers of Integers

Find the following sums.

a. $\displaystyle\sum_{n=1}^{7} n^3 = 1^3 + 2^3 + 3^3 + 4^3 + 5^3 + 6^3 + 7^3$ **b.** $\displaystyle\sum_{n=1}^{4} (6n - 4n^2)$

Solution

a. Using the formula for the sum of the cubes of the first n positive integers, you obtain

$$\sum_{n=1}^{7} n^3 = 1^3 + 2^3 + 3^3 + 4^3 + 5^3 + 6^3 + 7^3$$

$$= \frac{7^2(7 + 1)^2}{4} = \frac{49(64)}{4} = 784.$$

b. $\displaystyle\sum_{n=1}^{4} (6n - 4n^2) = \sum_{n=1}^{4} 6n - \sum_{n=1}^{4} 4n^2$

$$= 6\sum_{n=1}^{4} n - 4\sum_{n=1}^{4} n^2$$

$$= 6\left[\frac{4(4 + 1)}{2}\right] - 4\left[\frac{4(4 + 1)(8 + 1)}{6}\right]$$

$$= 6(10) - 4(30)$$

$$= 60 - 120$$

$$= -60$$

Example 5 ▶ Proving an Inequality by Mathematical Induction

Prove that $n < 2^n$ for all positive integers n.

Solution

1. For $n = 1$ or 2, the statement is true, because

$$1 < 2^1 \quad \text{and} \quad 2 < 2^2.$$

2. Assuming that

$$k < 2^k$$

you need to show that $k + 1 < 2^{k+1}$. For $n = k$, you have

$$2^{k+1} = 2(2^k) > 2(k) = 2k. \qquad \text{By assumption}$$

Because $2k = k + k > k + 1$ for all $k > 1$, it follows that

$$2^{k+1} > 2k > k + 1$$

or

$$k + 1 < 2^{k+1}.$$

Therefore, $n < 2^n$ for all integers $n \geq 1$.

To check a result that you have proved by mathematical induction, it helps to list the statement for several values of n. For instance, in Example 5, you could list

$$1 < 2^1 = 2, \qquad 2 < 2^2 = 4, \qquad 3 < 2^3 = 8,$$
$$4 < 2^4 = 16, \qquad 5 < 2^5 = 32, \qquad 6 < 2^6 = 64.$$

From this list, your intuition confirms that the statement $n < 2^n$ is reasonable.

Pattern Recognition

Although choosing a formula on the basis of a few observations does *not* guarantee the validity of the formula, pattern recognition *is* important. Once you have a pattern or formula that you think works, you can try using mathematical induction to prove your formula.

Finding a Formula for the *n*th Term of a Sequence

To find a formula for the *n*th term of a sequence, consider these guidelines.

1. Calculate the first several terms of the sequence. It is often a good idea to write the terms in both simplified and factored forms.

2. Try to find a recognizable pattern for the terms and write a formula for the *n*th term of the sequence. This is your *hypothesis* or *conjecture*. You might try computing one or two more terms in the sequence to test your hypothesis.

3. Use mathematical induction to prove your hypothesis.

Example 6 ▶ Finding a Formula for a Finite Sum

Find a formula for the finite sum and prove its validity.

$$\frac{1}{1 \cdot 2} + \frac{1}{2 \cdot 3} + \frac{1}{3 \cdot 4} + \frac{1}{4 \cdot 5} + \cdots + \frac{1}{n(n + 1)}$$

Solution

Begin by writing out the first few sums.

$$S_1 = \frac{1}{1 \cdot 2} = \frac{1}{2} = \frac{1}{1 + 1}$$

$$S_2 = \frac{1}{1 \cdot 2} + \frac{1}{2 \cdot 3} = \frac{4}{6} = \frac{2}{3} = \frac{2}{2 + 1}$$

$$S_3 = \frac{1}{1 \cdot 2} + \frac{1}{2 \cdot 3} + \frac{1}{3 \cdot 4} = \frac{9}{12} = \frac{3}{4} = \frac{3}{3 + 1}$$

$$S_4 = \frac{1}{1 \cdot 2} + \frac{1}{2 \cdot 3} + \frac{1}{3 \cdot 4} + \frac{1}{4 \cdot 5} = \frac{48}{60} = \frac{4}{5} = \frac{4}{4 + 1}$$

From this sequence, it appears that the formula for the kth sum is

$$S_k = \frac{1}{1 \cdot 2} + \frac{1}{2 \cdot 3} + \frac{1}{3 \cdot 4} + \frac{1}{4 \cdot 5} + \cdots + \frac{1}{k(k + 1)} = \frac{k}{k + 1}.$$

To prove the validity of this hypothesis, use mathematical induction, as follows. Note that you have already verified the formula for $n = 1$, so you can begin by assuming that the formula is valid for $n = k$ and trying to show that it is valid for $n = k + 1$.

$$S_{k+1} = \left[\frac{1}{1 \cdot 2} + \frac{1}{2 \cdot 3} + \frac{1}{3 \cdot 4} + \frac{1}{4 \cdot 5} + \cdots + \frac{1}{k(k + 1)} \right] + \frac{1}{(k + 1)(k + 2)}$$

$$= \frac{k}{k + 1} + \frac{1}{(k + 1)(k + 2)}$$

$$= \frac{k(k + 2) + 1}{(k + 1)(k + 2)}$$

$$= \frac{k^2 + 2k + 1}{(k + 1)(k + 2)}$$

$$= \frac{(k + 1)^2}{(k + 1)(k + 2)}$$

$$= \frac{k + 1}{k + 2}$$

So, the hypothesis is valid.

Finite Differences

The **first differences** of a sequence are found by subtracting consecutive terms. The **second differences** are found by subtracting consecutive first differences. The first and second differences of the sequence 3, 5, 8, 12, 17, 23, . . . are as follows.

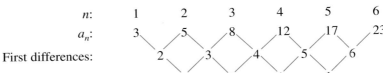

For this sequence, the second differences are all the same. When this happens, the sequence has a perfect *quadratic* model. If the first differences are all the same, the sequence has a *linear* model. That is, it is arithmetic.

Example 7 ▶ Finding a Quadratic Model

Find the quadratic model for the sequence

3, 5, 8, 12, 17, 23, ...

Solution

You know from the second differences shown above that the model is quadratic and has the form

$$a_n = an^2 + bn + c.$$

By substituting 1, 2, and 3 for n, you can obtain a system of three linear equations in three variables.

$a_1 = a(1)^2 + b(1) + c = 3$ Substitute 1 for n.

$a_2 = a(2)^2 + b(2) + c = 5$ Substitute 2 for n.

$a_3 = a(3)^2 + b(3) + c = 8$ Substitute 3 for n.

You now have a system of three equations in a, b, and c.

$$\begin{cases} a + b + c = 3 & \text{Equation 1} \\ 4a + 2b + c = 5 & \text{Equation 2} \\ 9a + 3b + c = 8 & \text{Equation 3} \end{cases}$$

Using the techniques discussed in Chapter 6, you can find the solution to be $a = \frac{1}{2}$, $b = \frac{1}{2}$, and $c = 2$. So, the quadratic model is

$$a_n = \frac{1}{2}n^2 + \frac{1}{2}n + 2.$$

Try checking the values of a_1, a_2, and a_3.

8.4 Exercises

In Exercises 1–4, find P_{k+1} for the given P_k.

1. $P_k = \dfrac{5}{k(k+1)}$

2. $P_k = \dfrac{1}{2(k+2)}$

3. $P_k = \dfrac{k^2(k+1)^2}{4}$

4. $P_k = \dfrac{k}{3}(2k+1)$

In Exercises 5–18, use mathematical induction to prove the formula for every positive integer n.

5. $2 + 4 + 6 + 8 + \cdots + 2n = n(n+1)$

6. $3 + 7 + 11 + 15 + \cdots + (4n-1) = n(2n+1)$

7. $2 + 7 + 12 + 17 + \cdots + (5n-3) = \dfrac{n}{2}(5n-1)$

8. $1 + 4 + 7 + 10 + \cdots + (3n-2) = \dfrac{n}{2}(3n-1)$

9. $1 + 2 + 2^2 + 2^3 + \cdots + 2^{n-1} = 2^n - 1$

10. $2(1 + 3 + 3^2 + 3^3 + \cdots + 3^{n-1}) = 3^n - 1$

11. $1 + 2 + 3 + 4 + \cdots + n = \dfrac{n(n+1)}{2}$

12. $1^2 + 2^2 + 3^2 + 4^2 + \cdots + n^2 = \dfrac{n(n+1)(2n+1)}{6}$

13. $1^3 + 2^3 + 3^3 + 4^3 + \cdots + n^3 = \dfrac{n^2(n+1)^2}{4}$

14. $\left(1 + \dfrac{1}{1}\right)\left(1 + \dfrac{1}{2}\right)\left(1 + \dfrac{1}{3}\right)\cdots\left(1 + \dfrac{1}{n}\right) = n + 1$

15. $\displaystyle\sum_{i=1}^{n} i^5 = \dfrac{n^2(n+1)^2(2n^2 + 2n - 1)}{12}$

16. $\displaystyle\sum_{i=1}^{n} i^4 = \dfrac{n(n+1)(2n+1)(3n^2 + 3n - 1)}{30}$

17. $\displaystyle\sum_{i=1}^{n} i(i+1) = \dfrac{n(n+1)(n+2)}{3}$

18. $\displaystyle\sum_{i=1}^{n} \dfrac{1}{(2i-1)(2i+1)} = \dfrac{n}{2n+1}$

In Exercises 19–28, find the sum using the formulas for the sums of powers of integers.

19. $\displaystyle\sum_{n=1}^{15} n$

20. $\displaystyle\sum_{n=1}^{30} n$

21. $\displaystyle\sum_{n=1}^{6} n^2$

22. $\displaystyle\sum_{n=1}^{10} n^3$

23. $\displaystyle\sum_{n=1}^{5} n^4$

24. $\displaystyle\sum_{n=1}^{8} n^5$

25. $\displaystyle\sum_{n=1}^{6} (n^2 - n)$

26. $\displaystyle\sum_{n=1}^{20} (n^3 - n)$

27. $\displaystyle\sum_{i=1}^{6} (6i - 8i^3)$

28. $\displaystyle\sum_{j=1}^{10} \left(3 - \tfrac{1}{2}j + \tfrac{1}{2}j^2\right)$

In Exercises 29–34, find a formula for the sum of the first n terms of the sequence.

29. $1, 5, 9, 13, \ldots$

30. $25, 22, 19, 16, \ldots$

31. $1, \dfrac{9}{10}, \dfrac{81}{100}, \dfrac{729}{1000}, \ldots$

32. $3, -\dfrac{9}{2}, \dfrac{27}{4}, -\dfrac{81}{8}, \ldots$

33. $\dfrac{1}{4}, \dfrac{1}{12}, \dfrac{1}{24}, \dfrac{1}{40}, \ldots, \dfrac{1}{2n(n+1)}, \ldots$

34. $\dfrac{1}{2 \cdot 3}, \dfrac{1}{3 \cdot 4}, \dfrac{1}{4 \cdot 5}, \dfrac{1}{5 \cdot 6}, \ldots, \dfrac{1}{(n+1)(n+2)}, \ldots$

In Exercises 35–40, prove the inequality for the indicated integer values of n.

35. $n! > 2^n, \quad n \ge 4$

36. $\left(\tfrac{4}{3}\right)^n > n, \quad n \ge 7$

37. $\dfrac{1}{\sqrt{1}} + \dfrac{1}{\sqrt{2}} + \dfrac{1}{\sqrt{3}} + \cdots + \dfrac{1}{\sqrt{n}} > \sqrt{n}, \quad n \ge 2$

38. $\left(\dfrac{x}{y}\right)^{n+1} < \left(\dfrac{x}{y}\right)^n, \quad n \ge 1$ and $0 < x < y$.

39. $(1 + a)^n \ge na, \quad n \ge 1$ and $a > 0$

40. $2n^2 > (n+1)^2, \quad n \ge 3$

In Exercises 41–49, use mathematical induction to prove the given property for all positive integers n.

41. $(ab)^n = a^n b^n$

42. $\left(\dfrac{a}{b}\right)^n = \dfrac{a^n}{b^n}$

43. If $x_1 \ne 0, x_2 \ne 0, \ldots, x_n \ne 0$, then
$$(x_1 x_2 x_3 \cdots x_n)^{-1} = x_1^{-1} x_2^{-1} x_3^{-1} \cdots x_n^{-1}.$$

44. If $x_1 > 0, x_2 > 0, \ldots, x_n > 0$, then
$$\ln(x_1 x_2 x_3 \cdots x_n) = \ln x_1 + \ln x_2 + \ln x_3 + \cdots + \ln x_n.$$

45. Generalized Distributive Law:

$$x(y_1 + y_2 + \cdots + y_n) = xy_1 + xy_2 + \cdots + xy_n$$

46. $(a + bi)^n$ and $(a - bi)^n$ are complex conjugates for all $n \geq 1$.

47. A factor of $(n^3 + 3n^2 + 2n)$ is 3.

48. A factor of $(2^{2n-1} + 3^{2n-1})$ is 5.

49. A factor of $(9^n - 8n - 1)$ is 64 for all $n \geq 2$.

In Exercises 50–53, write the first five terms of the sequence.

50. $a_0 = 1$

$a_n = a_{n-1} + 2$

51. $a_0 = 10$

$a_n = 4a_{n-1}$

52. $a_0 = 4$

$a_1 = 2$

$a_n = a_{n-1} - a_{n-2}$

53. $a_0 = 0$

$a_1 = 2$

$a_n = a_{n-1} + 2a_{n-2}$

In Exercises 54–63, write the first five terms of the sequence. Then calculate the first and second differences of the sequence. Does the sequence have a linear model, a quadratic model, or neither?

54. $a_1 = 0$

$a_n = a_{n-1} + 3$

55. $a_1 = 2$

$a_n = n - a_{n-1}$

56. $a_1 = 3$

$a_n = a_{n-1} - n$

57. $a_2 = -3$

$a_n = -2a_{n-1}$

58. $a_0 = 0$

$a_n = a_{n-1} + n$

59. $a_0 = 2$

$a_n = (a_{n-1})^2$

60. $a_1 = 2$

$a_n = a_{n-1} + 2$

61. $a_1 = 0$

$a_n = a_{n-1} + 2n$

62. $a_0 = 1$

$a_n = a_{n-1} + n^2$

63. $a_0 = 0$

$a_n = a_{n-1} - 1$

In Exercises 64–67, find a quadratic model for the sequence with the indicated terms.

64. $a_0 = 3, \ a_1 = 3, \ a_4 = 15$

65. $a_0 = 7, \ a_1 = 6, \ a_3 = 10$

66. $a_0 = -3, \ a_2 = 1, \ a_4 = 9$

67. $a_0 = 3, \ a_2 = 0, \ a_6 = 36$

Synthesis

True or False? In Exercises 68–71, determine whether the statement is true or false. Justify your answer.

68. If the statement P_1 is true but the true statement P_6 does *not* imply that the statement P_7 is true, then P_n is not necessarily true for all positive integers n.

69. If the statement P_k is true and P_k implies P_{k+1}, then P_1 is also true.

70. If the second differences of a sequence are all zero, then the sequence is arithmetic.

71. A sequence with n terms has $n - 1$ second differences.

72. *Writing* In your own words, explain what is meant by a proof by mathematical induction.

73. *Think About It* What conclusion can be drawn from the information about the sequence of statements P_n?

(a) P_3 is true and P_k implies P_{k+1}.

(b) $P_1, P_2, P_3, \ldots, P_{50}$ are all true.

(c) $P_1, P_2,$ and P_3 are all true, but the truth of P_k does not imply that P_{k+1} is true.

(d) P_2 is true and P_{2k} implies P_{2k+2}.

Review

In Exercises 74–77, solve the system of equations.

74. $\begin{cases} y = x^2 \\ -3x + 2y = 2 \end{cases}$

75. $\begin{cases} x - y^3 = 0 \\ x - 2y^2 = 0 \end{cases}$

76. $\begin{cases} x - y = -1 \\ x + 2y - 2z = 3 \\ 3x - y + 2z = 3 \end{cases}$

77. $\begin{cases} 2x + y - 2z = 1 \\ x - z = 1 \\ 3x + 3y + z = 12 \end{cases}$

In Exercises 78–81, find the product.

78. $(2x^2 - 1)^2$

79. $(2x - y)^2$

80. $(5 - 4x)^3$

81. $(2x - 4y)^3$

8.5 The Binomial Theorem

▶ **What you should learn**

- How to use the Binomial Theorem to calculate binomial coefficients
- How to use Pascal's Triangle to calculate binomial coefficients
- How to use binomial coefficients to write binomial expansions

▶ **Why you should learn it**

You can use binomial coefficients to model and solve real-life problems. For instance, in Exercise 69 on page 624, you are asked to use binomial coefficients to find the probability that a baseball player will get three hits during the next 10 times at bat.

Binomial Coefficients

Recall that a binomial is a polynomial that has two terms. In this section, you will study a formula that gives a quick method of raising a binomial to a power. To begin, let's look at the expansion of $(x + y)^n$ for several values of n.

$$(x + y)^0 = 1$$
$$(x + y)^1 = x + y$$
$$(x + y)^2 = x^2 + 2xy + y^2$$
$$(x + y)^3 = x^3 + 3x^2y + 3xy^2 + y^3$$
$$(x + y)^4 = x^4 + 4x^3y + 6x^2y^2 + 4xy^3 + y^4$$
$$(x + y)^5 = x^5 + 5x^4y + 10x^3y^2 + 10x^2y^3 + 5xy^4 + y^5$$

There are several observations you can make about these expansions.

1. In each expansion, there are $n + 1$ terms.

2. In each expansion, x and y have symmetrical roles. The powers of x decrease by 1 in successive terms, whereas the powers of y increase by 1.

3. The sum of the powers of each term is n. For instance, in the expansion of $(x + y)^5$, the sum of the powers of each term is 5.

$$4 + 1 = 5 \qquad 3 + 2 = 5$$
$$(x + y)^5 = x^5 + 5x^4y^1 + 10x^3y^2 + 10x^2y^3 + 5x^1y^4 + y^5$$

4. The coefficients increase and then decrease in a symmetric pattern.

The coefficients of a binomial expansion are called **binomial coefficients.** To find them, you can use the **Binomial Theorem.** A proof of this theorem is given in Appendix A.

The Binomial Theorem

In the expansion of $(x + y)^n$

$$(x + y)^n = x^n + nx^{n-1}y + \cdots + {}_nC_r\, x^{n-r} y^r + \cdots + nxy^{n-1} + y^n$$

the coefficient of $x^{n-r} y^r$ is

$$_nC_r = \frac{n!}{(n - r)!\,r!}.$$

The symbol $\binom{n}{r}$ is often used in place of $_nC_r$ to denote binomial coefficients.

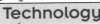
Example 1 ▶ Finding Binomial Coefficients

Find the binomial coefficients.

a. $_8C_2$ **b.** $\binom{10}{3}$ **c.** $_7C_0$ **d.** $\binom{8}{8}$

Solution

a. $_8C_2 = \dfrac{8!}{6! \cdot 2!} = \dfrac{(8 \cdot 7) \cdot 6!}{6! \cdot 2!} = \dfrac{8 \cdot 7}{2 \cdot 1} = 28$

b. $\binom{10}{3} = \dfrac{10!}{7! \cdot 3!} = \dfrac{(10 \cdot 9 \cdot 8) \cdot 7!}{7! \cdot 3!} = \dfrac{10 \cdot 9 \cdot 8}{3 \cdot 2 \cdot 1} = 120$

c. $_7C_0 = \dfrac{7!}{7! \cdot 0!} = 1$ **d.** $\binom{8}{8} = \dfrac{8!}{0! \cdot 8!} = 1$

When $r \neq 0$ and $r \neq n$, as in parts (a) and (b) above, there is a simple pattern for evaluating binomial coefficients.

$$\overset{\text{2 factors}}{\overbrace{}} \qquad\qquad \overset{\text{3 factors}}{\overbrace{}}$$

$$_8C_2 = \underset{\underset{\text{2 factors}}{\underbrace{}}}{\dfrac{8 \cdot 7}{2 \cdot 1}} \qquad \text{and} \qquad \binom{10}{3} = \underset{\underset{\text{3 factors}}{\underbrace{}}}{\dfrac{10 \cdot 9 \cdot 8}{3 \cdot 2 \cdot 1}}$$

Example 2 ▶ Finding Binomial Coefficients

Find the binomial coefficients.

a. $_7C_3$ **b.** $\binom{7}{4}$ **c.** $_{12}C_1$ **d.** $\binom{12}{11}$

Solution

a. $_7C_3 = \dfrac{7 \cdot 6 \cdot 5}{3 \cdot 2 \cdot 1} = 35$

b. $\binom{7}{4} = \dfrac{7 \cdot 6 \cdot 5 \cdot 4}{4 \cdot 3 \cdot 2 \cdot 1} = 35$

c. $_{12}C_1 = \dfrac{12}{1} = 12$

d. $\binom{12}{11} = \dfrac{12!}{1! \cdot 11!} = \dfrac{(12) \cdot 11!}{1! \cdot 11!} = \dfrac{12}{1} = 12$

It is not a coincidence that the results in parts (a) and (b) of Example 2 are the same and that the results in parts (c) and (d) are the same. In general, it is true that

$$_nC_r = {_nC_{n-r}}.$$

This shows the symmetric property of binomial coefficients that was identified earlier.

Pascal's Triangle

There is a convenient way to remember the pattern for binomial coefficients. By arranging the coefficients in a triangular pattern, you obtain **Pascal's Triangle.** This triangle is named after the famous French mathematician Blaise Pascal (1623–1662).

$$
\begin{array}{ccccccccccccccc}
&&&&&&& 1 &&&&&&& \\
&&&&&& 1 && 1 &&&&&& \\
&&&&& 1 && 2 && 1 &&&&& \\
&&&& 1 && 3 && 3 && 1 &&&& \\
&&& 1 && 4 && 6 && 4 && 1 &&& \quad 4+6=10 \\
&& 1 && 5 && 10 && 10 && 5 && 1 && \\
& 1 && 6 && 15 && 20 && 15 && 6 && 1 & \\
1 && 7 && 21 && 35 && 35 && 21 && 7 && 1 \quad 15+6=21
\end{array}
$$

The first and last numbers in each row of Pascal's Triangle are 1. Every other number in each row is formed by adding the two numbers immediately above the number. Pascal noticed that numbers in this triangle are precisely the same numbers that are the coefficients of binomial expansions, as follows.

$$(x + y)^0 = 1$$

$$(x + y)^1 = 1x + 1y$$

$$(x + y)^2 = 1x^2 + 2xy + 1y^2$$

$$(x + y)^3 = 1x^3 + 3x^2y + 3xy^2 + 1y^3$$

$$(x + y)^4 = 1x^4 + 4x^3y + 6x^2y^2 + 4xy^3 + 1y^4$$

$$(x + y)^5 = 1x^5 + 5x^4y + 10x^3y^2 + 10x^2y^3 + 5xy^4 + 1y^5$$

$$(x + y)^6 = 1x^6 + 6x^5y + 15x^4y^2 + 20x^3y^3 + 15x^2y^4 + 6xy^5 + 1y^6$$

$$(x + y)^7 = 1x^7 + 7x^6y + 21x^5y^2 + 35x^4y^3 + 35x^3y^4 + 21x^2y^5 + 7xy^6 + 1y^7$$

The top row in Pascal's Triangle is called the *zero row* because it corresponds to the binomial expansion

$$(x + y)^0 = 1.$$

Similarly, the next row is called the *first row* because it corresponds to the binomial expansion $(x + y)^1 = 1(x) + 1(y)$. In general, the *nth row* in Pascal's Triangle gives the coefficients of $(x + y)^n$.

Example 3 ▶ Using Pascal's Triangle

Use the seventh row of Pascal's Triangle to find the binomial coefficients.

$$_8C_0, \; _8C_1, \; _8C_2, \; _8C_3, \; _8C_4, \; _8C_5, \; _8C_6, \; _8C_7, \; _8C_8$$

Solution

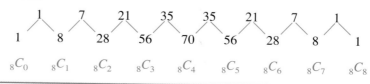

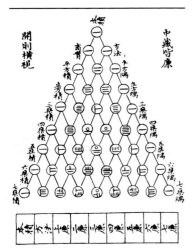

Historical Note
Precious Mirror "Pascal's" Triangle and forms of the Binomial Theorem were known in Eastern cultures prior to the Western "discovery" of the theorem. A Chinese text entitled *Precious Mirror* contains a triangle of binomial expansions through the eighth power.

 A computer animation of this concept appears in the *Interactive* CD-ROM and *Internet* versions of this text.

Binomial Expansions

As mentioned at the beginning of this section, when you write out the coefficients for a binomial that is raised to a power, you are **expanding a binomial.** The formulas for binomial coefficients give you an easy way to expand binomials, as demonstrated in the next four examples.

Example 4 ▶ Expanding a Binomial

Write the expansion for the expression

$$(x + 1)^3.$$

Solution

The binomial coefficients from the third row of Pascal's Triangle are

1, 3, 3, 1.

Therefore, the expansion is as follows.

$$(x + 1)^3 = (1)x^3 + (3)x^2(1) + (3)x(1^2) + (1)(1^3)$$
$$= x^3 + 3x^2 + 3x + 1$$

To expand binomials representing *differences* rather than sums, you alternate signs. Here are two examples.

$$(x - 1)^3 = x^3 - 3x^2 + 3x - 1$$
$$(x - 1)^4 = x^4 - 4x^3 + 6x^2 - 4x + 1$$

Example 5 ▶ Expanding a Binomial

Write the expansion for each expression.

a. $(2x + 3)^4$ **b.** $(2x - 3)^4$

Solution

a. The binomial coefficients from the fourth row of Pascal's Triangle are

1, 4, 6, 4, 1.

Therefore, the expansion is as follows.

$$(2x + 3)^4 = (1)(2x)^4 + (4)(2x)^3(3) + (6)(2x)^2(3^2) + (4)(2x)(3^3) + (1)(3^4)$$
$$= 16x^4 + 96x^3 + 216x^2 + 216x + 81$$

b. The binomial coefficients from the fourth row of Pascal's Triangle are

1, 4, 6, 4, 1.

Therefore, the expansion is as follows.

$$(2x - 3)^4 = (1)(2x)^4 - (4)(2x)^3(3) + (6)(2x)^2(3^2) - (4)(2x)(3^3) + (1)(3^4)$$
$$= 16x^4 - 96x^3 + 216x^2 - 216x + 81$$

Example 6 ▶ Expanding a Binomial

Write the expansion for $(x - 2y)^4$.

Solution

Use the fourth row of Pascal's Triangle, as follows.

$$(x - 2y)^4 = (1)x^4 - (4)x^3(2y) + (6)x^2(2y)^2 - (4)x(2y)^3 + (1)(2y)^4$$
$$= x^4 - 8x^3y + 24x^2y^2 - 32xy^3 + 16y^4$$

Example 7 ▶ Expanding a Binomial

Write the expansion for $(x^2 + 4)^3$.

Solution

Use the third row of Pascal's Triangle, as follows.

$$(x^2 + 4)^3 = (1)(x^2)^3 + (3)(x^2)^2(4) + (3)x^2(4^2) + (1)(4^3)$$
$$= x^6 + 12x^4 + 48x^2 + 64$$

Example 8 ▶ Finding a Term in a Binomial Expansion

Find the sixth term of $(a + 2b)^8$.

Solution

For the first term of the binomial expansion, you would use $n = 8$ and $r = 0$ to get ${}_8C_0 a^{8-0}(2b)^0$. For the second term of the binomial expansion, you would use $n = 8$ and $r = 1$ to get ${}_8C_1 a^{8-1}(2b)^1$. So, for the sixth term of this binomial expansion, use $n = 8$ and $r = 5$ to get

$${}_8C_5 a^{8-5}(2b)^5 = 56 \cdot a^3 \cdot (2b)^5$$
$$= 56(2^5)a^3b^5$$
$$= 1792a^3b^5.$$

Writing ABOUT MATHEMATICS

Error Analysis   You are a math instructor and receive the following solutions from one of your students on a quiz. Find the error(s) in each solution. Discuss ways that your student could avoid the error(s) in the future.

a. Find the second term in the expansion of $(2x - 3y)^5$.

$$5(2x)^4(3y)^2 = 720x^4y^2 \quad \times$$

b. Find the fourth term in the expansion of $\left(\frac{1}{2}x + 7y\right)^6$.

$${}_6C_4\left(\frac{1}{2}x\right)^2(7y)^4 = 9003.75x^2y^4 \quad \times$$

8.5 Exercises

In Exercises 1–10, find the binomial coefficients.

1. $_5C_3$

2. $_8C_6$

3. $_{12}C_0$

4. $_{20}C_{20}$

5. $_{20}C_{15}$

6. $_{12}C_5$

7. $\begin{pmatrix} 10 \\ 4 \end{pmatrix}$

8. $\begin{pmatrix} 10 \\ 6 \end{pmatrix}$

9. $\begin{pmatrix} 100 \\ 98 \end{pmatrix}$

10. $\begin{pmatrix} 100 \\ 2 \end{pmatrix}$

In Exercises 11–14, evaluate using Pascal's Triangle.

11. $\begin{pmatrix} 8 \\ 5 \end{pmatrix}$

12. $\begin{pmatrix} 8 \\ 7 \end{pmatrix}$

13. $_7C_4$

14. $_6C_3$

In Exercises 15–34, use the Binomial Theorem to expand and simplify the expression.

15. $(x + 1)^4$

16. $(x + 1)^6$

17. $(a + 6)^4$

18. $(a + 5)^5$

19. $(y - 4)^3$

20. $(y - 2)^5$

21. $(x + y)^5$

22. $(c + d)^3$

23. $(r + 3s)^6$

24. $(x + 2y)^4$

25. $(3a - b)^5$

26. $(2x - y)^5$

27. $(1 - 2x)^3$

28. $(5 - 3y)^3$

29. $(x^2 + 5)^4$

30. $(x^2 + y^2)^6$

31. $\left(\dfrac{1}{x} + y\right)^5$

32. $\left(\dfrac{1}{x} + 2y\right)^6$

33. $2(x - 3)^4 + 5(x - 3)^2$

34. $3(x + 1)^5 - 4(x + 1)^3$

In Exercises 35–38, expand the binomial using Pascal's Triangle to determine the coefficients.

35. $(2t - s)^5$

36. $(3 - 2z)^4$

37. $(x + 2y)^5$

38. $(2v + 3)^6$

In Exercises 39–46, find the coefficient a of the term in the expansion of the binomial.

Binomial	Term
39. $(x + 3)^{12}$	ax^5
40. $(x^2 + 3)^{12}$	ax^8
41. $(x - 2y)^{10}$	ax^8y^2
42. $(4x - y)^{10}$	ax^2y^8
43. $(3x - 2y)^9$	ax^4y^5
44. $(2x - 3y)^8$	ax^6y^2
45. $(x^2 + y)^{10}$	ax^8y^6
46. $(z^2 - t)^{10}$	az^4t^8

In Exercises 47–50, use the Binomial Theorem to expand and simplify the expression.

47. $\left(\sqrt{x} + 3\right)^4$

48. $\left(2\sqrt{t} - 1\right)^3$

49. $(x^{2/3} - y^{1/3})^3$

50. $(u^{3/5} + 2)^5$

In Exercises 51–54, expand the binomial in the difference quotient and simplify.

$$\dfrac{f(x + h) - f(x)}{h} \qquad \text{Difference quotient}$$

51. $f(x) = x^3$

52. $f(x) = x^4$

53. $f(x) = \sqrt{x}$

54. $f(x) = \dfrac{1}{x}$

In Exercises 55–60, use the Binomial Theorem to expand the complex number. Simplify your result.

55. $(1 + i)^4$

56. $(2 - i)^5$

57. $(2 - 3i)^6$

58. $\left(5 + \sqrt{-9}\right)^3$

59. $\left(-\dfrac{1}{2} + \dfrac{\sqrt{3}}{2}i\right)^3$

60. $\left(5 - \sqrt{3}i\right)^4$

Approximation **In Exercises 61–64, use the Binomial Theorem to approximate the given quantity accurate to three decimal places. For example, in Exercise 61, use the expansion**

$$(1.02)^8 = (1 + 0.02)^8 = 1 + 8(0.02) + 28(0.02)^2 + \cdots .$$

61. $(1.02)^8$

62. $(2.005)^{10}$

63. $(2.99)^{12}$

64. $(1.98)^9$

Graphical Reasoning **In Exercises 65 and 66, use a graphing utility to graph f and g in the same viewing window. What is the relationship between the two graphs? Use the Binomial Theorem to write the polynomial function g in standard form.**

65. $f(x) = x^3 - 4x, \quad g(x) = f(x + 4)$

66. $f(x) = -x^4 + 4x^2 - 1, \quad g(x) = f(x - 3)$

▦ 67. *Graphical Reasoning* Use a graphing utility to graph the functions in the given order and in the same viewing window. Compare the graphs. Which two functions have identical graphs, and why?

(a) $f(x) = (1 - x)^3$

(b) $g(x) = 1 - 3x$

(c) $h(x) = 1 - 3x + 3x^2$

(d) $p(x) = 1 - 3x + 3x^2 - x^3$

Probability **In Exercises 68–71, consider *n* independent trials of an experiment in which each trial has two possible outcomes: "success" or "failure." The probability of a success on each trial is *p*, and the probability of a failure is $q = 1 - p$. In this context, the term $_nC_k\, p^k q^{n-k}$ in the expansion of $(p + q)^n$ gives the probability of *k* successes in the *n* trials of the experiment.**

68. A fair coin is tossed seven times. To find the probability of obtaining four heads, evaluate the term

$$_7C_4 \left(\frac{1}{2}\right)^4 \left(\frac{1}{2}\right)^3$$

in the expansion of $\left(\frac{1}{2} + \frac{1}{2}\right)^7$.

69. The probability of a baseball player getting a hit during any given time at bat is $\frac{1}{4}$. To find the probability that the player gets three hits during the next 10 times at bat, evaluate the term

$$_{10}C_3 \left(\frac{1}{4}\right)^3 \left(\frac{3}{4}\right)^7$$

in the expansion of $\left(\frac{1}{4} + \frac{3}{4}\right)^{10}$.

70. The probability of a sales representative making a sale with any one customer is $\frac{1}{3}$. The sales representative makes eight contacts a day. To find the probability of making four sales, evaluate the term

$$_8C_4 \left(\frac{1}{3}\right)^4 \left(\frac{2}{3}\right)^4$$

in the expansion of $\left(\frac{1}{3} + \frac{2}{3}\right)^8$.

71. To find the probability that the sales representative in Exercise 70 makes four sales if the probability of a sale with any one customer is $\frac{1}{2}$, evaluate the term

$$_8C_4 \left(\frac{1}{2}\right)^4 \left(\frac{1}{2}\right)^4$$

in the expansion of $\left(\frac{1}{2} + \frac{1}{2}\right)^8$.

72. *Life Insurance* The average amount of life insurance per household $f(t)$ (in thousands of dollars) from 1980 through 1996 can be approximated by

$$f(t) = 0.0348t^2 + 5.1083t + 41.0250, \quad 0 \le t \le 16$$

where $t = 0$ represents 1980 (see figure). You want to adjust this model so that $t = 0$ corresponds to 1990 rather than 1980. To do this, you shift the graph of f 10 units *to the left* and obtain

$$g(t) = f(t + 10).$$

(Source: American Council of Life Insurance)

(a) Write $g(t)$ in standard form.

▦ (b) Use a graphing utility to graph f and g in the same viewing window.

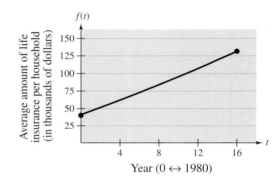

73. *Health Maintenance Organizations* The number of people $f(t)$ (in millions) enrolled in health maintenance organizations in the United States from 1976 through 1997 can be approximated by the model

$$f(t) = 0.0834t^2 + 0.07657t + 5.3680, \quad 0 \le t \le 21$$

where $t = 0$ represents 1976 (see figure). You want to adjust this model so that $t = 0$ corresponds to 1980 rather than 1976. To do this, you shift the graph of f 4 units *to the left* and obtain

$$g(t) = f(t + 4).$$

(Source: Group Health Association of America, Interstudy)

(a) Write $g(t)$ in standard form.

▦ (b) Use a graphing utility to graph f and g in the same viewing window.

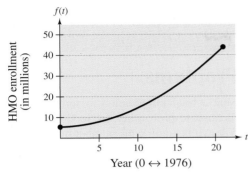

FIGURE FOR 73

Synthesis

True or False? **In Exercises 74 and 75, determine whether the statement is true or false. Justify your answer.**

74. The Binomial Theorem could be used to produce each row of Pascal's Triangle.

75. A binomial that represents a difference cannot always be accurately expanded using the Binomial Theorem.

76. *Writing* In your own words, explain how to form the rows of Pascal's Triangle.

77. Form the first nine rows of Pascal's Triangle.

78. *Think About It* How many terms are in the expansion of $(x + y)^n$?

79. *Think About It* How do the expansions of $(x + y)^n$ and $(x - y)^n$ differ?

In Exercises 80–83, prove the given property for all integers r and n where $0 \le r \le n$.

80. $_nC_r = {_nC_{n-r}}$

81. $_nC_0 - {_nC_1} + {_nC_2} - \cdots \pm {_nC_n} = 0$

82. $_{n+1}C_r = {_nC_r} + {_nC_{r-1}}$

83. The sum of the numbers in the nth row of Pascal's Triangle is 2^n.

Review

In Exercises 84–87, describe the relationship between the graphs of f and g.

84. $g(x) = f(x) + 8$ **85.** $g(x) = f(x - 3)$

86. $g(x) = f(-x)$ **87.** $g(x) = -f(x)$

In Exercises 88–91, the graph of $y = g(x)$ is shown. Graph f and use the graph to write an equation for the graph of g.

88. $f(x) = x^2$ **89.** $f(x) = x^2$

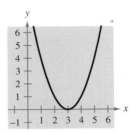   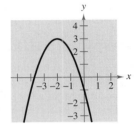

90. $f(x) = \sqrt{x}$ **91.** $f(x) = \sqrt{x}$

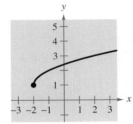   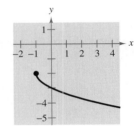

8.6 Counting Principles

▶ **What you should learn**

- How to solve simple counting problems
- How to use the Fundamental Counting Principle to solve counting problems
- How to use permutations to solve counting problems
- How to use combinations to solve counting problems

▶ **Why you should learn it**

You can use counting principles to solve counting problems that occur in real life. For instance, in Exercises 53 and 54 on page 634, you are asked to use counting principles to determine the number of possible ways of selecting the numbers on a lottery ticket.

Dion Ogust/The Image Works

Simple Counting Problems

This section and Section 8.7 present a brief introduction to some of the basic counting principles and their application to probability. In Section 8.7 you will see that much of probability has to do with counting the number of ways an event can occur.

Example 1 ▶ Selecting Pairs of Numbers at Random

Eight pieces of paper are numbered from 1 to 8 and placed in a box. One piece of paper is drawn from the box, its number is written down, and the piece of paper is *replaced in the box*. Then, a second piece of paper is drawn from the box, and its number is written down. Finally, the two numbers are added together. How many different ways can a total of 12 be obtained?

Solution

To solve this problem, count the different ways that a total of 12 can be obtained using two numbers from 1 to 8.

First number	4	5	6	7	8
Second number	8	7	6	5	4

From this list, you can see that a total of 12 can occur in five different ways.

Example 2 ▶ Selecting Pairs of Numbers at Random

Eight pieces of paper are numbered from 1 to 8 and placed in a box. Two pieces of paper are drawn from the box *at the same time*, and the numbers on the pieces of paper are written down and totaled. How many different ways can a total of 12 be obtained?

Solution

To solve this problem, count the different ways that a total of 12 can be obtained *using two different numbers* from 1 to 8.

First number	4	5	7	8
Second number	8	7	5	4

So, a total of 12 can be obtained in four different ways.

The difference between the counting problems in Examples 1 and 2 can be expressed by saying that the random selection in Example 1 occurs **with replacement,** whereas the random selection in Example 2 occurs **without replacement,** which eliminates the possibility of choosing two 6's.

The Fundamental Counting Principle

Examples 1 and 2 describe simple counting problems in which you can *list* each possible way that an event can occur. When it is possible, this is always the best way to solve a counting problem. However, some events can occur in so many different ways that it is not feasible to write out the entire list. In such cases, you must rely on formulas and counting principles. The most important of these is the **Fundamental Counting Principle.**

Fundamental Counting Principle

Let E_1 and E_2 be two events. The first event E_1 can occur in m_1 different ways. After E_1 has occurred, E_2 can occur in m_2 different ways. The number of ways that the two events can occur is $m_1 \cdot m_2$.

The Fundamental Counting Principle can be extended to three or more events. For instance, the number of ways that three events E_1, E_2, and E_3 can occur is $m_1 \cdot m_2 \cdot m_3$.

Example 3 ▶ Using the Fundamental Counting Principle

How many different pairs of letters from the English alphabet are possible?

Solution

There are two events in this situation. The first event is the choice of the first letter, and the second event is the choice of the second letter. Because the English alphabet contains 26 letters, it follows that the number of two-letter pairs is $26 \cdot 26 = 676$.

Example 4 ▶ Using the Fundamental Counting Principle

Telephone numbers in the United States currently have 10 digits. The first three are the *area code* and the next seven are the *local telephone number.* How many different telephone numbers are possible within each area code? (Note that at this time, a local telephone number cannot begin with 0 or 1.)

Solution

Because the first digit cannot be 0 or 1, there are only eight choices for the first digit. For each of the other six digits, there are 10 choices.

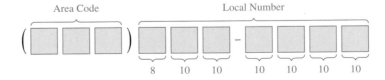

So, the number of local telephone numbers that are possible within each area code is $8 \cdot 10 \cdot 10 \cdot 10 \cdot 10 \cdot 10 \cdot 10 = 8{,}000{,}000$.

Permutations

One important application of the Fundamental Counting Principle is in determining the number of ways that n elements can be arranged (in order). An ordering of n elements is called a **permutation** of the elements.

Definition of Permutation

A **permutation** of n different elements is an ordering of the elements such that one element is first, one is second, one is third, and so on.

Example 5 ▶ Finding the Number of Permutations of n Elements

How many permutations are possible for the letters A, B, C, D, E, and F?

Solution

Consider the following reasoning.

> *First position:* Any of the *six* letters
> *Second position:* Any of the remaining *five* letters
> *Third position:* Any of the remaining *four* letters
> *Fourth position:* Any of the remaining *three* letters
> *Fifth position:* Any of the remaining *two* letters
> *Sixth position:* The *one* remaining letter

So, the numbers of choices for the six positions are as follows.

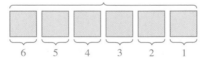

Permutations of six letters

The total number of permutations of the six letters is

$$6! = 6 \cdot 5 \cdot 4 \cdot 3 \cdot 2 \cdot 1$$
$$= 720.$$

Number of Permutations of n Elements

The number of permutations of n elements is

$$n \cdot (n - 1) \cdots 4 \cdot 3 \cdot 2 \cdot 1 = n!.$$

In other words, there are $n!$ different ways that n elements can be ordered.

Eleven thoroughbred racehorses hold the title of Triple Crown winner for winning the Kentucky Derby, the Preakness, and the Belmont Stakes in the same year. Forty horses have won two out of the three races.

Occasionally, you are interested in ordering a *subset* of a collection of elements rather than the entire collection. For example, you might want to choose (and order) *r* elements out of a collection of *n* elements. Such an ordering is called a **permutation of *n* elements taken *r* at a time.**

Example 6 ▶ Counting Horse Race Finishes

Eight horses are running in a race. In how many different ways can these horses come in first, second, and third? (Assume that there are no ties.)

Solution

Here are the different possibilities.

> Win (first position): *Eight* choices
> Place (second position): *Seven* choices
> Show (third position): *Six* choices

Using the Fundamental Counting Principle, multiply these three numbers together to obtain the following.

Different orders of horses

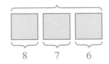

So, there are $8 \cdot 7 \cdot 6 = 336$ different orders.

Permutations of *n* Elements Taken *r* at a Time

The number of permutations of *n* elements taken *r* at a time is

$$_nP_r = \frac{n!}{(n-r)!}$$

$$= n(n-1)(n-2) \cdots (n-r+1).$$

Using this formula, you can rework Example 6 to find that the number of permutations of eight horses taken three at a time is

$$_8P_3 = \frac{8!}{5!}$$

$$= \frac{8 \cdot 7 \cdot 6 \cdot 5!}{5!}$$

$$= 336$$

which is the same answer obtained in the example.

Remember that for permutations, order is important. So, if you are looking at the possible permutations of the letters A, B, C, and D taken three at a time, the permutations (A, B, D) and (B, A, D) are counted as different because the *order* of the elements is different.

Suppose, however, that you are asked to find the possible permutations of the letters A, A, B, and C. The total number of permutations of the four letters would be $_4P_4 = 4!$. However, not all of these arrangements would be *distinguishable* because there are two A's in the list. To find the number of distinguishable permutations, you can use the following formula.

Distinguishable Permutations

Suppose a set of n objects has n_1 of one kind of object, n_2 of a second kind, n_3 of a third kind, and so on, with

$$n = n_1 + n_2 + n_3 + \cdots + n_k.$$

Then the number of **distinguishable permutations** of the n objects is

$$\frac{n!}{n_1! \cdot n_2! \cdot n_3! \cdot \cdots \cdot n_k!}.$$

Example 7 ▶ Distinguishable Permutations

In how many distinguishable ways can the letters in BANANA be written?

Solution

This word has six letters, of which three are A's, two are N's, and one is a B. So, the number of distinguishable ways the letters can be written is

$$\frac{n!}{n_1! \cdot n_2! \cdot n_3!} = \frac{6!}{3! \cdot 2! \cdot 1!}$$

$$= \frac{6 \cdot 5 \cdot 4 \cdot 3!}{3! \cdot 2!}$$

$$= 60.$$

The 60 different distinguishable permutations are as follows.

AAABNN	AAANBN	AAANNB	AABANN	AABNAN	AABNNA
AANABN	AANANB	AANBAN	AANBNA	AANNAB	AANNBA
ABAANN	ABANAN	ABANNA	ABNAAN	ABNANA	ABNNAA
ANAABN	ANAANB	ANABAN	ANABNA	ANANAB	ANANBA
ANBAAN	ANBANA	ANBNAA	ANNAAB	ANNABA	ANNBAA
BAAANN	BAANAN	BAANNA	BANAAN	BANANA	BANNAA
BNAAAN	BNAANA	BNANAA	BNNAAA	NAAABN	NAAANB
NAABAN	NAABNA	NAANAB	NAANBA	NABAAN	NABANA
NABNAA	NANAAB	NANABA	NANBAA	NBAAAN	NBAANA
NBANAA	NBNAAA	NNAAAB	NNAABA	NNABAA	NNBAAA

Combinations

When you count the number of possible permutations of a set of elements, *order* is important. As a final topic in this section, you will look at a method of selecting subsets of a larger set in which order is *not* important. Such subsets are called **combinations of *n* elements taken *r* at a time.** For instance, the combinations

$$\{A, B, C\} \quad \text{and} \quad \{B, A, C\}$$

are equivalent because both sets contain the same three elements, and the order in which the elements are listed is not important. So, you would count only one of the two sets. A common example of how a combination occurs is a card game in which the player is free to reorder the cards after they have been dealt.

Example 8 ▶ Combinations of *n* Elements Taken *r* at a Time

In how many different ways can three letters be chosen from the letters A, B, C, D, and E? (The order of the three letters is not important.)

Solution

The following subsets represent the different combinations of three letters that can be chosen from the five letters.

$$\{A, B, C\} \qquad \{A, B, D\}$$
$$\{A, B, E\} \qquad \{A, C, D\}$$
$$\{A, C, E\} \qquad \{A, D, E\}$$
$$\{B, C, D\} \qquad \{B, C, E\}$$
$$\{B, D, E\} \qquad \{C, D, E\}$$

From this list, you can conclude that there are 10 different ways that three letters can be chosen from five letters.

Combinations of *n* Elements Taken *r* at a Time

The number of combinations of *n* elements taken *r* at a time is

$$_nC_r = \frac{n!}{(n-r)!r!}.$$

Note that the formula for $_nC_r$ is the same one given for binomial coefficients. To see how this formula is used, let's solve the counting problem in Example 8. In that problem, you are asked to find the number of combinations of five elements taken three at a time. So, $n = 5$, $r = 3$, and the number of combinations is

$$_5C_3 = \frac{5!}{2!3!} = \frac{5 \cdot \overset{2}{4} \cdot 3!}{2 \cdot 1 \cdot 3!} = 10$$

which is the same answer obtained in the example.

Example 9 ▶   Counting Card Hands

A standard poker hand consists of five cards dealt from a deck of 52. How many different poker hands are possible? (After the cards are dealt, the player may reorder them, and therefore order is not important.)

Solution

You can find the number of different poker hands by using the formula for the number of combinations of 52 elements taken five at a time, as follows.

$$_{52}C_5 = \frac{52!}{47!5!}$$

$$= \frac{52 \cdot 51 \cdot 50 \cdot 49 \cdot 48 \cdot 47!}{5 \cdot 4 \cdot 3 \cdot 2 \cdot 1 \cdot 47!}$$

$$= 2{,}598{,}960$$

Example 10 ▶   The Number of Subsets of a Set

Find the total number of subsets of a set that has 10 elements.

Solution

Begin by considering the number of subsets with 0 elements, the number with 1 element, the number with 2 elements, and so on.

Number of subsets	=	Subsets with 0 elements	+	Subsets with 1 element	+	Subsets with 2 elements	+ · · · +	Subsets with 10 elements

$$= \binom{10}{0} + \binom{10}{1} + \binom{10}{2} + \cdots + \binom{10}{10}$$

By comparing this expression with the binomial expansion of $(1 + 1)^{10}$, you see that they are the same.

$$(1 + 1)^{10} = \binom{10}{0}1^{10}1^0 + \binom{10}{1}1^91^1 + \binom{10}{2}1^81^2 + \cdots + \binom{10}{10}1^01^{10}$$

$$= \binom{10}{0} + \binom{10}{1} + \binom{10}{2} + \cdots + \binom{10}{10}$$

This implies that the total number of subsets of a set of 10 elements is

$$(1 + 1)^{10} = 2^{10}$$

$$= 1024.$$

The result of Example 10 can be generalized to conclude that the total number of subsets of a set of n elements is 2^n.

8.6 Exercises

Random Selection **In Exercises 1–8, determine the number of ways a computer can randomly generate one or more such integers from 1 through 12.**

1. An odd integer

2. An even integer

3. A prime integer

4. An integer that is greater than 9

5. An integer that is divisible by 4

6. An integer that is divisible by 3

7. Two integers whose sum is 8

8. Two *distinct* integers whose sum is 8

9. *Entertainment Systems* A customer can choose one of three amplifiers, one of two compact disc players, and one of five speaker models for an entertainment system. Determine the number of possible system configurations.

10. *Computer Systems* A customer in a computer store can choose one of four monitors, one of three keyboards, and one of five computers. If all the choices are compatible, determine the number of possible system configurations.

11. *Job Applicants* A college needs two additional faculty members: a chemist and a statistician. In how many ways can these positions be filled if there are five applicants for the chemistry position and three applicants for the statistics position?

12. *Course Schedule* A college student is preparing a course schedule for the next semester. The student may select one of two mathematics courses, one of three science courses, and one of five courses from the social sciences and humanities. How many schedules are possible?

13. *True-False Exam* In how many ways can a six-question true-false exam be answered? (Assume that no questions are omitted.)

14. *True-False Exam* In how many ways can a 12-question true-false exam be answered? (Assume that no questions are omitted.)

15. *Toboggan Ride* Three people are lining up for a ride on a toboggan, but only two of the three are willing to take the first position. With that constraint, in how many ways can the three people be seated on the toboggan?

16. *Aircraft Boarding* Eight people are boarding an aircraft. Two have tickets for first class and board before those in the economy class. In how many ways can the eight people board the aircraft?

17. *License Plate Numbers* In a certain state, each automobile license plate number consists of three letters followed by a four-digit number. How many distinct license plate numbers can be formed?

18. *License Plate Numbers* In a certain state, each automobile license plate number consists of two letters followed by a four-digit number. To avoid confusion between "O" and "zero" and between "I" and "one," the letters "O" and "I" are not used. How many distinct license plate numbers can be formed?

19. *Three-Digit Numbers* How many three-digit numbers can be formed under the following conditions?
    (a) The leading digit cannot be zero.
    (b) The leading digit cannot be zero and no repetition of digits is allowed.
    (c) The leading digit cannot be zero and the number must be a multiple of 5.
    (d) The number is at least 400.

20. *Four-Digit Numbers* How many four-digit numbers can be formed under the following conditions?
    (a) The leading digit cannot be zero.
    (b) The leading digit cannot be zero and no repetition of digits is allowed.
    (c) The leading digit cannot be zero and the number must be less than 5000.
    (d) The leading digit cannot be zero and the number must be even.

21. *Combination Lock* A combination lock will open when the right choice of three numbers (from 1 to 40, inclusive) is selected. How many different lock combinations are possible?

22. *Combination Lock* A combination lock will open when the right choice of three numbers (from 1 to 50, inclusive) is selected. How many different lock combinations are possible?

23. *Concert Seats* Four couples have reserved seats in a given row for a concert. In how many different ways can they be seated if
    (a) there are no seating restrictions?
    (b) the two members of each couple wish to sit together?

24. *Single File*   In how many orders can four girls and four boys walk through a doorway single file if

(a) there are no restrictions?

(b) the girls walk through before the boys?

In Exercises 25–30, evaluate $_nP_r$.

25. $_4P_4$

26. $_5P_5$

27. $_8P_3$

28. $_{20}P_2$

29. $_5P_4$

30. $_7P_4$

In Exercises 31 and 32, solve for *n*.

31. $14 \cdot {}_nP_3 = {}_{n+2}P_4$

32. $_nP_5 = 18 \cdot {}_{n-2}P_4$

In Exercises 33–38, evaluate using a calculator.

33. $_{20}P_5$

34. $_{100}P_5$

35. $_{100}P_3$

36. $_{10}P_8$

37. $_{20}C_5$

38. $_{10}C_7$

In Exercises 39–42, find the number of distinguishable permutations of the group of letters.

39. A, A, G, E, E, E, M

40. B, B, B, T, T, T, T, T

41. A, L, G, E, B, R, A

42. M, I, S, S, I, S, S, I, P, P, I

43. Write all permutations of the letters A, B, C, and D.

44. Write all the permutations of the letters A, B, C, and D if the letters B and C must remain between the letters A and D.

45. Write all the possible selections of two letters that can be formed from the letters A, B, C, D, E, and F. (The order of the two letters is not important.)

46. Write all the possible selections of three letters that can be formed from the letters A, B, C, D, E, and F. (The order of the three letters is not important.)

47. *Posing for a Photograph*   In how many ways can five children line up in a row?

48. *Riding in a Car*   In how many ways can six people sit in a six-passenger car?

49. *Choosing Officers*   From a pool of 12 candidates, the offices of president, vice-president, secretary, and treasurer will be filled. In how many different ways can the offices be filled?

50. *Assembly Line Production*   There are four processes involved in assembling a certain product, and these processes can be performed in any order. The management wants to test each order to determine which is the least time-consuming. How many different orders will have to be tested?

51. *Forming an Experimental Group*   In order to conduct a certain experiment, five students are randomly selected from a class of 20. How many different groups of five students are possible?

52. *Test Questions*   You can answer any 10 questions from a total of 12 questions on an exam. In how many different ways can you select the questions?

53. *Lottery Choices*   There are 40 numbers in a particular state lottery. In how many ways can a player select six of the numbers?

54. *Lottery Choices*   There are 50 numbers in a particular state lottery. In how many ways can a player select six of the numbers?

55. *Number of Subsets*   How many subsets of four elements can be formed from a set of 100 elements?

56. *Number of Subsets*   How many subsets of five elements can be formed from a set of 80 elements?

57. *Geometry*   Three points that are not on a line determine three lines. How many lines are determined by seven points, no three of which are on a line?

58. *Defective Units*   A shipment of 10 microwave ovens contains three defective units. In how many ways can a vending company purchase four of these units and receive (a) all good units, (b) two good units, and (c) at least two good units?

59. *Job Applicants*   An employer interviews eight people for four openings in the company. Three of the eight people are women. If all eight are qualified, in how many ways can the employer fill the four positions if (a) the selection is random and (b) exactly two selections are women?

60. *Poker Hand*   You are dealt five cards from an ordinary deck of 52 playing cards. In how many ways can you get a full house? (A full house consists of three of one kind and two of another. For example, A-A-A-5-5 and K-K-K-10-10 are full houses.)

61. *Forming a Committee* Four people are to be selected at random from a group of four couples. In how many ways can this be done under the following conditions?

(a) There are no restrictions.

(b) The group must have at least one couple.

(c) Each couple must be represented in the group.

62. *Interpersonal Relationships* The complexity of the interpersonal relationships increases dramatically as the size of a group increases. Determine the number of different two-person relationships in a group of people of size (a) 3, (b) 8, (c) 12, and (d) 20.

In Exercises 63–66, find the number of diagonals of the polygon. (A line segment connecting any two nonadjacent vertices is called a diagonal of the polygon.)

63. Pentagon

64. Hexagon

65. Octagon

66. Decagon (10 sides)

Synthesis

True or False? In Exercises 67 and 68, determine whether the statement is true or false. Justify your answer.

67. The number of letter pairs that can be formed from any of the first 13 letters in the alphabet (A–M) is an example of a permutation.

68. The number of permutations of n elements can be determined by using the Fundamental Counting Principle.

69. What is the relationship between $_nC_r$ and $_nC_{n-r}$?

70. Without calculating the numbers, determine which of the following is greater. Explain.

(a) The combinations of 10 elements taken six at a time

(b) The permutations of 10 elements taken six at a time

In Exercises 71–74, prove the identity.

71. $_nP_{n-1} = {_nP_n}$

72. $_nC_n = {_nC_0}$

73. $_nC_{n-1} = {_nC_1}$

74. $_nC_r = \dfrac{_nP_r}{r!}$

75. *Think About It* Can your calculator evaluate $_{100}P_{80}$? If not, explain why.

76. *Writing* Explain in words the meaning of $_nP_r$.

Review

In Exercises 77–80, evaluate the function at the specified values of the independent variable.

77. $f(x) = 3x^2 + 8$

(a) $f(3)$ (b) $f(0)$ (c) $f(-5)$

78. $g(x) = \sqrt{x-3} + 2$

(a) $g(3)$ (b) $g(7)$ (c) $g(x+3)$

79. $f(x) = -|x-5| + 6$

(a) $f(-5)$ (b) $f(-1)$ (c) $f(11)$

80. $f(x) = \begin{cases} x^2 - 2x + 5, & x \le -4 \\ -x^2 - 2, & x > -4 \end{cases}$

(a) $f(-4)$ (b) $f(-1)$ (c) $f(-20)$

In Exercises 81–84, use the Binomial Theorem to expand and simplify the expression.

81. $(x+1)^5$

82. $(y-2)^6$

83. $(x^2 + 2y)^5$

84. $(x^2 - y^2)^4$

8.7 Probability

▶ **What you should learn**

- How to find the probability of an event
- How to find the probabilities of mutually exclusive events
- How to find the probabilities of independent events
- How to find the probability of the complement of an event

▶ **Why you should learn it**

You can use probability to solve a variety of problems that occur in real life. For instance, in Exercise 33 on page 644, you are asked to use probability to help analyze the age distribution of employees who find work through temporary help agencies.

Telegraph Colour Library/FPG

The Probability of an Event

Any happening for which the result is uncertain is called an **experiment.** The possible results of the experiment are **outcomes,** the set of all possible outcomes of the experiment is the **sample space** of the experiment, and any subcollection of a sample space is an **event.**

For instance, when a six-sided die is tossed, the sample space can be represented by the numbers 1 through 6. For this experiment, each of the outcomes is *equally likely.*

To describe sample spaces in such a way that each outcome is equally likely, you must sometimes distinguish between or among various outcomes in ways that appear artificial. Example 1 illustrates such a situation.

Example 1 ▶   Finding the Sample Space

Find the sample space for each of the following.

a. One coin is tossed.

b. Two coins are tossed.

c. Three coins are tossed.

Solution

a. Because the coin will land either heads up (denoted by H) or tails up (denoted by T), the sample space is

$$S = \{H, T\}.$$

b. Because either coin can land heads up or tails up, the possible outcomes are as follows.

HH = heads up on both coins

HT = heads up on first coin and tails up on second coin

TH = tails up on first coin and heads up on second coin

TT = tails up on both coins

So, the sample space is

$$S = \{HH, HT, TH, TT\}.$$

Note that this list distinguishes between the two cases HT and TH, even though these two outcomes appear to be similar.

c. Following the notation of part (b), the sample space is

$$S = \{HHH, HHT, HTH, HTT, THH, THT, TTH, TTT\}.$$

To calculate the probability of an event, count the number of outcomes in the event and in the sample space. The *number of outcomes* in event E is denoted by $n(E)$, and the number of outcomes in the sample space S is denoted by $n(S)$. The probability that event E will occur is given by $n(E)/n(S)$.

The Probability of an Event

If an event E has $n(E)$ equally likely outcomes and its sample space S has $n(S)$ equally likely outcomes, the **probability** of event E is

$$P(E) = \frac{n(E)}{n(S)}.$$

Because the number of outcomes in an event must be less than or equal to the number of outcomes in the sample space, the probability of an event must be a number between 0 and 1. That is,

$$0 \le P(E) \le 1.$$

If $P(E) = 0$, event E *cannot occur*, and E is called an impossible event. If $P(E) = 1$, event E *must occur*, and E is called a certain event.

Example 2 ▶ Finding the Probability of an Event

a. Two coins are tossed. What is the probability that both land heads up?

b. A card is drawn from a standard deck of playing cards. What is the probability that it is an ace?

Solution

a. Following the procedure in Example 1(b), let

$$E = \{HH\}$$

and

$$S = \{HH, HT, TH, TT\}.$$

The probability of getting two heads is

$$P(E) = \frac{n(E)}{n(S)}$$

$$= \frac{1}{4}.$$

b. Because there are 52 cards in a standard deck of playing cards and there are four aces (one in each suit), the probability of drawing an ace is

$$P(E) = \frac{n(E)}{n(S)}$$

$$= \frac{4}{52}$$

$$= \frac{1}{13}.$$

FIGURE **8.8**

 A computer simulation of this example appears in the *Interactive* CD-ROM and *Internet* versions of this text.

Example 3 ▶   **Finding the Probability of an Event**

Two six-sided dice are tossed. What is the probability that the total of the two dice is 7? (See Figure 8.8.)

Solution

Because there are six possible outcomes on each die, you can use the Fundamental Counting Principle to conclude that there are $6 \cdot 6$ or 36 different outcomes when two dice are tossed. To find the probability of rolling a total of 7, you must first count the number of ways in which this can occur.

First die	1	2	3	4	5	6
Second die	6	5	4	3	2	1

So, a total of 7 can be rolled in six ways, which means that the probability of rolling a 7 is

$$P(E) = \frac{n(E)}{n(S)}$$

$$= \frac{6}{36}$$

$$= \frac{1}{6}.$$

You could have written out each sample space in Examples 2 and 3 and simply counted the outcomes in the desired events. For larger sample spaces, however, you should use the counting principles discussed in Section 8.6.

Example 4 ▶   **Finding the Probability of an Event**

Twelve-sided dice, as shown in Figure 8.9, can be constructed (in the shape of regular dodecahedrons) such that each of the numbers from 1 to 6 appears twice on each die. Prove that these dice can be used in any game requiring ordinary six-sided dice without changing the probabilities of different outcomes.

Solution

For an ordinary six-sided die, each of the numbers 1, 2, 3, 4, 5, and 6 occurs only once, so the probability of any particular number coming up is

$$P(E) = \frac{n(E)}{n(S)} = \frac{1}{6}.$$

For one of the 12-sided dice, each number occurs twice, so the probability of any particular number coming up is

$$P(E) = \frac{n(E)}{n(S)} = \frac{2}{12} = \frac{1}{6}.$$

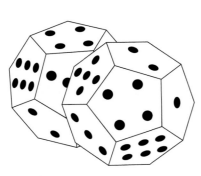

FIGURE **8.9**

Although popular in the early 1800s, lotteries were banned in more and more states until, by 1894, no state allowed lotteries. In 1964, New Hampshire became the first state to reinstitute a state lottery. Today, lotteries are conducted by almost all states.

Example 5 ▶ The Probability of Winning a Lottery

In a state lottery, a player chooses six different numbers from 1 to 40. If these six numbers match the six numbers drawn by the lottery commission, the player wins (or shares) the top prize. What is the probability of winning?

Solution

To find the number of elements in the sample space, use the formula for the number of combinations of 40 elements taken six at a time.

$$n(S) = {}_{40}C_6$$

$$= \frac{40 \cdot 39 \cdot 38 \cdot 37 \cdot 36 \cdot 35}{6 \cdot 5 \cdot 4 \cdot 3 \cdot 2 \cdot 1}$$

$$= 3{,}838{,}380$$

If a person buys only one ticket, the probability of winning is

$$P(E) = \frac{n(E)}{n(S)}$$

$$= \frac{1}{3{,}838{,}380}.$$

Example 6 ▶ Random Selection

The numbers of colleges and universities in various regions of the United States in 1996 are shown in Figure 8.10. One institution is selected at random. What is the probability that the institution is in one of the three southern regions? (Source: U.S. National Center for Education Statistics)

Solution

From the figure, the total number of colleges and universities is 3696. Because there are $607 + 265 + 298 = 1170$ colleges and universities in the three southern regions, the probability that the institution is in one of these regions is

$$P(E) = \frac{n(E)}{n(S)} = \frac{1170}{3696} \approx 0.317.$$

FIGURE 8.10

Mutually Exclusive Events

Two events A and B (from the same sample space) are **mutually exclusive** if A and B have no outcomes in common. In the terminology of sets, the intersection of A and B is the empty set, which is expressed as

$$P(A \cap B) = 0.$$

For instance, if two dice are tossed, the event A of rolling a total of 6 and the event B of rolling a total of 9 are mutually exclusive. To find the probability that one or the other of two mutually exclusive events will occur, you can *add* their individual probabilities.

Probability of the Union of Two Events

If A and B are events in the same sample space, the probability of A *or* B occurring is given by

$$P(A \cup B) = P(A) + P(B) - P(A \cap B).$$

If A and B are mutually exclusive, then

$$P(A \cup B) = P(A) + P(B).$$

Example 7 ▶ The Probability of a Union of Events

One card is selected from a standard deck of 52 playing cards. What is the probability that the card is either a heart or a face card?

Solution

Because the deck has 13 hearts, the probability of selecting a heart (event A) is

$$P(A) = \frac{13}{52}.$$

Similarly, because the deck has 12 face cards, the probability of selecting a face card (event B) is

$$P(B) = \frac{12}{52}.$$

Because three of the cards are hearts *and* face cards (see Figure 8.11), it follows that

$$P(A \cap B) = \frac{3}{52}.$$

Finally, applying the formula for the probability of the union of two events, you can conclude that the probability of selecting a heart or a face card is

$$P(A \cup B) = P(A) + P(B) - P(A \cap B)$$

$$= \frac{13}{52} + \frac{12}{52} - \frac{3}{52} = \frac{22}{52} \approx 0.423.$$

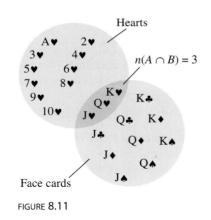

Hearts

$n(A \cap B) = 3$

Face cards

FIGURE **8.11**

Example 8 ▶   **Probability of Mutually Exclusive Events**

The personnel department of a company has compiled data on the numbers of employees who have been with the company for various periods of time. The results are shown in the table.

Years of service	Number of employees
0–4	157
5–9	89
10–14	74
15–19	63
20–24	42
25–29	38
30–34	37
35–39	21
40–44	8

If an employee is chosen at random, what is the probability that the employee has 9 or fewer years of service?

Solution

To begin, add the number of employees and find that the total is 529. Next, let event A represent choosing an employee with 0 to 4 years of service and let event B represent choosing an employee with 5 to 9 years of service. Then

$$P(A) = \frac{157}{529} \quad \text{and} \quad P(B) = \frac{89}{529} \cdot$$

Because A and B have no outcomes in common, you can conclude that these two events are mutually exclusive and that

$$P(A \cup B) = P(A) + P(B)$$

$$= \frac{157}{529} + \frac{89}{529}$$

$$= \frac{246}{529}$$

$$\approx 0.465.$$

So, the probability of choosing an employee who has 9 or fewer years of service is about 0.465.

Independent Events

Two events are **independent** if the occurrence of one has no effect on the occurrence of the other. For instance, rolling a total of 12 with two six-sided dice has no effect on the outcome of future rolls of the dice. To find the probability that two independent events will occur, *multiply* the probabilities of each.

> **Probability of Independent Events**
>
> If A and B are independent events, the probability that both A and B will occur is
>
> $$P(A \text{ and } B) = P(A) \cdot P(B).$$

Example 9 ▶ Probability of Independent Events

A random number generator on a computer selects three integers from 1 to 20. What is the probability that all three numbers are less than or equal to 5?

Solution

The probability of selecting a number from 1 to 5 is

$$P(A) = \frac{5}{20}$$

$$= \frac{1}{4}.$$

So, the probability that all three numbers are less than or equal to 5 is

$$P(A) \cdot P(A) \cdot P(A) = \left(\frac{1}{4}\right)\left(\frac{1}{4}\right)\left(\frac{1}{4}\right)$$

$$= \frac{1}{64}.$$

Example 10 ▶ Probability of Independent Events

In 1997, 58% of the population of the United States was 30 years old or older. Suppose that in a survey, 10 people were chosen at random from the population. What is the probability that all 10 were 30 years old or older? (Source: U.S. Bureau of the Census)

Solution

Let A represent choosing a person who was 30 years old or older. Because the probability of choosing a person who was 30 years old or older was 0.58, you can conclude that the probability that all 10 people were 30 years old or older is

$$[P(A)]^{10} = (0.58)^{10}$$

$$\approx 0.0043.$$

Exploration

You are in a class with 22 other people. What is the probability that at least two out of the 23 people will have a birthday on the same day of the year?

The complement of the probability that at least two people have the same birthday is the probability that all 23 birthdays are different. So, first find the probability that all 23 people have different birthdays and then find the complement.

Now, determine the probability that in a room with 50 people at least two people have the same birthday.

The Complement of an Event

The **complement of an event** A is the collection of all outcomes in the sample space that are *not* in A. The complement of event A is denoted by A'. Because $P(A \text{ or } A') = 1$ and because A and A' are mutually exclusive, it follows that $P(A) + P(A') = 1$. Therefore, the probability of A' is

$$P(A') = 1 - P(A).$$

For instance, if the probability of *winning* a certain game is

$$P(A) = \frac{1}{4}$$

the probability of *losing* the game is

$$P(A') = 1 - \frac{1}{4}$$

$$= \frac{3}{4}.$$

Probability of a Complement

Let A be an event and let A' be its complement. If the probability of A is $P(A)$, the probability of the complement is

$$P(A') = 1 - P(A).$$

Example 11 ▶ Finding the Probability of a Complement

A manufacturer has determined that a certain machine averages one faulty unit for every 1000 it produces. What is the probability that an order of 200 units will have one or more faulty units?

Solution

To solve this problem as stated, you would need to find the probabilities of having exactly one faulty unit, exactly two faulty units, exactly three faulty units, and so on. However, using complements, you can simply find the probability that all units are perfect and then subtract this value from 1. Because the probability that any given unit is perfect is 999/1000, the probability that all 200 units are perfect is

$$P(A) = \left(\frac{999}{1000}\right)^{200}$$

$$\approx 0.8186.$$

Therefore, the probability that at least one unit is faulty is

$$P(A') = 1 - P(A)$$

$$\approx 0.1814.$$

8.7 Exercises

In Exercises 1–6, determine the sample space for the experiment.

1. A coin and a six-sided die are tossed.

2. A six-sided die is tossed twice and the sum of the points is recorded.

3. A taste tester has to rank three varieties of yogurt, A, B, and C, according to preference.

4. Two marbles are selected from a sack containing two red marbles, two blue marbles, and one black marble. The color of each marble is recorded.

5. Two county supervisors are selected from five supervisors, A, B, C, D, and E, to study a recycling plan.

6. A sales representative makes presentations of a product in three homes per day. In each home there may be a sale (denote by S) or there may be no sale (denote by F).

Heads or Tails In Exercises 7–10, find the probability for the experiment of tossing a coin three times. Use the sample space $S = \{HHH, HHT, HTH, HTT, THH, THT, TTH, TTT\}$.

7. The probability of getting exactly one tail

8. The probability of getting a head on the first toss

9. The probability of getting at least one head

10. The probability of getting at least two heads

Drawing a Card In Exercises 11–14, find the probability for the experiment of selecting one card from a standard deck of 52 playing cards.

11. The card is a face card.

12. The card is not a face card.

13. The card is a red face card.

14. The card is a 6 or less.

Tossing a Die In Exercises 15–20, find the probability for the experiment of tossing a six-sided die twice.

15. The sum is 4.

16. The sum is at least 7.

17. The sum is less than 11.

18. The sum is 2, 3, or 12.

19. The sum is odd and no more than 7.

20. The sum is odd or prime.

Drawing Marbles In Exercises 21–24, find the probability for the experiment of drawing two marbles (without replacement) from a bag containing one green, two yellow, and three red marbles.

21. Both marbles are red.

22. Both marbles are yellow.

23. Neither marble is yellow.

24. The marbles are of different colors.

In Exercises 25–28, you are given the probability that an event *will* happen. Find the probability that the event *will not* happen.

25. $p = 0.7$

26. $p = 0.36$

27. $p = \dfrac{1}{3}$

28. $p = \dfrac{5}{6}$

In Exercises 29–32, you are given the probability that an event *will not* happen. Find the probability that the event *will* happen.

29. $p = 0.15$

30. $p = 0.84$

31. $p = \dfrac{13}{20}$

32. $p = \dfrac{87}{100}$

33. *Graphical Reasoning* In 1997 there were approximately 1.3 million temporary help agency workers in the United States. The figure gives the age profile of these workers. (Source: U.S. Bureau of Labor Statistics)

Age of U.S. Workers

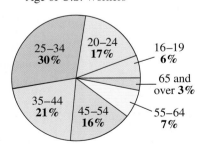

(a) Estimate the number of temporary help agency workers in the age category 16–19.

(b) What is the probability that a person selected at random from the population of temporary agency workers is in the 25–34 age group?

(c) What is the probability that a person selected at random from the population of temporary agency workers is in the 35–54 age group?

34. *Graphical Reasoning* In 1997 there were approximately 111 million workers in the civilian labor force in the United States. The figure gives the educational attainment of these workers. (Source: U.S. Bureau of Labor Statistics)

Educational Attainment
of Workers

(a) Estimate the number of workers whose highest education level was a high school education.

(b) A person is selected at random from the civilian work force. What is the probability that the person has 1 to 3 years of college education?

(c) A person is selected at random from the civilian work force. What is the probability that the person has more than a high school education?

35. *Data Analysis* A study of the effectiveness of a flu vaccine was conducted with a sample of 500 people. Some in the study were given no vaccine, some were given one injection, and others were given two injections. The results of the study are listed in the table.

	No vaccine	One injection	Two injections	Total
Flu	7	2	13	22
No flu	149	52	277	478
Total	156	54	290	500

A person is selected at random from the sample. Find the specified probability.

(a) The person had two injections.

(b) The person did not get the flu.

(c) The person got the flu and had one injection.

36. *Data Analysis* One hundred college students were interviewed to determine their political party affiliations and whether they favored a balanced-budget amendment to the Constitution. The results of the study are listed in the table.

	Favor	Not favor	Unsure	Total
Democrat	23	25	7	55
Republican	32	9	4	45
Total	55	34	11	100

A person is selected at random from the sample. Find the probability that the described person is selected.

(a) A person who doesn't favor the amendment

(b) A Republican

(c) A Democrat who favors the amendment

37. *Alumni Association* A college sends a survey to selected members of the class of 2000. Of the 1254 people who graduated that year, 672 are women, of whom 124 went on to graduate school. Of the 582 male graduates, 198 went on to graduate school. If an alumnus member is selected at random, what is the probability that the person is (a) female, (b) male, and (c) female and did not attend graduate school?

38. *Post–High School Education* In a high school graduating class of 72 students, 28 are on the honor roll. Of these, 18 are going on to college, and of the other 44 students, 12 are going on to college. If a student is selected at random from the class, what is the probability that the person chosen is (a) going to college, (b) not going to college, and (c) on the honor roll, but not going to college?

39. *Winning an Election* Taylor, Moore, and Jenkins are candidates for public office. It is estimated that Moore and Jenkins have about the same probability of winning, and Taylor is believed to be twice as likely to win as either of the others. Find the probability of each candidate winning the election.

40. *Winning an Election* Three people have been nominated for president of a class. From a poll, it is estimated that the first has a 37% chance of winning and the second has a 44% chance of winning. What is the probability that the third candidate will win?

In Exercises 41–52, the sample spaces are large and you should use the counting principles discussed in Section 8.6.

41. *Preparing for a Test*   A class is given a list of 20 study problems, from which 10 will be part of an upcoming exam. If a given student knows how to solve 15 of the problems, find the probability that the student will be able to answer (a) all 10 questions on the exam, (b) exactly eight questions on the exam, and (c) at least nine questions on the exam.

42. *Preparing for a Test*   A class is given a list of eight study problems, from which five will be part of an upcoming exam. If a given student knows how to solve six of the problems, find the probability that the student will be able to answer (a) all five questions on the exam, (b) exactly four questions on the exam, and (c) at least four questions on the exam.

43. *Letter Mix-Up*   Four letters and envelopes are addressed to four different people. If the letters are randomly inserted into the envelopes, what is the probability that (a) exactly one will be inserted in the correct envelope and (b) at least one will be inserted in the correct envelope?

44. *Payroll Mix-Up*   Five paychecks and envelopes are addressed to five different people. If the paychecks are randomly inserted into the envelopes, what is the probability that (a) exactly one will be inserted in the correct envelope and (b) at least one will be inserted in the correct envelope?

45. *Game Show*   On a game show you are given five digits to arrange in the proper order to form the price of a car. If you are correct, you win the car. What is the probability of winning, given the following conditions?

   (a) You guess the position of each digit.

   (b) You know the first digit and guess the positions of the others.

46. *Game Show*   On a game show you are given five digits in the price of a car. The first digit is 1, and you are given the other four digits to arrange in the correct order to win the car. What is your probability of winning given the following conditions?

   (a) You guess the position of each digit.

   (b) You know the second digit but guess the others.

47. *Drawing Cards from a Deck*   Two cards are selected at random from an ordinary deck of 52 playing cards. Find the probability that two aces are selected, given the following conditions.

   (a) The cards are drawn in sequence, with the first card being replaced and the deck reshuffled prior to the second drawing.

   (b) The two cards are drawn consecutively, without replacement.

48. *Poker Hand*   Five cards are drawn from an ordinary deck of 52 playing cards. What is the probability that the hand drawn is a full house? (A full house is a hand that consists of two of one kind and three of another kind.)

49. *Defective Units*   A shipment of 12 microwave ovens contains three defective units. A vending company has ordered four of these units, and because each is identically packaged, the selection will be random. What is the probability that (a) all four units are good, (b) exactly two units are good, and (c) at least two units are good?

50. *Defective Units*   A shipment of 20 compact disc players contains four defective units. A retail outlet has ordered five of these units. What is the probability that (a) all five units are good, (b) exactly four units are good, and (c) at least one unit is defective?

51. *Random Number Generator*   Two integers from 1 through 30 are chosen by a random number generator. What is the probability that (a) the numbers are both even, (b) one number is even and one is odd, (c) both numbers are less than 10, and (d) the same number is chosen twice?

52. *Random Number Generator*   Two integers from 1 through 40 are chosen by a random number generator. What is the probability that (a) the numbers are both even, (b) one number is even and one is odd, (c) both numbers are less than 30, and (d) the same number is chosen twice?

53. *Backup System*   A space vehicle has an independent backup system for one of its communication networks. The probability that either system will function satisfactorily during a flight is 0.985. What is the probability that during a given flight (a) both systems function satisfactorily, (b) at least one system functions satisfactorily, and (c) both systems fail?

54. *Backup Vehicle* A fire company keeps two rescue vehicles. Because of the demand on the vehicles and the chance of mechanical failure, the probability that a specific vehicle is available when needed is 90%. If the availability of one vehicle is *independent* of the availability of the other, find the probability that (a) both vehicles are available at a given time, (b) neither vehicle is available at a given time, and (c) at least one vehicle is available at a given time.

55. *Making a Sale* A sales representative makes a sale at approximately one-fourth of all calls. If, on a given day, the representative contacts five potential clients, what is the probability that a sale will be made with (a) each of the five contacts, (b) none of the contacts, and (c) at least one contact?

56. *A Boy or a Girl?* Assume that the probability of the birth of a child of a particular sex is 50%. In a family with four children, what is the probability that (a) all the children are boys, (b) all the children are the same sex, and (c) there is at least one boy?

57. *Flexible Work Hours* In a survey, people were asked if they would prefer to work flexible hours—even if it meant slower career advancement—so they could spend more time with their families. The results of the survey are shown in the figure. Suppose that three people from the survey were chosen at random. What is the probability that all three people would prefer flexible work hours?

Flexible Work Hours

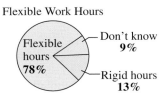

58. *Will That Be Cash or Charge?* Suppose that the methods used by shoppers to pay for merchandise are as shown in the circle graph. If two shoppers are chosen at random, what is the probability that both shoppers paid for their purchases only in cash?

How Shoppers Pay for Merchandise

59. *Geometry* You and a friend agree to meet at your favorite fast-food restaurant between 5:00 and 6:00 P.M. The one who arrives first will wait 15 minutes for the other, and then will leave (see figure). What is the probability that the two of you will actually meet, assuming that your arrival times are random within the hour?

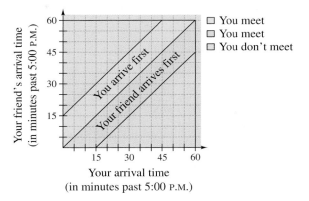

Your arrival time
(in minutes past 5:00 P.M.)

60. *Estimating* π A coin of diameter d is dropped onto a paper that contains a grid of squares d units on a side (see figure).

(a) Find the probability that the coin covers a vertex of one of the squares on the grid.

(b) Perform the experiment 100 times and use the results to approximate π.

Synthesis

True or False? **In Exercises 61 and 62, determine whether the statement is true or false. Justify your answer.**

61. If A and B are independent events with nonzero probabilities, then A can occur when B occurs.

62. Rolling a number less than 3 on a normal six-sided die has a probability of $\frac{1}{3}$. The complement of this event is to roll a number greater than 3, and its probability is $\frac{1}{2}$.

63. *Pattern Recognition and Exploration*   Consider a group of n people.

(a) Explain why the following pattern gives the probabilities that the n people have distinct birthdays.

$$n = 2: \quad \frac{365}{365} \cdot \frac{364}{365} = \frac{365 \cdot 364}{365^2}$$

$$n = 3: \quad \frac{365}{365} \cdot \frac{364}{365} \cdot \frac{363}{365} = \frac{365 \cdot 364 \cdot 363}{365^3}$$

(b) Use the pattern in part (a) to write an expression for the probability that $n = 4$ people have distinct birthdays.

(c) Let P_n be the probability that the n people have distinct birthdays. Verify that this probability can be obtained recursively by

$$P_1 = 1 \ \text{ and } \ P_n = \frac{365 - (n-1)}{365} P_{n-1}.$$

(d) Explain why $Q_n = 1 - P_n$ gives the probability that at least two people in a group of n people have the same birthday.

(e) Use the results of parts (c) and (d) to complete the table.

n	10	15	20	23	30	40	50
P_n							
Q_n							

(f) How many people must be in a group so that the probability of at least two of them having the same birthday is greater than $\frac{1}{2}$? Explain.

64. *Think About It*   A weather forecast indicates that the probability of rain is 40%. What does this mean?

Review

In Exercises 65–68, find all real solutions of the polynomial equation.

65. $6x^2 + 8 = 0$

66. $4x^2 + 6x - 12 = 0$

67. $x^3 - x^2 - 3x = 0$

68. $x^5 + x^3 - 2x = 0$

In Exercises 69–74, find all real solutions of the rational equation.

69. $\dfrac{12}{x} = -3$

70. $\dfrac{32}{x} = 2x$

71. $\dfrac{2}{x - 5} = 4$

72. $\dfrac{3}{2x + 3} - 4 = \dfrac{-1}{2x + 3}$

73. $\dfrac{3}{x - 2} + \dfrac{x}{x + 2} = 1$

74. $\dfrac{2}{x} - \dfrac{5}{x - 2} = \dfrac{-13}{x^2 - 2x}$

In Exercises 75–80, find all real solutions of the exponential equation.

75. $e^x = 27$ **76.** $e^x + 7 = 35$

77. $e^{2x} - 4e^x + 3 = 0$ **78.** $e^{2x} - 7e^x + 12 = 0$

79. $200e^{-x} = 75$ **80.** $800e^{-x} = 250$

In Exercises 81–84, find all real solutions of the logarithmic equation.

81. $\ln x = 8$ **82.** $3 - 4 \ln x = 6$

83. $4 \ln 6x = 16$ **84.** $5 \ln 2x - 4 = 11$

Chapter Summary

What did you learn?

	Review Exercises
Section 8.1	
☐ How to use sequence notation to write the terms of a sequence	1–4
☐ How to use factorial notation	5–8
☐ How to use summation notation to write sums	9–16
☐ How to find the sum of an infinite series	17–20
☐ How to use sequences and series to model and solve real-life problems	21, 22
Section 8.2	
☐ How to recognize and write arithmetic sequences	23–30
☐ How to find an nth partial sum of an arithmetic sequence	31–36
☐ How to use arithmetic sequences to model and solve real-life problems	37, 38
Section 8.3	
☐ How to recognize and write geometric sequences	39–46
☐ How to find the nth partial sum of a geometric sequence	47–56
☐ How to find the sum of an infinite geometric series	57–62
☐ How to use geometric sequences to model and solve real-life problems	63, 64
Section 8.4	
☐ How to use mathematical induction to prove a statement	65–68
☐ How to find the sums of powers of integers	69–72
☐ How to recognize patterns and write the nth term of a sequence	73–76
☐ How to find finite differences of a sequence	77–80
Section 8.5	
☐ How to use the Binomial Theorem to calculate binomial coefficients	81–84
☐ How to use Pascal's Triangle to calculate binomial coefficients	85–88
☐ How to use binomial coefficients to write binomial expansions	89–94
Section 8.6	
☐ How to solve simple counting problems	95, 96
☐ How to use the Fundamental Counting Principle to solve counting problems	97, 98
☐ How to use permutations to solve counting problems	99, 100
☐ How to use combinations to solve counting problems	101, 102
Section 8.7	
☐ How to find the probability of an event	103, 104
☐ How to find the probabilities of mutually exclusive events	105, 106
☐ How to find the probabilities of independent events	107, 108
☐ How to find the probability of the complement of an event	109, 110

Review Exercises

8.1 **In Exercises 1–4, write the first five terms of the sequence. (Assume that n begins with 1.)**

1. $a_n = 2 + \dfrac{6}{n}$

2. $a_n = \dfrac{5n}{2n - 1}$

3. $a_n = \dfrac{72}{n!}$

4. $a_n = n(n - 1)$

In Exercises 5–8, evaluate the factorial expression.

5. $5!$

6. $3! \cdot 2!$

7. $\dfrac{3! \cdot 5!}{6!}$

8. $\dfrac{7! \cdot 6!}{6! \cdot 8!}$

In Exercises 9–14, find the sum.

9. $\displaystyle\sum_{i=1}^{6} 5$

10. $\displaystyle\sum_{k=2}^{5} 4k$

11. $\displaystyle\sum_{j=1}^{4} \dfrac{6}{j^2}$

12. $\displaystyle\sum_{i=1}^{8} \dfrac{i}{i + 1}$

13. $\displaystyle\sum_{k=1}^{10} 2k^3$

14. $\displaystyle\sum_{j=0}^{4} (j^2 + 1)$

In Exercises 15 and 16, use sigma notation to write the sum.

15. $\dfrac{1}{2(1)} + \dfrac{1}{2(2)} + \dfrac{1}{2(3)} + \cdots + \dfrac{1}{2(20)}$

16. $\dfrac{1}{2} + \dfrac{2}{3} + \dfrac{3}{4} + \cdots + \dfrac{9}{10}$

In Exercises 17–20, find the sum of the infinite series.

17. $\displaystyle\sum_{i=1}^{\infty} \dfrac{5}{10^i}$

18. $\displaystyle\sum_{i=1}^{\infty} \dfrac{3}{10^i}$

19. $\displaystyle\sum_{k=1}^{\infty} \dfrac{2}{100^k}$

20. $\displaystyle\sum_{k=2}^{\infty} \dfrac{9}{10^k}$

21. *Job Offer*   The starting salary for a job is $34,000 with a guaranteed salary increase of $2250 per year. Determine (a) the salary during the fifth year and (b) the total compensation through 5 full years of employment.

22. *Baling Hay*   In the first two trips baling hay around a large field, a farmer obtains 123 bales and 112 bales, respectively. Because each round gets shorter, the farmer estimates that the same pattern will continue. Estimate the total number of bales made if there are another six trips around the field.

8.2 **In Exercises 23–26, determine whether the sequence is arithmetic. If it is, find the common difference.**

23. $5, 3, 1, -1, -3, \ldots$

24. $0, 1, 3, 6, 10, \ldots$

25. $\frac{1}{2}, 1, \frac{3}{2}, 2, \frac{5}{2}, \ldots$

26. $\frac{9}{9}, \frac{8}{9}, \frac{7}{9}, \frac{6}{9}, \frac{5}{9}, \ldots$

In Exercises 27–30, find a formula for a_n for the arithmetic sequence.

27. $a_1 = 7, d = 12$

28. $a_1 = 25, d = -3$

29. $a_1 = y, d = 3y$

30. $a_1 = -2x, d = x$

In Exercises 31–34, find the sum.

31. $\displaystyle\sum_{j=1}^{10} (2j - 3)$

32. $\displaystyle\sum_{j=1}^{8} (20 - 3j)$

33. $\displaystyle\sum_{k=1}^{11} \left(\tfrac{2}{3}k + 4\right)$

34. $\displaystyle\sum_{k=1}^{25} \left(\dfrac{3k + 1}{4}\right)$

35. Find the sum of the first 100 positive multiples of 5.

36. Find the sum of the integers from 20 to 80 (inclusive).

37. *Running*   The first time you run a 5-mile distance course, it takes you 49 minutes. If you run the same course 30 seconds faster each week, how fast can you run the 5-mile course after 12 weeks?

38. *Running*   On the first day of a new training schedule you run 2 miles. If you increase your distance by one-half mile every day, how many total miles will you run in 14 days?

8.3 **In Exercises 39–42, write the first five terms of the geometric sequence.**

39. $a_1 = 4, r = -\frac{1}{4}$

40. $a_1 = 2, r = 2$

41. $a_1 = 9, a_3 = 4$

42. $a_1 = 2, a_3 = 12$

In Exercises 43–46, write an expression for the nth term of the geometric sequence and find the sum of the first 20 terms of the sequence.

43. $a_1 = 16, a_2 = -8$ **44.** $a_3 = 6, a_4 = 1$

45. $a_1 = 100, r = 1.05$ **46.** $a_1 = 5, r = 0.2$

In Exercises 47–52, find the sum.

47. $\sum\limits_{i=1}^{7} 2^{i-1}$ **48.** $\sum\limits_{i=1}^{5} 3^{i-1}$

49. $\sum\limits_{i=1}^{4} \left(\frac{1}{2}\right)^{i}$ **50.** $\sum\limits_{i=1}^{6} \left(\frac{1}{3}\right)^{i-1}$

51. $\sum\limits_{i=1}^{5} (2)^{i-1}$ **52.** $\sum\limits_{i=1}^{4} 6(3)^{i}$

▦ In Exercises 53–56, use a graphing utility to find the sum.

53. $\sum\limits_{i=1}^{10} 10\left(\frac{3}{5}\right)^{i-1}$ **54.** $\sum\limits_{i=1}^{15} 20(0.2)^{i-1}$

55. $\sum\limits_{i=1}^{25} 100(1.06)^{i-1}$ **56.** $\sum\limits_{i=1}^{20} 8\left(\frac{6}{5}\right)^{i-1}$

In Exercises 57–62, find the sum of the infinite series.

57. $\sum\limits_{i=1}^{\infty} \left(\frac{7}{8}\right)^{i-1}$ **58.** $\sum\limits_{i=1}^{\infty} \left(\frac{1}{3}\right)^{i-1}$

59. $\sum\limits_{i=1}^{\infty} (0.1)^{i-1}$ **60.** $\sum\limits_{i=1}^{\infty} (0.5)^{i-1}$

61. $\sum\limits_{k=1}^{\infty} 4\left(\frac{2}{3}\right)^{k-1}$ **62.** $\sum\limits_{k=1}^{\infty} 1.3\left(\frac{1}{10}\right)^{k-1}$

63. *Depreciation*   A company buys a machine for $120,000. During the next 5 years it will depreciate at a rate of 30% per year. (That is, at the end of each year the depreciated value will be 70% of what it was at the beginning of the year.)

(a) Find the formula for the nth term of a geometric sequence that gives the value of the machine t full years after it was purchased.

(b) Find the depreciated value of the machine at the end of 5 full years.

64. *Total Compensation*   A job pays a salary of $32,000 the first year. During the next 39 years, suppose there is a 5.5% raise each year. What would be the total amount earned over the 40-year period?

8.4 In Exercises 65–68, use mathematical induction to prove the formula for every positive integer n.

65. $1 + 4 + \cdots + (3n - 2) = \dfrac{n}{2}(3n - 1)$

66. $1 + \dfrac{3}{2} + 2 + \dfrac{5}{2} + \cdots + \dfrac{1}{2}(n + 1) = \dfrac{n}{4}(n + 3)$

67. $\sum\limits_{i=0}^{n-1} ar^{i} = \dfrac{a(1 - r^{n})}{1 - r}$

68. $\sum\limits_{k=0}^{n-1} (a + kd) = \dfrac{n}{2}[2a + (n - 1)d]$

In Exercises 69–72, find the sum using the formulas for the sums of powers of integers.

69. $\sum\limits_{n=1}^{30} n$ **70.** $\sum\limits_{n=1}^{10} n^{2}$

71. $\sum\limits_{n=1}^{7} n^{4}$ **72.** $\sum\limits_{n=1}^{6} n^{5}$

In Exercises 73–76, find a formula for the sum of the first n terms of the sequence.

73. $9, 13, 17, 21, \ldots$ **74.** $68, 60, 52, 44, \ldots$

75. $1, \frac{3}{5}, \frac{9}{25}, \frac{27}{125}, \ldots$ **76.** $12, -1, \frac{1}{12}, -\frac{1}{144}, \ldots$

In Exercises 77–80, find the first five terms of the sequence. Then calculate the first and second differences of the sequence. Does the sequence have a linear model, a quadratic model, or neither?

77. $a_1 = 5$ **78.** $a_1 = -3$
$a_n = a_{n-1} + 5$ $a_n = a_{n-1} - 2n$

79. $a_1 = 16$ **80.** $a_0 = 0$
$a_n = a_{n-1} - 1$ $a_n = n - a_{n-1}$

8.5 In Exercises 81–84, use the Binomial Theorem to calculate the binomial coefficient.

81. $_6C_4$ **82.** $_{10}C_7$

83. $_8C_5$ **84.** $_{12}C_3$

In Exercises 85–88, use Pascal's Triangle to calculate the binomial coefficient.

85. $\dbinom{7}{3}$ **86.** $\dbinom{9}{4}$

87. $\dbinom{8}{6}$

88. $\dbinom{5}{3}$

In Exercises 89–94, use the Binomial Theorem to expand the binomial. Simplify your answer. (Remember that $i = \sqrt{-1}$.)

89. $\left(\dfrac{x}{2} + y\right)^4$

90. $\left(\dfrac{2}{x} - 3x\right)^6$

91. $(a - 3b)^5$

92. $(3x + y^2)^7$

93. $(5 + 2i)^4$

94. $(4 - 5i)^3$

8.6 **95.** *Dice* In how many different ways can a pair of dice be rolled to obtain a total of 10?

96. *Numbers in a Hat* If slips of paper numbered 1 through 14 are placed in a hat, in how many ways could you draw two numbers with replacement that total 12?

97. *Telephone Numbers* The same three-digit prefix is used for all of the telephone numbers in a small town. How many different telephone numbers are possible by changing only the last four digits?

98. *Telephone Numbers* A telephone number in another town can use any one of five different three-digit prefixes. How many different telephone numbers are possible in this town?

99. *Bike Race* There are 10 bicyclists entered in a race. In how many different orders could these 10 bicyclists finish?

100. *Bike Race* If 10 bicyclists are entered in a race, in how many different ways could the top three places be decided?

101. *Travel Wardrobe* You have eight different suits to choose from to take on a trip. How many combinations of three suits could you take?

102. *Neckties* If you have 20 different neckties in your wardrobe, how many combinations of three ties could you choose?

8.7 **103.** *Matching Socks* A man has five pairs of socks of which no two pairs are the same color. If he randomly selects two socks from a drawer, what is the probability that he gets a matched pair?

104. *Bookshelf Order* A child returns a five-volume set of books to a bookshelf. The child is not able to read, and so cannot distinguish one volume from another. What is the probability that the books are shelved in the correct order?

105. *Students by Class* At a particular high school, the numbers of students in each class are broken down by the following percents.

Freshmen	31%
Sophomores	26%
Juniors	25%
Seniors	18%

If a single student is picked randomly by lottery for a cash scholarship, what is the probability that the scholarship winner is

(a) a junior or senior?

(b) a freshman, sophomore, or junior?

106. *Data Analysis* A sample of college students, faculty, and administration were asked whether they favored a proposed increase in the annual activity fee to enhance student life on campus. The results of the study are listed in the table.

	Students	Faculty	Admin.	Total
Favor	237	37	18	292
Oppose	163	38	7	208
Total	400	75	25	500

A person is selected at random from the sample. Find the specified probability.

(a) The person is not in favor of the proposal.

(b) The person is a student.

(c) The person is a faculty member and is in favor of the proposal.

107. *Roll of the Dice* If a six-sided die is rolled three times, what is the probability of a 6 on each roll?

108. *Roll of the Dice* A six-sided die is rolled six times. What is the probability that each side appears exactly once?

109. *Picking a Card* You randomly select a card from a 52-card deck. What is the probability that the card is *not* a club?

110. *Tossing a Coin* Find the probability of obtaining at least one tail when a coin is tossed five times.

Synthesis

True or False? **In Exercises 111–114, determine whether the statement is true or false. Justify your answer.**

111. $\dfrac{(n + 2)!}{n!} = (n + 2)(n + 1)$

112. $\displaystyle\sum_{i=1}^{5}(i^3 + 2i) = \sum_{i=1}^{5}i^3 + \sum_{i=1}^{5}2i$

113. $\displaystyle\sum_{k=1}^{8}3k = 3\sum_{k=1}^{8}k$

114. $\displaystyle\sum_{j=1}^{6}2^j = \sum_{j=3}^{8}2^{j-2}$

115. An infinite sequence is a function. What is the domain of the function?

116. How do the two sequences differ?

(a) $a_n = \dfrac{(-1)^n}{n}$

(b) $a_n = \dfrac{(-1)^{n+1}}{n}$

117. In your own words, explain what makes a sequence (a) arithmetic and (b) geometric.

118. The graphs of two sequences are shown below. Identify each sequence as arithmetic or geometric. Explain your reasoning.

(a)

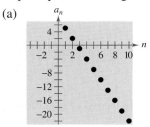

(b)

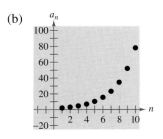

119. Explain what a recursion formula is.

120. Explain why the terms of a geometric sequence decrease when $0 < r < 1$.

In Exercises 121–124, match the sequence or sum of a sequence with its graph without doing any calculations. Explain your reasoning. [The graphs are labeled (a), (b), (c), and (d).]

(a)

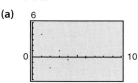

(b)

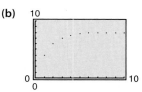

(c)

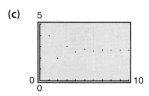

(d)

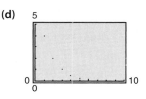

121. $a_n = 4\left(\frac{1}{2}\right)^{n-1}$

122. $a_n = 4\left(-\frac{1}{2}\right)^{n-1}$

123. $a_n = \displaystyle\sum_{k=1}^{n}4\left(\frac{1}{2}\right)^{k-1}$

124. $a_n = \displaystyle\sum_{k=1}^{n}4\left(-\frac{1}{2}\right)^{k-1}$

125. How do the expansions of $(x + y)^n$ and $(x - y)^n$ differ?

126. The probability of an event must be a real number in what interval? Is the interval open or closed?

127. The probability of an event is $\frac{2}{3}$. What is the probability that the event does not occur? Explain.

128. A weather forecast indicates that the probability of rain is 60%. Explain what this means.

Chapter Project ▶ Exploring Difference Quotients

In this project, you will explore difference quotients and see how they can be used to find average rates.

Example ▶ Finding an Average Speed

You are driving from Atlanta to Miami. The trip is about 700 miles and takes you 12 hours. Your average speed is

$$\text{Average speed} = \frac{\text{distance}}{\text{time}} = \frac{700}{12} \approx 58.3 \text{ miles per hour.}$$

The concept of average rate of change can be generalized as follows. Let f be a function defined on the interval $[a, b]$. The **average rate of change of** f from a to b is

$$\text{Average rate of change} = \frac{f(b) - f(a)}{b - a}.$$

This expression is called a **difference quotient.**

Chapter Project Investigations

1. (a) Calculate the average rate of change of $f(x) = 3x + 4$ on the interval $[2, 6]$. Select any other interval and show that you obtain the same average rate of change.

   (b) Calculate the average rates of change of $f(x) = x^2$ on the intervals $[1, 3]$ and $[4, 6]$. Are they equal? Explain.

   (c) During a 1-hour trip, your average speed is 50 miles per hour. Discuss the relationship between this speed and the speeds shown on your speedometer.

2. The data in the table gives the costs of first-class postage in the United States for selected years from 1971 to 1999, where $t = 71$ corresponds to 1971.   (Source: U.S. Postal Service)

t	71	74	75	78	81
Postage	$0.08	$0.10	$0.13	$0.15	$0.20

t	85	88	91	95	99
Postage	$0.22	$0.25	$0.29	$0.32	$0.33

   (a) Find the average rate of change of the cost of postage between each two adjacent time intervals.

   (b) When was the average rate of change largest? When was it smallest?

   (c) Are there intervals over which the average rate of change was zero? What does this mean?

Take this test as you would take a test in class. After you are done, check your work against the answers given in the back of the book.

1. Write the first five terms of the sequence $a_n = \dfrac{(-1)^n}{3n + 2}$.

2. Write an expression for the nth term of the sequence.

$$\frac{3}{1!}, \frac{4}{2!}, \frac{5}{3!}, \frac{6}{4!}, \frac{7}{5!}, \cdots$$

3. Find the fifth partial sum of the series.

$$6 + 17 + 28 + 39 + \cdots$$

4. The fifth term of an arithmetic series is 5.4, and the 12th term is 11.0. Find the 30th term.

5. Write the first five terms of the sequence $a_n = 5(2)^{n-1}$.

6. Find the sum of the finite series $\displaystyle\sum_{i=1}^{50} (2i^2 + 5)$.

7. Find the sum of the infinite series $\displaystyle\sum_{i=1}^{\infty} 4\left(\tfrac{1}{2}\right)^i$.

8. Use mathematical induction to prove the formula.

$$5 + 10 + 15 + \cdots + 5n = \frac{5n(n + 1)}{2}$$

9. Use the Binomial Theorem to expand the expression $(x + 2y)^4$.

10. To dress for a party, you can choose from six different pairs of slacks, 10 different shirts, and three different pairs of shoes. How many possible outfit combinations do you have?

In Exercises 11 and 12, evaluate the expression.

11. (a) $_9P_2$ (b) $_{70}P_3$ **12.** (a) $_{11}C_4$ (b) $_{66}C_4$

13. Eight people are going for a ride in a boat that seats eight people. The owner of the boat will drive, and only three of the remaining people are willing to ride in the two bow seats. How many seating arrangements are possible?

14. You attend a karaoke night and hope to hear your favorite song. The karaoke song book has 300 different songs. Assuming that the singers are equally likely to pick any song and no song is repeated, what is the probability that your favorite song is one of the 20 that you hear?

15. You are with seven of your friends at a party. Names of all of the 60 guests are placed in a hat and drawn randomly to award eight door prizes. If each guest is limited to one prize, what is the probability that you and your friends win all eight of the prizes?

16. The weather report calls for a 75% chance of rain. According to this report, what is the probability that it will not rain?

Cumulative Test for Chapters 6–8

Take this test to review the material from earlier chapters. After you are done, check your work against the answers given in the back of the book.

In Exercises 1–4, solve the system by the specified method.

1. Substitution

$$\begin{cases} y = 3 - x^2 \\ 2(y - 2) = x - 1 \end{cases}$$

2. Elimination

$$\begin{cases} x + 3y = -1 \\ 2x + 4y = 0 \end{cases}$$

3. Elimination

$$\begin{cases} -2x + 4y - z = 3 \\ x - 2y + 2z = -6 \\ x - 3y - z = 1 \end{cases}$$

4. Gauss-Jordan Elimination

$$\begin{cases} x + 3y - 2z = -7 \\ -2x + y - z = -5 \\ 4x + y + z = 3 \end{cases}$$

In Exercises 5 and 6, sketch the graph of the solution set of the system of inequalities.

5. $\begin{cases} 2x + y \geq -3 \\ x - 3y \leq 2 \end{cases}$

6. $\begin{cases} x - y > 6 \\ 5x + 2y < 10 \end{cases}$

7. Sketch a graph of the solution of the constraints and maximize the objective function $z = 3x + 2y$ subject to the constraints.

$$\begin{aligned} x + 4y &\leq 20 \\ 2x + y &\leq 12 \\ x &\geq 0 \\ y &\geq 0 \end{aligned}$$

8. A custom-blend bird seed is to be mixed from seed mixtures costing $0.75 per pound and $1.25 per pound. How many pounds of each seed mixture are used to make 200 pounds of custom-blend bird seed costing $0.95 per pound?

9. Find the equation of the parabola $y = ax^2 + bx + c$ passing through the points $(0, 4)$, $(3, 1)$, and $(6, 4)$.

$$\begin{cases} -x + 2y - z = 9 \\ 2x - y + 2z = -9 \\ 3x + 3y - 4z = 7 \end{cases}$$

SYSTEM FOR 10 AND 11

In Exercises 10 and 11, use the system of equations at the left.

10. Write the augmented matrix corresponding to the system of equations.

11. Solve the system using the matrix and Gauss-Jordan elimination.

In Exercises 12–15, use the following matrices.

$$A = \begin{bmatrix} 4 & 0 \\ -1 & 2 \end{bmatrix}, \quad B = \begin{bmatrix} -1 & 3 \\ 1 & 0 \end{bmatrix}$$

12. Find $A - B$.

13. Find $-2B$.

14. Find $A - 2B$.

15. Find AB, if possible.

16. Find the inverse (if it exists): $\begin{bmatrix} 1 & 2 & -1 \\ 3 & 7 & -10 \\ -5 & -7 & -15 \end{bmatrix}$.

17. The percents (by age group) of the total amounts spent on three types of shoes in 1996 are shown in the matrix. The total amounts (in millions) spent by each age group on the three types of shoes were $518.97 (14–17 age group), $336.16 (18–24 age group), and $753.37 (25–34 age group). How many dollars worth of gym shoes, jogging shoes, and walking shoes were sold in 1996? (Source: National Sporting Goods Association)

	Gym shoes	Jogging shoes	Walking shoes
Age group 14–17	0.14	0.13	0.03
Age group 18–24	0.05	0.10	0.04
Age group 25–34	0.10	0.19	0.11

MATRIX FOR **17**

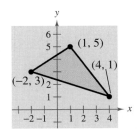

FIGURE FOR **20**

In Exercises 18 and 19, use Cramer's Rule to solve the system of equations.

18. $\begin{cases} 8x - 3y = -52 \\ 3x + 5y = 5 \end{cases}$

19. $\begin{cases} 5x + 4y + 3z = 7 \\ -3x - 8y + 7z = -9 \\ 7x - 5y - 6z = -53 \end{cases}$

20. Find the area of the triangle in the figure.

21. Write the first five terms of the sequence $a_n = \dfrac{(-1)^{n+1}}{2n+3}$.

22. Write an expression for the nth term of the sequence.

$$\frac{2!}{4}, \frac{3!}{5}, \frac{4!}{6}, \frac{5!}{7}, \frac{6!}{8}, \cdots$$

23. Sum the first 20 terms of the arithmetic sequence $8, 12, 16, 20, \ldots$.

24. The sixth term of an arithmetic series is 20.6, and the ninth term is 30.2. Find the 20th term.

25. Write the first five terms of the sequence $a_n = 3(2)^{n-1}$.

26. Find the sum: $\displaystyle\sum_{i=0}^{\infty} 3\left(\tfrac{1}{2}\right)^i$.

27. Use mathematical induction to prove the formula

$$3 + 7 + 11 + 15 + \cdots + (4n - 1) = n(2n + 1).$$

28. Use the Binomial Theorem to expand and simplify $(z - 3)^4$.

In Exercises 29–32, evaluate the expression.

29. $_7P_3$ **30.** $_{25}P_2$ **31.** $\dbinom{8}{4}$ **32.** $_{10}C_3$

33. A personnel manager has 10 applicants to fill three different positions. In how many ways can this be done, assuming that all the applicants are qualified for any of the three positions?

34. On a game show, the digits 3, 4, and 5 must be arranged in the proper order to form the price of an appliance. If the digits are arranged correctly, the contestant wins the appliance. What is the probability of winning if the contestant knows that the price is at least $400?

Appendix A   **Proofs of Selected Theorems**

SECTION P.7, PAGE 64

The Midpoint Formula

The midpoint of the segment joining the points (x_1, y_1) and (x_2, y_2) is given by the Midpoint Formula

$$\text{Midpoint} = \left(\frac{x_1 + x_2}{2}, \frac{y_1 + y_2}{2}\right).$$

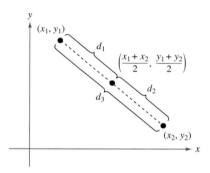

Midpoint Formula

Proof

Using the figure, you must show that

$$d_1 = d_2 \quad \text{and} \quad d_1 + d_2 = d_3.$$

By the Distance Formula, you obtain

$$d_1 = \sqrt{\left(\frac{x_1 + x_2}{2} - x_1\right)^2 + \left(\frac{y_1 + y_2}{2} - y_1\right)^2} = \frac{1}{2}\sqrt{(x_2 - x_1)^2 + (y_2 - y_1)^2}$$

$$d_2 = \sqrt{\left(x_2 - \frac{x_1 + x_2}{2}\right)^2 + \left(y_2 - \frac{y_1 + y_2}{2}\right)^2} = \frac{1}{2}\sqrt{(x_2 - x_1)^2 + (y_2 - y_1)^2}$$

$$d_3 = \sqrt{(x_2 - x_1)^2 + (y_2 - y_1)^2}.$$

So, it follows that $d_1 = d_2$ and $d_1 + d_2 = d_3$.

SECTION 3.3, PAGE 282

The Remainder Theorem

If a polynomial $f(x)$ is divided by $x - k$, the remainder is

$$r = f(k).$$

Proof

From the Division Algorithm, you have

$$f(x) = (x - k)q(x) + r(x)$$

and because either $r(x) = 0$ or the degree of $r(x)$ is less than the degree of $x - k$, you know that $r(x)$ must be a constant. That is, $r(x) = r$. Now, by evaluating $f(x)$ at $x = k$, you have

$$f(k) = (k - k)q(k) + r = (0)q(k) + r = r.$$

SECTION 3.3, PAGE 282

The Factor Theorem

A polynomial $f(x)$ has a factor $(x - k)$ if and only if $f(k) = 0$.

Proof

Using the Division Algorithm with the factor $(x - k)$, you have

$$f(x) = (x - k)q(x) + r(x).$$

By the Remainder Theorem, $r(x) = r = f(k)$, and you have

$$f(x) = (x - k)q(x) + f(k)$$

where $q(x)$ is a polynomial of lesser degree than $f(x)$. If $f(k) = 0$, then

$$f(x) = (x - k)q(x)$$

and you see that $(x - k)$ is a factor of $f(x)$. Conversely, if $(x - k)$ is a factor of $f(x)$, division of $f(x)$ by $(x - k)$ yields a remainder of 0. So, by the Remainder Theorem, you have $f(k) = 0$.

SECTION 3.4, PAGE 288

Linear Factorization Theorem

If $f(x)$ is a polynomial of degree n, where $n > 0$, then f has precisely n linear factors

$$f(x) = a_n(x - c_1)(x - c_2) \cdots (x - c_n)$$

where $c_1, c_2, \ldots, c_n$ are complex numbers.

Proof

Using the Fundamental Theorem of Algebra, you know that f must have at least one zero, c_1. Consequently, $(x - c_1)$ is a factor of $f(x)$, and you have

$$f(x) = (x - c_1)f_1(x).$$

If the degree of $f_1(x)$ is greater than zero, you again apply the Fundamental Theorem to conclude that f_1 must have a zero c_2, which implies that

$$f(x) = (x - c_1)(x - c_2)f_2(x).$$

It is clear that the degree of $f_1(x)$ is $n - 1$, that the degree of $f_2(x)$ is $n - 2$, and that you can repeatedly apply the Fundamental Theorem n times until you obtain

$$f(x) = a_n(x - c_1)(x - c_2) \cdots (x - c_n)$$

where a_n is the leading coefficient of the polynomial $f(x)$.

SECTION 3.4, PAGE 292

Factors of a Polynomial

Every polynomial of degree $n > 0$ with real coefficients can be written as the product of linear and quadratic factors with real coefficients, where the quadratic factors have no real zeros.

Proof

To begin, you use the Linear Factorization Theorem to conclude that $f(x)$ can be *completely* factored in the form

$$f(x) = d(x - c_1)(x - c_2)(x - c_3) \cdots (x - c_n).$$

If each c_i is real, there is nothing more to prove. If any c_i is complex ($c_i = a + bi$, $b \neq 0$), then, because the coefficients of $f(x)$ are real, you know that the conjugate $c_j = a - bi$ is also a zero. By multiplying the corresponding factors, you obtain

$$(x - c_i)(x - c_j) = [x - (a + bi)][x - (a - bi)]$$
$$= x^2 - 2ax + (a^2 + b^2)$$

where each coefficient is real.

SECTION 4.4, PAGE 351

Standard Equation of a Parabola (Vertex at Origin)

The standard form of the equation of a parabola with vertex at $(0, 0)$ and directrix $y = -p$ is

$$x^2 = 4py, \qquad p \neq 0. \qquad \text{Vertical axis}$$

For directrix $x = -p$, the equation is

$$y^2 = 4px, \qquad p \neq 0. \qquad \text{Horizontal axis}$$

The focus is on the axis p units (directed distance) from the vertex.

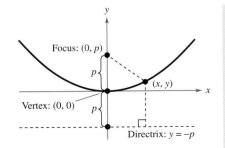

Proof

Because the two cases are similar, a proof will be given for the first case only. Suppose the directrix ($y = -p$) is parallel to the x-axis. In the figure, you assume that $p > 0$, and because p is the directed distance from the vertex to the focus, the focus must lie above the vertex. Because the point (x, y) is equidistant from $(0, p)$ and $y = -p$, you can apply the Distance Formula to obtain

$$\sqrt{(x - 0)^2 + (y - p)^2} = y + p$$
$$x^2 + (y - p)^2 = (y + p)^2$$
$$x^2 + y^2 - 2py + p^2 = y^2 + 2py + p^2$$
$$x^2 = 4py.$$

SECTION 5.3, PAGE 404

Properties of Logarithms

Let a be a positive number such that $a \neq 1$, and let n be a real number. If u and v are positive real numbers, the following properties are true.

1. $\log_a(uv) = \log_a u + \log_a v$ 1. $\ln(uv) = \ln u + \ln v$

2. $\log_a \dfrac{u}{v} = \log_a u - \log_a v$ 2. $\ln \dfrac{u}{v} = \ln u - \ln v$

3. $\log_a u^n = n \log_a u$ 3. $\ln u^n = n \ln u$

Proof

To prove Property 1, let

$$x = \log_a u \quad \text{and} \quad y = \log_a v.$$

The corresponding exponential forms of these two equations are

$$a^x = u \quad \text{and} \quad a^y = v.$$

Multiplying u and v produces $uv = a^x a^y = a^{x+y}$. The corresponding logarithmic form of $uv = a^{x+y}$ is $\log_a(uv) = x + y$. So, $\log_a(uv) = \log_a u + \log_a v$.

SECTION 8.1, PAGE 583

Properties of Sums

1. $\displaystyle\sum_{i=1}^{n} ca_i = c\sum_{i=1}^{n} a_i,$ c is any constant.

2. $\displaystyle\sum_{i=1}^{n} (a_i + b_i) = \sum_{i=1}^{n} a_i + \sum_{i=1}^{n} b_i$

3. $\displaystyle\sum_{i=1}^{n} (a_i - b_i) = \sum_{i=1}^{n} a_i - \sum_{i=1}^{n} b_i$

Proof

Each of these properties follows directly from the Associative Property of Addition, the Commutative Property of Addition, and the Distributive Property of multiplication over addition. For example, note the use of the Distributive Property in the proof of Property 1.

$$\sum_{i=1}^{n} ca_i = ca_1 + ca_2 + ca_3 + \cdots + ca_n$$

$$= c(a_1 + a_2 + a_3 + \cdots + a_n) = c\sum_{i=1}^{n} a_i$$

SECTION 8.2, PAGE 592

The Sum of a Finite Arithmetic Sequence

The sum of a finite arithmetic sequence with n terms is

$$S_n = \frac{n}{2}(a_1 + a_n).$$

Proof

Begin by generating the terms of the arithmetic sequence in two ways. In the first way, repeatedly add d to the first term to obtain

$$S_n = a_1 + a_2 + a_3 + \cdots + a_{n-2} + a_{n-1} + a_n$$
$$= a_1 + [a_1 + d] + [a_1 + 2d] + \cdots + [a_1 + (n-1)d].$$

In the second way, repeatedly subtract d from the nth term to obtain

$$S_n = a_n + a_{n-1} + a_{n-2} + \cdots + a_3 + a_2 + a_1$$
$$= a_n + [a_n - d] + [a_n - 2d] + \cdots + [a_n - (n-1)d].$$

If you add these two versions of S_n, the multiples of d cancel and you obtain

$$\overbrace{2S_n = (a_1 + a_n) + (a_1 + a_n) + (a_1 + a_n) + \cdots + (a_1 + a_n)}^{n \text{ terms}}$$
$$= n(a_1 + a_n).$$

So, you have $S_n = \dfrac{n}{2}(a_1 + a_n).$

SECTION 8.3, PAGE 601

The Sum of a Finite Geometric Sequence

The sum of the geometric sequence

$$a_1, \; a_1 r, \; a_1 r^2, \; a_1 r^3, \; a_1 r^4, \; \ldots, a_1 r^{n-1}$$

with common ratio $r \neq 1$ is given by $S_n = a_1\left(\dfrac{1 - r^n}{1 - r}\right).$

Proof

Begin by writing out the nth partial sum.

$$S_n = a_1 + a_1 r + a_1 r^2 + \cdots + a_1 r^{n-2} + a_1 r^{n-1}$$

Multiplication by r yields

$$rS_n = a_1 r + a_1 r^2 + a_1 r^3 + \cdots + a_1 r^{n-1} + a_1 r^n.$$

Subtracting the second equation from the first yields

$$S_n - rS_n = a_1 - a_1 r^n.$$

So, $S_n(1 - r) = a_1(1 - r^n)$, and, because $r \neq 1$, you have $S_n = a_1\left(\dfrac{1 - r^n}{1 - r}\right).$

SECTION 8.5, PAGE 618

The Binomial Theorem

In the expansion of $(x + y)^n$

$$(x + y)^n = x^n + nx^{n-1}y + \cdots + {}_nC_r\, x^{n-r}y^r + \cdots + nxy^{n-1} + y^n$$

the coefficient of $x^{n-r}y^r$ is

$${}_nC_r = \frac{n!}{(n-r)!r!}.$$

Proof

The Binomial Theorem can be proved quite nicely using mathematical induction. The steps are straightforward but look a little messy, so only an outline of the proof is presented.

1. If $n = 1$, you have

$$(x + y)^1 = x^1 + y^1 = {}_1C_0 x + {}_1C_1 y$$

and the formula is valid.

2. Assuming that the formula is true for $n = k$, the coefficient of $x^{k-r}y^r$ is

$${}_kC_r = \frac{k!}{(k-r)!r!} = \frac{k(k-1)(k-2)\cdots(k-r+1)}{r!}.$$

To show that the formula is true for $n = k + 1$, look at the coefficient of $x^{k+1-r}y^r$ in the expansion of

$$(x + y)^{k+1} = (x + y)^k(x + y).$$

From the right-hand side, you can determine that the term involving $x^{k+1-r}y^r$ is the sum of two products.

$$\left({}_kC_r x^{k-r}y^r\right)(x) + \left({}_kC_{r-1}x^{k+1-r}y^{r-1}\right)(y)$$

$$= \left[\frac{k!}{(k-r)!r!} + \frac{k!}{(k-r+1)!(r-1)!}\right]x^{k+1-r}y^r$$

$$= \left[\frac{(k+1-r)k!}{(k+1-r)!r!} + \frac{k!r}{(k+1-r)!r!}\right]x^{k+1-r}y^r$$

$$= \left[\frac{k!(k+1-r+r)}{(k+1-r)!r!}\right]x^{k+1-r}y^r$$

$$= \left[\frac{(k+1)!}{(k+1-r)!r!}\right]x^{k+1-r}y^r$$

$$= {}_{k+1}C_r x^{k+1-r}y^r$$

So, by mathematical induction, the Binomial Theorem is valid for all positive integers n.

Appendix B   Concepts in Statistics

B.1   Representing Data

Line Plots

Statistics is the branch of mathematics that studies techniques for collecting, organizing, and interpreting data. In this section, you will study several ways to organize data. The first is a **line plot,** which uses a portion of a real number line to order numbers. Line plots are especially useful for ordering small sets of numbers (about 50 or less) by hand.

Example 1 ►   Constructing a Line Plot

Use a line plot to organize the following test scores. Which number occurs with the greatest frequency?

93, 70, 76, 67, 86, 93, 82, 78, 83, 86, 64, 78, 76, 66, 83,
83, 96, 74, 69, 76, 64, 74, 79, 76, 88, 76, 81, 82, 74, 70

Solution

Begin by scanning the data to find the smallest and largest numbers. For this data, the smallest number is 64 and the largest is 96. Next, draw a portion of a real number line that includes the interval [64, 96]. To create the line plot, start with the first number, 93, and enter an × above 93 on the number line. Continue recording ×'s for each number in the list until you obtain the line plot shown in Figure B.1. From the line plot, you can see that 76 had the greatest frequency.

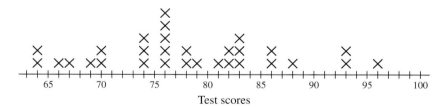

FIGURE B.1

Test Scores

93, 70, 76, 58, 86, 93, 82, 78,
83, 86, 64, 78, 76, 66, 83, 83,
96, 74, 69, 76, 64, 74, 79, 76,
88, 76, 81, 82, 74, 70

Stems	Leaves
5	8
6	4 4 6 9
7	0 0 4 4 4 6 6 6 6 6 8 8 9
8	1 2 2 3 3 3 6 6 8
9	3 3 6

Stem-and-Leaf Plots

Another type of plot that can be used to organize sets of numbers by hand is a **stem-and-leaf plot.** A set of test scores and the corresponding stem-and-leaf plot are shown at the left.

Note that the *leaves* represent the units digits of the numbers and the *stems* represent the tens digits. Stem-and-leaf plots can also be used to compare two sets of data, as shown in the following example.

Example 2 ▶   Comparing Two Sets of Data

Use a stem-and-leaf plot to compare the test scores given on page A7 with the following test scores. Which set of test scores is better?

90, 81, 70, 62, 64, 73, 81, 92, 73, 81, 92, 93, 83, 75, 76,
83, 94, 96, 86, 77, 77, 86, 96, 86, 77, 86, 87, 87, 79, 88

Solution

Begin by ordering the second set of scores.

62, 64, 70, 73, 73, 75, 76, 77, 77, 77, 79, 81, 81, 81, 83,
83, 86, 86, 86, 86, 87, 87, 88, 90, 92, 92, 93, 94, 96, 96

Now that the data has been ordered, you can construct a *double* stem-and-leaf plot by letting the leaves to the right of the stems represent the units digits for the first group of test scores and letting the leaves to the left of the stems represent the units digits for the second group of test scores.

Leaves (2nd Group)	Stems	Leaves (1st Group)
	5	8
4 2	6	4 4 6 9
9 7 7 7 6 5 3 3 0	7	0 0 4 4 4 6 6 6 6 6 8 8 9
8 7 7 6 6 6 6 3 3 1 1 1	8	1 2 2 3 3 3 6 6 8
6 6 4 3 2 2 0	9	3 3 6

By comparing the two sets of leaves, you can see that the second group of test scores is better than the first group.

Example 3 ▶   Using a Stem-and-Leaf Plot

The table at the left above shows the percent of the population of each state and the District of Columbia that was at least 65 years old in 1997. Use a stem-and-leaf plot to organize the data.   (Source: U.S. Bureau of the Census)

Solution

Begin by ordering the numbers, as shown below.

5.3, 8.7, 9.9, 10.1, 10.1, 11.1, 11.2, 11.2, 11.3, 11.3, 11.4, 11.5, 11.5, 11.5,
12.1, 12.1, 12.2, 12.3, 12.3, 12.4, 12.5, 12.5, 12.5, 12.5, 12.5, 12.9, 13.0,
13.2, 13.2, 13.2, 13.2, 13.3, 13.4, 13.4, 13.4, 13.5, 13.7, 13.7, 13.7, 13.9,
13.9, 14.1, 14.3, 14.3, 14.4, 14.4, 15.0, 15.1, 15.8, 15.8, 18.5

Next construct the stem-and-leaf plot using the leaves to represent the digits to the right of the decimal points, as shown at the left. From the stem-and-leaf plot, you can see that Alaska has the lowest percent and Florida has the highest percent.

AK	5.3	MT	13.2
AL	13.0	NC	12.5
AR	14.3	ND	14.4
AZ	13.2	NE	13.7
CA	11.1	NH	12.1
CO	10.1	NJ	13.7
CT	14.4	NM	11.2
DC	13.9	NV	11.5
DE	12.9	NY	13.4
FL	18.5	OH	13.4
GA	9.9	OK	13.4
HI	13.2	OR	13.3
IA	15.0	PA	15.8
ID	11.3	RI	15.8
IL	12.5	SC	12.1
IN	12.5	SD	14.3
KS	13.5	TN	12.5
KY	12.5	TX	10.1
LA	11.4	UT	8.7
MA	14.1	VA	11.2
MD	11.5	VT	12.3
ME	13.9	WA	11.5
MI	12.4	WI	13.2
MN	12.3	WV	15.1
MO	13.7	WY	11.3
MS	12.2		

Stems	Leaves
5.	3
6.	
7.	
8.	7
9.	9
10.	1 1
11.	1 2 2 3 3 4 5 5 5
12.	1 1 2 3 3 4 5 5 5 5 5 9
13.	0 2 2 2 2 3 4 4 4 5 7 7 7 9 9
14.	1 3 3 4 4
15.	0 1 8 8
16.	
17.	
18.	5

Histograms and Frequency Distributions

With data such as that given in Example 3, it is useful to group the numbers into intervals and plot the frequency of the data in each interval. For instance, the **frequency distribution** and **histogram** shown in Figure B.2 represent the data given in Example 3.

Frequency Distribution

Interval	Tally
$[5, 7)$	I
$[7, 9)$	I
$[9, 11)$	III
$[11, 13)$	LHT LHT LHT LHT I
$[13, 15)$	LHT LHT LHT LHT
$[15, 17)$	IIII
$[17, 19)$	I

Histogram

FIGURE B.2

A histogram has a portion of a real number line as its horizontal axis. A histogram is similar to a bar graph, except that the rectangles (bars) in a bar graph can be either horizontal or vertical and the labels of the bars are not necessarily numbers.

Another difference between a bar graph and a histogram is that the bars in a bar graph are usually separated by spaces, whereas the bars in a histogram are not separated by spaces.

Interval	Tally
100–109	LHT III
110–119	I
120–129	III
130–139	III
140–149	LHT II
150–159	LHT
160–169	LHT III
170–179	LHT I
180–189	II
190–199	LHT

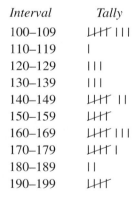

Example 4 ▸ Constructing a Histogram

A company has 48 sales representatives who sold the following numbers of units during the first quarter of 2000. Construct a grouped frequency distribution for this data.

107	162	184	170	177	102	145	141
105	193	167	149	195	127	193	191
150	153	164	167	171	163	141	129
109	171	150	138	100	164	147	153
171	163	118	142	107	144	100	132
153	107	124	162	192	134	187	177

Solution

To begin constructing a grouped frequency distribution, you must first decide on the number of groups. There are several ways to group this data. However, because the smallest number is 100 and the largest is 195, it seems that 10 groups of 10 each would be appropriate. The first group would be 100–109, the second group would be 110–119, and so on. By tallying the data into the 10 groups, you obtain the distribution shown at the left above. A histogram for the distribution is shown in Figure B.3.

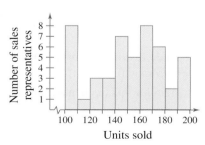

FIGURE B.3

B.1 Exercises

1. *Gasoline Prices*   The line plot shows a sample of prices of unleaded regular gasoline from 25 different cities.

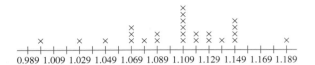

(a) What price occurred with the greatest frequency?

(b) What is the range of prices?

2. *Livestock Weights*   The line plot shows the weights (to the nearest hundred pounds) of 30 head of cattle sold by a rancher.

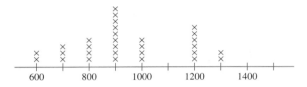

(a) What weight occurred with the greatest frequency?

(b) What is the range of weights?

Quiz and Exam Scores   **In Exercises 3–6, use the following scores from a math class of 30 students. The scores are for two 25-point quizzes and two 100-point exams.**

Quiz #1   20, 15, 14, 20, 16, 19, 10, 21, 24, 15, 15, 14, 15, 21, 19, 15, 20, 18, 18, 22, 18, 16, 18, 19, 21, 19, 16, 20, 14, 12

Quiz #2   22, 22, 23, 22, 21, 24, 22, 19, 21, 23, 23, 25, 24, 22, 22, 23, 23, 23, 23, 22, 24, 23, 22, 24, 21, 24, 16, 21, 16, 14

Exam #1   77, 100, 77, 70, 83, 89, 87, 85, 81, 84, 81, 78, 89, 78, 88, 85, 90, 92, 75, 81, 85, 100, 98, 81, 78, 75, 85, 89, 82, 75

Exam #2   76, 78, 73, 59, 70, 81, 71, 66, 66, 73, 68, 67, 63, 67, 77, 84, 87, 71, 78, 78, 90, 80, 77, 70, 80, 64, 74, 68, 68, 68

3. Construct a line plot for each quiz. For each quiz, which score occurred with the greatest frequency?

4. Construct a line plot for each exam. For each exam, which score occurred with the greatest frequency?

5. Construct a stem-and-leaf plot for Exam #1.

6. Construct a double stem-and-leaf plot to compare the scores for Exam #1 and Exam #2. Which set of scores is higher?

7. *Insurance Coverage*   The table shows the total numbers of persons (in thousands) without health insurance coverage in the 50 states and the District of Columbia in 1996. Use a stem-and-leaf plot to organize the data. (Source: U.S. Bureau of the Census)

AK	89	AL	550	AR	566	AZ	1159	CA	6514
CO	644	CT	368	DC	80	DE	98	FL	2722
GA	1319	HI	101	IA	335	ID	196	IL	1337
IN	600	KS	292	KY	601	LA	890	MA	766
MD	581	ME	146	MI	857	MN	480	MO	700
MS	518	MT	124	NC	1160	ND	62	NE	190
NH	109	NJ	1317	NM	412	NV	255	NY	3132
OH	1292	OK	570	OR	496	PA	1133	RI	93
SC	634	SD	67	TN	841	TX	4680	UT	240
VA	811	VT	65	WA	761	WI	438	WV	261
WY	66								

8. *Snowfall*   The data below shows the seasonal snowfall (in inches) at Lincoln, Nebraska for the years 1968 through 1997 (the amounts are listed in order by year). How would you organize this data? Explain your reasoning. (Source: University of Nebraska–Lincoln)

39.8, 26.2, 49.0, 21.6, 29.2, 33.6, 42.1, 21.1, 21.8, 31.0, 34.4, 23.3, 13.0, 32.3, 38.0, 47.5, 21.5, 18.9, 15.7, 13.0, 19.1, 18.7, 25.8, 23.8, 32.1, 21.3, 21.8, 30.7, 29.0, 44.6

9. *Retirement Contributions*   The employees of a company must contribute 7% of their monthly salaries to a company-sponsored retirement plan. The contributed amounts (in dollars) for the company's 35 employees are as follows.

100, 200, 130, 136, 161, 156, 209, 126, 135, 98, 114, 117, 168, 133, 140, 124, 172, 127, 143, 157, 124, 152, 104, 126, 155, 92, 194, 115, 120, 136, 148, 112, 116, 146, 96

(a) Construct a frequency distribution using groups of 20. The first group should be 90–109.

(b) Construct a histogram for this frequency distribution.

▶ **What you should learn**

- How to find and interpret the mean, median, and mode of a set of data
- How to determine the measure of central tendency that best represents a set of data
- How to find the standard deviation of a set of data
- How to use box-and-whisker plots

▶ **Why you should learn it**

Measures of central tendency and dispersion provide a convenient way to describe and compare sets of data. For instance, in Exercise 36 on page A19, the mean and standard deviation are used to analyze the price of gold for the years 1978 through 1997.

Mean, Median, and Mode

In many real-life situations, it is helpful to describe data by a single number that is most representative of the entire collection of numbers. Such a number is called a **measure of central tendency.** The most commonly used measures are as follows.

1. The **mean,** or **average,** of n numbers is the sum of the numbers divided by n.

2. The **median** of n numbers is the middle number when the numbers are written in order. If n is even, the median is the average of the two middle numbers.

3. The **mode** of n numbers is the number that occurs most frequently. If two numbers tie for most frequent occurrence, the collection has two modes and is called **bimodal.**

Example 1 ▶   Comparing Measures of Central Tendency

On an interview for a job, the interviewer tells you that the average annual income of the company's 25 employees is $60,849. The actual annual incomes of the 25 employees are shown below. What are the mean, median, and mode of the incomes? Was the person telling you the truth?

$17,305,	$478,320,	$45,678,	$18,980,	$17,408,
$25,676,	$28,906,	$12,500,	$24,540,	$33,450,
$12,500,	$33,855,	$37,450,	$20,432,	$28,956,
$34,983,	$36,540,	$250,921,	$36,853,	$16,430,
$32,654,	$98,213,	$48,980,	$94,024,	$35,671

Solution

The mean of the incomes is

$$\text{Mean} = \frac{17{,}305 + 478{,}320 + 45{,}678 + 18{,}980 + \cdots + 35{,}671}{25}$$

$$= \frac{1{,}521{,}225}{25} = \$60{,}849.$$

To find the median, order the incomes as follows.

$12,500,	$12,500,	$16,430,	$17,305,	$17,408,
$18,980,	$20,432,	$24,540,	$25,676,	$28,906,
$28,956,	$32,654,	$33,450,	$33,855,	$34,983,
$35,671,	$36,540,	$36,853,	$37,450,	$45,678,
$48,980,	$94,024,	$98,213,	$250,921,	$478,320

From this list, you can see that the median (the middle number) is $33,450. From the same list, you can see that $12,500 is the only income that occurs more than once. So, the mode is $12,500. Technically, the person was telling the truth because the average is (generally) defined to be the mean. However, of the three measures of central tendency *Mean:* $60,849 *Median:* $33,450 *Mode:* $12,500 it seems clear that the median is most representative. The mean is inflated by the two highest salaries.

Choosing a Measure of Central Tendency

Which of the three measures of central tendency is the most representative? The answer is that it depends on the distribution of the data *and* the way in which you plan to use the data.

For instance, in Example 1, the mean salary of $60,849 does not seem very representative to a potential employee. To a city income tax collector who wants to estimate 1% of the total income of the 25 employees, however, the mean is precisely the right measure.

Example 2 ▶   Choosing a Measure of Central Tendency

Which measure of central tendency is the most representative of the data given in each of the following frequency distributions?

a. Number	Tally		b. Number	Tally		c. Number	Tally
1	7		1	9		1	6
2	20		2	8		2	1
3	15		3	7		3	2
4	11		4	6		4	3
5	8		5	5		5	5
6	3		6	6		6	5
7	2		7	7		7	4
8	0		8	8		8	3
9	15		9	9		9	0

Solution

a. For this data, the mean is 4.23, the median is 3, and the mode is 2. Of these, the mode is probably the most representative.

b. For this data, the mean and median are each 5 and the modes are 1 and 9 (the distribution is bimodal). Of these, the mean or median is the most representative.

c. For this data, the mean is 4.59, the median is 5, and the mode is 1. Of these, the mean or median is the most representative.

Variance and Standard Deviation

Very different sets of numbers can have the same mean. You will now study two **measures of dispersion,** which give you an idea of how much the numbers in a set differ from the mean of the set. These two measures are called the *variance* of the set and the *standard deviation* of the set.

Definitions of Variance and Standard Deviation

Consider a set of numbers $\{x_1, x_2, \ldots, x_n\}$ with a mean of $\bar{x}$. The **variance** of the set is

$$v = \frac{(x_1 - \bar{x})^2 + (x_2 - \bar{x})^2 + \cdots + (x_n - \bar{x})^2}{n}$$

and the **standard deviation** of the set is $\sigma = \sqrt{v}$ (σ is the lowercase Greek letter *sigma*).

The standard deviation of a set is a measure of how much a typical number in the set differs from the mean. The greater the standard deviation, the more the numbers in the set *vary* from the mean. For instance, each of the following sets has a mean of 5.

$$\{5, 5, 5, 5\}, \qquad \{4, 4, 6, 6\}, \qquad \text{and} \qquad \{3, 3, 7, 7\}$$

The standard deviations of the sets are 0, 1, and 2.

$$\sigma_1 = \sqrt{\frac{(5-5)^2 + (5-5)^2 + (5-5)^2 + (5-5)^2}{4}}$$

$$= 0$$

$$\sigma_2 = \sqrt{\frac{(4-5)^2 + (4-5)^2 + (6-5)^2 + (6-5)^2}{4}}$$

$$= 1$$

$$\sigma_3 = \sqrt{\frac{(3-5)^2 + (3-5)^2 + (7-5)^2 + (7-5)^2}{4}}$$

$$= 2$$

Example 3 ▶ Estimations of Standard Deviation

Consider the three sets of data represented by the bar graphs in Figure B.4. Which set has the smallest standard deviation? Which has the largest?

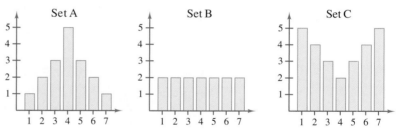

FIGURE B.4

Solution

Of the three sets, the numbers in set *A* are grouped most closely to the center and the numbers in set *C* are the most dispersed. So, set *A* has the smallest standard deviation and set *C* has the largest standard deviation.

Example 4 ▶   **Finding Standard Deviation**

Find the standard deviation of each set shown in Example 3.

Solution

Because of the symmetry of each bar graph, you can conclude that each has a mean of $\bar{x} = 4$. The standard deviation of set A is

$$\sigma = \sqrt{\frac{(-3)^2 + 2(-2)^2 + 3(-1)^2 + 5(0)^2 + 3(1)^2 + 2(2)^2 + (3)^2}{17}}$$

$$\approx 1.53.$$

The standard deviation of set B is

$$\sigma = \sqrt{\frac{2(-3)^2 + 2(-2)^2 + 2(-1)^2 + 2(0)^2 + 2(1)^2 + 2(2)^2 + 2(3)^2}{14}}$$

$$= 2.$$

The standard deviation of set C is

$$\sigma = \sqrt{\frac{5(-3)^2 + 4(-2)^2 + 3(-1)^2 + 2(0)^2 + 3(1)^2 + 4(2)^2 + 5(3)^2}{26}}$$

$$\approx 2.22.$$

These values confirm the results of Example 3. That is, set A has the smallest standard deviation and set C has the largest.

The following alternative formula provides a more efficient way to compute the standard deviation.

Alternative Formula for Standard Deviation

The standard deviation of $\{x_1, x_2, \ldots, x_n\}$ is

$$\sigma = \sqrt{\frac{x_1^2 + x_2^2 + \cdots + x_n^2}{n} - \bar{x}^2}.$$

Because of messy computations, this formula is difficult to verify. Conceptually, however, the process is straightforward. It consists of showing that the expressions

$$\sqrt{\frac{(x_1 - \bar{x})^2 + (x_2 - \bar{x})^2 + \cdots + (x_n - \bar{x})^2}{n}}$$

and

$$\sqrt{\frac{x_1^2 + x_2^2 + \cdots + x_n^2}{n} - \bar{x}^2}$$

are equivalent. Try verifying this equivalence for the set $\{x_1, x_2, x_3\}$ with $\bar{x} = (x_1 + x_2 + x_3)/3$.

AK	66	MT	62
AL	40	NC	42
AR	39	ND	47
AZ	51	NE	63
CA	62	NH	59
CO	69	NJ	77
CT	80	NM	45
DC	94	NV	49
DE	44	NY	73
FL	50	OH	55
GA	46	OK	47
HI	80	OR	70
IA	55	PA	61
ID	53	RI	56
IL	61	SC	41
IN	47	SD	49
KS	51	TN	53
KY	53	TX	47
LA	45	UT	66
MA	74	VA	54
MD	68	VT	57
ME	47	WA	68
MI	62	WI	65
MN	67	WV	43
MO	53	WY	52
MS	37		

Example 5 ▶ Using the Alternative Formula

Use the alternative formula for standard deviation to find the standard deviation of the following set of numbers.

5, 6, 6, 7, 7, 8, 8, 8, 9, 10

Solution

Begin by finding the mean of the set, which is 7.4. So, the standard deviation is

$$\sigma = \sqrt{\frac{5^2 + 2(6^2) + 2(7^2) + 3(8^2) + 9^2 + 10^2}{10} - (7.4)^2}$$

$$= \sqrt{\frac{568}{10} - 54.76}$$

$$= \sqrt{2.04}$$

$$\approx 1.43.$$

You can use the statistical features of a graphing utility to check this result.

A well-known theorem in statistics, called *Chebychev's Theorem*, states that at least

$$1 - \frac{1}{k^2}$$

of the numbers in a distribution must lie within k standard deviations of the mean. So, 75% of the numbers in a collection must lie within two standard deviations of the mean, and at least 88.9% of the numbers must lie within three standard deviations of the mean. For most distributions, these percentages are low. For instance, in all three distributions shown in Example 3, 100% of the numbers lie within two standard deviations of the mean.

Example 6 ▶ Describing a Distribution

The table at the left above shows the number of dentists (per 100,000 people) in each state and the District of Columbia. Find the mean and standard deviation of the numbers. What percent of the numbers lie within two standard deviations of the mean? (Source: American Dental Association)

Solution

Begin by entering the numbers into a graphing utility that has a standard deviation program. After running the program, you should obtain

$$\bar{x} \approx 56.76 \qquad \text{and} \qquad \sigma = 12.14.$$

The interval that contains all numbers that lie within two standard deviations of the mean is

$$[56.76 - 2(12.14), 56.76 + 2(12.14)] \qquad \text{or} \qquad [32.48, 81.04].$$

From the histogram in Figure B.5, you can see that all but one of the numbers (98%) lie in this interval—all but the number that corresponds to the number of dentists (per 100,000 people) in Washington, DC.

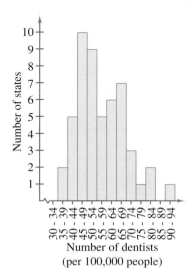

FIGURE B.5

Box-and-Whisker Plots

Standard deviation is the measure of dispersion that is associated with the mean. **Quartiles** measure dispersion associated with the median.

> ## Definition of Quartiles
>
> Consider an ordered set of numbers whose median is m. The **lower quartile** is the median of the numbers that occur before m. The **upper quartile** is the median of the numbers that occur after m.

Example 7 ▶ Finding Quartiles of a Set

Find the lower and upper quartiles for the following set.

34, 14, 24, 16, 12, 18, 20, 24, 16, 26, 13, 27

Solution

Begin by ordering the set.

12, 13, 14, 16, 16, 18, 20, 24, 24, 26, 27, 34

1st 25%   2nd 25%   3rd 25%   4th 25%

The median of the entire set is 19. The median of the six numbers that are less than 19 is 15. So, the lower quartile is 15. The median of the six numbers that are greater than 19 is 25. So, the upper quartile is 25.

Quartiles are represented graphically by a **box-and-whisker plot,** as shown in Figure B.6. In the plot, notice that five numbers are listed: the smallest number, the lower quartile, the median, the upper quartile, and the largest number. Also notice that the numbers are spaced proportionally, as though they were on a real number line.

12   15   19   25   34

FIGURE B.6

The next example shows how to find quartiles when the number of elements in a set is not divisible by 4.

EXAMPLE 8 ▶ **Sketching Box-and-Whisker Plots**

Sketch a box-and-whisker plot for each of the given sets.

a. 27, 28, 30, 42, 45, 50, 50, 61, 62, 64, 66
b. 82, 82, 83, 85, 87, 89, 90, 94, 95, 95, 96, 98, 99
c. 11, 13, 13, 15, 17, 18, 20, 24, 24, 27

Solution

a. This set has 11 numbers. The median is 50 (the sixth number). The lower quartile is 30 (the median of the first five numbers). The upper quartile is 62 (the median of the last five numbers).

b. This set has 13 numbers. The median is 90 (the seventh number). The lower quartile is 84 (the median of the first six numbers). The upper quartile is 95.5 (the median of the last six numbers).

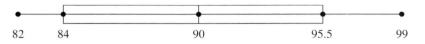

c. This set has 10 numbers. The median is 17.5 (the average of the fifth and sixth numbers). The lower quartile is 13 (the median of the first five numbers). The upper quartile is 24 (the median of the last five numbers).

B.2 Exercises

In Exercises 1–6, find the mean, median, and mode of the set of measurements.

1. 5, 12, 7, 14, 8, 9, 7

2. 30, 37, 32, 39, 33, 34, 32

3. 5, 12, 7, 24, 8, 9, 7

4. 20, 37, 32, 39, 33, 34, 32

5. 5, 12, 7, 14, 9, 7

6. 30, 37, 32, 39, 34, 32

7. *Reasoning* Compare your answers for Exercises 1 and 3 with those for Exercises 2 and 4. Which of the measures of central tendency is sensitive to extreme measurements? Explain your reasoning.

8. *Reasoning*

(a) Add 6 to each measurement in Exercise 1 and calculate the mean, median, and mode of the revised measurements. How are the measures of central tendency changed?

(b) If a constant k is added to each measurement in a set of data, how will the measures of central tendency change?

9. *Electric Bills* A person had the following monthly bills for electricity. What are the mean and median of the collection of bills?

January	$67.92	February	$59.84
March	$52.00	April	$52.50
May	$57.99	June	$65.35
July	$81.76	August	$74.98
September	$87.82	October	$83.18
November	$65.35	December	$57.00

10. *Car Rental* A car rental company kept the following record of the numbers of miles a rental car was driven. What are the mean, median, and mode of this data?

Monday	410	Tuesday	260
Wednesday	320	Thursday	320
Friday	460	Saturday	150

11. *Six-Child Families* A study was done on families having six children. The table gives the numbers of families in the study with the indicated numbers of girls. Determine the mean, median, and mode of this set of data.

Number of girls	0	1	2	3	4	5	6
Frequency	1	24	45	54	50	19	7

12. *Baseball* A baseball fan examined the records of a favorite baseball player's performance during his last 50 games. The numbers of games in which the player had 0, 1, 2, 3, and 4 hits are recorded in the table.

Number of hits	0	1	2	3	4
Frequency	14	26	7	2	1

(a) Determine the average number of hits per game.

(b) Determine the player's batting average if he had 200 at-bats during the 50-game series.

13. *Think About It* Construct a collection of numbers that has the following properties. If this is not possible, explain why it is not.

Mean = 6, median = 4, mode = 4

14. *Think About It* Construct a collection of numbers that has the following properties. If this is not possible, explain why it is not.

Mean = 6, median = 6, mode = 4

15. *Test Scores* A professor records the following scores for a 100-point exam.

99, 64, 80, 77, 59, 72, 87, 79, 92, 88,
90, 42, 20, 89, 42, 100, 98, 84, 78, 91

Which measure of central tendency best describes these test scores?

16. *Shoe Sales* A salesman sold eight pairs of a certain style of men's shoes. The sizes of the eight pairs were as follows: $10\frac{1}{2}$, 8, 12, $10\frac{1}{2}$, 10, $9\frac{1}{2}$, 11, and $10\frac{1}{2}$. Which measure (or measures) of central tendency best describes the typical shoe size for this data?

In Exercises 17–24, find the mean ($\bar{x}$), variance (v), and standard deviation (σ) of the numbers.

17. 4, 10, 8, 2

18. 3, 15, 6, 9, 2

19. 0, 1, 1, 2, 2, 2, 3, 3, 4

20. 2, 2, 2, 2, 2, 2

21. 1, 2, 3, 4, 5, 6, 7

22. 1, 1, 1, 5, 5, 5

23. 49, 62, 40, 29, 32, 70

24. 1.5, 0.4, 2.1, 0.7, 0.8

In Exercises 25–30, use the alternative formula to find the standard deviation of the numbers.

25. 2, 4, 6, 6, 13, 5

26. 10, 25, 50, 26, 15, 33, 29, 4

27. 246, 336, 473, 167, 219, 359

28. 6.0, 9.1, 4.4, 8.7, 10.4

29. 8.1, 6.9, 3.7, 4.2, 6.1

30. 9.0, 7.5, 3.3, 7.4, 6.0

In Exercises 31 and 32, line plots of sets of data are given. Determine the mean and standard deviation of each set.

31. (a)

(b)

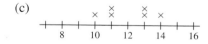

(c)

(d)

32. (a)

(b)

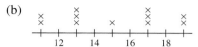

(c)

(d)

33. *Reasoning* Without calculating the standard deviation, explain why the set {4, 4, 20, 20} has a standard deviation of 8.

34. *Reasoning* If the standard deviation of a set of numbers is 0, what does this imply about the set?

35. *Test Scores* An instructor adds five points to each student's exam score. Will this change the mean or standard deviation of the exam scores? Explain.

36. *Price of Gold* The following data represents the average prices of gold (in dollars per fine ounce) for the years 1978 to 1997. Use a computer or calculator to find the mean, variance, and standard deviation of the data. What percent of the data lies within two standard deviations of the mean? (Source: U.S. Bureau of Mines)

194,	308,	613,	460,	376,
424,	361,	318,	368,	448,
438,	383,	385,	363,	345,
361,	385,	386,	389,	333

37. *Think About It* The histograms represent the test scores of two classes of a college course in mathematics. Which histogram has the smaller standard deviation?

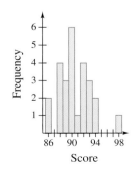

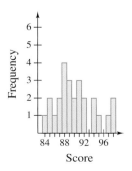

In Exercises 38–41, sketch a box-and-whisker plot for the data without the aid of a graphing utility.

38. 23, 15, 14, 23, 13, 14, 13, 20, 12

39. 11, 10, 11, 14, 17, 16, 14, 11, 8, 14, 20

40. 46, 48, 48, 50, 52, 47, 51, 47, 49, 53

41. 25, 20, 22, 28, 24, 28, 25, 19, 27, 29, 28, 21

In Exercises 42–45, use a graphing utility to create a box-and-whisker plot for the data.

42. 19, 12, 14, 9, 14, 15, 17, 13, 19, 11, 10, 19

43. 9, 5, 5, 5, 6, 5, 4, 12, 7, 10, 7, 11, 8, 9, 9

44. 20.1, 43.4, 34.9, 23.9, 33.5, 24.1, 22.5, 42.4, 25.7, 17.4, 23.8, 33.3, 17.3, 36.4, 21.8

45. 78.4, 76.3, 107.5, 78.5, 93.2, 90.3, 77.8, 37.1, 97.1, 75.5, 58.8, 65.6

46. *Product Lifetime* A company has redesigned a product in an attempt to increase the lifetime of the product. The two sets of data list the lifetimes (in months) of 20 units with the original design and 20 units with the new design. Create a box-and-whisker plot for each set of data, and then comment on the differences between the plots.

Original Design

15.1	78.3	56.3	68.9	30.6
27.2	12.5	42.7	72.7	20.2
53.0	13.5	11.0	18.4	85.2
10.8	38.3	85.1	10.0	12.6

New Design

55.8	71.5	25.6	19.0	23.1
37.2	60.0	35.3	18.9	80.5
46.7	31.1	67.9	23.5	99.5
54.0	23.2	45.5	24.8	87.8

B.3 Least Squares Regression

▶ **What you should learn**

- How to use the sum of squared differences to determine a least squares regression line
- How to find a least squares regression line for a set of data
- How to find a least squares regression parabola for a set of data

▶ **Why you should learn it**

The method of least squares provides a way of creating mathematical models for a set of data, which can then be analyzed.

In many of the examples and exercises in this text, you have been asked to use the regression capabilities of a graphing utility to find mathematical models for sets of data. Another way to find a mathematical model for a set of data is to use the **method of least squares.** As a measure of how well a model fits a set of data points

$$\{(x_1, y_1), (x_2, y_2), (x_3, y_3), \ldots, (x_n, y_n)\}$$

you can add the squares of the differences between the actual y-values and the values given by the model to obtain the **sum of the squared differences.** For instance, the table shows the heights x (in feet) and the diameters y (in inches) of eight trees. The table also shows the values of a linear model $y^* = 0.69x - 40$ for each x-value. The sum of squared differences for the model is 53.3.

x	70	72	75	76	85	78	77	80
y	8.3	10.5	11.0	11.4	12.9	14.0	16.3	18.0
y^*	8.3	9.68	11.75	12.44	18.65	13.82	13.13	15.2

The model that has the least sum of squared differences is the **least squares regression line** for the data. The least squares regression line for the data in the table is $y \approx 0.43x - 20.3$. The sum of squared differences is 43.32.

To find the least squares regression line $y = ax + b$ for the points $\{(x_1, y_1), (x_2, y_2), (x_3, y_3), \ldots, (x_n, y_n)\}$ algebraically, you need to solve the following system for a and b.

$$\begin{cases} nb + \left(\sum_{i=1}^{n} x_i\right)a = \sum_{i=1}^{n} y_i \\ \left(\sum_{i=1}^{n} x_i\right)b + \left(\sum_{i=1}^{n} x_i^2\right)a = \sum_{i=1}^{n} x_i y_i \end{cases}$$

In the system,

$$\sum_{i=1}^{n} x_i = x_1 + x_2 + \cdots + x_n$$

$$\sum_{i=1}^{n} y_i = y_1 + y_2 + \cdots + y_n$$

$$\sum_{i=1}^{n} x_i^2 = x_1^2 + x_2^2 + \cdots + x_n^2$$

$$\sum_{i=1}^{n} x_i y_i = x_1 y_1 + x_2 y_2 + \cdots + x_n y_n.$$

Example 1 ▶ **Finding a Least Squares Regression Line**

Find the least squares regression line for the points $(-3, 0)$, $(-1, 1)$, $(0, 2)$, and $(2, 3)$.

Solution

Begin by constructing a table like that shown below.

x	y	xy	x^2
-3	0	0	9
-1	1	-1	1
0	2	0	0
2	3	6	4
$\displaystyle\sum_{i=1}^{n} x_i = -2$	$\displaystyle\sum_{i=1}^{n} y_i = 6$	$\displaystyle\sum_{i=1}^{n} x_i y_i = 5$	$\displaystyle\sum_{i=1}^{n} x_i^2 = 14$

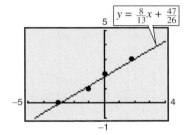

$$y = \tfrac{8}{13}x + \tfrac{47}{26}$$

FIGURE **B.7**

Applying the system for the least squares regression line with $n = 4$ produces

$$\begin{cases} nb + \left(\displaystyle\sum_{i=1}^{n} x_i\right)a = \displaystyle\sum_{i=1}^{n} y_i \\ \left(\displaystyle\sum_{i=1}^{n} x_i\right)b + \left(\displaystyle\sum_{i=1}^{n} x_i^2\right)a = \displaystyle\sum_{i=1}^{n} x_i y_i \end{cases} \implies \begin{cases} 4b - 2a = 6 \\ -2b + 14a = 5 \end{cases}.$$

Solving this system of equations produces $a = \frac{8}{13}$ and $b = \frac{47}{26}$. So, the least squares regression line is $y = \frac{8}{13}x + \frac{47}{26}$, as shown in Figure B.7.

The least squares regression parabola $y = ax^2 + bx + c$ for the points

$$\{(x_1, y_1), (x_2, y_2), (x_3, y_3), \ldots, (x_n, y_n)\}$$

is obtained in a similar manner by solving the following system of three equations in three unknowns for a, b, and c.

$$\begin{cases} nc + \left(\displaystyle\sum_{i=1}^{n} x_i\right)b + \left(\displaystyle\sum_{i=1}^{n} x_i^2\right)a = \displaystyle\sum_{i=1}^{n} y_i \\ \left(\displaystyle\sum_{i=1}^{n} x_i\right)c + \left(\displaystyle\sum_{i=1}^{n} x_i^2\right)b + \left(\displaystyle\sum_{i=1}^{n} x_i^3\right)a = \displaystyle\sum_{i=1}^{n} x_i y_i \\ \left(\displaystyle\sum_{i=1}^{n} x_i^2\right)c + \left(\displaystyle\sum_{i=1}^{n} x_i^3\right)b + \left(\displaystyle\sum_{i=1}^{n} x_i^4\right)a = \displaystyle\sum_{i=1}^{n} x_i^2 y_i \end{cases}$$

Fortunately, graphing utilities have built-in least squares regression capabilities.

Looking for solutions to your math problems? These study aids offer more than just the answers!

Two technology options offer complete solutions to every odd-numbered problem, help you practice and assess your skills, and provide problem-solving tools.

Interactive College Algebra 2.0 CD-ROM

- Enhance your understanding with step-by-step solutions to all odd-numbered exercises in the text.
- Assess your skill levels with diagnostic pre- and post-tests for each chapter.
- Strengthen your skills with tutorial exercises accompanied by examples and diagnostics.
- Visualize and graph with a built-in Meridian Graphing Calculator Emulator.
- Enjoy learning with additional interactive features.

Internet College Algebra 1.0

A subscription to this web site offers all of the CD-ROM features listed above, plus:
- Chat rooms for peer support
- Bulletin boards for sharing ideas and insights on each chapter

To purchase Interactive College Algebra 2.0 CD-ROM:

- Visit Houghton Mifflin's College Store at **college.hmco.com** or contact your campus bookstore.

To subscribe online to Internet College Algebra 1.0:

- Visit Houghton Mifflin's College Division web site at **college.hmco.com** and select mathematics.

Use these print supplements for added practice and convenient support.

Study and Solutions Guide to accompany
College Algebra, 5th Edition

• Work through step-by-step solutions for all odd-numbered exercises in the text.
• Test your skills by taking practice tests with accompanying solutions.
• Find useful study strategies designed to help you succeed.

Student Success Organizer
Ask your instructor about this new study aid.

• Use its practical format to guide you step-by-step through difficult concepts.
• Enhance your organizational skills for approaching problems and assignments.

To purchase these print supplements:

• Visit Houghton Mifflin's College Store at **college.hmco.com**
  or contact your campus bookstore.

Larson • Hostetler

Answers to Odd-Numbered Exercises and Tests

Chapter P

Section P.1   *(page 9)*

1. (a) $5, 1, 2$ (b) $-9, 5, 0, 1, -4, 2, -11$

(c) $-9, -\frac{7}{2}, 5, \frac{2}{3}, 0, 1, -4, 2, -11$ (d) $\sqrt{2}$

3. (a) 1 (b) $-13, 1, -6$

(c) $2.01, 0.666\ldots, -13, 1, -6$ (d) $0.010110111\ldots$

5. (a) $\frac{6}{3}, 8$ (b) $\frac{6}{3}, -1, 8, -22$

(c) $-\frac{1}{3}, \frac{6}{3}, -7.5, -1, 8, -22$ (d) $-\pi, \frac{1}{2}\sqrt{2}$

7. 0.625 **9.** $0.\overline{123}$ **11.** $\frac{41}{10}$

13. $\frac{92}{9}$ **15.** $-\frac{1811}{900}$ **17.** $-1 < 2.5$

19. $-4 > -8$ **21.** $\frac{3}{2} < 7$

23. $\frac{5}{6} > \frac{2}{3}$

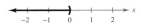

25. $x \le 5$ denotes the set of all real numbers less than or equal to 5. Unbounded

27. $x < 0$ denotes the set of all negative real numbers. Unbounded

29. $x \ge 4$ denotes the set of all real numbers greater than or equal to 4. Unbounded

31. $-2 < x < 2$ denotes the set of all real numbers greater than -2 and less than 2. Bounded

33. $-1 \le x < 0$ denotes the set of all negative real numbers greater than or equal to -1. Bounded

35. $\frac{127}{90}, \frac{584}{413}, \frac{7071}{5000}, \sqrt{2}, \frac{47}{33}$ **37.** $-2 < x \le 4$ **39.** $y \ge 0$

41. $10 \le t \le 22$ **43.** $W > 65$

45. This interval consists of all real numbers greater than or equal to 0 and less than 8.

47. This interval consists of all real numbers greater than -6.

49. 10 **51.** If $x \le 3, 3 - x$. If $x > 3, x - 3$. **53.** -1

55. -1 **57.** -1 **59.** $|-3| > -|-3|$

61. $-5 = -|5|$ **63.** $-|-2| = -|2|$ **65.** 4

67. 51 **69.** $\frac{5}{2}$ **71.** $\frac{128}{75}$ **73.** (a) Negative (b) Negative

75. $|7 - 18| = 11$ miles **77.** $|60 - 23| = 37°$

79. $|x - 5| \le 3$ **81.** $|y| \ge 6$

83. $|\$113{,}356 - \$112{,}700| = \$656 > \$500$

$0.05(\$112{,}700) = \$5635$

Because the actual expenses differ from the budget by more than \$500, there is failure to meet the "budget variance test."

85. $|\$37{,}335 - \$37{,}640| = \$305 < \$500$

$0.05(\$37{,}640) = \$1882$

Because the difference between the actual expenses and the budget is less than \$500 and less than 5% of the budgeted amount, there is compliance with the "budget variance test."

87. $|92.5 - 92.2| = 0.3$

There was a surplus of \$0.3 billion.

89. $|1032.0 - 1253.2| = 221.2$

There was a deficit of \$221.2 billion.

91. $7x$ and 4 are the terms; 7 is the coefficient.

93. $\sqrt{3}x^2, -8x,$ and -11 are the terms; $\sqrt{3}$ and -8 are the coefficients.

95. $4x^3, x/2,$ and -5 are the terms; 4 and $\frac{1}{2}$ are the coefficients.

97. (a) -10 (b) -6 **99.** (a) 14 (b) 2

101. (a) Division by 0 is undefined. (b) 0

103. Commutative Property of Addition

105. Multiplicative Inverse Property

107. Distributive Property

109. Multiplicative Identity Property

111. Associative and Commutative Properties of Multiplication

113. $\frac{1}{2}$ **115.** $\frac{3}{8}$ **117.** 48 **119.** $\frac{5x}{12}$

121. -2.57 **123.** 1.56

125. (a)

n	1	0.5	0.01	0.0001	0.000001
$5/n$	5	10	500	50,000	5,000,000

(b) The value of $5/n$ approaches infinity as n approaches 0.

127. (a) No. If one variable is negative while the other is positive, the expressions are unequal.

(b) $|u + v| \le |u| + |v|$

The expressions are equal when u and v have the same sign. If u and v differ in sign, $|u + v|$ is less than $|u| + |v|$.

129. The only even prime number is 2, because its only factors are itself and 1.

131. False. The denominators cannot be added when adding fractions.

133. Yes. $|a| = -a$ if $a < 0$.

Section P.2   *(page 21)*

1. $(8 \times 8 \times 8 \times 8 \times 8)$

3. $-(0.4 \times 0.4 \times 0.4 \times 0.4 \times 0.4 \times 0.4)$ **5.** 4.9^6

7. $(-10)^5$ **9.** (a) 27   (b) 81

11. (a) 729   (b) -9 **13.** (a) $\frac{243}{64}$ (b) -1

15. (a) $\frac{5}{6}$ (b) 4   **17.** -1600 **19.** 2.125

21. -24 **23.** 6   **25.** -54 **27.** 1

29. (a) $-125z^3$ (b) $5x^6$ **31.** (a) $24y^{10}$ (b) $3x^2$

33. (a) $\frac{7}{x}$ (b) $\frac{4}{3}(x + y)^2$ **35.** (a) 1   (b) $\frac{1}{4x^4}$

37. (a) $-2x^3$ (b) $\frac{10}{x}$ **39.** (a) $\frac{a^6}{64b^9}$ (b) $\frac{1}{625x^8y^8}$

41. (a) 3^{3n} (b) $\frac{b^5}{a^5}$ **43.** $9^{1/2} = 3$ **45.** $\sqrt[5]{32} = 2$

47. $\sqrt{196} = 14$ **49.** $(-216)^{1/3} = -6$ **51.** $\sqrt[3]{27^2} = 9$

53. $81^{3/4} = 27$

55. (a) 3   (b) 2   **57.** (a) 3   (b) $\frac{1}{2}$

59. (a) -125 (b) 3   **61.** (a) $\frac{1}{8}$ (b) $\frac{27}{8}$

63. (a) -4 (b) 2   **65.** (a) 7.550   (b) -7.225

67. (a) 14.499   (b) 0.528

69. (a) -0.011 (b) 0.005

71. (a) $2\sqrt{2}$ (b) $2\sqrt[3]{3}$ **73.** (a) $6x\sqrt{2x}$ (b) $\frac{18}{z\sqrt{z}}$

75. (a) $2x\sqrt[3]{2x^2}$ (b) $\frac{5|x|\sqrt{3}}{y^2}$ **77.** 625   **79.** $\frac{2}{x}$

81. $\frac{1}{x^3}$, $x > 0$ **83.** (a) $\frac{\sqrt{3}}{3}$ (b) $4\sqrt[3]{4}$

85. (a) $\frac{x(5 + \sqrt{3})}{11}$ (b) $3(\sqrt{6} - \sqrt{5})$

87. (a) $\frac{2}{\sqrt{2}}$ (b) $\frac{3}{\sqrt[3]{75}}$

89. (a) $\frac{2}{3(\sqrt{5} - \sqrt{3})}$ (b) $-\frac{1}{2(\sqrt{7} + 3)}$

91. (a) $\sqrt{3}$ (b) $\sqrt[3]{(x + 1)^2}$

93. (a) $2\sqrt[4]{2}$ (b) $\sqrt[8]{2x}$

95. (a) $34\sqrt{2}$ (b) $22\sqrt{2}$ **97.** (a) $2\sqrt{x}$ (b) $4\sqrt{y}$

99. (a) $13\sqrt{x + 1}$ (b) $18\sqrt{5x}$

101. $\sqrt{5} + \sqrt{3} > \sqrt{5 + 3}$

103. $5 > \sqrt{3^2 + 2^2}$ **105.** 5.73×10^7 **107.** 8.99×10^{-5}

109. 604,800,000   **111.** 0.0000000000000000001602

113. (a) 50,000   (b) 200,000

115. (a) 954.448   (b) 3.077×10^{10}

117. (a) 67,082.039   (b) 39.791

119. When any positive integer is squared, the units digit is 0, 1, 4, 5, 6, or 9. Therefore, $\sqrt{5233}$ is not an integer.

121. $\frac{\pi}{2} \approx 1.57$ seconds   **123.** 13.29 seconds   **125.** 0.280

127. True. When dividing variables, you subtract exponents.

129. $a^0 = 1$, $a \ne 0$, using the property $\frac{a^m}{a^n} = a^{m-n}$:
$\frac{a^m}{a^m} = a^{m-m} = a^0 = 1$.

131. No. A number written in scientific notation has the form $c \times 10^n$, where $1 \le c < 10$ and n is an integer. In true scientific notation, the number 52.7×10^5 is 5.27×10^6.

Section P.3   *(page 30)*

1. d   **3.** b   **5.** f

7. $-2x^3 + 4x^2 - 3x + 20$ (Answers will vary.)

9. $-15x^4 + 1$ (Answers will vary.)

11. Degree: 2; Leading coefficient: 2

13. Degree: 5; Leading coefficient: 1

15. Degree: 5; Leading coefficient: -4

17. Degree: 5; Leading coefficient: 1

19. Polynomial: $-3x^3 + 2x + 8$

21. Not a polynomial because of the operation of division

23. Polynomial: $-y^4 + y^3 + y^2$ **25.** $-2x - 10$

27. $3x^3 - 2x + 2$ **29.** $8.3x^3 + 29.7x^2 + 11$

31. $12z + 8$ **33.** $3x^3 - 6x^2 + 3x$ **35.** $-15z^2 + 5z$

37. $-4x^4 + 4x$ **39.** $7.5x^3 + 9x$ **41.** $-\frac{1}{2}x^2 - 12x$

43. $4x^3 - 2x^2 + 4$ **45.** $5x^2 - 4x + 11$

47. $-30x^3 + 57x^2 + 25x - 12$

49. $x^4 - x^3 + 5x^2 - 9x - 36$ **51.** $x^4 + x^2 + 1$

53. $x^2 + 7x + 12$ **55.** $6x^2 - 7x - 5$

57. $4x^2 + 12x + 9$ **59.** $4x^2 - 20xy + 25y^2$

61. $x^2 - 100$ **63.** $x^2 - 4y^2$ **65.** $m^2 - n^2 - 6m + 9$

67. $x^2 + 2xy + y^2 - 6x - 6y + 9$ **69.** $4r^4 - 25$

71. $x^3 + 3x^2 + 3x + 1$ **73.** $8x^3 - 12x^2y + 6xy^2 - y^3$

75. $16x^6 - 24x^3 + 9$ **77.** $\frac{1}{4}x^2 - 3x + 9$

79. $\frac{1}{9}x^2 - 4$ **81.** $1.44x^2 + 7.2x + 9$

83. $2.25x^2 - 16$ **85.** $2x^2 + 2x$ **87.** $u^4 - 16$

89. $x - y$ **91.** $x^2 - 2\sqrt{5}x + 5$

93. (a) $x^2 - 1$ (b) $x^3 - 1$ (c) $x^4 - 1$

The result is $x^5 - 1$.

95. $(3 + 4)^2 = 49 \neq 25 = 3^2 + 4^2$.

If either x or y is zero, then $(x + y)^2 = x^2 + y^2$.

97. $P = \$548$

99. (a) $1200r^3 + 3600r^2 + 3600r + 1200$

(b)

r	2%	3%	$3\frac{1}{2}$%
$1200(1 + r)^3$	\$1273.45	\$1311.27	\$1330.46

r	4%	$4\frac{1}{2}$%
$1200(1 + r)^3$	\$1349.84	\$1369.40

(c) The amount increases with increasing r.

101. $V = \frac{1}{2}(6x^3 - 135x^2 + 675x)$

x (cm)	3	5	7
V (cm³)	486	375	84

103. (a) $6x^2 - 3x$ (b) $42x^2$ **105.** $2x^2 + 46x + 252$

107. (a) Estimates will vary.

(b) The difference decreases in magnitude.

109. $(x + a)(x + a) = x(x + a) + a(x + a)$

Distributive Property

111. False. $(4x + 3) + (-4x + 6)$ **113.** n

$= 4x + 3 - 4x + 6$

$= 3 + 6$

$= 9$

115. (a) Yes.

$(6x^3 + 3x + 1) + (5x^4 + 2x^2 + x + 2) =$

$5x^4 + 6x^3 + 2x^2 + 4x + 3$

(Examples will vary.)

(b) No. When third- and fourth-degree polynomials are added, the fourth-degree term of the polynomial will be in the sum.

(c) No. The sum will be of fourth degree. The *product* of the two polynomials would be of seventh degree.

Section P.4   *(page 39)*

1. 30   **3.** $6x^2y$ **5.** $3(x + 2)$ **7.** $2x(x^2 - 3)$

9. $(x - 1)(x + 6)$ **11.** $(x + 3)(x - 1)$ **13.** $\frac{1}{2}(x + 8)$

15. $\frac{1}{2}x(x^2 + 4x - 10)$ **17.** $\frac{2}{3}(x - 6)(x - 3)$

19. $(x + 6)(x - 6)$ **21.** $(4y + 3)(4y - 3)$

23. $\left(4x + \frac{1}{3}\right)\left(4x - \frac{1}{3}\right)$ **25.** $(x + 1)(x - 3)$

27. $(3u + 2v)(3u - 2v)$ **29.** $(x - 2)^2$ **31.** $(2t + 1)^2$

33. $(5y - 1)^2$ **35.** $(3u + 4v)^2$ **37.** $\left(x - \frac{2}{3}\right)^2$

39. $-5(x^2 - 5)$ **41.** $-2(t^3 - 2t - 3)$

43. $(x + 2)(x - 1)$ **45.** $(s - 3)(s - 2)$

47. $-(y + 5)(y - 4)$ **49.** $(x - 20)(x - 10)$

51. $(3x - 2)(x - 1)$ **53.** $(5x + 1)(x + 5)$

55. $-(3z - 2)(3z + 1)$ **57.** $(x - 2)(x^2 + 2x + 4)$

59. $(y + 4)(y^2 - 4y + 16)$ **61.** $(2t - 1)(4t^2 + 2t + 1)$

63. $(u + 3v)(u^2 - 3uv + 9v^2)$ **65.** $(x - 1)(x^2 + 2)$

67. $(2x - 1)(x^2 - 3)$ **69.** $(3 + x)(2 - x^3)$

71. $(3x^2 - 1)(2x + 1)$ **73.** $(x + 2)(3x + 4)$

75. $(2x - 1)(3x + 2)$ **77.** $(3x - 1)(5x - 2)$

79. $6(x + 3)(x - 3)$ **81.** $x^2(x - 4)$ **83.** $(x - 1)^2$

85. $(1 - 2x)^2$ **87.** $-2x(x + 1)(x - 2)$

89. $(9x + 1)(x + 1)$ **91.** $\frac{1}{81}(x + 36)(x - 18)$

93. $(3x + 1)(x^2 + 5)$ **95.** $x(x - 4)(x^2 + 1)$

97. $\frac{1}{4}(x^2 + 3)(x + 12)$ **99.** $(t + 6)(t - 8)$

101. $(x + 2)(x + 4)(x - 2)(x - 4)$

103. $5(x + 2)(x^2 - 2x + 4)$ **105.** $(3 - 4x)(23 - 60x)$

107. $5(1 - x)^2(3x + 2)(4x + 3)$

109. $(x - 2)^2(x + 1)^3(7x - 5)$

111. $3(x^6 + 1)^4(3x + 2)^2(33x^6 + 20x^5 + 3)$

113. $3w^2[15w(9w + 1)^4 + (2w + 1)^5]$ **115.** c   **117.** d

119.

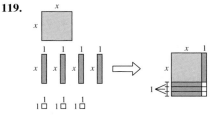

121.

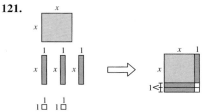

123. $r^2(4 - \pi)$ **125.** $\frac{5}{8}(x + 7)(x - 1)$

127. $-51,\ 51,\ -15,\ 15,\ -27,\ 27$

129. $-25,\ 25,\ -14,\ 14,\ -11,\ 11,\ -10,\ 10$

131. Two possible answers: 3, -8

133. Two possible answers: 9, 7

135. No. $3(x - 2)(x + 1)$ **137.** $kx(Q - x)$

139. $(x^n + y^n)(x^{2n} - x^n y^n + y^{2n})$

141. True. $(x^2 - 16) = (x + 4)(x - 4)$

143. A polynomial is in factored form when it is written as a product.

Section P.5   *(page 49)*

1. All real numbers   **3.** All nonnegative real numbers

5. All real numbers x such that $x \neq 2$

7. All real numbers x such that $x \geq -1$

9. $3x,\ x \neq 0$ **11.** $x - 2,\ x \neq 2$ **13.** $x,\ x \neq 0$

15. $\dfrac{3x}{2},\ x \neq 0$ **17.** $\dfrac{3y}{y + 1},\ x \neq 0$ **19.** $\dfrac{-4y}{5},\ y \neq \dfrac{1}{2}$

21. $-\dfrac{1}{2},\ x \neq 5$ **23.** $y - 4,\ y \neq -4$

25. $\dfrac{x(x + 3)}{x - 2},\ x \neq -2$ **27.** $\dfrac{y - 4}{y + 6},\ y \neq 3$

29. $\dfrac{-(x^2 + 1)}{(x + 2)},\ x \neq 2$ **31.** $z - 2$

33.

x	0	1	2	3	4	5	6
$\dfrac{x^2 - 2x - 3}{x - 3}$	1	2	3	Undef.	5	6	7
$x + 1$	1	2	3	4	5	6	7

The expressions are equivalent except at $x = 3$.

35. The expression cannot be simplified.

37. $\dfrac{\pi}{4},\ r \neq 0$ **39.** $\dfrac{1}{5(x - 2)},\ x \neq 1$

41. $\dfrac{x - 3}{(x + 2)^2},\ x \neq -5$ **43.** $\dfrac{r + 1}{r},\ r \neq 1$

45. $\dfrac{t - 3}{(t + 3)(t - 2)},\ t \neq -2$ **47.** $\dfrac{x - y}{x(x + y)^2},\ x \neq -2y$

49. $\dfrac{3}{2},\ x \neq -y$ **51.** $\dfrac{(x^2 + 16)(x + 2)(x - 2)}{(x + 4)^2}$

53. $\dfrac{(x + 6)(x + 1)}{x^2},\ x \neq 6$ **55.** $\dfrac{x + 5}{x - 1}$ **57.** $\dfrac{6x + 13}{x + 3}$

59. $-\dfrac{2}{x - 2}$ **61.** $\dfrac{x - 4}{(x + 2)(x - 2)(x - 1)}$

63. $-\dfrac{x^2 + 3}{(x + 1)(x - 2)(x - 3)}$ **65.** $\dfrac{2 - x}{x^2 + 1},\ x \neq 0$

67. $\dfrac{x^7 - 2}{x^2}$ **69.** $\dfrac{3x^2 - 2}{x^{1/2}}$ **71.** $\dfrac{-1}{(x^2 + 1)^5}$

73. $\dfrac{2x^3 - 2x^2 - 5}{(x - 1)^{1/2}}$ **75.** $\dfrac{-2(x - 6)}{x + 2}$ **77.** $\dfrac{1}{2},\ x \neq 2$

79. $x(x + 1),\ x \neq -1,\ 0$ **81.** $\dfrac{1}{x},\ x \neq -1$

83. $\dfrac{(x + 3)^3}{2x(x - 3)},\ x \neq -3$ **85.** $-\dfrac{2x + h}{x^2(x + h)^2},\ h \neq 0$

87. $\dfrac{2x - 1}{2x},\ x > 0$ **89.** $\dfrac{3x - 1}{3}$

91. $-\dfrac{1}{x^2(x + 1)^{3/4}}$ **93.** $\dfrac{1}{\sqrt{x + 2} + \sqrt{x}}$

95. (a) $\dfrac{1}{16}$ minute   (b) $\dfrac{x}{16}$ minute(s)

     (c) $\dfrac{60}{16} = \dfrac{15}{4}$ minutes

97. $\dfrac{11x}{30}$ **99.** (a) 9.09%   (b) $\dfrac{288(MN - P)}{N(MN + 12P)}$, 9.09%

101. (a)

t	0	2	4	6	8	10
T	75	55.9	48.3	45	43.3	42.3

t	12	14	16	18	20	22
T	41.7	41.3	41.1	40.9	40.7	40.6

   (b) The model is approaching a T-value of 40.

103. $\dfrac{x}{3(x + 3)}$ **105.** $\dfrac{7}{6x(x + 5)}$

107. False. In order for the simplified expression to be equivalent to the original expression, the domain of the simplified expression needs to be restricted. If n is even, $x \neq -1,\ 1$. If n is odd, $x \neq 1$.

109. False. The least common denominator of several fractions consists of the product of all *prime factors* in the denominators, with each factor given the highest power of its occurrence in any denominator.

111. No. $\dfrac{ax - b}{b - ax} = \dfrac{a\cancel{x} \cancel{- b}}{-(a\cancel{x} \cancel{- b})} = -1,\ x \neq \dfrac{b}{a}$

Section P.6   *(page 58)*

1. Change all signs when distributing the minus sign.

     $2x - (3y + 4) = 2x - 3y - 4$

3. Change all signs when distributing the minus sign.

     $\dfrac{4}{16x - (2x + 1)} = \dfrac{4}{14x - 1}$

5. z occurs twice as a factor.

$$(5z)(6z) = 30z^2$$

7. The fraction as a whole is multiplied by a, not the numerator and denominator separately.

$$a\left(\frac{x}{y}\right) = \frac{ax}{y}$$

9. The exponent applies to the denominator also.

$$\left(\frac{x}{y}\right)^3 = \frac{x \cdot x \cdot x}{y \cdot y \cdot y} = \frac{x^3}{y^3}$$

11. $\sqrt{x + 9}$ cannot be simplified.

13. $\dfrac{6x + y}{6x - y}$ cannot be simplified.

15. The negative exponent is on a term of the denominator, not a factor.

$$\frac{1}{x + y^{-1}} = \frac{y}{xy + 1}, \quad y \neq 0$$

17. Exponents are applied before multiplying.

$$x(2x - 1)^2 = x(4x^2 - 4x + 1) = 4x^3 - 4x^2 + x$$

19. Radicals apply to every factor of the radicand.

$$\sqrt[3]{x^3 + 7x^2} = \sqrt[3]{x^2(x + 7)} = \left(\sqrt[3]{x^2}\right)\left(\sqrt[3]{x + 7}\right)$$

21. To add fractions, first find a common denominator.

$$\frac{3}{x} + \frac{4}{y} = \frac{3y + 4x}{xy}$$

23. $3x + 2$ **25.** $2x^2 + x + 15$ **27.** $\frac{1}{4}$ **29.** $-\frac{1}{4}$

31. $\dfrac{1}{2}$ **33.** $\dfrac{1}{2x^2}$ **35.** $\dfrac{25}{9}, \dfrac{49}{16}$ **37.** $1, 2$

39. $1 - 5x$ **41.** $1 - 7x$ **43.** $3x - 1$ **45.** $\dfrac{16}{x} - 5 - x$

47. $4x^{8/3} - 7x^{5/3} + \dfrac{1}{x^{1/3}}$ **49.** $\dfrac{3}{\sqrt{x}} - 5x^{3/2} - x^{7/2}$

51. $\dfrac{-7x^2 - 4x + 9}{(x^2 - 3)^3(x + 1)^4}$ **53.** $\dfrac{27x^2 - 24x + 2}{(6x + 1)^4}$

55. $\dfrac{-1}{(x + 3)^{2/3}(x + 2)^{7/4}}$ **57.** $\dfrac{4x - 3}{(3x - 1)^{4/3}}$ **59.** $\dfrac{x}{x^2 + 4}$

61. $\dfrac{(3x - 2)^{1/2}(15x^2 - 4x + 45)}{2(x^2 + 5)^{1/2}}$

63. (a) Answers will vary.

(b)

x	-2	-1	$-\frac{1}{2}$	0	1	2	$\frac{5}{2}$
y_1	-8.7	-2.9	-1.1	0	2.9	8.7	12.5
y_2	-8.7	-2.9	-1.1	0	2.9	8.7	12.5

65. $y_1(0) = 0$ and $y_2(0) = 2$ so, $y_1 \neq y_2$.

$$y_2 = \frac{2x - 3x^3}{\sqrt{1 - x^2}}. \text{ (Answers will vary.)}$$

67. False. Cannot move term-by-term from denominator to numerator.

69. False. $x^2 - 9$ does not factor into $\left(\sqrt{x} + 3\right)\left(\sqrt{x} - 3\right)$.

71. There is no error.   **73.** There is no error.

Section P.7   *(page 66)*

1.   **3.**

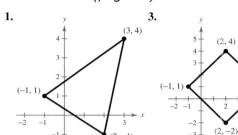

5. $A: (2, 6), \ B: (-6, -2), \ C: (4, -4), \ D: (-3, 2)$

7. $(-3, 4)$ **9.** $(-5, -5)$

11. Quadrant IV   **13.** Quadrant II

15. Quadrants III and IV   **17.** Quadrant III

19. Quadrants I and III   **21.** $(0, 1), \ (4, 2), \ (1, 4)$

23. $(-3, 6), (2, 10), (2, 4), (-3, 4)$

25.

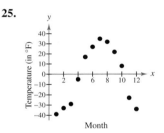

27. 1996 : $1.65 per one-half gallon   **29.** 166.67%

31. 1990s   **33.** 65   **35.** 8   **37.** 5

39. (a) $4, 3, 5$ (b) $4^2 + 3^2 = 5^2$

41. (a) $10, 3, \sqrt{109}$ (b) $10^2 + 3^2 = \left(\sqrt{109}\right)^2$

43. (a)

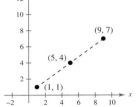

(b) 10

(c) $(5, 4)$

45. (a)

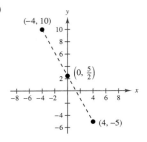

(b) 17

(c) $\left(0, \frac{5}{2}\right)$

47. (a)

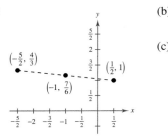

(b) $2\sqrt{10}$

(c) $(2, 3)$

49. (a)

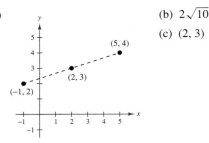

(b) $\dfrac{\sqrt{82}}{3}$

(c) $\left(-1, \frac{7}{6}\right)$

51. (a)

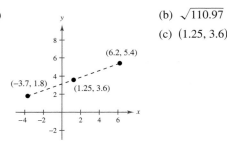

(b) $\sqrt{110.97}$

(c) $(1.25, 3.6)$

53. (a)

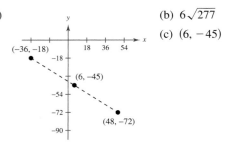

(b) $6\sqrt{277}$

(c) $(6, -45)$

55. $630,000

57. $\left(\sqrt{5}\right)^2 + \left(\sqrt{45}\right)^2 = \left(\sqrt{50}\right)^2$

59. Opposite sides have equal lengths of $2\sqrt{5}$ and $\sqrt{85}$.

61. $(2x_m - x_1,\ 2y_m - y_1)$

63. $\left(\dfrac{3x_1 + x_2}{4}, \dfrac{3y_1 + y_2}{4}\right), \left(\dfrac{x_1 + x_2}{2}, \dfrac{y_1 + y_2}{2}\right),$

$\left(\dfrac{x_1 + 3x_2}{4}, \dfrac{y_1 + 3y_2}{4}\right)$

65. $5\sqrt{74} \approx 43$ yards

67.

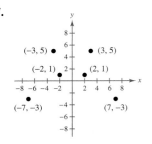

(a) The point is reflected through the y-axis.

(b) The point is reflected through the x-axis.

(c) The point is reflected through the origin.

69. $1002.6 million

71. False. The Midpoint Formula would be used 15 times.

73. Point on x-axis: $y = 0$; Point on y-axis: $x = 0$

75. Use the Midpoint Formula to prove that the diagonals of the parallelogram bisect each other.

$$\left(\frac{b + a}{2}, \frac{c + 0}{2}\right) = \left(\frac{a + b}{2}, \frac{c}{2}\right)$$

$$\left(\frac{a + b + 0}{2}, \frac{c + 0}{2}\right) = \left(\frac{a + b}{2}, \frac{c}{2}\right)$$

77. c   **79.** a

Review Exercises   *(page 71)*

1. (a) 11   (b) $11, -14$

(c) $11, -14, -\frac{8}{9}, \frac{5}{2}, 0.4$ (d) $\sqrt{6}$

3. (a) $0.8\overline{3}$ (b) 0.875

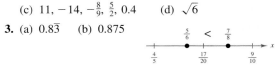

$\frac{5}{6} < \frac{7}{8}$

5. The set consists of all real numbers less than or equal to 7.

7. 155   **9.** $|x - 7| \geq 4$ **11.** $|y + 30| < 5$

13. (a) -7 (b) -19 **15.** (a) -1 (b) -3

17. Associative Property of Addition

19. Multiplicative Inverse Property

21. -11 **23.** $\frac{1}{12}$ **25.** -144

27. (a) $\dfrac{3u^5}{v^4}$ (b) m^{-2} **29.** 3.3674×10^{10}

31. 483,600,000   **33.** (a) 9 (b) 343

35. (a) $\sqrt[6]{7}$ (b) $\sqrt[4]{19}$ **37.** (a) $2x^2$ (b) $\dfrac{3|u|\sqrt{2b}}{b^2}$

39. (a) $2\sqrt{2}$ (b) $26\sqrt{2}$

41. Radicals cannot be combined unless the index and the radicand are the same.

43. $2 + \sqrt{3}$ **45.** 64   **47.** $6x^{9/10}$ **49.** $16^{1/2}$

51. $-11x^2 + 3$ **53.** $-12x^2 - 4$ **55.** $-3x^2 - 7x + 1$

57. $15x^2 - 27x - 6$ **59.** $4x^2 - 12x + 9$ **61.** 41

63. (a)

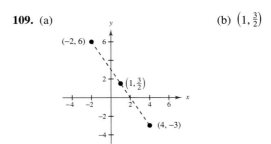

(b) $168\pi \approx 527.79$ in.^{2}

Explanations will vary.

65. $x(x + 1)(x - 1)$ **67.** $(5x + 7)(5x - 7)$

69. $(x - 4)(x^2 + 4x + 16)$ **71.** $(x + 10)(2x + 1)$

73. $(x - 1)(x^2 + 2)$ **75.** All real numbers except $x = -6$

77. $\dfrac{x - 8}{15}, x \neq -8$ **79.** $\dfrac{1}{x^2}, x \neq \pm 2$ **81.** $\dfrac{4x^2 - 16x + 3}{2(x - 4)}$

83. $\dfrac{3x}{(x - 1)(x^2 + x + 1)}$ **85.** $\dfrac{3ax^2}{(a^2 - x)(a - x)}$

87. The multiplication in parentheses comes first.
$10(4 \cdot 7) = 10(28) = 280$

89. The exponent applies to the coefficient also. $(2x)^4 = 16x^4$

91. Multiply exponents when raising a power to a power.
$(3^4)^4 = 3^{16}$

93. Add the numbers in parentheses before squaring.
$(5 + 8)^2 = 13^2 \neq 5^2 + 8^2$

95. $16x^2 - 9x + 20$ **97.** $-5x^2 + 2x + 15$

99. $\dfrac{2(x + 1)}{(x + 2)^{1/2}}$ **101.**

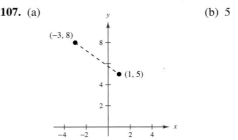

$\left(\sqrt{185}\right)^2 + \left(\sqrt{185}\right)^2 = \left(\sqrt{370}\right)^2$

103. Quadrant IV   **105.** Quadrant IV

107. (a) (b) 5

109. (a) (b) $\left(1, \frac{3}{2}\right)$

111. 80°F

113. False. There is also a cross-product term when a binomial sum is squared.
$(x + a)^2 = x^2 + 2ax + a^2$

Chapter Test   *(page 75)*

1. $-\frac{10}{3} > -|-4|$ **2.** 9.15

3. (a) -18 (b) $\frac{4}{27}$

4. (a) $-\frac{27}{125}$ (b) $\frac{8}{729}$

5. (a) 25 (b) 6

6. (a) 1.8×10^5 (b) 2.7×10^{13}

7. (a) $12z^8$ (b) $(u - 2)^{-7}$ (c) $\dfrac{3x^2}{y^2}$

8. (a) $15z\sqrt{2z}$ (b) $-10\sqrt{y}$ (c) $\dfrac{2\sqrt[3]{2v}}{v^2}$

9. $2x^2 - 3x - 5$

10. $x^2 - 5$ **11.** $8, x \neq 3$ **12.** $\dfrac{x - 1}{2x}, x \neq \pm 1$

13. (a) $x^2(2x + 1)(x - 2)$ (b) $(x - 2)(x + 2)^2$

14. (a) $4\sqrt[3]{4}$ (b) $-3\left(1 + \sqrt{3}\right)$

15. $\dfrac{1}{420}$ minute; $\dfrac{1}{4200}$ minute; $\dfrac{x}{4200}$ minute(s)

16. $545

17.

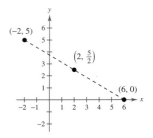

Midpoint: $\left(2, \frac{5}{2}\right)$; Distance: $\sqrt{89}$

18. $\frac{5}{6}\sqrt{3}\,x^2$

Chapter 1

Section 1.1   *(page 85)*

1. (a) Yes   (b) Yes   **3.** (a) No   (b) Yes

5.

x	-1	0	1	2	$\frac{5}{2}$
y	7	5	3	1	0

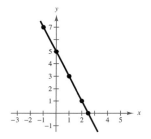

7.

x	-1	0	1	2	3
y	4	0	-2	-2	0

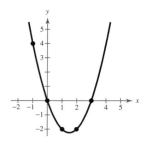

9. y-axis symmetry   **11.** Origin symmetry

13. Origin symmetry   **15.** x-axis symmetry

17.

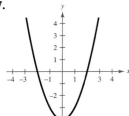

19.

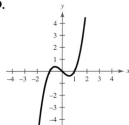

21. c   **23.** b

25. x-intercepts: $(\pm 2, 0)$
y-intercept: $(0, 16)$

27. x-intercepts: $(0, 0)$, $\left(\frac{5}{2}, 0\right)$
y-intercept: $(0, 0)$

29.

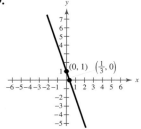

31.

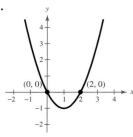

33.

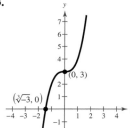

35.

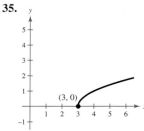

37.

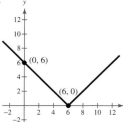

39.

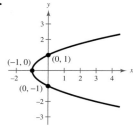

41.

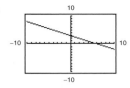

Intercepts: $(6, 0)$, $(0, 3)$

43.

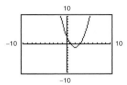

Intercepts: $(3, 0)$, $(1, 0)$, $(0, 3)$

45.

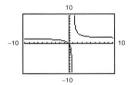

Intercept: $(0, 0)$

47.

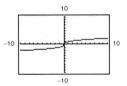

Intercept: $(0, 0)$

49.

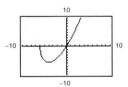

Intercepts: $(0, 0)$, $(-6, 0)$

51.

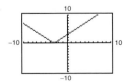

Intercepts: $(-3, 0)$, $(0, 3)$

53. $x^2 + y^2 = 16$

55. $(x - 2)^2 + (y + 1)^2 = 16$

57. $(x + 1)^2 + (y - 2)^2 = 5$

59. $(x - 3)^2 + (y - 4)^2 = 25$

61. Center: $(0, 0)$; Radius: 5

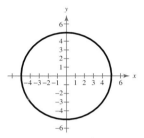

63. Center: $(1, -3)$; Radius: 3

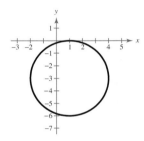

65. Center: $\left(\frac{1}{2}, \frac{1}{2}\right)$; Radius: $\frac{3}{2}$

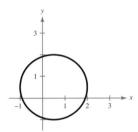

67.

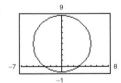

Circle

69.

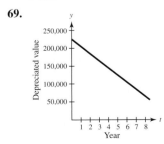

71. (a)

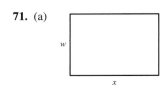

(b) Answers will vary.

(c)

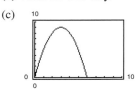

(d) $x = 3$, $w = 3$

73. (a) and (b)

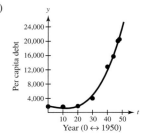

(b) The curve seems to be a good fit for the data.

(c) 2002: \$27,142; 2004: \$30,589

75. 3.9 ohms

77. True. All linear equations of the form $y = mx + b$, which excludes vertical lines, cross the y-axis one time.

79. (a) $a = 1$, $b = 0$ (b) $a = 0$, $b = 1$

81. False. $\dfrac{1}{3 \cdot 4^{-1}} = \dfrac{4}{3}$ **83.** $9x^5$, $4x^3$, -7

85. $2\sqrt{2x}$ **87.** $\dfrac{10\sqrt{7x}}{x}$ **89.** $\sqrt[3]{|t|}$

Section 1.2   *(page 94)*

1. (a) No   (b) No   (c) Yes   (d) No

3. (a) Yes   (b) Yes   (c) No   (d) No

5. (a) Yes   (b) No   (c) No   (d) No

7. (a) No   (b) No   (c) No   (d) No

9. (a) Yes   (b) No   (c) Yes   (d) No

11. Identity   **13.** Conditional equation   **15.** Identity

17. Conditional equation   **19.** Conditional equation

21. Original equation
Subtract 32 from both sides.
Simplify.
Divide both sides by 4.
Simplify.

23. 6   **25.** 7   **27.** 4   **29.** -9 **31.** 5

33. 9   **35.** No solution   **37.** -4 **39.** $-\frac{6}{5}$

41. 9   **43.** $\frac{7}{3}$ **45.** 13

47.

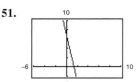

49.

$x = 3$ $x = 10$

51.

$x = \frac{7}{5}$

53. No solution. The x-terms sum to zero.

55. 10   **57.** 4   **59.** 3

61. No solution. The variable is divided out.

63. $\frac{5}{3}$ **65.** No solution. The solution is extraneous.

67. 5   **69.** No solution. The solution is extraneous.

71. 0   **73.** All real numbers

75. $\dfrac{1}{3 - a}$, $a \neq 3$ **77.** $\dfrac{5}{4 + a}$, $a \neq -4$

79. $\dfrac{18}{36 + a}$, $a \neq -36$ **81.** $\dfrac{-17}{10 - 2a}$, $a \neq 5$

83. 138.889   **85.** 62.372   **87.** 19.993

89. (a) 6.46   (b) $\dfrac{1.73}{0.27} \approx 6.41$; Yes

91. (a) 1.00   (b) $\dfrac{6.01}{5.98} \approx 1.01$; Yes

93. (a)

x	-1	0	1	2	3	4
$3.2x - 5.8$	-9	-5.8	-2.6	0.6	3.8	7

(b) $1 < x < 2$. The expression changes from negative to positive in this interval.

(c)

x	1.5	1.6	1.7	1.8	1.9	2
$3.2x - 5.8$	-1	-0.68	-0.36	-0.04	0.28	0.6

(d) $1.8 < x < 1.9$. To improve accuracy, evaluate the expression in this interval and determine where the sign changes.

95. 61.2 inches   **97.** $T = 10,000 + \frac{1}{2}E$, $0 \leq E \leq 20,000$

99. \$7600   **101.** $x = 10$ centimeters

103. (a)

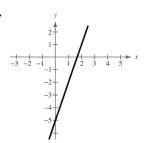

(b) 1994

105. 23,437.5 miles

107. False. $x(3 - x) = 10$

$$3x - x^2 = 10$$

The equation cannot be written in the form $ax + b = 0$.

109. Equivalent equations have the same solution set, and one is derived from the other by steps for generating equivalent equations.

$$2x = 5, \ 2x + 3 = 8$$

111. $4x^3, \ x \neq 0$ **113.** $\dfrac{x - 4}{2x - 1}, \ x \neq -9$

115. $\dfrac{-5x + 31}{x - 6}$

117.

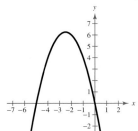

119.

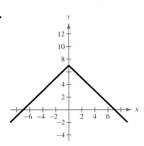

121.

Section 1.3   *(page 105)*

1. A number increased by 4   **3.** A number divided by 5

5. A number decreased by 4 is divided by 5.

7. Negative 3 is multiplied by a number increased by 2.

9. 12 is multiplied by a number and that product is multiplied by the number decreased by 5.

11. $n + (n + 1) = 2n + 1$

13. $(2n - 1)(2n + 1) = 4n^2 - 1$

15. $50t$ **17.** $0.20x$ **19.** $6x$ **21.** $25x + 1200$

23. $0.30L$ **25.** $N = p(500)$ **27.** $4x + 8x = 12x$

29. 262, 263   **31.** 37, 185   **33.** $-5, -4$

35. 13.5   **37.** 27%   **39.** 2400   **41.** $22,316.98

43. Income tax: $738 billion; Corporation taxes: $182 billion; Social Security taxes: $540 billion; Other: $120 billion

45. January: $71,590.00; February: $85,908.00

47. 22% decrease   **49.** 46% decrease

51. (a)

(b) $l = 1.5w; \ p = 5w$

(c) 7.5 meters × 5 meters

53. 97   **55.** 3 hours

57. 3 hours at 58 miles per hour; 2 hours and 45 minutes at 52 miles per hour

59. (a) 3.8 hours, 3.2 hours   (b) 1.1 hours

(c) 25.6 miles

61. $66\frac{2}{3}$ kilometers per hour   **63.** 1.29 seconds

65. 91.4 feet   **67.** 4.36 feet   **69.** $16,666.67

71. Compact: $450,000; Midsize: $150,000

73. ≈ 32.1 gallons   **75.** 50 pounds of each kind

77. 8064 units   **79.** $x = 6$ feet from the 50-pound child

81. $\dfrac{2A}{b}$ **83.** $\dfrac{S}{1 + R}$ **85.** $A\left(1 + \dfrac{r}{n}\right)^{-nt}$ **87.** $\dfrac{360A}{\pi r^2}$

89. $\dfrac{3V}{4\pi a^2}$ **91.** $\dfrac{Fr^2}{\alpha m_1}$ **93.** $\dfrac{CC_2}{C_2 - C}$

95. $\dfrac{2S - n(n - 1)d}{2n}$ **97.** $\sqrt[3]{\dfrac{4.47}{\pi}} \approx 1.12$ inches

99. False. The expression should be $\dfrac{z^3 - 8}{z^2 - 9}$.

101. (a) Negative   (b) Positive; answers will vary.

103. $5st\sqrt{6t}$ **105.** $\dfrac{-2x^2 + 30x - 45}{x(x + 3)(x - 3)}$

107. $\sqrt{10} + 2$ **109.** $\dfrac{42\sqrt{10} + 14}{89}$

Section 1.4   *(page 119)*

1. $2x^2 + 8x - 3 = 0$ **3.** $x^2 - 6x + 6 = 0$

5. $3x^2 - 90x - 10 = 0$ **7.** $0, -\frac{1}{2}$ **9.** $4, -2$

11. -5 **13.** $3, -\frac{1}{2}$ **15.** $2, -6$ **17.** $-\frac{20}{3}, -4$

19. $-a$ **21.** $\pm 7; \pm 7.00$ **23.** $\pm\sqrt{11}; \pm 3.32$

25. $\pm 3\sqrt{3}; \pm 5.20$ **27.** $8, 16; 8.00, 16.00$

29. $-2 \pm \sqrt{14}; 1.74, -5.74$ **31.** $\dfrac{1 \pm 3\sqrt{2}}{2}; 2.62, -1.62$

33. $2; 2.00$ **35.** $0, 2$ **37.** $4, -8$ **39.** $-3 \pm \sqrt{7}$

41. $1 \pm \dfrac{\sqrt{6}}{3}$ **43.** $2 \pm 2\sqrt{3}$ **45.** $\dfrac{1}{(x + 1)^2 + 4}$

47. $\dfrac{1}{\left(x + \frac{1}{2}\right)^2 - 1}$ **49.** $\dfrac{1}{\sqrt{9 - (x - 3)^2}}$

51. 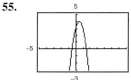   **53.**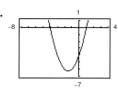

$x = -1, -5$ $x = 3, 1$

55.   **57.**

$x = -\frac{1}{2}, \frac{3}{2}$ $x = 1, -4$

59. No real solution   **61.** Two real solutions

63. No real solution   **65.** Two real solutions

67. $\frac{1}{2}, -1$ **69.** $\frac{1}{4}, -\frac{3}{4}$ **71.** $1 \pm \sqrt{3}$

73. $-7 \pm \sqrt{5}$ **75.** $-4 \pm 2\sqrt{5}$ **77.** $\dfrac{2}{3} \pm \dfrac{\sqrt{7}}{3}$

79. $-\frac{4}{3}$ **81.** $-\frac{1}{2} \pm \sqrt{2}$ **83.** $\frac{2}{7}$

85. $2 \pm \dfrac{\sqrt{6}}{2}$ **87.** $6 \pm \sqrt{11}$ **89.** $-\dfrac{3}{8} \pm \dfrac{\sqrt{265}}{8}$

91. $0.976, -0.643$ **93.** $1.355, -14.071$

95. $1.687, -0.488$ **97.** $-0.290, -2.200$

99. $1 \pm \sqrt{2}$ **101.** $6, -12$ **103.** $\frac{1}{2} \pm \sqrt{3}$

105. $-\dfrac{1}{2}$ **107.** $\dfrac{3}{4} \pm \dfrac{\sqrt{97}}{4}$

109. (a) and (b) $x = -5, -\frac{10}{3}$

(c) The method used in part (a) reduces the number of algebraic steps.

111. $x^2 + 15x + 44 = 0$ **113.** $30x^2 + 7x - 2 = 0$

115. $x = 7.5$ m and $y = 23\frac{1}{3}$ m or $x = 17.5$ m and $y = 10$ m

117. 14 centimeters × 14 centimeters

119. 9 seats per row

121. (a) $s = -16t^2 + 1368$ (b) 1112 feet

(c) ≈ 9.25 seconds

(d)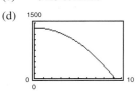

123. (a)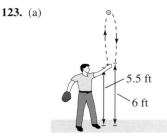

(b) $S = -16t^2 + 45t + 5.5$

(c) 24 feet

(d) ≈ 2.8 seconds

(e)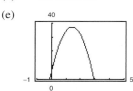

125. $\dfrac{20\sqrt{3}}{3} \approx 11.55$ inches

127. (a)

(b) $x^2 + 15^2 = l^2$

$30\sqrt{6} \approx 73.5$ feet

129. 3761 units or 146,239 units   **131.** 100 units

133. 48 units   **135.** (a) $16.8°$ (b) 2.5

137. 259 kilometers, 541 kilometers

139. False. The product must equal zero for the Zero-Factor Property to be used.

141. (a) Neither   (b) Both   (c) Quadratic   (d) Neither

143. Associative Property of Multiplication

145. Additive Inverse Property   **147.** $\dfrac{5}{6u^2v^3}$

149. $-(x+10),\ x \neq 10$ **151.** $x^2(x-3)(x^2+3x+9)$

153. $(x^2-2)(x+5)$ **155.** $x^2-3x-18$

157. x^3+3x^2-2x+8 **159.** $6x^4-3x^3-2x^2+3x-1$

Section 1.5   *(page 129)*

1. $a=-10,\ b=6$ **3.** $a=6,\ b=5$ **5.** $4+3i$

7. $2-3\sqrt{3}\,i$ **9.** $5\sqrt{3}\,i$ **11.** 8   **13.** $-1-6i$

15. $0.3i$ **17.** $11-i$ **19.** 4   **21.** $3-3\sqrt{2}\,i$

23. $-14+20i$ **25.** $\frac{1}{6}+\frac{7}{6}i$ **27.** $-2\sqrt{3}$ **29.** -10

31. $5+i$ **33.** $12+30i$ **35.** 24   **37.** $-9+40i$

39. -10 **41.** $6-3i,\ 45$ **43.** $-1+\sqrt{5}\,i,\ 6$

45. $-2\sqrt{5}\,i,\ 20$ **47.** $\sqrt{8},\ 8$ **49.** $-5i$

51. $\frac{8}{41}+\frac{10}{41}i$ **53.** $\frac{4}{5}+\frac{3}{5}i$ **55.** $-5-6i$

57. $-\frac{120}{1681}-\frac{27}{1681}i$ **59.** $-\frac{1}{2}-\frac{5}{2}i$ **61.** $\frac{62}{949}+\frac{297}{949}i$

63. $1\pm i$ **65.** $-2\pm\frac{1}{2}i$ **67.** $-\frac{3}{2},\ -\frac{5}{2}$

69. $2\pm\sqrt{2}\,i$ **71.** $\dfrac{5}{7}\pm\dfrac{5\sqrt{15}}{7}$

73. (a) 1   (b) i (c) -1 (d) $-i$

75. $-4+2i$ **77.** i **79.** -8 **81.** $\frac{1}{8}i$

83. (a) 16   (b) 16   (c) 16   (d) 16

85. False. If the complex number is real, the number equals its conjugate.

87. False.

$i^{44}+i^{150}-i^{74}-i^{109}+i^{61}=1-1+1-i+i=1$

89. Answers will vary.   **91.** $-x^2-3x+12$

93. $3x^2+\frac{23}{2}x-2$ **95.** -31 **97.** $\frac{27}{2}$

99. $a=\dfrac{\sqrt{3V\pi b}}{2\pi b}$ **101.** 1 liter

Section 1.6   *(page 139)*

1. $0,\ \pm\dfrac{3\sqrt{2}}{2}$ **3.** $\pm3,\ \pm3i$ **5.** $-6,\ 3\pm3\sqrt{3}\,i$

7. $-3,\ 0$ **9.** $3,\ 1,\ -1$ **11.** $\pm1,\ \dfrac{1}{2}\pm\dfrac{\sqrt{3}}{2}i$

13. $\pm\sqrt{3},\ \pm1$ **15.** $\pm\frac{1}{2},\ \pm4$

17. $1,\ -2,\ 1\pm\sqrt{3}\,i,\ -\dfrac{1}{2}\pm\dfrac{\sqrt{3}i}{2}$

19. $-\frac{1}{5},\ -\frac{1}{3}$ **21.** $\frac{1}{4}$ **23.** $1,\ -\frac{125}{8}$

25. (a)

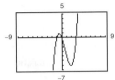

(b) $(0,0),\ (3,0),\ (-1,0)$

(c) $x=0,\ 3,\ -1$

(d) The x-intercepts and the solutions are the same.

27. (a)

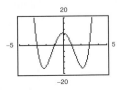

(b) $(\pm3,0),\ (\pm1,0)$

(c) $x=\pm3,\ \pm1$

(d) The x-intercepts and the solutions are the same.

29. 50   **31.** 26   **33.** -16 **35.** $2,\ -5$

37. 0   **39.** 9   **41.** $\frac{101}{4}$ **43.** 14   **45.** 9

47. $-3\pm16\sqrt{2}$ **49.** $\pm\sqrt{14}$ **51.** 1

53. (a)

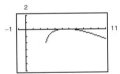

(b) $(5,0),\ (6,0)$

(c) $x=5,\ 6$

(d) The x-intercepts and the solutions are the same.

55. (a)

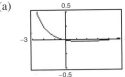

(b) $(0,0),\ (4,0)$

(c) $x=0,\ 4$

(d) The x-intercepts and the solutions are the same.

57. $4,\ -5$ **59.** $\dfrac{-3\pm\sqrt{21}}{6}$ **61.** $2,\ -\dfrac{3}{2}$ **63.** $1,\ -3$

65. $3,\ -2$ **67.** $\sqrt{3},\ -3$ **69.** $3,\ \dfrac{-1-\sqrt{17}}{2}$

71. (a)

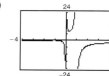

(b) $(-1, 0)$

(c) $x = -1$

(d) The x-intercept and the solution are the same.

73. (a)

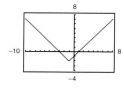

(b) $(1, 0), (-3, 0)$

(c) $x = 1, -3$

(d) The x-intercepts and the solutions are the same.

75. ± 1.038 **77.** 16.756 **79.** $x^2 - 3x - 10 = 0$

81. $21x^2 + 31x - 42 = 0$ **83.** $x^3 - 4x^2 - 3x + 12 = 0$

85. $x^4 - 1$ **87.** 34 students   **89.** 400 miles per hour

91. 4%   **93.** $x = 6, -4$ **95.** $y = \pm 15$

97. 26,250 passengers   **99.** 500 units

101. $x = 2$ miles or 0.382 mile

103. (a)

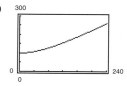

173

(b)

h	160	165	170	175	180	185
d	188.7	192.9	197.2	201.6	205.9	210.3

$d = 200$ when h is between 170 and 175 feet.

(c) 173.2

(d) Solving graphically or numerically yields an approximate solution. An exact solution is obtained algebraically.

105. $7 + \sqrt{65} \approx 15$ hours

$9 + \sqrt{65} \approx 17$ hours

107. $g = \dfrac{\mu s v^2}{R}$ **109.** False. See Example 7 on page 135.

111. $a = 9, b = 9$ **113.** $a = 4, b = 24$ **115.** $\dfrac{25}{6x}$

117. $\dfrac{-3z^2 - 2z + 4}{z(z + 2)}$ **119.** $\dfrac{2(x^2 - 4x - 4)}{5(2 - x)^2}$

121. 11   **123.** $5, -45$

Section 1.7 *(page 149)*

1. $-1 \le x \le 5$. Bounded   **3.** $11 < x < \infty$. Unbounded

5. $-\infty < x < -2$. Unbounded

7. b   **9.** d   **11.** e

13. (a) Yes   (b) No   (c) Yes   (d) No

15. (a) Yes   (b) No   (c) No   (d) Yes

17. (a) Yes   (b) Yes   (c) Yes   (d) No

19. $x < 3$

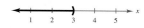

21. $x > \frac{3}{2}$

23. $x \ge 12$

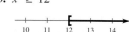

25. $x > 2$

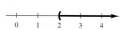

27. $x \ge \frac{2}{7}$

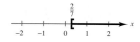

29. $x < 5$

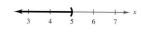

31. $x \ge 4$

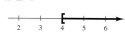

33. $x \ge 2$

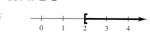

35. $x \ge -4$

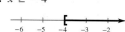

37. $-1 < x < 3$

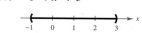

39. $-\frac{9}{2} < x < \frac{15}{2}$ **41.** $-\frac{3}{4} < x < -\frac{1}{4}$

43. $10.5 \le x \le 13.5$

45. $-6 < x < 6$

47. $x < -10, x > 10$

49. No solution

51. $14 \leq x \leq 26$

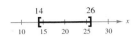

53. $x \leq -\frac{3}{2}, x \geq 3$

55. $x \leq -7, \ x \geq 13$

57. $4 < x < 5$

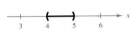

59. $x \leq -\frac{29}{2}, \ x \geq -\frac{11}{2}$

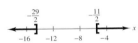

61.   **63.**

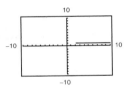   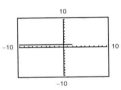

$x > 2$ $x \leq 2$

65.   **67.**

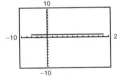

$-6 \leq x \leq 22$ $x \leq -\frac{27}{2}, x \geq -\frac{1}{2}$

69.

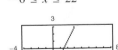

(a) $x \geq 2$

(b) $x \leq \frac{3}{2}$

71.

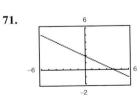

(a) $4 \geq x \geq -2$

(b) $x \leq 4$

73.

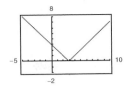

(a) $1 \leq x \leq 5$

(b) $x \leq -1, \ x \geq 7$

75. $[5, \infty)$ **77.** $[-3, \infty)$ **79.** $\left(-\infty, \frac{7}{2}\right]$

81. All real numbers within 8 units of 10

83. $|x| \leq 3$ **85.** $|x - 7| \geq 3$ **87.** $|x - 12| < 10$

89. $|x + 3| > 5$ **91.** More than 400 miles

93. $r > 3.125\%$ **95.** $x \geq 36$

97. (a) and (b)   (c) $x \geq 186$

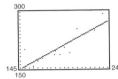

(d) An athlete's weight is not a particularly good indicator of the athlete's maximum bench press weight. Other factors, such as muscle tone and exercise habits, influence maximum bench press weight.

99. 2002   **101.** 6¢   **103.** $20 \leq h \leq 80$

105. False. c has to be greater than zero.   **107.** b

109. $5\sqrt{5}; \left(-\frac{3}{2}, 7\right)$ **111.** $2\sqrt{65}; (-1, -1)$

113. 11   **115.** 10   **117.** $-\frac{1}{2}, 10$ **119.** $\frac{1}{7}, -\frac{1}{2}$

121. $(-3, 10)$ **123.** 13

Section 1.8   *(page 160)*

1. (a) No   (b) Yes   (c) Yes   (d) No

3. (a) Yes   (b) No   (c) No   (d) Yes

5. $2, -\frac{3}{2}$ **7.** $\frac{7}{2}, 5$

9. $[-3, 3]$ **11.** $(-7, 3)$

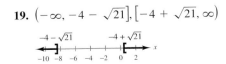

13. $(-\infty, -5], [1, \infty)$ **15.** $(-3, 2)$

17. $(-3, 1)$

19. $\left(-\infty, -4 - \sqrt{21}\right], \left[-4 + \sqrt{21}, \infty\right)$

21. $(-1, 1), (3, \infty)$ **23.** $[-3, 2], [3, \infty)$

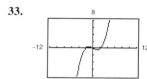

25. $(-\infty, 0), \left(0, \frac{3}{2}\right)$ **27.** $[-2, 0], [2, \infty)$ **29.** $[-2, \infty)$

31.

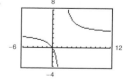

(a) $x \le -1, x \ge 3$

(b) $0 \le x \le 2$

33.

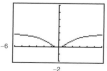

(a) $-2 \le x \le 0$,

$2 \le x < \infty$

(b) $x \le 4$

35. $(-\infty, -1), (0, 1)$

37. $(-\infty, -1), (4, \infty)$

39. $(5, 15)$ **41.** $\left(-5, -\frac{3}{2}\right), (-1, \infty)$

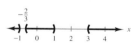

43. $\left(-\frac{3}{4}, 3\right), [6, \infty)$ **45.** $(-3, -2], [0, 3)$

47. $(-\infty, -1), \left(-\frac{2}{3}, 1\right), (3, \infty)$

49.

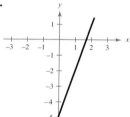

(a) $0 \le x < 2$

(b) $2 < x \le 4$

51.

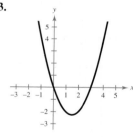

(a) $2 \le |x|$

(b) $-\infty < x < \infty$

53. $[-2, 2]$ **55.** $(-\infty, 3] \cup [4, \infty)$

57. $(-5, 0], (7, \infty)$ **59.** $(-3.51, 3.51)$

61. $(-0.13, 25.13)$ **63.** $(2.26, 2.39)$

65. (a) $t = 10$ seconds (b) 4 seconds $< t < 6$ seconds

67. 13.8 meters $\le L \le 36.2$ meters **69.** $r > 4.88\%$

71. 1999 **73.** $R_1 \ge 2$

75. (a) $L = 6 - y + 2\sqrt{16 + y^2}$

(b) $0 \le y \le 6$

$y = 0: L = 14$

$y = 6: L = 4\sqrt{13} \approx 14.4$

Decrease

(c)

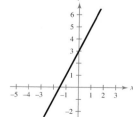

(d) $\frac{5}{3} < y < 3$

77. True. The y-values are greater than zero for all values of x.

79. $(-\infty, \infty)$ **81.** $\left(-\infty, -2\sqrt{10}\right] \cup \left[2\sqrt{10}, \infty\right)$

83. 0 **85.** $(2x + 5)^2$ **87.** $(x + 3)(x + 2)(x - 2)$

89. $2x^2 + x$ **91.** $22x$

Review Exercises (page 164)

1.

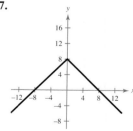

3.

5.

7.

9.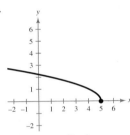

11.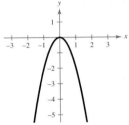

13. x-intercept: $\left(\frac{9}{2}, 0\right)$

y-intercept: $(0, -9)$

15. x-intercept: $(-1, 0)$

y-intercept: $(0, 1)$

17. x-intercepts: $(0, 0), \left(\pm 2\sqrt{2}, 0\right)$

y-intercept: $(0, 0)$

19. x-intercepts: $(0, 0), (\pm 3, 0)$

y-intercept: $(0, 0)$

21. x-intercept: $(8, 0), (0, 0)$

y-intercept: $(0, 0)$

23.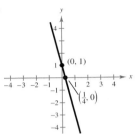

25.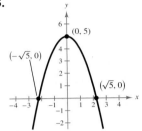

27.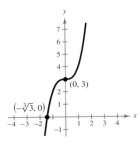

29.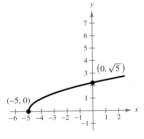

31.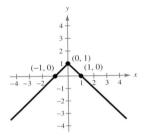

33. Center: $(0, 0)$; Radius: 5

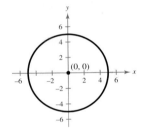

35. Center: $(-2, 0)$; Radius: 4

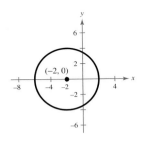

37. Center: $\left(\frac{1}{2}, -1\right)$; Radius: 6

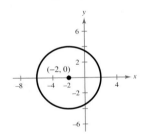

39. $(x - 2)^2 + (y + 3)^2 = 13$

41. (a)

x	0	4	8	12	16	20
F	0	5	10	15	20	25

(b)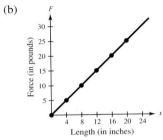

(c) 12.5 pounds

43. Identity   **45.** Identity

47. (a) No   (b) Yes   (c) Yes   (d) No

49. 20   **51.** $-\frac{1}{2}$ **53.** -5 **55.** 9

57. $h = 10$ inches

59. September: $325,000; October: $364,000

61. 2π meters   **63.** Nine   **65.** $2\frac{6}{7}$ liters

67. $r = \dfrac{\sqrt{3V\pi h}}{\pi h}$ **69.** $\frac{2}{3}$ hour   **71.** $-\frac{5}{2}, 3$ **73.** $\pm\sqrt{2}$

75. $-4 \pm 3\sqrt{2}$ **77.** $6 \pm \sqrt{6}$ **79.** $-\dfrac{5}{4} \pm \dfrac{\sqrt{241}}{4}$

81. (a) $x = 0, 20$

(b)

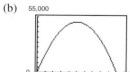

(c) $x = 10$

83. $6 + 2i$ **85.** $-1 + 3i$ **87.** $3 + 7i$

89. $40 + 65i$ **91.** $-4 - 46i$ **93.** $\frac{23}{17} + \frac{10}{17}i$

95. $\frac{21}{13} - \frac{1}{13}i$ **97.** $\pm\frac{\sqrt{3}}{3}i$ **99.** $1 + 3i$ **101.** $0, \frac{12}{5}$

103. $\pm\sqrt{2}, \pm\sqrt{3}$ **105.** 5   **107.** $-3 \pm 4i$

109. No solution   **111.** $-124, 126$

113. $-2 \pm \dfrac{\sqrt{95}}{5}, -4$ **115.** $-5, 15$ **117.** $1, 3$

119. $143,203$ units   **121.** $-7 < x \leq 2$. Bounded

123. $-\infty < x \leq -10$. Unbounded   **125.** $(-\infty, 12]$

127. $\left[\frac{32}{15}, \infty\right)$ **129.** $\left(-\frac{2}{3}, 17\right)$ **131.** $[-4, 4]$

133. $(-\infty, -1), (7, \infty)$ **135.** 36 units

137. $(-3, 9)$ **139.** $\left(-\frac{4}{3}, \frac{1}{2}\right)$ **141.** $[-5, -1), (1, \infty)$

143. $[-4, -3], (0, \infty)$ **145.** 4.9%

147. False. $\sqrt{-18}\sqrt{-2} = \left(3\sqrt{2}\,i\right)\left(\sqrt{2}\,i\right) = 6i^2 = -6$
and $\sqrt{(-18)(-2)} = \sqrt{36} = 6$

149. Some solutions to certain types of equations may be extraneous solutions, which do not satisfy the original equation. So, a check is crucial.

151. Answers will vary.

Chapter Test *(page 169)*

1.

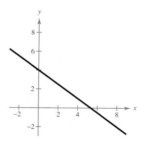

2.

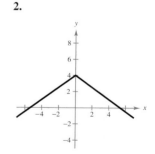

3.

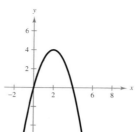

4.

5.

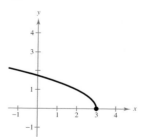

6.

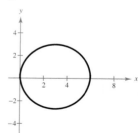

7. $\frac{128}{11}$ **8.** $-4, 5$ **9.** No solution

10. $\pm\sqrt{2}, \pm\sqrt{3}i$ **11.** 4   **12.** $-2, \frac{8}{3}$

13. $-\frac{11}{2} \leq x < 3$

14. $x < -6$ or $0 < x < 4$

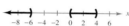

15. $x < -4$ or $x > \frac{3}{2}$

16. $x \le 10$ or $x \ge 20$

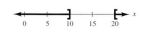

17. (a) $-3 + 5i$ (b) 7 (c) $2 - i$

18.

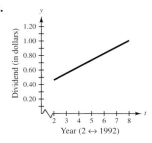

19. $93\frac{3}{4}$ kilometers per hour   **20.** $a = 80, b = 20$ **21.** 5%

Chapter 2

Section 2.1   *(page 181)*

1. (a) L_2 (b) L_3 (c) L_1

3.

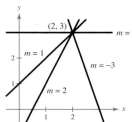

5. $\frac{8}{5}$ **7.** 0 **9.** -4

11. $m = 2$

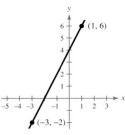

13. m is undefined.

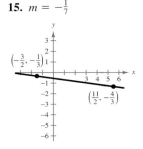

15. $m = -\frac{1}{7}$

17. $m = 0.15$

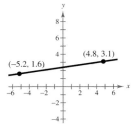

19. $(0, 1), (3, 1), (-1, 1)$ **21.** $(6, -5), (7, -4), (8, -3)$

23. $(-8, 0), (-8, 2), (-8, 3)$

25. $(-4, 6), (-3, 8), (-2, 10)$

27. $(9, -1), (11, 0), (13, 1)$ **29.** Perpendicular

31. Parallel

33. (a) Sales increasing 135 units per year

(b) No change in sales

(c) Sales decreasing 40 units per year

35. (a) Greatest increase: 1990 to 1991 and 1996 to 1997

Greatest decrease: 1997 to 1998

(b) $m = 0.037$

(c) Each year, the earnings per share increase by $0.037.

37. (a) and (b)

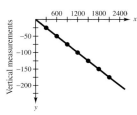

(c) $y = -\frac{1}{12}x$

(d) For every 12 horizontal measurements, the vertical measurement decreases by 1.

(e) "8.3% grade"

39. 12 feet

41. $m = 1$; Intercept: $(0, -10)$

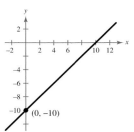

43. $m = 0$; Intercept: $\left(0, -\frac{5}{3}\right)$ **45.** $m = -\frac{2}{3}$; Intercept: $(0, 3)$

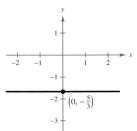

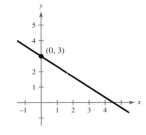

47. $y = -x + 10$

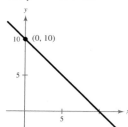

49. $y = 4x$

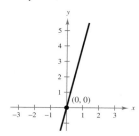

51. $y = \frac{3}{4}x - \frac{7}{2}$

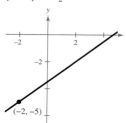

53. $y = 4$

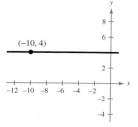

55. $y = -3x$

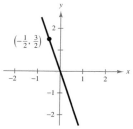

57. $y = -2.5x - 2.75$

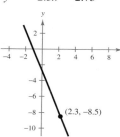

59. $y = \frac{7}{8}x - \frac{1}{2}$

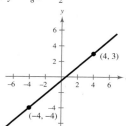

61. $y = 4$

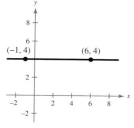

63. $y = -\frac{1}{3}x + \frac{4}{3}$

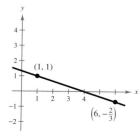

65. $y = -\frac{3}{25}x + \frac{159}{100}$

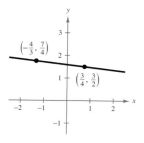

67. $y = 0.3x - 1.8$

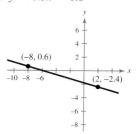

69. $4x - 3y + 12 = 0$ **71.** $3x - y - 2 = 0$

73. $x + y - 1 = 0$

75. (a) $y = -x - 1$ (b) $y = x + 5$

77. (a) $y = -\frac{5}{3}x + \frac{53}{24}$ (b) $y = \frac{3}{5}x + \frac{9}{40}$

79. (a) $x = 2$ (b) $y = 5$

81. (a) $y = -3x - 13.1$ (b) $y = \frac{1}{3}x - 0.1$

83. (a) is parallel to (c). (b) is perpendicular to (a) and (c).

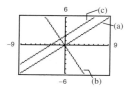

85. (a) is parallel to (b). (c) is perpendicular to (a) and (b).

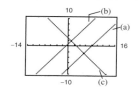

87. $V = 4.5t + 151.5$

89. c; The slope is 2, which represents the hourly wage per unit produced.

91. d; The slope is -100, which represents the decrease in the value of the word processor each year.

93. $y = -\frac{5}{13}x - \frac{2}{13}$ **95.** $y = -\frac{16}{21}x - \frac{13}{56}$

97. $y = 54t + 3927$; 1999: 4089 stores, 2000: 4143 stores

99.

C	$-17.8°$	$-10°$	$10°$	$20°$	$32.2°$	$177°$
F	$0°$	$14°$	$50°$	$68°$	$90°$	$350.6°$

101. 3014 students   **103.** $V = 25{,}000 - 2300t$

105. $W = 0.75x + 11.50$

107. (a) $x = -\frac{1}{15}p + \frac{266}{3}$ (b) 45   (c) 49

109. $W = 0.07S + 2500$

111. (a) $12{,}000 - x$ (b) $y = -0.015x + 480$

(c)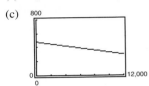

(d) As the amount invested at the lower interest rate increases, the annual interest decreases.

113. (a) and (b)

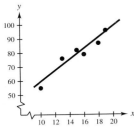

(c) $y = 4x + 19$ (d) 87

(e) Vertical shift 4 units upward

115. False. The slope of the first line is $\frac{2}{7}$ and the slope of the second line is $-\frac{11}{7}$.

117. The slope of a vertical line is undefined because division by zero is undefined.

119. The line with a slope of -4 is steeper. The slope with the greatest magnitude corresponds to the steepest line.

121. Yes. The rate of change remains the same on a line.

123. d   **125.** a   **127.** -1 **129.** $\frac{7}{2}, 7$

131. No solution

Section 2.2 *(page 195)*

1. Yes   **3.** No

5. Yes, each input value has exactly one output value.

7. No, the input value of 7 has two output values, 6 and 12.

9. (a) Function

(b) Not a function, because the element 1 in A corresponds to two elements, -2 and 1, in B.

(c) Function

(d) Not a function, because not every element in A is matched with an element in B.

11. Each is a function. For each year there corresponds one and only one circulation.

13. Not a function   **15.** Function   **17.** Function

19. Not a function   **21.** Function

23. (a) 4   (b) 0   (c) $4x$ (d) $(x + c)$

25. (a) -1 (b) -9 (c) $2x - 5$

27. (a) 36π (b) $\frac{9}{2}\pi$ (c) $\frac{32}{3}\pi r^3$

29. (a) 1   (b) 2.5   (c) $3 - 2|x|$

31. (a) $-\frac{1}{9}$ (b) Undefined   (c) $\dfrac{1}{y^2 + 6y}$

33. (a) 1   (b) -1 (c) $\dfrac{|x - 1|}{x - 1}$

35. (a) -1 (b) 2   (c) 6

37.

x	-2	-1	0	1	2
$f(x)$	1	-2	-3	-2	1

39.

t	-5	-4	-3	-2	-1
$h(t)$	1	$\frac{1}{2}$	0	$\frac{1}{2}$	1

41.

x	-2	-1	0	1	2
$f(x)$	5	$\frac{9}{2}$	4	1	0

43. 5   **45.** $\frac{4}{3}$ **47.** ± 3 **49.** $0, \pm 1$

51. $2, -1$ **53.** $3, 0$ **55.** All real numbers x

57. All real numbers $t \neq 0$

59. $y \geq 10$ **61.** $-1 \leq x \leq 1$

63. All real numbers $x \neq 0, -2$ **65.** $s \geq 1, s \neq 4$

67. All real numbers $x \neq 0$

69. $\{(-2, 4), (-1, 1), (0, 0), (1, 1), (2, 4)\}$

71. $\{(-2, 0), (-1, 1), (0, \sqrt{2}), (1, \sqrt{3}), (2, 2)\}$

73. $g(x) = -2x^2$ **75.** $r(x) = \dfrac{32}{x}$ **77.** $3 + h, h \neq 0$

79. $3x^2 + 3xc + c^2, c \neq 0$

81. $3, x \neq 3$ **83.** $\dfrac{\sqrt{5x} - 5}{x - 5}$ **85.** $A = \dfrac{p^2}{16}$

87. $A = 8\sqrt{s^2 - 64}$

89. (a)

Height x	Width	Volume V
1	$24 - 2(1)$	$1[24 - 2(1)]^2 = 484$
2	$24 - 2(2)$	$2[24 - 2(2)]^2 = 800$
3	$24 - 2(3)$	$3[24 - 2(3)]^2 = 972$
4	$24 - 2(4)$	$4[24 - 2(4)]^2 = 1024$
5	$24 - 2(5)$	$5[24 - 2(5)]^2 = 980$
6	$24 - 2(6)$	$6[24 - 2(6)]^2 = 864$

Maximum when $x = 4$

(b)

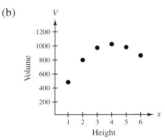

V is a function of *x*.

(c) $V = x(24 - 2x)^2, 0 < x < 12$

91. $A = \dfrac{x^2}{2(x-2)}, x > 2$

93. 1978: \$15,198; 1988: \$25,558;
1993: \$30,800; 1997: \$41,008

95. (a) $C = 12.30x + 98,000$

(b) $R = 17.98x$

(c) $P = 5.68x - 98,000$

97. (a) $R = \dfrac{240n - n^2}{20}, n \geq 80$

(b)

n	90	100	110	120	130	140	150
R(n)	\$675	\$700	\$715	\$720	\$715	\$700	\$675

The revenue is maximum when 120 people take the trip.

99. (a)

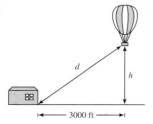

(b) $h = \sqrt{d^2 - 3000^2}, d \geq 3000$

101. Yes, the ball will be at a height of 6 feet.

103. True. Each *x*-value corresponds to one *y*-value.

105. The domain is the set of inputs of the function, and the range is the set of outputs.

107. $\frac{15}{8}$ **109.** $-\frac{1}{5}$ **111.** $2x - 3y - 11 = 0$

113. $10x + 9y + 15 = 0$

Section 2.3   *(page 209)*

1. Domain: all real numbers
Range: all real numbers

3. Domain: all real numbers
Range: $(-\infty, 1]$

5. Domain: $(-\infty, -1], [1, \infty)$
Range: $[0, \infty)$

7. Domain: $[-4, 4]$
Range: $[0, 4]$

9. Function   **11.** Not a function   **13.** Function

15. $-\frac{5}{2}, 6$ **17.** 0   **19.** $0, \pm\sqrt{2}$ **21.** $\pm\frac{1}{2}, 6$

23. $-\frac{5}{3}$ **25.** $-\frac{11}{2}$

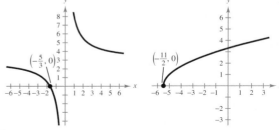

27. $\frac{1}{3}$

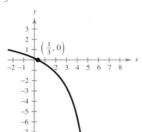

29. (a) Increasing on $(-\infty, \infty)$ (b) Odd

31. (a) Increasing on $(-\infty, 0)$ and $(2, \infty)$
Decreasing on $(0, 2)$

(b) Neither even nor odd

33. (a)

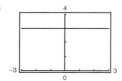

Constant on $(-\infty, \infty)$

(b)

x	-2	-1	0	1	2
f(x)	3	3	3	3	3

35. (a)

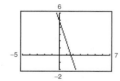

Decreasing on $(-\infty, \infty)$

(b)

x	-2	-1	0	1	2
f(x)	11	8	5	2	-1

37. (a)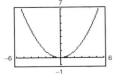

Decreasing on $(-\infty, 0)$; Increasing on $(0, \infty)$

(b)

s	-4	-2	0	2	4
$g(s)$	4	1	0	1	4

39. (a)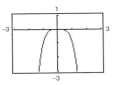

Increasing on $(-\infty, 0)$; Decreasing on $(0, \infty)$

(b)

t	-2	-1	0	1	2
$f(t)$	-16	-1	0	-1	-16

41. (a)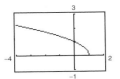

Decreasing on $(-\infty, 1)$

(b)

x	-3	-2	-1	0	1
$f(x)$	2	$\sqrt{3}$	$\sqrt{2}$	1	0

43. (a)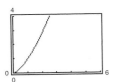

Increasing on $(0, \infty)$

(b)

x	0	1	2	3	4
$f(x)$	0	1	2.82	5.2	8

45. (a)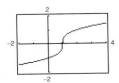

Increasing on $(-\infty, \infty)$

(b)

t	-2	-1	0	1	2
$g(t)$	-1.44	-1.26	-1	0	1

47. (a)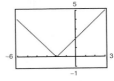

Decreasing on $(-\infty, -2)$; Increasing on $(-2, \infty)$

(b)

x	-6	-4	-2	0	2
$f(x)$	4	2	0	2	4

49. (a)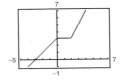

Increasing on $(-\infty, 0)$ and $(2, \infty)$; Constant on $(0, 2)$

(b)

x	-2	-1	0	1	2	3	4
$f(x)$	1	2	3	3	3	5	7

51.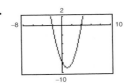

Relative minimum: $(1, -9)$

53.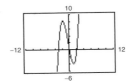

Relative maximum: $(-1.79, 8.21)$

Relative minimum: $(1.12, -4.06)$

55.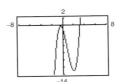

Relative maximum: $(-0.33, -0.30)$

Relative minimum: $(2, -13)$

57.

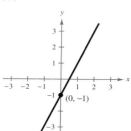

59.

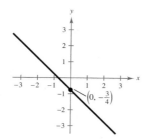

61.

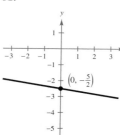

63.

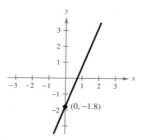

65. $f(x) = -2x + 6$ **67.** $f(x) = -3x + 11$

69. $f(x) = \frac{2}{5}x - 3$ **71.** $f(x) = \frac{6}{7}x - \frac{45}{7}$

73.

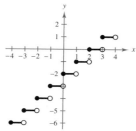

Vertical shift 2 units downward

75.

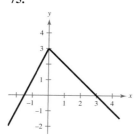

77.

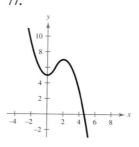

79. $(-\infty, 4]$

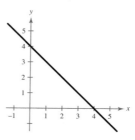

81. $(-\infty, -3], [3, \infty)$

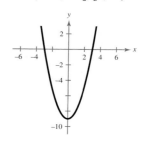

83. $[-1, 1]$

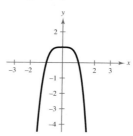

85. $(-\infty, \infty)$

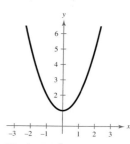

87. $f(x) < 0$ for all x.

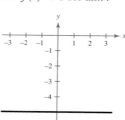

89. $(-2, 8]$

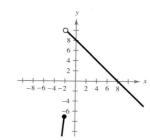

91.

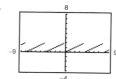

Domain: $(-\infty, \infty)$

Range: $[0, 2)$

Sawtooth pattern

93. Even   **95.** Odd   **97.** Neither even nor odd

99. (a) $\left(\frac{3}{2}, 4\right)$ (b) $\left(\frac{3}{2}, -4\right)$

101. (a) $(-4, 9)$ (b) $(-4, -9)$

103. (a) C_2 is the appropriate model, because the cost does not increase until after the next minute of conversation has started.

(b)

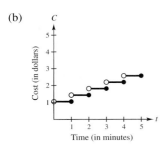

$7.89

105.

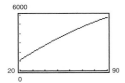

30 watts

107. $h = 3 - 4x + x^2$ **109.** $h = 2 - \sqrt[3]{x}$

111. $L = 2 - \sqrt[3]{2y}$ **113.** $L = \dfrac{2}{y}$

115. (a) $A = 64 - 2x^2,\ 0 \le x \le 4$

(b)

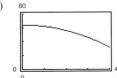

$32 \le A \le 64$

(c) Square with sides $4\sqrt{2}$ m

117.

Interval	Inside Pipe	Drainpipe 1	Drainpipe 2
$[0, 5]$	Open	Closed	Closed
$[5, 10]$	Open	Open	Closed
$[10, 20]$	Closed	Closed	Closed
$[20, 30]$	Closed	Closed	Open
$[30, 40]$	Open	Open	Open
$[40, 45]$	Open	Closed	Open
$[45, 50]$	Open	Open	Open
$[50, 60]$	Open	Open	Closed

119. False. A piecewise-defined function is a function that is defined by two or more equations over a specified domain. That domain may or may not include x- and y-intercepts.

121. Answers will vary.

123. Yes. For each value of y there corresponds one and only one value of x.

125. (a)

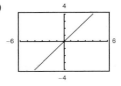

(b)

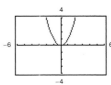

(c)

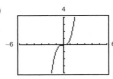

(d)

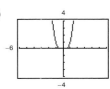

(e)

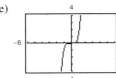

(f)

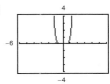

All the graphs pass through the origin. The graphs of the odd powers of x are symmetric with respect to the origin, and the graphs of the even powers are symmetric with respect to the y-axis. As the powers increase, the graphs become flatter in the interval $-1 < x < 1$.

127. $0, 10$ **129.** $0, \pm 1$

131. (a) 37 (b) -28 (c) $5x - 43$

133. (a) -9 (b) $2\sqrt{7} - 9$ (c) $-9 + 3\sqrt{2}i$

135. $h + 4,\ h \ne 0$

Section 2.4   *(page 220)*

1. (a)

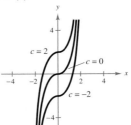

(b)

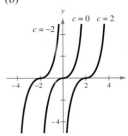

3. (a)

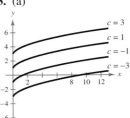

(b)

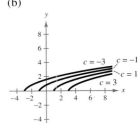

(c)

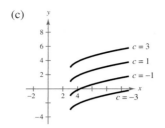

5. (a)

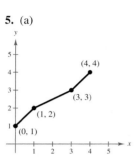

(b)

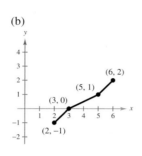

(c)

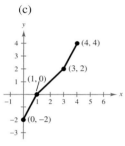

(d)

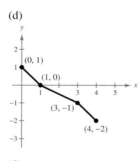

(e)

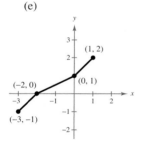

(f)

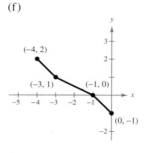

7. (a)

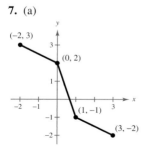

(b)

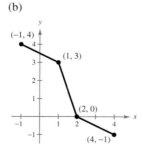

(c)

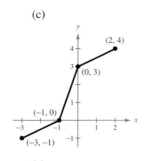

(d)

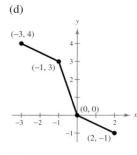

(e)

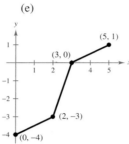

(f)

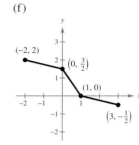

9. (a) $y = x^2 - 1$ (b) $y = 1 - (x + 1)^2$
   (c) $y = -(x - 2)^2 + 6$ (d) $y = (x - 5)^2 - 3$

11. (a) $y = |x| + 5$ (b) $y = -|x + 3|$
    (c) $y = |x - 2| - 4$ (d) $y = -|x - 6| - 1$

13. Horizontal shift of $y = x^3$; $y = (x - 2)^3$

15. Reflection in the x-axis of $y = x^2$; $y = -x^2$

17. Reflection in the x-axis and vertical shift of $y = \sqrt{x}$; $y = 1 - \sqrt{x}$

19. Reflection of $f(x) = x^2$ in x-axis and vertical shift of 12 units upward

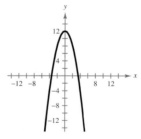

21. Vertical shift of $f(x) = x^3$ seven units upward

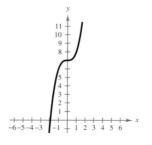

23. Reflection of $f(x) = x^2$ in x-axis, vertical shift of 2 units upward, and horizontal shift of 5 units to the left

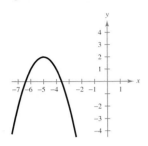

25. Vertical shift of $f(x) = x^3$ 2 units upward and horizontal shift of 1 unit to the right

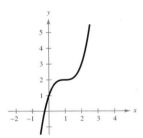

27. Reflection of $f(x) = |x|$ in x-axis and vertical shift of 2 units downward

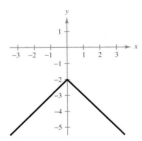

29. Reflection of $f(x) = |x|$ in x-axis, vertical shift of 8 units upward, and horizontal shift of 4 units to the left

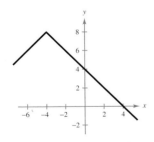

31. Horizontal shift of $f(x) = \sqrt{x}$ 9 units to the right

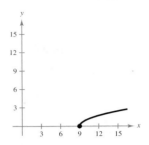

33. Reflection of $f(x) = \sqrt{x}$ in y-axis, vertical shift of 2 units downward, and horizontal shift of 7 units to the right

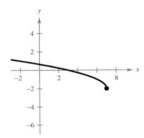

35. $f(x) = (x - 2)^2 - 8$ **37.** $f(x) = (x - 13)^3$

39. $f(x) = -|x| - 10$ **41.** $f(x) = -\sqrt{-x} + 6$

43. (a) $y = -3x^2$ (b) $y = 4x^2 + 3$

45. (a) $y = -\frac{1}{2}|x|$ (b) $y = 3|x| - 3$

47. Vertical stretch of $y = x^3$; $y = 2x^3$

49. Reflection in x-axis and vertical shrink of $y = x^2$; $y = -\frac{1}{2}x^2$

51. Reflection in y-axis and vertical shrink of $y = \sqrt{x}$; $y = \frac{1}{2}\sqrt{-x}$

53. $y = -(x - 2)^3 + 2$ **55.** $y = -\sqrt{x} - 3$

57. (a)   (b)

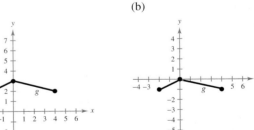

(c)   (d)

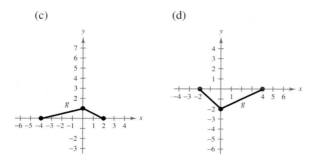

59. (a) Vertical shrink of 0.04 and vertical shift of 20.46 units upward

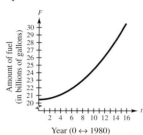

Year (0 ↔ 1980)

(b) $f(t) = 20.46 + 0.04(t + 10)^2$. By shifting the graph 10 units to the left, you obtain $t = 0$ represents 1990.

61. True. $|-x| = |x|$

63. (a) $g(t) = \frac{3}{4} f(t)$ (b) $g(t) = f(t) + 10{,}000$
   (c) $g(t) = f(t - 2)$

65. $(-2, 0), (-1, 1), (0, 2)$

67. $\dfrac{4}{x(1 - x)}$ **69.** $\dfrac{3x - 2}{x(x - 1)}$ **71.** $\dfrac{(x - 4)\sqrt{x^2 - 4}}{x^2 - 4}$

73. $5(x - 3), x \neq -3$

75. (a) 38   (b) $\frac{57}{4}$ (c) $x^2 - 12x + 38$

77. All real numbers $x \neq 11$ **79.** $-9 \leq x \leq 9$

Section 2.5 *(page 230)*

1.   **3.**

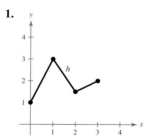   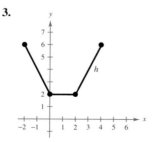

5. (a) $2x$ (b) 4   (c) $x^2 - 4$ (d) $\dfrac{x + 2}{x - 2}; \ x \neq 2$

7. (a) $x^2 - x + 2$ (b) $x^2 + x - 2$
   (c) $2x^2 - x^3$ (d) $\dfrac{x^2}{2 - x}; \ x \neq 2$

9. (a) $x^2 + 6 + \sqrt{1 - x}$ (b) $x^2 + 6 - \sqrt{1 - x}$
   (c) $(x^2 + 6)\sqrt{1 - x}$ (d) $\dfrac{(x^2 + 6)\sqrt{1 - x}}{1 - x}; \ x < 1$

11. (a) $\dfrac{x + 1}{x^2}$ (b) $\dfrac{x - 1}{x^2}$ (c) $\dfrac{1}{x^3}$ (d) $x; \ x \neq 0$

13. 3   **15.** 5   **17.** $9t^2 - 3t + 5$ **19.** 74

21. 26   **23.** $\frac{3}{5}$

25.   **27.**

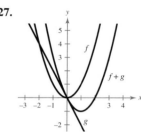

29.

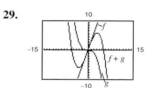

$f(x), g(x)$

31. $T = \frac{3}{4}x + \frac{1}{15}x^2$

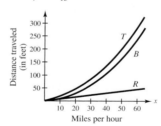

33. $y_1 = -0.59t^2 + 7.66t - 144.90$
   $y_2 = 16.58t + 245.06$
   $y_3 = 1.85t + 21.88$

35. (a) For each time t there corresponds one and only one temperature T.
   (b) $60°, \ 72°$
   (c) All the temperature changes would occur 1 hour later.
   (d) The temperature would be decreased by 1 degree.

37. (a) $(x - 1)^2$ (b) $x^2 - 1$ (c) x^4

39. (a) $20 - 3x$ (b) $-3x$ (c) $9x + 20$

41. (a) $\sqrt{x^2 + 4}$ (b) $x + 4$

Domain of f and $g \circ f$: $x \geq -4$;

Domain of g and $f \circ g$: all real numbers

43. (a) $x - \frac{8}{3}$ (b) $x - 8$

Domain of $f, g, f \circ g$, and $g \circ f$: all real numbers

45. (a) x^{16} (b) x^{16}

Domain of $f, g, f \circ g$, and $g \circ f$: all real numbers

47. (a) $|x + 6|$ (b) $|x| + 6$

Domain of $f, g, f \circ g$, and $g \circ f$: all real numbers

49. (a) $\dfrac{1}{x + 3}$ (b) $\dfrac{1}{x} + 3$

Domain of f and $g \circ f$: all real numbers $x \neq 0$

Domain of g: all real numbers

Domain of $f \circ g$: all real numbers $x \neq -3$

51. (a) 3   (b) 0   **53.** (a) 0   (b) 4

55. $f(x) = x^2$, $g(x) = 2x + 1$

57. $f(x) = \sqrt[3]{x}$, $g(x) = x^2 - 4$

59. $f(x) = \dfrac{1}{x}$, $g(x) = x + 2$

61. $f(x) = \dfrac{x + 3}{4 + x}$, $g(x) = -x^2$

63. (a) $r(x) = \dfrac{x}{2}$ (b) $A(r) = \pi r^2$

(c) $(A \circ r)(x) = \pi\left(\dfrac{x}{2}\right)^2$; $(A \circ r)(x)$ represents the area of the circular base of the tank on the square foundation with side length x.

65. $(C \circ x)(t) = 3000t + 750$; $(C \circ x)(t)$ represents the cost after t production hours.

67. True. The range of g must be a subset of the domain of f for $(f \circ g)(x)$ to be defined.

69. Answers will vary.

71. 3   **73.** $\dfrac{-4}{x(x + h)}$

75. $3x - y - 10 = 0$ **77.** $3x + 2y - 22 = 0$

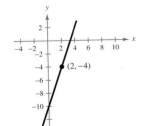

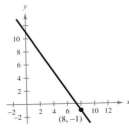

Section 2.6   *(page 239)*

1. c   **3.** a   **5.** $f^{-1}(x) = \frac{1}{6}x$ **7.** $f^{-1}(x) = x - 9$

9. $f^{-1}(x) = \dfrac{x - 1}{3}$ **11.** $f^{-1}(x) = x^3$

13. (a) $f(g(x)) = f\left(\dfrac{x}{2}\right) = 2\left(\dfrac{x}{2}\right) = x$

$g(f(x)) = g(2x) = \dfrac{(2x)}{2} = x$

(b)

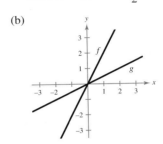

15. (a) $f(g(x)) = f\left(\dfrac{x - 1}{5}\right) = 5\left(\dfrac{x - 1}{5}\right) + 1 = x$

$g(f(x)) = g(5x + 1) = \dfrac{(5x + 1) - 1}{5} = x$

(b)

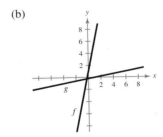

17. (a) $f(g(x)) = f\left(\sqrt[3]{x}\right) = \left(\sqrt[3]{x}\right)^3 = x$

$g(f(x)) = g(x^3) = \sqrt[3]{x^3} = x$

(b)

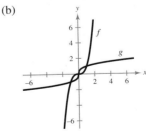

19. (a) $f(g(x)) = f(x^2 + 4)$, $x \geq 0$

$= \sqrt{(x^2 + 4) - 4} = x$

$g(f(x)) = g\left(\sqrt{x - 4}\right)$

$= \left(\sqrt{x - 4}\right)^2 + 4 = x$

(b)

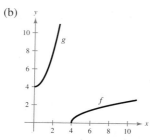

21. (a) $f(g(x)) = f(\sqrt{9-x}),\ x \le 9$

$\quad = 9 - (\sqrt{9-x})^2 = x$

$\quad g(f(x)) = g(9 - x^2),\ x \ge 0$

$\quad = \sqrt{9 - (9 - x^2)} = x$

(b)

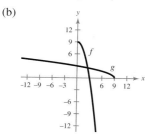

23. (a) $f(g(x)) = f\left(-\dfrac{5x+1}{x-1}\right) = \dfrac{-\left(\dfrac{5x+1}{x-1}\right)-1}{-\left(\dfrac{5x+1}{x-1}\right)+5}$

$\quad = \dfrac{-5x - 1 - x + 1}{-5x - 1 + 5x - 5} = x$

$\quad g(f(x)) = g\left(\dfrac{x-1}{x+5}\right) = \dfrac{-5\left(\dfrac{x-1}{x+5}\right)-1}{\dfrac{x-1}{x+5}-1}$

$\quad = \dfrac{-5x + 5 - x - 5}{x - 1 - x - 5} = x$

(b)

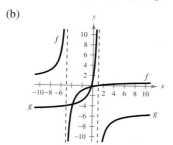

25. No

27.

x	-2	0	2	4	6	8
$f^{-1}(x)$	-2	-1	0	1	2	3

29. Yes **31.** No

33.

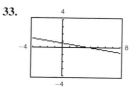
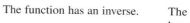

The function has an inverse.

35.

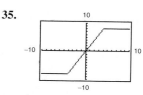

The function does not have an inverse.

37.

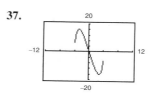

The function does not have an inverse.

39. $f^{-1}(x) = \dfrac{x+3}{2}$ **41.** $f^{-1}(x) = \sqrt[5]{x+2}$

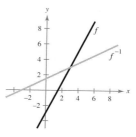
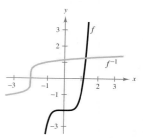

43. $f^{-1}(x) = x^2,\ x \ge 0$

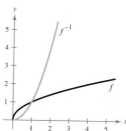

45. $f^{-1}(x) = \sqrt{4 - x^2},\ 0 \le x \le 2$

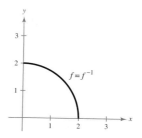

47. $f^{-1}(x) = \dfrac{4}{x}$

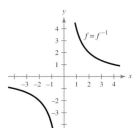

49. $f^{-1}(x) = \dfrac{2x + 1}{x - 1}$

51. $f^{-1}(x) = x^3 + 1$

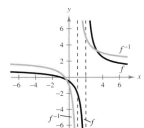 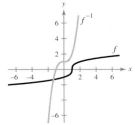

53. $f^{-1}(x) = \dfrac{5x - 4}{6 - 4x}$

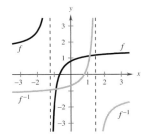

55. No inverse   **57.** $g^{-1}(x) = 8x$ **59.** No inverse

61. $f^{-1}(x) = \sqrt{x} - 3$ **63.** No inverse

65. $h^{-1}(x) = \dfrac{1}{x}$ **67.** $f^{-1}(x) = \dfrac{x^2 - 3}{2}, \; x \geq 0$ **69.** 32

71. 600   **73.** $2\sqrt[3]{x + 3}$ **75.** $\dfrac{x + 1}{2}$ **77.** $\dfrac{x + 1}{2}$

79. (a) $y = \dfrac{x - 8}{0.75}$

(b) $y = $ number of units produced; $x = $ hourly wage

(c) 19 units

81. (a) $y = \sqrt{\dfrac{x - 245.50}{0.03}}, \; 245.5 < x < 545.5$

$x = $ degrees Fahrenheit; $y = \%$ load

(b) 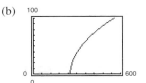   **(c)** $0 < x < 92.11$

83. No. The function would not pass the Horizontal Line Test.

85. (a) Yes. f^{-1} yields the year for a given per capita regular soft drink consumption.

(b) 5

87. True. If $f(x) = x - 6$ and $f^{-1}(x) = x + 6$, then the y-intercept of f is $(0, -6)$ and the x-intercept of f^{-1} is $(-6, 0)$.

89. False. $f(x) = \dfrac{1}{x} = f^{-1}(x)$.

91.

x	-2	-1	1	3
y	-5	-2	2	3

x	-5	-2	2	3
$f^{-1}(x)$	-2	-1	1	3

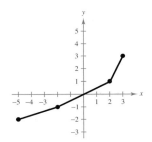

93.

x	-4	-2	0	3
y	3	4	0	-1

The graph of f does not pass the Horizontal Line Test, so $f^{-1}(x)$ does not exist.

95. ± 8 **97.** $\dfrac{3}{2}$ **99.** $3 \pm \sqrt{5}$ **101.** $5, -\dfrac{10}{3}$

103. All real numbers   **105.** All real numbers $x \neq 0, 4$

107. 16, 18   **109.** $b = h = 2\sqrt{5}$ feet

Review Exercises   *(page 244)*

1. (a) L_2 (b) L_3 (c) L_1 (d) L_4

3.

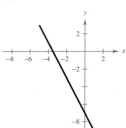

5.

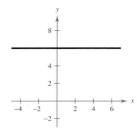

7.

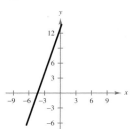

9.

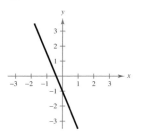

11. $t = \frac{7}{3}$ **13.** $(6, 0),\ (10, 1),\ (-2, -2)$

15.

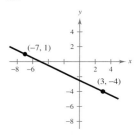

17.

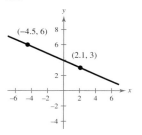

$m = -\frac{1}{2}$ $m = -\frac{5}{11}$

19. $x = 0$ **21.** $4x + 3y - 8 = 0$

23. $3x - 2y - 10 = 0$ **25.** $x + 2y - 4 = 0$

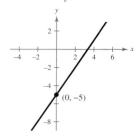

27. (a) $5x - 4y - 23 = 0$ (b) $4x + 5y - 2 = 0$

29. $V = 850t + 12{,}500,\ 0 \le t \le 5$

31. \$210,000

33. (a) Not a function because 20 in the domain corresponds to two values in the range.

(b) Function

(c) Function

(d) Not a function because 30 in A is not matched with any element in B.

35. No   **37.** Yes

39. (a) 5   (b) 17   (c) $t^4 + 1$ (d) $-x^2 - 1$

41. $-5 \le x \le 5$ **43.** All real numbers $s \ne 3$

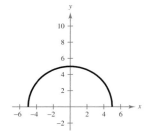
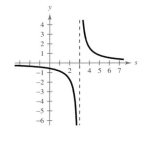

45. All real numbers $x \ne 3, -2$

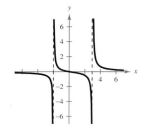

47. (a) 16 feet per second   (b) 1.5 seconds

(c) -16 feet per second

49. (a) $A = x(12 - x)$

(b) $0 < x < 12$

51. Function   **53.** Not a function

55. $\frac{7}{3}, 3$ **57.** $-\frac{3}{8}$

59.

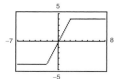

Increasing on $(-2, 2)$;

Constant on $(-\infty, -2)$ and $(2, \infty)$

61.
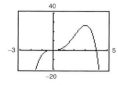

Increasing on $(-\infty, 3)$;
Decreasing on $(3, \infty)$

63. $f(x) = -3x$ **65.** $f(x) = \frac{5}{3}x + \frac{10}{3}$

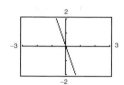   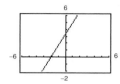

67.
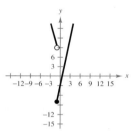

69. Neither even nor odd   **71.** Odd

73. Vertical shift of 4 units upward and horizontal shift of 4 units to the left of $y = x^3$

75. Horizontal shift of 6 units to the right of $y = |x|$

77. Vertical shift of 9 units downward

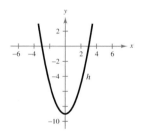

79. Horizontal shift of 7 units to the right

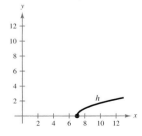

81. Reflection in x-axis, horizontal shift of 3 units to the left, and vertical shift of 1 unit upward

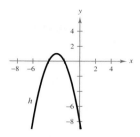

83. Reflection in x-axis, horizontal shift of 1 unit to the left, and vertical shift of 9 units upward

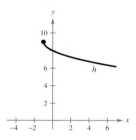

85. Reflection in x-axis, horizontal shift of 1 unit to the left, vertical shift of 3 units downward, and vertical stretch

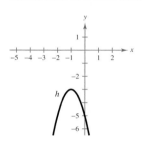

87. Reflection in x-axis, vertical stretch, and horizontal shift of 4 units to the right

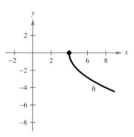

89. -7 **91.** 5   **93.** 23   **95.** $\sqrt{7}$

97. $y_1 = 0.80t^2 + 3.34t + 24.23$
$y_2 = -0.43t^2 + 18.14t + 62.89$

99. $f^{-1}(x) = \dfrac{x}{6}$

$f(f^{-1}(x)) = 6\left(\dfrac{x}{6}\right) = x$

$f^{-1}(f(x)) = \dfrac{6x}{6} = x$

101. $f^{-1}(x) = x + 7$

$f(f^{-1}(x)) = x + 7 - 7 = x$

$f^{-1}(f(x)) = x - 7 + 7 = x$

103.

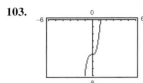

105.

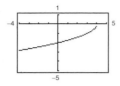

The function has an inverse.

The function has an inverse.

107.

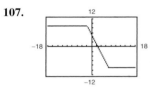

The function does not have an inverse.

109. (a) $f^{-1}(x) = 2x + 6$

(b)

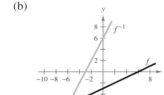

(c) $f^{-1}(f(x)) = 2\left(\tfrac{1}{2}x - 3\right) + 6 = x - 6 + 6 = x$

$f(f^{-1}(x)) = \tfrac{1}{2}(2x + 6) - 3 = x + 3 - 3 = x$

111. (a) $f^{-1}(x) = x^2 - 1,\ x \geq 0$

(b)

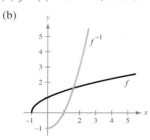

(c) $f^{-1}(f(x)) = f^{-1}\left(\sqrt{x + 1}\right)$

$= (x + 1) - 1$

$= x$

$f(f^{-1}(x)) = f(x^2 - 1),\ x \geq 0$

$= \sqrt{x^2 - 1 + 1}$

$= x$

113. $x \geq 4;\ f^{-1}(x) = \sqrt{\dfrac{x}{2} + 4}$

115. False. The graph is reflected in the x-axis, shifted 9 units to the left, and then shifted 13 units downward.

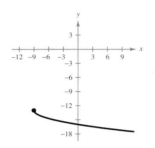

117. A function from a set A to a set B is a relation that assigns to each element x in the set A exactly one element y in the set B.

Chapter Test *(page 249)*

1. $2x + y - 1 = 0$

2. $17x + 10y - 59 = 0$

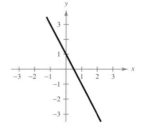

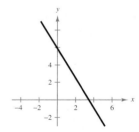

3. (a) $4x - 7y + 44 = 0$ (b) $7x + 4y - 53 = 0$

4. (a) -9 (b) 1 (c) $|x - 4| - 15$

5. (a) $-\dfrac{1}{8}$ (b) $-\dfrac{1}{28}$ (c) $\dfrac{\sqrt{x}}{x^2 - 18x}$

6. $-10 \leq x \leq 10$ **7.** All real numbers

8. (a)

(b) Increasing on $(-0.31, 0)$, $(0.31, \infty)$

Decreasing on $(-\infty, -0.31)$, $(0, 0.31)$

(c) Even

9. (a)

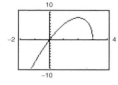

(b) Increasing on $(-\infty, 2)$

Decreasing on $(2, 3)$

(c) Neither even nor odd

10. (a)

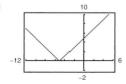

(b) Increasing on $(-5, \infty)$

Decreasing on $(-\infty, -5)$

(c) Neither even nor odd

11.

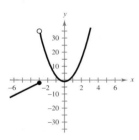

12. Reflection in x-axis and vertical shift of $y = x^3$

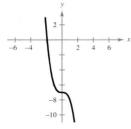

13. Reflection in x-axis, horizontal shift, and vertical shift of $y = \sqrt{x}$

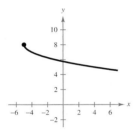

14. Vertical shrink, horizontal shift, and vertical shift of $y = |x|$

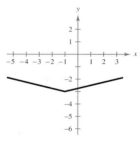

15. -2 **16.** 12 **17.** -35 **18.** 5

19. $f^{-1}(x) = \sqrt[3]{x - 8}$ **20.** No inverse

21. $f^{-1}(x) = \left(\frac{8}{3}x\right)^{2/3}$, $x \geq 0$ **22.** $\$153$

Cumulative Test for Chapters P–2
(page 250)

1. $\dfrac{4x^3}{15y^5}$, $x \neq 0$ **2.** $2x^2y\sqrt{6y}$ **3.** $5x - 6$

4. $x^3 - x^2 - 5x + 6$ **5.** $\dfrac{s - 1}{(s + 1)(s + 3)}$

6. $(x + 3)(7 - x)$ **7.** $x(x + 1)(1 - 6x)$

8. $2(3 - 2x)(9 + 6x + 4x^2)$

9.

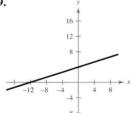

10.

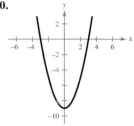

11.

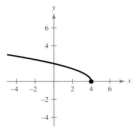

12. $4x^2 + 12x$ **13.** $\frac{3}{2}x^2 + 8x + \frac{5}{2}$ **14.** $1, 3$

15. $2 \pm \sqrt{10}$ **16.** ± 4

17. $\dfrac{-5 \pm \sqrt{97}}{6}$ **18.** $-\dfrac{3}{2} \pm \dfrac{\sqrt{69}}{6}$

19. ± 8 **20.** $0, -12, \pm 2i$ **21.** $0, 3$ **22.** ± 8

23. 6 **24.** $-5, 9$ **25.** No solution

26. (a) Not a solution (b) Not a solution (c) Solution
(d) Solution

27. (a) Not a solution (b) Not a solution (c) Solution
(d) Solution

28. (a) Not a solution (b) Solution (c) Not a solution
(d) Not a solution

29. $-7 \le x \le 5$

30. $x < -\frac{3}{2}, \; x > -\frac{1}{4}$

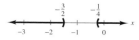

31. $x \le -\frac{7}{5}, \; x \ge -1$

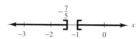

32. $x < \dfrac{1 - \sqrt{17}}{2}, \; x > \dfrac{1 + \sqrt{17}}{2}$

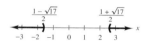

33. $2x - y + 2 = 0$

34. For some values of x there correspond two values of y.

35. (a) $\frac{3}{2}$ (b) Division by 0 is undefined. (c) $\dfrac{s + 2}{s}$

36. (a) Vertical shrink by $\frac{1}{2}$
(b) Vertical shift of 2 units upward
(c) Horizontal shift of 2 units to the left

37. (a) $5x - 2$ (b) $-3x - 4$ (c) $4x^2 - 11x - 3$
(d) $\dfrac{x - 3}{4x + 1}$; Domain: all real numbers except $x = -\frac{1}{4}$

38. (a) $\sqrt{x - 1} + x^2 + 1$ (b) $\sqrt{x - 1} - x^2 - 1$
(c) $x^2\sqrt{x - 1} + \sqrt{x - 1}$ (d) $\dfrac{\sqrt{x - 1}}{x^2 + 1}$; Domain: $x \ge 1$

39. (a) $2x + 12$ (b) $\sqrt{2x^2 + 6}$

40. (a) $|x| - 2$ (b) $|x - 2|$

41. $h^{-1}(x) = \frac{1}{5}(x + 2)$ **42.** 9 people

43. (a) $R(n) = n[8 - 0.05(n - 80)], \; n \ge 80$

(b)

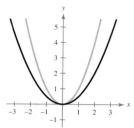

120 passengers

Chapter 3

Section 3.1 *(page 260)*

1. g   **3.** b   **5.** f   **7.** e

9. (a)   (b)

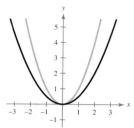

Vertical shrink

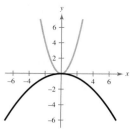

Vertical shrink and
reflection in the x-axis

(c)   (d)

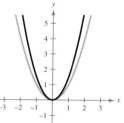

Vertical stretch

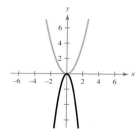

Vertical stretch and
reflection in the x-axis

11. (a)   (b)

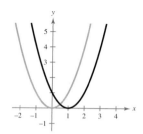

Horizontal translation

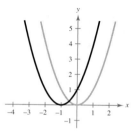

Horizontal translation

(c)

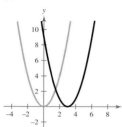

(d)

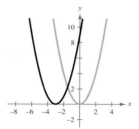

Horizontal translation

Horizontal translation

13.

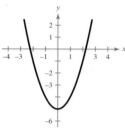

Vertex: $(0, -5)$

x-intercepts: $\left(\pm\sqrt{5}, 0\right)$

15.

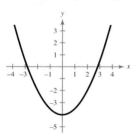

Vertex: $(0, -4)$

x-intercepts: $\left(\pm 2\sqrt{2}, 0\right)$

17.

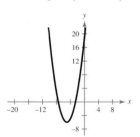

Vertex: $(-5, -6)$

x-intercepts: $\left(-5 \pm \sqrt{6}, 0\right)$

19.

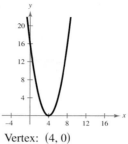

Vertex: $(4, 0)$

x-intercept: $(4, 0)$

21.

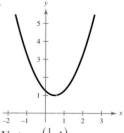

Vertex: $\left(\frac{1}{2}, 1\right)$

No x-intercept

23.

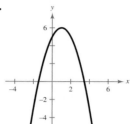

Vertex: $(1, 6)$

x-intercepts: $\left(1 \pm \sqrt{6}, 0\right)$

25.

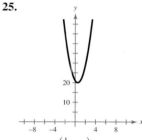

Vertex: $\left(\frac{1}{2}, 20\right)$

No x-intercept

27.

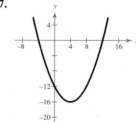

Vertex: $(4, -16)$

x-intercepts: $(-4, 0), (12, 0)$

29.

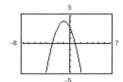

Vertex: $(-1, 4)$

x-intercepts: $(1, 0), (-3, 0)$

31.

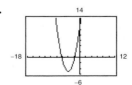

Vertex: $(-4, -5)$

x-intercepts: $\left(-4 \pm \sqrt{5}, 0\right)$

33.

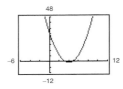

Vertex: $(4, -1)$

x-intercepts: $\left(4 \pm \frac{1}{2}\sqrt{2}, 0\right)$

35.

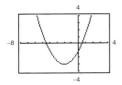

Vertex: $(-2, -3)$

x-intercepts: $\left(-2 \pm \sqrt{6}, 0\right)$

37. $y = (x - 1)^2$ **39.** $y = -(x + 1)^2 + 4$

41. $y = -2(x + 2)^2 + 2$ **43.** $f(x) = (x + 2)^2 + 5$

45. $f(x) = -\frac{1}{2}(x - 3)^2 + 4$ **47.** $f(x) = \frac{3}{4}(x - 5)^2 + 12$

49. $f(x) = -\frac{24}{49}\left(x + \frac{1}{4}\right)^2 + \frac{3}{2}$ **51.** $f(x) = -\frac{16}{3}\left(x + \frac{5}{2}\right)^2$

53. $(\pm 4, 0)$. The x-intercepts and solutions of the equation are the same.

55. $(5, 0), (-1, 0)$. The x-intercepts and solutions of the equation are the same.

57.

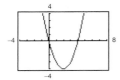

$(0, 0), (4, 0)$

59.

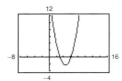

$(3, 0), (6, 0)$

61.

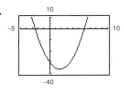

$\left(-\frac{5}{2}, 0\right), (6, 0)$

63.

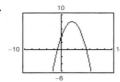

$(7, 0), (-1, 0)$

65. $f(x) = x^2 - 2x - 3$
$g(x) = -x^2 + 2x + 3$

67. $f(x) = x^2 - 10x$
$g(x) = -x^2 + 10x$

69. $f(x) = 2x^2 + 7x + 3$
$g(x) = -2x^2 - 7x - 3$

71. 55, 55   **73.** 12, 6

75. (a) $A = x(50 - x), 0 < x < 50$

(b)

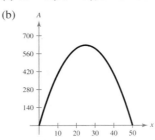

(c) 25 feet $\times$ 25 feet

77. (a)

x	y	Area
2	$\frac{1}{3}[200 - 4(2)]$	$(2)(2)\left(\frac{1}{3}\right)[200 - 4(2)] = 256$
4	$\frac{1}{3}[200 - 4(4)]$	$(2)(4)\left(\frac{1}{3}\right)[200 - 4(4)] \approx 491$
6	$\frac{1}{3}[200 - 4(6)]$	$(2)(6)\left(\frac{1}{3}\right)[200 - 4(6)] = 704$
8	$\frac{1}{3}[200 - 4(8)]$	$(2)(8)\left(\frac{1}{3}\right)[200 - 4(8)] = 896$
10	$\frac{1}{3}[200 - 4(10)]$	$(2)(10)\left(\frac{1}{3}\right)[200 - 4(10)] \approx 1067$
12	$\frac{1}{3}[200 - 4(12)]$	$(2)(12)\left(\frac{1}{3}\right)[200 - 4(12)] = 1216$

(b)

x	y	Area
20	$\frac{1}{3}[200 - 4(20)]$	$(2)(20)\left(\frac{1}{3}\right)[200 - 4(20)] = 1600$
22	$\frac{1}{3}[200 - 4(22)]$	$(2)(22)\left(\frac{1}{3}\right)[200 - 4(22)] \approx 1643$
24	$\frac{1}{3}[200 - 4(24)]$	$(2)(24)\left(\frac{1}{3}\right)[200 - 4(24)] = 1664$
26	$\frac{1}{3}[200 - 4(26)]$	$(2)(26)\left(\frac{1}{3}\right)[200 - 4(26)] = 1664$
28	$\frac{1}{3}[200 - 4(28)]$	$(2)(28)\left(\frac{1}{3}\right)[200 - 4(28)] \approx 1643$
30	$\frac{1}{3}[200 - 4(30)]$	$(2)(30)\left(\frac{1}{3}\right)[200 - 4(30)] = 1600$

$x = 25$ ft, $y = 33\frac{1}{3}$ ft

(c) $A = \dfrac{8x(50 - x)}{3}$

(d)

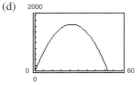

$x = 25$ feet; $y = 33\frac{1}{3}$ feet

(e) $A = -\frac{8}{3}(x - 25)^2 + \frac{5000}{3}$

79. 4500 units   **81.** 20 fixtures   **83.** 350,000 units

85. (a) 4 feet   (b) 16 feet   (c) 25.86 feet   **87.** 16 feet

89. (a)

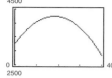

(b) 4242; Yes   (c) 9703 annually; 27 daily

91. (a) and (c)

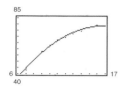

(b) $y = -0.35t^2 + 11.8t - 21$

(d) No. The model decreases and eventually becomes negative.

93. True. The vertex of $f(x)$ is $\left(-\frac{5}{4}, \frac{53}{4}\right)$ and the vertex of $g(x)$ is $\left(-\frac{5}{4}, -\frac{71}{4}\right)$.

95. Conditions (a) and (d) are preferable because profits would be increasing.

97. Answers will vary.   **99.** $y = -\frac{1}{3}x + \frac{5}{3}$

101. $y = \frac{5}{4}x + 3$ **103.** 27   **105.** $-\frac{1408}{49}$ **107.** 109

Section 3.2   (page 274)

1. c   **3.** h   **5.** a   **7.** d

9. (a)

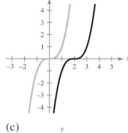

(b)

(c)

(d)

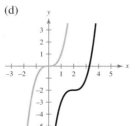

11. (a)

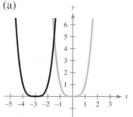

(b)

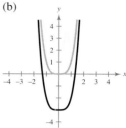

(c)

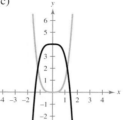

(d)

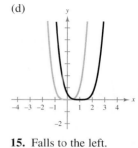

13. Falls to the left.
Rises to the right.

15. Falls to the left.
Falls to the right.

17. Rises to the left.
Falls to the right.

19. Rises to the left.
Falls to the right.

21. Falls to the left.
Falls to the right.

23.

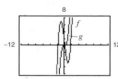

25.

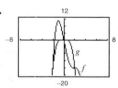

27. ± 5 **29.** 3   **31.** $1, -2$ **33.** $2 \pm \sqrt{3}$

35. $2, 0$ **37.** ± 1 **39.** $0, \pm \sqrt{3}$ **41.** No real zeros

43.

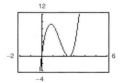

45.

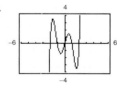

$(0, 0), \left(\frac{5}{2}, 0\right)$ $(0, 0), (\pm 1, 0), (\pm 2, 0)$

47. $f(x) = x^2 - 10x$ **49.** $f(x) = x^2 + 4x - 12$

51. $f(x) = x^3 + 5x^2 + 6x$

53. $f(x) = x^4 - 4x^3 - 9x^2 + 36x$

55. $f(x) = x^2 - 2x - 2$ **57.** $f(x) = x^2 + 4x + 4$

59. $f(x) = x^3 + 2x^2 - 3x$ **61.** $f(x) = x^3 - 3x$

63. $f(x) = x^4 + x^3 - 15x^2 + 23x - 10$

65. $f(x) = x^5 + 16x^4 + 96x^3 + 256x^2 + 256x$

67.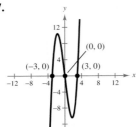

69.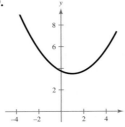

71.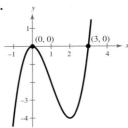

73.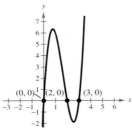

75.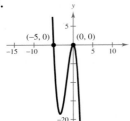

77.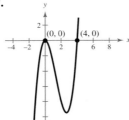

79.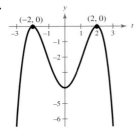

81.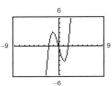

Zeros: $0, \pm 2$, all of odd multiplicity

83.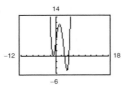

Zeros: -1, even multiplicity; $3, \frac{9}{2}$, odd multiplicity

85. $(-1, 0), (1, 2), (2, 3)$ **87.** $(-2, -1), (0, 1)$

89. (a) and (b)

Box height	Box width	Box volume
1	$36 - 2(1)$	$1[36 - 2(1)]^2 = 1156$
2	$36 - 2(2)$	$2[36 - 2(2)]^2 = 2048$
3	$36 - 2(3)$	$3[36 - 2(3)]^2 = 2700$
4	$36 - 2(4)$	$4[36 - 2(4)]^2 = 3136$
5	$36 - 2(5)$	$5[36 - 2(5)]^2 = 3380$
6	$36 - 2(6)$	$6[36 - 2(6)]^2 = 3456$
7	$36 - 2(7)$	$7[36 - 2(7)]^2 = 3388$

6 in. $\times$ 24 in. $\times$ 24 in.

(c) Domain: $0 < x < 18$

(d)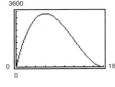

$x = 6$

91. $x = 200$

93. False. A fifth-degree polynomial can have at most four turning points.

95. (a) Degree: 3; Leading coefficient: positive

(b) Degree: 2; Leading coefficient: positive

(c) Degree: 4; Leading coefficient: positive

(d) Degree: 5; Leading coefficient: positive

97. (a)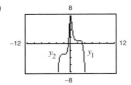

y_1 is decreasing. y_2 is increasing.

(b) Either always increasing or always decreasing. The behavior is determined by a.

(c)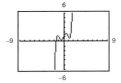

Because $H(x)$ is not always increasing or always decreasing, H cannot be written in the form $H(x) = a(x - h)^5 + k$.

99. $-\frac{2}{3}, 8$ **101.** -12 **103.** $4 \pm \sqrt{14}$

105. $\dfrac{-2 \pm \sqrt{31}}{3}$ **107.** $x(6x - 1)(x - 10)$

109. $(y + 6)(y^2 - 6y + 36)$

Section 3.3   *(page 285)*

1. Answers will vary.   **3.** Answers will vary.

5.

7. $2x + 4$ **9.** $x^2 - 3x + 1$ **11.** $x^3 + 3x^2 - 1$

13. $7 - \dfrac{11}{x + 2}$ **15.** $3x + 5 - \dfrac{2x - 3}{2x^2 + 1}$

17. $x^2 + 2x + 4 + \dfrac{2x - 11}{x^2 - 2x + 3}$

19. $x + 3 + \dfrac{6x^2 - 8x + 3}{(x - 1)^3}$ **21.** $3x^2 - 2x + 5$

23. $4x^2 - 9$ **25.** $-x^2 + 10x - 25$

27. $5x^2 + 14x + 56 + \dfrac{232}{x - 4}$

29. $10x^3 + 10x^2 + 60x + 360 + \dfrac{1360}{x - 6}$

31. $x^2 - 8x + 64$

33. $-3x^3 - 6x^2 - 12x - 24 - \dfrac{48}{x - 2}$

35. $-x^3 - 6x^2 - 36x - 36 - \dfrac{216}{x - 6}$

37. $4x^2 + 14x - 30$

39. $f(x) = (x - 4)(x^2 + 3x - 2) + 3, \quad f(4) = 3$

41. $f(x) = \left(x + \frac{2}{3}\right)(15x^3 - 6x + 4) + \frac{34}{3},$
$f\left(-\frac{2}{3}\right) = \frac{34}{3}$

43. $f(x) = \left(x - \sqrt{2}\right)\left[x^2 + \left(3 + \sqrt{2}\right)x + 3\sqrt{2}\right] - 8,$
$f\left(\sqrt{2}\right) = -8$

45. $f(x) = \left(x - 1 + \sqrt{3}\right)\left[-4x^2 + \left(2 + 4\sqrt{3}\right)x + \left(2 + 2\sqrt{3}\right)\right],$
$f\left(1 - \sqrt{3}\right) = 0$

47. (a) 1   (b) 4   (c) 4   (d) 1954

49. (a) 97   (b) $-\frac{5}{3}$ (c) 17   (d) -199

51. $(x - 2)(x + 3)(x - 1)$; Zeros: 2, -3, 1

53. $(2x - 1)(x - 5)(x - 2)$; Zeros: $\frac{1}{2}$, 5, 2

55. $\left(x + \sqrt{3}\right)\left(x - \sqrt{3}\right)(x + 2)$; Zeros: $-\sqrt{3}, \sqrt{3}, -2$

57. $(x - 1)\left(x - 1 - \sqrt{3}\right)\left(x - 1 + \sqrt{3}\right)$;
Zeros: 1, $1 + \sqrt{3}$, $1 - \sqrt{3}$

59. (a) Answers will vary.   (b) $2x - 1$
(c) $f(x) = (2x - 1)(x + 2)(x - 1)$ (d) $\frac{1}{2}, -2, 1$
(e)

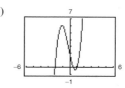

61. (a) Answers will vary.   (b) $(x - 1), (x - 2)$
(c) $f(x) = (x - 1)(x - 2)(x - 5)(x + 4)$
(d) $1, 2, 5, -4$ (e)

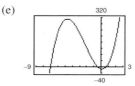

63. (a) Answers will vary.   (b) $x + 7$
(c) $f(x) = (x + 7)(2x + 1)(3x - 2)$ (d) $-7, -\frac{1}{2}, \frac{2}{3}$
(e)

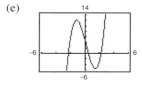

65. (a) Answers will vary.   (b) $\left(x - \sqrt{5}\right)$
(c) $f(x) = \left(x - \sqrt{5}\right)\left(x + \sqrt{5}\right)(2x - 1)$ (d) $\pm\sqrt{5}, \frac{1}{2}$
(e)

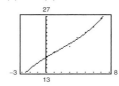

67. (a) Zeros are 2 and $\approx\pm 2.236$.
(b) $f(x) = (x - 2)\left(x - \sqrt{5}\right)\left(x + \sqrt{5}\right)$

69. (a) Zeros are -2, ≈ 0.268, and ≈ 3.732.
(b) $h(t) = (t + 2)\left[t - \left(2 + \sqrt{3}\right)\right]\left[t - \left(2 - \sqrt{3}\right)\right]$

71. $2x^2 - x - 1, \quad x \neq \frac{3}{2}$ **73.** $x^2 + 2x - 3, \quad x \neq -1$

75. $x^2 + 3x, \quad x \neq -2, -1$

77. (a) and (b)

$R = 0.01326t^3 - 0.0677t^2 + 1.231t + 16.68$

(c)

t	-2	-1	0	1	2
R	13.84	15.37	16.68	17.86	18.98

t	3	4	5	6	7
R	20.12	21.37	22.80	24.49	26.53

(d) $44.62. No, because the model will approach infinity quickly.

79. False. $-\frac{4}{7}$ is a zero of f.　　**81.** $x^{2n} + 6x^n + 9$

83. The remainder is 0.　　**85.** $c = -210$

87. 0; $x + 3$ is a factor of f.

89. (a) $f(x) = (x - 2)x^2 + 5 = x^3 - 2x^2 + 5$

(b) $f(x) = -(x + 3)x^2 + 1 = -x^3 - 3x^2 + 1$

91. $\pm\dfrac{\sqrt{21}}{4}$　　**93.** $\dfrac{5}{4}, \dfrac{3}{2}$　　**95.** $\dfrac{-3 \pm \sqrt{21}}{2}$

97. $f(x) = x^2 + 5x - 6$

(Answer is not unique.)

99. $f(x) = x^4 - 3x^3 - 5x^2 + 9x - 2$

(Answer is not unique.)

Section 3.4 *(page 298)*

1. $0, 6$　　**3.** $2, -4$　　**5.** $-6, \pm i$　　**7.** $\pm 1, \pm 3$

9. $\pm 1, \pm 3, \pm 5, \pm 9, \pm 15, \pm 45, \pm\frac{1}{2}, \pm\frac{3}{2}, \pm\frac{5}{2}, \pm\frac{9}{2}, \pm\frac{15}{2}, \pm\frac{45}{2}$

11. $1, 2, 3$　　**13.** $1, -1, 4$　　**15.** $-1, -10$

17. $\frac{1}{2}, -1$　　**19.** $-2, 3, \pm\frac{2}{3}$　　**21.** $-1, 2$　　**23.** $-6, \frac{1}{2}, 1$

25. (a) $\pm 1, \pm 2, \pm 4$

(b)

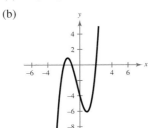

(c) $-2, -1, 2$

27. (a) $\pm 1, \pm 3, \pm\frac{1}{2}, \pm\frac{3}{2}, \pm\frac{1}{4}, \pm\frac{3}{4}$

(b)

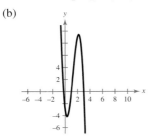

(c) $-\frac{1}{4}, 1, 3$

29. (a) $\pm 1, \pm 2, \pm 4, \pm 8, \pm\frac{1}{2}$

(b)

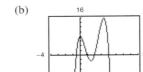

(c) $-\frac{1}{2}, 1, 2, 4$

31. (a) $\pm 1, \pm 3, \pm\frac{1}{2}, \pm\frac{3}{2}, \pm\frac{1}{4}, \pm\frac{3}{4}, \pm\frac{1}{8}, \pm\frac{3}{8}, \pm\frac{1}{16}, \pm\frac{3}{16}, \pm\frac{1}{32}, \pm\frac{3}{32}$

(b)

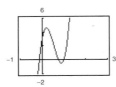

(c) $1, \frac{3}{4}, -\frac{1}{8}$

33. (a) $\pm 1, \approx\pm 1.414$

(b) $f(x) = (x + 1)(x - 1)(x + \sqrt{2})(x - \sqrt{2})$

35. (a) $0, 3, 4, \approx\pm 1.414$

(b) $h(x) = x(x - 3)(x - 4)(x + \sqrt{2})(x - \sqrt{2})$

37. $x^3 - x^2 + 25x - 25$　　**39.** $x^3 + 4x^2 - 31x - 174$

41. $3x^4 - 17x^3 + 25x^2 + 23x - 22$

43. (a) $(x^2 + 9)(x^2 - 3)$

(b) $(x^2 + 9)(x + \sqrt{3})(x - \sqrt{3})$

(c) $(x + 3i)(x - 3i)(x + \sqrt{3})(x - \sqrt{3})$

45. (a) $(x^2 - 2x - 2)(x^2 - 2x + 3)$

(b) $(x - 1 + \sqrt{3})(x - 1 - \sqrt{3})(x^2 - 2x + 3)$

(c) $(x - 1 + \sqrt{3})(x - 1 - \sqrt{3})(x - 1 + \sqrt{2}i)$
$(x - 1 - \sqrt{2}i)$

47. $-\frac{3}{2}, \pm 5i$　　**49.** $\pm 2i, 1, -\frac{1}{2}$　　**51.** $-3 \pm i, \frac{1}{4}$

53. $2, -3 \pm \sqrt{2}i, 1$　　**55.** $\pm 5i; (x + 5i)(x - 5i)$

57. $2 \pm \sqrt{3}; (x - 2 - \sqrt{3})(x - 2 + \sqrt{3})$

59. $\pm 3, \pm 3i; (x + 3)(x - 3)(x + 3i)(x - 3i)$

61. $1 \pm i; (z - 1 + i)(z - 1 - i)$

63. $2, 2 \pm i; (x - 2)(x - 2 + i)(x - 2 - i)$

65. $-2, 1 \pm \sqrt{2}i; (x + 2)(x - 1 + \sqrt{2}i)(x - 1 - \sqrt{2}i)$

67. $-\frac{1}{5}, 1 \pm \sqrt{5}i; (5x + 1)(x - 1 + \sqrt{5}i)(x - 1 - \sqrt{5}i)$

69. $2, \pm 2i; (x - 2)^2(x + 2i)(x - 2i)$

71. $\pm i, \pm 3i; (x + i)(x - i)(x + 3i)(x - 3i)$

73. $-10, -7 \pm 5i$　　**75.** $-\frac{3}{4}, 1 \pm \frac{1}{2}i$　　**77.** $-2, -\frac{1}{2}, \pm i$

79. No real zeros　　**81.** No real zeros

83. One positive zero　　**85.** One or three positive zeros

87. Answers will vary.   **89.** Answers will vary.

91. $1, -\frac{1}{2}$ **93.** $-\frac{3}{4}$ **95.** $\pm 2, \pm \frac{3}{2}$ **97.** $\pm 1, \frac{1}{4}$

99. d   **101.** b

103. $f(x) = -2x^3 + 3x^2 + 11x - 6$

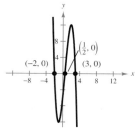

(Equations and graphs will vary.) There are infinitely many possible functions for f.

105. (a) $-2, 1, 4$

(b) The graph touches the x-axis at $x = 1$.

(c) The least possible degree of the function is 4 because there are at least four real zeros (1 is repeated) and a function can have at most the number of real zeros equal to the degree of the function. The degree cannot be odd by the definition of multiplicity.

(d) Positive. From the information in the table, you can conclude that the graph will eventually rise to the left and to the right.

(e) $f(x) = x^4 - 4x^3 - 3x^2 + 14x - 8$
(Answer is not unique.)

(f)

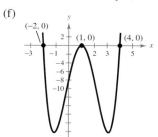

107. (a)

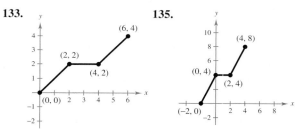

(b) $V = x(9 - 2x)(15 - 2x)$

Domain: $0 < x < \frac{9}{2}$

(c)

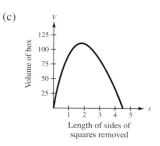

$1.82 \text{ cm} \times 5.36 \text{ cm} \times 11.36 \text{ cm}$

(d) $\frac{1}{2}, \frac{7}{2}, 8$; 8 is not in the domain of V.

109. $x \approx 38.4$, or \$384,000   **111.** $x \approx 40$

113. No. Setting $h = 64$ and solving the resulting equation yields imaginary roots.

115. False. The most complex zeros it can have is two and the Linear Factorization Theorem guarantees that there are three linear factors, so one zero must be real.

117. r_1, r_2, r_3 **119.** $5 + r_1, 5 + r_2, 5 + r_3$

121. The zeros cannot be determined.

123. (a) $0 < k < 4$ (b) $k = 4$ (c) $k < 0$ (d) $k > 4$

125. (a) $x^2 + b$ (b) $x^2 - 2ax + a^2 + b^2$

127. $-11 + 9i$ **129.** $20 + 40i$ **131.** i

133.   **135.**

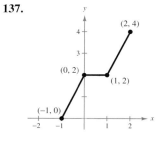

137.

Section 3.5   *(page 309)*

1.

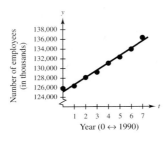

The model is a "good fit" for the actual data.

3. Inversely

5.

x	2	4	6	8	10
$y = kx^2$	4	16	36	64	100

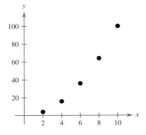

7.

x	2	4	6	8	10
$y = kx^2$	2	8	18	32	50

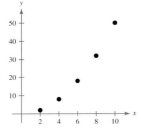

9.

x	2	4	6	8	10
$y = k/x^2$	$\frac{1}{2}$	$\frac{1}{8}$	$\frac{1}{18}$	$\frac{1}{32}$	$\frac{1}{50}$

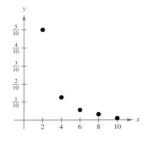

11.

x	2	4	6	8	10
$y = k/x^2$	$\frac{5}{2}$	$\frac{5}{8}$	$\frac{5}{18}$	$\frac{5}{32}$	$\frac{1}{10}$

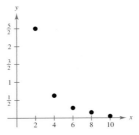

13. $y = \dfrac{5}{x}$ **15.** $y = -\dfrac{7}{10}x$ **17.** $y = \dfrac{12}{5}x$

19. $y = 205x$ **21.** $I = 0.035P$

23. Model: $y = \frac{33}{13}x$

Inches	5	10	20	25	30
Centimeters	12.7	25.4	50.8	63.5	76.2

25. $y = 0.0368x$; $7360

27. (a) 0.05 meter   (b) $176\frac{2}{3}$ newtons   **29.** 39.47 pounds

31. $A = kr^2$ **33.** $y = \dfrac{k}{x^2}$ **35.** $F = \dfrac{kg}{r^2}$

37. $P = \dfrac{k}{V}$ **39.** $F = \dfrac{km_1 m_2}{r^2}$

41. The area of a triangle is jointly proportional to its base and height.

43. The volume of a sphere varies directly as the cube of its radius.

45. Average speed is directly proportional to the distance and inversely proportional to the time.

47. $A = \pi r^2$ **49.** $y = \dfrac{28}{x}$ **51.** $F = 14rs^3$

53. $z = \dfrac{2x^2}{3y}$ **55.** ≈ 0.61 mile per hour   **57.** 506 feet

59. 400 feet   **61.** No. The 15-inch pizza is the best buy.

63. (a) The velocity is increased by one-third.

(b) The velocity is decreased by one-fourth.

65. (a)

(b) Yes. $k = 0.575$

(c) 5 pounds

67. (a)

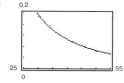

(b) 0.2857 microwatt per square centimeter

69. (a) $y = 127.4t + 218.4$

(b)

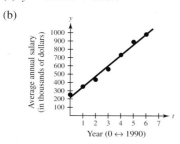

(c) 1997: $1110 thousand
1998: $1238 thousand
1999: $1365 thousand

(d) Answers will vary.

71. (a) $y = 489.58t + 5628.4$

(b)

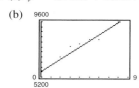

(c) $10,035 million   (d) Answers will vary.

73. (a) $y = -0.0186x + 689$

(b)

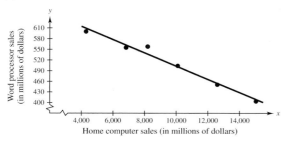

(c) $\approx$ $354 million

75. False. E is jointly proportional to the mass of an object and the square of its velocity.

77. Poor approximation   **79.** Good approximation

81.

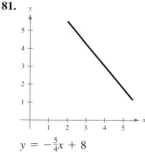

$y = -\frac{5}{4}x + 8$

83.

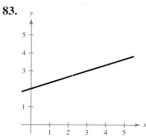

$y = \frac{1}{3}x + 2$

85. $x \leq 4, x \geq 6$

87. $x > 5$

89. (a) $-\frac{5}{3}$ (b) $-\frac{7}{3}$ (c) 21   **91.** All real numbers

93. All real numbers $x \neq -7$

Review Exercises   *(page 316)*

1. $f(x) = -\frac{1}{2}(x - 4)^2 + 1$ **3.** $f(x) = (x - 1)^2 - 4$

5. (a)

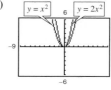

Vertical stretch

(b)

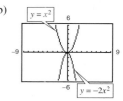

Vertical stretch and reflection in the x-axis

(c)

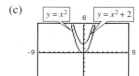

Vertical translation

(d)

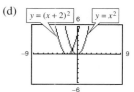

Horizontal translation

7. $g(x) = (x - 1)^2 - 1$ **9.** $f(x) = (x + 4)^2 - 6$

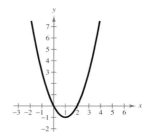

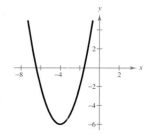

11. $f(t) = -2(t-1)^2 + 3$ **13.** $h(x) = 4\left(x + \frac{1}{2}\right)^2 + 12$

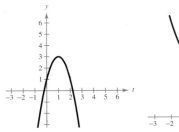
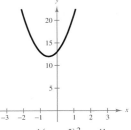

15. $h(x) = \left(x + \frac{5}{2}\right)^2 - \frac{41}{4}$ **17.** $f(x) = \frac{1}{3}\left(x + \frac{5}{2}\right)^2 - \frac{41}{12}$

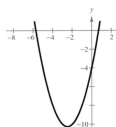
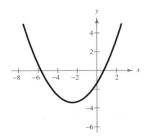

19. (a)

x	y	Area
1	$4 - \frac{1}{2}(1)$	$(1)[4 - \frac{1}{2}(1)] = \frac{7}{2}$
2	$4 - \frac{1}{2}(2)$	$(2)[4 - \frac{1}{2}(2)] = 6$
3	$4 - \frac{1}{2}(3)$	$(3)[4 - \frac{1}{2}(3)] = \frac{15}{2}$
4	$4 - \frac{1}{2}(4)$	$(4)[4 - \frac{1}{2}(4)] = 8$
5	$4 - \frac{1}{2}(5)$	$(5)[4 - \frac{1}{2}(5)] = \frac{15}{2}$
6	$4 - \frac{1}{2}(6)$	$(6)[4 - \frac{1}{2}(6)] = 6$

(b) $x = 4$, $y = 2$ (c) $A = x\left(\dfrac{8-x}{2}\right)$, $0 < x < 8$

(d)

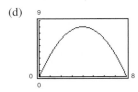

$x = 4$, $y = 2$

(e) $A = -\frac{1}{2}(x-4)^2 + 8$; $x = 4$, $y = 2$

21. \$1020 **23.** 25

25.

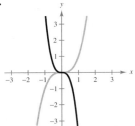

27.

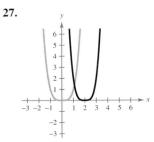

29.

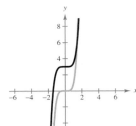

31.

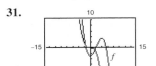

33.

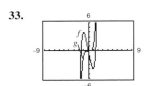

35. Falls to the left. **37.** Rises to the left.
Rises to the right.   Falls to the right.

39. $0, -3$ **41.** $0, 8$

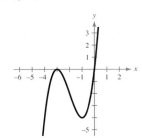
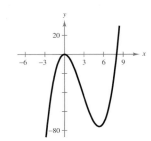

43. $0, 2, -1$

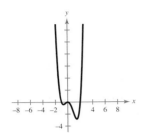

45. $(-5, -4)$ **47.** $(-2, -1), (-1, 0)$

49. $\dfrac{4}{3} + \dfrac{29}{3(3x - 2)}$ **51.** $3x^2 + 3 + \dfrac{3}{x^2 - 1}$

53. $3x^2 + 5x + 8 + \dfrac{10}{2x^2 - 1}$

55. $0.1x^2 + 0.8x + 4 + \dfrac{19.5}{x - 5}$ **57.** $3x^2 + 11x - 4$

59. (a) Yes (b) No (c) Yes (d) No

61. (a) -3276 (b) 0

63. (a) Answers will vary. (b) $(2x + 5), (x - 3)$
(c) $f(x) = (2x + 5)(x - 3)(x + 6)$ (d) $-\frac{5}{2}, 3, -6$
(e)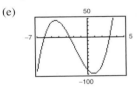

65. (a) Answers will vary. (b) $(x - 1), (x - 3)$
(c) $f(x) = (x - 1)(x - 3)(x - 2)(x - 5)$ (d) $1, 2, 3, 5$
(e)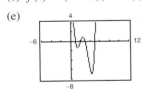

67. $0, 2$ **69.** $8, 1$ **71.** $-4, 6, \pm 2i$

73. $\pm 1, \pm 3, \pm 5, \pm 15, \pm\frac{1}{2}, \pm\frac{3}{2}, \pm\frac{5}{2}, \pm\frac{15}{2}, \pm\frac{1}{4}, \pm\frac{3}{4}, \pm\frac{5}{4}, \pm\frac{15}{4}$

75. $-1, -3, 6$ **77.** $1, 8$ **79.** $-4, 3$

81. $3x^4 - 14x^3 + 17x^2 - 42x + 24$

83. $1, \frac{3}{4}$ **85.** $-1, \frac{3}{2}, 3, \frac{2}{3}$

87. (a)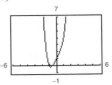
(b) Two zeros
(c) $-1, -0.54$

89. (a)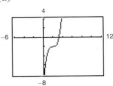
(b) One zero
(c) 3.26

91. Two or no positive real zeros, one negative real zero

93. (a)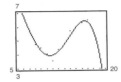
The model is a fairly "good fit."
(b) Recession; yes
(c) $0.3 billion; more
(d) $-$3.47 billion. The model is not accurate in predicting future sales because the sales eventually become negative.

95. 2438.7 kilowatts   **97.** $y = \dfrac{49.5}{x}$

99. (a) $y_1 = 0.414t + 13.213$
$y_2 = 0.437t + 12.904$
(b)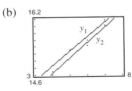
(c) The slope represents the average increase of hourly wages per year.
(d) Mining: $18.18; construction: $18.15

101. True. If y is directly proportional to x, then $y = kx$, so $x = (1/k)y$. Therefore, x is directly proportional to y.

Chapter Test *(page 321)*

1. (a) Reflection in the x-axis followed by a vertical translation
(b) Horizontal translation

2. Vertex: $(-2, -1)$;
Intercepts: $(0, 3), (-3, 0), (-1, 0)$

3. $y = (x - 3)^2 - 6$

4. (a) 50 feet

(b) 5. Yes, changing the constant term results in a vertical translation of the graph and therefore changes the maximum height.

5. Rises to the left.

Falls to the right.

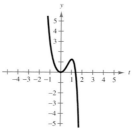

6. $3x + \dfrac{x - 1}{x^2 + 1}$ **7.** $2x^3 + 4x^2 + 3x + 6 + \dfrac{9}{x - 2}$

8. $(4x - 1)(x - \sqrt{3})(x + \sqrt{3})$;

Real solutions: $\frac{1}{4}, \pm \sqrt{3}$

9. $\pm 1, \pm 2, \pm 3, \pm 4, \pm 6, \pm 8, \pm 12, \pm 24, \pm \frac{1}{2}, \pm \frac{3}{2}$

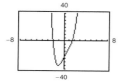

$-2, \frac{3}{2}$

10. $\pm 1, \pm 2, \pm \frac{1}{3}, \pm \frac{2}{3}$

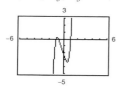

$\pm 1, -\frac{2}{3}$

11. $-0.819, 1.380$ **12.** $-1.414, -0.667, 1.414$

13. $f(x) = x^4 - 9x^3 + 28x^2 - 30x$

14. $f(x) = x^4 - 6x^3 + 16x^2 - 24x + 16$

15. $v = 6\sqrt{s}$ **16.** $A = \dfrac{25}{6}xy$ **17.** $b = \dfrac{48}{a}$

Chapter 4

Section 4.1   *(page 329)*

1. (a)

x	$f(x)$	x	$f(x)$	x	$f(x)$
0.5	-2	1.5	2	5	0.25
0.9	-10	1.1	10	10	$0.\overline{1}$
0.99	-100	1.01	100	100	$0.\overline{01}$
0.999	-1000	1.001	1000	1000	$0.\overline{001}$

(b) Vertical asymptote: $x = 1$

Horizontal asymptote: $y = 0$

(c) Domain: all real numbers $x \neq 1$

3. (a)

x	$f(x)$	x	$f(x)$	x	$f(x)$
0.5	4	1.5	12	5	5
0.9	36	1.1	44	10	$4.\overline{4}$
0.99	396	1.01	404	100	$4.\overline{04}$
0.999	3996	1.001	4004	1000	$4.\overline{004}$

(b) Vertical asymptote: $x = 1$

Horizontal asymptotes: $y = \pm 4$

(c) Domain: all real numbers $x \neq 1$

5. (a)

x	$f(x)$	x	$f(x)$	x	$f(x)$
0.5	-1	1.5	5.4	5	3.125
0.9	-12.79	1.1	17.29	10	$3.\overline{03}$
0.99	-147.8	1.01	152.3	100	$3.\overline{0003}$
0.999	-1498	1.001	1502	1000	3

(b) Vertical asymptotes: $x = \pm 1$

Horizontal asymptote: $y = 3$

(c) Domain: all real numbers $x \neq \pm 1$

7. Domain: all real numbers $x \neq 0$

Vertical asymptote: $x = 0$

Horizontal asymptote: $y = 0$

9. Domain: all real numbers $x \neq 2$

Vertical asymptote: $x = 2$

Horizontal asymptote: $y = -1$

11. Domain: all real numbers $x \neq \pm 1$

Vertical asymptotes: $x = \pm 1$

13. Domain: all real numbers

Horizontal asymptote: $y = 3$

15. d   **17.** f   **19.** e

21. (a) Domain of f: all real numbers $x \neq -2$;

Domain of g: all real numbers

(b) Vertical asymptote: none

(c)

x	-4	-3	-2.5	-2	-1.5	-1	0
$f(x)$	-6	-5	-4.5	Undef.	-3.5	-3	-2
$g(x)$	-6	-5	-4.5	-4	-3.5	-3	-2

(d) The functions differ only where f is undefined.

23. (a) Domain of f: all real numbers $x \neq 0, 2$;

   Domain of g: all real numbers $x \neq 0$

   (b) Vertical asymptote: $x = 0$

   (c)

x	-1	-0.5	0	0.5	2	3	4
$f(x)$	-1	-2	Undef.	2	Undef.	$\frac{1}{3}$	$\frac{1}{4}$
$g(x)$	-1	-2	Undef.	2	$\frac{1}{2}$	$\frac{1}{3}$	$\frac{1}{4}$

   (d) The functions differ only where f is undefined and g is defined.

25. 1   **27.** 6

29. (a) 4   (b) Less than   (c) Greater than

31. (a) 2   (b) Greater than   (c) Less than

33.

M	200	400	600	800	1000
t	0.472	0.596	0.710	0.817	0.916

M	1200	1400	1600	1800	2000
t	1.009	1.096	1.178	1.255	1.328

The greater the mass, the more time required per oscillation.

35. (a) 28.33 million dollars

   (b) 170 million dollars

   (c) 765 million dollars

   (d) No. The function is undefined at $p = 100$.

37. (a) 333 deer, 500 deer, 800 deer   (b) 1500 deer

39. (a)

n	1	2	3	4	5	6
P	0.50	0.74	0.82	0.86	0.89	0.91

n	7	8	9	10
P	0.92	0.93	0.94	0.95

   P approaches 1 as n increases.

   (b) 100%

41. (a)

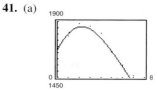

   (b) 817 thousand

   (c) No, because there may be an unexpected rise in military personnel.

43. False. Polynomials do not have vertical asymptotes.

45. $f(x) = \dfrac{1}{x^2 + x - 2}$ (There are many correct answers.)

47. $f(x) = \dfrac{2x^2}{x^2 + 1}$ (There are many correct answers.)

49. $f(x) = \dfrac{1}{x^2 + 2}; f(x) = \dfrac{1}{x - 20}$ (Answers are not unique.)

51. $f^{-1}(x) = \dfrac{x + 7}{8}$

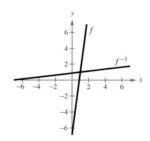

53. $f^{-1}(x) = -\frac{1}{2}(5x - 7)$

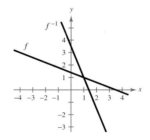

55. $x + 9 + \dfrac{42}{x - 4}$ **57.** $2x - 9 + \dfrac{34}{x + 5}$

Section 4.2 *(page 338)*

1.

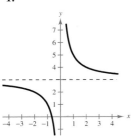

3.

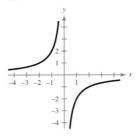

5.

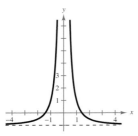

7.

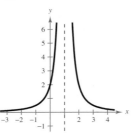

9.

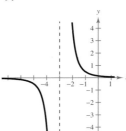

11.

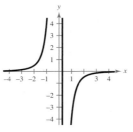

13. (a) y-intercept: $\left(0, \frac{1}{2}\right)$

    (b) Vertical asymptote: $x = -2$

         Horizontal asymptote: $y = 0$

    (c) No axis or origin symmetry

    (d)

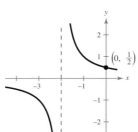

15. (a) y-intercept: $\left(0, -\frac{1}{2}\right)$

    (b) Vertical asymptote: $x = -2$

         Horizontal asymptote: $y = 0$

(c) No axis or origin symmetry

(d)

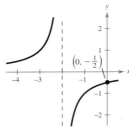

17. (a) x-intercept: $\left(-\frac{5}{2}, 0\right)$

        y-intercept: $(0, 5)$

    (b) Vertical asymptote: $x = -1$

         Horizontal asymptote: $y = 2$

    (c) No axis or origin symmetry

    (d)

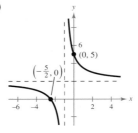

19. (a) x-intercept: $\left(-\frac{5}{2}, 0\right)$

        y-intercept: $\left(0, \frac{5}{2}\right)$

    (b) Vertical asymptote: $x = -2$

         Horizontal asymptote: $y = 2$

    (c) No axis or origin symmetry

    (d)

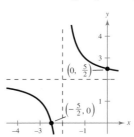

21. (a) Intercept: $(0, 0)$

    (b) Horizontal asymptote: $y = 1$

    (c) y-axis symmetry

(d)

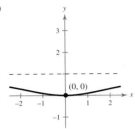

23. (a) Intercept: $(0, 0)$

(b) Vertical asymptotes: $x = \pm 3$

Horizontal asymptote: $y = 1$

(c) y-axis symmetry

(d)

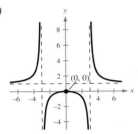

25. (a) Intercept: $(0, 0)$

(b) Horizontal asymptote: $y = 0$

(c) Origin symmetry

(d)

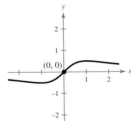

27. (a) x-intercept: $(-1, 0)$

(b) Vertical asymptotes: $x = 0$, $x = 4$

Horizontal asymptote: $y = 0$

(c) No axis or origin symmetry

(d)

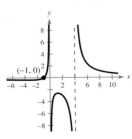

29. (a) Intercept: $(0, 0)$

(b) Vertical asymptotes: $x = -1$, $x = 2$

Horizontal asymptote: $y = 0$

(c) No axis or origin symmetry

(d)

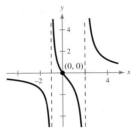

31. (a) Intercept: $(0, 0)$

(b) Vertical asymptotes: $x = -2$, $x = 7$

Horizontal asymptote: $y = 0$

(c) No axis or origin symmetry

(d)

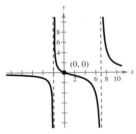

33. (a) x-intercept: $\left(-\frac{1}{2}, 0\right)$

y-intercept: $\left(0, -\frac{1}{2}\right)$

(b) Vertical asymptotes: $x = 2$, $x = -1$

Horizontal asymptote: $y = 0$

(c) No axis or origin symmetry

(d)

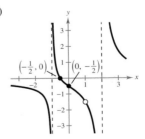

35. (a) Domain of f: all real numbers $x \neq -1$;

Domain of g: all real numbers

(b) Vertical asymptote: none

(c)

x	-3	-2	-1.5	-1	-0.5	0	1
$f(x)$	-4	-3	-2.5	Undef.	-1.5	-1	0
$g(x)$	-4	-3	-2.5	-2	-1.5	-1	0

(d)

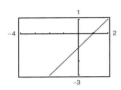

(e) Because there are only a finite number of pixels, the utility may not attempt to evaluate the function where it does not exist.

37. (a) Domain of f: all real numbers $x \neq 0, 2$;

   Domain of g: all real numbers $x \neq 0$

(b) Vertical asymptote: $x = 0$

(c)

x	-0.5	0	0.5	1	1.5	2	3
$f(x)$	-2	Undef.	2	1	$\frac{2}{3}$	Undef.	$\frac{1}{3}$
$g(x)$	-2	Undef.	2	1	$\frac{2}{3}$	$\frac{1}{2}$	$\frac{1}{3}$

(d)

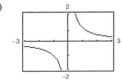

(e) Because there are only a finite number of pixels, the utility may not attempt to evaluate the function where it does not exist.

39.

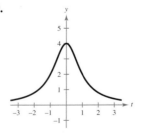

Domain: all real numbers

$y = 0$

41.

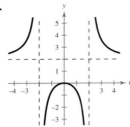

Domain: all real numbers $t \neq \pm 2$

$t = \pm 2, \ y = 2$

43.

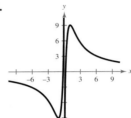

Domain: all real numbers $x \neq 0$

$x = 0, \ y = 0$

45. (a) No intercepts

(b) Vertical asymptote: $x = 0$

   Slant asymptote: $y = 2x$

(c) Origin symmetry

(d)

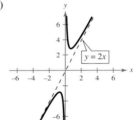

47. (a) No intercepts

(b) Vertical asymptote: $x = 0$

   Slant asymptote: $y = x$

(c) Origin symmetry

(d)

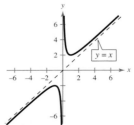

49. (a) Intercept: $(0, 0)$

(b) Vertical asymptotes: $x = \pm 1$

Slant asymptote: $y = x$

(c) Origin symmetry

(d)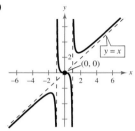

51. (a) y-intercept: $(0, -1)$

(b) Vertical asymptote: $x = 1$

Slant asymptote: $y = x$

(c) No axis or origin symmetry

(d)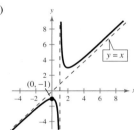

53.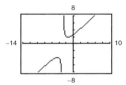

Domain: all real numbers $x \neq -3$

Vertical asymptote: $x = -3$

Slant asymptote: $y = x + 2$

$y = x + 2$

55.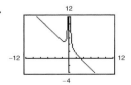

Domain: all real numbers $x \neq 0$

Vertical asymptote: $x = 0$

Slant asymptote: $y = -x + 3$

$y = -x + 3$

57. $(-1, 0)$ **59.** $(1, 0), (-1, 0)$ **61.** $(-4, 0)$

63. $(3, 0), (-2, 0)$

65. (a) Answers will vary.   (b) $[0, 950]$

(c)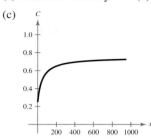

Increases more slowly; 0.75

67. (a) Answers will vary.   (b) $(4, \infty)$

(c)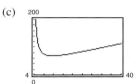

11.75 inches $\times$ 5.87 inches

69. 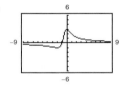   Minimum: $(-2, -1)$

Maximum: $(0, 3)$

71.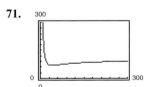

$x \approx 40.45$, or 4045 components.

73. 12.8 inches $\times$ 8.5 inches

75. (a) Answers will vary.

(b) Vertical asymptote: $x = 25$

Horizontal asymptote: $y = 25$

(c)

x	30	35	40	45	50	55	60
y	150	87.5	66.7	56.3	50	45.8	42.9

(d) Yes. You would expect the average speed for the round trip to be the average of the average speeds for the two parts of the trip.

(e) No. At 20 miles per hour you would use more time in one direction than is required for the round trip at an average speed of 50 miles per hour.

77. False. The graph of $f(x) = \dfrac{x}{x^2 + 1}$ crosses $y = 0$, which is a horizontal asymptote.

79.

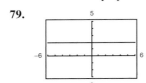

The fraction is not reduced.

81. No. $f(x) = \dfrac{1}{x^2 + 1}$ does not have a vertical asymptote.

83. $(3x - 4)(x + 9)$ **85.** $(x + 6)\left(x + \sqrt{2}\right)\left(x - \sqrt{2}\right)$

87. $x < 0$ **89.** $x \le -\dfrac{13}{2},\, x \ge \dfrac{7}{2}$

Section 4.3   *(page 348)*

1. b   **3.** d

5. $\dfrac{A}{x} + \dfrac{B}{x - 14}$ **7.** $\dfrac{A}{x} + \dfrac{B}{x^2} + \dfrac{C}{x - 10}$

9. $\dfrac{A}{x - 5} + \dfrac{B}{(x - 5)^2} + \dfrac{C}{(x - 5)^3}$

11. $\dfrac{A}{x} + \dfrac{Bx + C}{x^2 + 10}$ **13.** $\dfrac{A}{x} + \dfrac{Bx + C}{x^2 + 1} + \dfrac{Dx + E}{(x^2 + 1)^2}$

15. $\dfrac{1}{2}\left(\dfrac{1}{x - 1} - \dfrac{1}{x + 1}\right)$

17. $\dfrac{1}{x} - \dfrac{1}{x + 1}$ **19.** $\dfrac{1}{x} - \dfrac{2}{2x + 1}$ **21.** $\dfrac{1}{x - 1} - \dfrac{1}{x + 2}$

23. $-\dfrac{3}{x} - \dfrac{1}{x + 2} + \dfrac{5}{x - 2}$ **25.** $\dfrac{3}{x} - \dfrac{1}{x^2} + \dfrac{1}{x + 1}$

27. $\dfrac{3}{x - 3} + \dfrac{9}{(x - 3)^2}$ **29.** $-\dfrac{1}{x} + \dfrac{2x}{x^2 + 1}$

31. $\dfrac{-1}{x - 1} + \dfrac{x + 2}{x^2 - 2}$

33. $\dfrac{1}{3(x^2 + 2)} - \dfrac{1}{6(x + 2)} + \dfrac{1}{6(x - 2)}$

35. $\dfrac{1}{8(2x + 1)} + \dfrac{1}{8(2x - 1)} - \dfrac{x}{2(4x^2 + 1)}$

37. $\dfrac{1}{x + 1} + \dfrac{2}{x^2 - 2x + 3}$

39. $1 - \dfrac{2x + 1}{x^2 + x + 1}$

41. $2x - 7 + \dfrac{17}{x + 2} + \dfrac{1}{x + 1}$

43. $x + 3 + \dfrac{6}{x - 1} + \dfrac{4}{(x - 1)^2} + \dfrac{1}{(x - 1)^3}$

45. $\dfrac{3}{2x - 1} - \dfrac{2}{x + 1}$ **47.** $\dfrac{2}{x} - \dfrac{1}{x^2} - \dfrac{2}{x + 1}$

49. $\dfrac{1}{x^2 + 2} + \dfrac{x}{(x^2 + 2)^2}$ **51.** $2x + \dfrac{1}{2}\left(\dfrac{3}{x - 4} - \dfrac{1}{x + 2}\right)$

53. $\dfrac{3}{x} - \dfrac{2}{x - 4}$

$y = \dfrac{x - 12}{x(x - 4)}$ $y = \dfrac{3}{x},\, y = -\dfrac{2}{x - 4}$

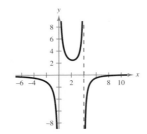   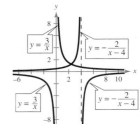

The vertical asymptotes are the same.

55. $\dfrac{3}{x - 3} + \dfrac{5}{x + 3}$

$y = \dfrac{2(4x - 3)}{x^2 - 9}$ $y = \dfrac{3}{x - 3},\, y = \dfrac{5}{x + 3}$

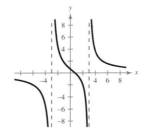   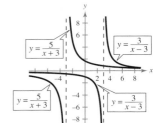

The vertical asymptotes are the same.

57. (a) $\dfrac{2000}{7 - 4x} - \dfrac{2000}{11 - 7x},\quad 0 < x \le 1$

(b) $\text{Ymax} = \left|\dfrac{2000}{7 - 4x}\right|$

$\text{Ymin} = \left|\dfrac{-2000}{11 - 7x}\right|$

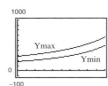

59. False. The expression is an improper rational expression, so you must first divide before applying partial fraction decomposition.

61. $\dfrac{1}{a}\left(\dfrac{1}{x} - \dfrac{1}{x + a}\right)$ **63.** $\dfrac{1}{a + 1}\left(\dfrac{1}{x + 1} + \dfrac{1}{a - x}\right)$

65.
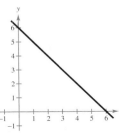

67.
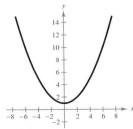

69.
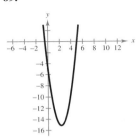

71.
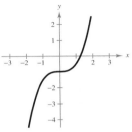

73.
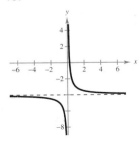

75.
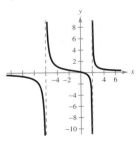

77.
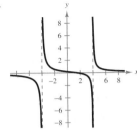

Section 4.4   (page 359)

1. Not shown   **3.** e   **5.** Not shown   **7.** f

9. Not shown

11. Vertex: $(0, 0)$
Focus: $\left(0, \frac{1}{2}\right)$

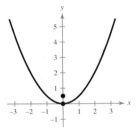

13. Vertex: $(0, 0)$
Focus: $\left(-\frac{3}{2}, 0\right)$

15. Vertex: $(0, 0)$
Focus: $(0, -2)$

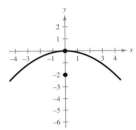

17. $y^2 = 8x$ **19.** $x^2 = -6y$ **21.** $x^2 = 4y$

23. $y^2 = -12x$ **25.** $y^2 = 9x$

27. $x^2 = \frac{3}{2}y$; Focus: $\left(0, \frac{3}{8}\right)$ **29.** $y^2 = \frac{9}{5}x$; Focus: $\left(\frac{9}{20}, 0\right)$

31. $y = \dfrac{1}{14}x^2$ **33.** (a) $y = \dfrac{x^2}{12{,}288}$ (b) 22.6 feet

35. Center: $(0, 0)$
Vertices: $(0, \pm13)$

37. Center: $(0, 0)$
Vertices: $(\pm2, 0)$

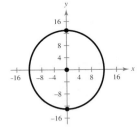

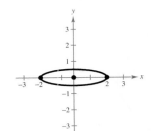

39. Center: $(0, 0)$
Vertices: $(0, \pm8)$

41. Center: $(0, 0)$
Vertices: $(\pm3, 0)$

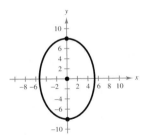

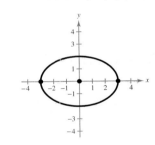

43. 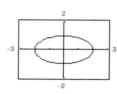   **45.**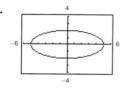

47. $\dfrac{x^2}{25} + \dfrac{y^2}{36} = 1$ **49.** $\dfrac{x^2}{49} + \dfrac{y^2}{49/4} = 1$

51. $\dfrac{x^2}{48} + \dfrac{y^2}{64} = 1$ **53.** $\dfrac{x^2}{16} + \dfrac{y^2}{12} = 1$ **55.** $\dfrac{x^2}{4} + \dfrac{y^2}{16} = 1$

57. (a)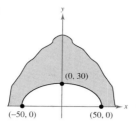

    (b) $y = \frac{3}{5}\sqrt{2500 - x^2}$ (c) 13.08 feet

59. 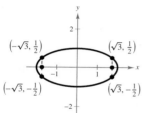   **61.**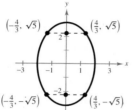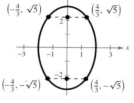

63. Center: $(0, 0)$ **65.** Center: $(0, 0)$
Vertices: $(\pm 1, 0)$ Vertices: $(0, \pm 1)$

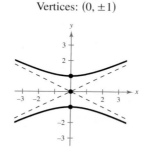

67. Center: $(0, 0)$ **69.** Center: $(0, 0)$
Vertices: $(0, \pm 5)$ Vertices: $\left(0, \pm\frac{1}{2}\right)$

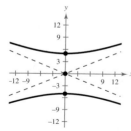 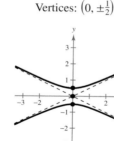

71. $\dfrac{y^2}{4} - \dfrac{x^2}{12} = 1$ **73.** $\dfrac{x^2}{1} - \dfrac{y^2}{9} = 1$

75. $\dfrac{17y^2}{1024} - \dfrac{17x^2}{64} = 1$ **77.** $\dfrac{y^2}{9} - \dfrac{x^2}{9/4} = 1$

79. $\left(12\left(\sqrt{5} - 1\right), 0\right) \approx (14.83, 0)$

81. False. The equation represents a hyperbola:
$$\dfrac{x^2}{144} - \dfrac{y^2}{144} = 1.$$

83. False. If the graph crossed the directrix, there would exist points nearer the directrix than the focus.

85. No. If it were an ellipse, the equation would have to be second degree.

87. The shape continuously changes from an ellipse with a vertical major axis of length 8 and minor axis of length 2 to a circle with a diameter of 8 and then to an ellipse with a horizontal major axis of length 16 and minor axis of length 8.

89. Answers will vary.

91. 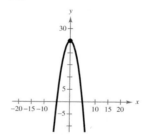   **93.**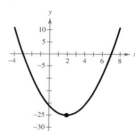

Vertex: $(0, 25)$ Vertex: $(2, -25)$

95. $\pm 5i, \dfrac{3}{2}$ **97.** $\dfrac{1}{9}\left(\dfrac{-2}{x + 6} + \dfrac{11}{x - 3}\right)$

99. $\dfrac{1}{2}\left(\dfrac{3}{x - 3} - \dfrac{1}{x - 5}\right)$

Section 4.5   *(page 368)*

1. Center: $(0, 0)$ **3.** Center: $(-3, 8)$ **5.** Center: $(1, 0)$
    Radius: 7   Radius: 4   Radius: $\sqrt{10}$

7. $(x - 1)^2 + (y + 3)^2 = 1$ **9.** $\left(x + \frac{3}{2}\right)^2 + (y - 3)^2 = 1$
    Center: $(1, -3)$ Center: $\left(-\frac{3}{2}, 3\right)$
    Radius: 1   Radius: 1

11. Vertex: $(1, -2)$
Focus: $(1, -4)$
Directrix: $y = 0$

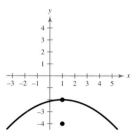

13. Vertex: $\left(5, -\frac{1}{2}\right)$
Focus: $\left(\frac{11}{2}, -\frac{1}{2}\right)$
Directrix: $x = \frac{9}{2}$

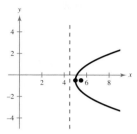

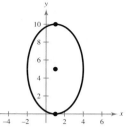

15. Vertex: $(1, 1)$
Focus: $(1, 2)$
Directrix: $y = 0$

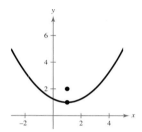

17. Vertex: $(-2, -3)$
Focus: $(-4, -3)$
Directrix: $x = 0$

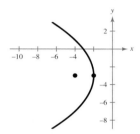

37. Center: $(-2, -4)$
Foci: $\left(\dfrac{-4 \pm \sqrt{3}}{2}, -4\right)$
Vertices: $(-3, -4)$, $(-1, -4)$

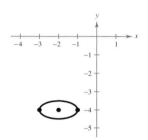

19. Vertex: $(-2, 1)$
Focus: $\left(-2, -\frac{1}{2}\right)$
Directrix: $y = \frac{5}{2}$

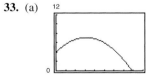

21. Vertex: $\left(\frac{1}{4}, -\frac{1}{2}\right)$
Focus: $\left(0, -\frac{1}{2}\right)$
Directrix: $x = \frac{1}{2}$

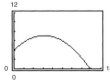

39. Center: $(-2, 3)$
Foci: $\left(-2, 3 \pm \sqrt{5}\right)$
Vertices: $(-2, 6)$, $(-2, 0)$

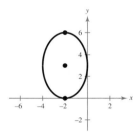

23. $(x - 3)^2 = -(y - 1)$ **25.** $(y - 2)^2 = -8(x - 3)$

27. $x^2 = 8(y - 4)$ **29.** $(y - 2)^2 = 8x$

31. (a) $17{,}500\sqrt{2}$ miles per hour
(b) $x^2 = -16{,}400(y - 4100)$

33. (a)

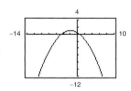

(b) $(6.25, 7.125)$; 15.69 feet

35. Center: $(1, 5)$
Foci: $(1, 9)$, $(1, 1)$
Vertices: $(1, 10)$, $(1, 0)$

41. Center: $(1, -1)$
Foci: $\left(\frac{7}{4}, -1\right)$, $\left(\frac{1}{4}, -1\right)$
Vertices: $\left(\frac{9}{4}, -1\right)$, $\left(-\frac{1}{4}, -1\right)$

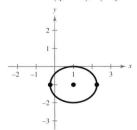

43. $\dfrac{(x - 4)^2}{9} + \dfrac{y^2}{16} = 1$ **45.** $\dfrac{(x - 2)^2}{4} + \dfrac{(y - 2)^2}{1} = 1$

47. $\dfrac{x^2}{48} + \dfrac{(y - 4)^2}{64} = 1$ **49.** $\dfrac{(x - 3)^2}{9} + \dfrac{(y - 5)^2}{16} = 1$

51. $\dfrac{x^2}{16} + \dfrac{(y-4)^2}{12} = 1$ **53.** $\dfrac{x^2}{25} + \dfrac{y^2}{16} = 1$

55. 2,756,832,000 miles; 4,575,168,000 miles

57. $e = \dfrac{c}{a} \approx 0.054$

59. Center: $(1, -2)$
  Vertices: $(3, -2), (-1, -2)$
  Foci: $\left(1 \pm \sqrt{5}, -2\right)$

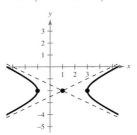

61. Center: $(2, -6)$
  Vertices: $(2, -5), (2, -7)$
  Foci: $\left(2, -6 \pm \sqrt{2}\right)$

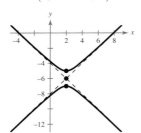

63. Center: $(2, -3)$
  Vertices: $(3, -3), (1, -3)$
  Foci: $\left(2 \pm \sqrt{10}, -3\right)$

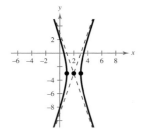

65. The graph of this equation is two lines intersecting at $(-1, -3)$.

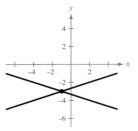

67. Center: $(1, -1)$
  Vertices: $(-2, -1), (4, -1)$
  Foci: $\left(1 \pm \sqrt{13}, -1\right)$

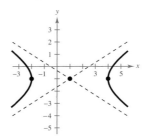

69. $(y-1)^2 - x^2 = 1$ **71.** $\dfrac{(x-4)^2}{4} - \dfrac{y^2}{12} = 1$

73. $\dfrac{(y-5)^2}{16} - \dfrac{(x-4)^2}{9} = 1$ **75.** $\dfrac{y^2}{9} - \dfrac{4(x-2)^2}{9} = 1$

77. $\dfrac{(x-3)^2}{9} - \dfrac{(y-2)^2}{4} = 1$

79. Circle **81.** Hyperbola

83. Ellipse **85.** Parabola

87. True. The equation in standard form is
$$\dfrac{(x-3)^2}{1/3} + \dfrac{(y-4)^2}{1/2} = 1.$$

89. Additive Inverse Property **91.** Distributive Property

93. $f^{-1}(x) = -\dfrac{x-10}{7}$

Review Exercises (page 372)

1. Domain: all real numbers $x \neq -12$

3. Domain: all real numbers $x \neq 6, 4$

5. Vertical asymptote: $x = -3$
  Horizontal asymptote: $y = 0$

7. Vertical asymptotes: $x = -2, x = 2$
  Horizontal asymptote: $y = 1$

9. $0.50 is the horizontal asymptote of the function.

11. $x \approx 0.346$ inch

13. No intercepts
    y-axis symmetry
    Vertical asymptote: $x = 0$
    Horizontal asymptote: $y = 0$

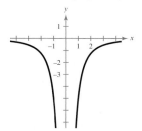

15. x-intercept: $(-2, 0)$
    y-intercept: $(0, 2)$
    No axis or origin symmetry
    Vertical asymptote: $x = 1$
    Horizontal asymptote: $y = -1$

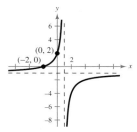

17. Intercept: $(0, 0)$
    y-axis symmetry
    Horizontal asymptote: $y = 1$

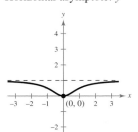

19. Intercept: $(0, 0)$
    Origin symmetry
    Horizontal asymptote: $y = 0$

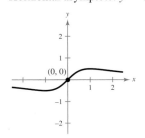

21. Intercept: $(0, 0)$
    y-axis symmetry
    Horizontal asymptote: $y = -6$

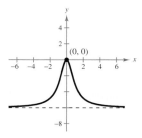

23. Intercept: $(0, 0)$
    Origin symmetry
    Vertical asymptotes: $x = \pm 1$
    Horizontal asymptote: $y = 0$

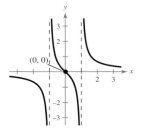

25. Slant asymptote: $y = 2x$

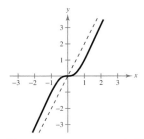

27. Slant asymptote: $y = x + 1$

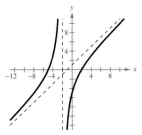

29. (a)

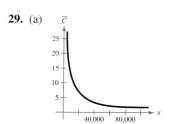

(b) $100.90, $10.90, $1.90

(c) $0.90 is the horizontal asymptote of the function.

31.

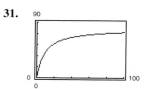

80.3 milligrams per square decimeter per hour

33. $\dfrac{A}{x-7} + \dfrac{B}{x+4}$ **35.** $\dfrac{A}{x} + \dfrac{Bx+C}{x^2+2} + \dfrac{Dx+E}{(x^2+2)^2}$

37. $\dfrac{1}{x+1} - \dfrac{2}{x+2}$ **39.** $\dfrac{1}{2}\left(\dfrac{3}{x-3} - \dfrac{3}{x+3}\right)$

41. $\dfrac{4}{3(x-1)} + \dfrac{2}{3(x-1)^2}$ **43.** $2\left(\dfrac{1}{x-1} + \dfrac{x+1}{x^2+1}\right)$

45. Ellipse   **47.** Ellipse   **49.** Circle   **51.** Ellipse

53. $x^2 = -8y$ **55.** $x^2 = 12y$

57. (a)

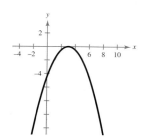

(b) $6737 million

59. $\dfrac{x^2}{16} + \dfrac{y^2}{100} = 1$ **61.** $\dfrac{x^2}{49} + \dfrac{y^2}{13} = 1$

63. $\dfrac{x^2}{1} - \dfrac{y^2}{4} = 1$ **65.** $\dfrac{y^2}{1} - \dfrac{x^2}{8} = 1$

67. $(x-3)^2 = -2y$ **69.** $(1, 2)$

Parabola   Degenerate conic (a point)

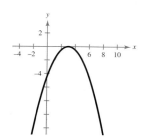

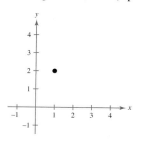

71. $\dfrac{(x+5)^2}{9} + (y-1)^2 = 1$ **73.** $(x-4)^2 - \dfrac{(y-4)^2}{9} = 1$

Ellipse   Hyperbola

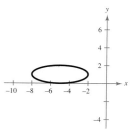

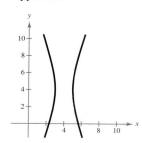

75. $(x+6)^2 = -9(y-4)$ **77.** $(x-4)^2 = -8(y-2)$

79. $(y-2)^2 = -4x$ **81.** $\dfrac{(x-5)^2}{25} + \dfrac{(y-3)^2}{9} = 1$

83. $\dfrac{(x-2)^2}{25} + \dfrac{y^2}{21} = 1$ **85.** $\dfrac{(x-2)^2}{4} + (y-1)^2 = 1$

87. $\dfrac{x^2}{36} - \dfrac{(y-7)^2}{9} = 1$ **89.** $\dfrac{(x+2)^2}{64} - \dfrac{(y-3)^2}{36} = 1$

91. $\dfrac{5(x-4)^2}{16} - \dfrac{5y^2}{64} = 1$ **93.** $8\sqrt{6}$ meters

95. 10.29 centimeters   **97.** True

Chapter Test   *(page 377)*

1. Domain: all real numbers $x \neq 4$
    Vertical asymptote: $x = 4$
    Horizontal asymptote: $y = 0$

2. Domain: all real numbers
    Vertical asymptote: None
    Horizontal asymptote: $y = -1$

3. Domain: all real numbers $x \neq 2$
    Vertical asymptote: $x = 2$
    Slant asymptote: $y = x + 4$

4.

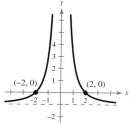

Vertical asymptote: $x = 0$
Horizontal asymptote: $y = -1$

5.

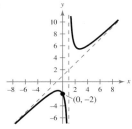

Vertical asymptote: $x = 1$

Slant asymptote: $y = x + 1$

6. $f(x) = \dfrac{3x^2}{x^2 - 4}$

7. (a) Answers will vary.

(b) $A = \dfrac{x^2}{2(x - 2)}, \quad x > 2$

(c)

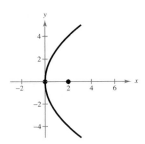

$A = 4$

8. $\dfrac{3}{x - 2} - \dfrac{1}{x + 1}$ **9.** $\dfrac{2}{x^2} - \dfrac{3}{x - 2}$

10. $-\dfrac{5}{x} + \dfrac{3}{x - 1} + \dfrac{3}{x + 1}$ **11.** $-\dfrac{2}{x} + \dfrac{3x}{x^2 + 2}$

12. Vertex: $(0, 0)$ **13.** Vertex: $(1, 0)$

Focus: $(2, 0)$ Focus: $(2, 0)$

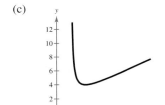

14. Vertices: $(\pm 1, 0)$ **15.** Vertices: $(0, 0), (4, 0)$

Foci: $\left(\pm \sqrt{5}, 0 \right)$ Foci: $\left(2 \pm \sqrt{5}, 0 \right)$

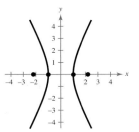   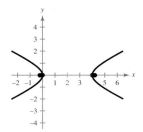

16. $\dfrac{(x - 4)^2}{16} + \dfrac{(y - 2)^2}{4} = 1$ **17.** $\dfrac{y^2}{9} - \dfrac{x^2}{4} = 1$

18.

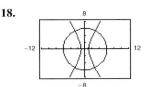

$(\pm 2.83, \pm 5.29)$

19. Smallest distance: $\approx 364{,}224$ kilometers

Greatest distance: $\approx 404{,}582$ kilometers

Chapter 5

Section 5.1 *(page 388)*

1. 946.852   **3.** 7.352   **5.** 0.006   **7.** 673.639

9. 0.472   **11.** d   **13.** a

15. Shift the graph of f 4 units to the right.

17. Shift the graph of f 5 units upward.

19. Reflect f in the x-axis and shift f 4 units to the left.

21. Reflect f in the x-axis and shift f 5 units upward.

23.

x	-2	-1	0	1	2
$f(x)$	4	2	1	0.5	0.25

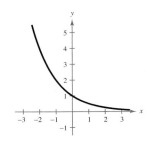

25.

x	-2	-1	0	1	2
$f(x)$	0.25	0.5	1	2	4

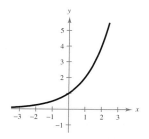

27.

x	-2	-1	0	1	2
$f(x)$	0.125	0.25	0.5	1	2

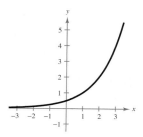

29.

x	-2	-1	0	1	2
$f(x)$	0.135	0.368	1	2.718	7.389

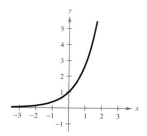

31.

x	-8	-7	-6	-5	-4
$f(x)$	0.055	0.149	0.406	1.104	3

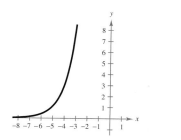

33.

x	-2	-1	0	1	2
$f(x)$	4.037	4.100	4.271	4.736	6

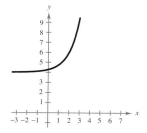

35.

x	-1	0	1	2	3
$f(x)$	3.003	3.016	3.063	3.25	4

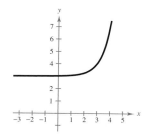

37.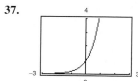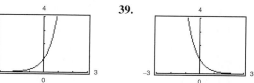

39.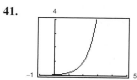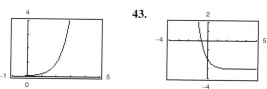

41.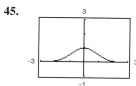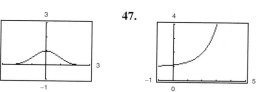

43.

45.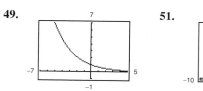

47.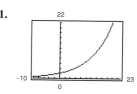

49.

51.

53.

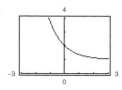

55.

n	1	2	4
A	\$5397.31	\$5477.81	\$5520.10

n	12	365	Continuous
A	\$5549.10	\$5563.36	\$5563.85

57.

n	1	2	4
A	\$11,652.39	\$12,002.55	\$12,188.60

n	12	365	Continuous
A	\$12,317.01	\$12,380.41	\$12,382.58

59.

t	1	10	20
A	\$12,999.44	\$26,706.49	\$59,436.39

t	30	40	50
A	\$132,278.12	\$294,390.36	\$655,177.80

61.

t	1	10	20
A	\$12,805.91	\$22,986.49	\$44,031.56

t	30	40	50
A	\$84,344.25	\$161,564.86	\$309,484.08

63. \$222,822.57

65. (a) The steeper curve represents the investment earning compound interest, because compound interest earns more than simple interest.

(b) $A = 500(1.07)^t$

$A = 500(0.07)t + 500$

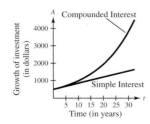

67. \$35.45 **69.** (a) 100 (b) 300 (c) 900

71. (a) 25 units (b) 16.30 units

(c)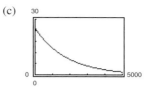

73. (a)

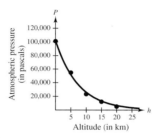

(b)

h	0	5	10	15	20
P	102,303	51,570	25,996	13,104	6606

(c) 34,190 pascals (d) ≈ 11.6 kilometers

75. True. As $x \to -\infty$, $f(x) \to 0$ but never reaches 0.

77. $f(x) = h(x)$ **79.** $f(x) = g(x) = h(x)$

81. (a) $x < 0$ (b) $x > 0$

83. (a) 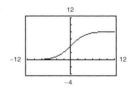 Horizontal asymptotes: $y = 0, y = 8$

(b) 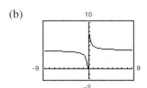 Horizontal asymptote: $y = 4$

Vertical asymptote: $x = 0$

85.

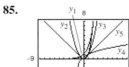

$y = e^x$

87. It usually implies rapid growth.

89. The value of $f(x)$ approaches e^r.

x	1	10	100	200	500
$[1 + (1/x)]^x$	2	2.5937	2.7048	2.7115	2.7156

x	1100	10,000
$[1 + (1/x)]^x$	2.7170	2.718

91. $1 < \sqrt{2} < 2$
$2^1 < 2^{\sqrt{2}} < 2^2$

93. $y_4 = 1 + \dfrac{x}{1!} + \dfrac{x^2}{2!} + \dfrac{x^3}{3!} + \dfrac{x^4}{4!}$

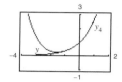

As more terms are added, the polynomial approaches e^x.

95. $y = \frac{1}{7}(2x + 14)$ **97.** $y = \pm\sqrt{25 - x^2}$

99.

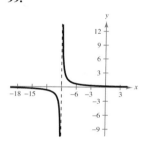

101.

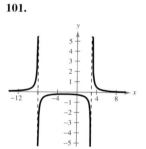

103. Ellipse

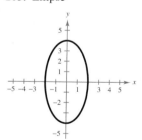

105. Hyperbola

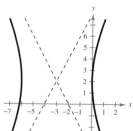

Section 5.2   (page 399)

1. $4^3 = 64$ **3.** $7^{-2} = \frac{1}{49}$ **5.** $32^{2/5} = 4$

7. $e^0 = 1$ **9.** $\log_5 125 = 3$ **11.** $\log_{81} 3 = \frac{1}{4}$

13. $\log_6 \frac{1}{36} = -2$ **15.** $\ln 20.0855\ldots = 3$

17. $\ln 1 = 0$ **19.** 4   **21.** $\frac{1}{2}$ **23.** 0

25. -2 **27.** $\frac{5}{3}$ **29.** 3   **31.** 2

33. 2.538   **35.** -0.097 **37.** 2.913   **39.** -3.418

41. 1.005   **43.** -0.405 **45.** c   **47.** d   **49.** b

51. Domain: $(0, \infty)$
Intercept: $(1, 0)$
Vertical asymptote: $x = 0$

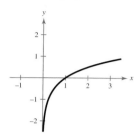

53. Domain: $(0, \infty)$
Intercept: $(9, 0)$
Vertical asymptote: $x = 0$

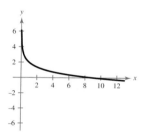

55. Domain: $(-2, \infty)$
Intercept: $(-1, 0)$
Vertical asymptote: $x = -2$

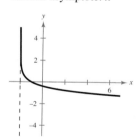

57. Domain: $(0, \infty)$
Intercept: $(5, 0)$
Vertical asymptote: $x = 0$

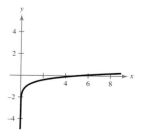

59. Domain: $(2, \infty)$

Intercept: $(3, 0)$

Vertical asymptote: $x = 2$

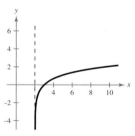

61. Domain: $(-\infty, 0)$

Intercept: $(-1, 0)$

Vertical asymptote: $x = 0$

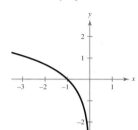

63. 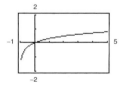   **65.**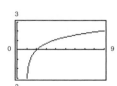

67.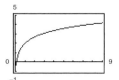

69. (a) 80   (b) 68.1   (c) 62.3

71. (a)

r	0.005	0.01	0.015	0.02	0.025	0.03
t	138.6	69.3	46.2	34.7	27.7	23.1

(b) Answers will vary.

73.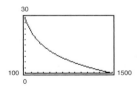

17.66 cubic feet per minute

75. 21,357 foot-pounds   **77.** 30 years

79. Total amount: $396,234

Interest: $246,234

81. (a)

x	1	5	10	10^2
$f(x)$	0	0.322	0.230	0.046

x	10^4	10^6
$f(x)$	0.00092	0.0000138

(b) 0

(c)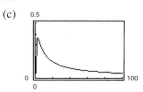

83. True. $\log_3 27 = 3 \Longrightarrow 3^3 = 27$

85. 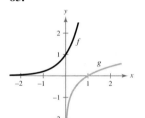   **87.**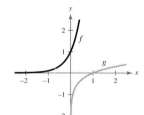

$g = f^{-1}$ $g = f^{-1}$

89. (a) False   (b) True   (c) True   (d) False

91. $y_4 = (x - 1) - \frac{1}{2}(x - 1)^2 + \frac{1}{3}(x - 1)^3 - \frac{1}{4}(x - 1)^4$

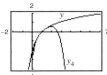

93. (a)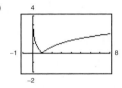

(b) Increasing: $(1, \infty)$

Decreasing: $(0, 1)$

(c) Relative minimum: $(1, 0)$

95. (a) 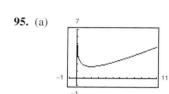 (b) Increasing: $(2, \infty)$
Decreasing: $(0, 2)$
(c) Relative minimum:
$\left(2, 1 - \ln \frac{1}{2}\right)$

97. $8n - 3$ **99.** $83.95 + 37.50t$

101. Vertical asymptote: $x = -8$
Horizontal asymptote: $y = 0$

103. Vertical asymptotes: $x = -3, x = \frac{5}{2}$
Horizontal asymptote: $y = 0$

105. 403.429   **107.** 0.018

Section 5.3   *(page 407)*

1. 1.771   **3.** -2.000 **5.** -0.417 **7.** 2.633

9. (a) $\dfrac{\log_{10} x}{\log_{10} 5}$ (b) $\dfrac{\ln x}{\ln 5}$ **11.** (a) $\dfrac{\log_{10} x}{\log_{10} \frac{1}{5}}$ (b) $\dfrac{\ln x}{\ln \frac{1}{5}}$

13. (a) $\dfrac{\log_{10} \frac{3}{10}}{\log_{10} x}$ (b) $\dfrac{\ln \frac{3}{10}}{\ln x}$

15. (a) $\dfrac{\log_{10} x}{\log_{10} 2.6}$ (b) $\dfrac{\ln x}{\ln 2.6}$

17. $f(x) = \dfrac{\log_{10} x}{\log_{10} 2} = \dfrac{\ln x}{\ln 2}$ **19.** $f(x) = \dfrac{\log_{10} x}{\log_{10} \frac{1}{2}} = \dfrac{\ln x}{\ln \frac{1}{2}}$

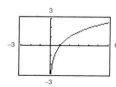 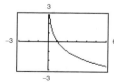

21. $f(x) = \dfrac{\log_{10} x}{\log_{10} 11.8} = \dfrac{\ln x}{\ln 11.8}$

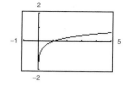

23. $\log_{10} 5 + \log_{10} x$ **25.** $\log_{10} 5 - \log_{10} x$

27. $4 \log_8 x$ **29.** $\frac{1}{2} \ln z$ **31.** $\ln x + \ln y + \ln z$

33. $\frac{1}{2}\ln(a - 1)$ **35.** $\ln z + 2 \ln(z - 1)$

37. $\frac{1}{3} \ln x - \frac{1}{3} \ln y$ **39.** $4 \ln x + \frac{1}{2} \ln y - 5 \ln z$

41. $2 \log_b x - 2 \log_b y - 3 \log_b z$

43. $\ln 3x$ **45.** $\log_4 \dfrac{z}{y}$ **47.** $\log_2(x + 4)^2$

49. $\log_3 \sqrt[4]{5x}$ **51.** $\ln \dfrac{x}{(x + 1)^3}$ **53.** $\ln \dfrac{x - 2}{x + 2}$

55. $\ln \dfrac{x}{(x^2 - 4)^4}$ **57.** $\ln \sqrt[3]{\dfrac{x(x + 3)^2}{x^2 - 1}}$

59. $\ln \dfrac{\sqrt[3]{y(y + 4)^2}}{y - 1}$ **61.** $\ln \dfrac{9}{\sqrt{x^2 + 1}}$

63. $\log_2 \frac{32}{4} = \log_2 32 - \log_2 4$; Property 2

65. 2   **67.** 2.4   **69.** -9 is not in the domain of $\log_3 x$.

71. 2   **73.** -3 **75.** 0 is not in the domain of $\log_{10} x$.

77. 4.5   **79.** $\frac{3}{2}$ **81.** $-3 - \log_5 2$ **83.** $6 + \ln 5$

85. (a) 90   (b) 77   (c) 73   (d) 9 months
(e) $90 - \log_{10}(t + 1)^{15}$
(f)

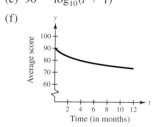

87. False. $\ln 1 = 0$ **89.** False. $\ln(x - 2) \neq \ln x - \ln 2$

91. False. $u = v^2$ **93.** Answers will vary.

95.

97.

$f(x) = h(x)$; Property 2

99. $\dfrac{3x^4}{2y^3}, x \neq 0$ **101.** $1, x \neq 0, y \neq 0$

103. 2502.655   **105.** 0.002   **107.** 4.325

109. 1.415   **111.** 2.361

Section 5.4   *(page 416)*

1. (a) Yes   (b) No

3. (a) No   (b) Yes   (c) Yes, approximate

5. (a) Yes, approximate   (b) No   (c) Yes   **7.** 2

9. 4   **11.** -2 **13.** -5 **15.** 3   **17.** 4

19. 2   **21.** $\ln 2 \approx 0.693$ **23.** $e^{-1} \approx 0.368$

25. 64   **27.** 100   **29.** $\frac{1}{10}$ **31.** $(3, 8)$ **33.** $(9, 2)$

35. x^2 **37.** $x - 2, x > 2$ **39.** $7x + 2$

41. $5x + 2, x > -\frac{2}{5}$ **43.** $2x - 1$ **45.** $\ln 10 \approx 2.303$

47. 0   **49.** $\dfrac{\ln 12}{3} \approx 0.828$ **51.** $\ln \dfrac{5}{3} \approx 0.511$

53. $\ln 5 \approx 1.609$ **55.** $2 \ln 75 \approx 8.635$

57. $\log_{10} 42 \approx 1.623$ **59.** $\dfrac{\ln 80}{2 \ln 3} \approx 1.994$

61. 2   **63.** $\dfrac{\ln 8 - \ln 565}{\ln 2} \approx -6.142$

65.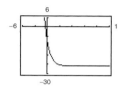

-0.427

67.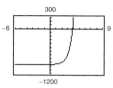

3.847

69.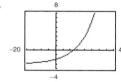

12.207

71.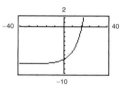

16.636

73. $\dfrac{1}{3} \log_{10} \left(\dfrac{3}{2}\right) \approx 0.059$ **75.** $1 + \dfrac{\ln 7}{\ln 5} \approx 2.209$

77. $\dfrac{\ln 4}{365 \ln\left(1 + \frac{0.065}{365}\right)} \approx 21.330$

79. $\dfrac{\ln 2}{12 \ln\left(1 + \frac{0.10}{12}\right)} \approx 6.960$ **81.** $\dfrac{\ln 1498}{2} \approx 3.656$

83. $e^{-3} \approx 0.050$ **85.** $\dfrac{e^{2.4}}{2} \approx 5.512$

87. $\dfrac{e^{10/3}}{5} \approx 5.606$ **89.** $e^2 - 2 \approx 5.389$

91. $e - 1 \approx 1.718$ and $-e - 1 \approx -3.718$

93. $1 + \sqrt{1 + e} \approx 2.928$ **95.** No solution

97. $10^2 + 3 = 103$ **99.** $2(3^{11/6}) \approx 14.988$

101. $\dfrac{-1 + \sqrt{17}}{2} \approx 1.562$ **103.** 2

105. $\dfrac{725 + 125\sqrt{33}}{8} \approx 180.384$

107. 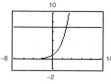   **109.**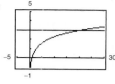

$(2.807, 7)$
$(20.086, 3)$

111. 8.2 years   **113.** 12.9 years

115. (a) 1426 units   (b) 1498 units

117. (a)

(b) $V = 6.7$. Yield will approach 6.7 million cubic feet per acre.

(c) 29.3 years

119. (a) $y = 100$ and $y = 0$; the range falls between 0% and 100%.

(b) Males: 69.71 inches   Females: 64.51 inches

121. (a) $T = 20$; Room temperature   (b) ≈ 0.81 hour

123. $\log_b uv = \log_b u + \log_b v$

True by Property 1 in Section 5.3.

125. $\log_b(u - v) = \log_b u - \log_b v$

False. $1.95 \approx \log_{10}(100 - 10) \neq \log_{10} 100 - \log_{10} 10 = 1$

127. For $rt < \ln 2$ years, double the amount you invest. For $rt > \ln 2$ years, double the interest rate or double the number of years, because either of these will double the exponent in the exponential function.

129. No. It is dependent on the interest rate.

131. $4\sqrt{2} - 10$ **133.** $\dfrac{1}{2}\sqrt{10} + 1$ **135.** $t = \dfrac{k}{s^3}$

137. $x = \dfrac{k}{b - 3}$ **139.** 1.226   **141.** -5.595

Section 5.5   *(page 427)*

1. c   **3.** b   **5.** d

Initial Investment	Annual % Rate	Time to Double	Amount After 10 years
7. $1000	12%	5.78 yr	$3,320.12
9. $750	8.9438%	7.75 yr	$1,834.37
11. $500	11.0%	6.3 yr	$1,505.00
13. $6376.28	4.5%	15.4 yr	$10,000.00

15. $112,087.09

17. (a) 6.642 years   (b) 6.330 years
(c) 6.302 years   (d) 6.301 years

19.

r	2%	4%	6%	8%	10%	12%
t	54.93	27.47	18.31	13.73	10.99	9.16

21.

r	2%	4%	6%	8%	10%	12%
t	55.48	28.01	18.85	14.27	11.53	9.69

23.

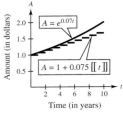

Continuous compounding

Isotope	Half-life (Years)	Initial Quantity	Amount After 1000 Years
25. ^{226}Ra	1620	10 g	6.52 g
27. ^{14}C	5730	2.26 g	2 g
29. ^{239}Pu	24,360	2.16 g	2.1 g

31. $y = e^{0.7675x}$ **33.** $y = 5e^{-0.4024x}$

35. 2023   **37.** $k = 0.0112$; 2796

39. (a) Croatia: $y = 5e^{-0.0018t}$; 4.7 million

Mali: $y = 9.9e^{0.0314t}$; 27.9 million

Singapore: $y = 3.5e^{0.0090t}$; 4.7 million

Sweden: $y = 8.9e^{0.0028t}$; 9.8 million

(b) The greater the rate of growth, the greater the value of b.

(c) b determines whether the population is increasing ($b > 0$) or decreasing ($b < 0$).

41. 61.16 hours   **43.** 15,683 years

45. (a) $V = -750t + 2000$ (b) $V = 2000e^{-0.6931t}$

(c) 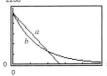   Exponential

(d) 1 year:
Straight-line, $1250; Exponential, $1000
3 years:
Straight-line, $-$250; Exponential, $250

(e) Decreases $750 per year

47. (a) $S = \dfrac{500{,}000}{1 + 0.6e^{0.053t}}$ (b) 280,771 units

49. $583,275

51. (a)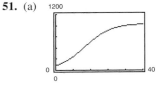

Asymptotes: $y = 0, y = 1000$. The population size will approach 1000 as time increases.

(b) 203 animals

(c) 13 months

53. (a) 398,107,171   (b) 5,011,872   (c) 50,118,723

55. (a) 30 decibels   (b) 85 decibels
(c) 90 decibels   (d) 115 decibels

57. 97%   **59.** 4.95

61. $10^{-3.2} \approx 6.3 \times 10^{-4}$ moles per liter   **63.** 10

65. (a)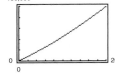

(b) ≈ 21 years; Yes

67. 7:30 A.M.

69. False. A logistics growth function never has an x-intercept.

71. (a) Logarithmic   (b) Logistic   (c) Exponential
(d) Linear   (e) None of the above   (f) Exponential

73. Answers will vary.

75. $8x^2 - 24x + 18$

77. $x^3 - 5x^2 + 25x - 128 + \dfrac{641}{x + 5}$

79. 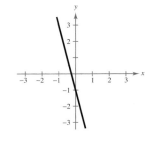   **81.**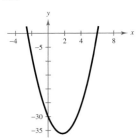

83.

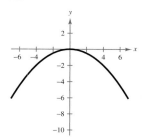

85.

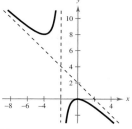

87.

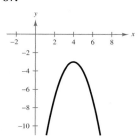

89.

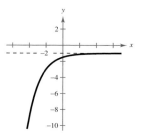

91.

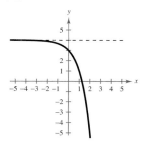

Review Exercises *(page 434)*

1. 76.699   **3.** 0.337   **5.** 1201.845   **7.** c   **9.** a

11.

x	-1	0	1	2	3
$f(x)$	8	5	4.25	4.063	4.016

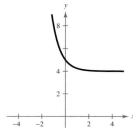

13.

x	-2	-1	0	1	2
$f(x)$	-0.377	-1	-2.65	-7.023	-18.61

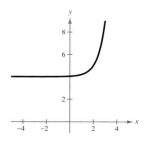

15.

x	-1	0	1	2	3
$f(x)$	4.008	4.04	4.2	5	9

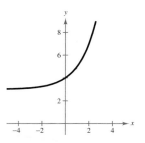

17.

x	-2	-1	0	1	2
$f(x)$	3.25	3.5	4	5	7

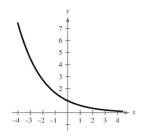

19. 2980.958   **21.** 0.183

23.

x	-2	-1	0	1	2
$h(x)$	2.72	1.65	1	0.61	0.37

25.

x	-3	-2	-1	0	1
$f(x)$	0.37	1	2.72	7.39	20.09

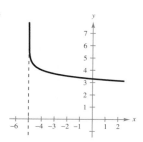

27.

n	1	2	4	12
A	\$6569.98	\$6635.43	\$6669.46	\$6692.64

n	365	Continuous
A	\$6704.00	\$6704.39

29.

t	1	10	20
P	\$184,623.27	\$89,865.79	\$40,379.30

t	30	40	50
P	\$18,143.59	\$8152.44	\$3663.13

31. (a) 0.154 (b) 0.487 (c) 0.811

33. (a) \$1,069,047.14 (b) 7.9 years

35. $\log_4 64 = 3$ **37.** 3 **39.** -3

41.

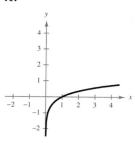

Vertical asymptote: $x = 0$

43.

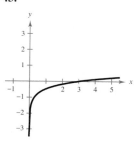

Vertical asymptote: $x = 0$

45.

Vertical asymptote: $x = -5$

47. 3.118 **49.** -12 **51.** 2.034

53.

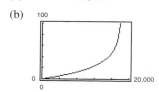

Vertical asymptote: $x = 0$

55.

Vertical asymptote: $x = 0$

57. 27.16 miles **59.** 2.132 **61.** -1.159

63. and 65. Answers will vary.

67. $1 + 2\log_5|x|$ **69.** $\log_{10} 5 + \frac{1}{2}\log_{10} y - 2\log_{10}|x|$

71. $\log_2 5x$ **73.** $\ln \dfrac{\sqrt{|2x - 1|}}{(x + 1)^2}$

75. (a) $0 \le h < 18{,}000$

(b)

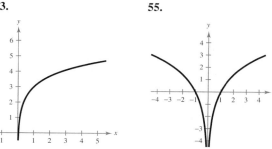

Vertical asymptote: $h = 18{,}000$

(c) Climbing at a slower rate, so the time required increases.

(d) 5.46 minutes

77. 6 **79.** 6 **81.** 3 **83.** $\dfrac{\ln 25}{3} \approx 1.073$

85. $\frac{1}{3}(\ln 40 - 2) \approx 0.563$ **87.** $\ln 20 \approx 2.996$

89. $\dfrac{\ln 95}{\ln 12} \approx 1.833$ **91.** $\ln 2 \approx 0.693, \ln 4 \approx 1.386$

93. $-7.04, -1.53$ **95.** 0.68 **97.** $\frac{1}{5}e^{7.2} \approx 267.886$

99. $\frac{1}{3}e^{15/4} \approx 14.174$ **101.** $e^6 - 8 \approx 395.429$

103. $5e^4 \approx 272.991$ **105.** $-2 + \sqrt{6} \approx 0.449$

107. -104 **109.** $0, 0.42, 13.63$ **111.** $-3.99, 1.48$

113. 21.3 years   **115.** e   **117.** f   **119.** a

121. 2025   **123.** (a) 13.8629%   (b) $11,486.98

125. $y = \frac{1}{2}e^{0.4605x}$

127. (a) 7.7 weeks   (b) 13.3 weeks

129. (a) 251,188,643   (b) 7,079,458   (c) 1,258,925,412

131. True by properties of exponents.

$$e^{x-1} = e^x \cdot e^{-1} = \frac{e^x}{e}$$

133. False. $\ln(x \cdot y) = \ln x + \ln y \neq \ln(x + y)$

135. False. The domain of $f(x) = \ln x$ is $(0, \infty)$.

Chapter Test   *(page 439)*

1. 1123.690   **2.** 687.291   **3.** 0.497   **4.** 22.198

5.

x	-1	$-\frac{1}{2}$	0	$\frac{1}{2}$	1
$f(x)$	10	3.162	1	0.316	0.1

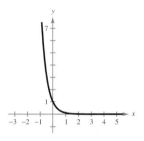

6.

x	-1	0	1	2	3
$f(x)$	-0.005	-0.028	-0.167	-1	-6

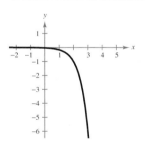

7.

x	-1	$-\frac{1}{2}$	0	$\frac{1}{2}$	1
$f(x)$	0.865	0.632	0	-1.718	-6.389

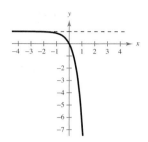

8. (a) -0.89 (b) 9.2

9.

x	$\frac{1}{2}$	1	$\frac{3}{2}$	2	4
$f(x)$	-5.699	-6	-6.176	-6.301	-6.602

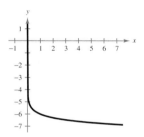

Vertical asymptote: $x = 0$

10.

x	5	7	9	11	13
$f(x)$	0	1.099	1.609	1.946	2.197

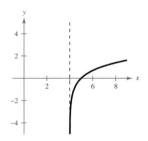

Vertical asymptote: $x = 4$

11.

x	-5	-3	-1	0	1
$f(x)$	1	2.099	2.609	2.792	2.946

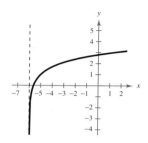

Vertical asymptote: $x = -6$

12. 1.945 **13.** 0.115 **14.** 1.328

15. $\log_2 3 + 4 \log_2 a$ **16.** $\ln 5 + \frac{1}{2}\ln x - \ln 6$

17. $\log_3 13y$ **18.** $\ln \dfrac{x^4}{y^4}$ **19.** $\dfrac{\ln 197}{4} \approx 1.321$

20. $\frac{800}{501} \approx 1.597$ **21.** $y = 2745 e^{0.1570x}$ **22.** 55%

23. (a)

x	$\frac{1}{4}$	1	2	4	5	6
H	58.720	75.332	86.828	103.43	110.59	117.38

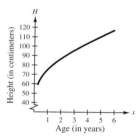

(b) 103 centimeters; 103.43 centimeters

Cumulative Test for Chapters 3–5
(page 440)

1. $y = -\frac{3}{4}(x + 8)^2 + 5$

2.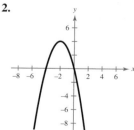

3.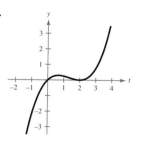

4.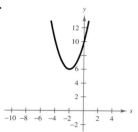

5. $-2, \pm 2i$ **6.** $-7, 0, 3$ **7.** $3x - 2 - \dfrac{3x - 2}{2x^2 + 1}$

8. $2x^3 - x^2 + 2x - 10 + \dfrac{25}{x + 2}$ **9.** 1.20

10. $x^4 + 3x^3 - 11x^2 + 9x + 70$

11.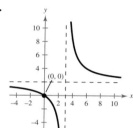

Vertical asymptote: $x = 3$

Horizontal asymptote: $y = 2$

12.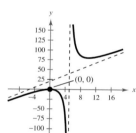

Vertical asymptote: $x = 5$

Slant asymptote: $y = 4x + 20$

13.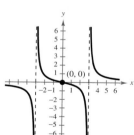

Vertical asymptotes: $x = \pm 3$

Horizontal asymptote: $y = 0$

14. $\dfrac{1}{5}\left(\dfrac{4}{x - 7} - \dfrac{4}{x + 3}\right)$ **15.** $\dfrac{5}{x - 4} + \dfrac{20}{(x - 4)^2}$

16.

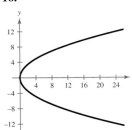

17.

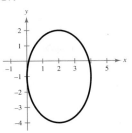

18. $(x - 3)^2 = \frac{3}{2}(y + 2)$ **19.** $\frac{(y - 2)^2}{\frac{4}{5}} - \frac{x^2}{\frac{16}{5}} = 1$

20. Reflect f in the x-axis and y-axis, and shift f 3 units to the right.

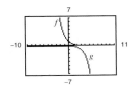

21. Reflect f in the x-axis, and shift f 4 units upward.

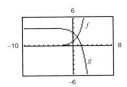

22. 1.991   **23.** -0.067 **24.** 1.717   **25.** 0.281

26. 0.302   **27.** -1.733 **28.** -4.087

29. $\ln(x + 4) + \ln(x - 4) - 4 \ln x, \; x > 4$

30. $\ln \dfrac{x^2}{\sqrt{x + 5}}, \; x > 0$ **31.** $\dfrac{\ln 12}{2} \approx 1.242$ **32.** $\dfrac{64}{5}$

33.

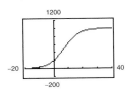

Asymptotes: $y = 0, \; y = 1000$

34. $2000

35. (a) and (b)

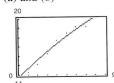

(c) No, because the model will eventually become negative.

36. $16,302.05   **37.** 6.3 hours

Chapter 6

Section 6.1   *(page 451)*

1. d   **3.** b   **5.** $(2, 2)$ **7.** $(2, 6), (-1, 3)$

9. $(0, -5), (4, 3)$ **11.** $(0, 0), (2, -4)$

13. $(0, 1), (1, -1), (3, 1)$ **15.** $(5, 5)$ **17.** $\left(\frac{1}{2}, 3\right)$

19. $(1, 1)$ **21.** $\left(\frac{20}{3}, \frac{40}{3}\right)$ **23.** No solution

25. $(-2, 4), (0, 0)$ **27.** $(0, 0), (-1, -1), (1, 1)$

29. $(4, 3)$ **31.** $\left(\frac{5}{2}, \frac{3}{2}\right)$ **33.** $(2, 2), (4, 0)$

35. $(1, 4), (4, 7)$ **37.** $\left(4, -\frac{1}{2}\right)$

39. No solution   **41.** $(4, 3), (-4, 3)$

43.

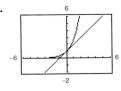

$(0, 1)$

45.

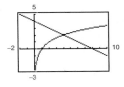

$(4, 2)$

47.

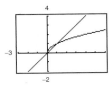

$(0, 0), (1, 1)$

49.

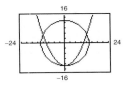

$(0, -13), (\pm 12, 5)$

51. $(1, 2)$ **53.** $(-2, 0), \left(\frac{29}{10}, \frac{21}{10}\right)$ **55.** No solution

57. $(0.287, 1.75)$ **59.** $(-1, 0), (0, 1), (1, 0)$

61. $\left(\frac{1}{2}, 2\right), \left(-4, -\frac{1}{4}\right)$ **63.** 192 units   **65.** 3133 units

67. (a) 6400 units   (b) 8800 units

69. (a) $\begin{cases} x + y = 25{,}000 \\ 0.06x + 0.085y = 2{,}000 \end{cases}$

(b)

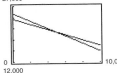

Decreases; interest is fixed.

(c) $5000

71. More than $11,666.67

73. (a)

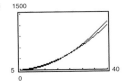

(b) 24.7 inches   (c) Doyle Log Rule

75. 6×9 meters   **77.** 9×12 inches

79. 8×12 kilometers

81. (a) $f(t) = 1.53t + 38.94$

$g(t) = -0.325t^2 + 4.455t + 32.765$

(b)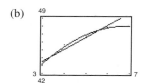

(c) $(3.382, 44.114), (5.618, 47.536)$

83. False. The system can have at most four solutions because a parabola and a circle can intersect at most four times.

85. For a linear system the result will be a contradictory equation such as $0 = N$, where N is a nonzero real number. For a nonlinear system there may be an equation with imaginary solutions.

87. (a) $y = 2x$ (b) $y = 0$ (c) $y = x - 2$

89. $2x + 7y - 45 = 0$ **91.** $y - 3 = 0$

93. $30x - 17y - 18 = 0$

95. Domain: All real numbers $x \neq 6$

Horizontal asymptote: $y = 0$

Vertical asymptote: $x = 6$

97. Domain: All real numbers $x \neq \pm 4$

Horizontal asymptote: $y = 1$

Vertical asymptotes: $x = \pm 4$

99. 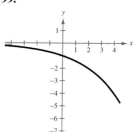   **101.**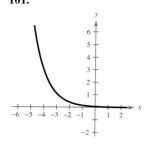

Section 6.2 *(page 463)*

1. $(2, 1)$ **3.** $(1, -1)$ **5.** No solution

7. $\left(a, \frac{3}{2}a - \frac{5}{2}\right)$ **9.** $\left(\frac{1}{3}, -\frac{2}{3}\right)$ **11.** $\left(\frac{5}{2}, \frac{3}{4}\right)$

13. $(3, 4)$ **15.** $(4, -1)$ **17.** $\left(\frac{12}{7}, \frac{18}{7}\right)$

19. No solution   **21.** $\left(\frac{18}{5}, \frac{3}{5}\right)$ **23.** $\left(a, \frac{5}{6}a - \frac{1}{2}\right)$

25. $\left(\frac{90}{31}, -\frac{67}{31}\right)$ **27.** $\left(-\frac{6}{35}, \frac{43}{35}\right)$ **29.** $(5, -2)$

31. 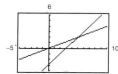   **33.**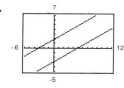

Consistent, one solution   Inconsistent

35. 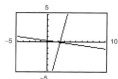   **37.**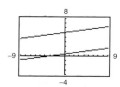

Consistent, one solution   Inconsistent

39. 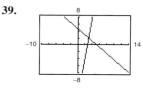   **41.**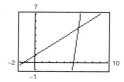

$(3, 2)$ $(6, 5)$

43. 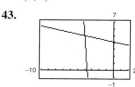   **45.**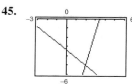

$(-4, 5)$ $(2, -5)$

47. $(4, 1)$ **49.** $(2, -1)$ **51.** $(6, -3)$ **53.** $\left(\frac{43}{6}, \frac{25}{6}\right)$

55. $(80, 10)$ **57.** $(2,000,000, 100)$

59. 550 miles per hour, 50 miles per hour

61. (a) $\begin{cases} x + y = 10 \\ 0.2x + 0.5y = 3 \end{cases}$

(b)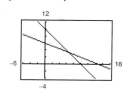

Decreases

(c) 20% solution: $6\frac{2}{3}$ liters

50% solution: $3\frac{1}{3}$ liters

63. $6000   **65.** 400 adults, 1035 students

67. Machine 1: 1134 containers

Machine 2: 630 containers

69. $y = 0.97x + 2.10$ **71.** $y = 0.318x + 4.061$

73. $y = -2x + 4$

75. (a) and (b) $y = -240x + 685$

(c) 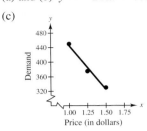 (d) 349 units

77. False. Two lines that coincide have infinitely many points of intersection.

79. True

81. $\begin{cases} x + y = 6 \\ 2x - y = 18 \end{cases}$ **83.** $\begin{cases} 3x + 3y = -32 \\ 3x + 6y = -62 \end{cases}$

85. $(300, 315)$. It is necessary to change the scale on the axes to see the point of intersection.

87. (a) $\begin{cases} x + y = 10 \\ x + y = 20 \end{cases}$ (b) $\begin{cases} x + y = 3 \\ 2x + 2y = 6 \end{cases}$

89. $k = -2$

91. $x > 1$

93. $\frac{4}{3} \le x < \frac{16}{3}$

95. All real numbers x

97. $x < -4, x > 0$

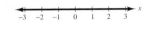

99. $\frac{1}{2}\left(-\frac{6}{x} + \frac{3}{x+1} + \frac{3}{x-1}\right)$ **101.** $\ln \frac{x}{(x+3)^5}$

103. $\log_6 \sqrt[4]{3x}$ **105.** $\left(\frac{3}{2}, \frac{3}{10}\right)$

Section 6.3 *(page 475)*

1. d **3.** c **5.** $(1, -2, 4)$ **7.** $(1, 2, -2)$

9. $\left(\frac{1}{2}, -2, 2\right)$

11. $\begin{cases} x - 2y + 3z = 5 \\ y - 2z = 9 \\ 2x \quad\quad - 3z = 0 \end{cases}$

First step in putting the system in row-echelon form

13. $(1, 2, 3)$ **15.** $(-4, 8, 5)$ **17.** $(5, -2, 0)$

19. No solution **21.** $\left(-\frac{1}{2}, 1, \frac{3}{2}\right)$

23. $(-3a + 10, 5a - 7, a)$ **25.** $(-a + 3, a + 1, a)$

27. $(2a, 21a - 1, 8a)$ **29.** $\left(-\frac{3}{2}a + \frac{1}{2}, -\frac{2}{3}a + 1, a\right)$

31. $(1, 1, 1, 1)$ **33.** No solution **35.** $(0, 0, 0)$

37. $(9a, -35a, 67a)$

39. $y = \frac{1}{2}x^2 - 2x$

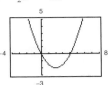

41. $y = x^2 - 6x + 8$

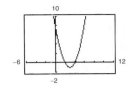

43. $x^2 + y^2 - 4x = 0$

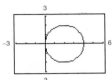

45. $x^2 + y^2 + 6x - 8y = 0$

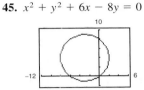

47. $s = -16t^2 + 144$ **49.** $s = -16t^2 - 32t + 500$

51. 8 touchdowns, 6 extra points, 6 field goals

53. \$300,000 at 8%
\$400,000 at 9%
\$75,000 at 10%

55. $250,000 - \frac{1}{2}s$ in certificates of deposit,
$125,000 + \frac{1}{2}s$ in municipal bonds,
$125,000 - s$ in blue-chip stocks,
s in growth stocks

57. 20 liters of spray X,
18 liters of spray Y,
16 liters of spray Z

59. Use four medium trucks or use two large, one medium, and two small trucks. Other answers are possible.

61. $t_1 = 96$ pounds
$t_2 = 48$ pounds
$a = -16$ feet per second squared

63. $\frac{1}{2}\left(-\frac{2}{x} + \frac{1}{x-1} + \frac{1}{x+1}\right)$

65. $\frac{1}{2}\left(\frac{1}{x} - \frac{1}{x-2} + \frac{2}{x+3}\right)$

67. $y = -\frac{5}{24}x^2 - \frac{3}{10}x + \frac{41}{6}$ **69.** $y = x^2 - x$

71. (a) $y = 0.165x^2 - 6.55x + 103$

(b)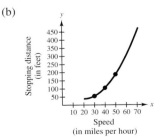

(c) 453 feet

73. $x = 5$
$y = 5$
$\lambda = -5$

75. $x = \pm\sqrt{2}/2$ or $x = 0$
$y = \frac{1}{2}$ $y = 0$
$\lambda = 1$ $\lambda = 0$

77. False. Equation 2 does not have a leading coefficient of 1.

79. No. Answers will vary.

81. There will be a row representing a contradictory equation such as $0 = N$, where N is a nonzero real number.

83. $\begin{cases} x + y + z = -6 \\ -2x - y + 3z = 15 \\ x + 4y - z = -14 \end{cases}$

85. $\begin{cases} 2x - y + 3z = -28 \\ -6x + 4y + z = 18 \\ -4x - 2y - 3z = 19 \end{cases}$

87. 150%   **89.** 275

91. (a) $\pm 2, 0$
(b)
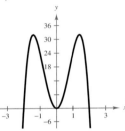

93. (a) $-\frac{1}{2}, \frac{1}{3}, 5$
(b)
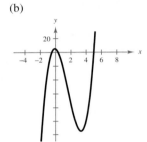

95.

x	-2	-1	0	1	2
y	11.625	2.25	-1.5	-3	-3.6

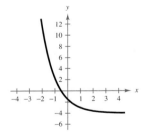

97.

x	$-\frac{1}{2}$	0	$\frac{1}{2}$	1	2
y	28.918	18.25	12.548	9.5	7

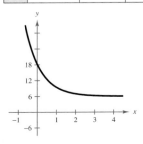

99. $\left(\frac{1}{2}, 0\right)$

Section 6.4 *(page 487)*

1.
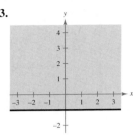

3.
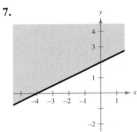

5.
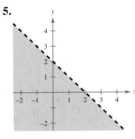

7.
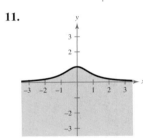

9.
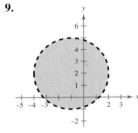

11.

13.
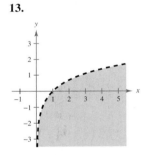

15.
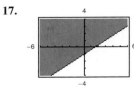

17.
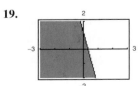

19.
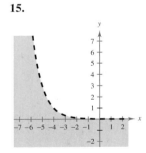

21.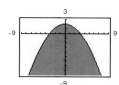

23.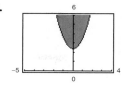

51.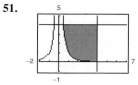

25. $y \leq \frac{1}{2}x + 2$ **27.** $y \geq -\frac{2}{3}x + 2$

29. c and d **31.** a, c, and d

53. $\begin{cases} y \leq 4 - x \\ x \geq 0 \\ y \geq 0 \end{cases}$ **55.** $\begin{cases} y \geq 4 - x \\ y \geq 2 - \frac{1}{4}x \\ x \geq 0,\ y \geq 0 \end{cases}$

33.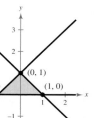

35.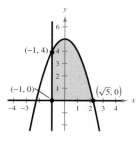

57. $\begin{cases} x^2 + y^2 \leq 16 \\ x \geq 0 \\ y \geq 0 \end{cases}$ **59.** $\begin{cases} 2 \leq x \leq 5 \\ 1 \leq y \leq 7 \end{cases}$ **61.** $\begin{cases} y \leq \frac{3}{2}x \\ y \leq -x + 5 \\ y \geq 0 \end{cases}$

63. Consumer surplus: 1600

  Producer surplus: 400

65. Consumer surplus: 40,000,000

  Producer surplus: 20,000,000

37.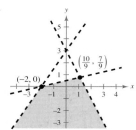

39. No solution

67. $\begin{cases} x + \frac{3}{2}y \leq 12 \\ \frac{4}{3}x + \frac{3}{2}y \leq 15 \\ x \qquad \geq 0 \\ \qquad y \geq 0 \end{cases}$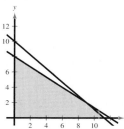

41.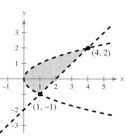

43.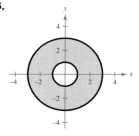

69. $\begin{cases} x + y \leq 20{,}000 \\ \qquad y \geq 2x \\ x \qquad \geq 5{,}000 \\ \qquad y \geq 5{,}000 \end{cases}$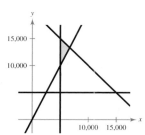

45.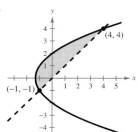

71. $\begin{cases} 55x + 70y \leq 7500 \\ x \qquad\quad \geq 50 \\ \qquad y \geq 40 \end{cases}$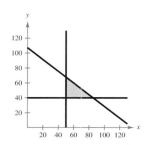

47.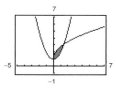

49.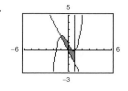

73. $\begin{cases} xy \geq 500 \\ 2x + \pi y \geq 125 \\ x \qquad \geq 0 \\ \qquad y \geq 0 \end{cases}$

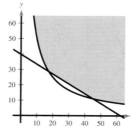

75. False. The graph shows the solution of the system

$\begin{cases} y < 6 \\ -4x - 9y < 6. \\ 3x + y^2 \geq 2 \end{cases}$

77. b   **79.** a

81. (a) $\begin{cases} \pi y^2 - \pi x^2 \geq 10 \\ \qquad y > x \\ \qquad x > 0 \end{cases}$

(b)

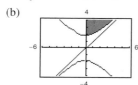

(c) The line is an asymptote to the boundary. The larger the circles, the closer the radii can be and the constraint still be satisfied.

83. The graph is a half-line on the real number line; on the rectangular coordinate system, the graph is a half-plane.

85. $x + 11y + 8 = 0$ **87.** $2x - y + 1 = 0$

89. $60x - 35y + 113 = 0$ **91.** 52.619 **93.** 0.022

95. 0.064 **97.** $(-4, 3, -7)$

Section 6.5   *(page 497)*

1. Minimum at $(0, 0)$: 0

   Maximum at $(5, 0)$: 20

3. Minimum at $(0, 0)$: 0

   Maximum at $(0, 5)$: 40

5. Minimum at $(0, 0)$: 0

   Maximum at $(3, 4)$: 17

7. Minimum at $(0, 0)$: 0

   Maximum at $(4, 0)$: 20

9. Minimum at $(0, 0)$: 0

   Maximum at $(60, 20)$: 740

11. Minimum at $(0, 0)$: 0

   Maximum at any point on the line segment connecting $(60, 20)$ and $(30, 45)$: 2100

13.

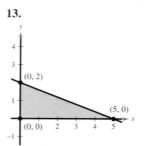

Minimum at $(0, 0)$: 0

Maximum at $(5, 0)$: 30

15.

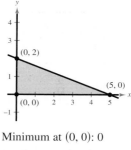

Minimum at $(0, 0)$: 0

Maximum at $(0, 2)$: 48

17.

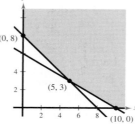

Minimum at $(5, 3)$: 35

No maximum

19.

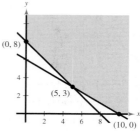

Minimum at $(10, 0)$: 20

No maximum

21.

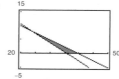

Minimum at $(24, 8)$: 104

Maximum at $(40, 0)$: 160

23.

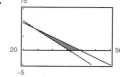

Minimum at $(36, 0)$: 36

Maximum at $(24, 8)$: 56

25. Maximum at $(3, 6)$: 12

27. Maximum at $(0, 10)$: 10

29. Maximum at $(0, 5)$: 25

31. Maximum at $\left(\frac{22}{3}, \frac{19}{6}\right)$: $\frac{271}{6}$

33. 750 units of model A

   1000 units of model B

   Maximum profit: $83,750

35. 0 units of the $150 model

   200 units of the $200 model

   Maximum profit: $8000

37. Three bags of brand X

   Six bags of brand Y

   Minimum cost: $195

39. 4 audits

   32 tax returns

   Maximum revenue: $17,600

41.

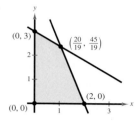

The maximum, 5, occurs at any point on the line segment connecting $(2, 0)$ and $\left(\frac{20}{19}, \frac{45}{19}\right)$.

43.

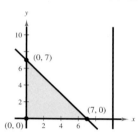

The constraint $x \leq 10$ is extraneous. Maximum at $(0, 7)$: 14

45.

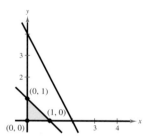

The constraint $2x + y \leq 4$ is extraneous. Maximum at $(0, 1)$: 4

47. True. The objective function has a maximum value at any point on the line segment connecting the two vertices.

49. (a) $t \geq 9$ (b) $\frac{3}{4} \leq t \leq 9$ **51.** $z = x + 5y$

53. $z = 4x + y$

55. $\dfrac{9}{2(x + 3)}, \; x \neq 0$ **57.** $\dfrac{x^2 + 2x - 13}{x(x - 2)}, \; x \neq \pm 3$

59.

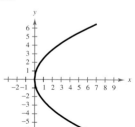

61.

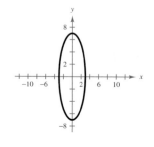

63.

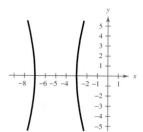

65. $\ln 3 \approx 1.099$ **67.** $4 \ln 38 \approx 14.550$

69. $\frac{1}{3}e^{12/7} \approx 1.851$

Review Exercises   *(page 502)*

1. $(5, 4)$ **3.** $(0, 0), (2, 8), (-2, 8)$ **5.** $(4, -2)$

7. $(0, -2)$ **9.** $(1.41, -0.66), (-1.41, 10.66)$

11. 10,417 units   **13.** 96×144 meters   **15.** $\left(\frac{3}{10}, \frac{2}{5}\right)$

17. $\left(\frac{1}{3}, -\frac{1}{2}\right)$ **19.** $(-3, 7)$ **21.** No solution

23.

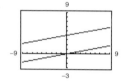

25.

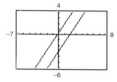

Inconsistent   Inconsistent

27. 400 at \$9.95; 250 at \$14.95   **29.** $\left(\dfrac{500{,}000}{7}, \dfrac{159}{7}\right)$

31. $(2, -4, -5)$ **33.** $\left(\frac{24}{5}, \frac{22}{5}, -\frac{8}{5}\right)$

35. $(3a + 4, 2a + 5, a)$ **37.** $y = 2x^2 + x - 5$

39. $x^2 + y^2 - 4x + 4y - 1 = 0$ **41.** $y = 1.01x + 1.54$

43. $(a - 1, a, a + 3)$

45. 10 gallons of spray X
   5 gallons of spray Y
   12 gallons of spray Z

47. $y = 0.114x^2 + 0.800x + 1.591$

49.

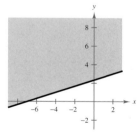

51.

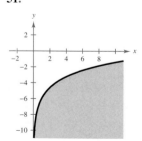

53.

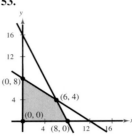

55.

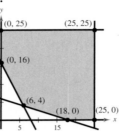

57.

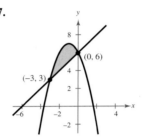

59.

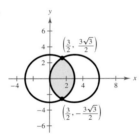

61. $\begin{cases} 20x + 30y \le 24{,}000 \\ 12x + 8y \le 12{,}400 \\ \quad x \qquad \ge \quad 0 \\ \qquad y \ge \quad 0 \end{cases}$

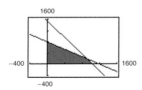

63. Consumer surplus: $4,000,000

Producer surplus: $6,000,000

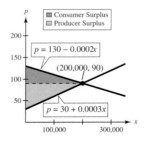

65. Minimum at $(25, 50)$: 600

67. Maximum at $(500, 500)$: 60,000

69. Five units of product A
Two units of product B
Maximum profit: $138

71. $\frac{1}{3}$ gallon of type A
$\frac{2}{3}$ gallon of type B
Minimum cost: $1.45

73. False. An objective function can have no maximum value, one point where a maximum value occurs, or an infinite number of points where a maximum value occurs.

75. $\begin{cases} x - y = 9 \\ 3x + y = 11 \end{cases}$

77. $\begin{cases} -x + 4y = 10 \\ 3x - 8y = -21 \end{cases}$

79. $\begin{cases} x - 2y + z = -7 \\ 2x + y - 4z = -25 \\ -x + 3y - z = 12 \end{cases}$

81. $\begin{cases} 4x + y - z = -7 \\ 8x + 3y + 2z = 16 \\ 4x - 2y + 3z = 31 \end{cases}$

83. For a linear system, the result will be a contradictory equation such as $0 = N$, where N is a nonzero real number. For a nonlinear system, there may be an equation with imaginary roots.

85. (a) One   (b) Two   (c) Four

Chapter Test   *(page 507)*

1. $(-3, 4)$ **2.** $(0, -1), (1, 0), (2, 1)$

3. $(8, 4), (2, -2)$ **4.** $(3, 2)$ **5.** $(-3, 0), (2, 5)$

6. $(1, 12), (0.034, 8.619)$ **7.** $(1, 5)$ **8.** $(2, -3, 1)$

9. $\begin{cases} 3x - y = 9 \\ 6x + y = 3 \end{cases}$

(Answer is not unique.)

10. $y = -\frac{1}{2}x^2 + x + 6$

11. $\begin{cases} 2x - y - 4z = -2 \\ 4x + 3y + 8z = 6 \\ -6x + y - 12z = 24 \end{cases}$

(Answer is not unique.)

12.

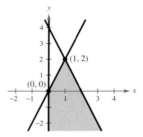

13.

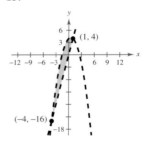

14.

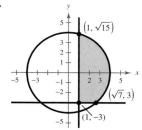

15. $(12, 0)$; $z = 240$

16. \$275 model: 160 units

\$400 model: 140 units

Maximum profit: \$19,300

17. 0 units of model I

5300 units of model II

Maximum profit: \$212,000

Chapter 7

Section 7.1 *(page 520)*

1. 1×2 **3.** 3×1 **5.** 2×2

7. $\begin{bmatrix} 4 & -3 & \vdots & -5 \\ -1 & 3 & \vdots & 12 \end{bmatrix}$

9. $\begin{bmatrix} 1 & 10 & -2 & \vdots & 2 \\ 5 & -3 & 4 & \vdots & 0 \\ 2 & 1 & 0 & \vdots & 6 \end{bmatrix}$

11. $\begin{bmatrix} 7 & -5 & 1 & \vdots & 13 \\ 19 & 0 & -8 & \vdots & 10 \end{bmatrix}$

13. $\begin{cases} x + 2y = 7 \\ 2x - 3y = 4 \end{cases}$

15. $\begin{cases} 2x \quad\quad + 5z = -12 \\ \quad\quad y - 2z = \quad 7 \\ 6x + 3y \quad\quad = \quad 2 \end{cases}$

17. $\begin{cases} 9x + 12y + 3z \quad\quad = \quad 0 \\ -2x + 18y + 5z + 2w = \quad 10 \\ x + 7y - 8z \quad\quad = -4 \\ 3x \quad\quad + 2z \quad\quad = -10 \end{cases}$

19. Reduced row-echelon form

21. Not in row-echelon form

23. $\begin{bmatrix} 1 & 4 & 3 \\ 0 & 2 & -1 \end{bmatrix}$

25. $\begin{bmatrix} 1 & 1 & 4 & -1 \\ 0 & 5 & -2 & 6 \\ 0 & 3 & 20 & 4 \end{bmatrix}$

$\begin{bmatrix} 1 & 1 & 4 & -1 \\ 0 & 1 & -\frac{2}{5} & \frac{6}{5} \\ 0 & 3 & 20 & 4 \end{bmatrix}$

27. Add 5 times Row 2 to Row 1.

29. Interchange Row 1 and Row 2. Add 4 times new Row 1 to Row 3.

31. (a) $\begin{bmatrix} 1 & 2 & 3 \\ 0 & -5 & -10 \\ 3 & 1 & -1 \end{bmatrix}$ (b) $\begin{bmatrix} 1 & 2 & 3 \\ 0 & -5 & -10 \\ 0 & -5 & -10 \end{bmatrix}$

(c) $\begin{bmatrix} 1 & 2 & 3 \\ 0 & -5 & -10 \\ 0 & 0 & 0 \end{bmatrix}$ (d) $\begin{bmatrix} 1 & 2 & 3 \\ 0 & 1 & 2 \\ 0 & 0 & 0 \end{bmatrix}$

(e) $\begin{bmatrix} 1 & 0 & -1 \\ 0 & 1 & 2 \\ 0 & 0 & 0 \end{bmatrix}$

The matrix is in reduced row-echelon form.

33. $\begin{bmatrix} 1 & 1 & 0 & 5 \\ 0 & 1 & 2 & 0 \\ 0 & 0 & 1 & -1 \end{bmatrix}$ **35.** $\begin{bmatrix} 1 & -1 & -1 & 1 \\ 0 & 1 & 6 & 3 \\ 0 & 0 & 0 & 0 \end{bmatrix}$

37. $\begin{bmatrix} 1 & 0 & 0 \\ 0 & 1 & 0 \\ 0 & 0 & 1 \end{bmatrix}$ **39.** $\begin{bmatrix} 1 & 2 & 0 & 0 \\ 0 & 0 & 1 & 0 \\ 0 & 0 & 0 & 1 \\ 0 & 0 & 0 & 0 \end{bmatrix}$

41. $\begin{bmatrix} 1 & 0 & 3 & 16 \\ 0 & 1 & 2 & 12 \end{bmatrix}$

43. $\begin{cases} x - 2y = \quad 4 \\ \quad\quad y = -3 \end{cases}$ **45.** $\begin{cases} x - y + 2z = \quad 4 \\ \quad\quad y - z = \quad 2 \\ \quad\quad\quad z = -2 \end{cases}$

$(-2, -3)$ $(8, 0, -2)$

47. $(3, -4)$ **49.** $(-4, -10, 4)$ **51.** $(3, 2)$

53. $(-5, 6)$ **55.** $(-1, -4)$ **57.** Inconsistent

59. $(4, -3, 2)$ **61.** $(7, -3, 4)$ **63.** $(-4, -3, 6)$

65. $(2a + 1, 3a + 2, a)$

67. $(4 + 5b + 4a, 2 - 3b - 3a, b, a)$ **69.** Inconsistent

71. $(0, 2 - 4a, a)$ **73.** $(1, 0, 4, -2)$

75. $(-2a, a, a, 0)$ **77.** Yes; $(-1, 1, -3)$ **79.** No

81. $\begin{bmatrix} 1 & 3 & \frac{3}{2} & \vdots & 4 \\ 0 & 1 & \frac{7}{4} & \vdots & -\frac{3}{2} \\ 0 & 0 & 1 & \vdots & 2 \end{bmatrix}, \begin{bmatrix} 1 & 3 & 1 & \vdots & 3 \\ 0 & 1 & 2 & \vdots & -1 \\ 0 & 0 & 1 & \vdots & 2 \end{bmatrix}$

83. $\dfrac{4x^2}{(x + 1)^2(x - 1)} = \dfrac{1}{x - 1} + \dfrac{3}{x + 1} - \dfrac{2}{(x + 1)^2}$

85. \$100,000 at 9%   **87.** $y = -x^2 + 2x + 8$

\$250,000 at 10%

\$150,000 at 12%

89. (a) $y = -128.5t^2 + 1587.5t - 4304.0$

(b)

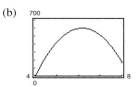

(c) -1279. The estimate is not reasonable because it is a negative number.

91. (a) $x_1 = 500 - s - t,\ x_2 = -200 + s + t,\ x_3 = s,$

$x_4 = 350 - t,\ x_5 = t$

(b) $x_1 = 100,\ x_2 = 200,\ x_3 = 50,\ x_4 = 0,\ x_5 = 350$

(c) $x_1 = 150,\ x_2 = 150,\ x_3 = 0,\ x_4 = 0,\ x_5 = 350$

93. False. A matrix is in reduced row-echelon form if (1) all rows consisting entirely of zeros occur at the bottom of the matrix, (2) for each row that does not consist entirely of zeros, the first nonzero entry is 1, (3) for two successive nonzero rows, the leading 1 in the higher row is farther to the left than the leading 1 in the lower row, and (4) every column that has a leading 1 has zeros in every position above and below its leading 1.

95.
$$\begin{cases} x + y + 7z = -1 \\ x + 2y + 11z = 0 \\ 2x + y + 10z = -3 \end{cases}$$

97. Interchange two rows.
Multiply a row by a nonzero constant.
Add a multiple of a row to another row.

99. A matrix in reduced row-echelon form has zeros above the leading 1s. A matrix in row-echelon form may have any real number above the leading 1s.

101.

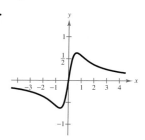

Horizontal asymptote: $y = 0$

103. Parabola   **105.** Hyperbola   **107.** Circle

109.

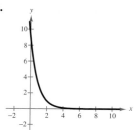

111.

Section 7.2   *(page 535)*

1. $x = -4,\ y = 22$ **3.** $x = 2,\ y = 3$

5. (a) $\begin{bmatrix} 3 & -2 \\ 1 & 7 \end{bmatrix}$ (b) $\begin{bmatrix} -1 & 0 \\ 3 & -9 \end{bmatrix}$ (c) $\begin{bmatrix} 3 & -3 \\ 6 & -3 \end{bmatrix}$

(d) $\begin{bmatrix} -1 & -1 \\ 8 & -19 \end{bmatrix}$

7. (a) $\begin{bmatrix} 7 & 3 \\ 1 & 9 \\ -2 & 15 \end{bmatrix}$ (b) $\begin{bmatrix} 5 & -5 \\ 3 & -1 \\ -4 & -5 \end{bmatrix}$ (c) $\begin{bmatrix} 18 & -3 \\ 6 & 12 \\ -9 & 15 \end{bmatrix}$

(d) $\begin{bmatrix} 16 & -11 \\ 8 & 2 \\ -11 & -5 \end{bmatrix}$

9. (a) $\begin{bmatrix} 3 & 3 & -2 & 1 & 1 \\ -2 & 5 & 7 & -6 & -8 \end{bmatrix}$

(b) $\begin{bmatrix} 1 & 1 & 0 & -1 & 1 \\ 4 & -3 & -11 & 6 & 6 \end{bmatrix}$

(c) $\begin{bmatrix} 6 & 6 & -3 & 0 & 3 \\ 3 & 3 & -6 & 0 & -3 \end{bmatrix}$

(d) $\begin{bmatrix} 4 & 4 & -1 & -2 & 3 \\ 9 & -5 & -24 & 12 & 11 \end{bmatrix}$

11. (a), (b), and (d) not possible

(c) $\begin{bmatrix} 18 & 0 & 9 \\ -3 & -12 & 0 \end{bmatrix}$

13. $\begin{bmatrix} -8 & -7 \\ 15 & -1 \end{bmatrix}$ **15.** $\begin{bmatrix} -24 & -4 & 12 \\ -12 & 32 & 12 \end{bmatrix}$

17. $\begin{bmatrix} 10 & 8 \\ -59 & 9 \end{bmatrix}$ **19.** $\begin{bmatrix} -17.143 & 2.143 \\ 11.571 & 10.286 \end{bmatrix}$

21. $\begin{bmatrix} -1.581 & -3.739 \\ -4.252 & -13.249 \\ 9.713 & -0.362 \end{bmatrix}$

23. $\begin{bmatrix} -6 & -9 \\ -1 & 0 \\ 17 & -10 \end{bmatrix}$ **25.** $\begin{bmatrix} 3 & 3 \\ -\frac{1}{2} & 0 \\ -\frac{13}{2} & \frac{11}{2} \end{bmatrix}$

27. (a) $\begin{bmatrix} 0 & 15 \\ 6 & 12 \end{bmatrix}$ (b) $\begin{bmatrix} -2 & 2 \\ 31 & 14 \end{bmatrix}$ (c) $\begin{bmatrix} 9 & 6 \\ 12 & 12 \end{bmatrix}$

29. (a) $\begin{bmatrix} 0 & -10 \\ 10 & 0 \end{bmatrix}$ (b) $\begin{bmatrix} 0 & -10 \\ 10 & 0 \end{bmatrix}$ (c) $\begin{bmatrix} 8 & -6 \\ 6 & 8 \end{bmatrix}$

31. (a) $\begin{bmatrix} 7 & 7 & 14 \\ 8 & 8 & 16 \\ -1 & -1 & -2 \end{bmatrix}$ (b) $[13]$ (c) Not possible

33. Not possible   **35.** $\begin{bmatrix} 3 & -4 \\ 10 & 16 \\ 26 & 46 \end{bmatrix}$ **37.** $\begin{bmatrix} 3 & 0 & 0 \\ 0 & -4 & 0 \\ 0 & 0 & -10 \end{bmatrix}$

39. $\begin{bmatrix} 0 & 0 & 0 \\ 0 & 0 & 0 \\ 0 & 0 & 0 \end{bmatrix}$ **41.** $\begin{bmatrix} 41 & 7 & 7 \\ 42 & 5 & 25 \\ -10 & -25 & 45 \end{bmatrix}$

43. $\begin{bmatrix} 151 & 25 & 48 \\ 516 & 279 & 387 \\ 47 & -20 & 87 \end{bmatrix}$ **45.** Not possible

47. $\begin{bmatrix} 5 & 8 \\ -4 & -16 \end{bmatrix}$ **49.** $\begin{bmatrix} -4 & 10 \\ 3 & 14 \end{bmatrix}$

51. (a) $\begin{bmatrix} -1 & 1 \\ -2 & 1 \end{bmatrix}\begin{bmatrix} x_1 \\ x_2 \end{bmatrix} = \begin{bmatrix} 4 \\ 0 \end{bmatrix}$ (b) $\begin{bmatrix} 4 \\ 8 \end{bmatrix}$

53. (a) $\begin{bmatrix} -2 & -3 \\ 6 & 1 \end{bmatrix}\begin{bmatrix} x_1 \\ x_2 \end{bmatrix} = \begin{bmatrix} -4 \\ -36 \end{bmatrix}$ (b) $\begin{bmatrix} -7 \\ 6 \end{bmatrix}$

55. (a) $\begin{bmatrix} 1 & -2 & 3 \\ -1 & 3 & -1 \\ 2 & -5 & 5 \end{bmatrix}\begin{bmatrix} x_1 \\ x_2 \\ x_3 \end{bmatrix} = \begin{bmatrix} 9 \\ -6 \\ 17 \end{bmatrix}$ (b) $\begin{bmatrix} 1 \\ -1 \\ 2 \end{bmatrix}$

57. (a) $\begin{bmatrix} 1 & -5 & 2 \\ -3 & 1 & -1 \\ 0 & -2 & 5 \end{bmatrix}\begin{bmatrix} x_1 \\ x_2 \\ x_3 \end{bmatrix} = \begin{bmatrix} -20 \\ 8 \\ -16 \end{bmatrix}$ (b) $\begin{bmatrix} -1 \\ 3 \\ -2 \end{bmatrix}$

59. Not possible   **61.** Not possible   **63.** 2×2

65. Not possible   **67.** 2×3 **69.** $\begin{bmatrix} 84 & 60 & 30 \\ 42 & 120 & 84 \end{bmatrix}$

71. $BA = [\$1037.50 \quad \$1400 \quad \$1012.50]$

The entries represent the profits from both products at each of the three outlets.

73. $\begin{bmatrix} \$15,770 & \$18,300 \\ \$26,500 & \$29,250 \\ \$21,260 & \$24,150 \end{bmatrix}$

The entries are the wholesale and retail inventory values of the inventories at the three outlets.

75. $\begin{bmatrix} \$18.10 & \$15.40 \\ \$29.80 & \$25.40 \\ \$51.20 & \$43.80 \end{bmatrix}$

The entries are labor costs at each plant for each size of boat.

77. True. The sum of two matrices of different orders is undefined.

79. False.

$\begin{bmatrix} -2 & 4 \\ -3 & 0 \\ 6 & 1 \end{bmatrix}\begin{bmatrix} 1 & 1 \\ 1 & 1 \end{bmatrix} = \begin{bmatrix} 2 & 2 \\ -3 & -3 \\ 7 & 7 \end{bmatrix}$

81. $AB = \begin{bmatrix} 0 & 0 \\ 0 & 0 \end{bmatrix}$

$AB = O$ and neither A nor B is O.

83. $A^2 = \begin{bmatrix} 1 & 0 \\ 0 & 1 \end{bmatrix} = I$, the identity matrix.

85. $-8, \dfrac{4}{3}$ **87.** $0, \dfrac{-5 \pm \sqrt{37}}{4}$ **89.** $4, \pm\dfrac{\sqrt{15}}{3}i$

91. $\left(7, -\dfrac{1}{2}\right)$ **93.** $(3, -1)$

Section 7.3   *(page 545)*

1–9. $AB = I$ and $BA = I$

11. $\begin{bmatrix} \frac{1}{2} & 0 \\ 0 & \frac{1}{3} \end{bmatrix}$ **13.** $\begin{bmatrix} -3 & 2 \\ -2 & 1 \end{bmatrix}$ **15.** $\begin{bmatrix} 1 & -1 \\ 2 & -1 \end{bmatrix}$

17. Does not exist   **19.** Does not exist

21. $\begin{bmatrix} 1 & 1 & -1 \\ -3 & 2 & -1 \\ 3 & -3 & 2 \end{bmatrix}$ **23.** $\begin{bmatrix} 1 & 0 & 0 \\ -\frac{3}{4} & \frac{1}{4} & 0 \\ \frac{7}{20} & -\frac{1}{4} & \frac{1}{5} \end{bmatrix}$

25. $\begin{bmatrix} -\frac{1}{8} & 0 & 0 & 0 \\ 0 & 1 & 0 & 0 \\ 0 & 0 & \frac{1}{4} & 0 \\ 0 & 0 & 0 & -\frac{1}{5} \end{bmatrix}$ **27.** $\begin{bmatrix} -175 & 37 & -13 \\ 95 & -20 & 7 \\ 14 & -3 & 1 \end{bmatrix}$

29. $\begin{bmatrix} -1.5 & 1.5 & 1 \\ 4.5 & -3.5 & -3 \\ -1 & 1 & 1 \end{bmatrix}$ **31.** $\begin{bmatrix} -12 & -5 & -9 \\ -4 & -2 & -4 \\ -8 & -4 & -6 \end{bmatrix}$

33. $\begin{bmatrix} 0 & -1.\overline{81} & 0.\overline{90} \\ -10 & 5 & 5 \\ 10 & -2.\overline{72} & -3.\overline{63} \end{bmatrix}$ **35.** Does not exist

37. $\begin{bmatrix} 1 & 0 & 1 & 0 \\ 0 & 1 & 0 & 1 \\ 2 & 0 & 1 & 0 \\ 0 & 1 & 0 & 2 \end{bmatrix}$ **39.** $\begin{bmatrix} \frac{3}{19} & \frac{2}{19} \\ -\frac{2}{19} & \frac{5}{19} \end{bmatrix}$

41. Does not exist   **43.** $\begin{bmatrix} \frac{16}{59} & \frac{15}{59} \\ -\frac{4}{59} & \frac{70}{59} \end{bmatrix}$ **45.** $(5, 0)$

47. $(-8, -6)$ **49.** $(3, 8, -11)$ **51.** $(2, 1, 0, 0)$

53. $(2, -2)$ **55.** No solution   **57.** $\left(3, -\dfrac{1}{2}\right)$

59. $(-4, -8)$ **61.** $(-1, 3, 2)$

63. $\left(\dfrac{5}{16}a + \dfrac{13}{16}, \dfrac{19}{16}a + \dfrac{11}{16}, a\right)$ **65.** $(2a - 1, -3a + 2, a)$

67. $(5, 0, -2, 3)$

69. $7000 in AAA-rated bonds
\$1000 in A-rated bonds
\$2000 in B-rated bonds

71. $9000 in AAA-rated bonds
\$1000 in A-rated bonds
\$2000 in B-rated bonds

73. $I_1 = -3$ amperes
$I_2 = 8$ amperes
$I_3 = 5$ amperes

75. True. If B is the inverse of A, then $AB = I = BA$.

77. True. If A is of order $m \times n$ and B is of order $n \times m$ (where $m \neq n$), the products AB and BA are of different orders and so cannot be equal to each other.

79. The inverse matrix can be calculated once and used for more than one exercise.

81. $x = \dfrac{2 \ln 315}{\ln 3} \approx 10.47$ **83.** $x = 2^{6.5} \approx 90.51$

85. $\begin{bmatrix} 12 & -18 \\ -6 & 24 \\ -3 & -36 \end{bmatrix}$ **87.** $\begin{bmatrix} 6 & -1 \\ -27 & 19 \end{bmatrix}$

Section 7.4 *(page 553)*

1. 5 **3.** 5 **5.** 27 **7.** 0 **9.** 6 **11.** -9

13. 72 **15.** $\frac{11}{6}$ **17.** -0.002 **19.** -4.842 **21.** 0

23. (a) $M_{11} = -5, M_{12} = 2, M_{21} = 4, M_{22} = 3$

  (b) $C_{11} = -5, C_{12} = -2, C_{21} = -4, C_{22} = 3$

25. (a) $M_{11} = -4, M_{12} = -2, M_{21} = 1, M_{22} = 3$

  (b) $C_{11} = -4, C_{12} = 2, C_{21} = -1, C_{22} = 3$

27. (a) $M_{11} = 3, M_{12} = -4, M_{13} = 1, M_{21} = 2, M_{22} = 2,$
    $M_{23} = -4, M_{31} = -4, M_{32} = 10, M_{33} = 8$

  (b) $C_{11} = 3, C_{12} = 4, C_{13} = 1, C_{21} = -2, C_{22} = 2,$
    $C_{23} = 4, C_{31} = -4, C_{32} = -10, C_{33} = 8$

29. (a) $M_{11} = 30, M_{12} = 12, M_{13} = 11, M_{21} = -36,$
    $M_{22} = 26, M_{23} = 7, M_{31} = -4, M_{32} = -42, M_{33} = 12$

  (b) $C_{11} = 30, C_{12} = -12, C_{13} = 11, C_{21} = 36, C_{22} = 26,$
    $C_{23} = -7, C_{31} = -4, C_{32} = 42, C_{33} = 12$

31. (a) -75 (b) -75

33. (a) 96 (b) 96

35. (a) 170 (b) 170

37. 0 **39.** 0

41. -9 **43.** -58 **45.** -30 **47.** -168 **49.** 0

51. 412 **53.** -126 **55.** 0 **57.** -336 **59.** 410

61. (a) -3 (b) -2 (c) $\begin{bmatrix} -2 & 0 \\ 0 & -3 \end{bmatrix}$ (d) 6

63. (a) -8 (b) 0 (c) $\begin{bmatrix} -4 & 4 \\ 1 & -1 \end{bmatrix}$ (d) 0

65. (a) -21 (b) -19 (c) $\begin{bmatrix} 7 & 1 & 4 \\ -8 & 9 & -3 \\ 7 & -3 & 9 \end{bmatrix}$ (d) 399

67. (a) 2 (b) -6 (c) $\begin{bmatrix} 1 & 4 & 3 \\ -1 & 0 & 3 \\ 0 & 2 & 0 \end{bmatrix}$ (d) -12

69.–73. Answers will vary. **75.** $-1, 4$ **77.** $-1, -4$

79. $8uv - 1$ **81.** e^{5x} **83.** $1 - \ln x$

85. True. If an entire row is zero, then each cofactor in the expansion is multiplied by zero.

87. Answers will vary.

89. A square matrix is a square array of numbers. The determinant of a square matrix is a real number.

91. $(y - 3)^2 = 8x$ **93.** $\dfrac{x^2}{64} + \dfrac{y^2}{28} = 1$

95.

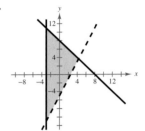

97. $\begin{bmatrix} \frac{1}{4} & \frac{1}{4} \\ 2 & 1 \end{bmatrix}$

99. Does not exist

Section 7.5 *(page 565)*

1. $(2, -2)$ **3.** $\left(\frac{32}{7}, \frac{30}{7}\right)$ **5.** $(-1, 3, 2)$

7. $(-2, 1, -1)$ **9.** $\left(0, -\frac{1}{2}, \frac{1}{2}\right)$ **11.** $(1, 2, 1)$ **13.** 7

15. 14 **17.** $\frac{33}{8}$ **19.** $\frac{5}{2}$ **21.** 28 **23.** $\frac{16}{5}$ or 0

25. -3 or -11 **27.** 250 square miles **29.** Collinear

31. Not collinear **33.** Collinear **35.** $x = -3$

37. $3x - 5y = 0$ **39.** $x + 3y - 5 = 0$

41. $2x + 3y - 8 = 0$

43. Uncoded: $[20\ 18\ 15], [21\ 2\ 12], [5\ 0\ 9], [14\ 0\ 18],$
    $[9\ 22\ 5], [18\ 0\ 3], [9\ 20\ 25]$

  Encoded: $-52\ 10\ 27\ -49\ 3\ 34\ -49\ 13\ 27$
    $-94\ 22\ 54\ 1\ 1\ -7\ 0\ -12\ 9$
    $-121\ 41\ 55$

45. $-6\ -35\ -69\ 11\ 20\ 17\ 6\ -16\ -58\ 46\ 79\ 67$

47. $-5\ -41\ -87\ 91\ 207\ 257\ 11\ -5\ -41\ 40\ 80$
    $84\ 76\ 177\ 227$

49. HAPPY NEW YEAR **51.** CLASS IS CANCELED

53. SEND PLANES **55.** MEET ME TONIGHT RON

57. True. If the determinant of the coefficient matrix is zero, then the solution of the system would result in division by zero, which is undefined.

59. Answers will vary. **61.** $\left(5, -\frac{1}{2}\right)$ **63.** $(2, -2, 5)$

65.

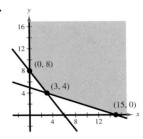

Minimum at $(3, 4)$: 46

No maximum

67. -13.78 **69.** -115

Review Exercises   *(page 569)*

1. 3×1 **3.** 1×1 **5.** $\begin{bmatrix} 3 & -10 & \vdots & 15 \\ 5 & 4 & \vdots & 22 \end{bmatrix}$

7. $\begin{cases} 5x + y + 7z = -9 \\ 4x + 2y = 10 \\ 9x + 4y + 2z = 3 \end{cases}$

9. $\begin{bmatrix} 1 & 0 & 0 \\ 0 & 1 & 0 \\ 0 & 0 & 1 \end{bmatrix}$

11. Consistent, infinitely many solutions

13. Consistent, one solution   **15.** $(10, -12)$

17. $(-0.2, 0.7)$ **19.** $(5, 2, -6)$

21. $\left(-2a + \frac{3}{2}, 2a + 1, a\right)$ **23.** $(1, 0, 4, 3)$

25. $\dfrac{8}{x+1} - \dfrac{8}{x+2} - \dfrac{7}{(x+2)^2}$ **27.** $(2, -3, 3)$

29. $(2, 3, -1)$ **31.** $(2, 6, -10, -3)$

33. $x = 12, y = -7$ **35.** $x = 1, y = 11$ **37.** Yes

39. Not possible because matrices are not of the same order.

41. $\begin{bmatrix} 17 & -17 \\ 13 & 2 \end{bmatrix}$ **43.** $\begin{bmatrix} 54 & 4 \\ -2 & 24 \\ -4 & 32 \end{bmatrix}$

45. $\begin{bmatrix} 48 & -18 & -3 \\ 15 & 51 & 33 \end{bmatrix}$ **47.** $\begin{bmatrix} -14 & -4 \\ 7 & -17 \\ -17 & -2 \end{bmatrix}$

49. $\frac{1}{3}\begin{bmatrix} 9 & 2 \\ -4 & 11 \\ 10 & 0 \end{bmatrix}$ **51.** Yes   **53.** Yes

55. $\begin{bmatrix} 14 & -2 & 8 \\ 14 & -10 & 40 \\ 36 & -12 & 48 \end{bmatrix}$ **57.** $\begin{bmatrix} 44 & 4 \\ 20 & 8 \end{bmatrix}$

59. $\begin{bmatrix} 24 & -8 \\ 36 & -12 \end{bmatrix}$ **61.** $\begin{bmatrix} 1 & 17 \\ 12 & 36 \end{bmatrix}$

63. $\begin{bmatrix} 14 & -22 & 22 \\ 19 & -41 & 80 \\ 42 & -66 & 66 \end{bmatrix}$

65. $\begin{cases} 5x + 4y = 2 \\ -x + y = -22 \end{cases}$

67. (a) $[\$274{,}150 \quad \$303{,}150]$

The merchandise shipped to warehouse 1 is worth $\$274{,}150$ and the merchandise shipped to warehouse 2 is worth $\$303{,}150$.

(b) $A_n = \begin{bmatrix} 10{,}250 & 9{,}250 \\ 8{,}125 & 12{,}250 \\ 6{,}750 & 6{,}000 \end{bmatrix}$

$BA_n = [\$342{,}687.5 \quad \$378{,}937.5]$

69. and 71. $AB = I$ and $BA = I$

73. $\begin{bmatrix} 3 & 5 \\ -2 & -3 \end{bmatrix}$ **75.** $\begin{bmatrix} 1 & 11 & 8 \\ -1 & -7 & -5 \\ -1 & -14 & -10 \end{bmatrix}$

77. $\begin{bmatrix} 0.0434\ldots & 0.2173\ldots \\ -0.0869\ldots & 0.0652\ldots \end{bmatrix}$ **79.** Does not exist

81. $\begin{bmatrix} \frac{3}{2} & -2 \\ -\frac{7}{2} & 5 \end{bmatrix}$ **83.** $\begin{bmatrix} -\frac{2}{3} & -\frac{5}{8} \\ \frac{1}{5} & -\frac{3}{16} \end{bmatrix}$ **85.** $(2, -3)$

87. $(-2, 1)$ **89.** $(-2, 4, 3)$ **91.** $(-3, 5, 0)$

93. $(5, 6)$ **95.** $(4, -2, 1)$ **97.** -41 **99.** 78

101. (a) $M_{11} = -4, M_{12} = 5, M_{21} = 6, M_{22} = 3$

(b) $C_{11} = -4, C_{12} = -5, C_{21} = -6, C_{22} = 3$

103. (a) $M_{11} = 19, M_{12} = -24, M_{13} = 26, M_{21} = 2,$

$M_{22} = 32, M_{23} = 20, M_{31} = -47, M_{32} = -96,$

$M_{33} = 22$

(b) $C_{11} = 19, C_{12} = 24, C_{13} = 26, C_{21} = -2,$

$C_{22} = 32, C_{23} = -20, C_{31} = -47, C_{32} = 96,$

$C_{33} = 22$

105. -117 **107.** -255 **109.** $(3, -2)$

111. $(6, 8, 1)$

113. 40 liters of 75% solution; 60 liters of 50% solution

115. 16,667 units   **117.** 24   **119.** $\frac{25}{8}$

121. $3x + 2y - 16 = 0$ **123.** $10x - 5y + 9 = 0$

125. Uncoded: $[18 \ 5 \ 20], [21 \ 18 \ 14], [0 \ 20 \ 15],$

$[0 \ 2 \ 1], [19 \ 5 \ 0]$

Encoded: $66 \ 28 \ 10 \ -24 \ -59 \ -22 \ -75 \ -90$

$-25 \ -9 \ -10 \ -3 \ 8 \ -11 \ -10$

127. MAY THE FORCE BE WITH YOU

129. True. Answers will vary.

131. If A is a square matrix, the cofactor C_{ij} of the entry a_{ij} is $(-1)^{i+j}M_{ij}$, where M_{ij} is the determinant obtained by deleting the ith row and jth column of A. The determinant of A is the sum of the entries of any row or column of A multiplied by their respective cofactors.

133. The part of the matrix corresponding to the coefficients of the system reduces to a matrix in which the number of rows with nonzero entries is the same as the number of variables.

Chapter Test *(page 575)*

1. $\begin{bmatrix} 1 & 0 & 0 \\ 0 & 1 & 0 \\ 0 & 0 & 1 \end{bmatrix}$ **2.** $\begin{bmatrix} 1 & 0 & -1 & 2 \\ 0 & 1 & 0 & -1 \\ 0 & 0 & 0 & 0 \\ 0 & 0 & 0 & 0 \end{bmatrix}$

3. $\begin{bmatrix} 4 & 3 & -2 & \vdots & 14 \\ -1 & -1 & 2 & \vdots & -5 \\ 3 & 1 & -4 & \vdots & 8 \end{bmatrix}, \left(1, 3, -\frac{1}{2}\right)$

4. $y = -\frac{1}{2}x^2 + x + 2$

5. (a) $\begin{bmatrix} 1 & 5 \\ 0 & -4 \end{bmatrix}$ (b) $\begin{bmatrix} 15 & 12 \\ -12 & -12 \end{bmatrix}$

(c) $\begin{bmatrix} 7 & 14 \\ -4 & -12 \end{bmatrix}$ (d) $\begin{bmatrix} 4 & -5 \\ 0 & 4 \end{bmatrix}$

6. $\begin{bmatrix} \frac{1}{2} & \frac{2}{5} \\ 1 & \frac{3}{5} \end{bmatrix}$ **7.** $\begin{bmatrix} -\frac{5}{2} & 4 & -3 \\ 5 & -7 & 6 \\ 4 & -6 & 5 \end{bmatrix}$

8. $(13, 22)$ **9.** -196 **10.** 29 **11.** $(-3, 5)$

12. $(-2, 4, 6)$ **13.** 7

14. Uncoded: $[11\ 14\ 15], [3\ 11\ 0], [15\ 14\ 0], [23\ 15\ 15],$
$[4\ 0\ 0]$

Encoded: $115\ -41\ -59\ 14\ -3\ -11\ 29\ -15$
$-14\ 128\ -53\ -60\ 4\ -4\ 0$

15. 75 liters of 60% solution
25 liters of 20% solution

Chapter 8

Section 8.1 *(page 585)*

1. $4, 7, 10, 13, 16$ **3.** $2, 4, 8, 16, 32$

5. $-2, 4, -8, 16, -32$ **7.** $3, 2, \frac{5}{3}, \frac{3}{2}, \frac{7}{5}$

9. $3, \frac{12}{11}, \frac{9}{13}, \frac{24}{47}, \frac{15}{37}$ **11.** $0, 1, 0, \frac{1}{2}, 0$ **13.** $\frac{5}{3}, \frac{17}{9}, \frac{53}{27}, \frac{161}{81}, \frac{485}{243}$

15. $1, \frac{1}{2^{3/2}}, \frac{1}{3^{3/2}}, \frac{1}{8}, \frac{1}{5^{3/2}}$ **17.** $3, \frac{9}{2}, \frac{9}{2}, \frac{27}{8}, \frac{81}{40}$

19. $-1, \frac{1}{4}, -\frac{1}{9}, \frac{1}{16}, -\frac{1}{25}$ **21.** $\frac{2}{3}, \frac{2}{3}, \frac{2}{3}, \frac{2}{3}, \frac{2}{3}$

23. $0, 0, 6, 24, 60$ **25.** -73

27. 0.000282 **29.** $\frac{44}{239}$

31.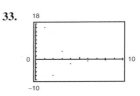

33.

35.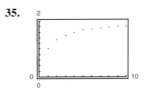

37. c **39.** d **41.** $a_n = 3n - 2$ **43.** $a_n = n^2 - 1$

45. $\dfrac{(-1)^n(n+1)}{n+2}$ **47.** $a_n = \dfrac{n+1}{2n-1}$ **49.** $a_n = \dfrac{1}{n^2}$

51. $a_n = (-1)^{n+1}$ **53.** $a_n = 1 + \dfrac{1}{n}$

55. $28, 24, 20, 16, 12$ **57.** $3, 4, 6, 10, 18$

59. $6, 8, 10, 12, 14$ **61.** $81, 27, 9, 3, 1$

$a_n = 2n + 4$ $\qquad a_n = \dfrac{243}{3^n}$

63. $\dfrac{1}{30}$ **65.** 90 **67.** $n + 1$ **69.** $\dfrac{1}{2n(2n+1)}$

71. 35 **73.** 40 **75.** 30 **77.** $\frac{9}{5}$ **79.** 88

81. 30 **83.** 81 **85.** $\frac{47}{60}$

87. $\displaystyle\sum_{i=1}^{9} \frac{1}{3i}$ **89.** $\displaystyle\sum_{i=1}^{8} \left[2\left(\frac{i}{8}\right) + 3\right]$ **91.** $\displaystyle\sum_{i=1}^{6}(-1)^{i+1}3i$

93. $\displaystyle\sum_{i=1}^{20} \frac{(-1)^{i+1}}{i^2}$ **95.** $\displaystyle\sum_{i=1}^{5} \frac{2^i - 1}{2^{i+1}}$ **97.** $\dfrac{75}{16}$ **99.** $-\dfrac{3}{2}$

101. $\frac{2}{3}$ **103.** $\frac{7}{9}$

105. (a) $A_1 = \$5100.00$, $A_2 = \$5202.00$, $A_3 = \$5306.04$,
$A_4 = \$5412.16$, $A_5 = \$5520.40$, $A_6 = \$5630.81$,
$A_7 = \$5743.43$, $A_8 = \$5858.30$
(b) $\$11,040.20$

107. $a_{-1} = \$627.23$, $a_0 = \$696.00$, $a_1 = \$760.03$,
$a_2 = \$819.32$, $a_3 = \$873.87$, $a_4 = \$923.68$,
$a_5 = \$968.75$, $a_6 = \$1009.08$

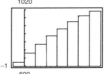

109. $\$23,660.88$ million

111. $1, 1, 2, 3, 5, 8, 13, 21, 34, 55, 89, 144$

$1, 2, \frac{3}{2}, \frac{5}{3}, \frac{8}{5}, \frac{13}{8}, \frac{21}{13}, \frac{34}{21}, \frac{55}{34}, \frac{89}{55}$

113. $500.95   **115.** Answers will vary.

117. True by the Properties of Sums

119. (a) $\begin{bmatrix} 8 & 1 \\ -3 & 7 \end{bmatrix}$ (b) $\begin{bmatrix} -26 & 1 \\ 15 & -24 \end{bmatrix}$

  (c) $\begin{bmatrix} 18 & 9 \\ 18 & 0 \end{bmatrix}$ (d) $\begin{bmatrix} 0 & 6 \\ 27 & 18 \end{bmatrix}$

121. (a) $\begin{bmatrix} -3 & -7 & 4 \\ 4 & 4 & 1 \\ 1 & 4 & 3 \end{bmatrix}$ (b) $\begin{bmatrix} 10 & 25 & -10 \\ -12 & -11 & 3 \\ -3 & -9 & -8 \end{bmatrix}$

  (c) $\begin{bmatrix} -2 & 7 & -16 \\ 4 & 42 & 45 \\ 1 & 23 & 48 \end{bmatrix}$ (d) $\begin{bmatrix} 16 & 31 & 42 \\ 10 & 47 & 31 \\ 13 & 22 & 25 \end{bmatrix}$

123. 26   **125.** -194

Section 8.2   *(page 595)*

1. Arithmetic sequence, $d = -2$

3. Not an arithmetic sequence

5. Arithmetic sequence, $d = -\frac{1}{4}$

7. Not an arithmetic sequence

9. Not an arithmetic sequence

11. 8, 11, 14, 17, 20

  Arithmetic sequence, $d = 3$

13. 7, 3, -1, -5, -9

  Arithmetic sequence, $d = -4$

15. $-1, 1, -1, 1, -1,$

  Not an arithmetic sequence

17. $-3, \frac{3}{2}, -1, \frac{3}{4}, -\frac{3}{5}$

  Not an arithmetic sequence

19. 15, 19, 23, 27, 31; $d = 4$; $a_n = 4n + 11$

21. 200, 190, 180, 170, 160; $d = -10$; $a_n = -10n + 210$

23. $\frac{5}{8}, \frac{1}{2}, \frac{3}{8}, \frac{1}{4}, \frac{1}{8}$; $d = -\frac{1}{8}$; $a_n = -\frac{1}{8}n + \frac{3}{4}$

25. 5, 11, 17, 23, 29   **27.** $-2.6, -3.0, -3.4, -3.8, -4.2$

29. 2, 6, 10, 14, 18   **31.** $-2, 2, 6, 10, 14$

33. $a_n = 3n - 2$ **35.** $a_n = -8n + 108$

37. $a_n = 2xn - x$ **39.** $a_n = -\frac{5}{2}n + \frac{13}{2}$

41. $a_n = \frac{10}{3}n + \frac{5}{3}$ **43.** $a_n = -3n + 103$

45. b   **47.** c

49.

51.

53. 620   **55.** 17.4   **57.** 265   **59.** 4000

61. 1275   **63.** 30,030   **65.** 355   **67.** 160,000

69. 520   **71.** 2725   **73.** 10,120

75. (a) $40,000   (b) $217,500

77. 2340 seats   **79.** 405 bricks   **81.** 490 meters

83. True. Given a_1 and a_2, $d = a_2 - a_1$ and
  $a_n = a_1 + (n - 1)d$.

85. Because $a_n = dn + c$, the geometric pattern is linear.

87. 10,000

89. (a) 4, 9, 16, 25, 36   (b) n^2

  (c) $\dfrac{n}{2}[1 + (2n - 1)] = n^2$

91. 4   **93.** $\left(\frac{3}{4}a, -\frac{3}{2}a, a\right)$ **95.** $(0, 0, 0)$

97. $(4, 1)$ **99.** $(1, -2, 3)$

Section 8.3   *(page 604)*

1. Geometric sequence, $r = 3$

3. Not a geometric sequence

5. Geometric sequence, $r = -\frac{1}{2}$

7. Geometric sequence, $r = 2$

9. Not a geometric sequence

11. 2, 6, 18, 54, 162   **13.** $1, \frac{1}{2}, \frac{1}{4}, \frac{1}{8}, \frac{1}{16}$

15. $5, -\frac{1}{2}, \frac{1}{20}, -\frac{1}{200}, \frac{1}{2000}$ **17.** $1, e, e^2, e^3, e^4$

19. $2, \dfrac{x}{2}, \dfrac{x^2}{8}, \dfrac{x^3}{32}, \dfrac{x^4}{128}$

21. 64, 32, 16, 8, 4; $r = \frac{1}{2}$; $a_n = 128\left(\frac{1}{2}\right)^n$

23. 7, 14, 28, 56, 112; $r = 2$; $a_n = \frac{7}{2}(2)^n$

25. $6, -9, \frac{27}{2}, -\frac{81}{4}, \frac{243}{8}$; $r = -\frac{3}{2}$; $a_n = -4\left(-\frac{3}{2}\right)^n$

27. $\dfrac{1}{128}$ **29.** $-\dfrac{2}{3^{10}}$ **31.** $100e^{8x}$ **33.** 1082.372

35. 9   **37.** -2 **39.** a   **41.** b

43.

45.

47. 511   **49.** 43   **51.** 29,921.311   **53.** 2092.596

55. 6.400   **57.** $\displaystyle\sum_{n=1}^{7} 5(3)^{n-1}$ **59.** $\displaystyle\sum_{n=1}^{7} 2\left(-\frac{1}{4}\right)^{n-1}$

61. $\displaystyle\sum_{n=1}^{6} 0.1(4)^{n-1}$ **63.** 2   **65.** $\frac{2}{3}$ **67.** $\frac{16}{3}$ **69.** $\frac{5}{3}$

71. -30 **73.** 32   **75.** Undefined   **77.** $\frac{4}{11}$ **79.** $\frac{7}{22}$

81.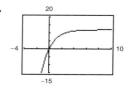

Horizontal asymptote: $y = 12$

Corresponds to the sum of the series

83. (a) $1790.85
(b) $1806.11
(c) $1814.02
(d) $1819.40
(e) $1822.03

85. $22,689.45   **87.** $7011.89

89. Answers will vary.

91. (a) $26,198.27   **93.** (a) $637,678.02
(b) $26,263.88   (b) $645,861.43

95. Answers will vary.   **97.** 126 square inches

99. (a) $5,368,709.11   **101.** (a) 152.42 feet
(b) $10,737,418.23   (b) 19 seconds
(c) $21,474,836.47

103. False. $a_n = a_1 r^{n-1}$

105. Given a_1 and a_2, $r = a_2/a_1$ and $a_n = a_1 r^{n-1}$.

107. $3x + 4$ **109.** $9x^2 + 24x + 15$

111. Does not factor   **113.** $4x^2(2 + x)(2 - x)$

115. $\frac{1}{3}, x \neq 2, -7$ **117.** $1, x \neq 3, 5$

119. $\dfrac{7x^2 + 21x - 53}{(x - 1)(x + 4)}$

59. 2, 4, 16, 256, 65,536
First differences: 2, 12, 240, 65,280
Second differences: 10, 228, 65,040
Neither

61. 0, 4, 10, 18, 28
First differences: 4, 6, 8, 10
Second differences: 2, 2, 2
Quadratic

63. 0, -1, -2, -3, -4
First differences: -1, -1, -1, -1
Second differences: 0, 0, 0
Linear

65. $a_n = n^2 - 2n + 7$ **67.** $a_n = \frac{7}{4}n^2 - 5n + 3$

69. False. P_1 must be proven to be true.

71. False. A sequence with n terms has $n - 2$ second differences.

73. (a) P_n is true for integers $n \geq 3$.
(b) P_n is true for integers $1 \leq n \leq 50$.
(c) $P_1, P_2,$ and P_3 are true.
(d) P_{2n} is true for any positive integer n.

75. (8, 2) and (0, 0)   **77.** (4, -1, 3)

79. $4x^2 - 4xy + y^2$ **81.** $8x^3 - 48x^2y + 96xy^2 - 64y^3$

Section 8.4   *(page 616)*

1. $\dfrac{5}{(k + 1)(k + 2)}$ **3.** $\dfrac{(k + 1)^2(k + 2)^2}{4}$

5.–17. Answers will vary.

19. 120   **21.** 91   **23.** 979   **25.** 70   **27.** -3402

29. $S_n = n(2n - 1)$ **31.** $S_n = 10 - 10\left(\frac{9}{10}\right)^n$

33. $S_n = \dfrac{n}{2(n + 1)}$ **35.–49.** Answers will vary.

51. 10, 40, 160, 640, 2560   **53.** 0, 2, 2, 6, 10

55. 2, 0, 3, 1, 4
First differences: -2, 3, -2, 3
Second differences: 5, -5, 5
Neither

57. $\frac{3}{2}, -3, 6, -12, 24$
First differences: $-\frac{9}{2}, 9, -18, 36$
Second differences: $\frac{27}{2}, -27, 54$
Neither

Section 8.5   *(page 623)*

1. 10   **3.** 1   **5.** 15,504   **7.** 210   **9.** 4950

11. 56   **13.** 35   **15.** $x^4 + 4x^3 + 6x^2 + 4x + 1$

17. $a^4 + 24a^3 + 216a^2 + 864a + 1296$

19. $y^3 - 12y^2 + 48y - 64$

21. $x^5 + 5x^4y + 10x^3y^2 + 10x^2y^3 + 5xy^4 + y^5$

23. $r^6 + 18r^5s + 135r^4s^2 + 540r^3s^3 + 1215r^2s^4 +$
$1458rs^5 + 729s^6$

25. $243a^5 - 405a^4b + 270a^3b^2 - 90a^2b^3 + 15ab^4 - b^5$

27. $1 - 6x + 12x^2 - 8x^3$

29. $x^8 + 20x^6 + 150x^4 + 500x^2 + 625$

31. $\dfrac{1}{x^5} + \dfrac{5y}{x^4} + \dfrac{10y^2}{x^3} + \dfrac{10y^3}{x^2} + \dfrac{5y^4}{x} + y^5$

33. $2x^4 - 24x^3 + 113x^2 - 246x + 207$

35. $32t^5 - 80t^4s + 80t^3s^2 - 40t^2s^3 + 10ts^4 - s^5$

37. $x^5 + 10x^4y + 40x^3y^2 + 80x^2y^3 + 80xy^4 + 32y^5$

39. 1,732,104   **41.** 180   **43.** $-326,592$ **45.** 210

47. $x^2 + 12x^{3/2} + 54x + 108x^{1/2} + 81$

49. $x^2 - 3x^{4/3}y^{1/3} + 3x^{2/3}y^{2/3} - y$ **51.** $3x^2 + 3xh + h^2$

53. $\dfrac{1}{\sqrt{x+h} + \sqrt{x}}$ **55.** -4 **57.** $2035 + 828i$

59. 1   **61.** 1.172   **63.** 510,568.785

65.

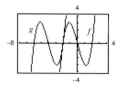

g is shifted 4 units to the left of f.

$g(x) = x^3 + 12x^2 + 44x + 48$

67.

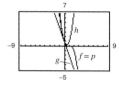

$p(x)$ is the expansion of $f(x)$.

69. 0.2503   **71.** 0.273

73. (a) $g(t) = 0.0834t^2 + 0.74377t + 7.00868$

(b)

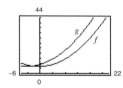

75. False. Expanding binomials that represent differences is accurate. The coefficients have alternating signs.

77.
```
              1
            1   1
          1   2   1
        1   3   3   1
      1   4   6   4   1
    1   5  10  10   5   1
  1   6  15  20  15   6   1
1   7  21  35  35  21   7   1
1  8 28 56  70  56 28  8   1
```

79. The signs of the terms in the expansion of $(x - y)^n$ alternate from positive to negative.

81. and 83. Answers will vary.

85. $g(x)$ is shifted 3 units to the right of $f(x)$.

87. $g(x)$ is the reflection of $f(x)$ in the x-axis.

89.

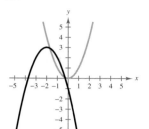

$g(x) = -(x + 2)^2 + 3$

91.

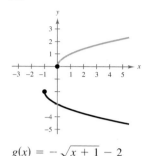

$g(x) = -\sqrt{x+1} - 2$

Section 8.6   *(page 633)*

1. 6   **3.** 5   **5.** 3   **7.** 7   **9.** 30   **11.** 15

13. 64   **15.** 4   **17.** 175,760,000

19. (a) 900   (b) 648   (c) 180   (d) 600

21. 64,000   **23.** (a) 40,320   (b) 384

25. 24   **27.** 336   **29.** 120   **31.** $n = 5$ or $n = 6$

33. 1,860,480   **35.** 970,200   **37.** 15,504

39. 420   **41.** 2520

43. ABCD, ABDC, ACBD, ACDB, ADBC, ADCB, BACD, BADC, CABD, CADB, DABC, DACB, BCAD, BDAC, CBAD, CDAB, DBAC, DCAB, BCDA, BDCA, CBDA, CDBA, DBCA, DCBA

45. AB, AC, AD, AE, AF, BC, BD, BE, BF, CD, CE, CF, DE, DF, EF

47. 120   **49.** 11,880   **51.** 15,504   **53.** 3,838,380

55. 3,921,225   **57.** 21

59. (a) 70   (b) 30

61. (a) 70   (b) 54   (c) 16

63. 5   **65.** 20

67. False. It is an example of a combination.

69. They are the same.   **71. and 73.** Answers will vary.

75. No. For some calculators the number is too great.

77. (a) 35   (b) 8   (c) 83

79. (a) -4 (b) 0   (c) 0

81. $x^5 + 5x^4 + 10x^3 + 10x^2 + 5x + 1$

83. $x^{10} + 10x^8y + 40x^6y^2 + 80x^4y^3 + 80x^2y^4 + 32y^5$

Section 8.7   *(page 644)*

1. $\{(H, 1), (H, 2), (H, 3), (H, 4), (H, 5), (H, 6),$
$(T, 1), (T, 2), (T, 3), (T, 4), (T, 5), (T, 6)\}$

3. $\{ABC, ACB, BAC, BCA, CAB, CBA\}$

5. {(A, B), (A, C), (A, D), (A, E), (B, C),

    (B, D), (B, E), (C, D), (C, E), (D, E)}

7. $\frac{3}{8}$ **9.** $\frac{7}{8}$ **11.** $\frac{3}{13}$ **13.** $\frac{3}{26}$ **15.** $\frac{1}{12}$

17. $\frac{11}{12}$ **19.** $\frac{1}{3}$ **21.** $\frac{1}{5}$ **23.** $\frac{2}{5}$ **25.** 0.3

27. $\frac{2}{3}$ **29.** 0.85   **31.** $\frac{7}{20}$

33. (a) 78,000   **35.** (a) 58%   **37.** (a) $\frac{112}{209}$

    (b) 30%   (b) 95.6%   (b) $\frac{97}{209}$

    (c) 37%   (c) 0.4%   (c) $\frac{274}{627}$

39. $P(\{\text{Taylor wins}\}) = \frac{1}{2}$

    $P(\{\text{Moore wins}\}) = P(\{\text{Jenkins wins}\}) = \frac{1}{4}$

41. (a) $\frac{21}{1292}$ **43.** (a) $\frac{1}{3}$ **45.** (a) $\frac{1}{120}$

    (b) $\frac{225}{646}$ (b) $\frac{5}{8}$ (b) $\frac{1}{24}$

    (c) $\frac{49}{323}$

47. (a) $\frac{1}{169}$ **49.** (a) $\frac{14}{55}$ **51.** (a) $\frac{1}{4}$

    (b) $\frac{1}{221}$ (b) $\frac{12}{55}$ (b) $\frac{1}{2}$

                   (c) $\frac{54}{55}$ (c) $\frac{9}{100}$

                                (d) $\frac{1}{30}$

53. (a) 0.9702   **55.** (a) $\frac{1}{1024}$

    (b) 0.9998   (b) $\frac{243}{1024}$

    (c) 0.0002   (c) $\frac{781}{1024}$

57. 0.4746   **59.** $\frac{7}{16}$

61. True. Two events are independent if the occurrence of one has no effect on the occurrence of the other.

63. (a) As you consider successive people with distinct birthdays, the probabilities must decrease to take into account the birth dates already used. Because the birth dates of people are independent events, multiply the respective probabilities of distinct birthdays.

    (b) $\dfrac{365}{365} \cdot \dfrac{364}{365} \cdot \dfrac{363}{365} \cdot \dfrac{362}{365}$

    (c) Answers will vary.

    (d) Q_n is the probability that the birthdays are *not* distinct, which is equivalent to at least two people having the same birthday.

    (e)

n	10	15	20	23	30	40	50
P_n	0.88	0.75	0.59	0.49	0.29	0.11	0.03
Q_n	0.12	0.25	0.41	0.51	0.71	0.89	0.97

    (f) 23

65. No real solution   **67.** $0, \dfrac{1 \pm \sqrt{13}}{2}$ **69.** -4

71. $\frac{11}{2}$ **73.** -10 **75.** $\ln 27 \approx 3.296$

77. $\ln 1 = 0, \ln 3 \approx 1.099$ **79.** $\ln \frac{8}{3} \approx 0.981$

81. $e^8 \approx 2980.958$ **83.** $\dfrac{e^4}{6} \approx 9.100$

Review Exercises   (page 650)

1. $8, 5, 4, \frac{7}{2}, \frac{16}{5}$ **3.** $72, 36, 12, 3, \frac{3}{5}$ **5.** 120   **7.** 1

9. 30   **11.** $\dfrac{205}{24}$ **13.** 6050   **15.** $\displaystyle\sum_{k=1}^{20} \dfrac{1}{2k}$

17. $\frac{5}{9}$ **19.** $\frac{2}{99}$

21. (a) \$43,000   (b) \$192,500

23. Arithmetic sequence, $d = -2$

25. Arithmetic sequence, $d = \frac{1}{2}$ **27.** $a_n = 12n - 5$

29. $a_n = 3ny - 2y$ **31.** 80   **33.** 88   **35.** 25,250

37. 43 minutes   **39.** $4, -1, \frac{1}{4}, -\frac{1}{16}, \frac{1}{64}$

41. $9, 6, 4, \frac{8}{3}, \frac{16}{9}$ or $9, -6, 4, -\frac{8}{3}, \frac{16}{9}$

43. $a_n = 16\left(-\frac{1}{2}\right)^{n-1}, 10.67$

45. $a_n = 100(1.05)^{n-1}, 3306.60$ **47.** 127   **49.** $\frac{15}{16}$

51. 31   **53.** 24.849   **55.** 5486.45   **57.** 8

59. $\frac{10}{9}$ **61.** 12

63. (a) $a_t = 120,000(0.7)^t$

    (b) \$20,168.40

65. and 67. Answers will vary.

69. 465   **71.** 4676

73. $S_n = n(2n + 7)$ **75.** $S_n = \frac{5}{2}\left[1 - \left(\frac{3}{5}\right)^n\right]$

77. 5, 10, 15, 20, 25

    First differences: 5, 5, 5, 5

    Second differences: 0, 0, 0

    Linear

79. 16, 15, 14, 13, 12

    First differences: $-1, -1, -1, -1$

    Second differences: 0, 0, 0

    Linear

81. 15   **83.** 56   **85.** 35   **87.** 28

89. $\dfrac{x^4}{16} + \dfrac{x^3 y}{2} + \dfrac{3x^2 y^2}{2} + 2xy^3 + y^4$

91. $a^5 - 15a^4 b + 90a^3 b^2 - 270a^2 b^3 + 405ab^4 - 243b^5$

93. $41 + 840i$ **95.** 3   **97.** 10,000

99. 3,628,800   **101.** 56   **103.** $\frac{1}{9}$

105. (a) 43%   (b) 82%   **107.** $\frac{1}{216}$ **109.** $\frac{3}{4}$

111. True. $\dfrac{(n + 2)!}{n!} = \dfrac{(n + 2)(n + 1)n!}{n!} = (n + 2)(n + 1)$

113. True by Properties of Sums   **115.** Natural

117. (a) Each term is obtained by adding the same constant (common difference) to the previous term.

(b) Each term is obtained by multiplying the same constant (common ratio) by the previous term.

119. Each term of the sequence is defined in terms of the previous term.

121. d   **123.** b

125. The terms of the expansion of $(x + y)^n$ are positive; the signs of the terms in the expansion of $(x - y)^n$ alternate.

127. $\frac{1}{3}$. The probability that an event does not occur is 1 minus the probability that it does occur.

Chapter Test   *(page 655)*

1. $-\frac{1}{5}, \frac{1}{8}, -\frac{1}{11}, \frac{1}{14}, -\frac{1}{17}$ **2.** $a_n = \dfrac{n + 2}{n!}$ **3.** 140

4. 25.4   **5.** 5, 10, 20, 40, 80   **6.** 86,100   **7.** 4

8. Answers will vary.

9. $x^4 + 8x^3y + 24x^2y^2 + 32xy^3 + 16y^4$ **10.** 180

11. (a) 72   (b) 328,440   **12.** (a) 330   (b) 720,720

13. 720   **14.** $\frac{1}{15}$ **15.** 3.908×10^{-10} **16.** 25%

Cumulative Test for Chapters 6–8
(page 656)

1. $(1, 2), \left(-\frac{3}{2}, \frac{3}{4}\right)$ **2.** $(2, -1)$

3. $(4, 2, -3)$ **4.** $(1, -2, 1)$

5.   **6.**

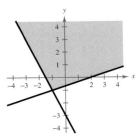

7.

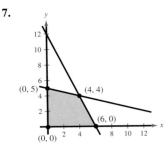

Maximum at $(4, 4)$: $z = 20$

8. 120 pounds, 80 pounds   **9.** $y = \frac{1}{3}x^2 - 2x + 4$

10. $\begin{bmatrix} -1 & 2 & -1 & \vdots & 9 \\ 2 & -1 & 2 & \vdots & -9 \\ 3 & 3 & -4 & \vdots & 7 \end{bmatrix}$ **11.** $(-2, 3, -1)$

12. $\begin{bmatrix} 5 & -3 \\ -2 & 2 \end{bmatrix}$ **13.** $\begin{bmatrix} 2 & -6 \\ -2 & 0 \end{bmatrix}$ **14.** $\begin{bmatrix} 6 & -6 \\ -3 & 2 \end{bmatrix}$

15. $\begin{bmatrix} -4 & 12 \\ 3 & -3 \end{bmatrix}$ **16.** $\begin{bmatrix} -175 & 37 & -13 \\ 95 & -20 & 7 \\ 14 & -3 & 1 \end{bmatrix}$

17. Gym shoes: \$164.80 million

Jogging shoes: \$244.22 million

Walking shoes: \$111.89 million

18. $(-5, 4)$ **19.** $(-3, 4, 2)$ **20.** 9

21. $\frac{1}{5}, -\frac{1}{7}, \frac{1}{9}, -\frac{1}{11}, \frac{1}{13}$ **22.** $a_n = \dfrac{(n + 1)!}{n + 3}$

23. 920   **24.** 65.4   **25.** 3, 6, 12, 24, 48   **26.** 6

27. Answers will vary.

28. $z^4 - 12z^3 + 54z^2 - 108z + 81$ **29.** 210   **30.** 600

31. 70   **32.** 120   **33.** 720   **34.** $\frac{1}{4}$

Appendix B

Section B.1   *(page A10)*

1. (a) \$1.109   (b) [\$0.999, \$1.189]

3. Quiz 1:

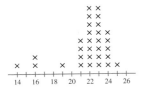

Quiz 2:

22 and 23

5.

Stems	Leaves
7	0 5 5 5 7 7 8 8 8
8	1 1 1 1 2 3 4 5 5 5 5 7 8 9 9 9
9	0 2 8
10	0 0

7.

Stems	Leaves
0	89 66 67 65 80 98 62 93
1	09 01 46 24 96 90
2	92 55 40 61
3	68 35
4	12 96 80 38
5	81 18 50 70 66
6	44 00 34 01
7	61 66 00
8	11 57 41 90
9	
10	
11	60 59 33
12	92
13	19 17 37
27	22
31	32
46	80
65	14

9. (a)

Interval	Tally
90–109	⊥⊥⊥⊥⊥
110–129	⊥⊥⊥⊥⊥ ⊥⊥⊥⊥⊥ l
130–149	⊥⊥⊥⊥⊥ llll
150–169	⊥⊥⊥⊥⊥ l
170–189	l
190–209	lll

(b)

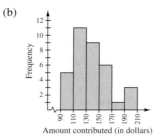

Amount contributed (in dollars)

Section B.2 (page A17)

1. Mean: 8.86; median: 8; mode: 7

3. Mean: 10.29; median: 8; mode: 7

5. Mean: 9; median: 8; mode: 7

7. The mean is sensitive to extreme values.

9. Mean: \$67.14; median: \$65.35

11. Mean: 3.07; median: 3; mode: 3

13. One possibility: $\{4, 4, 10\}$

15. The median gives the most representative description.

17. $\bar{x} = 6,\ v = 10,\ \sigma = 3.16$

19. $\bar{x} = 2,\ v = \frac{4}{3},\ \sigma = 1.15$ **21.** $\bar{x} = 4,\ v = 4,\ \sigma = 2$

23. $\bar{x} = 47,\ v = 226,\ \sigma = 15.03$ **25.** 3.42

27. 101.55   **29.** 1.65

31. (a) $\bar{x} = 12;\ \sigma = 2.83$ (b) $\bar{x} = 20;\ \sigma = 2.83$
(c) $\bar{x} = 12;\ \sigma = 1.41$ (d) $\bar{x} = 9;\ \sigma = 1.41$

33. $\bar{x} = 12$ and $|x_i - 12| = 8$ for all x_i

35. It will increase the mean by 5, but the standard deviation will not change.

37. First histogram

39.

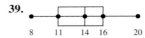

8   11   14 16   20

41.

19   21.5   25   28 29

43.

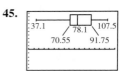

4 5 7 9   12

45.

37.1   78.1   107.5
70.55   91.75

Index of Applications

Index

FORMULAS FROM GEOMETRY

Triangle:

$h = a \sin \theta$

$\text{Area} = \dfrac{1}{2}bh$

(Laws of Cosines)

$c^2 = a^2 + b^2 - 2ab \cos \theta$

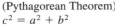

Right Triangle:

(Pythagorean Theorem)

$c^2 = a^2 + b^2$

Equilateral Triangle:

$h = \dfrac{\sqrt{3}s}{2}$

$\text{Area} = \dfrac{\sqrt{3}s^2}{4}$

Parallelogram:

$\text{Area} = bh$

Trapezoid:

$\text{Area} = \dfrac{h}{2}(a + b)$

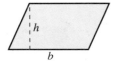

Circle:

$\text{Area} = \pi r^2$

$\text{Circumference} = 2\pi r$

Sector of Circle:

(θ in radians)

$\text{Area} = \dfrac{\theta r^2}{2}$

$s = r\theta$

Circular Ring:

(p = average radius,

w = width of ring)

$\text{Area} = \pi(R^2 - r^2)$

$\qquad = 2\pi p w$

Sector of Circular Ring:

(p = average radius,

w = width of ring,

θ in radians)

$\text{Area} = \theta p w$

Ellipse:

$\text{Area} = \pi a b$

$\text{Circumference} \approx 2\pi \sqrt{\dfrac{a^2 + b^2}{2}}$

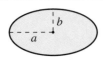

Cone:

(A = area of base)

$\text{Volume} = \dfrac{Ah}{3}$

Right Circular Cone:

$\text{Volume} = \dfrac{\pi r^2 h}{3}$

$\text{Lateral Surface Area} = \pi r \sqrt{r^2 + h^2}$

Frustum of Right Circular Cone:

$\text{Volume} = \dfrac{\pi(r^2 + rR + R^2)h}{3}$

$\text{Lateral Surface Area} = \pi s(R + r)$

Right Circular Cylinder:

$\text{Volume} = \pi r^2 h$

$\text{Lateral Surface Area} = 2\pi r h$

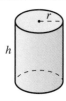

Sphere:

$\text{Volume} = \dfrac{4}{3}\pi r^3$

$\text{Surface Area} = 4\pi r^2$

Wedge:

(A = area of upper face,

B = area of base)

$A = B \sec \theta$

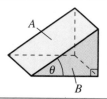